P9-DGK-509

PHYSIOLOGY AND BIOCHEMISTRY OF EXERCISE

PHYSIOLOGY AND BIOCHEMISTRY OF EXERCISE

Roy J. Shephard, M.D., Ph.D.

Professor of Applied Physiology
Dept. of Preventive Medicine and Biostatistics
University of Toronto

Director, School of Physical and Health Education
University of Toronto

Professor, Institute of Medical Sciences
University of Toronto

PRAEGER SPECIAL STUDIES • PRAEGER SCIENTIFIC

Library of Congress Cataloging in Publication Data

Shephard, Roy J.
Physiology and biochemistry of exercise.

Bibliography: p.
Includes index.
1. Exercise—Physiological aspects. 2. Biological chemistry. I. Title.
QP301.S494 612'.04 81-1833
ISBN 0-03-059289-5 AACR2

Published in 1982 by Praeger Publishers
CBS Educational and Professional Publishing
a Division of CBS Inc.
521 Fifth Avenue, New York, New York 10175 U.S.A.

3456789 147 98765432

Printed in the United States of America

CONTENTS

PREFACE

Many writers spend long and lonely hours contemplating the first blank sheet of paper that will set the course of their book. Once the topic has been broached, ideas gather momentum, but choosing the point of departure can be a fateful and time-consuming decision. Is the author of a systematic textbook spared such problems? It could be argued that both the scope of his work and the recognized approach to a given discipline are determined by hallowed tradition. Thus, textbooks of exercise physiology are supposed to start with chapters devoted to muscle, heart, and lungs, and then proceed to consider more specific applications in sports medicine or physical education. Although the required material is undoubtedly covered in this manner, the student then fails to grasp a vision of the integrated response that translates the stored energy of electrons into human movement.

There would be little point in writing another book that merely wallowed in the wake of predecessors launched by other authors. Nor would I wish to develop in any detail material already covered in other volumes from this laboratory. The detailed theory of human fitness and its methodology are already available in *Endurance Fitness* and *Fundamentals of Exercise Testing*. Anthropological aspects of human performance are discussed in *Human Physiological Work Capacity* and specific features of the top competitor are reviewed in *The Fit Athlete*. Unfortunately, other points of departure for a thorough discussion of the physiology and biochemistry of human movement are beset by uncharted shoals, gaps in the reader's knowledge that must be left unexplored until subsequent chapters are reached. But this is, perhaps, an inevitable consequence of our topic; the function of each link in the energy-transducing system is closely interrelated through a series of complicated feedback loops, involving not only chemical messengers and an omnipresent nerve network, but also mechanical and hydraulic linkages. As a medical undergraduate, I confronted this same dilemma. Where should my study of human function begin? I surmised that my professors had abandoned all hope of finding a good starting point to instruct a person such as myself, with no knowledge of biological science. They simply required all students to muddle through a 3-month survey course in anatomy, physiology, and biochemistry, subsequently returning to "square one" for a more serious and systematic study of their individual disciplines.

The present book breaks with tradition by starting from within the active muscle fiber, as the stored chemical energy of the phosphate bond is translated into a tension that can perform external work. The comfortable status quo of the fiber is disturbed, and the reader is challenged to explore the closely integrated compensatory mechanisms that are called into play to restore the initial state and allow work to continue. The life-sustaining process of homeostasis is thus approached from the perspective of the working cell rather than from the secondary reactions that develop in gross organs such as the heart and lungs. The cellular emphasis seems appropriate to our present day because most new knowledge of the biology of work is now being won at cellular and even subcellular levels of organization.

To minimize the array of unfamiliar concepts, a brief review is first made of the various biological factors influencing human performance—the energy balance of the body, stores of energy, and maximum power output. The energy transducers of the body are next examined, and the book then proceeds to a systematic study of internal transport systems, noting their role in maintaining the constant *milieu interieur* demanded by the transducers. Subsequent chapters consider the coordinating role of the central nervous system and humoral messengers, and the impact of physical activity upon other body systems. This knowledge is then applied to specific categories of individuals: the young, the aged, females, the obese, and patients with cardiovascular disease. The influence upon homeostasis of unusual environments—heat, cold, underwater, and high altitudes—is considered. A final section examines briefly such topics as diurnal rhythms of performance, fatigue, influence of ergogenic and socially abused drugs, and energy costs of specific occupational and recreational activities.

The general plan of writing has been to combine scientific rigor with a minimum of unexplained technical jargon. Simplicity of style is particularly important in discussing the biology of work, for this is an interdisciplinary subject. The material covered has application to the task of the physician who is unaware of the finer points of physical education. It is equally vital knowledge for a physical educator who is unfamiliar with complex medical terminology and to an ergonomist whose primary degree may be in engineering or industrial psychology. Unfortunately, there are limits to possible simplification of a substantial textbook. Although there are no formal prerequisites, it is assumed that the reader has stayed awake in at least a proportion of high school physics and mathematics classes. The book is also followed more easily after undertaking the equivalent of brief introductory courses in anatomy and physiology. The student can begin use of the book as a senior undergraduate, rereading it for greater detail as a graduate or practitioner. The student will find his task facilitated by a detailed index and a comprehensive bibliography of more than 5000 references drawn from recent literature in all parts of the world.

Although I have claimed sole authorship, in a real sense this book is a team effort because it incorporates the fruits of stimulating discussions with professional colleagues in every corner of the globe. Where possible, the sources of these ideas have been identified by appropriate references, but if any such debt lacks formal acknowledgment, I apologize here for the omission. A second group of contributors have been my pupils at the University of Toronto, both undergraduates taking advanced course work in the physiology of physical activity and a succession of graduate students who have shared in the vigorous exploration of almost every facet of human physical endeavor. Last, I must praise the fortitude of the "home team," my wife and two daughters. The writing of yet another major textbook has meant many long evenings enlivened only by the pounding of a typewriter and the occasional excitement of a page of copy being torn to shreds. My family have met this challenge with serenity, and by their generous understanding and loving support have greatly helped in the completion of my task.

Toronto, 1981

PHYSIOLOGY AND BIOCHEMISTRY OF EXERCISE

—CHAPTER 1

Energy Balance in Humans

Movement and Energy Homeostasis
Units of Measurement
Laws of Thermodynamics
Force, Energy, and Power
Energy Storage in Humans
Body Energy Requirements
Human Energy Transducers
Efficiency and External Work

MOVEMENT AND ENERGY HOMEOSTASIS

> Living organisms . . . must pay for all their activities in the currency of metabolism (Baldwin, 1967).

Motion of any object, animate or inanimate, demands the transformation of energy. Humanity is not exempt from this principle. The cost of movement must be met from a body store (such as the elastic energy of a stretched tendon) or an external resource (such as the biochemical energy of ingested food). Life itself depends on the balance between income and expenditure of energy. It is thus convenient to discuss human movement from the starting point of homeostasis, that is, maintenance of the status quo with respect to chemical and physical stores of energy. The famous French physiologist Claude Bernard (1878) noted how the body strove to maintain the constancy of its internal environment, or *milieu interieur*, thereby protecting humankind against sudden and potentially dangerous changes in the external world. For example, the rate of many biochemical reactions would be doubled by a 10°C rise of tissue temperature (the Law of Arrhenius). However, a rather precise regulation of local muscle temperature ensures a consistent response to stimulation despite wide variations of environmental temperature. Similar advantages stem from a close regulation of oxygen pressure, acidity, and food reserves at both cellular and subcellular levels of organization (Robin, 1977).

Homeostasis can be disturbed by a gross change of external environment (as when a subject flies to a high-altitude resort) or by a sudden upsurging of energy expenditure within the body (as when vigorous exercise is undertaken). If the disturbance is sustained or repeated, adaptation normally occurs; the subject becomes acclimatized to high altitude or trained to withstand a high intensity of physical activity. However, on occasion the disturbance of homeostasis is too severe for adaptive mechanisms, and cellular changes become irreversible (as in high-altitude edema and the heat deaths that affect long-distance cyclists, runners, and football players) (Shephard, 1976b).

UNITS OF MEASUREMENT

Older textbooks present a confusing array of units for the measurement of energy, work, and power. Fortunately, internationally agreed recommendations (G. Ellis, 1971) are now gaining wide acceptance. The standard system uses metric units of length and mass (meters and kilograms) and is independent of variations in gravitational acceleration (Table 1.1).

Gravitational forces vary slightly with latitude and altitude (Grombach, 1960). This reflects two physical circumstances: (1) the earth is not a perfect sphere and (2) the gravitational attraction varies inversely as the square of the distance separating a body from the earth's center of mass. Variations of gravity are too small to be of serious consequence in most physiological experiments. However, gravity-independent units are essential when dealing with such problems as the piloting of high-performance aircraft and the exploration of space. An astronaut on the moon's surface, for example, encounters only a sixth of the normal gravitational force; he slips when trying to run, and finds that the most effective method of moving about is to adopt a bouncing pattern of motion (Margaria, 1971).

LAWS OF THERMODYNAMICS

Sir Isaac Newton propounded three fundamental laws of motion. These may be related to human activity when presented as follows:

LAW 1. INERTIA AND CONSERVATION OF MOMENTUM. A body segment remains in a state of rest or of uniform motion unless acted upon by some force external to the segment. If at rest, there is inertial resistance proportional to the mass of the segment and any attached load. If in motion, there is momentum proportional to the product of mass and the velocity of the part.

The inertia of a 100-kg football player is twice that of a 50-kg coxswain. Thus the force needed to accelerate the football player is twice that required for a similar acceleration of the coxswain. If the coxswain decided to take up football he could develop a momentum equal to that of a 100-kg opponent by the tactic of moving at twice the speed of the other player.

One important metabolic corollary of Newton's first law is that matter (or its energy equivalent) can be neither created nor destroyed. Painstaking research has shown that this principle applies to the human machine. Heat generated by subjects living inside a metabolic chamber coincides closely with the heat that would have been liberated had ingested food been burnt in a bomb calorimeter (Perkins, 1964; Durnin and Passmore, 1967). Specifically, the energy content of the food eaten is exactly balanced by the sum of any work accomplished, changes in the energy stores of the body, and losses as heat and excreta.

The change in heat content (*enthalpy*) of a system ($\Delta H'$) depends on the change of *internal energy* ($\Delta E'$) and any accompanying volume change (the *dimensional component* $P'\Delta V$). If heat is evolved during a reaction, enthalpy is lost; the change of enthalpy thus equals the heat produced, but has a negative sign (heat = $-\Delta H'$). If the products of a reaction have a larger volume than the reactants, then work is performed against atmospheric pressure P′, and the change of enthalpy is smaller than the change of internal energy:

$$-\Delta H' = -\Delta E' - P'\Delta V \qquad (1.1)$$

Applying these concepts to the release of energy in muscle, we find that all reactions occur in solution, and changes in the volume of the system are negligible. The term $P'\Delta V$ can thus be ignored, and for practical purposes $-\Delta H'$ and $-\Delta E'$ are equal. However, not all of the enthalpy change is available for performance of external work; part is used to increase the *entropy* or randomness of molecular structure in the chemical constituents. If T′ is the absolute temperature and $\Delta S'$ is the change in entropy, then

$$-\Delta H' = -\Delta G' - T'\Delta S' \qquad (1.2)$$

where $-\Delta G'$ is the *free energy change*. When food is broken down in the body, the free energy can per-

TABLE 1.1
SI Units of Work

Force	=	newton (N) (1 kg • m • sec^{-2}; a mass of 1 kg exerts a force of 9.81 N in a standard gravitational field)				
Distance	=	meter (m)				
Work	=	force × distance	=	newton • meter	=	joule (J)
Power	=	work ÷ time	=	joule • second^{-1}	=	watt (W)
Pressure	=	force ÷ area	=	newton • meter^{-2}	=	pascal (Pa)

form useful work, but may also appear as heat; the entropy change necessarily appears as heat. Thus,

$$h = -\Delta G' - T'\Delta S' - W' \quad (1.3)$$

where h is the heat production and W′ is the useful external work that has been performed.

Biochemical reactions proceed readily if there is a liberation of heat (an exothermal reaction). However, endothermal reactions can also occur, provided that the increase in heat content or enthalpy of the system (ΔH′) is balanced by an even greater increase of entropy (−T′ ΔS′).

LAW 2. ACCELERATION AND APPLIED FORCE. Any change in motion of a body is proportional to the external force and takes place in the direction of the applied force. Thus,

$$F = ma \quad (1.4)$$

where F is the applied force, m is the displaced mass, and a is the linear or angular acceleration. If m is measured in g and a in cm • sec^{-2}, then F is expressed in dynes. Again, the law applies to the human body. The acceleration of an individual limb may be induced by the normal force of gravity, an external force such as a gust of wind or a mechanical impact, or a force generated by muscular activity in an adjacent body segment. If a sprinter attains a velocity of 5 m • sec^{-1} within 1 sec, his acceleration is 5 m • sec^{-2}, and given a body weight of 70 kg, the minimum force needed to develop this speed is 350×10^5 dynes, or 350 newtons (N). Once any part of the body has been set in motion, an opposing force (such as the contraction of opposing, "antagonistic" muscles) must be exerted for it to be brought to rest. In a "ballistic" movement (such as the swing of a cricket bat), little sustaining action is needed from the muscles initiating the movement (the "agonists"); however, if the bat is brought to rest in a specific position such as the "crease," then a burst of activity can be seen as the movement is checked (see Fig. 4.18).

Newton's second law implies a constraint on the direction of energy change. Natural processes are associated with an increase of entropy. Thus physical activity inevitably involves a liberation of heat. If the temperature of the working muscle is to remain uniform, this heat must be dissipated to the external environment. Heat loss is essential not only to homeostasis, but also to the functioning of the body as a machine.

LAW 3. REACTION. Every action induces an equal and opposing reaction, thereby conserving the energy of the system. The practical consequences of this law are best appreciated when a worker is trying to turn a lever under water. An opposing torque spins his body in the opposite direction to the motion he is attempting to induce. The same task is performed quite easily on dry land because the twisting force acting through the shoes is opposed by friction at the ground surface.

FORCE, ENERGY, AND POWER

Force

The traditional metric unit of force has been the *dyne*. Since force is the product of mass and acceleration [equation (1.4)], it follows that in a standard gravitational field of 981 cm • sec^{-2}, a mass of 1 g exerts a force of 981 dynes. While a unit of this size is appropriate for precise physical experiments, it is far too small for the exercise physiologist. One suggested alternative has been the *kilopond*, the force exerted by a mass of 1 kg in a standard gravitational field (1 kilopond, kp, = 9.81×10^5 dynes). Unfortunately, there is a potential for confusion between the pond and the imperial pound. The international committee (G. Ellis, 1971) thus recommended adoption of the newton and the *kilonewton* (1 kilonewton, kN, = 10^8 dynes).

Devices such as bicycle ergometers and handgrip dynamometers frequently have scales that are calibrated in kg. Given a standard gravitational acceleration, 1 kg force = 1 kp = 9.81 N.

Work and Energy

Work (W′) is the product of force (F′) and the distance through which it acts (S′):

$$W' = F'S' \quad (1.5)$$

In current metric units, if a force of 1 kN is sustained over a distance of 1 m, the amount of work performed is 1 *kilonewton-meter* (kN • m). Returning to the example of the sprinter, if the applied force is 350 N and the acceleration is 5 m • sec^{-2}, the distance covered in the first second of running is 2.5 m. The work performed is thus $350 \times 2.5 = 875$ N • m.*

*This calculation ignores a substantial additional amount of work performed in raising and lowering the body mass and accelerating and decelerating individual body segments.

Other units accepted by the international committee are based on electrical work (the *watt-minute*, W • min) and thermal work (the *kilojoule*, kJ). The kilocalorie (kcal), kilogram-meter (kg • m), and kilopond-meter (kp • m) have been discarded.

$$1 \text{ kN} \cdot \text{m} = 16.7 \text{ W} \cdot \text{min} = 1 \text{ kJ} = 0.239 \text{ kcal} = 101.9 \text{ kg} \cdot \text{m} = 101.9 \text{ kp} \cdot \text{m}$$

Although dimensionally equivalent to work, the term *energy* implies a store available to perform work. In a mechanical system, possible forms of energy include potential, pressure, and kinetic energy.

Potential Energy

Potential energy is developed by the displacement of a mass above some arbitrary reference point. It is calculated as the product of force and distance [equation (1.5)].

For example, a man weighing 70 kg is sitting on a teeter-totter. If he raises his center of mass 1 m from the floor, he has accumulated 687 N • m (70 × 9.81 × 1) of potential energy. Assuming that the teeter-totter is well oiled, much of this energy can be recovered during his subsequent descent; for example, work can be performed in lifting a partner on the other end of the device. However, efficiency is less than 100%. Some energy is lost in the bearings of the teeter-totter, and if the partner also weighs 70 kg he will be lifted less than 1 m.

Pressure Energy

The exercise physiologist is commonly concerned with gas pressures in the lungs and hydrostatic pressures in the circulation. The traditional unit has been the equivalent column of mercury, with or without standardization of gravitational field (torr, mmHg). However, the current recommendation (G. Ellis, 1971) is for use of the kilopascal (kPa). In unit gravitational field,

$$1 \text{ kPa} = 7.52 \text{ torr} = 7.52 \text{ mmHg} = 1 \text{ kN} \cdot \text{m}^{-2}$$

When studying mechanical problems, such as the pressure exerted on the soles of the feet, it is convenient to use units of kN • m^{-2}. If a man weighs 70 kg, his body mass exerts a force of 70 × 9.81, or 687 N. Given that his soles cover an area of 160 cm^2 or 160 × 10^{-4} m^2, the resultant pressure is 687 × $10^{-3}/160 \times 10^{-4}$, or 42.9 kN • m^{-2}.

Usage of pressure energy (work, W') can be calculated as the product of pressure (P') and change of volume (ΔV):

$$W' = P' \Delta V \qquad (1.6)$$

Let us suppose that an average pressure of 100 mm Hg is developed during the forced expiration of 5000 cm^3 of air from the lungs. P' is 100/7.52 = 13.3 kN • m^{-2} and W' is 13.3 × $5000/100^3$ = 66.5 N • m. We may note that since pressure is force per unit area [F' • $(s')^{-2}$] and volume is distance times area [s' • $(s')^2$], equation (1.6) is a special case of the force-distance product of equation (1.5):

$$W' = F'(s')^{-2} \cdot s'(s')^2 \qquad (1.7)$$

Kinetic Energy

Kinetic energy is possessed by virtue of linear or angular motion. It is calculated from the mass of the body (m) and the velocity of movement (v):

$$W' = \frac{mv^2}{2} \qquad (1.8)$$

If m is measured in kg and v in m • sec^{-1}, then the kinetic energy is expressed in N • m. A 100-kg football player moving with a velocity of 5 m • sec^{-1} thus has a kinetic energy of 100 × $5^2/2$ N • m, or 1.25 kN • m.

Since force is the product of mass and acceleration (mvt^{-1}), and the distance covered in time t is equal to vt/2, we are again dealing with a special case of the force-distance product:

$$mvt^{-1} \frac{vt}{2} = \frac{mv^2}{2} \qquad (1.9)$$

Power

Power is the rate of working ($W't^{-1}$). Currently accepted metric units are kN • m • min^{-1}, and kJ • min^{-1}. Outmoded units include kcal • min^{-1}, kg • m • min^{-1}, and kp • m • min^{-1}.

Interchange of Energy

In the absence of external losses, it follows from the laws of thermodynamics that a decrease in one form of energy within a system must be accompanied by an equivalent increase in other forms of energy. This concept was recognized many years ago by Bernouilli. He stated the theorem that the sum of potential, pressure, and kinetic energy stores remained constant within a frictionless hydraulic system.

There are many examples of energy exchange in the body at both the molecular and the gross structural levels. In the circulation, for example, a conversion of kinetic to pressure energy allows a filling of the atrial chambers even if intramural pressures are greater than in the veins. Again, the potential energy of a runner diminishes as his center of gravity falls; kinetic energy is developed during the descent, and a part of this energy is stored as elastic energy when the leg tendons are stretched by the impact of the foot on the ground (Cavagna et al., 1964a; Thys et al., 1972; Komi and Bosco, 1978). However, energy exchange is never perfect. Entropy must always increase, and some heat must be lost if the system is to function.

ENERGY STORAGE IN HUMANS

Physical Stores

On occasion, the body may store substantial amounts of energy in some physical form. A cyclist who pushes his machine to the summit of a 1000-m pass accumulates ~1000 kN • m of *potential energy* (mass of machine plus rider = 100 kg, equivalent force = 0.981 kN, work = force × distance = 0.981 × 1000 kN • m). Since the external work required for steady cycling on level ground is about 10 kN • m • min^{-1}, sufficient energy has been accumulated to coast for 100 min during the subsequent descent to sea level.

A pirouetting skater has a considerable store of *kinetic energy* by virtue of her speed of rotation. If she lowers her outstretched arms, their angular velocity diminishes, and the kinetic energy is transferred to the body as a whole, increasing its speed of rotation. We have noted that a sprinting 100-kg American football player can accumulate a kinetic energy of 1.25 kN • m; if 25% of this energy were applied to the displacement of an equally massive opponent, he could be knocked 30 cm off the ground (cost 0.294 kN • m). Smaller amounts of energy are stored through the stretching of elastic structures. Thus, with maximum inspiration the chest is displaced ~3000 cm^3 from its resting position, at a pressure of 130 mmHg (17.2 kN • m^{-2}) and energy equivalent to 17.2 × 3000/100^3 kN • m or 51.6 N • m accumulates in the elastic elements of the respiratory pump.

Some energy may be stored through an increase of *body temperature* (ΔT'). The specific heat of the body, the energy needed to raise unit mass of tissue 1°C, is ~0.2 J • g^{-1}. Thus, in a subject of mass m, the work performed is:

$$W' = 0.2m\,\Delta T' \quad (1.10)$$

If m is expressed in kg and ΔT' in °C, then W' is given in kJ. The 4°C rise of core temperature encountered during marathon racing (Pugh et al., 1967; Wyndham and Strydom, 1972; Kavanagh and Shephard, 1977a) is equivalent to the storage of ~56 kJ (56 kN • m) of energy. This is ~0.5% of the 12 megajoules (MJ) of heat liberated during the event.

Under normal circumstances, a subject can recoup little of his stored heat. However, if he were to jump suddenly into an ice-cold lake, a longer than normal interval would elapse before he had to augment body heat production by shivering and/or vigorous swimming.

Chemical Stores

All chemical molecules have within their structure vast reserves of energy consequent upon their ordered structure. However, the body does not have at its disposal a nuclear reactor that can liberate this resource in its entirety. Interest thus attaches to simpler rearrangements of chemical structures that lead to changes of free energy [−ΔG', equation (1.3)]. Given a suitable transducer, such reactions can be used to perform external work.

The Transducer

One important energy transducer within the human body is the actin/myosin system of skeletal and cardiac muscle. Actin and myosin are long-chained protein molecules. Under appropriate local environmental conditions, these molecules can accept energy, with a consequent shortening of their joint dimensions (formation of actomyosin). A tension is developed in the tendinous attachments of the muscle protein filaments and the resultant elastic energy can be used for performance of external work (Fig. 1.1).

Phosphagen Stores

The primary energy store for muscle (and many other body systems) is the phosphate bond of the adenosine triphosphate (ATP) molecule. Energy is liberated as the ATP is broken down to adenosine diphosphate (ADP) and a phosphate radical. The yield is ~46 kJ • mol^{-1}. Given a total muscle mass of

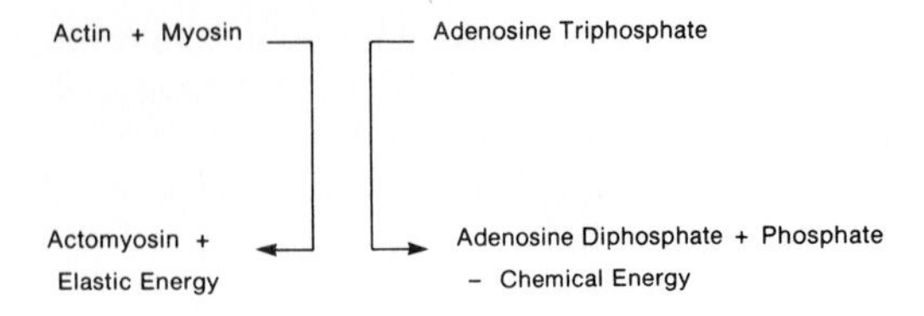

FIG. 1.1. The energy-transducing system of human muscle.

28 kg and a muscle ATP content of 6.9 mmol • kg^{-1}, the total energy stored in this form is ~8.9 kJ.

One mechanism for the regeneration of ATP is the breakdown of ADP to adenosine monophosphate (AMP):

$$2ADP \xrightarrow{\text{myokinase}} ATP + AMP \quad (1.11)$$

This reaction is catalyzed by the enzyme myokinase (adenylate kinase). A more usual process in muscle cytoplasm is interaction with a second high-energy phosphate compound, creatine phosphate (CP), which is broken down to creatine (C) and phosphate radical (P):

$$ADP + CP \xrightarrow{\text{creatine phospho-transferase}} ATP + C + P \quad (1.12)$$

This reaction is catalyzed by creatine phosphotransferase. Inevitably, the transfer of energy is incomplete, but the loss of free energy is only ~6.7 kJ • mol^{-1} (di Prampero, 1971a, b). The total resource of ATP, CP, and smaller amounts of other high-energy phosphate compounds is sometimes described as the phosphagen store. It has been set at 32 mmol • kg^{-1}; given an energy yield of 46 kJ • mol^{-1} this amounts to ~41 kJ distributed over 28 kg of muscle (McGilvery, 1975). Depletion is usually incomplete. The usable fraction is probably ~30 mmol • kg^{-1}, or 38.7 kJ.

Phosphagen stores are small relative to the maximum rate of working of the muscle transducer (250–300 kJ • sec^{-1}). For various reasons, the body cannot attain the maximum steady rate of phosphagen depletion instantaneously. Nevertheless, phosphagen energy reserves are exhausted by less than 10 sec of sprinting or other all-out energy expenditure.

The main bases for replenishment of stored ATP are the anerobic breakdown of glycogen and glucose to lactate within the cytoplasm (Fig. 1.2) and the aerobic breakdown of glycogen and fatty acids to carbon dioxide and water within the mitochondria (Fig. 1.3). Assuming a free supply of oxygen, the half-time of the recuperative process has been set at 22 sec (di Prampero, 1971a, b).

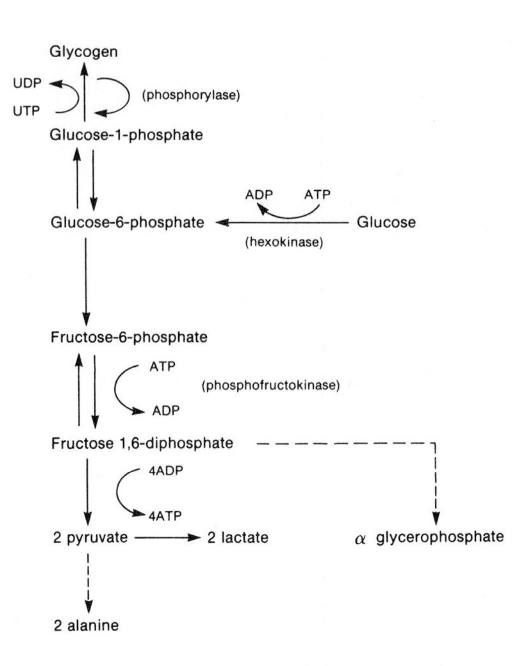

FIG. 1.2. Anaerobic mechanisms for the resynthesis of phosphagen. Note that these reactions proceed in the cytoplasm. UTP, uridine triphosphate; ATP, adenosine triphosphate; UDP, uridine diphosphate; ADP, adenosine diphosphate

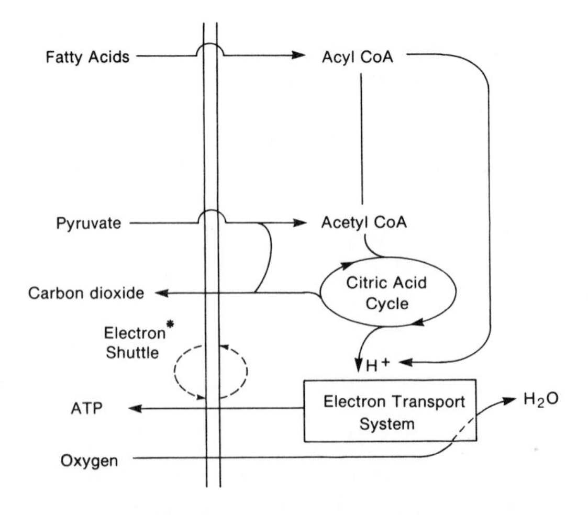

FIG. 1.3. Aerobic mechanisms for the resynthesis of phosphagen. Note that these reactions proceed within the mitochondrion.
*See Fig. 2.8.

Anerobic Sources of Energy

In the absence of oxygen, glycogen or glucose is broken down to pyruvate, and there is an accumulation of intermediate metabolites (some alanine and α-glycerol phosphate, but mainly lactic acid). If glucose is used as the fuel, this must first be converted to glucose-6-phosphate at the expense of one ATP molecule (Fig. 1.2). The net yield under anerobic conditions is thus 2 mol of ATP per mol of glucose, or 3 mol of ATP per mol of glycogen-derived glucose-6-phosphate.

The factor limiting energy release by anerobic mechanisms is not normally the glycogen content of the cytoplasm, but rather the accumulation of acid metabolites within the active muscle fibers. The limiting local concentration of lactate is 25–40 mmol • kg^{-1}; given 20 kg of active muscle, this amounts to 500–800 mmol of lactate. The free energy yielded by formation of lactate from glycogen is ~0.92 kJ • g^{-1}, or 82.8 kJ • mol^{-1} (di Prampero, 1971a, b). Since there is also some dispersion of lactate into other less active parts of the body, the maximum possible anerobic energy release is ~65–70 kJ, attained within 40 sec of exhausting effort.

Part of the accumulating lactate may be used as a fuel during exercise; it diffuses to other, better oxygenated muscle fibers, or is transported by the blood to other tissues with an adequate partial pressure of oxygen. A larger fraction of the accumulated lactate is oxidized to carbon dioxide and water during the recovery period, while most of the remainder is reconverted to glucose and glycogen within the liver. The half-time for disappearance of lactate is 10–15 min (di Prampero, 1971a, b). Several heats of an anerobic event can thus be run on the same day, providing that the tissue reserves of glycogen have not been exhausted.

Energy Stores Under Aerobic Conditions

If oxygen is freely available, energy can be derived from the breakdown of carbohydrate, fat, protein, and alcohol. The principal end-products of aerobic metabolism are carbon dioxide and water, although if a large part of the energy is derived from fat, this may undergo incomplete breakdown to ketone bodies (acetone, acetoacetate, and β-hydroxybutyrate), while if the energy is derived from protein there is necessarily some urea formation.

The average energy yield of the several classes of foodstuff was first estimated by Atwater and Benedict (1899). They proposed the now familiar scale of 4, 9, and 4 kcal • g^{-1} for the metabolism of carbohydrate, fat, and protein, respectively. Such values were based on the heat of combustion, as determined by burning the food in a bomb calorimeter, with adjustments for (1) digestibility and (2) excretion of urea (an energy loss of 1.25 kcal • g^{-1} of protein). McCance and Widdowson (1960) suggested a slight revision of these figures (Table 1.2). Their estimates made no adjustment for digestibility and were based on glucose rather than starch carbohydrate.

Mayer (1972) stressed that the energy yield varies from 15.1 to 17.2 kJ • g^{-1} (3.6–4.1 kcal • g^{-1}) for different types of carbohydrate, from 35.0–37.7 kJ • g^{-1} (8.35–9.0 kcal • g^{-1}) for different fats, and from 12.2 to 18.2 kJ • g^{-1} (2.90–4.35 kcal • g^{-1}) for different proteins. If energy expenditures are to be estimated accurately from food intake data, it is necessary to record not only the broad categories of ingested foods, but also the specific characteristics of the carbohydrate, fat, and protein that have been metabolized.

CARBOHYDRATE. The main carbohydrate stores of the human body are glycogen and glucose. Taking the data of McCance and Widdowson (1960), we would anticipate an energy yield of 2.83 MJ • mol^{-1} of glucose. However, not all of this energy can be applied to the resynthesis of ATP. According to current views, 38 mol of ATP is formed during the aerobic breakdown of 1 glucose molecule, but 2 molecules of ATP are used in the associated transfer of electrons from the cytoplasm to the mitochondria. The net yield is thus 36 mol of ATP, ~59% of the bomb calorimetry estimate. The residual 41% appears as heat. If the fuel is glycogen, it is not necessary to form glucose-6-phosphate as a preliminary to oxidation; 39 rather than 38 mol of ATP is generated with the breakdown of 1 glucose equivalent, and after allowance for electron transport between the cytoplasm and mitochondria, the net yield is 37 mol of ATP, ~60% of the stored energy as indicated by bomb calorimetry.*

A typical resting blood glucose concentration is 6.7 mmol • l^{-1} (120 mg • dl^{-1}). Given a blood volume of 5 l, there is thus a total of ~34 mmol of glucose in the bloodstream. About half of this glucose can be used before serious symptoms of hypoglycemia (glucose lack) appear. The equivalent energy reserve is thus 2.83 × 34/2 or ~48 kJ.

The glycogen content of resting muscle is

*While most of the stored glycogen is converted to glucose-6-phosphate, ~8–10% may appear as free glucose requiring phosphorylation (Field, 1966).

TABLE 1.2
Energy Yield of Principal Classes of Foodstuffs and Alcohol

	Atwater and Benedict (1899)			McCance and Widdowson (1960)	
Class	Bomb calorimetry (kcal • g^{-1})	Digestibility (%)	Corrected (kcal • g^{-1})	(kcal • g^{-1})	(kJ • g^{-1})
Protein	5.65	92	4.0*	4.1*	17.2
Fat	9.4	95	9.0	9.3	39.0
Carbohydrate	4.1	99	4.0	3.75	15.7
Alcohol	7.1	99	7.0	7.0	29.3

*Corrected for loss of 1.25 kcal as urea.

~1.5 g per 100 g of wet tissue, a reserve of ~420 g in 28 kg of muscle. During intense effort, the body can also draw upon ~100 g of glycogen stored in the liver. The total glycogen store is thus just over 500 g, equivalent to ~7.9 MJ of energy. With vigorous effort, this resource can be exhausted in as little as 100 min (Hultman, 1971). It is difficult to carry out anerobic work until the store has been replenished; unfortunately, this can take as long as 24 hr (Saltin and Hermansen, 1967).

FAT. The fat found in adipose tissue is the main energy reserve of the body. The energy yield in vivo (29.3 kJ • g^{-1}, 7.0 kcal • g^{-1}) is lower than the bomb calorimeter value because part of the weight loss that occurs during depletion of adipose tissue stores is the result of factors other than the combustion of pure triglyceride.

Complete oxidation of a typical fatty acid molecule to carbon dioxide and water yields 138 mol of ATP. However, the yield of ATP per mole of oxygen is somewhat poorer than for glycogen (5.63 rather than 6.17 mol). Thus, if oxygen transport is the factor limiting performance, then the rate of working can be increased ~10% by combustion of carbohydrate rather than fat.

Depending on the age, sex, and habitual activity of a subject, fat can account for 5–40% of body mass. The endurance athlete carries only 3–4 kg of fat, but in a sedentary young man a burden of 10–14 kg is common. The sedentary subject can lose as much as 75% of his store without ill effect, giving an effective energy reserve of up to 300 MJ. Assuming a daily energy usage of 7 MJ, this would allow 43 days of survival without food. Incomplete oxidation of fat to ketone bodies shortens the possible period of starvation. The likely ceiling of energy losses through ketosis can be deduced from studies of diabetics; such patients waste up to 1.7 MJ of energy per day through the excretion of ketone bodies. Replenishment of fat stores typically proceeds slowly over a period of months or even years.

PROTEIN. Tissue protein is not used as a source of energy except during dieting and starvation. In the latter situation, reserves of carbohydrate are exhausted within a day, and since glucose cannot be formed from fat (Figs. 1.4 and 1.5), protein breakdown becomes essential to meet the blood sugar requirements of carbohydrate-metabolizing tissues (particularly the red cells and the central nervous system).

There is a labile reserve of protein in all tissues, especially the liver. This organ provides a physiological store of ~300 g of protein. Once the labile reserve has been exhausted, there is inevitably tissue wasting, most obvious in the skeletal muscles.

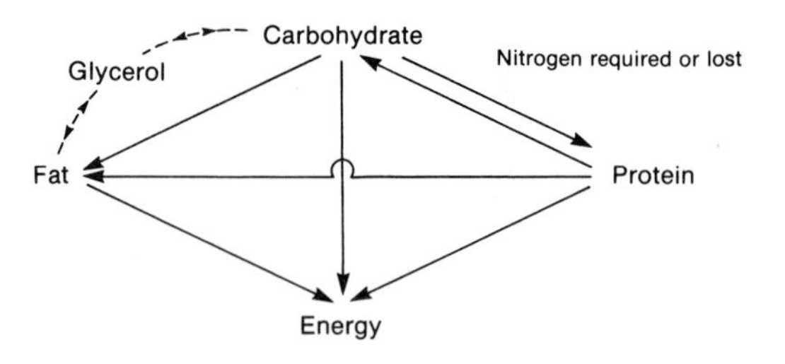

FIG. 1.4. Interconversion of energy stored as protein, fat, and carbohydrate. Note particularly that fat cannot be converted directly to either protein or carbohydrate. For detail, see Fig. 1.5.
Notes:
1. A certain minimum of carbohydrate is needed to avoid incomplete combustion of fat (ketosis).
2. Certain amino acids (essential amino acids) cannot be synthesized in the body.
3. According to some authors, the body also needs small quantities of certain essential fatty acids.

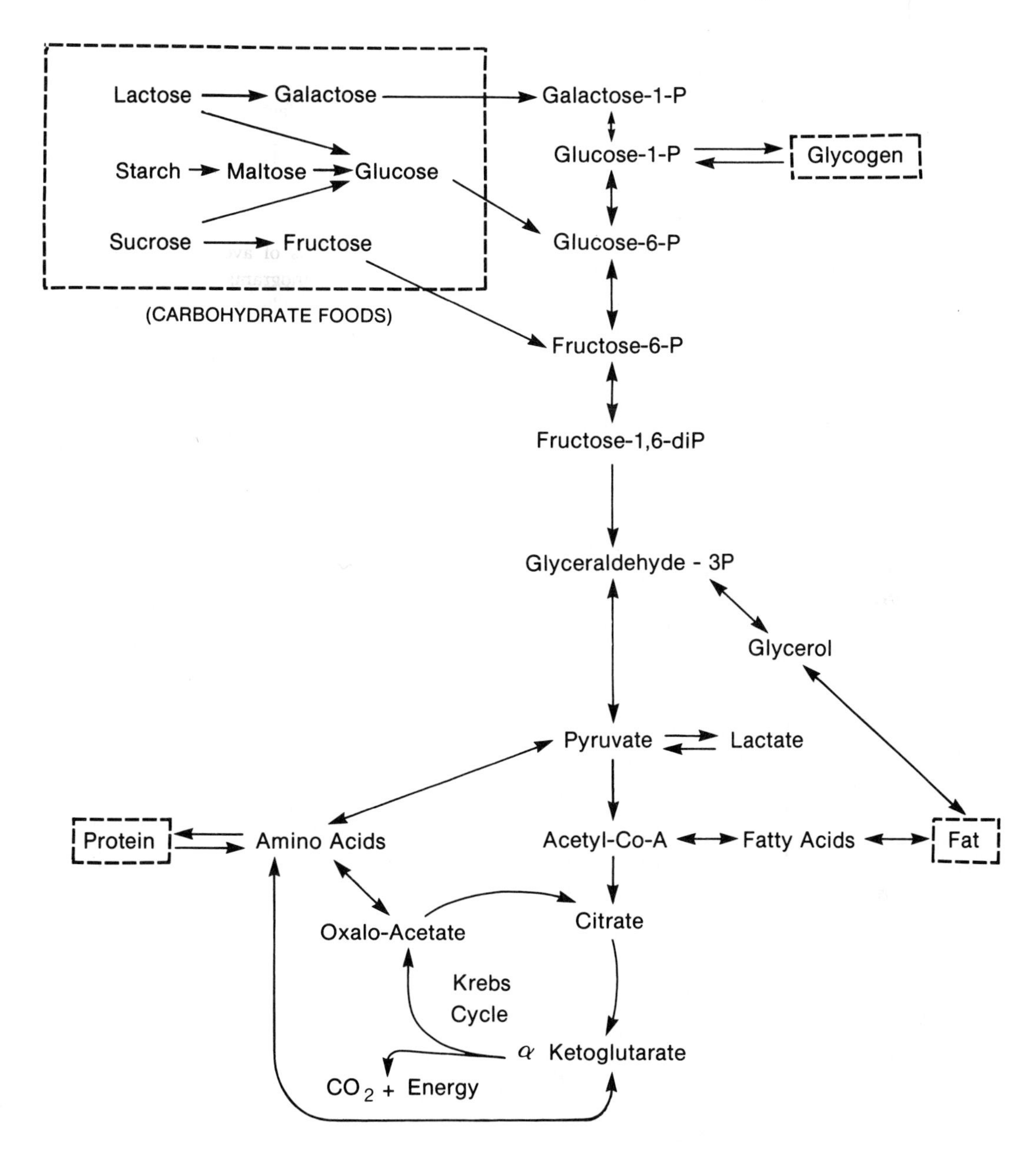

FIG. 1.5. Interrelationships of carbohydrate, fat, and protein metabolism. Note that carbohydrate and nonessential amino acids are interconvertible, but because of the point of entry into the Krebs cycle (Acetyl-CoA-Citrate stage), fat cannot be converted to carbohydrate or protein.

During the first few days of starvation, protein usage accounts for ~15% of energy needs, 60 g • day^{-1} in a young man, but there is a progressive increase in the rate of protein combustion as the body stores of fat become depleted. On the figures of McCance and Widdowson (1960), the physiological store corresponds to an energy reserve of ~5.2 MJ, while the maximum protein usage compatible with survival provides ~40 MJ of energy.

Alcohol. Energy can finally be derived from the oxidation of alcohol. The energy yield is similar to that of depot fat (29.3 kJ • g^{-1}). Some people ingest a substantial proportion of their daily energy needs in the form of alcohol. However, alcohol metabolism can occur only in the liver, and the maximum rate of usage (7 g • hr^{-1}) is relatively slow. Furthermore, there is little possibility of storing energy as alcohol because symptoms of intoxication appear with blood concentrations as low as 80 mg • dl^{-1}.

BODY ENERGY REQUIREMENTS

Homeostasis

Even if observations are made at rest or under basal conditions, the body has a finite consumption of energy. This reflects the costs of homeostasis at molecular, cellular, and higher levels of organization.

Molecular

Although the structure of the adult appears to be relatively constant, many body proteins are continuously being broken down to their constituent amino acids and then resynthesized. In muscle, the half-time of the synthetic reaction is 7.2 days for fibrillar protein and 2.8 days for other protein (Millward, 1970). In some tissues, such as the liver, the turnover is even faster, while in others it is much slower. As much as 450 g of muscle and a total of 600 g of protein is broken down and resynthesized each day. Energy losses associated with this cycle undoubtedly make a major contribution to basal metabolism.

Cellular

At the cellular level, substantial amounts of work are performed to maintain gradients of osmotic pressure and electrical potential across membranes permeable to charged ions and osmotically active compounds. Work is also involved in transporting molecules and electrons across intracellular boundaries. For example, the malate/aspartate shuttle transports electrons from the cytoplasm to active sites of aerobic metabolism within the mitochondria; the process uses ~5.3% of the ATP that is generated (2 of every 38 ATP molecules, McGilvery, 1975).

Organ and Whole-Body

Energy is needed to maintain the status quo of the body as a whole. Work is performed in pumping air into the chest and blood around the circulation. Furthermore, heat is lost from the body in most environments, and this loss must be made good if the body temperature is to be sustained.

Heat loss is closely related to the surface area of the body. E. F. DuBois (1927) showed many years ago that a person's surface area could be approximated by the following formula:

$$BSA = (M'^{0.425})(H'^{0.725})(71.84 \times 10^{-4}) \tag{1.13}$$

where BSA is the surface area in m^2, M' is the unclothed mass of the body in kg, and H' is the standing height in cm. Nomograms are available for making this calculation (Fig. 1.6) or, alternatively, an approximate solution can be programmed for a small computer (Shephard, 1970b). The DuBois formula works well for persons of average build, although more complicated nomograms are available for individuals of extreme body type (Sendroy and Cecchini, 1954).

The main factors influencing resting metabolism are probably maintenance of the protein pool (proportional to lean body mass) and counteraction of heat loss (proportional to $M'^{0.425}$). The resting energy requirement is thus approximated by

$$W' = a + b(M')^{0.75} \tag{1.14}$$

where M' is the body mass and a and b are constants (J. R. Brown, 1966).

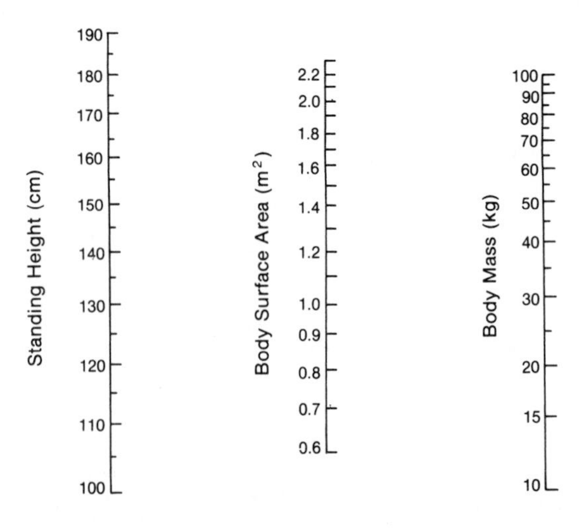

FIG. 1.6. The relationship of body surface area to standing height and body mass. (Based on the nomogram of E. F. DuBois: BSA = $M^{0.425}$ • $H^{0.725}$ • 10^{-4}.)

Estimation of basal or resting metabolism is a necessary first step in calculating the net efficiency of any type of physical activity. It is possible to determine such energy expenditures from measurements of basal or resting gas exchange, but often a predicted value based on body mass or surface area is of sufficient accuracy for the exercise physiologist. In a young man, the basal metabolism is ~2.8 $kJ \cdot min^{-1} \cdot m^{-2}$. In children, the figure is about 25% higher, while in women and older men it is some 10% lower.

Growth and Hypertrophy

Energy is required for the growth of new tissue not only in infants, children, and pregnant women, but also in athletes who engage in a muscle-building program. The cost of such growth cannot be less than the potential energy of the protein that is accumulated (17.2 $kJ \cdot g^{-1}$; Table 1.2) and is probably considerably larger than this (W. J. O'Hara et al., 1977b, 1979). One estimate can be derived from the normal protein turnover. Setting this at 600 $g \cdot day^{-1}$ and assuming it to be responsible for 50% of basal metabolism (that is, ~3600 $kJ \cdot day^{-1}$), the cost of protein synthesis is 6 $kJ \cdot g^{-1}$. Adding this figure to the accumulated potential energy, the total cost becomes 23 $kJ \cdot g^{-1}$ of new protein.

Movement and Tension

Two main types of muscular activity are distinguished: isotonic and isometric contraction. During an isotonic effort, the muscle shortens with no increase of tension, while during isometric effort tension is developed without shortening. Most of the activities encountered in real life involve a mixture of isotonic and isometric effort; however, a rhythmic activity such as running or cycling is mainly isotonic in character, while the support of a heavy weight depends largely upon isometric effort.

Energy is expended in both types of contraction. The "useful" component of isotonic work can be calculated as the product of the force exerted by a muscle and the distance over which it shortens. This approach is readily applicable to a frog nerve-muscle preparation, but in humans it is more convenient to measure external work—the electrical energy generated on a bicycle ergometer or the potential energy accumulated during the ascent of a staircase. During isometric effort, there is no obvious shortening of the muscle because contraction of the fibers is masked by an equivalent lengthening of the elastic elements to which they are attached.

As in any machine, the performance of muscular work is associated with a substantial heat production. Classical studies distinguished four phases:

1. **Activation heat.** This is associated with the transfer of energy from ATP to myosin and the development of an "active state" within the muscle fibers (formation of actomyosin).
2. **Shortening heat.** Internal (viscous) work is performed as individual filaments of actin and myosin slide over one another and the muscle shortens.
3. **Relaxation heat.** Let us suppose that muscle shortening has been used to lift either an external load or the body mass. If this same load is now lowered carefully as the muscle relaxes, potential energy stored in the load becomes liberated as heat within the muscle. This phase of heat loss is one reason that overall efficiency is lower for a step test than for bicycle ergometry.
4. **Delayed heat.** Resynthesis of high-energy phosphate bonds (ATP and CP) gives a final "delayed" phase of heat production. A more detailed consideration of muscle heat production follows.

In theory, the energy costs of movement and tension development should be determined by performing the work inside a metabolic chamber and measuring the resultant heat production. This approach was used by a number of classical investigators, but the necessary chambers are both cumbersome and expensive. Furthermore, they usually have a protracted response time, so that satisfactory results are obtained only if work is performed steadily over a long period. The much simpler technique of *indirect calorimetry* is commonly substituted. This approach exploits the fairly consistent relationship between oxygen consumption and energy usage (19.9 $kJ \cdot l^{-1}$ of oxygen for fat combustion, 21.2 $kJ \cdot l^{-1}$ of oxygen if carbohydrate is the fuel). The relative usage of fat and carbohydrate is estimated from the ratio of carbon dioxide output to oxygen intake (the respiratory quotient, RQ). With fat combustion, the RQ is 0.70, while with carbohydrate usage it is 1.00. An exactly intermediate value (RQ = 0.85) implies an equal usage of fat and carbohydrate. More precisely, the total work performed (W') is given by

$$W' (kJ) = 15.8 (V_{O_2}) + 4.86 (V_{CO_2}) - 12.5 (U_n) \quad (1.15)$$

where V_{O_2} is the volume of oxygen consumed in l, V_{CO_2} is the volume of carbon dioxide produced in l, and U_n is the mass of urinary nitrogen in g.

Applying this equation to a 1-min period of rest, we would find an oxygen consumption of ~0.25 l • min^{-1}, a CO_2 output of about 0.2 l • min^{-1}, and a nitrogen excretion of about 11 mg • min^{-1}. Thus,

$$W' = 15.8 (0.25) + 4.86 (0.20) - 12.5 (0.011) \quad (1.16)$$

or 4.59 kJ • min^{-1}. In many applications, the correction for urinary nitrogen is so small that it can be ignored. In the context of exercise, it must be stressed that equation (1.15) applies only to steady-state conditions. All work must be performed aerobically, and the gas exchange measured at the mouth must equal that occurring within the active tissues. At the beginning and end of vigorous physical activity, calculations can be upset by quite large changes in body stores of oxygen and carbon dioxide.

The average young man has a maximum aerobic energy expenditure of 60–65 kJ • min^{-1}. The corresponding figure for a well-trained endurance athlete can be as great as 120–130 kJ • min^{-1}, while in a sedentary senior citizen a value of 30–35 kJ • min^{-1} is to be anticipated. A much more rapid usage of energy can be developed for brief periods at the expense of a depletion of phosphagen reserves and an accumulation of acid metabolites. Thus, a young man who runs up a staircase can reach work rates that would require a steady aerobic energy expenditure of 230–240 kJ • min^{-1}.

HUMAN ENERGY TRANSDUCERS

While many of the energy needs of the body can be met temporarily by a depletion of physical or chemical stores, homeostasis depends ultimately on the liberation of chemical energy from ingested food. As noted above, in the absence of oxygen carbohydrates are broken down to pyruvate in the cytoplasm, while if oxygen is freely available fatty acids and pyruvate are oxidized to carbon dioxide and water within the mitochondria.

Cytoplasm

The anerobic mechanism of energy release has already been sketched briefly (Figs. 1.1 and 1.2). The enzyme myosin ATPase, firmly bound to the contractile protein myosin, converts ATP to ADP. Some 49% of the change in free energy can be applied to the development of muscle force through the bonding of actin and myosin, while the remaining 51% (23.5 kJ • mol^{-1}) appears as heat. Anerobic regeneration of ATP uses only a small proportion of the potential energy of glucose or glycogen. Despite a bomb calorimeter heat yield of 2.83 MJ • mol^{-1}, the anerobic metabolism of 1 mol of glucose provides energy to resynthesize only 2 mol of ATP (92 kJ, 3.2% of the potential energy indicated by combustion); the corresponding figure for glycogen (3 mol of ATP resynthesized) is 138 kJ, 4.9% of the potential energy being exploited (Fig. 1.7).

According to traditional thinking (Margaria et al., 1933), ~90% of the pyruvate formed during anerobic activity was reconverted to glycogen during the recovery period; the necessary energy was derived from oxidation of the residual 10% to carbon dioxide and water. In effect, each molecule of glucose or glycogen was utilized 10 times, so that ultimately 33% of the glucose energy and 49% of the glycogen energy was liberated. Bearing in mind the 49% efficiency of energy transfer from ATP to actomyosin, the overall efficiency from food store to actomyosin bonding would then be

For glucose: 49% × 33% = 16.2%
For glycogen: 49% × 49% = 24.0%

There are several difficulties with Margaria's interpretation of glycogen resynthesis. Two merit emphasis in the present context:

1. While some ATP is undoubtedly needed for the regeneration of glycogen, there is no strong evidence that exactly 9 molecules of glycogen are resynthesized for every one that is oxidized.
2. The rate of resynthesis of glycogen is relatively slow; 1–2 days is needed for the full replenishment of muscle stores (Saltin and Hermansen, 1967). On the other hand, the accumulated lactate disappears with a half-time of 10–15 min (di Prampero, 1971a, b).

In practice, some of the accumulated lactate is used elsewhere in the body during exercise, and most

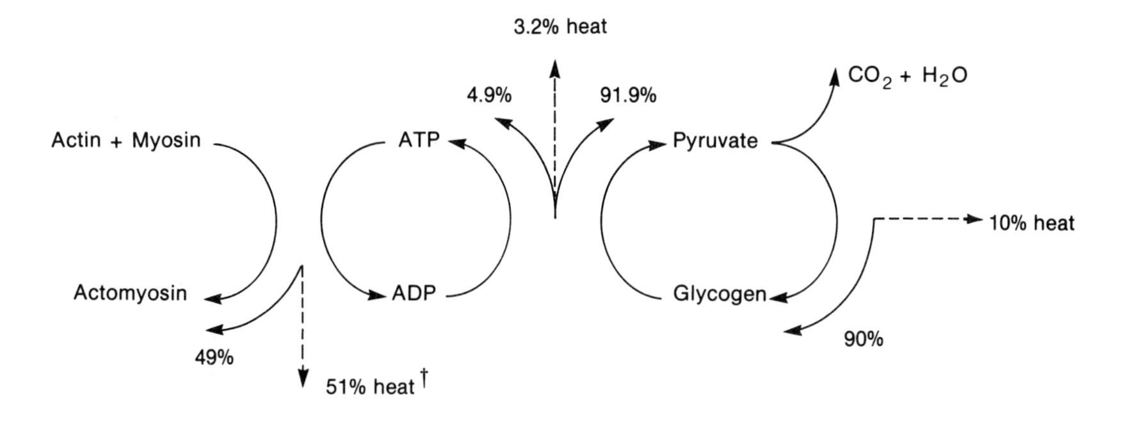

FIG. 1.7. Efficiency of the energy-transducing system under anerobic conditions. (Reutilization of pyruvate occurs during aerobic recovery.)

*If creatine phosphate (CP) forms an intermediate step in the process, the transfer of energy from creatine phosphate to ATP has an efficiency of 85%. The efficiency of actomyosin formation is thus (49) 85/100 = 41.7% and the over all efficiency is (41.7/100) • (4.9/100) • (100/10) = 20.4% if the glycogen is used in CP resynthesis, or (41.7/100) • (3.3/100) • (100/10) = 13.8% if glucose radicals are used for this purpose.

†The efficiency of energy transfer from ATP to actomyosin is not absolutely constant, being less at both low and high rates of contraction (Kushmerick & Davies, 1969). A much higher efficiency (~70%) can be calculated if allowance is made for such items as the work performed in pumping calcium ions (Infante et al., 1964).

of the remainder is burnt in place of ingested glucose over the immediate recovery period. However, glucose is subsequently taken from the bloodstream to rebuild muscle glycogen stores. In terms of energy exchange, the ultimate effect is thus approximately as suggested by Margaria et al. (1933).

Mitochondria

Under aerobic conditions, the process of energy transfer proceeds as illustrated in Fig. 1.8. Approximately 58.6% of the chemical energy of glucose and 60.3% of the energy content of an equivalent glycogen molecule is transferred to the ADP/ATP reaction. The overall efficiency of the transducer with respect to the combustion of glycogen is thus:

$$49\% \times 60.3\% = 29.5\%$$

This value sets a ceiling to the mechanical efficiency of any form of physical activity.

Muscle

Once the coupling of actin and myosin has occurred, external work can be performed. This is defined by the product of the force developed and the muscle shortening that occurs [equation (1.5)]. The working potential of a muscle is indicated by the area of its length tension diagram (Fig. 1.9).

Unfortunately, one cannot always exploit the entire diagram during physical activity. For example, maximum shortening can occur only if the muscle first undergoes passive elongation (as in some

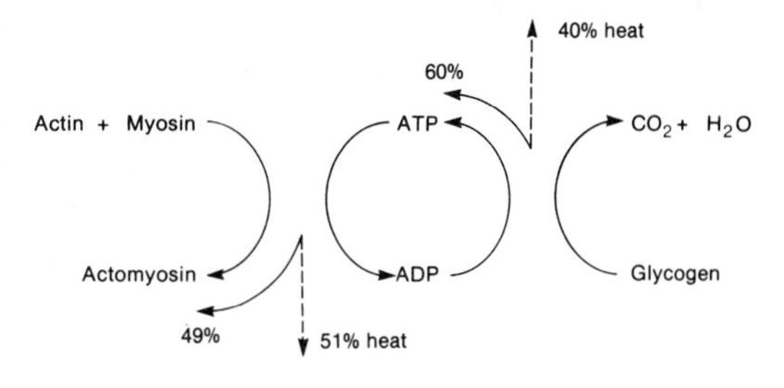

FIG. 1.8. Efficiency of the energy-transducing system under aerobic conditions.

*If creatine phosphate forms an intermediate step in the process, the transfer of energy from CP to ATP has an efficiency of 85%. The efficiency of actomyosin formation is then (49) 85/100 = 41.7%, and the overall efficiency is (29.5) 85/100 = 25.2%.

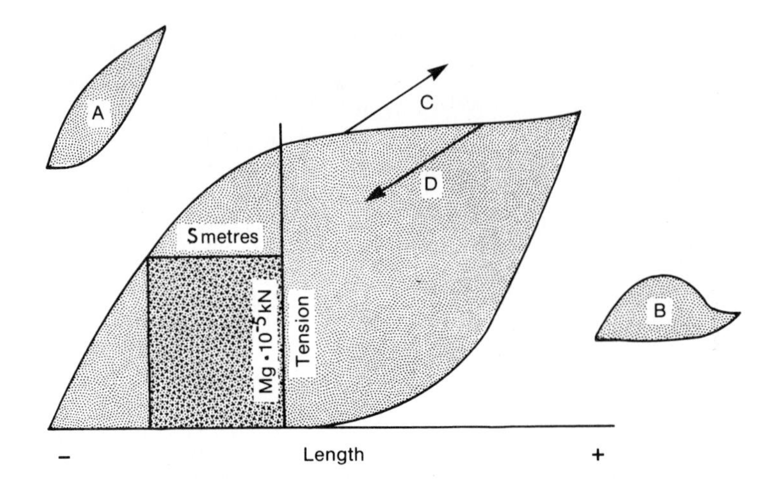

FIG. 1.9. Schematic length/tension diagram for a muscle. The total potential of the muscle for work is indicated by the area of the diagram (light shading). In order to lift a mass M grams through S meters, a tension Mg • 10^{-8} kN is developed (where g is the acceleration due to gravity). A shortening of S meters then occurs. The work performed (Mg • S • 10^{-8} kN-m) is indicated by the darker-shaded portion of the length/tension diagram. Inserts A and B illustrate changes in the shape of the length/tension diagram in muscles with much and little related connective tissue, respectively. Line C illustrates the response to combined lengthening and maximum contraction (eccentric activity). Line D illustrates the response to combined forcible shortening and maximum contraction (concentric activity).

rhythmic activities, such as cycling). Starting from normal relaxation, the possible extent of shortening is greatest at low tensions, but the area of the length-tension diagram and thus the work that can be performed is greatest if shortening is initiated after development of ~60% of the maximum isometric tension.

The darker-shaded area of Fig. 1.9 indicates the simplest possible pattern of behavior during the lifting of a heavy object such as a suitcase. Given a mass of M kg, a tension equal to 9.81M N is first developed by the arm muscles. Shortening then occurs through a distance of s meters, the total work performed being 9.81Ms N • m. In actual practice, the response of the body is more complicated because both the necessary tension and the associated muscle shortening are influenced by changing leverage as the movement develops.

Heart

Actin/myosin combination occurs in cardiac muscle much as in skeletal muscle. However, the tension developed is converted to a hydraulic pressure within the ventricular cavities. The relationship between tension and pressure is given by the law of Laplace:

$$P'_v = T' \frac{1}{R'_1} + \frac{1}{R'_2} \qquad (1.17)$$

where P'_v is the ventricular pressure, T′ is the tension in the ventricular wall, and R'_1 and R'_2 are the principal radii of the ventricle.* For example, if the ventricular pressure is 100 mmHg (≈ 130 g • cm^{-2}) and the two radii are each 5 cm, then 130 = T′(2/5), and T′ = 325 g • cm^{-1}. Much of the work of the heart is normally performed against this wall tension.

The external work performed by the heart is equal to the product of pressure and volume change [equation (1.6)]. The maximum work per beat available for pumping blood around the circulation is thus indicated by the pressure-volume diagram of ventricular muscle (Fig. 1.10). As with skeletal muscle, a portion of the diagram usually remains unexploited because the ventricular chambers are not completely emptied of blood when the heart contracts. If the end-diastolic volume is 160 ml, the end-systolic volume is 80 ml, and the average pressure during systole is 15 kPa (15 kN • m^{-2}), the work performed per beat is 80 × 10^{-6} m^3 × 15 kN • m^{-2}, or 1.2 N • m. Given a heart rate of 100 beats per min, the power output is 120 N • m • min^{-1}.

Uncoupling

The normal objective of the body is to transform chemical energy into external or internal work with

*Within the major blood vessels, R'_2 approximates to infinity and the formula reduces to $P' = T'/R'_1$.

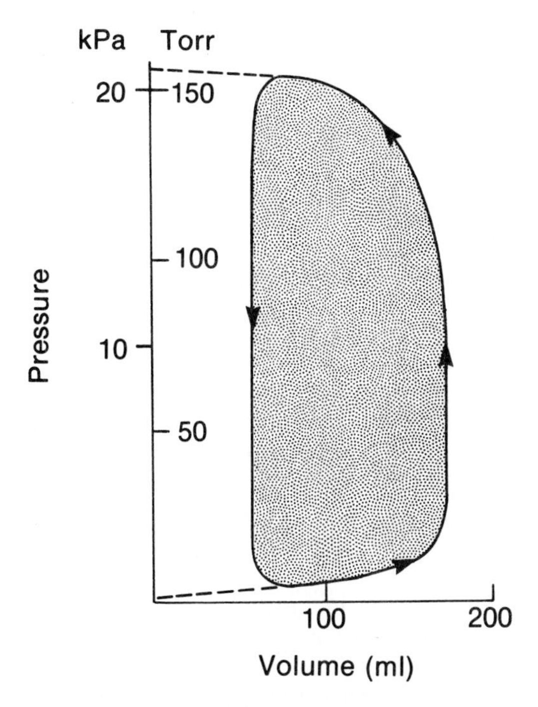

FIG. 1.10. Pressure–volume diagram for left ventricle (schematic). The shaded area indicates the work that can be performed during vigorous effort. Note that the heart is not completely emptied at systole—a systolic reserve of 70 to 80 ml is retained within the ventricular cavity.

the maximum possible efficiency. In this connection, heat production for a given work rate is reduced if the muscle is kept under tension (eccentric work, Fig. 1.9, Bigland-Ritchie and Woods, 1976). However, on occasion, homeostasis is challenged by a falling body temperature. It is then useful to minimize efficiency, thereby increasing the output of body heat.

There are three short-term mechanisms that can increase heat production:

1. Augmentation of voluntary physical activity (for example, the commuter pacing up and down at a cold bus stop).
2. A general increase of muscle "tone" (increased electrical activity and a rise of tension in "resting" muscle fibers).
3. Sudden involuntary contractions of trunk and limb muscles (shivering). An almost simultaneous contraction of agonistic muscles and their antagonists causes the limbs to oscillate without producing purposeful movement. Violent shivering can increase resting energy expenditure by a factor of 5.

A longer term expedient is to increase the production of thyroid hormone. This compound stimulates protein turnover and thus resting heat production. In large doses it also "uncouples" the phosphagen/carbohydrate reaction, so that less food energy is transferred to high-energy phosphate bonds and more energy appears as heat. There is some evidence that thyroid secretion is increased when men are exposed to a cold environment for several months (Itoh, 1974; J. LeBlanc, 1975).

EFFICIENCY AND EXTERNAL WORK

Definition of Efficiency

The mechanical efficiency ϵ of a system is given by the ratio of work output to energy expended:

$$\epsilon = \frac{\text{work output}}{\text{energy expended}} \tag{1.18}$$

In biochemical terms, efficiency is equal to work output divided by free energy change:

$$\epsilon = \frac{\text{work output}}{-\Delta G'} \tag{1.19}$$

From equation (1.3), the residue of the free energy change ($-\Delta G' - W'$) is converted to heat, with the creation of additional entropy $\Delta S'$:

$$\Delta S' = \frac{-\Delta G' - W'}{T'} \tag{1.20}$$

The exercise physiologist usually distinguishes gross efficiency [equation (1.18)] from net efficiency (where an allowance has been made for basal energy expenditure):

$$\text{Net } \epsilon = \frac{\text{work output}}{\text{total} - \text{basal energy expenditure}} \tag{1.21}$$

Because of practical difficulties in measuring the basal energy expenditure, a resting value is often substituted; if the total energy expenditure is large, this does not change the net efficiency greatly:

$$\text{Net } \epsilon' = \frac{\text{work output}}{\text{total} - \text{resting energy expenditure}} \tag{1.22}$$

For example, a 70-kg man climbs a 50-cm bench 20 times per min. The work performed = force × distance = 687 N • 10 m = 6.87 kN • m or 6.87 kJ • min^{-1}. The oxygen consumption is 2.3 l • min^{-1}, and since 1 l of oxygen liberates approximately 20.95 kJ of energy, the total cost of the work is 48.2 kJ • min^{-1}. The gross efficiency is thus 6.87/48.2 = 14.3%. The basal energy expenditure is approximately 2.8 kJ • m^2 of body surface min^{-1}, or in a man of 1.8 m^2 surface area 5.0 kJ • min^{-1}. The net cost is thus 48.2 − 5.0 = 43.2 kJ • min^{-1} and the net efficiency is 6.87/43.2 = 15.9%. The resting energy expenditure is ~6.0 kJ • min^{-1}, so that the net efficiency ϵ' is given by 6.87/(48.2 − 6.0) = 16.3%.

Even the resting energy expenditure can be quite variable in an anxious subject. There have thus been suggestions that efficiency should be calculated as the ratio of the increase in work output over the corresponding increase of energy expenditure, both measured relative to data for light activity:

$$\text{Net } \epsilon'' = \frac{\Delta \text{ work output}}{\Delta \text{ energy expenditure}} \qquad (1.23)$$

A final possibility is to plot work output W′ against energy expenditure E′ (Fig. 1.11); the slope of the line then indicates an efficiency ϵ''' and the intercept on the abscissa is the resting energy expenditure E'_r:

$$W' = \epsilon'''(E') + E'_r \qquad (1.24)$$

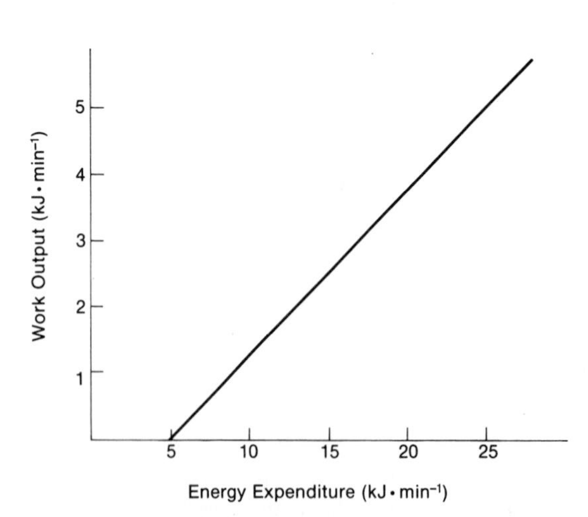

FIG. 1.11. Relationship between total energy expenditure and work performed. The slope of the line is one measure of mechanical efficiency, whereas the intercept on the abscissa corresponds to the resting energy expenditure.

In the specific example of Fig. 1.11, a Δ work output of 5 kJ • min^{-1} requires a Δ energy expenditure of 20 kJ • min^{-1}. Net ϵ''' thus equals 5/20, or 25%. Equation (1.24) may be written as follows:

$$W' = 0.25\,(E') + 5$$

so that

$$\epsilon''' = 25\% \text{ and } E'_r = 5 \text{ kJ} \bullet \text{min}^{-1}$$

Occasionally, the several methods of calculating efficiency lead to differing conclusions about the optimal operating conditions for a machine. The relationship of bicycle ergometer loading to the several indices of efficiency is illustrated in Fig. 1.12.

Cellular Efficiency

We have already noted that the efficiency with which a person can operate a machine is ultimately limited by the efficiency of the cellular energy transducer. Under aerobic conditions (Whipp and Wasserman, 1969; Gaesser and Brooks, 1975) values as high as 29.5% are possible, and 25.2% is usual (Fig. 1.8). With anerobic effort, a lower efficiency is inevitable (Christensen and Högberg, 1950b; Agnevik et al., 1969); even after resynthesis of glycogen stores, the

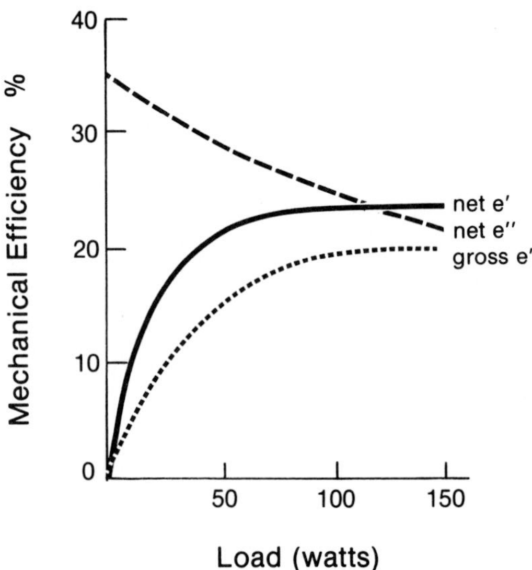

FIG. 1.12. The influence of bicycle ergometer loading on several estimates of mechanical efficiency. Gross e', gross efficiency; net e', net efficiency; net e", Δ work output/Δ energy expenditure. (Based on data of C. M. Hesser et al., 1977.)

overall efficiency cannot be better than 24.0% and figures of 13.8–20.4% are more usual (Fig. 1.7).

Organ Efficiency

The efficiency of the muscle as a whole is necessarily a little poorer than that of the cellular transducer. Sources of energy loss in the whole organ include (1) viscous work performed as the muscle fibers slide over one another, (2) energy lost in series and parallel elastic elements, and (3) energy lost through angulation of the muscle fibers.

Viscous Work

Hill (1938) described a hyperbolic relationship between the speed of muscle shortening V′ and its loading:

$$V' = \frac{(P'_0 - P')b}{P' + a} \qquad (1.25)$$

Here P'_0 is the maximum isometric tension, P′ the observed tension, a is a constant with the dimensions of force, and b is a constant with the dimensions of velocity. Useful external work is proportional to the product P′V′, whereas wasted internal work is proportional to a V′. Further, since (P′ + a)V′ decreases in proportion to the applied load $P'_0 - P'$, the constant b describes the relationship between work performance and external loading.

We can see intuitively that if the contraction is performed against a rigid isometric lever, V′ = 0 and the observed tension P′ becomes equal to the maximum isometric force. Conversely, under lightly loaded conditions, P′ is almost zero, and a large proportion of the work performed is due to component aV′ of equation (1.25) (internal work). The maximum rate of external working is achieved between these extremes, at about 30% of maximum isometric tension. Well-designed machines (such as the bicycle) demand forces of this order.

F. D. Carlson and Wilkie (1974) presented a more formal analysis of the equations of motion for a muscle. If F′ is the applied force, M′ the mass, and s′ the speed of displacement of an external load, then the tension P′ is given by

$$P' = F' + M' \frac{\partial s'}{\partial t} \qquad (1.26)$$

Further, if c_i and c_e are compliances* associated with internal and external elastic elements, then the speed of shortening (V′) can be deduced as

$$V' = s' + (c_i + c_e)\frac{\partial P'}{\partial t} \qquad (1.27)$$

Under isometric conditions, the muscle shortens by no more than 3%. Thus, for practical purposes, s′ = 0, P′ = F′, and $V' = (c_i + c_e)(\partial P'/\partial t)$.

Elastic Work

The contractile elements of a muscle shorten against the compliance of internal parallel and series elastic elements and external series elements (Fig. 1.13). If the structures concerned were perfectly elastic, the area of the length/tension diagram (Fig. 1.9) would be independent of the elastic content of the tissue, and there would be a possibility that energy expended against internal or external elastic elements could later be recovered as useful external work. Some recovery of energy is observed in the chest, where a portion of the elastic work performed during inspiration can be applied to the work of expiration. However, in practice the length-tension curve for the relaxation of elastic tissue does not exactly match that for its extension; in other words, there is some

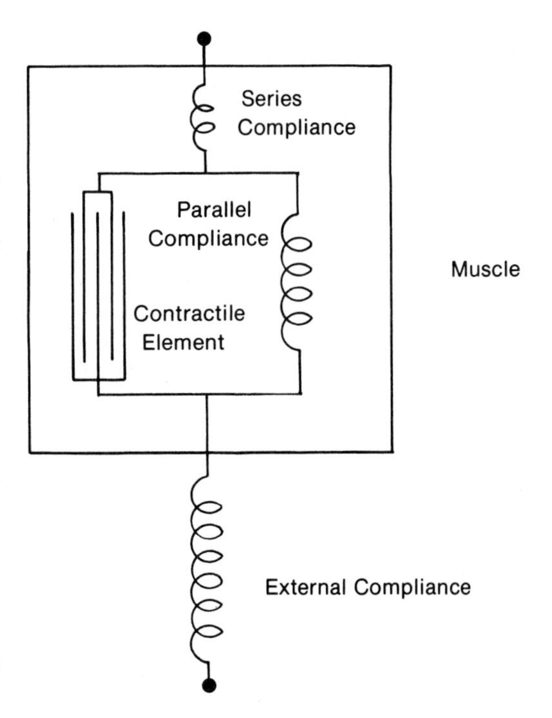

FIG. 1.13. Schematic diagram of muscle to show series and parallel elastic elements (compliances).

*Compliance is the reciprocal of elasticity.

hysteresis, and consequently a part of the elastic energy is dissipated as heat. Further, in many circumstances the direction and/or the magnitude of elastic forces are such that not all of the stored energy can be applied to external work when the muscle relaxes. For example, quite vigorous breathing is needed to make maximum use of elastic energy accumulated during expansion of the chest.

Fiber Angulation

Not all muscle fibers are aligned exactly with the direction of pull of the corresponding tendon. Consider a bipennate arrangement (Fig. 1.14), whereby fibers are inserted at an angle on both sides of a tendon. The force F′ developed by a given fiber can be resolved into a component $F' \cdot (\cos \theta)$ along the line of pull and a second component $F' \cdot (\sin \theta)$ at right angles to this. The second component is counteracted by other fibers that develop an opposing force on the opposite side of the tendon; work is performed against the elastic elements, and this appears as heat.

Whole-Body Efficiency

There is a further deterioration of efficiency when calculations are based on the energy balance for the entire body, because a substantial part of muscular effort is applied to factors other than the performance of useful external work. Energy is expended to stabilize and support the body during almost every type of activity. Costs are also incurred through increased activity of the heart and chest muscles. Other more specific sources of energy loss may be illustrated for the tasks of stepping, riding a bicycle ergometer, and running on a treadmill.

Stepping

During stepping, useful work is calculated as an accumulation of potential energy; thus, if a subject of mass M′ kg climbs a total vertical height of H′, the work is $9.81M'H'$ N • m. On an escalator (M. Richardson, 1966) or a ladder mill (Kamon and Pandolf, 1972), the subject continues moving in an upward direction throughout the test period. However, in the more usual form of laboratory step test, an equal number of descents are made, and an additional amount of work (up to a third of the cost of ascent) (Nagle et al., 1965) is required to control the descent of the body. If the rate of stepping is very slow (less than 50 paces per min), further energy is lost holding the limb poised and awaiting the beat of the metronome (Shephard and Olbrecht, 1970).

Bicycle Ergometer Work

Useful work on a bicycle ergometer is usually measured in terms of the friction imposed by a belted flywheel or the electrical energy generated in a dynamo. G. R. Cumming and Alexander (1968) pointed out sizable unmeasured energy losses in the friction of pedal bearings and chain drive (typically ~8% of the total work performed). Allowance can be made for such losses if the ergometer is calibrated by coupling a torque generator to the pedals rather than the flywheel. Some energy is also dissipated in motion of the legs, as their centers of mass are raised, lowered, accelerated, and decelerated. Finally, unmeasured work is performed in stabilizing the shoulder girdle by gripping the handlebars.

Treadmill Running

Margaria (1971) made a detailed analysis of the energy balance during walking and running. While people generally find it useful to walk or run on level ground, by convention work is defined as the accumulation of potential energy by moving uphill ($9.81M'H'$ N • m). When walking or running on the level, there is no gain of potential energy, and efficiency is thus zero. Nevertheless, work is continually being performed in raising and lowering the center of mass of the body and in accelerating and decelerating the limbs.

If the treadmill is adjusted so that the subject must run uphill, efficiency rises progressively with the slope because more potential energy is generated. On the other hand, the calculated efficiency is negative for downhill running. Potential energy is lost from the system, and at the same time the subject must perform work to control the rate of descent.

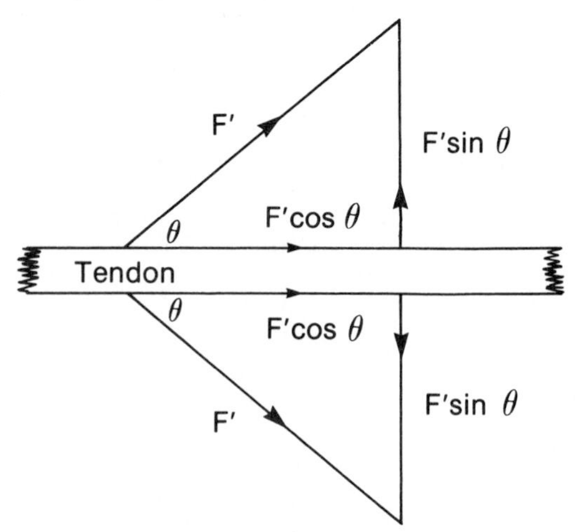

FIG. 1.14. Loss of efficiency with angular insertion of muscle fibers into tendon. The force F′ developed by a given fiber must be resolved into a useful component $F' \cos \theta$ in the line of motion, and a wasted component $F' \sin \theta$ at right angles to the line of motion.

CHAPTER 2
Resources of Energy, Fluid, and Minerals

Having sketched briefly the energy balance of the body, it is now necessary to examine energy resources in more detail, considering limitations of performance imposed by the magnitude of food, fluid, and mineral stores within active tissues, the rate of transfer of nutrients from other body depots, and the rate of ingestion of further food supplies.

PHOSPHAGEN

Energy Yield

The heat arising from the direct combustion of a few molecules of glucose would be sufficient to destroy a living cell. The high-energy phosphate compounds such as adenosine triphosphate (ATP) and creatine phosphate (CP) thus serve two important functions. Not only do they provide a small immediate reserve of energy, but they also allow a controlled release of the larger energy reserves available in ingested food.

Early studies indicated an energy yield of ~ 50 kJ for the hydrolysis of both ATP (Meyerhof and Lohmann, 1932) and CP (Meyerhof and Lohmann, 1928). If the reactants are present in molar concentrations, the free energy of the reaction [$-\Delta G'$ of equation (1.2)] lies between 29 and 33 kJ • mol^{-1} (Morales et al., 1955; George and Rutman, 1960; Karlson, 1966; Banister, 1971). However, under physiological conditions, concentrations are several orders smaller than this, and simultaneous measurements of work, heat production, and CP depletion have set the free energy of the phosphate bond within the active muscle fibers at 42–50 kJ • mol^{-1} (F. D. Carlson and Wilkie, 1974, Gower and Kretzschmar, 1976).

Phosphagen Equilibrium

Lohmann (1934) first demonstrated that in muscle extracts ATP and CP were held in equilibrium by an enzyme now known as creatine phosphotransferase. Equation 1.12 may be written in the form

$$\frac{[ATP]}{[ADP]} = K \frac{[CP]}{[C]} \qquad (2.1)$$

where K is the equilibrium constant for the reaction, varying from 20 to 100. The implication is that CP stores are depleted much more rapidly than ATP. In practive (Fig. 2.1), the reaction summarized by equation (2.1) involves not only the fully ionized forms of the high-energy phosphate compounds, but also magnesium complexes (chelates) and undissociated acids (Kuby and Noltmann, 1962). The relative proportions of the several reactants is thus

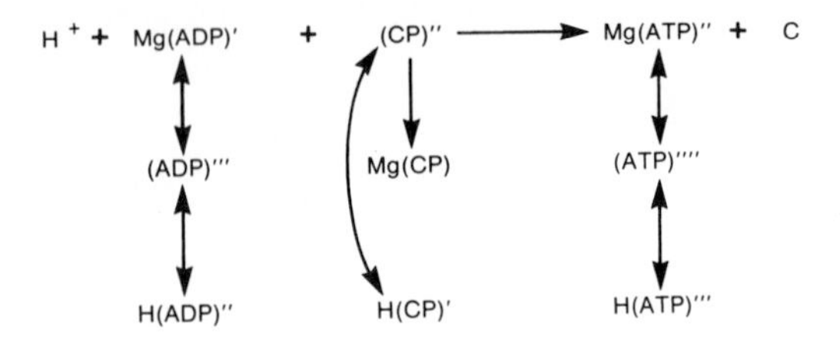

FIG. 2.1. Details of regeneration of adenosine triphosphate (ATP) from adenosine diphosphate (ADP) and phosphocreatine (CP). (After McGilvery, 1975.)

influenced by tissue levels of free magnesium and hydrogen ions, offering potential "feedback" mechanisms for the control of phosphagen breakdown. Under resting conditions, the CP/P ratio is ~2 and the ATP/ADP ratio is ~8 (Bergström, Harris et al., 1971). The latter ratio is less than would have been predicted from in vitro calculations, because a substantial part of the ADP measured in vivo is bound to the proteins actin (Perry, 1952) and myosin (Marston and Tregear, 1972).

Adenosine Triphosphate (ATP)

The ATP content of resting muscle has been estimated from biopsy specimens (Bergström, 1967; Hultman et al., 1967). The observed result (6.9 mmol • kg^{-1}) could be biased downward by difficulty in arresting tissue reactions after biopsy (C. Gilbert et al., 1971). Similar values have been derived from the maximum speed of ascent of a staircase (Margaria et al., 1964), although McGilvery (1975) calculated a somewhat larger store from the height of a standing jump. Taking the 6.9-mmol figure, and assuming a muscle mass of 28 kg, the total ATP content of the muscle would be 193 mmol; at 46 kJ • mol^{-1}, this would represent a reserve of 8.9 kJ of energy.

The reaction of equation (2.1) leads to a rapid regeneration of ATP during physical activity. Application of the equilibrium constants of Kuby and Noltmann (1962) suggests that ADP and AMP do not accumulate until phosphagen reserves are almost exhausted (Fig. 2.2).* However, it is possible to demonstrate that ATP hydrolysis is the first step in energy release if the enzyme creatine phosphotransferase is inhibited by a large dose of 1-fluor-2,4-dinitrobenzene (FDNB). Contraction subsequently depends on the breakdown of ATP (Cain et al., 1962). As an emergency measure, accumulating AMP is removed by deamination to IMP (inosine monophosphate), thus helping forward the regeneration of ATP from ADP [equation (1.11)]. Physical activity ceases when the muscle ATP content has dropped to about 1 mmol • kg^{-1}, suggesting that there is some compartmentalization of ATP within the active fibers. Certainly, it is localized to specific sites (the isotropic I bands; see Chap. 4).

Muscle biopsy data do not agree completely with the predictions of McGilvery (1975) (Fig. 2.2). ATP levels show some fall with moderate work (Hultman et al., 1967; Karlsson et al., 1971), and in maximum effort as much as 40% of muscle ATP may be utilized (Gollnick and Hermansen, 1973). It is not yet clear as to how far these discrepancies are attributable to compartmentalization of ATP and CP in vivo, and how far they reflect breakdown of ATP subsequent to biopsy.

*Newsholme (1977) suggests that one reason ADP concentrations are kept low is that it acts as a competitive inhibitor in many of the reactions for which ATP is a substrate.

Creatine Phosphate

McGilvery (1975) set the CP content of resting skeletal muscle at 20 mmol • kg^{-1}. Other authors generally agree with these estimates (Lohmann, 1937; Hohorst et al., 1962; Danforth, 1965), although they describe differences between muscle fibers. Specifically, "fast-twitch" white fibers (with a high glycolytic capacity) contain more CP than the "slow-twitch" red fibers (Kirsten et al., 1966).

Assuming a muscle mass of 28 kg, the total CP content of muscle would be 560 mmol. At 46 kJ • mol^{-1}, this represents a reserve of 25.8 kJ of energy. Knuttgen and Saltin (1972) observed little depletion of CP when subjects were working at less than 60%

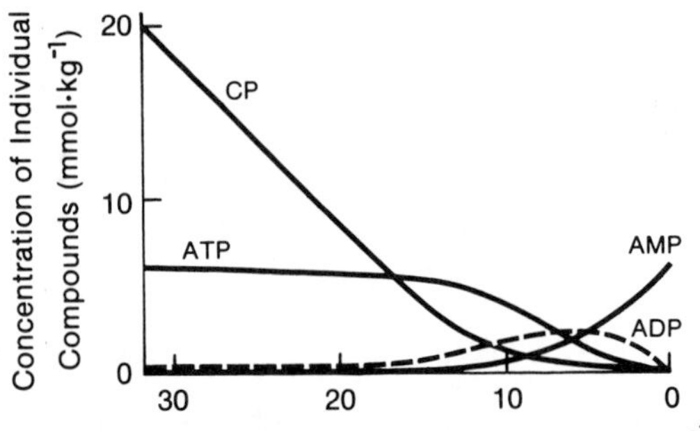

FIG. 2.2. Concentration of individual high-energy phosphate compounds in relation to total store of phosphagen. (Based on calculations of McGilvery, 1975, for human muscle.)

of their maximum oxygen intake. Under such circumstances, energy demands are presumably satisfied by aerobic metabolism. A small increase of effort (to 75% of maximum oxygen intake) gave 80% depletion of CP, but Knuttgen and Saltin found little further usage of CP at still higher rates of working. In contrast, Hultman et al. (1967) reported CP levels <3% of resting figures at exhaustion.

Phosphagen Capacity and Power

Because of the complex interactions between the various high-energy phosphorus compounds (Fig. 2.1) and puzzling discrepancies between theoretical equilibria and in vivo data, it is convenient to consider jointly the overall "capacity" (energy store) and power of the ATP and CP reactions.

Capacity

It is possible to use at least 30 of the 32 mmol • kg^{-1} of stored phosphagen. With a muscle mass of 28 kg, the total reserve amounts to 840 mmol, or 38.7 kJ of energy. On the assumption that 1 l of oxygen liberates 21 kJ, this is equivalent to 1.85 l of oxygen. However, we must qualify this calculation in two respects: (1) in most vigorous activities the mass of active muscle is <20 kg (27.6 kJ, 1.32 l of oxygen) and (2) the energy yielded by phosphagen is used more efficiently than that derived from aerobic metabolism.

Power

In very brief maximal activity, the crucial factor is not the size of the phosphagen store, but rather the power of the reaction, the rate at which free energy can be transferred from CP to ATP and thus the actin/myosin interaction. There is a brief initial lag associated with enzyme activation (a half-time of perhaps 0.5 sec; McGilvery, 1975) and thereafter the determinant is the maximum velocity of the creatine phosphotransferase reaction. McGilvery (1975) calculated that in a vertical jump a power of 6 mmol • kg^{-1} • sec^{-1} was sustained for 0.5 sec.

Margaria and his associates (1964) had subjects run up a short staircase, timing the vertical velocity for the period 2–5 sec* (Fig. 2.3); they

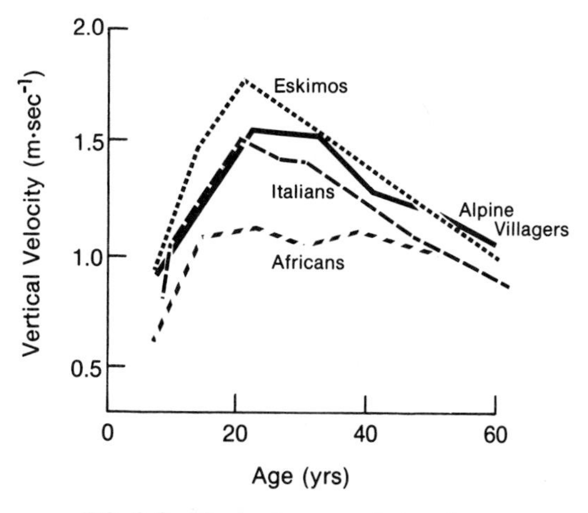

FIG. 2.3. Maximal power of phosphagen reaction, as assessed by velocity of ascent of a staircase. If a mechanical efficiency of 25% is assumed, the peak rate of energy usage in the best performed (a 20-year-old Eskimo) is some 4.7 kJ • sec^{-1}. On the less conventional (but more reasonable) assumption of cumulative phosphagen depletion, efficiency should be 41.7% (Fig. 1.7), and the energy expenditure of the Eskimo should be 2.8 kJ • sec^{-1}. (Based on data of di Prampero, 1971a.)

estimated a maximum power of ~3 mmol • kg^{-1} • sec^{-1} (di Prampero, 1971a,b). Given 28 kg of muscle, this would be equivalent to an energy usage of 3.87 kJ • sec^{-1} (Fig. 2.3).† A more reasonable ceiling, based upon the observed rate of working and an efficiency of 41.7%, is 2.8 kJ • sec^{-1}. If 1 l of oxygen liberates 21 kJ, and this is used with 25% efficiency, a phosphagen breakdown of 2.8 kJ • sec^{-1} would be equivalent to a steady oxygen transport of ~11.1 l • min^{-1} (158 ml • kg^{-1} • min^{-1} in a standard 70-kg man). If there has been preceding vigorous exercise, the maximum vertical velocity is reduced. The loss of speed amounts to ~15% with a previous net oxygen consumption of 40 ml • kg^{-1} • min^{-1} (1.78 mmol • kg^{-1} • min^{-1}).

Given a maximum rate of phosphagen depletion of 2–6 mmol • kg^{-1} • sec^{-1}, it follows that body stores can be exhausted by 5–15 sec of maximal

*They assumed there was no lactate formation until phosphagen reserves were exhausted. Karlsson (1971b) criticized this concept on the basis of muscle biopsy studies. Hermansen (1971) also deduced from blood lactate measurements that there was substantial lactate production in the first 10 sec of exercise. Unfortunately, there is a possibility that phosphagen/lactate exchange occurs between the end of exercise and the completion of muscle biopsy or blood sampling. For this reason the issue is still not finally resolved.

†di Prampero and his associates apparently assumed an efficiency of 25%, more appropriate to sprint plus recovery. They thus overestimated the power of the phosphagen reaction from measurements of the speed of ascent of the staircase. However, the useful work (and thus the oxygen equivalent) was measured correctly.

work. The half-time of the recuperative process has been estimated at 22 sec (di Prampero, 1971a). This figure is based on a partitioning of the oxygen debt and also includes time needed for the replenishment of tissue oxygen stores (Table 2.1).

Control of the Reaction

The enzyme responsible for the hydrolysis of ATP and initiation of muscular contraction (myosin ATPase) seems to be one part of the myosin molecule (heavy meromyosin). Under resting conditions, its enzymic properties are inhibited by divalent cations (calcium and magnesium). The high rate of ATP breakdown during exhausting work reflects a reversible binding of the inhibiting cations to another muscle protein, troponin.

The simplest description of myosin ATPase activity would be a reaction with myosin (M′) of the type

$$M' + ATP \underset{}{\overset{k_1}{\rightleftharpoons}} M' \cdot ATP \rightarrow M' + ADP + P' \tag{2.2}$$

However, the reaction does not conform to the first-order kinetics predicted for such a system (Fig. 2.4). There is an early burst of rapid hydrolysis rather than a steady linear breakdown with time. The currently postulated sequence (E. W. Taylor, 1972) thus involves the formation of intermediate com-

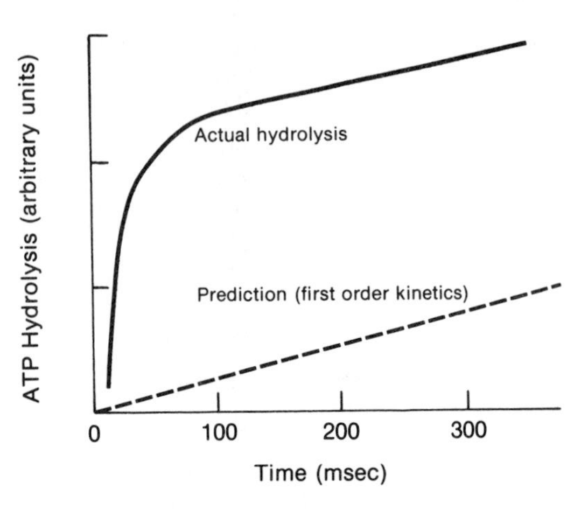

FIG. 2.4. A comparison of the observed hydrolysis of actomyosin/ATP with predictions for a simple first order reaction (after F. D. Carlson & Wilkie, 1974). The data of di Prampero (1971b) also implied departure from a first order reaction, in that the speed of a staircase sprint (Fig. 2.3) was almost constant from the second to the sixth sec.

TABLE 2.1
Components of the Oxygen Debt in a Sedentary Young Man

Component	Tissue Weight	Change of Concentration	Contribution to Oxygen Debt
Lactate debt			
Muscle	20 kg	40 mmol • l⁻¹	800 mmol = 5.38 l*
Alactate debt			
Depletion of oxygen stores			
Venous blood	4 kg	−100 ml • l⁻¹	0.40 l
Myoglobin	20 kg	−10 ml • l⁻¹	0.20 l†
Tissue fluids	38 kg	−0.5 ml • l⁻¹	0.02 l
Phosphagen	20 kg	30 mmol • l⁻¹	600 mmol = 2.18 l‡
Total			8.18 l

Note: It is assumed that exhausting exercise is performed by 20 kg of body muscles.

*The oxygen *debt* is larger than the oxygen equivalent of the corresponding anerobic energy release since resynthesis of glycogen and phosphagen is not 100% efficient. For the purpose of this calculation we have accepted the premise of di Prampero (1971a) that oxidation of 10% of the accumulated lactate provides energy to reconvert the remaining 90% to glycogen. On this basis, 1 mol of lactic acid is equivalent to ~6.73 l of oxygen. Likewise, resynthesis of phosphagen from glycogen has an efficiency of ~60% (Fig. 1.8); thus, 1 mol of ATP usage is equivalent to 1/37th mol of glucose-6-phosphate and an oxygen debt of ~3.63 l. The size of the oxygen debt naturally varies somewhat according with the consumptions that are made.

†P. O. Åstrand and Rodahl (1977) quote a myoglobin store as large as 0.43 l.

‡Karlsson (1971b) estimated an oxygen deficit of 1.3 l on the basis of muscle biopsy studies; this would require a repayment of 2.14 l.

plexes with both myosin (M′) and actomyosin (AM′):

$$\begin{array}{ccccc} AM'+ATP \rightleftharpoons AM'\cdot ATP & & AM'\cdot ADP\cdot P \rightleftharpoons AM'+ADP+P & & \\ \updownarrow +A & & \updownarrow +A & & (2.3) \\ M'+ATP \rightleftharpoons M'\cdot ATP \rightleftharpoons & & M'\cdot ADP\cdot P \rightleftharpoons M'+ADP+P & & \end{array}$$

Free myosin associates rapidly with ATP to form M′ • ATP, and this in turn dissociates to the relatively stable M′ • ADP • P. Addition of actin reduces the lifetime of M′ • ADP • P from 20 sec to $\sim 10^{-1}$ sec, and the AM′ • ADP • P thus formed is rapidly hydrolyzed to AM′, ADP, and P with concurrent contraction of the muscle.

The reverse reaction, muscle relaxation, involves the release of calcium and magnesium ions from troponin and consequent inhibition of linkages between the actin and myosin molecules (Chap. 4). Replenishment of ATP reserves is regulated by the reaction of equation (2.1). The equilibrium constant K and thus the ratio of ATP to ADP is increased by both a decrease of tissue pH and an increase in the concentration of divalent cations such as magnesium (Kuby and Noltmann, 1962):

	Values of K at 3 Concentrations of Free Mg^{2+} (mmol • l^{-1})		
pH	**0.2**	**0.5**	**1.0**
7.40	13.5	21.2	29.4
7.00	30.1	45.9	54.2
6.60	63.6	90.6	125.8

LACTATE FORMATION

In the absence of oxygen, CP is regenerated by anerobic pathways (Fig. 1.2) using energy derived from the conversion of glycogen and glucose to lactate.

Capacity and Power

The capacity and power of this reaction have classically been studied in terms of blood lactate concentrations. Originally, venous samples were used (Margaria et al., 1933), but more recently arterial blood or arterialized capillary specimens have been preferred (Mohme-Lundholm et al., 1965; Shephard, Allen et al., 1968a,b). If work is being performed with the legs, it is plainly unsatisfactory to collect blood from the antecubital vein of the forearm; there is a substantial circulation time from artery to vein, and there is also likely to be a loss of lactate into the tissues of the hand during circulation.

Even arterial samples often show quite low concentrations relative to intramuscular conditions. If an average young man performs a single brief bout of exhausting exercise, the arterial lactate concentration reaches a peak of 11–13 mmol • l^{-1} 1–2 min after ceasing exercise (Shephard, Allen et al., 1968a). Somewhat higher concentrations are found in trained distance runners. However, a blood level of some 32 mmol • l^{-1} was observed in an experiment in which five short bursts of treadmill running were spaced over 20 min (Hermansen, 1971); presumably, this protocol allowed better equilibration of lactate between muscle and blood. Intramuscular lactate concentrations of 20–25 mmol • kg^{-1} are usual during exhausting bicycle ergometry (Karlsson, 1971a,b; Saltin and Essén, 1971), and values as high as 39 mmol • kg^{-1} have been developed in bursts of exercise lasting <1 min. During treadmill exercise it seems likely that concentrations can reach 40 mmol • kg^{-1}, with an intracellular pH of 6.5 or less (Bergström et al., 1971).

C. L. Evans (1930) suggested that blood and muscle concentrations of lactate were approximately the same within 30 sec of a short burst of muscle contraction, but more recent research (Diamant et al., 1968) indicates a much slower equilibration, over 5–10 min.* The dynamics of the exchange are illustrated in Fig. 2.5. The concentration gradient between active muscle and blood is more than 3 times the arteriovenous lactate gradient across the working muscles. The effective conductance of lactate across the muscle membrane must thus be about a third of that imposed by circulatory removal of the metabolite. Given a muscle perfusion of 20 l • min^{-1}, the membrane conductance would be 6 l • min^{-1}, and the initial lactate flux for a muscle/blood gradient of 30 mmol • l^{-1} would be 180 mmol • min^{-1}. Benade and Heisler (1978) commented on a 14- to 50-fold difference in the rate of transport of hydrogen ions and lactate across the muscle membrane.

If 20 kg of muscle are active, the initial lactate content of the muscle amounts to ~ 800 mmol. Assuming the flux is correctly calculated at 180 mmol • min^{-1}, equilibration with blood and tissue water cannot take less than 4–5 min. The free energy yield of 1 mol of lactate is ~ 83.0 kJ (di Prampero, 1971a,b), so that 800 mmol is equivalent to 66 kJ, or an oxygen reserve of about 3.17 l. If fatigue develops over 40 sec, the power of the reaction is 1.65 kJ • sec^{-1}, equivalent to a steady oxygen transport of 4.76 l • min^{-1}.

*Given a "progressive" maximal test format, peak values will be reached sooner than 5 min after maximal exercise.

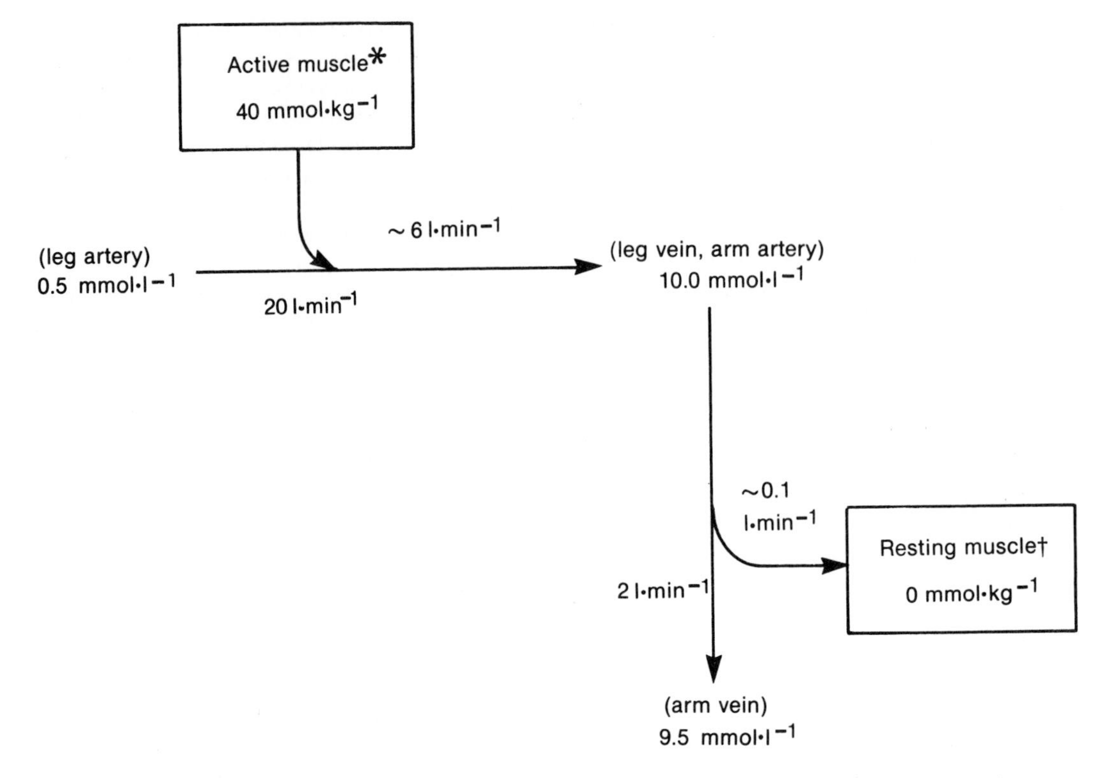

FIG. 2.5. Schematic diagram of lactate flux during exhausting exercise. Transfer of lactate between muscle and blood is indicated as a conductance (reciprocal of resistance). Although most early authors assumed a free passage of lactate between muscle and blood (Knuttgen, 1970), some authors now postulate the need for an active transport mechanism (Kubler et al., 1965; Tranquade, 1966).

*The concentration shown is intracellular. About one-quarter of muscle is extracellular water, so that concentrations in wet biopsy specimens are appreciably lower than this figure (Karlsson, 1971b).

†Low concentrations of lactate (~1.4 mmol • kg⁻¹) may be found in resting muscle (Karlsson, 1971b).

Lactate Formation in Submaximum Work

The overall course of glycogen breakdown in the cytoplasm can be summarized as

$$\frac{1}{n}(C_6H_{10}O_5)_n + 2NAD^+ + 3ADP + 3P_i \rightarrow$$

$$2NADH + 2H^+ + 3ATP + \underset{\text{pyruvic acid}}{2(C_3H_4O_3)} \qquad (2.4)$$

where NAD is the coenzyme or hydrogen carrier nicotinamide dinucleotide. In the oxidized state, NAD exists as the cation NAD^+, while in the reduced state it forms $NADH + H^+$. If oxygen is freely available, the NADH transfers its H^+ and 2 extra electrons to the cytochrome chain of the mitochondrion, while the pyruvic acid also crosses the mitochondrial membrane to enter the Krebs cycle (Fig. 1.3). However, pyruvate is also in equilibrium with lactate [see equation (2.5)]. Thus, in the absence of oxygen, formation of lactate prevents an excessive accumulation of pyruvate and restores the NAD^+ needed to help forward the breakdown of glycogen.

From the foregoing, a large increase of pyruvate could of itself favor the accumulation of lactate. In practice, pyruvate levels are no more than doubled during effort (Karlsson, 1971a,b). A second possible reason for lactate buildup might be that the activity of lactate dehydrogenase was increased by physical effort. However, this explanation can also be dis-

$$\text{Pyruvate} + NADH + H^+ \xrightleftharpoons{\text{lactic dehydrogenase}} \text{lactate} + NAD^+ \qquad (2.5)$$

counted, since lactate production never uses more than a quarter of the potential activity of the enzyme. A buildup of lactate during effort thus reflects an increase of NADH and H^+ secondary to an inadequate supply of oxygen to the mitchondrion.

According to classical theory, the blood lactate did not rise except (1) with time lags in circulatory adaptation at the beginning of exercise and (2) if the steady rate of working exceeded the overall potential of the body for aerobic energy release. Thus, Saiki et al. (1967) found no lactate accumulation during treadmill running at 60–70% of $\dot{V}O_{2(max)}$, but a substantial accumulation at 90–100% of $\dot{V}O_{2(max)}$. Subsequent studies at intermediate workloads (Fig. 2.6) have shown that there is some accumulation of lactate in submaximum effort, particularly when a large proportion of the required effort is sustained by relatively small muscle groups (Asmussen, Von Döbeln, and Nielsen, 1948; Shephard, Allen et al., 1968b). There are at least two reasons for this. First, not all muscle fibers reach 100% of their aerobic power simultaneously. Slowly contracting fibers are recruited preferentially in light work, and such fibers may reach their aerobic limit before the subject has attained his overall maximum oxygen intake. Second, contraction of some muscles is sufficiently intense to cause a local restriction of blood flow before maximum oxygen intake is reached.

With moderate intensities of work (50–85% of maximum oxygen intake), the blood lactate slowly increases for 5–10 min, but then plateaus or returns toward its resting value (Bang, 1936; P. O. Åstrand, Hällback et al., 1963; P. Harris et al., 1968; Fig. 2.7). Various factors contribute to this secondary adjustment. Circulation "catches up" with metabolic demand, a rise of blood pressure gives better perfusion of vigorously contracting muscles, patterns of fiber recruitment change, and peripheral utilization of lactate may increase (Minaire and Forichion, 1975). With heavier exercise (>90% maximum oxygen intake), the buildup of lactate continues until work is stopped by local muscular fatigue (C. Kay and Shephard, 1969). Lactate can also be formed during resynthesis of phosphagen molecules; Cerretelli, Ambrosoli et al. (1975) provided one example of this in an experiment in which subjects undertook a brief burst of exercise at 2.5 times their maximum oxygen intake and continued to run at their maximum oxygen intake during the "recovery" period.

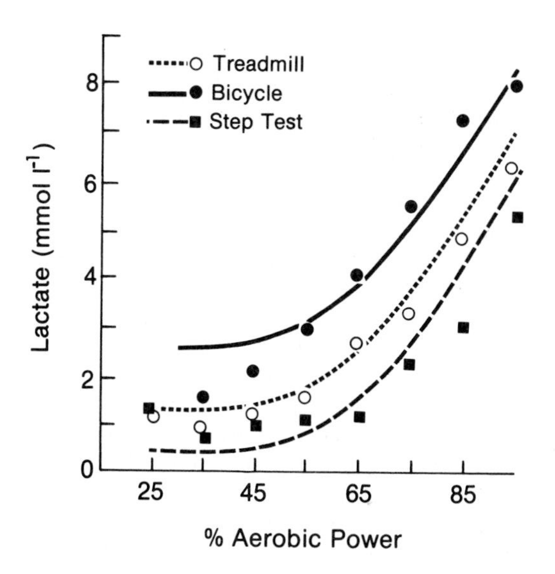

FIG. 2.6. The lactic acid content of arterial blood two minutes after performance of three types of submaximum exercise. (Based on data of Shephard, Allen et al., 1968b; illustration from author's book *Endurance Fitness* [University of Toronto Press, 2nd edition, 1977a] by permission of the publishers.)

Lactate Elimination

Following exhausting exercise, lactic acid is cleared from the blood in a roughly exponential manner. Physiologists have usually required complete rest during the recovery period. The half-time of the process is then ~10–15 min (di Prampero, 1971a,b), although faster in trained than in untrained subjects (G. V. Mann and Garrett, 1978). For the athlete, warm-down activity is more realistic. Hermansen and Stensvold (1972) thus compared lactate clearance rates between a resting recovery and various intensities of warm-down exercise. At rest, the rate of removal of lactic acid from the blood was 2–3 mg • dl^{-1} • min^{-1} or, on the by no means proven assumption of uniform distribution throughout 75% of body water (30 l volume), a total clearance of 0.75 g • min^{-1} (8.3 mmol • min^{-1}). Continued physical activity increased the clearance rate to a maximum of 8 mg • dl^{-1} • min^{-1} at a workload demanding 63% of the maximum oxygen intake, with a slower clearance at higher workloads. Belcastro and Bonen (1975), Weltman et al. (1977), and W. J. Ryan et al. (1979) also found some advantage from moderate exercise during recovery. In the experiments of Bonen and Belcastro (1976), lactate removal was fastest with free jogging at some 32% of maximum oxygen intake; there was also a positive correlation between the

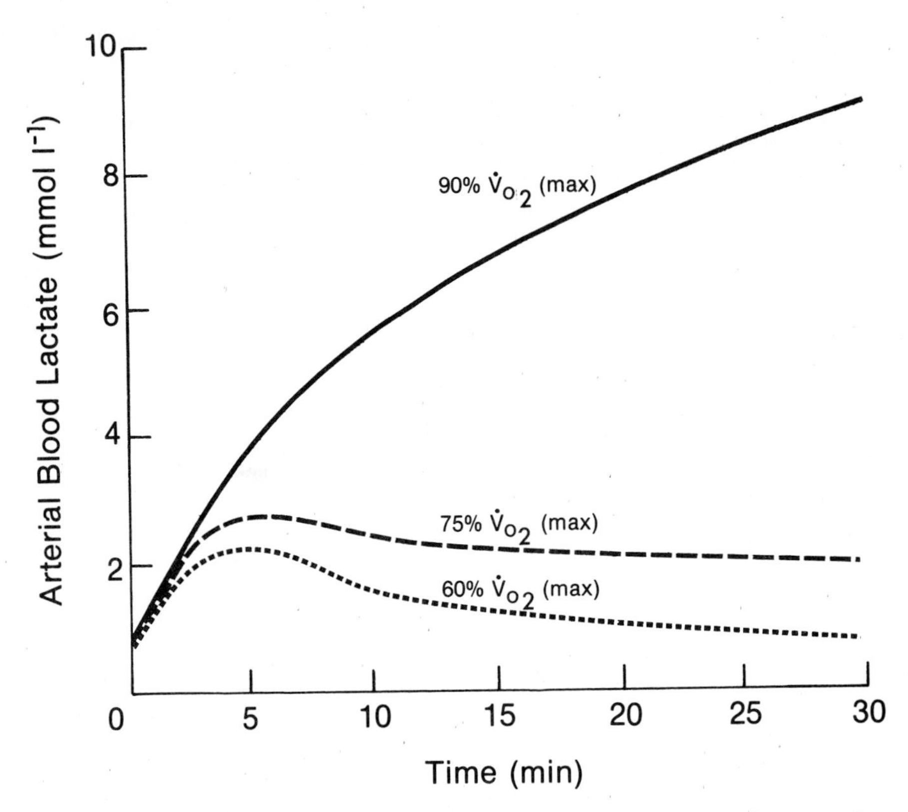

FIG. 2.7. Time course of arterial blood lactate at various intensities of effort, all expressed as a percentage of maximum oxygen intake.

percentage of slow-twitch fibers in the active muscle and the rate of lactate removal (Bonen et al., 1978).

Among possible fates for the lactic acid, we may note reconversion to glucose in the liver;* oxidation by the heart, kidney, and resting skeletal muscle; oxidation by the exercising muscles; and elimination in urine and sweat.

According to classical theory, a large fraction of the lactate was reconverted to glycogen in the liver (and perhaps also the kidneys) (Krebs and Woodford, 1965) using energy derived from oxidation of the remainder. A. V. Hill (1925) and Meyerhof (1920) argued from frog muscle experiments and thermodynamic considerations that 80% was resynthesized. Margaria, Edwards, and Dill (1933) and di Prampero (1971a) compared total body lactate and excess oxygen consumption, and concluded that 90% was resynthesized. There are three major objections to these views. First, biopsy studies suggest that muscle glycogen replenishment is not complete for 1–2 days, while lactate disappears in 0.5–1 hr. Second, maximum hepatic blood flow and hepatic arteriovenous lactate differences (Rowell, Kraning et al., 1966) would account for a removal of only 0.1–0.2 $g \cdot min^{-1}$, 4–8% of the observed rate of lactate clearance. Finally, if [^{14}C] lactate is administered, much more than 10% of the radioactive carbon appears as $^{14}CO_2$ (Drury and Wick, 1965; G. A. Brooks, Brauner, and Cassens, 1973).† In an experiment on dogs, Issekutz, Shaw, and Issekutz (1976) found that with a resting recovery period, 50% of accumulated lactate was oxidized and 18% was converted to glucose. Exercise recovery gave a threefold increase in the rate of lactate clearance, but still 55% was oxidized and 25% converted to glucose. Conversion to other compounds such as glycerol and glycogen is usually less than 20% (Jorfeldt, 1971).

The limiting assumption for the heart, kidneys, and resting muscle is that all of their meta-

*During severe starvation, the kidney also produces some glucose (Owen et al., 1969).

†Cohen and Woods (1976) caution that substantial quantities of labelled CO_2 can appear without any net conversion of lactate into CO_2.

bolic needs, ~6–7 kJ • min^{-1} during the recovery period, are met by oxidation of lactate;* the corresponding clearance, 0.3–0.4 g • min^{-1} (33–44 mmol • min^{-1}), is only about 15% of that observed during recovery with continuing moderate exercise. Excretion of lactate in urine and sweat also accounts for less that 5% of the total (Jervell, 1928; R. E. Johnson and Edwards, 1937; Ström, 1949; Hermansen, Maehlum et al., 1975). It is thus clear that a major part of the lactate is oxidized within the working muscles.

Even during exhausting exercise, not all of the active muscle fibers are contracting at 100% of their aerobic potential. There is thus some possibility for lactate to diffuse from anerobically active to aerobically active fibers during maximal effort. If a high rate of blood flow is sustained by continued activity during the recovery period, the potential lactate conductance of 6 l • min^{-1} (Fig. 2.5) operates in the reverse direction, from the arterial blood into the working muscles.† Given an arterial lactate of 10 mmol • l^{-1}, the total flux could amount to 60 mmol • min^{-1}, faster than the *average* clearance usually observed during the recovery period. R. D. Cohen and Iles (1977) suggest that passage across the cell membrane involves the active transport of lactic acid or lactate and/or a carrier-facilitated diffusion of lactic acid rather than a passive diffusion. This might explain both the relatively slow transfer of lactate and differences in rates for inward and outward transport.

Excess Lactate

Accumulation of lactic acid during exercise is widely assumed to be an indication of a deficient oxygen supply to the mitochondria. However, because of the equilibrium indicated by equation (2.5), a part of any increase in lactate could theoretically reflect an accumulation of pyruvate. Thus, a rise of blood lactate concentration immediately prior to exercise has been attributed to catecholamine secretion and an increase of glycogen breakdown (Nowacki et al., 1969). Likewise, Jöbsis and Stainsby (1968) and Keul, Doll, and Keppler (1972) argued strongly for an accumulation of pyruvate during exercise, although this is not borne out by the measurements of Karlsson (1971a).

*Keul, Doll et al. (1966) cite the example of a well-trained cyclist who met 89% of the metabolic needs of his heart by the oxidation of lactate.

†Not all of the lactate diffusing out of the bloodstream necessarily undergoes immediate metabolism—the muscles may be serving in part as a lactate reservoir.

In an attempt to obtain a fairer measure of cellular oxygen lack, Huckabee (1958a) proposed the calculation of an "excess lactate":

$$\text{Excess L}' = (L'_e - L'_r) - (P'_e - P'_r)\frac{L'_r}{P'_r} \tag{2.6}$$

or, more simply,

$$\text{Excess L}' = (L'_e - P'_e)\frac{L'_r}{P'} \tag{2.7}$$

where L′ is lactate, P′ is pyruvate, and the subscripts e and r refer to exercise and resting conditions, respectively. On the basis of such calculations, Huckabee (1958a) claimed that all of the oxygen debt could be explained by lactate accumulation.

Within a given body compartment, the L′/P′ ratio corresponds closely with the NADH/NAD status (Bücher and Rüssmann, 1963). However, there are several objections to more general application of the excess lactate concept (Olson, 1963; Keul, Keppler, and Doll, 1967; R. L. Hughes et al., 1968). There is no guarantee of tissue/blood equilibrium for either lactate or pyruvate during vigorous exercise. Lactate and pyruvate may themselves take some time to equilibrate (Wasserman, 1967b). Cytoplasmic L′/P′ ratios do not necessarily indicate NADH/NAD ratios and thus oxygen available to the respiratory chains within the mitochondria. The relationship of excess lactate to total oxygen debt is not stoichiometric, but varies with workload (Wasserman, 1967b; C. T. M. Davies, 1967a). Increases of blood lactate levels are closely correlated with changes in lactate/pyruvate ratios and excess lactate (C. T. M. Davies, 1967a), and because of such correlations little new information is generated by calculation of these values. Finally, all authors are now agreed that for a complete understanding of the oxygen debt it is necessary to consider factors other than lactate accumulation, whether expressed as such or as an L′/P′ ratio.

Control of Glycolysis

There is a 1000-fold increase in the activity of the glycolytic pathway during the first 20 sec of vigorous activity. It might be supposed that this was triggered from the mitochondrion through an accumulation of NADH and a depletion of pyruvate [equation (2.5)]. However, this is not the case because the mitochondrial membrane is virtually impermeable to NADH; the only transfer of electrons between the cytoplasm and the interior of the mitochondrion occurs via the glycerol-3-phosphate shuttle of the white muscle fibers and the malate-aspartate shuttle of the red fibers and cardiac muscle (Fig. 2.8). The rapid on-transient thus reflects modulation of key enzymes that lie much earlier in the chain of glycolysis. The principal bottlenecks are phosphorylase and phosphofructokinase (Fig. 1.2). The latter enzyme is present in particularly limited amounts, as can be deduced from the accumulation of fructose-6-phosphate and other hexose phosphates during vigorous exercise.

Phosphorylase

Mechanisms controlling phosphorylase activity have been explored in some detail (Fig. 2.9). The enzyme is normally present in the inactive b form, but through a complex cascade of protein kinases it is converted to the active a form soon after a muscle has been stimulated. The reaction is helped forward by two products of contraction: calcium ions and AMP (Holzer, 1969; Brostrom et al., 1971). All of the phosphorylase can be converted into the a form in as little as 3 sec (Danforth et al., 1962). Adrenaline secretion accelerates the reaction by increasing the production of cyclic-3,5-AMP from ATP. On the other hand, an accumulation of glucose mops up 5-AMP, preventing the phosphorylase a from attaining its full potential activity. Both 3,5-AMP and adrenaline inhibit glycogen synthetase, which reconverts glucose-1-phosphate to glycogen.

Phosphofructokinase

Phosphofructokinase (PFK) also exists in active and inactive forms. It is activated by 5-AMP, 3,5-AMP, ADP, hexose monophosphate, and inorganic phosphate; however, the key substance seems to be AMP (Newsholme, 1971). Very small amounts of AMP (25–100 μmol • l^{-1}) (Shen et al., 1968) activate the enzyme, thus speeding the formation of fructose-1,

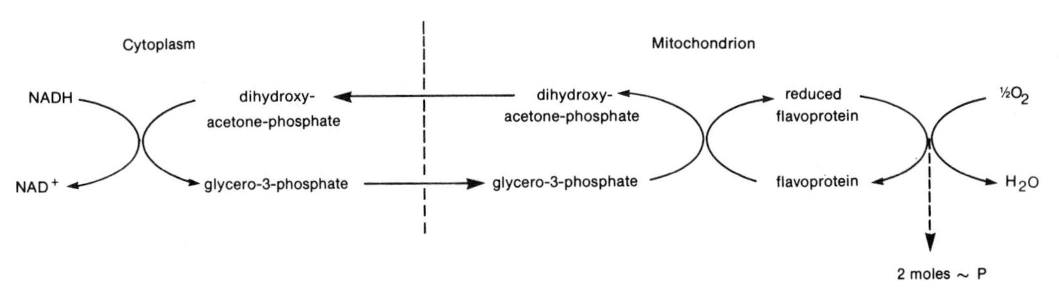

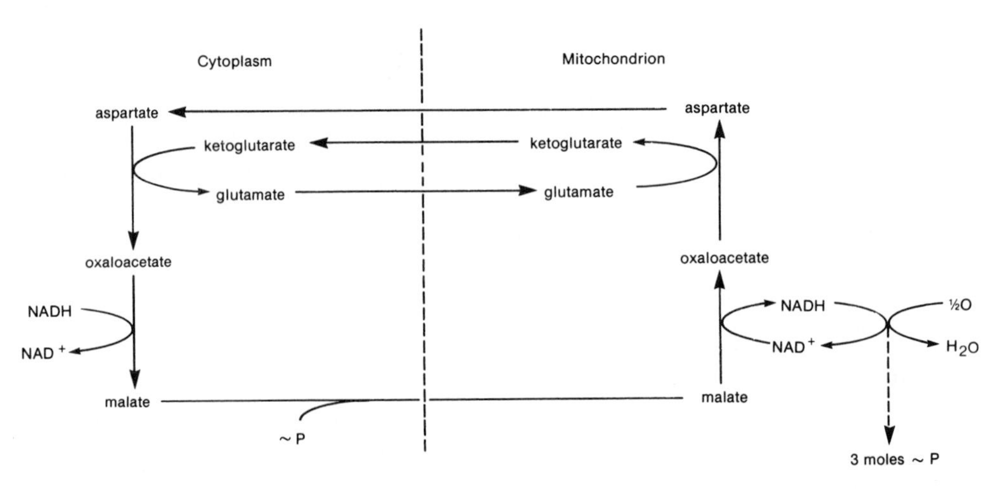

FIG. 2.8. Shuttles for transferring electrons from the cytoplasm to the mitochondria. The glycerol 3-phosphate shuttle is active in white muscle fibers, and the malate-aspartate shuttle operates in red muscle and cardiac muscle fibers. (After McGilvery, 1975.)

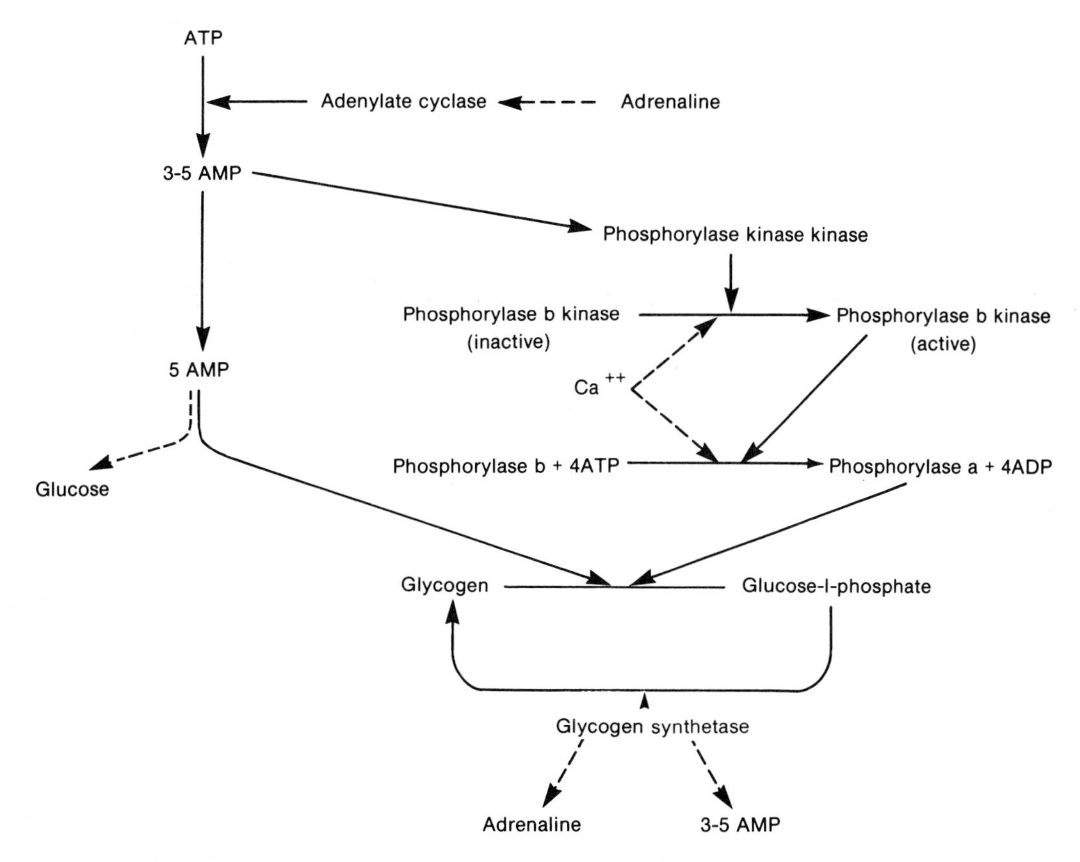

FIG. 2.9. Mechanisms for activation of phosphorylase. Phosphorylase b is inhibited by physiological levels of ATP and glucose-6-phosphate; phosphorylase a is not. High levels of phosphate decrease the need for AMP during activation.

6-diphosphate from fructose-6-phosphate. Even lower concentrations of AMP (0.15–0.35 μmol • l^{-1}) inhibit fructose diphosphatase, which catalyzes the reverse reaction (Newsholme, 1977). Creatine phosphate inhibits PFK until the concentration of the former is less than 2–3 mmol • l^{-1} (Krzanowski and Matschinsky, 1969); ATP and citrate also inhibit PFK (Essén, 1977; Newsholme, 1977).

Other Controls

Under resting conditions, one factor limiting ATP synthesis is the availability of ADP. At the two stages of glucose breakdown at which ATP is synthesized (Fig. 2.10), the maximum rates of reaction (K_m) are reached only when 40% of the total phosphagen and 60% of the CP have been used, with a corresponding increase of ADP concentration.

Creatine phosphate inhibits glycogen breakdown through actions on the enzymes phosphoenol pyruvate kinase (R. G. Kemp, 1973) and glyceraldehyde-3-phosphate dehydrogenase (R. G. Kemp, 1973, Fig. 2.10).

Effect of pH

The buildup of acid within the muscle fibers soon slows the course of glycolysis, with an accumulation of hexose monophosphate (Bergström, Harris et al., 1971). Mechanisms of action of the hydrogen ions include an inhibition of PFK (Trivedi and Danforth, 1966; Bergström, Harris et al., 1971), restraint of phosphorylase activation (Danforth, 1965), and an alteration of the Michaelis constant for PFK (Hofer and Pette, 1968). Glycolysis is halted when the pH has dropped to ~6.3 (A. V. Hill, 1955/56).

OXYGEN DEBT

The oxygen consumption increases relatively slowly at the onset of vigorous exercise, due mainly

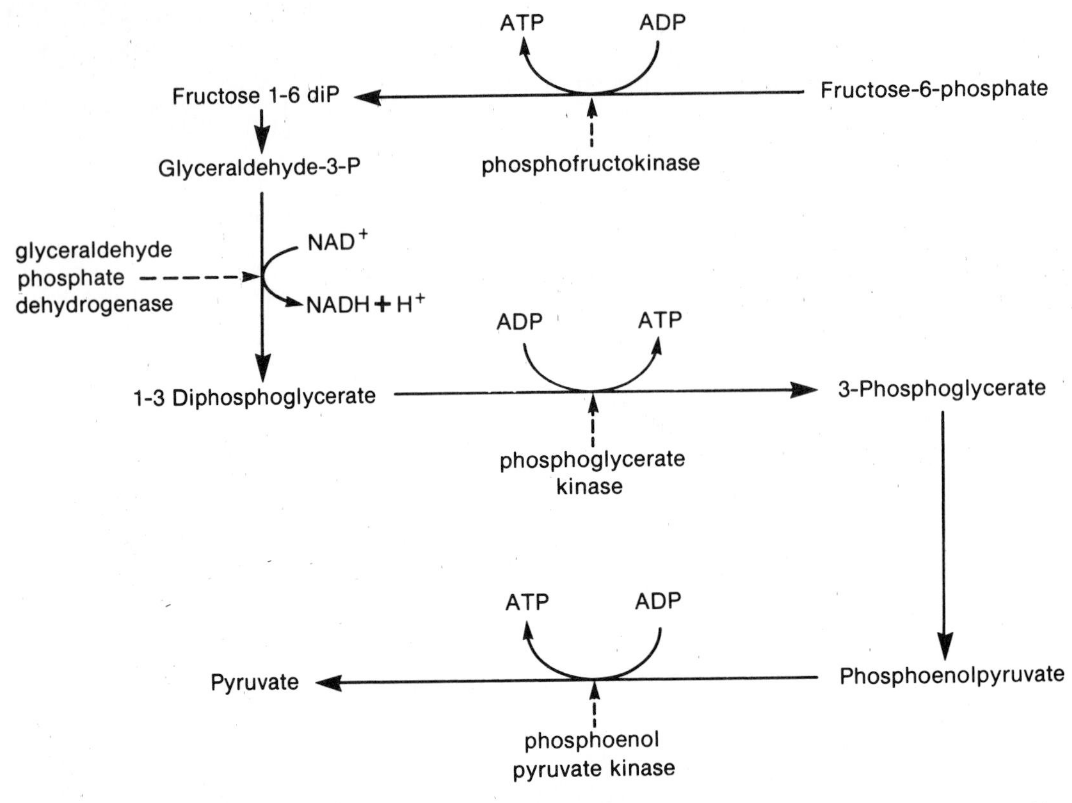

FIG. 2.10. Anerobic reactions involved in ATP production. Each mole of hexose phosphate provides energy for synthesis of three mol of ATP. Under aerobic conditions, the two mol of NADH yield a further six mol of ATP (Fig. 2.13), two of which are used in operation of the NADH shuttle (Fig. 2.8).

(Meakins and Long, 1927; Huckabee, 1958a) but perhaps not entirely (de Coster, 1971) to the time required for circulatory adaptations in the working muscle. Several recent studies of healthy young adults have found oxygen consumption to follow a single-exponent on-transient curve with a half-time of 30–45 sec (C. T. M. Davies, di Prampero, and Cerretelli, 1972; Shindell et al., 1977; Whipp et al., 1977). In consequence, tissue oxygen needs are not fully satisfied, and an oxygen debt is accumulated (A. V. Hill et al., 1924/25) (Table 2.1). The debt comprises work performed by glycolysis (the "lactate" component) and an "alactate" component associated with depletion of phosphagen and body oxygen stores (de Coster, 1971; di Prampero, 1971a,b).

Modern technology allows the measurement of oxygen consumption on a breath-by-breath basis. If oxygen consumption reaches a plateau before exhaustion, it is quite possible to calculate the lag in response to a steady workload (the oxygen deficit). Despite the complication of alactate debt, several authors have claimed a good quantitative relationship between the size of this deficit and muscle or blood lactate concentrations (Karlsson, 1971b; Margaria, Cerretelli, and Mangili, 1964). However, much older work was based on a supposed repayment of the oxygen debt during the recovery period. A. V. Hill et al. (1924/25), and Royce (1962) reported a close match between the incurred debt and its repayment, but others (Lukin and Ralston, 1962; de Coster, 1971) have not confirmed their findings. There are two serious difficulties with debt repayment calculations. First, part of the lactate is oxidized to meet the normal metabolic demands of other tissues. This fraction fails to generate any excess oxygen consumption during the recovery period. Second, various factors (the oxygen cost of respiratory work, cardiac work, and restoration of electrolyte balance; liberation of catecholamines and other hormones; and an increase of body temperature) conspire to keep the oxygen consumption above its preexercise value even though much or all of the lactate has been oxidized (Knuttgen, 1962; Stainsby and Barclay, 1970; G. A. Brooks, Hittelman et al.,

1971a,b; Claremont et al., 1975). Estimates of oxygen debt repayment thus vary widely, depending on the rigor with which a return to the preexercise oxygen consumption has been sought.

The present author has suggested arbitrarily restricting examination of the oxygen recovery curve to the first 15 min after exercise. Excess oxyen consumption for this period is considered the total oxygen debt. The residual debt at any given instant is plotted semilogarithmically against time (Fig. 2.11). A linear regression is fitted to points from 2 to 15 min after effort. Extrapolation back to the ordinate gives the volume of the lactate debt. During the first two minutes of recovery, excess oxygen consumption is greater than would be anticipated from the extrapolated semilog plot. The additional oxygen cost is attributable to the alactate debt. This is estimated by subtraction of observed data from the first line, differences being replotted in a semilogarithmic manner. Actual values (Wright et al., 1976) for a team of well-trained oarsmen (maximum oxygen intake 4.85 ± 0.32 l • min^{-1}, 217 ± 14 mmol • min^{-1}) were a total debt of 8.32 ± 1.35 l (371 ± 60 mmol) and a lactate debt of 5.67 ± 1.33 l (253 ± 59 mmol). Both total debt and lactate debt exceeded predictions for sedentary subjects. Athletes attain a larger lactate debt than sedentary subjects, mainly by virtue of a larger muscle mass; intramuscular lactate concentrations at exhaustion are sometimes but not always higher in athletic individuals.

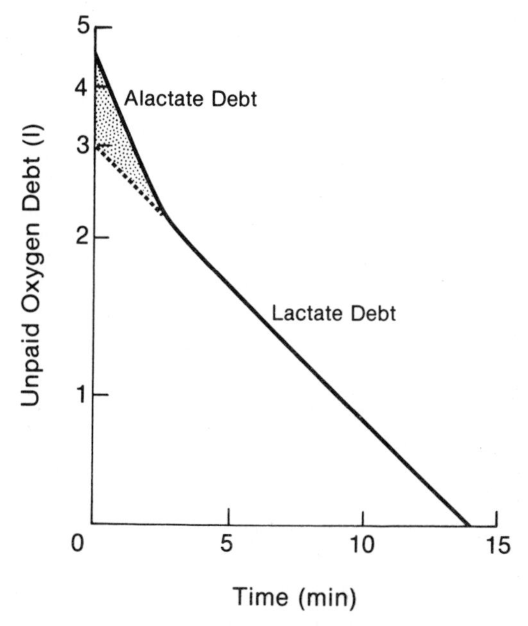

FIG. 2.11. Semilogarithmic plot of oxygen debt following exhausting exercise. A linear regression is fitted to data for 2 to 15 min after effort. The intercept on the ordinate indicates the lactate debt. By subtraction, a series of data points is obtained for the first two min of the recovery process; these differences represent repayment of the alactate debt, and may be plotted in a similar semilogarithmic fashion.

AEROBIC ENERGY RELEASE

Overview

Mechanisms for the aerobic release of energy from fatty acids and pyruvate are sketched in Fig. 1.3 and are illustrated in greater detail in Figs. 2.12 and 2.13. After entering the mitochondrion, the substrate (which we may represent as RH_2) combines with coenzyme A and then passes into the Krebs cycle, where it is progressively broken down to carbon dioxide and hydrogen. The latter combines with NAD^+, its phosphorylated derivative $NADP^+$, or in one instance directly with flavoprotein, forming the corresponding reduced compounds (for example, $NADH + H^+$; Fig. 2.12):

$$NAD^+ + RH_2 \rightarrow NADH + H^+ + R \tag{2.8}$$

The NADH or NADPH is then passed to the chain of cytochrome pigments (Fig. 2.13). Note that 1 mol of ATP is regenerated by a reaction that involves the intervention of another high-energy phosphate compound, guanosine phosphate (Fig. 2.12).

The cytochrome chain comprises several similarly structured compounds, arranged in an orderly fashion from one side of the mitochondrial membrane to the other. Each member of the chain can oxidize (or accept an electron from) the compound on its right (Fig. 2.13). Substrates such as NADH are received at the right-hand side of the chain and are reconverted to NAD^+ with the generation of 2 electrons (e) and a proton (H^+):

$$NADH \rightarrow NAD^+ + H^+ + 2e \tag{2.9}$$

Because NADH generates electrons, it has a negative oxidation/reduction (redox) potential. At the opposite end of the cytochrome chain, the protons of reactions (2.8) and (2.9) and the corresponding electrons react with molecular oxygen to form water:

$$O_2 + 4H^+ + 4e \rightarrow 2H_2O \tag{2.10}$$

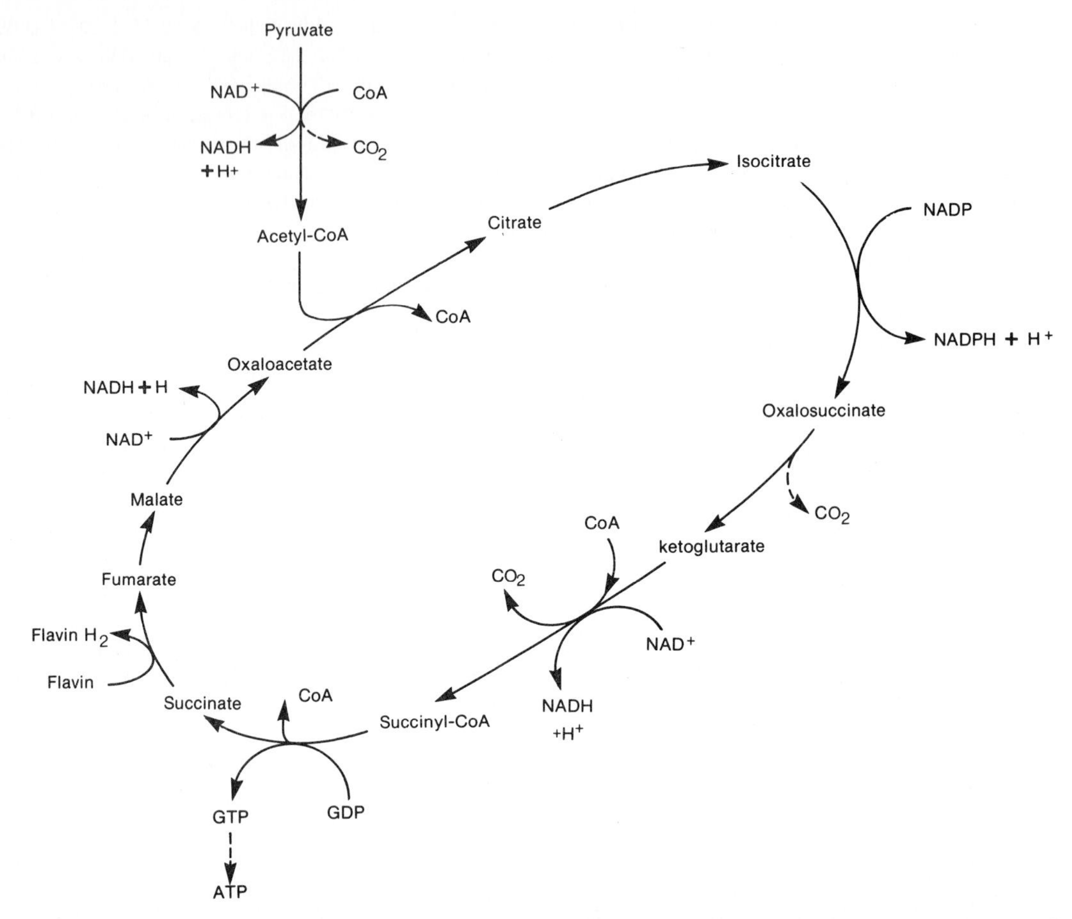

FIG. 2.12. The citric acid cycle of Hans Krebs. Note that pyruvate first combines with coenzyme A, and is then progressively broken down to CO_2 and H_2; the latter combines with NAD^+, its phosphorylated derivative $NADP^+$, or flavoprotein (flavin adenine dinucleotide, FAD). Fifteen mol of ATP are generated for each mol of pyruvate that enters the citric acid cycle. Guanosine diphosphate (GDP) is also an intermediary for the conversion of succinyl CoA to succinate.

Because oxygen eliminates electrons from the system, it is said to have a positive redox potential. The reaction of equation (2.10) could lead to the formation of toxic intermediaries if protons were accepted singly:

$$O_2 + H^+ + e \rightarrow {}^\bullet HO_2 \quad (2.11)$$
$$^\bullet HO_2 + H^+ + e \rightarrow H_2O_2 \quad (2.12)$$
$$H_2O_2 + H^+ + e \rightarrow {}^\bullet HO' + H_2O \quad (2.13)$$
$$^\bullet HO + H^+ + e \rightarrow H_2O \quad (2.14)$$

However, the current view is that two protons are transferred simultaneously:

$$O_2 + 2H^+ + 2e \rightarrow H_2O_2 \quad (2.15)$$
$$H_2O_2 + 2H^+ + 2e \rightarrow 2H_2O \quad (2.16)$$

While the electrons of equation (2.10) are known to move along the cytochrome chain, the precise pathway for movement of the protons has yet to be established.

The difference of redox potential between the NADH and oxygen ends of the cytochrome chain is ~1.2 volts (V) (Table 2.2). The passage of a single electron against a potential difference of 1 V (1 eV) should yield 96 kJ of free energy per mol. Since each oxygen atom reacts with 2 electrons, the total energy liberated by the cytochrome mechanism should be $2 \times 1.2 \times 96 = 230$ kJ • mol^{-1}. In fact, 3 mol of ATP are resynthesized for each atom of oxygen that is used. The energy conserved is thus ~138 kJ, and the efficiency of ATP resynthesis is ~60% (Fig. 1.8).

If a subject performs exhausting work such as running at maximum speed, the proportion of the

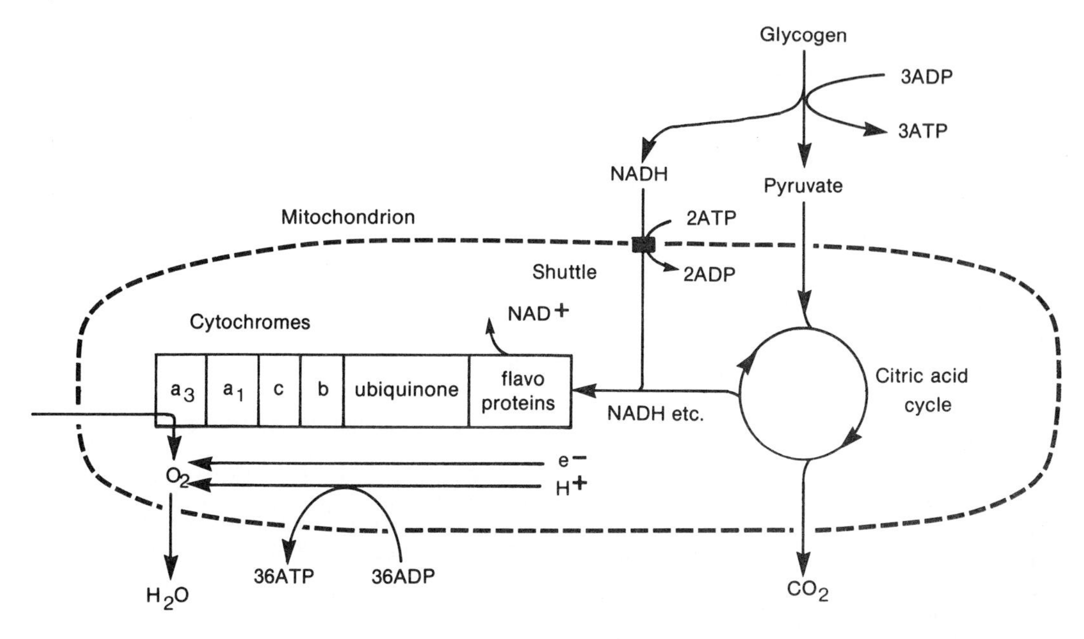

FIG. 2.13. Regeneration of ATP by aerobic energy release. The ATP generation is calculated per mol of hexose; three mol of ATP are generated for each NADH reaching the cytochrome chain. Details of the individual cytochrome constituents are given by Ohnishi (1976). Major increases of redox potential, and thus energy for ATP resynthesis, occur between NADH and ubiquinone, between ubiquinone and cytochrome c, and between cytochrome a_3 and oxygen (Table 2.2; Fig. 4.4).

work undertaken by aerobic mechanisms increases with the total duration of the task (Fig. 2.14). In the first minute, as much as two-thirds of the energy may be derived from anerobic sources, but with a 5-min run about four-fifths of the total energy usage has an aerobic basis. Unfortunately, the acid metabolites associated with anerobic metabolism cause a distressing increase of ventilation. Thus, an experienced distance runner attempts to operate within aerobic limits for most of a race, conserving his anerobic reserves for the final sprint (A. V. Hill, 1925); the optimum technique involves a gradual tapering of speed so that the physiological stress is held constant (Ariyoshi et al., 1979a,b).

TABLE 2.2

Approximate Redox Potentials at Various Points in the Chain of Food Oxidation

Reaction	Redox Potential (mV)
Acetaldehyde/acetic acid	−0.60
$NADH/NAD^+ + H^+$	−0.32
$FAD/FADH_2$	−0.04
Ubiquinone/dihydro-ubiquinone	+0.09
Cytochrome b ox/red	+0.07
Cytochrome c ox/red	+0.23
Cytochrome a ox/red	+0.22
Cytochrome a_3 ox/red	+0.38
$H_2O/\frac{1}{2}O_2$	+0.82

Note: The free energy between any two points in the chain is given by the product of the difference in redox potential, the number of electrons involved (2 per atom of oxygen), and a constant (the Faraday, 96 kJ • mol^{-1}). Thus, between $NADH/NAD^+ + H^+$ and cytochrome c ox/red the free energy is 0.55 × 2 × 96 = 106 kJ • mol^{-1}. With 100% efficiency of transfer, this would be more than the energy needed to generate 2 mol of ATP (92 kJ).

Control

We have already noted that during sustained effort, a well-trained endurance athlete can increase his overall aerobic energy release as much as 12-fold, from a resting figure of about 5 kJ • min^{-1} to more than 60 kJ • min^{-1}. F. D. Carlson and Wilkie (1974) suggested that a 20-fold increase of aerobic activity is possible within a given muscle. In humans, an even greater range of change can be deduced from blood flow and oxygen extraction measurements. Local blood flow is increased at least 30- to 40-fold, and oxygen extraction is also augmented 3- to 4-fold, implying a 90- to 160-fold increase of aerobic metabolism during maximum effort. Mottram (1971) set the oxygen consumption of resting muscle at only 3 ml • kg^{-1}. If 6 l • min^{-1} is consumed by 20 kg of muscle during maximum work, this also indicates a

100-fold increase of aerobic energy release.

Unlike glycolysis, control of the aerobic reaction seems independent of enzyme activation. Critical factors are the availability of substrate (often NADH), oxygen, and small amounts of ADP and phosphate. We have seen that ADP concentrations rise relatively little during physical activity (Fig. 2.2). Nevertheless, the increase seems sufficient to sustain the oxidative reaction until all of the CP reserves have been restored. It also seems to be the drop in ADP concentration that halts the reaction when phosphagen has finally been regenerated.

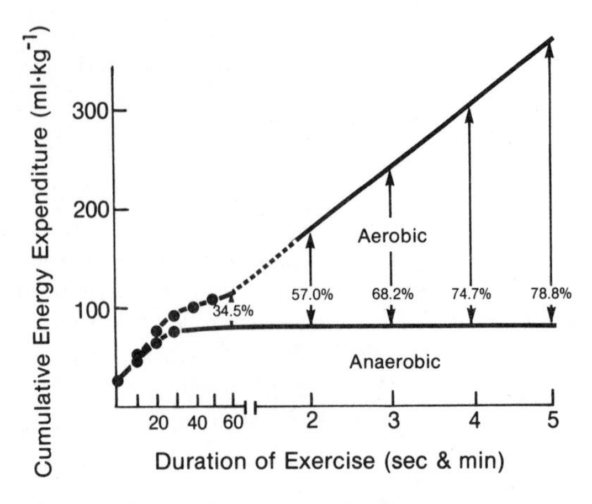

FIG. 2.14. The relationship between the duration of steady, exhausting activity and the proportion of the energy expenditure attributable to a steady transport of oxygen. (From Shephard, 1970a, by permission of the *Journal of Sports Medicine and Physical Fitness*.)

Fuel

Protein

Classical biochemists maintained that protein was not used as fuel during exercise unless the subject was either dieting or starving (Chauveau, 1896; Krogh and Lindhard, 1920). The main evidence for this view was that nitrogen excretion did not rise appreciably during effort (Crittenden, 1904), even if activity was sustained until exhaustion of muscle glycogen (Hedman, 1957).

It is now questioned as to whether urinary nitrogen excretion provides a sufficiently sensitive index of protein breakdown. Shephard and Kavanagh (1975a) found an increase of blood urea nitrogen equivalent to the combustion of 50 g of protein over the course of a 5-hr "postcoronary" marathon run. Assuming their subjects usually had an energy expenditure of 12.6 MJ • day^{-1} with a diet that provided 12% protein, the normal protein metabolism would have averaged 3.8 g • hr^{-1}. This implies that the marathon event led to the breakdown of an additional 30 g of protein. Others (Refsum and Stromme, 1974; F. Cerny, 1975; Rougier and Babin, 1975) have had similar findings. Refsum and Stromme (1974) saw a 60–80% increase of urea production during sustained skiing. While some energy is necessarily liberated by protein breakdown, more important functions of the reaction (F. Cerny, 1975) are (1) pyruvate removal as alanine (Fig. 2.15; Poortmans and Delisse, 1977; Felig, 1977), (2) glucose formation (gluconeogenesis), and (3) replenishment of intermediate compounds in the citric acid cycle (since these tend to leak from the mitochondria during vigorous exercise). In support of the first mechanism, Felig and Wahren (1971) demonstrated a substantial release of alanine from exercising muscle.

Not only is protein burned during acute bouts of strenuous activity, but many classes of athlete seek to increase their muscle bulk during a conditioning program. Sometimes 10 kg or more of lean tissue may be developed over a 4-month period of heavy training. One recent report set the cost of the synthetic process at 21 kJ • g^{-1} (N. J. Smith, 1976). Our calculations suggest a figure of up to 23 kJ • g^{-1} (an efficiency of 74%). Taking the latter estimate, and assuming that 25% of the lean tissue is protein, synthesis would demand 28 g of first-class protein per day, or 0.35 g • kg^{-1} day^{-1} in an 80-kg athlete. Adding a resting requirement of 1.0 g • kg^{-1} • day^{-1} (Nutrition Canada, 1973) and a modest combustion of protein during exercise (0.4 g • kg^{-1} • day^{-1}), the total need of the athlete could easily reach 1.7–1.8 g • day^{-1} of good-quality protein (Consolazio, Johnson et al., 1975; Celejowa, 1978; Laritcheva et al., 1978).*

While this conclusion may be anathema to some expert committees on nutrition (Durnin, 1978), it should be stressed that the subjects they have considered were neither exercising hard nor undergoing rapid muscle hypertrophy. In support of the present author's estimate of protein needs, Yamaji (1951) and Yoshimura (1970) noted that unless athletes in training were given at least 2 g • kg^{-1} • day^{-1} (14% of a 16.7 MJ • day^{-1} diet), there were decreases in both hemoglobin level and blood protein concentra-

*The quality of protein is determined by its content of "essential" amino acids that the body is unable to synthesize.

tions. This might be dismissed as an expansion of plasma volume. On the one hand, Celejowa and Homa (1970) found that a protein intake of 145 g • day^{-1} was necessary to maintain a slight positive nitrogen balance in Polish weight lifters who were attending a training camp, whereas Stucke et al. (1972) found an intake of 2 g • kg^{-1} • day^{-1} boosted muscle mass. Nevertheless, if minimal protein needs are met, the overall energy balance is more important to the maintenance of a positive nitrogen balance than is the precise proportion of protein ingested (Callaway, 1975).

Fat

The early work of Krogh and Lindhard (1920) and Christensen and Hansen (1939a) established that both fat and carbohydrate are metabolized during exercise. The relative usage of the two fuels is often assessed from the ratio of carbon dioxide output to oxygen consumption (the respiratory quotient, RQ). Ignoring the complications of protein usage [the term U_n of equation (1.15)] and the 7% lower energy yield per mol of oxygen when fat is the fuel, the proportion of energy derived from fat can be approximated by the simple ratio $(1 - RQ)/0.3$. For example, an RQ of 0.85 implies that $[(1 - 0.85)/0.3] \times 100$, or 50% of the energy, is derived from fat. Factors influencing the relative combustion of fat and carbohydrate include the intensity and duration of work (Bock et al., 1928; Christensen and Hansen, 1939a), blood levels of fatty acids (Bremer, 1967; P. Paul, 1970; Wahren, 1977a), and the state of training of the individual (Holloszy, 1973). In vigorous activity (70–80% of maximum oxygen intake), carbohydrate provides at least 75% of the fuel (Pruett, 1970a; Hultman, 1971), and in near maximal work almost all of the energy seems to be released from carbohydrate stores (C. T. M. Davies and Barnes, 1972). On the other hand, with more moderate work (<50% of maximum oxygen intake), as much as 50–60% of the energy may be derived from fat (C. Williams et al., 1975; P. Paul, 1975). Intermittent effort with brief rest pauses favors fat usage. Local stores of phosphagen and oxygen may then meet peak energy needs without recourse to anerobic work, while some metabolic product (perhaps citrate) inhibits glycogen breakdown (Essén, Hagenfeldt, and Kaijser, 1977).

If steady, rhythmic activity, such as cycling, is prolonged, the proportion of fat used by the skeletal muscles increases progressively, mainly because local stores of glycogen have been depleted (Hultman, 1971; Bergström, Hultman, and Saltin, 1973b; Gollnick, Piehl, and Saltin, 1974; P. Paul, 1975; Wahren, 1977a). However, if the work is severe enough to tax the oxygen debt mechanism in some or all of the active muscle fibers, such anerobic activity must be sustained by carbohydrate breakdown. In the heart, the relative usage of fat and carbohydrate does not change with prolonged exercise (Kaijser, Lassers et al., 1972).

The influence of diet upon the performance of sustained work was shown in the classical experiments of Christensen and Hansen (1939a). When carbohydrate provided only 5% of the total energy intake, subjects could work at a standard load (10.6 kN • m • min^{-1}) for a period of 1 hr, 70–99%

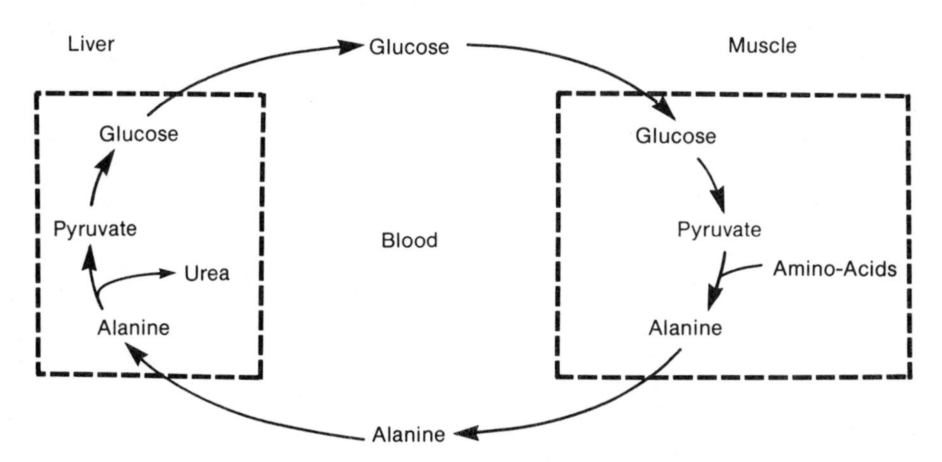

FIG. 2.15. Suggested pyruvate-removing role of alanine in vigorous effort. (After Felig & Wahren, 1971.)

of the energy being obtained from the breakdown of fat (Fig. 2.16). On the other hand, with a 90% carbohydrate diet, the same exercise was tolerated for 4 hr; in these circumstances, the proportion of energy derived from carbohydrate was initially 70–75% and was still 40% at the end of the experiment.

The relative advantages of fat and carbohydrate as fuels are summarized in Table 2.3. Carbohydrate is best suited to brief bouts of intense work. It can be used anerobically, and the energy yield per mol of oxygen is slightly higher than for fat. Further, since most of the carbohydrate is stored as glycogen, substantial reserves of water associated with the glycogen molecule (2.7 g • g^{-1}, Weis-Fogh, 1967) are liberated during its metabolism. In contrast, fat has several advantages as a long-term food store (Dole, 1964). Perhaps most important from the viewpoint of locomotion, fat is energy-dense; a given reserve has a much lower mass than has an equivalent amount of an alternative fuel. Fat consists mainly of carbon and hydrogen atoms, and it is for this reason that combustion yields more energy per g than could be derived from either protein or carbohydrate (Tabel 1.1). The absence of associated water molecules further enhances the energy density of fat relative to carbohydrate. Finally, fat is stable, it can be molded readily to fit the various storage spaces in the body, yet it can be mobilized rapidly and transported to the active tissues at minimal energy cost.

For practical purposes, the capacity of the fat stores (300 MJ) can be regarded as infinite. However, the usable fat content of the muscle cytoplasm is relatively small (about a third of the energy reserve due to glycogen; Keul, 1975). During sustained physical activity, additional amounts of lipid are thus transported from the adipose tissue, mainly as a complex of free fatty acid (FFA) and albumin. While FFA accounts for only 2–3% of the fat content of the plasma, it has a rapid turnover (25% per min; L. A. Carlson, 1967a,b), so that several hundred grams of FFA can be transported to the working tissues over the space of 24 hr.

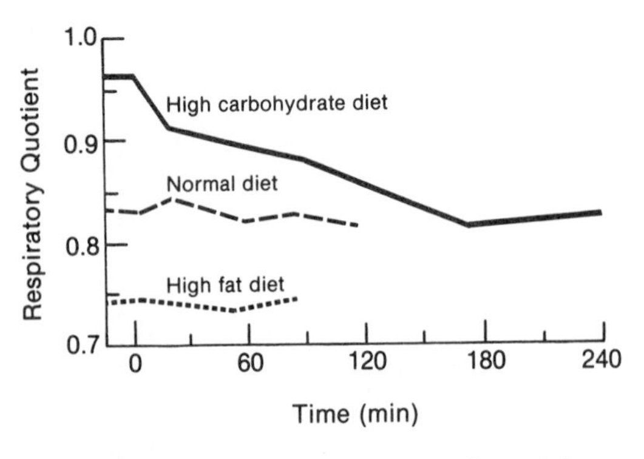

FIG. 2.16. Effect of diet on the respiratory quotient and the effort tolerance of a subject working at a constant rate of 10.6 kN • m • min^{-1}. When provided with a high-fat diet, the subject was exhausted in 90 min; with a high-carbohydrate diet, the activity could be sustained for four hours. (Based on an experiment of Christensen & Hansen, 1939a.)

TABLE 2.3
Relative Advantages of Fat and Carbohydrate as Fuels

Fat	Carbohydrate
High-energy density	Can be used anerobically
Stable	Liberates water during breakdown
Readily mobilized	Higher energy yield per mol of oxygen
Low cost of storage and transport	

Transfer of FFA across the mitochondrial membrane is facilitated by the enzyme acyl-transferase (Fig. 2.17). Once within the mitochondrion, the FFA combines with coenzyme A. Thereafter, oxidative enzymes on the inner aspect of the mitochondrial membrane split from the FFA chain a sequence of 2-carbon acetyl-CoA units which enter the citric acid cycle. Every 2 carbon atoms that are split from the original fatty acid chain generate 1 mol of NADH and 1 mol of reduced flavin adenine nucleotide (FADH). This provides energy for the resynthesis of 6 mol of ATP from ADP [equations (2.9) and (2.10)]. After entry into the citric acid cycle (Fig. 2.12), there is potential for the formation of an additional 12 mol of ATP as each acetyl-CoA unit is broken down to CO_2 and water. If fat is being used very rapidly, the capacity of the citric acid cycle may be exceeded. Fatty acids then accumulate as 4-carbon "ketone bodies" (β-hydroxybutryate and acetoacetate); these can later be reused in the fatty acid oxidation cycle of cardiac and skeletal muscle, but may also be excreted in the breath and the urine with a resultant energy loss.

Starvation and diabetes are two commonly cited reasons for ketosis. A "cold ketosis" is well recognized in arctic travelers. We found a heavy ketone loading of the urine when untrained troops were required to undertake a rigorous military exercise in the arctic (energy expenditure 15 MJ • day^{-1}, W. O'Hara et al., 1977b,1979). Body composition studies showed a rapid breakdown of body fat to meet the energy demands of vigorous work, muscle hypertrophy, and maintenance of body temperature.

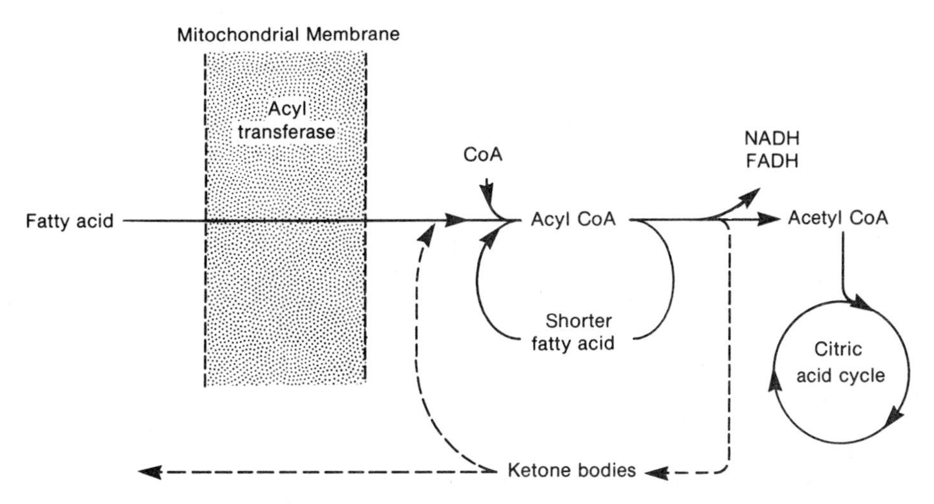

FIG. 2.17. Fatty acid metabolism within mitochondrion. Six mol of ATP are regenerated by the breakdown of Acyl-CoA to Acetyl-CoA, and a further 12 mol of ATP are regenerated with entry of the Acetyl-CoA into the citric acid cycle.

B. B. Lloyd (1966) estimated the power of fat transporting mechanisms from athletic records. Running speeds for events lasting 30 min to 2 hr were equivalent to a steady energy expenditure of ~85.5 kJ • min^{-1} or (allowing for a substantial proportion of fat combustion) an oxygen consumption of ~205 mmol • min^{-1}, 4.6 l • min^{-1}. During the period 2–5 hr, when glycogen stores were presumably exhausted, running speeds were substantially slower, equivalent to a steady energy usage of 59.1 kJ • min^{-1}. If this was derived mainly from the metabolism of fat (energy yield 39.0 kJ • g^{-1}, Table 1.1) this would imply a fat transport of ~1.5 g • min^{-1}. A plot of distance against record times yielded a positive intercept of 13.4 km (corresponding to 3.64 MJ, presumably a 230-g store of glycogen).

Somewhat similar rates of fat transport can be derived from blood concentrations of FFA during vigorous exercise (Rodahl, Miller, and Issekutz, 1964). Given a muscle flow of 20 l • min^{-1} and a blood FFA concentration of 1 mmol • l^{-1}, the total delivery to the tissues is 20 mmol • min^{-1}. Assuming a 25% clearance (L. A. Carlson, 1967a,b), the effective transport of FFA is 5 mmol • min^{-1} or, for an average molecular weight of 270 g, 1.35 g • min^{-1}.

Carbohydrate

Carbohydrate provides an ever increasing proportion of the metabolic fuel as maximum effort is approached. Reliance on carbohydrate reflects in part selective fiber recruitment. Other significant factors influencing carbohydrate utilization are the maximum rate of fat mobilization and an oxygen pressure in some muscle fibers that is less than the threshold needed to sustain aerobic metabolism. Available carbohydrate stores include blood glucose, muscle glycogen, and liver glycogen.

Blood Glucose. In theory, the blood glucose reserve is quite limited. The blood volume of ~5 l contains ~6 g of glucose (33.3 mmol, 94 kJ).* Furthermore, the brain depends on glucose for its function, so that only about a half of this store (16 mmol) can be used before problems arise from hypoglycemia.

Some authors have argued that sustained exercise can be halted by hypoglycemic fatigue. However, the immediate effect of vigorous exercise is a small increase of blood glucose (Wahren, 1977a,b). We have observed well-maintained or even elevated concentrations of blood glucose after 4 or 5 hr of running (Shephard and Kavanagh, 1975a). In exhausting bicycle work, on the other hand, 4-hr readings have fallen as low as 3 mmol • l^{-1} (Wahren, Felig et al., 1975), and in such experiments hypoglycemia could conceivably be limiting effort.

The glucose uptake of the exercising limbs increases at least 10- to 20-fold with exercise (Jorfeldt and Wahren, 1970), to as much as 4 mmol • min^{-1}, 11.3 kJ • min^{-1} (Wahren, 1977a,b). This is matched by a roughly proportional output of glucose from the liver. Wahren (1977a,b) measured the hepatic output at 4 mmol • min^{-1}, while Rowell (1971) set the max-

*Using the figures of McCance and Widdowson (1960), Table 1.1.

imum rate of liver glucose formation at 5.6 mmol • min^{-1} (20.4 kJ • min^{-1}). Under resting conditions, 25–30% of hepatic glucose is derived from gluconeogenesis (glucose formation from lactate and amino acids, with small contributions from glycerol and pyruvate). During exercise of 40 min duration, Wahren (1977a,b) found increasing reliance on glycogen stores (16% of glucose output due to gluconeogenesis at a load of 65 W, 11% at 135 W, and 6% at 200 W). However, if moderate work was sustained for 240 min, gluconeogenesis finally accounted for 45% of hepatic glucose output.

The glucose uptake of the active muscles depends upon the hexokinase level in the cell membrane (H. E. Morgan et al., 1964). The enhanced activity of this enzyme and thus the greater glucose uptake of an exercising subject is not due to an augmented release of insulin or a decrease of circulatory glucagon (Fig. 2.18); indeed, insulin concentrations fall and glucagon levels rise during prolonged effort. However, it is possible that the enhanced muscle blood flow increases the delivery of insulin to the active tissues, while the increased blood concentrations of glucagon may favor gluconeogenesis in the liver (Wahren, Felig et al., 1975).

Quite a number of endurance athletes ingest strong (for example, 30%, 1.67 mol • l^{-1}) solutions of glucose while exercising. Such solutions may counter immediate hypoglycemia, but have other less beneficial effects. As blood glucose rises, the blood level of free fatty acids and thus the usage of fat falls (Felig, 1977). At the same time, insulin concentrations rise, glucagon concentrations fall, and the liver uptake of the gluconeogenic metabolites lactate, glycerol, and alanine is reduced (Ahlborg and Felig, 1976; Wahren, 1977a,b).

MUSCLE GLYCOGEN. Glycogen provides a convenient energy reserve for muscle because quite large quantities of fuel can be stored in this form without an excessive rise of intracellular osmotic pressure. We have already noted the associated water of hydration, ~2.7 g for each gram of glycogen. This water is liberated as the glycogen is broken down,

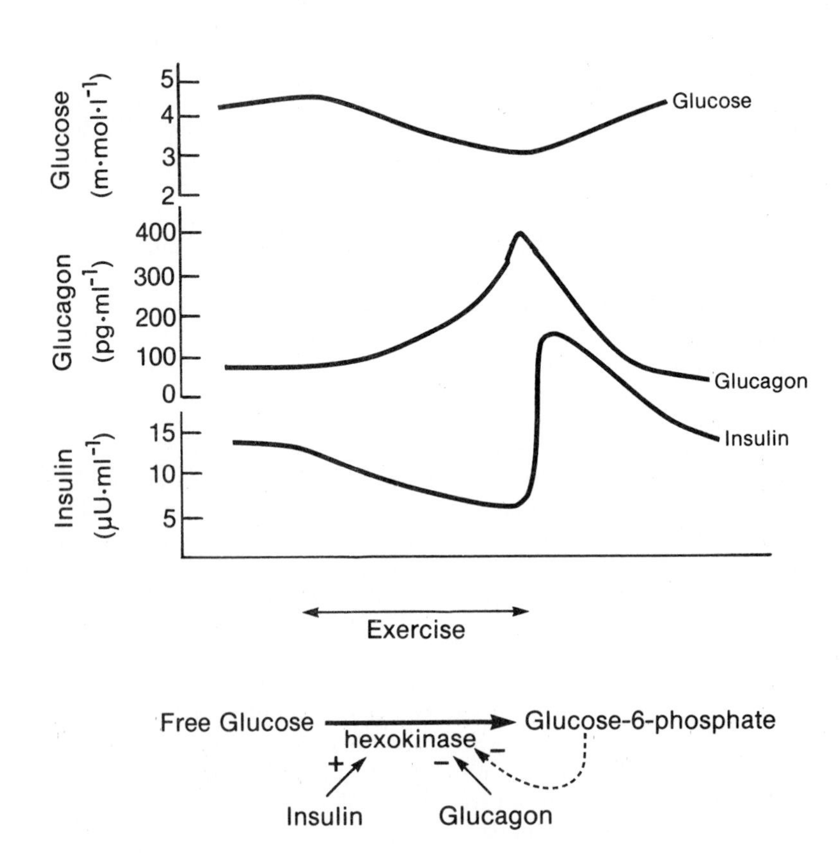

FIG. 2.18. Schematic representation of changes in blood concentrations of glucose, insulin, and glucagon during and following sustained vigorous exercise (After Wahren, 1977a,b.)

and it can make an important contribution to the fluid needs of a sustained effort such as a marathon race (Shephard and Kavanagh, 1975a).

Early views concerning the relative usage of carbohydrate and fat during exercise were based upon observations of the respiratory quotient (Fig. 2.16). More recent data on glycogen metabolism has been obtained by the technique of needle biopsy (Bergström, 1962). Unfortunately, there is a limit to the number of biopsies that one subject will tolerate, and information is thus restricted to small segments of a few muscles. The first biopsies of the quadriceps femoris (Hermansen, Hultman, and Saltin, 1967; Hultman, 1971) set the resting glycogen content at about 1.4 g per 100 g of wet tissue, with little age or sex variation. Given a total muscle mass of 28 kg, the muscle glycogen reserve of a young man was thus estimated at 390 g, 6.1 MJ.

Reserves were only depleted 30–40% after a week of starvation but were apparently exhausted almost completely by 75–100 min of vigorous cycling. Hultman (1971) described a triphasic curve of glycogen usage. There was a brief initial phase of anerobic glycolysis, triggered by mechanisms that have already been discussed (Fig. 2.9), including the products of the initial muscle contraction, adrenaline release, and possibly a drop in local oxygen tension. In the second phase, circulation adapted to the workload, and a steady aerobic usage of glycogen developed. This increased disproportionately with the intensity of work:

Intensity of Aerobic Power (%)	Rate of Glycogen Breakdown (mg per kg of muscle per min)
25	50
54	140
78	260

During this phase of activity, glycogen was the preferred carbohydrate substrate, and—perhaps because of limitations on the rate of transfer from blood to muscle—the rate of depletion of intramuscular glycogen could not be reduced by giving intravenous glucose. The rate-controlling step during the second phase was the enzyme phosphofructokinase. In the final phase, carbohydrate utilization was greatly slowed by a relative lack of glycogen. Not only did this deprive the glycolytic reaction of substrate, but it tended to inactivate phosphorylase b kinase (Fig. 2.9). There was thus increased usage of both free fatty acids and liver-derived glucose. Hultman (1971) found final muscle glycogen concentrations as low as 0.1 g per 100 g of muscle. Assuming that 20 kg of muscle are active during bicycle ergometry, glycogen usage would total 260 g, an energy yield of 40–50 kJ • min^{-1}, sustained for 75–100 min of effort. Such estimates coincide rather closely with rates of carbohydrate consumption calculated from the oxygen consumption and the respiratory quotient.

There are differences in the glycogen content of individual muscles. For example, figures for the resting deltoid average 0.98 rather than 1.40 g per 100 g (Hultman, 1971). Initial glycogen concentrations apparently reflect in part the proportions of the several types of fiber present in individual muscles. Slow-twitch, aerobic fibers contain little glycogen, while fast-twitch fibers (whether anerobic or aerobic) contain much larger quantities (Piehl, 1974; Saltin, 1973; Saltin, Henriksson et al., 1977). The extent of glycogen usage thus depends not only on the oxygen supply to the muscle, but also on the pattern of fiber recruitment. As maximum effort is approached, more fast-twitch fibers are brought into play, and the rate of glycogen depletion is accelerated. Unfortunately, humans do not have as clearly identifiable fiber types as some animals. Nevertheless, there is evidence that in humans, also, glycogen usage varies from one type of activity to another. Thus glycogen is lost from the human quadriceps more rapidly during bicycle ergometry than during running (Costill, Sparks et al., 1971; Gollnick, Piehl et al., 1975). However, if the run is uphill and of sufficient duration, fairly complete glycogen depletion can develop in both the gastrocnemius and soleus muscles (Costill, Jansson et al., 1974; Gollnick, Piehl et al., 1975). During marathon running, effort is often limited to ~75% of maximum aerobic power. Such activity is sustained mainly by slow-twitch fibers. Selective depletion of glycogen in these fibers thus tends to fatigue and exhaustion, at a time when muscle biopsy specimens still show substantial glycogen reserves in the fast-twitch fibers. Nevertheless, some anerobic activity can continue in the slow-twitch fibers if lactate is accepted from the fast-twitch fibers and from the bloodstream (Essén, Pernow et al., 1975; Klausen, Piehl, and Saltin, 1975).

Glycogen resynthesis is regulated by the enzyme glycogen synthetase. This exists in I and D forms, the latter being inactive in the presence of physiological concentrations of phosphate. The I form is converted to the inactive D form in the presence of adrenaline, high concentrations of glycogen, and the by-products of muscular contraction. However, it is reactivated when glycogen levels are low, as in maximum exercise to exhaustion (A. W. Taylor, 1975). Replenishment of glycogen stores is thus largely under local control; if a subject performs one-legged exercise, both depletion and resynthesis are localized to the active limb (Hultman, 1971). In the rat, some resynthesis of glycogen stores can be seen as soon as 15 min after vigorous activity (Terjung, Baldwin et al., 1974; McCafferty and Edington, 1975). In humans, replenishment seems a slower process; ~40% is replaced in 5 hr (Klausen, Piehl, and Saltin, 1975) and complete recovery may take a week (Hultman, 1971; Piehl, 1974). Fast-twitch muscle fibers are usually replenished more rapidly than the slow-twitch variety (Terjung et al., 1974). The rate of resynthesis varies with the diet, being speeded 10-fold by a high-carbohydrate regimen (Hultman, 1971; Maehlum, 1978). The optimum nutritional plan for an endurance athlete (Fig. 2.19) seems a period of hard work to exhaust muscle reserves, 2–3 days on a fat and protein diet, and then 2 days on a high-carbohydrate diet. With such a technique, the glycogen reserves of active fibers and possibly of the liver can be boosted to several times the normal resting figure (up to 6 g • 100 g^{-1}; Saltin and Hermansen, 1967; Hultman, 1971).*

The practical importance of glycogen reserves has been demonstrated by both laboratory and field experiments. In the laboratory, the tolerance of bicycle ergometer work at 75% of aerobic power is closely proportional to the initial glycogen content of the quadriceps (Bergström, Hermansen et al., 1967). In the field, the performance of many athletes flags after the first hour of running, but deterioration is most obvious in contestants with a low initial muscle glycogen (Saltin, 1970a). A final sprint also produces a lower blood lactate concentration than could have been developed earlier during exhausting exercise; this phenomenon has been well documented for ice hockey players (H. J. Green, Bishop et al., 1976), and it can be attributed simply to a depletion of glycogen stores (Karlsson, Diamant, and Saltin, 1968).

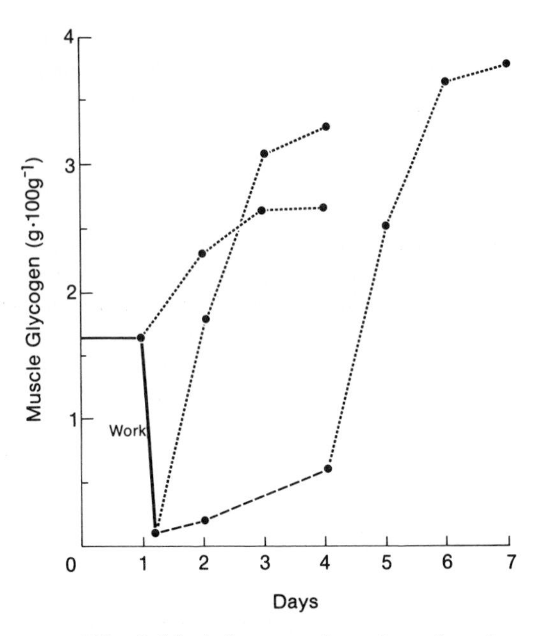

FIG. 2.19. Influence of work and various dietetic regimens on muscle glycogen content. (After Saltin & Hermansen, 1967.)
——— mixed diet; – – – fat and protein; ----carbohydrate

The resting hepatic output of glucose is about 90 mg • min^{-1} (0.5 mmol • min^{-1}, 1.41 kJ • min^{-1}). Rowell, Masoro, and Spencer (1965) found that if exhausting work was performed at normal room temperatures, output rose to 2.5–3.5 mmol • min^{-1}, while with exercise in the heat values were 14 times as large as the resting figure (up to 7 mmol • min^{-1}, 20.4 kJ • min^{-1}). More recently, Wahren (1977a,b) reported a glucose output of 4 mmol • min^{-1} with exhausting work at room temperatures. We have already commented that gluconeogenesis accounts for 6–45% of the hepatic glucose release.

Liver Glycogen. The glycogen content of the resting liver ranges from 1.5 to 8.0 g • 100 g^{-1} (average 4.4 g • 100 g^{-1}) wet wt. Given a liver weight of 1.5 kg, the total hepatic store is thus ~66 g (1.04 MJ). However, not all of this reserve can be utilized. At exhaustion, hepatic concentrations range from 0.6 to 3.6 g • 100 g^{-1} (average 2.0 g • 100 g^{-1}). Thus, the available liver store is ~36 g or ~566 kJ (Hultman, 1971). One or two meals is sufficient to restore liver reserves (Costill, 1977a).

*One recent report (Tremblay et al., 1978) failed to confirm the superiority of a carbohydrate diet; there is as yet no clarification of reasons for this divergent result.

ROLE OF VITAMINS

Athletes frequently take vitamin supplements in an attempt to increase the energy available to them for competition. However, there do not seem strong biochemical grounds for eating more than a normal, well-balanced diet.

Admittedly, a number of the B vitamins play a significant role in carbohydrate metabolism. Thus *pantothenic acid* is an essential constituent of coenzyme A (Fig. 2.12), while *thiamine* and *lipoic acid* are essential for the conversion of pyruvic acid to acetyl-coenzyme A and α-ketoglutaric acid to succinyl-coenzyme A (Fig. 2.12). *Nicotinamide* (niacin) is a constituent of the hydrogen acceptors NAD and NADP (Figs. 2.12 and 2.13). *Riboflavin* is a constituent of flavin mononucleotide (FMN) and flavin adenine nucleotide (FAD), cofactors in the cytochrome chain (Figs. 2.12 and 2.13). It has been suggested (but never proven) that ascorbic acid (*vitamin C*) is also an essential cofactor for oxidation-reduction reactions (Staudinger et al., 1961). Finally, there have been reports that a lack of *vitamin E* (α-tocopherol) leads to partial inactivation of the cytochrome reductase system, this vitamin serving as a cofactor or catalyst in the electron transport chain just prior to cytochrome c (Kobayashi, 1974; Shephard, 1978d).

Clinical syndromes can arise from a lack of thiamine, nicotinamide, riboflavin, or ascorbic acid. Large doses of vitamin C are also reputed to reduce the period of disability from upper respiratory infections (T. W. Anderson, Suranyi, and Beaton, 1974).

It is arguable that the endurance athlete may require more of some vitamins than the average citizen (Bourne, 1948; Vytchikova, 1958). The recommended daily intake of the B vitamins is linked to energy expenditure (FAO/WHO, 1967; Horstman, 1972; Nutrition Canada, 1973):

Thiamine	0.4 mg • 1000 kcal^{-1} (4.18 MJ)
Nicotinamide	6.6 mg • 1000 kcal^{-1} (4.18 MJ)
Riboflavin	0.55 mg • 1000 kcal^{-1} (4.18 MJ)

There is also some loss of water-soluble vitamins in the sweat (Dluzniewska et al., 1965; Strydom, Rogers et al., 1977), 0.2–1.0% of the daily intake for thiamine, and 2–5% of the daily intake for riboflavin (Shephard, Kavanagh, and Moore, 1978). Finally, there is some evidence that injury increases the demand for vitamin C. The athlete is naturally at an increased risk of gross injuries (MacIntosh et al., 1972), and the distance runner may also develop periodic microtraumata (Stanesçu, 1971), thereby augmenting his ascorbic acid requirements. On the other hand, most of the vitamins under discussion are so widely prevalent in natural foods that a deficiency is unlikely while eating a well-balanced diet. Any increase of demand in the athlete is inevitably satisfied by his greater overall food consumption.

The main argument advanced by proponents of vitamin therapy has been kinetic: substantial amounts of the various water-soluble vitamins can be administered to an athlete before his body becomes "saturated" and excretion equals intake (Komarevtsev, 1975; Jirka, 1975). Anecdotal evidence has linked saturation treatment with superior performance, but there are not good double-blind trials to back up such claims. If the intake of thiamine is greatly curtailed, there is a fairly long interval before the appearance of deficiency symptoms and an associated deterioration of physical performance. Timing varies with the rigor of the diet, the extent of initial body stores, and the intensity of daily work; however, even with extreme deprivation significant findings are rare during the first week (R. D. Williams et al., 1940; Keys, Henschel et al., 1943,1944, 1945). Some early field reports claimed that cycling, swimming, and shuttle running performance were all improved by the administration of large doses of thiamine. Controlled laboratory experiments have generally shown no gains of endurance, efficiency, or strength with such treatment (Keys and Henschel, 1942; Horstman, 1972), although one double-blind trial in fencers claimed small gains of reaction time, hit frequency, and neuromuscular responses (Van Dam, 1978).

Vitamin C deprivation leads to a reduction of work performance, although the time needed to see such an effect is at least 2 months (R. E. Johnson, Darling et al., 1945). Russian authors, such as Yakovlev (cited by Horstman, 1972), have attributed at least a part of the success of their athletes to the use of vitamin C supplements (75 mg • day^{-1} for the recovery period following speed and strength exercises, 300 mg • day^{-1} for those undertaking prolonged endurance effort). The main rationale for such vitamin therapy seems the quantity of ascorbic acid needed to saturate body stores. Hoogerwerf and Hoitink (1963) also claimed that subjects receiving vitamin C supplements were able to perform better, the effect being similar to that of physical training, while Van Huss (1966) commented on a speeding of the recovery process in those of his subjects who were given the vitamin. In contrast, Keys (1943) found no improvement of performance with vitamin C supplements of 270 mg • day^{-1}, while D. A. Bailey, Carron et al. (1970a,b) reported no change in ventilation or the oxygen cost of a standard exercise test after administering a daily dose of 2 g ascorbic acid.

The possible influence of vitamin E supplements upon performance has been reviewed recently (Shephard, 1978d). Several well-controlled trials (Sharman et al., 1971,1975; Shephard, Campbell et al., 1974) saw no gain of performance relative to placebo groups, although Shephard, Campbell et al., (1974) commented that their experimental group suffered less loss of muscle strength than the controls during 3 months of rigorous endurance training. One as yet unpublished report (Kobayashi, 1974) indicated a gain of maximum oxygen intake, a decrease of oxygen consumption in submaximum work, and a reduction of oxygen debt when vitamin E was administered to athletes exercising at high altitudes.

RESOURCES NEEDED FOR LONG-DURATION WORK

In exercise of very long duration, body reserves of fluid and energy can become exhausted, and performance is then determined by the ability to ingest necessary water, minerals, and metabolites.

Fluid

Lean tissue contains 72–73% water. The total body water content of a typical young man is thus 40–45 l. Under resting conditions, daily water losses are approximately 2.5 l (urine 1200–1500 ml, "insensible" water loss at the skin 500 ml, respiration 400 ml, feces 150–200 ml). During vigorous exercise, urine formation may be reduced, but cutaneous water loss is greatly augmented by the onset of frank sweating. Given adverse environmental conditions, sweat production may amount to 2 l in 1 hr and 6 l or more in 4 hr. Even in a temperate climate, the cumulative sweat loss for a marathon event is often 3–4 l (Wyndham and Strydom, 1969,1972; Costill, 1972a,b; Kavanagh, Shephard, and Pandit, 1974; Kavanagh, and Shephard, 1975a; Shephard, Kavanagh, and Moore, 1978). Under warm and humid conditions, little water vapor is added to expired gas, and respiratory water loss is not greatly increased by the hyperventilation of effort. However, in an arctic environment vigorous exercise can induce a respiratory water loss of up to 1.5 l • day^{-1} (Brebbia et al., 1957; O'Hara, Allen et al., 1977b).

Water reserves are normally maintained by (1) ingestion of fluids (~1000 ml • day^{-1}), (2) ingestion of water as a constituent of foods (~1200 ml • day^{-1}), and (3) production of water during the metabolism of foods (~300 ml • day^{-1}). A further significant resource is the water of hydration associated with glycogen, ~2.7 g • g^{-1}; depending on the extent of glycogen reserves, this can amount to 1000–1600 ml of water. In an endurance event such as a marathon race, much of the glycogen may be burnt, and the metabolic production of water is also greatly increased. Thus, a weight loss of as much as 2 kg may be incurred without significant dehydration (Shephard and Kavanagh, 1975a; Table 2.4).

The maximum safe level of dehydration is still discussed. Much depends upon (1) the amount of water liberated from glycogen and (2) the ease of heat dissipation. In a temperate environment, a weight loss of 7–8% can be well tolerated (Shephard and Kavanagh, 1975a), but margins are likely to be smaller if exercise has been lighter and the environment is hot and humid.

The water balance of a distance runner or team sportsman can be improved if he is preloaded with up to 500 ml of fluid 15–30 min before exercise commences (Kavanagh and Shephard, 1975a). Further small quantities of fluid should be taken at regular intervals as exercise proceeds. It is important that the fluid be not only drunk, but also absorbed. The pyloric sphincter of the stomach is influenced by the osmotic pressure (or possibly the energy content; Hunt and Stubbs, 1975) of the gastric contents. Under resting conditions, Hunt and Pathak (1960) and Hunt (1961) noted that gastric emptying proceeded most rapidly when subjects ingested saline solutions of approximately 250 mosmol • l^{-1}. Both potassium ions and glucose molecules slowed emptying relative to pure water. It remains uncertain as to how far such experiments can be extrapolated to endurance exercise. Hunt's subjects also ingested a single and relatively large volume of fluid (750 ml), thus ruling out the possibility of adjustments in tonicity by subsequent gastric secretion. Furthermore, most authors have found that vigorous physical activity modifies both gastric secretion and gastric emptying (J. M. H. Campbell et al., 1928; Hellebrandt and Tepper, 1934; Fordtran and Saltin, 1967). Costill (1972b) persuaded young runners to drink 100 ml of fluid every 5 min. After 100 minutes, they were uncomfortably distended, but nevertheless had realized a fluid intake of 800–1000 ml • hr^{-1}, absorption being marginally greater for water than for weak glucose solutions (4–5 g • dl^{-1}, 0.22–0.28 mol •

TABLE 2.4
Water Balance Calculations for Middle-Aged Persons Participating in a Marathon Race[a]

1. Average weight loss	1.95 kg
2. Weight loss due to oxidation of food	0.20 kg
3. Weight loss due to sweat loss (1 − 2)	1.75 kg
4. Water gained from oxidation of food	0.30 kg
5. Water liberated by breakdown of glycogen	1.05 kg
6. Net dehydration [3 − (4 + 5)]	0.40 kg
7. Water intake	2.19 l
8. Urine secretion	0.50 l
9. Net fluid intake (7 − 8)	1.69 l
10. Total sweat loss (3 + 9)	3.44 l

[a]Water was provided ad libitum.
Source: Based on data of Kavanagh and Shephard, 1977a.

l^{-1}). Kavanagh and Shephard (1977a) followed middle-aged men over a 4- to 5-hr run; their group sustained a fluid intake of up to 630 ml • hr^{-1}, the largest volumes again being ingested by those who were provided with pure water (Table 2.5).

Minerals

Substantial quantities of various minerals and water-soluble vitamins are excreted in sweat. However, tissue pools of the items in question are quite large relative to potential daily losses, and body needs are usually satisfied by normal feeding between bouts of exercise. Cumulative deficits only develop in exceptional circumstances.

Sodium Ions

The Na^+ concentration found in sweat (20–80 meq • l^{-1}) is lower than that in plasma (~140 meq • l^{-1}) and interstitial fluid (145 meq • l^{-1}). Thus, there is a rise of plasma sodium ion concentration over the course of an event in which sweat is being secreted, irrespective of the fluid available to the participant (Table 2.6). At this stage, a salt-rich drink plainly has no advantage (other than flavor) relative to water (Ladell, 1955).

The total body pool of exchangeable sodium ions ranges from 1880 to 3140 meq, of which ~2175 meq lies in the extracellular fluid (Dancaster and Whereat, 1971; Davidson et al., 1972). The normal daily intake of salt is 5–20 g of NaCl, 85–340 meq of Na^+. The maximum recorded sweat production seems ~141 • day^{-1} (Davidson et al., 1972). Subjects who sweat this heavily normally secrete a dilute sweat. Nevertheless, a cutaneous sodium loss of up to 160 meq • hr^{-1} or 1000 meq • day^{-1} is theoretically possible. Urinary excretion of Na^+ is normally 100–150 meq • day^{-1}, but in severe salt depletion an increased output of adrenal cortical hormones reduces urinary sodium losses almost to zero. There are also small losses of Na^+ in the feces (~5 meq • day^{-1}). Plainly, a cumulative deficiency of sodium is likely if a subject sweats heavily for several days and fails to increase his salt intake. Deficiency syndromes are well recognized among workers in hot industries. Symptoms have also been described in athletes who are training and/or competing in hot climates (Wyndham and Strydom, 1972); under such circumstances, it is a wise precaution to provide competitors with additional sodium, either in a beverage or through the deliberate heavy salting of salads and vegetables.

Potassium Ions

Many proprietary drinks for the athlete contain potassium ions. Under resting conditions, the potassium content of the sweat is quite variable. Haralambie (1975) reported norms of 4–6 meq • l^{-1}, while Ahlman et al. (1952) noted a range of 5–16 meq • l^{-1}, with a mean value of 9 meq • l^{-1}. During prolonged exercise, values of 6.3 meq • l^{-1} (Körge and Seene, 1973), 6.5 meq • 1^{-1}, 4–5 meq • 1^{-1} (Costill, 1977a,b), and 6.6 meq • l^{-1} (MacAraeg, 1975) have been observed. Such figures slightly exceed the normal potassium content of plasma and extracellular fluid (~5 meq • l^{-1}), so that a fall of plasma potassium might be anticipated in response to heavy sweating. A drop of plasma potassium has sometimes

TABLE 2.5
Fluid Balance over the Course of a Marathon Race

Weight Loss (kg)	Fluid Intake (l)	Urine Loss (l)	Sweat Loss (l)	Potential Dehydration (l)	Δ Rectal Temperature (°C)
Special solution (n = 4)					
2.33	1.25	0.29	3.01	0.70	2.3
Erg (n = 3)					
2.33	1.72	0.32	3.53	0.78	2.6
Water (n = 2)					
1.95	2.19	0.50	3.43	0.40	2.5

Note: Averaged data of Kavanagh and Shephard (1977a) for subjects receiving a "special" isotonic solution, the proprietary solution Erg, or water. The special fluid provided before and during the race contained Na^+ 21 meq • l^{-1} and glucose 4.1 g • dl^{-1} (228 mmol • l^{-1}); total osmolarity was ~270 mosmol • l^{-1}. After the race, a second mixture was substituted; this contained Na^+ 20 meq • l^{-1}, K^+ 4.7 meq • l^{-1}, and glucose 4.1 g • dl^{-1} with a total osmolarity of ~278 mosmol • l^{-1}. Erg is a proprietary hyperosmolar solution recommended by its makers for use in marathon events. Analysis shows a composition Na^+ 19 meq • l^{-1}, K^+ 10.7 meq • l^{-1}, and glucose 5.3 g • dl^{-1} (294 mmol • l^{-1}); total osmolarity is approximately 354 mosmol • l^{-1}.

TABLE 2.6
Average Changes of Plasma Mineral Ion Concentrations over the Course of Marathon Events

			Weight	Sweat	Mineral Ion Changes				
Event	n	Fluid	Loss (kg)	Loss (l)	Na^+ (meq • l^{-1})	K^+ (meq • l^{-1})	Fe^{2+} (g • dl^{-1})	Ca^{2+} (mg • dl^{-1})	Mg^{2+} (mg • dl^{-1})
Boston 1973	8	Lucozade, etc.	4.00	4.57	+6.5	+0.3			
Hawaii 1974	3	Erg	1.27	2.89	+0.3	+0.5	−27	+0.6	
Toronto 1974	3	Erg	2.30	3.53	+2.0	+0.4			
	4	Special	2.25	3.01	+3.0	+0.3			
	2	Water	1.95	3.43	+1.5	+0.6			
Boston 1975	6	Saline (50 meq • l^{-1})	2.81	4.51	+4.0	+0.0	+14		−0.22

Note: Data of Kavanagh, Shephard, and associates (1973–1975) for middle-aged men, mostly postcoronary patients.

been seen at the end of marathon races (Ulmeanu et al., 1958), but readings of >5.5 meq • l^{-1} are more common (Peracino et al., 1965; McKechnie et al., 1967; Rougier and Babin, 1969, Table 6). This reflects a leakage of potassium ions from the active muscles, plus some hemolysis of circulating red cells. If effort is continued, a plateau of plasma K^+ concentration develops, 15–20% higher than the resting level, and at this stage K^+ begins to reenter the active muscles (Costill, 1977a,b). Again, there seems no justification for adding potassium to the fluids administered while a person is exercising.

The body pool of potassium is quite large; 80 meq is found in extracellular fluid and ~3500 meq in the tissues, 80% of the latter being in muscle. The daily intake ranges from 50 to 150 meq, a typical figure being 65 meq (Davidson et al., 1972). This closely matches the normal loss in the urine (60 meq • day^{-1}) and feces (10 meq • day^{-1}). An imbalance can arise if there is also vigorous sweating (Malhotra, Sridharan et al., 1976; Lane et al., 1978), or aldosterone-mediated renal excretion of potassium in an attempt to conserve sodium ions (Knochel, 1977). The maximum possible skin loss, with a sweat production of 14 l, would be 77 meq • day^{-1}. Table salt usually contains potassium; thus an increased salt intake usually meets potassium needs. Good natural sources of potassium are dried fruit and nuts. In practice, it is rare to encounter hypokalemia in athletes (Haralambie, 1975), except when training is undertaken in a very hot climate (Knochel et al., 1972; Knochel, 1977). Long-term consequences of potassium deficiency could include impaired glycogen storage (Blackley et al., 1974) and muscle damage due to lack of exercise vasodilation (Knochel, 1977).

Magnesium Ions

In a series of 43 samples collected over a 5-mile run, the magnesium content of sweat ranged from 0.2 to 2.2 mg • dl^{-1}, with a median value of 0.6 mg • dl^{-1} (0.25 meq • l^{-1}). Stromme, Stensvold et al. (1975) also reported initial concentrations of 0.2–0.8 meq • l^{-1}, but after 30–60 min of activity, concentrations dropped to an average of 0.04 meq • l^{-1}. On the other hand, Costill (1977b) noted values of 1.5–5.0 meq • l^{-1}. Such results may be compared with the plasma magnesium content of 1.5–2.0 meq • l^{-1}. Theoretically, sweating might be expected to cause an increase of plasma magnesium ion concentrations. However, in practice plasma magnesium levels decrease during and immediately subsequent to protracted work (Haralambie and Keul, 1970a; L. I. Rose et al., 1970; Refsum, Meen et al., 1973; Haralambie, 1975; Stromme, Stensvold et al., 1975; Table 2.5). This is presumably the result of an exchange of magnesium between the plasma and the other body compartments, including muscle (Costill, 1977b; I. Cohen and Zimmerman, 1978). Bergström, Guarneri, and Hultman (1971) found no changes of intracellular magnesium, but their exercise bout lasted for only 20 min. Stromme, Stensvold et al. (1975) noted a rise of erythrocyte magnesium as plasma magnesium fell; furthermore, plasma concentrations were restored within 2 hr of ceasing exercise, even if no specific attempt was made to remedy the deficiency.

Magnesium is an essential constituent of many

of the enzyme systems associated with energy transfer; for example, both cocarboxylase and coenzyme A function only in the presence of traces of this mineral. The critical level for the appearance of electrocardiographic abnormalities and muscular hypoexcitability is a plasma concentration of ~1.6 meq • l^{-1}. Figures below this threshold are commonly encountered with sustained exercise (Haralambie, 1975).

The total body store of magnesium is ~2000 meq, a large part of this being present in bone as phosphate and bicarbonate. The usual diet provides 17-35 meq of magnesium per day, much of it in the chlorophyll of green vegetables. Urinary losses are normally about 0.7 meq • hr^{-1} but are reduced to 0.22-0.26 meq • hr^{-1} during prolonged activity. Shephard, Kavanagh, and Moore (1978) estimated the total sweat loss over a marathon race at 1.3 meq. Even if their average magnesium concentration is applied to a sweat production of 14 l, daily losses would not exceed 3.5-4.0 meq. Many days of activity would thus be needed to deplete body stores. Strømme, Stensvold et al. (1975) saw no cumulative loss over two ski races of 90 and 70 km, respectively, although others have described low resting magnesium levels in endurance athletes (Haralambie, 1975).

Iron

The total body content of iron is ~4 g; 2.5 g is in the form of hemoglobin, and 1 g is stored as ferritin in the reticuloendothelial cells (Davidson et al., 1972). Dietary intake is 12-14 mg • day^{-1}, but only about 1 mg of this is absorbed, the remainder being eliminated in the feces. Urinary losses are negligible (50-150 μg • day^{-1}), but substantial quantities of iron are eliminated as desquamated epithelial cells (0.5-1.0 mg • day^{-1}). In women, menstruation leads to further loss averaging 28 mg of iron per menstrual period.

There has been much discussion as to the iron content of sweat because it is difficult to collect specimens that are not contaminated by epithelium. Hussain and Patwardhan (1959) quoted an average concentration of 1.6 mg Fe^{2+} • l^{-1}, but more recent estimates have been lower (200 μg • l^{-1}, Vuori et al., 1966; 450 μg • l^{-1}, Shephard, Kavanagh, and Moore, 1978). Accepting the present author's values, a sweat loss of 1.8 mg Fe^{2+} is incurred over a marathon event, and the maximum daily sweat loss could exceed 6 mg Fe^{2+}.

The acute response to sustained exercise is commonly a rise of serum iron (Haralambie et al., 1975; Shephard, Kavanagh, and Moore, 1978). Possibly, there is an increase of both serum iron and its transport protein transferrin to allow an augmented synthesis of iron containing enzymes and tissue metalloproteins (Haralambie, 1975). However, if the subject is well trained, the rise of serum iron does not occur (Haralambie, 1975), and indeed if hydration is well maintained a decrease of serum iron may be detected (Kavanagh and Shephard, 1975a).

Do athletes become anemic (Gilligan et al., 1943; Ekblom and Hermansen, 1968; de Wijn et al., 1971; Clement et al., 1977; Wirth et al., 1978) as a long-term consequence of iron losses in the sweat and, if so, should they be given iron supplements? Certainly, the distribution of serum iron values is nongaussian, ~20% of athletes having values of less than 80 μg • dl^{-1} (Haralambie, 1975). Among our "postcoronary" runners, two out of nine fell into this category (Shephard, Kavanagh, and Moore, 1978; Kavanagh and Shephard, 1975a). Relative to track officials, an increased percentage of athletes also show serum iron depletion (saturation less than 20%) or a latent iron deficit (saturation less than 15%) (de Wijn et al., 1971). However, a variety of factors other than sweat loss contribute to this situation. An increased formation and destruction of red cells and an increased synthesis of other iron-containing tissues augments the demand for iron, while at the same time a large total energy intake is achieved by boosting the fat content of the diet, a maneuver that reduces the proportion of iron absorbed from the intestines.

Other Ions

The *calcium* loss in sweat is about 2 mg • dl^{-1} (Verde et al., unpublished). This has a negligible effect on calcium balance unless the exercise is repeated for many days. The body pool of miscible calcium is 4-7 g, dietary absorption is about 200 mg • day^{-1}, and urinary losses are 100-350 mg • day^{-1} (Davidson et al., 1972).

Manganese is a trace element which activates or is a component of many enzymes, including constituents of the Krebs cycle and enzymes concerned with the metabolism of connective tissue. In animals, an experimental deficiency of manganese leads to weakness of the Achilles tendon, a common complaint of human athletes. Sweat levels of manganese are ~0.15 μg • dl^{-1} (Haralambie, 1975). The dietary intake is 5-10 mg • day^{-1}, so that it is unlikely a deficiency could arise from sweating.

The sweat content of *copper* is 5-12 μg • dl^{-1} (Haralambie, 1975). The normal dietary intake is at least 2 mg • day^{-1} (Davidson et al., 1972), and there is no evidence that a deficiency can arise in humans; indeed, athletes have higher resting levels than normal subjects (Haralambie and Keul, 1970b).

Zinc is a constituent of many vital enzyme systems. Sweat concentrations are about 35 $\mu g \cdot dl^{-1}$. The body pool is 1–2 g, and the dietary intake (10–15 $mg \cdot day^{-1}$) is quite large relative to sweat losses, although much of the ingested zinc passes into the feces without absorption. The immediate effect of endurance exercise is a dramatic increase of serum zinc (Hetland et al., 1975). This seems the result of an escape of zinc-containing enzymes from the active muscles.

Energy Intake

We have seen already that the body contains a sufficient reserve of energy to live for ~4 weeks without additional nutrients. Thus the provision of fluid normally has priority over the intake of food during bouts of endurance work. However, there are occasional protracted feats, such as mountaineering expeditions and transcontinental runs, when priorities must be reserved.

The mainland segment of the Trans-Canada run covers a distance of ~6768 km. If a man completes this in 10 weeks, he must make daily runs of 100 km, dropping to distances of 60 km on mountainous sections of the route. For a 60-kg runner, the daily energy expenditure is just under 28.5 MJ (Shephard, Conway et al., 1977). Further, up to 8 hr a day are spent in running, so that the equivalent of at least one large meal must be ingested while exercising. The drink often recommended to endurance athletes (4–5% glucose) provides a fluid volume of 600–800 $ml \cdot hr^{-1}$, but the corresponding glucose intake is at most 40 g (222 mmol) or 637 $kJ \cdot hr^{-1}$. Fat emulsions have not usually been considered for athletes because they are reputed to slow gastric emptying. However, in practice they provide quite an effective basis of energy supply for the ultralong-distance performer (Shephard, Conway et al., 1977; O'Hara, Allen et al., 1977a,b,c; Table 2.7).

TABLE 2.7
Intake of Energy and Fluid Attained During Physical Exercise (Distance Running) Using Various Preparations

Preparation	Ingested Mass ($g \cdot hr^{-1}$)	Percentage Moisture (%)	Energy Yield ($kJ \cdot g^{-1}$)	Ingested Energy ($kJ \cdot hr^{-1}$)	Ingested Water ($ml \cdot hr^{-1}$)
Water [a]	930	100	0	0	930
Mushroom soup plus milk [a]	740	83.3	4.44	3290	615
Tea, syrup plus sugar [a]	625	81.9	3.18	1995	510
Beef broth [a]	520	96.2	0.34	175	500
Ice cream plus syrup and milk [a]	615	65.6	7.50	4620	405
Instant breakfast [b]	306	85.0	3.01	922	260
Celery soup, skim milk, milk powder [b]	283	91.9	1.66	469	260
10% glucose [b]	286	90.9	1.45	415	260
5% glucose [c]	630	95.2	0.75	471	600
Ice cream, corn syrup, milk, chicken bouillon [b]	288	90.3	1.73	499	260
Ovaltine, condensed milk, sugar [b]	324	80.2	3.54	1148	260
Ice cream, condensed milk, choc. syrup [b]	349.5	74.4	5.04	1760	260
Instant breakfast plus condensed milk [b]	350.5	74.2	4.36	1529	260
Ice cream, skimmed milk, corn syrup [b]	316.5	82.1	3.57	1131	260

[a] Subject drank each of five fluids *ad libitum* in rotation every 12 min.
[b] Subject drank 130 ml of each fluid every 30 min.
[c] Based on experiments of Costill, 1972b.
Source: Based on data of Shephard, Conway et al., 1977 and W. J. O'Hara, Allen et al., 1977a.

CHAPTER 3
Limitations Upon the Rate of Working

In the previous chapter, we sketched the energy resources available to humans and indicated various mechanisms that could limit the rate of energy release. We now examine the nature of these limitations in more detail.

MAXIMUM HUMAN POWER OUTPUT

Several authors have estimated the maximum human power output from the dynamics of body movement. Fletcher and Lewis (1959,1960) calculated a power of 1.49 kJ • sec^{-1} for a champion weight lifter and 0.45 kJ • sec^{-1} for a pole vaulter, while Cavagna, Margaria, and Arcelli (1965) estimated that a champion sprinter developed an initial power of 2.22 kJ • sec^{-1}. C. T. M. Davies (1971a) based calculations for the vertical jump on recordings from a force platform (Fig 3.1); he found an average power in men of 3.90 kJ • sec^{-1} and in women 2:35 kJ • sec^{-1}. McGilvery (1975) based his calculations on the height of a vertical jump; he obtained a lower figure of 2.08 kJ • sec^{-1} A simplistic analysis supports the McGilvery figure. The average vertical jump for a young man is ~55 cm, or for a body weight of 70 kg a work of 0.55 × 70 × 9.81 = 378 N • m; if this is performed over 0.2 sec, as in the tracings of C. T. M. Davies (1971a), the rate of working is 1.89 kJ • sec^{-1}. One reason that Davies obtained higher figures is that his calculation (quite properly) took account of the initial descent of the center of gravity that precedes a vertical jump. In a staircase sprint (Fig. 1.16) C. T. M. Davies (1971a) found a power of 0.98 kJ • sec^{-1} for men and 0.73 kJ • sec^{-1} for women.

There is a rapid drop of maximum power output as the duration of activity is extended (Fig. 3.2). If we assume that the curve has a simple exponential form, the half-time is ~10 sec (C. T. M. Davies, 1971a).

From the biochemical point of view, muscular contraction depends upon the transfer of energy from the creatine phosphate molecule to actin/myosin cross-bridges. The ultimate limitation of physical performance is thus the power of the creatine phosphotransferase reaction (Fig. 1.7). Expressing data

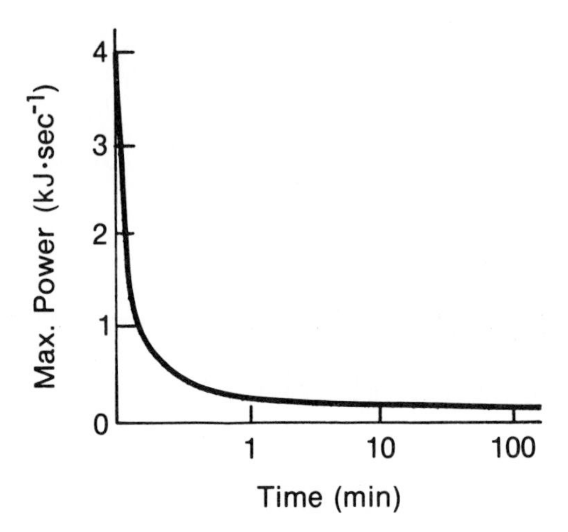

FIG. 3.1. Relationship between maximum power output of human subjects and duration of activity. (After D. R. Wilkie, 1960, and C. T. M. Davies, 1971a.)

per kg of muscle tissue, and allowing for differences between in vitro and in vivo temperatures, the maximum power of this process has been set at 6 mmol • kg^{-1} • sec^{-1}, or 277 J • kg^{-1} • sec^{-1} (Kleine, 1967; McGilvery, 1975). Until stores of phosphagen are exhausted, it is only necessary to consider the efficiency of the primary transfer processes. ATP yields its energy with an efficiency of ~49% (Fig. 1.7), while the creatine phosphotransferase reaction has an efficiency of ~85%. Neglecting delays imposed by a possible on-transient, the usable portion of the initial power thus amounts to about 116 J • kg^{-1} • sec^{-1}. Applying this result to the activities already discussed, the work demands of weight lifting would be met by the phosphagen stores found in 12.8 kg of active muscle; corresponding figures for pole vaulting and sprinting would be 3.9 and 8.5 kg, respectively.* Davies' estimate of the work involved in a vertical jump would imply 33.8 kg of active muscle in men and 20.2 kg in women, while McGilvery's estimate would need only 18.0 kg.† C. T. M. Davies argues strongly that it is unrealistic to expect the simultaneous and effective contraction of all the body musculature during a vertical jump. Since the average man has only 28 kg of muscle, C. T. M. Davies explains his 33.8-kg figure on the basis that a preliminary stretching of the muscles increases their potential energy through an increase of ATP stores (Abbott et al., 1951) or a stretching of series elastic elements (Cavagna, Saibene, and Margaria, 1964 a,b; Cavagna, 1970). Davies noted that the center of gravity was lowered by ~30 cm preparatory for a vertical jump. If all of this potential energy were stored, a 70-kg person could perform an additional 206 N • m of work; dispersed over 0.2 sec, as in a typical jump, this would contribute a power of 1.03 kJ • sec^{-1}, reducing the mass of active muscle to the more realistic figure of 24.7 kg.

PROPERTIES OF INDIVIDUAL MUSCLE FIBERS

Inherent Contractile Properties

The sustained muscular contraction demanded by an activity such as a vertical jump is built from a sequence of individual twitches each lasting a fraction of a second. Thus, while the factor limiting the power output of the body as a whole is the rate of breakdown of creatine phosphate by creatine phospho-

*This last value may need to be augmented, because the staircase sprint has to be sustained for several seconds.

†McGilvery (1975) actually speaks of 30 kg of active muscle, but this is because he assumes a mechanical efficiency of 25%, a value that is unnecessarily low for a single maximum effort with the rebuilding of phosphagen stores during later leisure.

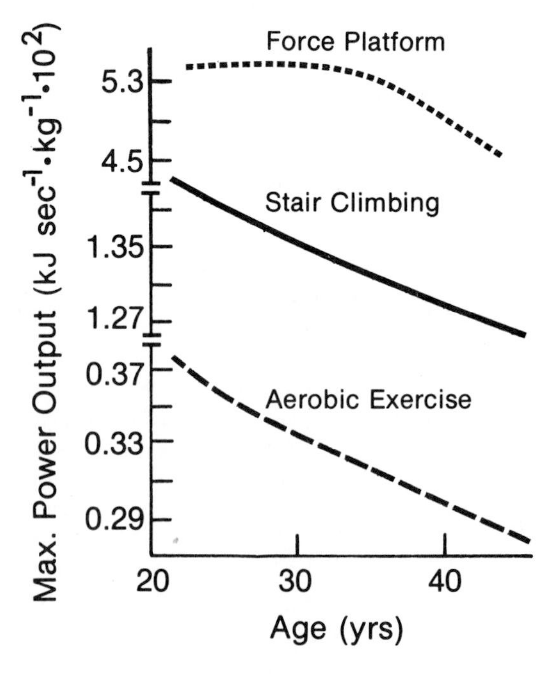

FIG. 3.2. Influence of age upon maximum power output. (Based on data of C. T. M. Davies, 1971a.)

transferase, the performance of the individual fiber depends upon its myosin ATPase content (Bárány, 1967; Barnard, Edgerton et al., 1971).

Animal studies distinguish three fiber types (Edgerton, 1976). Fast, fatiguable fibers reach their peak tension in ~30 msec; they contain large amounts of myosin ATPase and moderate amounts of acid ATPase. Fast, fatigue-resistant fibers also reach their peak tension in ~30 msec; they have a fast half-time of relaxation and contain large amounts of myosin ATPase but small amounts of acid ATPase. Slow fibers take at least 80 msec to reach peak tension, and relax more slowly; they stain lightly for myosin ATPase, but relatively darkly for acid ATPase. Some authors maintain that human muscles contain only fast fatigable and slow-twitch fibers. However, the species difference seems mainly a matter of degree, and recent investigators have described two or even three types of fast-twitch fibers (types IIa, IIb, IIc) in human muscle preparations (Kaijser and Jansson, 1977).

Operating Temperature

As with most biochemical processes (the Law of Arrhenius), the speed of the myosin ATPase reaction is increased 2.0–2.5 fold by a 10°C rise of local temperature. Under resting conditions, the limbs are often cooler than the core temperature of 37°C. However, local heat production during warm-up exercises brings the intramuscular temperature to 41–43°C within a few minutes (Asmussen and Bøje, 1945; Edington and Edgerton, 1976), thereby augmenting maximum power by at least 50%.

An investigator may deliberately cool a muscle preparation. This slows the speed of the various biochemical reactions, increasing the resolving power of any apparatus used to study muscular contraction. However, it is difficult to extrapolate the information thus obtained to normal body temperatures, since it is unlikely that temperature coefficients are comparable for all of the various processes involved.

Changes in muscle force at a given temperature also reflect changes in the entropy or thermoelastic heat of the system [equations (1.2) and (1.3)]. If S′ is entropy, L′ is muscle length, F′ is muscle force, and T′ is temperature, then

$$\left(\frac{\partial S'}{\partial L'}\right)_{T'} = -\left(\frac{\partial F'}{\partial T'}\right)_{L'} \tag{3.1}$$

The entropy change for a given change of length at a given temperature can thus be calculated from the variation of force with temperature at a given muscle length. Under resting conditions, the muscle behaves rather like rubber. Passive stretching of the fibers restricts the movement of their lightly bonded molecules; entropy is decreased, and the tissue is warmed. If stretching is continued, the point is ultimately reached where the tissue is supported not by the actin and myosin molecules, but by parallel fibers of elastic connective tissue. The entropy effect then changes its sign, and the system behaves like other firmly bonded molecules such as metals.

If a muscle fiber is put under tension while it is contracting, the actin and myosin molecules are more firmly bonded to one another; in this situation, stretching leads to absorption rather than release of heat. From studies of isometric contractions in the frog sartorius, F. D. Carlson and Wilkie (1974) set the change of entropy at 3.35 mJ • g^{-1}, or 94 J for 28 kg of muscle. It is hardly surprising that this small component of the total heat exchange remained undiscovered until 1953 (Woledge, 1961).

Patterns of Stimulation

Muscle shows an all-or-none response. If a stimulus exceeds a certain threshold, a given motor unit (the group of muscle fibers supplied by a single motoneuron) always contracts. The intensity of response is unchanged by any further increase in the strength of the stimulus, but can be modified by local operating conditions such as temperature, fatigue, and initial fiber length. Rapid repetition of stimulation may augment the tension that is produced (Milner-Brown et al., 1973). This probably reflects facilitation of actin/myosin bonding through an incomplete return of calcium ions to their storage sites in the sarcoplasmic reticulum (Chap. 4; Edington and Edgerton, 1976).

If a second suprathreshold stimulus is applied within a few seconds of the first, there is no additional response: the muscle fiber is said to be in a "refractory" state. However, it is possible to select a slightly longer stimulus interval, or to stimulate an adjacent fiber, so that the second response is enhanced by the after-effect of the initial contraction (summation, Fig. 3.3). This reflects the fact that series elastic elements (Fig. 1.13) are already under tension, and dissipate less of the force generated by the second muscle twitch.

If a series of suprathreshold stimuli are applied in close succession, a tremulous contraction or subtetanus is seen. If the interstimulus interval is further reduced, fusion of the individual twitches becomes

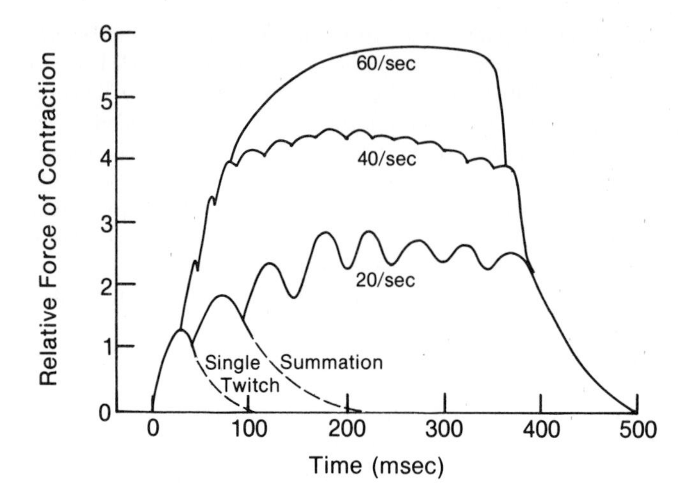

FIG. 3.3. Summation and tetanus in a fast-twitch muscle. The response to a second suprathreshold stimulus is increased if it follows soon after the first. With repetitive stimuli (30 to 40 Hz), a partial tetanus is observed. Complete tetanus develops at frequencies of about 60 Hz. Rather lower frequencies (~20 Hz) are needed for fusion of contractions in slow-twitch muscles.

more complete, until at a frequency of ~60 stimuli per sec (higher for fast- than for slow-twitch fibers), a fully developed tetanus is seen. The force now developed is 3 or 4 times as great as during a single twitch. In some muscles, the tension associated with a partial tetanus can also be augmented for a substantial period by a single shortened interstimulus interval (the "catch" effect, R. E. Burke and Edgerton, 1975).

Tetanic contractions are frequently used in performing normal physical activities. However, a tremulous pattern of movement is rarely seen because the motor units fire asynchronously and individual tremors cancel one another out.

Length/Tension Characteristics

We have noted previously some characteristics of the muscle length/tension diagram (Fig. 1.9). Resting tension (indicated by the lower margin of the diagram) does not vary linearly with muscle length, but rather shows a progressive increase as the mesh of connective tissue within the muscle belly (the parallel elastic element of Fig. 1.13) is tautened. The vertical separation of the upper and lower borders of the length/tension diagram indicates the maximum active tension that the muscle can develop at a fixed length (isometric force), while the horizontal separation indicates the maximum potential shortening at a fixed tension (isotonic contraction). The idealized diagram of Fig. 1.9 can readily be reproduced in frog muscle experiments. In the human body, on the other hand, the external forces that can be generated at any given muscle length are influenced greatly by considerations of leverage (Figs. 3.4 and 3.5). Further, the anatomical characteristics of many joints restrict the lengthening and shortening of muscles to a 20–30% range about their resting lengths.

Isometric force is usually expressed per cm^2 of muscle cross-section. Mammalian muscles typically can realize forces of 15–40 N • cm^2. Maximum tension is usually developed when a contraction is initiated at ~20% above resting length. Athletes thus

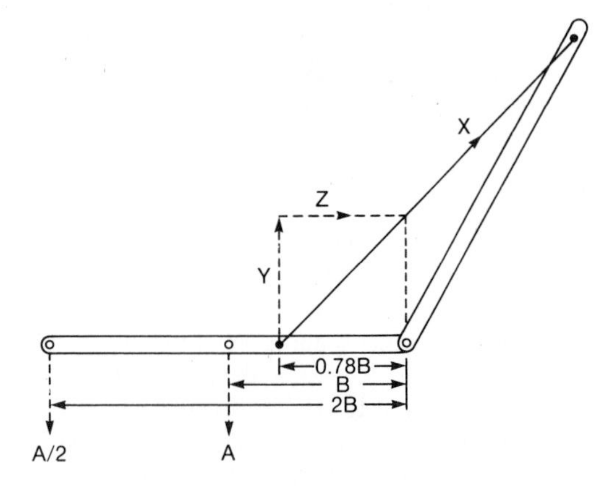

FIG. 3.4. The principle of leverage. A muscle exerts a pull of X kN in the direction indicated by the arrow, drawing the lower (mobile) lever arm toward the upper lever arm (fixed by appropriate accessory muscles). The force X can be resolved into two smaller components, Y at right angles to the lower lever, and Z in parallel with the lower lever. The distance from the fulcrum of the lever to the lower insertion of the muscle is $0.7B$ cm. The torque developed is thus $0.7YB$ kN • cm. The force developed is recorded by two cable tensiometers, harnessed at right angles to the lower lever, B and $2B$ cm from the fulcrum. The torque at both sites is AB kN • cm = $0.7YB$ kN • cm, but the forces indicated by the tensiometer are A and $A/2$ kN, respectively.

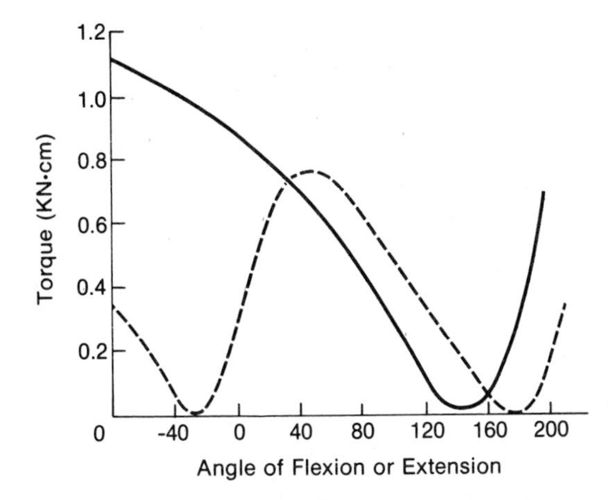

FIG. 3.5. The influence of the angle of flexion or extension on the isometric torque developed by the two parts of the deltoid muscle. (Adapted from a study by Hvorslev, 1928). Similar curves can be drawn for isotonic contraction (Asmussen, Hansen & Lammert, 1965).

tend to place a muscle under a slight stretch prior to making a maximal contraction.

From a mechanical point of view, the optimum length of an individual sarcomere (the portion of a muscle fiber lying between two adjacent Z lines) (Fig. 4.1), is 2.0–2.5 μm (Fig. 3.6). If such a unit is stretched to >2.5 μm, this reduces the overlap of the actin and myosin filaments, thereby limiting the potential for cross-bridge formation during contraction. No contraction is possible if the sarcomere length is more than 3.5 μm (A. F. Huxley, 1962). We may presume that with such extension, the actin and myosin filaments no longer overlap, and no cross-bridges can be formed. The reduction of maximum tension with compression of the sarcomere is more difficult to explain. Possibly the tubules of the sarcoplasmic reticulum are distorted sufficiently to hamper the release of calcium ions (a key step in initiating muscular contraction). There may also be mechanical disruption of the actomyosin; and an

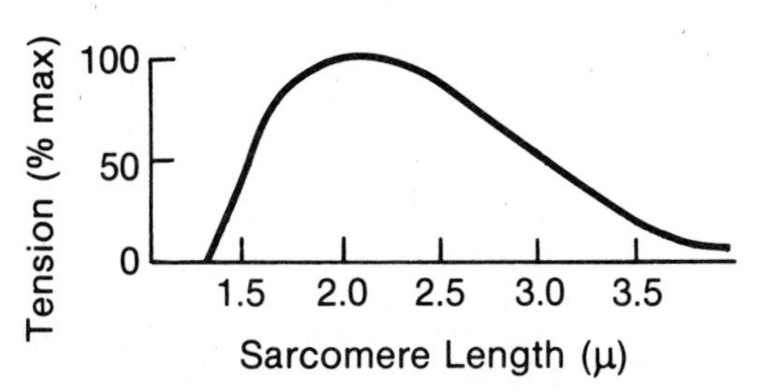

FIG. 3.6. Influence of sarcomere length on muscle tension. (After A. M. Gordon et al., 1966.)

overlap of actin filaments may restrict cross-bridge formation. Last, increased rigidity of the tissue may absorb much of any force that is developed (F. D. Carlson and Wilkie, 1974; Edington and Edgerton, 1976).

A muscle usually shortens as it contracts ("concentric" effort). However, it can undergo lengthening during tetanic contraction, as when the biceps is used to control the rate of descent of the body following a chin-up ("negative" work). The tension developed during such an "eccentric" effort (line c, Fig. 1.9) is typically greater than during concentric contraction. This reflects partly a change in the contractile properties of the stretched myofibrils (optimum overlap of actin and myosin filaments) and partly reversible storage of energy (an increase of entropy) plus a stretching of the series elastic elements. In other situations, a muscle may be forcibly shortened as it contracts. The tension developed (line d, Fig. 1.9, and Fig. 3.7) is then considerably less than that predicted from fiber length.

Because the force developed by a given motor unit is greater for eccentric than concentric contraction, fewer motor units are needed to sustain a given tension during eccentric activity. This point can be confirmed by electromyography (Bigland and Lippold, 1954a; Bouisset, 1973). Normally, there is a linear relationship between force and motor unit activity as expressed by the integrated electromyogram (Fig. 3.8). If a muscle is weak, more motor units are recruited to develop a given tension than if it is strong. Equally, more units are recruited under concentric than under eccentric conditions.

Because of differences in the number of active motor units, oxygen consumption is less and efficiency is higher for eccentric than for concentric effort. One convenient laboratory form of eccentric work is downhill treadmill running. Margaria (1971) calculated an efficiency of −120% for subjects who were running downhill; in other words, potential energy was dissipated as heat, and the muscles added a further 20% to this work to control descent of the subject. In terms of chemical energy usage, the cost was about 1/4.8 times that of uphill running. Asmussen (1953) had his subjects perform similar experiments, riding a bicycle both up and down a treadmill. He found that the cost of positive (uphill) work was 5.9 times that of downhill (negative) work.

Speed of Contraction

We noted [equation (1.25)] the classical hyperbolic relationship between the speed of contraction of a

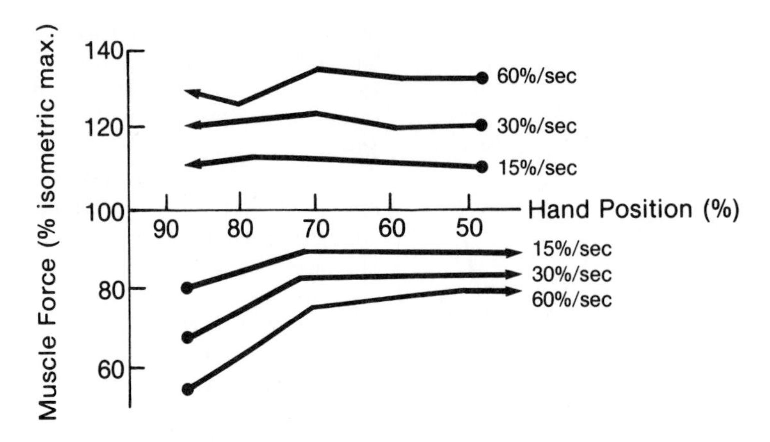

FIG. 3.7. Isometric force developed while arm is being extended (upper panel) or forcibly shortened (lower panel). Ordinates are muscle force, expressed as percentage of isometric maximum for a concentric contraction. Abscissae are hand position, expressed as percentage of arm length for full arm extension. Rates of lengthening and shortening are also expressed as percentage of arm length (change per second). (Based on experiments of Asmussen, Hansen & Lammert, 1965.)

muscle and its loading. Useful external work is proportional to the term P′V′, while the term aV′ is the fraction of energy dissipated in overcoming internal (viscous) resistance. The total work performed (P′ + a)V′ decreases with increasing load. However, with very rapid contractions a large part of the available energy is dissipated against internal resistance. The power output per contraction is thus maximal at an intermediate loading (Fig. 3.9), perhaps 50–60% of maximum isometric tension (Fig. 1.9). Nevertheless, a faster and less forceful contraction allows more repetitions of a given effort in unit time. The power of rhythmic work is thus greatest when individual contractions develop ~30% of maximum isometric force. Well-designed machines such as the bicycle require a force of this order (E. W. Banister and Jackson, 1967). The force/velocity relationship is displaced toward the right by an increase of muscle temperature (Sargeant and Jones, 1978).

The several phases of muscle heat production

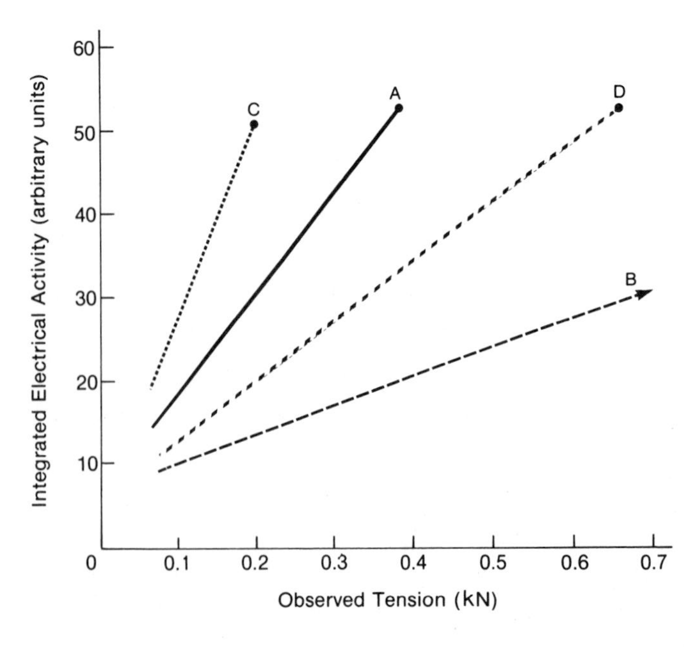

FIG. 3.8. Relationship between observed tension and integrated electrical activity as seen in the electromyogram. Line A is the integrated EMG for muscle developing maximum concentric tension of 0.37 kN. Line B is the same muscle, contracting eccentrically (maximum > 0.7 kN). Line C is the concentric contraction of weak muscle (maximum concentric tension 0.19 kN). Line D is the concentric contraction of strong muscle (maximum concentric tension 0.66 kN).

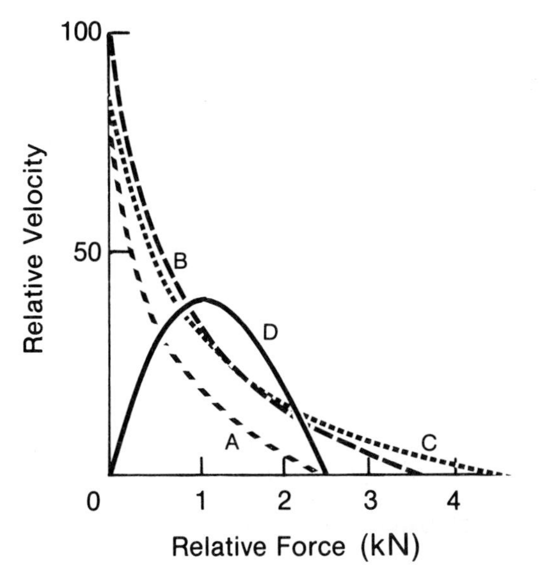

FIG. 3.9. Force/velocity relationship for muscle. Curve A is the response under cool conditions. Curve B is the response under warm conditions. Curve C is the response under warm conditions after hypertrophy due to strength training (Binkhorst & van't Hof, 1973). Curve D is the power output. Note that the form of the curve differs, depending upon the controlled variable (loading or speed).

have already been considered briefly. Typical events during an isometric contraction are illustrated in Fig. 3.10. Available data refer to toad muscle at 0°C. *Activation heat* first appears 10–15 msec after stimulation, reaches a peak at 20–30 msec, and then declines exponentially with a half-time of ~50 msec (A. V. Hill, 1953). The maximum rate for this phase of heat production is ~69 mJ • g^{-1} • sec^{-1}, and the total output is ~4 mJ • g^{-1}. About a third of the heat is attributable to internal shortening and thermoelastic effects [equation (3.1)], while the remainder is probably due to the release of calcium and its subsequent interaction with the contractile proteins (F. D. Carlson and Wilkie, 1974). A second component of "labile heat production" is now distinguished (curve A, Fig. 3.10). This has a half-time of ~1.5 sec. The rate and total production of heat for this phase are ~5 mJ • g^{-1} • sec^{-1} and 10 mJ • g^{-1}, respectively. The labile heat varies with the speed of operation of the muscle; it seems to arise from some chemical process not directly associated with contraction and relaxation. A third component of "stable heat production" is also described (line B, Fig. 3.10). This reflects inefficiencies in the transfer of energy from ATP to the actin/myosin bonding; it appears throughout tetanus at a rate of ~5 mJ • g^{-1} • sec^{-1}.

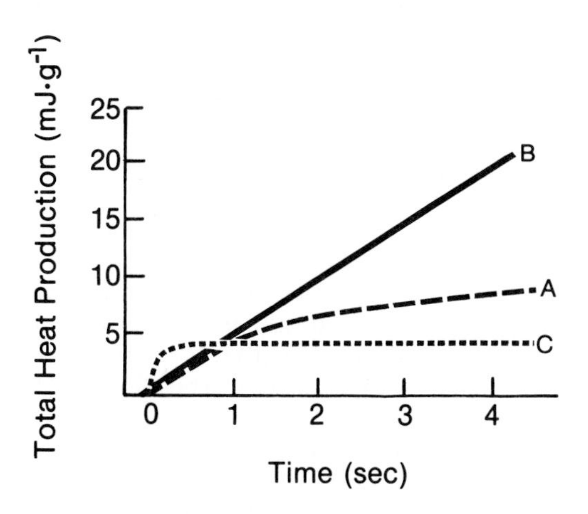

FIG. 3.10. Relationship of total heat production (toad muscle, isometric contraction at 0°C) to time. A, labile heat; B, stable heat; C, activation heat. (Based on F. D. Carlson & Wilkie, 1974.)

If the muscle is allowed to shorten, additional heat is produced. The maximum *heat of shortening* is ~21 mJ • g^{-1} • sec^{-1}. The extra heat is roughly proportional to the speed of shortening, or a constant times the distance shortened, so that

$$\text{Shortening heat} = \alpha \times \text{distance shortened} \tag{3.2}$$

The constant α has the dimensions of a force, and is analogous to but not identical with the constant a of equation (1.25). It is equal to ~25% of maximum isometric tension.

While the rate of heat production is faster if a muscle is allowed to shorten, the continuing stable heat production of an isometric twitch causes a very similar total heat production for the two forms of contraction (Fig. 3.11). This implies a differing total output of energy in the two types of experiment, for the isotonic contraction has also accomplished external work, whereas the isometric twitch has not. Fenn (1923) was the first to notice that shortening caused an increased output of energy roughly equal to the external work that was performed. This phenomenon has since been termed the Fenn effect. Its precise explanation remains uncertain. Heat production is further complicated if shortening is associated with a tetanic contraction. Individual sarcomeres may then spend a considerable part of the total contraction period at such a short length that there is interference with both cross-bridge formation and associated heat production (Aubert, 1956).

Subsequent to isotonic contraction, a muscle

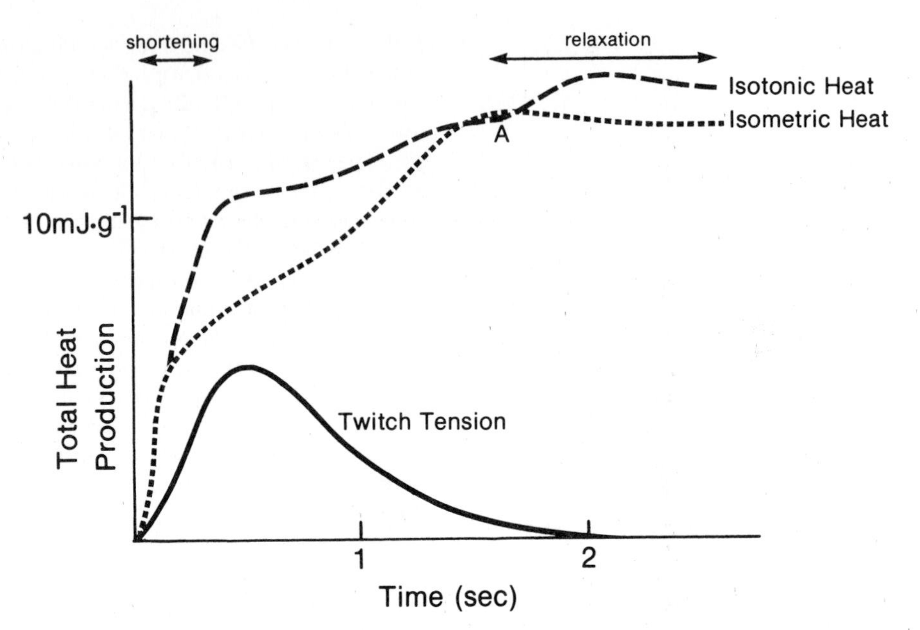

FIG. 3.11. Comparison of heat production in isometric twitch, and isotonic shortening and relaxation with a light load. At point A, the total heat production is similar for the two forms of contraction. Subsequently, the isotonic heat is boosted by relaxation heat, as the load that has been lifted is lowered. (Based on an experiment of R. C. Woledge, cited by F. D. Carlson & Wilkie, 1974.)

may perform further eccentric work in controlling the descent of the initial load. This accounts for the *relaxation heat*, seen after point A in Fig. 3.11.

Finally, there is a phase of *recovery heat*. This normally reflects inefficiencies in the aerobic processes that restore the initial state of the muscle. It is closely correlated with both the oxygen consumption of the recovery period and the total energy expended to this time (heat + work). Measurements of creatine phosphate resynthesis have been made in muscles perfused by oxygenated Ringer's solution at 0°C; the half-time of recovery from a tetanic contraction is then ~11.5 min (Dydynska and Wilkie, 1966). This figure agrees quite well with classical estimates of oxygen debt repayment in humans (di Prampero, 1971a). Recent work suggests that some recovery heat can appear even under anerobic conditions (F. D. Carlson and Wilkie, 1974); however, the precise basis of this phenomenon has yet to be elucidated.

SPECIAL PROPERTIES OF CARDIAC MUSCLE

Problems of Experimentation

The histological appearance of cardiac muscle is very similar to that of skeletal muscle. However, it is difficult to investigate the mechanical properties of cardiac muscle, since there are problems in isolating small groups of parallel fibers from this tissue. Attempts to slice strips out of the heart wall leave many cut and damaged fibers, while the complex geometry of the heart hampers the conversion of pressure and volume measurements for the whole heart into corresponding length and tension data for individual fibers [equation (1.17)]. Experiments are often conducted on the small papillary muscles of the mitral and tricuspid valves, although there is no certain proof that the behavior of these elements is typical of the heart as a whole.

Patterns of Stimulation

It is not possible to induce a tetanus in cardiac muscle. Nevertheless, the response to supramaximal stimulation is very labile (Fig. 3.12). On changing from a low to a high frequency of stimulation, the twitch size increases progressively for many contractions, while there is an exponential decline back toward the original pattern of contraction if the slower speed of stimulation is resumed.

Mechanical Characteristics

The mechanical behavior of cardiac muscle corresponds to the familiar model of a contractile element

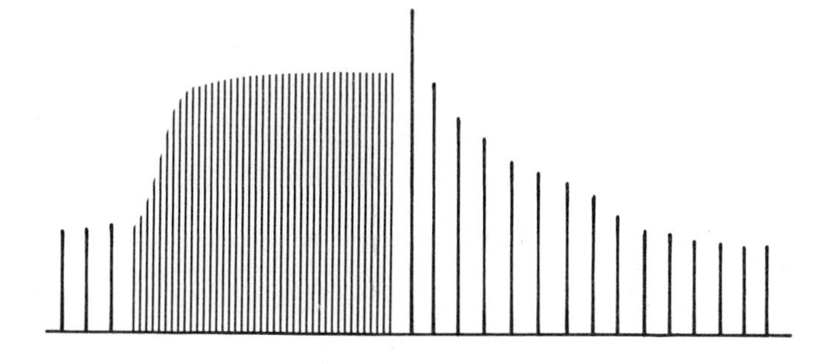

FIG. 3.12. The effect of stimulus interval on twitch force in cardiac muscle. (Based on an experiment of Blinks & Koch-Weser, 1963.)

arranged in series with an elastic component (Fig. 1.13). However, the elastic component is more compliant than that of skeletal muscle; at maximum isometric tension, the elongation of elastic tissue is 10% of muscle length, compared with 3% in skeletal muscle.

The form of the force/velocity curve is much as for skeletal muscle (Fig. 3.9), but unlike skeletal muscle the heart wall has an appreciable resting tension, and the contractile state develops slowly after stimulation. The velocity of shortening thus depends on both the force to be developed and the time that has elapsed since stimulation.

PHOSPHAGEN EXHAUSTION

While events such as a vertical jump or a pole vault depend upon the rate at which phosphagen energy can be transferred to the contractile proteins, slightly more protracted activities are limited by exhaustion of phosphagen stores. The limits of phosphagen-based activity have been judged from staircase sprints, from the negative intercept of lactate formation during graded "supramaximal" work (Margaria, Cerretelli, and Mangili, 1964), and from a resolution of the oxygen debt repayment into alactate and lactate components.

Staircase Sprints

Margaria and his associates calculated that the maximum speed of ascent of a staircase in a young man was equivalent to a sustained phosphagen usage of 3 mmol • kg^{-1} • sec^{-1} (di Prampero, 1971a,b). This estimate apparently assumed the participation of all skeletal muscles (28 kg), with a mechanical efficiency of 25%. We have already noted that while the reaction is limited to usage rather than replenishment of phosphagen stores, an efficiency of 41.7% is more appropriate. Furthermore, if the maximum power of the creatine phosphotransferase reaction could be sustained throughout the task, the necessary energy could be delivered by a mere 8.5 kg of muscle, with exhaustion in 5 sec. Effort slows after 5–6 sec of either a vertical or a horizontal sprint; if the reaction is carried to exhaustion of phosphagen, it is reasonable to suggest that the average rate of energy transfer is about a half of the maximum (that is, 3 mmol • kg^{-1} • sec^{-1} rather than 6 mmol • kg^{-1} • sec^{-1}). On this basis, 17 kg of muscle would be activated, and phosphagen reserves would be exhausted in a minimum of 10 sec. Di Prampero (1971a,b) is thus correct in his estimate of the duration of phosphagen-based activity, although he apparently arrived at this figure from a combination of an overstated muscle volume and an understated efficiency.

Negative Intercepts of Lactate Production

Shortly before the staircase sprint was introduced as a test of anerobic power, Margaria, Cerretelli, and Mangili (1964) suggested that phosphagen could be deduced from a plot of lactate production against the duration of supramaximal work. The negative lactate intercept obtained by backward extrapolation of the graph to the onset of a given work bout indicated the work performed at the expense of phosphagen breakdown.

The limitations of this approach were substantial. It was necessary to assume that no significant energy was supplied from lactate formation or aerobic processes during the first 5 sec of activity and that venous lactate concentrations were representative of intramuscular lactate concentrations in both timing and magnitude. Further, it was necessary to correct the overall estimate of "alactate" energy release by a figure that allowed for usage of oxygen stored in hemoglobin and myoglobin.

The total alactate resource was estimated at 281 J • kg^{-1} of body weight, or 23.0 kJ in 28 kg of muscle (821 J • kg^{-1}). According to di Prampero (1971a,b), this corresponds to 29 mmol of phosphagen per kg, a value agreeing well with biochemical estimates. Unfortunately, the agreement again seems fortuitous, since it was based on the assumption that phosphagen was resynthesized with a mechanical efficiency of 25%. The true biochemical estimate of phosphagen storage is 38.7 kJ.

Margaria, Cerretelli, and Mangili (1964) pointed out that only about a half of the 821 J • kg^{-1} phosphagen store was usable. This presumably reflects the fact that only 15–20 kg of muscle are recruited in most common activities, since the CP content of the active fibers drops to <3% of the resting level at exhaustion (Hultman et al., 1967).

Oxygen Debt

We have noted that the maximum phosphagen component of the oxygen debt is equivalent to an oxygen *delivery* of 1.85 l. However, experimental measurements are usually made in terms of debt repayment, so that if 20 kg of muscle is involved in a particular task, the phosphagen debt drops to ~ 1.32 l (Table 1.2). Alactate repayment also includes replenishment of 0.62 l taken from the oxygen stores, for a total of 1.94 l.

Intermittent Work

Saltin and Essén (1971) examined changes of phosphagen levels during vigorous intermittent exercise. With a cycle of 10 sec exercise and 20 sec recovery, they found little change of intramuscular ATP; on the other hand, CP was ~50% depleted during the first work bout and showed little recovery until exercise was stopped. Muscle lactate concentrations increased only slightly over the experiment, and it was unclear as to why CP resynthesis did not occur during the rest pauses. Nevertheless, it was plain that the main basis of energy release was aerobic glycolysis, replenishment of oxygen stores during the rest pauses making an important contribution to continuation of the required effort (I. Åstrand, P. O. Åstrand et al., 1960a,b).

If work and rest intervals were extended, both ATP and CP stores were progressively depleted; with a 1-min exercise, 2 min rest cycle 60–70% of ATP and >90% of CP were eventually utilized.

Isometric Work

The majority of investigators have considered phosphagen depletion in the context of rhythmic, isotonic work. However, Bergström, Harris et al. (1971) studied muscle phosphagen levels during sustained isometric contractions.

At 100% of maximum voluntary force, subjects became exhausted in ~20 sec. Over this period, almost 70% of CP and 10–15% of ATP stores were utilized. At 40% of maximum effort, the contraction could be held for 90 sec. Phosphagen usage was then 3–4 times slower, although the final depletion was similar to that observed with a maximum contraction. In neither instance could exhaustion be blamed upon a lack of phosphagen. However, if several sustained isometric contractions were made with brief (20 sec) recovery intervals, there was a cumulative usage of phosphagen (R. H. T. Edwards, Nordesjø et al., 1971). If the circulation remained uninterrupted, ATP stores were well maintained, but CP dropped to 37% of control at the beginning and 15% of control at the end of each contraction. If the circulation was occluded (as usually occurs during vigorous isometric effort), ATP dropped rapidly to 60% of control and CP to 7% of control. Plainly, in the latter situation phosphagen lack could have contributed to final exhaustion of the active muscle.

LIMITING INTRACELLULAR pH

Various feedback mechanisms govern the rate of anerobic glycolysis by modulating the activity of the enzymes phosphorylase and phosphofructokinase. However, the principal factor limiting exhausting effort of 10–60 sec duration is the intracellular accumulation of lactic acid. This restricts not only glycolysis, but also the immediate process of muscle contraction (Fuchs et al., 1970; Katz, 1970).

Limits of Intracellular pH

The majority of estimates of intramuscular pH have been based on tissue homogenates (Hermansen and Osnes, 1972; Sahlin et al., 1976). Such results are based on a mixture of intracellular and interstitial fluid; however, Sahlin et al. (1976) recently used a CO_2 method to estimate true intracellular pH, and at least under resting conditions their values agree well with those obtained from homogenates.

We may thus conclude that the limiting intracellular pH is ~6.3

Limiting Lactate Concentration

The limiting intramuscular concentration of lactate varies with the duration of activity. Karlsson (1971b) tested work loads that were exhausting in 2.5, 6, and 16 min (130, 100, and 90% of maximum oxygen intake). In all three experiments, similar values were noted for the accumulated oxygen deficit (~5 l), ATP usage (20–30%), and CP depletion (70–80%). However, the terminal concentrations of intramuscular lactate were higher in the first two experiments (~16 mmol • kg^{-1}) than in the third (~12 mmol • kg^{-1}); furthermore, even higher lactate concentrations (up to 39 mmol • kg^{-1}) were encountered in very brief (< 1 min) bursts of supramaximal activity.

Karlsson suggested two explanations of his findings: (1) with a slower approach to exhaustion, lactate was distributed more uniformly throughout body fluids and (2) some additional factor such as a reduction of tissue buffering capacity sensitized the muscle to lactate accumulation during prolonged exercise. It is hard to see how the first mechanism could influence intramuscular lactate concentrations at exhaustion, unless the limiting factor was related to blood rather than intramuscular lactate concentration. In an unfit subject who stops exercise because of breathlessness, the limit might indeed be the respiratory response to blood lactate. In most subjects, the second explanation seems the more likely. Wasserman (1967a), Gaisl and Harnoncourt (1975), and Kindermann and Keul (1977) have all drawn attention to a rapid decrease of buffering during lactate acidosis. The arterial bicarbonate concentration falls by 16–18 mmol • l^{-1} in as little as 15 min of heavy work (Wasserman, 1967a), the change being almost a mirror image of lactate accumulation (8–9 mmol • l^{-1} in the same experiments). Except in very heavy work, elimination of carbon dioxide proceeds rapidly enough to maintain a relatively constant intravascular pH.

Proof that the ultimate limitation is intramuscular rather than intravascular pH was provided by Hermansen (1971). He had his subjects perform five bursts of exhausting work, each separated by 4-min rest periods. The blood lactate concentrations rose progressively over the five bouts of exercise, reaching a ceiling of 31–32 mmol • l^{-1}. At this stage, blood and muscle lactate concentrations were very similar. In another series of experiments, a subject performed 13 bouts of exercise producing exhaustion in 20- 600 sec; the rate of lactate production naturally varied widely from one experiment to another, but the final blood lactate concentration in all 13 experiments (~18 mmol • l^{-1}) was lower than when rest pauses allowed equilibration between muscle and blood.

Accumulation of blood lactate is often used as a criterion of a "good" effort during the direct measurement of a subject's maximum oxygen intake. If there is a plateauing of oxygen consumption as work load is increased, the additional work must be performed anerobically. With the usual progressive test protocol, final blood lactate readings for a healthy young adult are 11–13 mmol • l^{-1} (Shephard, Allen et al., 1968a). Readings for young children and for the elderly are often lower (~9 and ~7 mmol • l^{-1}). This is partly a question of motivation, and in one series of elderly subjects (aged 60–83 years) many lactate readings >9 mmol • l^{-1} were observed (Sidney and Shephard, 1977a). Another important variable is the ratio of muscle mass to blood volume; in children and elderly subjects this ratio is less than in young adults, so that the blood lactate concentration is inevitably lower for a given intramuscular reading.

Oxygen Debt

We have noted discussion regarding the size of the lactate component of the oxygen debt. In athletic subjects we estimated the lactate component at ~5.7 l, while in sedentary subjects we found figures as low as 3–4 l. In contrast, some observers have reported figures of 15–20 l. Theoretical calculations suggest that the lower estimates are more reasonable. One mole of lactate is equivalent to an oxygen debt repayment of 6.73 l (Table 2.1). Thus, our figures of 3.5–5.7 l are equivalent to a total lactate accumulation of 500–800 mmol. Lactate is not distributed through all of the body water. For example, it does not penetrate the cerebrospinal fluid (Hermansen, 1971). However, if it entered a fluid compartment of only 40 l (most of both the extracellular and the intracellular water), the smaller oxygen debt figures would reflect equilibrium lactate concentrations of 12.5–20 mmol • l^{-1}, well up to the directly measured blood lactate readings.

Carbon Dioxide Stores

Lactate accumulation rapidly leads to hyperventilation, with a compensatory reduction of CO_2 stores.

There is thus a possibility of estimating lactic acid accumulation from the respiratory gas exchange ratio (Issekutz, Birkhead, and Rodahl, 1962; Naimark, Wasserman, and McIlroy, 1964). Unfortunately, a part of the CO_2 produced during effort is not expired immediately, but rather accumulates in body stores. To circumvent this problem, it was suggested that a carbon dioxide balance sheet should be drawn up (Shephard, 1955c; Clode and Campbell, 1969; N. L. Jones, Campbell et al., 1975). This has four components as follows:

$$\begin{aligned} &\text{Total } CO_2 \text{ output (1)} = \\ &\quad CO_2 \text{ produced by aerobic metabolism (2)} \\ &\quad \pm\ CO_2 \text{ moving in or out of body stores (3)} \\ &\quad +\ CO_2 \text{ resulting from lactate production (4)} \end{aligned} \qquad (3.3)$$

Item (1) is readily measured. If the experiment is of 5 min duration and CO_2 production is 2000 ml • min^{-1}, (1) = 10,000 ml. Item (2) is calculated from the measured oxygen consumption by assuming an appropriate respiratory quotient for the active muscle. A value between 0.85 and 1.00 is selected, depending on the severity of the exercise. For example, if the oxygen consumption is 2100 ml • min^{-1}, 70% of aerobic power, an RQ of 0.9 may be assumed. Then (2) = 2100 × 5 × 0.9 = 9450 ml. Item (3) is estimated as the product of the change in mixed venous carbon dioxide pressure (torr) and the body mass (kg); given a 5-torr increase of venous PCO_2 and a body mass of 70 kg, item (3) = 350 ml. Item (4) is now estimated by difference:

$$\begin{aligned} &CO_2 \text{ from lactate production} = \\ &\quad 10{,}000 - 9450 + 350 = \\ &\quad 900 \text{ ml (40.2 mmol)} \end{aligned}$$

This is equivalent to the production of 40.2 mmol of lactic acid. N. L. Jones, Campbell et al. (1975) assumed in their calculations that the lactate would be distributed throughout body water; however, dispersion through 70–75% of body water (~30 l) seems more realistic (Alpert and Root, 1954). In our example, the aqueous concentration of lactate will then be 40.2/30.0 = 1.34 mmol • l^{-1}. If we further assume that blood is 80% water, the blood concentration will be 0.8 × 1.34, or 1.07 mmol • l^{-1}.

Clode and Campbell (1969) claimed that this method is accurate to ±1 mmol • l^{-1}; if so, it has ~50% of the accuracy of the more usual enzymatic measurements of blood lactate concentration.

Intermittent Work

For a given performance of external work, lactate production is usually larger with intermittent than with continuous effort (R. H. T. Edwards, Nordesjø et al., 1971). However, intermittent work may yield either high or low lactate readings, depending upon the frequency of the work/rest cycles and the duration of recovery pauses. The metabolic needs of rapid cycles (for example, 10 sec work, 20 sec rest) are largely satisfied by "alactate" stores, and little lactate is produced (Saltin and Essén, 1971). Much larger accumulations of lactate are seen if the same total amount of work is performed with a 30/60-sec or a 60/120-sec activity/rest cycle.

The athlete who chooses an interval training plan can stress his alactate, lactate, or aerobic mechanisms by suitable adjustment of the work/recovery schedule. The postcoronary patient with angina of effort must avoid anerobic activity; his training prescription should require bursts of activity of no longer than 1 min, with relatively long recovery intervals (Kavanagh and Shephard, 1975b).

Isometric Work

Endurance of isometric effort depends greatly upon "central fatigue" (Bigland-Ritchie et al., 1978), including motivation (Shephard, 1974a). Nevertheless, there is an almost exponential curve relating force (expressed as a percentage of maximum effort) to the endurance of a given contraction (Fig. 3.13). An effort of less than 10% (Björksten and Jonsson, 1977) or 15% (Rohmert, 1968a) of maximum force can be held almost indefinitely. With stronger contractions, endurance decreases rapidly to a minimum of ~20 sec at 100% of maximum effort (Grose, 1958; Rohmert, 1960; Monod and Scherrer, 1957; Kogi and Hakamada, 1962; Bowie and Cumming, 1971; Shephard, 1974a). In subjective terms, effort is halted by weakness and pain as in exhausting isotonic work, and we may presume that lactate accumulation is limiting performance in both situations. Certainly, blood flow to the active muscles is impeded by vigorous isometric effort (Lind and McNicol, 1967). Occlusion of the arterial flow by a sphygmomanometer cuff makes no difference to the tolerance of maximum effort, but it shortens markedly the endurance of less intense contractions (Merton, 1954; Royce, 1958; Shephard, 1974a). From such experiments, it has been deduced that intramuscular pressure first restricts local blood flow when contractions reach 15% of maximum voluntary force. At about 70% of

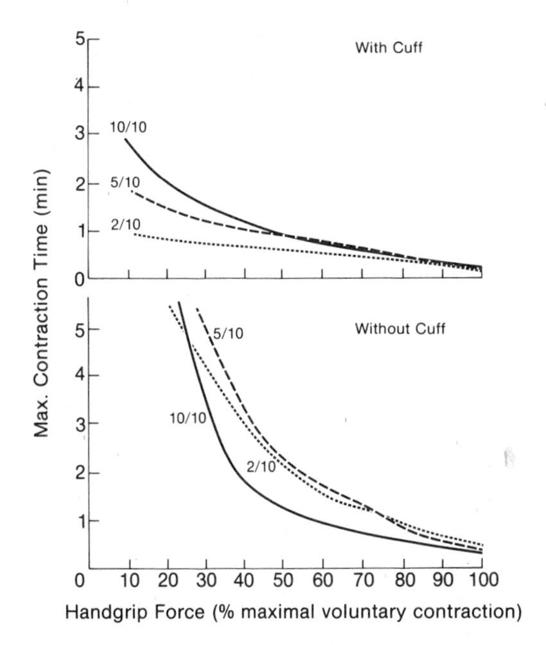

FIG. 3.13. Endurance of isometric handgrip in relation to force exerted (percentage of maximum voluntary contraction), occlusion of circulation by cuff, and intermittency of effort (2 sec of 10 sec, 5 sec of 10 sec, continuous; after Shephard, 1974a.)

maximum force (a higher percentage in women than in men, Heyward and McCreary, 1978), intramuscular pressure exceeds systolic pressure, and arterial flow is halted. The resultant discomfort can be related to an accumulation of lactate and other metabolically linked phenomena such as an escape of potassium ions from the active tissues and a rise of local temperature.*

Irrespective of mechanisms, the endurance time T is dependent upon the accumulation of a metabolic factor A, at a rate related to the restriction of local blood flow and the intensity of muscle metabolism (both functions of effort I), the size of a local reservoir S (for instance, oxygen and phosphagen stores), and any diffusion of the limiting substance to other regions within the muscle (Shephard, 1974a). Thus we may write

$$A = IT - S + BI \quad (3.4)$$

or

$$T = \frac{1}{I}(A + S) - B \quad (3.5)$$

where B is a constant describing the on-transient of isometric contraction.† If effort is submaximal, there is a possibility that contraction may alternate between different muscle groups, with intermittent restoration of blood flow. Finally, if the required force can be sustained for more than a few seconds, compensatory increases of systemic blood pressure (Lind and McNicol, 1967) help to overcome the restriction of blood flow in the active tissues.

There is little possibility of fat combustion during vigorous isometric activity. Furthermore, anerobic effort has only ~1/13th of the efficiency of aerobic work. Nevertheless, it is unlikely that the glycogen reserves of a muscle would be exhausted by less than 5-7 min of isometric effort, that is, 10-12 maximal contractions (Shephard, 1974a). Josenhans (1967a) noted the accumulation of a 1.7 l oxygen debt when both legs developed a force of 40 kg for 2½ min. With the more usual laboratory type of isometric contraction (a sustained handgrip), the muscle mass is only ~0.5 kg and the increase of blood lactate (~1 mmol • l^{-1}) is close to the accuracy of analysis (Shephard, 1974a).

Direct measurements of intramuscular lactate concentrations have been made during deliberate isometric exercise of the quadriceps (Bergström, Harris et al., 1971) and after maximal skiing (Tesch, Larsson et al., 1978). Lactate levels were at least as high as those encountered during rhythmic work to exhaustion (>40 mmol • kg^{-1} wet wt in the laboratory experiments), and phosphofructokinase inhibition was deduced from a dramatic decrease in the ratio of fructose-1-diphosphate to fructose-6-phosphate (Bergström, Harris, et al., 1971; see Fig. 1.2).

Some 20% of submaximal endurance was restored with 5 sec of rest and recovery was 87% complete after 40 min (Stull and Kearney, 1978).

LIMITING OXYGEN PRESSURE

When oxygen pressure drops below a certain critical value, aerobic metabolism is halted, and anerobic

*Lind and McNicol (1967) argue that changes of lactate and hydrogen ion concentration are maximal after exercise has ceased and thus cannot be responsible for pain, discomfort, and the associated reflex increase of heart rate and blood pressure. However, the phase lag which they discuss applies only to the intravascular compartment. If intramuscular conditions are considered, lactate remains a good candidate for the causation of these phenomena.

†The on-transient comprises such factors as the stretching of series elastic elements within the muscle and (in a mechanical dynamometer) the work performed in approximating the grip plates.

glycolysis then provides the sole basis for generation of NAD^+ and thus ATP formation [equation (2.5)]. The critical oxygen pressure is reasonably well established, but there is more discussion as to whether this pressure is reached in the muscles during endurance exercise. Some authors support the traditional view that performance is limited by oxygen delivery to the tissues (Warburg, 1923; Kawashiro et al., 1978), while others argue that the real limitation is the ability of the cells to utilize the oxygen that they receive.

Critical Oxygen Pressure

The critical oxygen pressure is so small that it is difficult to deduce from the oxygen saturation of hemoglobin or even polarographic (oxygen electrode) measurements of tissue oxygen tension. One practical alternative is to examine the oxygen/reduction status of respiratory chain components such as $NADH/NAD^+$ and cytochrome c (Chance, 1957; Chance et al., 1964; Chance and Pring, 1968; Starlinger and Lübbers, 1973; Granger et al., 1975; Chance, 1977), using the techniques of fluorometry and spectrophotometry, respectively (Lübbers, 1977a,b; Wodick, 1977; Mandel et al., 1977). The disadvantage of using such biological indicators is that their oxidation/reduction status is influenced not only by the local oxygen pressure, but also by the metabolic rate and the pH of the tissue.

In isolated mitochondria, Chance and Pring (1968) observed a 50% oxidation of NADH at an oxygen pressure of 0.01 mmHg (1.3 Pa). However, if the mitochondria were metabolically active (Fig. 3.14), the critical pressure increased to 0.05 mmHg (6.7 Pa) for $NADH/NAD^+$ and 0.18 mmHg (23.9 Pa) for cytochrome c (Starlinger and Lübbers, 1973; Chance, 1977).

Substantially higher oxygen pressures (~1 mmHg, 0.13 kPa) are necessary for 50% oxidation of NADH in the whole cell (Chance, 1957; Granger et al., 1975). This probably reflects mainly the pressure gradient needed for diffusion of oxygen within the muscle cells, although it is also conceivable that mitochondria behave differently under in vivo conditions of temperature, pH, and metabolism.

Tissue Oxygen Pressure

The oxygen pressure within active tissues has been deduced from models of capillary diffusion, from observations of myoglobin pigment, from oxygen electrode determinations, and from measurements of venous oxygen pressures.

Models of Capillary Diffusion

Krogh (1929a,b) introduced the concept that each capillary could be considered as supplying oxygen to a surrounding cylinder of tissue (Fig. 3.15). The diffusion equation for this model is

$$\dot{V}_{O_2} = \frac{\alpha \dot{D}_{O_2}}{C_{R,r}} \Delta P_{R,r} \tag{3.6}$$

where $\dot{V}_{O_2}$ is the oxygen consumption of the tissue, α is the solubility of oxygen in the tissue, $\dot{D}_{O_2}$ is the diffusion coefficient for oxygen, $\Delta P_{R,r}$ is the difference of oxygen pressure necessary to sustain metabolism, and $C_{R,r}$ is a factor describing the capillary geometry.

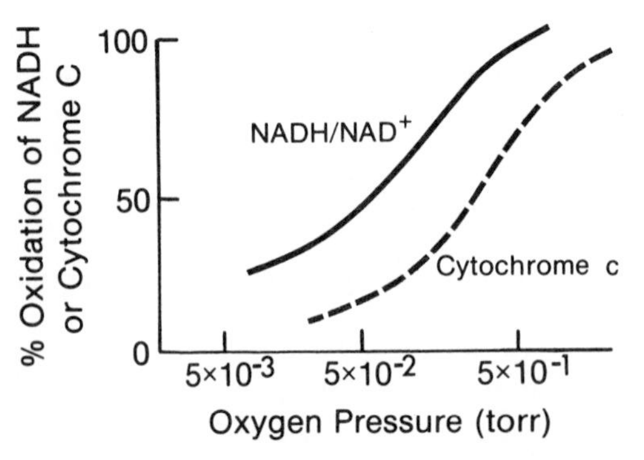

FIG. 3.14. Mitochondrial oxygen pressure and oxidation/reduction status of $NADH/NAD^+$ and cytochrome c. Based on data of B. Chance (1977), obtained in metabolically active state 3 mitochondria at 23°C.

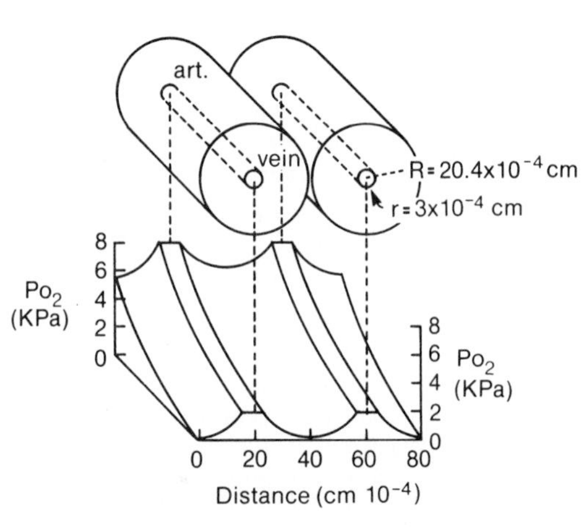

FIG. 3.15. Krogh cylinder model of tissue oxygen supply, but applied to maximally active skeletal muscle. (After Opitz & Schneider, 1950.)

$$C_{R,r} = \frac{1}{4}\left[R^2 \ln\left(\frac{R}{r}\right)^2 - (R^2 - r^2)\right]^* \qquad (3.7)$$

where R is the radius of the tissue cylinder and r the radius of the capillary in cm.

Given a maximum oxygen intake of 3 l • min^{-1} and an active muscle mass of 20 kg, we may set the average muscle oxygen consumption at 0.15 ml • g^{-1} • min^{-1}. A typical capillary count for exercising muscle is 600 mm^{-2}, implying an intercapillary distance of 40.9×10^{-4} cm, while a typical capillary radius is 3×10^{-4} cm.† We thus find

$$C_{R,r} = \frac{1}{4}\left[20.4^2 \ln\left(\frac{20.4}{3.0}\right)^2 - (20.4^2 - 3^2)\right] 10^{-8} = 296.8 \times 10^{-8} \qquad (3.8)$$

The diffusion factor $\alpha\dot{D}$ is $\sim 1.6 \times 10^{-5}$ ml • cm^{-1} • min^{-1} per atmosphere, so that

$$\Delta P_{R,r} = \frac{296.8 \times 10^{-8} \times 0.15}{1.6 \times 10^{-5}} = 2.78 \times 10^{-2} \text{ atmospheres} \qquad (3.9)$$

Given that 1 atmosphere of dry gas at body temperature has a pressure of 101 kPa (760 mmHg), the average pressure gradient $\Delta P_{R,r}$ needed to sustain metabolism at the points most distant from the capillary would be 2.81 kPa (21.1 mmHg).

As we shall see, the answers yielded by this traditional model threaten parts of the working muscle with severe oxygen lack. In practice, several factors limit the likelihood of such an occurrence:

1. The classical estimate of $\alpha\dot{D}$ may be too low. Some authors (for example, Grunewald, 1973; Hutten et al., 1973) set the diffusion factor at 2.7×10^{-5} ml • cm^{-1} • min^{-1} per atmosphere. This would immediately decrease the terminal oxygen pressure gradient to 1.66 kPa, 12.4 mmHg. The diffusion of oxygen may be further facilitated by displacement of the myoglobin molecule (Scholander, 1960). The diffusion coefficient for myoglobin is about 1/25th that for oxygen (R. E. Forster, 1964), as would be predicted from their respective molecular weights. Given a capillary oxygen pressure of 1.6 kPa (12 mmHg), a mitochondrial oxygen pressure of 13.3 Pa (0.1 mmHg), and an oxygen solubility coefficient of 0.024, the concentration gradient for dissolved oxygen, from capillary to mitochondrion, would be

$$\left(\frac{1.600 - 0.013}{101}\right)0.024 = 377 \times 10^{-6} \text{ ml} \cdot \text{ml}^{-1} \qquad (3.10)$$

The concentration of myoglobin in muscle is typically ~ 1.2 mg • g^{-1}. Since the molecular weight of myoglobin is 16,400 and each mole can combine with 22,400 ml of oxygen, the oxygen capacity is

$$\frac{0.0012 \times 22{,}400}{16{,}400} = 164 \times 10^{-5} \text{ ml} \cdot \text{ml}^{-1} \qquad (3.11)$$

If we assume a 50% gradient of myoglobin saturation between the capillaries and a typical point in the cytoplasm, the gradient of oxymyoglobin concentration is 82×10^{-5} ml • ml^{-1} and the relative flux of the two forms of oxygen is given by the respective products of concentration gradient and diffusion coefficient:

$$\frac{\text{Oxymyoglobin flux}}{\text{Oxygen flux}} = \frac{1 \times 82 \times 10^{-5}}{25 \times 377 \times 10^{-6}} = \frac{8.7}{100} \qquad (3.12)$$

The myoglobin concentration can sometimes be as great as 7 mg • g^{-1}; in such a situation, the myoglobin molecules augment oxygen flux by $\sim 50\%$. The oxygen-storing capacity of myoglobin is very limited. Given the normal myoglobin concentration of 1.2 mg × g^{-1}, the maximum volume of oxygen that can be stored as oxymyoglobin is only 16×10^{-4} ml per ml of muscle. Expressing maximum aerobic power per unit mass of muscle (0.15 ml • g^{-1} • min^{-1}), it can be seen that this reserve would be exhausted within 0.64 sec.

2. Autoregulation (Bruley et al., 1973) opens up new capillaries in hypoxic areas of muscle, thus reducing the average intercapillary distance and the radius of the corresponding tissue cylinder R of equation (3.7). According to Otis (1963), application of the Krogh equation suggests that a capillary density of only 30 mm^{-2} is sufficient to prevent hypoxia under resting conditions. However, exercise sufficient to induce a muscle oxygen consumption of 0.04 ml • g^{-1} • min^{-1} demands a capillary density of 350 mm^{-2} (Fig. 3.16). The Otis calculation assumes a resting gradient ($\Delta P_{R,r}$) of 35 mmHg (4.67 kPa) and an exercise gradient of 15 mmHg (2.0 kPa).

*The current formula (Lübbers, 1977b) differs somewhat from that described by Otis (1963).

†The average capillary is slightly narrower than a red cell.

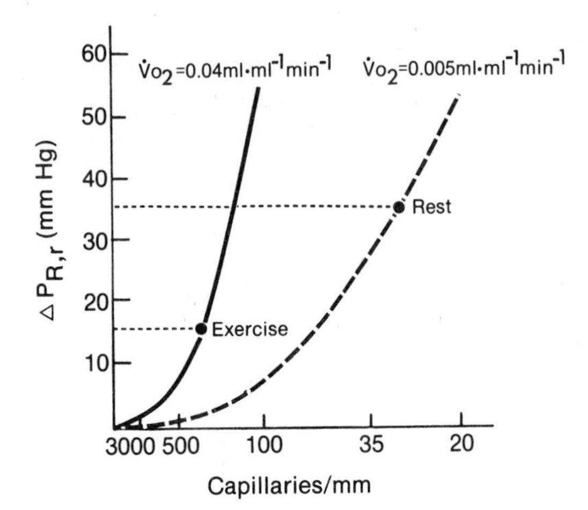

FIG. 3.16. Autoregulation in the muscle capillaries. Application of the Krogh equation (3.6) suggests that with a resting oxygen consumption of 0.005 ml • ml^{-1} • min^{-1} and a capillary-tissue oxygen pressure gradient of 35 mm Hg (4.67 kPa), a capillary density of 30 per mm^2 suffices. However, in moderate exercise (oxygen consumption 0.04 ml • ml^{-1} • min^{-1}, $\Delta P_{R,r}$ = 15 mm Hg, 2.0 kPa), the capillary density must be increased to 350 per mm^2.

3. Krogh's cylindrical tissue model fails to take account of the possibility that a muscle fiber may be supplied from several capillaries (Hutten et al., 1973). It also neglects three-dimensional flow (Metzger, 1973), diffusion of oxygen along the length of the capillary (Grunewald, 1973), and possible asymmetric or countercurrent arrangements of the capillaries (Grunewald, 1973), all of which reduce the likelihood of tissue hypoxia.

Observations of Myoglobin Pigment

According to the classical studies of Theorell (1934), the red pigment of the muscle cytoplasm is 50% saturated with oxygen at a temperature of 37°C and a pressure of 0.43 kPa (3.26 mmHg). Commencing with Millikan (1937), a number of authors (Landis and Pappenheimer, 1963; H. Barcroft, 1963; Coburn and Mayers, 1971) thus used the absorption spectra of myoglobin to indicate oxygen pressures within the muscle cytoplasm. Estimates ranged from 0.4 kPa (3 mmHg) to 0.66 kPa (5 mmHg).

Tissue Oxygen Pressures

It might seem a simple matter to insert a needle electrode (Longmuir, 1964; Cater, 1964) into an active muscle and make direct determinations of tissue oxygen pressure. In practice there are several technical difficulties. Movement of the electrode through the semisolid medium disturbs both the reduction process at the cathode surface and diffusion within the electrode membrane (Schuchhardt and Losse, 1973). Further, the electrode is large relative to a typical capillary. The pressure recorded is therefore the average for a substantial volume of tissue, and local pockets of hypoxia may be overlooked (Lübbers, 1973).

Despite these problems, there have been a number of applications of the oxygen electrode technique. Whalen et al. (1973) reported pressures ranging from 0 to 4.13 kPa (31 mmHg) in the beating cat heart. The mean pressure was 0.92 kPa (6.9 mmHg). Although many individual readings came closer to the critical value of 0.13 kPa, especially in the deeper layers of the heart wall, the authors concluded that autoregulation generally maintained oxygen pressure above hypoxic levels. Whalen and Nair (1967) reported resting intracellular oxygen pressures of 0.13–0.52 kPa (1–4 mmHg) for the gracilis muscle of the guinea pig. Somewhat higher values have been observed for human skeletal muscle (Fig. 3.17).

Oxygen pressure histograms for skeletal muscle cannot be explained simply in terms of parallel (concurrent) capillary flow. Computer simulations suggest that the observed distribution of oxygen pressures is best explained by a model with a mixture of concurrent flow (61%), countercurrent flow (15%), and asymmetric flow (24%). An asymmetric countercurrent arrangement of the capillaries gives the most efficient and a simple countercurrent arrange

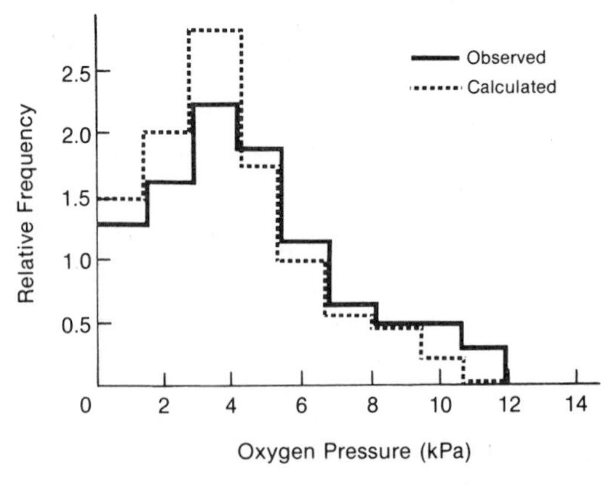

FIG. 3.17. Relationship between observed (Kunze, 1969) and calculated (Lübbers, 1977b) oxygen pressure in resting human skeletal muscle (blood flow 4.5 • 10^{-2} ml • g^{-1} • min^{-1}). The calculation assumes 61% concurrent, 15% countercurrent, and 24% asymmetric arrangement of the capillaries.

ment the least efficient transfer of oxygen from the capillaries to the active tissues (Lübbers, 1977b).

Venous Oxygen Pressure

Animal experiments (Fig. 3.18) have shown that muscle consuming oxygen at a rate equivalent to a whole-body oxygen consumption of 3 l • min^{-1} has a venous oxygen pressure of 0.8–1.3 kPa (6–10 mmHg; Stainsby, 1966). If hypoxia is induced by rebreathing, the critical venous oxygen pressure at which the oxygen consumption of the leg falls is ~4 kPa (30 mmHg) under resting conditions and ~1.33 kPa (10 mmHg) when an oxygen consumption of 0.04 ml • g^{-1} • min^{-1} has been induced by nerve stimulation.

Femoral venous oxygen pressures as high as 20–22 mmHg (2.7–2.9 kPa) have been reported during maximum leg exercise in man (Doll, Keul, and Maiwald, 1968; Pirnay, Lamy et al., 1971; Keul and Doll, 1973). This is somewhat surprising since the oxygen content of *mixed* venous blood drops to ~30-40 ml • l^{-1} during maximum exercise. Blood leaving the active muscles is likely to have an oxygen content of only 6–20 ml • l^{-1} (Shephard, 1968a; Hartley and Saltin, 1969), and from standard oxygen dissociation curves this would be equivalent to an oxygen pressure of 1.1–1.9 kPa (8–14 mmHg). It is possible that the femoral venous blood specimens examined by Keul and his associates were contaminated with substantial quantities of oxygen-rich blood coming from skin and inactive muscles.

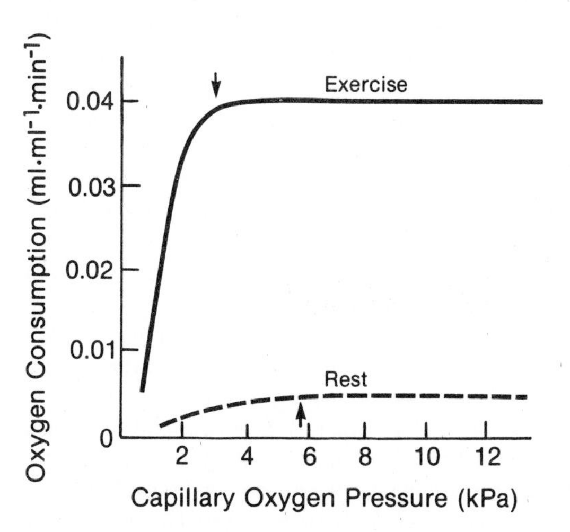

FIG. 3.18. Critical oxygen pressures for sustaining metabolism in the dog limb (after Otis, 1963). When exercising, oxygen consumption begins to diminish at an arterial oxygen pressure of 4.67 kPa (35 mm Hg), a capillary oxygen pressure of 2.67 (20 mm Hg), and a venous oxygen pressure of 1.33 kPa (10 mm Hg).

Overview of Tissue Oxygen Supply

We are now in a position to review briefly why some authors have suggested there is a peripheral limitation of metabolism and to make a reasoned rebuttal of this concept.

Arguments for a Peripheral Limitation of Metabolism

Kaijser (1970) advanced six main arguments in favor of a peripheral limitation of metabolism during sustained effort, as follows.

VARIATIONS OF MAXIMUM OXYGEN INTAKE WITH THE MODE OF EXERCISE. Kaijser (1970) reasoned that if maximum oxygen intake was limited by the performance of the cardiorespiratory system, then it should have a unique value, independent of the method of measurement. In fact, there are small differences between the results obtained on the treadmill, the bicycle ergometer, and the step test (Table 3.1; Shephard, 1977a), and the forearm ergometer often yields an oxygen consumption plateau that is only ~70% of that obtained by uphill treadmill running (P. O. Åstrand and Saltin, 1961a; R. C. G. Simmons and Shephard, 1971a,b; Pirnay, Deroanne et al., 1971; Secher et al., 1974).

The cardiac stroke volume during arm ergometer exercise is also only ~70% of that attained during treadmill effort, and the failure of the body to call upon its known cardiac potential does suggest some peripheral limitation with this type of activity. Such a view is supported by complaints of local muscular fatigue, with the accumulation of lactate at quite low workloads (Fig. 2.6). On the other hand, the oxygen intake attained during exhausting uphill treadmill running is a true maximum, and the plateau value cannot be increased by deliberate addition of arm work (H. L. Taylor, Buskirk, and Henschel, 1955; Stenberg et al., 1967; Pirnay, Deroanne et al., 1971; Secher et al., 1974; Gleser, Horstman, and Mello, 1974).

If we assume that the maximum value for uphill treadmill running is limited by the oxygen conductance of the cardiorespiratory system, then the lower result for arm work can be explained by the addition of a peripheral resistance ~43% of that imposed by the cardiorespiratory system (Shephard, 1976a). This resistance does not necessarily lie in the

TABLE 3.1
Comparison of the Maximum Oxygen Intake Obtained by Treadmill, Step, and Bicycle Ergometer Tests for Healthy Young Men

Mode of Exercise	Maximum Oxygen Intake (l • min⁻¹ STPD)	Maximum Oxygen Intake (mmol • min⁻¹)	Heart Rate (mmol • min⁻¹)	Blood Lactate (mmol • l⁻¹)
Treadmill	3.81 ± 0.76	170 ± 34	190 ± 5	13.6 ± 2.3
Step test	3.68 ± 0.71	164 ± 32	188 ± 6	11.7 ± 2.9
Bicycle ergometer	3.56 ± 0.71	159 ± 32	187 ± 9	12.4 ± 1.7

Source: After Shephard, Allen et al., 1968a.

enzyme systems; indeed, it is probably a vascular resistance related to difficulty in perfusing the active muscles and/or a limitation of venous return associated with a pooling of blood in the veins of the inactive limbs. In other words, the performance of the heart is modified by its pre- and afterloading. Viewed in such terms, the distinction between central and peripheral circulatory factors becomes fine, and perhaps impossible to justify in physiological terms (P. O. Åstrand, 1952).

VARIATIONS OF MAXIMUM OXYGEN INTAKE WITH AMBIENT PRESSURE. Kaijser (1970) further argued that if circulatory delivery of oxygen was the limiting factor, then exposure to oxygen at 3 atmospheres pressure should augment performance. Perhaps because his subjects were suffering from oxygen poisoning, Kaijser did not find any increase of endurance time when bicycle ergometer exercise was carried out in oxygen at a pressure of 3 atmospheres. However, many other authors have shown gains of maximum oxygen intake from quite modest increases of oxygen pressure (Shephard, 1977a), while there is now universal agreement that even a modest decrease of ambient pressure (for example, an altitude of 2000–2500 m) leads to a decrement of maximum oxygen intake (Goddard, 1967; Kollias and Buskirk, 1974).

Kaijser's own experiments speak against a peripheral limitation of oxygen transport. When breathing oxygen at a pressure of 3 atmospheres (240 kPa), he found that the oxygen pressure in venous blood was only 8.6 kPa; thus, the major part of the pressure drop, and by inference the major resistance, was in the cardiorespiratory system rather than in the tissues.

SPECIFICITY OF TRAINING. Clausen (1970) claimed that a program of forearm training had relatively little effect on the heart rate response to subsequent leg ergometry, while leg training had little influence on the subsequent response to arm work (Chap. 11). Kaijser interpreted these findings as further evidence of a peripheral limiting system, susceptible to training.

When Clausen (1973) repeated his experiments with direct measurement of maximum oxygen intake, he obtained less categoric results; leg training increased the aerobic power during arm work by 57% of the gain found during leg work. However, even if this additional evidence had not been forthcoming, the original experiments of Clausen should not have been accepted as proof that training was occurring predominantly in tissue enzyme systems. The specificity of training could equally well have been explained by a strengthening of the local musculature, with a consequent facilitation of local perfusion.

CARDIAC OUTPUT AND MAXIMUM OXYGEN INTAKE. A fourth argument of the "peripheralists" was that if cardiac output limited oxygen transport, then cardiac output should reach a fixed maximum value when a plateau of oxygen consumption was attained. In fact, a higher cardiac output was seen during a long (6–10 min) than a brief (3–5 min) measurement of maximum oxygen intake (P.O. Åstrand and Saltin, 1961b), while β-blocking drugs such as propranolol reduced cardiac output without changing maximum oxygen intake (P. O. Åstrand, Ekblom, and Goldberg, 1971).

Again, the explanation of these observations is probably a peripheral restriction of cardiac performance. The heart reaches a maximum output for any given level of peripheral resistance, but if a subcutaneous "shunt" is opened up by the rise of body temperature during extended maximum effort, a further increase of flow becomes possible. Likewise, the β-blocking drugs close off shunts not concerned with flow to the nutrient capillaries of the muscles.

MAXIMUM OXYGEN INTAKE AND MUSCLE TEMPERATURE. A fifth "proof" advanced by Kaijser (1970) concerned the influence of muscle temperature upon maximum oxygen intake. Since the rate of biochemical reactions is halved by a 10°C fall of tissue temperature, Kaijser reasoned that a decrease of maximum oxygen intake with local cooling would show that tissue enzyme systems were limiting performance.

The maximum heart rate and the maximum arteriovenous oxygen differences were both reduced when the limbs were cooled. However, the decrease of maximum oxygen intake (< 1 l • min^{-1} for a 10°C decrease of tissue temperature) was less than would have been predicted from the Law of Arrhenius. Possible explanations of the experimental results other than a limitation imposed by tissue enzyme systems include (1) a greater relative contribution to the "mixed" venous specimens of blood flowing from regions other than the cooled limb and (2) an increase of the diffusion pathway to the active muscles consequent upon cold-induced vasoconstriction within the active muscles.

Tissue Enzyme Responses to Training. Teleological arguments for the peripheral limitation of effort are based on (1) close correlations between fiber type, tissue enzyme levels, and maximum oxygen intake (Howald, 1975) and (2) an increase of key enzymes such as phosphofructokinase and succinic dehydrogenase with training (Saltin, 1973b; Holloszy, 1973; Holloszy, Booth et al., 1975). The teleogists suggest a causal link between gains of maximum oxygen intake and increases of enzyme activity.

One counterargument is that much larger changes in the activity of many enzyme systems can be induced by isometric and anerobic patterns of training, without altering maximum oxygen intake (Gollnick and Hermansen, 1973). Again, the enzymes of female subjects show a lower oxidative potential than would be anticipated in male subjects having a comparable maximum oxygen intake (Hedberg and Jansson, 1976; Costill, Daniels et al., 1976). If training is stopped in either sex, increases of enzyme activity are reversed much more quickly than the corresponding gain of maximum oxygen intake (Saltin et al., 1977). Finally, it is difficult to attribute the widening of arteriovenous oxygen difference with training to an increased oxygen extraction within the working muscles. Hartley and Saltin (1969), for example, found only 6 ml of oxygen per l of venous blood during maximum activity, and it is difficult to see how such high rates of oxygen extraction could be increased further by the development of muscle enzymes. Doll, Keul, and Maiwald (1968) commented that femoral venous oxygen pressures did not differ between athletes and sedentary subjects. A more convincing explanation of why the arteriovenous oxygen difference is widened in trained subjects is that a larger proportion of their total cardiac output is directed to the active muscles. This reflects a greater total cardiac output, a greater reduction of visceral blood flow, and a reduction of subcutaneous blood flow secondary to earlier sweating and a loss of subcutaneous fat (Simmons and Shephard, 1971b).

What explanation can we then offer for the increase of tissue enzyme activity during training? One possibility may be compensation for an increased diffusion pathway. In Saltin's experiments (1973b), a 15% gain of maximum oxygen intake was associated with a 37% increase of muscle fiber area. In the absence of other adaptations, there would thus have been a drop of oxygen pressure at the center of the active fibers. Often, hypertrophy is accompanied by a compensatory increase of capillarity, but an increase of tissue enzymes could plainly serve a similar function. A second advantage of an increased enzyme activity is that less phosphagen depletion is needed to "turn on" mechanisms of energy release at the beginning of exercise. Last, with more active enzymes, there is a greater steady-state utilization of fat, conserving stored glycogen for occasional bouts of anerobic work (Holloszy, 1973; Moesch and Howald, 1975).

The Oxygen Pressure Gradient

The strongest argument against a peripheral limitation of performance comes from an examination of the oxygen pressure gradient between the atmosphere and the working tissues. If the enzymes were the limiting factor, we would anticipate a buildup of oxygen waiting to enter enzyme systems within the tissues. However, we have seen that the mitochondria continue to function well until their oxygen pressure is reduced to ~ 0.0067 kPa (0.05 mmHg). Treating the oxygen delivery system as a series of linked conductances (Shephard, 1977a), we find the pressure gradient between the atmosphere and the mitochondrion is 20.093 kPa, while the gradient imposed by the enzyme systems is only 0.0067 kPa! Plainly, a doubling or even a tripling of enzyme activity would have a negligible effect on oxygen transport.

One possible counterargument is that tissue metabolism was insufficient to induce the muscle vasodilatation necessary to bring about a large buildup of oxygen pressure in the tissues. However, such a suggestion can immediately be rejected on the

TABLE 3.2
Cumulative Estimate of Minimum Oxygen Pressure Needed to Sustain Aerobic Metabolism

	Cumulative Minimum Pressure		Gradient from Adjacent Site	
Site	kPa	mmHg	kPa	mmHg
Mitochondrion	0.0067	0.05	0.0067	0.05
Mitochondrial surface	0.0200	0.15	0.0133	0.10
Cell surface	0.1133	0.85	0.0933	0.70
Capillary surface	1.73–2.93	13–22	1.60–2.80	12–21
Capillary blood	1.75–3.60	13.1–27.0	0.0133–0.613	0.1–4.6

basis of capillary counts. We have seen that the number of open capillaries during maximum exercise is ~600 mm^{-2}, and careful counts show that the maximum potential number of capillaries is only slightly larger than this (782 mm^{-2} in the dog gastrocnemius) (Schmidt-Nielsen and Pennycuik, 1961).

It is finally instructive to review the minimum pressure gradient needed to sustain aerobic metabolism within an active mitochondrion (Table 3.2). The required local oxygen pressure is 6.7 Pa (0.05 mmHg). Taking typical dimensions, R. E. Forster (1964) calculated that a gradient of 13.3 Pa (0.10 mmHg) is needed to maintain diffusion from the surface to the interior of the mitochondrion. Similar calculations for a typical cell suggest the need for a total intracellular gradient of 93.3 Pa (0.7 mmHg). The Krogh cylinder data, previously reviewed, indicate a gradient of 1.60–2.80 kPa (12–21 mmHg) from the capillary wall to the surface of the most distant cell. Finally, there is a pressure loss of 13.3–163 Pa (0.1–4.6 mmHg) associated with delays in the dissociation of oxyhemoglobin within the tissue capillaries (R. E. Forster, 1964). The total pressure drop from capillary to a distant mitochondrion is thus 1.73–3.08 kPa (13.0–26.5 mmHg). This estimate agrees quite well with the observed oxygen pressure of venous blood leaving the active muscles.

Oxygen Delivery and Physical Performance

The practical importance of oxygen delivery to the performance of moderate duration exercise (1–60 min) is supported by substantial coefficients of correlation between the results of athletic contests and the directly measured maximum oxygen intake of the individual (Tables 3.3 and 3.4).

In well-trained sportsmen, the correlation is less than perfect because (1) the laboratory test employed for the determination of maximum oxygen intake (usually uphill running on the treadmill) requires a different pattern of activity from that needed during performance of the sport in question, (2) there is only a small spread of data between successful and less successful performers, and (3) many contests are determined as much by psychological as by physiological factors (Shephard, 1978c). Sometimes, poor coefficients of correlation have also been found when maximum oxygen intake has been related to the running speed of sedentary subjects; in such groups, difficulties arise from an inappropriate choice of running pace, a limited motivation for all-out effort, and possibly the carriage of inert mass (body fat, K. J. Cureton, Sparling et al., 1978).

TABLE 3.3
Coefficients of Correlation Between Maximum Oxygen Intake (Direct Treadmill Measurements) and Athletic Performance

Subjects	n	Coefficient of Correlation	Author
Whitewater paddlers	10	0.84	Sidney and Shephard (1973)
Distance swimmers	6	0.84	Shephard, Godin, and Campbell (1973)
Sailors (high wind)	10	0.75	Niinimaa, Wright et al. (1977)
Cross-country skiers	10	0.40	Niinimaa, Shephard, and Dyon (1979)

TABLE 3.4
Some Coefficients of Correlation Between Maximum Oxygen Intake and Running Speed

Subjects	n	Authors	Coefficients of Correlation with $\dot{V}_{O_2,max}$
402 m			
College males	35	Wiley and Shaver (1972)	−0.22
College males	11	Ribisl and Kachadorian (1969)	−0.31
549 m			
Boys, 10 years	20	Larivière et al. (1974)	−0.58
Boys 12–13 years	30	Metz and Alexander (1970)	−0.67
Boys 14–15 years	30	Metz and Alexander (1970)	−0.27
Boys, grade 9	9	Doolittle and Bigbee (1968)	−0.62
Faculty and staff	87	Falls et al. (1966)	−0.64
Sedentary men	141	V. Drake et al. (1968)	−0.27
805 m			
Boys 10 years	20	Larivière et al. (1974)	−0.37
College males	11	Ribisl and Kachadorian (1969)	−0.67
Phys. ed. majors	10	Kearney and Byrnes (1974)	−0.30
College males	11	Byrnes and Kearney (1974)	−0.73
Phys. ed. majors	11	Byrnes and Kearney (1974)	−0.04
Cross-country runners	11	Byrnes and Kearney (1974)	−0.42
1610 m			
College males	25	Wiley and Shaver (1972)	−0.29
College males	11	Ribisl and Kachadorian (1969)	−0.79
College males	11	Byrnes and Kearney (1974)	−0.72
Phys. ed. majors	11	Byrnes and Kearney (1974)	−0.25
Cross-country runners	11	Byrnes and Kearney (1974)	−0.51
3220 m			
College males	11	Ribisl and Kachadorian (1969)	−0.85
College males	25	Wiley and Shaver (1972)	−0.47
Older males	24	Ribisl and Kachadorian (1969)	−0.86
4830 m			
College males	35	Wiley and Shaver (1972)	−0.43
8050 m			
Cross-country runners	17	Kearney and Byrnes (1974)	−0.38
9-min run			
Boys 9–12 years	25	Coleman (1974)	0.82
Girls 9–12 years	25	Coleman (1974)	0.71
12-min run			
Boys 9–12 years	25	Coleman (1972)	0.82
Boys 10 years	20	Larivière et al. (1974)	0.44
Boys 11–14 years	17	Maksud and Coutts (1971)	0.65
Boys grade 9	9	Doolittle and Bigbee (1968)	0.90
Girls 9–12 years	25	Coleman (1974)	0.71
College males	7	Kearney and Byrnes (1974)	0.80
Phys. ed. majors	10	Kearney and Byrnes (1974)	0.64
Cross-country runners	17	Kearney and Byrnes (1974)	0.28
Adult males	115	K. H. Cooper (1968a)	0.90
College females	30	Burris (1970)	0.74
College females	36	F. I. Katch et al. (1973)	0.67

Source: Based partly on data collected by Disch et al., 1975.

Maximum Oxygen Delivery

Maximum oxygen intake is usually measured by having the subject exercise at a progressively increasing work rate until a plateau of oxygen consumption is reached (Shephard, 1977a,1978a). Some subjects reach a clear limiting value, but others do not; an arbitrary criterion of a plateauing is thus adopted, such as a change in oxygen consumption of less than 6.7 mmol • min^{-1} (150 ml • min^{-1}) or 89 μmol • kg^{-1} • min^{-1} (2 ml • kg^{-1} • min^{-1}).

Results may be expressed in absolute units or

TABLE 3.5
Highest Reported Results for the Oxygen-Transporting Power of Contestants in Various Sports

Sport	Oxygen Transport* Men	Women
Cross-country skiing	82	63
1000–10,000 m running	82	
Long-distance cycling	80	
Speed skating	79	53
Orienteering	77	60
100–1500 m running	77	
Marathon	76	
Pentathlon	74	
Diathlon	73	
Distance walking	71	
Canoeing	70	
Rowing	70	
Swimming	70	58
Downhill skiing	68	57
400 m running	67	56
Handball	62	
Association football	61	
Gymnastics	60	43
Jumping	59	
Basketball	59	
Ice hockey	58	
Decathlon	58	
Weight lifting	56	
Wrestling	56	
Sprinting	56	45
Badminton	55	
Throwing	55	38
Boxing	55	
Baseball	52	
Volleyball	52	33
Rugby football	50	
Dinghy sailing	50	
Judo	49	
Archery	—	40
Field hockey	63	
Tennis	61	
Whitewater paddling	60	49
American football	60	
Table tennis	59	43

*Milliliters of oxygen transported per min per kg of body mass.
Source: Based on data of Shephard, 1978c.

per kg of body mass. Since the energy cost of most tasks varies with body mass, the relative units are the more popular. Typical maximum values for a sedentary young man lie between 1.8 and 2.2 mmol • kg^{-1} • min^{-1}, or in a subject of 70 kg, ~ 140 mmol • min^{-1} (Fig. 3.19). Resting metabolism accounts for perhaps 11 mmol • min^{-1} of this total, leaving 129 mmol • min^{-1} for the performance of external work. Aerobic energy release ranges from 441 to 479 kJ per mol of oxygen, being least for the combustion of fat and greatest for the combustion of carbohydrate [equation (1.5)]. If we assume a net mechanical efficiency of 25%, the aerobic performance of external work cannot exceed a rate of 14.2–15.5 kJ • min^{-1}. During maximum work the principal fuel is usually carbohydrate, so that the upper of these two figures is the more realistic.

The relative maximum oxygen intake is somewhat greater in a child of 10 years than in a sedentary adult, and aerobic power declines by a further 25% over the span of working life (Fig. 3.19). At all ages after puberty, relative values for women are ~25% smaller than those for men (Fig. 3.20). Endurance athletes have a relative maximum oxygen intake up to twice as large as that of a sedentary adult (Table 3.5). Heavy athletes, such as players of American football, have a substantial absolute maximum oxygen intake, but their scores are poor when expressed per unit of body mass. In some sports (such as rowing and swimming) the body mass is largely supported, and the absolute oxygen delivery is then a more important determinant of performance than the relative figure.

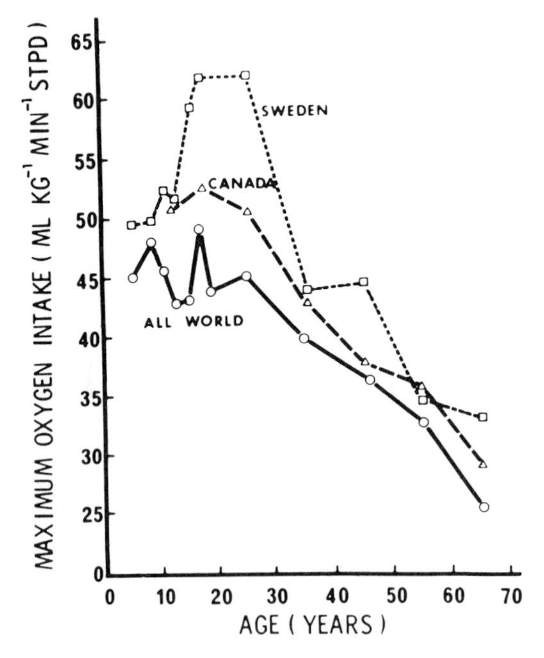

FIG. 3.19. The maximum oxygen intake of 6633 nonathletic men from a wide range of developed countries, compared with data reported for Sweden and author's data for 505 Torontonians. (Cross-sectional data, from Shephard, 1977a, by permission of University of Toronto Press.)

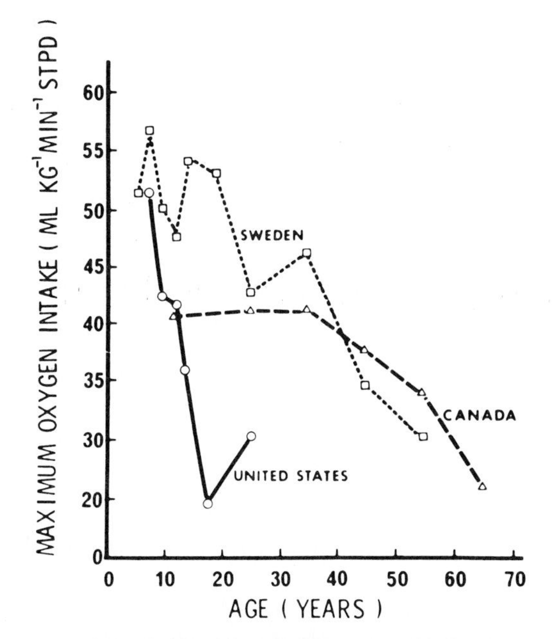

FIG. 3.20. The maximum oxygen intake of women, including author's data for 156 Torontonians, reported data for 286 Scandinavians, and reported data for 211 U.S. citizens. The U.S. data include 58 subjects with very low maximum oxygen intakes (K. Rodahl et al., 1961). (Cross-sectional data, from Shephard, 1977a, by permission of University of Toronto Press.)

FLUID AND MINERAL BALANCE

Water depletion is usually acute, as in the football player or distance runner who sweats profusely. Subacute dehydration occurs when an athlete trains for several days in a very hot environment or attempts to "make weight" (Table 3.6); wrestlers, in particular, may decrease their body mass 3–20% by food deprivation, water deprivation, and such tricks as the use of rubber suits, laxatives, and diuretics (Table 3.7). Chronic dehydration can impair the performance of those exposed for weeks or months to very high altitudes and/or an arctic climate. Finally, if a high rate of sweating is sustained for several weeks, a progressive deficit of key mineral ions can develop.

Acute and Subacute Dehydration

Dehydration in Endurance Sports

U.S. football teams have long recognized their need of fluid replenishment. The Rome Olympic games stimulated interest in the thermoregulatory problems of track competitors. R. H. Fox (1960) pointed out that the rate of sweating and/or the evaporative capacity could be critical to homeostasis in an endurance runner. S. Robinson (1963) reported rectal temperatures of 41.1°C in two record holders after runs of only 14 and 30 min. The weather was hot on this occasion, but Pugh, Corbett, and Johnson (1967) found a rectal reading of 41.1°C in the winner of one marathon contest when the ambient temperature was only 23°C.

Under such circumstances, it is hardly surprising that the body attaches more importance to the regulation of core temperature by sweating than it does to the maintenance of fluid volumes. The rate of sweating can exceed 2 l • hr^{-1}, particularly in an ac-

TABLE 3.6
Techniques Used by Wrestlers for Losing Weight (%)

Method	Used Percent	Not Used Percent	No Reply Percent
Eating less food	82.2	17.0	0.8
Drinking less fluid	74.9	23.3	2.0
Performing more exercise	87.0	12.1	0.9
Exercise in heat	74.1	22.0	3.9
Exercise in rubber suit	40.3	56.1	3.6
Exercise, rubber suit, heat	31.1	63.4	5.5

Source: Based on data of Tcheng and Tipton (1973) for 747 high school wrestlers.

TABLE 3.7
Loss of Body Weight in 747 High School Wrestlers in the 17 Days Prior to Certification

Weight Category (kg)	n	Percentage Weight Loss
Less than 43.1	55	6.1
43.5–46.7	55	6.7
47.2–50.8	73	5.6
51.3–54.4	81	5.4
54.9–57.6	73	5.2
58.1–60.3	78	5.2
60.8–62.6	60	5.5
63.0–65.8	80	4.9
66.2–69.9	69	5.0
70.3–74.8	57	3.7
75.3–79.4	21	3.5
Heavy weight	45	1.4
All categories	747	4.9

Source: Based on data of Tcheng and Tipton (1973).

TABLE 3.8
Expected Decreases of Body Mass (kg) in a 90-kg Man Playing American Football Under Environmental Conditions Specified

Temperature	Relative Humidity (%)			
(°C)	<40	40-60	60-80	80-100
37.8	2 7	3.1	3.3	3.5
32.2	2.3	2.6	2.8	3.1
26.7	1.8	2.2	2.4	2.6
21.1	1.4	1.7	1.9	2.2
15.6	0.3	0.5	0.7	0.9

climatized subject (W. C. Adams, Fox et al., 1975b), and there is a rapid decrease of body mass (Table 3.8) approximately proportional to energy expenditure (Costill, 1977a,b). Even if the rules of a contest permit fluid ingestion, the absorbed volume is unlikely to be more than 0.8 l • hr^{-1}. Given an initial hyperhydration of 0.5 l and a metabolic water release of ~ 1.9 l (Table 2.4), hypohydration is likely if near maximum exercise is sustained for more than 2 hr.

Site of Water Loss

The site of the water loss remains controversial. Some authors have argued that plasma loss is characteristic of a pure heat exposure, but that during exercise water is lost mainly from the intracellular space (Adolph, 1947a; Pugh, Corbett, and Johnson, 1967; Kozlowski and Saltin, 1973; Shephard, Kavanagh, and Moore, 1978; but not Costill and Fink, 1974; or Costill and Saltin, 1975). Discrepant results arise from differences in the severity of heat and exercise stress and the timing of observations relative to the exercise bout. Laboratory analysis is further complicated by a dehydration of the red cells, and by bulk movements of protein and fluid between the tissues and the plasma; the first factor tends to invalidate hematocrit estimations of plasma volume (Van Beaumont, 1973; Costill and Fink, 1974) while the second excludes use of many traditional markers of fluid volumes (Senay and Christensen, 1968; Wyndham, 1973; Shephard, Kavanagh, and Moore, 1978). One advantage of the trained subject may be his ability to keep within the blood vessels protein that enters this body compartment during the early stages of exercise (Senay, 1972). Costill, Coté, and Fink, (1976) made muscle biopsy studies 30 min after exposure to a combined exercise and heat stress; they found that plasma and muscle water declined an average of 2.4 and 1.2%, respectively, for each 1% decrease of body mass. Initially, a large loss occurred from interstitial fluid, but intracellular water was better maintained; Costill suggested this might reflect the influence of water liberated by glycogen breakdown.

Water Loss and Endurance Performance

Acute heat exposure has little effect upon the performance of brief bouts of maximal work, particularly if the subject is in good physical condition. With the exception of Klausen, Dill et al. (1967), most authors found no immediate decrease of maximum oxygen intake (C. G. Williams, Bredell et al., 1962; Saltin, 1964a; Rowell, Marx et al., 1966; Saltin, Gagge et al., 1972). On the other hand, a decrease of maximum oxygen intake is seen if the body is preheated (Rowell, Brengelmann et al., 1969a,b; Pirnay, Deroanne, and Petit, 1970), and Pirnay, Petit, Deroanne et al. (1968) also found a reduction of aerobic power in the period following heat dehydration. The on-transient of oxygen intake develops similarly in normal and hypohydrated subjects (Fig. 3.21), but the endurance of maximum effort is curtailed in the latter (Buskirk, Iampietro, and Bass, 1958; Kozlowski, 1966). The time to exhaustion at a fixed percentage of maximum effort is also shortened (Fig. 3.22). Ladell (1965) postulated the existence of at least 2 l of free circulating water; in his experiments, a cumulative deficit of ~2.5 l was needed before symptoms of heat intolerance developed. Most of Ladell's experiments involved quite vigorous stepping exercise, and it is likely that a fair part of his free circulating water was liberated by glycogen breakdown. Saltin (1964b) maintained that up to 5% of body mass (~3.5 l of fluid) could be lost without change of maximum oxygen intake, cardiac output,

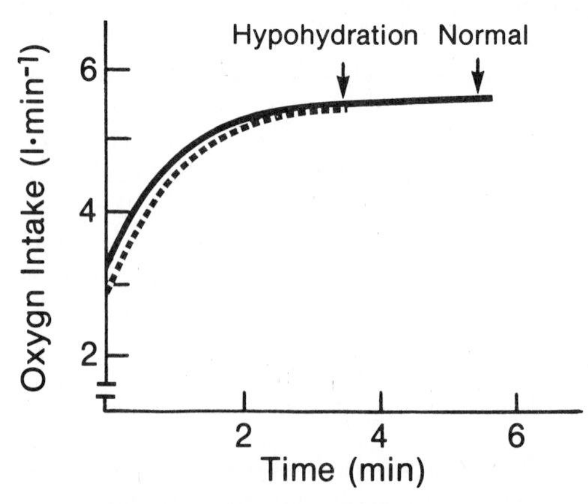

FIG. 3.21. Reduction in tolerance of maximum aerobic work after dehydration (broken line). (After Saltin, 1964a.)

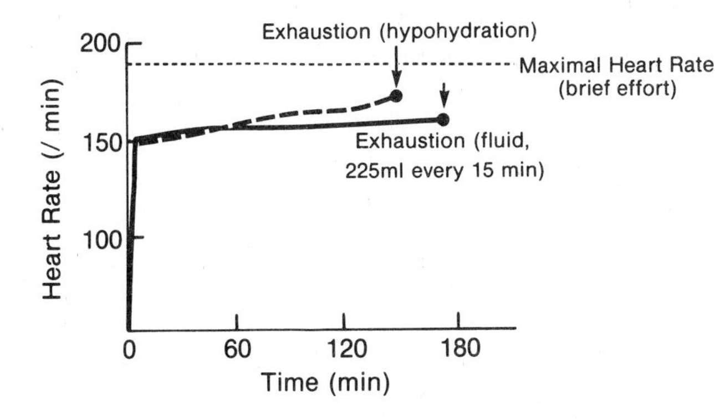

FIG. 3.22. Heart rate during treadmill exercise at 70% of maximum aerobic power. Note that the time to exhaustion was substantially prolonged when hypohydration was avoided by drinking 225 ml of water every 15 min. (After Staff & Nilsson, 1971.)

or stroke volume. Buskirk and Beetham (1960) also pointed out that the pace of the marathon runner was well sustained during his event; despite a 2.5–7.4% decrease of body mass, many competitors were still capable of a final sprint.

Cardiac Responses to Dehydration

The stroke volume of the heart is usually reduced by hypohydration. The effect is larger in the upright than in the supine position (Saltin, 1964a,b), and it is also larger in women, perhaps because of a less ready exchange of osmotically active particles between the tissues and the capillaries (Senay and Fortney, 1975). There is some associated increase of heart rate (Staff and Nilsson, 1971), but compensation is usually incomplete, leading to a widening of the arteriovenous oxygen difference (Saltin, 1964a,b; Sproles et al., 1976). The circulation adapts well to a 4–5% decrease of body mass, but less well to a 7% loss; a parallel may perhaps be drawn with the response to hemorrhage, a sudden fall of blood pressure occurring when a critical portion (10%) of the circulating blood volume has been lost.

Factors contributing to the decrease in stroke volume include (1) a depletion of central blood volume, with reduced diastolic filling of the heart, and (2) peripheral pooling due to a rise of body temperature and associated cutaneous vasodilatation. Well-trained subjects are less affected by hypohydration than sedentary individuals (Saltin, 1964a), but heat acclimatization apparently does not protect against hypohydration.

Neuromuscular Effects

Dehydration does not change the excitability of the muscle membrane (Costill and Fink, 1974). Nevertheless, maximum isometric strength may be somewhat reduced (Bosco et al., 1968; but not Tuttle, 1943; or Saltin, 1964a). This presumably reflects water loss from the muscle cytoplasm and associated electrolyte disturbances. The smaller fluid pool also restricts the possible accumulation of lactate (Saltin, 1964a).

The wrestler is sometimes concerned with the rate at which performance is restored during rehydration. Under laboratory conditions, recovery may be complete within 5 hr (Ribisl and Herbert, 1970), but this is less likely during actual competition. In one study, wrestlers were allowed free access to fluids, but at competition they remained more than 2% underweight, with an associated deficit of performance (Herbert and Ribisl, 1972; American College of Sports Medicine, 1976; American Medical Association, 1976; Sproles et al., 1976).

Dehydration and Core Temperature

The problem of fluid balance is closely intertwined with that of temperature control. If adequate fluid is provided, the rise of rectal temperature with effort is appreciably reduced (Fig. 3.23). A relationship thus exists between the rise of rectal temperature and the loss of mass incurred through sweating (Adolph, 1974a; Buskirk and Beetham, 1960; Gisolfi and Copping, 1974). Equally, the rate of cooling of a subject following exertion is influenced by his state of hydration (G. R. Walder et al., 1975). One reason sweat loss causes a rise of temperature is that the body heat store becomes redistributed through a smaller tissue volume. If the body core is initially $\Delta T'$°C hotter than the environment, the initial heat store $\Delta H'$ relative to ambient conditions is

$$\Delta H' = \Delta T' \, M' \times 3.48 \qquad (3.13)$$

where M' is the body mass in kg and 3.48 is the specific heat of the body ($kJ \cdot kg^{-1}$).* Let us assume that

*The specific heat ranges from 2.93 to 3.56 $kJ \cdot kg^{-1}$, depending on body composition (Hardy, Gagge, and Stolwijk, 1970).

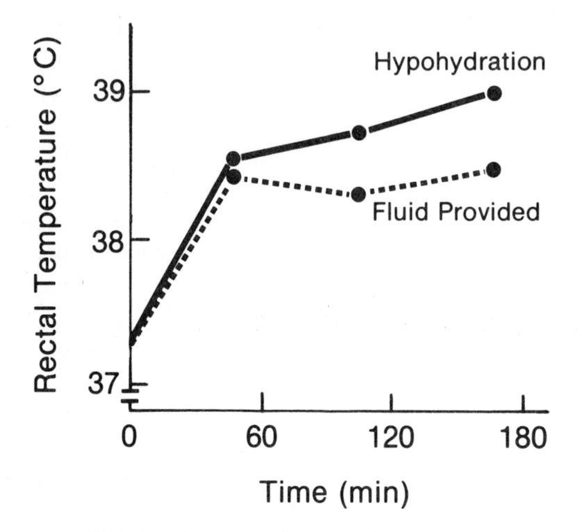

FIG. 3.23. The influence of hypohydration on the course of rectal temperature during treadmill exercise at 70% of aerobic power. In the experiment with broken line, 225 ml of water was provided every 15 min. (After Staff & Nilsson, 1971.)

in the experiment of Fig. 3.23, the environmental temperature was 20°C and the initial body mass was 70 kg. Then:

$$\Delta H' = 17.3 \times 70 \times 3.48 = 4214 \text{ kJ} \qquad (3.14)$$

The total volume of water ingested over the 165-min experiment was 2475 ml. Not all of this fluid was necessarily absorbed. Costill, Kammer, and Fisher (1970) found that as much as two-thirds of the fluid ingested by a runner could remain unabsorbed in the stomach. However, let us assume that the water did replace sweat loss. The subjects would then be dehydrated by 2475 ml in the experiment where no drink was provided, and the corresponding decrease of body mass would necessarily increase $\Delta T'$ from 17.3 to 17.9°C. Heat redistribution could thus account for all of the difference between experiments with and without fluid replacement.

If the 2475 ml of ingested fluid was chilled to 10°C below ambient temperatures, the body heat store would be reduced by $\Delta T' \, M' \times 3.48 = 10 \times 2.475 \times 3.48 = 86.1$ kJ, enough to lower $\Delta T'$ from 17.3 to 16.9°C.

Adolph (1947a) previously reported that body temperature rose by 0.55°F for a 1% decrease of body mass, a change of about 0.44°C l^{-1}. One reason for the smaller change in Fig. 3.23 might be that not all of the ingested volume of 2475 ml was absorbed. There may also be a threshold for the dehydration effect, although this would be easier to explain as a circulatory failure than as a consequence of heat redistribution. Wyndham and Strydom (1969,1972) found that temperatures rose only if body mass decreased by more than 3%; if the loss of mass was more than 4%, the rise of temperature was liable to be excessive.

Dehydration and Sweating

If exercise in a hot environment is prolonged, the priorities of the body change. The need becomes one of conserving fluid at the expense of temperature regulation, and sweat production falls (Greenleaf and Castle, 1971). Sweat production is less in a hypohydrated than in a normally hydrated or hyperhydrated subject (Gerking and Robinson, 1946; F. P. Ellis, Ferres, and Lind, 1954; Moroff and Bass, 1965; Greenleaf and Castle, 1971). Some authors have suggested that at least 2 l of sweat must be lost before secretion diminished, but Ekblom, Greenleaf et al. (1970) found a decrease with only 1% hypohydration. The principal signal for the reduction of sweating seems a change of plasma sodium ion concentration (Sargent, 1962; Senay, 1968). Ladell (1964) suggested that there is also a mechanical factor, the sweat pores becoming blocked by maceration of the skin. Finally, the sweat gland response itself is depressed as the skin becomes wet (Nadel and Stolwijk, 1973).

Assessing Dehydration

The simplest method of assessing acute dehydration is serial weighing. Sweat-sodden clothing and shoes must be removed, the bladder emptied, and allowances made for food ingested and any feces passed. Some authors have recommended the use of an accurate beam balance, but this is hardly necessary if several liters of sweat have been lost; much can be learned from the use of a carefully calibrated clinical scale. If an athletic team is spending several weeks in a hot climate, there should be a daily check of body mass as well as urine composition and flow. A progressive decrease of body mass is a harbinger of inadequate salt and/or water intake. When examining wrestling contestants, the observed body mass should be matched with predictions based on body shape (Wilmore and Behnke, 1969; Katch and Michael, 1971; Tcheng and Tipton, 1973; K. S. Clarke, 1974; Sinning, 1974). It is also useful to check a wrestler's "weight" in previous contests and to examine his urine for evidence of dehydration (concentrated specimens, with a high specific gravity).

Acceptable Levels of Dehydration

There is still disagreement as to what is an acceptable level of dehydration. The working capacity sometimes deteriorates with a 1-2% decrease of body mass (Pitts et al., 1944; Adolph, 1947a; Ladell, 1955; Saltin, 1964a,b; Gisolfi and Copping, 1974). However, in endurance sports, the situation is complicated by the release of water linked to glycogen breakdown (Table 2.5); if all of the glycogen reserves are utilized, up to 2 kg of body mass can be lost without dehydration. Wyndham and Strydom (1969,1972) suggested that a 2% loss would cause thirst, and that with a 3% loss the rise of rectal temperature would be greater than in a normally hydrated individual exercising at the same intensity. With a 4% weight loss, the temperature rise became excessive, with 6% loss there was oliguria, an ill appearance, irritability, and aggressiveness, and with more than 6% loss there was a marked impairment of physical and mental performance. Davidson et al. (1972) suggested that severe symptoms developed when more than 10% of body mass was lost. Presumably, much depends on the initial hydration of the subject, the proportion of the thermal load created by body metabolism, the speed with which dehydration develops, and the extent of compensation by glycogen breakdown.

Chronic Dehydration

Physiologists have recognized for many years that some subjects adapt poorly to sustained physical activity at high altitudes (Hultgren and Lundberg, 1968; Peñazola et al., 1971; Arias-Stella, 1971). In those affected by the syndrome of high-altitude deterioration, the working capacity declines, the appetite becomes poor, sleep is lost, and the individual becomes progressively more lethargic. The causes are multiple, including an inadequate intake of food, intense physical and mental stress, biting cold, lack of sleep, and illness. Nevertheless, progressive dehydration seems a major contributory factor. Vigorous ventilation in the dry mountain air and sweating induced by solar radiation create a fluid need of 2-3 l • day^{-1}. It is surprisingly difficult to satisfy this need when all water must be obtained by melting snow; primitive stoves are liable to set a climber's tent on fire, and the boiling point of any fluid produced is too low to make acceptable tea. High-altitude deterioration can be prevented by insistence on an adequate intake of food and fluids, but once the condition has become established, the only remedy seems evacuation to a lower altitude (Chap. 12).

Dehydration is also seen when vigorous physical activity is undertaken in the arctic. O. Wilson (1960) reported a 3-kg decrease of body mass during 40 days of sledging, this deficit being made good within 24 hr of return to base. Likewise, Shephard, Hatcher, and Rode (1973) commented on the substantial dehydration of Eskimo hunters tested soon after their return from an arctic journey. The extra-urinary water losses of an arctic traveler can amount to 2.1-3.3 l • day^{-1} (Table 3.9). Given a urine flow of perhaps 1.2 l • day^{-1}, the total fluid requirement is thus 3.3-4.5 l • day^{-1}.

O'Hara, Allen et al. (1977a,b,1979) simulated an arctic expedition in a climatic chamber. During the first 2 days, the fluid intake was even less than the urine flow, so that there was an early decline of body mass not matched by a decrease of skinfold thickness (Fig. 3.24). Part of this change could be attributed to a usage of glycogen (with an associated liberation of the water of hydration); however, measurements of skin thicknesses on the dorsum of the hand showed that there was also a substantial dehydration. This remained uncorrected until the cold exposure was terminated. Unfortunately, attempts to increase fluid intake in the cold lead to a greater urine flow rather than a rehydration of the tissues.

Wyndham and Strydom (1972) stressed that persons who are training hard in a hot climate can also accumulate a progressive water deficit. This is usually linked to a deficient intake of salt, and while it can be detected by regular early morning "weight" checks, it cannot be corrected by drinking more fluid unless salt is also provided.

TABLE 3.9
Estimates of Extraurinary Fluid Loss During Exercise in a Cold Environment

Route of Loss	Volume of Water Loss (ml • day^{-1})	Authors
Feces	123	Masterson et al. (1956)
Insensible perspiration	550-567	Welch et al. (1959)
Sweating	700-1097	Welch et al. (1959)
Respiratory tract	700-800	T. A. Rogers et al. (1964)
	1000-1500	Brebbia et al. (1957)
Total	2073-3287	

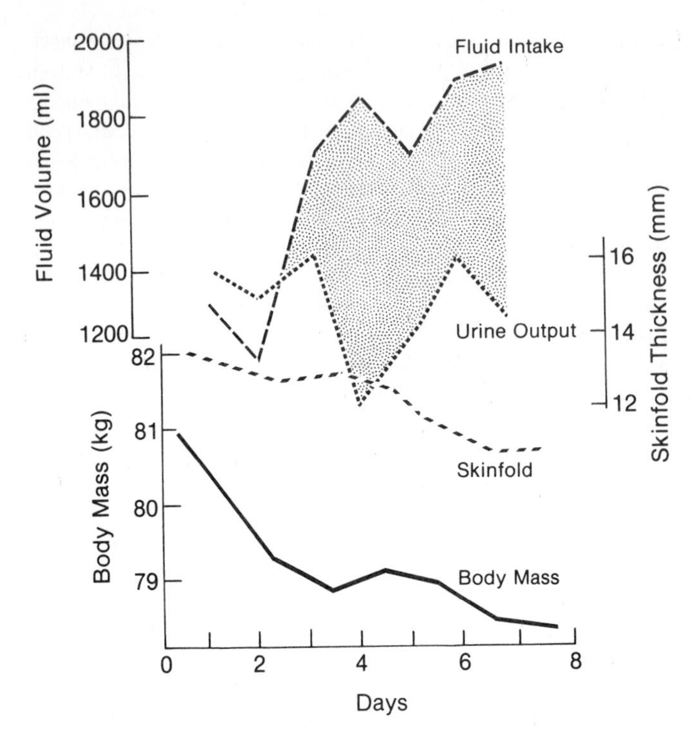

FIG. 3.24. Changes of body mass, skinfold (average of three thicknesses), urine output, and fluid intake during a one-week simulated arctic expedition (W. J. O'Hara, Allen et al., 1977c).

Mineral Deficits

Body pools of most minerals are sufficiently large that a deficit is unlikely to result from a single bout of vigorous work. Nevertheless, cumulative deficiencies can develop when physical activity is carried out repeatedly in a hot climate (Wyndham and Strydom, 1972; Malhotra, Sridharan et al., 1976).

Sodium Ions

A deficiency of sodium ions would lead to an associated water loss, with many of the sequelae already discussed under dehydration. The most obvious physiological manifestation is a liability to painful muscular cramps (the stoker's cramps of old-time steamships). The condition may also present as a chronic weakness, with lassitude, irritability, giddiness, and fainting spells. On other occasions the symptoms are more acute, with marked weight loss, constipation, a scanty urine, headache, and (if dehydration is advanced) nausea and vomiting, sunken eyes, an inelastic skin, and circulatory failure. Once vomiting is established, the condition can be fatal, but symptoms are corrected rapidly if a normal tissue and plasma composition is restored.

The subject most at risk is the newcomer to a hot environment, since with repeated exercise in the heat renal reabsorption of sodium is increased and the sodium content of the sweat is decreased (Smiles and Robinson, 1971). However, the likelihood of developing a sodium deficit is not very great, given an adequate dietary intake. Costill, Coté et al. (1975) had subjects exercise to a 3% weight loss on 5 successive days. In some experiments, a proprietary fluid containing 23 meq of sodium ions per liter was given between exercise bouts, and in other experiments pure water was provided. Both groups of subjects *accumulated* small amounts of sodium, with expansion of their plasma volume, the effect being larger for the water (392 meq) than for the electrolyte (334 meq) experiments. Smiles and Robinson (1971) produced small sodium deficits (140–320 meq) over 7 days of exercise in the heat, but their subjects were restricted to a sodium intake of 30 meq • day^{-1}. After a period in the tropics, most subjects learn to increase their salt intake above the temperate average of 12 g • day^{-1} (204 meq • day^{-1}).

Other Mineral Ions

Much less is known about possible cumulative deficits of other minerals. A deficiency of potassium ions could alter the charge across cell membranes, with consequent alterations in tissue excitability and physical performance; it could also limit muscle vasodilatation, predisposing to local ischemic damage during exercise (MacAraeg, 1974). Knochel et al. (1972) reported that the total body potassium decreased by 349 meq over the first 4 days of military training, but

Costill (1977b) argued that since sweat contains only about 6 meq • l^{-1}, this would be impossible unless the subjects had produced about 20 l of sweat per day! Costill (1977b) had a group of distance runners exercise for 4 days on a diet that provided only 25 meq of potassium per day (about 25% of the normal intake); nevertheless, there was little change of whole-body or muscle potassium; the total deficit for the 4 days was 80 meq, ~2.5% of the body potassium pool. When the experiment was repeated with a normal dietary potassium, the subjects remained in positive potassium balance throughout. Costill noted that some authors inferred a potassium deficiency from a low serum potassium (K. D. Rose, 1975); however, this could be an expression of sodium retention and plasma volume expansion.

Calcium and magnesium deficiencies would have a direct effect upon mechanisms for the initiation of muscular contraction. Substantial amounts of calcium and magnesium are lost in the sweat. According to Costill (1977b), repeated days of distance running do not alter the body content of either ion, but Cohen and Zimmerman (1978) noted low prerace magnesium levels in ultralong-distance runners.

Implications for Fluid and Mineral Replacement

The ideal fluid for the exercising athlete continues to be debated. Palatability is plainly a key issue. Costill (1977b) stressed that preferences are altered during exercise and that under such circumstances a bland fluid such as water is more acceptable than a syrupy citrus drink. We have further noted that the plasma concentration of most mineral ions rises during effort, while the energy yield of any added glucose is not significant in events lasting less than 2 hr. There thus seems little advantage in providing anything more complicated than fresh water during most types of contest (Coyle et al., 1978).

There are advantages to cooled drinks. Henschel (1965) suggested that the optimum temperature was 15 ± 5°C; he considered that iced water often caused vomiting and gastric distress. Costill and Saltin (1974) noted that cold drinks could lower the gastric temperature by 7–18°C, not only increasing gastric motility, but also inducing electrocardiographic changes. They suggested that some increase in the activity of the stomach was an advantage, since this speeded gastric emptying. We have also noted the effect of cool fluids upon body heat content.

In the postcompetition period, there is still little advantage in taking expensive drinks fortified with various mineral ions. The needs of the body are well met by adjustments in the composition of sweat and urine and an increased salting of food. If a salt deficiency is suspected, it is particularly undesirable to give salt tablets, as these provoke a temporary peak of plasma sodium ion concentration, with a further loss of body water (Wyndham et al., 1973a).

THERMAL BALANCE

Individual cells can withstand temperatures ranging from −1°C to +45°C. At the lower extremity of this range, structural damage is caused by freezing of the tissue, while at the upper extremity vital proteins undergo irreversible heat change. Physical and mental performance deteriorates if the temperature of the body core drops lower than 35°C or rises higher than 41°C. Man's tolerance of the environment covers much broader limits, determined largely by his skill in creating a comfortable microclimate through the use of protective clothing and other devices. Thus physical activity can continue over environmental temperatures ranging from −50°C or lower to +100°C or higher.

Under temperate conditions, the exercise-induced rise of body temperature is insufficient to limit endurance performance (Buskirk and Beetham, 1960; Wyndham and Strydom, 1969; Costill, 1972b; Maron et al., 1977; Tanaka et al., 1978), unless there is considerable solar radiation (Pugh, Corbett, and Johnson, 1967). However, in a warm environment the accumulation of body heat can limit performance in a very direct manner, causing cardiovascular collapse, impairment of coordination, loss of consciousness, and even death. This type of reaction was shown by Dorando during the Olympic marathon of 1908. He entered the stadium in a dazed condition, turned in the wrong direction, collapsed twice, and remained in coma for 2 days after the event (Shephard, 1976b). More commonly, core temperature rises to a dangerous level without the subject being aware of his problem. Occasionally (for example, hill walks in wet weather, swimming in cold water, dinghy sailing, and cave exploration) metabolic heat production is less than heat loss. Performance may then be limited by difficulty in sustaining body temperature.

Excessive Heat

Limiting Conditions

In extremely hot industries, tolerance of the environment is limited by overstimulation of pain receptors in the skin. Leithead and Lind (1964) found that if the air was dry, subjects could withstand temperatures of 120°C for ~10 min and 200°C for ~2 min.

More commonly, difficulty arises from an accumulation of metabolic heat. Since most tasks are performed with a mechanical efficiency of 20% or less, ~80% of the free energy of metabolism appears as heat; heat production is even greater if the subject is unfit and has recourse to anerobic metabolism for a part of his energy needs (Wyndham, 1977). The accumulating heat must be dissipated to the environment by a combination of conduction, convection, radiation, and sweating. Under cool conditions, the convective heat loss h_c (kJ • m^{-2} • $°C^{-1}$) is proportional to the relative air velocity V_a (m • sec^{-1}) (Nishi and Gagge, 1970):

$$h_c = 0.516V_a^{0.531} \qquad (3.15)$$

Thus, a runner moving at 5 m • sec^{-1} would be able to dissipate 2.3 kJ • min^{-1} • m^{-2} of body surface area for each 1°C gradient of temperature between his skin surface and the environment. Given a skin temperature of 30°C, an air temperature of 23°C, and a body surface area of $1.8m^2$*, the convective loss in still air could amount to 29.6 kJ • min^{-1}. Under hot conditions, the main method of heat loss is sweating. Each liter of sweat that is evaporated dissipates 2.43 MJ of heat.

Let us suppose that a 70-kg subject with a maximum oxygen intake of 5 l • min^{-1} (223 mmol • min^{-1}) is engaged in an endurance event that lasts for several hours. His oxygen consumption is then likely to be ~75% of maximum (3.75 l • min^{-1}, 167 mmol • min^{-1}); given also an energy usage of 20.9 kJ • l^{-1} and a mechanical efficiency of 20%, his heat production will be 3.75 × 20.9 × 0.8 = 63 kJ • min^{-1}, or 3.76 MJ • hr^{-1}. A sweat evaporation of 1.5 l • hr^{-1} would be necessary to maintain a constant body temperature. Sweat secretion can exceed 2 l • hr^{-1}, but this is not necessarily sufficient for thermal equilibrium, since much of the sweat rolls to the ground unevaporated.

Industrial physiologists have recommended that the rectal temperature of workers should not rise above 39.2°C (Fig. 3.25). However, figures of 41°C are fairly commonplace among endurance athletes (S. Robinson, 1963; Pugh, Corbett, and Johnson, 1967; W. C. Adams, Fox et al., 1975b). The simplest calculation suggests that the quantity of heat needed

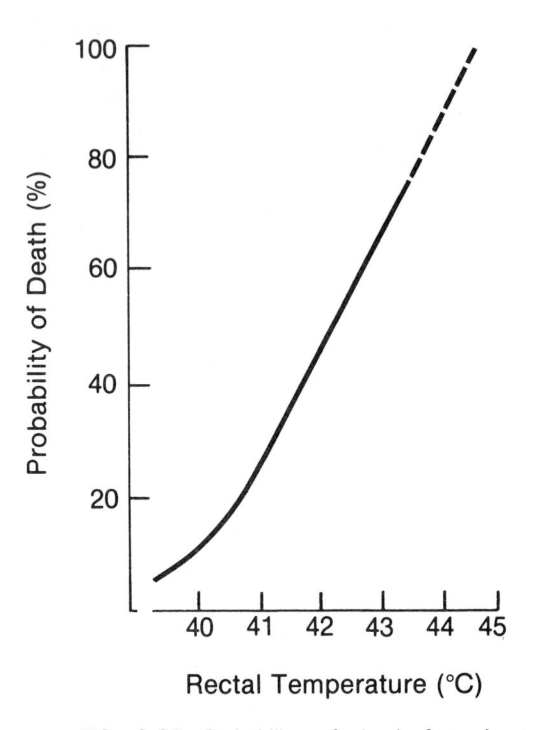

FIG. 3.25. Probability of death from heat stroke in relation to rectal temperature on admission to hospital. (Based on data of Wyndham, 1974.)

to raise the body temperature of a 70-kg man from 37 to 41°C is 4 × 3.48 × 70 = 974 kJ, where 3.48 is the specific heat of the body. In practice, the estimate is complicated by dehydration and a redistribution of heat between the deep and the peripheral tissues.

A. C. Burton (1935) suggested that the mean body temperature T' was given by

$$T' = 0.65T'_r + 0.35T'_s \qquad (3.16)$$

where T'_r is the rectal temperature and T'_s is the average temperature recorded from a number of representative skin sites (Table 3.10). The temperature gradient from core to surface depends largely upon constriction of the skin vessels. Under resting conditions, T'_s can be 10–20°C lower than T'_r. During vigorous physical activity in a cool climate, the skin temperatures may still be only 27–28°C (Pugh, 1970), but in a warm or a humid environment (W. C. Adams, 1977) the corresponding core surface gradient drops to as little as 1–4°C. Taking an initial gradient of 20°C, and assuming a decrease to 1.2°C (as in an example cited by P. O. Åstrand and Rodahl, 1977), heat redistribution would accommodate 18.8 × 3.48 × 70.0 × 0.35 = 1604 kJ. The

*The constants in the formula of Nishi and Gagge (1970) take account of the fact that the projected area of the subject is less than his true surface area.

TABLE 3.10
Weighting Factors Used in Calculating the Mean Skin Temperature

Body Site	Weighting Factor
Head	0.07
Arms	0.14
Hands	0.05
Feet	0.07
Legs	0.13
Thighs	0.19
Trunk	0.35
Total	1.00

Source: After Hardy and DuBois, 1938.

total heat production needed to reach a rectal temperature of 41°C would thus be 974 + 1604 = 2578 kJ. Given the heat production typical of an endurance runner (63 kJ • min^{-1}), this limit would be reached in 41 min [rather less after allowance for sweating, according to equation (3.13)].

Actual data support these theoretical calculations. S. Robinson (1963) found temperature of 41.1°C in both Rice and Lash after races lasting 14 and 30 min, respectively. The weather was sunny, hot (29.5°C), and humid when his observations were made. It is thus likely that initial skin temperatures were higher than 17°C, reducing the potential for an accumulation of heat in the superficial tissues. On a cooler but clear day, outdoor sportspersons may gain appreciable quantities of heat from solar radiation. Pugh, Corbett, and Johnson (1967) calculated a solar heat load of 400 kJ • hr^{-1}, while Blum (1945) noted a figure of over 1000 kJ • hr^{-1} under desert conditions. Much depends upon the time of day. D. H. K. Lee (1972) reported a minimum radiant loading of 360 kJ • hr^{-1} when the sun was directly overhead and a maximum of 900 kJ • hr^{-1} in the afternoon when the sun was lower and the ground was hotter. Pugh's experiments concerned marathon runners in southern England. Here, solar heating was offset by low-temperature radiation and convective exchange amounting to over 2000 kJ •hr^{-1}. Nevertheless, the contestants still needed to dissipate ~2.40 MJ • hr^{-1} by sweating and other forms of evaporative water loss in order to achieve homeostasis. Thermal equilibrium could have been maintained by evaporating 1 l of water per hr. The actual sweat secretion of the winner was 1.8 l • hr^{-1}, but his final temperature after 165 min of running was 41.1°C. Despite a cool environment (23°C dry bulb, 17°C wet bulb), less than a half of the sweat had evaporated. Kerslake suggested (1963,1972) that sweat starts to drip from the body when about a third of the maximum evaporative capacity is being used (Fig. 3.26). The extent of evaporation depends upon skin temperature, local and general humidity, and wind speed (Gagge, 1972); in the example of Fig. 3.22, evaporative heat loss could not exceed 1 MJ • hr^{-1} even if skin temperature rose to 37°C. Skin temperature is a linear function of environmental temperature and is relatively independent of the intensity of exercise (B. Nielsen, 1969).

Effects on Performance

Preliminary exposure to heat can improve the performance of brief activities through a warm-up effect. However, the accumulation of heat is a disadvantage in longer events, and times are generally poorer when races must be run under warm conditions. B. B. Lloyd (1966) noted that world record speeds for

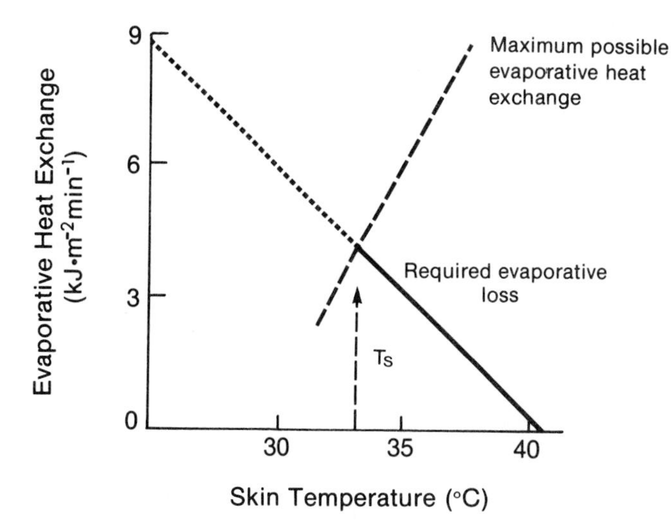

FIG. 3.26. Maximum evaporative heat exchange for specified conditions (air-speed 0.25 m • sec^{-1}; water vapor pressure 4.0 kPa) in relation to skin temperature. T_s is the minimum skin temperature that will permit the required heat exchange. (After Kerslake, 1972.)

runners were 6% poorer for 30-min than for 4-min events. Many factors obviously contribute to a slowing of pace over the longer distances, but Lloyd suggested that one major consideration was a buildup of body heat, with diversion of blood flow from the active muscles to the skin.

In industry, a rise of body temperature can lead to a loss of productivity, accidents, and impairment of consciousness. Many years ago, Vernon et al. (1927) studied conditions in the Cornish tin mines. Accidents (particularly minor accidents) were more frequent in parts of the mines where temperatures were greater than 80°F (26.7°C), and in such shafts productivity was only 74% of that found in cooler and better ventilated areas. Wyndham (1974) likewise noted a heat-related loss of productivity in Bantu miners (Fig. 3.27), although perhaps on account of greater initial acclimatization and differing patterns of labor relations, the critical temperature was substantially higher than in the Cornish population. Wyndham suggested that work became increasingly difficult if rectal temperature rose over 38°C and that heat was excessive if the rectal reading was greater than 39.2°C. Laboratory studies have also shown losses of dexterity, coordination, visual acuity, and vigilance (Pepler, 1963) as body temperature rises. During acute exposures, Blockley and Taylor (1949) found that a heat storage of 210 kJ • m^{-2} led to appreciable discomfort and that an accumulation of 335 kJ • m^{-2} usually resulted in impairment of consciousness. Such quantities of heat are apparently much smaller than those stored by the endurance athlete, yet the majority of distance competitors do not suffer any obvious disturbance of consciousness. Possible explanations of the divergent behavior of the athlete include a slow onset of heat loading, relative youth, a high level of personal fitness, and a measure of heat acclimatization.

Acute Physiological Consequences

Oxygen delivery to the muscles is usually reduced by the increase of skin blood flow, which in some regions is enormous. The blood flow to the fingers, for example, can increase from 0.2 to 120 ml • min^{-1} • 100 ml^{-1} on moving from a cold to a warm room (S. Robinson, 1963). The maximum overall effect is seen when subjects perform moderate exercise in the supine position, with skin temperatures substantially elevated by water-filled suits. Under such circumstances, cardiac output may be increased by 7–10 l • min^{-1} relative to cool conditions (Koroxenidis et al., 1961; Folkow et al., 1965; Rowell, Brengelmann et al., 1969a; Detry et al., 1972). With moderate exercise (oxygen consumption 0.9–1.5 l • min^{-1}, 40–67 mmol • min^{-1}) and a skin temperature of 39°C, cardiac output is still 3 l • min^{-1} greater than under cool conditions (Rowell, Murray et al., 1969; Rowell, Brengelmann et al., 1969a).

It is less certain as to how far maximum oxygen delivery is curtailed by the blood flow needs of the skin. During moderate activity, exercise has little influence upon the relationship between core temperature and skin blood flow (Nadel, Wenger et al., 1977), but there seems to be a progressive reduction in cutaneous vasodilatation as maximum effort is approached (Fig. 3.28). The priority of the body in

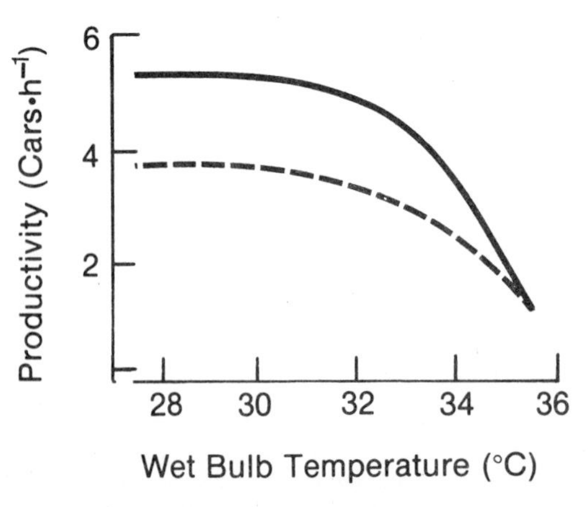

FIG. 3.27. The influence of ambient conditions on the productivity of Bantu miners. Airspeed 0.4 m • sec^{-1}, good supervision (solid line), no supervision (broken line). (Based on data of Wyndham, 1974.)

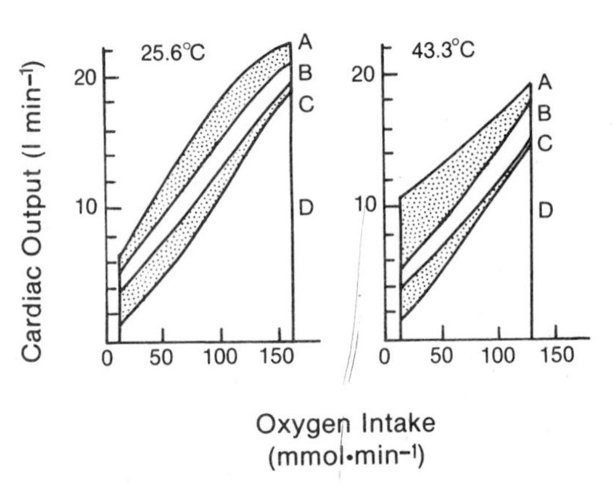

FIG. 3.28. Distribution of cardiac output in cool (25.6°C) and hot (43.3°C) environments. A, skin flow; B, flow to heart and brain; C, flow to viscera; D, muscle flow. (After Wade & Bishop, 1962, and Rowell, 1974.)

effect switches from heat regulation to maintenance of the cerebral circulation (Dill, Edwards et al., 1931; Asmussen, 1940; H. L. Taylor, Henschel, and Keys, 1943). In theory, a lowering of peripheral resistance might permit a subject to develop a larger maximum cardiac output, thereby satisfying the demands of both oxygen delivery and heat exchange, but in practice the maximum cardiac output is often no greater in a hot environment (C. G. Williams, Bredell et al., 1962; Rowell, Marx et al., 1966; Rowell, Kraning et al., 1967). Muscle flow is thus conserved by a greater than normal reduction of visceral blood flow (Radigan and Robinson, 1949; Rowell, Blackmon et al., 1967; Rowell, 1971).

While some authors have found a decrease of maximum oxygen intake in the heat (Brouha, Smith et al., 1960; F. N. Craig and Cummings, 1966; Klausen, Dill et al., 1967; Pirnay, Deroanne, and Petit, 1970), others have not (C. G. Williams, Bredell et al., 1962; Rowell, Blackmon et al., 1965; Saltin, Gagge et al., 1972). Presumably, much depends on the duration of heat exposure and the extent of dehydration that has developed in the various body compartments.

The tolerance of exhausting tasks is substantially reduced in the heat (Saltin, 1964a; MacDougall et al., 1974), particularly when the heat exposure has been relatively prolonged (Pirnay, Deroanne, and Petit, 1970). The reduced endurance reflects the rise of body temperature and associated dehydration, with a considerable contribution from other factors limiting performance. At any given level of submaximal work, the blood lactate is often augmented (Rowell, Blackmon et al., 1965; MacDougall et al., 1974; but not Klausen, Dill et al., 1967; or Rowell, Murray et al., 1969). The heart rate is also higher (F. N. Craig and Froehlich, 1968; Pirnay et al., 1968; C. G. Williams, Bredell et al., 1962) and cardiac output is increased (Drinkwater, Denton et al., 1976a,b). Finally, some authors have found an increase in the oxygen cost of a given activity (for references, see Rowell, 1974).

Lactate accumulates partly because a falling systolic pressure leads to difficulty in perfusing the active muscles (C. G. Williams, Bredell et al., 1962) and partly because a reduction of visceral flow curtails the hepatic metabolism of lactate (Rowell, Brengelmann et al., 1968a). The high heart rate during submaximal work compensates in part for a declining stroke volume and is in part a direct response to the increase of core temperature (Pirnay, Deroanne, and Petit, 1970). In a cool environment, a decrease of stroke volume presages the end of an exhausting work bout. Under hot conditions, the decline occurs earlier (MacDougall et al., 1974); an increase in the compliance of the cutaneous venules encourages peripheral pooling of blood (Rowell, Murray et al., 1969), while an increase of local temperature overrides the peripheral venous constriction that normally accompanies exercise (Bevegård and Shepherd, 1966a; Zitnik et al., 1971). Factors that tend to increase the oxygen cost of work include (1) a decrease in the efficiency of phosphorylative energy transfer in the active muscles (Brooks et al., 1971a,b), (2) an increase in the oxygen consumption of inactive tissues through the Law of Arrhenius (H. W. Burton et al., 1978), (3) progressive fatigue, with a loss of coordination and a decrease of mechanical efficiency (Rowell, Brengelmann et al., 1969b), and (4) the metabolic cost of increases in cardiac output, ventilation, and sweat gland activity (MacDougall et al., 1974). On the other hand, in the early stages of exercise a warming of the muscles decreases their viscosity, and thus the extent of internal work. Conflicting results in the literature (Rowell, 1974) may thus reflect differences in the duration of exercise and/or heat exposure.

Pathological Rise of Temperature

EXTENT OF RISE. Rectal temperature normally rises to a new steady level over a 60-min bout of exercise (M. Nielsen, 1938). The shift of core temperature is apparently an active regulatory process and, although influenced by personal variables such as age, sex, size, and body composition (Wyndham and Strydom, 1972), its main determinant is the intensity of metabolic activity. In any given individual, the relationship between work rate and rectal temperature is sustained over a wide range of ambient conditions (5–30°C). According to B. Nielsen (1968), it also applies to intermittent work (30 sec exercise, 30 sec rest); in contrast, Ekblom, Greenleaf et al. (1971) noted higher temperatures during interval work.

Some authors maintain that the rise of rectal temperature depends more on relative work load (percentage of maximal oxygen intake) than on the absolute intensity of exercise (I. Åstrand, 1960; Saltin and Hermansen, 1966; B. Nielsen, 1969; Kok et al., 1972). This view is supported by the response to combined exercise and heat stress after carbon monoxide poisoning (B. Nielsen, 1971), but there is a less consistent increase of the rectal temperature response when the maximum oxygen intake is reduced by simulated high-altitude exposure (Greenleaf, Greenleaf et al., 1969; Saltin, 1970a) or the use of arm rather than leg exercise (B. Nielsen, 1968).

C. T. M. Davies, Barnes, and Sargeant (1971) suggested that the response was to relative work load when the subject was first seen, but that after habituation to work and acclimatization to heat, the relationship became linked to the absolute rate of working. Some authors reported a linear increase of rectal temperature with work load, but C. T. M. Davies, Barnes, and Sargeant (1971) observed a curvilinear relationship at high work rates.

The body compensates for a moderate rise of ambient temperature by an increased conductance of heat from the core to the skin and an increased sweat rate; while the tolerance of prolonged activity may deteriorate, there is no immediate danger to the subject. However, if a critical combination of work rate and environmental loading is exceeded, adaptive reactions are checked by a combination of dehydration, peripheral pooling of blood, and a large demand for blood flow in the active muscles. The rectal temperature is then forced to a higher plateau or may continue to rise to an ever more dangerous level (Fig. 3.29; Wyndham, Bouwer et al., 1953; Lind, 1963; W. C. Adams, Fox et al., 1975b). As the critical point is approached, there are feelings of strain, and ventilation and heart rate rise.

Discomfort seems related to skin rather than deep-body temperature (Ekblom, Greenleaf et al., 1970,1971). Nevertheless, the spiraling temperature is associated with a reduction of cutaneous blood flow and a decreased conduction of heat to the skin surface (W. C. Adams, Fox et al., 1975b); in one experiment, the decrease was from 6.5 kJ • min^{-1} • °C^{-1} to 5.6 kJ • min^{-1} • °C^{-1}. Depression of sweating leads to a rise of skin temperature and a diminished thermal gradient from the deeper tissues, further impeding the elimination of body heat.

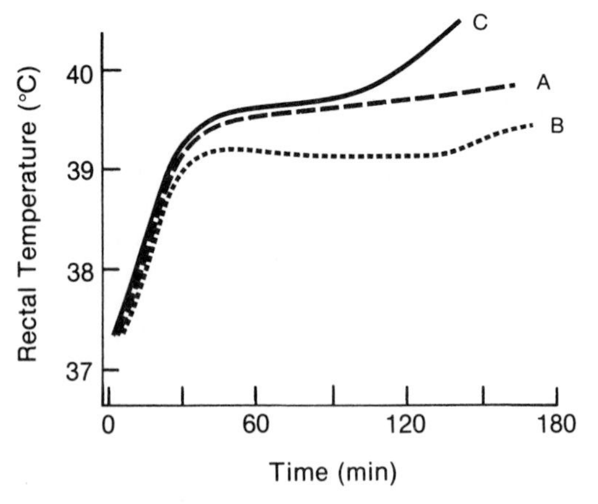

FIG. 3.29. The influence of acclimatization and environment upon the course of rectal temperature during a marathon run. A, moderate environment before acclimatization; B, moderate environment after acclimatization; C, hot environment after acclimatization. (Based on data of W. C. Adams, Fox et al., 1975b.)

During an athletic contest, all subjects work at a roughly comparable proportion of their aerobic power. Vulnerability to an excessive rise of body temperature is thus associated mainly with lack of heat acclimatization. In industry, age and lack of personal fitness are also important considerations.

Heat tolerance is much poorer in older subjects (Robinson, 1963; Leithead and Lind, 1964). Indeed, 70% of industrial heat stroke victims are over 60 years of age (Minard and Copman, 1963). If work is paced, a greater relative load is inevitably imposed on those with low levels of fitness. In a study of Belgian miners, Lavenne and Belayew (1966) concluded that a maximum oxygen intake of 40 ml • kg^{-1} • min^{-1} (17.8 mmol • min^{-1}) was the "dividing line between those who are able to tolerate a high temperature and those who are adversely affected." Likewise, Wyndham, Strydom et al. (1967) found that rectal temperatures >40°C were most likely in those of their mine rescue workers who had a poor aerobic power:

$\dot{V}_{O_2 max}$		Probability of Rectal Temperature >40°C	
l • min^{-1}	mmol • min^{-1}	Unacclimatized	Acclimatized
<2.0	89	0.17	No. est. possible
2.0–2.5	89–112	0.045	0.03
2.5–3.0	112–134	0.018	0.0003
>3.0	134	0.009	<0.0001

Vulnerable Bantu miners were noted to be older and very light in weight (Strydom, Wyndham, and Benade, 1971), although it is difficult to separate these characteristics from a poor absolute aerobic power.

Limiting Emergencies. Fatalities are well documented in endurance sports. One cyclist in the 1959 Tour de France died of heat stroke after reaching a rectal temperature of 43°C, and there were three nonfatal episodes in Danish cyclists at the Rome Olympics (Shibolet et al., 1976). Equally, many North American football seasons have been marred by three or four heat deaths (Murphy and Ashe, 1965; Spickard, 1968; Buskirk, 1968). Dancaster et al. (1969) saw two cases of nephropathy in marathon runners. One was anuric for 48 hr, with a blood pressure of 200/130 mmHg (26.6/17.3 kPa) and a blood urea of 186 mg • dl^{-1} (31.0 mmol • l^{-1}). Most patients surviving the acute episode make a complete recovery, although ~10% develop a chronic nephritis (Kew, Bersohn et al., 1970).

The use of amphetamines is particularly dangerous under hot conditions, since such drugs curtail skin blood flow and diminish subjective feelings of fatigue (Wyndham, Rogers et al., 1971). A number of the fatalities in cyclists have allegedly involved the use of amphetamines.

DETERIORATION OF ORGAN AND CELLULAR FUNCTION. An excessive rise of core temperature leads to a deterioration of both organ and cellular function. Specific changes have been described in the kidney (Kew, Abrahams et al., 1967; Schrier, Henderson et al., 1967), the heart (Kew, Tucker et al., 1969), the liver (Kew, Bersohn, et al., 1970), and the clotting mechanisms of the blood (Schrier, Hano et al., 1970; Shibolet et al., 1967).

There are very large increases in serum enzymes such as lactic dehydrogenase (LDH) (McKechnie et al., 1967); since the -3, -4, and -5 isozyme fractions of LDH are particularly increased, it seems that much of the circulating enzyme has come from muscle. Some leakage occurs in all subjects who participate in endurance activity; blood levels are higher if the rectal temperature exceeds 39°C (Wyndham, 1977), and are 5–10 times larger in individuals who develop heat stroke (Kew, Abrahams et al., 1967). Such increments of serum LDH seem pathological rather than physiological, but it is less clear as to how the blame should be apportioned between the high temperature itself and associated anoxia and acidosis.

There may also be some increase of LDH -1 and -2 isozymes in subjects suffering from heat stroke; this indicates that cellular damage is occurring in the heart and the kidneys as well as the skeletal muscle.

Chronic Reactions to Heat

During more protracted heat exposure, work may be limited by adverse psychological reactions, including aggression, hysteria, and apathy. Wyndham and Strydom (1972) describe the reactions of one psychologist as follows:

> In the first half hour, I worked easily, and thought that the severity of the stress I had been told about had been exaggerated to me. Towards the end of the hour, the position began to change. I felt hot, my face was flushed, I became aware of my breathing . . . my body felt remarkably weak and tired, but there was no anatomical region where I could locate this feeling . . . the control of my social inhibitions began to slip about this time . . . I realized that I was hyper-reacting to trivial irritations but could not control myself. My entire world narrowed down to the single activity of continuing with the task . . . doing so seemed to take all my mental effort.

Such symptoms, Wyndham and Strydom noted, were greatly reduced by a preliminary period of acclimatization.

Excessive Cold

If normal heat loss is exaggerated by high winds, loss of insulation in clothing, or water immersion, the energy liberated during physical activity may be insufficient to sustain body temperature. The oxygen cost of submaximal work is then increased by shivering and performance deteriorates, with eventual loss of consciousness and fatal hypothermia.

Heat Loss in a Cold Environment

If the environment is cool, the initial skin temperature is likely to be low, so that core temperature cannot be conserved at the expense of further peripheral cooling. The heat loss needed to lower core temperature from 37°C to the critical level of 35°C is then given simply by the product of temperature difference, specific heat, and body mass, in a typical young man: $2 \times 3.48 \times 70 = 487$ kJ. The main route of heat loss in the cold is usually by convection. Substantial protection is gained from the thin film of still air in immediate contact with the body. The thickness of this "barrier layer," and thus the rate of heat transfer, is markedly influenced by relative wind speed. Other variables affecting heat loss are environmental temperature, skin temperature, and extent of clothing.

Given normal indoor clothing, a skin temperature of 35°C, an air temperature of 0°C, and a wind speed of 5 m • sec^{-1}, convective heat loss could reach 75 kJ • min^{-1}. This figure is much greater than the likely heat production of a hill walker (34 kJ • min^{-1} if walking at an oxygen consumption of 2 l • min^{-1}, with a gross mechanical efficiency of 20%). The hill walker gains some protection from a decrease of his skin temperature, the preferred value of 29–31°C being less than at rest (Pugh, 1969b). On the other hand, wind speeds much higher than 5 m • sec^{-1} are often encountered in hill walking, and much higher relative air velocities can be developed in other sports, such as snow mobiling, downhill skiing, speed skating, cycling, and dinghy sailing. It is thus easy to see how dangerous heat losses can accumulate in the athlete who is exposed to cool conditions.

Additional heat is lost by evaporation of water. Insensible perspiration causes a minimum loss of 20 ml • hr^{-1} at the skin surface, and the respiratory loss amounts to a further 10–300 ml • hr^{-1} (depending on the absolute water content of the ambient air and the respiratory minute volume). In cold, dry air, unavoidable water loss may lead to an evaporative cooling of ~13 kJ • min^{-1} (J. W. Mitchell, Nadel, and Stolwijk, 1972).

The insulation needed to sustain core temperature is an inverse function of the intensity of physical activity (Fig. 3.30). Clothing is assessed in CLO units. Under resting conditions, 11 CLO units (the equivalent of a double layer of caribou skin) is required to ensure thermal equilibrium at a temperature of −40°C, even in the absence of wind. A person wearing standard arctic military clothing (4 CLO units) loses ~2 kJ • min^{-1} of heat in this environment, and he must develop an energy expenditure of 15 kJ • min^{-1} in order to maintain his body temperature.

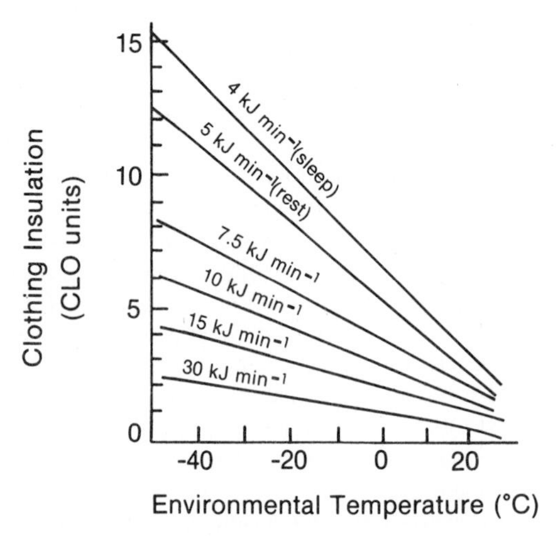

FIG. 3.30. Relationship between activity level and insulation required by subjects when protected from the wind. The CLO unit of insulation = 0.75 $(T_s' - T_a')$ m^2 • hr • kJ^{-1}, where T_a' is the ambient temperature, and m^2 is the surface area of the body in square meters. The indoor clothing of a British businessman provides an insulation of about one CLO unit. (After A. C. Burton & Edholm, 1969.)

It is necessary to do more than make good the resting heat loss, since the insulation of clothing deteriorates when activity is undertaken. Pugh (1967) estimated that the dry clothing of hill walkers provided an initial insulation of 2.58 CLO; this was reduced to 1 13 CLO when walking, and when wet the protection obtained by an active man was only 0.39 CLO. Very fit subjects were able to counter the heat loss by a rapid rate of walking, but those with a poorer working capacity could not develop a sufficient energy expenditure to avoid dangerous hypothermia.

Oxygen Cost of Activity

Cold exposure sometimes reduces physical performance because a given amount of work costs more to perform. This has been demonstrated for bicycle ergometry in a cold chamber (Hart, 1967; Pugh, 1972), and for bicycle ergometry (Craig and Dvorak, 1968,1969; McArdle et al., 1976) and swimming (Costill et al., 1967; Nadel et al., 1974) in cold water. The increase of oxygen consumption relative to warm conditions is 0.25–0.5 l • min^{-1} STPD, 11.2–22.3 mmol • min^{-1}, a value almost as large as the potential energy cost of the shivering it replaces. In a cold environment, much of the potential heat gain from physical activity is dissipated by vasodilation, the performance of external work (Craig and Dvorak, 1968), and increased heat loss at the skin surface.

A cold subject is in effect denied the advantages of an athletic warm-up. Part of the increased oxygen cost of activity could reflect a greater viscosity of the muscles (and of the water, in the case of the swimmer). An added work load might also be anticipated from the weight of winter clothing. However, in practice such effects are sometimes offset by a decrease of oxygen consumption in the inactive tissues, so that in the absence of frank shivering, the energy expenditure during moderate activity may be much as in a temperate environment (Godin and Shephard, 1973a; O'Hara, Allen et al., 1977b,c,1979).

The major influence of shivering and a related increase of muscle tone is suggested by the fact that the augmentation of oxygen consumption is greatest at light work loads. On the bicycle ergometer, the temperate oxygen cost is approximated once the energy expenditure exceeds the minimum required for thermal equilibrium (Fig. 3.31). Data for swimming have been obtained on tethered subjects (Costill, Cahill, and Eddy, 1967) and in the "flume," a

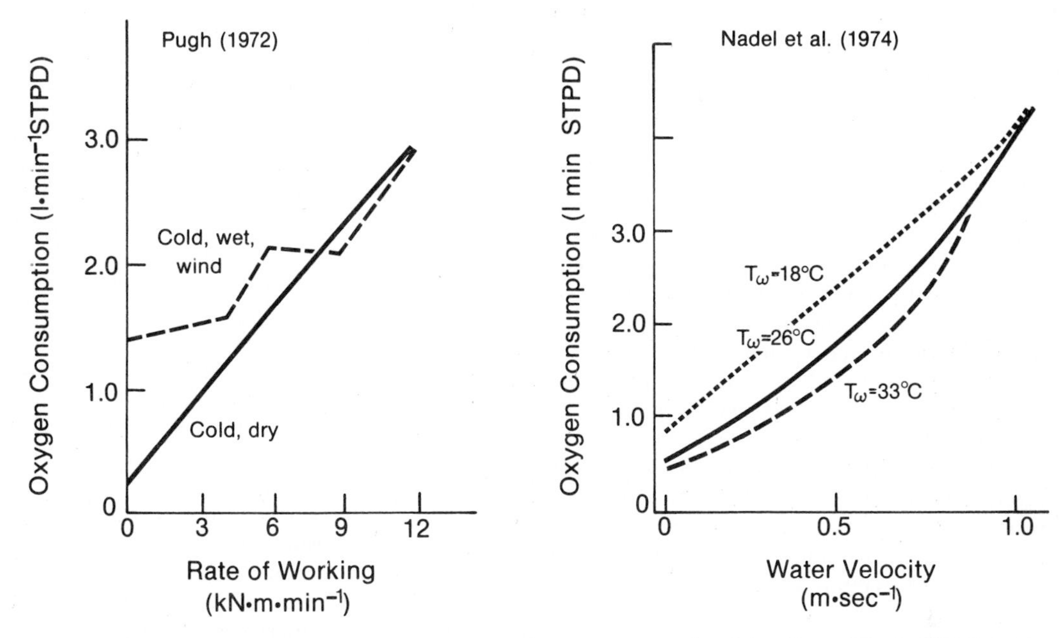

FIG. 3.31. The influence of environment upon the oxygen cost of activity. Data of Pugh (1972) for bicycle ergometry and Nadel, Holmér et al. (1974) for swimming in a flume.

form of millrace developed in Sweden (Holmér, 1974); again, there is a tendency for the added cost to diminish at higher work loads, although in the flume the problem is complicated by an increased convective heat loss from the body at high water speeds.

Effects on Performance

Local Changes. The rate of most cellular reactions is depressed by cooling, as would be predicted from the Law of Arrhenius. The velocity of nerve conduction decreases 15 $m \cdot sec^{-1}$ for each 10°C decrease of local temperature, complete nerve block developing at 8–10°C (Vangaard, 1975). At 20°C, the pressure and touch receptors of the skin have only a sixth of the sensitivity found at 35°C (Irving, 1966), and at 5°C they no longer react to stimuli. The sensitivity of the muscle spindles is increased by a slight fall of local temperature, but 50% of their responsiveness is lost at 27°C, and their function ceases at 15–20°C (Stuart et al., 1963). Below 27°C, there is a rapid decrease in the maximum force of muscle contraction (R. S. J. Clarke, Hellon, and Lind, 1958).

These various changes lead to a slowing of reactions and clumsy performance (J. Le Blanc, 1975), especially if fine coordinated movements are required. The critical loss of body heat for the deterioration of psychomotor skills is about 300 kJ. Associated symptoms include muscle weakness, stiffness, stumbling, and cramp (Pugh, 1972).

Overall Performance. Cold exposure has some beneficial effects during submaximal work; less of the cardiac output is shunted to the skin, heart rates are lower for a given work load, central blood volume is increased, and there may be a parallel increase of stroke volume (A. Craig and Dvorak, 1969; Rennie, di Prampero et al., 1971). Nevertheless, maximal oxygen intake is diminished and the performance of heavy work deteriorates (Holmér and Bergh, 1974; C. T. M. Davies, Ekblom et al., 1975; but not Horstman, 1977). Nadel, Holmér et al. (1974) found that their two leanest subjects were able to attain no more than 85% of maximal aerobic power when swimming in water at 18°C. Body temperatures dropped to less than 36°C, maximum heart rates were reduced, and at the limiting speed of swimming subjects complained of inability to contract their muscles and tiredness rather than cardiorespiratory distress.

Adee (1953) suggested that performance of a 50-yard swim was optimal at a water temperature of 28–34°C, while the best temperature for a 1500-m race was 23–26°C.

Thermally Neutral Temperature. We have noted the combinations of insulation and physical activity needed to avoid heat loss at various air temperatures (Fig. 3.30). Water has ~20 times the thermal conductivity of air, and the neutral temperature is thus much higher. Under resting conditions, figures of 36.4–36.8°C (Wick, 1894) and 35.0–35.5°C (A. Craig and Dvorak, 1968) have been suggested. On the other hand, the rectal temperature rises to 39.5°C when a 1500-m race is swum in water at 34.5°C (Adee, 1953). A. Craig and Dvorak (1968) reported a neutral water temperature of 34°C for a subject cycling at 2.5 times the resting energy expenditure, and the neutral reading dropped to 29°C when the energy expenditure of the cyclist was increased to 3.4 times the resting value. Costill, Cahill, and Eddy (1967) found a slight rise of rectal temperature when subjects engaged in tethered swimming developed an oxygen consumption of 3 l • min^{-1} (134 mmol • min^{-1}) at a water temperature of 17°C; however, most studies of free swimmers have shown a fall of body temperature when the water was cooler than 20°C.

Dangerous Hypothermia

EXHAUSTING HYPOTHERMIA. A hill walker normally sets a pace he can sustain throughout the day, 50–60% of his maximum oxygen intake (Pugh, 1972). However, under cold conditions, shivering may cause a 15% increase of energy expenditure. The time to exhaustion is then halved, and the walker may lack the endurance to reach his destination. If excessive energy expenditure is sensed, the pace may be slowed to accommodate the cost of shivering, but this inevitably extends the period of heat loss; in consequence, the destination may not be reached before nightfall, food supplies become exhausted, and hypothermia is compounded by hypoglycemia. Accidents of this type are particularly probable if a party of hill walkers are not well matched in terms of their initial effort tolerance; however, difficulties can arise even if fitness levels are well matched, since subjects also differ in their cold tolerance. One person may be just above his shivering threshold and will experience little additional fatigue, but another with less effective clothing, different body type, less subcutaneous fat, or a greater sensitivity to cold (Strømme, Andersen, and Elsner, 1963) will experience a considerable increase of oxygen consumption and early exhaustion.

Similar phenomena develop during laboratory exposures to cold. Adolph and Molnar (1946) found that subjects working in the nude became exhausted within 1 hr at 0°C, whereas they had no difficulty in maintaining the same rate of working for 4 hr in a temperate environment. In these experiments, pain and discomfort were considered important factors hastening the onset of exhaustion.

HYPOTHERMIAL CONFUSION. A decrease of cerebral temperature causes confusion, irrational behavior, and eventual loss of consciousness. The function of the brain begins to deteriorate at a rectal temperature <35°C, although some cross-channel champions have continued swimming to readings of 34–34.5°C (Pugh, Edholm et al., 1960). When the core temperature drops below 34°C, the shivering response passes off, and the rate of body cooling is accelerated (Burton and Edholm, 1969). Other mechanisms of heat conservation fail at about 32°C; the skin becomes vasodilated, giving a sensation of warmth, and with the associated confusion clothing may be removed (Wedin, 1976). The victims of the infamous Dachau "experiments" reputedly became unconscious when their rectal temperatures had dropped to 30–31°C (Alexander, 1946). Some cases of hypothermia are still conscious at rectal temperatures of 27°C, but others lose consciousness at 31°C (K. E. Cooper et al., 1964). Variables influencing the time of useful consciousness in the cold include the development of ketosis (Pugh, 1969b), hypoglycemia, and (particularly in snowmobile accidents) alcohol ingestion.

Rate of Cooling

The rate of body cooling in air is so dependent on relative wind speed, solar radiation, precipitation, and the quantity and quality of clothing that it is difficult to quote average tolerance times for a given environment and rate of working.

Forced convection from swimming greatly reduces the survival times in water (Keatinge, 1961; Bullard and Rapp, 1970), and unless the shore is close at hand the safest tactic for a sailor who is tipped into an icy lake or ocean is to remain crouching in a fetal position until he is rescued (Hayward, 1975). The obese subject has a substantial advantage over a thinner person when immersed in cold water (Pugh, Edholm et al., 1960). The heat flow H′ across an insulating layer of unit thickness and area is given by the temperature difference (ΔT′) divided by the thermal insulation (I′):

$$H' = \frac{\Delta T}{I'} \quad (3.17)$$

The thermal conductivity of human fat (the reciprocal of its thermal insulation) is approximately 1.22

kJ • min^{-1} • $°C^{-1}$ for a thickness of 1 cm (Hatfield and Pugh, 1951). Under resting conditions, the heat transfer across the fat is perhaps 2.1 kJ • m^{-2} • min^{-1}, so that the 1-cm layer of subcutaneous fat found in a typical cross-channel swimmer (Pugh, Edholm et al., 1960) will support a temperature gradient of 1.7°C. However, during exercise the heat flow may be 10 times as large, so that the same 1-cm layer of fat will support a thermal gradient of 17°C. In other words, fat offers much greater protection while swimming than under resting conditions.* The active swimmer also obtains useful protection from a 1-mm layer of grease. Assuming a 10-fold increase over resting heat transfer, thermal gradients for three types of grease would be

Form of Grease	Thermal Conductivity (kJ • m^{-2} • min^{-1} • $°C^{-1}$)	Thermal Gradient (°C • mm^{-1})
Lanoline anhydrous	1.51	1.4
Lanoline hydrous	1.38	1.5
Vaseline	1.23	1.7

R. E. G. Sloan and Keatinge (1973) suggested that the rate of cooling $\partial\theta/\partial t$ of an actively swimming subject was given by an equation of the type:

$$\frac{d\theta}{dt} = \frac{2R'A'K'\,\Delta T'}{W'S'} - \frac{M'}{W'S'} \qquad (3.18)$$

where K′ is the thermal conductivity of unit area of fat (0.0122 kJ • m^{-1} • min^{-1} • $°C^{-1}$), ΔT′ is the difference of temperature between the body core and water (16.6°C in their example), R′ is the mean reciprocal of fat thickness (~100 m^{-1} in their example), A′ is the body surface area in m^2, W′ is the body mass in kg, S′ is the average specific heat of the body (3.48 kJ • kg^{-1} • $°C^{-1}$), and M′ is the metabolic rate.

Putting A′ = 1.8 m^2, W′ = 70 kg, and M′ = 35 kJ • min^{-1}, then

$$\frac{d\theta}{dt} = \frac{2 \times 100 \times 1.8 \times 0.0122 \times 16.6}{70 \times 3.48} - \frac{35}{70 \times 3.48} = 0.299 - 0.144°C \cdot min^{-1} \qquad (3.19)$$

*These calculations led Pugh to the view that it was an advantage for a shipwrecked sailor to continue swimming in cold water. However, in practice, the greater thermal gradient across the fat is usually offset by loss of the thin film of still water in immediate contact with the skin surface.

The example of R. E. G. Sloan and Keatinge was based on a water temperature of 20.3°C. According to their calculations, consciousness should have been impaired within 13 min of immersion (2°C decrease of core temperature), death occurring within 30–40 min (4.5–6.2°C decrease of core temperature). In practice, many of their subjects (a party of schoolboys) were able to continue swimming for the full 40-min class period, and the average rate of cooling was only about a quarter of that predicted by the above equations. Most of the group must thus have obtained additional insulation, either from still water trapped in the bathing costume or from tissues other than fat. However, the fattest subjects lost heat faster than predicted, suggesting that in their case the potential insulation of a thick layer of subcutaneous fat was bypassed by a substantial cutaneous blood flow. Presumably, cooling of the obese subjects was insufficient to induce a full cutaneous vasoconstriction.

LIMITATIONS OF FOOD SUPPLY

Carbohydrate Reserves

Lack of carbohydrate normally impairs performance through a direct effect on the metabolism of skeletal muscle. However, in some circumstances, a deficiency of glucose in the cerebral circulation can contribute to the reduction of working capacity.

Glycogen Lack

The glycogen reserves of skeletal muscle and the liver can be almost exhausted by 75–100 min of vigorous endurance work. Thereafter, effort depends on fat mobilization plus the limited gluconeogenic power of the liver. B. B. Lloyd (1966) illustrated the consequent limitation of effort by an analysis of world track records. Winning times for events lasting 2–5 hr could be described by a steady rate of running, equivalent to a fat combustion of 59.1 kJ • min^{-1}, and a zero intercept of 13.4 km, equivalent to a glycogen combustion of 3.64 MJ, ~230 g. Total body glycogen stores are ~500 g, but the intercept of 230 g seems realistic for that fraction of the body musculature involved in distance running.

In a sport such as soccer or ice hockey, where there are frequent short sprints, the deterioration of performance appears in less than 2 hr. The player tires during the final period, although at this stage his blood lactate levels are lower than earlier in the match (H. J. Green, Bishop et al., 1976). The problem is particularly acute if the sport continues to demand intermittent bursts of anerobic work, since fat cannot replace glycogen as a fuel in the absence of oxygen. A cross-country cyclist who has exhausted his glycogen reserves finds little difficulty in sustaining a reasonable pace on the level, but he is no longer able to tackle quite modest gradients, as the more intense contractions of the quadriceps that are then needed would interrupt muscle blood flow, forcing a reliance on anerobic metabolism.

One suggested short-term remedy is to administer a glucose-containing drink immediately before (Brooke, 1978a) or while activity is proceeding. Uncontrolled experiments with soccer players have suggested that glucose administration at half-time gives an increased rate of scoring in the second period (Muckle, 1973). However, the total absorption of glucose during physical activity is unlikely to exceed 30 g • hr^{-1}, boosting existing carbohydrate reserves by no more than 7–8%. It is difficult to give glucose in double-blind fashion without gastric intubation, and any benefit gained from the usual glucose drinks may be psychological rather than physiological. If the glucose has a direct physiological effect, it could be supporting cerebral rather than muscular metabolism (see below).

If several days are available for dietary preparation, the glycogen reserves of specific muscles can be increased substantially (Fig. 2.19). High initial levels of glycogen in the active muscles prevent the "fading" of performance in events such as a 30-km cross-country run (Fig. 3.32; Essén, 1977).

Glucose Lack

In our experiments, the blood glucose was well maintained over as much as 4–5 hr of slow marathon running (Shephard and Kavanagh, 1975a). Gollnick, Karlsson et al. (1974) and Gollnick, Piehl, and Saltin (1974) also reported that blood glucose levels remained high until liver glycogen stores were almost depleted. Nevertheless, others found a fall of blood glucose (Brooke, 1978a), particularly in activities lasting >90 min (D. R. Young et al., 1967; Ahlborg et al., 1974; Udassin et al., 1977). There have been suspicions that cerebral function declines over the course of endurance activities such as orienteering (P. O. Åstrand and Rodahl, 1977) and bicycle ergometry (Marsh and Murlin, 1928). Certainly, the glucose needs of the brain continue unabated during prolonged effort (Ahlborg and Wahren, 1972), and although the blood insulin falls during activity, this does not reduce the muscle glucose uptake on account of the increased perfusion of the active fibers (Wahren, 1977b).

An adequate cerebral glucose supply is particularly important in sports that require much tactical thought. The dinghy sailor, for example, is playing the analog of a vast chess game, each competitor seeking to anticipate many future moves by his opponents. The crew of a small boat are particularly vulnerable to a fall in blood glucose, since sustained isometric contractions of the leg and abdominal muscles are needed when counterbalancing the vessel. It is thus not surprising that there is a positive correlation between the resting blood glucose levels of dinghy sailors and the team captain's rating of performance under light-wind conditions (Niinimaa, Wright et al., 1977).

Fat Supply

Local Fat Reserves

Essén (1977) has set the local triglyceride content of the muscle fibers at 5–15 mmol per kg of wet tissue, a

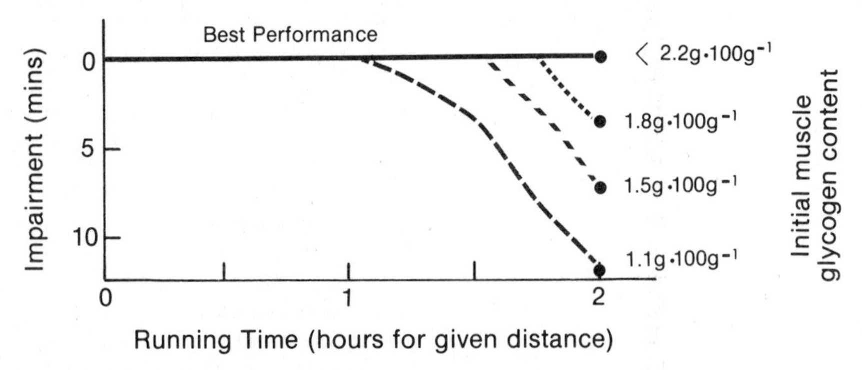

FIG. 3.32. Impairment of running speed in 30-km race in relation to initial glycogen content of muscle. All data are plotted relative to best performance for the distance. (Based on an experiment of B. Saltin, cited by P. O. Åstrand & Rodahl, 1977.)

total of 200–300 g (~10,000 kJ). Type I fibers contain 2–3 times more fat (~207 mmol • kg^{-1} dry wt) than type II fibers (74 mmol • kg^{-1} dry wt). Much of this fat seems usable, and a long ski race can reduce fiber stores by 50%.

Fat Uptake

The maximum rate of working declines dramatically when glycogen reserves have been exhausted. B. B. Lloyd (1966) calculated that the oxygen intake of champion runners was 4.6 l • min^{-1} (205 mmol • min^{-1}) in events lasting 30–120 min, but dropped to 3.3 l • min^{-1} (147 mmol • min^{-1}) in events of 2–5 hr duration. The equivalent energy usage dropped from 85.5 to 59.1 kJ • min^{-1}. B. B. Lloyd suggested that delivery of substrate to the cell now restricted performance, although the size of intramuscular fat reserves (sufficient for 120 min of fat-based activity) seems rather against this hypothesis.

The mobilization of free fatty acids (FFA) is increased by stimuli such as noradrenaline (Issekutz, 1964) and possibly growth hormone (Sidney and Shephard, 1978a), while being inhibited by lactate accumulation (Issekutz and Miller, 1962; Fredholm, 1969; Issekutz, Shaw, and Issekutz, 1975). The net effect of prolonged work is therefore an increase rather than a decrease of blood FFA levels, and it seems clear that the limiting factor is not the delivery of fat to the active muscles, but rather the rate at which they can use local reserves and can clear FFA from the bloodstream (L. A. Carlson, 1967a,b). Holloszy (1977) commented that fatty acids could not enter the cell fast enough to support more than ~50% of the maximum oxygen intake; in his view, the ultimate limitation was set by the mitochondrial utilization of FFA. Others suggested that the problem is due to a slow diffusion of FFA from the capillaries to the active fibers (Saltin et al., 1977; Gollnick, 1977), but this is hard to reconcile with (1) the substantial intramuscular reserves of triglyceride and (2) the fact that the oxygen equivalent of fat and glucose delivered to the active muscles far exceeds the observed oxygen intake after exhaustion of local glycogen reserves (Gollnick, 1977).

Depot Fat

The depot fat of the average sedentary subject is usually regarded as infinite. This is a permissible approximation in the case of the average city dweller. However, an ultralong-distance runner (such as a Trans-Canada contestant, Shephard, Conway et al., 1977) may occasionally find that performance is limited by fat depletion and a subsequent excessive breakdown of protein. Such subjects usually commence a race with a low percentage of body fat (perhaps 5% of 60 kg, 88 MJ), yet they may have an energy expenditure of as much as 25 MJ • day^{-1} (Shephard, Conway et al., 1977). Ketosis can waste a further 1 MJ • day^{-1}, and because there is little time or inclination for heavy eating, an energy deficit of 5–10 MJ • day^{-1} must be met from fat stores. The labile fraction of their body fat (75% of 88 MJ = 66 MJ) is thus exhausted in 7–13 days, much less than the time needed to run across Canada.

The remedy seems to maximize energy ingestion while running. Menus that serve this purpose are indicated in Table 2.7.

Protein Intake

It is generally held that an athlete receiving a well-balanced diet needs no additional protein (Durnin, 1978). This is probably true of brief steady-state conditions where the usage of amino acids is minimal. However, in prolonged heavy work substantial quantities of alanine are converted to glucose. About 90% of circulating alanine is then extracted from the splanchnic vessels (Ahlborg et al., 1974), and as much as 5–6% of the energy requirements of a marathon run are met by protein catabolism (Shephard and Kavanagh, 1975a).

If the daily energy expenditure includes 6 MJ of sustained, deliberate activity, 300 kJ of this total may be met from protein (~17 g); with 15 MJ of deliberate energy expenditure (possible in a distance cyclist, a cross-country runner, or a football player), 750 kJ may come from protein (~44 g). We have noted that in the football player who is seeking to increase his body mass, the total protein need may reach 1.8 g • kg^{-1} • day^{-1}, or 180 g • day^{-1} in a 100-kg contestant (see also Stucke et al., 1972).

Over a 10-day period, the maximal oxygen intake is not reduced by a protein intake as low as 4 g • day^{-1} (Rodahl, Horvath et al., 1962), while performance is not improved if the protein intake is increased to 160 g • day^{-1} (Darling et al., 1944). Nevertheless, the long-term importance of a substantial protein intake to the productivity of heavy workers, such as miners, was recognized by both the British and the Germans during World War II. Under current economic conditions, a protein intake of 180 g • day^{-1} (~12% of the energy content of a 25-MJ diet) is quiet conceivable if the athlete eats at the training table of a well-endowed university. However, it is less likely when an impoverished student is feeding himself, and is improbable when the contestant comes from one of the less developed nations of the world.

Other Nutrients

Arguments for and against the provision of mineral and vitamin supplements are given in Chap. 2. Provided that the diet is well balanced, such supplements do not improve an individual's performance.

Timing of Meals

The cross-Canada runner or the ultralong-distance cyclist may need to take food and fluid while he is exercising. However, most classes of athlete should take a final light meal at least 2–3 hr prior to competition. Subsequent feeding is reputed to cause a diversion of blood flow from the muscles to the intestines, with a corresponding impairment of performance. There is also a danger of vomiting, which can be dangerous for a swimmer.

OTHER DETERMINANTS OF PERFORMANCE

When account has been taken of all the factors noted above, a large part of the variation in most types of human performance remains unexplained. Possible residual determinants include neuromuscular performance, motivation, and technique.

Neuromuscular Performance

Specific problems relating to vision, balance, and the central processing of information are deferred to Chap. 7. We here consider reaction time, skill and agility, coordination and kinesthetic sense, flexibility, and warm-up.

Reaction Time

Sprint events may be completed in less than 10 sec. Thus a fast reaction to the starter's pistol has an appreciable influence on the recorded performance.

The simplest reactions depend solely on the length of the reflex arc and the number of intervening synapses. However, responses of athletic interest are more complex, involving elements of both decision and movement. They vary considerably from person to person, being improved by training and an appropriate level of arousal, "mobilized readiness" (Genov, 1970a,b), or expectancy (Lueft, 1970).

Athletes have faster reaction times than nonathletes, the advantage being particularly marked for contestants in sprint events. Reaction times are also specific to a given body part; thus an individual may react quickly with his arms, but slowly with his legs (F. M. Henry and Rogers, 1960; Lotter, 1960; H. H. Clarke and Glines, 1962). However, it is less clear how far such differences are inherited and how far they are a reflection of practice. Reaction times reach a minimum between the ages of 20 and 30 years (Welford, 1951; Birren, 1959; Fig. 3.33) and are shorter in men than in women (F. M. Henry, 1961; Hodgkins, 1963; G. R. Wright and Shephard, 1978a).

At one time, athletes awaiting the starter's gun were trained to think of the initial movements that they would make (the technique of "motor set"). However, most individuals react more quickly if they direct attention to the initiating signal. With such a "sensory set," they rely upon learned responses, in contrast to the motor set, where responses may be slowed by an attempt to reprogram the sequence of movements (F. M. Henry, 1960).

Skill and Agility

In some common activities, such as walking, running, and cycling, the mechanical efficiency of effort varies little from one person to another. However,

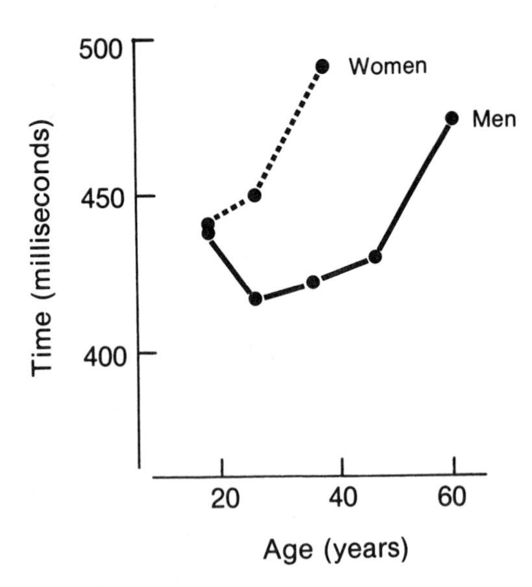

FIG. 3.33. Influence of age and sex upon braking time; the braking time comprises almost equal components of light-reaction time and leg-movement time. (Based on data of G. R. Wright & Shephard, 1978a.)

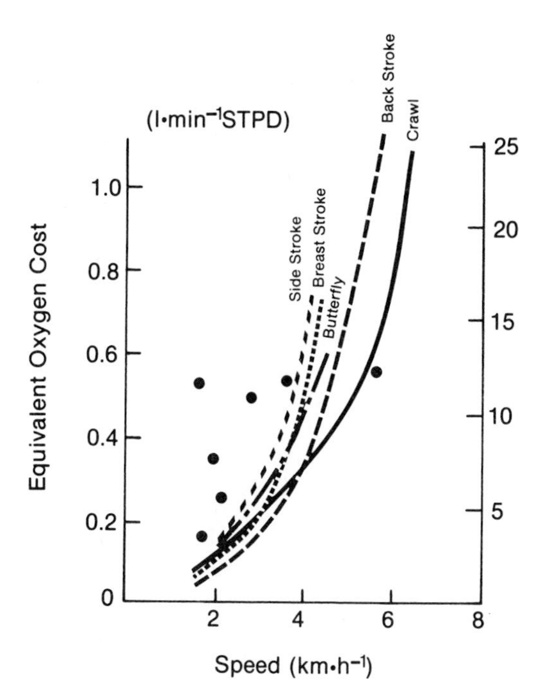

FIG. 3.34. Energy cost of swimming. Solid circles illustrate energy expenditure of inexperienced swimmers performing the crawl. Lines are for good swimmers. (Based on data of Karpovich & Millman, 1944.)

there are wide interindividual differences in the oxygen cost of many sports; for example, there is a fivefold difference of energy expenditure between a novice and an experienced swimmer at a given water speed (Fig. 3.34).

Secrets of improved mechanical efficiency in the skillful athlete include

1. Reduction of postural work through the development of balance and coordination
2. Elimination of unnecessary movements
3. Reduction of kinetic energy losses through adoption of a more uniform speed and a more appropriate direction of limb movements
4. Replacement of controlled movements by ballistic strokes that require much less sustained muscular contractions (A. W. Hubbard, 1960)
5. Reduction of activity in antagonistic muscles, so that opposition to a movement is minimized, and the least necessary force is exerted to halt displacement of the body part.

Thus an inexperienced swimmer allows his arm speed to vary widely; in contrast, the expert moves his arms forward in a smooth and graceful curve, maintaining an almost constant velocity. Likewise, it is easy to distinguish the well-trained from the indifferent runner; the former brings his thighs forward at close to the maximum speed permitted by the resonant frequency of the legs (~3Hz), taking a strike 20 cm longer than that of the novice. The increased speed of thigh movement requires a more intense contraction of the agonist muscles, while the longer stride is facilitated by a greater relaxation of the antagonists. The well-trained runner also maintains a better dynamic posture; in particular, he permits less lateral oscillation of his hips and trunk. In many other sports, the learning of optimum movement patterns is important to peak performance. While a novice may attempt to compensate for his lack of skill by making more forceful muscle contractions, this is an exhausting and largely ineffective tactic.

Certain aspects of skill and agility depend upon body build, and are thus less amenable to training. A tall person usually has a high center of gravity, with resultant difficulty in maintaining his balance (Travis, 1944; but not Singer, 1970). Height is a handicap in many sports; in gymnastics, for example, North Americans are consistently outscored by their shorter Asiatic competitors (Shephard, 1978a,c). However, height is of some advantage in running. Stride length is roughly proportional to leg length L, whereas the natural frequency of oscillation varies as $\sqrt{1/L}$; a tall runner can therefore take longer if slightly slower strides. In basketball, also, top performers now have a height of 205–210 cm (Shephard, 1978c); the associated long reach enables them to guide the ball through a substantial part of its trajectory. A heavy person is usually less skillful than someone who is lighter. Subcutaneous fat may physically impede movement; further, the turning moment tending to cause loss of balance is directly proportional to body mass. On the other hand, the size of the muscles and thus maximum voluntary force is highly correlated with lean body mass. The ideal weight for a given task thus depends on the relative demands of skill and force during peak performance.

Coordination and Kinesthetic Sense

Coordination and kinesthetic (position) sense make important contributions to skill. If complex muscle movements are to be both effective and economical, individual muscles must contract in the correct sequence, and at appropriate intensities. As a task is learned, a program describing both sequence and intensity of contraction is developed and stored in the brain. The body can then initiate the movement as a whole rather than as a series of isolated muscle contractions (see further, Chap. 7). Control shifts from the motor cortex of the cerebrum (where individual muscles are represented) to the cerebellum (where information on the sensory consequences of a given movement is stored and interpreted in terms of limb position; Granit, 1970,1972,1975; P. Gilbert, 1975).

When first playing tennis, a novice keeps his eyes on the racquet, making it difficult to watch either the ball or its target. With practice, the discharge of the muscle spindle and joint receptors corresponding to a given racquet position (P. B. C. Matthews, 1972) is impressed upon the cerebellum; execution of the movement is now much less dependent on immediate sensory input (Eccles, 1973; G. I. Allen and Tsukahara, 1974); and the eyes can concentrate on following the ball. However, individuals vary widely in both their kinesthetic sensitivity and their ability to store movement sequences in the brain (G. M. Scott, 1955). Much probably depends upon early training, since basic movement patterns are established by 3 years of age. If the reflexes concerned with body balance, for example, have been poorly developed as a child, it will take much perseverance to become even a moderate skater at the age of 35 or 40 years.

Flexibility

Any movement must be performed against viscous resistance offered by the displaced tissues. Efficiency in the performance of external work should thus be influenced by flexibility although, to date, experiments to test this hypothesis have yielded essentially negative results (deVries, 1963; Dintiman, 1964).

Static flexibility is simply the range of possible movement at any given articulation (K. F. Wells and Dillon, 1952; J. Leighton, 1955). At some joints, movement is limited by contact between opposing soft tissues (for example, elbow flexion) or bony structures (for example, elbow extension). At other joints, limitation is imposed by the elastic resistance of the muscle sheath, tendon, joint capsule, or supporting ligaments, together with the overlying skin (Table 3.11). Flexibility is generally greater in girls than in boys, is improved by warmth and habitual exercise, and deteriorates with age (Krahenbuhl and Martin, 1977). While it is debatable how important static flexibility is to the average citizen, a wide potential range of joint movement is essential to many athletic skills.

TABLE 3.11
Relative Contribution of Various Tissues to Resistance of a Joint for Wrist Flexion and Extension in the Cat

Tissue	Extension − 48° (% total torque)	Flexion + 48° (% total torque)
Muscle	42.4	36.9
Joint capsule	35.2	38.8
Tendon	11.2	33.0
Skin	11.2	−8.7

Source: After Johns and Wright, 1962; deVries, 1974.

Dynamic flexibility is an inverse function of the resistance encountered when moving a joint through its normal operating range. It depends largely upon the elasticity and resistance to plastic deformation of the joint tissues, with inertia, viscosity, and frictional resistance making a negligible contribution (V. Wright and Johns, 1960). It is likely that dynamic flexibility is increased by habitual activity and deteriorates with age.

Warm-up

An increase of muscle and/or body temperature can theoretically improve physical performance in several ways (D. Frank, 1976; Jensen, 1977). Muscle viscosity is reduced, increasing both the speed of a single twitch and the force developed at a given velocity of contraction (Fig. 3.9). The antagonists are more completely relaxed (thereby reducing the resistance to movement and the associated risk of injury). The mechanical efficiency is also improved, and this seems a specific effect of the increased intramuscular temperature, since it cannot be reproduced by passive stretching exercises (deVries, 1963). An increase of muscle blood flow improves the local oxygen supply during submaximal work, while a rightward shift of the oxyhemoglobin dissociation curve allows a greater and a more rapid release of oxygen at a given partial pressure. The power of local metabolic reactions is increased in accordance with the Law of Arrhenius. Finally, if the warm-up involves a mild or moderate form of the intended exercise, performance may be improved by recent practice of the required skill.

Needle thermocouple observations show that the intramuscular temperature rises 2–3°C over the first 5–10 min of moderate activity, thereafter remaining relatively constant. The extent of general body heating depends upon the environmental temperature and the state of training of the subject, but usually amounts to no more than 0.5–1.0°C, attained over the first 30 min of exercise. Most authors consider the beneficial effects of warm-up as related more to muscle than to deep body temperature (Asmussen and Bøje, 1945; Carlile, 1956); however, Muido (1946) claimed that benefit persisted for more than an hour, when muscle temperatures had returned to normal but core temperatures remained elevated.

Some comparisons have shown a slight increase of both physical working capacity (Simonson, Teslenko, and Gorkin, 1936; Asmussen and Bøje, 1945) and maximum oxygen intake (H. L. Taylor, Buskirk, and Henschel, 1955; Pirnay, Petit et al., 1966; B. J. Martin et al., 1975) after warm-up. In forms of exercise such as bicycle ergometry and one-legged ergometry (Stamford et al., 1978), where endurance is limited by local metabolic change within the active muscles, the warm-up could conceivably improve performance by speeding the circulatory on-transient, thereby reducing the early accumulation of anerobic metabolites. However, an excessive warm-up can cause the metabolic changes it is designed to prevent, leading to an accumulation of lactate and/or an exhaustion of food reserves (Richards, 1968; Bailin and Stewart, 1976). A nice judgment must thus be used in regard to both the intensity and the duration of warm-up (R. K. Burke, 1957). An athletic subject may find it useful to warm up for longer and at a higher absolute work intensity than a person who is unfit. Because of the interaction between muscle strength and lactate accumulation, the athlete may also find it advantageous to work at a larger percentage of his aerobic power than a sedentary person.

Performance times for speed events are generally improved by a warm-up (Kaufman and Ware, 1977), although it is difficult to separate a physiological effect from psychological consequences of the athlete's attitude toward a warm-up (Genov, 1970b; Karpovich, 1971). Accuracy, movement time, range of motion, and strength may all be improved, but there is little change in reaction time (D. Frank, 1976), rating of perceived exertion, or state of anxiety (Aronchick and Burke, 1977). Asmussen and Bøje (1945) found a close parallel between the rise of intramuscular temperature and the decrease in time needed for a bicycle ergometer sprint (Fig. 3.35). In their experiments, the maximum gain in performance was about 10%, irrespective of the mode of warm-up (submaximal exercise, diathermy, or warm showers). With physical exercise, most of the improvement was seen in the first 10 min. Högberg and Ljunggren (1947) found a better response to a 15-min than to a 5-min warm-up. There was no further significant improvement with 30 min of preliminary activity, but nevertheless they advocated a 15- to 30-min warm-up at a high rate of energy expenditure (3.0–3.4 l • min^{-1}, 134–152 mmol • min^{-1}); this regimen yielded a 3–6% improvement in performance over all distances from a 100-m dash to an 800-m race.

Strength is increased by general (Asmussen and Bøje, 1945; R. K. Burke, 1957), but not by local, warm-up (R. S. J. Clarke et al., 1958; Grose, 1958). Activities requiring dynamic or explosive strength (such as throwing and vertical jumping) show performance gains of 2.6–20% (Merlino, 1959; Pacheco, 1957,1959; Richards, 1968), ballistic stretching being more effective than passive stretching.

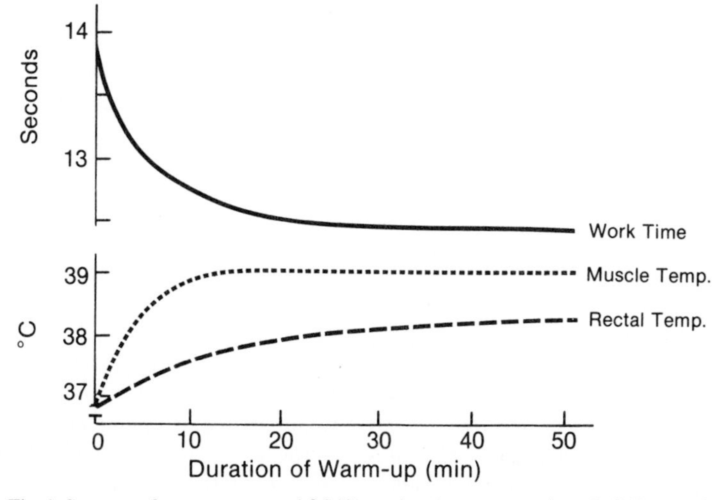

FIG. 3.35. The influence of a warm-up at 160 W on the time to complete 9.3 N • m of bicycle ergometer work, with corresponding data for muscle and rectal temperatures. (After Asmussen & Bøje, 1945.)

Perhaps because the temperature of most swimming baths is held below the thermally neutral figure, a warm-up also offers an advantage of 1–4% in swimming contests of 1–5 min duration (Muido, 1946; Carlile, 1956). However, a warm-up contributes little to the performance of more long-lasting or endurance-type activities (R. S. J. Clarke et al., 1958; Grose, 1958; Sedgewick et al., 1964).

Occasionally, a body part may be warmed by the local application of heat rather than by muscular activity. A large proportion of the available blood flow may then be diverted from muscle to the overlying skin (Nukada, 1955), with an earlier onset of fatigue and a lessened work output (Grose, 1958; R. S. J. Clarke et al., 1958).

Muscle tears are commonly attributed to a forceful contraction of the antagonists while the agonist muscles are still under tension. A warm-up should theoretically reduce the dangers of this type of injury by speeding the relaxation process. Sucec (1967) noted less episodes of muscle soreness when a warm-up was included in a longitudinal training program, but in general there is little experimental evidence regarding the effects of warm-up on the incidence of muscle injuries (D. Frank, 1976). A sudden burst of exercise can occasionally provoke ventricular fibrillation and sudden death (Shephard, 1979b). Barnard, Gardner et al. (1973) suggested that the incidence of electrocardiographic abnormalities can be lessened by a previous warm-up.

Cooling of the body following a warm-up may occupy an hour or more, and some authors find that the beneficial effects on performance also persist (Muido, 1946; Nukada, 1955). The implication is that an athlete could benefit from a warm-up, even if it proved necessary for this to be carried out some distance from the site of competition. A gradual warm-down is said to reduce the incidence of soreness and stiffness following exercise. Gentle activity during the recovery process also assists in the removal of lactate from the tissues and the return of fluid to the central circulation. A too-sudden cessation of activity can predispose to circulatory collapse, particularly in a warm environment.

Psychological Factors

Many observers would argue that the ultimate limit of athletic performance is set by psychological rather than by physiological factors. Considerations of psychology determine whether an athlete trains to his physiological potential and whether he realizes this potential during international competition (Sidney and Shephard, 1973). Space precludes more than a brief examination of three relevant topics: motivation, central inhibition, and cortical arousal.

Motivation

Motivation is perhaps the most important of all factors limiting an athlete. A ceiling of activity is reached not because a physiological system is fully taxed, but rather because a competitor perceives he is working to his limit (Gelfand, 1964; L. E. Smith, 1970; Borg, 1971; Borg and Noble, 1974). The complaint is of muscle pain, weakness, a "stitch," extreme breathlessness, or nausea rather than loss of consciousness.

The broad heading of motivation also covers acceptance of a long and arduous training program, a desire to excel in international competition, willingness to sacrifice safety to speed, persistence in the face of discouragement and physical discomfort, a preparedness to face and match aggressive opponents, and (in the team player) a subordination of personal glory to group objectives (Veit, 1970; Shephard, 1978c).

Central Inhibition

Physiological capabilities are normally restrained by inhibitory impulses arising from the cerebral cortex (Lehmann et al., 1939; Alles and Feigen, 1942; Wilmore, 1968). Sometimes (Roush, 1951; Steinhaus, 1963), but not invariably (W. R. Johnson, 1961), a greater part of the true potential can be revealed under hypnosis or in the stress of a sudden emergency. Athletes as a class are achievement-oriented (Cratty, 1968; Hegg, 1972; Rókusfalvy, 1972; Shephard, 1978c), and for this reason carry themselves much closer to their physiological potential than the average person. Nevertheless, most competitors find that the pressure of tough competition is still necessary to counteract inhibitions and achieve maximum performance (Strong, 1963). Psychological demands increase rapidly as the ultimate physiological limit is approached, and for this reason the psychological capabilities of a person may influence peak performance even if the physiological systems are not fully taxed at exhaustion.

Cortical Arousal

There is an inverted U-shaped relationship between the level of arousal or wakefulness of an individual and his performance (Malmo, 1957; Duffy, 1957; E. D. Ryan, 1962; Corcoran, 1965; Poulton, 1970; Genov, 1970a,b; Epuran et al., 1970; Cratty, 1971; Martens, 1974; Fig. 3.36). A moderate level of arousal ensures a brisk cardiac and respiratory anticipation of effort (Shephard and Callaway, 1966), a brief reaction time (Garcia-Austt et al., 1964; Cratty, 1968), and an adequate background of muscle tone for the development of maximum muscular force (Granit and Kaada, 1952; Gellhorn, 1960; Ikai and Steinhaus, 1961; Shephard, 1974a). However, premature arousal can lead to sleeplessness and general exhaustion of a contestant, while excessive arousal during a performance can lead to jerky, inefficient, and poorly coordinated movements. The appropriate level of arousal depends in part upon the personality of the athlete (extroverts tolerate more arousal than introverts; Bakan et al., 1960) and in part upon the level of skill required for the task (E. D. Ryan, 1962; Carron, 1968); thus, the desirable level of arousal is lower for an activity such as putting or pistol shooting than for the dull repetitive work of a long road race.

It is normally the responsibility of the coach to manipulate the arousal of his team members to an optimum level (Vañek, 1970; Cratty, 1971; M. J. Ellis, 1978). The physiologist can sometimes assist this process through autoconditioning procedures; an athlete is shown a display of his heart rate, blood pressure, or muscle action potentials, and learns to control these variables in the face of unpleasant, stressful stimuli (Brener and Kleinman, 1970).

Equipment and Technique

The efficiency of conversion of biochemical energy into external work depends not only upon the human machine, but also upon the design of ancillary equipment and the skill with which this is used. Dramatic gains of athletic performance can thus arise from changes in equipment and techniques (Shephard, 1978c). A prime example is the ski jump, where distances have increased 376% since 1900; in addition to changes in ski construction, competitors now choose more streamlined clothing and hold their bodies in a more aerodynamic posture during the jump, so that air resistance is drastically reduced (Jokl and Jokl, 1968a; Shephard, 1978c). The design of racing bicycles has also changed greatly over the present century; the use of lighter materials, narrower wheels, lower handlebars, and altered gear ratios all contribute to the observed 12% increase of speed in the paced racer. The height attained in pole vaulting increased steadily over a number of Olympic contests, but there was a sudden surge of performance in 1964, associated with the introduction of the fiberglass pole. Lastly, more than a second was pruned from the women's record for 100-m free-style swimming when this event was performed by an East German competitor without the encumbrance of a swimsuit.

SUMMARY OF LIMITING FACTORS

The principal factors limiting activities of various durations are summarized in Table 3.12.

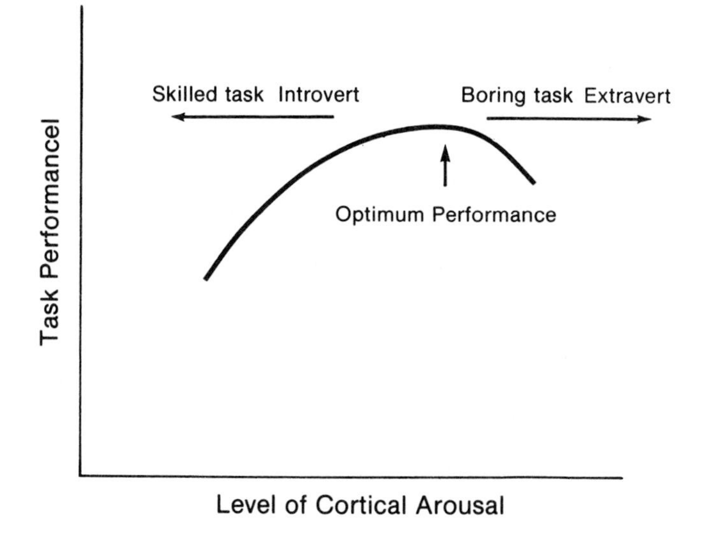

FIG. 3.36. The relationship between arousal and performance. The optimum level of arousal is moved leftward for a skilled task and an introverted subject (Martens, 1974). Most authors now reject the alternative hypothesis ("drive theory," Spence, 1971) that performance of a well-learned task improves progressively with arousal.

TABLE 3.12
Factors Limiting Performance in Relation to Duration of Event

0–10 sec	10–60 sec	1–60 min	60–120 min	2–5 hr	>5 hr
← Motivation, release of inhibition, arousal →					
Anerobic power	Anerobic capacity	$\dot{V}_{O_2(max)}$	Fluid and mineral loss	Glycogen stores	Fat mobilization
Reaction time	Strength	Strength	Heat elimination	Fluid and mineral loss	Fat stores
Strength	Skill		$\dot{V}_{O_2(max)}$	Heat elimination	Food intake
Skill				Fat mobilization	Protein reserves
Flexibility				$\dot{V}_{O_2(max)}$	Bone and joint strength

CHAPTER 4
Energy Transducers

We have already looked at the extent of available energy stores and the rate at which these can be utilized. It is now necessary to return to the energy transducers (Chap. 1), considering in more detail the mechanisms whereby energy is applied to the performance of external work, at both the cellular and the organ level.

STRUCTURE OF THE MUSCLE CELL

Because a major part of ingested food is metabolized within the muscles, it is convenient to examine biological energy transducers within skeletal muscle, recognizing that cells in other parts of the body function in a rather similar fashion. A final portion of this section reviews specific differences between skeletal and cardiac muscle.

Overall Arrangement

Skeletal muscle cells take the form of elongated fibers, 1–40 mm in length. Individual fibers are embedded in a fine connective tissue (the *endomysium*), hundreds or even thousands of such elements being grouped into a muscle bundle or fasciculus. The bundles are bounded in turn by a connective sheath termed the *perimysium*, while an outer fascia or *epimysium* encloses the entire muscle. The number and thickness of the connective tissue septae vary from one individual to another, and there is some evidence that muscular training leads to an increase of septal tissue. There is also a variable amount of fat within the muscle belly. Account must be taken of this "dilution" by fat and connective tissue if muscle bulk is to be estimated from measurements of external circumferences or from soft-tissue radiographs.

The perimysium and epimysium blend with the tendon. In muscles such as the hip and shoulder girdle, where power is more important than speed or range of movement, fibers are inserted at an angle to the tendon (the so-called pennate arrangement). In other muscles, such as those producing rapid arm movements, power requirements are subordinate to the demands of speed. The fibers are than arranged longitudinally (the so-called fusiform arrangement). Specific parts of the skeletal muscle fiber are illustrated in Fig. 4.1.

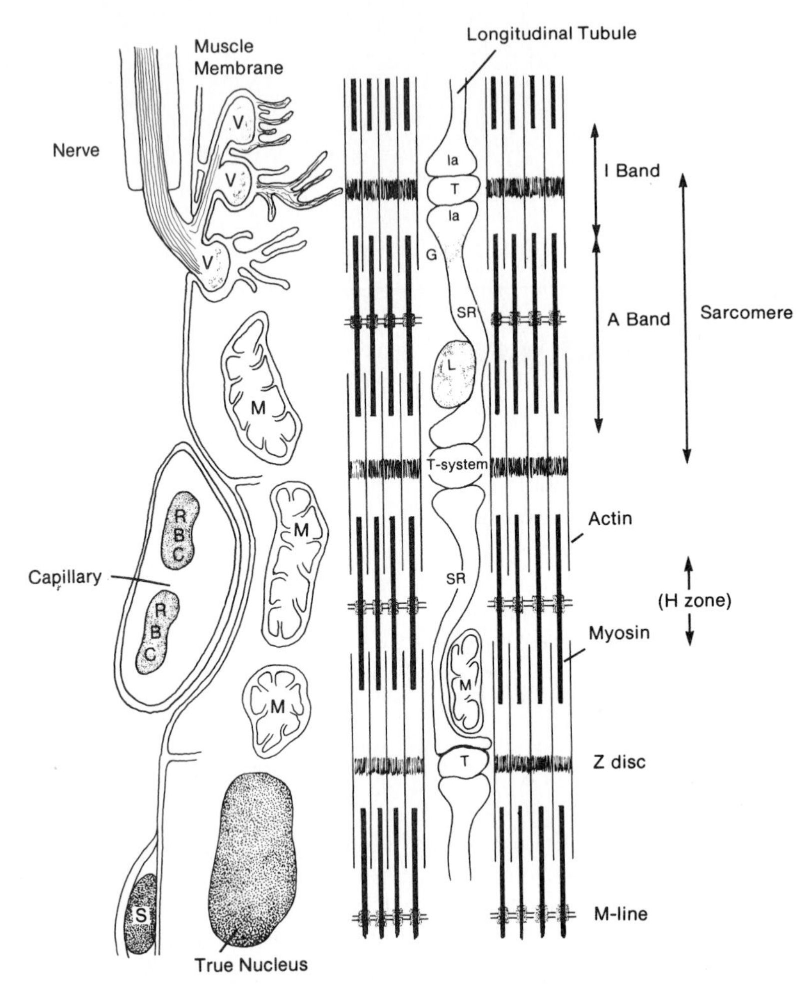

FIG. 4.1. Schematic drawing of a frog muscle fiber. RBC, red cell; M, mitochondrion; L, lipid; G, glycogen; SR, sarcoplasmic reticulum; S, satellite cell nucleus; T-system, triad of central and two lateral sacs (la). Note that in humans the triad may be displaced towards A/I junctions (Fahrenbach, 1965). V, terminal knobs of neural axon, lying in synaptic groove, filled with synaptic vesicles and mitochondria. Note infolding of sarcolemma beneath knobs.

Cell Membrane

The cell membrane plays an active role in regulating the intracellular environment, transporting some materials into the cell and excluding others. It may also be concerned in the synthesis of ATP and the transduction of energy.

A typical membrane has two principal layers. The *outer sheath* comprises a layer of fine reticular fibers, ~100 Å (1 Å = 10^{-8} cm) thick, bounding a basement membrane of striated collagen that is ~300 Å in thickness; this loosely structured mixture of protein and polysaccharide merges rather freely with the surrounding connective tissue, septae, and tendons (Mauro and Adams, 1961). An amorphous layer separates the outer sheath from the inner membrane, or sarcolemma, which is probably more vital to the life of the cell. The sarcolemma is 90–100 Å thick, consisting of cholesterol and a double layer of phospholipids such as lecithin (Davson and Danielli, 1952; J. D. Robertson, 1964; Dervichian, 1964), sandwiched between an outer layer of glycoproteins or mucopolysaccharides and an inner layer of extended polypeptides (Fig. 4.2).

Calcium ions apparently play an important role in the structure of the sarcolemma, bonding both lipid and protein components, and packing the lipids. Individual phospholipid molecules are arranged in polar fashion, with hydrophilic (water-soluble) heads on the inner and outer surfaces, and hydrophobic

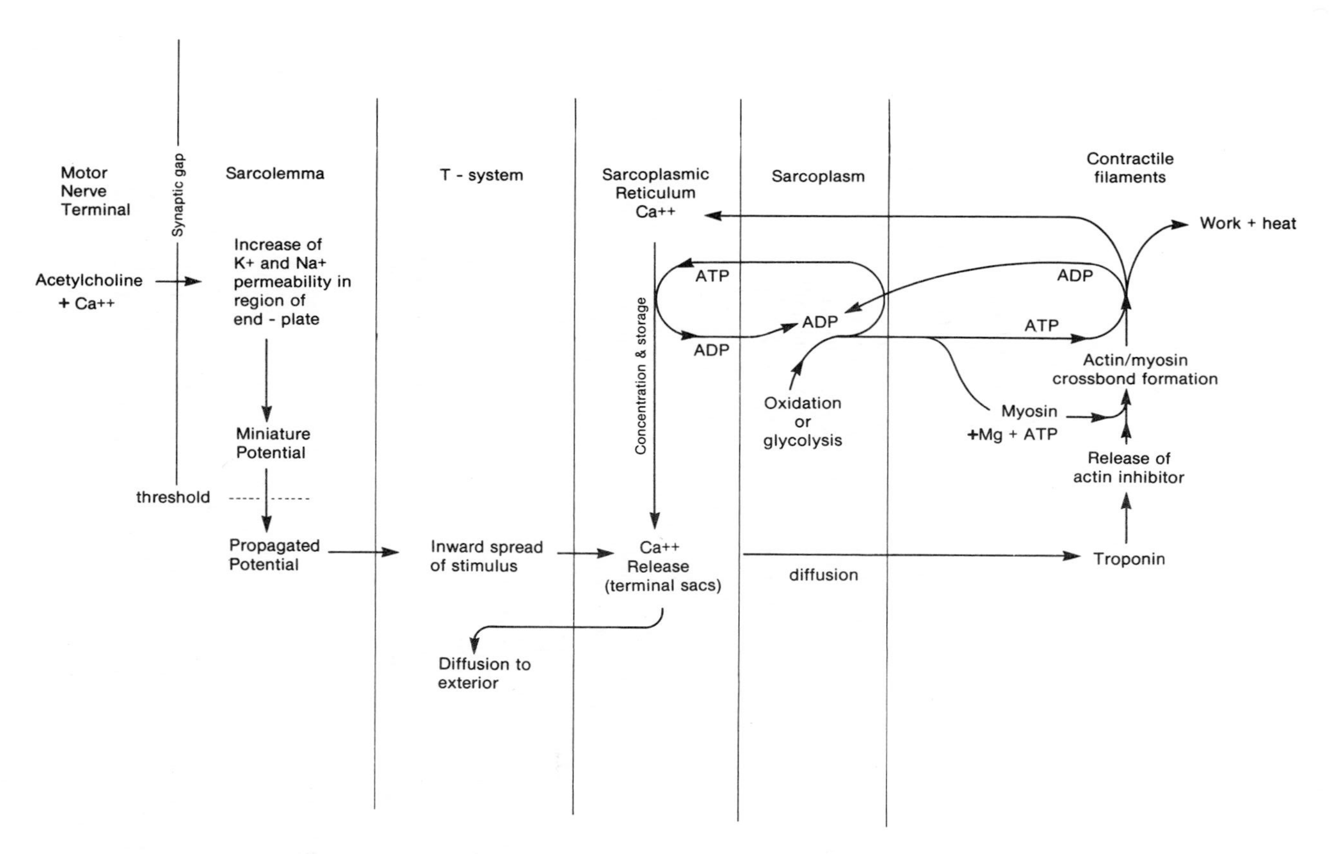

FIG. 4.2. The sequence of events associated with a muscle twitch. (Adapted from F. D. Carlson & Wilkie, 1974.)

tails (with a poor water affinity) meeting at the center of the membrane (J. D. Robertson, 1964).

Because of its characteristic structure, the sarcolemma has the electrical properties of a capacitance and a resistance, arranged in parallel. Water is an integral part of the membrane, probably forming a continuous phase throughout all of its layers. Plugs of nonlipid material, possibly water covered with a protein monolayer (Danielli, 1958), penetrate the membrane at intervals. Such "oval bodies" have a diameter of ~600 Å. They help to stabilize the membrane and allow the diffusion of mineral ions and electrically polar substances across it.

T System

The T system of transverse tubules is a regular series of invaginations of the sarcolemma (K. R. Porter and Palade, 1957; Franzini-Armstrong and Porter, 1964). In many species (although apparently not in humans), they are set at the level of the Z discs that separate the individual units or sarcomeres of the muscle fiber (Fig. 4.1). Tubules are rectangular or elliptical in shape, and run at right angles to the longitudinal axis of the muscle fiber; it is not yet certain whether they end blindly or continue to the opposite side of the fiber. In some muscles, they apparently form the central element of a *triad*, comprising a *central sac* and two *lateral sacs* connected with the sarcoplasmic reticulum, as described below.

Although small in volume (~0.3% of the total fiber volume), the tubules offer a large total surface, sometimes greater than that of the exterior of the fiber (Peachey, 1965). Their function is to conduct depolarizing waves of electrical activity from the surface of the sarcolemma into the interior of the fiber.

Sarcoplasmic Reticulum

The sarcoplasmic reticulum consists of a network of longitudinally oriented membranous sacs. These generally extend between successive systems of transverse tubules and drain to the lateral sacs. However, in some instances the reticulum of adjacent sarcomeres may be linked, and the lining of the sacs may also become continuous with the sarcolemma or the nuclear membrane (Peachey, 1965). The surface area of the sacs is approximately ($27 \times 10^3 \times a$) cm^2 per cm^2 of fiber surface, where a is the fiber radius in cm (Peachey and Adrian, 1973). For a typical fiber (a = 50 μm), the reticular surface is thus 135 times the external surface of the fiber, or 54×10^3 cm^2 per cm^3 of muscle. The volume of fluid within the reticulum accounts for ~ 12.7% of the wet weight of the muscle (Zierler, 1973).

Parts of the reticulum have a "rough" appearance because of the attachment of ribosome granules. The ribosomes contain some 85% of the cellular ribonucleic acid (RNA) and provide the framework upon which proteins are synthesized. Other segments of the reticulum bind glycogen and the related enzymes glycogen synthetase and phosphorylase.

The electrical signal that passes through the T system is communicated to the sarcoplasmic reticulum via the lateral sacs. Here, studies with a calcium-sensitive dye (Jobsis and O'Connor, 1966) and a protein that fluoresces in the presence of free calcium ions (C. C. Ashley and Ridgeway, 1968) have shown that the electrical change induces a release of calcium, amounting to approximately 10^{-7} mol of Ca^{2+} per cm^3 of muscle (Peachey and Adrian, 1973). Muscle contraction in turn is initiated by diffusion of the calcium ions to receptor sites on the contractile proteins. Relaxation of the muscle depends on (1) release of calcium from the contractile proteins and (2) reabsorption of calcium ions by the middle segment of the reticulum. Both of these processes consume ATP energy.

At slow rates of stimulation, the reticulum uses approximately 1 mol of ATP for every 2 mol of Ca^{2+} that are taken up (Needham, 1973), a rate of 0.05 μmol per stimulus per g of muscle. However, with more rapid stimulation, the ATP usage per stimulus decreases, finally reaching a steady level of ~0.25 $\mu mol \cdot g^{-1} \cdot sec^{-1}$ at 50 impulses per sec (R. E. Davies, 1973). Depression of Ca^{2+} transport after exhausting exercise may impair the contractile characteristics of both skeletal and cardiac muscle (Hashimoto, Sembrowich, and Gollnick, 1978).

Mitochondria

The mitochondria (Lehninger, 1964) are widely scattered throughout the cytoplasm (Fig. 4.1); in white muscle, they are concentrated in the I bands of individual fibers while in red and cardiac muscle they are found mainly in the A bands (E. Holmgren, 1910). Howald (1975) estimated that the central (those closely related to the contractile filaments) mitochondria have a density of 4.1–8.6% per unit volume of myofibril, interindividual differences mirroring differences in the maximum oxygen intake of the subjects. The volume densities of the peripheral mitochondria (those lying immediately beneath the sarcolemma) show an even larger range of interindividual differences, average values being 1.4 times larger in men than in women and 3.2 times larger in trained than in untrained men. When calculated per unit volume of myofibril, ratios range from ≃ 5% in untrained women to ≃ 11% in trained men, figures being larger for slow- than for fast-twitch fibers.

Each mitochondrion is a cylindrical-shaped body with a double membrane similar to that described for the sarcolemma (De Haan et al., 1973). The outer membrane is permeable to small molecules and ions. Many substances are actively transported across the inner membrane (see, for example, the electron shuttle of Fig. 2.8); however, water and some uncharged fat-soluble molecules can diffuse freely across this barrier. As with the sarcoplasmic reticulum, the surface created by the convoluted membranes of the mitochondrion is vast. Outer membranes have an area of 0.66–1.14 m^2 per ml of tissue, while inner membranes present a surface ranging from 1.35 $m^2 \cdot ml^{-1}$ in women to 2.77 $m^2 \cdot ml^{-1}$ in well-trained male long-distance runners.

Closer examination of the mitochondrion reveals a number of *cristae*. These are generally assumed to be inward folds of the inner membrane (Robertson, 1964), although Fernández-Morán (1962) and D. E. Green and Goldberger (1967) suggested that they comprise regularly repeating elementary tripartite particles (Fig. 4.3). While mito-

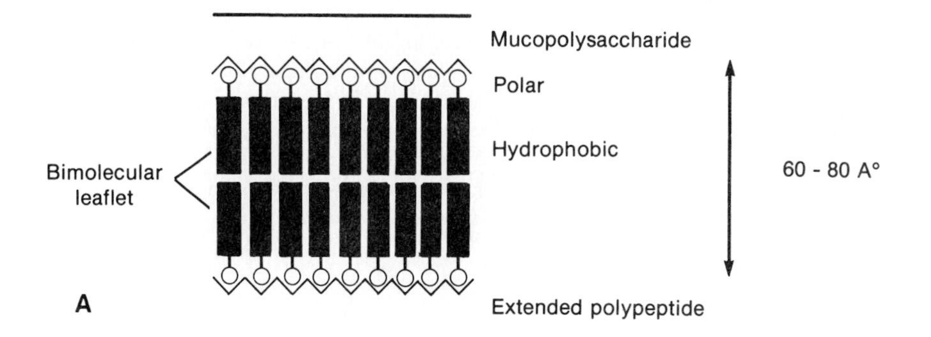

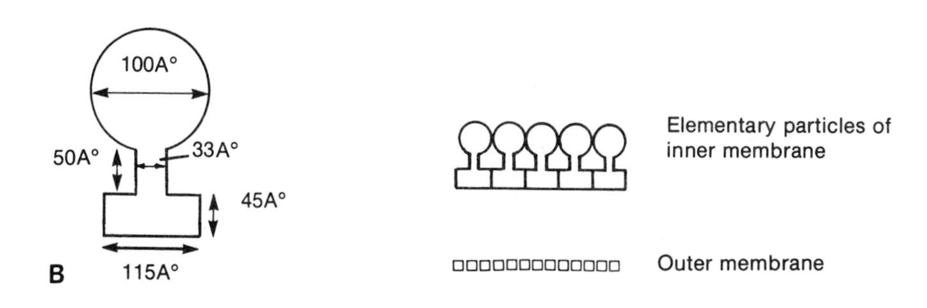

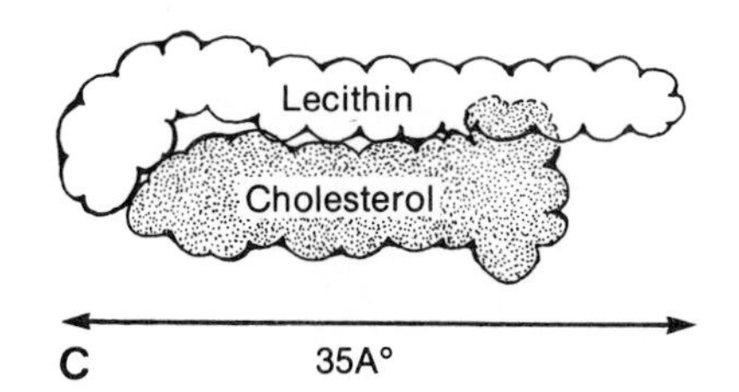

FIG. 4.3. A. Detail of cell membrane (after J. D. Robertson, 1964). **B.** The elementary particle of D. E. Green & Goldberger (1967). According to their hypothesis, the cell membrane is formed by a sequence of such particles. **C.** Interrelation of cholesterol and lecithin in cell membrane (after Zachar, 1971).

chondria are larger in trained than in untrained subjects, the increase of size does not seem to affect such details of structural organization as the cristal surface density.

We have already seen that the mitochondrion is the major site of ATP synthesis, so that the association between the volume density of the mitochondria and maximum oxygen intake is not surprising. Enzymes are attached to mitochondrial membranes in a regular sequence. The various components of the citric acid cycle, fatty acid oxidation, oxygen transport, and protein synthesis are all found at specific sites. The oxidation/reduction proteins (Fig. 2.13) are grouped in four distinct complexes (Fig. 4.4), the flow of electrons between these four segments being ensured by two mobile components (the coenzyme Q, or ubiquinone, and cytochrome c). Likewise, the citric acid cycle is split into five complexes, only four of which lie on the outer membrane of the mitochondrion (D. E. Green and Goldberger, 1967). A typical enzyme complex has a molecular weight of ~300,000; ~36% is lipid and 64% protein, a half of the latter being actual enzyme. A single mitochondrion contains up to 10^9 enzyme molecules, each of which can carry out several million catalytic cycles per min (Lehninger, 1973).

Contractile Mechanism

The contractile portion of the cytoplasm comprises a series of myofibrils. Each is no more than 1 μm in diameter. Many hundreds are bunched together to form a single muscle fiber.

When viewed under the light microscope, a typical myofibril consists of alternate light and dark structures (Fig. 4.1). These have been named on the basis of their optical properties the I (isotropic) and A (antisotropic) bands (Høncke, 1947). The *I band* is plainly birefringent when viewed in polarized light, but the *A band* is only slightly birefringent. At the midpoint of each I band is a darker *Z disc* (the *Zwischenscheibe*, or inserted disc), which marks off individual *sarcomeres*. In the middle of the A band is a lighter *H zone* (*hellige* = light), with a narrow *M line* (*Mittelscheibe*) at its center.

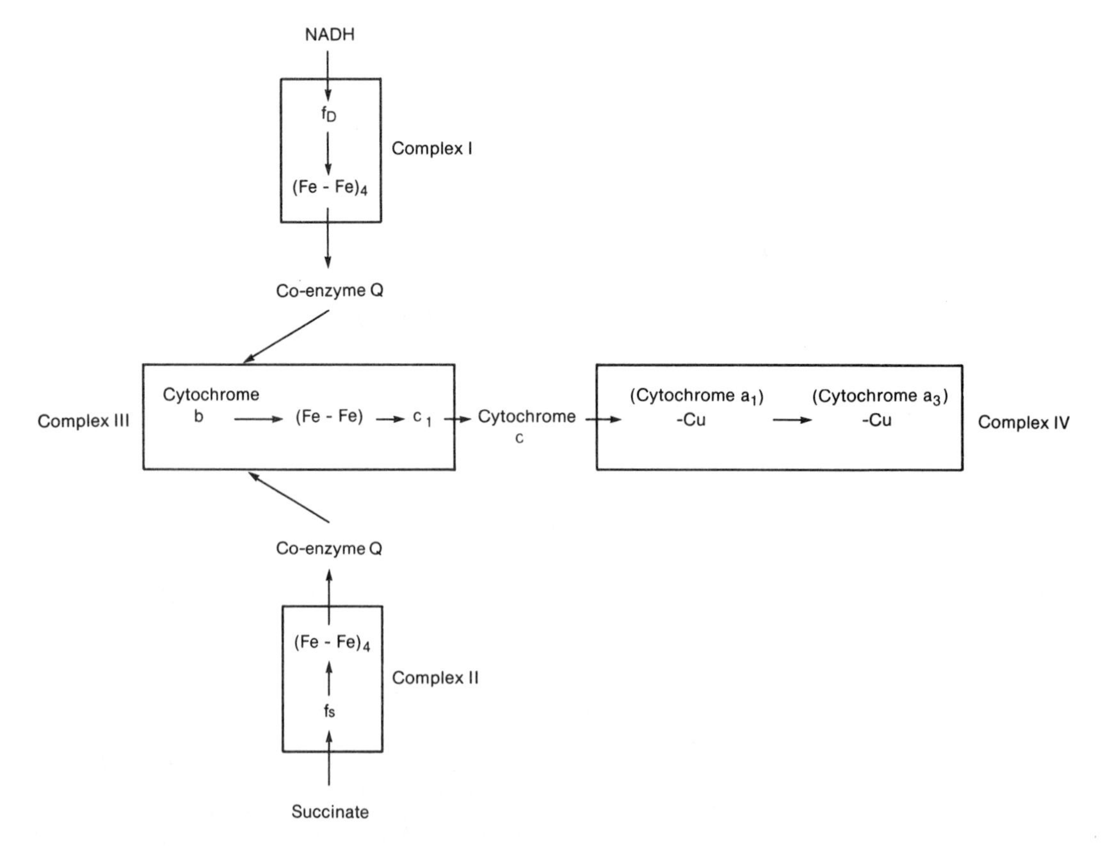

FIG. 4.4. Distribution of the oxidation–reduction proteins in the mitochondrial membrane.

The optical implication of anisotropy is that rodlike particles running parallel to the longitudinal axis are embedded in a medium of differing refractive index. Electron microscopy confirms this supposition, demonstrating an interdigitation of thick and thin filaments built from long-chained proteins, predominantly *myosin* and *actin*, respectively. When the muscle is activated, the actin filaments slide over the myosin and toward each other (A. F. Huxley and Niedergerke, 1954) and there is also a structural bonding of the two types of filament (the formation of *actomyosin*; Szent Gyorgyi, 1953).

During growth, muscle length is increased by the formation of additional sarcomeres; however, the length of individual sarcomeres remains constant. An increase in the cross-section of a fiber is brought about by the longitudinal splitting of myofibrils when they reach a certain critical size (G. Goldspink, 1970). Splitting seems a more common occurrence in animals than in humans. It can sometimes give the impression that the number of fibers has increased (hyperplasia); however, in reality, the fiber population seems set at an early point in development, and thereafter muscle cross-section can increase only by *hypertrophy*.

Structural proteins make up ~90% of the dry weight of the myofibril. Chemical analysis has identified eight proteins (Table 4.1). Their location within the myofibril has been deduced by comparing appearances before and after selective extraction, and by developing specific reactions to fluorescent antibodies (Pepe, 1966).

TABLE 4.1
Constituent Proteins of Skeletal Myofibril[a] and Their Relative Stability as Determined by Rate of Incorporation of Radioactive Amino Acids

Protein	Percent of Total Structural Protein	Relative Rate of Incorporation of Amino Acids
Myosin	54–60	73
Actin	20–25	30
Tropomyosin	4.5–11	100[b]
Troponin	2–5	202
α-Actinin	2–10	99
β-Actinin	0.5–2	—
M protein	0.5–2	—
C protein	2	—

[a]About 60% of the proteins of skeletal muscle are contained in myofibrillar structures.
[b]The average rate for soluble proteins is 124.
Source: After J. Hanson and Huxley, 1953; Perry and Corsi, 1958; Ebashi and Nonomura, 1973; F. D. Carlson and Wilkie, 1974; and Edington and Edgerton, 1976.

Myosin molecules comprise two heavy-coiled chains (α helices), intertwined but separated at one end, and four light chains (Fig. 4.5). Two of the latter are associated with the ATPase activity of the molecule (Engelhardt and Ljubimova, 1939; Szent Gyorgi, 1953) and differ in their structure according to fiber type (Tonomura, 1973). The straight parts of individual myosin molecules fit together and slightly overlap each other; the direction of placement undergoes reversal in the H zone (where the molecules are arranged tail to tail). The projecting "heads" of the molecules (Fig. 4.5) are lightly hinged to the tails (H. E. Huxley, 1969); one "hinge" (between the elements HMM-1 and HMM-2) allows outward rotation and cross-bridge formation with actin during contraction (Lowey et al., 1969; Pepe, 1971), while the second hinge (between HMM-2 and the light meromyosin component) generates the force of contraction. Because the orientation of the cross-bridges is reversed in the H zone, the two halves of a myosin filament work cooperatively when developing a muscular contraction.

Actin in vitro exists in a globular form (G-actin), with one molecule of ATP bound to each molecule of G-actin. However, in the presence of magnesium ions, it polymerizes to form a fibrous chain (F-actin); ADP is incorporated into the molecule in this process and phosphate is liberated (Martonosi et al., 1960). In muscle, actin is present only as the polymer. This forms a double-stranded right-handed helix (Depue and Rice, 1965) with a characteristic pitch, one complete turn occupying 710–740 Å (Fig. 4.6).

Tropomyosin (Perry and Corsi, 1958) is distributed along the length of actin filaments (H. E. Huxley, 1960; Pepe, 1966), lying in the groove between the two coils of the F-actin (Fig. 4.6). Tropomyosin molecules are arranged as an α helix and are ~400 Å in length (Ebashi and Nonomura, 1973). The main function of tropomyosin seems to be to inhibit the interaction between actin and myosin. It apparently achieves this by undergoing a slight lateral displacement within its groove; inhibition is released when it moves back to the center of the groove (H. E. Huxley, 1972). According to some authors, tropomyosin is also a constituent of the Z discs.

Troponin (Ebashi and Kodama, 1965) is a small globular protein that recurs every 400 Å along the actin filaments (Fig. 4.6). It has a high affinity for tropomyosin, and is normally attached about one-third of the distance from the end of the latter molecule. Troponin is the calcium receptor of the myofi-

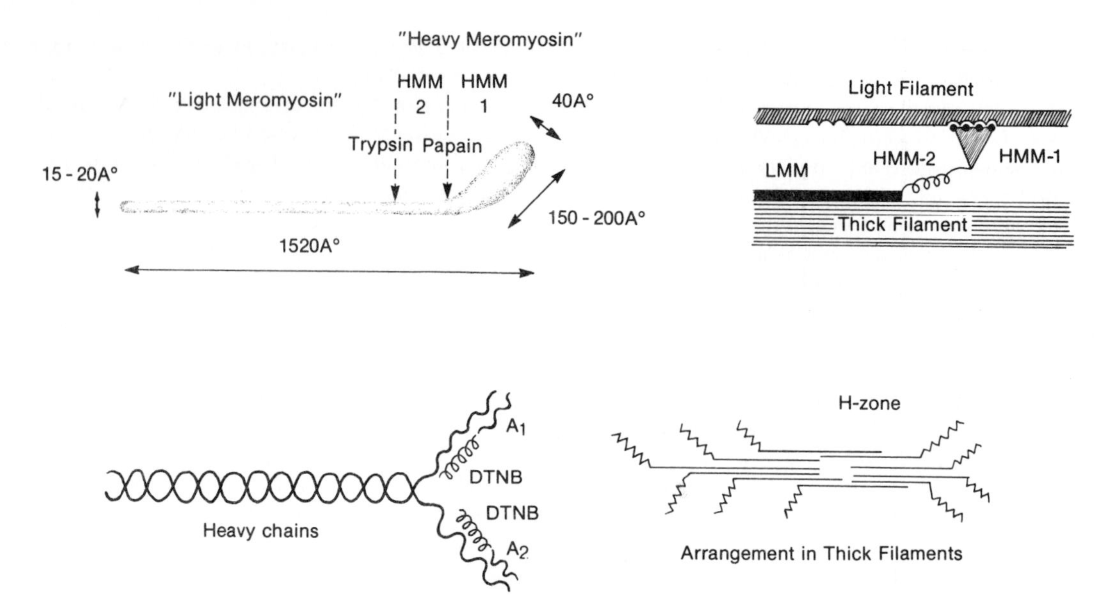

FIG. 4.5. Myosin molecules. The "tail" of the molecule comprises two heavy chains. These separate at the "head" to enclose two light chains (A_1 and A_2) with ATPase activity and two other light chains (DTNB) that can be separated by dinitrobenzoic acid. The molecule can be split by trypsin and papain at the sites indicated, forming light meromyosin and two fragments of heavy meromyosin. HMM-1 can rotate to interact with the adjacent thin filament, while HMM-2 is probably concerned with the generation of the muscular force. The myosin molecules overlap slightly in thick filaments. (After Lymm and Huxley, 1972, and F. D. Carlson & Wilkie, 1974.)

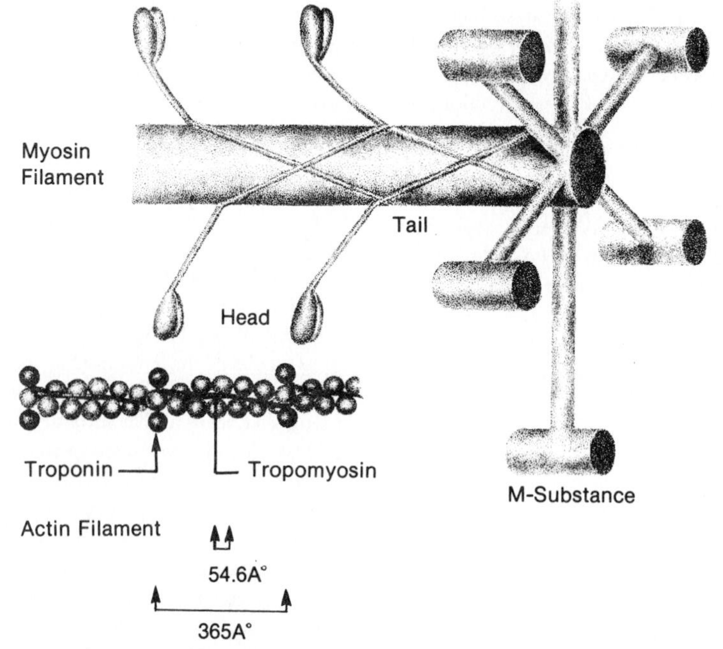

FIG. 4.6. Interrelation of protein molecules. The myosin molecules are wrapped around each other and spaced as a regular hexagon by the M-substance. The heads of each molecule are closely related to a troponin molecule, carried on the actin filament. There is one molecule of troponin and one longitudinally oriented tropomyosin molecule for each seven units of G-actin making up the actin filament. (After Ebashi et al., 1969.) The dimensions of the actin filament are based on X-ray diffraction studies (J. Hanson & Lowy, 1963; H. E. Huxley & Brown, 1967).

bril. Each molecule of troponin can bind two calcium ions, thereby undergoing a structural change that displaces the tropomyosin molecule, thereby reversing inhibition of the actin/myosin interaction (Ebashi, 1968; Svent-Gyorgi, 1975). One molecule of troponin and one molecule of tropomyosin are associated with every seven units of G-actin.

Recently, troponin has been split into several subfragments (Gergely, 1973). Troponin A (TN-C, TpC) contains the calcium-binding structures, troponin B (TN-I, TpT) inhibits the enzyme actomyosin ATPase, and the third fragment, troponin T (TN-T, TpT), possesses a high affinity for tropomyosin.

α-Actinin (Ebashi, Ebashi, and Maruyama, 1964) is found particularly in the region of the Z discs. Here, it is thought to cement the actin filaments to the Z disc and to each other. It is also found elsewhere in the myofibril and may contribute in some way to the actin/myosin interaction (Ebashi, 1968).

Electron micrographs of the Z disc region have been interpreted as showing a "square net" (Fig. 4.7). Each actin filament apparently gives rise to four I filaments that connect it with other actin filaments in adjacent sarcomeres (Knappeis and Carlsen, 1962). However, the Z discs do not seem firmly connected to the sarcolemma or the tendinous sheath, and the manner in which the muscular force is transmitted from the actin filaments to the tendon remains a mystery.

β-Actinin (Maruyama, 1965) is present in quite small quantities. It can modify the length of the actin filaments by preventing further polymerization of the actin molecule, but its precise function is unknown.

M protein has been isolated from the region of the M line (Morimoto and Harrington, 1972). The M substance forms a hexagonal, bridgelike structure (Fig. 4.5) that helps to maintain a regular orientation of the myosin filaments (Franzini-Armstrong and Porter, 1964). It also contains the enzyme creatine phosphokinase (D. C. Turner et al., 1973).

C protein is a part of the thick filament, being organized along its stripes (Offer, 1972; Pepe, 1972). Its functions are not clearly understood at present.

Yet more detailed analysis of ultrastructure is possible using the technique of small-angle X-ray diffraction. According to C. Cohen (1966), the central part of the thick filaments contains a number of proteins other than myosin. The myosin molecules are coiled as four cables around a central core. Each cable is formed from four intertwined myosin molecules. In consequence, the cross-bridges are also arranged in helical fashion along the length of the thick

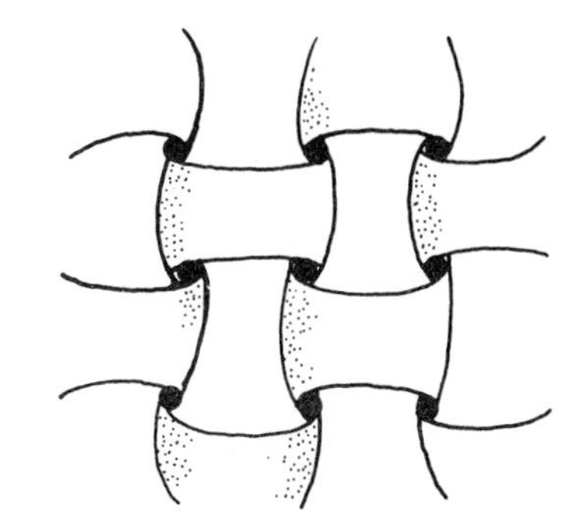

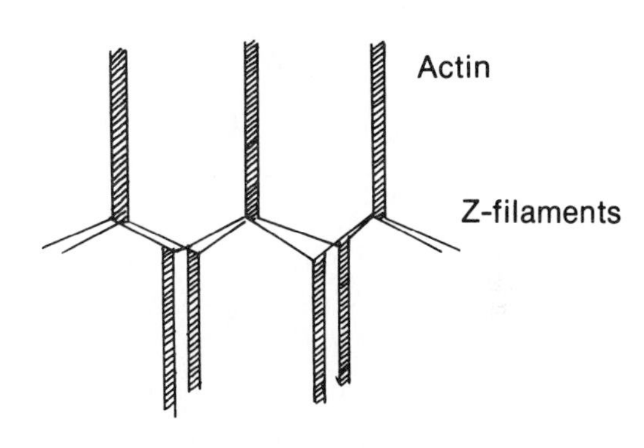

FIG. 4.7. Possible interpretation of Z-disc structure from electron micrographs. (After Reedy, 1964.)

filaments (Fig. 4.8). Successive pairs of bridges are rotated by 60° with respect to their predecessors and are set at a distance of 143 Å along the longitudinal axis. According to H. E. Huxley (1960), there are ~108 bridges to each half of a thick filament.

When viewed in transverse section, the filaments are arranged as a hexagonal lattice. The separation between individual filaments is 350–450 Å, depending on sarcomere length (G. F. Elliot et al., 1963; Fig. 4.9). The cross-bridges extend out to a radius of ~130 Å, almost making contact with the corresponding actin filaments at the trigonal points of the lattice (J. Hanson and Huxley, 1953; H. E. Huxley, 1960). Individual thick filaments are rotated by 120 or 240° with respect to their nearest neighbors, thus creating a superlattice with dimensions $\sqrt{3}$ times that of the basic hexagonal lattice (Fig. 4.9).

The actin filaments yield an X-ray diffraction pattern suggestive of a double helix with a pitch of 2 × 365 Å (J. Hanson and Lowy, 1963). Each subunit occupies 54.6 Å (as would be predicted for G-actin) and there are 13 or 14 subunits per turn.

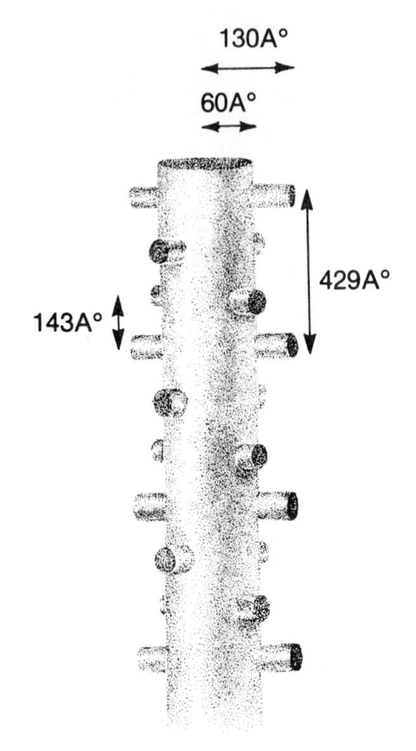

FIG. 4.8. 6/2 helical arrangement of myosin cross-bridges. There are a total of some 108 cross-bridges on each half of a thick filament. (Based on X-ray diffraction studies of the relaxed frog sartorius muscle, after H. E. Huxley & Brown, 1967.)

The X-ray diffraction patterns show other reflections, including one associated with the I filaments that repeats every 385 Å along the longitudinal axis. However, the structures responsible for these appearances have yet to be worked out.

Nucleus

Each muscle fiber contains many cylindrical-shaped nuclei. Each is bounded by a double nuclear membrane. Constituents include the nucleolus, the nuclear fluid, and strands of chromatin material—the deoxyribonucleic acids (DNA) which control inherited characteristics.

One of the main functions of the nucleus is to direct the synthesis of structural, enzymatic, and contractile proteins. Although such anabolic activity is maximal during periods of growth and hypertrophy, sedentary adults also show a continuous breakdown and resynthesis of cell protein. DNA is acquired through the mitotic division of the cells. A regular sequence of nucleotides encodes all the necessary information for the synthesis of body proteins. However, the DNA is normally surrounded by histone proteins that inhibit most or all of this function, and a specific segment of the DNA molecule must be exposed for synthesis to commence. Growth, exercise, and hormone secretions can all initiate the synthetic process, although the way in which they unmask the DNA is not yet clear.

Depending on which portion of the DNA molecule is exposed, four types of ribonucleic acid can be transcribed:

Nuclear RNA is normally concentrated in the nucleolus.
Messenger RNA contains a coded sequence of nucleotides to synthesize a specific protein.
Ribosomal RNA may adhere to the "rough" parts of the sarcoplasmic reticulum. It binds messenger RNA and builds onto this template sequences of amino acids brought to it by specific *transfer RNAs*. There is a separate transfer RNA for each of the 26 amino acids in the body.

Nuclei are sometimes found lying between the sarcolemma and the outer membrane of the muscle

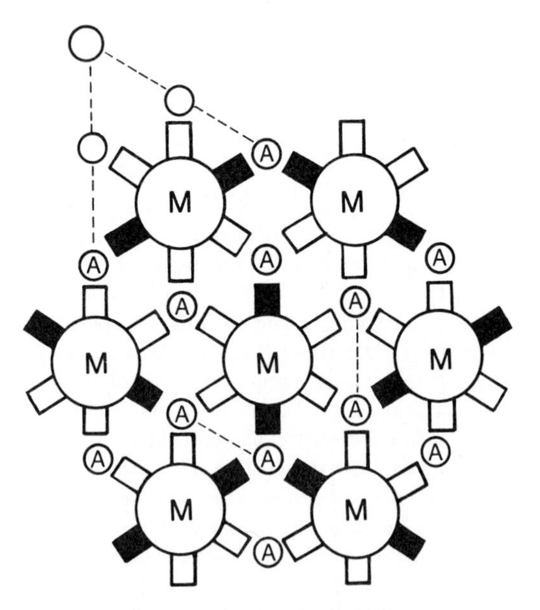

FIG. 4.9. Spatial orientation of myosin filaments, cross-bridges, and actin filaments. Note that each myosin filament is rotated by 120° or 240° with respect to its neighbors. This rotation imposes a superlattice (interrupted lines) upon the basic hexagonal arrangement. (After H. E. Huxley & Brown, 1967.)

fiber. Careful examination shows a small volume of associated cytoplasm. The function of such "satellite cells" is uncertain. There have been suggestions that they are dormant "myoblasts," which can be incorporated into the muscle fiber with suitable stimulation. If so, they could play a role in muscle regeneration, and perhaps also in the training response.

Cytoplasm

Strictly speaking, the cytoplasm includes all elements of the cell bounded by the sarcolemma, with the exception of the nucleus. However, we have already discussed the sarcoplasmic reticulum, the mitochondria, and the contractile filaments. In addition, the cytoplasm contains the pigment *myoglobin* and certain soluble enzymes.

The role of *myoglobin* in oxygen transport and storage has been noted. The muscle pigment has a structure similar to that of hemoglobin, but contains only one atom of iron and one heme molecule. In consequence (Millikan, 1936; Antonini, 1964), the two compounds have very different oxygen-binding properties (Fig. 4.10). The hemoglobin curve is sigmoid in shape, with half-saturation at an oxygen pressure of 3.2 kPa (24 mmHg), whereas the myoglobin curve is a hyperbola, with half-saturation at an oxygen pressure of 0.4–0.8 kPa (3–6 mmHg; Theorell, 1934).

Certain enzymic reactions occur within the cytoplasm (Fig. 2.13). Traditionally, at least one form of creatine phosphokinase and the enzymes concerned with glycolysis were thought to be "soluble" in the cytoplasm, although there is now some evidence that they are bound to the sarcoplasmic reticulum (D. E. Green and Goldberger, 1967) and/or the contractile elements (Edington and Edgerton, 1976).

Other Elements

Other important formed elements within the muscle are the terminals of the motor sensory nerves.

Structure of Cardiac Muscle

Cardiac muscle has certain structural differences from skeletal muscle (Leyton, 1974). Instead of a system of longitudinally oriented fibers, there is a branching network (syncytium) of interconnected muscle "cells." These have a diameter of 10–20 μm (narrow relative to a skeletal muscle fiber of 50–100 μm) and a length of 50–100 μm. The orientation of the cells is nearly vertical at both the outer (epicardial) and inner (endocardial) surfaces, but there is a smooth transition to a circumferential arrangement in the middle third of the wall (Jean et al., 1972).

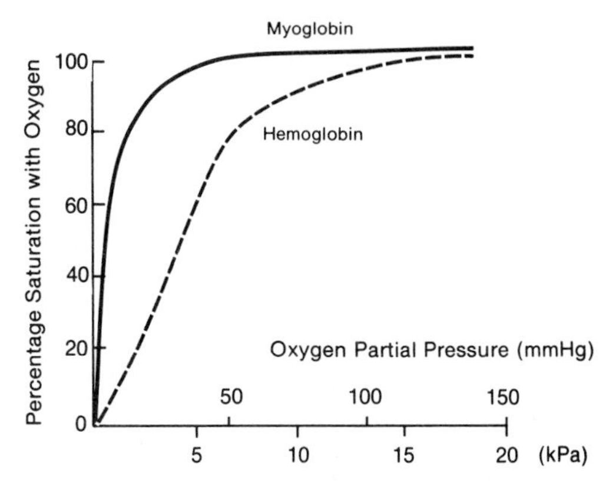

FIG. 4.10. A comparison of the oxygen-binding properties of myoglobin and hemoglobin.

Individual muscle cells are bounded by a sarcolemma membrane. This contains perhaps 30 myofibrils (each built from a sequence of up to 10 sarcomeres), a tubular system and sarcoplasmic reticulum similar to that found in skeletal muscle, mitochondria, a large central nucleus, and a number of capillaries. The extremities of each cell are attached to *intercalated discs*. These comprise the cell membranes of two adjacent cells and a small interposed cellular space. A large number of interdigitating projections gives great functional contact between adjoining cells. The electrical impedance is particularly low across what is termed the *fascia occludens* portion of the disc; here, the interposed space measures no more than 18 Å. This arrangement is important in allowing the electrical impulse to spread from one segment of the heart to the next, thereby forming a functional as well as a structural syncytium (Sjöstrand et al., 1958).

Mitochondria are larger and more frequent than in skeletal muscle, taking up as much as 30–50% of cell volume. They are arranged systematically between and in close proximity to the parallel rows of sarcomeres (Spiro and Sonnenblick, 1965).

ENERGY-TRANSDUCING MECHANISMS

Sliding Filament Model of Skeletal Muscle

Much early research on the interdigitation of thick and thin muscle filaments was carried out by standard light microscopy. Refraction by the cylindrical fibers made the interpretation of appearances difficult, and contradictory hypotheses were developed. However, more clear-cut information became available with the use of a special type of interference microscope.

The length of the thick filaments was estimated from the width of the A band (Fig. 4.1), while the length of the thin filaments was deduced from the distance separating the Z line from the edge of the H zone. Passive stretching increased the length of the I band but did not change the width of the A band (A. F. Huxley and Niedergerke, 1954). It was thus deduced that the length of the thick filaments remained unchanged. Similarly, during isotonic contraction there was a narrowing of the I band, but the length of the A band remained constant at ~1.5 μm.

Subsequent observations on isolated myofibrils established that the length of the thin filaments also remained unchanged with both passive stretching and isotonic contraction. It was thus inferred that the thin filaments slid between the thick filaments during muscular contraction. If the muscle was allowed to shorten by more than 65% of its original length, the borders of the A band impinged on the Z line and contraction bands formed around the Z line, probably due to pressure of the thin filaments on the Z line.

Further details became apparent with the introduction of low-angle X-ray diffraction procedures (H. E. Huxley and Brown, 1967). By synchronizing electrical stimulation of the muscle with the operation of a camera shutter, it was possible to build up a composite X-ray diagram, based on ~600–1200 muscle twitches carried out over the course of 10–20 hr. During contraction, the 143-Å reflections corresponding to the actin/myosin cross-bridges remained the same distance apart as at rest, but their intensity was substantially reduced, suggesting that their orientation was changed (H. E. Huxley and Brown, 1967). Since the reflections diminished rather than increased in intensity, it was inferred that cross-bridge formation did not occur synchronously along the muscle filament, and that at any given moment only a small proportion of the cross-bridges were united. Passive stretching to 125% of the resting length did not change the reflection pattern, but with more severe stretching, changes were observed that could be interpreted as a less accurate alignment of the thick and thin filaments.

We have already seen (Fig. 3.6) that the tension a skeletal muscle fiber can develop depends upon sarcomere length. The maximum force is realized when the distance between two adjacent Z lines is 2.0–2.5 μm. No contraction is possible if the thick and thin filaments do not overlap (sarcomere length > 3.5 μm). Muscle force is also weakened if there is too much overlap of the two types of filament (sarcomere length < 1.65 μm); possibly this increases the separation of actin and myosin filaments (Elliott, 1967). On the other hand, the maximum velocity of contraction is relatively independent of sarcomere length (Edman, 1978).

When adjusted for differences in cross-section and in the proportions of contractile material, cardiac muscle develops a peak tension similar to that of skeletal muscle (Sonnenblick, Parmley et al., 1968). However, the resting tension needed to develop peak tension is greater in cardiac than in skeletal muscle, perhaps because there is a larger amount of connective tissue between the fibers. As in skeletal muscle, the peak tension is developed at a sarcomere length of 2.2 μm; however, the decrease of tension with compression of the sarcomere is more dramatic, and no tension can be developed at unit lengths of less than 1.5–1.6 μm (Sonnenblick, Spotnitz, and Spiro, 1964). In theory, cardiac sarcomeres could shorten from 2.2 to 1.5 μm, but under normal circumstances the arterial pressure against which the ventricle must contract (afterloading) is such that shortening is limited to 12–18% of initial length. This corresponds to ejection of 50–65% of the ventricular contents (ejection fraction), as observed in radio-opaque (angiographic) studies of cardiac dynamics.

Molecular Mechanisms in Muscle

Cross-Bridge Formation

The common denominator to all hypotheses of muscle contraction seems the alternate formation and rupture of bonds between thick and thin muscle filaments.

J. Hanson and Huxley (1955) suggested that sliding of the filaments was initiated by a shortening of the cross-bridges, associated with the binding or breakdown of ATP. The bonds then dissociated, allowing the thin filaments to move forward the distance separating them from the next receptor sites.

This basic concept was elaborated by A. F. Huxley (1957). Thermal agitation caused an oscillation of the myosin bridges around their equilibrium position (Fig. 4.11). When contact was made with a receptor site on a thin filament, spontaneous bonding occurred and the receptor site on the actin molecule was drawn toward the equilibrium position of the M

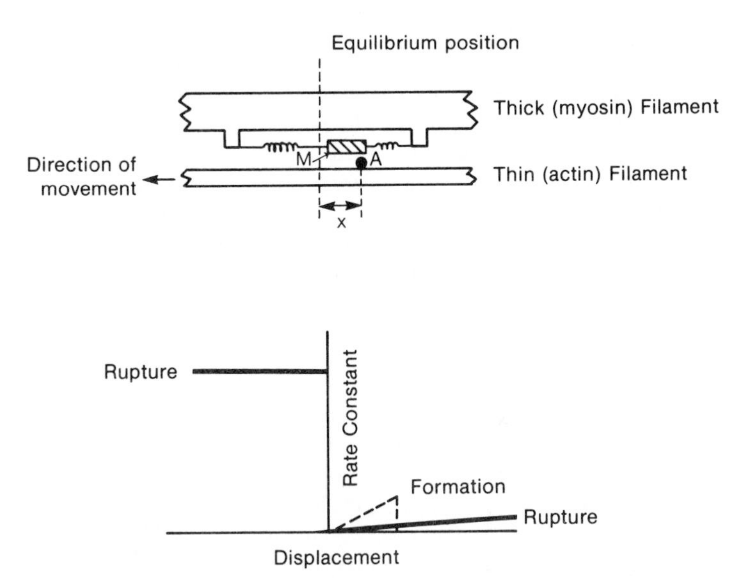

FIG. 4.11. Sliding filament model of muscle contraction. The rate-constants for formation and rupture of bonds between the bridges of the thick filaments (M) and receptor sites on the thin filaments (A) depend on the relative positions of the two elements. Bonding is a spontaneous process that tends to pull the thin filament to the left, restoring M to its equilibrium position. Rupture of the bond then becomes probable, preventing A from returning to its initial position (see also Fig. 4.5). (After A. F. Huxley, 1957.)

bridge. Once the equilibrium point had been passed, the bond ruptured rapidly, and recombination of this particular cross-bridge was not possible until it came within range of the next receptor site further along the thin filament.

The relative motion of the two types of filament might arise from a slight and temporary difference in the separation of bonding sites (J. Hanson and Huxley, 1955). For example, a segment of actin could undergo elongation and subsequently revert to a shorter form, drawing the remainder of the filament forward by an equivalent distance. The structure of the thin filaments offers some support for such a suggestion. The tropomyosin skeleton could conserve the basic form of the F-actin chain (Fig. 4.6), while allowing ATP-induced changes in lateral bonding and thus displacement of the receptor sites (Oosawa et al., 1966). A second possibility is a change in form of the HMM-2 component of the cross-bridge (Fig. 4.4).

Worthington (1964) proposed an electrical variant of these hypotheses. A wave of negativity originated at the Z disc. As it reached successive receptor sites on a thin filament, there was a displacement of 25 Å (the distance separating the receptor site from the next cross-bridge). Propagation of the wave through all 10 sites in a half sarcomere gave a displacement of ~250 Å, so that several waves of negativity were needed to account for the full possible range of motion within the sarcomere. Murukami (1960) linked this concept to previous views by suggesting that ATP was involved in generating the force of electrical attraction.

A final possibility is that the muscle force may be developed through expansion of some structure. Thus Ullrick (1967) suggested that ATP induced an expansion of the Z disc. However, this hypothesis would not explain how contraction occurs in animals that lack a Z disc.

Conversion of Chemical to Mechanical Energy

The minimum "quantum" of movement is that necessary for contact between a cross-bridge and the center of an actin monomere, often a distance of less than 10 Å (Fig. 4.9). At the molecular level, movement is relatively slow (perhaps 100 Å • sec^{-1}), but because there are ~10 bridges per half-sarcomere, arranged in series, the overall speed per sarcomere is about 2^9 times as fast as the movement in a single bridge. The sarcomeres are also arranged in series with one another, so that the overall velocity of the myofibril may reach 10^{10}–10^{11} Å • sec^{-1}.

It is generally agreed that ATP plays an essential part in the reaction, although it is still not completely resolved whether the phosphagen is needed for the formation of cross-bridges, for their rupture, or for both reactions (Ebashi and Nonomura, 1973). On the rupture hypothesis, the basic resource would be the potential energy of the elastic processes (Fig. 4.11), and ATP would provide the energy needed to restore initial conditions. Recent hypotheses [equation (2.3)] involve the formation of intermediate compounds with both myosin and actomyosin.

We have previously set the energy store of the ATP molecule at ~ 46 kJ • mol^{-1}, with a 49% efficiency of transfer to the actin/myosin bonding (Fig. 1.8). The usable energy is thus 22.5 kJ • mol^{-1} or 22.5×10^{10} ergs • g-mol^{-1} (1 kJ = 10^{10} ergs). Since by Avogadro's hypothesis each g-mol contains 6.02×10^{23} molecules, a single molecule of ATP yields 3.7×10^{-13} ergs (1 erg = 1 dyne • cm) of useful work. In one study of frog muscle (R. E. Davies, 1963), a cross-section containing 6.5×10^{12} cross-bridges per cm^2 yielded a force of 2 kg • cm^{-2}, or $\sim 3 \times 10^{-7}$ dynes per bond. On this basis, one molecule of ATP would provide sufficient energy to displace an actin filament by $3.7 \times 10^{-13}/3 \times 10^{-7}$ cm, or $\sim$ 123 Å. A. F. Huxley (1957) reached the similar figure of 153 Å displacement per molecule of ATP.

When a muscle is lightly loaded, the movement per molecule of ATP seem to be somewhat greater. H. E. Huxley (1960) found a heat production of $\sim 1.75 \times 10^{6}$ ergs • sec^{-1} • cm^{-3} at a half of maximum loading. Assuming the same density of cross-bridges as before and given an interbridge distance of 143 Å, there would be 4.8×10^{16} bridges per cm^3. The rate of working at a single bridge would then be 3.6×10^{-11} ergs • sec^{-1}. The total heat yield of each molecule of ATP is 7.6×10^{-13} ergs. H. E. Huxley's experiment thus implies the breakdown of 47 molecules of ATP per sec at each cross-bridge. The observed speed of interdigitation of the filaments at half-loading was 18,400 Å • sec^{-1}. This in turn indicates that one molecule of ATP was broken down for each 390 Å of movement. The discrepancy from the 123–153 Å movement previously cited can be explained on the basis that with rapid rates of sliding, not all potential bonds are formed. This is why muscle force depends upon the rate of shortening [equation (1.25)]. According to A. F. Huxley (1957), the maximum isometric tension is directly related to the number of cross-bridges per cm^2 of muscle, to sarcomere length, and to the work performed per bridge, but is inversely proportional to the distance between receptor sites on the actin filaments. A typical figure is ~ 40 N • cm^{-2} of muscle cross-section.

Studies in cardiac muscle have emphasized the dependence of contractile force upon calcium delivery (Repke and Katz, 1972). A concentration of 10^{-6} M Ca^{2+} is needed to occupy one-half of the receptor sites on the troponin molecules. Since cross-bridge formation is contingent upon calcium binding of the troponin, it can be calculated that each kg of wet ventricular tissue requires ~ 60 μmol of Ca^{2+} in order to develop its maximum possible force. With less adequate delivery of calcium, some of the potential actin/myosin bonds remain unformed, and systolic power is lost.

In the short term, an increased calcium delivery (and thus an increase in the contractility of the heart) is mediated by the release of calcium ions from binding sites in the glycocalyx, a 50-nm film external to the lipid bilayer of the cell membrane that penetrates the sarcotubular network (Langer, 1978). In the longer term, there is a liberation of calcium stored as an insoluble oxalate in the membrane vesicles (A. M. Katz, 1974). Disease may reduce cardiac contractility by altering the structure of the myosin molecule; if its ATPase activity is reduced, the number of actin/myosin bonds may remain unchanged, but their rate of interaction, and thus the maximum velocity of contraction, is diminished.

Osmotic Work

There is an ionic gradient across most cell membranes, nerve and muscle included, with high intracellular concentrations of potassium ions and high extracellular concentrations of sodium ions. A slow exchange of intracellular ions occurs under resting conditions, and in muscle the process is greatly accelerated during contraction. In order to restore the initial composition of the intracellular environment, osmotic work must be performed. Sodium ions must be pumped out of the cell, and potassium ions must be returned to the intracellular fluid. There are many other situations in the body where molecules are moved against osmotic gradients; two other examples in muscle are the return of calcium ions to their storage sites following contraction and the transfer of glucose from the blood to the interior of the cell.

It is difficult to study the conversion of chemical to osmotic work in muscle cells, since energy used in this way is overshadowed by that required for mechanical work. Much of our present knowledge of the "sodium pump" is thus based on studies of red cells (Ussing, 1949; Whittam, 1964). Some observers have claimed that sodium ion transfer accounts for at least 30% of resting cell metabolism (Ismail-Beigi and Edelman, 1970). One estimate (R. E. Davies,

1973) suggested that the sodium pump of skeletal muscle was using 0.01 μmol of ATP per g of tissue following a single twitch. Given 50 impulses per sec, this could account for an ATP breakdown of 30 μmol • g^{-1} • min^{-1}, or with 20 kg of active muscle 0.6 mol • min^{-1} (27.7 kJ • min^{-1}).

Conway (1960) proposed that energy was transferred directly from the electron transport chain (Fig. 2.13) to the ion-exchange process, the "redox-pump" hypothesis. However, the majority of authors believe that ATP plays an intermediary role. Each molecule of ATP provides sufficient energy to expel 1 ion against an electrochemical gradient of ~400 mV (Hodgkin, 1964). The average resting potential is about −80 mV, so that 5 ions could be transported across the membrane for each molecule of ATP that is broken down. The molecular mechanism of transfer is still unclear. Most suggestions involve a carrier, either a phospholipid or a phosphoprotein, which rotates or moves using energy derived directly or indirectly from ATP (Glynn and Karlish, 1975). As an example, we may cite the model of Opit and Charnock (1965). They visualize a surface-oriented anionic protein, with which the sodium and potassium ions can form associated ion pairs. The related enzyme is (Na^+ + K^+) ATPase, an enzyme activated by sodium, potassium, and magnesium ions (Skou, 1965). It has two separate activation sites. That on the inside of the membrane is activated by sodium ions, while that on the outside is activated by potassium ions. Let us suppose that the sodium concentration rises on the inside of the membrane (Fig. 4.12). This activates the ATPase, and as a result the ATP forms an intermediate phosphorylation compound with the inner protein layer. Elongation causes the protein film to rotate about its outer counterpart. The absorption energy changes with translocation, and ATPase is further activated, leading to hydrolysis of the phosphate radical. The protein then shortens and rotates back to its original position, carrying the potassium ions with it into the interior of the cell. Details of the hypothesis have yet to be confirmed, but there is now good evidence that muscle fibers contain a (Na^+ + K^+) ATPase (Rogus et al., 1969). It seems to be located mainly in the sarcoplasmic reticulum; sodium ions may thus be returned to the extracellular fluid by extrusion into the sarcoplasmic reticulum, with subsequent transport into the tubular system.

DIFFERING TYPES OF MUSCLE

Fiber Classifications

For many years observers have distinguished white (fast-twitch) and red (slow-twitch) types of skeletal

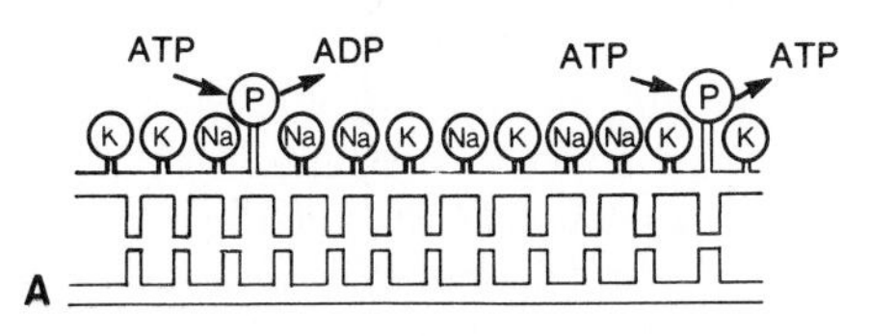

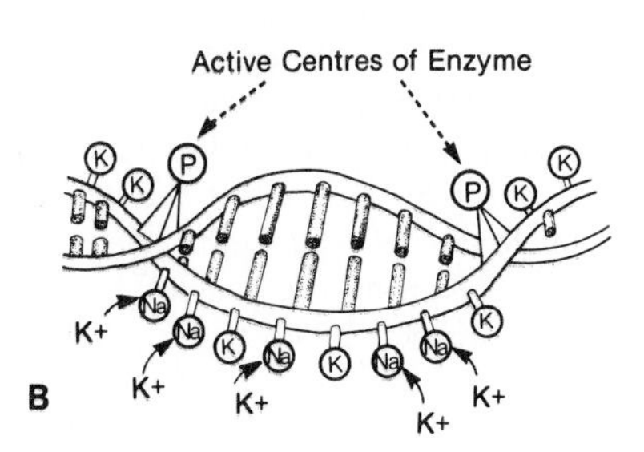

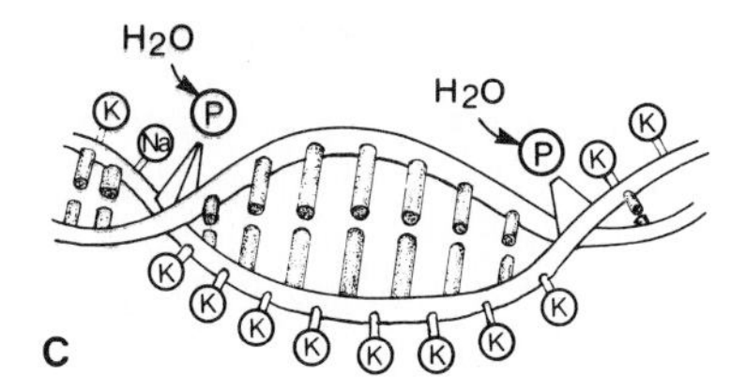

FIG. 4.12. Model of membrane transport. In **A**, an accumulation of sodium ions on the inner protein layer leads to a phosphorylation of active sites in the membrane. **B.** Elongation of the inner layer causes it to rotate about the outer layer, and an exchange of sodium and potassium ions occurs. **C.** This renders the phosphorus bond vulnerable to hydrolysis; the membrane then twists back to **A** carrying potassium ions into the interior of the cell. (After Opit and Charnock, 1965.)

muscle fibers on the basis of their myoglobin content (Needham, 1926; Lawrie, 1953). The red fibers (Table 4.2) have a good capillary supply; a large content of both fat (Schmalbruch and Kamieniecka, 1974; Essén, 1977) and oxidative enzymes (Batelli and Stern, 1912; Barnard et al., 1971; Essén, Jansson et al., 1975), but a limited content of glycolytic enzymes (Barnard, Edgerton et al., 1971; Essén, Jansson et al., 1975). In contrast, the white fibers have limited fat stores and low concentrations of certain oxidative enzymes but a compensatory increase in the concentrations of glycolytic enzymes, to the point where both classes of fiber have a similar maximum rate of ATP generation (Pette and Staudte, 1973).

Not only are there fewer mitochondria in white than in red fibers, but there are also characteristic differences of enzyme content (Spurway, 1978). For example, glycerol phosphate dehydrogenase (GPD) and cytochrome a_3 are present in almost equal concentrations (~ 1 U • g^{-1} wet wt) in white fiber mitochondria whereas in red fiber mitochondria there is a much greater concentration of cytochrome a_3 (~ 10 U • g^{-1}) than of GPD (~ 0.2 U • g^{-1}). On the other hand, both types of mitochondria have similar *relative* concentrations of the enzymes concerned with the citric acid cycle and with fatty acid oxidation (Pette and Staudte, 1973).

As a consequence of these various characteristics, the red fibers are well adapted to aerobic metabolism, with a capacity for sustained, fatigue-resistant activity, while the white fibers are particularly suited to brief bouts of anerobic effort, showing a high power output but early fatigue. R. E. Davies (1973) noted that in using 1 μmol of ATP per g of tissue, various muscles developed tension/time products ranging from 30,800 g • sec for the biceps brachii to 540,000 g • sec for the latissimus dorsi. Readiness of the white fibers for powerful but short-lasting activity is further indicated by large stores of creatine phosphate and a high concentration of the corresponding enzyme creatine phosphotransferase (Ogata, 1960). A large sarcoplasmic reticulum also favors a rapid uptake of Ca^{2+} and thus quick reactivation of this type of fiber.

W. K. Engel (1962) suggested an alternative system of fiber classification based on concentrations of the enzymes myosin and actomyosin ATPase. Type I (slow-twitch) fibers stain lightly for these enzymes, while type II (fast-twitch) fibers color darkly with the same stain. Apparently, both types of fiber have a similar content of actin and of myosin, but slight differences in the structure of the proteins modify their ATPase activity (Bárány, Bárány et al., 1965). In this respect, cardiac tissue seems an extreme form of red muscle, having an extremely low ATPase activity (A. M. Katz, 1974). While there is a fair parallel between the ATPase activity of a fiber and its speed of contraction, identical ATPase staining may be associated with a twofold difference of twitch speed (R. E. Burke and Tsairis, 1974). Certainly, a causal association has yet to be proven. It remains conceivable that a high ATPase activity is found in a form of myosin that for some other reason has a high velocity of contraction, or indeed that the rate-limiting step lies elsewhere, such as the rate of release and uptake of calcium from the sarcoplasmic reticulum (Gauthier, 1969; Close, 1972).

With the widespread application of needle biopsy and associated techniques of histochemistry, an intermediate (IIa) type of skeletal muscle fiber has been recognized in some species (Close, 1972; R. H. Fitts et al., 1973; W. K. Engel, 1974; Maxwell et al., 1977). The IIa fibers are characterized by a high actomyosin ATPase content, moderate amounts of myosin ATPase and myoglobin, more mitochondria and oxidative enzymes even than the type I fibers (Peter et al., 1972), at least moderate amounts of fat, and less glycolytic activity than type IIb fibers. While some authors maintain that the three types of fiber can be identified in humans (Edgerton, Smith, and Simpson, 1975; Prince et al., 1976; Garnett et al., 1978), others (Saltin, Henriksson et al., 1977) fail to find a consistent subdivision of type II fibers in human subjects.

M. H. Brooke and Kaiser (1970) proposed subdivision on the basis of differences in ATPase staining with alkaline (pH 10.3) and acid (pH 4.6–4.8) preincubation. Unfortunately, their method has increased rather than diminished the confusion, since it seems to identify more type IIa and fewer type IIb fibers than the other methods of classification.

Unequivocal identification of the several fiber types could presumably be achieved through an understanding of associated differences in the structure of the myosin molecule. The ultimate resolution of fiber typing thus lies with the protein chemists (M. H. Brooke and Kaiser, 1970).

Quantitative Studies in Humans

Early work was based on the staining characteristics of biopsy specimens and was at best semiquantitative. Subsequently, there have been more precise measurements of both the contractile characteristics and enzyme concentrations (Table 4.3). Calcium-

TABLE 4.2
Classification of Fiber Type

Fiber Type	Myoglobin Content	Capillary Supply (per fiber)	Fiber Area	Motor Neuron and Axon Size	Typical Innervation Ratio (gastrocnemius)	Axon Conduction Velocity ($m \cdot sec^{-1}$)	Liability to Accommodation	Mitochondrial Enzymes	Glycolytic Enzymes	Fat Content
Red (Slow-twitch, type I)	+++	4	++	+	540/unit	85	+	++	+	+++
Intermediate (Fast-twitch a, type IIa)	++	4	++++	++	440/unit	100	++	+++	++	++
White (Fast-twitch b, type IIb)	(+)	3	+++	+++	750/unit	100	++	(+)	+++	+

Fiber Type	Myofibrillar ATPase	ATPase after Acid Preincubation[a]	Time to Peak Tension (msec)	Tension Developed	Fatigue Resistance	Sag in Unfused Tetani	Liability to Recruitment	Function
Red (Slow-twitch, type I)	+	–	80	+	+++[b]	0	+++	Slow twitch, low tension (postural and endurance work)
Intermediate (Fast-twitch a, type IIa)	+++	(+)	40[c]	++	++	+	++	Fast twitch, moderate tension (medium endurance work)
White (Fast-twitch b, type IIb)	+++	++	30	+++	(+)	+	+	Fast twitch, high tension (rapid, powerful movements)

[a] It is not certain that classification by acid preincubation gives the same basis of subdivision as the other criteria listed.
[b] The fatigue resistance of intermediate fibers is less than that of slow-twitch fibers, despite more mitochondria; this may reflect the relative efficiences of isotonic and isometric contraction in the two types of fiber (Goldspink, 1970).
[c] According to Prince et al. (1976), the intermediate fibers contract faster than the white fibers in humans.

TABLE 4.3
Characteristics of Human Muscle Fibers

<table>
<tr><th>Substance</th><th>Slow-twitch (type I)</th><th>Fast-twitch a (type IIa)</th><th>Fast-twitch b (type IIb)</th></tr>
<tr><td>Ca^{2+} actomyosin ATPase (mmol • min^{-1} per mg of myosin)</td><td>0.16</td><td colspan="2">0.48</td></tr>
<tr><td>Mg^{2+} actomyosin ATPase (mmol • min^{-1} per g of protein)</td><td>0.30</td><td colspan="2">0.84</td></tr>
<tr><td>Time to peak tension (msec)</td><td>80</td><td colspan="2">30</td></tr>
<tr><td>Creatine phosphokinase (mmol • min^{-1} per g of protein)</td><td>13.1</td><td colspan="2">16.6</td></tr>
<tr><td>Phosphofructokinase (mmol • kg^{-1} • min^{-1} wet wt)</td><td>9.4</td><td>14.0</td><td>20.0</td></tr>
<tr><td>Succinate dehydrogenase (mmol • kg^{-1} • min^{-1} wet wt)</td><td>11.5</td><td>9.0</td><td>6.5</td></tr>
<tr><td>Cross-sectional area (μm^2)</td><td>5310</td><td>6110</td><td>5600*</td></tr>
</table>

*In female subjects, the average area of the fast-twitch b fibers is much smaller than that for the other two categories.
Source: Based on reports collected by Saltin, Henriksson et al. (1977), mainly for young men.

activated ATPase concentrations accord well with differences of contraction speed (J. S. Hanson, 1974). The usually selected enzymatic representatives of mitochondrial oxidation (succinic dehydrogenase) and glycolysis (phosphofructokinase) also show anticipated differences of activity between type I and type II fibers. On the other hand, all categories of fiber show sufficient enzyme concentrations to accommodate high levels of both aerobic and anerobic metabolism. Further, endurance training can enhance the aerobic activity of type II fibers until their oxidative potential exceeds that of type I fibers in untrained subjects (Essén, Jansson et al., 1975).

The relative proportions of the several fiber types are generally gauged from small-needle biopsy specimens, an approach that carries a danger of substantial sampling error (Elder, 1977; Gandy et al., 1978). The most frequently biopsied muscle is the vastus lateralis. Swedish authors report ~52% type I, 33% type IIa, and 14% type IIb fibers in average men and women (Hedberg and Jansson, 1976; Nygaard and Gøricke, 1976), whereas Edgerton, Smith et al. (1975) find respective percentages of 38, 37, and 25% for the three fiber types. Presumably, the mixed fiber pattern of the vastus reflects a need for both static and dynamic activity in this muscle. A similar pattern of fiber distribution is seen in many other limb muscles, including the gastrocnemius, rectus femoris, deltoid, and biceps, all of which contain ~50% type I fibers. On the other hand, the soleus, as might be predicted from its predominant postural role, contains 75–90% type II fibers.

The total number of fibers in a given muscle group is probably established by the fourth or fifth month of intrauterine life (MacCallum, 1898). At this stage of development (Tomanekh and Colling-Saltin, 1977), the several fiber types begin to emerge from a common stock of undifferentiated (type IIc) fibers in response to neurotrophic influences (M. Brown, Cotter et al., 1975). There is a reversion to the fetal fiber type if denervation occurs (Needham, 1973). Possibly, the messenger RNA that controls development is produced or transmitted via the motor nerve (Guth and Watson, 1967). Both the contractile and the enzymatic characteristics of differentiated fibers can be modified by cross-innervation (Bárány and Close, 1971; Luff, 1975), for example, an experimental exchange of motor nerves supplying the soleus and the fast digitorum longus muscles (Buller et al., 1960; Romanul and van der Meulen, 1966; Romanul and Pollock, 1969). Such procedures lead to changes in the myofibrillar proteins (particularly myosin and troponin) and in the architecture of the sarcoplasmic reticulum (Amphlett et al., 1975). Neural stimulation of type II fibers at a frequency appropriate to the type I variety (5–10 Hz) also increases fatigue resistance (M. Brown, Cotter et al., 1975). The response of the muscles to systematic electrical stimulation was initially interpreted as a change of fiber type, and reported increases of myoglobin content and aerobic enzyme activity (Pette and Staudte, 1973) opened up a vista of possible "neural doping" of endurance competitors. However, more recent studies suggest that much of the observed response can be attributed to an increase of capillary density, without postulating an increase of aerobic enzyme activity or other changes in fiber characteristics (Hudlicka and Cotter, 1977). To the discomfiture of coaches and trainers, there is little evidence that the relative proportions of type I and type II fibers in a given muscle can be modified either by event-specific athletic training (Roy et al., 1977; Kaijser and Jansson, 1977) or by faradic stimulation (A. W. Taylor et al., 1978). Thus identical twins show very similar fiber populations, regardless of their respective levels of habitual activity (Komi et al., cited by Sjodin, 1976). On the other hand, the proportions of the different fiber types found in a given muscle vary quite widely within a total population, suggesting the possibility that well-endowed candidates for a specific sport could be selected by performing muscle biopsies at an early age.

Differences of fiber population are much as anticipated between sprinters (where 20–30% of the vastus lateralis fibers are type I) and long-distance competitors (where 65–95% of the vasti are type I fibers) (Costill, Fink, and Pollock, 1976). Downhill skiers have a high proportion of type II fibers (Thorstenson et al., 1977). More surprisingly, throwers, jumpers, and weight lifters have characteristics similar to those of an average person (40–60% type I) (Gollnick, Piehl et al., 1975; Roy et al., 1977). Saltin et al. (1977) suggest that in sports in which a single vigorous effort is required, synchrony of contraction may be more important than fatigue resistance, and that synchrony is more readily established in type I than in type II fibers. Thorstenson et al. (1977) link type II fibers with the ability to develop a high peak torque.

Event-specific training cannot change type I into type II fibers, or vice versa, but there is some evidence that it can convert type IIb to type IIa fibers (Kaijser and Jansson, 1977); perhaps for this reason orienteers show a high normal content (18%) of type IIb fibers in the deltoid muscle, but a very low content (2–3%) in leg muscles such as the vasti and the gastrocnemii (Thorstenson et al., 1977). The cross-sectional area of individual fibers shows some increase in young men from age 16 to 20–30 (P. Andersen, 1975). On the other hand, young women show a decrease of fiber size over the same period, presumably because growth has ceased and their physical activity is also declining. In the 20- to 30-year-old age group, the type IIb fibers of women (average cross-section = 2235 μm^2) are much smaller than either type I (3948 μm^2) or type IIa (3637 μm^2) fibers. In general, the dimensions of type II fibers respond more readily to variations of habitual activity than do type I fibers. Costill et al. (1976b) reported that in elite cross-country runners type I fibers were 22% larger than type II fibers; his data refer to a mixture of IIa and IIb fibers, and thus do not differ greatly from findings for the general population.

The activity of aerobic enzymes varies widely with habitual activity. In sedentary subjects, typical succinic dehydrogenase readings per unit mass of wet muscle are 6–7 mmol • kg^{-1} • min^{-1} (Costill, Daniels et al., 1976). With extreme inactivity, values of 3–4 mmol • kg^{-1} • min^{-1} are possible (Saltin, Henriksson et al., 1977), while in well-trained distance athletes figures of 21 mmol • kg^{-1} • min^{-1} are likely, equally in type I and type II fibers. In contrast, the activity of many glycolytic enzymes is the same or even slightly less in athletes than in sedentary individuals (Costill, Daniels et al., 1976; Costill, Fink et al., 1976).

In sedentary subjects, the average number of capillaries per fiber is ~4 for types I and IIa fibers, and 3 for type IIb fibers (P. Andersen, 1975). Some authors find that the capillary/fiber ratio is closely correlated with maximum oxygen intake per unit of body mass, and Saltin, Henriksson et al. (1977) argued that the number of capillaries per fiber may be the most realistic measure of blood supply to a muscle; on the other hand, Maxwell (1978) found only a threefold difference of capillary density when comparing species in which the oxidative capacity of the muscles varied 40-fold.

We have previously reviewed Krogh's concept of an average diffusion path length in skeletal muscle [equation (3.6)], noting that fiber cross-section enters into this calculation of tissue oxygen supply. The precision of the classical concept is limited by variations in such factors as the number and distribution of mitochondria within the muscle fiber (Howald, 1975). Further, a given type of activity involves highly selective recruitment of muscle fibers (see below). Nevertheless, the basic soundness of Krogh's hypothesis (with its indication of a potential for local hypoxia) is supported by the characteristics of successful endurance sportspersons such as swimmers and orienteers—a combination of a high capillary density (5–6 vessels per fiber) and small muscle fibers (Nygaard and Gøricke, 1976; Nygaard, 1976; Brodal et al., 1977; Ingjer and Brodal, 1978).

NEUROMUSCULAR COUPLING

The Motor Unit

Anterior Horn Cells

Muscle fibers are called into action by electrical impulses from the anterior horn cells of the spinal cord (Fig. 4.13). A typical anterior horn cell has a diameter of 70 μm, and gives rise to many branching processes or dendrites. The majority of the dendrites are

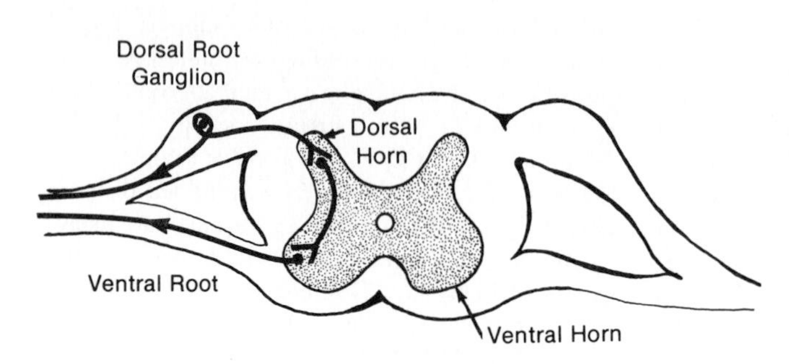

FIG. 4.13. Segmental reflex in spinal cord. Sensory fibers enter via the dorsal root, and their cell bodies are located in the dorsal root ganglion. Motor neurons are located in the ventral or anterior horn of the spinal cord, and the motor fibers emerge via the ventral root. In the example illustrated, there are two synapses, with a connecting *internuncial neuron*. This is typical of the flexor reflex that follows contact with a painful stimulus. The stretch reflex is monosynaptic, the internuncial neuron being omitted.

quite short, but one (the axon) passes out of the ventral root of the spinal cord, traveling up to 90 cm to reach the corresponding skeletal muscle. Each axon terminates in a series of motor end-plates, one per muscle fiber. A fatty insulating sheath of myelin covers the main length of the axon, with the exception of periodic nodes (the nodes of Ranvier). Electrical impulses travel the length of the nerve by jumping from one node to another (saltatory transmission), an arrangement that gives a high conduction velocity.

The terminals of axons from other nerves, sometimes several thousand in number, cover the surface of the anterior horn cell and its dendritic processes. Terminals related to the dendrites have an excitatory function, while those reaching the cell body are generally inhibitory in type. Eccles (1973) commented that this plan gives inhibitory circuits the last word on whether a command should be transmitted to the muscles.

The terminals, or synaptic knobs, are separated from the adjacent dendrites or cell body by a narrow synaptic cleft (width ~20 nm) and two cell membranes. Chemical neurotransmitters ensure passage of the signal across this gap.

The Motor End-plate

Motor end-plates are found in one or more troughs of the sarcolemma, about halfway along individual muscle fibers (Fig. 4.1). In these structures, the terminal membrane of the neural axon comes into intimate contact with the basal membrane of the muscle. The nerve emerges from its myelin sheath, forming a number of terminal knobs. Each knob has a diameter of ~2 μm. It contains numerous vesicles and mitochondria. The vesicles apparently store quantal packets of the neurotransmitter substance acetylcholine. The sarcolemmal surface is greatly extended in the end-plate region, through a complicated pattern of folds that are arranged perpendicular to the terminal knobs (Fig. 4.1). Structural details of the end-plate vary with the muscle type (Ogata et al., 1967; Nyström, 1968b; Padykula and Gauthier, 1970). The terminal nerve axons in fast-twitch fibers are long, smooth, and wide spreading, while the slow-twitch fibers have tightly packed, complex terminals, with a greater number of vesicles and many postsynaptic folds.

Neurotransmitter Substances

The sole neurotransmitter at the motor end-plate is acetylcholine. However, in other parts of the nervous system a variety of chemical agents (including noradrenaline and serotonin) serve to carry impulses across the synaptic gap. Acetylcholine facilitates passage of a nerve impulse, but within the brain and spinal cord inhibitory compounds are also liberated. One well-recognized cerebral inhibitor substance is γ-aminobutyric acid.

The process of synaptic transmission has been studied most closely at the motor end-plate. Principles are similar in the central nervous system, although the quantities of transmitter involved are much smaller. Under resting conditions, the terminal

$$\underbrace{CH_3-\overset{\overset{CH_3}{|}}{\underset{\underset{CH_3}{|}}{N^+}}-CH_2-CH_2-O-\overset{\overset{O}{\|}}{C}-CH_3}_{\text{acetylcholine}} + H_2O \xrightleftharpoons{\text{cholinesterase}} \underbrace{CH_3-\overset{\overset{CH_3}{|}}{\underset{\underset{CH_3}{|}}{N^+}}-CH_2-CH_2-OH}_{\text{choline}} + \underbrace{CH_3COOH}_{\text{acetic acid}} \tag{4.1}$$

knobs release occasional packets, each containing $\sim 2 \times 10^4$ molecules of acetylcholine. Individual occurrences follow a Poisson distribution, that is to say, the likelihood that a quantum will be released at any instant is a constant. In the resting frog sartorius muscle, for example, a packet escapes about once every second (F. D. Carlson and Wilkie, 1974). The event can be recognized by the appearance of a "miniature potential" in the end-plate region, with an associated alteration in the distribution of potassium and sodium ions across the muscle membrane. Subsequently, the acetylcholine is hydrolyzed by the enzyme cholinesterase, large amounts of which are present at the neuromuscular junction [see equation (4.1)]. The activity of the enzyme is greater in fast- than in slow-twitch fibers (Nyström, 1968a; Crockett and Edgerton, 1975), reflecting the greater frequency of miniature end-plate potentials in the former (J. J. McArdle and Albuquerque, 1973). Within the central nervous system, the glial cells also have a high content of cholinesterase, thus providing a form of chemical insulation between adjacent synapses.

Resting Membrane Potential

Under resting conditions, the interior of the muscle fiber has a negative potential of ~80–100 mV with respect to the extracellular fluid. Some (Yonemura, 1967; Campion, 1974) but not all authors (Albuquerque et al., 1972; Luff and Atwood, 1972) find a higher potential in fast- than in slow-twitch fibers. The difference of potential across the cell membrane is sustained by metabolic activity. It reflects the retention of potassium ions within the cell (the potassium pump), plus the active extrusion of sodium ions (the sodium pump; Eccles, 1965).

The membrane potential V′ can be calculated from the internal and external concentrations of potassium ions $[K^+]_i$ and $[K^+]_o$ according to the Nernst equation:

$$V' = \frac{RT}{F} \log \frac{[K^+]_o}{[K^+]_i} \qquad (4.2)$$

or, more precisely (allowing for some permeability of the membrane to sodium ions):

$$V' = \frac{RT}{F} \log \frac{[K^+]_o + 0.01[Na^+]_o}{[K^+]_i} \qquad (4.3)$$

where R is the gas constant, T is the absolute temperature, and F is Faraday's constant. At 37°C, RT/F is ~60, so that if $[K^+]_o$ = 140 meq • l^{-1} and $[K^+]_i$ = 150 meq • l^{-1}, V′ = −82 mV.

Membrane Depolarization

Given an adequate concentration of calcium ions in the presynaptic terminal knobs, acetylcholine alters the properties of the sarcolemma so that it becomes freely permeable to both sodium and potassium ions (B. Katz and Miledi, 1972; Sakmann, 1978). The negative potential across the membrane is progressively reduced; depolarization is said to have occured.

Arrival of a nerve impulse at the motor end-plate induces a large but temporary increase in the probability that quanta of acetylcholine will be released. The transmitter substance diffuses across the 1-μm synaptic gap, and in consequence the permeability of the sarcolemma is increased. The amounts of acetylcholine released per nerve impulse are larger in a fast-twitch than in a slow-twitch fiber. In the former, the near simultaneous delivery of ~100 quanta (~10^6 molecules) of acetylcholine in 1 msec produces a sufficient local concentration (~10^{-15} Molar) to induce not only depolarization but also a propagated action potential that travels the length of the muscle fiber. The critical stimulus to development of a propagated disturbance is a 30- to 40-mV decrease of the local membrane potential. Depending on the type of muscle fiber and the local temperature, the wave of electrical activity traverses the fiber (Fig. 4.14) at a speed of 0.5–5.0 m • sec^{-1} (Fatt and Katz, 1951); attenuation of the response is related only to the physical characteristics of the muscle fiber ("cable effect"). During the recovery period, acetylcholine is removed from the neuromuscular junction by a combination of diffusion and hydrolysis. The choline residue is taken up by the nerve terminals for further synthesis of acetylcholine (Axelrod, 1974), while the sodium pump restores the local ionic gradient across the sarcolemma.

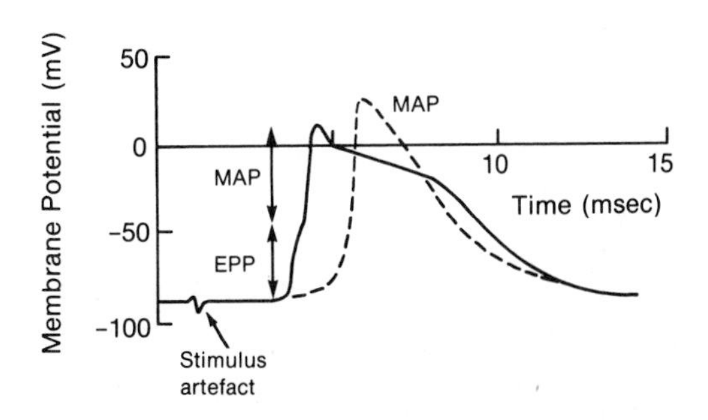

FIG. 4.14. Changes in membrane potential in the region of a motor end-plate following electrical stimulation. Solid line at end-plate. Interrupted line 2 mm distant from end-plate. MAP, muscle action potential; EEP, end-plate potential. (After Fatt & Katz, 1951.)

Other parts of the muscle membrane do not show the acetylcholine sensitivity of the end-plate region, except after denervation. It has thus been postulated that the nerve terminals have a trophic function, facilitating the synthesis of an acetylcholine-sensitive protein in the underlying sarcolemma. Fast fibers have ~20% more of the specific acetylcholine receptor sites than slow-twitch fibers (Edington and Edgerton, 1976).

Various factors can modify neuromuscular transmission. If the cholinesterase activity of the junction is depressed (for example, by a poisonous dose of an organophosphorus insecticide), the acetylcholine persists for an excessive time, giving a facilitation of response to nerve impulses or (with a larger buildup of the transmitter substance) a continuous fibrillar twitching of the affected muscle. In contrast, the response can be inhibited if receptor proteins on the sarcolemma are blocked (for example, by the use of a muscle relaxant such as d-tubocurarine). A facilitation or a depression of response may also occur with rapidly repeated neural stimuli, depending on the interstimulus interval (del Castillo and Katz, 1956).

Similar considerations influence impulse transmission at the anterior horn cell and elsewhere within the central nervous system. A stimulus may be *subliminal*, of insufficient strength to depolarize the postsynaptic membrane. Nevertheless, it gives rise to a local "excitatory postsynaptic potential," with release of a transmitter substance that temporarily brings the synapse closer to a situation where transmission can occur (Fig. 4.15). This is described as "facilitation." It offers a possibility for the *temporal summation* of closely succeeding stimuli and the *spatial summation* of activity in closely adjacent synapses. Such effects are short-lived because the transmitter is hydrolyzed within 1–10 msec.

Once a discharge has occurred, the postsynaptic membrane is unable to make a further response until a negative potential of approximately −50 mV has been restored. There is a brief absolute refractory period, but ~90% of normal excitability is restored within 1 msec, allowing nerves to transmit impulses to rates of 1000 • sec^{-1}. Potassium ion permeability remains increased for 15–100 msec after each discharge, giving rise to a phase of "after hyperpolarization," with an associated 5-mV augmentation of the resting membrane potential.

Inhibitory transmitter substances permit an outward diffusion of potassium ions without an associated inward movement of sodium ions. The result is a hyperpolarization of the postsynaptic membrane. Latency is longer than for the excitatory process. The inhibitory postsynaptic potential peaks in about 1.5–2.0 msec, and at this stage a stronger than normal stimulus is needed to activate the synapse. Restoration of normal sensitivity requires activity of the potassium pump (Fig. 4.16).

Motor Unit Characteristics

A motor unit comprises all muscle fibers supplied by a single anterior horn cell. It is thus the smallest functional unit that can be activated in any given muscle. A single motor nerve may control as many as 2000 fibers in some of the large limb muscles, but in the small external muscles of the eye only a single fiber may be activated (McComas et al., 1973). In all, the human spinal cord contains more than 200,000 motor units.

Kugelberg (1973) suggested that a motor unit sometimes comprises both slow and fast fibers. However, homogeneity of innervation conforms better with concepts of fiber differentiation by neural trophic factors (Guth, 1968) or neural impulse patterns

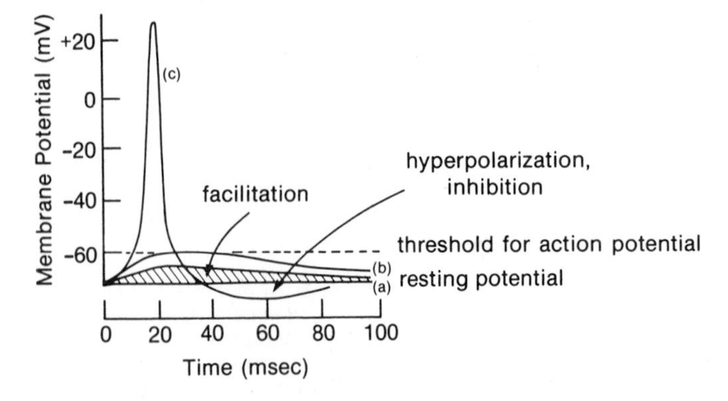

FIG. 4.15. Facilitation, with spatial summation. **A.** Excitatory postsynaptic potential (EPSP) with stimulation of single afferent. **B.** EPSP with stimulation of two closely adjacent afferents; potential still below threshold for action potential. **C.** EPSP with stimulation of three closely adjacent afferents. Note phase of hyperpolarization following action potential. (After Eccles, 1965.)

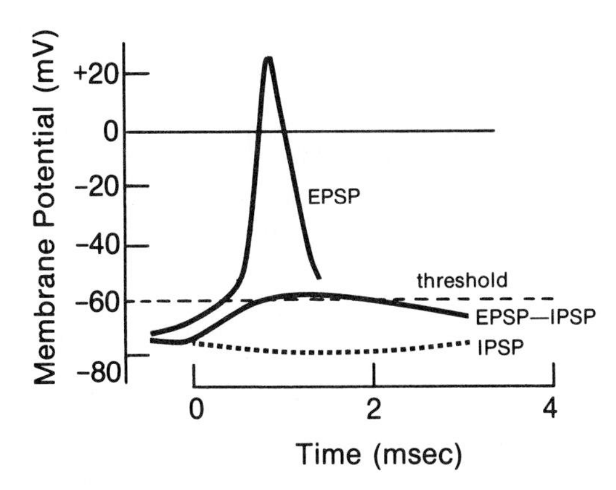

FIG. 4.16. Interaction of excitatory and inhibitory circuits. EPSP, excitatory post-synaptic potential induced by stimulation of excitatory circuit. Note that the action potential threshold is surpassed. IPSP, inhibitory post-synaptic potential induced by stimulation of inhibitory circuit. Note hyperpolarization. EPSP − IPSP = effect of combined stimulation; the resultant change in the membrane is insufficient to induce an action potential. (After Eccles, 1965.)

(Salmons and Vrbova, 1969). The number of fibers per motor unit is rather similar for large α and small α motoneurons, but the size of the fibers supplied is much greater in the case of large α motoneurons. The force developed may thus be 12 times as large for a large as for a small α motoneuron unit (R. E. Burke and Edgerton, 1975).

The fibers activated by a given motoneuron are scattered quite widely through a muscle belly, and any one portion of muscle may contain fibers belonging to 20–50 motor units (R. E. Burke and Tsairis, 1973; Stålberg and Ekstedt, 1973). This anatomical arrangement probably has advantages if there is asynchronous activation of motor units (Rack and Westbury, 1969) during a partial tetanus (Fig. 3.3). It also undoubtedly influences signal detection by the muscle spindles and Golgi tendon organs.

Motor units are activated both by simple reflex arcs (Fig. 4.13, receptors in the skin and muscle) and by volitional impulses traveling from the motor cortex and cerebellum via pyramidal and extrapyramidal pathways (for details, see Chap. 7). Granit (1970) distinguished two basic types of motor unit—the tonic and the phasic. The tonic was characterized by a low stimulation threshold, a slow rate of firing, and prolonged activity. The nerve fibers concerned had a relatively slow rate of conduction (50–80 m • sec^{-1}), innervating muscles that comprised mainly slow-twitch fibers. Phasic motor units showed a higher threshold, a faster rate of discharge, and only transient bursts of activity. The nerve fibers concerned had a rapid rate of conduction ($>$90 m • sec^{-1}), innervating muscles with a high proportion of fast-twitch fibers.

Subsequent studies distinguished three populations of motor units, corresponding to the three muscle fiber types described above (Stephens et al., 1973).

Electromechanical Coupling

Calcium ions are essential to the development of a propagated end-plate potential at the neuromuscular junction. At one time it was thought that calcium ions diffused directly from the motor end-plate to active sites on the contractile filaments. However, A. V. Hill (1948) showed that the rate of fiber tension development was too rapid to be explained by such a lengthy diffusion pathway. The T system was found to play a vital role, conducting the signal to the terminal sacs, where calcium reservoirs lie in much closer relationship with the contractile elements (A. F. Huxley and Taylor, 1958). The precise nature of the signal within the T system is still discussed, but it is probably analogous to the propagated action potential that can be recorded at the muscle surface.

Immediately upon stimulation, a muscle may show a slight "latency relaxation." This is followed by a phase of increasing rigidity, the active state. The externally recorded tension is developed more slowly (Fig. 4.17) because elastic elements are arranged in series with the muscle fibers. We have seen that a minimum electrical disturbance of 30–40 mV is needed to cause a propagated action potential. This seems the threshold stimulus necessary to initiate a release of calcium ions within the tubular system. A larger electrical disturbance accelerates calcium release, but the mechanism "saturates" when the negative potential across the sarcolemma has decreased to ~25 mV. The latent period for development of the active state in a muscle fiber, ~4 msec, comprises ~1 msec for initiation of calcium release plus a further 2–3 msec for it to accumulate and diffuse to active sites on the troponin molecule.

Neuromuscular Fatigue

Animal Preparations

If an isolated nerve muscle preparation is stimulated repeatedly, the maximum shortening of the muscle shows a progressive diminution. This is due largely to an incomplete relaxation between contractions; a

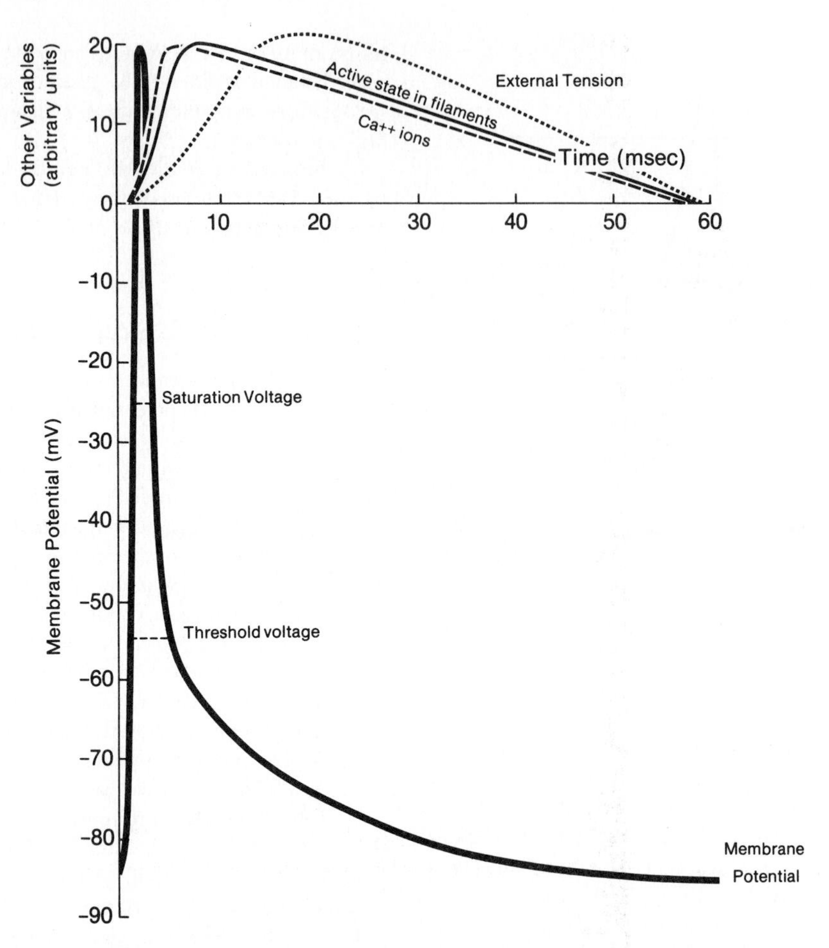

FIG. 4.17. The sequence of events during electromechanical coupling. The disturbance of membrane potential must pass a threshold to initiate calcium ion release, and further time is needed for diffusion of calcium ions to active sites on the contractile filaments. Because of series elastic elements, the externally recorded tension lags st'll further behind these events.

state of "contracture" is said to have developed. The immediate cause is usually a lack of the ATP needed to restore the initial status of the actin/myosin system (Simonson, 1971; R. H. T. Edwards, Hill, and Jones, 1975a). This can be traced in turn to a lack of oxygen for aerobic metabolism and/or an inhibition of glycolysis by an excessive decrease of intramuscular pH.

Animal experiments have shown that fatigue can also affect the mechanism of electromechanical coupling. A normal action potential may still traverse the sarcolemma, but no significant tension is developed; nevertheless, one or more maximal contractions of the muscle can be induced by application of a local chemical stimulus such as 0.1 M potassium chloride. Possibly, in such circumstances a lack of ATP has led to a depletion of calcium stores and thus problems in activation of the actin/myosin bonding.

Isotonic Fatigue in Humans

When a person makes repeated maximum isotonic contractions on a weight-lifting machine, a superficially similar phenomenon can be seen, with a progressive dimunition in the magnitude of muscle contractions. Again, chemical factors—oxygen lack, exhaustion of local stores of phosphagen, inhibition of glycolysis, and, ultimately, depletion of glycogen reserves—contribute to the phenomenon. However, there is an important psychological overlay, since the full initial contraction can be restored temporarily by suitable incentives such as cheering or a sudden emergency. It is possible that normal performance is reestablished by the recruitment of motor units not initially involved in the activity. However, there is also reason to believe that a part of the fatigue is due to central inhibition, originating in the brain rather than in the local reflex arc.

If the effort is submaximal, the time to fatigue can be extended by an alternation of activity between adjacent motor units. However, it would seem that individual muscle fibers develop progressively less tension as exhaustion is approached, since there is a progressive increase in the total electrical activity of the muscle (A. J. Lloyd et al., 1970; Simonson, 1971).

Isometric Fatigue in Humans

During the early stages of a maximal isometric contraction, one can apparently develop transmission fatigue at the neuromuscular junction. Over the first minute of maximal activity, high-frequency motoneurons cease firing and low-frequency units are recruited in their place (A. J. Lloyd, 1971; Stephens and Taylor, 1972). The relative activity of low- and high-frequency units is closely correlated with sensations of pressure, fatigue, pain, and loss of muscle force (Kogi and Hakamada, 1962; A. J. Lloyd et al., 1970). If the contraction is sustained for more than a few seconds, the physiological basis of the fatigue changes to become an inadequate regeneration of ATP within the active fibers (R. H. T. Edwards, Hill, and Jones, 1975a) or other pH-related changes (Karlsson, Funderburk et al., 1975). One practical consequence remains a recruitment of additional fibers to make good the declining tension developed by those elements that were first activated (Simonson, 1971). Because of incomplete relaxation, less energy is needed to sustain tension (R. H. T. Edwards, Hill, and Jones, 1975b). Glycogen exhaustion is not normally a factor in isometric fatigue (Ahlborg, Bergström et al., 1972).

Effects of Prior Exercise

If the subject has previously engaged in several hours of hard physical work, the time to fatigue may be shortened for both isometric and isotonic effort. The main problem in such circumstances is a lack of glycogen, seen first in the slow-twitch and later in the fast-twitch fibers (Piehl, 1974; R. E. Burke and Edgerton, 1975). Other possible contributory factors are an increase of intracellular fluid (Bergström, Guarnieri, and Hultman, 1973a), a loss of potassium ions (Nöcker, 1964), and a loss of cellular buffers (particularly bicarbonate; Wasserman, 1967a).

Recovery

Recovery from fatiguing work generally follows the time course of phosphagen regeneration, lactate removal, and glycogen replenishment. However, complete restoration of excitation/contraction coupling can be a slow process (R. H. T. Edwards, Hill et al., 1977).

DYNAMICS OF MUSCULAR CONTRACTION

Patterns of Fiber Recruitment

The order of fiber recruitment is more or less stereotyped for a given movement (Wyman et al., 1974; Desmedt and Godaux, 1977), although it is influenced by skin stimulation (Kanda et al., 1977; Buller et al., 1978), limb position (Wagman et al., 1965), and the speed of movement required. Fast-twitch fibers are likely to be called into play with rapid, forceful movements, while slow-twitch fibers are more likely to be activated during endurance-type work (Hannerz, 1974; L. Grimby and Hannerz, 1976). The tension yielded per mol of ATP is greater for slow- than for fast-twitch fibers.

With a mild isometric effort, the slow-twitch fibers are the first to be recruited, contracting at a frequency of 5–10 Hz. If the tension is increased, the contraction rate of these fibers is augmented until tetanus develops (20–30 Hz). Further increases of tension lead to an activation of fast-twitch fibers. These commence to contract at ~30 Hz and can increase their frequency to a maximum of 60–70 Hz.

Other factors being equal, the slow-twitch fibers are the most likely to be recruited in both reflex (Buchthal and Schmalbruch, 1970; R. E. Burke, Rymer, and Walsh, 1973) and voluntary movements (Milner-Brown et al., 1973). This reflects early activity of those motoneurons with small cell bodies (Henneman et al., 1965; Granit, 1970; Milner-Brown et al., 1973; Freund et al., 1975). However, there is no magic property in a given cell size (R. E. Burke, 1973; Zucker, 1973); rather, there are both qualitative and quantitative characteristics of the synaptic input which favor activation of the smaller cells, particularly a dense excitatory input from spindle afferents (Grimby and Hannerz, 1976). Recruitment of specific fibers can be inferred from a local depletion of glycogen reserves (Bosley et al., 1976), although in drawing conclusions, account must be taken of (1) glycogen usage from apparently minimal work (R. E. Burke and Edgerton, 1975), (2) interfiber differences in the efficiency of isotonic work, and (3) the simultaneous use of fat by the oxidative fibers, particularly at low work loads (Essén, 1977). At efforts demanding 30–85% of maximum oxygen intake, glycogen first disappears from type I fibers; however, as exercise continues depletion also develops in type II fibers (Gollnick, Armstrong et al., 1973a). If the intensity of effort is increased (for example, by performing the same amount of work in interval fashion), type II fibers are activated in preference to type I, and glycogen depletion is seen first in type IIa fibers (Essén, 1977). Mild isometric contractions lead to glycogen loss from type I fibers, while vigorous isometric efforts are associated with glycogen depletion in type II fibers (Gollnick, Piehl, and Saltin 1974).

Glycogen depletion leads to the fatigue of individual muscle fibers. Edgerton (1976) argued that with prolonged muscular effort there may also be a fatigue of neuronal output. Cetainly, the integrated électrical activity of the muscle (integrated electromyograph, EMG) falls during prolonged effort, the dimunition of potential being particularly marked for fast glycolytic (type IIb) fibers (Stephens and Taylor, 1972). The observed change could reflect a decrease in the rate of firing of individual muscle units (Person and Kudina, 1972), a drop in the amplitude of individual muscle potentials (Stephens and Taylor, 1972), or a slowing of conduction of the potential over the surface of the muscle (Lindström et al., 1970).

Regulation of Muscular Force

We have seen (Chap. 3) that a given motor unit responds to stimulation in an all-or-none fashion. However, various tactics are adopted by the body to develop the finely graded intensities of muscular force encountered in skilled movements:

1. Number of motor units recruited is varied.
2. Type of motor unit recruited is varied (R. E. Burke and Edgerton, 1975).
3. Frequency of steady activation of individual motor units is varied up to and including the development of tetanus (Fig. 3.3).
4. Force of contraction is increased by a single, short interstimulus interval (the catch effect; Norris and Gasteiger, 1955; R. E. Burke, Rudomin, and Zajac, 1970).
5. Energy is stored and later released from series-elastic elements.
6. Externally measured force is modified by leverage (Fig. 3.4), release of central inhibition (Steinhaus, 1963; Ikai et al., 1967), loading (Fig. 3.9), synchronization of motor unit activity (Stein and Milner-Brown, 1973), muscle temperature, and oxygen supply.

The first two mechanisms (varying the number and type of fibers recruited) are generally more important that the "rate" coding of mechanisms (3) and (4) (Bigland and Lippold, 1954b; Clamann, 1970; Desmedt and Godaux, 1978). There may be a 10-fold difference of force between recruitment of a fast and a slow motor unit, this reflecting differences of innervation ratio, fiber cross-section, and the force developed per unit cross-section (Table 4.1; R. E. Burke and Tsairis, 1973). While some increase of firing rate can occur in fibers initially activated at less than 25% of maximum voluntary force, fibers recruited at greater tensions show a progressively narrower band of firing frequencies (Clamann, 1970). One exception to this rule seems the small muscles of the fingers; here, perhaps because of a limited number of available motor units, fiber recruitment is complete at ~50% of maximum force, and further increases of tension depend on a faster rate of firing in units that are already active (Milner-Brown et al., 1973). The advantage of recruitment over rate coding is that the metabolic load on individual fibers is kept low (R. B. Stein, 1974); there is then less likelihood of fatigue secondary to an occlusion of the local blood supply.

The fifth mechanism (energy storage) is encountered in the experienced sprinter (Cavagna, Dusman, and Margaria, 1968; Cavagna, 1970; Cavagna, Komarek, and Mazzoleni, 1971). In such subjects, the measured power output at speeds above 6–7 m • sec^{-1} is greater than can be accounted for by the force/velocity curve of the leg muscles. Cavagna (1971) and his associates showed that the extensor muscles start contracting before the foot hits the ground. The prime function of such an action is to prevent an injury from excessive plantar flexion of the foot (Carlsöo and Johansson, 1962; G. M. Jones and Watt, 1971; Grillner, 1975). Nevertheless, the extensor activity also improves the efficiency of running. The extensor fibers are forcibly elongated while they are contracting, and a part of the energy of foot impact is stored in the series elastic elements, to be reused as extension develops (Fig. 4.18).

Forcible stretching of the fibers may also improve the transfer of chemical energy from the phosphagen molecules to the actin/myosin cross-bridges (Cavagna, Citterio, and Jacini, 1975) by altering the arrangement of the contractile components; one might envisage a difference in filament overlap, filament separation, or the number of cross-bridges formed. It could also be argued that a finite time is needed to develop maximum neural stimulation of the muscle and that a preliminary stretching provides opportunity for completion of this on-transient; how-

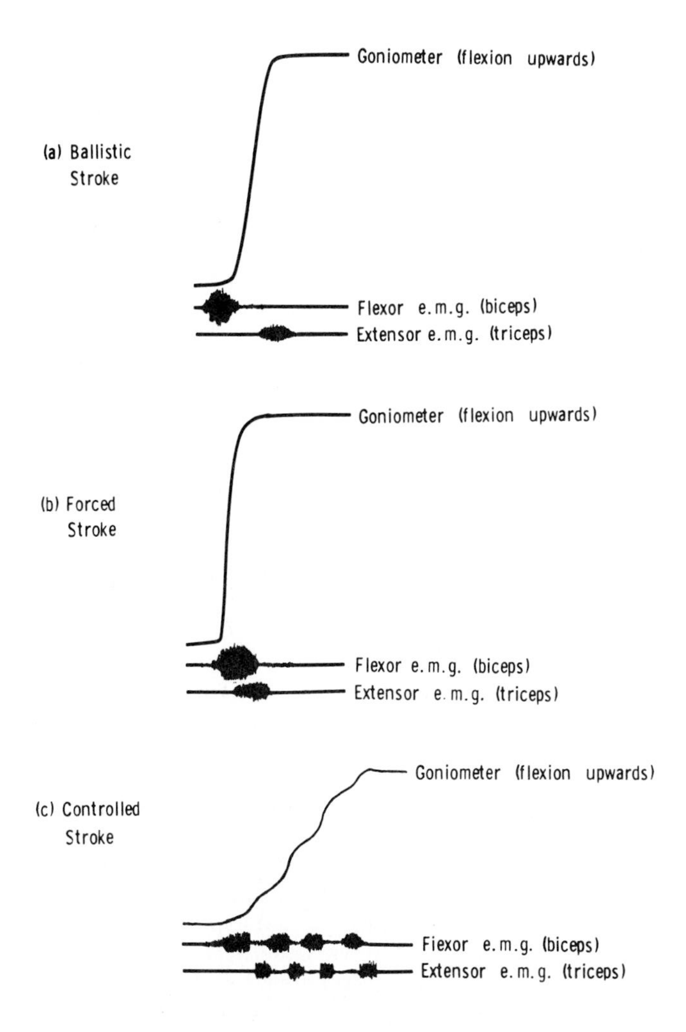

FIG. 4.18. A comparison of ballistic, forced, and controlled movements (electromyographic record of muscle contraction, goniometer record of limb displacement). Note that in the ballistic movement, the limb is set in motion by the vigorous activity of the agonists, or prime movers, and then travels at a speed determined largely by the natural frequency of the part until movement is finally checked or reversed by a carefully programmed burst of activity in the antagonists (Hallett et al., 1975). In contrast with the forced movement, the agonistic fibers remain active throughout most of the limb displacement, whereas in the controlled movement, displacement of the part is slowed by periodic bursts of activity in the antagonistic muscles (V. B. Brooks et al., 1973). With learning, the movement becomes smoother, and antagonist activity is reduced (Marsden et al., 1977).

ever, in practice maximum neural activity develops very rapidly (Asmussen and Sorenson, 1971).

A similar phenomenon of energy sparing or storage can occur during jumping (Asmussen and Bonde-Petersen, 1974a,b). By bouncing, subjects store and recover up to 23% of the negative work associated with the initial descent of the center of gravity of the body. A high jump performed from a squat position thus yields an 11% poorer performance than a jump initiated by a quick drop to the squat position (D. J. Morton, 1952). The counterforce exerted on the floor is correspondingly greater in the second type of jump (F. Andersen, 1967; Thys et al., 1972).

Under the final group of mechanisms (6), we may contrast the ballistic movement (A. W. Hubbard, 1960), where a limb moves freely at a speed determined by the natural frequency of the part and a controlled movement (where displacement is periodically braked by contraction of the antagonist muscles; Fig. 4.19).

Leverage

Because of problems associated with leverage (Fig. 3.4), it is often convenient to think in terms of the torque (force × leverage) rather than the force developed at a given joint. Leverage is commonly optimal at the midpoint of operation of a joint. Thus, the biceps has an optimal leverage when the elbow is held at 90°. In general, the lever systems of the human body are designed for rapid rather than for powerful movements. To achieve this objective, the distance between the fulcrum and the muscle insertion is kept short relative to the leverage exerted by the mass of the part or any external load. Thus, the olecranon provides a very short lever arm for the triceps; the resultant arm movement is fast, but lacking in

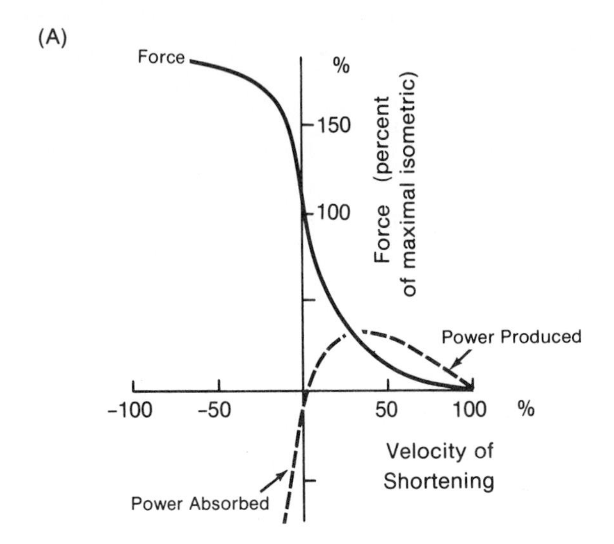

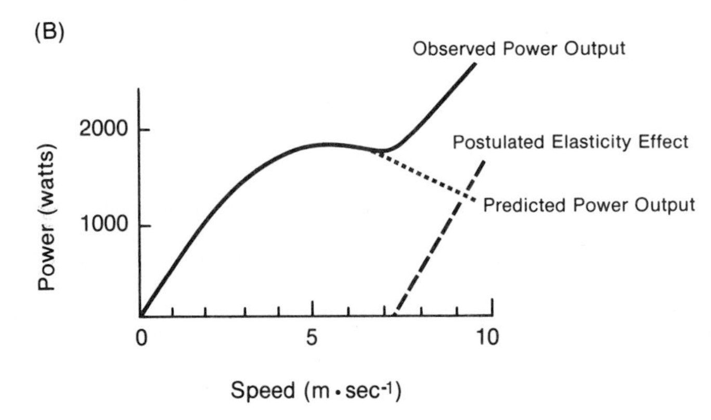

FIG. 4.19. A. Extension of the classical force/velocity curve to show a situation where power is being absorbed by muscle. (After P. O. Åstrand & Rodahl, 1977.) **B.** Power output of extensor muscles of an experienced sprinter as a function of running speed. (After Cavagna, Komarek & Mazzoleni, 1971.)

power. By way of contrast we may look at the ankle joint. Here, the calcaneum provides the gastrocnemius with a substantial lever, since this muscle is often required to lift the entire body mass.

The Athletic Wind-up

We have noted that individual muscle fibers develop their maximum isometric tension when stretched about 20% beyond their resting length (Fig. 1.9). The same principle holds for an entire muscle, and if it is necessary to develop a large force in the course of an athletic event, it would seem reasonable to put the muscle concerned under slight tension. The wind-up movement adopted by many competitors in throwing, jumping, and lifting events (Asmussen and Sorensen, 1971) is based partly on this concept and partly on factors associated with the storage of elastic energy. In practice, it is not always an advantage to tense a muscle. Sometimes the resultant posture results in an unfavorable leverage. The best results are obtained through an intelligent combination of the length/tension diagram (Fig. 1.9) and the principles of leverage (Fig. 3.4). In many events, account must also be taken of the force/velocity relationship (Fig. 3.9). The optimum speed of muscle shortening depends on the relative requirements of power and speed of movement.

Techniques of Lifting

The technique frequently recommended for the lifting of heavy loads has been criticized on the grounds of poor leverage (P. R. Davis et al., 1965). Workers are trained to maintain a rigid spine and to lift by extension of the knees (Chap. 7). However, if the load is lying on the floor or a low platform, the knees must initially be fully flexed, and the average person is unable to develop a sufficient force to lift legally permitted loads (40–50 kg for men, 15–20 kg for women), at least if the worker adheres to the recommended technique. One obvious remedy is to raise the height of the loading platform, thereby improving leverage. Momentum may also be used to carry the load past positions where the spine is under heavy stress (the so-called dynamic technique of lifting; D. Jones, 1969; J. R. Brown, 1971).

Inertial effects are minimized by relatively slow lifting. If an excess of kinetic energy is imparted to a load, effort is expended both in producing the acceleration and also in stopping its motion when the intended displacement has been realized. Minimization of inertia is important not only in the movement of external loads, but also in the displacement of body parts. The avoidance of sudden accelerations and decelerations is an important component of skill.

ASSESSMENT OF MUSCLE PERFORMANCE

Muscle Dimensions

Since there is a general relationship between the cross-section of a muscle and the force developed, it should be possible to form an impression of the strength of an individual from a measurement of muscle dimensions. This approach has the important advantage that little cooperation is required from the subject, and the result obtained is uninfluenced by test learning or motivation.

The simplest procedure is to measure the circumference, either at the broadest region of the muscle belly or at a fixed distance from some convenient bony protuberance. For instance, clinicians commonly measure the circumference of the thigh 12.5 cm above the condyles of the knee. The weakness of this method of assessment is that a broad limb may reflect large muscles, large bones, or an accumulation of fat; scores thus show only a limited correlation with tests of running, throwing, jumping ability, and muscle force (Malina, 1975). A better indication of muscularity is obtained if corrections are applied for the thickness of subcutaneous fat (as estimated by skinfold calipers; Weiner and Lourie, 1969) and the breadth of the long bones (as determined by intercondylar dimensions).

The same type of procedure can be carried out more accurately if soft-tissue radiographs are taken in the posteroanterior and lateral planes (Tanner, 1965; Fried and Shephard, 1970). Precise positioning of the limb with respect to the X-ray machine is important, and it must not be forgotten that the radiographic bulk of a muscle can be increased by edema, intramuscular fat, changes of muscular tone, and other technical factors.

In children, there are moderate correlations (r = 0.4–0.6) between radiographic measures of muscle breadth and isometric strength (Y. Dempsey, 1955; Rarick and Thompson, 1956). In adults, there is a useful correlation between the size of the muscle X-ray shadow and explosive force (Adamson and Cotes, 1967), and some authors (Malina, 1975) have also found significant correlations with isometric force. Presumably, much depends on (1) motivation during the force measurements and (2) the range of data available within a given sample; some laboratories find little or no relationship between the isometric strengths of adults and their lean leg volumes (L. E. Smith and Royce, 1963; Adamson and Cotes, 1967; F. I. Katch and Michael, 1971, Table 4.4). Discrepancies are particularly likely in the period following injury to a limb (Stoboy et al., 1968; Fried and Shephard, 1970); under such conditions, there may be a quite rapid restoration of muscle forces, but a much slower restoration of muscle bulk (Table 4.5).

Ikai and Fukunaga (1968) used ultrasound to determine muscular dimensions. The advantages and limitations of this method are similar to those of soft-tissue radiography. In particular, the dimensions recorded with both techniques are greatly influenced by limb position. Ikai and Fukunaga noted specifically that the cross-sectional area of the arm flexors was increased by 34% when the arm was held in a flexed position.

An overall assessment of muscularity can be obtained by calculating lean body mass. In children, this is quite closely correlated with grip strength (Forbes, 1965). In adults, there is a fair correlation between a measure of lean mass such as body potassium and explosive force (Adamson and Cotes,

TABLE 4.4
Coefficients of Correlation Between Quadriceps Bulk (as Seen in Soft-tissue Radiographs) and Isometric Knee Extension Strength at 115°*

	Laboratory A	Laboratory B
Before rehabilitation	0.10 ± 0.29	0.12 ± 0.29
After rehabilitation	0.34 ± 0.27	0.39 ± 0.27

*Workmen tested before and after rehabilitation for leg injuries.
Source: Based on data of Fried and Shephard, 1969.

TABLE 4.5
Changes in Muscle Force and Muscle Dimensions over the Course of 4–6 Weeks Lower Limb Rehabilitation

	Initial Value	Change	% Change
R. handgrip force (N)	464 ± 64	11 ± 49	2.3
L. handgrip force (N)	446 ± 80	13 ± 38	2.8
Knee extension force			
Injured leg (N)	473 ± 21	59 ± 136	12.5
Uninjured leg (N)	551 ± 20	24 ± 136	4.3
Muscle bulk (injured leg, cm^2)	108.6	3.5	2.3

Source: Based on data of Fried and Shephard, 1969.

1967), but again the correlation with isometric force is disappointingly small (r ~ .5; Laubach and McConville, 1969; Leedy et al., 1965). Part of the problem may lie in the whole-body potassium determination; training decreases "screening" from subcutaneous fat and may increase the potassium content of the lean tissues (Womersley et al., 1976).

Explosive Force

Simple field tests of explosive force are the vertical jump and reach (Shephard, 1978e) and the standing broad jump (AAHPER, 1965; Hayden and Yuhasz, 1966). Scores for both of these tests are greatly influenced by standing height, body mass, motivation, familiarity with the test (Ikai and Steinhaus, 1961; Berger, 1967; B. L. Johnson and Nelson, 1967), and environmental conditions. Espenschade and Meleney (1961) found that over a 24-year period, Californian boys and girls improved their jump and reach scores, but showed a deterioration of standing broad jump results; they attributed the discordant findings to opportunities for practice of the broad jump, present in 1934 but absent in 1958. Despite potential difficulties in the interpretation of field test scores, it was found that vertical jump height had correlations of .64 and .71 with static and dynamic leg strengths, respectively. Laboratory test measurements are based on the force platform (F. Andersen, 1967; C. T. M. Davies, 1971a; Asmussen and Bonde-Petersen, 1974a,b; Ramey, 1975). This was developed originally for studies of normal gait (Elftman, 1938) and prostheses (D. M. Cunningham and Brown, 1952), but has since been applied to analyses of industrial tasks (Lauru, 1957), jumping, running, hurdling, shot putting, and weight lifting (Payne et al., 1968; Cavagna, Komarek, and Mazzoleni, 1971; Ramey, 1972). Force transducers that have been used include (1) combinations of mechanical springs and pointers (Elftman, 1938), (2) crystals that generate a piezoelectric potential when deformed (Lauru, 1957), (3) differential transformers with a sensing core displaced by the applied force (J. H. Greene and Morris, 1958), and (4) strain gauges (D. M. Cunningham and Brown, 1952). One commercially available device (Kristal Instrument Corporation) has a stack of three piezoelectric crystals mounted at right angles to one another at each corner of the force plate. Practical problems of existing systems include (1) resolving the opposing demands of sensitivity to static forces and resonance with rapid movement and (2) eliminating "cross-talk" between sensors for the three axes of movement.

We have already discussed the potential power output of the body in the context of ATP hydrolysis. Typical forces developed during a vertical jump amount to 1.7 kN in a young man and 1.3 kN in a young woman.

Isometric Force

If several joints are involved, it may take 3–4 sec to develop a true maximum isometric force (Kroemer and Howard, 1970). Isometric force is thus best defined as a contraction that can be sustained for at least 3 sec (Chaffin, 1975). The strength of individual muscle groups is conveniently measured by a series of *dynamometers* and *tensiometers* (Hunsicker and Donelly, 1955). When using a dynamometer, the subject exerts maximum force (for example, handgrip, back lift, or leg extension) against a strong spring-loaded plate (D. K. Mathews, 1963), and displacement of the plate is recorded either electrically (by means of a strain gauge) or mechanically (by the movement of a unidirectional pointer over a suitable scale). The mechanical systems are simpler and more rugged, but electrical devices (Asmussen et al., 1959; Bäcklund and Nordgren, 1968; Höök and Tornvall, 1969) have the advantage that a contraction can be more completely isometric, a smaller movement of the muscle being necessary to produce a gauge reading. A continuous recording of contraction force can also be obtained from an electrical strain gauge, thus reducing the possibility of errors from undetected tremors or jerking movements.

The usual arrangement with a tensiometer is that one end of a stout steel cable is attached to the body by a leather harness, while the other end of the cable is hooked to a rigid support. The cable is tensed by action of the muscle group under study, and its tension is determined either electrically (by attaching a strain gauge) or mechanically (by fitting around the cable a "tensiometer" originally designed to measure the tension in aircraft control cables). The reading obtained is essentially an expression of torque, the recorded force depending on the angulation of the cable to the limb and the distance separating the cable harness from the axis of rotation of the joint (Fig. 3.4).

Both dynamometer and tensiometer readings are greatly dependent on the conditions of measurement, for example, the precise angulation of the joint as the tension is developed, the extent of immobilization of other body segments, jerking and inertial effects, the motivation of the subject, and the urging of the investigator. Results for some muscle groups show a poor reproducibility (Tornvall, 1963; Simonson, 1971), with considerable test learning (Table 4.6; Shephard, Lavallée et al., 1977d; 1978b) Because of such difficulties and the number of muscle groups available for study, some authors content themselves with measuring the handgrip force as an index of general muscularity (Bookwalter, 1950). If the separation of the grip-plates is adjusted to a convenient length for the palm of the subject, the dominant hand is chosen, and several practice attempts are allowed, the handgrip dynamometer gives reasonably consistent answers. Obviously, it is not representative of general muscularity in professions that call for special development of the wrist muscles (such as an anesthetist or a tennis player), and some authors find that grip strength bears little relationship to forces that can be developed elsewhere in the body (Asmussen, Hansen, and Lammert, 1965; O. Lambert, 1965). However, H. H. Clarke (1966) reported that in the ordinary sedentary population there is a correlation of 0.69 between handgrip force and general strength as assessed from tensiometer readings on a number of other muscle groups.

Data for Toronto children are summarized in Table 4.7. With repeated maximum isometric contractions, there is a progressive weakening of the tension that can be developed. Molbech (1963) found that after 2–6 min, efforts settled to a steady-state that could be sustained more or less indefinitely. At a frequency of 6 contractions per min, the steady state was 85% of maximum isometric force, and at 29 contractions per min it dropped to 60% of maximum force.

Isometric Endurance

Some field tests, such as the flexed arm hang (Hayden and Yuhasz, 1966), provide an index of isometric endurance. As a simple clinical test, Mertens et al. (1978) suggested timing the period for which the legs could be held in the horizontal plane when a standard mass was strapped to the ankles. Unfortunately, scores for such tests are strongly influenced by motivation and by the mass of the body part itself.

Laboratory tests are commonly based on the time for which a subject can hold a dynamometer or tensiometer at a predetermined reading. Exhaustion is rarely clear-cut, and the observer must therefore establish an objective criterion, for instance, the mo-

TABLE 4.6
Reproducibility of Muscle Force Measurements[a]

Muscle Group	Boys			Girls		
	Visit 1	Visit 2	Visit 3	Visit 1	Visit 2	Visit 3
Elbow flexion	195	192	201	174	177	186
Shoulder flexion	120	120	120	101	101	104
Hip flexion	218	229	206	187	204	208
Knee flexion	174	180	172	158	159	164
extension	209	203	191	208	213	214
Handgrip	159	148	150	131	124	131
Leg force (dynamometer)	134	149	141	97	138	142*
Back force (dynamometer)	53	60	60	38	52	55*

[a]Nine-year-old children without previous experience of tensiometry or dynamometry performed a series of three tests, each separated by several days. Results expressed in N. Significant learning effects are marked by asterisks.

Source: Based on data of Shephard, Lavallée et al., 1977d,1978b.

TABLE 4.7
Mean Values ($\bar{x}$), Coefficients of Variation (cv), and Coefficients of Correlation (r) Between Individual Measurements of Muscle Force and the Summated Strength Index

	Boys			Girls		
Muscle Group	$\bar{x}$ (N)	cv (%)	r	$\bar{x}$ (N)	cv (%)	r
Handgrip	211	21.0	0.61	193	28.1	0.90
Arm extension	181	26.9	0.84	150	34.7	0.85
Arm flexion	147	27.7	0.83	129	31.3	0.60
Trunk flexion	173	38.2	0.75	187	33.6	0.81
Trunk extension	193	31.2	0.80	177	40.9	0.80
Leg extension	381	27.6	0.70	379	33.4	0.90

Source: Based on data of Shephard, Allen et al. (1968c) for children aged 10–12 years.

ment when the subject first allows the gauge reading to drop 20 N below the required value.

Isotonic Force

Practical problems in measuring isotonic force include the need to control posture and loss of energy in the acceleration of body parts. Since there is a fair correlation (r ~ 0.8) between isotonic and isometric data (Fig. 4.20), the isotonic value is often inferred from an isometric measurement (Rasch and Pierson, 1963; Lammert, 1963; Asmussen, Hansen, and Lammert, 1965; Doss and Karpovich, 1965).

Performance tests such as speed sit-ups and pull-ups (F. R. Rogers, 1926; AAHPER, 1965; Hayden and Yuhasz, 1966) combine elements of both isotonic muscle force and isotonic endurance.

Laboratory measurements of isotonic force have traditionally been made by the tedious process of adjusting the loadings of pulley systems. Using such an approach, DeLorme and Watkins (1948) defined isotonic strength as the maximum load that could be lifted 10 times—a criterion that includes elements of force and endurance. Unfortunately, the trial and error technique used in such measurements inevitably leads to fatigue before the limiting value has been discovered. Further, even if only a single lift is attempted, the result is inevitably affected by the inertia of the load and the speed of contraction (Fig. 3.9). If several lifts are made in succession, the score becomes a measure of working capacity rather than isotonic force.

More elaborate torque generators (Thistle et al., 1967; Perrine, 1968; Thorstenson, 1976) now allow the measurement of isotonic strength under constant speed (isokinetic) conditions. Forces can be measured not only during shortening, but also when the muscles are extended beyond their resting length (Fig. 4.19). The main problems with this type of equipment are its expense and practical difficulties in aligning the measuring instrument with the axis of rotation of the joint under study.

Josenhans (1967b) suggested the possibility of determining a "one-repetition" maximal contraction by using an accelerometer to measure the force applied to a heavy external mass. The main difficulty with this approach is that an arbitrary allowance must be made for the inertia of the subject's arm.

Isotonic Endurance

Isotonic endurance is usually determined by a pulley system, the subject being required to lift a load that demands a known fraction of his maximum isotonic force. Endurance is expressed not as a time, but rather as the number of repetitions of a slow lift that can be undertaken. Other measures of isotonic strength and endurance examine the maximum amount of work that can be accomplished on a bicycle ergometer within a fixed time (30 sec, InBar and Bar-Or, 1977; or 45 sec, Edgren and Borg, 1975) and the number of times such work can be repeated (Edgren and Borg, 1975).

Quantitative Electromyography

Whether isometric or isotonic strength is to be measured, maximum effort tests are not always desirable. Performance may be limited by psychological rather than by physiological factors, especially in cases that are being assessed for compensation following industrial injury. The rise of blood pressure that accompanies maximum effort may also be dangerous for elderly patients and those recovering from myocardial infarction. We have noted previously (Fig. 3.8)

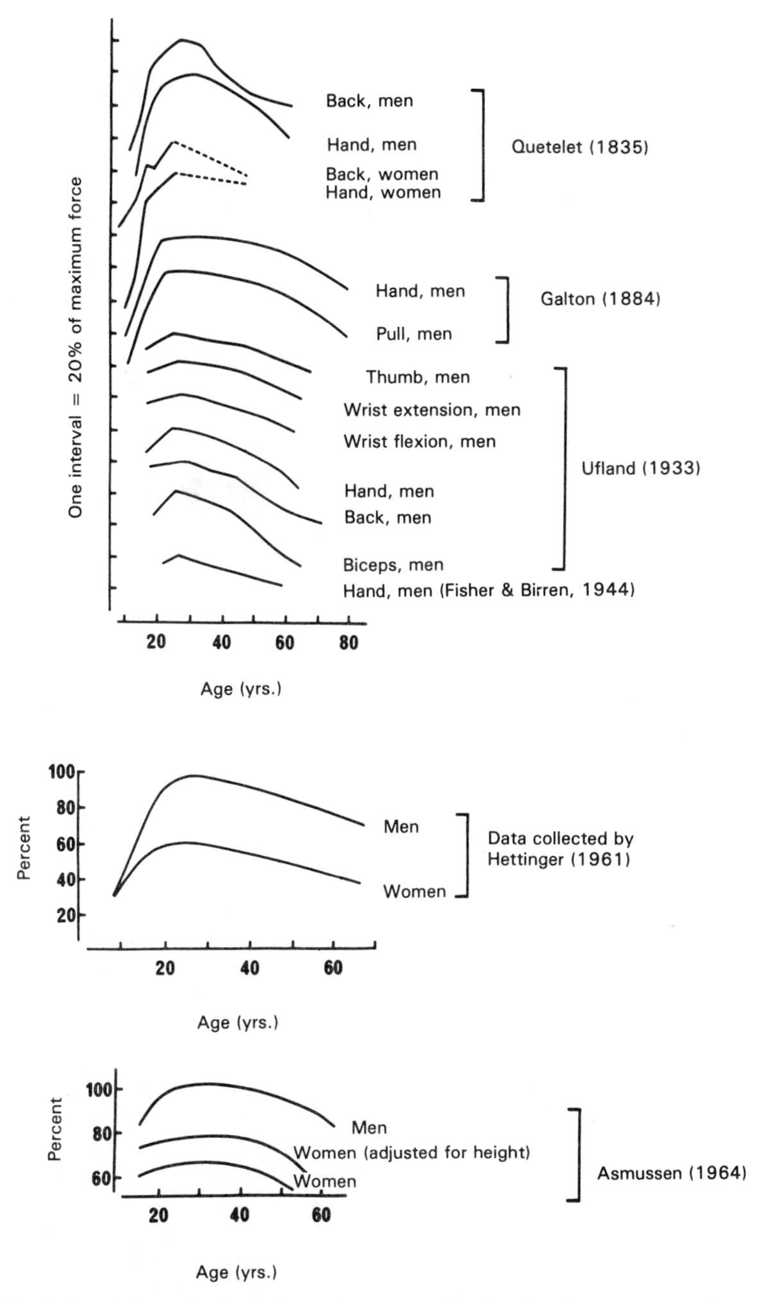

FIG. 4.20. Aging curves of selected muscle groups. (For details of sources, see Shephard, 1978b.)

the possibility of linking tensiometry with quantitative electromyography. Malingering may then be detected from the association between a poor muscle force and limited electrical activity.

The EMG is basically a display of action potential recorded over (plate electrodes) or within (needle electrodes) a given muscle belly (Basmajian, 1967). Needle electrodes provide information specific to a small segment of muscle; they are invaluable to the kinesiologist who wishes to identify the fibers active in a particular body movement. Plate electrodes give a better idea of the total muscular activity about a given joint, although results can be distorted by (1) an association between the electrical impedance of a muscle and its cross-section and (2) changes in the relative contributions of deep and superficial muscles to the recorded signal. For purposes of quantification, one may determine the mean height of the EMG, the total number of action potentials per sec as counted by a suitable scaling unit, the number of action potentials exceeding a certain threshold voltage, or the root mean square of the signal voltage (Haas, 1926; Ralston et al., 1947; J. L. Kennedy and Travis, 1947). There is a close relationship between tension and the amount of electrical activity at most intensities of submaximal effort (Bigland and Lippold, 1954a; Bigland-Ritchie and Woods, 1974), although there have been reports of an exponential increase in EMG activity with efforts requiring more than 90% of maximum voluntary contraction (Kuroda et al., 1970). The electrical activity associated with development of a given external tension varies directly with the strength of the subject (Fig. 3.8).

During maximal isometric efforts, the integrated EMG is independent of the speed of contraction (Komi, 1973; Rodgers and Berger, 1974); however, as would be predicted from oxygen costs, the electrical activity for a given tension is much greater in concentric than in eccentric activity (Bigland and Lippold, 1954a; Fig. 3.8). During isotonic contractions, there is a relationship between the total electrical activity (integrated over time) and the work performed (Bouisset, 1973; Bouisset and Goubel, 1973).

NORMAL VALUES

Influence of Laterality

There is remarkably little asymmetry of muscular strength. Heebøll-Nielsen (1964) commented that differences between the two arms increased until early adulthood, but thereafter remained constant at 5–11% for different muscle groups. Singh and Karpovich (1968) were unable to detect any differences of forearm strength between the preferred and the nonpreferred arms, while Hettinger and Hollman (1969) found no difference of forces between right and left arms.

Variations with Age and Sex

We have noted that strength is closely related to the cross-section of a muscle. If due account is taken of leverage, the force developed is 40–70 N • cm^{-2} (Ikai and Fukunaga, 1968; Howells and Jordan, 1978). On theoretical grounds, this relationship should hold for individuals of differing size, be they men and women or children and young adults; dimensional arguments suggest that muscle force should increase as the square of linear measurements such as standing height.

In practice, the strength of children increases faster than would be predicted from the square of standing height, a typical exponent being 2.9. In the boys, there is a particularly large increase of muscle force at puberty (M. B. Fisher and Birren, 1948; H. E. Jones, 1949; Asmussen, Heebøll-Nielsen, and Molbech, 1959; Asmussen, 1973a; H. H. Clarke, 1966). On the other hand, the strength of the girls reaches a plateau at the age of 12 or 13, and thereafter shows a slight decline. The sex difference is probably explicable in terms of differences of hormonal development and motivation; the pubertal strength spurt of the boys depends more on these factors than on an improvement of coordination (as postulated by Asmussen and Heebøll-Nielsen, 1955).

The European men studied by Ufland (1933) and Asmussen and Heebøll-Nielsen (1961; Fig. 4.20) grew in strength until they reached the age of 25–30 years, with a subsequent 15–20% decline to the age of 60. In North America, the average man reaches a peak of strength rather earlier (about 17 years of age), thereafter maintaining relatively constant readings to ~45 years of age.

The typical adult woman (Fig. 4.21) has 55–65% the isometric strength of a man (C. B. Morris, 1948; Asmussen and Heebøll-Nielsen, 1961; Hettinger and Hollmann, 1969), although wide discrepancies of sex differential can be found in data from different laboratories (Laubach, 1976). Laubach noted that the average of 13 reports for the arm muscles showed women with 55.8% of the male strength, while corresponding figures for trunk strength, leg strength, and dynamic strength were 41.9, 63.8 and 68.6%, respectively. Applying the dimensional analysis of Asmussen and Heebøll-Nielsen (1961) and Asmussen (1964), one would conclude that about a half of the relative weakness of the female is attributable to a shorter height. In support of this view, Costill, Daniels et al. (1976) noted some differences of fiber size between men and women; fast-twitch fibers averaged 85% and slow-twitch fibers 70% of the male cross-section in their female subjects. However, there were wide interindividual differences of microstructure, the fitter women having much larger fibers than the weaker men. In contrast to these various findings, Ikai and Fukunaga (1968) made ultrasound measurements of muscle dimensions and

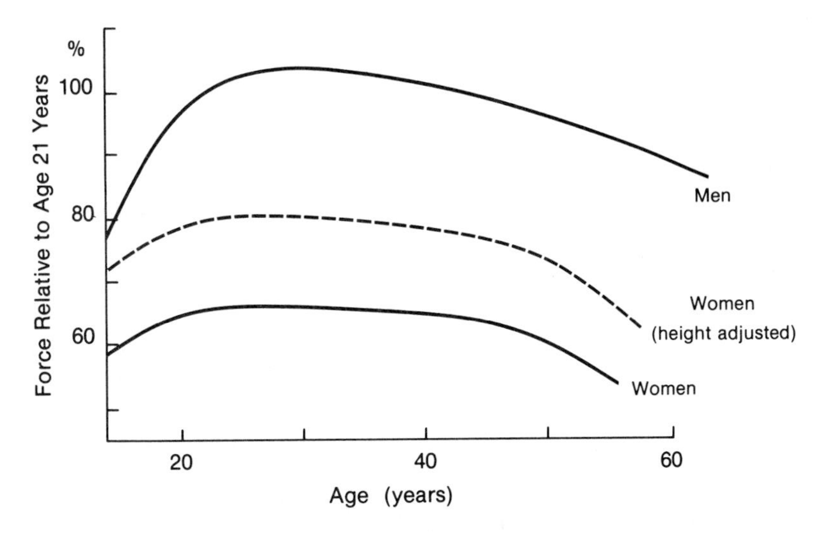

FIG. 4.21. Curves showing effect of age on muscle force of Danish men and women, and of women after adjustment for the shorter height of the latter. (After Asmussen & Heebøll-Nielsen, 1961.)

leverage in adolescents of both sexes and concluded that the maximum isometric force developed by the elbow flexors was a fixed proportion of cross-section, irrespective of the sex of the individual.

Body Build

Because muscle strength is a power function of stature, we might anticipate that tall people would be stronger than those who are short. In practice, this is not always the case, as tall people tend to have a lean, ectomorphic body build, with rather poor muscular development. Muscle strength is greatest in thick-set individuals with a mesomorphic body build.

In assessing the practical significance of the strength recorded by a tensiometer or a dynamometer, account must also be taken of (1) body mass (which influences the forces that must be developed in most tasks) and (2) differences of leverage afforded by long and short limbs.

Temporal Factors

Grip strength apparently shows a diurnal rhythm, being greatest during waking hours and least in the small hours of the night (Kleitman, 1963). Body temperature shows a rather similar diurnal pattern, and it is tempting to ascribe the changes of measured strength to alterations of muscular viscosity accompanying a warm-up of the tissues. Certainly, grip strength can be increased by immersion of the arm in hot water, and it is diminished by immersion in cold water. However, other factors such as variations in cortical arousal probably contribute to diurnal changes of strength.

Psychological Factors

We have noted the problem that the observed isometric strength is limited more by psychological than by physiological factors. Psychological factors probably account also for a fair part of the difference in readings between the sexes. Steinhaus (1963) demonstrated that the recorded strength could be increased 5–10% by various forms of encouragement and that even larger gains (~25%) could be achieved if central inhibition was relieved by hypnosis. In contrast, hypnotic suggestion of weakness reduced the recorded force by up to 30%.

Normal Standards

Normal values for the jump and reach test are fairly well established (Table 4.8). The explosive force developed in various types of jump depends greatly upon the percentage of the body musculature that is activated; it can further be augmented by a wind-up.

Various authors have published norms of isometric force (Table 4.9; Fig. 4.20). The reported results for some muscle groups show substantial differences from one laboratory to another, scores depending on the completeness of immobilization of accessory muscles and details of the measuring

TABLE 4.8
Normal Scores for Jump and Reach Test [a]

Percentile of Population	Age (Male Subjects, Years) 10	11	12	13	14	15	16	17	18–29	30–39
≥99.9	40.6	42.4	48.3	58.4	68.1	77.2	79.2	80.3	81.3	76.7
99.4–99.8	38.1	40.1	45.5	54.6	63.5	72.4	74.2	75.2	76.2	71.6
97–99.3	35.6	37.3	42.4	50.5	58.4	66.3	68.1	69.1	69.9	66.0
90–96	33.0	34.5	39.1	46.7	53.8	60.7	62.0	63.0	63.8	59.9
80–89	31.2	32.8	36.8	43.7	50.3	56.6	57.9	58.9	59.7	58.2
65–79	29.5	30.5	34.3	40.6	46.7	51.8	53.1	53.8	54.6	52.3
35–64	26.4	26.9	30.2	35.3	40.1	44.7	45.5	46.2	47.0	44.7
20–34	24.6	24.6	27.7	32.0	36.1	39.9	40.6	41.4	41.9	39.6
10–19	22.9	22.9	25.4	29.0	32.5	35.8	36.6	37.3	37.8	35.6
4– 9	20.3	20.1	22.1	24.9	27.7	30.2	30.7	31.2	31.8	30.2
≤3	20.1 or less	19.8 or less	21.8 or less	24.6 or less	27.4 or less	30.0 or less	30.5 or less	31.0 or less	31.5 or less	30.0 or less

Percentile of Population	Age (Female Subjects, Years) 10	11	12	13	14	15	16	17	18–39
≥99.0	43.2	47.5	50.8	55.9	62.2	67.3	68.6	68.6	68.6
99.4–99.8	40.1	43.7	47.0	51.3	57.9	62.2	63.5	63.5	63.5
97–99.3	36.6	39.4	42.7	47.0	52.6	56.1	57.4	57.4	57.4
90–96	32.8	35.1	38.4	42.7	47.2	49.8	51.1	51.1	51.1
80–89	30.2	31.8	35.1	39.4	43.7	45.7	47.0	47.0	47.0
65–79	27.4	28.4	31.8	36.1	39.6	40.6	41.9	41.9	41.9
35–64	22.9	22.9	26.2	30.5	33.0	33.0	34.3	34.3	34.3
20–34	20.1	20.1	23.4	27.7	29.0	27.9	29.2	29.2	29.2
10–19	17.5	17.5	20.8	25.1	25.4	23.9	25.1	25.1	25.1
4– 9	13.7	13.7	17.0	21.3	20.1	17.5	18.8	18.8	18.8
≤3	13.5 or less	13.5 or less	16.8 or less	21.1 or less	19.8 or less	17.3 or less	18.5 or less	18.5 or less	18.5 or less

[a]Values (cm) shown are minima for attaining a given percentile range.
Source: After Shephard, 1975a.

equipment. Nevertheless, it is possible to infer the general form of the growth and aging curves for both sexes (Asmussen and Heebøll-Nielsen, 1961). Handgrip force is one measurement whereby interlaboratory comparison seems possible. Many of the world's laboratories have tested this variable on similar dynamometers. The results apparently show a superiority of strength in Scandinavian subjects relative to their North American counterparts (Shephard, 1978a). However, it is dangerous to read too much into such comparisons until true random samples have been drawn from the various nations of the world.

Isometric endurance was discussed earlier in this chapter. Rohmert (1968a) made an extensive study of this question; he found that if data were expressed as a percentage of maximum isometric force, the endurance time was remarkably constant from one muscle group to another, irrespective of limb position, age, or sex of the subject (Fig. 3.13).

Little is known about normal values for isotonic force and endurance. Population studies have been carried out using simple field tests such as speed sit-ups (Table 4.10) and the flexed arm hang, but accurate objective standards must await the widespread use of isokinetic equipment.

SKELETAL SUPPORT SYSTEMS

The effectiveness of the muscular energy transducers ultimately depends not only upon their ability to generate substantial forces, but also upon the transmission of such forces through tendons and joints to the supporting structures of a rigid bone or cartilaginous skeleton.

Ligaments and Tendons

Ligaments and tendons are formed of white fibrous tissue, primarily long parallel fibers of collagen. The

TABLE 4.9
Muscle Force (N) Exerted by Danish Men (175 cm height) and Women (160 cm height) at Age 25 and 65

Movement	Men		Women		Movement	Men		Women	
	25	65	25	65		25	65	25	65
Trunk extension	857	752	538	453	Hip flexion	642	512	403	327
flexion	630	552	389	310	extension	499	386	314	236
Handgrip	588	455	355	301	abduction	459	362	321	281
Horizontal pull	482	421	274	247	adduction	538	468	338	262
push	323	294	191	166	Knee extension*	1715	1298	1122	883
Vertical pull (down)	563	523	333	299	flexion*	1388	983	894	633
Leg extension	304	213	208	149	Plantar flexion of foot*	1136	844	866	539
					Dorsiflexion of foot*	563	477	356	330

*Values cited as torque, kg • cm.
Source: Based on data for selected muscle groups, collected by Asmussen and Heebøll-Nielsen, 1961.

ligaments strengthen joint capsules by extending between adjacent bones, while the tendons connect muscle to bone. At the bone surface, the collagen merges with the fibrocartilage of its insertion, and according to some authors a proportion of the fibers continue into the core of the bone. Within the muscles, the collagen fibers of the tendons blend imperceptibly with the perimysium and endomysium.

The maximal tensile strength of a tendon is ~4 times the isometric strength of the associated muscle (Harkness, 1968). If a tendon is stretched forcibly, it may be lengthened permanently by as much as 4%, but it is doubtful as to whether such treatment weakens it significantly. Injuries usually occur at the point of insertion into bone (Viidik, 1972).

The careful animal experiments of Tipton and his associates have shown the effects of disuse and of training upon the strength of tendon insertions (Vailas et al., 1978a,b). Six to twelve weeks of disuse leads to a surface scalloping of the affected bone, infiltration by osteoclasts and fibroblasts (Laros et al., 1971), and a weakening of ligament attachments. Although the effects are most obvious after complete immobilization, confined caging of dogs is sufficient to initiate such changes. The affected ligaments show a thinning of the fiber bundles (Tipton et al., 1971), a decreased capillary volume (Rothman and Slogoff, 1967), greater extensibility (Tipton et al., 1971), and a possible reduction of collagen cross-linkages (Peacock, 1963), apparently without change of hydroxyproline concentration (Tipton, Tcheng, and Mergner, 1971). Conversely, training increases the strength of tendons (Viidik, 1967a; Kiiskinen and Heikkinen, 1975) and ligaments (Viidik, 1968). Structural changes include an increase of mucopolysaccharide and hydroxyproline content (Kiiskinen and Heikkinnen, 1975), with an increase of cellular nuclei and the thickness of collagen fibrils (Ingelmark, 1948), but a decrease in the number of collagen cross-linkages (Viidik, 1973). If a tendon has been injured, the rate of healing is enhanced by training (Tipton, Tcheng, and Mergner, 1971). However, it is less clearly established that physical activity increases the strength of bone attachments (Viidik, 1969; Zuckerman and Stull, 1973; Tipton, Matthes et al., 1975); conceivably the added movement does no more than reverse the harmful effects of caging.

The sensory innervation of tendons is in the form of Golgi organs, simple sprays of fine nerve terminals widely distributed throughout the connective tissue. Traditionally thought to have a high threshold, the Golgi receptors are nevertheless stimulated by a substantial increase of length, such as accompanies a general rise of static tension within a tendon. Through central connections in the gray matter of the spinal cord, the Golgi organs can initiate a relaxation of the corresponding muscle. This "antimyotatic reflex" has a protective function, reducing the risk of injury to both tendon and muscle.

Joints

There are three main categories of joint. The *fibrous joints*, such as those in the skull, allow almost no movement between adjacent bones *Cartilaginous joints*, such as the articulations of the vertebral column, are connected by a plate of white *fibrocartilage*; this can be deformed, allowing a modest range of movement. In *synovial joints*, the opposing bones are

TABLE 4.10
Normal Scores for Speed Sit-ups (Bent-Knee)*

Percentile of Population	Age (Male Subjects, Years) 10	11	12	13	14	15	16	17	18–29	30–39
≥99.9	50	55	55	56	57	58	59	59	59	53
99.4–99.8	47	52	51	53	54	55	55	55	56	52
97–99.3	43	48	46	49	51	51	51	50	53	50
90–96	39	42	42	45	45	47	47	46	50	45
80–89	35	38	38	41	41	43	43	42	45	41
65–79	31	34	33	37	36	38	39	38	42	36
35–64	23	26	26	29	28	29	31	31	33	29
20–34	19	22	21	23	24	23	25	25	30	25
10–19	14	18	18	19	20	19	21	19	26	20
4– 9	9	13	12	14	17	15	17	15	22	15
≤3	8 or less	12 or less	11 or less	13 or less	16 or less	14 or less	16 or less	14 or less	21 or less	14 or less

Percentile of Population	Age (Female Subjects, Years) 10	11	12	13	14	15	16	17	18–39
≥99.9	50	60	55	46	44	50	46	42	42
99.4–99.8	46	52	48	43	41	46	43	39	39
97–99.3	41	43	40	38	37	40	39	36	36
90–96	35	37	36	34	33	36	36	32	32
80–89	31	33	32	31	28	30	31	28	28
65–79	26	29	26	26	24	25	26	23	23
35–64	19	20	19	19	16	18	17	15	15
20–34	14	16	13	14	10	12	14	10	10
10–19	7	11	9	9	5	9	10	6	6
4– 9	2	5	4	4	4	5	5	3	3
≤3	1 or less	4 or less	3 or less	3 or less	3 or less	4 or less	4 or less	2 or less	2 or less

*Values shown are minima for attaining a given percentile range.
Source: After Shephard, 1975a.

separated by a fluid-filled synovial space; depending on the shape of the bones and the surrounding tissues, a wide range of movement is then possible.

White fibrocartilage is found in the attachment of ligaments and tendons to bone, in the intervetebral discs, and in the "cartilage" of the knee joints. It consists of a mixture of fibrous tissue and cartilage. The fibers of articular cartilage are characteristically arranged at right angles to the imposed stress. Because of this histological arrangement, the tension resistance is ~2 N • mm^{-2}, while resistance to compression is almost 10 times as great. The articular cartilages are repeatedly deformed during activity but normally regain their shape with rest. If other supporting mechanisms of the joint fail due to fatigue, previous injury, poor coordination, excessive external stress, or a combination of such factors, the deformation of the cartilage may exceed its elastic limit, and with many repetitions of the stress tearing may occur (Jenkins and Weightman, 1978; Fig. 4.22).

The articular surfaces of bones are covered by a layer of *hyaline cartilage*. This consists of an interlacing fabric of fine fibrils embedded in the mucoproteins chondroitin-4-sulfate and chondroitin-6-sulfate. It can be deformed, absorbing forces, and it plays a major role in ensuring smooth and relatively friction-free movement about a joint's axis of rotation.

Vigorous physical activity leads to an increase in the thickness of the cartilaginous layer, 12–13% within 10 min (Ingelmark and Ekholm, 1948). This has been attributed to a seeping of fluid into the cartilage from the underlying bone marrow. It helps to increase the contact area, and thus reduce unit pressure at the joint—a valuable by-product of a preexercise warm-up. Habitual activity also increases the thickness of articular cartilage, particularly in regions where it is initially quite thick (Holmdahl and Ingel-

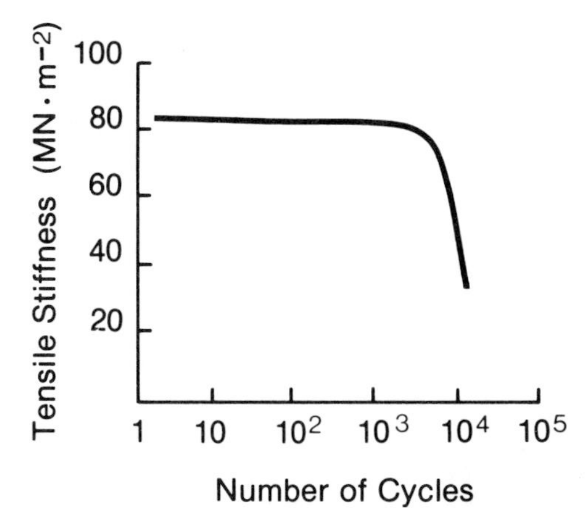

FIG. 4.22. Effect of repeated stress on the tensile stiffness of cartilage from the femoral head. (After Jenkins & Weightman, 1978.)

mark, 1948). There is some increase of cellular components and a 4 times larger increase of extracellular components. There is little evidence that chronic activity in itself is harmful to a healthy joint (Lanier, 1946), but trauma can cause a loss of the entire thickness of articular cartilage. Because an athlete is at greater risk of trauma, the ultimate condition of his joints may be worse than those of a sedentary counterpart. If bone is exposed through injury, it undergoes an abnormal proliferation (osteoarthritis), leading to stiff and painful joints. Most elderly people show some radiographic evidence of osteoarthritis, although the symptoms vary widely from one individual to another (see Shepard, 1978b).

The *synovial fluid* is secreted by the synovial membrane that lines the joint capsule. Normally, it forms a very thin viscous film that acts as a lubricant and shock absorber, but if the joint is injured or inflamed it may accumulate in excessive amounts. Four types of nerve endings are recognized in and around joints (Brodal, 1972). The first three types have specialized end organs. Type I receptors apparently indicate joint position and type II speed of movement, both being located in the capsule of the joint. Type III receptors register joint position and are found in the ligaments of a joint, while type IV receptors consist of freely branching pain-sensitive nerve endings. The extent of innervation varies from one joint to another, and at some articulations more reliance is placed upon muscle than upon joint receptors. However, the muscle receptors are concerned mainly with automatic movements, while the joint receptors play a larger role in controlling voluntary movements.

Bone

Bone consists of an organic phase (collagen fibers embedded in mucopolysaccharides, the glycosaminolysans) and an inorganic phase [apatite, crystals of $Ca_{10}(PO_4)_6(OH)_2$]; these components are combined to form a strong quasicrystalline matrix. The internal architecture of most bones presents an elegant array of reinforcing bars and braces that gives great strength with minimal mass. One author estimated the resistance of compact bone to tension, compression, and shearing forces at 12.2, 16.6, and 4.9 kN • cm^{-2}, respectively (Goss, 1973). Unlike most engineering structures, bone shows relatively little deterioration with age because remodeling occurs continuously. Studies with radioactive tracers indicate that at least 20% of the calcium and phosphorus content of bone has a short half-life (LeBlond and Greulich, 1956). One advantage of this lability is that the architecture of the bone can be adapted to meet changing patterns of daily activity.

The ratio of organic to inorganic matter in a typical bone decreases with age, being about 1:1 in a child, 1:4 in a young adult, and 1:7 in an old person. In consequence, bones become progressively more brittle as a person gets older. Calcium loss may also reduce the strength of bones in the elderly (see below). During the period of growth, the influence of physical activity upon the skeleton apparently depends upon the intensity of activity that is undertaken. Moderate effort leads to the development of longer and heavier bones (Saville and Whyte, 1969; Kiiskinen and Heikkinen, 1973,1975), while intense activity produces a shorter and lighter architecture (Kiiskinen and Heikkinen, 1973; Tipton, Matthes, and Maynard, 1972). Presumably, moderate compression of the epiphyses stimulates growth, but excessive pressure has an inhibitory effect (Steinhaus, 1933; Shephard, Lavallée et al., 1978a).

Regular light activity has no influence on the density of bones, but heavy activity encourages the formation of a dense cortex (Saville and Smith, 1966; King and Pengelly, 1973). Any increase of bone density reflects increases in both calcium (Heaney, 1962; Kiiskinen and Heikkinen, 1973; E. L. Smith and Reddan, 1976) and hydroxyproline content (Kiiskinen and Heikkinen, 1975; Chvapil, 1967). However, it is not yet agreed whether the latter reflects an enhanced collagen synthesis, a slower breakdown of existing collagen, or a combination of these two mechanisms (Booth and Gould, 1975). An increase of bone density increases mechanical strength (Saville and Smith, 1966; Kiiskinen and Heikkinen, 1975). In the event of fracture, the healing callus is apparently more rapidly impregnated with calcium and collagen in a previously active person than in a sedentary individual (Heikkinen et al., 1974).

Maintenance of a normal bone composition seems dependent on continued weight-bearing activity. Possibly, the piezoelectric potentials developed thereby regulate bone breakdown and resynthesis (Bassett, 1972). A progressive loss of both calcium and collagen occurs with bed rest, space travel, the sedentary habits of old age, and clinical problems such as muscular dystrophy, paraplegia, acute anterior poliomyelitis, and limb denervation (Hattner and McMillan, 1968; Mattsson, 1972). Usually, there is a close correlation between bone loss and atrophy of the related muscles (Gillespie, 1954; F. Doyle et al., 1970). Physical disuse is associated with an increased number of osteoclasts in the bone (Geiser and Trueta, 1958), and although the rate of bone formation is at least normal, the rate of resorption may rise to 2–3 times the usual figure (Heaney, 1962), often with an increase in the vascularity of the tissue (Geiser and Trueta, 1958). The strength of the affected bones is diminished, although it is not yet agreed as to whether elasticity is reduced (Gillespie, 1954; Haike et al., 1967).

Studies of subjects over 50 years of age have shown that increased weight bearing augments both the thickness and the mineral content of the bones (P. J. Atkinson et al., 1962; Eisenberg and Gordan, 1961; Sidney et al., 1977). Some authors have found that bedridden subjects benefit from the regular use of an ergometer, but most investigators now agree on the need for the specific longitudinal stresses of weight bearing to avoid a deterioration of bone structure (M. Jansen, 1920; Rodahl, Birkhead et al., 1966; Birge and Whedon, 1968; Rummell, Sawin et al., 1975).

CHAPTER 5

Cardiorespiratory Transport
1. The Respiratory System

BOTTLENECKS OF CELLULAR HOMEOSTASIS

The muscle cell can sustain brief bursts of physical activity by calling upon intracellular reserves of oxygen and food products. A limited accumulation of carbon dioxide, anerobic metabolites, and heat is also possible. However, continued release of energy depends upon an efficient matching of local metabolism with extramuscular mechanisms of homeostasis.

The replenishment of muscle nutrients hinges largely upon such factors as the mobilization of depot fat and glycogen, the ability of the liver to synthesize glucose, and the rate of transport of fat and glucose across cell membranes. The elimination of lactic acid also depends largely upon diffusion into and out of the vascular compartment.

In contrast, the exchange of oxygen and carbon dioxide is limited mainly by the power of the cardiorespiratory system. This may be represented schematically by a series of four conductances (Fig. 5.1), corresponding to alveolar ventilation, the interaction between pulmonary diffusion and blood flow, blood transport, and the interaction between tissue diffusion and blood flow, respectively (Shephard, 1971a;1977a). In most circumstances, the second and fourth conductances offer little impedance to gas transport. The overall cardiorespiratory conductance ($\dot{G}$) can thus be determined from the two remaining terms, representing alveolar ventilation ($\dot{V}_A$) plus the product of cardiac output ($\dot{Q}$) and a blood solubility factor λ:

$$\frac{1}{\dot{G}} = \frac{1}{\dot{V}_A} + \frac{1}{\lambda\dot{Q}} \qquad (5.1)$$

If a sedentary young man is performing maximum exercise, alveolar ventilation would be about 80 l • min^{-1} STPD,* with a corresponding cardiac

*When using the conductance equation, all values must be expressed under standard conditions of temperature and pressure, dry gas (STPD).

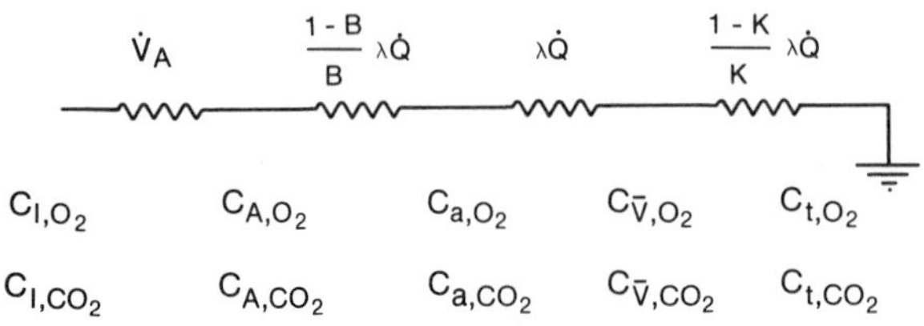

$\dot{V}_A$ $\frac{1-B}{B}\lambda\dot{Q}$ $\lambda\dot{Q}$ $\frac{1-K}{K}\lambda\dot{Q}$

C_{I,O_2} C_{A,O_2} C_{a,O_2} $C_{\bar{v},O_2}$ C_{t,O_2}

C_{I,CO_2} C_{A,CO_2} C_{a,CO_2} $C_{\bar{v},CO_2}$ C_{t,CO_2}

FIG. 5.1. Series conductances determining gas exchange between the tissues and the external environment. $\dot{V}_A$, alveolar ventilation; $B = e^{-\int \frac{1}{\lambda \dot{Q}} \dot{D}_L}$, where $\dot{D}_L$ = pulmonary diffusing capacity; λ, blood solubility factor; $\dot{Q}$ = cardiac output; K, $e^{-\int \frac{1}{\lambda \dot{Q}} \dot{D}_t}$, where D_t = tissue diffusing capacity; C, gas concentration (I, inspired; A, alveolar; a, arterial; $\bar{v}$, mixed venous; t, tissue)

output of ~25 l • min⁻¹. Solubility factors are normally ~1.2 for oxygen and 5 for carbon dioxide, so that we may write

$$\frac{I}{\dot{G}_{O_2}} = \frac{1}{80} + \frac{1}{1.2(25)} = \frac{1}{80} + \frac{1}{30} = \frac{1}{21.8} \quad (5.2)$$

and

$$\frac{1}{\dot{G}_{CO_2}} = \frac{1}{80} + \frac{1}{5(25)} = \frac{1}{80} + \frac{1}{125} = \frac{1}{49.4} \quad (5.3)$$

The circulatory system ($\lambda\dot{Q}$) normally offers the smallest conductance (and thus the largest impedance) to oxygen transport. On the other hand, alveolar ventilation offers rather more impedance to carbon dioxide elimination than does blood transport. Furthermore, the overall impedance is lower for carbon dioxide than for oxygen.

Let us suppose that there is an overall oxygen concentration gradient of 209 ml • l^{-1} from the atmosphere to active sites within the mitochondria of the working muscles. Oxygen transport should then equal the product of conductance and concentration gradient (21.8 × 209, or 4556 ml • min). In practice, the maximum oxygen transport of a sedentary young man is rather less than that predicted from the simple conductance equation (5.2), mainly because a part of the cardiac output is directed to tissues other than working muscles. From equation (5.3), a carbon dioxide transport of 4556 ml • min would require a CO_2 concentration gradient of 4556/49.4 or 92.2 ml • l^{-1} between the active tissues and inspired air.

A somewhat similar conductance equation can be developed for the exchange of heat between the muscles and the external environment. Here, the main series impedances (Fig. 5.2) are blood transport (from the skeletal muscles to the body core and from the core to the skin) and a subsequent convective transfer of heat or evaporated sweat across the barrier layer of still air that covers the skin surface. The driving pressure is a gradient of temperature rather than of gas concentration, and the solubility factor is the specific heat of the blood. Blood transport from the core to the skin is an important part of the total process of heat transfer, since it accounts for about half of the total thermal gradient. As with the transport of respiratory gases, cardiorespiratory conductance must thus be closely matched to metabolic demand.

In theory, the regulation of gas and heat transport could occur either locally, at the cellular level, or through a centrally mediated increase in the activity of respiratory and cardiac pumps. Local regulation inevitably carries a risk of jeopardizing homeostasis in other parts of the body, while central regulation could waste energy through an increase of cardiorespiratory transport that was excessive relative to local metabolic needs. The arrangement adopted by hu-

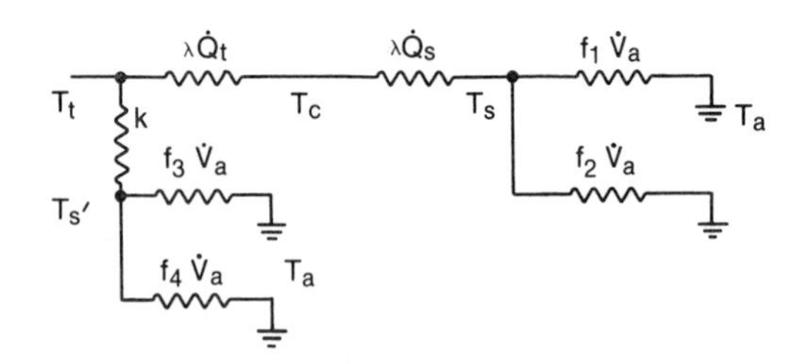

FIG. 5.2. Series conductances determining heat exchange between active tissues and environment. (Radiant heat exchange is assumed equal to zero.) λ, solubility factor (specific heat of blood); $\dot{Q}_t$, blood flow from active tissues; $\dot{Q}_s$, blood flow to skin; $f_1\dot{V}_a$, convective heat loss as function of air velocity; $f_2\dot{V}_a$, evaporative heat loss as function of air velocity for a given local temperature and humidity; $f_3\dot{V}_a$ and $f_4\dot{V}_a$, convective heat loss immediately over active tissues; K, conductivity of superficial tissues in active part; T, temperature (t, active part; c, body core; s, skin; s′, skin overlying active part; a, air).

mans seems a happy combination of these two options, with "fine tuning" of a centrally mediated response being obtained through a peripheral feedback from the active tissues.

OVERVIEW OF THE RESPIRATORY SYSTEM

The respiratory gases are drawn into and expelled from the chest in response to forces developed by the respiratory muscles and the elastic tissues of the thoracic cage and lungs. Work is performed against viscous and turbulent components of air flow resistance, viscous resistance to tissue displacement, and the inertia of both the gas molecules and heavy viscera such as the liver. Inspired air is warmed further to body temperature, saturated with water, and filtered to remove particulate matter and noxious vapors.

Ventilation is normally measured at the mouth. Unfortunately, not all of the external ventilation contributes to gas exchange. Part is wasted in ventilating the conducting airways (the anatomical dead space), and a further part is directed to lung regions that are either poorly perfused or have an inadequate diffusing capacity (the so-called alveolar dead space). The respiratory gas conductance of equations (5.2) and (5.3) thus depends upon the alveolar component of external ventilation.

A variable part of the oxygen delivered by the cardiorespiratory system is consumed by cardiac and respiratory muscle. Such oxygen is obviously not available for the performance of useful external work. Thus, in assessing the suitability of a given breathing pattern, we must consider not only the alveolar ventilation achieved, but also the impact of this tactic upon the oxygen consumption of the heart and respiratory muscles.

Strictly speaking, respiratory events continue through to the cellular level, where oxygen and carbon dioxide are exchanged at the mitochondrial surface (Chap. 3). However, in keeping with recent tradition, the final element considered under the rubric of the respiratory system will be the exchange of gas between the alveoli and the pulmonary capillaries (the pulmonary diffusing capacity).

IMPEDANCES IN THE RESPIRATORY CHAIN

As in an electrical system, the relative impedance offered by individual elements in the respiratory chain is indicated by their share of the total concentration gradient from the atmosphere to the active tissues (Shephard, 1971a,1977a). Thus, during maximum effort about a quarter of the overall oxygen concentration gradient ($C_{I,O_2} - C_{t,O_2}$) is dissipated between inspired air ($C_{I,O_2} = 209$ ml • l^{-1}) and alveolar gas ($C_{A,O_2} = 155$ ml • l^{-1}). In keeping with this estimate, we have seen from equation (5.2) that alveolar ventilation accounts for 30/(80 + 30), or 27% of the total impedance to oxygen intake.

The concentration gradient associated with blood transport is the arteriovenous difference ($C_{a,O_2} - C_{\bar{v},O_2}$). By a process of elimination, the gradient for pulmonary diffusion is apparently that from alveolar gas to arterial blood ($C_{A,O_2} - C_{a,O_2}$), while that for tissue diffusion is from mixed venous to tissue gas ($C_{\bar{v},O_2} - C_{t,O_2}$). Such gradients are at variance with traditional views on gas diffusion. The pulmonary diffusing capacity ($\dot{D}_L$) is normally expressed as the rate of gas transfer per unit of pressure gradient between alveolar gas and a hypothetical mean capillary pressure (Fig. 5.3). The mean capillary is so calculated that it would permit the observed gas exchange if it existed along the entire length of the capillary bed. The oxygen-diffusing capacity, $\dot{D}_{L,O_2}$, is then given by the equation

$$\dot{D}_{L,O_2} = \frac{\dot{V}_{O_2}}{P_{A,O_2} - P_{\overline{p.c.},O_2}} \tag{5.4}$$

The reason the gradient in our conductance model is $C_{A,O_2} - C_{a,O_2}$ rather than $C_{A,O_2} - C_{\overline{p.c.},O_2}$ is that we are studying not diffusion alone, but rather the interaction between blood transport and pulmonary diffusion. The corresponding impedance is a complex

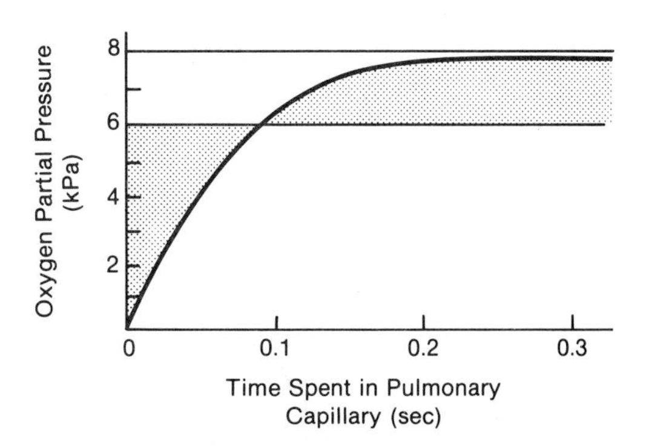

FIG. 5.3. The concept of a mean pulmonary capillary oxygen tension that will permit the observed oxygen exchange. The line is drawn so that the two shaded areas are of equal size. (Note that an analogous procedure can be carried out for the tissue capillaries.)

quantity $[(B/1 - B)(1/\lambda\dot{Q})]$, where B is an exponent based on the ratio of pulmonary diffusing capacity to blood transport, integrated over the length of the pulmonary capillary

$$(B = e^{-\int(\dot{D}_L/\lambda\dot{Q})}).$$

Similarly, in the tissues we are studying the interaction between blood transport and tissue diffusion. The relevant impedance is thus the complex quantity $[(K/1 - K)(1/\lambda\dot{Q})]$, where K is an exponent based on the ratio of tissue diffusing capacity $\dot{D}_t$ to blood transport, integrated over the length of the tissue capillary $(K = e^{-\int(\dot{D}_L/\lambda\dot{Q})})$

RESPIRATORY MINUTE VOLUME

Resting Conditions

The normal resting respiratory minute volume is ~4 l • min⁻¹ BTPS* per m² of body surface area. Thus, a ventilation of 7.2 l • min⁻¹ BTPS would be anticipated in a young man with a body surface area of 1.80 m². The breathing frequency of a sedentary young man is ~14 breaths per min. Given a ventilation of 7.2 l • min⁻¹, this would imply a tidal volume of 500 ml per breath. Athletes often have a slower and deeper pattern of respiration than sedentary subjects.

On-transients

Modern techniques for breath-by-breath recording of ventilation and oxygen consumption (R. H. T. Edwards, 1969; Fujihara et al., 1973a,b; Beaver et al., 1973; Reynolds and Milhorn, 1973) allow a detailed study of changes at the onset of vigorous exercise. The ventilatory response probably involves both a time delay and one or more exponential adjustments to the requirements of increased metabolism (Wigertz, 1970; Fujihara et al., 1973b).

Classical reports had described a rapid initial increase of ventilation, a slower secondary increase, commencing 20–30 sec later (Asmussen and Nielsen, 1948; Dejours, 1959,1964; Matell, 1963) and (in the case of severe work) a third very slow augmentation of ventilation (Knuttgen, 1970; Fig. 5.4). Transients were better seen in tidal volume than in respiratory rate (Dejours, 1967), the immediate response to leg exercise being an increase in the activity of the expiratory muscles.

According to Dejours, the rapid initial phase accounted for ~50% of the increase in ventilation. D'Angelo and Torelli (1971) found an even larger rapid component (60–100% of the steady-state increase in ventilation). Most (Asmussen and Nielsen, 1948; Dejours, Flandrois et al., 1961; D'Angelo and Torelli, 1971; Linnarson, 1974) but not all (Cerretelli, Sikand, and Farhi, 1966; Broman and Wigertz, 1971) authors find some variation of response, related to the intensity and type of work (Fig. 5.4). Indeed, Beaver and Wasserman (1968) suggested that the early increase of breathing is an inconsistent and learned response. Among naive subjects, an early exercise-induced increase of respiratory rate was often masked by a compensatory decrease of tidal volume (Beaver and Wasserman, 1970). Wigertz (1970) used sine-wave, step, and ramp-function workloads, and found a single on-transient (half-time 70 ± 6 sec) with no initial phase lag. Fujihara et al. (1973a,b) pointed out the need to use separate lag functions for the description of slow and fast responses; coupling this method of analysis with a square-wave increase of workload, they were able to reproduce both the initial rapid increase of ventilation and a slower adjustment that commenced after a 17-sec delay. Linnarson (1974) had similar findings.

Steady-State Response

The ventilatory response to a given intensity of rhythmic work is usually reported in terms of so-called steady-state data. A steady-state is assumed when the body has been allowed 4 or 5 min to adapt to the new level of metabolism and respiratory variables show no more than specified minor changes over the succeeding minute.

At moderate work loads, there is a linear relationship between the respiratory minute volume and the oxygen cost of a given type of activity, but as the intensity of effort is increased, a disproportionate hyperventilation develops (Fig. 5.5). This parallels and probably reflects the accumulation of anerobic metabolites in the blood. Some authors thus speak of an anerobic threshold (Wasserman et al., 1973). If

*Since oxygen intake usually exceeds carbon dioxide output, there are small (<1%) differences between inspired and expired volumes. Respiratory minute volumes are traditionally expressed as the volume *expired* at body temperature and pressure, saturated with water vapor (BTPS). The BTPS units are appropriate in terms of the chest movements sensed by the body, although in terms of gas exchange, the critical values are STPD volumes. Under normal conditions, 1 l STPD = ~1.2 l BTPS.

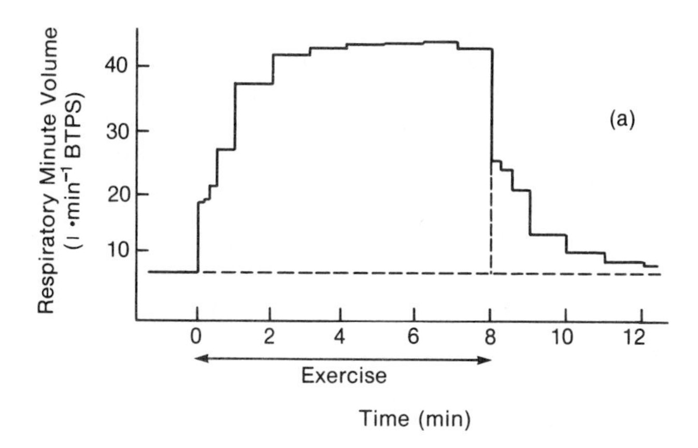

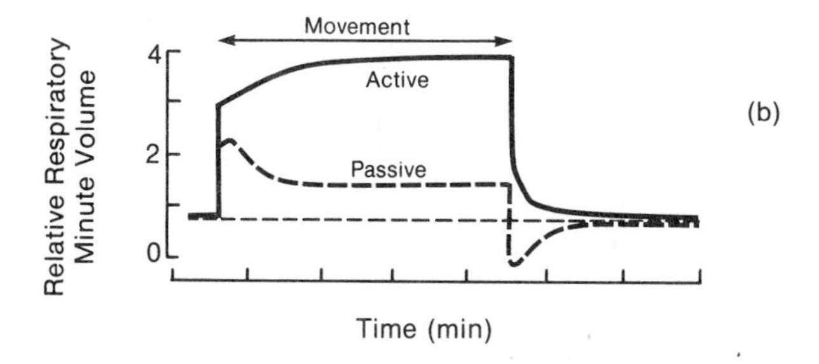

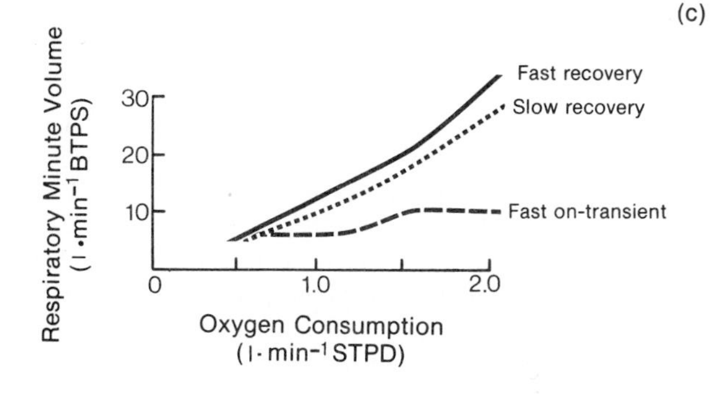

FIG. 5.4. A. Classical view of the time course of ventilation during exercise. **B.** Relative influence of active and passive limb movements. **C.** Fast and slow components of respiratory response in relation to oxygen consumption. (After Dejours, 1963.)

ventilation is plotted against absolute work load or absolute oxygen consumption, there is an obvious relationship between a low anerobic threshold and poor physical condition. Equally, the threshold is low if exercise is performed by a small group of muscles, as in forearm work. However, if the threshold is expressed in relative terms (as a percentage of the maximum attainable oxygen intake for a given individual performing a given type of work), differences between individuals and between modes of exercise become much smaller.

The anerobic threshold should not be overinterpreted, since blood lactate accumulation (and thus the extent of hyperventilation during a laboratory exercise test) depends on the pattern of effort that is required. The ventilation seen after 5 min of work at a steady rate of 9 kN • m • min^{-1}, for example, is very different from that observed if the same loading is attained by an increase of 1 kN • m • min^{-1} every minute for 9 min.

Maximum Ventilation

During maximum exercise, the average young man reaches a ventilation of 90–120 l • min^{-1} BTPS

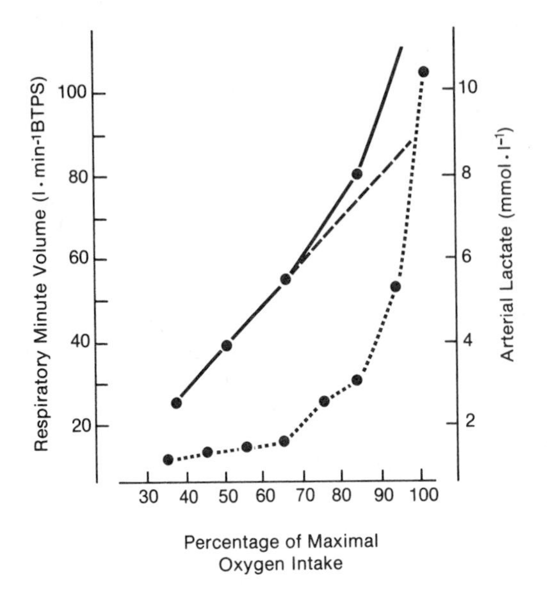

FIG. 5.5. Relationship between respiratory minute volume and intensity of stepping exercise, expressed as percentage of maximum oxygen intake. Note onset of disproportionate hyperventilation at loads above 70% of maximum oxygen intake, with associated increments of blood lactate. (After Shephard, 1967b.)

(Shephard, 1976a), although volumes as large as 160 $l \cdot min^{-1}$ are encountered in national class athletes (Saltin and Åstrand, 1967). The observed ventilation is generally less than the maximum ventilation that can be realized by voluntary effort (Table 5.1), and it is thus unlikely that the power of the thoracic pump limits performance of normal activity. The situation is less clear-cut when respiration is impeded by arm movement (as in a rower), intermittent submersion (as in a swimmer), or the use of respiratory equipment (as in a fireman). Holmér (1974a) noted that ventilation was less during maximal swimming than during all-out treadmill exercise, with a corresponding difference in the maximum attainable oxygen intake. Demedts and Anthonisen (1973) observed that when breathing through a high external resistance, both ventilation and work performance were limited (Fig. 5.6).

It is technically possible to breathe at 100–150 breaths per min, but in a young man vigorous physical effort rarely induces a faster rate than 40–50 breaths per min. Children attain maximum rates of 60–70 breaths per min during exercise.

The tidal volume increases progressively with the subject's power output; during maximum effort, it amounts to at least 50% of vital capacity (2500–3000 ml BTPS in a young man).

Coupling of Respiratory Frequency

There has been much recent discussion of a possible linkage between the rhythm of work and the breathing frequency or respiratory minute volume (Fried and Shephard, 1971; J. D. Kay et al., 1974; Gueli and Shephard, 1976; Bechbache and Duffin, 1977; Casaburi et al., 1978). In some sports, such as swimming and rowing, breathing rate is plainly set by the activity. If an oarsman adopts a rhythm of 36 strokes per min, he can satisfy his respiratory demand during the early part of the race by taking one breath per stroke, but as lactate accumulates it becomes necessary to introduce a quick inspiratory gasp during the power stroke.

Most studies of sedentary subjects have failed to show an association between pedal frequency and breathing rate during bicycle ergometry (J. D. Kay et al., 1974; Gueli and Shephard, 1976; Fig. 5.7). However, Bechbache and Duffin (1977) claimed that the breathing frequency corresponded with the exercise rhythm in 8 of 15 volunteers pedaling at 50 revo-

TABLE 5.1
Maximum Voluntary Ventilation Before, During, and After Exercise

Group	Respiratory Minute Volume during Exercise ($l \cdot min^{-1}$ BTPS)	Maximum Voluntary Ventilation ($l \cdot min^{-1}$ BTPS) (1) Before Exercise	(2) Final Minute of Exercise	(3) 3 Min after Exercise	Δ [(2) − (1)]
7 subjects exercising 5 min	60.7 ± 7.7	185.4 ± 14.0	211.3 ± 12.3	191.4 ± 17.1	25.9 ± 10.2
6 subjects exercising 10 min	70.2 ± 4.6	166.3 ± 6.6	175.6 ± 6.7	164.5 ± 12.9	9.3 ± 5.8
8 subjects exercising 20 min	69.5 ± 3.6	190.9 ± 8.2	184.1 ± 8.8	184.7 ± 12.6	−6.8 ± 5.9

Source: Based on data of Shephard, 1967a.

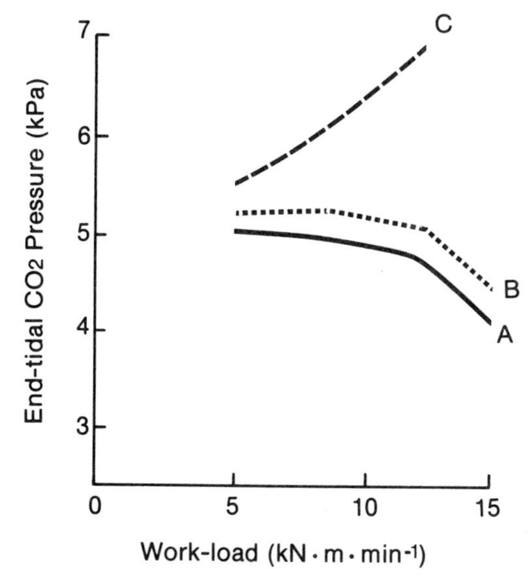

FIG. 5.6. Influence of an external resistance to breathing upon the end-tidal carbon dioxide pressure during bicycle ergometer work. A, no resistance; B, resistance doubling the work of breathing; C, large external resistance. Since carbon dioxide is a bronchodilator, the accumulation of CO_2 in situation C helps to minimize the work of breathing. (After Demedts & Anthonisen, 1973.)

lutions per min, and in 9 of 15 at 70 rpm; furthermore, they found entrainment of the exercise rhythm in 8 of 15 subjects while walking and 13 of 15 while running on a laboratory treadmill.

Ventilatory Equivalent

The overall efficiency of ventilation can be examined in terms of the ventilatory equivalent. This is the number of liters of ventilation (BTPS) required to supply each liter STPD of oxygen that is consumed. If the resting oxygen consumption is 0.25 l • min^{-1} STPD, then the ventilatory equivalent is 7.2/0.25, or 28.7 l • l^{-1}. The ventilatory equivalent remains at or near its resting level in moderate activity, but increases once the anerobic threshold has been surpassed. If a young man has a maximum oxygen intake of 3 l • min^{-1} STPD and a maximum exercise ventilation of 120 l • min^{-1} BTPS, his ventilatory equivalent for maximum effort is 120/3 or 40 l • l^{-1}.

Aging has little effect upon the ventilatory equivalent at a given relative work load (Tables 5.2 and 5.3); however, submaximal values are somewhat greater in children and smaller in athletes (P. O. Åstrand, 1952). If the external workload is held constant, the ventilatory equivalent is increased by small-muscle activity (Stenberg et al., 1967). On the other hand, the equivalent is reduced by both training and habituation (Table 5.4; C. T. M. Davies, Tuxworth, and Young, 1970; Girandola and Katch, 1976).

RESPIRATORY FUNCTION TESTS

Vital Capacity and Its Subdivisions

Definition

The vital capacity (Fig. 5.8a) is the volume of air that can be exhaled following a maximum inspiration. It is influenced somewhat by posture, but is essentially unchanged by exercise (Asmussen and Christensen, 1939a).

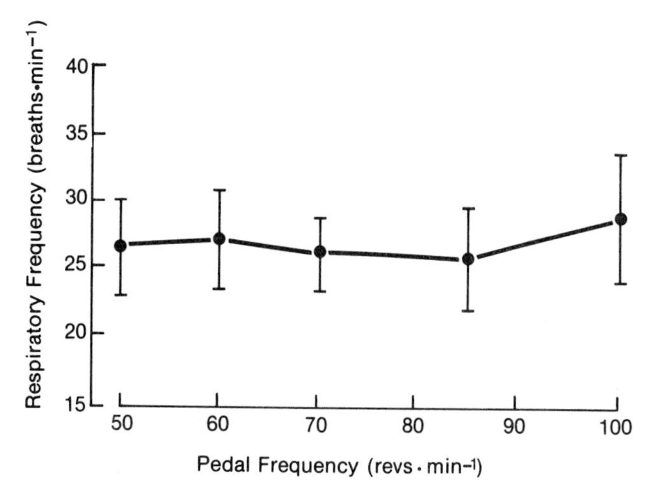

FIG. 5.7. The lack of relationship between pedal frequency (bicycle ergometer) and breathing frequency. (Data of Gueli & Shephard, 1976, for young men exercising to heart a rate of 140 to 145 beats • min^{-1}.)

TABLE 5.2
Ventilatory Equivalent at 50 and 75% of Maximum Oxygen Intake in Relation to Age and Sex*

Age (years)	Men (%) 50	Men (%) 75	Women (%) 50	Women (%) 75
9–11	30.8	33.4	29.8	28.6
12–13	27.9	26.4	28.8	29.0
14–16	26.0	27.4	26.0	27.5
17–19	24.0	27.7	26.1	28.2
20–29	24.0	25.8	26.0	27.4
30–39	23.4	25.3	27.1	29.7
40–49	24.5	25.2	27.8	21.0
50–59	24.8	25.1	25.3	27.1

*All data l BPTS per l STPD.
Source: Based on data of Rode and Shephard (1971) for Eskimo subjects.

Vital Capacity and Fitness

At one time, the vital capacity was considered an important measure of physical fitness (Dreyer, 1920; Dawber, Kannel, and Friedman, 1966). There is some correlation between a poor vital capacity and liability to ischemic heart disease, but this probably reflects a mutual association of the two variables with a high cigarette consumption. An apparent correlation between vital capacity and maximum oxygen intake can also be created by mixing results from children and adults (P. O. Åstrand and Rodahl, 1977), but such an association is spurious, reflecting a mutual dependence of the data on body size (Table 5.5; T. K. Cureton, 1936; Shephard, Lavallée et al., 1979b). If data are restricted to adults, there is little association between vital capacity and such measures of physical fitness as maximum oxygen intake (Ishiko, 1967; Table 5.6). Furthermore, endurance training does not augment the vital capacity of adults (Kollias, Boileau et al., 1972), although gains have been reported in teenagers (Ekblom, 1969a,b; Engström et al., 1971).

Correlations have been described between athletic performance and vital capacity (Ishiko, 1967; Table 5.6). Certainly, the measurement is influenced by muscular strength, and athletes who participate in sports that enhance development of the chest muscles (for instance, rowers and kayak paddlers) have a large vital capacity in relation to height (Sidney and Shephard, 1973; G. R. Wright, Bompa, and Shephard, 1976), with a positive correlation between vital capacity and competitive success. Medium-distance swimmers also gain a competitive advantage from the buoyancy associated with a large lung volume (Shephard, Godin, and Campbell, 1973). On the other hand, G. R. Cumming (1971) found no correlation between track performance and vital capacity.

TABLE 5.3
Ventilatory Equivalent at 100% of Maximum Oxygen Intake for Selected Groups of Torontonians*

Group	Ventilatory Equivalent	Authors
Rowers	26.0	G. R. Wright, Bompa, and Shephard (1976)
Skiers	34.5 (treadmill) 32.0 (skiing)	Niinimaa, Shephard, and Dyon (1979)
Swimmers	33.4	Shephard, Godin, and Campbell (1973)
Young men	35.4	Sidney and Shephard (unpublished)
Young women	41.8	Sidney and Shephard (unpublished)
Old men	39.0	Sidney and Shephard (1977a)
Old women	37.2	Sidney and Shephard (1977a)

*All data l BTPS per l STPD.

Normal Standards

The vital capacity decreases by 20–25% over the span of adult life (Shephard, 1971b,1978a). It is also 20–25% smaller in the female than in the male. Thus, if an elderly female subject wishes to reach the same absolute work rate as a young man, she must either breathe faster or utilize a larger portion of her vital capacity. Partly for this reason, effort is more commonly limited by breathlessness as a subject becomes older.

If account is taken of sex, age (A′, years), and height (H′, cm), a multiple-regression equation can be developed to predict an individual's vital capacity (VC, ml) with a standard deviation of about 10% (T. W. Anderson, Brown et al., 1968):

$$\text{Standing VC (men)} = 56.3H' - 17.4A' - 4210 \quad (5.5)$$

$$\text{Standing VC (women)} = 54.4H' - 10.5A' - 5210 \quad (5.6)$$

Thus a man with a height of 170 cm and an age of 30 years would have a predicted vital capacity of $56.3(170) - 17.4(30) - 42.0 = 4839$ ml BTPS.

TABLE 5.4
Changes in Ventilatory Cost of Exercise over 5 Days of Repeated Exercise, in Relation to Personality Type*

Total Extra Ventilation ($l \cdot min^{-1} \cdot W^{-1}$)	Personality Type: Normal	Anxious	Hysterical	Psychopathic	Extrovert
Day 1	0.30 ± 0.02	0.28 ± 0.02	0.30 ± 0.04	0.30 ± 0.02	0.27 ± 0.02
Δ Day 1-Day 5	−0.04 ± 0.01	−0.01 ± 0.004	−0.05 ± 0.02	−0.02 ± 0.01	−0.01 ± 0.01
Extra Ventilation during 15 Min Recovery Period ($l \cdot min^{-1}$)					
Day 1	6.0 ± 0.7	5.6 ± 0.8	6.9 ± 1.3	6.1 ± 0.8	4.6 ± 0.6
Δ Day 1-Day 5	−2.5 ± 0.5	1.7 ± 1.0	−5.2 ± 2.0	−2.2 ± 0.7	−1.1 ± 0.5

*As assessed by Maudsley Personality Inventory.
Source: Based on data of Shephard and Callaway (1966) and Shephard (1966c).

The equations imply that aging causes a functional loss of 17.4 ml • year^{-1} in men and 10.5 ml • year^{-1} in women.

The majority of authors have reported very similar rates of aging when analyzing cross-sectional data (Shephard, 1971b,1978a). In contrast, two longitudinal studies of physical education teachers have shown almost no loss of vital capacity with aging (Asmussen and Mathiasen, 1962; I. Åstrand, Åstrand et al., 1973a). These surprising findings probably reflect three peculiarities of the longitudinal samples: (1) continued physical activity, (2) a low proportion of cigarette smokers, and (3) a selective elimination of subjects affected by respiratory disease (T. W. Anderson, Brown et al., 1968; Shephard, 1976c). Certainly, respiratory disease greatly accelerates the loss of pulmonary function, and health-conscious nonsmokers preserve a superior lung function to the seventh decade of life (Niinimaa and Shephard, 1978a).

Tidal Volume

The resting tidal volume of 500 ml represents no more than 10% of the vital capacity of a young man. However, in maximum exercise, tidal volumes of 2500–3000 ml account for up to 60% of the vital capacity (Fig. 5.8a). The classical observations of Asmussen and Christensen (1939a) suggested that the increase of tidal volume involved a decrease of both inspiratory and expiratory reserve volumes. Incursion upon the expiratory reserve volume is particularly likely if the thorax is bent forward, as in the operation of a racing bicycle (Petit et al., 1959). With more normal posture, the exercise-induced increase of tidal volume takes place largely at the expense of the inspiratory reserve (Milic-Emili, Petit, and Deroanne, 1960; Vávra and Máček, 1968; Goldstein et al., 1975). Reliance upon the inspiratory reserve is particularly characteristic of the swimmer, where water pressure may reduce the resting expiratory reserve volume to 1 l or less (Holmér, 1974a).

As during the performance of a maximum voluntary ventilation test, a subject who is working maximally tends to breathe over the range 30–90% of vital capacity. There are several practical reasons for avoiding the final 30% of expiration. The maximum force that can be exerted by the expiratory muscles is limited at small alveolar volumes (Fig. 5.8b; Hyatt, 1972); the airways are also progressively narrowed as expiration proceeds (Shephard, 1959; Macklem and Mead, 1967); and the compliance of the lungs (the change of volume per unit change of pressure) is decreased by a progressive collapse of alveolar units in the basal regions of the lungs (Milic-Emili, Henderson et al., 1966). The final 5–10% of inspiration is also avoided because the inspiratory muscles can develop only a limited additional force when the chest has reached this degree of expansion (Olafsson and Hyatt, 1969), while the compliance of the system again becomes unfavorable as maximum distension is approached (Fig. 5.18; Goldstein et al., 1975).

Residual Volume

The residual volume is the quantity of gas remaining in the chest after a steady maximal forced expiration (Fig. 5.8a). It amounts to about 22% of total lung capacity in children, but increases to as much as 40% of the total lung capacity by the age of 60 (Shephard, 1971b; Niinimaa and Shephard, 1978a).

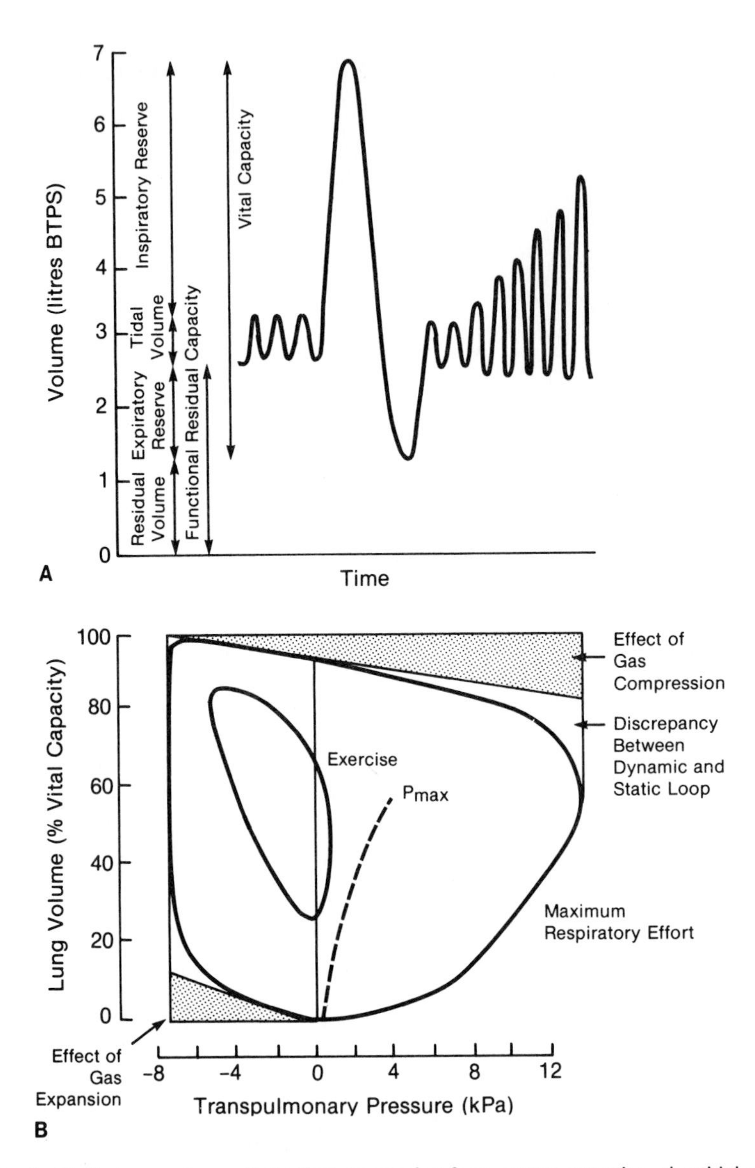

FIG. 5.8. **A.** Static lung volume. Note that on passing from rest to exercise, the tidal volume increases mainly at the expense of the inspiratory reserve, whereas the expiratory reserve remains relatively unchanged. **B.** Transpulmonary pressure–volume loops for exercise (small loop) and maximum respiratory effort (large loop). If the interrupted line (P_{max}) is exceeded, expiratory flow is limited by airway collapse (Fig. 5.15). (Based in part on an experiment of Hyatt, 1972.)

There are several possible methods of measurement. One approach is to note the dilution that occurs when a subject rebreathes from a spirometer circuit containing a known initial percentage of a relatively insoluble tracer gas such as helium. A variant of this procedure measures the volume of oxygen needed to flush nitrogen from the lungs. An alternative approach records changes in mouth and box pressures when a subject is enclosed in a rigid box and makes a forcible expiratory effort against a closed shutter. In general, the body box readings are larger than those obtained by gas dilution, since (1) equilibration of the tracer gas with poorly ventilated parts of the lung may be incomplete and (2) the body box method takes account of air trapped beyond points of airway collapse, along with the gas content of the

TABLE 5.5
Factors Contributing to Variance of Vital Capacity in High School Boys

Variable	Contribution to Variance of Vital Capacity (%)
Height	44.4
Weight	30.7
Rogers' Athletic Index	14.1
Age	6.9
Arm strength	2.1
Grip strength	1.2
Back strength	0.7
Leg strength	0.03

Source: Based on an analysis by T. K. Cureton, 1936.

stomach and intestines. In young subjects who are free of respiratory disease, a further possibility is to estimate the lung volume from chest radiographs (Fig. 5.9; Pratt and Klugh, 1967; Shephard and Seliger, 1969).

There are several important practical applications of residual volume measurements, as follows:

1. **Body fat.** An estimate of residual volume is needed when determining the percentage of body fat by underwater weighing. However, because of technical difficulties in making accurate helium dilution measurements while underwater, some authors prefer to predict the residual volume as a fixed percentage of vital capacity (28% in a young adult; Wilmore, 1969; Shephard, 1978e).
2. **Alveolar volume.** It is also necessary to determine residual volume when calculating the total volume of gas in the lungs (alveolar volume, functional residual capacity; Fig. 5.8a).

The alveolar gas volume influences the rate of change of gas tensions within the lungs during breath holding and the extent of oscillations of tension that occur over the course of a normal respiratory cycle (Shephard, 1968b; Fig. 5.10). The mean alveolar volume also influences the work of breathing, since the elastic forces developed by the chest and lungs are a nonlinear function of volume (Fig. 5.18).

A thoracic squeeze arises when an external force such as water pressure attempts to compress the lungs below their residual volume. Let us imagine a skin diver with a vital capacity of 5 litres and a residual volume of 1.2 l. If he commences a dive with a full inspiration and descends to approximately 40 m (5 atmospheres pressure), the gas content of the lungs is immediately compressed to residual volume and may attempt to shrink further with respiratory exchange.

The pulmonary vessels thus become overdistended with blood and may rupture into the lung; the likelihood of this pathology increases progressively as the external pressure is further increased.

During ascent, the diver must expel lung gas or absorb the excess gas pressure by expansion of the chest. If vital capacity is reached and the excess intrapulmonary pressure is still 11 kPa (80 mmHg) or more above atmospheric pressure, rupture of the lung may occur (Malhotra and Wright, 1960,1961; Chap. 12).

Dynamic Lung Volumes

Maximum Voluntary Ventilation (MVV)

The MVV may be defined as the maximum volume of air that a subject can respire in 15 sec. Typical figures for the average healthy young man are 160–200 l • min^{-1} BTPS, while values as high as 250 l • min^{-1} BTPS are not uncommon in athletes (Shephard, 1971b). Results for elderly and female subjects

TABLE 5.6
Coefficient of Correlation of Vital Capacity (ml, ml • kg^{-1}, and ml • cm^{-1}) with Selected Variables

Variable	Vital Capacity ml	Vital Capacity ml • kg^{-1}	Vital Capacity ml • cm^{-1}	Author
5000-m race time	−0.34	−0.47	—	Ishiko (1967)
Rowing performance	+0.71	+0.47	—	Ishiko (1967)
Maximum oxygen intake	−0.04	−0.01	—	Ishiko (1967)
Whitewater paddling performance	0.64	—	0.69	Sidney and Shephard (1973)
Sprint-swimming performance	−0.04	—	—	Shephard, Godin, and Campbell (1973)
Middle-distance-swimming performance	0.86	—	—	Shephard, Godin, and Campbell (1973)

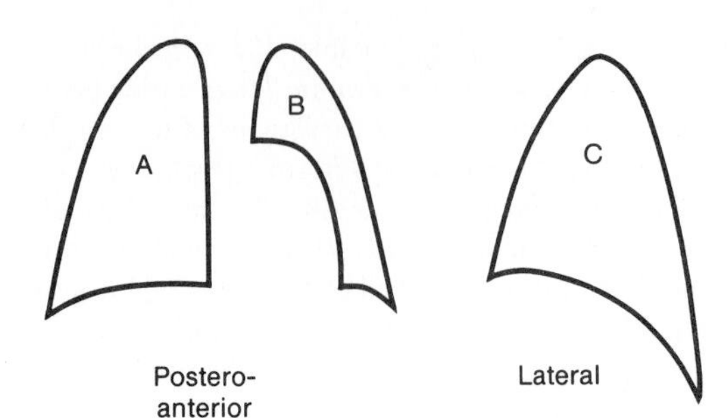

FIG. 5.9. Method of estimating the total lung capacity from chest radiographs. TLC = 0.67 ($^{4/3}\sqrt{AC}$ + $^{4/3}\sqrt{BC}$) + 320 ml. BTPS. (After Pratt & Klugh, 1967 and Shephard & Seliger, 1969.)

are 20–25% smaller than in the average young man (Tatai, 1957; Shephard, 1971b).

The prime physiological determinants of MVV are (1) ability of the chest muscles to convert chemical energy into ventilatory work and (2) a progressive increase in the cost of pulmonary ventilation at higher rates of gas flow. However, in practice the results are also influenced markedly by the dynamic properties of the recording apparatus, the motivation of the subject, and the breathing frequency that he adopts (Fig. 5.11; L. Bernstein et al., 1952; Shephard, 1957a).

Measurements are usually made while the subject is at rest, and under such conditions it is difficult to sustain a maximum ventilatory effort (Zocche et al., 1960; Tenney and Reese, 1968). Within 15 min, ventilation falls to ~53% of the 15-sec value, even if hypocapnia is avoided by supplying the subjects with carbon dioxide (Cournand and Richards, 1941; Zocche et al., 1960). If a similar decrease of MVV occurred during prolonged activity, it could seriously limit effort. However, a larger voluntary ventilation can be stimulated by vigorous pacing of breathing (S. Freedman, 1970), and during near maximal exercise subjects are able to sustain 75–80% of the 15-sec MVV (Shephard, 1967a). Several factors contribute to the better ventilatory performance during vigorous work (Table 5.1), including (1) a secretion of catecholamines leading to a decrease of airway resistance, (2) an increased sensory drive to the respiratory center, and (3) a warm-up of the chest muscles brought about by the general increase of body temperature.

Endurance training of the chest muscles augments MVV by 14% and brings subjects to a state where they can sustain 96% of the MVV over 15 min of exercise (Delhez et al., 1967–68; Leith and Bradley, 1976; Fig. 5.12). This gain has been attributed to a less ready fatigue of the inspiratory muscles, particularly the accessory muscles of the neck and shoulders.

Forced Expiratory Volume (FEV)

The MVV is an exhausting test for an elderly or sick person, and if there are some structural weaknesses

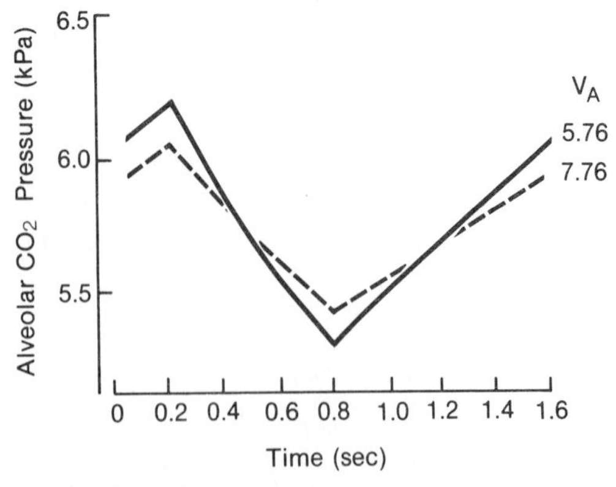

FIG. 5.10. Oscillations of alveolar gas composition over the course of a respiratory cycle. (After Shephard, 1968a.)

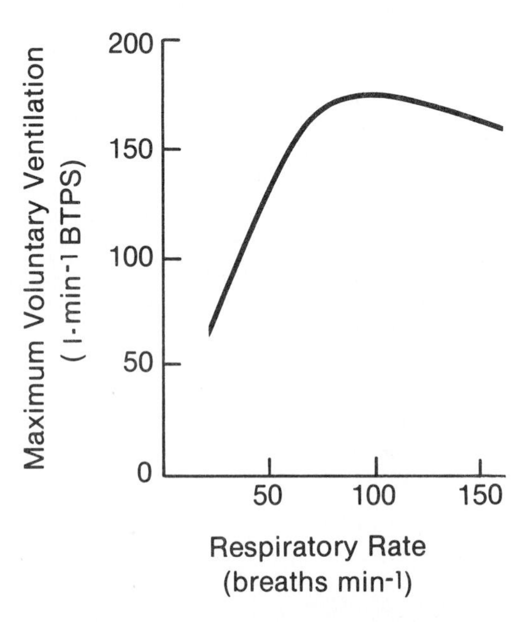

FIG. 5.11. Influence of respiratory rate upon maximum voluntary ventilation. (Based on data of Shephard, 1957a, for healthy young subjects breathing through a low resistance screen flowmeter.)

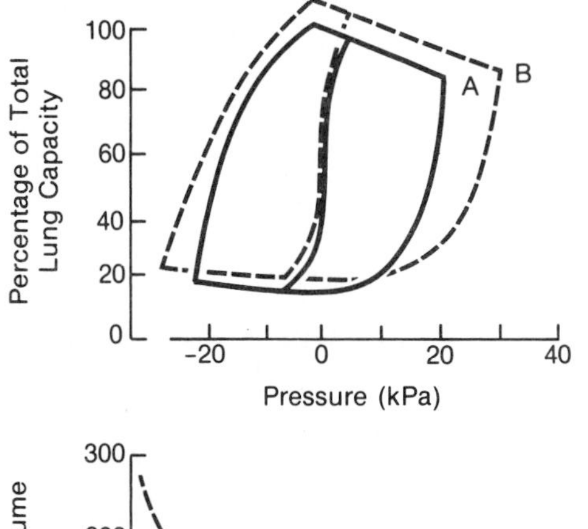

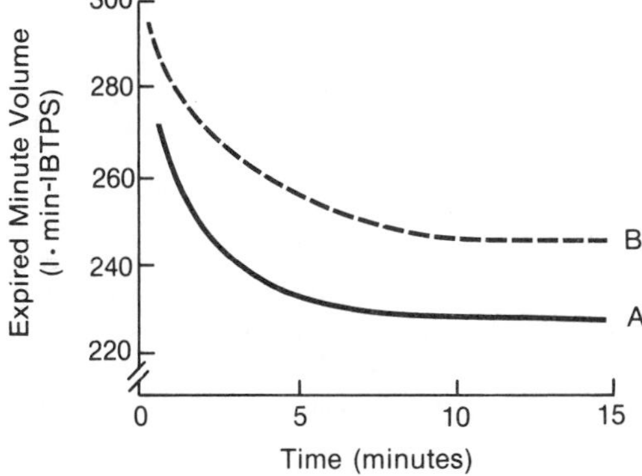

FIG. 5.12. Effects of training upon the respiratory muscles. Changes in pressure/volume diagram and sustained ventilatory capacity. A, before, and B, after five weeks of training. (After Leith & Bradley, 1976.)

of the lungs (for example, a partly calcified tuberculous cavity), there is a danger that this could be damaged by the forcible breathing. Many clinicians thus prefer to study the function of the respiratory pump in terms of a single forced expiration, initiated from a position of full inspiration (Fig. 5.13). Since the MVV expiration extends from 90% to perhaps 30% of vital capacity, the forced expiratory volume (FEV) and MVV curves overlie one another for much of their course.

Some laboratories still calculate an indirect MVV, multiplying the volume displaced in the first second of expiration ($FEV_{1.0}$) by a constant (usually 40). This practice is based on outdated views concerning the optimum breathing rate for the MVV test and, in any event, does not increase the information content of an $FEV_{1.0}$ reading. Current practice is thus to interpret the $FEV_{1.0}$ either as a number in its own right (Shephard, 1955b, 1971b) or as a percentage of the total forced vital capacity.

Equations for the prediction of absolute forced expiratory volumes are as follows (T. W. Anderson, Brown et al., 1968):

$$\text{Standing } FEV_{1.0}\text{(men)} = 37.5\,(H', \text{cm}) - 35.0\,(A', \text{year}) - 1290 \quad (5.7)$$

$$\text{Standing } FEV_{1.0}\text{(women)} = 4.0\,(H', \text{cm}) - 17.2\,(A', \text{year}) - 3130 \quad (5.8)$$

where H′ is the standing height (cm) and A′ is the age (years). The corresponding percentage values may be calculated by the equations:

$$\frac{FEV_{1.0}}{FVC\%}\text{(men)} = 128.95 - 0.201\,(H', \text{cm}) - 0.400\,(A', \text{year}) \quad (5.9)$$

$$\frac{FEV_{1.0}}{FVC\%}\text{(women)} = 93.03 - 0.253\,(A', \text{year}) \quad (5.10)$$

A young and healthy adult can expel 81–85% of his vital capacity in 1 sec. The percentage diminishes progressively with aging, amounting to ~70% of the forced vital capacity at 60 years. It is reduced acutely by exposure to irritant vapors (high concentrations of pollutants such as ozone; Folinsbee et al., 1977), finely suspended particulates (such as tobacco and industrial dusts), and cold air (Dautrebande, 1963).

Flow/Volume Curves and Airway Collapse

The detailed form of the forced expiratory flow curve is revealed by recording the pressure drop across a screen flowmeter (pneumotachograph) or by using an electronic spirometer. Flow rises rapidly to a peak of ~600 $l \cdot min^{-1}$, with a subsequent exponential fall (Fig. 5.14). The initial peak depends on the maximum force that can be exerted by the respiratory muscles, while the subsequent decline is brought about largely by a collapse of the air passages (Figs. 5.15 and 5.16). At the commencement of expiration, the pressure within the alveoli and smaller airways exceeds intrapleural pressure, and the equal pressure point is not reached until the expirate is traversing the major airways. In this region, the air passages are strengthened by cartilaginous plates, and collapse is unlikely to occur. However, the balance of forces changes as expiration proceeds. The elastic recoil of the lungs becomes smaller, and the airways are also narrowed, increasing the rate of pressure drop along

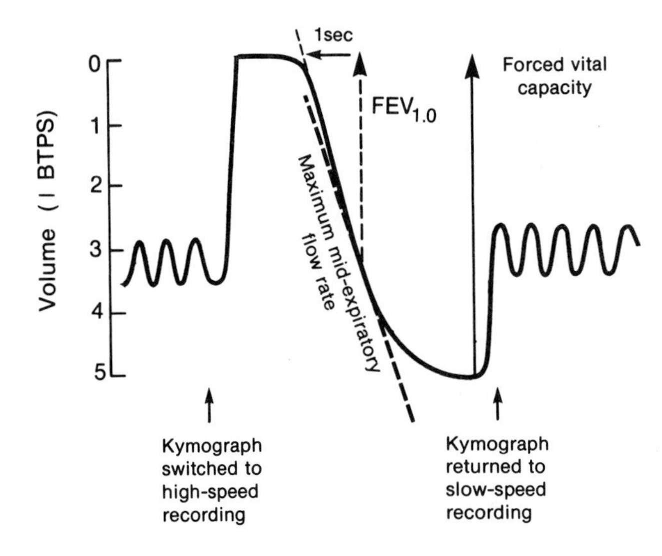

FIG. 5.13. A typical recording of forced vital capacity on a water-filled spirometer. Note the initial lag, due to inertia of the spirometer bell; this is ignored in calculating the one-second forced expiratory volume ($FEV_{1.0}$). The maximum mid-expiratory flow rate can be obtained by fitting a tangent to the FVC curve, although it is now more commonly estimated by plotting flow/volume loops.

their length. In consequence, the equal pressure point is progressively displaced toward the finer airways. These lack cartilaginous support, and collapse readily develops. Thereafter, attempts at more forcible expiration are fruitless. Collapse is worsened, and the expiratory flow is decreased rather than increased. The curve of Fig. 5.14 can thus be divided into an early portion, where flow depends upon muscle strength, and a later, effort-independent portion (Van de Woestijne and Zapletal, 1970; Fig. 5.16).

Healthy young adults do not reach the necessary conditions for airway collapse until at least 60% of their vital capacity has been expelled. Thus, the phenomenon is only marginally involved (Yamabayashi et al., 1970; Hyatt, 1972; Grimby, 1977) in the respiratory pattern of maximum exercise (ventilation from 90 to 30% of vital capacity). However, the equal pressure point is reached much earlier during expiration if the gradient along the airway is increased by the respiring of a dense gas, as during diving (L. D. H. Wood and Bryan, 1978). Other factors that exacerbate airway collapse include (1) a weakening of the cartilaginous plates, (2) a decrease in the elastic recoil of the lungs, and (3) spasm of the finer airways. Factors (1) and (2) are natural responses to aging, exaggerated by smoking and chronic chest disease. Factor (3) favors collapse by increasing the pressure drop from the alveolar spaces to the major bronchi.

An external expiratory resistance influences the maximum rate of expiratory airflow only when pressures upstream from it are high enough to prevent airway collapse (Mead, Turner et al., 1967; Fig. 5.16).

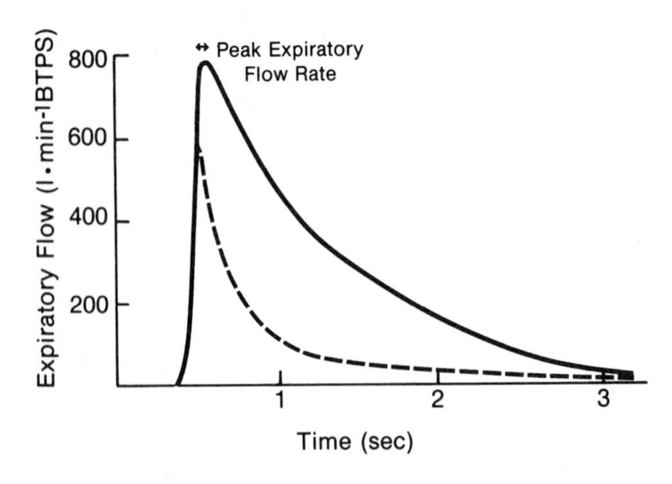

FIG. 5.14. Forced expiratory flow curves. Solid line, young adult male of above average physical fitness. Broken line, older subject with early collapse of airways. The peak expiratory flow rate is the maximum flow sustained for 10 msec.

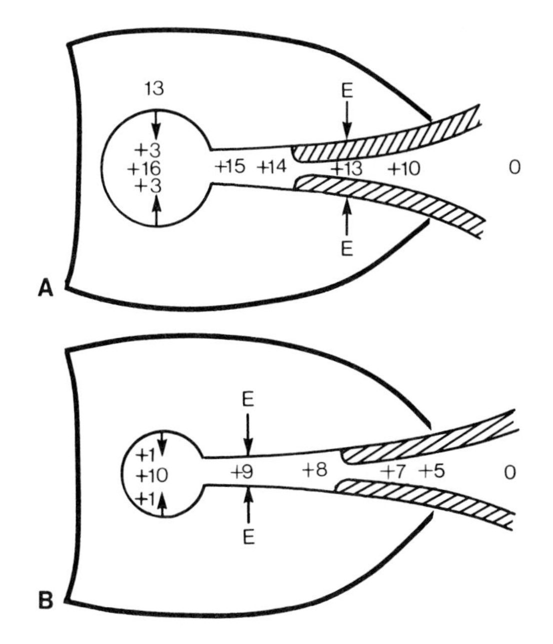

FIG. 5.15. The basis of airway collapse during a forced expiration. **A.** Commencement of expiration. The intrapleural pressure of 13 kPa is attributable to forces developed by the expiratory muscles and the elastic recoil of the thoracic cage. The intraalveolar pressure is boosted to 16 kPa by the elastic recoil of the lungs (3 kPa). The pressure within the airways does not drop to intrapleural pressure until the major airways are reached (equal pressure point, E). Here, cartilaginous plates prevent airway collapse. **B.** 60% of vital capacity expelled. The intrapleural pressure is now 9 kPa, and the elastic recoil of the lungs has decreased to 1 kPa, so that the intraalveolar pressure is 10 kPa. Further, the finer parts of the conducting airways are now narrower, and the pressure thus drops more quickly on passing along the airways. The equal pressure point is reached in the finer branches of the airway that lack stout walls, and collapse tends to occur. Attempts at more vigorous expiration merely increase airway collapse.

In some patients with respiratory disease, faster expiratory flows are accomplished during exercise than during forced vital capacity maneuvers (Pierce, Luterman et al., 1968). Presumably, there is less airway collapse during exercise. However, it is not clear as to whether this is due to less forceful respiratory efforts or to a change in airway properties induced by the physical activity.

PERCEPTION OF BREATHING

Breathlessness

The sensation of shortness of breath commonly occurs when ventilation exceeds 50% of the MVV (C. F. Schmidt and Comroe, 1944). On this basis, some workers have calculated a dyspnea index:

$$\text{Dyspnea index} = \frac{\text{observed limiting exercise ventilation}}{\text{MVV}} \times 100 \tag{5.11}$$

However, R. Gilbert and Auchincloss (1969) found a wide variation of the index in self-limited exercise (utilization of 33–75% of ventilatory reserve). In their view, breathlessness was more closely related to the total pressure change over the breathing cycle than to the fraction of the ventilatory reserve that was utilized.

Breathlessness rarely halts the activity of a person who is accustomed to vigorous exercise, but it may be the terminal complaint of a sedentary subject. For this reason, some authors (B. Noble et al., 1973; Bakers and Tenney, 1970) advocate using a scale of perceived exertion (Borg, 1971) to rate the se-

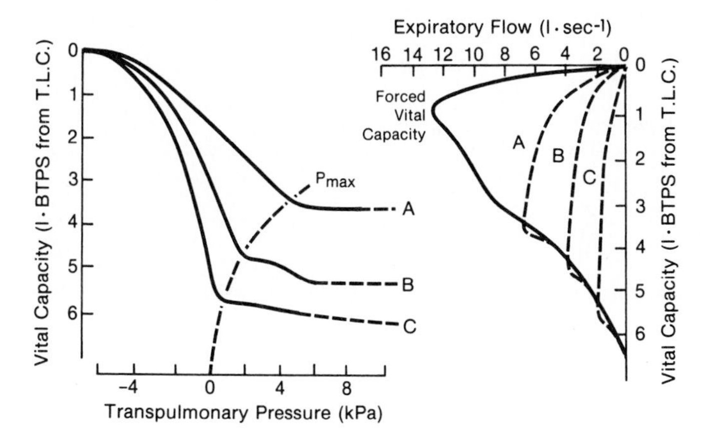

FIG. 5.16. Effects of external resistances upon forced expiratory flow–volume curves. Orifice A is 9 mm in diameter, B is 7 mm, and C is 5 mm. Note that in the latter part of expiration, flow is limited by airway characteristics rather than by the orifice. (After Olaffson and Hyatt, 1969.)

verity of respiratory symptoms. The subjective intensity of respiratory sensation seems related to the number (Halttunen, 1974) rather than the frequency (Viljanen, 1970,1972) of electrical impulses recorded from the respiratory muscles.

Discussion continues as to whether the breathlessness of vigorous exercise should be distinguished from the unpleasant sensation of labored breathing (dyspnea) encountered in certain forms of cardiorespiratory disease (J. B. L. Howell and Campbell, 1966). Kontos et al. (1964) noted that discomfort appeared at a 20% greater respiratory minute volume during exercise than during hypercapnia. Presumably, the minds of the subjects were diverted by their activity, and there may also have been a more direct modification of the sensory threshold by catecholamine secretion. Nevertheless, discussions with athletes and inspection of the sculptures of Tait MacKenzie (Fig. 5.17) suggest that the breathlessness of maximum effort is acutely unpleasant even for an experienced competitor.

E. J. M. Campbell (1966) suggested that the spindle receptors of the intercostal muscles provide the simplest and most complete explanation of respiratory sensations. Although their immediate projection is to spinal and cerebellar centers, the γ loop provides conscious proprioception of respiratory movement, allowing the body to distinguish elastic and viscous resistances to breathing (Pengelly, Rebuck, and Campbell, 1974). The γ loop also serves as a sounding board, allowing detection of inappropriate ventilatory effort. Fiber length and tension may be poorly matched in the intercostal muscles, ventilation may be disproportionate to physical activity, or the ventilation achieved may not match ventilatory demand. Since the tone of the diaphragm is largely unaffected by γ-loop activity (C. Von Euler, 1966a,b), Campbell's hypothesis would make dyspnea largely a phenomenon of the intercostal muscles. Further, the judgment of length/tension appropriateness would be essentially a learned response, capable of modification by habituation in both the

FIG. 5.17. Breathlessness (**left** and violent effort (**right**) of the athlete, as seen in the sculpture of Tait McKenzie. (''Violent Effort,'' by R. Tait McKenzie (for Plate VI, and ''Breathlessness,'' by R. Tait McKenzie (for Plate VII). From *R. Tait McKenzie: The Sculptor of Athletes*, by Andrew J. Kozar, reproduced by permission of The University of Tennessee Press. Copyright © 1975 by The University of Tennessee Press 37916.)

highly trained athlete and the patient with chronic respiratory or cardiac disease.

Pathological causes of intercostal length/tension disproportion could include a stiffening of the lung (pulmonary edema), stiffening of the rib cage (various bone and joint diseases), and narrowing of the airway (bronchospasm). An inappropriate sensation might equally arise in normal subjects through the use of equipment with a high respiratory resistance. It is unlikely that an accumulation of chemical stimuli makes a direct contribution to the feeling of breathlessness, although a buildup of lactate could be associated with dyspnea because it induces more intense respiratory effort.

"Second Wind"

Typical descriptions of "second wind" speak of a distress that develops after 30 sec to 1 min of heavy exercise, with relief appearing rather suddenly 1–3 min later. The predominant complaint is of breathlessness, but there may also be symptoms suggestive of an anxiety state, including rapid shallow breathing, a sense of constriction in the chest, throbbing or swimming in the head, muscle pains, and a fluttering, irregular pulse (Lefcoe and Yuhasz, 1971).

The relief of distress with continuation of exercise is controversial. There seem no changes of respiration or heart rates at the moment when athletes report second wind (Lefcoe and Yuhasz, 1971). In the treadmill experiment of Lefcoe and Yuhasz, 18 of 20 students noted some improvement of breathing, but the timing of relief varied from 2 to 18 min after the onset of exercise, 4 members of the group reporting relief on two separate occasions; 14 subjects also noted a lessening of leg soreness or fatigue, although in only 7 of these subjects was such relief concurrent with the improvement of respiratory symptoms.

Explanations of second wind (Shephard, 1974d,e) are far from satisfying. Ventilation may diminish concomitant with a decrease in blood levels of anerobic metabolites; at work loads demanding less than 70% of aerobic power, lactate accumulates in the first minute or so of exercise, but diminishes if the exercise continues (Fig. 2.7). Warm-up may also increase the efficiency of the respiratory muscles, with more rapid relaxation of the antagonists, and less stimulation of γ-loop receptors. The complaint of tightness in the chest is common in asthma, suggesting that the phenomenon may have a bronchospastic component. Certainly, exercise can induce bronchospasm (Shephard, 1977b). However, this is not a typical exercise response in healthy subjects (Table 5.1). Furthermore, in subjects who are liable to spasm, airway resistance reaches a maximum ~10 min following exercise. During activity, there is usually a diminution of airway resistance, perhaps related to the liberation of catecholamines (Shephard, 1966a). Nevertheless, it remains conceivable that an anxiety reaction could lead to hyperventilation, CO_2 washout, and hypocapnic bronchospasm in the early phases of exercise (Michaelis and Muller, 1942), with replenishment of carbon dioxide stores as effort continues (Issekutz and Rodahl, 1961). Another possibility is a metabolic fatigue of the intercostal muscles; this could arise from a slow adjustment of their regional circulation to the increased respiratory work load, or a subsequent diversion of blood flow from the chest wall to the active limbs, coupled with thoracic postural demands from the activity itself. Intercostal fatigue could account for both breathlessness (through an increased respiratory drive to the fatigued muscles) and an associated feeling of tightness in the chest. It would relate dyspnea rather closely to the diaphragmatic "stitch" discussed below. Certainly, there are situations in which dyspnea can be relieved by blocking the intercostal nerves (Gold and Nadel, 1966), and it is conceivable that local ischemia of the chest muscles could develop, to be relieved later as the systemic blood pressure rises. Respiratory muscle fatigue can plainly limit performance when exercise is carried out with external ventilatory loads (Fig. 5.6; Demedts and Anthonisen, 1973). However, if early intercostal fatigue developed with unimpeded ventilation, it would be difficult to explain how exercising subjects are able to augment their respiratory minute volumes voluntarily, without causing unpleasant symptoms.

Other explanations of dyspnea and its relief include airway collapse, a lack of surfactant, reactions to an increased pulmonary blood flow, and psychological sequelae of a precontest anxiety (Shephard, 1974d,e).

"Stitch"

A "stitch" is an unpleasant sensation associated with prolonged, vigorous effort. A sharp and rather severe pain is felt laterally over the lower part of the chest wall. Some authors have attributed the phenomenon to a spasm of the diaphragmatic muscles, while others have postulated that ischemia of the diaphragm is responsible. In support of the latter view, the symptom is common when circulatory adaptations to heavy work are slowed by a large, recent meal. On the other hand, it is quite possible to hyperventilate during near maximum effort (Shephard, 1967a), meeting the cost of such added respiratory activity without inducing a stitch.

MECHANICS OF BREATHING

Air Conditioning and the Role of the Nose

Most exercise physiologists use a mouthpiece and a noseclip to examine respiration. They are thus unable to determine the normal distribution of airflow between the mouth and the nose. Niinimaa and Cole et al. (1979) recently developed an apparatus to study this question. Total ventilation is measured by encasing the body from the neck downward in a rigid plethysmograph, while nasal flow is collected by a mask covering the upper half of the face. At light work loads, air is directed almost exclusively through the nose. However, the mouth opens at a ventilation of ~40 l • min^{-1} BTPS. The resistance of the nasal passages is relatively high (200 Pa • l^{-1} • sec^{-1}), and at higher work loads a large fraction of the respired air passes through the mouth.

The large surface of the nasal turbinate bones plays an important role in warming and humidifying inspired air. Animal experiments (Moritz, Henriques, and McLean, 1945; Moritz and Weisiger, 1945) suggest that air reaches the tracheobronchial tree at body temperature in the face of ambient conditions ranging from −100°C to +500°C. If work loads are sufficient to provoke oral respiration, the temperature and humidity of inspired gas are still regulated surprisingly effectively by the mouth and pharynx. During expiration, some of the heat and water vapor are recovered, but under cold, dry conditions respiratory water loss can amount to 300 ml • hr^{-1}.

Hard outdoor exercise in cold weather leads to oral respiration, with the impingement of cool, dry air on the tracheal and bronchial mucosae. The tracheal mucus is thickened, movements of the tracheal ciliae are slowed, and the tracheal nerve endings may be stimulated, precipitating both a local bronchospasm and a reflex spasm of the coronary vessels (Widdicombe, 1964). One method of circumventing such problems is to use a jogging mask; warm, moist air can then be inspired from within the subject's track suit (Kavanagh, 1970).

Another useful function of the nose is to filter out foreign matter. A combination of hairs around the external nares, turbulent airflow over the turbinate bones, and a large mucous surface allows impaction of large (>10 μm) particles (C. N. Davies, 1961) and absorption of up to 90% of noxious vapors, such as sulfur dioxide and anticholinesterases (Oberst, 1961). Smaller particles (1–5 μm) are deposited in the tracheobronchial tree by a combination of sedimentation and impaction; the cilia lining the airway carry them to the larynx, and they are then either expectorated or swallowed (P. E. Morrow et al., 1966; P. E. Morrow, 1970). The smallest particles penetrate to the alveoli, where they are ingested by wandering cells (alveolar macrophages); excretion is then via the lymphatics, blood stream, or respiratory tract (Newhouse et al., 1970).

Respiratory Workload

Ignoring for the moment complications presented by branching of the airway, turbulence, and non-sine-wave flow, the pressure P′ needed to move air in and out of the chest can be shown by the equation

$$P' = aV' + b\dot{V}' + c\ddot{V}' \qquad (5.12)$$

where V′ is the change of chest volume, $\dot{V}'$ is the rate of change of volume (velocity of air flow), $\ddot{V}'$ is the rate of change of velocity (acceleration), and a, b, and c are constants (Otis, Fenn, and Rahn, 1950). The first term (aV′) refers to the pressure developed against the elasticity of the rib cage, lungs, lining film of alveolar surfactant (Pattle, 1965), and any external forces (such as the difference between air and water pressure in a snorkel diver).

The potential energy stored in the elastic tissues of the expanded chest can perform viscous and inertial work during expiration. Utilization of this stored energy depends mainly upon the size of the second and third terms of equation (5.12). During vigorous exercise most of the elastic energy is recovered, but this is not possible during normal quiet breathing (Otis, 1964).

The second term ($b\dot{V}'$) relates to viscous work, performed in moving the air molecules over one another (airflow resistance) and in displacing the tissues (tissue resistance); it is by far the largest of the three terms during vigorous effort. The third term ($c\ddot{V}'$) refers to inertial work, performed in accelerating gas molecules and heavy viscera such as the liver and chest wall.

Compliance

A large surface tension might be anticipated at the air/tissue interface. This would have a major effect upon the mechanical properties of the lungs. However, in practice the alveolar macrophages secrete a thin (~5 nm) film of surfactant, a lipoprotein which lines the alveolar sacs; this reduces surface tension, averting both pulmonary collapse and the filling of the alveolar spaces with fluid (Pattle, 1965). The deep inspirations of vigorous exercise increase local reserves of alveolar surfactant, and thus decrease the likelihood that segments of the lung will collapse during expiration.

Because surfactant reduces alveolar forces, the elasticity of the thoracic pump is determined largely by the mechanical properties of the lungs and chest wall. Data are commonly expressed in compliance units. Compliance is the reciprocal of elasticity. The chest and lung compliance follow sigmoid curves (Otis, 1964; Fig. 5.18), and it is thus important to specify the lung volume at which measurements have been made.

Over the middle range, the compliance of the lungs (C_L) is ~2 l • kPa^{-1} (200 ml per cm H_2O) and the compliance of the chest wall (C_C) is of similar magnitude. The overall compliance of the chest and lungs (C_{C+L}) is obtained by reciprocal summation:

$$\frac{1}{C_{C+L}} = \frac{1}{C_C} + \frac{1}{C_L} \tag{5.13}$$

Under resting conditions, C_{C+L} thus amounts to ~1 l • kPa^{-1} (100 ml per cm H_2O).

During exercise, compliance could be modified by (1) an increase of mean alveolar volume (Fig. 5.18), (2) deeper breathing (with resultant opening of closed alveolar units, B. J. Ferris and Pollard, 1960), (3) more rapid breathing (the frequency dependence of compliance, only seen at very high breathing rates; Mead, Lindgren, and Gaensler, 1955; Macklem, 1970), and (4) increases of pulmonary blood flow and pulmonary blood volume (N. R. Frank, 1959). In practice, the resultant of these several factors leads to little change of compliance (Granath et al., 1959; Petit et al., 1960; Dejours, Bechtel-Labrousse et al., 1961; Chiang et al., 1965) other than effects attributable to the increase of tidal volume (R. Gilbert and Auchincloss, 1969).

The shape of the compliance curve has relevance not only to natural, but also to artificial, respiration. In order to produce a given tidal volume, a much larger pressure must be exerted during forcible deflation of the chest (back-pressure method) than during forcible inflation (arm lift, bellows, and mouth-to-mouth techniques). Considerable reductions of compliance may occur with pulmonary congestion and with interstitial or pleural fibrosis. A substantial positive pressure may thus be needed to produce even a small expansion of lungs congested by drowning. Rescuers who are being trained in the technique of mouth-to-mouth resuscitation should use mannikins that can simulate not only normal lung compliance, but also the reduced compliance of

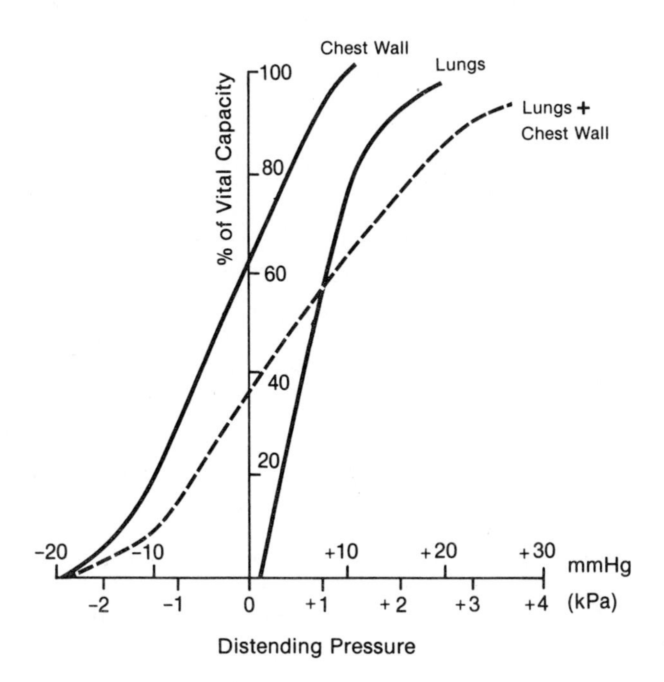

FIG. 5.18. The influence of lung volume upon the compliance of the chest wall, the lungs, and the lungs plus chest wall. (The compliance at any given lung volume is given by the ratio ΔV/ΔP.)

a typical casualty (W. R. Davies and Shephard, 1963; Shephard, 1966d).

A reduction of respiratory compliance may create problems when ventilation must be sustained by mechanical means. Some types of ventilatory equipment only deliver air until a certain cut-off pressure is reached. The purpose of this arrangement is to avoid damaging the lungs. However, if compliance is low, the limiting pressure (usually 2.7–4.0 kPa) may be developed before an adequate tidal volume has been delivered to the patient.

Hand-operated bellows resuscitators are also fitted with a relief valve to avoid lung damage. If the compliance of the chest and lungs is low, a vigorous compression stroke may cause most of the contents of such a bellows to be discharged via the relief valve instead of inflating the chest.

Resistance

Airflow Resistance

The resistance to air movement includes components related to laminar, turbulent, and orifice flow. During laminar flow, the gas molecules slip smoothly over one another; velocity is maximal at the center of the airway, and there is a thin film of still air adjacent to the wall. As air speed is increased, a critical velocity $\dot{V}'_c$ is reached where the orderly shearing process breaks down, and an irregular, turbulent pattern of flow develops (Fig. 5.19). The critical velocity can be calculated according to the equation

$$\dot{V}'_c = \frac{R\eta}{\delta r} \qquad (5.14)$$

where R is a constant (the Reynolds number, ~1000–2000 for a straight and smooth-walled tube), η is the viscosity of the gas, δ is its density, and r is the radius of the airway.

Flow is normally laminar in the smallest airways, but generalized turbulence develops in the nose, throat, and trachea, while more localized areas of turbulence are found at points of branching in the major bronchi.

Equation (5.12) is based simply on laminar flow. To a first approximation, the pressure drop associated with a steady laminar airflow is proportional to flow rate and gas viscosity, while the pressure drop for turbulent airflow is proportional to the square of the flow rate (Otis, Fenn, and Rahn, 1950). Classical theory summarized this information in the form of the equation (Rohrer, 1915):

$$P' = b_1(\dot{V}') + b_2(\dot{V}')^2 \qquad (5.15)$$

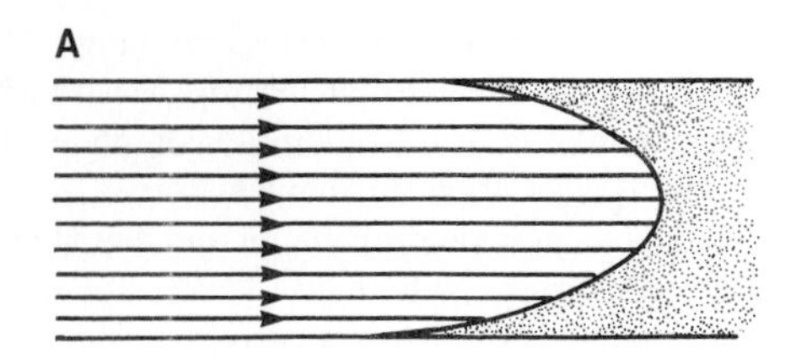

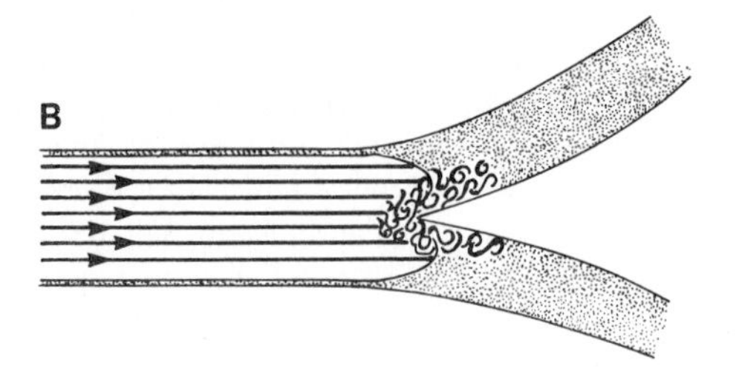

FIG. 5.19. Possible patterns of flow of a viscous fluid through a tube (air in the airways, blood in the arteries). **A.** Laminar flow. Still layer in contact with wall, maximum velocity at center of tube. **B.** Localized turbulence. Disturbances originating at branches of the tubular system. **C.** Generalized turbulence. Completely irregular pattern of fluid movement.

where P′ is the resistive pressure drop, V′ is the rate of airflow, and b_1 and b_2 are constants proportional to viscosity and density, respectively. Problems of analysis are posed by the complex waveform of respiratory flow (Varène and Jacquemin, 1970) and orifice effects (Jaeger and Mattys, 1970). There are thus some advantages to the adoption of an empirical nonlinear pressure/flow relationship of the type

$$P' = b(\dot{V}')^n \qquad (5.16)$$

where n is an exponent of ~1.66 (Ainsworth and Eveleigh, 1952; Shephard, 1959). Physiologists arbitrarily report resistance at a flow of 0.5 $l \cdot sec^{-1}$. When measured in this way, the resistance from the mouth to the alveoli is ~0.2 kPa (2 cm H_2O) per $l \cdot sec^{-1}$, while the resistance from the external nares to the oropharynx is of a similar order.

The bronchial tree comprises ~23 orders of bronchi (Fig. 5.28). At small lung volumes (25–50% of vital capacity), much of the airway resistance is attributable to the first three or four orders of the bronchial tree (Ainsworth and Shephard, 1961; Macklem, 1970). The resistance of this region is decreased markedly by further expansion of the lung (Macklem, 1970), but is increased by stimulation of tracheal nerve endings. Stimuli causing spasm of the large bronchi include carbon dioxide washout (Widdicombe, 1963; Coon and Kampine, 1977), cold air, particulate matter (Dautrebande, 1963), irritant gases, and the inhalation of vomit by an unconscious patient (Paintal, 1963,1973; Sampson et al., 1977). In contrast, the increase of airway resistance associated with allergic reactions (asthma and the anaphylactic shock induced by poison ivy and jellyfish stings) occurs mainly in the finest air passages.

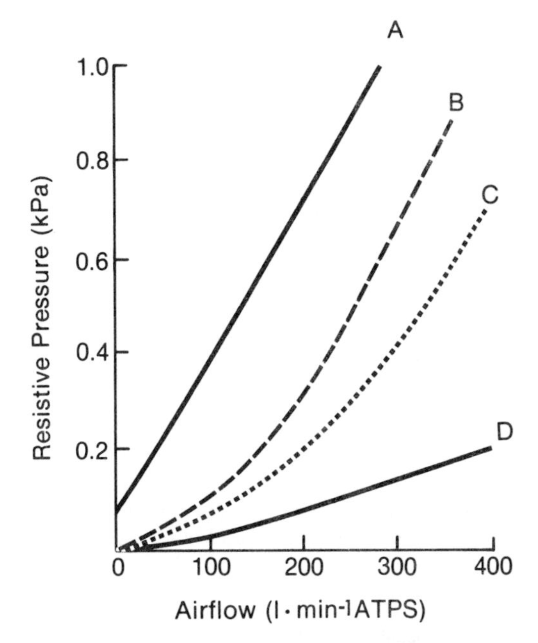

FIG. 5.20. Influence of an external resistance upon maximum voluntary ventilation. Line A, subjectively noticeable resistance. Line B, resistance leading to diminution of MVV. Line C, resistance tolerated without fall of MVV. Line D, ideal characteristics of respiratory equipment. (After Cotes, 1961.)

Exercise and Diving

During exercise, the increase of respiratory minute volume causes airflow turbulence to spread progressively (West, 1961) from the trachea as far as the seventh order of bronchial branching (Ainsworth and Shephard, 1961), and this in itself increases the effective airflow resistance. Turbulence becomes even more extensive when working at an increased ambient pressure (as in diving); in such situations, the work of breathing may become so large that the maximum voluntary ventilation drops to ~25 l • min BTPS, with a substantial resultant accumulation of carbon dioxide in the active tissues (Lanphier, 1969).

The use of low-density gas mixtures (such as 80% helium, 20% oxygen) is helpful to a diver from two points of view. Since the critical velocity is inversely proportional to gas density [equation (5.14)], the extent of turbulent airflow is reduced. Further, the coefficient of turbulent resistance [b_2 of equation (5.15)] decreases in proportion to the diminution of gas density. Unfortunately, helium has a somewhat greater viscosity than air, so that the laminar resistance coefficient b_1 rises by about 10%.

Added Resistance

During rest and light work, subjects tolerate a fourfold increase of airway resistance before they complain of difficulty in breathing. The smallest detectable resistive pressure is approximately a linear function of airflow (Fig. 5.20), whereas the pressures developed across an added resistance increase as a power function of airflow [equation (5.16)]. Bronchospasm is thus more readily detected during vigorous effort than when sitting at rest. A sedentary adult may not have a sufficiently large respiratory minute volume to notice the airway narrowing induced by smoking, but the endurance athlete knows that even the occasional cigarette seriously restricts his wind.

Military respirators (gas masks) are designed to allow vigorous physical activity and are thus required to have a total resistance of less than 1 kPa (10 cm H_2O) at an inspiratory flow of 85 l • min^{-1} (Shephard, 1961a). If this standard is not met, the subject is likely to complain of dyspnea. His respiration rate will be slowed (Zechman et al., 1957; S. Freedman, 1974) and physical effort will be limited by difficulty in breathing (Cotes, 1961; E. A. Cooper, 1961; Shephard, 1961a). In experiments with a respirator that met military standards, maximum effort on a bicycle ergometer was reduced by less than 5%, an effect roughly equivalent to that of adding 600 ml of external dead space (Table 5.7). However, the endurance time for an exhausting task was reduced from 9 to less than 4 min, and the time taken to cover an obstacle course was increased by 8–9% (Shephard, 1961a).

Some authors have suggested that inspiratory resistance is more unpleasant than an expiratory load. This may be true of moderate resistance, par-

TABLE 5.7
Maximum Effort of Subjects Exercised for 5 min on Bicycle Ergometer*

Condition	Maximum Work Performance (W)			
	Day 2	Day 3	Day 4	Δ Day 3
Control subjects (n = 6)	143 ± 5	148 ± 4	147 ± 5	+3.0 ± 0.8
Test subjects (n = 10)				
Respirator A day 3				
(immediate effect)	153 ± 2	146 ± 2	151 ± 2	−6.3 ± 2.3
(1 hr wear)	146 ± 3	143 ± 3	145 ± 3	−2.3 ± 2.2
Respiratory B day 3				
(immediate effect)	156 ± 4	152 ± 3	157 ± 4	−4.0 ± 1.3

*On day 3, two of the three groups of subjects were required to wear a military respirator.
Source: After Shephard, 1961a.

ticularly if expiratory flow is otherwise limited by airway collapse. However, the present author found that service volunteers disliked both inspiratory and expiratory resistances when working hard (Fig. 5.20). One practical implication of these observations is that an apparatus used in measuring ventilation should have a very low flow resistance (less than 0.15 kPa, 1.5 cm H_2O, at a steady flow of 100 l • min^{-1}; Shephard, 1977a).

Exercise-Induced Bronchospasm

Exercise is often accompanied by a small decrease of airway resistance, particularly if the air is cold (B. A. Wilson et al., 1978), but sensitive individuals subsequently show an increase of resistance (exercise-induced bronchospasm; Shephard, 1977b).

The initial bronchodilatation is due in part to release of catecholamines. These compounds produce an active dilatation of the bronchial muscles and also constrict blood vessels lining the airway (Capel and Smart, 1959; Bianco et al., 1974; Pierson and Bierman, 1975). A further factor reducing airflow resistance during exercise is an increase of mean alveolar volume (Fig. 5.21); this automatically causes a mechanical expansion of the airway (Shephard, 1959). Last, there may be some improvement of airway conductance due to a reopening of collapsed airways and an increased expectoration of mucus associated with the deeper breathing of exercise (Cropp, 1975).

Exercise-induced bronchospasm (EIB) has attracted recent interest, since several winners of inter-

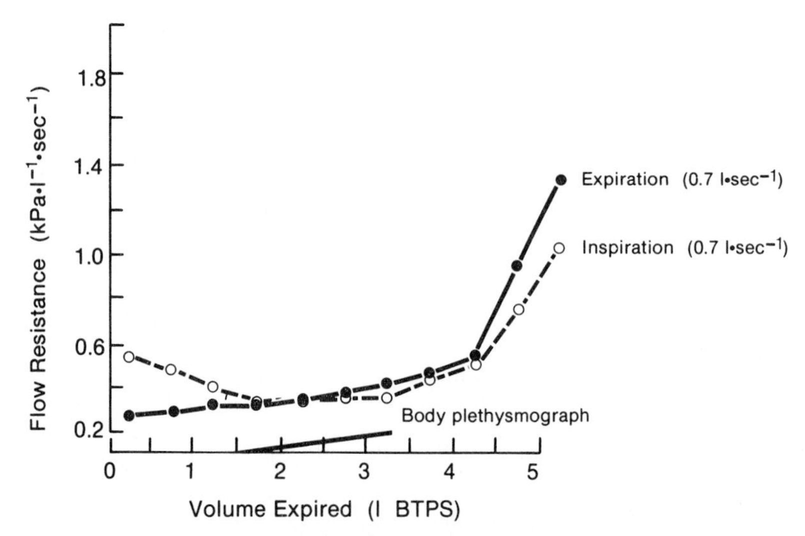

FIG. 5.21. Influence of lung volume upon airflow resistance. Note increase of flow resistance at lung volumes below the normal end-expiratory volume. (Subject of Shephard, 1959, inspiring or expiring at steady rate of 0.7 l • sec^{-1} BTPS; body plethysmograph data of Widdicombe, 1966.)

national athletic contests have been disqualified on the grounds that they took restricted sympathomimetic drugs to prevent or relieve EIB (Fitch, 1975b; Shephard, 1978c). Spasm typically appears a few minutes after effort, reaches a peak over the next 10–15 min, and resolves spontaneously within 40–60 min. The origin is multifactorial, the final common path being the smooth muscle of the airways (Shephard, 1977b; Bierman et al., 1977).

The bronchial muscle contracts in response to an increased concentration of cyclic guanosine monophosphate and relaxes in response to an increase of cyclic adenosine monophosphate (Middleton, 1975). The conversion of guanosine triphosphate to cyclic guanosine monophosphate is catalyzed by guanylate cyclase, an enzyme responsive to concentrations of calcium ions, acetylcholine, and prostaglandin F_2, while formation of cyclic adenosine monophosphate is catalyzed by adenylate cyclase, the so-called β receptor of the sympathetic nerve endings. Spasm can thus arise from an accumulation of acetylcholine (peripheral stimulation of vagal receptors by dust or dry air; central facilitation of vagal reflexes) or a decreased activity of sympathetic fibers. Prostaglandin F_2 is released in response to mechanical agitation of the lungs, hypoxia, respiratory alkalosis, and accumulation of acetylcholine, bradykinin, or histamine. Last, mast cells may be activated (by antigens, hypoxia, acidosis, changes in CO_2 tension, and various other humoral agents); the mast cells secrete histamine and slow-reacting substance A, both of which can cause bronchospasm. Many of these potential mechanisms contribute to EIB on different occasions.

Spasm is slight and difficult to demonstrate convincingly in normal individuals, but almost all asthmatic patients develop EIB if exercised under appropriate conditions (Silverman, 1973; Godfrey, 1974,1977). Some authors thus consider EIB as an attack of asthma induced by exercise (McFadden et al., 1977). Others find differences from asthma, EIB being characterized by a primary involvement of large airways (Buckley and Souhrada, 1975), a rapid reversal of signs and symptoms, and a different pattern of response to drugs (Oren, 1975).

EIB is most readily induced by running and arm cranking (S. D. Anderson, 1972), is less common after cycling, and is least readily induced by swimming (Godfrey et al., 1973; Godfrey, 1977). Effective treatments include selective β-sympathetic nerve agonists, theophylline, and cromolyn glycate. Use of the last of these drugs is permitted in international competition. It should be administered 1 hr prior to exercise (A. R. Morton and Fitch, 1974).

Exercise for Asthmatics

The response of asthmatics to training and exercise camps is generally favorable (Shephard, 1979a), although it is less clear whether changes are due to physical activity itself or the more general benefits of camp life. Given an appropriate choice of sport (such as swimming) and an adequate precontest warm-up, the asthmatic athlete can compete successfully at the highest levels of international competition (Fitch, 1975b).

Tissue Resistance

The tissue resistance is usually estimated as the difference between the total resistance to forced inflation of the chest and the airway resistance. Like most difference measurements, it is difficult to determine accurately (Shephard, 1966b). The normal value is ~50–100 Pa • l^{-1} • sec^{-1} (0.5–1.0 cm H_2O • l^{-1} • sec^{-1}). Part of the expiratory resistance is attributable to continued contraction of the inspiratory muscles (A. Taylor, 1960a); inspiratory activity in the early part of expiration builds up a store of elastic energy that can subsequently augment the power of expiration. In both phases of the respiratory cycle, relaxation of antagonistic muscles is usually incomplete; this component of tissue resistance can be reduced by either a local or a general warm-up of the tissues. A final component of tissue resistant reflects the rigidity of the thoracic cage; this increases markedly as a person gets older.

Inertia

The inertial work involved in acceleration of gas molecules and heavy abdominal viscera represents but a small part of the total energy cost of breathing. Typical values (Shephard, 1966b) are 1 Pa • l^{-1} • sec^{-2} (0.01 cm H_2O • l^{-1} • sec^{-2}) for both pulmonary gas and visceral inertance. Nevertheless, inertial effects can cause an apparent dependence of compliance upon breathing frequency (Dosman et al., 1975). Further, inertial forces operate most strongly in the narrower air passages, distal to the equal pressure point. Since the viscous resistance to airflow is low in this region, gas inertia may thus determine whether airway collapse occurs during expiration.

Inertia is important to the damping of oscillations within the airway (Shephard, 1966b). The normal bronchial tree is underdamped, and in consequence a distressing chatter may develop in the valves of any breathing equipment that is worn (Fig.

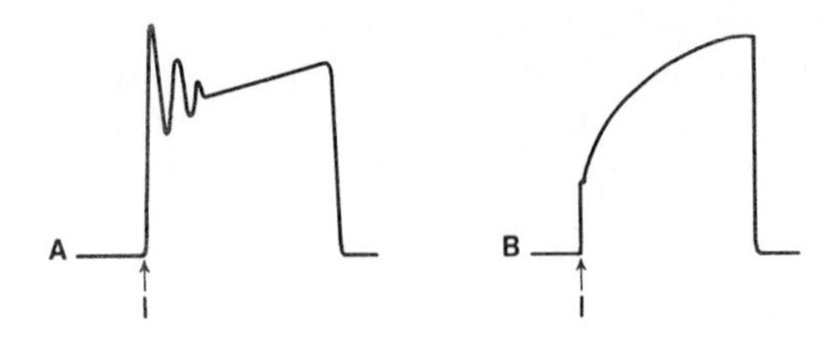

FIG. 5.22. Illustration of pressure changes at the mouth following the sudden interruption of airflow (I). **A.** Healthy young subject. Response is underdamped and overshoots the equilibrium value. **B.** Older, emphysematous subject. Response is overdamped, with a slow, exponential approach to the equilibrium value.

5.22). The damping ratio h for a simple breathing circuit is given by the formula

$$h = \frac{R'}{2}\sqrt{\frac{C'}{L'}} \tag{5.17}$$

where R′ is the resistance, C′ is the compliance, and L′ is the inertance. If h is less than unity, the system is underdamped and is liable to oscillate. Damping can be increased by a reduction of inertance (breathing a low-density gas mixture such as helium) or (less desirably) by an increase of R′ or C′. Conversely, the situation is worsened if the inertance is increased by breathing a dense gas mixture (as during diving).

The behavior of the human airway is more complex than this simple formula might suggest (Shephard, 1966b; Fig. 5.23). Nevertheless, damping is generally incomplete. Estimates based on oscillations of mouth pressure in young adults suggest that $h = 0.23 \pm 0.03$ for inspiration and 0.60 ± 0.03 for expiration (Shephard, 1966b).

Pressure/Volume Diagram

The total work performed per breathing cycle is indicated by the integral of pressure times volume. It is possible to plot a maximum pressure/volume diagram for the chest (Fig. 5.24), indicating the largest possible amount of work that can be performed by the respiratory muscles (Otis, 1964). The diagram is commonly drawn under static conditions, by having a subject draw varying fractions of the vital capacity into his lungs and then make a maximum inspiratory or expiratory effort against a closed valve. The volumes thus recorded are inevitably modified 5–15% by changes in pressure of the system during the maximum respiratory efforts. For this reason, it is not possible to determine maximum inspiratory or expiratory pressures at 100% of vital capacity.

During quiet breathing, only a very small fraction of the maximum pressure/volume diagram is utilized. The size of the pressure/volume loop increases with exercise, but even during maximum

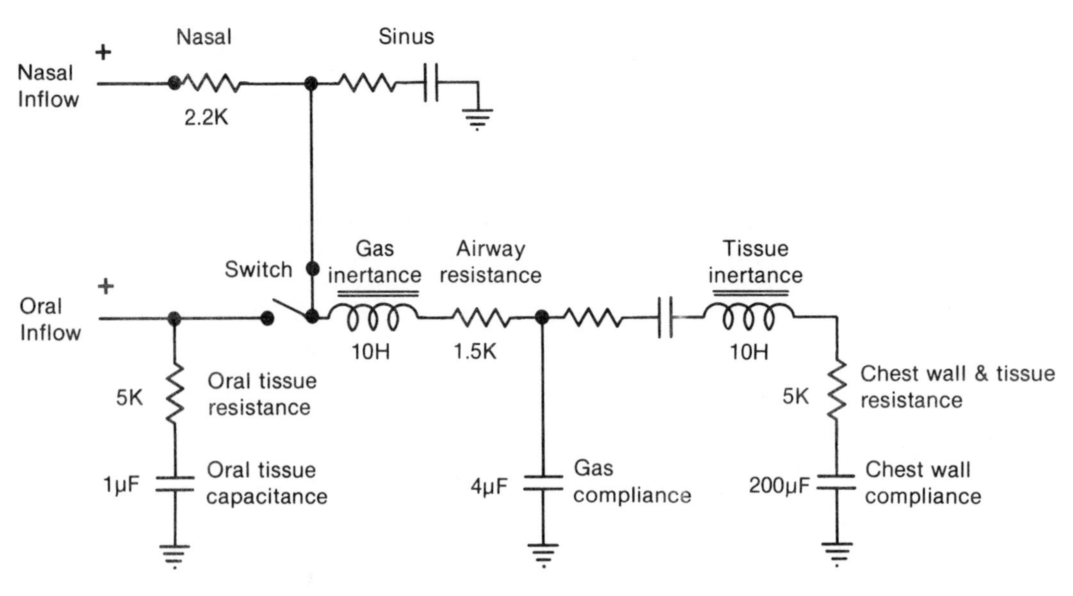

FIG. 5.23. Electrical analogue of impedance to gas flow in the respiratory tract. 1K resistance = 0.1 kPa • l^{-1} • sec^{-1} flow resistance. 1μF capacitance = 1 litre gas volume. 1H inductance = inertia of 0.1 kPa • l^{-1} • sec^{-2}. (After Shephard, 1966b.)

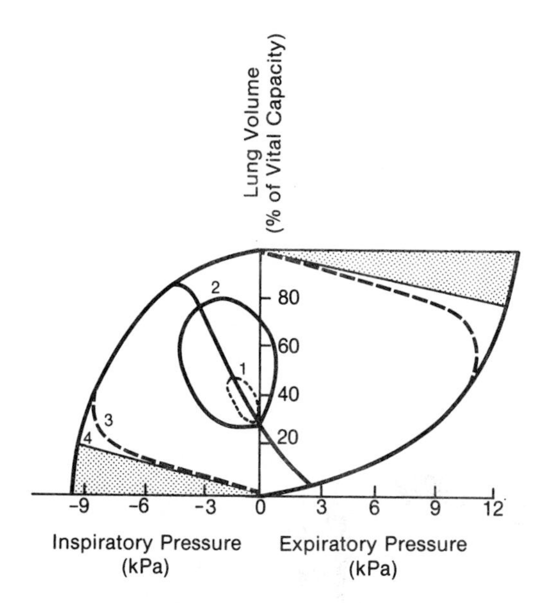

FIG. 5.24. Pressure-volume diagram for the chest. (1) Normal quiet breathing. (2) Maximum exercise. (3) Maximum voluntary ventilation. (4) Maximum static pressure-volume diagram. The shaded areas indicate the correction necessary for gas compression and expansion during development of the recorded pressures.

effort it rarely occupies more than a quarter of the static diagram (Hyatt, 1972). A maximum voluntary ventilation test yields a maximum dynamic pressure/volume diagram. The resultant surface is somewhat smaller than the static diagram, partly because volumes change before the muscles can develop their maximum tension and partly because limitations are imposed by the inherent force/velocity relationships of the respiratory muscles (Fig. 3.1).

Oxygen Cost of Breathing

The oxygen cost of breathing could be calculated theoretically from the work performed over a respiratory cycle (the area of the pressure/volume diagram). One difficulty with this approach is that the figures suggested for the mechanical efficiency of breathing range widely from 1 to 25% (Milic-Emili and Petit, 1959,1960). An alternative method is to increase ventilation (either voluntarily or by inhaling a carbon dioxide mixture) and note the resultant increase of oxygen consumption. The cost bears a nonlinear relationship to respiratory minute volume, being 0.5–1.0 ml of oxygen per l of ventilation during quiet breathing (Liljestrand, 1918; M. Nielsen, 1936; Shephard, 1955a; R. G. Bartlett et al., 1958; Otis, 1964), but increasing to as much as 5 ml • l^{-1} for the extra ventilation required during maximum exercise (Shephard, 1966a). In maximum effort, the chest muscles will consume 5–10% of the total oxygen intake.

One may ask whether the oxygen usage of the respiratory pump ever reaches a level at which a further increase of ventilation diminishes the amount of oxygen available to other body tissues (Fig. 5.25). Stated in mathematical terms, does the cost of a further increase in ventilation ($\Delta\dot{V}_{O_{2(R)}}/\Delta\dot{V}_E$) ever exceed the resulting gain in overall oxygen consumption ($\Delta\dot{V}_{O_2}/\Delta\dot{V}_E$)? In a healthy young person, such a situation is unlikely; $\Delta\dot{V}_{O_{2(R)}}/\Delta\dot{V}_E$ does not equal $\Delta\dot{V}_{O_2}/\Delta\dot{V}_E$ until the respiratory minute volume is 130–140 l • min^{-1} BTPS (Otis, 1964; Shephard, 1966a), whereas the maximum exercise ventilation is typically no more than 90–120 l • min^{-1} BTPS. Certainly, most authors agree that additional ventilation can be undertaken during maximum effort without impairing the performance of external work. The immediate problem at high ventilation rates is not so much an increase of $\Delta\dot{V}_{O_{2(R)}}/\Delta\dot{V}_E$ as a precipitous decrease of $\Delta\dot{V}_{O_2}/\Delta\dot{V}_E$ (Shephard, 1966a). Because of the shape of the oxygen dissociation curve of hemoglobin (Michel, 1974; Fig. 4.10), further augmentation of exercise ventilation does little to aid oxygen transport (Fig. 5.25).

Endurance athletes develop respiratory minute volumes as large as 160 l • min^{-1} BTPS, but because they have larger cardiac outputs, they still do not reach the critical point where $\Delta\dot{V}_{O_2}/\Delta\dot{V}_E$ falls precipitously. On the other hand, the effort tolerance of older subjects can be limited by the work of breathing, particularly if they suffer from emphysema; chronic chest disease sometimes causes a 10-fold increase in the oxygen cost of ventilation (Levison and Cherniak, 1968). Cigarette smoking also increases the oxygen cost of breathing, partly by causing an immediate bronchospasm and partly through more long-term effects such as the stimulation of an increased secretion of mucus. Typically, respiratory muscle oxygen usage of a smoker is twice that for a healthy nonsmoker of the same age (Rode and Shephard, 1971). In young adults, the effects of smoking are reversed by a few days of abstinence, but in middle-aged and older adults it may take months or even years for the oxygen cost to drop to that of a nonsmoker.

A final question is whether the respiratory muscles become fatigued by prolonged maximum effort. If this were to occur, respiratory work might limit endurance activity before the crossover point of Fig. 5.25 was reached.

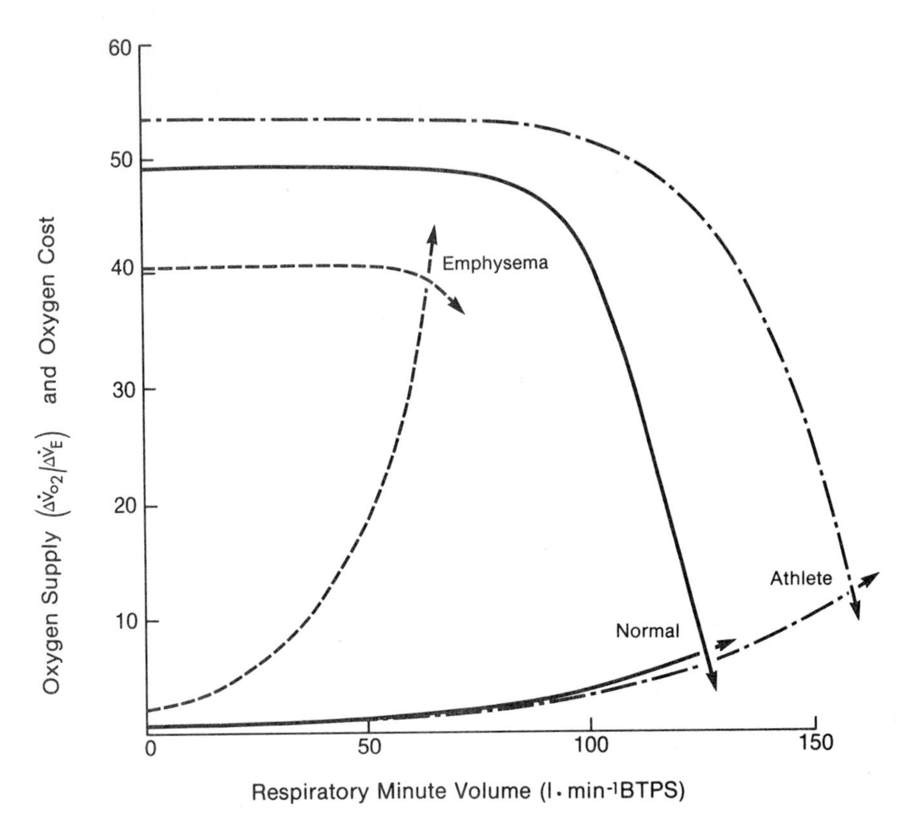

FIG. 5.25. The concept of a critical cost of breathing, where the oxygen consumed by the respiratory muscles ($\Delta\dot{V}_{O_{2(R)}}/\Delta\dot{V}_E$) exceeds the oxygen supplied by the additional ventilation ($\Delta\dot{V}_{O_2}/\Delta\dot{V}_E$).

Optimal Rate of Breathing

At slow breathing rates, a substantial amount of work is performed against the elastic forces of the lungs and chest, and not all of the stored energy is recovered during expiration. Furthermore, a decrease of end expiratory volume may cause narrowing and/or collapse of the airway, with an increase of viscous resistance (Shephard, 1959; Cheng et al., 1959; Fig. 5.21). In contrast, at fast breathing rates an excessive amount of work is performed against viscous forces. From a mechanical point of view, there is thus an optimum rate of breathing which varies with the intensity of physical activity (Yamashiro et al., 1975).

As the respiratory minute volume increases, so the optimum frequency of breathing rises (Milic-Emili, Petit, and Deroanne, 1960). However, the optimum rarely exceeds 40 breaths per min, even in maximum effort. Perhaps because costs are low, respiratory work is not always minimized under resting conditions. Nevertheless, during vigorous activity, the spontaneously selected breathing rate corresponds closely with the optimum over a wide range of respiratory minute volumes (Figs. 5.26 and 5.27; Grodins and Yamashiro, 1977).

Respiratory Muscles

Techniques

Studies of respiratory muscle function in humans have been based on anatomical inference and electromyography; in animals, added information has been obtained through microsphere determinations of blood flow distribution (C. H. Robertson et al., 1977a,b).

The Diaphragm

Electromyography suggests that in normal quiet breathing, the main responsibility for respiratory effort is borne by the diaphragm (M. D. Goldman, 1977). The efficiency of this muscle depends on its radius of curvature and the chest wall configuration (R. J. Marshall, 1962; Pengelly, Alderson, and Milic-Emili, 1971; M. Green, Mead, and Sears, 1974). The relative contributions of the diaphragm

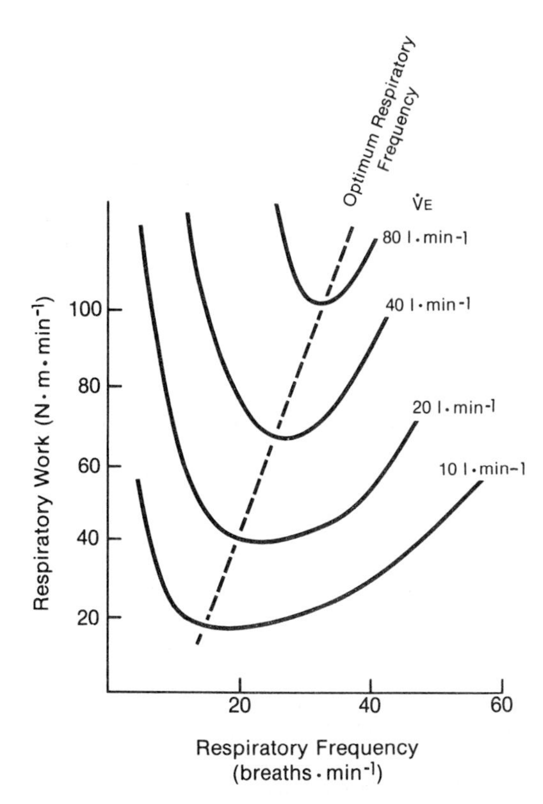

FIG. 5.26. The relationship between optimum frequency of breathing, respiratory minute volume, and work of breathing. Note that as respiratory minute volume increases, the acceptable zone of respiratory frequencies is narrowed. (Based in part on Milic-Emili & Petit, 1959.)

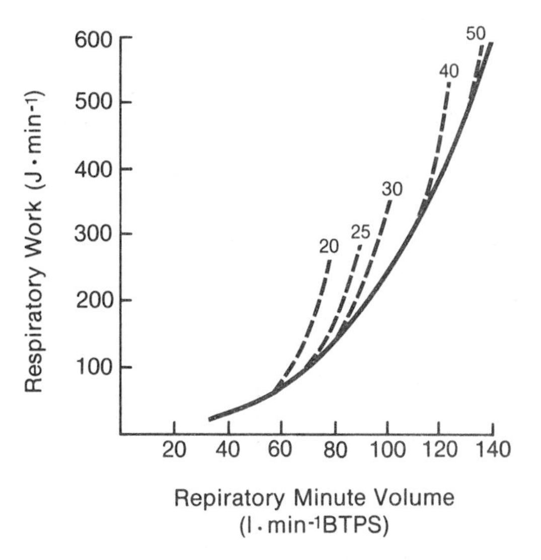

FIG. 5.27. Relationship between respiratory minute volume and respiratory work load. Solid line, spontaneously selected respiratory frequency; interrupted lines, controlled respiratory frequencies. (After Milic-Emili, Petit & Deroanne, 1960.)

and of other muscles thus vary with posture. If inspiration is produced by descent of the diaphragm, it is necessary to displace the liver and the other abdominal viscera in a downward and forward direction; this is most readily accomplished when a subject is standing erect or sitting with his abdominal muscles relaxed.

Intercostal Muscles

Electromyograms commonly show activity of the intercostal muscles during inspiration and of the internal intercostals during expiration (A. Taylor, 1960a; E. M. J. Campbell et al., 1970). The external intercostal muscles slope obliquely downward and forward; because leverage is greater for their lower insertions, contraction causes an elevation and outward rotation of the ribs, increasing the dimensions of the rib cage. The elastic forces of lung tissue, chest wall, and abdomen tend to empty the chest after the inspiratory muscles have relaxed. However, the internal intercostal fibers (which run at right angles to the external intercostals) can also produce a forcible deflation of the chest.

Under resting conditions, the practical importance of the intercostal contribution is uncertain. Much of the observed electrical activity may indicate mere stabilization of the chest or a passive adaptation to changing thoracic dimensions (D. S. Jones et al., 1953). C. S. Von Euler (1966b) described distinct efferent pathways from the intercostal fibers; one is linked to the respiratory center, while the other (with a tonic discharge) is linked to the cerebellum and is concerned with postural adjustments.

Abdominal and Accessory Muscles

As the respiratory minute volume is increased, various muscle groups are recruited for the task of breathing. The pattern of recruitment differs somewhat for CO_2 inhalation and external work (Remmers and Bartlett, 1977). When performing external work, the internal intercostals are activated first, and account for a major part of the energy cost of expiration at moderate loads (C. H. Robertson et al., 1977a,b). With more vigorous effort, the transversus abdominis, the internal obliques, and the external oblique muscles are successively brought into play. The abdominal muscles apparently operate at a mechanical disadvantage, working against the diaphragm (Agostoni and Rahn, 1960); nevertheless, they begin to make an active contribution to expiration at work loads of 50–100 W (M. D. Goldman, 1974,1977). At this stage various accessory muscles, such as the scalenes and the sternomastoids, contribute to inspiratory effort. The efficiency of the accessory muscles depends on the nature of the physical activity being undertaken. They function more effectively if there is external fixation of the shoulder girdle (as when riding a bicycle). However, the maximum contribution of the thoracic muscles to the ventilatory effort is relatively small, and if the thoracic cage is immobilized by a rigid binder, voluntary ventilation is reduced by no more than 20–30%.

ALVEOLAR VENTILATION

Measurement of Dead Space

A part of the respiratory effort is wasted in displacing gas from the dead space of the conducting airways (Fig. 5.28) and any external respiratory equipment, such as a snorkel tube. The extent of this inefficiency can be estimated if we calculate the ratio of dead space to tidal volume, using the classical Bohr equation:

$$\frac{V_D}{V_T} = \frac{P_{A,CO_2} - P_{E,CO_2}}{P_{A,CO_2} - P_{I,CO_2}} \qquad (5.18)$$

The partial pressures of carbon dioxide in inspired gas (P_{I,CO_2}) and expired gas (P_{E,CO_2}) are readily determined. The corresponding alveolar pressure (P_{A,CO_2}) presents a little more difficulty. The value required is an ideal figure that would permit the observed gas exchange; it is an appropriately weighted spatial and temporal average representative of the multitude of alveolar units over a typical breathing cycle.

Equilibrium of carbon dioxide across the alveolar/capillary membrane is fairly complete, and under many conditions the best measure of the ideal alveolar carbon dioxide tension is provided by specimens of arterial or arterialized capillary blood. A second possibility is to collect end-tidal gas specimens.* At rest and during moderate exercise, there is a close agreement between arterial and end-tidal gas pressures (Rahn, 1951; Suskind et al., 1951), the small discrepancy being due to temporal and spatial mismatching of ventilation and perfusion, the so-called alveolar component of dead space (Comroe et al., 1955; Farhi, 1966; J. B. West, 1966). During vigorous exercise, the end-tidal sample may have a higher CO_2 pressure than arterial blood (Matell, 1963; N. L. Jones, McHardy et al., 1966). This is due in part to larger oscillations of alveolar gas composition over the breathing cycle (Aitken and Clark-Kennedy, 1928; Shephard, 1968b; Nye, 1970) and in part to a more uniform distribution of ventilation and perfusion. Problems can also arise from the effect of core temperature upon blood gas tensions (C. A. Bradley et al., 1976). A 1°C rise of blood temperature causes a 4.4% rise of Pa,O_2 and a 6.6% rise of Pa,CO_2 (Holmgren and McIlroy, 1964), but this will not be detected if specimens are measured in a water bath at 37°C.

*The end-tidal sample is the final portion of a normal expirate. It may be collected mechanically or noted from a continuous record of expired gas concentrations.

Dead Space in Rest and Exercise

The V_D/V_T ratio of a resting subject is about 0.3, as can be seen from the corresponding gas pressures (P_{A,CO_2} = 5.3 kPa, 40 mmHg; P_{E,CO_2} = 3.7 kPa, 28 mmHg; P_{I,CO_2} ~0 kPa). The dead space of 150–180 ml is largely anatomical, an expression of the volume of the conducting airways; the alveolar component amounts to 30 ml or less (Shephard, 1959; West, 1974). During exercise, two main factors increase the anatomical dead space: a physical expansion of the airways and a reduction of the gas-mixing time (Fig. 5.29).

Inspiration leads to an increase in the volume of both alveolar units and conducting airways, expansion of a given element being roughly proportional to its initial volume. Let us consider an airway that comprises a relatively inelastic portion (mouth, throat, and larynx, volume 40 ml) and an elastic portion (trachea, bronchi, and bronchioles, volume 110 ml). If the initial lung volume is 3000 ml and a breath of 1000 ml is inhaled, then the conducting airway expands by 110 × (1000/3000), or 37 ml. In maximum exercise (tidal volume 3000 ml), physical expansion of the airway would thus add ~100 ml to the dead space.

A more significant problem is created by a shortening of the gas mixing time. Exchange between alveolar gas and newly inhaled air is dependent on the retrograde diffusion of mixed gas from the lungs through the smaller to the larger air passages. The postinspiratory phase must last for 2–3 sec (Fig. 5.29) if this diffusional exchange is to be reasonably complete (Roos et al., 1955; Shephard, 1956a; Shephard and Bar-Or, 1970). When the respiratory rate rises to

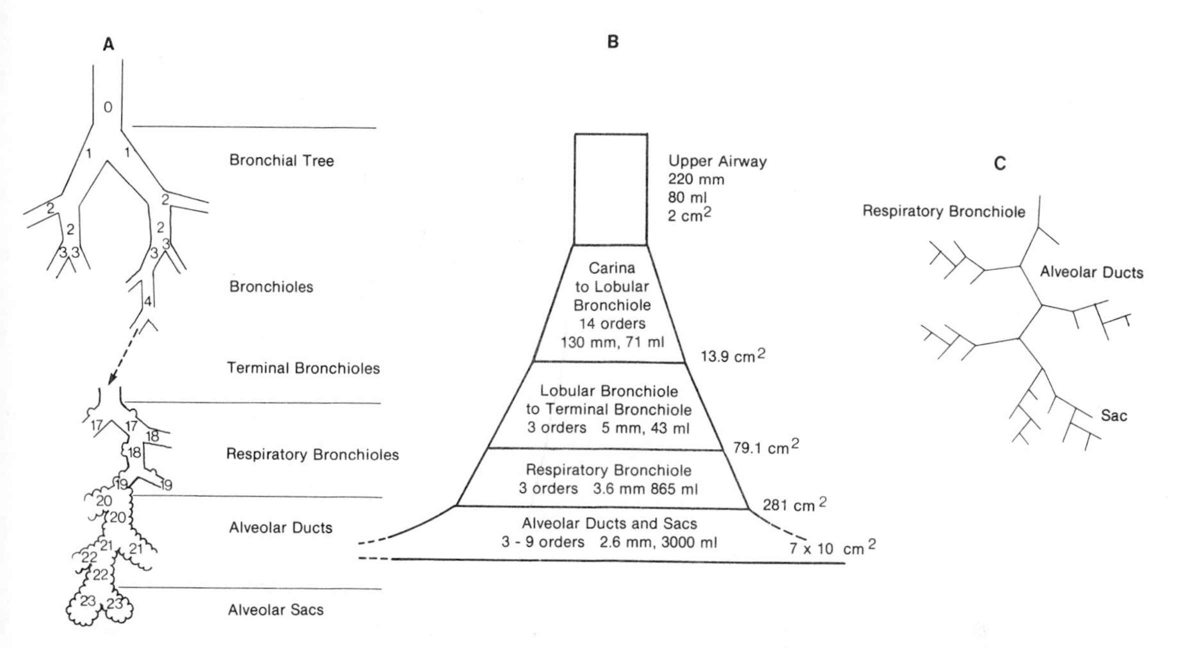

FIG. 5.28. **A.** A formalized anatomy of the bronchial tree (after Weibel, 1963). Weibel's schema assumes symmetric dichotomous branching. **B.** and **C.** A more recent model (G. Cumming, Horsfield & Preston, 1971; Thurlbeck & Wang, 1974; Yeates & Aspin, 1978) has allowed for asymmetry and a differing total number of branches between the trachea and the alveolar sacs. The number of alveoli ranges from 200 to 600 • 10^6, depending upon the individual's age and stature (Angus & Thurlbeck, 1972). At three-quarters of maximum inflation, 65% of the lung volume is contained within the alveolar sacs (Weibel, 1972).

35–40 breaths per min, such time is no longer available.

Computer calculations based on the passage of gas into a theoretical model of the bronchial tree (Fig. 5.28; Weibel, 1963; Wilson and Lin, 1970; LaForce and Lewis, 1970; Beeckmans and Shephard, 1971; G. Cumming et al., 1971; G. Cumming, 1974) suggest that during maximum effort the dead space increases to as much as 500 or 600 ml (Beeckmans and Shephard, 1971). In practice, volumes of 300–400 ml are more usual (Table 5.8), and it may be that diffusional mixing is supplemented by some mechanisms such as axial gas flow,* turbulent dispersion, or a physical massaging of the airways by the lungs and the heart (Beeckmans and Shephard, 1971; Hogg et al., 1972; Fukuchi et al., 1976; Mazzone et al., 1976; Ultman et al., 1978; Taulbee et al., 1978). Nevertheless, recent comparisons of helium and sulfur hexafluoride dead spaces stress that gas mixing is strongly dependent upon the time available for diffusion, both within the alveolar spaces and in the upper airway (Kawashiro et al., 1976; Sikand et al., 1976; Worth et al., 1977; Ultman and Blatman, 1977; Horsfield et al., 1977).

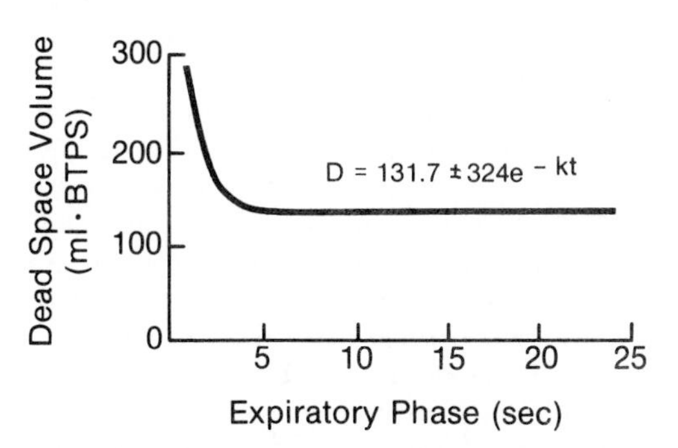

FIG. 5.29. Influence of expiratory phase duration upon volume and nitrogen dead space. (Data of Shephard, 1956a, for resting conditions.)

If the Bohr calculation is based on arterial rather than alveolar gas samples, the V_D estimate includes the alveolar component of dead space. The calculated V_D/V_T ratio for vigorous exercise then lies

*With axial or laminar flow, air passes through the center of a bronchiole, leaving a film of alveolar gas lining its wall (compare Fig. 5.12).

TABLE 5.8

Relationship Between Respiratory Frequency and Dead Space/Tidal Volume Ratio for Young Adult Men Performing Near Maximal Exercise[a]

	Observed V_D/V_T Ratio		
Respiratory Frequency[b]	Based on Arterialized Capillary Blood	Based on Mid-Tidal Samples	Based on End-Tidal Samples
25	0.232 ± 0.035	0.249 ± 0.042	0.294 ± 0.044
32*	0.225 ± 0.35	0.227 ± 0.039	0.279 ± 0.039
35	0.240 ± 0.047	0.237 ± 0.026	0.285 ± 0.033
45	0.293 ± 0.051	0.257 ± 0.010	0.298 ± 0.020
55	0.267 ± 0.062	0.252 ± 0.022	0.289 ± 0.024

[a]Mean ± S.D. of data.
[b]Controlled by metronome, with exception of spontaneously selected breathing frequency (marked by asterisk).
Source: From Shephard and Bar-Or, 1970.

in the range 0.20–0.25 (Table 5.8). The lowest ratio (and thus the most efficient ventilation from the viewpoint of gas exchange) is observed at or near the usual breathing frequency for this intensity of work (35–40 breaths per min).

Ventilation/Perfusion Ratios

Under resting conditions, a fall of alveolar oxygen pressure causes alveolar vasoconstriction while a fall of alveolar carbon dioxide pressure causes bronchoconstriction. Nevertheless, matching of ventilation and perfusion is far from complete (J. B. West, 1962a). Indeed, in a seated subject the upper part of the lung is grossly overventilated (ventilation/perfusion ratio of 3.3), while the lower part of the lung is somewhat overperfused (ventilation/perfusion ratio of 0.6).

Detailed knowledge of changes in alveolar dead space during exercise is lacking. In moderate exercise, a rise of pulmonary arterial pressure increases perfusion of the upper part of the lung (Dollery et al., 1960; Bake et al., 1968), and an increase of tidal volume increases the uniformity of alveolar ventilation (Fig. 5.30). The ventilation/perfusion gradient thus shows a much smaller gravitational gradient than at rest (Bryan et al., 1964; Harf et al., 1978). Considerable inequalities of ventilation and perfusion may persist at a given level within the lung (Gledhill et al., 1977), but the alveolar component becomes a minor fraction of the total dead space during vigorous exercise, unless, the subject deliberately hyperventilates; this immediately leads to a sharp worsening of ventilation/perfusion relationships, and creates a substantial alveolar dead space (Shephard and Bar-Or, 1970).

PULMONARY DIFFUSION

Principles of Measurement

The pulmonary diffusing capacity, or transfer factor as it is sometimes called, indicates the number of ml of a gas exchanged between the alveoli and the pulmonary capillary blood per min per unit of pressure gradient. The barriers to be traversed include the alveolar epithelium, the alveolar basement membrane, the interstitial tissue, the capillary basement membrane and endothelium, the plasma, and the red cell membrane. The harmonic mean thickness of the several layers of tissue is ~0.55 μm and the area of the exchange surface is ~80 m² (Weibel, 1970,1973; Scrimshire et al., 1973). As we have seen, the relevant pressure gradient is from alveolar gas to mean pulmonary capillary blood (Shephard, 1977a).

Carbon monoxide is commonly used to measure pulmonary diffusing capacity, since the mean pulmonary capillary pressure of oxygen is technically difficult to determine. On account of differences in molecular size and water solubility, the oxygen diffusing capacity ($\dot{D}_{L,O_2}$) is 1.23 times as great as that for carbon monoxide ($\dot{D}_{L,CO}$).

It was originally hoped that because carbon monoxide had a very great affinity for hemoglobin, the back-pressure of this gas in the pulmonary capillary blood could be ignored when calculating diffusing capacity. Unfortunately, this is not the case. An ordinary city dweller has ~1.0% carboxyhemoglobin in his blood (the combined result of endogenous carbon monoxide production and exposure to air pollutants, particularly car exhaust fumes). If the subject has a particularly heavy environmental exposure (for example, a traffic policeman or an attendant in a car-parking tower), his blood may contain 2 or 3% car-

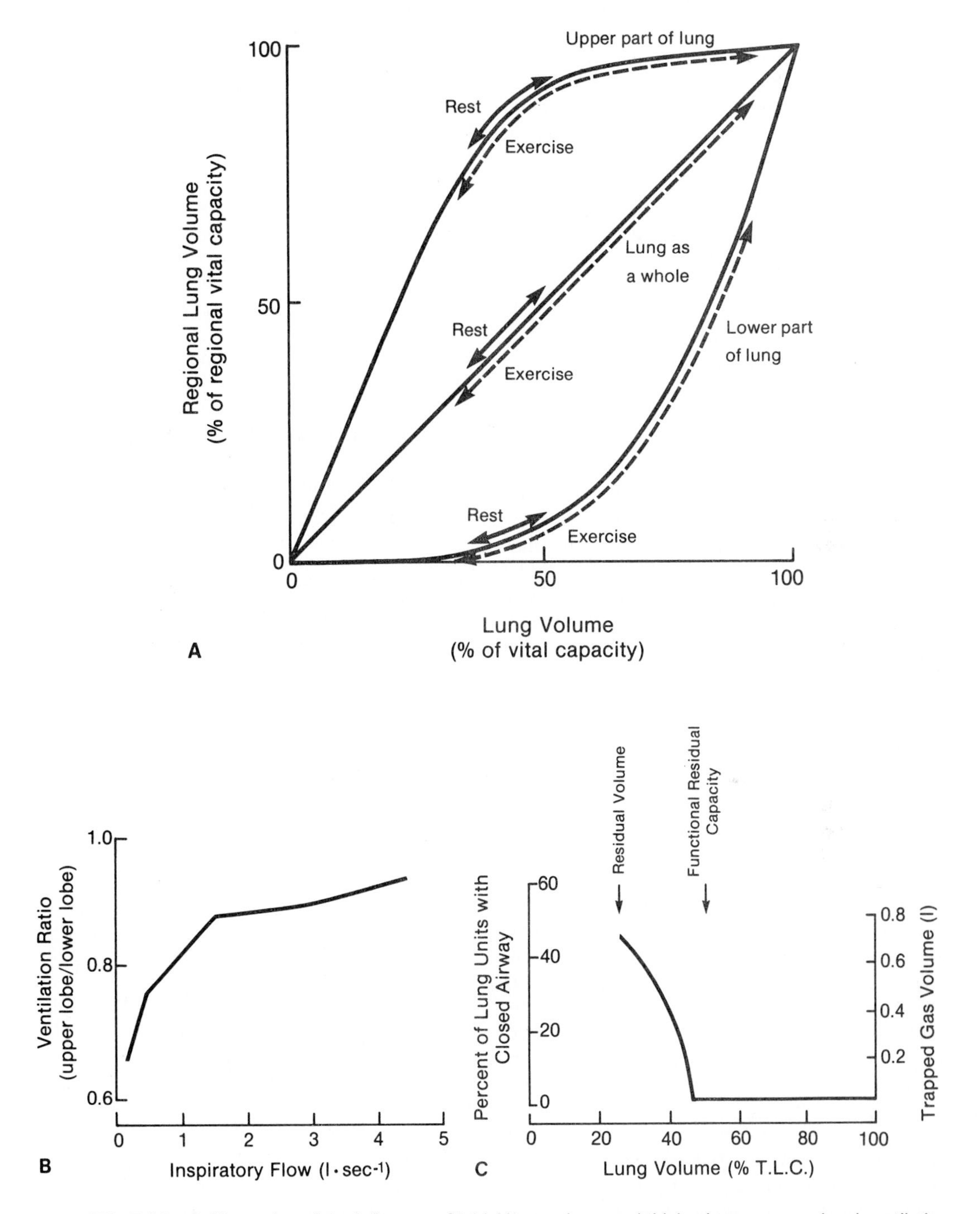

FIG. 5.30. A. Illustration of the influence of initial lung volume and tidal volume upon regional ventilation of the lungs. The upper part of the lung fills early during inspiration, but because of greater gravitational forces acting on the pleural fluid, the lower part of the lung does not fill until inspiration is nearing completion. The larger tidal volume of vigorous effort leads to ventilation of both upper and lower regions of the lung (based on experiments of Milic-Emili, Henderson et al., 1966). **B.** Influence of inspiratory flow rate upon relative ventilation of upper and lower lobes of the lungs (after Bake, 1977). **C.** Number of lung units with closed airways and trapped gas volume at various alveolar volumes. (After Milic-Emili, 1974.)

boxyhemoglobin (Ramsey, 1967; Godin, Wright, and Shephard, 1972; G. R. Wright, Jewczyk et al., 1975), and if he is a smoker, an average daytime value of 5-6% carboxyhemoglobin may be anticipated (Lawther, 1967; Rode, Ross, and Shephard, 1972). Unless account is taken of back-pressure, the resting diffusing capacity of a smoker is underestimated by 20–30% (T. W. Anderson and Shephard, 1968a).

There are two basic methods of measuring carbon monoxide diffusing capacity. In the single-breath method, a large volume of 0.1% carbon monoxide is inhaled, and the diffusing capacity is deduced from the change in alveolar carbon monoxide concentration over a 10-sec period of breath holding. The prolonged interruption of respiration is somewhat artificial even under resting conditions and is not practical during vigorous exercise. The alternative, steady-state procedure measures the uptake of carbon monoxide over a specified period of several minutes and relates this to the pressure gradient between alveolar gas and pulmonary capillary blood. It is then necessary to determine an appropriate mean alveolar carbon monoxide pressure. Some authors have done this by measuring the carbon monoxide content of the expired gas, assuming a figure for dead space volume, and applying the Bohr equation. In the resting state, the dead space can be predicted fairly accurately, but unfortunately there is much controversy concerning appropriate estimates for vigorous exercise. Others have assumed that the end-tidal samples are representative of alveolar gas concentration. This is not always the case; alveolar carbon monoxide levels vary widely over the breathing cycle, particularly if the respiratory rate is slow, and sometimes the average alveolar carbon monoxide pressure is closer to the mid-tidal than to the end-tidal reading. Reliance upon the end-tidal reading leads to overestimation of the pulmonary diffusing capacity in subjects with a slow and deep pattern of breathing (T. W. Anderson and Shephard, 1968a; Kindig and Hazlett, 1974), and this may account in part for the large reported diffusing capacity of athletes (Bates et al., 1955; Bannister et al., 1960; Newman et al., 1962), especially swimmers (Mostyn et al., 1963). A third approach is to measure CO_2 dead space by collecting arterial or arterialized capillary blood, and to assume that the CO and CO_2 dead spaces are identical. This is acceptable for quiet breathing but can lead to difficulties at the more rapid respiratory rates encountered in vigorous exercise (T. W. Anderson and Shephard, 1968a).

Response to Exercise

The resting CO diffusing capacity is ~150–190 ml • min^{-1} • kPa^{-1}, or 20–25 ml • min^{-1} • $mmHg^{-1}$. This increases progressively with exercise, reaching a maximum of perhaps 450–525 ml • min^{-1} • kPa^{-1} or 60–70 ml • min^{-1} • $mmHg^{-1}$ in a moderately active young man. Some authors have suggested that diffusing capacity plateaus before the maximum oxygen intake has been reached (Holmgren, 1965; Scherrer and Bitterli, 1970). Most reports have not confirmed this (R. L. Johnson et al., 1965; T. W. Anderson and Shephard, 1968c; di Prampero, Cerretelli, and Piiper, 1969; Pirnay, Fassotte et al., 1969; W. H. Lawson, 1970; Jebavy and Widimski, 1973; G. M. Andrew and Baines, 1974), although in some individuals (possibly those who are less well trained) the slope of the $\dot{D}_L$/heart rate line does become less steep between heart rates of 120 and 155 per min (Fig. 5.31; T. W. Anderson and Shephard, 1968c; K. L. Andersen and Magel, 1970).

The diffusing capacity ($\dot{D}_L$) samples two functional components, one related to the characteristics of the alveolar membrane ($\dot{D}_M$; Chinard, 1966) and the other to the rate of reaction (θ) between carbon monoxide and hemoglobin within the capillary blood volume (V_c). The two components behave as series conductances:

$$\frac{1}{\dot{D}_L} = \frac{1}{\dot{D}_M} + \frac{1}{\theta V_c} \qquad (5.19)$$

At rest, V_c lies in the range 50–100 ml, while the value of θ is ~0.6–0.7 (R. E. Forster, 1964). Thus the terms $\dot{D}_M$ and θV_c have an almost equal influence upon $\dot{D}_L$. Both $\dot{D}_M$ and θV_c increase with exercise, but because the measurements have a large experimental error, there is disagreement as to which of the two components increases the more.

At rest, only about a fifth of the alveolar surface of 80 m^2 is covered by patent capillaries (Weibel, 1970,1973). The increase of $\dot{D}_M$ reflects the expansion of existing capillaries to cover more of the available surface and/or the opening up of additional pulmonary capillaries. Both types of response also increase the value of V_c.

Practical Importance of Oxygen Diffusion

There has been much discussion as to whether athletes gain a competitive advantage from an above

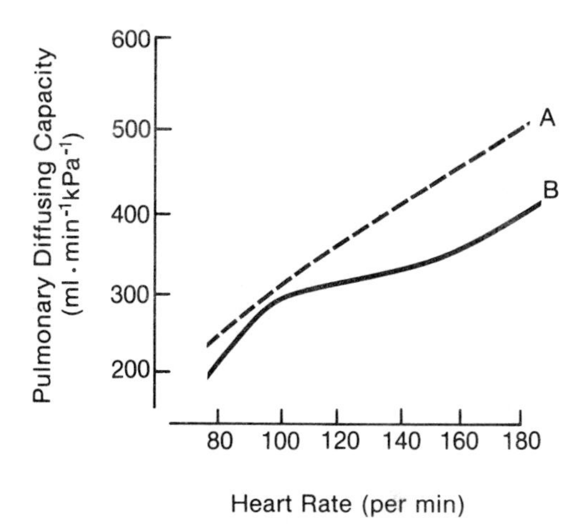

FIG. 5.31. The relationship between heart rate and pulmonary carbon monoxide diffusing capacity during exercise. In subject A, the relationship is relatively linear from light work to near maximum effort, but in subject B the slope is significantly flattened at heart rates between 120 and 155 • min^{-1}. (Based on data of T. W. Anderson & Shephard, 1968c.)

average pulmonary diffusing capacity. A part of their supposedly large D_L is a technical artifact associated with a slow rate of breathing (Table 5.9). Nevertheless, athletes often have a slightly larger diffusing capacity than sedentary individuals (Bannister et al., 1960; T. W. Anderson and Shephard, 1968b). Furthermore, although physical training does not change the diffusing capacity during submaximal work (T. W. Anderson and Shephard, 1968b; J. S. Hanson, 1969), it does lead to a small increase of maximum diffusing capacity, this being roughly proportional to the increase of cardiac output and maximum oxygen intake (T. W. Anderson and Shephard, 1968b; K. L. Andersen and Magel, 1970).

The purpose of oxygen diffusion is to saturate arterial blood with oxygen. The effectiveness of the system can thus be assessed from the alveolar/arterial oxygen tension gradient (P_{A,O_2} − Pa,O_2). We have already noted that in theory this gradient corresponds to the term $(B/1 - B)(1/\lambda\dot{Q})$ of the conductance equation, B being equal to $e - \int(\dot{D}_L/\lambda\dot{Q})$. Under normal conditions, this conductance accounts for but a small part of the total impedance to oxygen transport. In practice, two other variables contribute to the alveolar/arterial gradient inequalities of ventilation perfusion ratio and frank venous/arterial shunts through channels such as the thebesian vessels and the anterior cardiac veins.

At rest the alveolar/arterial gradient is typically ~1.3 kPa (10 mmHg), and is attributable largely to venous/arterial shunts. In vigorous exercise (Riley, 1974), the gradient widens to ~2.7 kPa (20 mmHg). The arterial oxygen tension changes little, but there is a substantial increase of P_{A,O_2} (Fig. 5.32). The percentage of shunted blood remains small (perhaps 1–2% of cardiac output), but because the oxygen content of venous blood is low during exercise, the shunt has a substantial effect on the tension of mixed arterial blood. The vertical matching of ventilation and perfusion is improved during effort, but there may be some increase of mismatching at a given level within the lung (Gledhill et al., 1977). The transit time through the average pulmonary capillary (Fig. 5.3; Staub and Schultz, 1968; Miyamoto and Moll, 1971; Wagner and West, 1972; E. P. Hill et al., 1973) decreases from its resting value of 0.75 sec to ~0.33 sec in maximum exercise (Roughton, 1945; Fishman, 1963; Frech et al., 1968). This still allows fairly complete equilibration, but in the shorter capillaries (where transit time is less than average) there may be some residual oxygen pressure gradient.

TABLE 5.9
Influence of Alveolar Gas-Sampling Artifacts upon the Diffusing Capacity of the Lungs*

Subject	Published Data (Mostyn et al., 1963): Normal Breathing (ml • min^{-1} • kPa^{-1})	Published Data (Mostyn et al., 1963): Breath Holding (ml • min^{-1} • kPa^{-1})	Recalculated Data (Shephard and Anderson, 1968): Computer Simulation of Alveolar Gas (ml • min^{-1} • kPa^{-1})	Recalculated Data (Shephard and Anderson, 1968): Modified Single Breath $\dot{D}_L$ (ml • min^{-1} • kPa^{-1})
C. B.	262	408	278	251
R. P.	337	702	380	366

*Published figures of Mostyn et al. (1963) recalculated using (1) a computer simulation of oscillations in alveolar carbon monoxide concentration and (2) a modified single-breath estimate of diffusing capacity.
Source: From Shephard and Anderson, 1968.

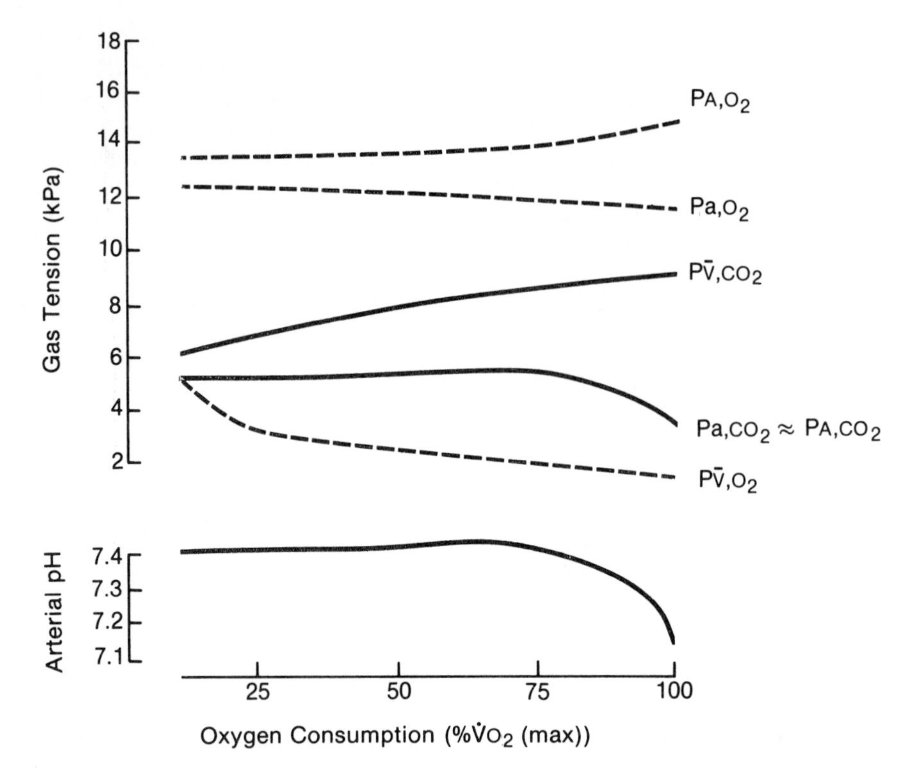

FIG. 5.32. Alveolar and blood gas pressures at various intensities of work.

Shephard (1971a) demonstrated that the pulmonary diffusing capacity was sufficient for a fairly complete equilibration of oxygen pressures during maximum effort under normal environmental conditions. Underwater (where the oxygen pressure is reduced by breath holding) and at high altitudes, pulmonary diffusing capacity becomes more critical; under such special circumstances, there may thus be a competitive advantage in having a large pulmonary diffusing capacity (Shephard, 1971a,1972b).

Carbon Dioxide Diffusion

Many physiology textbooks suggest that since carbon dioxide is very soluble in water, it has no problem in diffusing from the blood stream to alveolar gas. There are several fallacies in this argument, perhaps the most obvious being the composite nature of the pulmonary diffusing capacity [equation (5.19)]. The substantial water solubility of CO_2 ensures a large membrane diffusing capacity ($\dot{D}_M$), but does not augment the term θV_c. The latter is of the same order for CO_2 and for oxygen (Shephard, 1968b). $\dot{D}_L$ is thus only 2-4 times greater for CO_2 than for oxygen. Furthermore, the crucial factor in determining CO_2 exchange is not the magnitude of $\dot{D}_L$, but rather the ratio of $\dot{D}_L$ to $\lambda\dot{Q}$. Since λ is 5 times larger for CO_2 than for oxygen, $\dot{D}_L$ must also be 5 times as great in order to permit comparable equilibration.

RESPIRATORY CONTROL SYSTEM

Medullary Centers

The basic breathing control mechanism resides in the respiratory center or centers of the medulla. This region of the brainstem has an inherent respiratory rhythm even when it is isolated from the rest of the body (Hukuhara, 1978). Argument continues concerning details of organization (Merrill, 1970; Bianchi, 1971; Karczewski, 1974). Some authors regard the respiratory neurons as widely distributed throughout the brainstem, while others describe distinct clusters of neurons responsible for inspiration (inspiratory center) and expiration (expiratory center).

Regardless of anatomical localization, the concept of two centers facilitates description of respiratory control. During normal operation, we may view the inspiratory center as being activated by impulses arising in a somewhat higher apneustic center (Lumsden, 1924). The resultant central respiratory drive passes to the spinal cord, where a segmental integration of response occurs (C. Von Euler, 1966a,b; Sears, 1966; Aminoff and Sears, 1971; Newsom Davis, 1974; Fig. 5.33). Both the α and γ motoneurons of the inspiratory muscles are stimulated. At the same time, the motoneurons of the expiratory muscles are inhibited (Eklund et al., 1964) and there is feedback (C. Von Euler, 1974) to the pneumotaxic center of the pons (which functions

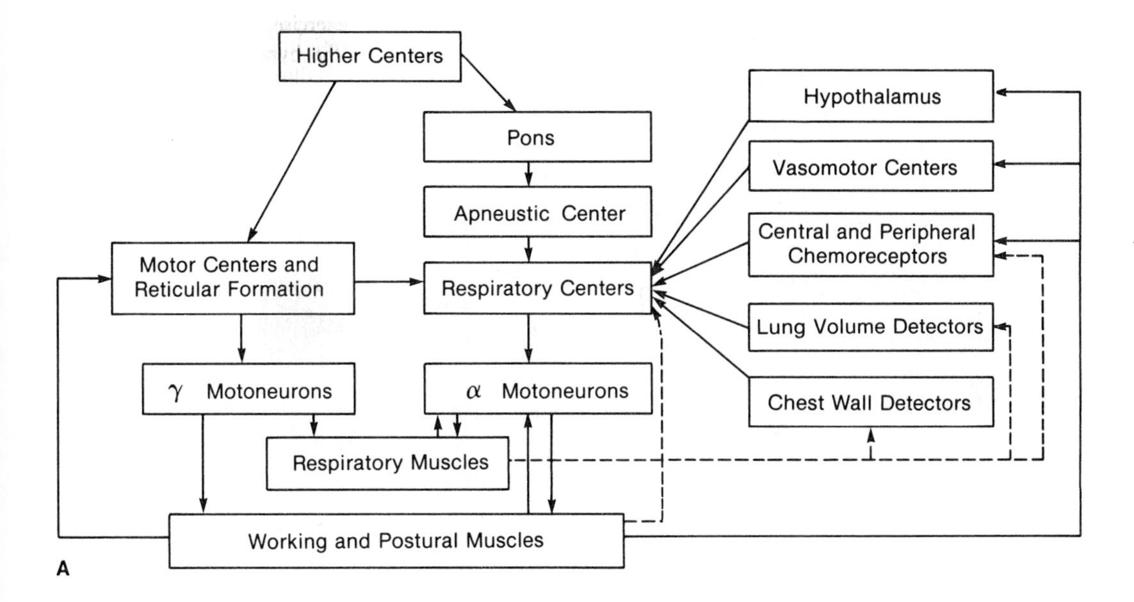

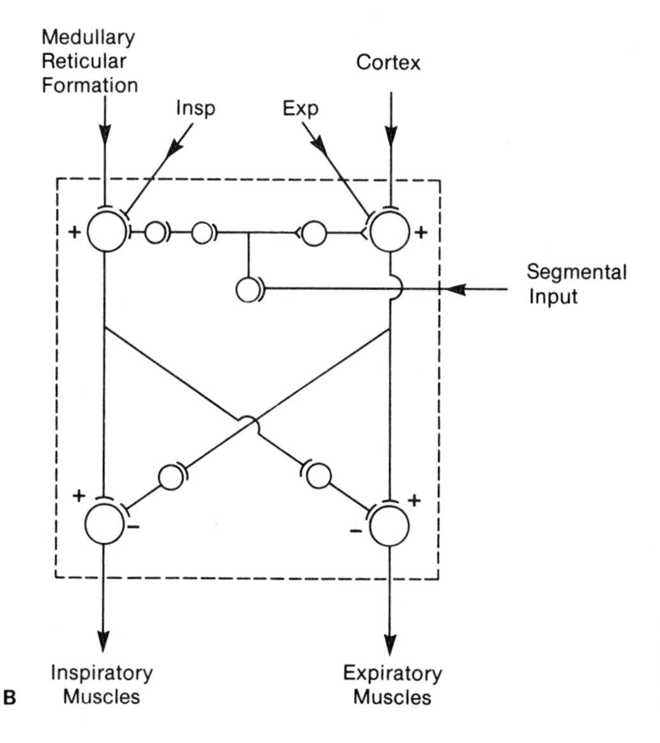

FIG. 5.33. A. Some of the systems involved in respiratory regulation during physical activity. **B.** Detail of segmental organization. (After Aminoff & Sears, 1971.)

as an off switch for the inspiratory center; G. W. Bradley et al., 1975). The respiratory motoneurons receive a second, independent input from both pyramidal and extrapyramidal fibers (Chap. 7), thereby allowing a partial voluntary control of breathing.

Despite the inherent rhythm of the medullary centers, inspiratory duration is normally set by a vagal feedback linked to the increase of lung volume (the Hering-Breuer reflex; C. Von Euler et al., 1970; Clark and Von Euler, 1972; Phillipson, 1974). A wide variety of other afferent impulses have the capacity to modulate the system (Remmers et al., 1974; Agostini and D'Angelo, 1976). However, many potential control mechanisms lack the negative feedback characteristic needed for effective regulation. Thus a rise of core temperature stimulates breathing, but in humans at least the increase of ventilation does little to restore body temperature to its initial value. Likewise, adrenaline stimulates respiration but an increase of minute volume has no influence upon blood catecholamine levels. Identification of feedback loops is complicated by a central neural process that maintains an increase of ventilation for a long period after the disturbing stimulus has passed (Eldridge, 1977). Details of control mechanisms thus are far from resolved (Dejours, 1964; Kao, 1972; Widdicombe, 1974).

Local Chemical Environment

The medulla is well supplied with blood vessels, and the respiratory center is sensitive to changes in the general chemical environment of the body. Exposure to moderate concentrations of carbon dioxide (2–5% CO_2 in air) increases the activity of the inspiratory center and also raises the threshold of the off-switch mechanisms (G. W. Bradley et al., 1975). However, higher concentrations of carbon dioxide (> 10%) depress respiratory function, with associated mental confusion. At one time, mixtures of carbon dioxide and oxygen were used for resuscitation, but it is now generally agreed that when an individual has stopped breathing, concentrations of CO_2 within the brain already approach narcotic levels; in such circumstances, it is plainly a disadvantage to add CO_2 to the breathing mixture.

In general, oxygen lack has a depressant effect on both respiratory and cardiovascular centers. However, if oxygen lack is severe, there is a rise of blood pressure, perhaps because medullary centers are stimulated by acidosis and release of catecholamines. At this stage, there may be some stimulation of respiration, with rapid shallow breathing.

Higher Centers

Other regions of the brain can modify the activity of the medullary centers (Planche and Bianchi, 1972). In humans, the highest areas of the cortex maintain a considerable control over breathing. Overbreathing (hyperventilation) is seen in acute stress (such as in an athlete prior to a contest) and in chronic anxiety. Examples of the voluntary arrest of breathing are provided by the diver, the player of wind instruments (Bouhuys, 1964), the singer (Bouhuys et al., 1966), and the soldier donning a gas mask. Conditioned reflexes may increase immediately before or coincident with the onset of vigorous physical activity (Shephard and McClure, 1965; Shephard and Callaway, 1966). Irradiation of impulses from the motor cortex also contributes to the increase of ventilation in the early stages of exercise.

Impulses from the hypothalamus, relayed to the medulla via the pons, initiate an increase of ventilation with a rise of deep-body temperature. In animals such as the dog, thermal panting is a useful cooling mechanism during and following vigorous exercise, but the importance of the thermal control of ventilation is less certain in humans. Finally, impulses could irradiate from the vasoregulatory centers of the medulla. A sudden rise of systemic blood pressure (due, for example, to a release of catecholamine or a strong isometric effort) causes a momentary depression of ventilation. However, recent work tends to minimize the importance of the baroreceptors during physical activity (McRitchie et al., 1976).

Specific Chemoreceptors

Specific areas of chemosensitive tissue are found in the lateral part of the fourth ventricle (R. D. Bradley and Semple, 1963; Mitchell, 1966; Loeschke, 1974) and in the carotid and aortic bodies (Biscoe, 1971).

Peripheral Chemosensors

If impulses are recorded from the glossopharyngeal nerve, the afferent pathway for the carotid bodies, action potentials can be detected under normal resting conditions. These are temporarily abolished a few seconds following a single breath of oxygen (Dejours, 1975), and there is an associated transient reduction of tidal volume (Fig. 5.34). We may thus conclude that the normal arterial oxygen pressure of 12–13 kPa (90–100 mmHg) provides some small stimulus to the receptors of the carotid body. However, the tension must be further reduced to 8–9 kPa (60–70 mmHg) before there is a major increase of either impulse

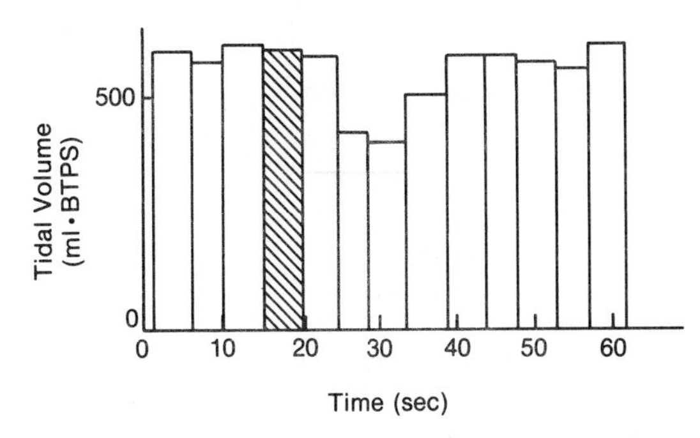

FIG. 5.34. Breath by breath changes of tidal volume in response to the inhalation of a single breath of oxygen. (Based on the concepts of Dejours, 1964.)

traffic in the glossopharyngeal nerve or ventilation (U.S. Von Euler, Liljestrand, and Zotterman, 1939; Witzleb, 1963; Fig. 5.35). Activity in the glossopharyngeal nerve is also increased by a sudden and substantial rise of arterial CO_2 tension (>1.3 kPa, 10 mmHg).

According to Torrance (1974), the peripheral chemoreceptor response R′ is described by a function of the type

$$R' = F'(Pa,O_2) \times (Pa,CO_2 - G') - H' \quad (5.20)$$

where F′, G′, and H′ are constants, and Pa,O_2 and Pa,CO_2 are the respective arterial partial pressures of oxygen and carbon dioxide. Possibly, CO_2 and a low pH potentiate the carotid body response to oxygen lack by enhancing changes of intracellular hydrogen ion concentration or by modifying local blood flow (Neil and Joels, 1963; Hornbein and Roos, 1963; Widdicombe, 1974; M. A. Hanson et al., 1978).

The local blood flow to the receptor organs seems an important variable. Vasoconstriction can be induced by catecholamines (adrenaline and noradrenaline) and by sympathetic nerve stimulation, with a corresponding increase in glossopharyngeal nerve impulse traffic (Floyd and Neil, 1952; Neil and Joels, 1963; K. D. Lee et al., 1964; Biscoe, 1971). In contrast, stimulation of the depressant sinus nerve efferents increases the local blood flow, reducing carotid body activity (Biscoe, 1971).

Central or Peripheral Response?

There may be a slight multiplicative interaction of central and peripheral chemoreceptor drive (Drysdale and Whipp, 1976). However, the CO_2 sensitivity of carotid and aortic chemoreceptors is normally much less than that of their central counterparts. Under steady-state conditions, the main responsibility for correcting an increase of arterial CO_2 tension is thus borne by the ventricular receptors.

With rapid changes, the situation is less clear-cut. The central chemoreceptors have a relatively long response time, since CO_2 must diffuse from the bloodstream into the cerebrospinal fluid (Lambertsen, 1963; Fig. 5.36). The carotid bodies respond much more briskly to transients of either oxygen or carbon dioxide tension, and they may well play an important role in immediate ventilatory responses to CO_2 (Dejours, 1962; A. M. S. Black, McCloskey, and Torrance 1971).

Response of Athletes

Most (Byrne-Quinn et al., 1971; Stegemann et al., 1975; Leitch et al., 1975; Miyamura et al., 1976; Saunders et al., 1976; Scoggin et al., 1978) but not all (Godfrey, Edwards et al., 1971) authors have described a reduced CO_2 responsiveness in the athlete.

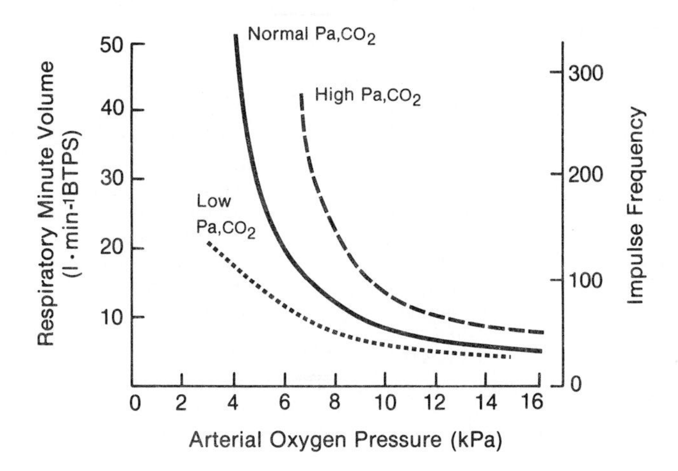

FIG. 5.35. The influence of arterial oxygen tension on respiratory minute volume and impulse frequency in chemoreceptor fibers (other factors held constant). (Based in part on data of U. S. Von Euler, Liljestrand & Zotterman, 1939; Witzleb, 1963; and B. B. Lloyd & Cunningham, 1963.)

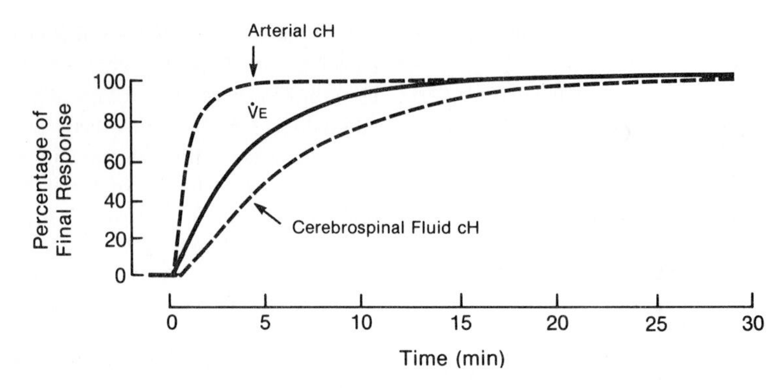

FIG. 5.36. Rate of change of ventilation ($\dot{V}_E$), arterial cH, and cerebrospinal fluid cH during exposure to 7% CO_2. (After Lambertsen, 1963.)

This apparently has a familial basis, with a stronger relationship to the responsiveness of the mothers than to that of the fathers. Under resting conditions, the response to hypoxia is also subnormal, but during exercise oxygen lack causes a normal increase of ventilation (Byrne-Quinn et al., 1971). The precise physiological basis of these changes has yet to be identified. One obvious possibility would be that the athlete has a greater than average perfusion of the carotid body, but there could equally well be an altered central integration of the afferent impulses.

Peripheral Reflexes

A wide range of peripheral receptors contributes afferent information to the medullary centers (Table 5.10). Vagal fibers from the trachea and bronchial tree are activated by normal distension of the lungs (the inspiration-inhibiting Hering-Breuer reflex; Fillenz and Widdicombe, 1971) and by forced distension of the lungs (as when the breathing pressure is increased without adequate counterpressure). They are also activated by exposure to cold, dry air, irritant vapors, and particulate matter (Dautrebande, 1963,1970; J. E. Mills et al., 1970; A. Davies, 1978; A. Davies and Kohl, 1978). Other fibers within the lung structure are stimulated by forced deflation (as may occur during diving); such fibers enhance the activity of inspiratory neurons within the medullary centers.

The muscles and joints both contain mechanoreceptors. Those from the joints (Comroe and Schmidt, 1943; Flandrois et al., 1966) and perhaps also those from the muscles (Besson et al., 1959; Gautier et al., 1964; but not Hodgson and Matthews, 1968 or Hornbein et al., 1969) induce an increase of ventilation when they are stimulated by either voluntary or passive movement of the limbs. Examples of ventilatory drive from passive move-

TABLE 5.10
Reflex Responses to Stimulation of Three Main Types of Lung Receptor

	Response Induced		
Effector System	**Stimulation of Stretch Receptors**	**Stimulation of Irritant Receptors**	**Stimulation of J Receptors***
Breathing	Inhibition	Hyperpnea	Apnea and rapid shallow breathing
Bronchial muscle	Relaxation	Contraction	? Contraction
Laryngeal caliber	Dilation	Constriction	Strong constriction
Heart	? Tachycardia	Unknown	Bradycardia
Vascular resistance	? Increase	Unknown	Bradycardia
Airway mucus	Unknown	Unknown	Depression
Spinal reflexes	Unknown	Unknown	Depression
Sensation	? No action	Unpleasant	Unpleasant

*The J receptors are alveolar nociceptive nerve endings (Paintal, 1970).
Source: After Widdicombe, 1974.

ment include a relaxed subject sitting on a bicycle ergometer that is driven by an electric motor and a pilot subjected to intense vibration while flying a low-level military aircraft. Receptors in the skin induce a brisk hyperventilation in response to cold (for instance, a 15°C shower or a plunge into an icy lake). Painful stimulation of the skin may also increase ventilation.

Respiratory Control in Exercise

Chemical Theories

We have noted a close correspondence between ventilation and metabolic effort, and it is thus tempting to explain the control of respiration during exercise in terms of changes in carbon dioxide tension, oxygen tension, and pH (Gray, 1950; Perkins, 1964). Large changes of blood gas tension are undoubtedly caused by vigorous exercise, but unfortunately, these occur on the venous side of the circulation (B. W. Armstrong et al., 1961; Riley, 1963; Wasserman et al., 1977). Attempts to demonstrate chemosensitive tissue either within the active muscles (Comroe and Schmidt, 1943) or in the great veins have not been successful (Cropp and Comroe, 1961; T. W. Lamb, 1966; Priban, 1966).

Early respiratory physiologists such as Haldane and Priestley (1935) suggested that since there was only a small increase of alveolar (and thus arterial) CO_2 tension during exercise, the receptors were extraordinarily sensitive to this particular gas. Such an argument may be tenable in moderate exercise (Fig. 5.32), where arterial CO_2 pressure increases (Dejours, 1964; Holmgren and McIlroy, 1964) from ~5.3 kPa (40 mmHg) to ~5.6 kPa (42 mmHg), but it is difficult to reconcile with the situation in maximum effort, where CO_2 pressure drops substantially below the resting value. By analogy with other sense organs, it is possible that the carotid body responds not to the mean gas tension but to some function of the rate of change of tension within the bloodstream (D. J. C. Cunningham, 1974; Yamamoto, 1977). Exercise could conceivably increase oscillations about the mean pressure (Black and Torrance, 1967; Fenner et al., 1968; R. C. Goode, Brown, et al., 1969), or alter the phase relationships between respiration and changes of gas tension within the peripheral and central receptors (A. M. S. Black and Torrance, 1971; D. J. C. Cunningham, Pearson et al., 1973; C. B. Wolff, 1977; Carruthers et al., 1978). While such hypotheses have yet to be categorically disproved (D. J. C. Cunningham, Elliott et al., 1965; Band et al., 1969; D. J. C. Cunningham, 1974), it seems unlikely that substantial oscillations of gas pressure can be transmitted from the lungs through the damping system offered by the heart, the blood vessels, and the chemosensitive tissues themselves (Kao, 1963).

The administration of oxygen causes a brisk (12–20%) decrease of ventilation, especially during heavy work (Asmussen and Nielsen, 1946; Bannister and Cunningham, 1954; Dejours, 1964; D. J. C. Cunningham, 1974). However, it is difficult to attribute the exercise-induced increase of ventilation to a decrease of arterial oxygen tension. Although the alveolar/arterial oxygen gradient is broadened by vigorous exercise (Fig. 5.32), the arterial oxygen pressure changes very little; certainly it does not fall to the level normally required for a brisk chemoreceptor response (Fig. 5.35).

It is possible that exercise increases the central or peripheral sensitivity to CO_2 and/or oxygen lack. Most authors have found a leftward displacement of the ventilatory response to CO_2 (decrease of zero intercept, without change of slope, Fig. 5.37; Asmussen and Nielsen, 1956; D'Angelo and Torelli, 1971), although there have been occasional reports of both an increased and a decreased slope. An increased peripheral sensitivity is suggested, since the chemoreceptor discharge is not silenced by administration of a gas mixture containing 35% oxygen during exercise

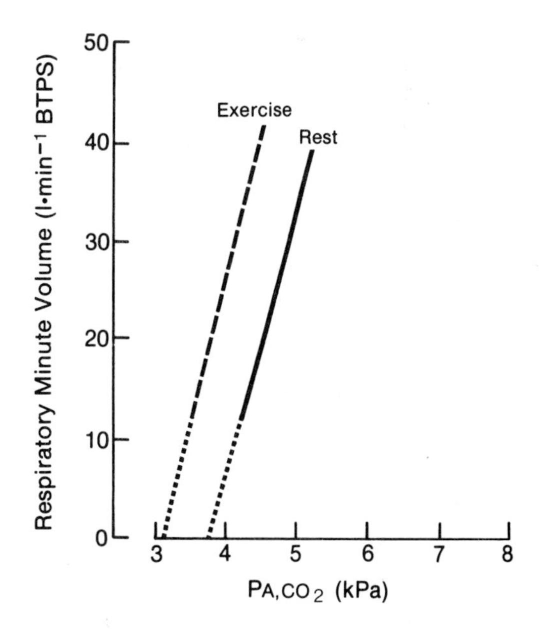

FIG. 5.37. The supposed effect of exercise upon the ventilatory response to alveolar CO_2—a decrease of zero intercept without change of slope.

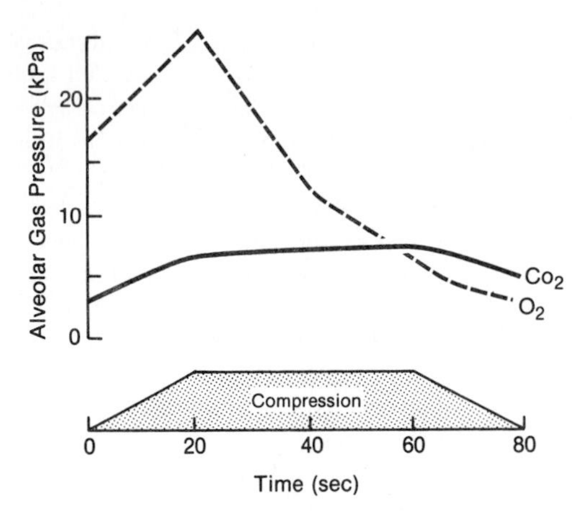

FIG. 5.38. Changes of alveolar gas pressure. Subject undertaking moderate exercise with initial hyperventilation, compression to two atmospheres pressure, and decompression over 80 sec breathhold. (After Rahn, 1963.)

(Kozlowski et al., 1971). Certainly, the sensitivity of the carotid bodies could be augmented through the action of the sympathetic nerves (Biscoe and Purves, 1965) or the secretion of catecholamines (Joels and White, 1968; Clancy et al., 1975), both responses decreasing perfusion of the glomus cells. Through this mechanism, respiration might be stimulated despite relatively normal arterial oxygen and CO_2 pressures (Neil and Joels, 1963; K. D. Lee et al., 1964; Whalen and Nair, 1975). Eisele et al. (1967) found no change of end-tidal PCO_2 or arterial pH after blocking the carotid body innervation with xylocaine. However, it may be that their nerve block was incomplete or that physical activity was too light to reveal the mechanism described by Biscoe and his associates.

In very vigorous exercise, there is a substantial production of lactic acid, and the increase of hydrogen ion concentration then contributes to the ventilatory response (Wasserman et al., 1973). But in most circumstances, it would seem that neural mechanisms initiate the increase of ventilation, with chemical factors playing no more than a secondary, feedback role.

Neural Mechanisms

As in the cardiovascular system, a substantial part of the early response to exercise seems due to an irradiation of impulses from the motor cortex (Krogh and Lindhard, 1913; Asmussen, 1967; Kao, 1972). The fibers of the intercostal muscles (and, to a lesser extent, those of the diaphragm) are activated in parallel with the skeletal muscles, modulating the force of inspiration in terms of both metabolic demand and any external resistance that may be encountered (Fig. 5.33).

The extent of ventilatory conditioned responses to exercise (F. N. Craig, Cummings, and Blevins, 1963; Shephard and McClure, 1965; Shephard and Callaway, 1966; Plum, 1970) is still discussed. Some authors think this mechanism is important in humans, although Flandrois et al. (1971) found no reaction to the noise of a treadmill motor in dogs, while Paulev (1971) was unable to develop a conditioned ventilatory response to an exercise metronome in high school students.

As exercise continues, the respiratory drive is probably augmented by a stimulation of peripheral mechanoreceptors (Kao, 1977), although it is difficult to disentangle the possible influence of increased activation of the motor cortex via the γ loop (Asmussen, 1967). The main mechanoreceptors are likely to be nerve endings in the active joints, although muscle receptors may also be involved (Asmussen and Nielsen, 1948; E. J. M. Campbell, 1964; C. Von Euler, 1974; Granit, 1975).

Overview of Exercise Response

Other possible ventilatory stimuli include catecholamine secretion (particularly if the exercise is stressful; Flandrois, Favier, and Pequignot, 1977) and hypothalamic stimulation (particularly if there is a rise of core temperature; Flandrois, Lacour, and Osman, 1971; Jennings and Macklin, 1972; J. G. Henry and Bainton, 1974).

Overventilation is avoided by a negative feedback from falling CO_2 pressures. If the systemic pressure rises, there may also be some irradiation of inhibitory impulses from the vasomotor centers.

Taken together, the interaction of these various stimuli seems sufficient to account for the magnitude of the ventilatory response to exercise. However, there remain several puzzles to challenge further research, including (1) the continued matching of ventilation with metabolism in situations where muscle tension is varied (as in negative and isometric effort; Asmussen, 1967) and (2) the similarity of ventilation during the active and resting phases of rapid intermittent work (Christensen, Hedman, and Saltin, 1960).

Breath Holding

Physiology of Apnea

The possible length of breath holding (apnea) varies markedly from one individual to another. At rest, the range is from 30 sec to several minutes, but in maximum exercise 5 sec is a common limit. The breaking point is reached when the sensations arising from CO_2 buildup, a decreasing oxygen tension, and the absence of respiratory movements cannot be counteracted by volun ary ınhibition of the respiratory centers (W S Fowler, 1954; Rahn, 1963). Typical limiting alveolar gas pressures are 5.3 kPa (40 mmHg) for carbon d'oxide and 6 0 kPa (45 mmHg) for oxygen Chemıcal changes normally play a major role in termınat ng a breath hold (Lin et al., 1974), and the duration of apnea can be extended appreciably by a prelim'nary perıod of hyperventilation or inhalation of oxygen.

At one time, the duration of breath holdıng (Kallfelz, 1962) and the somewhat related abılity to hold a column of mercury at a pressure of 40 mmHg, 5.3 kPa (Flack, 1920) were used as tests of fitness, particularly for aircrews. Unfortunately, tests of this type depend very much upon the motivation of the subject (D. Bartlett, 1977). By watching a clock driven at an unnaturally slow speed, the breath-holding time can often be extended to twice the best previous effort. The duration of breath holding is also increased if the lungs are fairly fully inflated. This is partly because changes of arterial gas tension develop more slowly and partly because the medullary centers are inhibited via the Herıng Breuer reflex. When the breaking point is reached, temporary relief can be obtained by swallow ng. The physical act of taking a breath further extends the period of apnea (Guy and Patrick, 1977) even if the gas mixture that is inhaled has no metabolic value (for example, a nitrogen/carbon dioxide mıx ure). Paralysis of the respiratory muscles by curare also pro ongs breath holding (S. Godfrey and Campbell, 1970).

Applications to Diving

A combination of vigorous physical activity and preliminary hyperventilation can allow the alveolar oxygen pressure of an apneic subject to drop low enough (~ 3 kPa) to cause convulsions or even fainting (P. O. Åstrand, 1960, Rahn, 1963). The diver is particularly vulnerable to this problem (Fig. 5.38) If he relies upon oxygen lack as a signal for surfacing, the margin between stımulat'on of the desire to breathe and a dangerous decrease of oxygen tension is very small. Furthermore the alveolar oxygen pressure inevitably falls rap d.y w th the decrease of water pressure during surfacing Othe adverse factors are a cerebral vasoconstriction (induced by the CO_2 washout), a reduction of cardiac output (due to the impairment of venous return by breath holding), and impairment of the diver's judgment by oxygen lack. It is scarcely surprising that there have been fatalities associated with loss of consciousness while underwater (A. Craig, 1961; J. H. Davis, 1961).

Intermittent Breathing

Intermittent breathing is a distressing problem of respiratory regulation encountered in the early stages of acclimatization to high altitudes. The tidal volume waxes and wanes, and respiration may cease altogether for short periods. The phenomenon seems an exaggeration of the normal hunting reaction, seen in the respiratory center as in other feedback-regulating systems. If the respiratory system is faced with a sudden (square-wave) challenge such as the inhalation of a 5% mixture of carbon dioxide in air, ventilation tends to overshoot the steady adjustment needed in the new environment, and there are further mınor oscillations before a new steady level of ventilatıon is reached (Defares, 1963; Lambertsen et al , 1965). The fact that the main CO_2 receptor responds to changes in the composition of cerebrospinal fluid gives a measure of damping to the system (Fig. 5.36), so that oscillations are rarely excessive. However, if a subject engages in vigorous physical activity on first arriving at high altitude, he may wash out so much carbon dioxide from his body that CO_2 no longer provides an effective respiratory stimulus. Ventilation now becomes dependent upon the response of the carotid bodies to hypoxia. The carotid receptors are in an intimate relationship with the arterial blood, so that little damping of their response occurs. When the oxygen tension falls, the glomus cells initiate vigorous ventilation, and this continues for several seconds until well-oxygenated blood from the lungs reaches the bifurcation of the carotid arteries. Ventilation is then suppressed until the situation is again reversed by the arrival of a further slug of poorly oxygenated blood.

Intermittent breathing was described among some of the athletes competing at the 1968 Olympic games in Mexico City. It is particularly likely to develop with frequent bouts of anerobic effort (for example, a sprint performer participating in repeated heats) and in conditions in which the circulation time from the lungs to the carotid bodies is prolonged (Milhorn and Guyton, 1965).

CHAPTER 6

Cardiorespiratory Transport 2. The Heart and Circulation

Overview of the Circulatory System
Cardiac Output
Energy and Work in the Cardiovascular System
Electrical Activity of the Heart
Control of the Circulation
Blood Flow to Specific Regions
Problems of Circulatory Regulation

OVERVIEW OF THE CIRCULATORY SYSTEM

As we have seen in Chapter 5, the heart and the circulation offer the main impedance to the transport of oxygen from the atmosphere to the working tissues. In this chapter we look at the function of the cardiac pump, the peripheral distribution of the available cardiac output, tissue blood flow, and venous return.

Under resting conditions, several body tissues are overperfused in terms of their oxygen needs. The skin receives ~10% of cardiac output, but accounts for only 5% of the total oxygen consumption. The kidneys, again, receive 25% of cardiac output, but account for only 10 percent of oxygen consumption. The explanation of such overperfusion is that blood flow meets demands other than oxygen consumption in these tissues—the elimination of heat in the skin, and of waste products in the kidney. In contrast, the resting muscle flow is low relative to the potential oxygen demand of this tissue.

During vigorous exercise, a local accumulation of metabolites causes widespread vasodilation in the active muscles. An increase of cardiac output with some redistribution of blood flow away from the kidneys and other viscera sustains the systemic blood pressure and helps to satisfy the increased local demand for transport of oxygen, carbon dioxide, and heat. The arteriovenous difference (and thus transport per unit flow) is also broadened for these several variables.

The skin blood flow usually increases in moderate exercise in order to carry the extra heat that must be eliminated from the body. However, a sudden decrease of cutaneous flow occurs as the maximum oxygen intake is approached. The body makes a last desperate search for blood, the demands of thermoregulation are overridden, and the face of the individual assumes an ashen gray pallor. This mechanism is appropriate only to very brief intense efforts; a sustained combination of exercise and cutaneous vasoconstriction can cause a dangerous hyperthermia.

CARDIAC OUTPUT

Principles of Measurement

The cardiac output, $\dot{Q}$, is the volume of blood pumped by each ventricle per min. It is given by the product of heart rate f_h and stroke volume Q_s:

$$\dot{Q} = f_h Q_s \tag{6.1}$$

Heart Rate

The heart rate is the number of ventricular beats per min, while the pulse rate is the frequency of pressure waves transmitted to a peripheral artery. Normally, the two values are identical, but in some types of dysrhythmia a proportion of ventricular contractions are too weak to produce a detectable pulse wave.

The resting heart rate can be counted by very simple techniques such as auscultation of the apex beat or palpation of a peripheral pulse; particularly if measurements are made from a short length of electrocardiograph record, due allowance must be made for respiratory variations of cardiac rhythm (the so-called sinus arrhythmia). During exercise, simple techniques are hampered by a rapid heart rate and physical movement of the subject. Expert observers can still obtain reliable data by strapping a stethoscopic diaphragm to the chest or palpating the carotid artery, but those with less experience are better advised to rely on an exercise electrocardiogram (thoracic leads) or an early postexercise pulse. If the pulse is counted in the first 10–15 sec after ceasing activity, the value observed agrees closely with the end-exercise result (F. S. Cotton and Dill, 1935; D. A. Bailey, Shephard, and Mirwald, 1976). However, care must be taken not to exert excessive pressure on the carotid arteries, as this can cause some slowing of the heart rate (J. R. White, 1977) and (rarely) loss of consciousness.

Cardiac Output

DIRECT FICK METHOD. The ideal way of measuring cardiac output in humans is to apply the Fick principle to samples of blood obtained at cardiac catheterization:

$$\dot{Q} = \frac{\dot{V}_{O_2}}{C_{a,O_2} - C_{\bar{v},O_2}} \tag{6.2}$$

where $\dot{V}_{O_2}$ is the oxygen consumption (measured under steady-state conditions), C_{a,O_2} is the oxygen content of arterial blood, and $C_{\bar{v},O_2}$ is the oxygen content of a specimen of mixed venous blood drawn from the pulmonary artery.

Under resting conditions, $\dot{V}_{O_2}$ might equal 250 ml • min^{-1}, C_{a,O_2} 195 ml • l^{-1}, and $C_{\bar{v},O_2}$ 155 ml • l^{-1}, so that $\dot{Q}$ = 250/(195 − 155) or 6.25 l • min^{-1}. Unfortunately, cardiac catheterization is needed to obtain the pulmonary arterial specimen, and the risks of this procedure do not justify its application except for diagnostic purposes; even in the absence of cardiac disease, there is a 2–8% incidence of serious complications (mainly abnormalities of cardiac rhythm, but including also cardiac perforations, knotting and fracture of catheters) (Corliss, 1979).

DYE INJECTION. A second popular clinical procedure requires the injection of a dye that is rapidly removed from the circulation (W. F. Hamilton, 1962). Indocyanine green is commonly used for this purpose. There are some problems from nonlinearity of the curves relating optical density and dye concentrations (Fig. 6.1; Simmons and She hard, 1971c), but discrepancies between dye and direct Fick estimates of cardiac output are said to average less than 10% (Holmgren and Pernow, 1959). Unfortunately, in order to realize this precision, int acardiac injection of the dye is needed with subse uent collection of serial blood samples by arterial puncture; attempts to

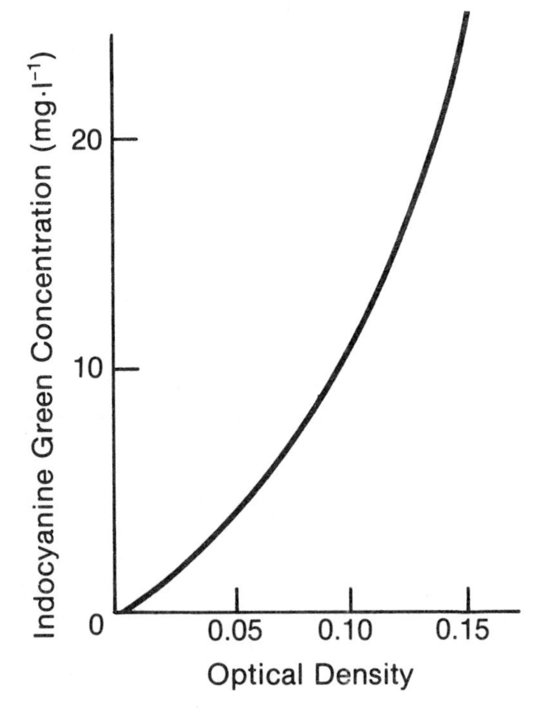

FIG. 6.1. Example of nonlinearity of optical density of indocyanine green. Varying concen rations of the dye were added to blood and subsequently centrifuged. Anticoagulant potassium oxalate, optical density measured at 800 nm. (Based on data o Simmons & Shephard, 1971c.)

inject the dye into a periphe a. vein and monitor bloodstream concentrations by means of an earpiece have not been particularly succe sful Like the direct Fick procedure, the dye method thus carries the hazards of cardiac catheterization and arterial puncture. While a medically qualified i ve tigator may choose to offer himself as a subject, it is difficult to suggest the general application of the dye method to normal individuals.

FOREIGN GAS METHOD. The foreign gas method applies the Fick equation to the uptake of a very soluble gas such as acetylene or nitrous oxide. Grollman (1931) required subjects to rebreathe from a small bag containing a 20% mixture o' acetylene in air. A preliminary series of deep breaths mixed the bag contents with alveolar air, and changes of gas concentration were then observed over a further period of rebreathing. Any decrement n volume of the lung/bag system was indicated by an increase of bag nitrogen, while decreases in oxygen and acetylene concentrations were used to calculate cardiac output according to the equation

$$\dot{Q} = \left(\frac{\dot{V}_{O_2}}{C_{a,C_2H_2}}\right)\left(\frac{\Delta C_2H_2}{\Delta O_2}\right) \qquad (6.3)$$

where $\dot{Q}$ was the cardiac output, $\dot{V}_{O_2}$ was the oxygen consumption determined a few seconds previously, C_{a,C_2H_2} was the arterial acetylene concentration (assumed equal to the average of concentrations found in the initial and final bag samples, multiplied by a factor describing the solubility of acetylene in arterial blood), ΔC_2H_2 was the change in acetylene concentration from the initial to the final sample (corrected for any decrement in volume of the lung/bag system), and ΔO_2 was the change of oxygen concentration over the same period.

We may note that Grollman used a special case of the Fick equation (6.2); the term $\Delta C_2H_2/\Delta O_2$ served to convert oxygen intake $\dot{V}_{O_2}$ to a corresponding acetylene absorption $\dot{V}_{C_2H_2}$, while the acetylene content of mixed venous blood was assumed to be zero. Given an oxygen consumption of 1500 ml • min^{-1}, initial and final lung/bag acetylene concentrations of 109 and 91 ml • l^{-1}, respectively, a solubility factor of 0.75, and a 30 ml • l^{-1} decrease of bag oxygen concentration, then

$$\dot{Q} = \left[\frac{1500}{\left(\frac{109 + 91}{2}\right)0.75}\right] \left[\frac{109 - 91}{30}\right]$$
$$= 12 \text{ l} \cdot \text{min}^{-1}$$

When the method was introduced by Grollman, the gas samples were analyzed chemically; carbon dioxide and oxygen were first absorbed, and then acetylene was burned to CO_2 and water vapor. The method was cumbersome, and high concentrations of acetylene were selected in an attempt to improve accuracy; the gas mixtures proved subjectively unpleasant, tended to be anesthetic, and had the unfortunate habit of exploding occasionally in the subject's lungs!

The figures that Grollman obtained for cardiac output are now recognized to have been rather low (Werkö et al., 1949; C. B. Chapman et al., 1950). Factors contributing to underestimation included use of an erroneous blood solubility coefficient for acetylene, solution of acetylene in the lung tissues (Jernerus et al., 1963), and continuation of rebreathing to the point where final gas samples were influenced by a recirculation of acetylene-contaminated blood. In maximum exercise, there is significant recirculation within 8–10 sec (Simmons and Shephard, 1971a). The necessary gas analyses have been greatly facilitated by such techniques as gas chromatography (Simmons and Shephard, 1971a) and mass spectrometry (Triebwasser et al., 1977), and it is now possible to use lower (1%), nonexplosive, and subjectively more pleasant acetylene concentrations. Adopting also revised solubility coefficients for acetylene and measuring gas concentrations as early as the fourth and seventh seconds of rapid rebreathing, systematic errors can be avoided (Asmussen and Nielsen, 1952); Klausen (1965a) and Simmons and Shephard (1971a) were able to obtain results that agreed well with other techniques (Fig. 6.2) and had a probable error of less than 3%.

While acetylene is the preferred foreign gas for cardiac output determinations (Cander and Forster, 1959) nitrous oxide can also be used in the rebreathing system, gas concentrations being determined by either infrared or chromatographic analysis (Rigatto, 1967; Ayotte et al., 1970). Some authors (Becklake, Varvis et al., 1962) have used an open-circuit "semi-steady-state" method with nitrous oxide. The gas mixture is then inhaled for a rather long period, and the procedure becomes open to one objection of the original Grollman technique, namely, that recirculation causes erroneously low cardiac output values; some published data using the Becklake method show arteriovenous oxygen differences as large as 200–220 ml • l^{-1}! More complicated open-circuit techniques allow for progressive saturation of the tissues with foreign gas (Hatch and Cook, 1955; Shephard, 1958a), but such procedures are cumbersome to apply during exercise.

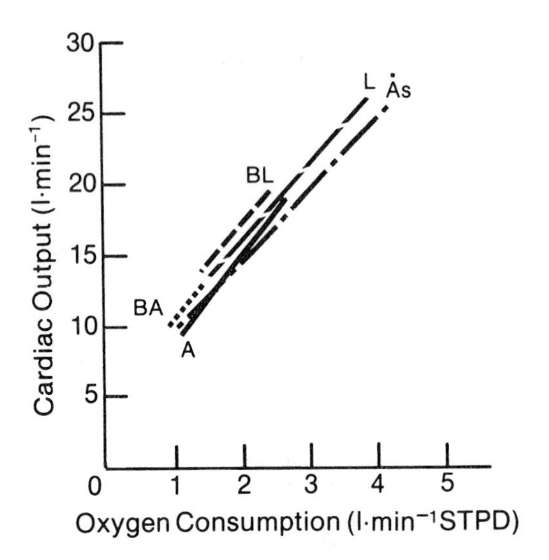

FIG. 6.2. A comparison of acetylene rebreathing cardiac output determinations (Simmons & Shephard, 1971a) for arm (A) and leg (L) ergometry with data obtained by other techniques on subjects of comparable cardiorespiratory fitness. (Data of P. O. Åstrand, Cuddy et al., 1964, for leg ergometry [Ȧs] and of Bevegard, Freyschuss & Strandell, 1966, for arm ergometry [BA] and leg ergometry [BL].)

CO_2 Rebreathing Methods. Carbon dioxide rebreathing methods apply the Fick principle to the exchange of carbon dioxide:

$$\frac{\dot{V}_{CO_2}}{C_{a,CO_2} - C_{\bar{v},CO_2}} \qquad (6.4)$$

where $\dot{V}_{CO_2}$ is the output of CO_2, C_{a,CO_2} is the arterial CO_2 concentration, and $C_{\bar{v},CO_2}$ is the concentration of CO_2 in mixed venous blood. The steady-state CO_2 output is measured by an open-circuit method, exercise continuing until the expired CO_2 concentration remains constant for two successive minutes. "Arterialized" capillary blood is most easily obtained by stabbing the finger tip or the ear lobe. If the finger tip is heated adequately, the arterioles become widely dilated, and the CO_2 tension of blood that emerges from a stab wound corresponds fairly closely with that of the arterial blood; nevertheless, some CO_2 is lost during filling of the capillary specimen tube (McEvoy and Jones, 1975), and it is thus advisable to adjust the observed readings upward by 0.06–0.07 kPa (Table 6.1). Alternatively, the arterial CO_2 tension can be estimated from a continuous record of expired CO_2 levels (Fig. 6.3). When the subject is at rest, the CO_2 pressure in the final portion of the expirate (the end-tidal sample) coincides rather closely with the arterial value; during exercise, variations of CO_2 over the course of the breathing cycle are larger (Shephard, 1968b), and commonly the best approximation to the arterial pressure is given by an average of mid-tidal and end-tidal readings (Matell, 1963). A third option is to assume a figure for dead space, calculating the alveolar CO_2 pressure and thus C_{a,CO_2} from the expired gas concentration using the Bohr formula [equation (5.18)]. There are uncertainties regarding the dead space during exercise, and one widely cited formula (Asmussen and Nielsen, 1956) is based on only four subjects breathing at an unspecified frequency. N. L. Jones, McHardy et al. (1966) proposed using the equation

$$V_D = 138.4 + 0.077\,(V_T) \qquad (6.5)$$

for the average adult male. Given a tidal volume of 3000 ml, this would imply a dead space of 138.4 + 0.077 (3000) = 369.4 ml. An alternative formula (Robertson et al., 1975) calculates the arterial CO_2 pressure from a combination of end-tidal and expired gas pressures:

TABLE 6.1
Some Comparisons of Arterial and Arterialized Capillary Blood Specimens[a]

Pa,CO_2 (kPa)	pH (Units)	Number of Samples	Author[b]
—	+0.0076	29	Gambino (1959)
−0.27	no difference	10	Anderson et al. (1960)
−0.061	—	21	Cooper and Smith (1961)
−0.018	+0.0023	22	Maas and van Heyst (1965)
—	+0.0100	75	McDonald et al. (1941)
−0.14	+0.0060	10, 12	Langlands and Wallace (1965)
−0.29	+0.0110	39, 46	Jung et al. (1966)

[a]All values shown as error of arterialized specimen.
[b]For details of references, see Shephard and Bar-Or (1970).

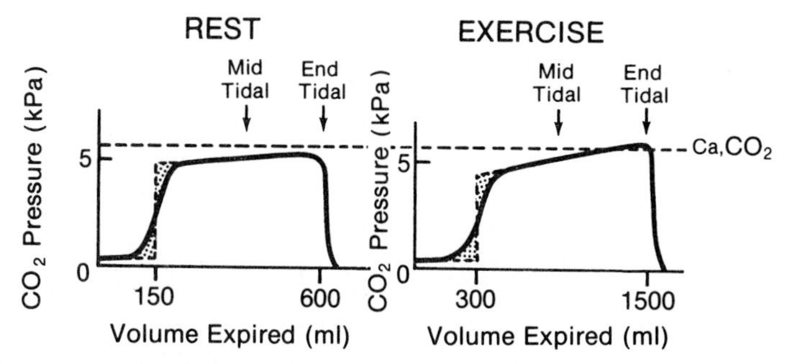

FIG. 6.3. The pressure of carbon dioxide over the course of a complete expiration. Note that at rest, the end-tidal sample approximates closely the composition of arterial blood, but that in exercise the arterial blood value lies midway between mid-tidal and end-tidal readings.

$$Pa, CO_2 = P_{ET,CO_2} + 0.59 - 0.00031V_T + 0.004f_R - 0.009P_{E,CO_2} \quad (6.6)$$

Given an end-tidal pressure (P_{ET,CO_2}) of 5.3 kPa, a tidal volume V_T of 3000 ml, a respiratory rate (f_R) of 45 breaths per min, and an expired gas pressure (P_{E,CO_2}) of 4.3 kPa, we find

$$\begin{aligned} Pa,CO_2 &= 5.3 + 0.59 - 0.00031\,(3000) \\ &\quad + 0.004\,(45) - 0.09\,(4.3) \\ &= 4.75 \text{ kPa} \end{aligned}$$

In this situation, Robertson's equation implies that the end-tidal reading is overestimating Pa,CO_2 by 0.55 kPa. Bar-Or and Shephard (1971) compared several possible approaches in preadolescent children and concluded that results based on the end-tidal reading were more realistic than those obtained from an assumed dead space (Table 6.2).

TABLE 6.2
Coefficients of Correlation Between $\dot{Q}$ and $\dot{V}_{O_2}$ with Three Different Methods of Estimating $Pa,_{CO_2}$

Variable Correlated with Oxygen Intake	Coefficient of Correlation		
	Girls	Boys	All Subjects
$\dot{Q}$ end-tidal	0.68 (n = 42)	0.79 (n = 36)	0.71 (n = 78)
$\dot{Q}$ Asmussen-Nielsen dead space	0.06 (n = 40)	0.40 (n = 36)	0.16 (n = 76)
$\dot{Q}$ revised dead space*	0.57 (n = 42)	0.77 (n = 36)	0.63 (n = 78)

*Arterialized capillary blood specimens were used to evolve a new formula for the prediction of dead space in preadolescent children ($V_D = 0.285V_T - 64$ ml).
Source: Based on data of Bar-Or and Shephard, 1971, for preadolescent boys and girls.

The mixed venous CO_2 reading is usually obtained by rebreathing from a bag containing 5–15% CO_2. In the Defares method (Defares, 1958; Amery et al., 1977), the subject takes one breath per sec, and a graph is plotted relating the CO_2 content of the Nth and the (N + 1)th breaths; this graph can be extrapolated to the line of identity, at which point the bag concentration has risen to the mixed venous level, and CO_2 elimination has ceased (Fig. 6.4). An alternative approach adopted by N. L. Jones and his associates (N. L. Jones, Campbell et al., 1975) is to preselect a gas mixture close to the anticipated mixed venous reading. If the "correct" mixture is chosen, the continuously recorded CO_2 tension shows a few oscillations as the gas in the rebreathing bag mixes with that within the lungs, and then a plateau is reached

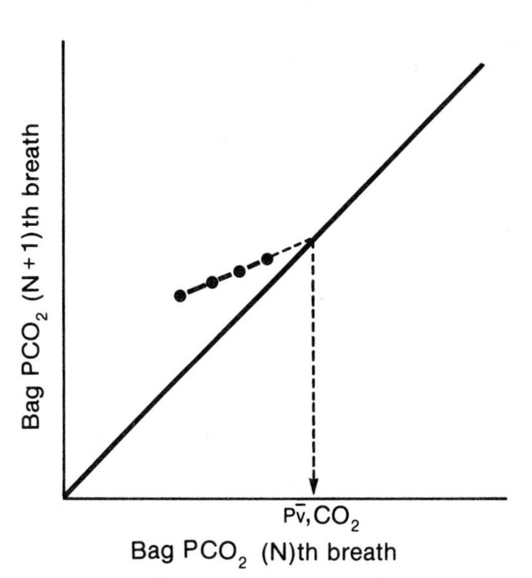

FIG. 6.4. Illustration of the Defares (1958) method for the estimation of mixed venous CO_2 pressure.

(Fig. 6.5). At this stage, CO_2 is neither being excreted nor absorbed, and a working assumption would be that the alveolar gas pressure equalled that in the mixed venous blood. The plateau is temporary in nature, and if rebreathing is continued for more than 12 or 15 sec, recirculation causes a secondary upward curving of the record. Rebreathing dams back CO_2 within the body, and blood with an increased CO_2 content quickly starts to appear at the lungs. During vigorous exercise, the recirculation time (and thus the period available to define the plateau) is only 8–10 sec (Simmons and Shephard, 1971a). For reasons that are still discussed (N. L. Jones, Campbell et al., 1969; Laszlo et al., 1971; Denison et al., 1971; Gurtner and Forster, 1977), most authors find that the equilibrium CO_2 pressure overestimates the true mixed venous value (N. L. Jones, Campbell et al., 1967; Denison et al., 1969; Amery et al., 1977). Jones et al. (1969) thus proposed "correcting" the bag value according to the equation

$$P_{\bar{v},CO_2} = P_{Bag,CO_2} - [(0.24P_{Bag,CO_2}) - 1.47] \quad (6.7)$$

The implication is that under resting conditions, (P_{Bag,CO_2} = 6.4 kPa), the mixed venous CO_2 ($P_{\bar{v},CO_2}$) is overestimated by 0.06 kPa, and that during vigorous exercise (P_{Bag,CO_2} = 10.0 kPa) the error increases to 0.93 kPa (more than 20% of the arteriovenous carbon dioxide difference). There are plainly problems in applying such large "corrections" to experimental data, and according to some authors (particularly those using the Defares method), uncorrected results yield cardiac outputs that are more consistent with values obtained by other methods (D. H. Paterson, 1972; S. Godfrey and Wolf, 1972; S. Godfrey, 1973; Heigenhauser and Faulkner, 1978).

The classical Defares method involves the rebreathing of 5% CO_2 in oxygen, and perhaps because the initial gas mixture differs markedly from the likely equilibrium value, results are less reproducible than those obtained with the plateau rebreathing method, especially during exercise (S. Godfrey, 1973). However, if a higher bag CO_2 concentration is selected, the Defares and Jones procedures have a similar accuracy (Rode and Shephard, 1973a; Amery et al., 1977).

Al-Dulymi and Hainsworth (1977) recently described a third option, where CO_2 is added to the inspirate in amounts that match findings in the expirate; they claim that their open-circuit method gives a faster approach to equilibrium than either of the traditional rebreathing techniques.

Irrespective of the method used to estimate mixed venous CO_2, the final stage in applying the Fick calculation is to convert the arteriovenous CO_2 pressure gradient to a corresponding CO_2 concentration gradient. At light work loads, an approximate conversion can be based upon the average slope of the oxygenated carbon dioxide dissociation curve (3.6 ml • kPa^{-1}). More precisely, individual blood CO_2 pressures can be converted to corresponding CO_2 contents, using the equation

$$\log_e C_{CO_2} = 0.874 (\log_e P_{CO_2}) + 2.4 \quad (6.8)$$

Further small corrections are necessary (1) if the hemoglobin (Hb) concentration is not 15 g • 100 ml^{-1} and (2) if the arterial oxygen saturation (Sa,O_2) is significantly less than 100% (N. L. Jones, Campbell et al., 1975):

$$\text{Hemoglobin correction} = (15 - Hb) \times 0.113 (P_{\bar{v},CO_2} - Pa,CO_2) \quad (6.9)$$

$$\text{Oxygen saturation correction} = (100 - Sa,O_2) \times 0.064 \quad (6.10)$$

RELATIVE MERITS OF FOREIGN GAS AND CO_2 METHODS. The main objection to the CO_2 re-

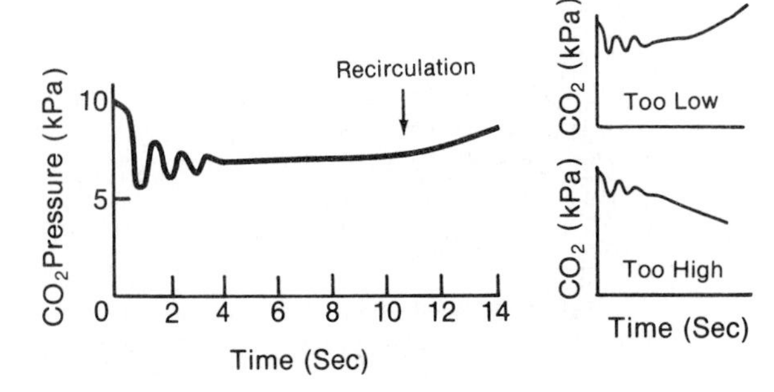

FIG. 6.5. Estimation of mixed venous CO_2 pressure by the rebreathing method. If the "correct" initial mixture is chosen, there is a plateau of some six seconds duration prior to the onset of recirculation. If the initial mixture contains too little CO_2, the plateau is replaced by an upward slope; if the concentration is too high, there is a downward slope.

breathing method under resting conditions is that the gradient of CO_2 partial pressure between arterial and mixed venous blood specimens is only ~0.8 kPa. This gradient tends to be overshadowed by measurement errors, since none of the available techniques indicate CO_2 pressures closer than ~0.2 kPa. In maximum exercise, the arteriovenous pressure difference increases to 4–5 kPa, but unfortunately errors in the estimation of P_{a,CO_2} and $P_{\bar{v},CO_2}$ also increase.

The foreign gas method works quite well in normal healthy men, but the analyses are more time consuming than those needed for the CO_2 rebreathing method.

Neither foreign gas nor CO_2 rebreathing procedures are satisfactory in patients with chronic chest disease because the permissible period of rebreathing is too short to establish an equilibrium between poorly ventilated regions of the lung and the rebreathing bag.

Stroke Volume and Arteriovenous Oxygen Difference

The stroke volume is normally calculated from the cardiac output and stroke volume, using equation (6.1). In cardiac patients, more direct estimates of stroke volume can be obtained by injection of a radioopaque dye and filming the cardiac chamber in the posterioanterior and lateral planes (cine-fluorographic biplane angiography). An ellipsoid formula is used to calculate ventricular volume:

$$\frac{4\pi}{3}\left(\frac{L_1}{2}+\frac{L_2}{2}+\frac{L_3}{3}\right) \qquad (6.11)$$

where L_1 is the length of the principal cardiac axis, and L_2 and L_3 are the minor axes, derived from planimetry of the dye shadow (J. W. Kennedy et al., 1966; Bartle and Sanmarco, 1966). More recently, computer technology has permitted very sophisticated three-dimensional analyses of angiographic videotapes (Heintzen et al., 1974; S. A. Johnson et al., 1974; Sandler and Dodge, 1974; Dodge and Sandler, 1974). However, such methods remain unsuitable for normal volunteers, since they involve the intracardiac injection of an occasionally irritant dye, followed by a substantial dose of X-irradiation.

Ultrasound (Edler, 1965; Gramiak and Shah, 1971; Feigenbaum, 1972,1974; Rennemann, 1974; D. H. Bennett and Evans, 1974; Laurenceau et al., 1979; Venco et al., 1979; Bubenheimer et al., 1979) has a potential for the estimation of stroke volume. High-frequency sound waves (2–10 MHz) are directed through the chest and reflected from the ventricular walls. The relative positions of the various cardiac structures can be deduced from the corresponding transit times of the sound waves (Table 6.3). At present, ultrasonic methods are used most commonly to detect portions of the ventricular wall with weakened or paradoxical motion following myocardial infarction (I. G. McDonald and Feigenbaum, 1972); however, the future may well see multiple ultrasound units with a three-dimensional computer analysis of the type already developed for angiography.

The arteriovenous oxygen difference is estimated from the cardiac output and the corresponding oxygen consumption using the classical Fick equation (6.2). Since the exercise on-transients for cardiac output and oxygen consumption have somewhat different time courses, it is important that subjects be close to a steady-state response to a given work rate if such calculations are to have validity.

Resting Data

The resting heart rate of a sedentary individual is commonly 70–80 beats per min, but values can vary

TABLE 6.3
Echocardiographic Comparison of Olympic Athletes and Sedentary Subjects

	Olympic Athletes	Sedentary Subjects
End-diastolic size (mm)	52.8 ± 5.1	46.3 ± 4.4
End-diastolic volume (index ml • m^{-2})	79.9 ± 19.8	60.3 ± 14.6
Posterior wall thickness (mm)	10.1 ± 1.4	8.4 ± 1.0
Left ventricular muscle mass index (g • m^{-2})	142.4 ± 32.0	96.9 ± 23.0
R.V./L.V. dimension ratio	0.39 ± 0.08	0.29 ± 0.06
Velocity of contraction (circ • sec^{-1})	1.09 ± 0.17	1.18 ± 0.24
Rate of thickening (normalized for post. wall thickness)	1.29 ± 0.43	1.75 ± 0.44

Note: Studies by Zoneraich et al. (1977) show also significant hypertrophy of the right ventricle and relative enlargement of the aortic root and left atrium. Morganroth and Maron (1977) note that in endurance athletes the tendency is to increase of ventricular mass and end-diastolic volume, whereas in strength training there is an increase of wall thickness and mass with a normal end-diastolic volume.

Source: Based on data of Laurenceau et al., 1979.

from less than 30 beats per min in a superb endurance athlete to as much as 100 beats per min in a middle-aged executive awaiting his annual medical examination. In general, readings are greater in women and in young children than in men. They are also increased by standing, smoking, and the recent ingestion of food. However, if such variables can be controlled, there is a fairly close inverse relationship between cardiorespiratory fitness and heart rate. Training increases the resting stroke volume without changing the cardiac output. Inevitably, there is thus a decrease of resting pulse rate.* The origin of the cardiac impulse may also be displaced. Sutton, Hughson et al. (1977) found a junctional rhythm in 4 of 12 marathon runners, although a sinus rhythm was restored when heart rate was increased by exercise or administration of atropine.

The resting stroke volume depends upon posture and cardiorespiratory fitness. When a sedentary subject is lying down, readings may amount to 120 ml per beat, but when standing or sitting values drop to ~80 ml (Wade and Bishop, 1962; Bevegärd and Shepherd, 1967). The reason for this is that when standing, blood pools in the veins of the legs, reducing end-diastolic filling and thus the stroke output of the heart.

Given a heart rate of 70–80 beats per min and a stroke volume of 80 ml, it follows that the resting cardiac output is typically 5.6–6.4 l • min^{-1}. Results are sometimes expressed as a "cardiac index" (flow per unit of body surface area). Since the body surface area of a typical man is $1.8 m^2$, it can be seen that the resting cardiac index is ~3.6 l • min^{-1} • m^{-2}.

On-transients

Heart Rate

At the beginning of exercise, heart rate increases with a latency of less than 0.5 sec (Petro et al., 1970); this rapid response is due to a decrease of vagal tone (Fagraeus and Linnarson, 1974). If the work load is moderate, a steady-state plateau is reached in 1–2 min, but if the exercise is more severe, the rate continues to increase (Broman and Wigertz, 1971), with a concomitant increase of cardiac output (Cerretelli, Sikand, and Farhi, 1966; W. B. Jones et al., 1970) until the subject is exhausted. As with respiration, the initial rapid increase is partly a conditioned reflex and/or a response to irradiation of impulses from the motor cortex, and it is lessened by previous test experience (Ninomiya and Wilson, 1966). Further increments of heart rate are a response to local vasodilation in the active limbs, and stimulation of proprioceptors in the joints and muscles. The final phase of slowly climbing heart rate during exhausting work reflects several additional factors, including (1) a rising deep-body temperature, (2) a peripheral sequestration of blood volume, (3) secretion of catecholamines, (4) recruitment of less efficient muscles as fatigue sets in, and (5) possible psychological reactions to the continuing activity.

Recovery after exercise follows an analogous two- or three-phase curve; there is an initial rapid slowing due to loss of cortical and proprioceptive stimulation, followed by slower adjustments as metabolites are being removed from the active tissues and body temperature is returning toward its normal resting value.

C. T. M. Davies (1968) advanced the concept of a "pulse deficit," first seen when the exercise on-transient was slowed by an early phase of anerobic work-rate. Data from our laboratory confirmed that there was a pulse deficit which increased with the intensity of work. Nevertheless, some deficit was seen at quite light loads, and the magnitude of the deficit had a low coefficient of correlation with blood lactate levels. Furthermore, most subjects failed to reach a constant heart rate within 6 min of exercise.

Detailed analysis of time constants shows that early adjustments of heart rate occur more quickly than the corresponding changes of ventilation (Broman and Wigertz, 1971; Linnarson, 1974) or stroke volume (Smuylan et al., 1965).

There is thus transient hypoventilation in the early stages of exercise (J. Dempsey, Hanson, and Masterbrook, 1978). The increase of heart rate involves a negligible time delay, followed by a two-component response. Wigertz (1970) found that with a step increase of loading, the respective time constants were 12 and 133 sec, the fast component accounting for 71% of the total response. Linnarson (1974) noted a very rapid change over the first 10–15 sec, a slower rise over the next 60–90 sec, and a continuing almost linear drift of heart rate. A formal two-component analysis yielded time constants of 11–22 sec and 367–1510 sec. Adjustments are generally slower at loads that surpass the anerobic threshold (Linnarson, 1974; Skinner et al., 1977). However, with exhausting work, the first component may quickly reach the maximum possible heart rate, giving no possibility for the detection of a second

*Frick (1967) argued that heart rate may determine stroke volume by varying the time available for diastolic filling; however, this seems unlikely, except in near maximal effort.

component (Linnarson, 1974). This could explain why P. O. Åstrand and Saltin (1961b) found a rapid adjustment to high work loads.

Oxygen Consumption

Similar types of analysis can be carried out for the oxygen consumption on-transient. In general, the response time seems longer than that for heart rate (Table 6.4; Shephard and Kavanagh, 1978a; Fardy and Hellerstein, 1978), although it is more difficult to be certain of details since most means of measuring oxygen consumption themselves have an inherent time lag (R. Gilbert and Auchincloss, 1971), and there are also inevitable short-term differences between the oxygen *uptake* of the pulmonary capillaries and the oxygen *intake* observed at the mouth (R. Gilbert, Baule, and Auchincloss, 1966).

During exhausting exercise (Fig. 6.6), the on-transient is apparently faster the heavier the applied work load (P. O. Åstrand and Saltin, 1961b). However, the comparison is not entirely fair, since the proportion of anerobic activity also increases with work load. If only submaximal intensities of effort are considered, the time required to reach a steady-state *increases* as work load is increased (W. B. Jones et al., 1970; Bason et al., 1973).

The response to any given submaximal load is slower in an unfit than in a fit subject (W. B. Jones et al., 1970; Weltman and Katch, 1977).

Practical Implications

The lag of the heart rate response to a given work load and the further lag of oxygen intake relative to a given heart rate have practical implications for exercise testing, since cardiorespiratory fitness is often deduced from the heart rate at a given submaximum work load or oxygen intake.

The standard submaximal test protocol recommended by the World Health Organisation (K. L. Andersen, Shephard et al., 1971), and the International Biological Programme (R. J. Shephard, Allen et al., 1968b; Weiner and Lourie, 1969) calls for 3 min at each state of a progressive exercise test, a sufficient period to bring the subject very close to a steady-state response (Shephard and Kavanagh, 1978a). However, some authors still use a protocol whereby the work load is increased every minute; the observed heart rate is then substantially less than the steady-state value, and fitness is overestimated (Table 6.5). Conversely, if cardiac output is being measured, as long as 6–8 min may be required per stage; the heart rate is then boosted by the consequences of prolonged exercise, and fitness is underestimated.

Steady-State Response

Concept

The concept of a steady-state response to exercise implies that an equilibrium has been reached, the variable in question (be it heart rate, ventilation, or oxygen intake) meeting the cellular demands of a given work load.

The oxygen intake induced by submaximum effort conforms closely to this pattern. An oxygen deficit accumulates over the first 1–2 min, but thereafter the oxygen cost of work is relatively stable until fatigue redistributes the task to less efficient muscle groups or glycogen exhaustion compels an increased use of fat as a metabolic substrate.

TABLE 6.4
Rate of Approach to Steady-State for Selected Physiological Variables*

		Respiratory Minute Volume	Respiratory Gas-Exchange Ratio	Heart Rate	Oxygen Consumption
Load 1	N	92.4	89.0	101.1	99.7
	C	103.9	105.0	100.2	96.2
Load 2	N	93.8	83.9	101.1	100.3
	C	93.9	98.7	99.0	93.4
Load 3	N	97.8	103.4	101.3	99.6
	C	94.8	100.0	98.1	97.0
Load 4	N	96.1	105.0	99.5	96.8
	C	90.1	100.0	95.5	94.4

*Results in third minute of progressive submaximal test protocol expressed as percent of equivalent data for 5 min exercise at same load. Data of Shephard (1967b) for healthy young men performing a stepping exercise (N) and of Shephard and Kavanagh (1978a) for newly recruited postcoronary patients exercised on bicycle ergometer (C).

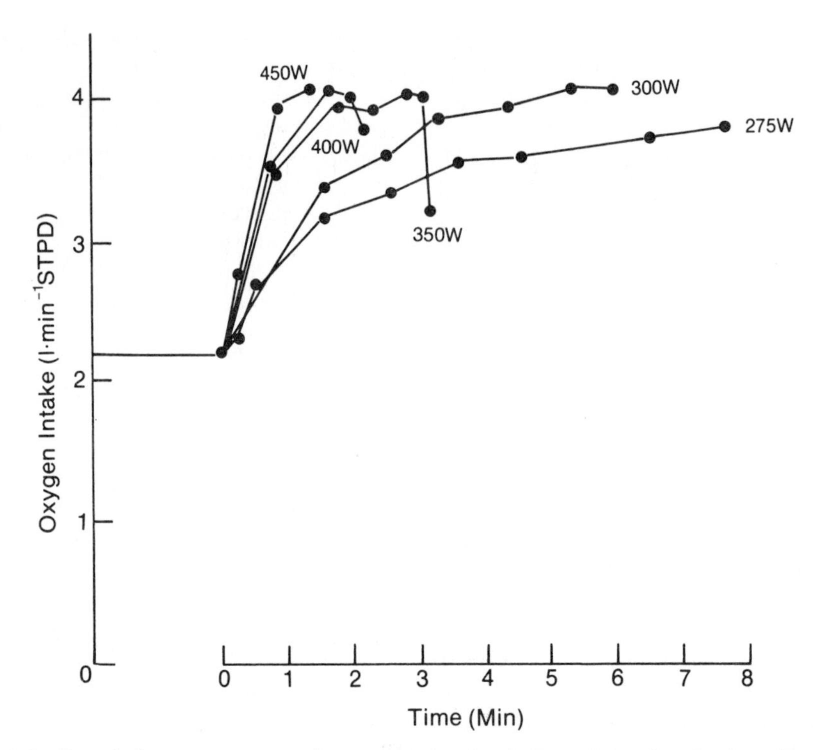

FIG. 6.6. Speed of oxygen consumption on transient in relation to intensity of exhausting work on bicycle ergometer. (After P.O. Åstrand & Saltin, 1961b.)

Heart rate reaches a much less clear-cut plateau, there being a slow upward drift of counts. Many clinicians are therefore content to accept a relative steady-state; for example, Sjöstrand (1967) accepts a heart rate that does not change by more than 10 beats per min from the second to the sixth minute of exercise.

Linearity of Response

The steady-state heart rate response to a given type of activity is linearly related to oxygen consumption over much of the working range. This relationship provides the basis for various predictions of maximum oxygen intake (Chap. 11) and for the use of heart rate in estimating industrial work loads. Unfortunately, departures from linearity occur at the light intensities of effort encountered in habitual daily activities; this reflects the increases of stroke volume and arteriovenous oxygen difference that develop during light work.

There has also been discussion of linearity at loads approaching maximum oxygen intake. Maritz et al. (1961) described an asymptotic relationship, the heart reaching a plateau before oxygen consumption. This type of response seems particularly likely when subjects are performing an unfamiliar pattern of laboratory exercise for the first time (C. T. M. Davies, 1968); when subjects are "habituated" to the experimenter and the required procedures, there do not seem to be significant departures from linearity (Shephard, 1967b).

Conditions of Work

At any given level of oxygen consumption, heart rates are higher for arm than for leg work (Table 6.6; Asmussen and Christensen, 1939b; Asmussen and

TABLE 6.5
Influence of Stage Duration upon the Apparent Fitness of Subjects as Estimated from Submaximal Exercise Tests*

	Predicted $\dot{V}_{O_2(max)}$	
Time per Test Stage (min)	Work Scale of Åstrand Nomogram (l • min⁻¹ STPD)	Oxygen Scale of Åstrand Nomogram (l • min⁻¹ STPD)
1	2.72 ± 0.70	—
3	2.38 ± 0.50	2.42 ± 0.47
6–8	2.06 ± 0.42	2.29 ± 0.49

*Data for 50 postcoronary patients.
Source: Based on data of Shephard and Kavanagh, 1978a.

TABLE 6.6
Comparison of Cardiac Function in Arm (A) and Leg (L) Ergometry

Oxygen Consumption ($l \cdot min^{-1}$ STPD)	Stroke Volume (ml)		Heart Rate (beats $\cdot$ min^{-1})		Cardiac Output ($l \cdot min^{-1}$)		Arteriovenous Oxygen Difference ($ml \cdot l^{-1}$)	
	A	L	A	L	A	L	A	L
1.5	96	125	132	96	12.3	13.2	123	120
2.0	101	131	155	117	14.9	15.9	133	127
2.5	108	138	174	142	18.5	18.5	139	135
Maximum (2.79, A; 3.70, L)	103	138	178	179	18.3	24.7	144	150

Source: Based on data of Simmons and Shephard, 1971a.

Nielsen, 1955; Vokac et al., 1975). Proportionality is only partially restored if heart rates are correlated with the relative work load (percentage of the corresponding maximum oxygen intake; Fig. 6.7). Heart rates are also higher for static than for rhythmic effort (Lind and McNicol, 1967) and are higher if exercise is performed in a hot or emotionally charged atmosphere rather than in a cool or psychologically "neutral" environment.

Sustained and Interval Work

If exercise is discontinuous, the extent of recovery between work bouts depends upon both the intensity of work and the length of the intervening rest periods. With a relatively light effort, there is complete recovery between each burst of exercise, and many repetitions can be undertaken. However, if the exercise is more severe, a cumulative displacement of the steady-state heart rate occurs. An athlete may value the progressive increase of heart rate as a means of cardiorespiratory conditioning (interval training), but in industry a rise of heart rate over the working day is regarded as evidence of excessive physical or thermal loading (Karrasch and Müller, 1951; Grandjean, 1971). Both the intensity of effort and the length of any rest pauses are thus adjusted so that heart rate

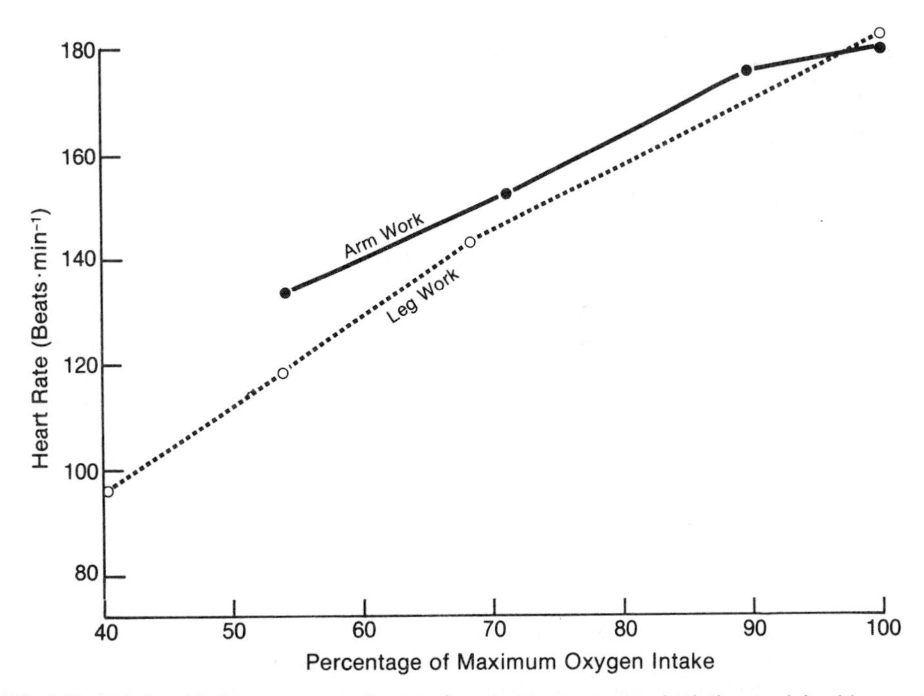

FIG. 6.7. Relationship betweeen steady-state heart rate response and relative work load (percentage of maximum oxygen intake). (Based on data of Simmons & Shephard, 1971a,b.)

and other physiological variables (respiratory gas-exchange ratio, blood lactate) remain fairly stable. On this basis, some French workers have defined a "*puissance maximale supportée*" (Chap. 11; Sadoul et al., 1966). Adaptations to prolonged work are considered later in this chapter.

Maximum Heart Rate

Transient heart rates of 250–300 beats per min can be developed for a few seconds during very demanding exercise, such as when making a difficult turn in slalom skiing (Christensen and Högberg, 1950a). However, it is unusual to find an adult who can maintain a rate of more than 200 beats per min over several minutes. The average "maximum," thus defined, is about 195 beats per min for a sedentary young man (P. O. Åstrand and Ryhming, 1954; I. Åstrand, 1960), while values for young women are marginally higher (198 beats per min).

There have been suggestions that physical conditioning decreases the maximum heart rate. Some training experiments show 5–10 beats per min decreases of maximum heart rate. Other evidence supporting such a trend includes a weak negative correlation between maximum oxygen intake and maximum heart rate (C. T. M. Davies, 1967b) and low maximum heart rates in some classes of athlete (Saltin and Åstrand, 1967). Low maxima seem likely if the test mode is similar to the sporting activity (for example, a runner tested on a treadmill). On the other hand, the average maximum for swimmers (also tested on a treadmill) is close to 195 beats per min (Shephard, Godin, and Campbell, 1973).

P. O. Åstrand (1952) described substantially higher maxima for children (210–215 beats per min) than for adults, but almost all other reports in the world literature (Shephard, 1971c) cite values of 195–200 beats per min for the preadolescent. Presumably, much depends on motivation. Similar problems arise in interpreting the decrease of maximum heart rate with aging. Early studies suggested a drop to 155–160 beats per min at age 65 (S. Robinson, 1938; Asmussen and Molbech, 1959), but recent work indicates that a sedentary North American who is exercised to a plateau of oxygen consumption develops a heart rate of at least 170 beats per min (Lester et al., 1968; S. M. Fox, 1969; Sidney and Shephard, 1977a; Shephard, 1978b).

The maximum heart rate is known to decrease at altitude (Pugh, 1962), but oxygen lack does not seem to be a factor in the low maxima of the elderly, since values typical of young adults cannot be restored by oxygen administration (I. Åstrand et al., 1959). Possible explanations of the age-related change include problems of diastolic filling associated with an increase of stiffness in the ventricular wall and more fundamental alterations in the sympathetic drive to the cardiac pacemaker. Certainly, problems of venous return set a heart rate ceiling of 100–150 beats per min for young adults during cardiac pacing (Kissling and Jacob, 1973), and stroke volume is poorly maintained when the elderly are exercised at high work rates.

Cardiac Emptying

Stroke Volume in Exercise

Determinants of stroke volume include (1) venous return, (2) ventricular distensibility, (3) contraction force, and (4) pressure in the aorta or pulmonary artery. Physical activity facilitates venous return, due to both compression of veins within the muscles (the "muscle pump") and increased variations of intrathoracic pressure (the "thoracic pump"). If the subject is recumbent, the legs are at or above heart level, and there is little tendency for blood to pool in the lower half of the body even before exercise is begun. Under these circumstances, the resting stroke volume is quite large and does not increase further with exercise. On the other hand, if leg work is performed in an upright or seated posture, stroke volume increases progressively from an initial value of 80 ml to a plateau of ~ 120 ml or more as the oxygen consumption increases to 50% or more of maximum. A young person sustains the plateau value and may even increase stroke volume to 100% of maximum oxygen intake (Simmons and Shephard, 1971a), but in an older adult the output falls at work loads above 70% of maximum oxygen intake (Niinimaa and Shephard, 1978b).

Stroke volume also varies with the type of activity (Table 6.7). Maximum values are greater on the treadmill than on the bicycle ergometer (Shephard, Allen et al., 1968a) and are particularly large during swimming (Saltin, 1973b). Very low stroke volumes are encountered during arm ergometry (Stenberg et al., 1967; Simmons and Shephard, 1971a); this has been attributed to peripheral venous pooling during arm work, but an additional consideration is probably the high peripheral resistance offered by the small mass of contracting muscle.

The maximum stroke volume is larger in the athlete (150–180 ml) than in the average young adult (Table 6.8). This is partly an expression of selection and partly a response to training. The stronger mus-

TABLE 6.7
Stroke Volume, Cardiac Output, and Arteriovenous Oxygen Difference for Eight Healthy Young Men Performing Maximum Exercise on the Treadmill and the Bicycle Ergometer (seated, leg work)*

	Stroke Volume (ml)	Cardiac Output ($l \cdot min^{-1}$)	Arterio-venous Oxygen Difference ($ml \cdot l^{-1}$)
Treadmill	150 ± 21	28.3 ± 4.7	140 ± 11
Bicycle ergometer	137 ± 20	25.6 ± 4.2	138 ± 8
Δ	13 ± 11	2.7 ± 1.9	2 ± 7

*Mean ± S.D. of data.
Source: After Shephard, Allen et al., 1968a.

cles of the athlete can develop a given rate of working at a smaller percentage of maximum force than in a sedentary subject, with a corresponding reduction in the perfusion resistance of vessels in the active tissues (Royce, 1958; Lind and McNicol, 1967). A further factor contributing to the large stroke volume of the athlete is a large blood volume. It follows that any factor leading to a decrease of blood volume and ventricular filling (excessive sweating, Saltin, 1964a,b; prolonged physical effort, Ekelund and Holmgren, 1964; Faulkner, Roberts et al., 1971; Bruce, Kusumi, et al., 1975; Sawka et al., 1978; or residence at high altitude, Vogel and Hansen, 1967; Greenleaf, Bernauer et al., 1978) is associated with a reduction of stroke volume. The vagus is sometimes implicated in changes of stroke volume; thus heart rate increases and stroke volume decreases with prolonged exercise in the supine position (Saltin and Stenberg, 1964), but such a reaction can be abolished by prior administration of atropine (Hartley, Pernow et al., 1970).

G. R. Cumming (1972) noted that with supine bicycle ergometer exercise, the stroke volume reached its highest value during the early part of the recovery period. Goldberg and Shephard (1980) recently demonstrated a similar phenomenon during upright bicycle ergometer work. In both cases, the peak recovery value is similar to that observed during maximum treadmill exercise. Particularly in the upright posture, it is hard to imagine that venous return is improved by the cessation of loaded pedaling; the most likely explanation of the data seems that cardiac output is increased once the high peripheral resistance of the actively contracting muscles is reduced.

End-diastolic Volume

The heart is not completely emptied with each beat. At rest, ~70–80 ml of blood remains in the left ventricle. Thus, the exercising heart can increase its stroke volume either by filling more fully during diastole or by emptying more completely during systole. Both mechanisms are probably operative.

The relationship between diastolic filling and stroke volume was first explored by Starling and his associates (S. W. Patterson et al., 1914). Using isolated dog heart-lung preparations, they produced beautiful curves showing that an increased rate of work (due to either an increase of arterial pressure or an increase of stroke volume) could be sustained by an increase of diastolic filling until a certain limit was reached; further loading led to the dangerous situation of a progressive fall in the work that could be performed (Fig. 6.8). The Starling curves are essentially analogous to the length/tension diagrams established for skeletal or cardiac muscle (Fig. 3.6); as E. H. Starling and his associates put it (S. W. Patterson et al., 1914):

> The law of the heart is therefore the same as that of skeletal muscle, namely that the mechanical energy set free on passage from

TABLE 6.8
Comparison of Maximum Exercise Test Data Between Average Men and Endurance Athletes

	Maximum Oxygen Intake ($ml \cdot kg^{-1} \cdot min^{-1}$) STPD	Stroke Volume (ml)	Heart Rate ($beats \cdot min^{-1}$)	Cardiac Output ($l \cdot min^{-1}$)	Arteriovenous Oxygen Difference ($ml \cdot l^{-1}$)
Average young men (n = 5)	48.3	140	184	25.7	133
Endurance athletes (n = 5)	66.6	172	184	31.6	153

Source: Based on data of Simmons, 1969.

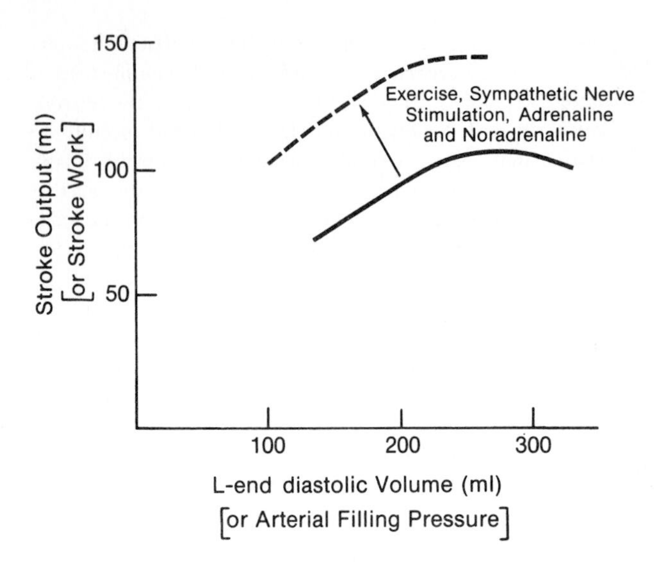

FIG. 6.8. The relationship between end diastolic volume (or atrial filling pressure) and stroke output (or stroke work) of the ventricle (Starling's law of the heart). Note the leftward displacement of this relationship (increase of contractility) with exercise, sympathetic nerve stimulation, and administration of adrenaline and noradrenaline.

the resting to the contracted state depends on the area of "chemically active surfaces", i.e. on the length of the muscle fibers.

Clinical Applications

Starling's "law of the heart" found early application in the analysis of cardiac valvular disease. While patients remain on the ascending limb of the Starling curve, a worsening of their valvular leakage or the demands of mild exercise can be met by an increase of stroke volume; the heart failure is said to be "compensated." However, exercise of unaccustomed severity or further deterioration of a diseased valve produces a situation whereby a further increase of diastolic volume causes a diminution of stroke volume. Heart failure is then "decompensated," and urgent treatment is required. The blood volume must be reduced (by diuretics and a low-salt diet) and/or the operating characteristics of the heart must be displaced to a more favorable portion of the Starling curve (by the use of drugs such as digitalis). It is unlikely that healthy people either can or will exercise to the point of decompensation. Certainly, subjective complaints and loss of the erect posture normally prevent this from occurring. However, in a diseased heart a less than anticipated increase of stroke volume and systemic blood pressure must be regarded as an ominous response to exercise, particularly if it is accompanied by severe breathlessness due to congestion of the pulmonary circulation.

Cardiac Contractility

While Starling's hypothesis is of considerable clinical value, it has proved difficult to reproduce his findings when chronic experiments are carried out on active, healthy dogs (Sarnoff and Mitchell, 1962; Mirsky and Parmley, 1974; V. Bishop and Horwitz, 1977). The probable explanation is that Starling's preparation was dissected free of all nerve supply; furthermore, it did not receive the normal delivery of circulating catecholamines such as adrenaline and noradrenaline. Under in vivo conditions such factors generate a whole family of Starling curves. During exercise, for example, a combination of increased heart rate, increased activity of the sympathetic nerves, decreased activity of the parasympathetic system and an outpouring of catecholamines moves the operating characteristics of the heart leftward to another of the possible curves (Braunwald et al., 1967; Randall and Smith, 1974; Vatner and Pagani, 1976). A larger stroke volume is produced for a given filling pressure, and at least initially the heart is emptied more completely with each beat. Part of the "reservoir" provided by the end-systolic volume is thus driven into the active part of the circulation.

The alteration in length/tension behavior of the heart is commonly termed a change in myocardial contractility. More precisely (Sonnenblick, 1974), it is an upward displacement of the three-dimensional surface relating length, tension, and contraction velocity (Fig. 6.9). The best available index of contractility in humans (Roskamm, 1973a; Mirsky, 1969; Mirsky, Pasternac et al., 1974) is a ratio based on pressure changes during isovolumic contraction of the ventricle $[(dp/dt)p]_{max}$. This is sensed by inserting a catheter into the brachial artery and passing it retrogradely into the left ventricle. An analog computer is then arranged to provide a continuous

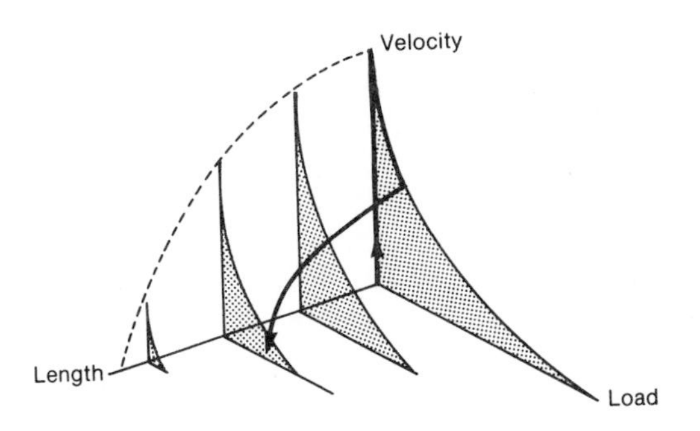

FIG. 6.9. Three-dimensional representation of length, load, and velocity for the myocardium. The heavy line indicates the behavior of a contractile element during a single contraction. An increase of contractility is associated with upward displacement of the curved surface traversed by the heavy line. (After Sonnenblick, 1974.)

calculation of the required ratio, and extrapolation back to zero pressure yields an estimate of $\dot{V}_{max}$, the maximum velocity of contraction (D. T. Mason et al., 1970). However, there are lingering doubts concerning the validity of extrapolation to zero pressure (Pollack, 1970), and the concept of ventricular catheterization is plainly unsuitable for use with normal volunteers.

Indirect assessments of cardiac contractility can be derived from echocardiography and angiocardiography. More simply, data may be taken from simultaneous recordings of the electrocardiogram, heart sounds (phonocardiogram), and the carotid pulse wave (Blumberger, and Sigisbert, 1959; G. R. Cumming and Edwards, 1963; Raab, 1966a; W. S. Harris, 1974). Possible measurements include the total period of ventricular systole (from the Q wave of the electrocardiogram to the second heart sound, QS_2), the ejection period (from the beginning of the carotid pulse wave to its dicrotic notch, LVET), and the preejection period (PEP = QS_2 − LVET). Since results vary with heart-rate-related changes of contractility, they are commonly "corrected" to a standard cardiac frequency (Van der Hoeven et al., 1977).

The resting PEP increases from ~80 msec at age 25 to 95–100 msec in a 65-year-old person (Gabbato and Media, 1956; T. R. Harrison et al., 1964); at least 10% of this difference persists after correction for age-related differences of heart rate. In contrast, the LVET is essentially independent of age. The best simple index of contractility seems the ratio PEP/LVET. This is generally reduced in athletes and well-trained individuals (Jordan et al., 1978) and lengthened in myocardial disease. Since it is difficult to record the carotid pulse wave during rhythmic exercise, assessments at increased cardiac loads are usually made either immediately following bicycle ergometer work, or during isometric hand grip contractions (Bloom and Vecht, 1978).

The Starling mechanism was originally thought responsible for the increased cardiac output of exercise. It was postulated that muscular activity increased the return of blood from the leg veins, thereby increasing diastolic volume and therefore the stroke output. Unfortunately for this hypothesis, the diastolic volume is not necessarily increased in exercise (Reindell, 1943; Kjellberg et al., 1949a,b; Holmgren and Ovenfors, 1960; W. J. Phillips et al., 1966; Horwitz et al., 1972; D. H. Bennett, Goldstein, and Leach, 1977), and it is now realized that the body has alternative mechanisms for the increase of cardiac output during work. Even the increase of cardiac contractility does not seem an essential feature of maximum performance, since the steady-state cardiac output is virtually unchange after administration of propranolol, a β-adrenergic blocking agent that causes an almost total inhibition of changes in cardiac contractility (Ekblom, Goldbarg et al., 1972a; Roskamm, 1973a).

The main functional role of the Starling mechanism is probably to equalize the output of the two sides of the heart. Let us suppose that in maximum exercise, the output of the right ventricle increased from 5 to 25 l • min^{-1}, while that of the left ventricle increased from 5 to 24 l • min^{-1}. In a mere 5 min, the entire blood volume would be waterlogging the lungs! Fortunately, the Starling mechanism permits a precise balancing of output between the two circulations, and such problems are avoided.

Heart Size

The volume of the heart is correlated not only with its stroke volume, but also with the total blood volume and the total hemoglobin content of the body (Kjellberg et al., 1949a; Sjöstrand, 1953; Mellerowicz, 1962). Estimates of cardiac volume can be derived from posteroanterior radiographs (Danzer, 1919; Nicogossian et al., 1976) or more precisely from a combination of such films and lateral views (Fig. 6.10; Blumchen et al., 1966; Reindell, König, and Roskamm, 1966; Roskamm and Reindell, 1972). If A, B, and C are the three principal dimensions, in cm, then the volume (ml) is given by

$$\text{Cardiac volume} = 0.4ABC \qquad (6.12)$$

The main weaknesses of the radiographic method are (1) development of cardiac muscle is not

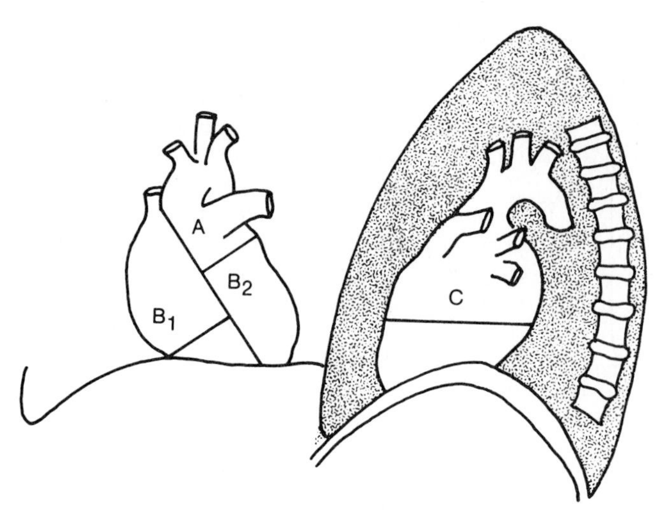

FIG. 6.10. Estimation of heart volume from posteroanterior and lateral chest radiographs.

distinguished from an accumulation of fat or fibrous tissue and (2) an increase of tissue mass may be confused with an increase of cardiac blood volume, as occurs in a thin, dilated, and failing heart. Clues to health or disease can sometimes be drawn from other features of the radiograph, such as congestion of the lung fields, but often the question must be resolved by relating heart volume to working capacity or maximum oxygen intake (Holmgren, Mossfeldt et al., 1964).

In a sedentary young man, a cardiac volume of 700–800 ml (10–11 ml • kg^{-1}) may be anticipated, while values of 900–1000 ml (~ 14 ml • kg^{-1}) are encountered in endurance athletes (Reindell, Kleipzig et al., 1960; Mellerowicz, 1962). It is generally assumed that hypertrophy of the cardiac muscle has occurred in the endurance performers in response to the demands of repeated exercise, although at least a part of the athletes' advantage could be a matter of genetic endowment. We know that hypertrophy of the heart wall occurs when it has to work against an abnormal valvular narrowing (as in pulmonary valvular or infundibular stenosis), but in such circumstances the "training stimulus" is present 24 hours per day. Significant gains in the mass and thickness of the ventricular wall have also resulted from as little as 12 weeks of endurance training in the dog (Wyatt and Mitchell, 1974). However, attempts to induce hypertrophy by the regular training of young men have had less uniform success (Roskamm and Reindell, 1972; Bruce, Kusumi et al., 1975; Roskamm, 1973b). Furthermore, longitudinal studies of ex-athletes indicate persistence of a large heart for some years after training has ceased (P. O. Åstrand, Engström et al., 1963), although there seems an eventual regression to more average values (Saltin and Grimby, 1968).

Cardiac Output in Exercise

There is normally a linear relationship between the steady-state exercise cardiac output and the corresponding oxygen consumption, and at any given submaximal oxygen intake rather similar cardiac output readings may be anticipated irrespective of the mode of exercise (walking, running, swimming, leg or arm ergometer exercise, single-leg work, etc.; Stenberg et al., 1967; Hermansen et al., 1970; Simmons and Shephard, 1971a,b; Holmér, 1974a; C. T. M. Davies and Sargent, 1974b; Clausen, 1977). However, the exercise cardiac output is smaller and the arteriovenous oxygen difference is wider when upright than when supine (Wade and Bishop, 1962).

P. O. Åstrand, Cuddy et al. (1964) argued that the slope of the increase in cardiac output with oxygen consumption became less steep between 70 and 100% of maximum oxygen intake. However, subsequent work from the same laboratory failed to demonstrate this feature (Grimby, Nilsson, and Saltin, 1966; Ekblom, Åstrand et al., 1968) in athletes and other subjects undergoing vigorous training. In our studies of young adults (Simmons and Shephard, 1971a,b), the cardiac output at maximum oxygen intake was marginally less than would be predicted from a linear extrapolation of the cardiac output/oxygen consumption line, but the discrepancy was not statistically significant. Presumably, much depends on the state of training of the subject and his ability to sustain a high stroke volume against the resistance offered by strongly contracting leg muscles. In support of this view, the maximum cardiac output is increased if cutaneous vasodilation accompanies vigorous exercise (for example, when an 8-min rather than a 3-min exercise protocol is adopted, Saltin, 1973b, or when maximum exercise is performed after ingestion of ethanol, Blomqvist et al., 1970).

During light physical activity, the cardiac output of an older subject is much as in a younger adult who is working at a comparable load (Becklake et al., 1965; Kilbom and Åstrand, 1971; Niinimaa and Shephard, 1978b). However, the stroke volume of an old person is less well maintained at high intensities of effort, so that in this age group the maximum cardiac output is substantially less than would be predicted from a linear extrapolation of the submaximal data.

Children have a low cardiac output in relation to oxygen consumption (Bar-Or, Shephard, and Allen, 1971; Rode, Shephard, and Bar-Or, 1973; Mocellin and Sebening, 1974); this reflects (1) small viscera, and thus a low blood flow requirement for tissues other than muscle; (2) a limited amount of subcutaneous fat, and thus less need for skin blood flow than in an adult; and (3) possible above-average tissue enzyme activities, with more complete extraction of oxygen in the working muscles.

Arteriovenous Oxygen Difference

The arteriovenous oxygen difference climbs quickly from its resting value of 40–50 ml • l^{-1} to ~120 ml • l^{-1} at 40% of maximum oxygen intake (Table 6.5). There is a further slow increase at higher work loads, to ~130–140 ml • l^{-1} in a sedentary young adult (Table 6.8), somewhat higher values in an athletic subject, and rather lesser readings in an elderly person (Niinimaa and Shephard, 1978b).

The maximum arteriovenous oxygen difference depends upon (1) the oxygen content of the arterial blood and (2) the completeness of oxygen extraction in the peripheral circulation. The arterial oxygen content varies with hemoglobin level, alveolar oxygen pressure, and the completeness of gas equilibration in the lungs (Chap. 5). When fully saturated, each g of hemoglobin can accept 1.34 ml of oxygen, and under normal ambient conditions a further 2 ml • l^{-1} of oxygen is dissolved in physical solution in the blood. The form of the oxygen dissociation curve (Fig. 4.10) is such that when breathing room air at sea level, arterial saturation is about 96%; in a man (average hemoglobin 156 g • l^{-1}), there will thus be an arterial oxygen content of ~203 ml • l^{-1}, while in a woman (average hemoglobin concentration 138 g • l^{-1}), the corresponding oxygen content is ~187 ml • l^{-1}. Many authors believe that oxygen saturation is well sustained in maximum effort, although others have suggested that there is a small decrease of arterial saturation at the highest work rates. Complicating factors during exercise include a hemoconcentration of 5–10%, a decrease of arterial pH, an increase of blood temperature, and an increase of blood CO_2 content. The last three of these variables displace the oxygen dissociation curve to the right, thus decreasing the oxygen saturation for a given partial pressure of oxygen (Haldane and Priestley, 1935; Ruch and Patton, 1974).

A modest rightward shift of the dissociation curve favors extraction of oxygen in the active tissues. P. O. Åstrand and Rodahl (1977) calculated the response for an oxygen pressure of 20 mmHg (2.67 kPa). If the pH fell from 7.4 to 7.2, and the blood temperature rose from 37 to 39°C, 26 ml of oxygen were liberated from each l of blood, improving the effective oxygen transport by ~12%. Further changes (to pH 7.0 and a temperature of 40°C) were less advantageous, since these were sufficient to modify not only the venous but also the arterial oxygen saturation.

The mixed venous oxygen content (Shephard, 1968a, 1977a) drops as muscle oxygen extraction increases. However, the minimum mixed venous content (and thus the maximum arteriovenous difference) depends largely upon the relative proportions of the cardiac output directed to muscle (where oxygen extraction is almost complete) and skin (where there is very little oxygen extraction). The large arteriovenous oxygen difference of the athlete is explained by a high ratio of muscular to extramuscular blood flow, and a low relative need for subcutaneous blood flow (since sweating is earlier and more vigorous, and there is less subcutaneous fat than in a sedentary subject).

Adaptations to Prolonged Exercise

The steady-state heart rate changes relatively little between 5 and 10 min of exercise, but if effort continues for 1 hr at ~75% of maximum oxygen intake there is a 15–20% increase of heart rate (~30 beats per min; Ekelund and Holmgren, 1964; Saltin, 1964a). This increment is greater in a warm than in a cool environment, and is less if the subject is well trained or the intensity of work is reduced (Ekblom, 1970). It reflects in part the rise of core temperature and in part a peripheral sequestration of blood with a reduction of plasma volume. The heart rate change is mediated via the autonomic nervous system. Plasma noradrenaline levels increase from ~2μg • l⁻¹ at 10 min to >4 μg • l⁻¹ at 60 min, while the atropine-induced reduction of heart rate decreases from ~15 beats per min to zero over the same period (Hartley, 1977). We may thus conclude that sympathetic nerve activity is increased, and the vagal influence on heart rate is abolished during prolonged work.

Cardiac output remains unchanged as heart rate rises (Ekelund and Holmgren, 1964; Saltin, 1964a). Stroke volume thus diminishes; indeed, some evidence suggests that the decline of stroke volume is the primary phenomenon. Filling pressures of both ventricles are reduced (Ekelund and Holmgren, 1964), while in animals at least there are changes of myocardial performance (lower peak tension, lower velocity of contraction; Maher et al., 1972) and alterations of cellular structure (King and Gollnick, 1970).

The mean systemic pressure drops by ~1.3 kPa (10 mmHg) over the first hour of work (Ekelund and Holmgren, 1964; Saltin, 1964a,b). Given the constancy of cardiac output, this implies a fall of peripheral resistance, possibly due to increasing cutaneous vasodilation.

Ninety minutes following the completion of exhausting work, Saltin (1964b) found that the heart rate response to submaximal exercise was still increased by 16 beats per min. Maximum oxygen intake was not significantly altered, but the endurance of maximum effort was reduced by 2.6 min, and the terminal blood lactate was 4.5 mmol • l⁻¹ less than before the prolonged work bout.

Hemoglobin Level

An increase of hemoglobin level increases the maximum possible arteriovenous oxygen difference, and thus the oxygen carriage per liter of cardiac output. Assuming that maximum cardiac output is not reduced by a concomitant increase of blood viscosity, an increase of hemoglobin should thus increase maximum oxygen intake.

While this proposition has been debated in the past, there is now good evidence that anemia restricts oxygen transport (Ekblom, Goldbarg, and Gullbring, 1972b; Woodson et al., 1978) and that an increase of hemoglobin level (as in "blood doping") increases both maximum oxygen intake and work capacity (Ekblom, Goldbarg, and Gullbring, 1972b; Buick et al., 1978).

Heart Sounds and Murmurs

"Functional" murmurs are a common feature of the normal heart, particularly in children; population studies of school children suggest a prevalence of ~20% (Shephard, Lavallée et al., 1979a), and some pediatric cardiologists claim that with very careful auscultation 45–90% of children show the phenomenon (Iliev and Velvev, 1977). The murmurs in question are soft in character, with a duration limited to early or mid-systole. A comparison between children with and without functional murmurs (Shephard, Lavallée et al., 1979a) reveals no differences of cardiorespiratory performance (Table 6.9). Nevertheless, misinterpreted murmurs are a frequent cause of unnecessary invalidism. A brief comment on the genesis of both heart sounds and murmurs thus seems appropriate.

Normally, the blood (like the respiratory gases; Fig. 5.19) has a laminar, streamline flow pattern that makes no sound detectable by a stethoscope. The sole cardiovascular noises are the two heart sounds (I, "lubb," and II, "dupp"). The lubb is initiated by vibration of the arterioventricular valves, and the dupp arises from a similar vibration of the aortic and pulmonary valves. Details of the two sounds can be studied by placing an electronic stethoscope over appropriate areas of the precordium. The second sound is often used in the timing of electromechanical systole (Q-S_2, a measurement in the noninvasive determination of myocardial contractility). Immediately postexercise a third loud and low-pitched heart sound is sometimes heard just prior to the P wave of the electrocardiogram. Zoneraich et al. (1977) recorded such a sound in 3 of 12 healthy marathon runners, although clinicians commonly associate it with a temporary ischemic weakening of left ventricular contractile force. A fourth heart sound may also develop, just prior to the QRS complex of the electrocardiogram. It is low in pitch and intensity and, although sometimes seen in normal subjects, is usually a warning of ischemic heart disease ("atrial gallop" rhythm; R. S. Ross, 1970; Niederberger, 1977; Siegel, 1978).

Under resting conditions, sensitive instruments show a slight disturbance of laminar flow just after the systolic peak, and it has been speculated that growth of the aortic lumen is regulated to avoid the development of more general turbulence (Boughner and Roach, 1971; Light, 1978). As the speed of blood

TABLE 6.9
Influence of "Functional" Cardiac Murmurs upon Cardiorespiratory Performance for a Large Population of Primary School Children (mean ± S.D.)

Variable	Boys: With Murmur	Boys: Without Murmur	Girls: With Murmur	Girls: Without Murmur
Maximum oxygen intake ($ml \cdot kg^{-1} \cdot min^{-1}$ STPD)	46.9 ± 8.2	48.4 ± 8.3	44.7 ± 5.9	44.1 ± 7.1
Physical working capacity (PWC_{170}, $kg \cdot m \cdot min^{-1}$ per kg of body mass)	14.5 ± 2.4	15.0 ± 3.5	11.1 ± 2.6	12.1 ± 2.9
50-yard run (sec)	9.0 ± 1.2	9.1 ± 1.1	9.5 ± 1.1	9.3 ± 1.2
300-yard run (sec)	73.2 ± 9.4	74.1 ± 8.9	76.6 ± 8.3	76.6 ± 17.1

Source: Based on data of Shephard, Lavallée et al., 1979a.

flow is increased by exercise,* the critical velocity is inevitably surpassed [equation (5.14)] and an irregular, turbulent pattern of flow spreads through the great vessels (Stehbens, 1959). Almost every athlete has an audible murmur over the aortic region of the precordium after severe exertion (Tunstall-Pedoe, 1970), and anxiety can bring blood flow to the critical velocity even while resting. If a physician reports that he has heard a resting heart murmur, without reassuring the patient, the person examined may be even more anxious at a subsequent clinical examination. In consequence, the resting cardiac output is further increased, turbulence spreads further along the aorta, and the murmur is heard over a larger area of the chest for a longer fraction of the cardiac cycle. A vicious circle of restricted activity, anxiety, and increasing murmur may thus be created. Anemia is another source of an "innocent" murmur; the tendency to turbulence is here brought about partly by a decrease of blood viscosity and partly by an increase of cardiac output in an attempt to compensate for the decreased oxygen-carrying capacity of the blood.

Pathological murmurs are localized areas of turbulence. The cause may be a narrowed, roughened, or leaky heart valve, or an abnormal communication between the pulmonary and systemic circulations (for example, atrial and ventricular septal defects, and persistent ductus arteriosus). The abnormal pattern of blood flow and the roughening of the ventricular wall lower the Reynold's number from the value of 1000–2000, typical for a straight and smooth-walled tube. Pathological murmurs may be localized to the region of the abnormality, but are often propagated over a wide area of the chest. They are typically louder than an innocent murmur and may extend into diastole. They are also usually audible at rest. However, on occasion exercise may induce an ischemic malfunction of the papillary muscles that support the mitral area of the precordium, thus leading to the appearance of a diastolic regurgitant murmur over the mitral valve during vigorous work (R. S. Ross, 1970).

If a pathological murmur is present, the work of the heart is generally increased by the associated abnormality. Nevertheless, it is increasingly recognized that many cardiac patients with quite loud murmurs benefit from an exercise prescription tailored to their condition and the demands of their daily life.

*Clinical readers may object that innocent murmurs are supposed to disappear with exercise or a change of posture. Murmurs that respond in this way probably arise from small defects in the interventricular septum. If the pressure in the pulmonary circulation is raised, the shunting of blood through an interventricular communication is diminished and the associated turbulent flow may no longer be audible.

ENERGY AND WORK IN THE CARDIOVASCULAR SYSTEM

Forms of Energy

As in any closed fluid circuit, the energy of the cardiovascular system exists in three forms: pressure energy, potential energy, and kinetic energy.

Pressure Energy

Pressure energy (P′) is largely self-explanatory. According to Pascal's law, the resting hydrostatic pressure increases with the depth below the free surface of a fluid. Other variables are the density of the fluid and the gravitational acceleration [compare equation (6.14)]. The current unit of measurement is the kilopascal (1 kPa = 7.51 mmHg). In most parts of the circulation, pressure varies over the cardiac cycle, and it may thus be useful to calculate a mean pressure energy. This is best accomplished by an electrical integration of the pressure signal, although the mean arterial pressure can also be approximated by summing the diastolic pressure (D′) and one-third of the difference between systolic (S′) and diastolic readings:

$$P' = D' + \frac{S' - D'}{3} \quad (6.13)$$

Potential Energy

A fluid has potential energy if it is stored above an arbitrary reference level. The quantity of energy involved depends on the density of the blood (δ_b), the gravitational acceleration (g), and the height (H′) above the reference level:

$$\text{Potential energy} = \delta_b g H' \quad (6.14)$$

Kinetic Energy

A fluid has kinetic energy by virtue of its motion. The quantity of energy stored in this form depends on the density of the blood and the local velocity of blood flow:

$$\text{Kinetic energy} = \tfrac{1}{2}\delta_b v^2 \quad (6.15)$$

Total Energy of the System

The total energy content at any point in the circulation (E′) is given by the sum of the three elements discussed above:

$$E' = P' + \delta_b g H' + \tfrac{1}{2}\delta_b v^2 \quad (6.16)$$

In the resting state, there is little kinetic energy at most points in the circulation, and the equation simplifies to

$$E' = P' + \delta_b g H' \quad (6.17)$$

During exercise, an increased proportion of the total energy is in kinetic form, particularly in the great veins and pulmonary circulation (where the pressure energy remains low). In a closed system, a decrease in one form of energy must be accompanied by an increase in some other form (Bernouilli's theorem, which is a special case of Newton's first law). A continuous interchange between the three types of energy thus occurs as the blood passes around the vascular system. For example, as blood is pumped to the head, pressure energy is converted to potential energy. Similarly, as blood enters the atria, flow is stopped, and kinetic energy is converted to pressure energy to meet the demands of diastolic filling (Fig. 6.8). The cardiovascular system is not completely closed, since some energy is degraded to heat. This loss is restored to the arterial side of the circulation by the activity of the cardiac pump.

Pressures in the Systemic Circulation

Reference Levels

Since the vascular system has no liquid surface, Pascal's law cannot be applied directly. The reference level usually chosen is that of the heart. In a seated subject, arterial pressures recorded by means of an arm cuff need little correction with respect to this reference. However, the valves of the venous system may distort local venous pressures from their anticipated hydrostatic values.

Resting Systemic Arterial Pressures

The standard clinical procedure for the measurement of systemic arterial pressure dates from the design of the mercury sphygmomanometer by Riva-Rocci (1896) and Korotkov's description (1905) of the sounds made by the blood as it pulses under an arm cuff (Fig. 6.11).

In order to appraise the accuracy of the clinical technique, it is necessary to record pressures simultaneously from the same limb by direct and indirect methods. Such an approach (Hunyor et al., 1978) suggests that the pressure at which sounds can first be heard is closely correlated with but slightly underestimates (Δ 1.5 kPa, 10 mmHg) the systolic pressure recorded from a catheter in the brachial artery. The stethoscopic estimate of diastolic pressure is less well correlated with the directly recorded value, whether the observer records the point where the sounds lose their tapping quality (fourth phase) or when they are no longer audible (fifth phase). Fabian et al. (1975) suggested that stethoscopic estimates were lower than values obtained by phonoarteriography, but Hunyor et al. (1978) found that the stethoscopic estimate of diastolic pressure systematically exceeded the directly measured value even if the fifth phase figure was recorded.

Both the reliability and the validity of indirect blood pressure measurements can be improved by careful technique. Practical suggestions include listening to standard tape recordings of the Korotkov sounds, instructing observers on the dangers of digit preference (some people will consistently record a pressure ending in 0 or 5 mmHg; G. A. Rose et al., 1964), choosing a suitable cuff (the inflatable bag

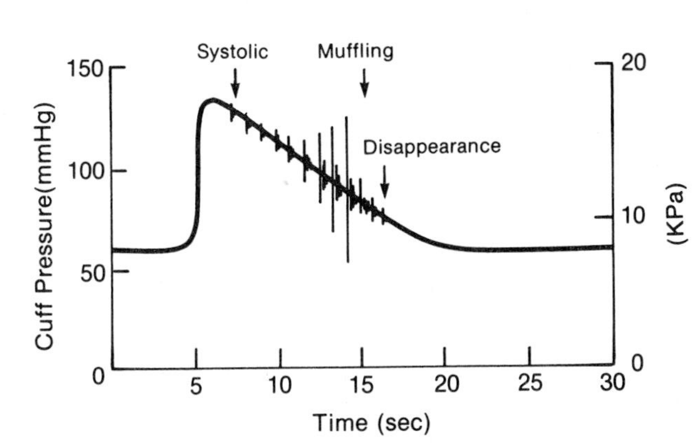

FIG. 6.11. Tracing obtained from simple form of automatic blood pressure recorder (Shephard, 1958b; Mastropaolo et al., 1964). The Korotkov sounds have been sensed over the brachial artery, and the corresponding signal is superimposed upon a pressure record obtained from a sphygmomanometer cuff. The systolic pressure is indicated by the first appearance of the Korotkov sounds, and the diastolic reading by their muffling (Phase IV) or, if desired, by their disappearance (Phase V). The cuff is automatically inflated to 20 kPa (150 mm Hg) twice every minute. The main difficulty during extended use of the technique is that physical activity displaces the microphone from the optimum site for sensing the Korotkov sounds.

should completely encircle the arm and the width should be 40% of arm circumference, *British Medical Journal*, 1975a; use of too large a cuff leads to overestimation of pressures), adopting a standard rate of cuff deflation (0.2–0.3 kPa • sec^{-1}), and using ultrasonic flow detectors.

Since exercise induces a substantial rise of systemic pressures, it is wise to exclude from unsupervised vigorous activity those individuals with a pathologically high blood pressure (M. M. Craig and Raftery, 1978; Nageotte and Kasch, 1978). However, there is no clear line of demarcation between normality and a pathological resting blood pressure (Stamler et al., 1967; Pickering, 1967), so that the precise level at which hypertension should be diagnosed remains a matter of dispute. The blood pressure is greatly increased by anxiety, and much thus depends on the degree of relaxation that is achieved during the measurement; readings may be high on first visiting a laboratory, but decline with the habituation of repeat visits. Pressures also rise over the course of the day, and the systolic pressure is increased following a heavy meal. Many physicians make measurements in the unsupported arm, forgetting that isometric muscular activity and dependency of the arm cause falsely elevated readings (Silverberg et al., 1977). Finally, both systolic and diastolic pressures increase by 1.3–2.0 kPa over the span of working life (Master et al., 1964).

The clinician usually attaches more significance to diastolic than to systolic pressure. A diastolic reading of 12 kPa (90 mmHg) is viewed with suspicion, and 13.3 kPa (100 mmHg) or more is indicative of hypertension. However, there is also much overdiagnosis. In one sample of 1130 adult Canadians, 100 persons had been told that their blood pressure was too high, but only 15 of the sample (including 12 of those diagnosed) had a pressure ≥12 kPa (Shephard, 1978j).

The long-term effects of activity upon the resting blood pressure are still disputed. Possible artifacts in experimental studies of this question include (1) a progressive habituation of the subjects to the observer and the test laboratory and (2) an improved fit of the sphygmomanometer cuff as subcutaneous fat is lost. Most (but not all) authors have reported that regular physical exercise produces a small (~1 kPa) decrease of systemic blood pressure in both hypertensive and normotensive individuals (for references, see Shephard, 1978b).

Systemic Arterial Pressure and Exercise

The error of indirectly measured blood pressures increases during exercise (Mastropaolo et al., 1964). Difficulty arises because (1) the kinetic component of the recorded pressure is increased by exercise, (2) reflection of pulse waves may augment the peripherally recorded pressures, (3) it is difficult to listen to the Korotkov sounds on a moving arm, and (4) the appearance and disappearance of the Korotkov sounds is less clear-cut during effort. Fortunately, the diastolic pressure changes little in the usual type of rhythmic exercise (Mellerowicz, 1962; P. O. Åstrand, Ekblom et al., 1965; Hanson, Tabakin, and Levy, 1968). For some purposes, it may thus be sufficient to estimate the systolic pressure during activity and to assume that the diastolic reading does not change from its resting value.

The risks of arterial puncture are such that direct intravascular recordings are not normally justified in applied physiology. Nevertheless, there have been comparisons between sphygmomanometer and intravascular readings in active, healthy subjects.

Cuff estimates of systolic pressure are commonly 1.1-2.0 kPa (8-15 mmHg) less than catheter readings during exercise but exceed the catheter figures by 2.1-5.1 kPa (16-38 mmHg) during recovery (Rowell, Brengelmann et al., 1968b; Kleinhauss and Franke, 1971). Note must be taken of the reference technique. The systolic pressure is substantially lower in the aorta than in a peripheral artery because of pulse wave reflection. Galichia et al. (1976) suggested that the central aortic pressure could be calculated as 4.9 kPa plus two-thirds of the brachial artery pressure. Further, much of the energy in the aorta is in kinetic form during exercise, and this energy is not detected by a lateral or distally directed pressure tapping (Marx et al., 1967). Nevertheless, there are substantial errors in indirect measurements during exercise, and such results must be interpreted with considerable caution.

The effect of rhythmic physical activity upon systolic pressure depends upon the balance that is struck between an increase of cardiac output, a general vasoconstriction, and a local vasodilation in the active muscles. If only the static, distending pressure is measured (Marx et al., 1967), there may be but a small increase during exertion. However, from the viewpoint of cardiac work load, it is important also to include the kinetic element. Experiments with centrally directed catheters and clinical blood pressure cuffs have both indicated a progressive rise of pressure during exercise (Table 6.10), although this varies in extent depending upon the type, intensity, and duration of effort and the condition of the myocardium (Mellerowicz, 1962; P. O. Åstrand, Ekblom et al., 1965; I. Åstrand, 1965; Hanson, Tabakin, and Levy, 1968). In submaximal effort, the response depends upon the proportion of the maximum oxygen intake that is utilized. Increases are larger in older subjects (Reindell, Kleipzig, et al., 1960; Hanson, Tabakin, and Levy, 1968; Gerstenblith et al., 1976) and in conditions whereby there is difficulty in perfusing the active muscle (use of small or weak muscles, particularly movements of the arms above the head; P. O. Åstrand, Ekblom et al., 1965; Bevegård, Freyschuss, and Strandell, 1966; I. Åstrand, 1971; Schwade et al., 1977; G. E. Adams et al., 1978). During sustained maximum effort (10-15 min of activity), peripheral systolic readings of 24-32 kPa (180-240 mmHg) may be anticipated (Masuda et al., 1967; Rowell, Brengelmann et al., 1968b). Low maxima are a warning of a deteriorating myocardium, for instance, in a patient who has sustained a myocardial infarction (Sheffield, 1974). A sudden drop of pressure during testing is an ominous sign, suggesting acute circulatory failure, and it is an urgent indication to halt exercise while sustaining venous return (Bruce et al., 1963; K. L. Andersen, Shephard et al., 1971).

Isometric exercise gives a large and rapid increase of both systolic and diastolic pressures, with an associated tachycardia (Lind and McNicol, 1967; G. E. Adams et al., 1978). The response to this type of work is generally proportional to the percentage of maximal muscle force exerted. Pressures first rise when the active muscles contract at more than 15% of their maximum force. At intermediate efforts (20-60% of maximum), the rate of rise of pressure varies with the intensity of effort, but the pressure at exhaustion is similar (Funderburk et al., 1974). A maximum effect is observed at 70% or more of maximum effort. There is a close parallel between the reduction of muscle blood flow by the isometric effort and the corresponding rise in mean systemic blood pressure. The hypertension seems an attempt by the body to compensate for the compression of the intramuscular vessels. The local stimulus is not precisely identified, but is probably related to vasodilator factors. The nerve pathway involves the sensory fibers of the dorsal column of the spinal cord. The response is thus absent in syringomyelia, a disease that damages this part of the spinal cord (Lind and McNicol, 1967).

A second factor that may contribute to the rise of blood pressure is a tendency to make a forced expiratory effort against a closed glottis (the Valsalva maneuver). This leads to a triphasic change of systemic blood pressure—a transient rise, an equally transient fall, and a more sustained rise, due to the effects of the increased intrathoracic pressure upon venous return. A Valsalva maneuver is common during weight lifting, arm movement being aided by a temporary fixation of the thoracic insertions of the pectoral muscles. Such expiratory efforts are unlikely to contribute to the hypertension of modest hand grip efforts, but they may develop if subjects find it difficult to sustain the required isometric force.

TABLE 6.10
Increments in Systolic Blood Pressure (kPa) in Relation to Age and Work Rate

Age	Work Rate (Mets*) and Syst. Pressure			
(Years)	4	6	8	10
20-29	3.3	4.8	6.3	7.8
30-39	3.3	5.3	7.2	8.9
40-49	3.5	5.7	6.8	8.1
50-59	3.9	6.4	8.5	10.5

*Met is the ratio to basal metabolic rate (i.e., 4 Mets equals 4 × basal metabolism).

Source: Based on data of S. M. Fox, personal communication to author.

The difference in blood pressure responses to rhythmic and isometric exercise has important application to the design of cardiac rehabilitation programs. Rhythmic exercise places little strain upon the heart unless it is so prolonged as to cause a large increase of systolic pressure and thus of tension work. On the other hand, a minute or so of isometric effort, weight lifting, or the supporting of body mass can give rise to a disastrous elevation of blood pressure and thus of cardiac work rate. The main hazard is probably the development of myocardial ischemia (oxygen lack), with a resultant increase in the risk of both electrical and mechanical failure of the left ventricle (Parker, et al., 1966; Shephard, 1974c; D. H. Paterson, Shephard, Youldon et al., 1979; Shephard, 1979h). Autopsy records also suggest an increased risk of bursting an aneurysm or some other weak point in the vessel walls (Jokl, 1958), although the Valsalva maneuver provides some protection in the intrathoracic vessels through a simultaneous increase of extramural pressures. If there is known myocardial ischemia, it is prudent to perform rhythmic exercise in an interval fashion (Kavanagh and Shephard, 1975b), allowing time for the dispersion of muscular metabolites, while any isometric contractions should be sustained for no more than a few seconds.

The Arterial Pulse Wave

Local recording of the arterial pulse wave at the wrist was a popular clinical technique until electrocardiograms became widely available. MacKenzie (1902) used a simple polygraph to distinguish three irregularities of rhythm: the "youthful" type, corresponding to sinus arrhythmia; the "adult" type, corresponding to ventricular extrasystoles; and the "dangerous" type, corresponding to atrial fibrillation. The heartometer (T. K. Cureton, 1947) also gives an indirect recording of the brachial artery pressure waveform. A standard sphygmomanometer cuff is fitted around the upper arm, inflated to 1.3 kPa (10 mmHg) above the local diastolic pressure, and coupled to a mechanical pressure recorder.

In the past, some workers have attached great significance to heartometer oscillations, calculating not only their amplitude and duration, but also the first and second differentials of the tracings (B. D. Franks, 1969). There is certainly a relationship between the amplitude of the primary pulse wave and cardiorespiratory fitness, since the pulse pressure varies with (stroke volume)/(arterial distensibility). Equally, the first and second derivatives are influenced by the vigor of cardiac contraction and thus myocardial contractility. However, quantitative interpretation of the records and complex mathematical treatment of the resultant data are hazardous. A large pulse pressure can reflect not only the large stroke volume of a fit subject, but also the rigid arteries of an older person. Sometimes, these possibilities can be distinguished from the form of the pulse record. If the arterial wall is rigid, both systolic and diastolic pressures are high, while the rise and fall of pressure both occur rapidly. Unfortunately, the heartometer tracing is also influenced by conditions in the tissue overlying the brachial artery, particularly the texture of the skin, the amount of subcutaneous fat, and the tension in the arm muscles. Training-induced "improvements" in the heartometer record may thus reflect alterations in these secondary factors rather than a true gain of cardiac performance.

Interest in the arterial pulse wave was renewed with recognition of the need for an indirect method of assessing cardiac contractility. The current approach (Cundiff and Corbun, 1969; B. D. Franks, 1969) is to make recordings from a sensitive electronic tambor or a lightly inflated cuff positioned over the carotid artery. Possible measurements include *electromechanical lag* (from the Q wave of the ECG to the first heart sound, S_1; T. R. Harrison et al., 1964); *ejection period* (times from the first to the second heart sound, $S_1 - S_2$; Weissler et al., 1961); *total systole* (QS_2; Weissler et al., 1961); *isovolumic contraction period* (mechanical systole minus ejection period; Rushmer, 1956); and *tension period* (electromechanical lag plus isovolumic contraction period; M. N. Frank and Kinlaw, 1962).

The figures most commonly used in the assessment of myocardial contractility are the preejection period (QS_2, ejection period) and the ejection period. The accuracy of such data is limited by the phase lag in transmission of the pulse wave from the heart to the recording site in the root of the neck. Although often assumed to be rapid and consistent in speed, the pulse wave velocity is in fact finite and inversely related to the distensibility of the great vessels, increasing from ~ 5 m • sec^{-1} (in a young person with elastic arteries) to ~ 10 m • sec^{-1} (in an older person with some vascular hardening). The rate of travel also depends on the mean systemic blood pressure; as this is increased, the arteries are put under stretch, their effective elasticity declines, and the pulse wave travels more rapidly. As in use of the heartometer, overlying tissues can cause problems for a neck tambor. The large size of the carotid artery and its superficial location favor recording, but it is still difficult to obtain good exercise tracings; readings have been obtained during bicycle ergometry (for example,

G. R. Cumming and Edwards, 1963), but it is more usual to collect data immediately following rhythmic exercise or during isometric hand grip efforts (Nutter et al., 1972).

Venous Pressures

The intravascular pressure decreases rapidly on passing through the arterioles, and venous pressures at heart level are close to zero. Because of the interchange between pressure and potential energy, the pressure in the neck veins is normally negative relative to the heart, and the vessels in this region are collapsed. The extent of body tilting necessary to induce venous collapse is used as a rough clinical measure of central venous pressure. In the lower part of the body, the potential energy of the vascular system is normally negative, but there is a substantial hydrostatic pressure relative to the heart (Sacdpraseuth, 1960); on the dorsum of the foot, this may amount to 11-12 kPa (80-90 mmHg).

Indirect recordings of venous pressure can be obtained by placing a very light mechanical or electrical tambor over one of the large veins. The central venous tracing usually shows three oscillations corresponding to atrial contraction (*a* wave), a pressure wave transmitted from the underlying artery (*c* wave), and a rise of venous pressure as the atria refill (*v* wave). Recent studies (Baerstschi and Gann, 1977) identified two distinct groups of mechanoreceptors in the atria. The first, or A group, fire during atrial contraction, and their function has yet to be elucidated. The B group fire during atrial refilling, and respond to both the absolute atrial volume and the rate of change of this volume.

Because the Starling curve is shifted to the left during exercise, an increased stroke volume can be developed with remarkably little rise of venous pressure (Fig. 6.8). Nevertheless, there is some association between atrial pressure and the heart rate at which cardiac output is maximal (Sugimoto et al., 1966). In certain forms of cardiac disease, patients may exercise to the point where the heart is failing. There is then a considerable elevation of venous pressure; the great veins are distended, and the *a* and *v* waves are transmitted much more forcibly to the recording tambor.

Potential Energy Within the Systemic Circulation

Arteries

Elevation of an artery above heart level leads to a local interchange of pressure and potential energy in accordance with the theorem of Bernouilli [equation (6.16)]. The influence upon the blood content of the vessel concerned is discussed below. Of greater practical importance is the modification of local perfusion pressure. This predisposes to fatigue when carrying out work above the head, for example, a man scraping or painting a ceiling (P. O. Åstrand, Ekblom et al., 1965; I. Åstrand, 1971). A house painter may find it advantageous to tie his brush to a long pole in such a situation. Although he must then lift a greater mass, his hands and arms can be kept at or below heart level.

Veins

In the venous system, an increase of potential energy is important primarily because it leads to an increase of pooling. Blood sequestered in this fashion is not available to meet the needs of homeostasis within the active tissues. Venous return is reduced, and (through the Starling mechanism and other reflexes) the stroke output of the heart is diminished.

The distensibility of the veins can be explored by noting (1) changes in the pressure of an isolated vein segment, (2) rate of rise of venous pressure and the associated change of segmental volume following venous occlusion, and (3) final volume of blood accumulating in a body segment with various pressures in the occluding cuff (Shepherd, 1963). Pressure/volume characteristics of typical large arteries and veins are compared in Fig. 6.12. The systemic arteries and arterioles typically contain ~800 ml of blood; they are relatively indistensible, and any increase of mean intravascular pressure has a small and rather uniform effect upon their blood content. In contrast, the systemic veins normally contain ~3000 ml of blood, more than 60% of the total blood volume. Further, they are readily distensible, with a "ceiling" of capacity set by fibrous tissue elements in the vessel wall (J. P. Henry, 1951). Thus a small increase of intravascular pressure is sufficient to fill an almost empty vein to near its maximum capacity.

The phenomenon is seen particularly well following a sudden passive shift from a supine to a vertical body posture (a "tilt-table" test), when a combination of poorly functioning valves in the great veins (Ludbrook 1966), venous pooling, and a decrease of hydrostatic pressure in the cerebral vessels may be sufficient to cause a brief loss of consciousness (Asmussen and Christensen, 1939a). As cardiac output and systemic blood pressure are reduced by the tilting (Damato et al., 1966; Sowton and Burkart, 1967), impulses from pressure receptors in the aortic arch and carotid sinus diminish and compensatory mechanisms are brought into play (Salzman and

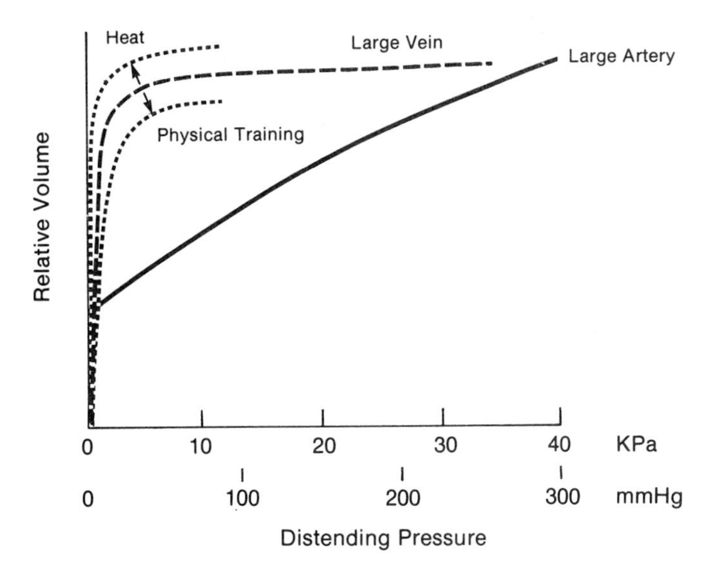

FIG. 6.12. A comparison of pressure/volume characteristics for a large systemic artery and a large systemic vein. Note the effects of heat and physical training upon the venous pressure/volume curve.

Leverett, 1956; Bartelstone, 1960; Ino, 1960; P. M. Stevens, 1966). Parasympathetic cardioinhibitory activity is decreased, vasodepressor neurons are inhibited, and there is an increase of adrenergic sympathetic nerve impulses. The heart rate is accelerated, with a vasoconstriction of resistance vessels in both the splanchnic area and the muscles (Rowell, 1974). The tone of the leg veins is also increased (Page et al., 1955; Sharpey-Schafer, 1961; C. A. Gilbert and Stevens, 1966), thereby diminishing their capacity at a given distending pressure. The extent of the postural disturbance thus depends upon (1) the available blood volume (Rushmer, 1976) and (2) the extent and speed of the various reflex adjustments; in general, adaptation is favored by vigorous habitual activity, a fact exploited in some older postural tests of cardiorespiratory fitness.

In a sedentary subject, blood pressure and pulse pressure fall for ~5 sec after tilting to the upright position (Sharpey-Schafer, 1961), but in a well-trained person there is a small increase of blood pressure. The individuals most prone to postural hypotension have avoided the stress of gravity for some time. Examples include cases of prolonged bed rest, astronauts engaged in space exploration, and athletes involved in prolonged swimming contests. It seems that if normal gravitational forces are not encountered periodically, the veins fail to constrict on standing. Manifestations of this loss of adaptation include tiredness, dizziness upon standing (Holmgren, 1967a; Fried and Shephard, 1969), and frank faints, all of these symptoms reflecting an inadequate blood supply to the brain.

If the postural response is poor, there may be an associated autonomic reaction, usually described as a "vasovagal attack" (McMichael and Sharpey-Schafer, 1944; Brigden et al., 1950; Greenfield, 1951). The skin becomes pale, moist, and clammy, the subject feels a wave of nausea, and the pulse rate which has been rapid suddenly slows. At the same time, arteries supplying the muscles dilate widely. The victim "bleeds into his muscles," the blood pressure falls dramatically, and consciousness is lost. If the individual concerned is allowed to lie down, pressure energy is reconverted to potential energy, and the hydrostatic load distending the veins is released. Blood floods back to the heart, increasing the stroke volume, and recovery is rapid. However, there have been incidents of misguided enthusiasm, particularly on the parade ground, where fainting guardsmen have been held erect by their colleagues with serious deprivation of blood supply to the brain.

Holmgren (1967a,1971) described a specific syndrome (vasoregulatory asthenia), characterized by a chronic deficiency of vascular adjustment and episodes of fainting; symptoms include precordial pain, palpitations, shortness of breath, and anxiety. Despite normal heart and blood volumes, the physical working capacity is low in both upright and supine positions, and the electrocardiogram shows evidence of increased sympathetic tone during both exercise and orthostatic (tilt-table) tests.

The capacity of the veins at any given distending pressure is increased by a rise of blood temperature (J. P. Henry, 1951). Unfortunately, body heating is also likely to deplete the total blood volume through sweating and an exudation of fluid from the vascular compartment. A person is thus particularly vulnerable to fainting in hot weather. An exercise-induced rise of rectal temperature is less hazardous, since exercise leads to a persistent increase of venous tone, and thus a diminution of venous pooling (Merritt and Weisler, 1959; Sharpey-Schafer, 1963; Bevegärd and Shepherd, 1965,1966a; Hanke et al., 1969). The tonic response seems restricted to the cutaneous veins (Zelis et al., 1969); however, contraction of the leg muscles also serves to reduce the effective hydrostatic pressure, forcibly pumping blood back toward the heart (Sacdpraseuth, 1960; Ludbrook, 1966), and this process is helped by the subatmospheric pressures generated within the chest (the thoracic pump).

During arm work, venous return is more difficult, especially if exercise is performed in an upright position. Under such conditions, the blood content of the leg veins may actually increase (Asmussen and Nielsen, 1955), so that it is not surprising to find a small maximum stroke volume during arm work (Simmons and Shephard, 1971a). A failure of venous return is also likely if a person remains standing following exercise. The body is hot and the arterioles are widely patent, permitting rapid filling of the veins. If it is necessary for a subject to remain upright after exercise, he should cease his efforts gradually (warm down); among other useful functions, this keeps both muscle and thoracic pumps operating at an adequate level to sustain venous return.

Venous pooling is increased if gravitational acceleration is augmented. Fighter pilots may have sudden 20- to 30-sec exposures to footward accelerations of 3–6 g when making tight turns in their aircraft; this leads to impairment of vision ("gray-out"), an increase in the effective density of the pleural fluid (with collapse of air spaces in the lower part of the lung), shunting of pulmonary blood flow through the poorly aerated bases of the lungs, and loss of consciousness. Tolerance of such maneuvers can be extended if local counterpressure is provided, for example, by bandaging the limbs (N. Lundgren, 1946; Arenander, 1960) or wearing water- or air-filled trousers (W. R. Franks, 1940; P. Howard, 1965). Headward acceleration is less common. Pilots may then note a "red-out" (unconsciousness, preceded by congestion of the eye). There is a risk of cerebral damage with accelerations exceeding 3 g.

Much greater accelerations are encountered during the launching of space vehicles. Astronauts minimize physiological disturbances by assuming a prone position. Much greater decelerations are encountered during the collision of either vehicles or competing sportsmen. However, the duration of such decelerations is short, and stresses are imposed on bones and ligaments rather than on the cardiovascular system.

Kinetic Energy of Circulating Blood

Under resting conditions, little of the energy content of the circulating blood is in the kinetic form except in the large veins and the atria. Filling of the ventricles depends on the conversion of kinetic energy into pressure energy, and diastolic pressures may actually be higher in the ventricles than in the atria.

During maximum exercise, blood flow is increased up to 6-fold, and since kinetic energy is proportional to the square of velocity [equation (6.15)] there is a 36-fold increase of the kinetic component. In the pulmonary artery, as much as a quarter of the energy of an exercising subject is in kinetic form, and even in the systemic arteries the kinetic component is sufficient to cause a substantial difference of readings between lateral and end-tapped pressure gauges. With end tapping, the kinetic energy is added to or substracted from the true local pressure; a change of flow and thus of kinetic energy may then be misinterpreted as a change of pressure (A. C. Burton, 1965a,b; Marx et al., 1967). The standard clinical sphygmomanometer cuff interrupts flow and thus necessarily adds the kinetic component to the systolic reading.

Work of the Heart

General Considerations

Frictional (viscous) and turbulent work are performed during each circulation of the blood; the resultant heat (plus a small component of sound) is dissipated to the environment, and the heart must work to restore the energy content of the system. Other sources of energy expenditure (O. Frank, 1898; A. V. Hill, 1938,1960; Sonnenblick, 1971; Blomqvist, 1974; Coulson, 1976; C. R. Jorgensen et al., 1977; Blick and Stein, 1977) are (1) electrochemical reactions associated with activation of the muscle fibers, including the displacement of sodium and calcium ions against concentration gradients (usually less than 1% of total energy expenditure), (2) the cost of developing and maintaining tension within the ventricular walls, (3) a small (sometimes immeasurable) component of internal work associated with fiber shortening (Urschel et al., 1968a), and (4) basal metabolism (usually 20-25% of the resting energy expenditure; McKeever et al., 1958).

The proportion of useful work, and thus the overall rate of working, is influenced markedly by changes of cardiac contractility and thus the rate of fiber shortening, $\dot{V}_{max}$ (Burns and Covell, 1972).

Useful Work

As in any mechanical pump [equation (1.6)], the useful work performed per stroke (W′) is given by the product of intraventricular pressure P'_v and volume change $\delta V'$, integrated over the ejection phase of the cardiac cycle (Fig. 1.10):

$$W' = \int_{V'_d}^{V'_s} P'_v \delta V' \qquad (6.18)$$

To a first approximation, this integral equals the product of the mean ventricular ejection pressure $\overline{P}'_v$ and the stroke volume Q'_s:

$$W' \approx \overline{P}'_v Q'_s \qquad (6.19)$$

Under resting conditions, both ventricles contain ~160 ml of blood at the end of diastole, and only 80 ml is expelled as the heart contracts. The remainder forms a reserve that may be drawn upon during exercise.

We have already seen (Chap. 1) that with a stroke volume of 80 ml and a mean ejection pressure of 15 kPa, the left ventricle performs ~1.2 N • m of useful work per beat. The mean ejection pressure of the right ventricle is about a fifth as great as that of the left, so that the useful work on the right side of the heart is ~0.24 N • m per beat.

Tension Work

The main determinant of cardiac work is the cost of developing and sustaining tension within the muscle fibers. The anatomical arrangement in the heart wall is complex, with both helical and spiral fibers (Streeter et al., 1969; Jean et al., 1972). Older authors (Burch et al., 1952; A. C. Burton, 1957) related tension T′ to intraventricular pressure P'_v according to the Law of Laplace [equation (1.17)]. The Laplace relationship is, strictly speaking, only valid for a thin-walled spherical structure; however, in the format cited it can be applied to a spheroid, pressures being adjusted by a dimensionless multiplier which is a function of the semimajor and semiminor axes. The equation implies that the wall tension associated with a given intraventricular pressure is augmented by an increase of heart size, whether due to dilation or hypertrophy of the ventricle. However, hypertrophy also increases wall thickness and thus reduces the stress per unit section of muscle. If a sphere has a finite thickness h, the Laplace relationship should be further modified to the form:

$$T' = \frac{P'_v}{h}\left[\frac{R_1 R_2}{R_1 + R_2}\right] \qquad (6.20)$$

Recent investigators have explored possible thick-shell formulas that allow for both radial stress and the transverse shear associated with bending (Sandler and Dodge, 1963; A. Y. K. Wong and Rautaharju, 1968; Mirsky, 1969,1974). Such calculations point to the conclusion that stresses in the ventricular muscle are far from uniform, tensions being largest in the inner (endocardial) region. This region is at a double disadvantage in terms of blood supply. Not only is the local intramuscular pressure high, but arteries reaching this part of the myocardium from the exterior have traversed a substantial thickness of the contracting ventricular wall. It is thus not surprising that myocardial infarctions frequently develop in subendocardial tissue.

Overall Equation for Cardiac Work

Components of cardiac work other than useful work [equation (6.18)] and tension work [equation (6.20)] account for less than 2% of the total energy consumed (Blomqvist, 1974).

The rate of working $\dot{W}'$ is thus approximated by

$$\dot{W}' = \left(\int_{V'_d}^{V'_s} P'_v \delta V' + \alpha \int_0^t T' \delta t\right) f_h \qquad (6.21)$$

where the first of the right hand terms corresponds to equation (6.18), the second term represents the wall tension integrated over the contraction phase of the cardiac cycle, α is a constant of proportionality between the tension/time integral and rate of working, and f_h is the heart rate. Since the first term is "useful" work, and the second term is "wasted" work, one index of cardiac efficiency E′ is given by the ratio

$$E' = \frac{\int_{V'd}^{V's} Pv' \delta V}{\int_{V'd}^{V's} Pv' \delta V + \alpha \int_0^t T \delta t} \qquad (6.22)$$

In practice, investigators have more commonly cited the ratio of useful work to the energy equivalent of the oxygen consumed. On the latter basis A. C. Burton (1965b) set the resting efficiency as low as 3%, although other calculations have yielded higher figures, particularly in trained athletes (Scheuer et al., 1974). With cardiac outputs of the order anticipated in exercise, efficiencies of 10–15% are seen (A. M. Katz, 1977). Even these last values are low relative to data for skeletal muscle, reflecting considerations of ventricular size and shape (A. C. Burton, 1957; Badeer, 1960; R. H. McDonald et al., 1966).

As would be predicted from equation (6.22), it has been found that an increase of external work was met more efficiently if induced by an increase of cardiac output than if due to a rise of systemic blood pressure. In humans, as in dogs, increase of the cardiac work load by a rise of blood pressure leads to a poor mechanical efficiency (Urschel et al., 1968a). On the other hand, moderate rhythmic activity gives a substantial increase of useful cardiac work, with little increase of mean ventricular pressure and a consequent improvement of efficiency. For this reason, it is better for the postcoronary patient to take moderate rhythmic exercise rather than sit at home with ever-increasing anxiety-induced hypertension. On the other hand, he must avoid work that causes a substantial rise of systemic blood pressure (sustained isometric contractions, the lifting of heavy weights, and prolonged near-maximal rhythmic exercise without rest pauses).

Various possible tactics for increasing cardiac output are reviewed later in this chapter. We may note here that at any given cardiac output, a rapid heart rate is less efficient than a slower rhythm. Since the duration of an individual contraction is similar whether a large or a small volume of blood is expelled from the heart, this conclusion can be anticipated from equation (6.22). For a given systemic ejection pressure, an increase of heart rate inevitably increases the useless component of cardiac work, the total tension work per min.

Oxygen Cost of Cardiac Work

The oxygen cost of cardiac work is commonly expressed per unit of ventricular mass. In this context, it must be noted that hypertrophy decreases tension per unit cross-section of ventricular wall, and thus the oxygen consumption per unit of ventricular mass.

Direct measurements of cardiac oxygen consumption can be made quite readily on the dog heart (T. P. Graham et al., 1968; Pool et al., 1968; Sonnenblick, Ross, and Braunwald, 1968). If contraction is halted, the oxygen consumption is 0.01 ml • min^{-1} per g of ventricle. Figures rise to 0.03 ml • min^{-1} • g^{-1} in the empty but actively beating heart (Britman and Levine, 1964), and when normally loaded, values range from 0.1 ml • min^{-1} • g^{-1} at rest to 0.6 ml • min^{-1} • g^{-1} in maximum effort (McKeever et al., 1958; D. E. Gregg and Fisher, 1964; G. G. Rowe et al., 1964; Van Citters and Franklin, 1969).

Measurements in humans involve the estimation of coronary blood flow (for example, by the nitrous oxide method; Bing et al., 1949; J. C. Scott, 1967) and the coronary arteriovenous oxygen difference. Most of the available data (Bing et al., 1949; Lombardo et al., 1953; Holmberg et al., 1971; C. R. Jorgensen, 1972; Ghista and Sandler, 1974) refer to clinical material observed under resting conditions. Values seem much as in the healthy dog (~0.1 ml • min^{-1} • g^{-1}).

Since a man's heart has a mass of ~350 g, its total resting oxygen consumption is ~35 ml • min^{-1}. Given an energy usage of 21 J = 21 N • m per ml of oxygen consumed, a total work rate of 0.74 kN • m • min^{-1} would be predicted. In fact, the useful work accomplished is ~1.44 N • m per beat, or with a heart rate of 65 beats per min, 94 N • m • min^{-1}. The efficiency indicated by these calculations is thus ~12.7%. During maximum exercise, both the mean systolic pressure and the mean stroke volume are increased by ~50%, while there is up to a three-fold increase of heart rate. Useful cardiac work is thus increased by a factor of 6, to ~560 N • m • min^{-1}. Assuming a 15% efficiency during exercise, energy usage would rise to 4.8 kN • m • min^{-1}, with an equivalent oxygen consumption of ~180 ml • min^{-1}.

Given a maximum cardiac output of 25 l • min^{-1}, the oxygen cost of pumping blood is ~6.4 ml of oxygen per l of cardiac output. Although a substantial charge upon the maximum oxygen intake, this is still much less than the equivalent oxygen transport (120 ml of oxygen per l of cardiac output).* There is thus little danger of reaching the point where the oxygen cost of pumping blood exceeds the extra oxygen introduced into the body thereby.

Clinical Estimation of Myocardial Oxygen Consumption

For clinical purposes, it is often useful to estimate the myocardial work load at which symptoms (for example, angina) or signs (for example, ECG ST seg-

*If the maximum oxygen intake is 3 l • min^{-1} and the cardiac output is 25 l • min^{-1}, then 120 ml of oxygen are transported per l of blood flow.

mental depression) first appear. While training usually increases the bicycle ergometer loading or treadmill slope at which an anginal patient develops ischemia, this is a reflection of an increase in the individual's maximum oxygen intake, and there is usually no change in the limiting myocardial work load.

The common noninvasive indices of myocardial oxygen consumption are (1) heart rate, (2) the product of systolic pressure and heart rate, sometimes called the tension/time index,* and (3) the triple product (systolic pressure × heart rate × duration of systole; Blomqvist, 1974).

It is plain from equation (6.21) that the heart rate is proportional to the cardiac work rate only if ventricular pressure, stroke volume, and cardiac contraction time all remain constant. Good correlations between heart rate and the directly measured myocardial oxygen consumption were reported in some experiments (C. R. Jorgensen, 1972; Y. Wang, 1972), but it was later realized that this arose from a correlation between heart rate and blood pressure under the conditions tested ($r = 0.73$); reanalysis of the data (C. R. Jorgensen, 1977) showed a dependence upon both heart rate (f_h) and mean systemic blood pressure (BP, kPa):

$$\text{Myocardial } \dot{V}_{O_2} = 0.24(f_h) + 1.20(\text{BP}) - 29.9 \qquad (6.23)$$

Given a heart rate of 135/min and a blood pressure of 17.5 kPa, this equation would predict an oxygen consumption of 23.5 ml • min^{-1} per 100 g of ventricular muscle.

The validity of estimates based upon the systolic-pressure-heart-rate product depends on the constancy of stroke volume and cardiac contraction time. During exercise, the second is more likely to be constant than the first; fortunately for the tension/time index, it is the cardiac contraction time that determines the second and larger component of cardiac work [equation (6.21)]. Perhaps for this reason, the systolic-pressure–heart-rate product is said to correlate quite well with myocardial oxygen consumption (Monroe and French, 1961; C. R. Jorgensen et al., 1977). The central aortic pressure should theoretically be used in the calculation, but in practice peripheral blood pressure measurements yield rate-pressure products that correlate satisfactorily with myocardial oxygen consumption (R. R. Nelson et al., 1974; R. J. Ferguson, Gauthier et al., 1975).

The triple product has the soundest theoretical basis, its relationship with cardiac oxygen consumption depending only upon the constancy of stroke volume. However, in practice its use is limited by problems of making a precise, noninvasive determination of cardiac contraction time, and its correlation with myocardial oxygen consumption is no better than that of the rate-pressure product (C. R. Jorgensen et al., 1977).

Cardiac Contractility, Oxygen Consumption, and Tactics for Increasing Cardiac Output

The three possible tactics for increasing cardiac output are (1) an increase of heart rate, (2) an increase of stroke volume induced by an increase of end-diastolic volume, and (3) an increase of stroke volume induced by an increase of cardiac contractility.

The first is inevitably the most expensive tactic in terms of myocardial metabolism, since it augments the nonproductive, isovolumic phase of the cardiac cycle (Pool et al., 1968) and in itself leads to an increase of cardiac contractility (Boerth et al., 1969).

An increase of end-diastolic volume is also costly, since the tension component of cardiac work is increased through the Laplace relationship [equation (6.20)]. Furthermore, as blood is expelled from a distended ventricle, the tension drops much less rapidly than would be the case at a smaller end-diastolic volume.

It might be thought that an increase of contractility would increase oxygen expenditure (Sonnenblick, 1971). This is true in isolated muscle, but *in vivo* the associated reduction of cardiac dimensions often more than compensates for the direct effect of increased contractility. As in skeletal muscle (Fig. 3.9), the useful work performed depends upon the velocity of contraction of the muscle fibers. The rate of muscle shortening is influenced somewhat by external loading and the properties of the series elastic elements, but the main determinant of contraction velocity and thus the performance of external work is cardiac contractility.

Long-term changes of contractility can arise from alterations in structure of the myosin molecule (with resultant alterations of ATPase activity). Both chronic ventricular overloading (Swynghedauw et al., 1976) and aging diminish ATPase activity. As in skeletal muscle, a slow velocity of contraction improves efficiency at heavy work loads, perhaps because the rate of hydrolysis of the actin/myosin cross-

*The true tension/time index is the area under the left ventricular pressure curve, usually measured as mmHg • sec (Sarnoff, Braunwald et al., 1958; Braunwald, Sarnoff et al., 1958).

bridges is slower during the phase of sustained tension. A reduction of contractility is therefore a useful adaptation to both the increased energy demands of chronic overloading and the impaired energy supply of old age.

Short-term increases of contractility are seen with extrasystoles, an increase of heart rate (the "staircase" phenomenon; Boerth et al., 1969), β-adrenergic activity, and administration of cardiac glycosides, such as digitalis (Covell et al., 1966). The pumping ability of the heart is then enhanced (Fig. 6.13) at the expense of an increase in cardiac oxygen consumption. Possible mechanisms for a temporary augmentation of contractility (Reuter, 1974; A. M. Katz, 1977) include (1) an increased rate of delivery of calcium ions to active sites on the troponin C molecule, (2) altered calcium ion binding by the troponin complex, (3) a prolongation of maximum calcium release, (4) a reduced rate of calcium ion removal, and (5) alterations in the levels of other intracellular constituents (pH, K^+ or ATP level). In the case of extrasystoles and the staircase phenomenon, electrochemical changes associated with the preceding contraction potentiate entry of calcium ions into the intracellular calcium pool (E. H. Wood et al., 1969), while the shorter interbeat interval allows less time for calcium ions to be pumped out of the cell. β-adrenergic agonists such as adrenaline activate adenylate cyclase found in the cardiac sarcolemma (Wollenberger, 1975); this converts ATP to cyclic AMP, which in turn seems to activate a protein kinase that phosphorylates the sarcolemma (A. M. Katz et al., 1975; Fig. 6.13), facilitating the inward movement of calcium ions during electrical activity. An excessive buildup of calcium ions is avoided by negative feedback, since the calcium ions inhibit adenylate cyclase and activate phosphodiesterase (an enzyme which converts cyclic AMP to AMP). Digitalis probably affects contractility by inhibiting the sodium pump, thereby increasing the intracellular sodium pool (Langer, 1972); contractility is influenced by the ratio $[Ca^{2+}]/[Na^+]^2$, since sodium ions compete with calcium at the pump that restores initial calcium levels (A. M. Katz, 1975).

Coronary Blood Flow and Cardiac Homeostasis

General Considerations

As in other tissues, homeostasis depends upon an adequate local blood flow. If a branch of the coronary arterial tree is occluded either by spasm of the vessel wall or lodgment of a blood clot, tissue distal to the occlusion dies, sometimes within as little as 5–10 min. The clinical picture is of myocardial infarction, the "heart attack" of popular parlance.

Under normal resting conditions, the coronary blood flow amounts to ~5% of cardiac output (80 ml • min^{-1} per 100 g of tissue, or 320 ml • min^{-1} in a 400-g heart). There is a correspondingly rich capillary blood supply (2500–3000 vessels per mm^2, compared with 200 per mm^2 in resting and 600 per mm^2 in exercising skeletal muscle).

Coronary venous blood drains largely to the coronary sinus, and blood samples collected from this site have a very low oxygen content (typically 50 ml • l^{-1}, but occasionally as low as 10–20 ml • l^{-1}, compared with the normal resting "mixed venous" oxygen content of ~150 ml • l^{-1}. Since arterial blood has an oxygen content of ~190 ml • l^{-1}, the resting heart is extracting 140–180 ml of oxygen from each l of blood, compared with the 40 ml • l^{-1} extracted by most other tissues (Varnauskas and Holmberg, 1971). The hemoconcentration of exercise gives some scope for augmenting oxygen extraction (Kitamura et al., 1972; R. R. Nelson et al., 1974), but it is plain that most of the increased oxygen needs of the

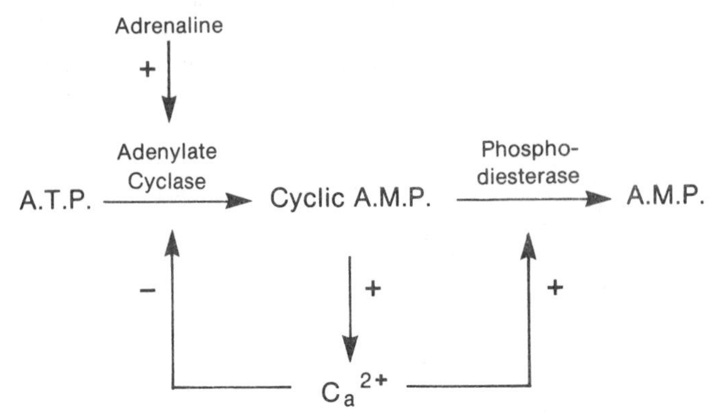

FIG. 6.13. Suggested mechanism for the increase of myocardial contractility by beta-adrenergic agonists such as adrenaline.

heart during physical activity must be met by an increase of coronary flow.

During vigorous exercise, the coronary flow of animals increases by a factor of 3–6 (Vatner et al., 1972; Ball et al., 1975; T. M. Sanders et al., 1975). In humans, exercise studies have been at fairly light work loads (Regan, Timmis et al., 1961; Messer et al., 1962; Kitamura et al., 1972). Nevertheless, the maximum likely response can be deduced from the effect of administering the vasodilator drug dipyridamole. This compound induces a fivefold increase of flow, from 80 to 400 ml • min^{-1} per 100 g of left ventricular mass (Tauchert et al., 1972). Since there is a sixfold increase of cardiac work rate, a small increase of oxygen extraction is needed to avoid local hypoxia. However, it is unlikely that coronary blood flow limits the performance of a healthy person. Maximum heart rate and blood pressure are well sustained if hypoxic gas mixtures are breathed during exercise (L. E. Lamb et al., 1969), and cardiac performance seems unaffected by the shift of blood flow from endocardial to epicardial regions during vigorous work (Ball et al., 1975; Sanders et al., 1975).

In subjects with extensive coronary atherosclerosis, the vasodilation necessary to meet the demands of exercise is hampered by fibrosis and calcification of the vessel walls, and the lumen of the major coronary arteries is narrowed by atheromatous plaques. Nevertheless, clinical evidence of ischemia (such as anginal pain, ST segmental depression, and limitation of maximum cardiac output) is not seen until some two-thirds of a major vessel is occluded. The explanation seems that while obstruction usually develops in the major coronary arteries, the main resistance to blood flow lies in much smaller vessels.

Cardiac Metabolism

Differences of metabolism between skeletal and cardiac muscle are quantitative rather than qualitative (Gertler and Leetma, 1973). As in other red muscles, the energy needs of the heart are normally met by aerobic metabolism. This is reflected in a high mitochondrial count. The preferred fuel is fatty acid (Neely et al., 1974). Glycogen, lactic acid, and certain amino acids can also be metabolized, although in practice protein is rarely a substrate (A. M. Katz, 1977). An accumulation of fatty acids (due, for example, to the release of adrenaline) inhibits the cardiac metabolism of glucose-6-phosphate, and this can apparently contribute to the development of cardiac arrhythmias (M. F. Oliver, 1972; Opie, 1972).

There is unfortunately little scope for developing an oxygen debt in the heart. The small reserves of oxygen stored in the red cell and muscle pigments (hemoglobin and myoglobin) are rapidly exhausted. The lactate dehydrogenase isozyme of the myocardium also has a low affinity for pyruvate, and little lactic acid is formed until the pyruvate concentration becomes very high (Most et al., 1969; Wildenthal, Morgan et al., 1976).

Under normal circumstances, lactate is removed from the blood during its passage through the coronary circulation (A. J. Drake, 1978), but in severe ischemia coronary venous samples may contain more lactate than arterial blood (Nakhjavan et al., 1975) and cardiac glycogen reserves become depleted (Barnard and Thorstensson, 1975). Oxygen lack leads rapidly to a reduction of myocardial contractility and pump failure (A. M. Katz, 1973); increased levels of ADP, AMP, and phosphate, with reduced levels of ATP and CP, give rise to a transient stimulation of glycolysis, but the reaction is soon halted by a lack of NAD (Fig. 2.12) and the inhibitory influence of hydrogen ions upon the enzyme phosphofructokinase (Fig. 2.10). At the time of failure, a fair amount of glycogen usually remains in the cardiac muscle fibers (Barnard and Thorstensson, 1975). The main problem is a lack of ATP; this hampers the dissociation of actin from myosin and halts operation of cellular sodium and calcium pumps, so that relaxation of the heart muscle cannot occur (Dunnett and Nayler, 1978).

External Compression of Coronary Vessels

Because the coronary vessels are embedded in the heart wall, they are subject to external pressure with each contraction of the ventricles. The left ventricular systolic pressure is normally 5 or 6 times as great as that in the right ventricle. In consequence, flow through the left coronary artery is completely stopped for much of systole (Fig. 6.14), whereas flow through the right coronary artery shows much smaller fluctuations over the cardiac cycle (D. E. Gregg, 1962). Increases of total coronary blood flow tend to increase intramyocardial pressure through an effect upon intramural fluid content and thus diastolic fiber length (Erbel et al., 1975).

During exercise, the left ventricular systolic pressure increases, but the right ventricular systolic pressure shows little change. The effective pressure available for perfusion of the right coronary artery is thus increased. However, the left ventricle is less fortunate. Its work load is increased by the rise of left ventricular systolic pressure, and the coronary vessels are also subjected to greater compression. If the diastolic pressure rises, there may be some increase in the

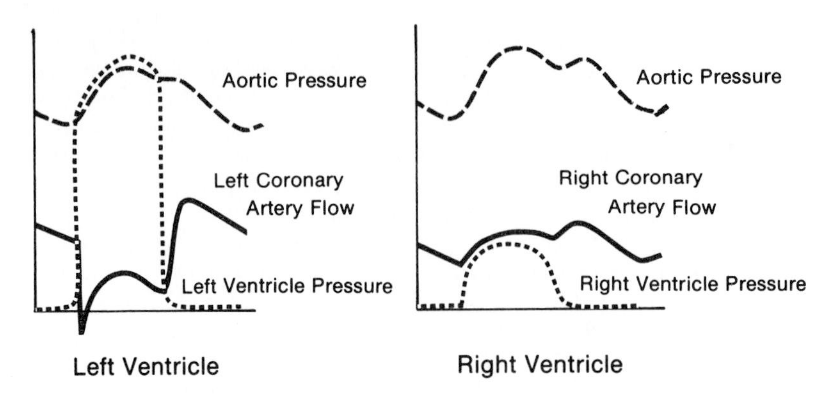

FIG. 6.14. Perfusion pressure and flow in the left and right coronary arteries over the course of a cardiac cycle. (Based on data presented by D. E. Gregg, 1962.)

pressure perfusing the left coronary artery. Nevertheless, flow remains restricted to the diastolic phase of the cardiac cycle, and as the intensity of exercise is increased the proportion of the cycle allocated to diastole is progressively shortened. For these reasons, it is usually the left ventricle that develops oxygen lack during vigorous exercise.

Since stress is greatest in the inner part of the ventricular wall (A. Y. K. Wong and Rautaharju, 1968; Mirsky, 1974), ischemia is particularly liable to develop in subendocardial tissue. Women generally have thinner heart walls than men, so that the tension per unit of cross-section is greater in the female during maximum exercise; this may explain the frequent electrocardiographic evidence of ischemia in elderly women despite a low incidence of atherosclerotic coronary narrowing (G. R. Cumming, Dufresne, and Samm, 1973; Sidney and Shephard, 1977b).

Measurement of Coronary Blood Flow

Until recently, there have been no good techniques for the noninvasive measurement of coronary blood flow. Myocardial oxygen lack has often been inferred from the development of a reversible horizontal or downward sloping ST segment of the exercise electrocardiogram. Injections of radioisotopes such as potassium-43 have provided estimates of overall coronary flow for a number of years (Zaret et al., 1973). More recently, sophisticated scanning and computing techniques (D. H. Schmidt, 1978) have permitted the mapping of regional coronary flow. However, the data thus obtained remain semiquantitative.

Regulation of Coronary Blood Flow

Since heart muscle lacks the ability to develop a large oxygen debt and oxygen extraction from blood perfusing the coronary vessels cannot greatly exceed that observed under resting conditions, the body needs a mechanism to make rapid and accurate adjustments of coronary flow to increases of cardiac work load.

The heart has a large sympathetic and parasympathetic nerve supply, and some coronary vasodilation can be induced by electrical stimulation of the stellate ganglion. However, there is no good evidence that the autonomic nerves play any significant role in the normal response to exercise (O. Lundgren and Jodal, 1975). Occasionally, they may be responsible for a reflex vascular spasm. One possible example of this is an attack of cardiac (anginal) pain induced by a brisk walk on a frosty morning. Exercise induces mouth breathing at a ventilation of ~35 $l \cdot min^{-1}$ (Niinimaa and Shephard, in preparation), and it has been suggested that the cold dry air stimulates vagal receptors in the air passages (Widdicombe, 1974), thus initiating a reflex contraction of both the bronchi and the coronary arteries. Others have argued that the angina arises from a poor coronary reserve rather than a specific effect of cold on either myocardial metabolism or the coronary vasomotor nerves (Neill et al., 1974).

The main stimulus to an increase of coronary blood flow is probably a local reduction of oxygen tension. The local accumulation of CO_2, adenosine, hydrogen ions, lactate, and other metabolites also contributes directly or indirectly to the vasodilation (O. Lundgren and Jodal, 1975; A. J. Drake et al., 1977). Adrenaline has some direct dilator action, with a stronger indirect dilator effect through (1) stimulation of metabolism in the vessel wall and (2) an increase of diastolic pressure in the systemic circulation. Noradrenaline has a weak dilator effect. Both amines are liberated into the circulation during sustained and vigorous exercise, and they could thus contribute to the increase of coronary flow during strenuous physical activity. Nicotine is a coronary vasoconstrictor, and one acute effect of smoking is thus to increase exercise-induced myocardial ischemia in patients with coronary atherosclerosis (G. R. Wright and Shephard, 1978b).

If an attack of angina occurs, the immediate treatment is commonly administration of an organic nitrite such as amyl nitrite or the longer acting glyceryl trinitrate. At one time, such compounds were thought to produce a beneficial effect by dilating the coronary vessels (Brunton, 1871). However, it is now accepted that they act by lowering systemic blood pressure, thus reducing the work of the heart [equation (6.21)] and easing external compression of the coronary arteries (Krantz et al., 1962).

The Collateral Circulation

Substantial anastomoses occur between the two main coronary arteries, particularly at the apex of the heart. These anastomoses are important in reducing the area of heart muscle subject to infarction if blockage of a major vessel should occur.

Some authors have suggested that regular physical activity encourages the development of collateral vessels, thereby reducing the liability to coronary attacks. It is postulated that exercise provides an anoxic stimulus that encourages the opening up of potential anastomotic pathways. Experiments in dogs exercised after surgically induced narrowing of the coronary vessels (Eckstein, 1957; Froehlicher, 1972) have sometimes (but not always shown such an effect. Regular daily exercise enlarges the coronary arterial tree of the rat (Tepperman and Pearlman, 1961; Stevenson, 1967), increases the capillary/fiber ratio in the heart of both the young rat (Tomanek, 1970) and the guinea pig (Rasmussen et al., 1978), and increases the survival rate of elderly rats after experimental infarction (Wexler and Greenberg, 1974). On the other hand, prior exercise does not seem to protect animals against subsequent occlusion of the coronary vessels (Burt and Jackson, 1965; T. M. Sanders et al., 1978).

In the average middle-aged man, there is usually some coronary narrowing before training is begun; the situation might therefore be thought analogous to the Eckstein experiment. Some authors have claimed that both vigorous exercise (D. E. Gregg, 1974) and the increased cardiac work loads of aortic stenosis (MacAlpin et al., 1973) lead to collateral development. However, most authors do not find such a response when a postcoronary patient undergoes exercise rehabilitation (Kattus and Grollman, 1972; R. J. Ferguson, Petitclerc et al., 1974; Ellerstad, 1975; Hellerstein, 1977; Semple, 1977; Wenger, 1977), except possibly when there is an extension of the disease process (Hellerstein, 1977). After training, a lessening of ST segmental depression can be seen in the exercise electrocardiogram, not only at a given work load (Hellerstein et al., 1973) but also at a given heart rate (Shephard and Kavanagh, 1975b) and rate-pressure product (Hellerstein, 1977). Nevertheless, such changes could reflect a lessening of cardiac work load and an improvement of relative coronary flow secondary to hypertrophy, a change of cardiac dimensions, or a reduction of myocardial contractility rather than a development of anastomotic vessels. Certainly, angiographic studies reveal little evidence of increased collateral vascularization among postcoronary patients who have participated in training programs (Kattus and Grollman, 1972; R. J. Ferguson, Petitclerc et al., 1974). It can be objected (1) that angiography only reveals vessels >100 μm, without reference to the flow achieved, and (2) that the intensity of training undertaken by the average postcoronary patient is insufficient to elicit the collateral response seen in dogs. However, Eckstein (1957) obtained positive results with quite mild exercise and the prime explanation of discrepant data in the coronary victim is probably that the vessels are already fibrosed or calcified, and therefore difficult to dilate.

Exercise and Sudden Death

There is some evidence that exercise may increase the immediate liability to both myocardial infarction and sudden death (Shephard, 1974c,1979b; Maron, 1979; Munscheck, 1979). Several explanations of the phenomenon may be advanced. First, narrowing of the vascular supply may cause transient, relative hypoxia in a patch of heart muscle, making it a potential focus for the appearance of abnormal rhythms, including ventricular fibrillation. Second, a more persistent oxygen lack may develop in a wedge-shaped segment of the ventricular wall because hardening of the coronary vessels prevents a normal vasodilator response to the oxygen demands of exercise. Ischemia may be sufficient to cause local death of the myocardium (infarction), with subsequent cardiac arrest, dysrhythmia, or pump failure (A. M. Katz, 1974). Third, a frank occlusion of a coronary vessel may be induced by vigorous and unaccustomed exercise (Shephard, 1974c). One suggested explanation notes the drop in pressure as blood flows past a fat plaque in the arterial wall. Exercise leads to an increase of coronary blood flow, and thus a greater pressure drop across the plaque. Hemorrhage tends to occur into the plaque from the upstream side, and this progressively occludes the coronary vessel. Alternatively, the increase of blood flow may dislodge a thrombus from the arterial wall and this can then block a smaller vessel. Finally, platelet aggregation is increased following maximum stress (Scheele, 1979), and this may cause partial occlusion of an artery to become complete. Attempts to identify the individual who will sustain an infarction or "electrical death" during exercise have not been very successful (Shephard and Kavanagh, 1978b; Shephard, 1979b). Possible clues are an aggressive, competitive personality, with reluctance to accept the slow progression of exercise necessary to a safe rehabilitation program, and the appearance of electrical abnormalities during laboratory exercise tests (premature ventricular contractions and ST segmental depression). In patients with a history of previous infarction, other adverse signs include the absence of a normal rise of blood pressure during exercise and difficulty in sustaining cardiac stroke volume at high work rates.

ELECTRICAL ACTIVITY OF THE HEART

Cardiac Action Potential

The electrical events that accompany the contraction of cardiac muscle are more complex than their counterparts in skeletal muscle; the sequence also has a longer duration, and varies from region to region within the heart.*

The upstroke of electrical potential (phase 0, Fig. 6.14) corresponds to depolarization of the skeletal muscle fiber. There is then a brief phase of early repolarization (phase I) and a sustained plateau (phase II) before repolarization (phase III) and restoration of the resting potential (phase IV). The cardiac sarcolemma is highly permeable to potassium ions, but relatively impermeable to sodium, calcium, and chloride ions. An ATP-fueled sodium pump keeps sodium ions out of the cardiac muscle but sustains a high intracellular concentration of potassium ions. In keeping with the Nernst equation, the equilibrium electrical potential V′ across the membrane is thus given by

$$V' = \frac{RT}{zF} \log_n \frac{C_0}{C_I} \qquad (6.24)$$

*There is now some evidence that mechanical stimulation of the ventricle can induce an electrical potential that may generate a dysrhythmia in an ischemic heart (Lab, 1978).

where R is the gas constant, T is the absolute temperature, z is the valence of the ion, F is Faraday's constant, and C_0 and C_I are ionic concentrations (more strictly activities) of ions outside and inside the cell membrane. For a monovalent ion, this simplifies to

$$V' = 61.5 \log_{10} \frac{C_0}{C_I} \qquad (6.25)$$

Given respective potassium ion concentrations of 4.0 and 140 mmol • l^{-1} and assuming free permeability to potassium, with impermeability to other ions, V′ is seen to approximate −95 mV.

Departures from the Nernst prediction occur at both high and low potassium concentrations. If extracellular potassium ion concentrations are less than 3.0 mmol • l^{-1}, the potassium permeability falls, and effects due to other ions become significant. Under resting conditions, potassium conductance is not infinite, and an increase of potassium permeability (caused, for example, by administration of acetylcholine) may lead to a slight increase of resting potential.

During the upstroke of the action potential, the permeability of the membrane toward potassium is greatly decreased, while sodium ion conductance is increased. The rate of depolarization and the amplitude of the action potential depend upon the voltage-dependent opening of a "gate" that temporarily allows sodium ions to penetrate the lipoprotein barrier of the cell membrane. The potential across the membrane becomes slightly positive (~20 mV), and the gate then abruptly closes. Reopening of the gate must await a recovery process that depends on the level of the membrane voltage and the time that has elapsed since the previous action potential. The prolonged plateau of persistent depolarization contributes to the refractory period of heart muscle. Drugs such as quinidine (which cause a partial depolarization of heart muscle) further extend the refractory period.

Early repolarization (phase I) is due mainly to a drop of sodium ion conductance, as the gate closes (Table 6.11). There is also a transient increase of chloride ion conductance. During the first part of the plateau (phase II), membrane conductance remains fairly high, but potential is kept constant by the neutralizing effects of inward and outward migration of ions. A "slow channel" opens up, allowing calcium ions to enter the cell, although the effect is less than would be predicted from respective extracellular and intracellular calcium ion concentrations; possibly much of the calcium passes to a special subsarcolemmal region related to the cisternae of the sarcoplasmic

TABLE 6.11
Ionic Movements Contributing to the Cardiac Action Potential

Phase	Electrical Current	Ion	Direction of Ion Movement
0	Inward	Na^+	Inward
I	Outward	Cl^-	Inward
		K^+	Outward
II	Inward	Ca^{2+}	Inward
	Outward	K^+	Outward
III	Outward	K^+	Outward

Source: After A. M. Katz, 1977.

reticulum (Chap. 4). In the latter part of the phase II plateau, membrane conductance is low. However, as the membrane voltage becomes more negative, potassium ion permeability is restored, the slow channels close, and the membrane is repolarized by an efflux of potassium from the cardiac cells (Fozzard and Das Gupta, 1975).

In skeletal muscle, it is normal to distinguish an absolute refractory period, when no stimulus will elicit a response, and a subsequent relative refractory period, when a stimulus of greater than the normal threshold strength will initiate a propagated response. In cardiac muscle, it is useful to distinguish an "effective refractory period"; stimuli that arrive in the latter part of this interval, while causing a local rather than a propagated response, can influence the behavior of subsequent action potentials. There is then a relative refractory period when stronger than normal stimuli are needed to induce a propagated response; the corresponding action potentials are characterized by a prolonged latency, a slow rise, and a low amplitude. There is finally a brief period of supranormal excitability before the normal resting state is restored (Fig. 6.14).

The clinical electrocardiogram bears little resemblance to the action potential described above since it arises largely from a mutual cancellation of electrical activity in different parts of the heart. The width of the P wave indicates the time needed for depolarization to spread over the atria. During the PR interval, the wave of depolarization passes through the atrioventricular node, the bundle of His, the bundle branches, and the Purkinje network, but no external electrical activity is seen because the mass of tissue involved is small.* The QRS complex reflects ventricular depolarization; the voltage change is large, but the duration is brief, since the electrical impulse is conducted rapidly through the Purkinje network. During the ST interval, all parts of the ventricle remain depolarized. The subsequent broad T wave reflects a slow and asynchronous repolarization of the ventricles.

*A sharp H wave can be recorded from intracardiac electrodes.

Lead Placement

Most clinicians like to examine a 12-lead resting electrocardiogram prior to exercise testing. The usual record comprises three standard bipolar limb leads (I, left arm relative to right arm; II, left leg to right arm; III, left leg to left arm) and nine unipolar leads in which the potential is measured relative to Wilson's central terminal (leads from the arms and the left leg, each joined to a common terminal through a 5000-ohm resistor).

The bipolar leads were first described by Einthoven (1907), who made the simplifying assumption that the cardiac action potential was generated at the center of an equilateral triangle formed by the three electrode placements (Fig. 6.15). Some authors have attempted to draw more precise triangles, allowing for such factors as the noncentral placement of the heart and the nonuniform conductivity of body tissues. However, such triangles differ from one person to another and for clinical purposes the Einthoven approximation is still preferred. In measuring potential differences between the limb leads, Einthoven arbitrarily arranged circuits so that the main deflections would normally be positive in all three leads.

There is no practical way of determining a true zero potential for the heart.† However, it is often assumed that repolarization is complete between the T and P waves, so that this is taken as the isoelectric reference point (Master, 1942). Unfortunately, during exercise the T wave begins to overlap the P wave, obscuring the isoelectric interval (Lepeschkin, 1969). Other authors (McHenry et al., 1968) use the PR interval as the isoelectric reference potential, again, this is satisfactory at rest, but during exercise an atrial T wave may give a downward slope to the PR interval. A third possibility is to take the potential at the onset of the QRS complex (Sheffield, 1974).

The unipolar leads are one attempt to obtain a zero reference potential, the resistors minimizing the

† Since potential diminishes as the square of distance from the heart, a close approximation to zero potential could theoretically be obtained by placing the reference electrode in the far corner of a saltwater swimming pool in which the subject was standing.

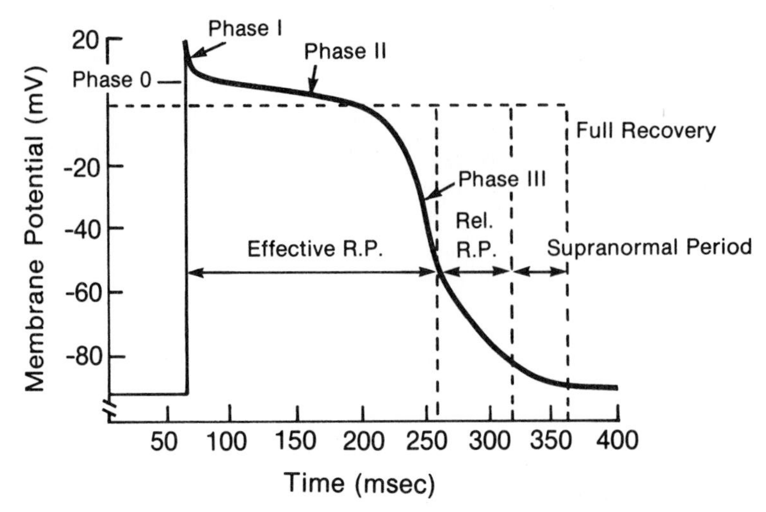

FIG. 6.15. Characteristics of cardiac action potential and refractory period (R. P.) as recorded from a Purkinje fiber. (After A. M. Katz, 1977.)

effects of varying skin impedances at the three electrode sites. Some clinicians use augmented leads whereby the recorded voltage is increased by disconnecting the lead under examination from the central terminal; thus lead aVF measures potentials between the left leg and a central terminal connected to left and right arm electrodes. Although of possible empirical value, such augmentation plainly undermines the theoretical concept of a zero reference potential.

It has been argued that electrical changes in the heart during the cardiac cycle can be represented by the movement of a small dipole (a conductor with a positive charge at one end and a negative charge at the other) in a three-dimensional space (E. Frank, 1956; Schaffer, 1956; Langner et al., 1958; Horan et al., 1963; Gelernter and Swihart, 1964; Willems et al., 1979).

It might thus seem that the entire information content of a 12-lead electrocardiogram could be obtained if potentials were recorded in three planes, X, Y, and Z, at right angles ("orthogonal") to each other (Table 6.12; E. Frank, 1956; Scher et al., 1960; Blomqvist, 1965); since measurements are necessarily made outside the active cells, the resting portion of the dipole would have a positive charge, while the active segment would assume a negative charge. Blackburn and Katizback (1964) found empirically that 89% of the information relating to behavior of the ST segment of the exercise ECG was contained in a single lead (V_5; Tables 6.13, 6.14). Nevertheless, pathways of impulse conduction within the heart are sufficiently complicated that multiple dipoles can coexist within the ventricles (Selvester et al., 1965), and additional information can therefore be obtained as unipolar exploring electrodes (F. N. Wilson et al., 1946; Taccardi, 1963) are moved over the chest wall (lead positions V_1 to V_6; Table 6.13). The voltages recorded reflect the characteristics of the immediately underlying myocardium; an abnormality may be apparent in some chest leads but not in others.

TABLE 6.12
Analysis of Frank Vectorcardiogram in Top Athletes

Differences of Athletes from Controls	Differences of Soccer Players from Endurance Athletes
(1) Decreased amplitude of initial rightward vector	(1) Larger amplitude of maximum QRS vector
(2) Increased duration of Q in Z lead	(2) Larger R in Z and Y leads
(3) Increased amplitude of R in Z lead	(3) Shorter PQ interval
(4) Increased duration of S in X lead	(4) QRS axis less horizontal
(5) Decreased spatial QRS-T gradient	(5) Terminal forces less oriented to right

Source: Based on findings of P. L. Williams et al., 1979.

Amplitude and Form

ECG potentials are small (peak 1-2 mV), and the amplitude of all waves is markedly dependent on impedances interposed between the heart and the re-

TABLE 6.13
Placement of Unipolar Chest Leads*

Lead	Placement of Exploring Electrode
V_1	Fourth intercostal space, just to right of sternum
V_2	Fourth intercostal space, just to left of sternum
V_3	Halfway between V_2 and V_4
V_4	Fifth intercostal space, at left mid-clavicular line
V_5	Anterior axillary line, at level of V_4
V_6	Mid-axillary line, at level of V_4

*Many laboratories now record leads V_2, V_4, and V_6 during exercise.

cording electrodes (Schaffer, 1956; Rautaharju and Karvonen, 1967). In endurance athletes, such as long-distance skiers, a large heart gives a low impedance, while in other subjects a high impedance may result from interposition of emphysematous, air-filled lung, avascular fat, or a poor skin contact. No particular significance can thus be attached to either a high- or a low-voltage signal.

Both the amplitude and the quality of ECG recordings can be improved if (1) the outer, poorly conducting layer of keratinized skin is removed from the sites of electrode placement using a dental burr (Tregear, 1965), (2) the distance between skin and electrodes is kept constant by a suitable electrode mounting, (3) the resultant signal is fed to an amplifier having an impedance at least 100 times as large as the source (Lewes, 1965; Blackburn, 1969; James and Patnoi, 1974), and (4) the combined response characteristics of amplifier and recorder give an error of less than 0.05 mV in the early part of the ST segment (0.05 Hz cutoff, 6 db per octave rolloff, ensuring a consistent response to both slow changes of potential and high-frequency impulses <100 Hz; Berson and Pipberger, 1966; American Heart Association, 1967).

TABLE 6.14
Sensitivity of Selected Lead Combinations in Detecting ST Segment Depression

Lead	Percent of ST Abnormalities Detected
V_5	89
$V_{5,6}$	91
$V_{4,5,6}$	93
$V_{3,4,5,6}$	94
II, $V_{3,4,5,6}$	96
II, $aVF_{3,4,5,6}$	100

Source: After Blackburn and Katizback, 1964.

An increase in amplitude or duration of the P wave, with or without notching, suggests left or right atrial enlargement; it is quite common in endurance athletes (Morganroth and Maron, 1977). Some authors have also claimed that a large upright T wave is associated with athletic endurance (Hoogerwerf, 1929; Beckner and Winsor, 1954; Beswick and Jordan, 1964) or a large stroke volume (P. M. Dawson, 1935; Ellerstad, 1975). This reflects in part the fact that fit individuals have less subcutaneous fat. Other factors such as a high serum potassium level also give rise to large upright T waves (Littman, 1972; Morganroth and Maron, 1977). On the other hand, intense training sometimes leads to flattening or even inversion of the T waves (T. K. Cureton, 1951; Plas, 1963, 1978; Gott et al., 1968; Morganroth and Maron, 1977; Kindermann et al., 1978). We must thus conclude that the association between T-wave size and stroke volume is not particularly close, and that there are easier noninvasive methods of estimating the latter variable.

The duration of the various portions of the ECG waveform is determined by the rate of conduction of electrical impulses through the myocardium (Table 6.15); prolongation of a given phase implies a conduction block in the corresponding portion of the myocardium.

Electrical Axis: Deviation and Hypertrophy

The electrical axis of the heart is normally determined from the projection of the QRS complex in the frontal plane, voltages for the three standard leads being represented in scalar fashion along the sides of the Einthoven triangle (Fig. 6.16). The electrical axis is influenced by the position of the heart in the chest and by ventricular hypertrophy. If lead I is taken as 0°, the normal range is from $-30°$ to $+90°$.

Left axis deviation (axis $-30°$ or less, with a

TABLE 6.15
Approximate Duration of Various Portions of the Electrocardiogram

Variable	Duration (msec)
P wave	80–100
QRS complex	60–100
PR interval	120–200
QT interval	300–400

Note: All times, but particularly the QT interval, are shortened by an increase of heart rate.

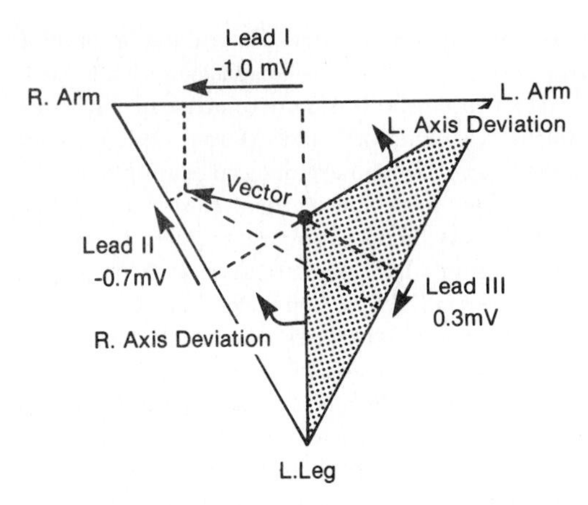

FIG. 6.16. Method of determining the electrical axis of the heart. QRS voltages for the three standard limb leads are plotted in scalar fashion along the sides of an equilateral (Einthoven) triangle. Perpendiculars are dropped to locate the equivalent dipole at the center of the triangle. Shaded area indicates normal range of axes for QRS complex. Vector illustrated shows marked right axis deviation.

large R wave in lead I and an S wave in lead III) is seen with the horizontally placed heart of an obese subject (Fig. 6.16), while right axis deviation (axis $+90°$ or more, with an S wave in lead I and a large R wave in lead III) is seen in a thin individual with a vertically placed heart. Hypertrophy of the left ventricle usually produces left axis deviation, but the voltage of the various waves is much greater than with a simple rotation of the heart, due to (1) increase of muscle mass and (2) slower passage of the wave of depolarization over the enlarged ventricle (A. M. Katz, 1977). Thus, the S wave in lead V_1 is >1.5 mV and the R wave in leads V_4, V_5, or V_6 is >2.5 mV. Hypertrophy of the right ventricle produces a right axis deviation, with tall R waves in leads V_1 and V_2 and conspicuous S waves in leads V_5 and V_6. Left ventricular hypertrophy (Fig. 6.17) may reflect the well-trained heart of an endurance athlete, in which case it is associated with a slow resting heart rate (Zoneraich et al., 1977; Morganroth and Maron, 1977). Alternatively, it may be due to some pathological abnormality such as a narrowing of the aortic valve (aortic stenosis). Right ventricular hypertrophy is often pathological (reflecting pulmonary stenosis or hypertension), but it also may occur in healthy athletes (Hanne-Paparo et al., 1976; Morganroth and Maron, 1977).

Contraindications to Exercise

What are danger signs in a preexercise electrocardiogram? Detailed clinical interpretation of the ECG is

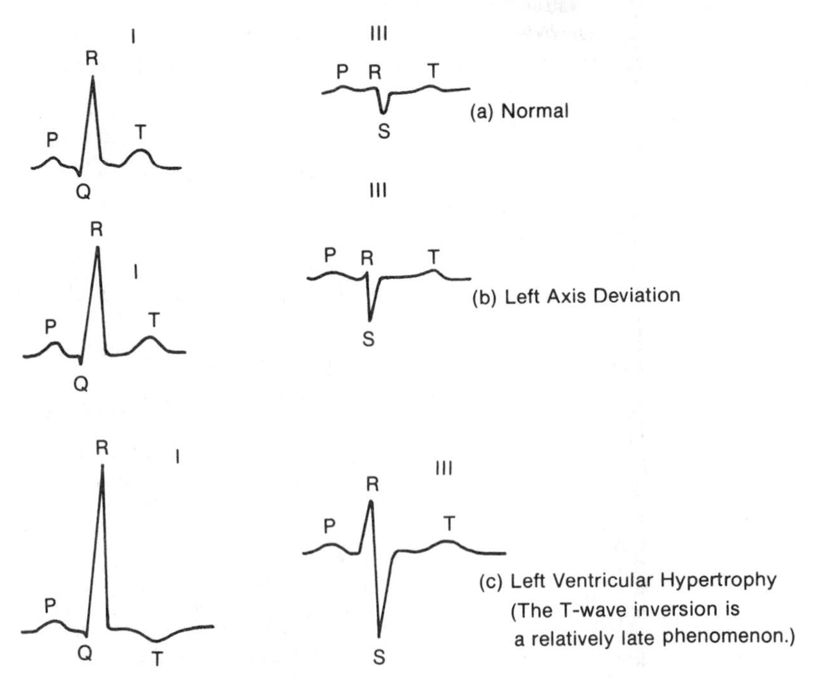

FIG. 6.17. The influence of left axis deviation and left ventricular hypertrophy upon the appearance of Leads I and III of the electrocardiogram.

the responsibility of the cardiologist, but nevertheless it is well to be warned against certain absolute contraindications to exercise that can be recognized in the resting tracing.

Myocardial Infarction and Acute Myocarditis

It is important to exclude imminent or recent myocardial infarction. This may be suspected if there is a prominent Q wave and elevation of the ST segment when an exploring electrode is placed immediately over the affected area of myocardium. However, both Q waves and ST elevation are sometimes seen in healthy endurance athletes (Zoneraich et al., 1977; Morganroth and Maron, 1977); the Q waves may then be a mechanical effect arising from rapid ventricular filling (Ferrero and Doret, 1954). If there is an infarct on the posterior surface of the heart, the ST segment may be depressed rather than elevated. In doubtful cases, it is helpful to show a progression of changes when electrocardiograms are repeated over the course of several days. Acute myocarditis is a second important contraindication to exercise; this may also cause depression of the ST segment and inversion of the T wave.

Pulmonary Embolism

Exercise should not be undertaken if there is a probability of recent pulmonary embolism. A large embolus leads to a rapid heart rate and ECG evidence of right heart strain (including inverted T waves when the exploring electrode is over the right ventricle). Smaller emboli may give no abnormal ECG appearances.

Conduction Disorders

Bursts of rapid ventricular rhythm (ventricular tachycardia) with independent P waves and broadened QRS complexes can progress to ventricular fibrillation, this being particularly likely to occur if the abnormal beats occur in the "vulnerable" early phase of repolarization (immediately following the peak of the T wave). The Wolff-Parkinson-White syndrome (S'Jongers et al., 1976) presents a superficially similar picture. In this condition, an accessory pathway bypasses the normal conduction delay of the atrioventricular node. There is a normal P wave, a P-R interval of less than 120 msec, and then an abnormal QRS complex (with a slowly rising or falling initial deflection). Since the abnormal beat ("preexcitation" of the ventricle) is followed by an electrical impulse following the standard route, there is a danger that abnormalities of rhythm such as atrial and ventricular tachycardia and ventricular fibrillation may develop (Chamberlain and Clark, 1977).

Relative Contraindications to Exercise

Atrial Flutter and Fibrillation

Care must be observed when exercising patients with atrial flutter or fibrillation. In atrial flutter, an irritable focus or a reentry of the electrical impulse into the atrium (Cranefield et al., 1973) gives a rapid succession of P waves (up to 250–300/min); every third or fourth impulse is transmitted to the ventricle. In atrial fibrillation, there are irregular and rapid f waves, while the ventricular contractions are slow and irregular in timing.

Conduction Disturbances

Indications for a cautious approach to exercise include sinoatrial block, marked atrioventricular block, and left bundle-branch block. In *sinuatrial block*, some of the impulses arising in the sinuatrial node fail to depolarize the atria; there are thus "pauses" when an entire ECG complex is missing. In the elderly, this may be a sign of coronary vascular disease, and a "sick sinus syndrome" can restrict maximum heart rate (Ferrer, 1973). In athletes (Table 6.16), hypertonia of the right vagal nerve is thought to be responsible.

Plas (1978) has observed competitors who al-

TABLE 6.16
Frequency of ECG Abnormalities in a Sample of 12,000 Athletes

	Per cent
Nodal rhythm	0.33
Coronary sinus rhythm	0.15
Premature systoles	1.38
Paroxysmal tachycardia	0.07
Atrioventricular block	
1st degree	6.17
2d degree	0.13
3d degree	0.02
Atrioventricular dissociation	0.12
Focal block	3.15
Right bundle-branch block	0.08
Ventricular preexcitation syndrome	0.16
Pseudoischemic abnormalities	0.55

Note: Of the 12,000 subjects, 37 were considered to have a cardiopathy. In 19 cases, this was incontravertible (4 atrial septal defect, 2 pulmonary stenosis, 3 pulmonary regurgitation, 2 aortic regurgitation, 4 mitral disease, 2 arterial hypertension, 1 aortic coarctation, 1 bicuspid aortic valve), but in some of the remaining 18 (8 right bundle-branch block, 7 paroxysmal tachycardia, 1 atrial flutter, 1 atrial fibrillation, 1 atrioventricular block) the disorder might be considered a normal response to the slow resting heart rate and the large heart of the endurance athlete.
Source: After Piovano et al., 1971a,b.

ternate between a sinus rhythm of 65–70 beats per min and a nodal rhythm of ~44 beats per min; on occasion, the alternation may be regular enough to be confused with sinus arrhythmia *First degree block* at the a-v node is shown by a lengthening of the PR interval to >200 msec. It can be associated with vigorous activity of the left vagus nerve, and as such is a common finding in endurance athletes (Zoneraich et al., 1977; Plas, 1978). The PR interval is also affected by the ionic composition of the plasma. An increase of potassium ions (as in freshwater drowning) lengthens the PR interval, while an increase of sodium ions (as in saltwater drowning) has the reverse effect. Lengthening of the PR interval is sometimes related to ischemia or the administration of an excessive dose of cardiac glycosides. In such circumstances, it serves mainly as a warning of possible progression to a more severe degree of block. In *second degree block*, only a proportion of impulses are transmitted to the ventricles; one or more ventricular contractions are "dropped" completely, and a QRS complex may sometimes appear without a preceding P wave ("ventricular escape"). Commonly, the PR interval gets progressively longer until a beat is dropped (the Wenckebach phenomenon, sometimes shown by healthy athletes; Morganroth and Maron, 1977). More rarely, the PR interval remains constant until a beat is suddenly dropped (Mobitz type II a-v block); this variety seems particularly prone to progress to *third degree block*, where the atria and ventricles beat independently of each other. Marked a-v block usually indicates some underlying abnormality of the myocardium, and there is danger of progression to a Stokes-Adams attack, with complete ventricular asystole. *Bundle-branch block* is indicated by a broadening of the QRS complex, and is a cause for concern mainly because it indicates underlying heart disease. Left bundle-branch block causes a leftward shift of the QRS vector (Fig. 6.16), with tall, broad R waves in lead I and deep, broad S waves in lead III; it is usually pathological in origin. Right bundle-branch block gives a rightward shift of the QRS vector, with broad, late S waves in leads I and II and a small QRS complex in lead III. A minor degree of right bundle-branch block is a common finding in athletes (Hanne-Paparo et al., 1976; Plas, 1978; Kindermann et al., 1978; Venco et al., 1979; de Andrade and de Rose, 1979).

Other Abnormalities

Sinus Bradycardia

Clinicians define sinus bradycardia as a resting heart rate of less than 60 beats per min, where normal QRS complexes follow normal P waves with a standard PR interval. The underlying mechanism is a vagally induced increase of potassium ion permeability in the sinoatrial node. The phenomenon is seen in endurance athletes (Chap. 11; Zoneraich et al., 1977) and in other normal subjects during vasovagal attacks.

Sinus Arrhythmia

Sinus arrhythmia is an innocent form of irregular cardiac rhythm. It is a common feature of the resting electrocardiogram, more apparent in young athletes than in older sedentary individuals (E. A. Hunt, 1963; Tunstall-Pedoe, 1970; Zoneraich et al., 1977; Morganroth and Maron, 1977). The heart rate quickens during inspiration and slows during expiration. The ECG has a normal form, and the rhythm becomes more regular as exercise is commenced. Some authors have attributed this type of arrhythmia to the Bainbridge reflex, the respiratory movements inducing a cyclic variation of venous return; others have suggested there is a primary increase of vagal tone and a decrease of sympathetic activity during inspiration, with the reverse change occurring during expiration.

Premature Systoles

These may arise in the sinus, atrium, a-v node, or ventricle. Although sometimes called extrasystoles, they may replace a normal beat, rather than adding to the total number of beats per min.

Persistent irregular atrial tachycardia is compatible with normal health and continued athletic performance (Fleischmann and Kellermann, 1969). Ventricular premature systoles are usually "ectopic," with a broadened and abnormal QRS waveform, but other types of premature systole arise within the normal conduction pathway. Occasionally, normal impulse conduction is blocked without appearance of a visible ECG wave ("concealed conduction").

One basis for an abnormal systole is the premature discharge of an irritated cell "below" the sinoatrial node; this could arise from any factor that causes a partial depolarization of the tissue, for example, an accelerated decline of potassium conductance, an increase of depolarizing current, or a decreased threshold for a propogated discharge (A. M. Katz, 1977). Such "ectopic beats" can apparently be caused by nicotine accumulation in a heavy smoker and by excessive sympathetic discharge in a chronically anxious individual. The added beats become less frequent as exercise is commenced, because there is an increased "drive" from the normal pacemaker. Although some authors have associated this type of dysrhythmia with an increased frequency of sudden death and myocardial infarction, the effect upon prognosis is probably small (F. D. Fisher and Tyroler, 1963; Goldschlager et al., 1973; Rodstein et al., 1971). A second source of premature systoles is an inhomogeneity of ventricular repolarization. Tissue that has remained depolarized then reexcites areas of heart muscle that have regained their normal membrane potential (Surawicz, 1975). The third and perhaps the most important cause for a premature contraction is the development of an area of tissue where decreased electrical conduction has progressed to cause a unidirectional conduction block. Fibrosis or ischemia leads to an asymmetric decrease of the normal action potential, so that impulses are only conducted in a reversed direction; the wave of electrical depolarization thus passes forward through healthy tissue, returns slowly through the diseased area, and reaches the healthy tissue when this is sufficiently repolarized to initiate a further discharge (Fig. 6.18). Mechanisms 2 and 3 are plainly indicative of cardiac disease. They tend to become worse with exercise and carry an adverse prognosis (R. H. Mann and Burchell, 1952; Rodstein et al., 1971; Vedin et al., 1972; Ellerstad, 1975).

Response to Exercise

The ST Segment

As the intensity of exercise is increased, the oxygen consumed by the heart muscle outstrips the oxygen supplied by the coronary vessels. Because of differences in the relative compression of the vessels and the work load to be performed, hypoxia is usually more severe in the left than the right ventricle. This leads to a progressive change in the appearance of the ST segment of the electrocardiogram (Feil and Siegal, 1928; Goldhammer and Scherf, 1932), most apparent in the unipolar chest lead V_4 (Wolferth et al., 1932). There is first a depression of the S-ST junction (Fig. 6.19) and then a horizontal or downward sloping depression of the entire ST segment.*

The underlying physiological mechanism is a nonuniform repolarization of the ventricles. This commonly reflects an ischemic malfunction of the sodium pump in certain parts of the ventricular wall, but it is important to note that ST depression is not specific evidence of myocardial ischemia (K. L. Andersen, Shephard et al., 1971). There are many other possible causes of the phenomenon (Table 6.17), including chemically induced malfunction of the sodium pump (caused by drugs such as digitalis) and altered plasma electrolyte concentrations (due, for example, to anxiety hyperventilation or the use of diuretics).

If the coronary vasculature is healthy, oxygen lack does not progress beyond a slight junctional depression even in maximum exercise (Bruce, Alexander et al., 1969). More severe changes are thus usually interpreted as evidence of myocardial ischemia. Some authors attach pathological significance to marked depression of the S-ST junction (Kurita et al., 1977), regarding the tracing as abnormal if (1) the junctional depression is >0.15 mV or (2) the ST segment is upward sloping yet remains 0.1 mV or more below the isoelectric (zero) potential at the origin of the T wave. However, false positive tests are less frequent if attention is directed simply to horizontal and downward sloping ST segments (Table 6.18; Sheffield and Reeves, 1965).

Horizontal ST segment depression forms the basis of several clinical tests of myocardial perfusion. The Master test (Master, 1969) has been widely used by physicians for ~ 50 years; initially observations were limited to recovery heart rates and blood pressures, but more recently (Master and Jaffe, 1941) the

*Occasionally, the hypoxic tissue may be so placed as to cause elevation of the ST segment ("Printzmetal angina") (Printzmetal et al., 1959).

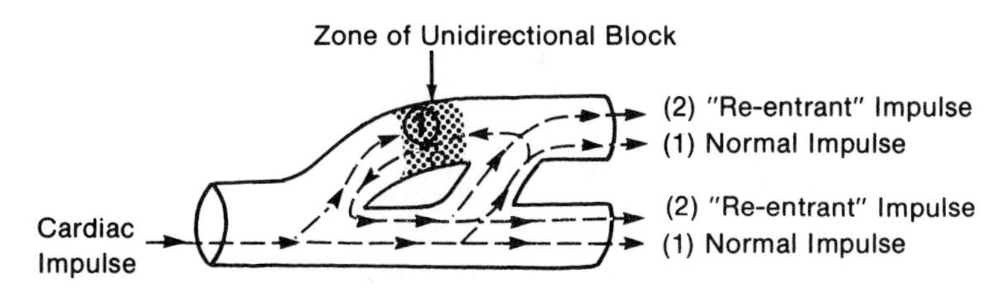

FIG. 6.18. The possible role of a unidirectional conduction block in the development of a premature systole. The shaded area can conduct only in the reverse direction. Impulses passing slowly through this zone reach normal tissue after this has repolarized, thereby initiating a further propagated action potential. (After A. M. Katz, 1977.)

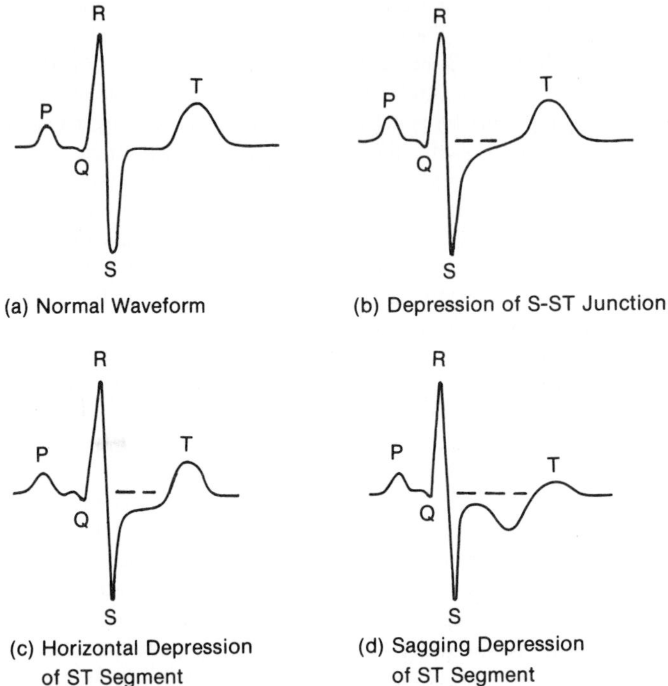

FIG. 6.19. The effects of progressively more severe hypoxia upon the waveform of the electrocardiogram.

recovery electrocardiogram has also been examined. The patient is required to climb backward and forward over a double 9-in. (22.9-cm) step for 1.5 min (single Master test) or 3.0 min (double Master test). The rate of climbing is varied somewhat with age and body mass, but generally the terminal pulse rate is ~120 beats per min. An abnormal response is reported if the ST segment of the recovery electrocardiogram is depressed by more than 0.1 mV at any time during the first 10 min of recovery. A patient showing such a response has a several fold increase in risk of premature death from ischemic heart disease (Mattingly, 1962; Rumball and Acheson, 1963; G. P. Robb and Marks, 1964; Kasser and Bruce, 1969; Aronow and Cassidy, 1975; G. P. Robb and Seltzer, 1975; Fig. 6.20). There are two main criticisms of the Master test: (1) the level of effort required of the average patient is rather mild and (2) the test imposes a greater strain on elderly than on young patients (since the maximum heart rate declines with age). The light work rate has the attraction of safety and also avoids false positive tests, but at the same time leads to a poor sensitivity, with some falsely negative reports.

A number of cardiology laboratories now carry out "symptom-limited" maximum tests on all patients submitted for electrocardiographic examination (P. D. S. Wood et al., 1950; Bruce, Alexander et al., 1969). The chance of provoking ventricular fibrilla-

TABLE 6.17
Factors Hampering Interpretation of ST Depression

False Positive Results	False Negative Results
Digitalis therapy	Nitroglycerine and other vasodilators
Antidysrhythmic drugs (e.g., procaine amide, quinidine)	Abnormalities of ventricular conduction
Abnormal stress on left ventricle	ST elevation at rest
Abnormalities of ventricular conduction (e.g., left ventricular bundle-branch block)	Insufficient exercise intensity
Glucose and carbohydrate loads	
ST depression at rest	
Cigarette smoking	
Hyperventilation	
Diuretics, potassium loss	

TABLE 6.18
Sensitivity and Specificity of Selected Exercise Test Protocols in Detecting Myocardial Ischemia

	Sensitivity (%)	Specificity (%)
(a) *Effect of test type*		
Master two-step test (recovery)	33	93
Progressive exercise test	59	94
(b) *Effect of test criterion*		
Junctional depression > 0.2 mV	60	50
Ischemic ST segment	62	91
Up-sloping ST segment	30	93

Note: Because ST changes have a continuous rather than a discontinuous distribution, Simoons and Hugenholtz (1977) suggested expressing results as a "likelihood ratio" rather than attempting to classify individuals as "normal" and "abnormal."
Source: After Ascoop, 1977.

tion by such a test is about 1 in 10,000 when dealing with a cardiac patient (Rochmis and Blackburn, 1971). There is some evidence that risks are greater for maximal than for submaximal testing (McDonough and Bruce, 1969). The symptom-limited endpoint is also, necessarily, subjective; if the confidence of the patient or his examiner allows a second test to be pushed to a higher work load, the misleading impression may be formed that ST depression has worsened.

A third option is to record the electrocardiogram when the patient is exercising at a fixed percentage, 75% (Shephard, 1971d) or 85% (I. Åstrand, 1967a), of his maximum aerobic power. The required work load can be defined adequately for clinical purposes by using an age-related target heart rate:

Age (year)	Target Heart Rate 75% Load	85% Load
25	160	170
35	150	160
45	140	150
55	130	140

In terms of striking an appropriate balance between sensitivity and specificity (Table 6.18, R. E. Mason et al., 1969), most authors now set the threshold for reporting ST depression at 0.1 mV. The proportion of any population diagnosed as having myocardial ischemia is thus larger if observations are made during exercise to a target heart rate rather than relying upon the recovery records of a traditional Master

TABLE 6.19
Yield of Abnormal ECG Records from Exercise and Recovery Tests

Exercise Only (%)	Recovery Only (%)	Exercise plus Recovery (%)	Author
41.5	12.1	46.4	G. R. Cumming (1972)
46.2	15.4	38.4	Sidney and Shephard (1977b)

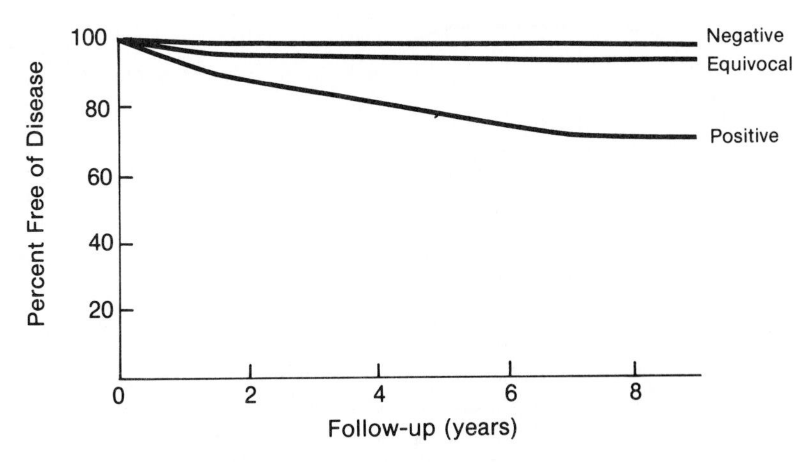

FIG. 6.20. Percentage of subjects remaining free of myocardial infarction over nine-year follow-up, in relation to results of exercise stress test on treadmill (positive = 1.0 mm ST depression 80 msec after J point). (After Ellerstad, 1975.)

two-step test (Table 6.19; G. R. Cumming, 1972). Some 10% of men over the age of 40 and at least 20% of men over 60 show significant ST depression with the more vigorous type of test (G. R. Cumming, Borysyk, and Dufresne, 1972; Froehlicher, Yanowitz et al., 1975); in women, the proportion of subjects with abnormal records is even higher (G. R. Cumming, Dufresne, and Samm, 1973; Sidney and Shephard, 1977b), although in their case it is less clearly established that ischemia is responsible for the abnormality. With any criterion, there is a substantial proportion of both false positive and false negative test results, and some authors maintain that the appearance of angina during exercise is a better indicator of an adverse prognosis than any interpretation of ST segment voltages (Cole and Ellerstad, 1978).

Accurate and consistent measurements of ST displacement are difficult during vigorous exercise (Seymour and Conway, 1969; Elgrishi et al., 1970). Blackburn et al. (1968) found that the percentage of abnormal records reported by 14 cardiologists ranged from 5 to 58% and that on different occasions the same physician would interpret the same ECG record differently. Reasons for such inconsistent interpretation included lack of objective criteria, uncertainty as to the significance of junctional depression, and the poor technical quality of many records. Technical problems such as interference and a wandering baseline are largely a reflection of varying electrode impedance. The quality of the ECG can be much improved by careful skin preparation (Tregear, 1965) and use of a high-input impedance amplifier (Lewes, 1965; Geddes and Baker, 1966). Subsequent treatment of the electrical signal by high-frequency filters (Von der Groeben et al., 1969), with specific ("common mode") rejection of 60-Hz noise, and analog or digital averaging of the residual input (Rautaharju et al., 1971; A. Pedersen and Andersen, 1971; Siegel, 1974; James and Patnoi, 1974; Simoons, 1976; Jansson et al., 1976) minimizes artifacts such as muscle noise (Blomqvist, 1965; Fig. 6.21), allowing precise objective measurements without interobserver variation (Pipberger et al., 1962; Blackburn, Blomqvist et al., 1968; Caceres and Hochberg, 1970; J. J. Bailey et al., 1974a–c). The signal improvement varies as the square root of the number of averaged cycles. However, if too many complexes are averaged, a brief ST displacement may be missed, while slight asynchrony of the QRS waves leads to a progressive rounding of the primary waveform (D. A. Winter, 1969; Rautaharju et al., 1971); 16 or 32 successive beats are thus used. Analog equipment presents the averaged signal as a visual display or as a series of perhaps 1024 (2^{10}) discrete voltage readings. Although this type of data reduction overcomes the problem of baseline drift, a single abnormal beat (such as a ventricular premature systole) causes a grossly inaccurate averaged signal (L. K. Jackson et al., 1969). A digital computer can be programmed to give a similar but more sophisticated solution of the problem. The 1024 measurements are made on each of 48 successive beats. The computer program then selects 16 of the 48 beats that have a mutually similar waveform and calculates an averaged tracing from these 16 beats. Problems still arise in some patients; in a series of 7084 tracings, Simoons (1977)

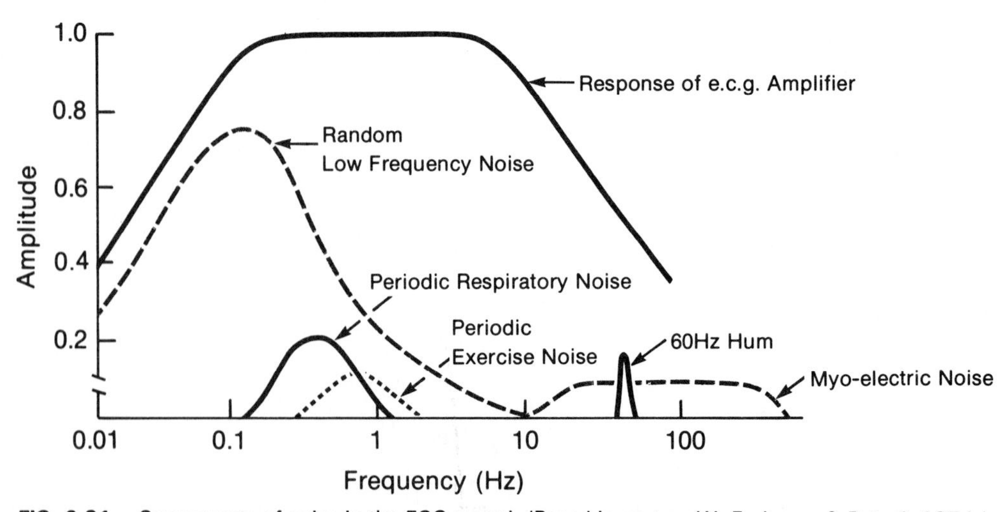

FIG. 6.21. Components of noise in the ECG record. (Based in part on W. E. James & Patnoi, 1974.)

noted incorrect averaging due to excessive drift or noise in 1.8%, incorrect QRS detection in 0.04%, P-wave detection errors in 6.5%, and T-wave detection errors in 0.6% of the sample. Many programs measure ST depression at a fixed interval, for example, 80 msec after the S-ST junction (Fig. 6.22) or a stated time after the nadir of the R or S waves. Unfortunately, such approaches do not allow for variations in duration of the QRS complex and the ST segment. Alternatively, the onset of the T wave can be determined by inspection of the averaged tracing or by electrical differentiation of the averaged signal. Other techniques of interpretation include a calculation of the ST slope (Lester, Sheffield, and Reeves, 1967; McHenry et al., 1968) or the ST integral below a baseline established at the onset of the Q wave (Sheffield, 1974; Fig. 6.23).

Some authors have used the behavior of the ST segment not only as an empirical guide to prognosis, but also as a more immediate indicator of myocardial oxygen supply (Table 6.18; Salzman et al., 1969; R. E. Mason et al., 1969; Siegel, 1978). The observed appearance depends somewhat upon the duration of activity at a given loading (Shephard and Kavanagh, 1978a); as acid metabolites accumulate, the coronary vessels dilate and ST depression may diminish. In order to make comparisons from one test or type of activity to another (I. Åstrand, 1972), it is necessary to equate cardiac work loads (in terms of heart rate, tension/time index, or triple product). Even this approach fails to allow for possible changes in heart volume and wall thickness (Sidney and Shephard, 1977b), so that it becomes quite difficult to assess the extent of myocardial ischemia from ST segment behavior (Tables 6.18, 6.20), particularly if the activity status of a subject has changed. Nevertheless, there have been suggestions that the appearance of ST depression in a patient previously known to have had a normal exercise ECG is a strong warning of progressing coronary vascular disease (J. T. Doyle and Kinch, 1970).

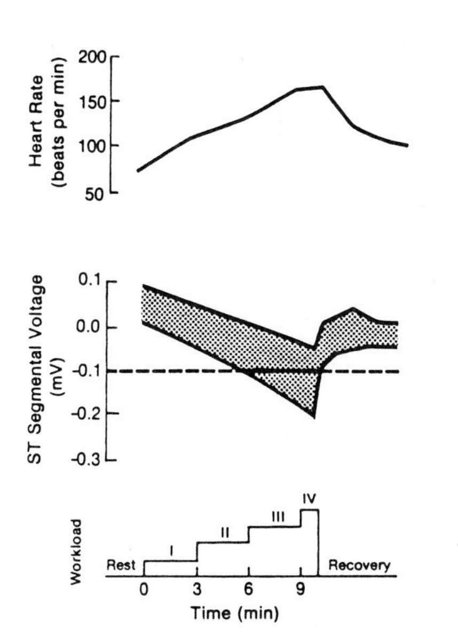

FIG. 6.22. Pattern of ST voltage change in 48 healthy middle-aged U.S. men performing progressive treadmill exercise. Voltages measured 50 to 59 msec after S nadir. Load I = 2.7 km • hr^{-1}, 10% slope; Load II = 4.0 km • hr^{-1}, 12% slope; Load III = 5.4 km • hr^{-1}, 14% slope; Load IV = 6.7 km • hr^{-1}, 16% slope. (After Bruce, Alexander et al., 1969).

Elevation of the ST segment has been described during very sustained effort (such as the 6-day Tour de France race; Plas, 1978; Venco et al., 1979). Again, this sign represents a disturbance of repolarization, in this instance associated with glycogen depletion of the cardiac muscle (Barnard and Thorstensson, 1975). It is important mainly because it must be distinguished from the ST displacement of myocardial infarction.

Other Changes in the Electrocardiogram

The premature systoles of the anxious patient commonly become less frequent with effort. However, an abnormality of rhythm that appears for the first time during exercise, whether a form of heart block or a series of extrasystoles, is generally regarded more seriously. Responsible factors include local hypoxia and an accumulation of catecholamines. Hypoxia impairs conduction, facilitating the reentry phenomenon discussed above (A. M. Katz, 1973). Circulating catecholamines (adrenaline and noradrenaline) may lower the threshold of abnormal pacemaking foci. They also speed both depolarization and repolarization (A. M. Katz, 1977), so that nonuniformities in response to these drugs can cause a reentry excitation. On occasion, catecholamines are probably responsible for sudden death during severe physical or emotional stress (Shephard, 1974c,1979b).

Although a less sensitive indicator than ST segmental depression, the onset of premature ventricular systoles has some value in diagnosing the coronary-prone patient; however, it is less clearly established that the abnormality of rhythm adds new evidence of risk to that based on behavior of the ST segment (Sheffield, 1974). Some authors maintain that premature systoles appearing during exercise carry a more adverse prognosis than occasional rest-

TABLE 6.20
Measures of Accuracy of the Exercise ECG in Diagnosing Myocardial Ischemia

(a) Percent of correct diagnoses in angina (relative to coronary angiography)

Type of Pain	Correct Diagnoses (%) History	Correct Diagnoses (%) Exercise ECG
Typical angina	90–95	70–75
Previous myocardial infarction	70–80	50
Atypical angina	60	50–70
Non-coronary chest pain	95	80

(b) Percent of correct diagnoses (relative to subsequent ischemic heart disease)

First Author*	Subjects No.	Follow-up Period (yrs)	Test Sensitivity (%)	Test Specificity (%)
Bruce (1969)	221	5	60	86
Aronow (1973)	100	1.5	75	77
Kattus (1971)	313	2.5	100	67
Cumming (1975)	510	3	58	75
Froehlicher (1974)	710	6	19	90
Bruce (1975)	1339	7	29	91

(c) Percent of correct diagnoses relative to coronary arteriogram.

First Author†	Subjects (No.)	Test Sensitivity (%)	Test Specificity (%)
Mason (1967)	84	77	89
Kassebaum (1968)	68	51	96
Roitman (1970)	46	80	87
Ascoop (1971)	96	59	94
Moran (1972)	140	65	91
Kelemen (1972)	61	54	94

(d) Percent of correct diagnoses relative to severity of angiographic lesion (Amsterdam & Mason, 1977).

Affected Vessels (No.)	Sensitivity (%)
1	42
2	66
3	80

*For details of references, see G. R. Cumming (1978).
†For details of references see Ascoop (1977).

Source: After Ascoop, 1977; Amsterdam & Mason, 1977 and G. R. Cumming, 1978.

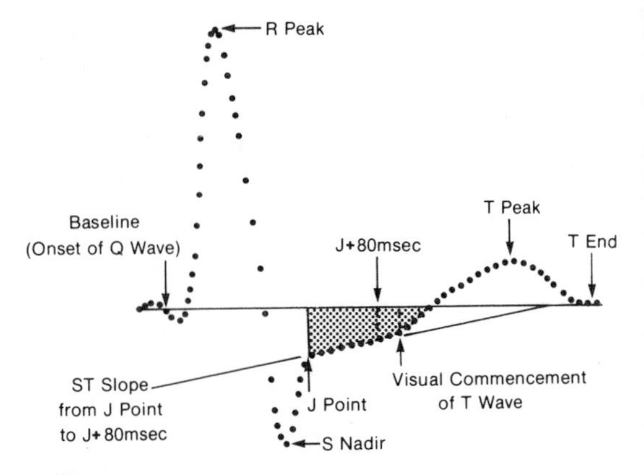

FIG. 6.23. Possible methods of interpreting ST depression in digitally averaged ECG complex. Shaded area indicates ST integral (normal value of integral J to J + 80 < − 10 μv • sec).

ing extrasystoles (McHenry, Fisch et al., 1972; Beard and Owen, 1973; Blackburn, Taylor et al., 1973; *British Medical Journal*, 1973). Others maintain that both types of premature beat carry an above average risk of developing overt ischemic heart disease (Chiang et al., 1969; Hinckle et al., 1969; Kohn et al., 1971; Sharma et al., 1974), so that detection can be based on 24-hr tape recordings of the ECG rather than an exercise stress test (Mogensen, 1977). Our own observations upon postcoronary patients suggest that polyfocal extrasystoles (indicated by differences of QRS waveform) increase the risk of a recurrence, whereas unifocal extrasystoles do not (Shephard and Kavanagh, 1979).

Other possible changes of ECG waveform include an increased prominence of the atrial P waves during activity and an increase of both P and T waves during recovery (Simoons and Hugenholtz, 1975). These changes may reflect an altered intracardiac blood volume and thus electrical conductivity for the cardiac impulse.

Indications To Halt a Test

Both ST segmental depression (>0.2 mV, horizontal or downward sloping) and premature ventricular systoles (more than 3 in 10 sec, especially if polyfocal and occurring during the vulnerable period), are indications to halt an exercise test. If ignored, there is a risk that ventricular tachycardia and even ventricular fibrillation will supervene.

ST segmental depression and disturbances of cardiac rhythm are often worse in the period immediately following exercise, and if there is any doubt about the condition of a patient, it is wise to stop a test. Various factors contribute to a worsening of myocardial status in the immediate postexercise period. These include a fall in the mean systemic blood pressure, a continuing decrease in the oxygen content of arterial blood, and a postexercise peak of acidity and catecholamine levels.

Ventricular Fibrillation

Ventricular fibrillation is an irregular, writhing contraction of the ventricles that is ineffective in expelling blood from the heart. Unless a normal rhythm is restored, death occurs rapidly. Irreversible changes in the brain begin after 4 min interruption of the cerebral circulation, and each succeeding minute of fibrillation reduces the likelihood of a complete resuscitation. Cardiac massage and defibrillation may still be technically successful, but it becomes increasingly probable that the patient will suffer permanent brain damage. In such circumstances, the restarting of the circulation must be regarded as an unwarranted interference with the natural course of life.

The first event leading to ventricular fibrillation seems an extrasystole (Moe et al., 1941; Milnor et al., 1958). This initiates a train of further extrasystoles, which either accelerate and become ventricular fibrillation, or slow and gradually cease (Wiggers, 1940). The essentials are thus (1) an ectopic focus, (2) a rapid burst of extrasystoles, and (3) an environment that favors the reentry phenomenon. A variety of factors such as electrical shock, freshwater drowning, myocardial infarction, and anticholinesterase poisoning are all possible triggers.

The combination of myocardial ischemia and an accumulation of catecholamines may induce an attack during vigorous exercise, particularly if there is associated anxiety (Shephard, 1974c; Shephard and Kavanagh, 1979). In some exercise-testing laboratories (McDonough and Bruce, 1969; Rochmis and Blackburn, 1971) and gymnasia (Pyfer et al., 1975), ventricular fibrillation occurs several times per year, but our experience to date has been more fortunate. After 10 years operation of a large postcoronary rehabilitation program, we have seen only three episodes in the gymnasium and none in the exercise laboratory. This probably reflects our care to avoid severe myocardial ischemia (ST depression > 0.2 mV) and trains of ventricular premature contractions (>3 in 10 sec) during either the test exercise or subsequent prescribed activity.

If the electrocardiogram is monitored continuously while testing, a clear diagnosis can be made immediately after a dysrhythmia develops. The treatment of ventricular fibrillation is to apply one or more electrical shocks to the chest wall using an external defibrillator. The intense burst of electrical energy leads to a complete depolarization of the myocardium, the hope being that with repolarization a normal rhythm will be restored.

Early designs of defibrillator used alternating current, with potentials of up to 750 V. There was then a danger that the rescuer himself might be sent into fibrillation by careless handling of the equipment, and the high voltage was liable to cause tissue damage (Shephard, 1961b; Peleška, 1963; Pantridge et al., 1975). Partly for these reasons and partly because of difficulty in timing brief shocks reliably (Shephard, 1961b), the currently accepted technique is to use a direct current defibrillator (Gurvich and Yuniev, 1947; Lown et al., 1962). A small condenser ($\sim 2.5\ \mu f$) is brought to a very high potential (5000 V). The charge Q is then given by

$$A = VF \tag{6.26}$$

where V is the potential and F is the capacity. In the example cited, the charge is (25×10^{-6}) 5000 or 0.125 C, and the stored energy is ½QV, or 312.5 W • sec. A series inductance ensures an optimum shock duration and avoids high peak currents that might otherwise cause tissue damage (Peleška, 1966; Tacker et al., 1968,1969). In the dog, pulse lengths of 0.018–27 msec will cause defibrillation, but energy requirements are minimal with a pulse duration of 1–8 msec (Geddes et al., 1970). Because energy losses in the system may be up to 50%, it is desirable that instruments show the *delivered* energy in W • sec. Modern devices incorporate an ECG monitoring oscilloscope and a synchronizing circuit that prevents administration of a shock to a normally beating ventricle during the early (vulnerable) phase of repolarization (Peleška, 1963; James and Patnoi, 1974). In clinical use, the initial setting is 100 W • sec, but the delivered energy is increased progressively if further shocks are needed, depending upon the age of the individual and the thickness of the skin and underlying fat (Detmer et al., 1964).

It is difficult to obtain reliable statistics on the success of defibrillation. In some institutions, all patients "brought in dead" are given resuscitative treatment in order to familiarize staff with procedures. Nevertheless, a properly used defibrillator has saved the lives of many patients who would otherwise have died (Table 6.21; H. J. Smith and Anthonisen, 1965). The success of the defibrillating procedure is markedly influenced by attendant hypoxia. In a healthy, well-oxygenated heart it is by no means easy

TABLE 6.21
Results of Cardiac Resuscitation in 1063 Patients

Diagnosis	Total	Temporary Success (%)	True Survival (%)	Death (%)
Myocardial infarction	357	22.4	15.1	62.5
Cardiovascular disease	220	22.3	6.4	71.3
Surgery (during or after)	60	35.0	20.0	45.0
Respiratory	82	25.6	7.3	67.1
Trauma	98	20.4	1.0	78.6
Neurological	101	28.7	2.0	69.3
Overdose of drugs	18	27.8	22.2	50.0
Other	127	21.3	0.0	78.7

Source: After Peatfield et al., 1977.

to induce persistent fibrillation, even by direct application of an electrical shock to the myocardium (Shephard, 1961b). But once coronary flow has stopped, fibrillation is readily induced and it becomes hard to restore a normal rhythm (Shephard, 1961b; Portal et al., 1963). Early treatment is thus vital. Because fibrillation is quite a rare event, it is debatable as to whether it is necessary to have a defibrillator on-site when the normal middle-aged adult is carrying out vigorous exercise. However, if an emergency does develop, defibrillation carried out by a well-trained paramedical team within 2–3 min is much more beneficial than skilled medical assistance provided after a longer time delay (Frey and Nolte, 1968; Murtomaa and Korttila, 1974; Hampton et al., 1977). High priority should thus be given to legislation permitting the training of specialist ambulance crews and first aid teams in cardiac resuscitation.

Cardiac Arrest

Ventricular contractions may cease (1) as a sequel to heart block, (2) as a secondary response to the various stimuli inducing ventricular fibrillation and attendant hypoxia, or (3) as a consequence of the increased plasma sodium ion concentration that accompanies saltwater drowning. The heart may also fail to restart following defibrillation.

If the ECG is being monitored, there will be a complete absence of electrical activity. If an electrocardiogram is not available, there will be no pulse palpable in the neck and no heart sounds audible over the precordium. Since irreversible brain damage is rapid (see above), treatment must be started immediately. Unfortunately, some medical licensing authorities still have not authorized widespread training in cardiac resuscitation, although there is good evidence that well-taught paramedical workers can administer such treatment safely and effectively (Genaud et al., 1965; L. B. Rose and Press, 1972; Briggs et al., 1976; N. M. White et al., 1973; Hampton et al., 1977). In Seattle, the cost of successful resuscitation by an ambulance team has been set at ~U.S. $2000 (Cobb et al., 1975), a very small fraction of the expense associated with successful medical treatment (Shephard, 1975b). Considerable interest attaches to the Seattle project, where cardiac resuscitation is now being taught to entire high school classes (Cobb et al., 1975).

Sometimes the heart can be restarted by a vigorous blow on the chest (Harwood-Nash, 1962), but if this fails, cardiac massage must be undertaken (Kouwenhoven et al., 1960; Jude et al., 1964). The patient is placed on a rigid surface, since much of the energy of a potential rescuer can be dissipated in the springs of a bed or mattress. A fist is used to apply very vigorous pressure over the sternum, ~60 times per min. The ribs and underlying viscera are protected by interposing the rescuer's free hand. If ventilation has ceased, a second person must apply mouth-to-mouth or some other form of artificial respiration.

In smaller mammals, such as laboratory dogs, very effective mean arterial pressures can be generated by external compression of the chest (Kouwenhoven et al., 1960). In humans, the thorax is more rigid and the resulting blood pressure is quite marginal for survival (Gurewich et al., 1961; Weale and Rothwell-Jackson, 1962; G. J. MacKenzie et al., 1964). Indeed, some observers maintain that an elderly victim with calcified costochondral joints cannot receive effective cardiac massage unless the thoracic cage is ruptured (Table 6.22). In younger individuals, systolic pressures of ~100 mmHg (13 kPa) have been recorded, but the pulse wave still has a spiky form, with a very low diastolic pressure and a

TABLE 6.22
Complications Observed in 126 Episodes of Cardiac Resuscitation

Complication	Percent
Ruptured liver	1.6
Ruptured spleen	0.8
Hemopericardium	1.6
Mediastinal emphysema	0.8
Pneumothorax	0.8
Pulmonary fat emboli	0.8
Fractured ribs or sternum	19.8

Source: After Klassen et al., 1963.

poor average reading (Gurewich et al., 1961; Weale and Rothwell-Jackson, 1962). Unfortunately, the victim usually has a high venous pressure, further reducing the gradient available for perfusion of the vital organs. Nevertheless, there have been many instances of successful cardiac massage. Manikins are now available to teach rescue workers the forceful efforts necessary to sustain an adequate arterial pressure.

Some authors still recommend that if the condition of the patient does not improve within a minute, the thorax should be opened by a broad incision through the fourth or fifth interspace, allowing internal massage of the heart. However, such a procedure is plainly restricted to licensed physicians. A doctor may administer sodium bicarbonate (45-90 meq every 5 min) to counter acidosis (J. S. S. Stewart, 1964) and pressor amines to increase myocardial contractility (particularly if the cardiac contractions are weak and ineffective). Calcium chloride may also be given (10 ml of 10% solution) if inadequate contractions persist after treatment with adrenaline or noradrenaline.

If resuscitation is successful, the patient must be watched very carefully for the next 48 hr. The heart remains abnormally irritable, and it is usual to administer procaine amide or lidocaine to reduce the risk that the aberrant rhythm will recurr (Klassen et al., 1963). The brain may swell as a late response to the period of hypoxia (cerebral edema). Measures adopted to minimize cerebral damage include a reduction of the circulating blood volume and deliberate induction of hypothermia.

CONTROL OF THE CIRCULATION

General Considerations

The primary function of the circulation is to maintain homeostasis in the cells of the brain. There is little possibility of varying vascular dimensions within the rigid walls of the cranium, so that cerebral homeostasis depends upon effective control of the central arterial pressure. Pressure is influenced by left ventricular output, systemic peripheral resistance, elasticity of the major arteries, blood volume, and blood viscosity. However, regulation normally occurs through appropriate adjustments in the first two of these variables.

The control system forms a typical cybernetic loop (Fig. 6.21). The input is by specialized pressure sensors in the aorta and at the bifurcation of the common carotid arteries (Heymans and Neil, 1958; Neil, 1960). Information is fed to controlling centers in the medulla. Some authors distinguish cardiac, vasoconstrictor, and vasodepressor centers, on the basis of specific responses to electrical stimulation (Oberholzer, Folkow et al., 1960). The vasoconstrictor area has its own spontaneous activity, suppressed or enhanced by impulses from the cortex and hypothalamus, and depressed by the activity of the vasodepressor center (Öberg, 1976). The vasodepressor center seems the comparator, where reported systemic pressures are matched to a predetermined "set point." Negative feedback is applied to the spontaneous activity of the vasoconstrictor and cardiac centers, and to sympathetic connector cells in the lateral horns of the spinal cord.

Information from many other sources (including the emotions, environmental temperature, and physical activity) is fed to the vasoconstrictor, and possibly also the vasodepressor center (Korner, 1971). In effect, such stimuli alter the set point of the medullary controller. In some instances, independent responses of cardiac output or local reactions of the blood vessels may override general control mechanisms (Fig. 6.24).

Sensory Input

Higher Centers

Impulses from the higher centers of the brain have a marked influence upon the setting of cardioregulatory and vasomotor centers (Delgado, 1960). Animal experiments (Hess, 1949; Lindgren, et al., 1956; Folkow, 1960; Uvnäs, 1960; V. C. Abrahams et al., 1960) have demonstrated that stimulation of the brainstem causes muscle vasodilation as part of an integrated fight-or-flight defense reaction involving the autonomic nervous system. Observations on human subjects confirm that anger, anxiety, and other emotional responses increase heart rate and blood pressure (Brod et al., 1959; O. A. Smith,

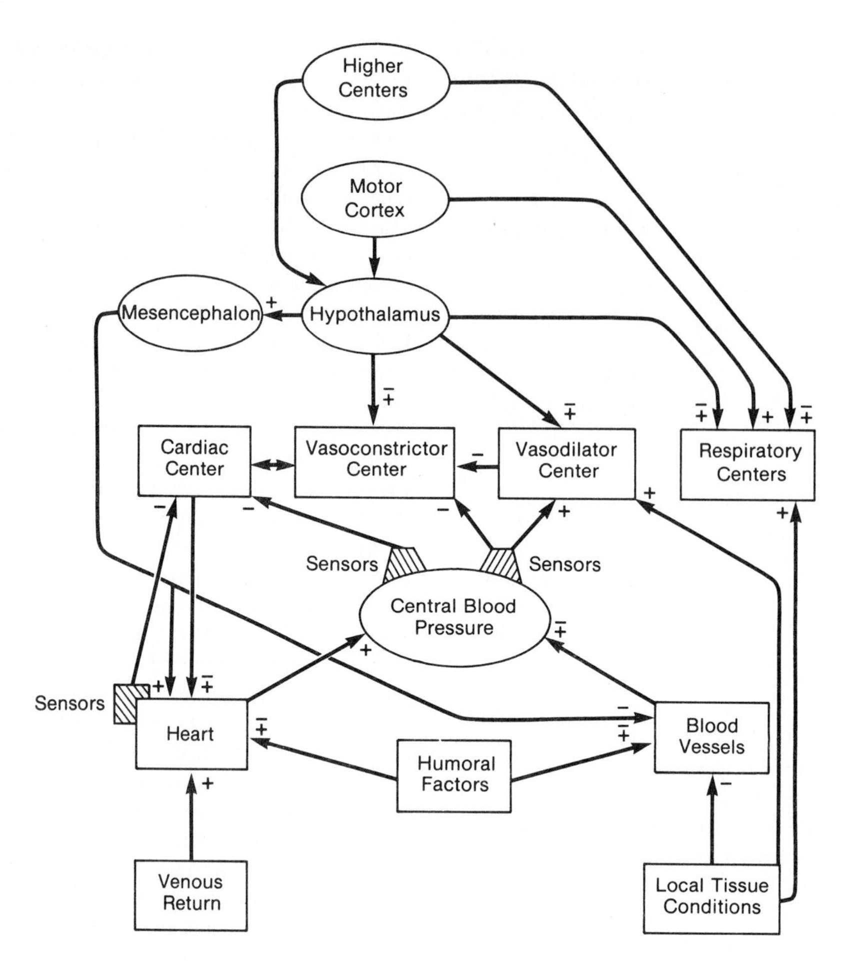

FIG. 6.24. Schematic representation of the control of the heart and circulation.

1974), sometimes throwing a heavier work load on the heart than vigorous exercise. Cardiac output is also increased, with a decrease of both stroke volume and arteriovenous oxygen differences (Hickam et al., 1948; Golenhofen and Hildebrandt, 1957; D. A. Blair et al., 1959a; Brod et al., 1959).

Conditioned reflexes from the cerebral cortex initiate similar changes immediately prior to a contest; racing drivers awaiting the starter's signal, for example, have heart rates of 180–200 beats per min (Lonne et al., 1968). Such anticipation of effort may be an advantage to the endurance athlete because it shortens the circulatory on-transient, reducing the size of the oxygen deficit at the beginning of exercise. On the other hand, anxiety-induced tachycardia exaggerates the heart rate response to submaximal exercise, increasing cardiac workload and leading to errors in the prediction of maximum oxygen intake from submaximal tests (Shephard, 1969; C. T. M. Davies, Tuxworth, and Young, 1970).

Irradiation from Motor Cortex

Information may irradiate from the motor cortex to the cardiac and vasomotor centers (Krogh and Lindhard, 1913). When voluntary activity is initiated, there is a tremendous increase in the neural discharge from the motor regions of the brain. The majority of impulses travel via the pyramidal fibers to the anterior horn cells of the spinal cord, whence they are relayed to the contracting muscles (Chap. 7). However, a small proportion of the neural signals reach the cardioregulatory centers of the medulla, thereby contributing to the early on-transient of heart rate.

In support of this view, the cardiac output response to dynamic leg exercise is related to the intensity of work even when the circulation to the active limbs is occluded (Asmussen, Christensen, and Nielsen, 1940), while the heart rate and blood pressure responses to a given work load are increased if muscle strength is reduced by administration of curare (Ochwadt et al., 1959), or the "central command" is increased by simultaneous stimulation of spindle afferents from antagonistic muscles (Chap. 7; Goodwin et al., 1972).

Hypothalamic Input

Centers regulating body temperature are found in the hypothalamus (Chap. 7; A. D. Keller, 1960; O. A. Smith et al., 1960). If body temperature rises as a result of exercise or exposure to a warm environment, impulses from the hypothalamus induce tachycardia, vasodilation of the cutaneous vessels, an increase in blood flow to the skin, and a reduction in blood flow to the viscera (Rowell, 1971). Nevertheless, the set point of systemic pressure seems remarkably constant (Marx et al., 1967).

Moderate cold exposure leads to a constriction of superficial arteries and veins, also through a hypothalamic reflex. However, if the local cold is severe enough to cause pain, there may be other effects (Chap. 12), including a rise of heart rate and systemic blood pressure (cold pressor response), alternation of intense vasoconstriction and vasodilation (a metabolic "hunting" reaction), and paradoxical vasodilation (due to paralysis of the nerves regulating the arteriovenous anastomoses).

Irradiation from Respiratory Centers

The medullary centers concerned with the circulation receive impulses from the respiratory centers (McAllen and Spyer, 1978). On exposure to a low partial pressure of oxygen, the respiratory centers are stimulated by impulses from the carotid bodies (J. M. Marshall, 1977), and irradiation of this information to the vasomotor center leads to a marked rise of blood pressure, although the sensitivity to other pressor stimuli remains unchanged (D. J. C. Cunningham, Strange-Petersen et al., 1972b). In artificially ventilated animals, secondary reactions to the increase of blood pressure include a slowing of the heart, with a decrease of myocardial contractility and cardiac output (M de B. Daly and Scott, 1958; M. de B. Daly, 1964; Greenwood et al., 1977). However, with spontaneous respiration, hyperventilation causes an increase of heart rate.

Peripheral Pressor Receptors

Information is transmitted to the medullary centers from a wide range of peripheral receptors. The most important are the pressor sensors of the carotid sinus and aortic arch (Heymans and Neil, 1958; Neil, 1960). These receptors send depressor impulses to the cardioregulatory centers when systemic pressure rises. They are stimulated by deformation of the vessel wall, being more receptive to the rate of pressure change and the frequency of distension than to the absolute pressure level. Their rate of discharge at a given mean arterial pressure is thus greater when the pulse pressure is widened (Sharpey-Schafer, 1956; Heymans and Neil, 1958; Neil, 1960; Gero and Gerova, 1965). Sensitivity is diminished by aging, since this causes the arteries to become more rigid (L. H. Peterson, 1967). The response of the sinus pacemaker depends also upon the phase of the cardiac cycle in which the pressor receptors are stimulated (Eckberg, 1976).

Most workers have found a rise of peripheral arterial pressure during upright exercise. This would play a useful role in sustaining perfusion of the vigorously contracting muscle. However, it would imply either local or central suppression of pressor receptor activity during physical effort (Heymans, 1950; Freyschuss, 1970; Gebber and Snyder, 1970; Korner, 1971). Marx et al. (1967) pointed out that the pressor receptors sense central rather than peripheral pressures. Given central recording and the use of adequate manometers, pressures do not rise during brief rhythmic activity, so that there is no need to postulate any adjustment of control settings. Neither dynamic (Bevegård and Shepherd, 1966b) nor isometric (Ludbrook et al., 1978) exercise influences the blood pressure response of the carotid receptors to external suction. On current evidence, an alteration of peripheral receptor sensitivity thus seems unlikely. However, the bradycardial response to neck suction is reduced by isometric effort (Mancia et al., 1978; Ludbrook et al., 1978) and given an adequate intensity of exercise and an appropriate method of analysis (pulse interval rather than pulse rate evaluation), a reduced response to drug-induced hypertension can also be demonstrated (more during isometric than during rhythmic work; D. J. C. Cunningham, Howson et al., 1970; D. J. C. Cunningham, Strange-Petersen et al., 1972a; Bristow et al., 1971; McRitchie et al., 1976). Possibly, the responsiveness of the vasomotor center is reduced because there is a high background level of stimulation from other receptors during exercise (Klevans and Gebber, 1970; Quest and Gebber, 1972).

Other Mechanoreceptors

Mechanoreceptors have been described in the great veins, atria, ventricles, and pulmonary arteries (Aviado and Schmidt, 1955; Coleridge and Kidd, 1960,1963; Paintal, 1973; T. C. Lloyd, 1975). The intraatrial and intraventricular receptors are occasionally stimulated by exposure to industrial irritants such as organic nitriles, giving rise to bradycardia, vasodilation, and systemic hypotension. Their normal function is uncertain. If the great veins or left atrium are deliberately distended, the receptors within their walls usually initiate an increase of heart rate (the Bainbridge reflex). However, this reflex is unlikely to be involved in the tachycardia of exercise, since the cardiac output of a normally active person is increased without any appreciable increase of venous filling pressure. Possibly, information on the filling of the atria relative to thoracic or total blood volume may influence the secretion of antidiuretic hormone and thus urine production, blood volume (Folkow and Neil, 1971), and sodium ion excretion (Eisele et al., 1978).

Muscle Receptors

Receptors have been postulated (but not yet demonstrated) in the muscles (Alam and Smirk, 1938; Lind, McNicol et al., 1968; Hollander and Bouman, 1975). A response can be elicited by stimulation of the afferent nerves from muscle, apparently due to fine myelinated and nonmyelinated fibers (types III and IV) that arise from pressure and pain receptors. Suggested stimuli to these endings include potassium ions and hyperosmolarity (Wildenthal et al., 1968, 1969). However, the importance of such receptors to the overall exercise response has yet to be established.

Cutaneous Receptors

Receptors in the skin and around the nose are probably responsible for the dramatic slowing of heart rate seen in many species during diving (Scholander et al., 1962; L. Irving, 1963). The contribution of these receptors to normal cardiovascular regulation is uncertain.

Cardiac Regulation

Normal Control Mechanism

The heart is slowed by impulses passing down the vagus nerve to the sinoatrial pacemaker. Conversely, it is speeded by a reduction of vagal discharge, with an increase in the activity of nerves supplying sympathetic β receptors. Parallel effects can be induced experimentally by injecting massive doses of the corresponding neurotransmitter substances (acetylcholine and noradrenaline, respectively).

There is little interaction between vagal and sympathetic activity at the sinuatrial node, and the response to vagal impulses is thus unaffected by the background level of sympathetic stimulation (Warner and Russell, 1969; M. N. Levy and Zieske, 1969). Both neurotransmitter substances are broken down rapidly through the action of appropriate enzymes (cholinesterase and amine oxidase, respectively). Transmitters produced elsewhere in the body do not normally affect the heart unless the corresponding enzyme is inhibited. One example of such inhibition is seen when cholinesterase is poisoned by an organophosphorus insecticide. The accumulation of acetylcholine is then sufficient to slow the sinus rhythm greatly, and heart block may develop at the atrioventricular node (Shephard, unpublished data).

Regulation in the Endurance Athlete

Trained athletes have a slow resting heart rate (30–60 beats per min) relative to untrained men (70–80 beats per min). The resting cardiac output is rather similar in the two groups, but the athlete has a larger stroke volume than the nonathlete. Literature from Central and Eastern Europe extols the virtues of "vegetative" adjustments and "vagotonia" (Raab, 1966b). Endurance training probably increases vagal drive and decreases sympathetic activity (Ekblom, Kilbom, and Soltysiak, 1973), as can be shown by administration of suitable blocking agents (Sutton, Hughson et al., 1977). This may have practical significance with respect to the genesis of dysrhythmias in athletes, although vagotonia scarcely merits the mystical significance with which it has sometimes been endowed (Raab, 1966b).

Exercise Response

Tactics for the increase of cardiac output during exercise have already been discussed. Augmentation of heart rate, stroke volume, and cardiac contractility normally involve an increase of sympathetic activity and a decrease of vagal discharge. If β-adrenergic activity is blocked by administration of propranolol, there is little or no increase of exercise heart rate, and the cardiac output response to supine submaximal work is reduced (J. M. Bishop and Segel, 1963; G. R. Cumming and Carr, 1966; S. E. Epstein, Robinson et al., 1965; Hamer and Sowton, 1965), with inhibition of the anticipated increase of myocardial contractility (Paley et al., 1965). However, there is surprisingly little change of maximum cardiac output or maximum oxygen intake in such circumstances (S. E. Epstein, Robinson et al., 1965; Ekblom, Goldbarg, and Gullbring, 1972b). If vagal activity is blocked by atropine, the heart rate is decreased in light work but remains unchanged in maximum effort (Ekblom, Goldbarg, and Gullbring,

1972b). After administration of both propranolol and atropine, the steady-state exercise response is much as in control subjects (G. R. Cumming and Carr, 1967; Ekblom, Goldbarg, and Gullbring, 1972b).

The versatility of regulating mechanisms is seen in patients in whom the heart rate is controlled by an artificial pacemaker. Compensatory adjustments of stroke volume ensure that a very normal cardiac output response to exercise is maintained in such individuals (Benchimol et al., 1964; Braunwald, Sonnenblick et al., 1967).

Experiments in dogs with denervated hearts (Donald and Shepherd, 1963; Ashkar, 1965a,b) confirm the wide range of alternative control loops. Provided that denervation is restricted to the cardiac nerves, working capacity is unaffected (D. E. Donald, Milburn, and Shepherd, 1964a). The on- and off-transients of heart rate are slower than normal, but the steady-state cardiac output at a given oxygen consumption is unchanged. With mild effort, the increase of cardiac output occurs largely through an increase of stroke volume, but with more severe exercise, an increase in heart rate and stroke volume contribute equally to the cardiac response. The increase of heart rate is not mediated by an increase of blood temperature (D. E. Donald and Shepherd, 1963). Furthermore, it persists after adrenalectomy or administration of propranolol and atropine (D. E. Donald and Samueioff, 1966). Presumably, if all other control mechanisms are blocked, cardiac regulation is still possible through an increase of venous return and Starling's law of the heart (D. E. Donald and Shepherd, 1964b).

Peripheral Regulation

General Considerations

Blood vessels supplying the viscera have both parasympathetic and sympathetic innervation. In contrast, the nerve supply of the intramuscular blood vessels is derived entirely from the sympathetic system, although it comprises both adrenergic fibers that release noradrenaline and cholinergic fibers that release acetylcholine.

Adrenergic Innervation

At least two types of adrenergic activity are observed: excitatory and inhibitory. Ahlquist (1948) explained this finding on the existence of two types of catecholamine receptor, α and β. The α receptors were predominantly excitatory in function and the β receptors predominantly inhibitory. The pattern of reaction in a given tissue thus depended on the relative numbers of α and β receptors and their sensitivity to the catecholamine in question. The α receptors were widely distributed in both resistance and capacity vessels of the vascular system. Their normal response was to noradrenaline. As this transmitter substance was liberated from the sympathetic nerve terminals, a strong general vasoconstriction was induced. Adrenaline had a similar effect when acting upon α receptors. The β receptors responded to adrenaline (secreted by the adrenal medulla) and the synthetic catecholamine isopropyl adrenaline, but were unaffected by noradrenaline. Such receptors were found in the resistance vessels of skeletal muscle, and possibly also of cardiac muscle; stimulation caused vasodilation. Adrenaline also caused a stimulation of local metabolism in the blood vessel wall, contributing to a phase of late vasodilation.

Recent research has clarified details of both α and β responses (Roddie and Shepherd, 1963; Shepherd, 1967; Van Houtte, 1978). Noradrenaline is stored in the terminal vesicles of the sympathetic nerves, having been taken up from extracellular fluid or synthesized locally from tyrosine (Fig. 6.25). Small amounts are lost by leakage, and larger amounts are displaced by sympathomimetic amines such as ephedrine and amphetamine. However, the normal stimulus to release of noradrenaline is an increase of neuroplasmic Ca^{2+} resulting from neuronal depolarization. The noradrenaline then binds to receptor sites on the vascular smooth muscle. Most of these are α-adrenergic in character, and a vasoconstriction is thus induced. Various other substances can act directly on the smooth muscle and facilitate or inhibit the release of noradrenaline (Fig. 6.25). The β receptors seem to be without innervation. In research with β-blocking drugs, it has become useful to distinguish two types of β receptor (β_1 and β_2; Furchgott, 1978). The objective of the pharmacologist is to block β receptors in one organ (such as the lungs) without influencing function elsewhere (for example, the heart). Thus, an asthmatic patient may be treated by salbutamol (an agent which relieves bronchospasm by blocking the β_2 receptors of the airways) with minimal disturbance of adrenergic function in the heart.

Cholinergic Fibers

Sympathetic cholinergic fibers (Uvñas, 1960; Folkow and Neil, 1971) have also been described. These fibers are strongly activated during physical activity. Their origin is in the motor cortex. With relay stations in the hypothalamus and mesencephalon, their output bypasses the vasomotor centers to reach sym-

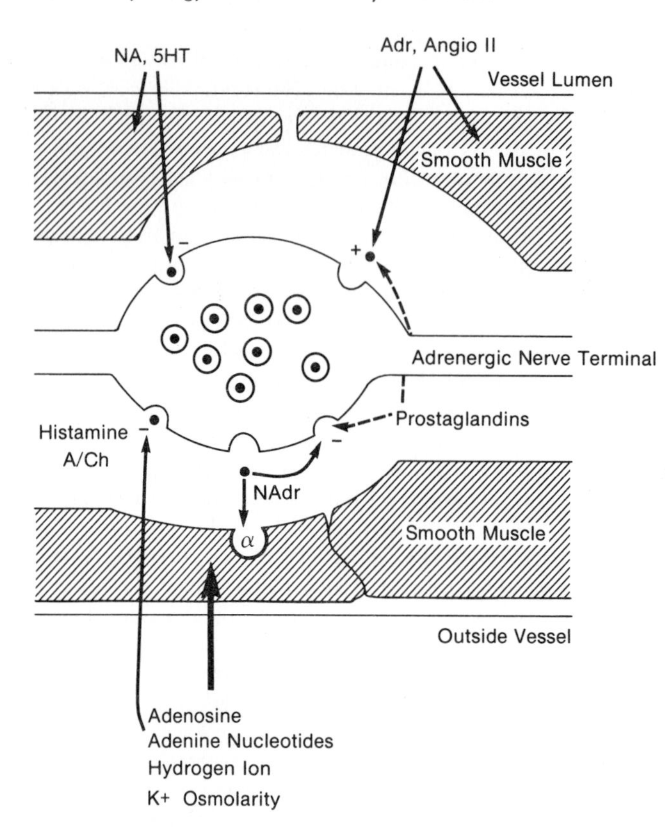

FIG. 6.25. Substances capable of acting upon both smooth muscle cells of vessel wall and adrenergic nerve terminals in blood vessel wall. Noradrenaline (NA), acetyl choline (A/Ch), histamine, 5-hydroxytryptamine (5-HT), adenosine, adenine nucleotides, H^+, and K^+ all inhibit the release of NA, while angiotension II (angio II) and adrenaline facilitate it. Prostaglandins may cause either facilitation or depression. (After J. Shepherd, 1978).

pathetic neurons in the lateral horn of the spinal cord. Dilation is induced in the precapillary resistance vessels of skeletal muscle, but vasoconstriction occurs in the skin and visceral regions; the cholinergic pathway thus ensures a substantial redistribution of blood flow during physical exercise (Folkow, 1960). In the skin, the cholinergic fibers stimulate the sweat glands, which in turn secrete an enzyme producing a vasodilator substance, either bradykinin or a close homolog of this compound (H. Barcroft, 1960).

Mechanical Factors

Some changes of vessel caliber have a mechanical basis. An increase of intravascular pressure stretches the blood vessel and [through the Laplace relationship, equation (1.17)] creates an unstable situation of increasing diameter and diminishing wall tension (A. C. Burton, 1954; Ashton, 1963). In reaction to this, there is an increase of tone in the smaller arteries, brought about by either a spontaneous contraction of the smooth muscle ("myogenic activity"; G. C. Patterson and Shepherd, 1954; Folkow and Neil, 1971) or a reflex initiated from the veins ("venivasomotor" response; Yamada and Burton, 1954). In the capacity veins, the opposite type of effect occurs; a reduction of transmural pressure secondary to arteriolar constriction leads to a collapse of the vessel wall, with passive displacement of the blood contents into more central venous reservoirs (J. H. Mitchell, et al., 1958; Braunwald and Kelly, 1960; Folkow and Neil, 1971).

Overall Response

Under resting conditions, the sum total of sympathetic activity leads to partial constriction in a substantial proportion of both arterioles and veins. This background of "vasomotor tone" is important for both the preloading of the heart (Bassenge et al., 1978) and the regulation of systemic blood pressure, since either the viscera or the skin plus muscles could readily sequester the entire blood volume (Marx et al., 1967). If local blood flow needs are increased, there is scope for vasodilation to occur through a release of vasomotor tone, whether centrally mediated or occurring as a local response (to an increase of temperature in the skin or an accumulation of metabolites in the active muscles; D. E. Donald, Rowlands, and Ferguson, 1970). In the skin and the viscera, release of preexisting tone yields a very large increase of perfusion, but in skeletal muscle it produces only a doubling of blood flow. Other factors that contribute to the exercise hyperemia of muscle include an active cholinergic vasodilation and (in severe stress) stimulation of β receptors.

Adaptation to Exercise

Overview

The total cardiovascular adaptation to exercise comprises an increase of cardiac output (discussed above), arteriolar vasodilation in the active muscles, vasoconstriction in the viscera, inactive muscles, and —depending upon skin temperature—the skin, with a reflex increase of tone in the venous capacity vessels (Shepherd, 1966; Hanke et al., 1969); this last change is reversed by a rise of local temperature (Rowell, Brengelmann et al., 1971; Zitnik et al., 1971).

Patterns of Flow Redistribution

The resting muscle blood flow is low (1–2 ml • min^{-1} per 100 ml of tissue, a total of ~500 ml • min^{-1}, or less than 10% of cardiac output). However, during maximum exercise, ~20 kg of active muscle accommodates a flow of up to 20 l • min^{-1} (80% of cardiac output, a flow of ~100 ml • min^{-1} per 100 ml of tissue). In contrast, we have noted that under resting conditions the skin accounts for ~5% of oxygen consumption while receiving 10% of cardiac output, while the kidneys account for 10% of oxygen consumption but receive 25% of cardiac output.

During vigorous exercise, the blood flow to the viscera is progressively reduced (Rowell, 1971). This change is particularly obvious if the environment is hot, and under such circumstances the visceral flow may fall to less than a third of its normal resting value.

The blood flow to the skin is not reduced in moderate exercise; indeed, it usually increases as more heat must be eliminated from the body. However, a rather sudden decrease occurs as the maximum oxygen intake is approached (Shephard, Allen et al., 1968a). Certain drugs, such as the amphetamines, produce cutaneous vasoconstriction at a lower intensity of exercise. Performance may possibly be improved thereby, but the penalty is an impairment of heat elimination that is reputed to have cost the life of at least one international competitor.

Neural Integration

During exercise, the initial "command" signal to the vasomotor centers probably originates in the cerebral cortex. It leads to a decrease of parasympathetic discharge and a simultaneous increase of sympathetic nerve activity. There is a short-lived vasodilation in all of the body musculature (Bevegård and Shepherd, 1967), but after ~10 sec vasoconstriction develops in the inactive muscles. This presumably implies a stimulation of adrenergic α receptors. In the working muscles, sympathetic activity is largely overridden (Rein, 1930) by responses to local metabolites (particularly potassium ions; Beaty and Donald, 1977). During moderate activity, there may be some response to sympathetic nerve stimulation (Strandell and Shepherd, 1967) but with more vigorous work the local blood flow becomes unresponsive to sympathetic nerve section, administration of noradrenaline or angiotensin (H. Barcroft and Swan, 1953; D. E. Donald, Rowlands, and Ferguson, 1970; Burcher and Garlick, 1975). Sensitivity can be restored by administration of Ca^{2+} ions, and it seems that contraction of the arteriolar smooth muscle is prevented by disturbance of the normal K^+/Ca^{2+} balance with consequent depletion of intracellular calcium (Beaty and Donald, 1977). As noted above, cholinergic activity is mainly responsible for constriction of the visceral and cutaneous vessels.

The contribution of baroreceptors to the exercise response is still debated. Some authors (for example, Asmussen and Nielsen, 1955) have held that at the beginning of exercise a local vasodilation in the active muscles unloads the baroreceptors, providing a prime stimulus to the cardiac and vasomotor centers. Others have suggested that the baroreflexes hold in check responses to enhanced sympathetic nervous activity (Bevegård and Shepherd, 1967). Since any fall of blood pressure at the onset of exercise is slight (Holmgren, 1956) and qualitatively unchanged by denervation of the pressor receptors (Krasney et al., 1974), the first mechanism seems unlikely. However, denervation does decrease the heart rate response to exercise, suggesting that the pressor receptors play more than a secondary "braking" role during physical activity.

The close relationship between cardiac output and oxygen intake (Fig. 6.7) implies some linkage between metabolism and the regulatory process. It is less clear as to whether maximum cardiac output is limited by a baroreceptor "brake." Differences of maximum cardiac output with varying degrees of cutaneous vasodilation (Saltin, 1973a,b) suggest that this may be the case.

BLOOD FLOW TO SPECIFIC REGIONS

Measurement of Peripheral Blood Flow

Venous occlusion plethysmography is the standard technique for measuring blood flow to skin and muscle. The method works best on the upper limb. A classical plethysmograph consists of a rigid metal or perspex cylinder with a tightly fitting rubber sleeve at either end. The arm is slipped through both sleeves, and the cylinder is filled with water at a "neutral" temperature (33–34°C). A change in the dimensions of the enclosed limb segment is transmitted by the water to a suitable recording device. The veins are first emptied by raising the limb above heart level. Venous return is then blocked by inflating a blood pressure cuff to 6.7–8.0 kPa (50–60 mmHg), and blood flow is deduced from the initial rate of swelling of the enclosed segment.

In some applications, both hand and forearm are enclosed by the plethysmograph. However, it is more usual to measure flow to the hand and the forearm separately. The hand is largely representative of skin blood flow, while the proximal part of the forearm reflects mainly muscle blood flow (Shepherd, 1963). For the forearm measurements, the return of blood from the hand is obstructed by inflating a second cuff, distal to the plethysmograph, to above arterial pressure.

The water-filled plethysmograph is at best heavy and cumbersome. It severely restricts movement of the limb and commonly discharges its water content at an inopportune moment. It is thus increasingly replaced by a series of two or more mercury-in-rubber strain gauges (Whitney, 1953). Fine silastic capillary tubes are filled with mercury. Their electrical resistance then varies with the extent to which they are stretched, so that if they are fitted around a limb, they can be used in the same manner as a classical plethysmograph. The strain gauges have several important advantages. In particular, they are light and can be worn during vigorous exercise without impeding normal heat loss. Unfortunately, the quality of records obtained during physical activity is poor, since muscular contractions themselves change limb dimensions. It is thus quite common to measure flow immediately following activity (Simmons and Shephard, 1971b). Recovery data give an idea of vascular tone during activity, but fail to take account of any restriction of perfusion by the contracting muscles (Lind and McNicol, 1967).

Methods that can be used on humans during exercise are remarkably few. Values based on plethysmography (J. E. Black, 1959), indicator dilution techniques (Agrifoglio et al., 1961), arteriovenous oxygen differences (Zierler, 1961; Mottram, 1971), and flow sensors inside or outside arteries or veins are generally "contaminated" by information relating to skin and bone blood flow, although the contribution of the skin vessels can be reduced by local applications of adrenaline (iontophoresis; Wahren, 1966). The rate of heat loss from an electrically warmed intramuscular probe gives a qualitative index of muscle flow (Grayson, 1952; Hensel et al., 1954; Perl and Cucinell, 1965; Levy et al., 1967; Golenhofen and Felix, 1972). However, the responses observed vary considerably with displacements of the exploring needle toward or away from major arteries (Shephard, 1957b). The rate of removal of radioactive materials (such as $^{24}NaCl$ or $^{133}Xenon$) indicates flow through the nutritive capillaries of a muscle (P. Howard et al., 1955; Clausen and Lassen, 1971); blood flowing through arteriovenous anastomoses does not contribute to the clearance of a radioactive depot. Disadvantages of the radioisotope technique are that counting must be averaged over a minute or more to overcome random fluctuations of radioactivity, and measurements are distorted if the relationship between the limb and the counting device is altered (as is likely during physical activity). Electrical impedance measurements have been used on occasion to examine both limb blood flow and thoracic blood content (Nyboer, 1959; Denniston et al., 1976). Problems arise from nonuniformity of tissue impedance and (during exercise) sweating and displacement of electrodes (Glaser and Shephard, 1963), although such difficulties can be reduced by multiple electrode systems (B. H. Brown et al., 1975). We may conclude that the ideal method of measuring blood flow to an active limb has yet to be devised.

Muscle Blood Flow

Overall Flow

The overall regulation of muscle flow has been discussed above. An early cholinergic vasodilation readies the muscle for activity; this is rapidly countered by an adrenergic discharge that prevents overperfusion of inactive muscle groups, although in the working muscles accumulation of metabolites soon produces a functional sympathectomy.

A single muscle twitch gives a local increase of flow in 0.5–1.0 sec (Corcondilas et al., 1964), but with repeated twitches vasodilation builds to a level proportional to the intensity of contraction over a period of 30–60 sec (Tønnesen, 1964; Piiper et al., 1968; Clausen and Lassen, 1971). The white muscle fibers (up to 80% of the total muscle mass in a sedentary subject) have a resting flow of 2–3 ml • min^{-1} per 100 ml of tissue, and in vigorous effort this can increase to 40–60 ml • min^{-1} per 100 ml of tissue. The red fibers have a much higher resting flow

(30–50 ml • min^{-1} per 100 ml), with a possibility of increase to ~150 ml • min^{-1} per 100 ml in vigorous work. Sympathetic nerve stimulation leads to an 80% reduction in the perfusion of white muscle, but only a 50% decrease in the perfusion of red fibers.

Vasodilator Metabolites

Discussion continues concerning the most important local vasodilator metabolite (Mellander and Johansson, 1968; Haddy and Scott, 1968; Verhaeghe et al., 1978). A number of likely candidates, such as bradykinin and lactate, have now been excluded, but advocates can still be found for many other possibilities, including a local decrease of oxygen tension (Granger et al., 1975; Stanbrook, 1978), pH (Haddy and Scott, 1975), and extracellular accumulation of potassium ions (Haddy and Scott, 1975; Beaty and Donald, 1977), changes of osmotic pressure (Lundvall, 1972), and an accumulation of adenosine triphosphate products (Clausen, 1973; Forrester and Hamilton, 1975).

Lorne Owen (personal communication) found that under the conditions of his experiments (pH 7.16, pO_2 2.5 kPa), muscle vasodilation was abolished if changes of pH and pO_2 were prevented. Others (Duling and Pittman, 1975) maintain that under normal conditions the change of pO_2 is insufficient to cause hyperemia. From Fig. 6.25, it would appear likely that several mechanisms are involved, for example, an early release of potassium ions and a later reaction to metabolic factors such as the accumulation of hydrogen ions (Haddy and Scott, 1975). Although potassium ions interfere with the transport of calcium ions to the arterial wall, a comparison of thresholds suggests that much of the observed effect is mediated by interference with noradrenaline release (Verhaege et al., 1978). Certainly, there is an early release of potassium ions proportional to the severity of exercise (Kjellmer, 1965a; N. S. Skinner and Costin, 1969), while if activity is sustained the K^+ concentration falls at the point where the arterioles are again beginning to respond to sympathetic nerve stimulation (Beaty and Donald, 1977). There may also be a local myogenic reaction to the increase of capillary and venous pressures (P. C. Johnson and Henrick, 1975), while a hyperosmolarity of the vessel contents could interfere with the normal vasoconstrictor response by causing a shrinkage of the muscle cells.

Local Mechanical Factors

Rhythmic Work. Despite vasodilation, the mechanical effects of rhythmic activity impair flow through the working muscles. Flow may increase several fold, but it fails to meet metabolic demand, as can be seen from a further increase of flow during intervals of relaxation (H. Barcroft and Dornhorst, 1949; Fig. 6.26). Folkow et al. (1970) suggested that during rhythmic exercise, the muscles functioned as a peripheral heart, receiving blood during diastole and forcibly expelling it during muscle contraction; calculations suggest that such a muscle pump could provide up to 30% of the energy needed by the circulation during vigorous work (Stegall, 1966). The degree of obstruction to arterial flow depends upon the type of work that is performed, the intensity of effort relative to maximum voluntary force, and the relative durations of contraction and relaxation phases. The optimum rhythm seems a contraction of 0.3 sec followed by a relaxation of 1.0 sec (Folkow et al., 1970). This is determined in part by local stores of ATP and oxymyoglobin, and in part by the desirability of minimizing local venous pressure. Most subjects choose such a rhythm spontaneously during running and cycling.

Clausen (1973) observed that the local flow to the vastus lateralis muscle increased regularly with bicycle ergometer loading to some 70% of the maximum tolerated work rate, but that at higher loads the blood flow remained constant and finally decreased (Fig. 6.27). Xenon-133 clearance was curtailed at a similar heart rate during arm ergometry and leg work, although the limiting absolute work rate was lower in the former case. Clausen argued that one of the factors limiting human performance was the ability to develop a sufficient blood pressure to sustain perfusion of actively contracting muscle. He noted that if maximum bicycle ergometer work was performed in the supine position, the flow was usually less than following vigorous isometric work (Clausen and Lassen, 1971; Koch, 1974). However, if the perfusion pressure was increased by sitting upright, the flow during rhythmic effort could exceed the postisometric hyperemia. Equally, the gain of arm maximum oxygen intake following training of the legs could be explained in part by the ability to sustain a greater rise of systemic blood pressure, and thus a greater perfusion of the arm muscles.

Wahren et al. (1974) reached similar conclusions from measurements of femoral arteriovenous oxygen differences during leg work. However, Bonde-Petersen et al. (1975a,b) compared ^{133}Xe clearance with arteriovenous differences, and found that whereas the former leveled off at 50–60% of maximum oxygen intake, the arteriovenous samples showed flow as increasing to 90–100% of maximum

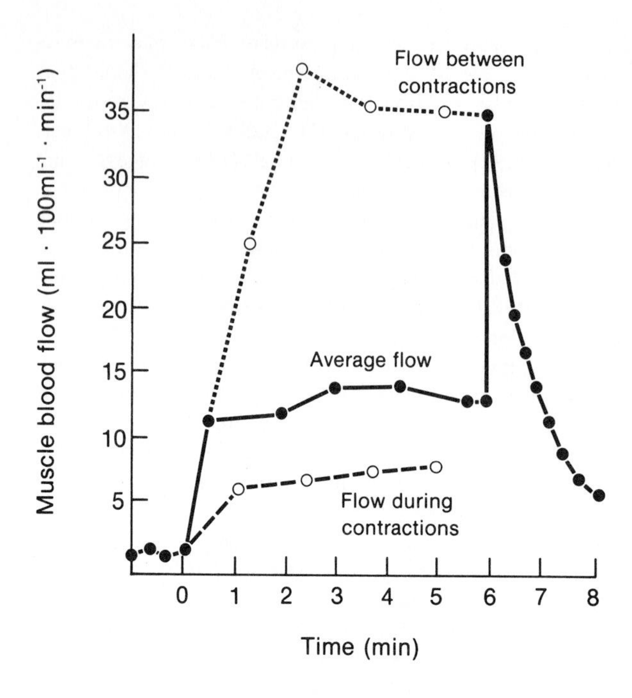

FIG. 6.26. The influence of rhythmic exercise on muscle blood flow. (After H. Barcroft & Dornhorst, 1949.)

oxygen intake. Bonde-Petersen and his associates thus speculated that ^{133}Xe clearance might be limited by diffusion rather than perfusion at high flow rates.

Isometric Work. Isometric contractions start to restrict local blood flow when 15–20% of maximum voluntary force is developed (Royce, 1958; Lind and McNicol, 1967; Maréchal et al., 1973; Bonde-Petersen et al., 1975a,b; Kilbom, 1976). Measurements of torque suggest that this intensity of effort is surpassed during bicycle ergometry (Hoes et al., 1968). Maximum flow is reached at <50% maximum voluntary force, and at 70–80% of maximum the vessels are completely occluded (Royce, 1958). The release of adrenaline and noradrenaline is much greater with isometric than with rhythmic work, presumably because there is a reflex activation of both α- and β-adrenergic systems (Kozlowski et al., 1973; Eklund and Kaijser, 1976). If repeated isometric contractions are held for 0.5 sec (Rodbard and Pragay, 1968) or 2 sec (Lind and Williams, 1977), there is little cumulative fatigue, ischemic pain, or postcontraction hyperemia. However, with a 4-sec contraction/8-sec recovery cycle, fatigue

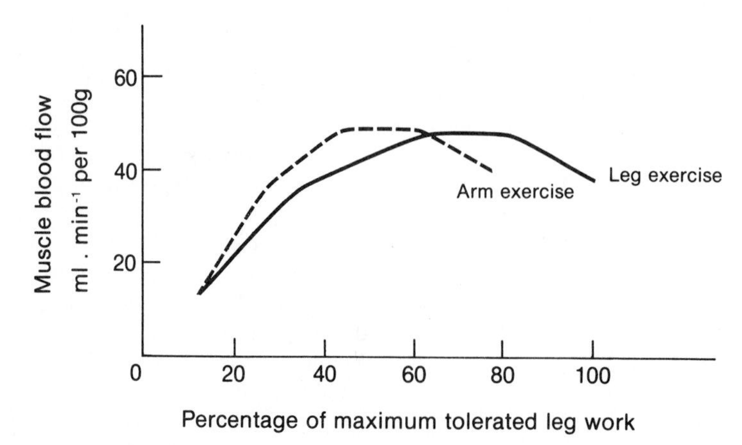

FIG. 6.27. Muscle blood flow (vastus lateralis) as estimated by ^{133}Xenon clearance during various intensities of bicycle ergometer work. (After Clausen, 1973, schematic.)

develops and postexercise blood flow becomes proportional to the number of contractions that have been made (Lind and Williams, 1977). Presumably, in the latter situation the depletion of O_2, ATP, and CP is more than can be made good during recovery intervals (R. C. Harris, Hultman et al., 1975). Measurements of regional perfusion during sustained isometric effort shows that elevation of systemic blood pressure increases blood flow to the skin and the muscles of the inactive limbs, but at best sustains flow to the working muscles (Kilbom and Brundin, 1976). Circulatory responses to isometric work are less well marked if the static contractions are initiated while dynamic exercise is in progress (Kilbom and Brundin, 1976).

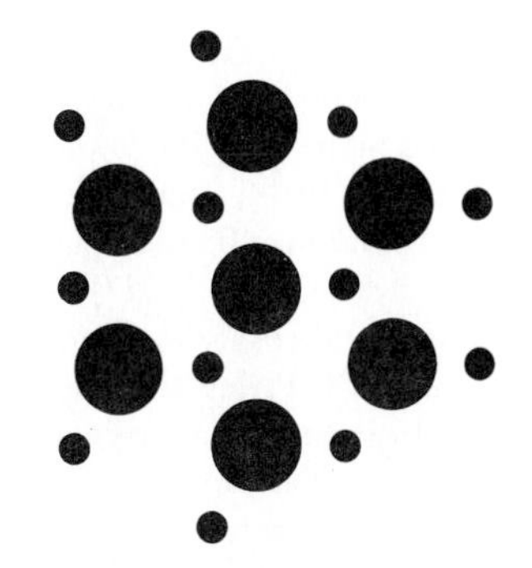

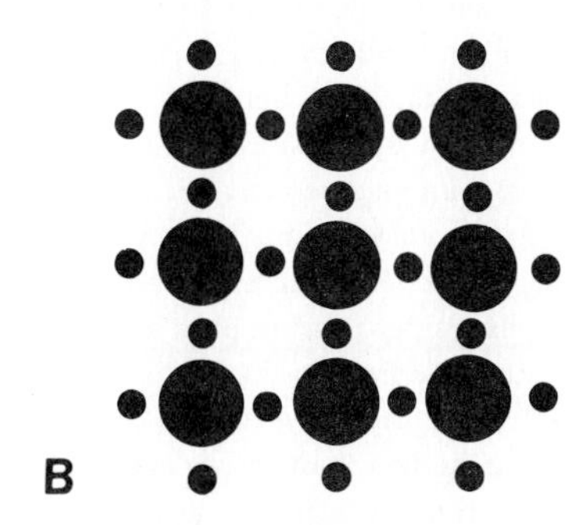

FIG. 6.28. Two possible interrelationships of capillaries and muscle fibers. (After Plyley & Groom, 1975).

Capillary Blood Supply

Anatomical Arrangement

The role of the arteriovenous anastomoses in bypassing the capillary microcirculation of the muscles will be discussed in a following section. We shall discuss here the behavior of the muscle capillaries proper, with particular reference to physical activity.

In some instances, the terminal arterioles give rise to capillary-like "preferential channels" (Elias and Pauly, 1960), but in other places the capillaries arise directly from the arterioles. The preferential channels of skeletal muscle are often guarded by a muscular precapillary sphincter. A typical channel supplies 8–10 true capillaries. The latter generally run parallel with the muscle fibers, having a length <1 mm (Honig et al., 1977) and a diameter of 3–20 μm. Lipid-soluble substances such as oxygen and carbon dioxide can diffuse through the thin endothelial cell membranes (Fig. 6.28) lining these vessels, but water-soluble materials can only pass through pores in the intercellular cement substance (Landis and Pappenheimer, 1963; Karnovsky, 1969; Trap-Jensen and Lassen, 1971). The cytoplasm of the endothelium contains numerous 20- to 30-mm vesicles that may facilitate the blood/tissue exchange of macromolecules (Zweifach, 1973).

Capillary Counts

The large increase of muscle blood flow during, and immediately following, vigorous exercise is matched by a large increase in the number of patent capillaries within the active muscles. August Krogh (1918/19, 1929b) described a 20-fold increase of capillary density (from 5 vessels per mm^2 at rest to >100 vessels per mm^2 during exercise). Krogh and other early investigators faced serious technical difficulties in making accurate counts. Reported maxima for the cat gastrocnemius muscle, for example, have ranged from 379 vessels per mm^2 to 2341 vessels per mm^2 (Plyley and Groom, 1975). Rapid freezing, with the use of high perfusion pressures and low viscosity perfusates, has greatly improved the precision of morphometric work, minimizing tissue distortion (Goldspink, 1961; M. J. Moore et al., 1971) and incomplete filling of capillaries (Plyley and Groom, 1975). Recent results have been discussed in relation to capillary/mitochondrial oxygen pressure gradients, fiber type (Plyley and Groom, 1975), and training (Reitsma, 1973). Plyley and Groom found little difference of vascularity between muscles that were predominantly composed of white or red fibers, and they commented that previously reported differences were based on either unacceptably low capillary counts for white muscle (incomplete perfusion) or inexplicably high counts for red muscle. Under resting conditions, a figure of ~200 vessels per mm² is likely (Honig et al., 1970), and while exercising 500–600 vessels per mm² may be anticipated. P. Andersen (1975) and Brodal et al. (1977) described a small effect of training; after allowance for technical problems (Reitsma, 1973; Plyley and Green, 1975), such as tissue shrinkage and the appearance of artificial spaces between capillaries, values of 325 and 460/mm² were noted for untrained and trained subjects, respectively.

Given 30 kg of muscle and 600 vessels per mm², the effective surface area of the muscle capillaries would amount to 300–600 m², accounting for the substantial maximum diffusing capacity of the tissues. Expansion of the capillary bed during exercise shortens the average distance between the vessel lumen and metabolically active sites within the muscle cells, thereby enabling a given oxygen and carbon dioxide exchange to occur with smaller terminal gradients of partial pressure. For example, an increase from 200 to 600 vessels per mm² implies a reduction of intercapillary distance from >70 μm to ~40 μm. In cardiac muscle, the capillary count (2000–4000 vessels per mm²) is higher than in skeletal muscle, and in consequence the intercapillary distance is decreased to 20–10 μm (Henquell et al., 1976).

The estimated proportion of the muscle cross-section occupied by capillaries depends on assumptions made regarding the size of these vessels. Given a diameter of 3 μm, they would occupy only 0.4% of the muscle cross-section, but at a diameter of 8 μm the proportion would rise to an unrealistically high figure of 2.9%.

Capillary/Fiber Ratios

A number of recent authors have expressed their data as capillary/fiber ratios or as capillary supply per muscle fiber. The latter value depends upon fiber architecture and the number of "shared" capillaries. Given a hexagonal fiber lattice, a capillary/fiber ratio of 1:1, and a three-way sharing of capillaries, each fiber would be surrounded by six capillary vessels (Fig. 6.29), while with a square lattice, a capillary/fiber ratio of 1:1, and a two-way sharing, each fiber would have a surrounding ring of four blood vessels.

Most recent reports find three to five vessels per fiber (Romanul, 1965; Mai et al., 1970; E. Eriksson and Myrhage, 1972; Plyley and Groom, 1975). The last authors observed a very constant capillary supply per fiber, irrespective of fiber type, and concluded that differences of capillary count per unit of cross-section reflected differences of muscle fiber size. More recently, Brodal et al. (1977) noted five to six vessels per fiber, with significantly higher counts in (1) trained subjects and (2) red muscles. They attribute differences from earlier results to use of an electron rather than a light microscope to identify individual capillaries.

Transit Times

Capillary transit times vary widely from 90 msec to 43 sec, with a mean of 4.29 sec and a median of 8 sec (Honig et al., 1977); the dispersion of these values is sufficient to upset many of the traditional calculations for the exchange of oxygen between the capillaries and the tissues (Middleman, 1972; C. P. Rose and Goresky, 1976).

Fluid Balance

Some fluid normally "leaks" into the tissues at the arterial end of the capillaries. The hydrostatic pressure in this region may be 4 kPa, compared with ~1 kPa in the tissues. There is thus a net hydrostatic pressure of ~3 kPa tending to produce exudation of fluid. This is opposed by osmotic forces. Given a colloid osmotic pressure of ~3.3 kPa for the plasma and

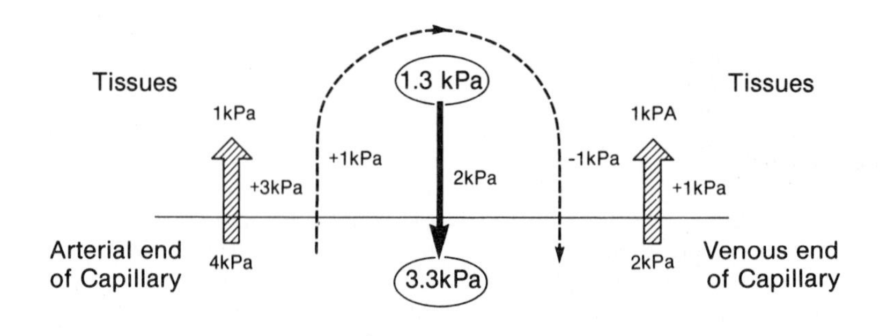

FIG. 6.29. Fluid balance within the capillaries.

~1.3 kPa for the extracellular fluid, there is a net osmotic gradient of ~2kPa favoring absorption of water and a resultant pressure (3 − 2 = 1 kPa) driving fluid from the capillaries into the tissues (Fig. 6.29). At the venous end of the capillaries, the hydrostatic pressure has dropped by at least 2 kPa, while osmotic forces are largely unchanged. Thus, there is a net pressure of at least 1 kPa favoring resorption of fluid at the venous end of the capillary.

Fluid circulates, leaving the vessels at the arterial end of the capillaries and being reabsorbed both at the venous end of widely patent vessels and along the length of vessels where the precapillary sphincter is closed. There is normally little tendency for the tissues to become waterlogged. However, the balance is a delicate one and can be upset by any factor causing an increase of hydrostatic pressure within the capillaries (for example, prolonged standing, incompetence of venous valves, heart failure, pressure breathing) or a reduction of the effective plasma osmotic pressure (starvation and changes of capillary permeability).

The smallest capillaries (3 μm) are smaller than the red cells (6–7 μm). Compression of the erythrocytes enables them to contribute to the extraction of fluid from the tissues (T. Hansen, 1961) and also facilitates gas exchange.

During exercise, several factors favor an increased formation of tissue fluid. The opening up of the precapillary sphincters causes the pressure within the muscle capillaries to rise by 1.0–1.5 kPa, while expansion of the capillary surface also increases the fluid filtration coefficient. The initial loss of fluid from the circulation has been set at 0.4 ml • min^{-1} per 100 ml of muscle or, if 20 kg of muscle are active, 80 ml • min^{-1}. If this rate of loss continued throughout sustained activity, it would represent a serious drain upon a total blood volume of 5 l. Fortunately, the exudation of fluid increases both tissue hydrostatic pressure and plasma colloid pressure, so that the process is rapidly self-limiting. Depending upon the intensity of work, Lundvall (1972) found a total displacement of 19–45 ml • kg^{-1} into the leg muscles over 6 min of bicycle ergometry. With heavy effort, they suggested the cumulative sequestration of fluid could amount to 1100 ml; however, the plasma volume diminished by only 600 ml, since the increased osmolarity of the plasma tended to draw fluid out of other tissues.

Skin Blood Flow

The blood flow to the skin varies widely between 1–150 ml • min^{-1} per 100 ml of tissue. Assuming a body surface area of 1.8 m^2 and a thickness averaging 2 mm, the maximum flow would be a little over 5 l • min^{-1}. In fact, flows of up to 10 l • min^{-1} seem possible in extreme heat (Brouha and Radford, 1960; Rowell, 1974). The increase of flow in a hot environment occurs quite rapidly through (1) a reduction of normal vasoconstrictor tone, (2) some local effect of warmth on the blood vessels, and (3) the formation of bradykinin (if there has been significant sweating). However, any response to cholinergic fibers is constrictor rather than dilator in type.

Much of the increased flow to the extremities bypasses the normal capillary circulation. Blood proceeds directly from the small arterioles to the venules through a series of thick-walled muscular arteriovenous anastomoses. When the anastomoses are open, a large blood flow is directed to the superficial veins and (depending on the environmental conditions) a substantial heat loss occurs from the venous reservoirs. Blood flowing via this "anatomical shunt" bypasses metabolically active tissues, and oxygen extraction is therefore minimal (Shephard, 1968a). The extent of such nonmetabolic flow largely determines the maximum arteriovenous oxygen difference that is reached in exhausting exercise. A well-trained subject loses a larger proportion of his heat production by sweating and less by pumping blood through the arteriovenous anastomoses. Mainly for this reason, he can develop an overall arteriovenous oxygen difference of 16–17 ml per 100 ml of blood compared with 13–14 ml per 100 ml in an untrained individual (Shephard, 1968a).

Under resting conditions, the command signal to the cutaneous circulation is influenced largely by core temperature. However, during exercise the basis of control changes (Table 6.23), both local and general skin temperature playing a much more important role (J. M. Johnson, 1977). It has been suggested that the vasoconstrictor pathway is more responsive to skin temperature, while the vasodilator pathway is more responsive to internal temperature (Wyss et al., 1974). This may explain why the rise of skin blood flow during submaximal exercise shows a lag of ~10 min. During the early phases of activity the normal response to a rising core temperature is countermanded by a strong vaosconstrictor discharge, and this effect is only reversed as skin temperature rises (Fig. 6.30). The increase of skin flow is probably a major factor in the drift of heart rate and stroke volume with sustained work (L. H. Hartley, 1977). Associated increases of wall compliance and intravascular pressure in the cutaneous veins lead to a peripheral sequestration of blood, with a reduction

TABLE 6.23
Influence of Exercise and Local Arm Temperature on Regression Coefficients Relating Forearm Flow to Esophageal Temperature (T_{es}) and Skin Temperature (T_s)

Treatment	Regression Coefficient (ml • min^{-1} • °C^{-1} per 100 ml) T_{es}	T_s
Rest, arm cool	13.05	0.07
arm warm	13.13	0.95
Exercise, arm cool	2.15	1.63
arm warm	2.78	2.34

Source: Based on data of J. M. Johnson (1977) for four subjects completing all treatment combinations.

of stroke volume and a compensatory increase of heart rate (Rowell, 1974).

When working in a hot environment, as much as 25% of the cardiac output may be directed to the skin, and performance is influenced substantially by the ability to shut off this flow as maximum effort is approached. During brief periods of maximum effort, cutaneous vasodilation may bring about a small increase of maximum cardiac output (Saltin, 1973 a,b), but even with the vigorous stimulus of a heated, water-filled suit, the potential skin flow of a vigorously exercising subject is no more than 2.0–2.5 l • min^{-1}. Under normal "warm" conditions, cutaneous vasoconstriction becomes fairly complete as maximum effort is approached since (1) maximum cardiac output is influenced little by environmental temperatures (J. M. Johnson, Rowel, and Brengelmann, 1974), (2) there is little scope for further redirection of blood flow from the viscera to the skin (Rowell, 1971,1974), and (3) maximum oxygen intake is generally unaffected by brief exposure to heat.

A decrease of blood flow to the extremities occurs in the cold. This reflects largely a shutting down of arteriovenous anastomoses, with a shifting of venous return to the deep veins that run alongside the major arteries. The reduction of cutaneous blood flow is particularly marked if physical activity involving intense vibration is performed in the cold. Spasm may affect not only the arteriovenous anastomoses, but also the vessels supplying the metabolic needs of the digits. Some individuals are particularly susceptible, and a distinct syndrome (Raynaud's disease) is recognized by clinicians (Allan, Barker, and Hines, 1962). The usual descriptions refer to men operating pneumatic drills, but the condition is also quite common in forestry workers operating chain saws. The critical frequency of vibration is high (~100 Hz), and attempts to eliminate such frequencies from mechanical equipment have not been particularly successful. Not all workers are affected; there seems to be an undue sensitivity to both vibration and cold in those who are susceptible to the phenomenon.

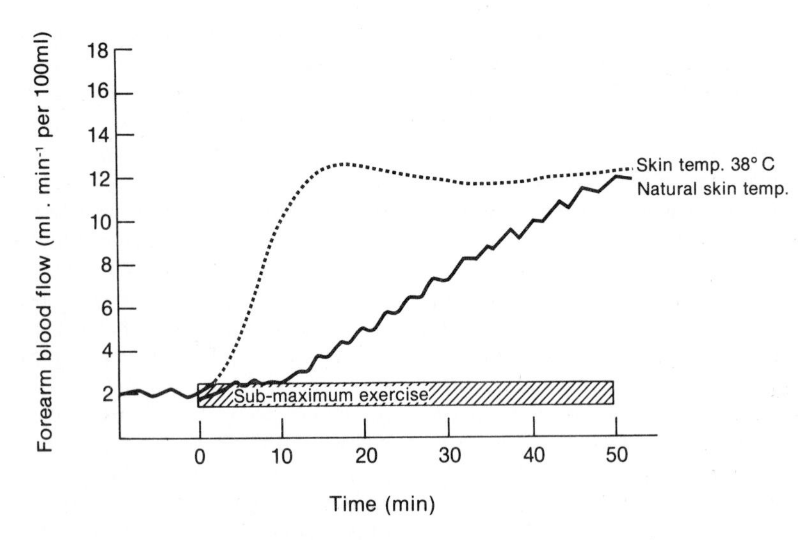

FIG. 6.30. Schematic representation of results of J. M. Johnson (1977). With the naturally occurring skin temperature, skin blood flow (indicated by forearm flow) rises after a lag of about 10 minutes. If the skin temperature is held at 38°C by use of a water-filled suit, the skin blood flow climbs rapidly and generally reaches a plateau value despite continuing increases of esophageal temperature.

Visceral Flow and Function

We have noted above that a number of the visceral organs receive more than their fair share of the total cardiac output at rest. The kidneys, for example, take 20% of the resting cardiac output (a flow of ~1.1 l • min^{-1}), with an arteriovenous oxygen difference as low as 15 ml • l^{-1}. When exercise is performed, the flow to all of the viscera is drastically reduced—particularly in the heat (Rowell, 1971, 1974). Renal flow may drop to 0.25–0.3 l • min^{-1}, and although a decrease of renal filtration reduces local oxygen consumption somewhat, the arteriovenous oxygen difference is increased by a factor of 3–4.

The changes of visceral flow are brought about mainly through an altered balance of parasympathetic and sympathetic nerve discharge. However, the catecholamines also have a direct constricting effect upon the visceral arterioles, and to the extent that blood levels of catecholamines are increased by physical activity, these hormones contribute to the regional redistribution of perfusion. Rowell (1977) has commented on a close inverse correlation between blood levels of noradrenaline and the reduction of visceral flow.

The maximum possible redistribution, perhaps 2–3 l • min^{-1}, 10% of the total exercise requirement of 20–30 l • min^{-1}, is less than the potential redistribution from dilated skin vessels. The extent of visceral flow redistribution at a given work load shows substantial interindividual differences, but variation is reduced if the exercise stress is expressed as a percentage of the individual's maximum oxygen intake.

The reduction of visceral flow inevitably leads to some reduction in the function of the organs concerned. In one 85-km ski race, Castenfors et al. (1967) saw a 30% reduction in the renal clearance of creatinine, with a consistent proteinuria. He explained the latter on the basis of (1) elevation of body temperature (since fever is often accompanied by albuminuria) and (2) a constriction of the efferent arterioles of the renal glomeruli with a resultant increase of filtration pressure and a stretching of "pores" in the capillary wall. Others have described a reduced renal excretion of paraamino hippuric acid and a reduction in the hepatic clearance of indocyanine green during vigorous activity (Rowell, 1971). However, the cellular metabolism of the liver seems unchanged (Rowell, 1977). In general, the effects of exercise-induced visceral ischemia are reversible, although enzymes from the renal and hepatic tissues are liberated into the bloodstream, and the urine may contain red cells and casts in addition to protein (Wyndham, Kew, et al., 1974; Castenfors, 1977; O'Donnell, 1977; Wyndham, 1977).

Flow to Other Organs

Brain

The cerebral blood flow is about 0.6 l • min^{-1}. The total figure varies little with rest, physical activity (Zobl et al., 1965), or mental activity (although it is conceivable that flow is redirected within the cranium to meet varying demands from specific regions of the brain). There have been occasional reports of gains in mental performance during and following vigorous physical activity, but these are probably mediated through an increase of arousal rather than an improvement of local blood flow.

Bone

The resting blood flow to bone is ~0.5 l • min^{-1}. Little is known concerning the flow during exercise. Reeve et al. (1977) noted that mechanical loading of rabbit bone increased local perfusion by up to 40%. However, moderate bicycle ergometer exercise (heart rate 110 beats per min) had no effect on the bone blood flow of human subjects.

Pulmonary Flow

Pressures

Resting pressures in the pulmonary artery are quite low (~2.4 kPa systolic, 1.0 kPa diastolic; Bevegärd, Holmgren, and Jonsson, 1960). The entire pulmonary arterial system is very distensible, and in theory a threefold increase of blood flow could be accommodated without a significant rise of pressure. In practice, a modest increase of mean pressure parallels increases of bicycle ergometer loading (Degré et al., 1972; Tartulier et al., 1972), but even in maximum exercise a young person does not exceed pressures of 3.3/1.3 kPa.

Reservoir Function

The pulmonary arteries and, to a greater extent, the pulmonary veins probably function as a variable reservoir for the left ventricle (Sjöstrand, 1953), although some authors have ascribed to the lesser circulation a permissive rather than a deterministic role in the regulation of left ventricular stroke volume (Giuntini et al., 1971). At rest, the lungs contain ~500 ml of blood (Sjöstrand, 1953; Yu, 1969). Some authors have suggested that this volume is increased up to three-fold during exercise (Sjöstrand, 1953; A. Farhi et al., 1968). However, R. J. Marshall and

Shepherd (1961) stressed that errors could arise when "central blood volume" was measured by the Stewart-Hamilton method due to changes in the extrathoracic component of central volume during exercise. They found "no important increase" in the blood content of the lungs or the left side of the heart when dogs were exercised (R. J. Marshall, Wang et al., 1961). Radiographic appearances suggest a permanent enlargement in the capacity of the upper lobe veins in endurance athletes (Reindell, Kleipzig et al., 1960; Rossi et al., 1977).

Pulmonary Capillaries

Little of the stored blood is in contact with alveolar gas. The blood content of the pulmonary capillaries is no more than 100 ml at rest, rising to perhaps 200 ml in vigorous exercise (Roughton and Forster, 1957). The cross-section of the pulmonary capillary bed undergoes a threefold expansion during vigorous activity, and since the cardiac output reaches about 6 times its resting value, the speed of flow through the capillaries is also increased during physical work. A typical red cell spends ~0.75 sec in the pulmonary capillaries of a resting man (Staub and Schulz, 1968), but only 0.5 sec during moderate work and 0.3 sec in maximum effort (Roughton, 1945; R. L. Johnson, Taylor, and Lawson, 1965; Fishman, 1963; Frech et al. 1968).

Some calculations suggest the transit time in maximum effort is rather brief for a full equilibration of alveolar gas with pulmonary capillary blood (Miyamoto and Moll, 1971). However, it is rare to find a decrease of arterial oxygen saturation during exercise. The explanations include (1) a facilitation of oxygen diffusion by rotation of the red cells, (2) displacement of red cells from the axial stream, and (3) a possible movement of hemoglobin and cytochrome molecules within the red cells (Kreuzer, 1970; E. P. Hill et al., 1973). Under steady-state conditions both the capillary blood volume and the pulmonary diffusing capacity are closely correlated with pulmonary blood flow, but at the beginning and end of exercise changes in diffusing capacity lag behind changes in cardiac output (R. L. Johnson, Taylor, and Lawson 1965).

Pulmonary Zones

When a man is standing upright, four zones of pulmonary perfusion can be recognized (J. B. West, 1977). Because of gravitational effects, the pulmonary arterial pressure falls by ~1.06 cm H_2O for each cm of vertical height. Thus, in the upper part of the lung (zone 1), both pulmonary arterial and pulmonary venous pressures are close to zero (Fig. 6.31). The alveolar gas pressure exceeds pulmonary arterial pressure, and with the possible exception of a few vessels in the angles between alveoli, all of the vasculature smaller than 30 μm is collapsed (Glazier et al., 1969; Rosenzweig et al., 1970). Red cells are trapped and distorted, and there is no significant perfusion of this zone. In zone 2, alveolar gas pressure is intermediate between pulmonary arterial and pulmonary venous pressures, and the system functions somewhat as a sluice (J. Banister and Torrance, 1960; Permutt, Bromberger-Barnea, and Bane, 1962). While venous pressure remains lower than the pericapillary pressure P_c, the flow $\dot{Q}$ is given by

$$\dot{Q} = \frac{P_a - P_c}{R_a + R_c} \tag{6.27}$$

where P_a is the pulmonary arterial pressure and R_a and R_c are the corresponding resistances of the arterial and capillary segments. However, once the venous pressure P_v exceeds pericapillary pressure, the equation becomes

$$\dot{Q} = \frac{P_a - P_v}{R_a + R_v} \tag{6.28}$$

The pericapillary pressure in these equations is normally close to alveolar pressure, although it is also affected by the surface tension of the alveolar lining. Forced inflation lowers the pericapillary pressure, while forced deflation increases it (T. C. Lloyd and Wright, 1960; Bruderman et al., 1964). The boundaries of zone 2 thus depend on the recent "volume history" of the lungs. Small variations of alveolar pressure (with respiration) or pulmonary arterial pressure (with the cardiac cycle; Wearn et al., 1934; G. de J. Lee and DuBois, 1955) also shift the limits of this zone. Flow within the zone increases progressively with distance from the apex of the lung (Fig. 6.31). In zone 3, both pulmonary arterial and pulmonary venous pressures exceed pericapillary pressure. However, flow continues to increase gradually with distance from the apex of the lung, because the rising intravascular pressure distends resistance vessels that are already open and recruits additional vascular pathways (K. T. Fowler, 1965; Maloney, 1965; Fung and Sobin, 1969), probably capillaries rather than arterioles (K. T. Fowler, 1965; Permutt, Caldini et al., 1969). Finally, there is a small zone 4 at the base of the lung. In this region, flow diminishes because some of the larger vessels (>100 μm) are influenced by interstitial pressure developed during

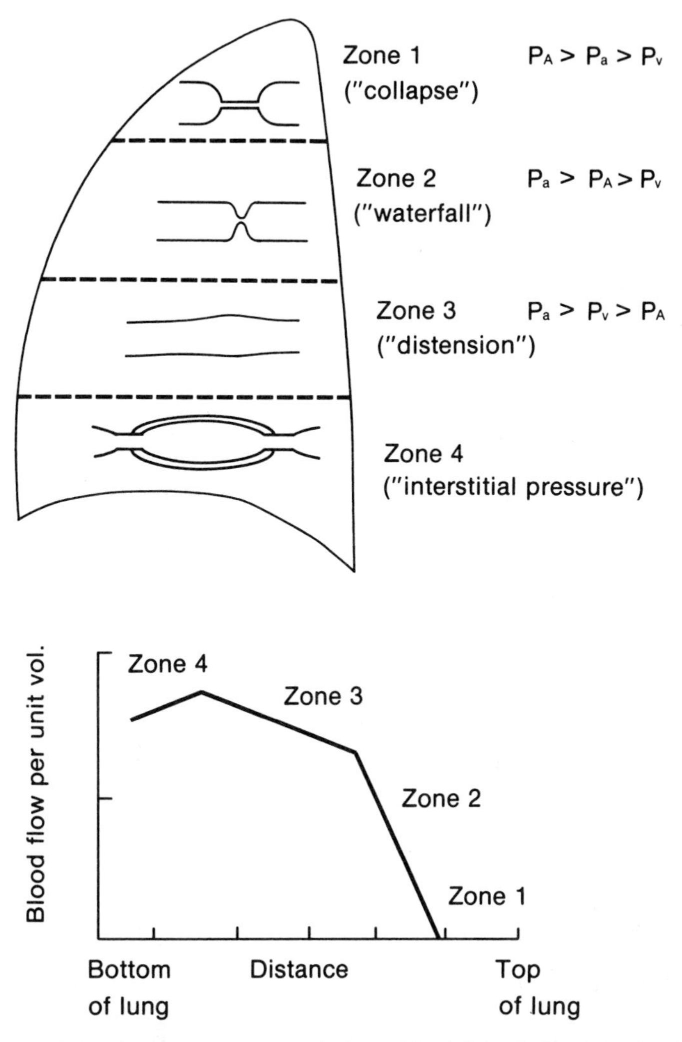

FIG. 6.31. Influence of alveolar pressure upon pulmonary blood flow. In Zone 1, alveolar pressure exceeds pulmonary arterial pressure, and the pulmonary capillaries are compressed throughout their length. In Zone 2, the alveolar pressure is less than the pulmonary arterial pressure, but exceeds the venous pressure; the capillaries are therefore compressed at the venous end. In Zone 3, the alveolar pressure is less than venous pressure, and the capillaries are open throughout their length. In Zone 4, the main flow resistance arises from larger vessels subject to interstitial rather than alveolar pressure. (After J. B. West, 1977.)

collapse of the alveoli (Mead and Whittenberger, 1964; Mead, Takishima, and Leith, 1970; Lambert and Wilson, 1973). The size of this zone naturally depends upon the degree of lung inflation in relation to closing volume (Fig. 6.32; Chap. 5).

During a normal cardiac cycle, ~30% of the fluctuation of pulmonary arterial pressure is transmitted to the pulmonary capillaries, and there is in consequence a minor (< 2 cm) displacement of interzone boundaries. Aging (Holland et al., 1968) and hypoxia (A. Dawson, 1969) both increase pulmonary arterial pressures, reducing the size of zone 1.

Exercise Response

Since exercise raises pulmonary arterial pressure, perfusion of the upper part of the lung is improved by vigorous physical activity (Bryan et al., 1964). The increase of pressure also distends existing vessels and, more important, recruits additional vessels in zone 2. An increase of mean alveolar volume lessens collapse of the lung, reducing the extent of zone 4, but at high levels of lung inflation an increase of tension in the alveolar walls leads to an increase of pulmonary vascular resistance (Roos et al., 1961). Mismatching of perfusion and ventilation also persists at

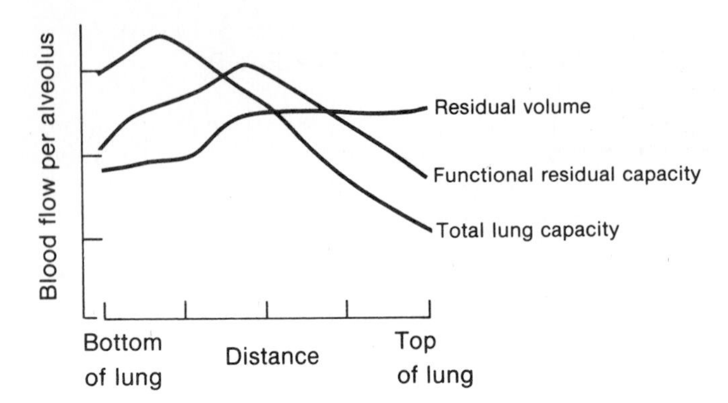

FIG. 6.32. Influence of lung distension upon pulmonary blood flow. (After J. M. B. Hughes et al., 1968.)

a given level within the lung, so that a fraction of the alveolar-arterial gradient remains attributable to an inappropriate spatial and temporal distribution of pulmonary blood flow (Gledhill, Froese, and Dempsey, 1977).

Fluid Balance

Normally, there is little tendency for fluid to escape from the pulmonary capillaries. However, the balance can be upset if "negative" pressure breathing is carried out when capillary permeability has been increased by prolonged oxygen lack (Greenleaf, Bernauer et al., 1978) or chemical poisoning. Oxygen lack also raises pressures in the pulmonary circulation, a factor that may contribute to the development of high-altitude edema (Chap. 12; R. Porter and Knight, 1971).

PROBLEMS OF CIRCULATORY REGULATION

Collapse

Collapse in the final lap or immediately following an exhausting race is by no means uncommon, particularly in a hot environment. The immediate symptoms are due to a failure of blood supply to the brain. Intense dilation of the arteries and veins within the limbs* is supplemented by a decrease of blood volume from sweating and extravascular accumulation of fluid. Cardiac output also falls due to a reduction of venous return (cessation of the muscle pump), and the situation is sometimes exacerbated by a frank vasovagal attack (M. A. Greene et al., 1961). Circulatory catecholamines produce a delayed vasodilation of resistance vessels, and there may be a delayed reaction of central regulatory mechanisms to withdrawal of exercise-induced hypertension (a situation analogous to that of the person who faints after a source of anxiety has been resolved; D. T. Graham et al., 1961). The blood pressure can no longer be maintained, and premonitory local symptoms such as mental confusion, incoordination, or blacking out of part of the visual field (Shephard, Allen et al., 1968a) are followed—often quite rapidly—by loss of consciousness.

No harm results if the patient falls to the ground, fluids are provided, and the body is cooled by tepid sponging (Wyndham and Strydom, 1972). However, more serious consequences may result if a contestant is urged to his feet and continues to struggle toward the finish line. Failure of the sweat glands, a progressive rise of body temperature, renal failure, and exhaustion of the adrenal glands may all occur, sometimes with fatal consequences (Wyndham and Strydom, 1972; Wyndham, 1977; Sulman et al., 1977; Castenfors, 1977).

Deaths from heat collapse and its complications are a recognized hazard of U.S. football games, particularly if the contestants wear nylon clothing that is impermeable to sweat (Murphy and Ashe, 1965; Buskirk, 1968; Spickard, 1968; Goodman, 1968). Problems can also arise in marathon races and long-distance cycling events, particularly if these are held in the heat of the day (Wyndham and Strydom, 1972; American College of Sports Medicine, 1975a; Shephard, 1976b). A more general increase of deaths occurs in the humid cities of the southern United States during hot spells (Schuman, 1972; F. P. Ellis et al., 1975). Heat collapse is also a danger among workers in deep underground mines (Vernon et al., 1927; Wyndham, 1974).

*This situation is in marked contrast to the normal vasovagal attack, where venous tone is increased (S. E. Epstein, Stampfer, and Beiser, 1968).

Shock

Shock was a frequent diagnosis in both world wars, particularly in casualties that had suffered severe hemorrhage (Blalock and Duncan, 1942; Rushmer et al., 1962; Fine, 1965). The condition is best defined as a loss of blood volume to which the body cannot readily adjust. Let us suppose that an experimental animal, such as a cat, is bled repeatedly. At first, the removal of 10–20 ml of blood gives only a transient fall of systemic blood pressure. However, if the bleeding is repeated at short intervals, adjustment becomes slower and incomplete in character. For a time, recovery is still possible if the animal is tended carefully, but ultimately the mean blood pressure is reduced to a critical value (between 5.2 and 6.5 kPa) whereby further removal of blood leads to an irreversible deterioration of condition. In addition, in humans the critical pressure is in the range 5.2–6.5 kPa.

Probable factors making the situation irreversible are an inadequate blood supply, oxygen lack, and loss of intracellular constituents such as purines from vital organs including the heart, brain, kidneys, liver, and intestines (Crowell, 1970). General relaxation of the precapillary sphincters (Mellander and Lewis, 1963), increased capillary permeability (Clarkson et al., 1960), and failure of the sodium pump cause a shift of fluid into the cells (Slonim and Stahl, 1968; R. E. Matthews and Douglas, 1969). Proteolytic enzymes, possibly liberated from the pancreas, react on plasma proteins to yield vasoactive peptides; these depress myocardial function (Gomez and Hamilton, 1964; Lefer, 1970) and cause peripheral vasodilation. The large intestine has an enormous bacterial population, and decline of its barrier function (Bounos et al., 1967; Crowell, 1970) leads to a massive circulatory infusion of microorganisms (septicemia). Under field conditions, infection is supplemented by local contamination of injuries. The situation may later be complicated by exhaustion of defense mechanisms such as the adrenal glands (Motsay et al., 1970) and poor distribution of blood flow within the capillary bed (Appelgren, 1972).

The "shocked" patient has a cold, pale, and sweaty skin, since there is an intense constriction of the subcutaneous blood vessels in an attempt to maintain the systemic blood pressure. At one time, it was common practice to "treat" the cold skin by heavy blankets and even radiant heat cradles. This inevitably led to a dilation of the superficial blood vessels, a further fall of blood pressure, and rapid death. The essential item of treatment is to restore blood pressure. Drugs to improve cardiac function, hypoxanthine to correct purine loss, and oxygen to reverse the causes of hypotension may (in late cases) be of more value than pressor amines and an increase in the volume of circulating fluid (Crowell, 1970). However, in the early stages of shock the best treatment is an appropriately matched blood transfusion (Hackel and Breitenecker, 1963; Rothe, 1970). If this is not immediately available, a physician can give an infusion of plasma or a glucose/dextran polymer mixture (Lepley et al., 1963). Drinks such as tea are of doubtful value. In the absence of blood protein or a high molecular weight polymer, the fluid is rapidly excreted, and if surgical treatment of an injury is required, the drink, may be vomited during induction of anesthesia. However, prevention of hemorrhage, elevation of the legs, and bandaging of the limbs are useful first aid measures.

Pressure Breathing

The term *pressure breathing* is something of a misnomer, since respiration would be impossible in the absence of pressure. What is implied is breathing against a pressure that is higher than normal (positive pressure breathing) or less than normal (negative pressure breathing). Exposure to the pressure may be intermittent (as in various mechanical respirators; Rahn et al., 1951; Elam, 1965; Bergman, 1967; Nunn, 1969) or continuous (positive pressure, as developed by a weight lifter during a Valsalva maneuver, Sharpey-Schafer, 1965, or supplied to an aviator at altitudes in excess of 12,000 m, Ernsting, 1965); negative pressure is encountered by a snorkel diver with a surface line.

Normally, alveolar pressure exceeds pulmonary arterial pressure toward the apex of the lung (Fig. 6.31), but if the alveolar pressure is increased, perfusion becomes impossible in a larger proportion of the lung, with a consequent wastage of ventilation. One remedy is to ensure that the expired gas pressure is low and that expiration occupies at least 50% of the respiratory cycle (Whittenberger, 1955). Alternatively, a respiratory pump may be fitted with a phase of negative pressure (Birnbaum and Thompson, 1942). This assists venous return to the chest and allows a more uniform blood flow distribution in the lung (Kilburn and Sieker, 1960). However, in practice these advantages are outweighed by the onset of collapse in the more dependent and better perfused parts of the lung. Theoretically, the negative pressure also increases the formation of edema fluid within the lung, but this disadvantage is hard to demonstrate experimentally.

The snorkel diver has his entire body surface exposed to a higher pressure than alveolar gas. He is thus engaged in a form of continuous negative pressure respiration. As with intermittent negative pressure, there is a risk of causing pulmonary collapse, and breathing is carried out at a relatively small alveolar volume (Agostoni, Gurtner et al., 1966; Hong, Cerretelli, et al., 1970). If the airway is closed and the external pressure is further increased, then the chest may be compressed to its minimum dimensions (residual volume), yet leaving the air pressure within the lungs less than the fluid pressure within the pulmonary capillaries. Hemorrhage into the lung tissues is then likely (the "thoracic squeeze" syndrome).

The Valsalva maneuver (forcible expiration against a closed glottis; Sharpey-Schafer, 1955) has a triphasic effect upon the circulation. Immediately, the pulmonary veins are emptied into the left side of the heart, causing an increase of cardiac stroke volume and systemic blood pressure. Venous return from the limbs is then checked until the peripheral vessels have filled to match the intrathoracic pressure. Cardiac output and blood pressure fall during this stage (Wolthuis et al., 1974). Finally, venous return is re-established; there may be some compensatory peripheral vasoconstriction (Blair et al., 1959b; McNamara et al., 1969; Rowell, Wyss, and Brengelmann, 1973) and systemic blood pressure rises to slightly above its resting level. A similar sequence of events may be traced if a local difference of pressure is created over parts of the body by wearing a partial pressure suit (as in military aviation). With such equipment, the breathing pressure is quite high (10–14 kPa) and substantial volumes of blood are dammed back in the veins (Fenn et al., 1947; Ernsting, 1965; Wolthuis et al., 1974). Peripheral arterial flow is briefly increased, but later there is vasoconstriction as during the Valsalva maneuver (P. Howard et al., 1955; Shephard, 1957b). Cardiac output falls (Parry, 1958, cited by Ernsting, 1965) as circulatory problems are compounded by mechanical stretching of the baroreceptors (Ernsting and Parry, 1957) and a rapid exudation of fluid from the distended capillaries (Sobel et al., 1959). A vasovagal type of collapse often develops within 1–5 min (Kaufman and Marbarger, 1956; McGuire et al., 1957; Ernsting, 1965). With more moderate stress, adaptation occurs via the renal/endocrine system, secretion of renin, antidiuretic hormone, and aldosterone reducing the excretion of both electrolytes and water (Wolthuis et al., 1974).

CHAPTER 7
Control Mechanisms: Neuromuscular System

PRECISION OF CONTROL

Occasionally, the body may find some advantage in a "wasteful" usage of energy. The sprinter, for example, learns to produce a forced oscillation of his legs at a rate that exceeds the frequency possible with an economical ballistic movement, while a man exposed to extreme cold heats his body by simultaneous contraction of agonistic and antagonistic muscles of the limbs (shivering). In normal circumstances, however, the energy released by the muscle transducers is closely matched to the minimum required to produce a given movement, and the initial energy balance is restored rather precisely in not more than 24–48 hr.

In this chapter we look at the phenomenon of coordinated movement and discuss some of the neuromuscular mechanisms that permit a precise performance of work. Attention is directed to techniques for the assessment of psychomotor performance. We next examine possibilities for control of the muscle at the cellular and spinal levels, and consider the additional scope for coordination offered by the cerebral cortex and cerebellum. We then study the operation of these various control loops in the regulation of muscle tone and the maintenance of body posture, concluding with a brief review of the function of the autonomic nervous system.

PHENOMENON OF COORDINATED MOVEMENT

Reflex and Voluntary Movements

Many authors distinguish "reflex" and "voluntary" movements. Others prefer a more functional classification of postural, locomotor, manipulative, and podal movements (Hess, 1954; Paillard, 1971; K. U. Smith and Smith, 1970). We will first explore the validity of the distinction between reflex and voluntary activity, deferring to the section on visual factors a specific discussion of postural movements.

A reflex movement is in essence an involuntary reaction to an immediate external stimulus (Posner and Davidson, 1978). We may take as an example the "knee jerk," the response to a brisk tap on the patellar tendon. The nature of the resultant movement is dependent upon the site and—with some reflexes—the intensity of the stimulus. A gentle stimulation of the sole of the foot gives an extensor thrust, part of the normal reflex of walking. More painful stimulation of the same area leads to an involuntary withdrawal of the limb.

Reflexes such as the knee jerk, extensor thrust, and flexor withdrawal are coordinated at the spinal level. Other reflexes involve the higher centers of the brain. An example of a centrally coordinated reflex would be the turning of the head toward a sudden loud noise. The reflex is then "conditioned" by the cumulative experience of the individual. If the noise is the slam of a door, and it is invariably followed by the appearance of a pretty girl wearing a deeply slit skirt, the reflex may well persist. Indeed, the sound has had an arousing effect upon the central mechanisms that underlie voluntary control (Posner, 1975). However, if the noise is usually the slam of a door closing, and there are never any interesting sequelae, the reflex may be progressively extinguished. Deconditioning or habituation has occurred.

The central component of reflex responses can be analyzed by a cost/benefit analysis (Posner and Davidson, 1978). Let us suppose that a subject is warned of the direction in which he must move his hand to catch a ball. On the occasions when the information is correct, the reaction time is speeded by the warning, but when the information is incorrect, the reaction time is lengthened. Movement of the eyes toward a centrally placed light is influenced much less by warnings of this type, since a large part of the response is a reflex rather than a "voluntary" reaction to the light stimulus.

In one sense, voluntary movement is also reflex, since there is invariably some identifiable external stimulus. However, the pattern of response depends not only upon the nature and intensity of the immediate stimulus, but also on information stored in the cortex and cerebellum as a result of a lifetime of favorable and unfavorable experiences. Unfamiliar voluntary movements require considerable concentration and thought. A person who is not accustomed to typing or to playing a piano must watch carefully where each finger is placed. But with repetition of the task, the appropriate movement patterns are "learned" and stored (N. Bernstein, 1967; Figs. 7.17 and 7.29). The movement becomes automatic or what some authors have called, confusingly, a "reflex act."

As Hughlings Jackson (1932) recognized, the distinction is not absolute; there are a hierarchy of movements ranging from the "most voluntary" (such as speech and singing) to the "most automatic" (such as respiration). Many automatic movements are undoubtedly "programs" built up by patient learning, but some authors have claimed there is also a significant genetic component (J. M. Cooper, 1971). This view is based on similarities of movement patterns between identical twins, even when learning has proceeded independently. It is likely that the "most automatic" actions, such as breathing, are inherited. Other activities, such as movements of the hand, seem "less automatic," but these also cannot proceed without the help of "more automatic" movements such as postural fixation of adjacent body parts (C. G. Phillips and Porter, 1977). Furthermore, the same muscles in different circumstances can play an automatic or a voluntary role; for example, the thumb muscles contribute to general postural reflex movements of the arm, but more discrete and localized movements cannot proceed without voluntary control, based on the feedback of sensation from the thumb to the sensory cortex of the brain (Marsden, Merton et al., 1976b; Gandevia and McCloskey, 1977a,b).

Description of Movement

The movement patterns needed to perform a given task have many characteristics that are common from one person to another. However, they also embody features peculiar to the individual and his specific situation at the time of observation; in effect, we are looking at a library of motor skills, accumulated over a lifetime and interpreted in the light of the immediate environment.

Careful observation and description of movement patterns is important to those engaged in the teaching of physical activities. Overall body movements may be described relative to a grid (Hutt et al., 1963), while displacement of individual parts can be studied by time-lapse photography (Wuellner et al., 1970). Several systems of "shorthand" (Roebuck, 1967; Jokl, Jokl-Ball, et al., 1970) are available to record movement characteristics. Aspects to be noted include the following:

Breadth of movement. Training broadens the vocabulary of movement, while injury induces a temporary restriction of vocabulary.

Quality of movement. Tense or inhibited movements reflect an increase of resting muscle tone. At other times, movement may be exuberant, with exertion excessive to the task, or forceful, with most of the available motor units called into action to meet an external resistance.

Skill of movement. Skill can improve markedly with training, but it deteriorates with fatigue. With movement of a major body part, note is taken of the general form of movement (clumsy or neat) and the overall mechanical efficiency (see Chap. 1), while with fine motor tasks specific assessments of accuracy may be made (see below).

Dimensions of movement. A correct *tempo* is very important to athletic success. The optimum tempo varies with external factors such as temperature, altitude, and fatigue. For example, if a distance swimmer competing in Mexico City adopts the same tempo of arm movement and the same rate of breathing that he has found successful in Toronto, he will accumulate an exhausting oxygen debt before the race is completed.

Other characteristics. Posture (see below) has an important bearing on both the energy cost of an activity and the effectiveness of "righting reflexes." Note should also be taken of the *size* of a movement; its *direction*, *speed*, and *force*; the *rhythm* (even or irregular); the *consistency* of pattern; and the *body parts* that are active.

Movement and Personality

Movement patterns often reflect the personality of an individual. This is most obvious with regard to the small muscles that control facial expression. However, overall body posture and movement are also influenced by a person's feelings about one's appearance, one's success in a particular situation, and one's apparent role in society (Layman, 1974).

The vigorous stride of the winning athlete, the dejected movements of the loser, the pompous bearing of the professional diplomat, and the slouch of the high school dropout all reflect the immediate or the long-term body image of the individual concerned. The same can be said of sexual differences in movement patterns that appear around puberty. Despite few physiological differences from her male contemporaries, the growing girl learns to feel "unladylike" if she runs to school. Her traditional role in many communities has been to cultivate a slow and languid droop, moving at two thirds the energy expenditure of a boy of comparable age (Durnin and Passmore, 1967).

Movement patterns are profoundly modified by temperament and culture. For example, the Latin people find it necessary to reinforce conversation with vigorous gestures, in marked contrast with the quietly restrained speech of a typical Anglo-Saxon.

There may be a more immediate interaction of personalities between performer and spectator (Layman, 1974; Ingham and Smith, 1974). A nervous apprentice suffers a temporary loss of acquired skills when watched by his supervisor. Equally, a hostile crowd can destroy the confidence of a skilled sportsman (R. N. Singer, 1965; A. R. da Silva, 1970). On the other hand a cheering audience is often needed to realize maximum performance in a top competitor (Steinhaus, 1963).

Inherent Limitations of Movement

Speed and precision of movement are highly specific to a given individual. One competitor is good at accelerating, while another has the ability to sustain speed. Acceleration is probably determined by the shape of the individual's force/velocity curve for his active muscles, and thus by muscle temperature and the proportion of type II (fast-twitch fibers; Chap. 4). Maintenance of speed depends more upon the efficiency of neuromuscular coordination, particularly an ability to restrict unnecessary activity in antagonistic muscles (Marsden, Merton et al., 1977). Speed also shows intraindividual specificity. One subject may be capable of very rapid arm movements, but his leg movements are relatively slow. Again, he may be able to make a rapid forward swing of his arm yet show a rather slow backward movement.

One might anticipate that if a person performed a task inefficiently, speed could be maintained through the use of greater force. In practice, there is little relationship between speed and static strength, unless the task involves movement of a heavy mass. However, there is a good correlation between speed and dynamic strength (the force exerted by a muscle group in accelerating an external load), while both isometric and isotonic training improve an individual's speed of performance.

One might also expect that an increase of flexibility would improve speed by reducing the internal resistance to movement. However, experiments conducted to date do not confirm this hypothesis. Women perform most movements more slowly than men. This reflects partly their lesser muscular strength (Asmussen, 1964), partly the shorter length of their limbs, and partly differences of motivation between the sexes (Durnin and Passmore, 1967).

Because of inertial effects, speed is greater for

smooth, curved trajectories than for sharp, jerky movements. Again because of inertia, speed is inversely related to the mass of the moving parts; adverse factors include a large limb and an excessively heavy bat or racquet. In general, a horizontal movement is performed more rapidly than a vertical one (Rubin et al., 1952), while flexion proceeds faster than extension (Glanville and Kreezer, 1937). Urging increases speed, but decreases accuracy; the mechanism seems an increase of arousal (Chap. 3), and thus of tension in the active part (see below). There is an optimum muscle tone for most activities, and excessive tension due to anxiety or other causes leads to an awkward, stiff, and jerky performance.

Short hand movements are made more precisely if the extremity moves away rather than toward the body, but this trend is reversed for movements of more than 10 cm (J. S. Brown et al., 1948). Precision is improved by the use of both hands rather than one hand and by an increase of external loading (Helson, 1949; W. L. Jenkins et al., 1951); presumably, greater inertia increases the stimulation of muscle spindle receptors (see below). With zero load, performance is uninfluenced by the direction of hand rotation (J. D. Reed, 1949), but if a force is to be exerted, both hands operate more precisely when moving toward the supine position; a right-handed movement is best made in a counterclockwise direction, while the reverse is true of a left-handed movement. Right-handed subjects also find it easier to turn their bodies counterclockwise than clockwise (Vanden Abeele, 1978). For these reasons, a right-handed cricketer prefers to hit the ball to the left rather than the right hand side of the pitch.

In jumping, the right-handed person uses the left leg for take-off. However, when kicking, the right leg is preferred (Vanden Abeele, 1978). This suggests that one limb specializes in lifting movements while the other develops its supporting role (Azémar, 1970).

Types of Movement

The limbs are lightly pivoted, underdamped, and relatively long levers. They thus have many of the features of a compound pendulum, including a natural frequency of oscillation that is proportional to the square root of the distance separating the center of gravity from the pivot (V. B. Brooks, Cooke, and Thomas, 1973).

The extended lower limb vibrates at 2–3 Hz. Lakie and Tsementzis (1978) cite a similar resonant frequency for the wrist. Finger tremor shows a major peak at 0.3 Hz (corresponding to respiratory movements), a 1-Hz peak coincident with heart frequency, and an 8- to 10-Hz peak related to electromyographic activity in the fingers. The extent of tremor is increased by (1) a synchronization of motor unit firing (for example, the forceful efforts associated with an "intentional tremor"; J. C. Joyce and Rack, 1974; Furness and Jessop, 1976; Jessop and Lippold, 1977) and (2) an increase of muscle tone (for example, the overaroused person who is "shaking with fright," and cases of Parkinson's disease whereby normal inhibition of the γ loop via the basal ganglia is weakened; Hagbarth et al., 1970; Gottlieb and Agarwal, 1973).

If the subject is working against a stiff spring, the limb may help to damp out oscillations inherent in the mechanical system (J. C. Joyce and Rack, 1974). However, in most circumstances the tendency to oscillation and thus the limiting frequency of movement is set by the natural frequency of the body parts. One older report (Amar, 1920) cited the following maximum rhythms for the upper extremity:

shoulder, 5–6 Hz; elbow, 8–9 Hz;
forearm, 3–4 Hz; wrist, 10–11 Hz;
fingers, 8–9 Hz.

Once the natural frequency is exceeded, a movement becomes very inefficient. Resonance thus limits the economical pace of running; if the natural frequency has been attained, greater speed can usefully be developed only by a lengthening of stride. The endurance performer respects these precepts, but a sprinter sometimes gains competitive advantage by initiating a fatiguing "forced vibration" at a somewhat higher stride frequency. Rapid movements such as tapping and shaking are initiated by a sudden rise of tension in the concentric muscles; this reveals the natural frequency of the part. A *ballistic movement* (Stetson and McGill, 1923; Desmedt and Godaux, 1978) can be viewed as a single oscillation of the relatively slow-moving pendulum formed by a limb. It is initiated by a sudden contraction of the concentric muscles (Fig. 4.18). One advantage of a ballistic movement is that little muscle shortening occurs during the active phase. For this reason, the muscle is able to develop almost its full isometric tension (Fig. 1.9). The main disadvantage is that the movement cannot be modified by peripheral feedback once it has been initiated. The entire act is controlled by a preexisting, "packaged" motor program (V. B. Brooks, 1974; R. A. Schmidt, 1975; Marteniuk and Sullivan, 1978) that specifies the required force, velocity, or displacement of the part (V. B. Brooks and Stoney, 1971).

A *reciprocating movement* (such as running) involves a series of ballistic strokes. Contraction of the antagonistic muscles continues for sufficient time to halt movement of the limb in one direction and initiate a fresh ballistic stroke in the opposite sense.

A *controlled movement* (such as careful manipulation of a lever) involves damping of the limb by a suitably graded and phased contraction of the antagonistic muscles (Fig. 4.18; Woodworth, 1899; Kozlovskaya et al., 1970; V. B. Brooks, Cooke, and Thomas, 1973). In essence, a slow controlled movement is a recognition task; the body part is kept moving forward until peripheral feedback indicates that a correct position has been reached (R. A. Schmidt, 1976). Vision (Posner, 1967), proprioceptive information (Marteniuk and Sullivan, 1978), and time judgments (Leuba, 1909) all contribute to successful performance (see below).

The electromyograph is a useful tool in deciphering the type of movement (ballistic or controlled). It also serves to identify active muscle groups (Basmajian, 1967) and exposes wasteful use of muscles when performing a particular movement. An increase in the amplitude of activity and an accompanying excess of electrical activity is a striking feature of the electromyogram in a man who has been performing heavy manual work and is suffering from physical fatigue (J. G. Wells, 1963; Gheron, 1970).

TECHNIQUES FOR ASSESSMENT OF PSYCHOMOTOR PERFORMANCE

In this section, we look briefly at some of the techniques used to assess psychomotor performance and then examine the specific contributions of such senses as vision and balance to overall achievement.

Gross Measures of Performance

Some tests assess the overall performance of a physical task. For example, a subject may be required to thread as many bolts as possible into a suitably tapped metal plate over a 1-min period. This task examines finger dexterity, responses being adversely affected by exposure to cold.

Alternatively, the requirement may be to transfer steel ball-bearings from a tray to a series of depressions in an inclined and rotating turntable, using a pair of forceps. This test also depends on manual dexterity and is adversely affected by a lack of hand steadiness. A third type of task requires a subject to match the movement of one pointer or dial with a second pointer or dial, using a control knob or lever. This is known as a "tracking" or "pursuit" test. The movement of the target pointer may be continuous or discontinuous. Performance can be displayed as the discrepancy between the two pointer readings (Fig. 7.1). With a discontinuous task, there is an initial "response time" before the subject notices that the target reading has changed, a period of "travel" when the control lever is being moved rapidly, and a third phase of "manipulation" as the subject attempts to match the target position (Shephard, 1956c). Finally, a certain level of accuracy is accepted. Slight depression of the central nervous system (mild oxygen lack, a small dose of alcohol) decreases cortical control of the movement. The initial response time is then shorter and travel is more rapid, but the matching of the two pointers is less accurate (Shephard, 1956d). Greater depression of the brain leads to a slowing of all phases of the reaction. Usually, the tracking pointer is controlled by manipulation of a knob or light lever, but in one form of apparatus developed by the Applied Psychology Unit at Cambridge University, the control lever is heavily weighted.

Interruption of light beams and a gridlike target can be used to measure both the speed and accuracy of throwing (Malina and Rarick, 1968; Pauwels, 1978).

Reaction Time

The contribution of reaction time to performance was reviewed in Chapter 3. As usually measured, it involves the entire chain of events between stimulating signal and response. For example, an electrical timer accurate to 0.01 sec is started coincident with a cue (such as a noise or light) and is stopped when the subject responds by touching a suitable key.

A good example of this type of test is the brake reaction time device available at many driving schools (G. R. Wright and Shephard, 1978a). The observed interval of 340–400 msec comprises roughly equal components of true reaction time and movement time. Some tasks also require an initial movement of the eyes to feed the signal to the central nervous system (Fujita, 1978). More complex reaction times may be measured, whereby the subject is presented with one of several possible cues and must make a choice before initiating his response (Baxter, 1942; S. N. Blair, 1970; Granit, 1973). This usually, but not always (Carrière, 1978), lengthens reaction time without increasing movement time. On the other hand, if the required response is made more complex, the reaction time is unchanged but the movement time is increased.

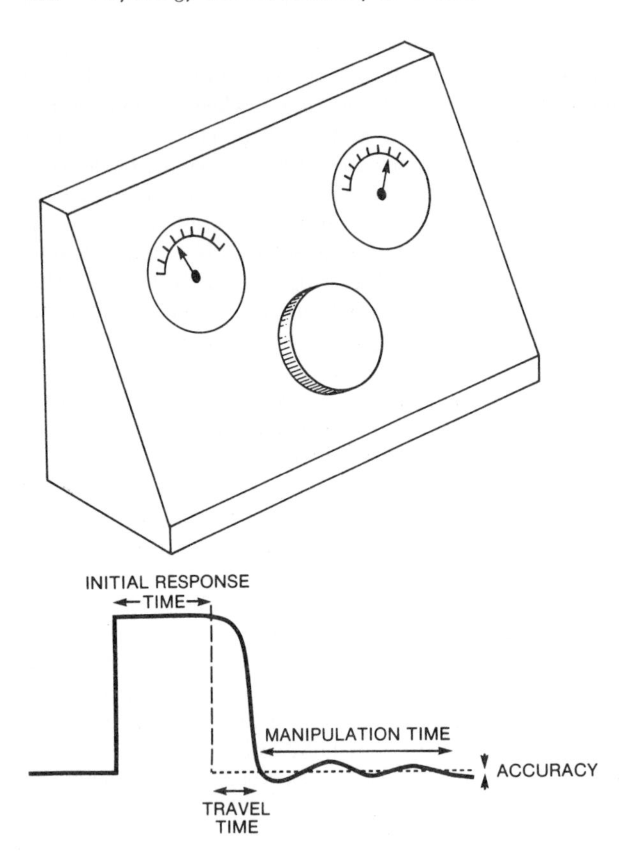

FIG. 7.1. Display of a discontinuous tracking task. The difference between two voltmeter readings that the subject is attempting to match gives a measure of initial response time, travel time, manipulation time, and final accuracy of matching of the two dials. (After Shephard, 1956c.)

Travel or Movement Time

Travel time can be distinguished by fitting a suitable recorder to the accelerator pedal of the brake reaction timing device (G. R. Wright and Shephard, 1978a).

The detailed course of large body movements can be followed by cinematography (Adrian, 1973; Atwater, 1973; Dillman, 1975), use of a stroboscope (Pechar and Nelson,1970), or the interruption of a series of light beams (N. R. Ellis and Pryer, 1959). The task may be "terminated' if it ends in striking an object, but usually the final timing mechanism is also operated by a light beam, so that "nonterminated" tasks with a "follow-through" can be studied.

Further subdivision of reaction time can be accomplished by simultaneous recording of brain potentials and electrical activity in the muscles (Granit, 1973; Evarts, 1973a). Evarts (1973a) found that neurons in the postcentral region of a monkey's brain showed a discharge 10 msec after presentation of a stimulus (movement of a control handle). Some neurons in the motor area of the brain were active at 14 msec, but the upper motoneurons (pyramidal cells) were slower to respond; their activity was depressed at 20 msec and not increased until at least 24 msec after stimulation. The first burst of electromyographic activity at 12 msec is probably a monosynaptic spinal reflex, but the second and third bursts (at 30-40 msec and ~80 msec, respectively) involve the motor cortex.

Visual Factors

Acuity

Visual acuity is usually assessed from the ability to read Snellen test type. Assuming that there is no error of refraction at the lens, a sport requiring visual skill is limited by the angle that can be distinguished at the nodal point of the eye (where light from the target is brought to a focus). The average individual can distinguish two objects that are separated by an angle of 1 min (equivalent to a separation of 4.5 μm at the retina).

The more distant an object, the smaller the angle it subtends; hence, performance can be enhanced by use of a nearby reference point (for example, the spot system in bowling). Many sports do not fully exploit normal visual acuity. For example, a cricket stump subtends an angle of ~15 min at a distance of 40 m.

Kinesthetic Skill

In many rapidly moving ball games, an important element of skill is the kinesthetic sensitivity of the neck and ocular muscles. Receptors in these muscles allow a sportsman to track and assess the speed of an oncoming ball (Sharp, 1978). The relative use of the retina and the muscle proprioceptors depends upon viewing time. If this is <240 msec, a player must rely upon retinal information. However, there is a dramatic improvement of performance with longer viewing times, when use can also be made of the proprioceptors (Dichgans and Brandt, 1972; Sharp, 1978; Fig. 7.2). Information derived from convergence of the eyes is pooled with other impressions such as the increasing size and clarity of the ball and movement of the head. The association areas of the sensory cortex then make what often seems a subconscious judgment, predicting the behavior of the target at the "preattentive" level of processing (Neisser, 1967) through contextual information and cumulative experience. Presumably because stored knowledge is greater, expert performers make fewer visual fixations than nonexperts (Bard and Fleury, 1978).

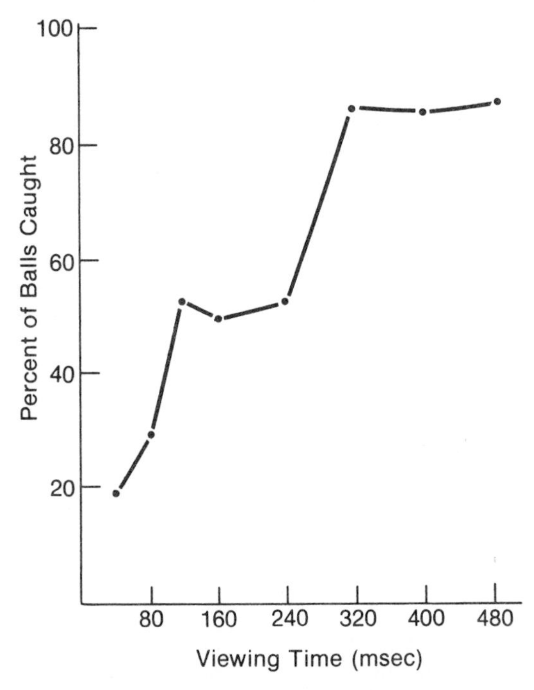

FIG. 7.2. Influence of viewing time upon performance of a ball-catching task. Note the sudden improvement of scores with viewing > 240 msec (After Sharp, 1978.)

Balance

Static Equilibrium

The ability to maintain a relatively fixed posture can be assessed by an *ataxiameter*. This device measures body sway while a subject is trying to stand still (Wapner and Witkin, 1950).

Dynamic Equilibrium

Dynamic equilibrium is tested by means of a *stabilometer* (Travis, 1944; Singer, 1970; Niinimaa, Wright et al., 1977). This is essentially a short board pivoted on a central roller. The subject stands astride the board and endeavors to prevent it from rotating. Scoring is based on the period for which it can be kept off the ground (Fig. 7.3) or the cumulative angular rotation in 30–60 sec (J. R. Richardson and Pew, 1970).

The results for adult men show a moderate coefficient of correlation (r ~ 0.6) with static equilibrium scores. Certain categories of athletes, such as water skiers and gymnasts, achieve much better results than sedentary subjects (Fig. 7.3).

Kinesthetic Sensitivity

Kinesthetic activity is measured in terms of the subject's ability to reproduce a given limb position, to

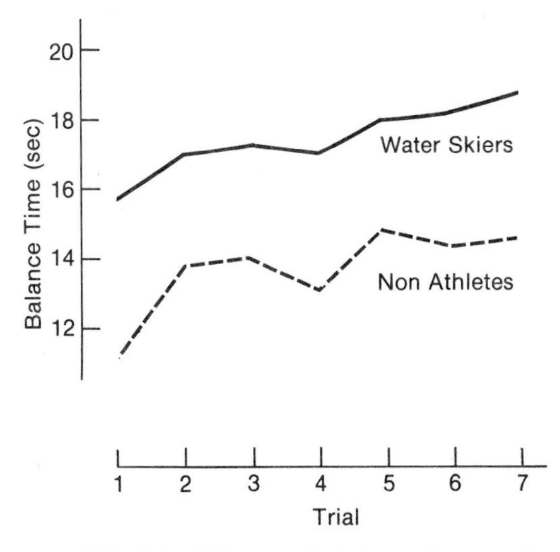

FIG. 7.3. Difference in balance time on stabilometer for waterskiers and nonathletes. (After Singer, 1970.)

detect the amplitude and direction of limb displacement, to duplicate a given muscle tension, or to judge speed of movement (G. M. Scott, 1955; Marteniuk and Sullivan, 1978). Mechanical or electrical goniometers (Adrian, 1973) may be used to record joint angles during such studies.

Hand Steadiness

Hand steadiness can be evaluated by passing a brass probe down a tapered slot (G. R. Wright, Randell, and Shephard, 1973). Note is taken of the distance traveled before electrical contact is made with the side walls of the slot.

CELLULAR AND SPINAL CONTROL MECHANISMS

Receptor Organs

The sensory supply of skeletal muscle has many functions, but perhaps the most important is to detect changes of tension within the muscle and its associated tendinous structures during contraction, relaxation, and forcible stretching movements. The muscles themselves contain spindle organs, free nerve endings, and a few Pacinian corpuscles. Golgi organs are found in tendons and ligaments, while the joint capsules contain Ruffini spray endings and free nerve endings (Eldred et al., 1967; Granit, 1970; P. B. C. Matthews, 1972).

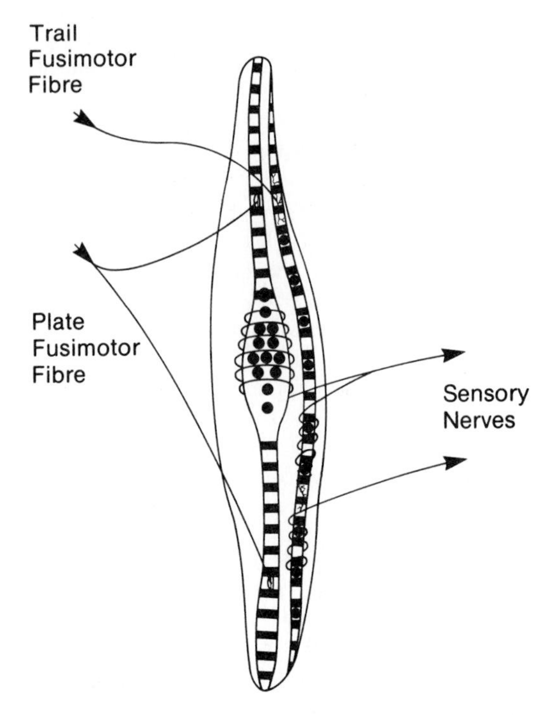

FIG. 7.4. Simplified representation of nuclear chain and nuclear bag fibers within a muscle spindle. For clarity, one nuclear bag and one nuclear chain fiber are shown. However, mammalian spindle receptors usually contain four or five nuclear chain fibers.

Muscle Spindle Receptors

The spindle organs are cigar-shaped structures a few mm in length, embedded between normal muscle fibers and functioning in parallel with them. A typical spindle organ has an outer connective tissue sheath enclosing two types of modified "intrafusal" muscle fibers, a single "nuclear bag" and four or five "nuclear chain" fibers (Boyd, 1962). *The nuclear bag* comprises striated polar regions and a central swelling; the latter contains up to 12 nuclei, as many as 2 or 3 lying abreast of one another (Figs. 7.4 and 7.5). Originally, it was held that myofibrils were confined to the polar regions of the receptor, but most authorities now believe that certain fibrils traverse the central swelling. Two types of nuclear bag have been distinguished on the basis of ultrastructural and histochemical differences (Barker, 1974). Neglecting this detail, we may note that the sarcoplasm of a typical nuclear bag differs from that of an extrafusal fiber in that it contains relatively few mitochondria, there is no M line at the center of the sarcomere, the H lines are poorly developed, and there are few triads (Adal, 1969; Banks et al., 1975). Large medullated afferent nerves send nonmedullated sensory spirals around the central ("equatorial") swelling ("primary" or "annulospiral" endings; P. B. C. Matthews, 1972). The polar regions are innervated by fine motor fibers, most of which end in motor-end plates. The p_1 plates have a typical motor-end plate structure and they are sometimes supplied by β nerves that also innervate extrafusal fibers. The p_2 plates (Barker et al., 1970) are about twice as long as the p_1 plates and show less infolding of the intrafusal membrane, with few soleplate nuclei; they are invariably supplied by γ motor fibers specific to the spindle organs.

The nuclear chain fibers have nuclei distributed along their length. The myofilaments are arranged less regularly than in the nuclear bag fibers; other distinguishing features are an appreciable amount of sarcoplasm, prominent mitochondria, well-developed M and H lines, and a moderate number of triads (Adal, 1969; Corvaja et al., 1969). Histochemical analysis shows a greater content of mitochondrial and myofibrillar ATPase than in the nuclear bag fibers

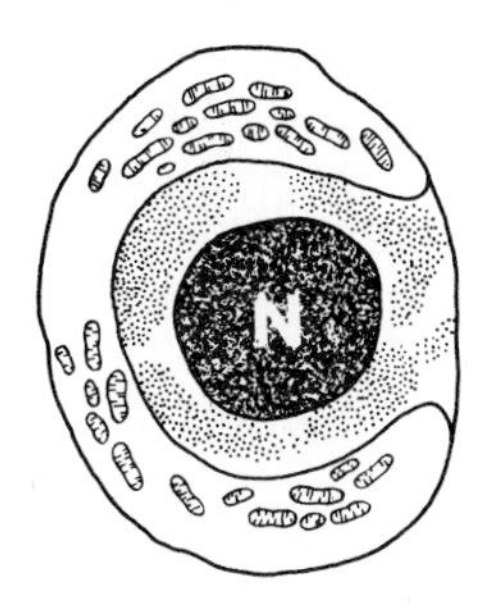

(a) Nuclear Chain

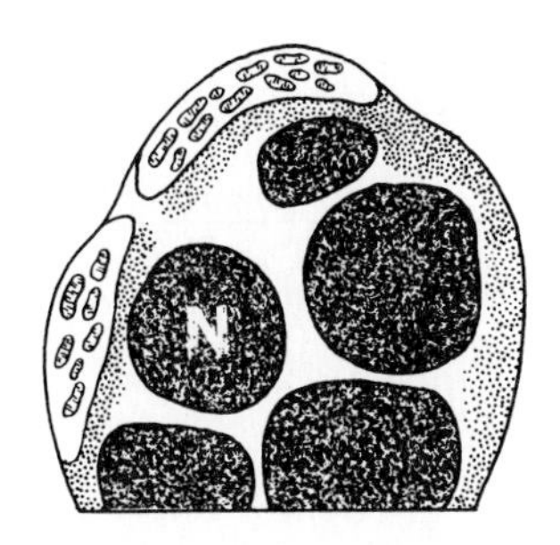

(b) Nuclear Bag (smaller scale)

FIG. 7.5 Microstructure of equatorial region of nuclear chain and nuclear bag fibers. The nuclei (N) are surrounded by myofibrils (more numerous in the nuclear chain fiber). Outside the myofibrils lies the sensory endings with prominent mitochondria. (After Adal, 1969.)

(Spiro and Beilin, 1969; Banks et al., 1975). The sensory receptors consist partly of distinct annulospiral endings and partly of finer "spray" ("secondary") branches (Boyd, 1962; Barker, 1967). There does not seem to be any ultrastructural difference between the two types of sensor, although the secondary endings are juxtaequatorial rather than equatorial in distribution. The motor supply of the chain fibers is from fine γ motor nerves, most of which terminate in "trail endings" (Barker and Ip, 1965). In general, the innervation is independent of that supplying the nuclear bag fibers.

Spindle Distribution

All somatic muscles contain spindles, with the possible exception of the small extraocular muscles in species such as the cat and dog. The number of spindle organs varies from one or two in the smallest muscles to several hundred in large muscles, the density (number per g) being particularly large in muscles concerned with fine movements (such as the muscles of the neck and hand, Tables 7.1 and 7.2; Barker, 1974).

Free Nerve Endings

Free endings are the most numerous type of muscle receptors, being found in close association with muscle fibers, fascia, fat, blood vessels, tendons, and muscle spindles (Stacey, 1969). The degree of specialization of such endings is unknown, but occasionally a single nerve may terminate in several different types of tissue.

Paciniform Corpuscles

The pacinian corpuscles are concerned with touch and are found in the skin. A few "paciniform" structures are found on the fascial sheets between muscles, at musculotendinous joints, and in the connective tissue around joints (Barker, 1967). The basic structure of the receptor is illustrated in Fig. 7.6. A medullated fiber loses its myelin and runs with uniform diameter inside a lamellated sheath.

Golgi Organs

The Golgi organs (Merrilees, 1962; Bridgman, 1968) are simple sprays of fine nerve terminals distributed in the connective tissue of the tendons at the origin and insertion of the muscles and in the intermuscular septa. They are usually connected to a single large afferent nerve (diameter 8–12 μm).

Joint Receptors

The connective tissue of the joint capsule contains free nerve endings, and the ligaments may show Golgi endings; specialized Ruffini and lamellated endings have also been described. The Ruffini ending (Fig. 7.6) is somewhat smaller than the Golgi organ. A medium-size afferent fiber gives rise to several spray-type terminals (Freeman and Wyke, 1967). Each spray is enclosed by a delicate capsule, sometimes containing a fine accessory nerve. The joint capsule contains some lamellated endings,

TABLE 7.1
Density of Muscle Spindles in Selected Muscle Groups

Muscle	Spindle Count	Spindle Density (Spindles per g Mass)
Obliquus capitis superior	141	42.7
Rectus capitis post. major	122	30.5
Abductor pollicis brevis	80	29.3
Opponens pollicis	44	17.3
Pectoralis major	450	1.5
Teres major	44	0.4

Source: Based on data collected by P. B. C. Matthews, 1972.

TABLE 7.2
Approximate Density of Afferent Nerve Endings in Selected Muscle Groupings of Cat

Muscle	Number of Endings per Muscle (a) Primary	(b) Secondary	(c) Golgi	(b)/(a)	(c)/(a)
Rectus femoris	102	100	78	98	76
Soleus	55	43	31	78	56
Semitendinosus	137	162	85	118	62
Medial flexor digitorum longus	51	47	17	92	33
Fifth pes interosseus	27	16	25	59	92
Fourth intercostal (internal + external)	49	67	17	137	35

Source: Based on data of D. Barker, 1962.

which seem in essence smaller versions of the paciniform corpuscles.

Skin

Touch and pressure receptors in the skin (Burgess and Perl, 1973) contribute to reflex responses, possibly by maintaining an adequate level of depolarization at one or more of the synapses that separate the other receptors from the brain (see Fig. 7.16).

Nerve Supply of Muscles

Afferent Fibers

Four categories of afferent fiber are described, although they are less clearly demarcated than the two main types of motor fiber to be discussed below (D. P. C. Lloyd, 1943a). Group I fibers (diameter 12–20 μm, conduction velocity 72–120 m • sec^{-1}) receive information from primary spindle receptors (type Ia fibers) and Golgi tendon organs (type Ib fibers), with some input from Golgi organs in joints and free endings. Group II fibers (diameter 4–12 μm, conduction velocity 24–72 m • sec^{-1}) arise mainly from secondary spindle receptors, with some contribution from paciniform corpuscles, free endings (both muscle and joint), and Ruffini endings (joint only). Group III fibers (diameter 1–4 μm) and group IV fibers (nonmedullated, diameter < 1 μm) are derived solely from free nerve endings in muscles and joints.

Study of the cat soleus (P. B. C. Matthews, 1972) shows an almost 1:1 relationship between receptors and afferent fibers; 50 primary spindle receptors give rise to 50 type Ia fibers, 45 Golgi tendon organs to 40 type Ib fibers, and 50 secondary spindle receptors to 50 type II fibers.

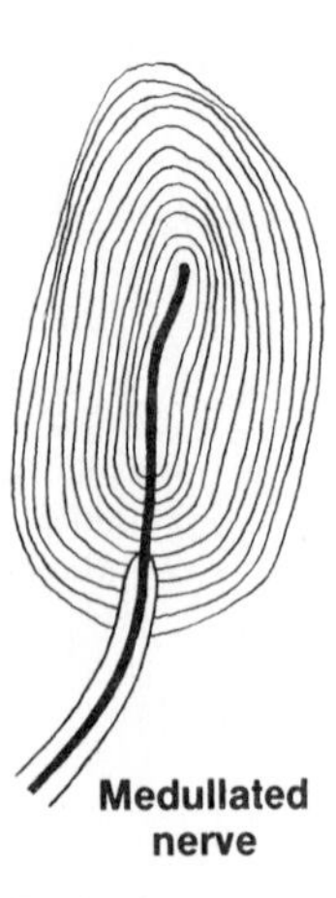

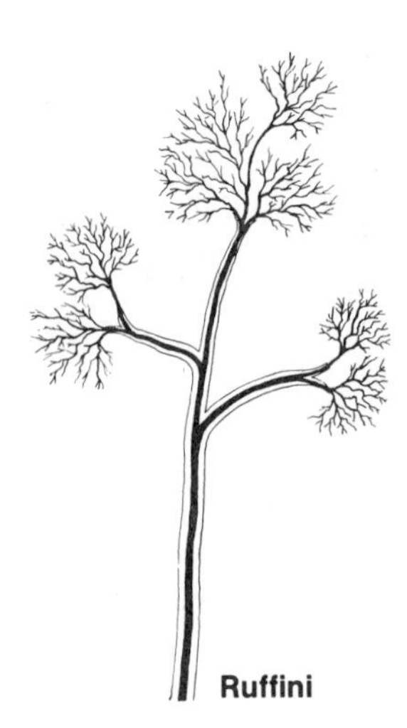

FIG. 7.6. Structure of paciniform and Ruffini type nerve endings.

Motor Fibers

The fibers of the large motor nerves fall rather precisely into two size categories (Eccles and Sherrington, 1930). The large α fibers (diameter 8–20 μm) are responsible for the peak of the sciatic nerve action potential and have a conduction velocity of 50–120 m • sec^{-1}. They innervate normal muscle, each nerve fiber dividing into branches that supply up to 150 extrafusal muscle fibers. The small γ fibers (diameter 2–8 μm) have a conduction velocity of 10–50 m • sec^{-1} (Leksell, 1945). They innervate only the muscle spindles (Langley, 1922), each 100 γ fibers being linked to perhaps 300 intrafusal fibers in 50 spindle organs. Attempts to distinguish γ_1 fibers (supplying the nuclear bag fibers) and γ_2 fibers (supplying the nuclear chain fibers) have yet to be substantiated (Adal and Barker, 1965; Laporte and Emonet-Dénand, 1973); however, there are functionally distinct static and dynamic γ fibers (see above). Nerve fibers that innervate both extra and intrafusal muscle are termed β fibers irrespective of their size.

Afferent Impulse Patterns

Early Concepts

Electrophysiological studies distinguished two types of afferent impulse from the working muscles. These were thought to reflect, respectively, stimulation of spindle receptors and Golgi tendon organs (B. H. C. Matthews, 1933). The discharge attributed to the spindle receptors ceased with muscle contraction, as though receptors and muscle fibers were arranged in parallel. In contrast, the discharge attributed to the Golgi tendon organs increased as the muscle shortened, suggesting a series arrangement of the receptors (Fulton and Pi-Suner, 1928). Furthermore, the threshold tension needed to stimulate the spindle receptors was low, and the rate of change of tension had more impact on the discharge rate than was typical of the impulses thought to be derived from the Golgi tendon organs.

The simple series/parallel arrangement envisaged by Fulton and Pi-Suner (1928) and B. H. C. Matthews (1933) has a number of complications in practice. Only part of a muscle may contract, so that distortion of the muscle could stimulate some of the spindles (Houk and Henneman, 1967); nevertheless, the spindles usually respond to tension developed in the corresponding motor unit rather than in the muscle as a whole (Binder et al., 1976). Simultaneous activation of α and γ motor nerves may cause a parallel shortening of muscle and spindle, so that no stimulation occurs, while γ-nerve activation may produce an afferent discharge in the absence of muscle shortening.

Comparison of Primary and Secondary Endings

Recent investigations have compared the sensitivity of primary and secondary endings within the muscle spindles. Sensitivity to static loading is relatively similar in the two types of receptor, but the primary endings respond much more readily to the rate of change of tension. Some investigators (Jansen and Matthews, 1962) express the latter type of sensitivity as a dynamic index (the decrease in discharge frequency in the first 0.5 sec after completing a stretch at constant velocity).

Other types of responsiveness are also described. The primary endings show what could be interpreted as an acceleration response (a brief burst of impulses as a movement is initiated; Lennerstrand and Thoden, 1968; Schäfer and Schäfer, 1969). The sensitivity to brief phasic stimuli (such as a tap on the related tendon) is about 8 times as great for primary as for secondary endings (Lundberg and Winsbury, 1960). The primary endings also have about 5 times as great a sensitivity as the secondary endings to vibration of 100–500 Hz (M. C. Brown, Engberg, and Matthews, 1967). Over an appropriate range, both primary and secondary endings show a linear increase of discharge frequency (impulses per sec) with an increase in the amplitude of regular sinusoidal stretching. The slope of this relationship (sensitivity) may in turn be plotted against the frequency of stretching (Fig. 7.7); the graph is flat from 0.1 to 1.0 Hz, but then begins to rise along a curve reflecting the vector sum of length and velocity of stretching, the turning point indicating the relative sensitivity of a receptor to length and velocity changes (P. B. C. Matthews and Stein, 1969).

The dynamic sensitivity of the primary endings depends upon rapid adaptation at some step within the receptor mechanism. In theory, the critical link could be mechanical transmission of the stimulus to the terminals of the receptor nerve, the conversion of the mechanical force into a generator potential at the neural membrane, or the resultant development of propogated action potentials in the afferent nerve (Fig. 7.8; P. B. C. Matthews, 1972). In practice, the major basis of adaptation is mechanical.

The anatomical characteristics of the nuclear bag fiber give its equatorial region elastic properties while the striated poles function as series viscous elements (B. H. C. Matthews, 1933; R. S. Smith, 1966; Boyd, 1971; Houk et al., 1973; Poppele, 1973).

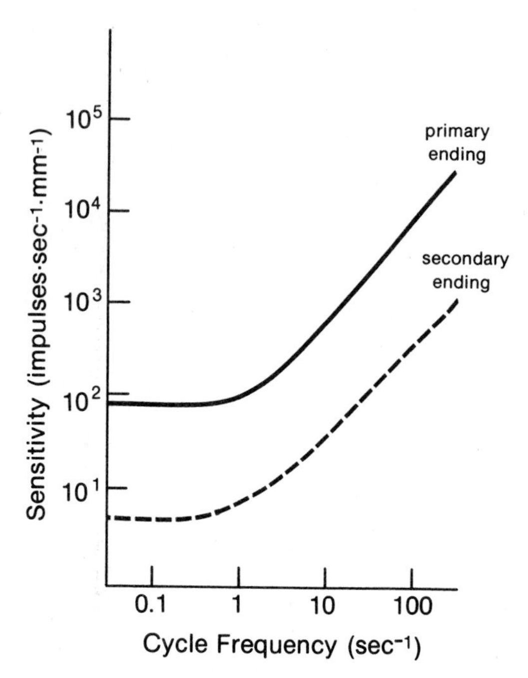

FIG. 7.7. Sensitivity of primary and secondary endings to sinusoidal stretching. The sensitivity is given by the discharge frequency divided by the amplitude of stretching (observations confined to linear range). At low frequencies, the response is to length (horizontal relationship), while at high frequencies the response is to velocity (diagonal slope). (After P. B. C. Matthews & Stein, 1969.)

When an elongating force is applied, the poles resist extension with a counterforce proportional to the velocity of extension. In contrast, the equatorial region stretches relatively easily, and in this zone the counterforce is proportional only to the amount of extension. As the dynamic stretch ceases the poles continue to elongate, allowing the equatorial region to shorten and giving an adaptation of the primary endings (Gottlieb et al., 1970; Rudjord, 1970). A further variable is the behavior of the encoding mechanism (Poppele, 1973). The afferent impulses are generated by a time-varying leak of cations across the receptor membrane; characteristics of the leak depend on the conductance and capacitance of the membrane, features which differ for primary and secondary endings.

The sensitivity of the primary endings to acceleration could also be attributed to mechanical factors, either a complicated arrangement of viscous and elastic elements (Angers, 1965; Gottlieb et al., 1970) or a rapid diminution in the viscosity of the polar regions secondary to a rupture of actomyosin cross-bridges by the applied force (D. K. Hill, 1968; Chap. 4). A similar phenomenon would explain the decrease in sensitivity of the primary endings with a change from small to large amplitudes of sinusoidal vibration.

The nuclear chain fibers have a more uniform internal structure than the nuclear bag fibers, and they would thus not be expected to show mechanical viscoelastic phenomena.

Tendon Organs

The tendon organs were originally thought to have a high threshold, being stimulated only when muscle tension reached a dangerous level (B. H. C. Matthews, 1933). More recent research suggests a low threshold when static tension is produced within the musculotendinous junction by active contraction of the related muscle. The threshold to passive stretching may thus be high simply because much of the applied tension is developed in the fascial sheath and fails to be transmitted to the tendon organ (Jansen and Rudjord, 1964).

The relationship of discharge frequency (D) to tension (T) is of the type

$$D = KT^{n} \tag{7.1}$$

where K is a constant and n is an exponent of 0.4–0.6. It has yet to be clarified as to whether individual

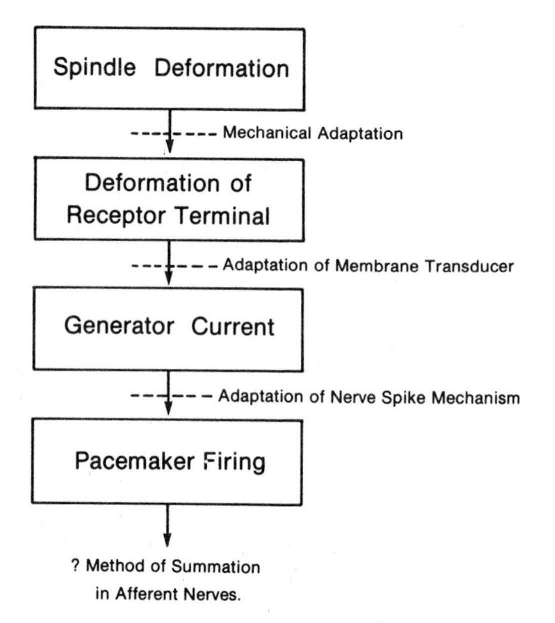

FIG. 7.8. Possible sites of adaptation of muscle stretch receptors. (After P. B. C. Matthews, 1972.)

tendon organs respond preferentially to fast- and slow-twitch muscle fibers (P. B. C. Matthews, 1972).

Other Receptors

The fine nonmedullated fibers arising from the free nerve endings are apparently uninfluenced by either active contraction or passive stretching, but respond to a variety of painful stimuli (including local ischemia; W. W. Douglas and Ritchie, 1957; Iggo, 1960).

In the joints, the Ruffini endings (Skoglund, 1956) respond to both position and angular velocity; sensitivity is increased by a tensing of the muscles, possibly because the joint capsule is deformed (Goodwin, 1976). B. L. Andrew and Dodt (1953) maintained that each ending sensed only a small portion of the total range of joint movement, although Burgess and Clark (1969) have attacked this traditional view. Golgi organs in the surrounding ligaments sense static position but not rate of movement, while paciniform corpuscles respond to both dynamic motion and vibration (Skoglund, 1956). In applying static tensions to the limbs, it is thus important to avoid using equipment with a tendency to vibrate (P. B. C. Matthews, 1972).

Functional Significance

Although afferent nerve discharges can be demonstrated for each of the different patterns of stimulation discussed, it has yet to be shown how far the central nervous system uses this information in the coordination of movement.

Efferent Impulse Patterns

α Motoneurons

The fibers from the α motoneurons innervate only the extrafusal muscle fibers. Anterior horn cells supplying type I (slow-twitch) muscle are readily excited by Ia afferent nerves, and are also powerfully inhibited by antagonistic group Ia afferents and cutaneous afferents (R. E. Burke, Rymer, and Walshe, 1973). Such motor units thus play an important role in postural and segmental reflexes. Motoneurons supplying type IIa (fast, fatigue-resistant) muscle also receive group Ia excitation and inhibition, but are less sensitive to cutaneous afferents. This type of unit can thus participate in reflexes but probably has additional functions. Type IIb (fast, fatigue-sensitive) motoneurons have much less input from group Ia afferents, and it is likely that they are used mainly for less stereotyped movements.

β Motoneurons

Beta nerves supply both normal muscle and spindle fibers (Besson et al., 1965; Adal and Barker, 1965), but their functional importance is uncertain.

γ Motoneurons

Gamma efferent fibers innervate only the muscle spindles (Diete-Spiff, 1967). Stimulation of such fibers increases spindle afferent activity, but there is no detectable contraction of the muscle as a whole, provided that spinal reflexes are interrupted by blockade of the large motor fibers.

Current research distinguishes *static* and *dynamic* fusimotor fibers (Tables 7.3 and 7.4). In general, the static γ efferents pass to the rapidly contracting and elastic fibers of nuclear chain elements, while the dynamic γ efferents innervate the viscous and slowly contracting elements of the nuclear bag. Nevertheless, the distinction between the two classes of efferent fiber is not absolute (LaPorte and Emonet-Dénand, 1973).

Stimulation of the static γ efferents depresses the dynamic response of primary endings. In contrast, stimulation of dynamic γ efferents increases the response of primary endings to dynamic stimulation. Both types of γ fiber increase the sensitivity of the primary spindle receptors to static stretching (M. C. Brown, Crowe, and Matthews, 1965). These changes are mediated by a change in the viscoelastic properties of the fiber (see above). Changes in polar stiffness are associated with a rapid turnover of actin/myosin bonds (that is, a rapidly contracting fiber); such changes influence position sensitivity. In contrast, changes of polar viscosity are associated with slowly contracting fibers and modify dynamic sensitivity (M. C. Brown and Matthews, 1966).

The secondary afferent endings are innervated only by static γ motor fibers. Stimulation of such efferents increases the sensitivity of the secondary endings to static stretching but leaves them largely unresponsive to dynamic stimulation (Appelberg et al., 1966). Two explanations may be advanced for the increase in static response without change of dynamic sensitivity. Possibly, the portion of the spindle beneath the secondary afferent ending is somewhat deficient in myofibrils, and can thus yield when the fiber as a whole contracts. Alternatively, the portion of the myofibril related to the ending fails to become fully depolarized.

The speed of nerve conduction tends to be slower in the static than in the dynamic γ fibers. However, presumably because of differences in the viscoelastic characteristics of the responding spindles

TABLE 7.3
Comparison of Responses to the Stimulation of Dynamic and Static γ Fibers

	Dynamic γ Fibers	Static γ Fibers
Static response of primary spindle receptor	Increase	Large increase
Static response of secondary spindle receptor	No effect	Increase
Dynamic response of primary receptor to ramp stretch	Increase	Decrease
Dynamic response of secondary receptor	No effect	Remains small
Response of primary receptor to low frequency sinusoidal stretching	Remains high	Decrease
Response of primary receptor to vibration	Increase	Increase
Speed of intrafusal contraction	Slow (fusion frequency $< 100 \cdot sec^{-1}$)	Fast (fusion frequency $> 100 \cdot sec^{-1}$)
Conduction velocity	15–50 $m \cdot sec^{-1}$	15–50 $m \cdot sec^{-1}$ (*but* higher proportion of slow fibers)
Responding fiber	Nuclear bag	Nuclear spindle

Source: After P. B. C. Matthews, 1972.

(see above), the change of afferent activity induced by stimulation of the fusimotor nerves develops more rapidly in those receptors supplied by static γ fibers (Besson et al., 1968; Emonet-Dénand and LaPorte, 1969).

Administration of adrenaline and stimulation of the sympathetic nerves produce both excitatory and depressant effects upon the primary spindle receptors. It is as yet unclear whether these actions are a direct response to catecholamines or a secondary asphyxial reaction to interference with the local circulation (Hodgson et al., 1969).

Interaction of Spindles

The functional importance of the two types of spindle, arranged in parallel, has yet to be fully explored.

TABLE 7.4
Comparison of Primary and Secondary Spindle Endings

Variable	Primary	Secondary
Site	Mid-equatorial bag and chain fibers	Juxta-equatorial chain fibers
Afferent nerve	Ia (12–20 μm, 70–120 $m \cdot sec^{-1}$)	II (4–12 μm, 20–70 $m \cdot sec^{-1}$)
Efferent control	γ_s and γ_D	γ_s
Ramp stretching		
Initial	Acceleration response	Linear response to stretch
Dynamic index (after 10 $mm \cdot sec^{-1}$)	40–120 impulses $\cdot sec^{-1}$	5–21 impulses $\cdot sec^{-1}$
Release	Abrupt silence	Progressive decrease
Tendon tap threshold	<50 μm	>500 μm
Vibration threshold	<50 μm at 100–300 Hz	>250 μm at 100 Hz
Variability of discharge	High	Low
Displacement sensitivity	High	Low

Source: After P. B. C. Matthews, 1972, for cat soleus.

It is possible that contraction of a nuclear bag fiber could "unload" a nuclear chain fiber (or vice versa). However, this is unlikely in practice, since both are attached to relatively unyielding extrafusal fibers. Electrical interaction is certainly necessary because signals from the two generators must be converted to a single train of impulses carried via a common afferent nerve; possibly, there is a simple summation of information at a unique pacemaker, or possibly there are independent pacemakers, priority being given to the pacemaker with the highest instantaneous discharge rate.

Interaction with Spinal and Cerebral Centers

Spinal Reflex Arcs

The various afferent pathways cited in the section on afferent impulse patterns can all influence the tone of corresponding muscle groups through simple spinal reflex arcs (Figs. 4.13, 7.9, and 7.10). These facilitate or inhibit the discharge of α motoneurons in the ventral horn of the spinal cord.

The most well-known reflex is the knee jerk. This was first attributed to a direct mechanical stimu-

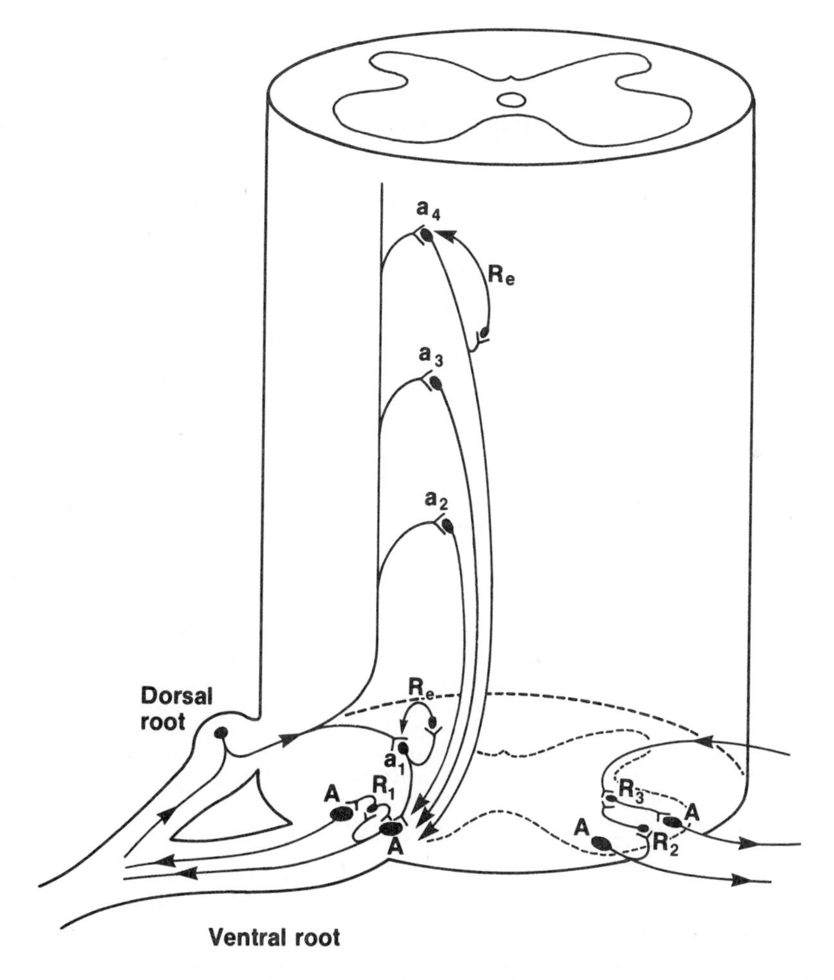

FIG. 7.9. Possible neuronal circuits within the spinal cord. The shortest loop passes through an internuncial neuron, a_1, in the same segment. However, other longer loops pass through internuncial neurons a_2 to a_4 in other segments. The signal can be maintained by closed-loop excitatory circuits (reverberators, R_e). Collateral branches from the anterior horn cells, A, may stimulate inhibitory Renshaw cells (R_1, R_2). R_1 inhibits both anterior horn cells it supplies. R_2 inhibits another inhibitory interneuron R_3, thus activating the corresponding anterior horn cell. (Based in part on Keele & Neil, 1971.)

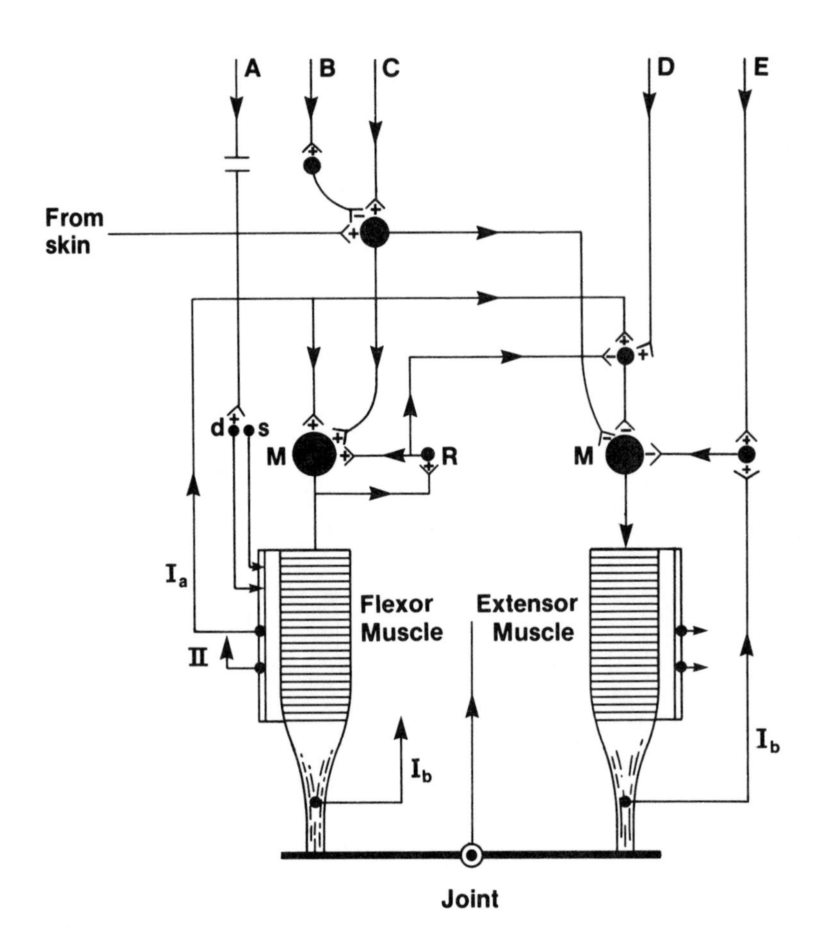

FIG. 7.10. Details of the segmental proprioceptive reflex network. A flexor and an extensor muscle act antagonistically about a joint; shortening one lengthens the other. M, motoneuron pool. The extensor muscle is inhibited by a disynaptic link from the corresponding tendon organ (Ib fibers). The flexor muscle receives monosynaptic excitation and the extensor muscle disynaptic inhibition (Ia fibers, mixed length and velocity signal). Circuits for type II fibers are still uncertain. R, Renshaw cells. These recurrently inhibit agonists and inhibitory interneurons of antagonists. d,s, dynamic and static fusimotor neurons. The cerebrospinal tract provides excitation to d and s (A), inhibits input from the skin (B), excites neurons that act on the agonist and inhibit the antagonist (C), reinforces reciprocal inhibition by the type I_a fibers (D), and reinforces inhibitory feedback by the tendon organs and type I_b fibers (E). (After C. G. Phillips & Porter, 1977.)

lation of muscle, since the response was thought too rapid for transmission of information via a reflex pathway (Waller, 1881). More accurate timing later established (Jolly, 1911) that impulses could travel via the spinal cord; nevertheless, there was insufficient time to allow transmission of the signal across more than one synapse (because of delays associated with the diffusion of transmitter substances, each synapse adds 0.5–1 msec to the overall conduction time).

Excitatory reflexes generally involve only a single synapse, but inhibitory circuits include an additional "internuncial" neuron (Eccles, Fatt, and Langdren, 1956; Eccles, 1969; Figs. 7.9 and 7.10). Such an anatomical arrangement permits the use of differing chemical transmitter substances for excitation and inhibition (Chap. 4). It also allows activation of the inhibitory neurons by extrasegmental circuits (Lundberg, 1970; for example, corticospinal, rubrospinal, and vestibulospinal tracts descending from the brain).

Inhibitory circuits act on both α and γ motoneurons (Ellaway, 1971; Fromm and Noth, 1975), serving to relax antagonistic muscles. They also help in the "focusing" of signals by suppressing discharge in pathways that have been only weakly excited (Eccles, 1973). During all-out activity, quite strong signals from the working muscles and joints can be kept from consciousness by this mechanism. Acupuncture, hypnosis, and yoga are all techniques that exploit the potential to inhibit normal sensory pathways.

The synapse shows little tendency to fatigue (Eccles, 1973), and signals are sometimes transmitted at rates as high as 1000 impulses per sec. One reason for fatigue resistance is that the release of the synaptic transmitter (acetylcholine) stimulates its resynthesis.

Group I Afferents

Afferent volleys associated with both a tendon jerk and high-frequency vibration (Coppin et al., 1970) travel with a conduction velocity of ~100 m • sec^{-1}. The pathway is thus the group I sensory fiber (see above) and the responsible receptors are the primary spindle endings (D. P. C. Lloyd, 1943b; Lundberg

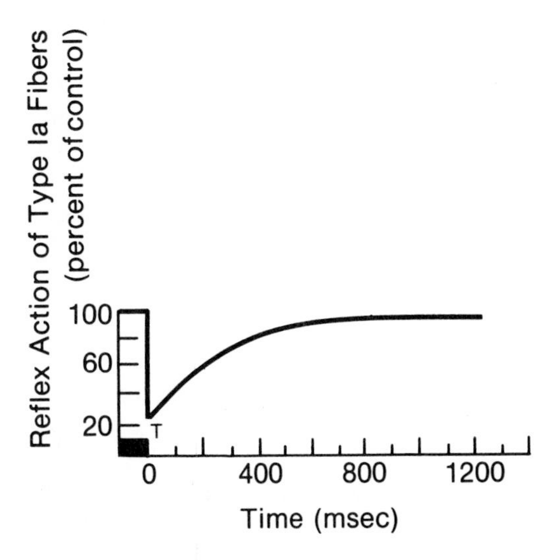

FIG. 7.11. Prolonged depression of reflex action of Type Ia fibers with presynaptic inhibition from a tetanic muscle contraction T. (After Eccles, 1969.)

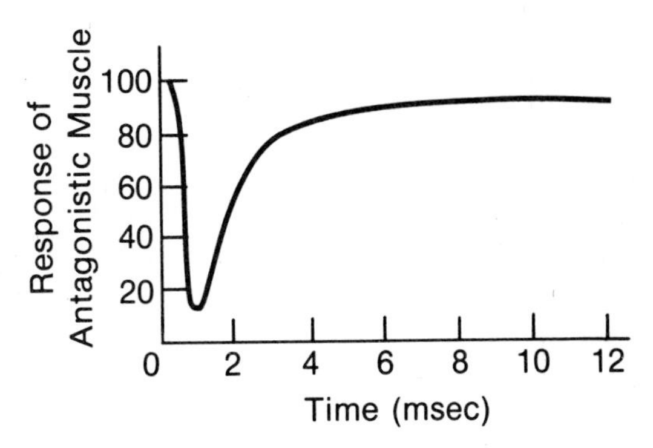

FIG. 7.12. The time course of inhibition of an antagonistic muscle by stimulation of Type Ia afferents. (After K. Bradley et al., 1953.)

and Winsbury, 1960). According to classical theory, there is an excitation of motoneurons acting on the muscle itself and its close synergists (Eccles, Eccles, and Lundberg, 1957a; R. F. Schmidt and Willis, 1963), with inhibition of the corresponding antagonistic muscles (Figs. 7.11 and 7.12), the response defining a functional whole (the "myotatic unit"; D. P. C. Lloyd, 1946). More recent studies confirm the notion of a functional entity, although this may embrace quite a wide range of muscles (Clough et al., 1968), individual group Ia afferents acting on as many as 300 motoneurons (Mendell and Henneman, 1971).

Stronger stimulation is usually (Laporte and Lloyd, 1952; K. Bradley and Eccles, 1953) but not always (Jack and MacLennan, 1971) needed to activate the tendon receptors (Fig. 7.13). These transmit information via group Ib afferent fibers, leading to an "inverse myotatic reflex" (inhibition of the agonists and stimulation of the antagonists). The reflex arc involves at least one and sometimes two internuncial neurons (Eccles, Eccles, and Lundberg, 1957a,b; Hongo et al., 1969).

Other Afferents

Activation of group II, III, and IV afferents leads to a general flexor response (Holmqvist and Lundberg, 1961), with extension of flexor motoneurons and inhibition of the extensors. Two or more internuncial neurons are involved. Other functions of the group II fibers have yet to be elucidated (Pacheco and Guzman, 1969); there is no conclusive evidence that joint afferents play any important role in the reflex control of movement (P. B. C. Matthews, 1972; Grillner, 1973). Group III fibers probably contribute to the respiratory and cardiac reflexes already discussed

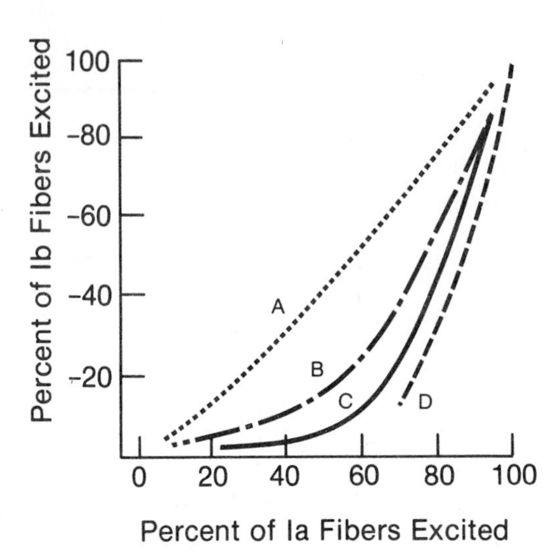

FIG. 7.13. To illustrate variations in the relative excitability of Type Ia and Type Ib fibers. A, peroneus longus; B, medial gastrocnemius; C, soleus; D, semitendinosus muscle. (After P. B. C. Matthews, 1972.)

(Chaps. 5 and 6; Wildenthal et al., 1968; Coote et al., 1971).

Presynaptic Inhibition

Certain terminals of the afferent fibers exert an inhibitory effect directly upon other afferent terminals (presynaptic inhibition; Conradi, 1970; Barnes and Pompeiano, 1970). Depolarization of the inhibited terminal reduces the number of quanta of neurotransmitter substance that it liberates (Kellerth, 1968). The group Ib fibers seem particularly prone to cause this type of action. It may serve as a form of "gain control" for a reflex arc, but the response is too slow and too generalized for it to have a more specific function in muscle control (P. B. C. Matthews, 1972).

Central Projections of Receptors

Afferent fibers from both muscles and joints project information to the contralateral sensory cortex of the brain (Oscarsson and Rosén, 1963; Landgren and Silfvenius, 1969; Fig. 7.14; also see below). Group III fibers from the joints provide conscious information of the position of the limbs (kinesthesis) and can also produce conditioned reflexes and arousal (E. Gardner and Hadded, 1953; Andersen et al., 1967). There is no strong evidence that group I afferents contribute to any of these phenomena (Pompeaino and Swett, 1962; Swett and Bourassa, 1967), nor do they modify the output of the motor cortex significantly (P. Anderson and Phillips, 1971). However, they may become involved in more complex forms of perception, such as the relationship between motoneuron output and muscle tension (length/tension appropriateness). Certainly, position can be sensed more readily with active than with passive displacement of a limb (Paillard and Brouchon, 1968), and alteration of spindle receptor discharge by vibration (Goodwin, McCloskey, and Matthews, 1972) or gravitational effects can impair judgments of limb position.

The spinocerebellar tracts (Fig. 7.15) carry impulses from group I afferents to the cerebellum (Oscarsson, 1967). The dorsal spinocerebellar tract transmits information from both group Ia and Ib fibers. The existence of a synapse on this pathway allows modification of the discharge, for example, by afferent impulses from other muscles. About 15 presynaptic inputs converge upon a single spinocerebellar neuron (Eide et al., 1969), leading to signal averaging, with some loss of detail in the information reaching the cerebellum (Walløe, 1970).

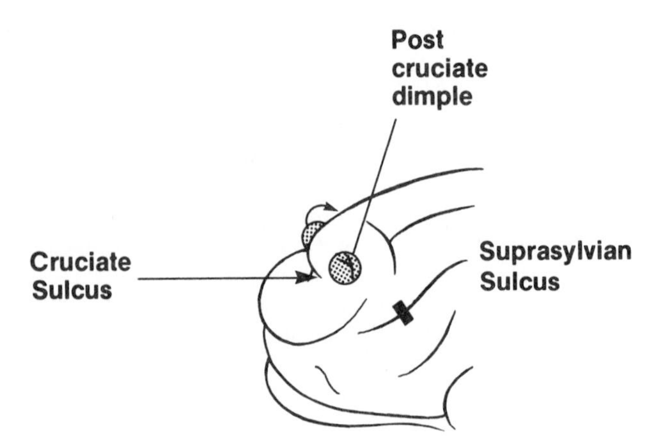

FIG. 7.14. Areas of the cat cerebral cortex where type 1 muscle afferents are known to project. Cruciate region is from the hindlimbs, the postcruciate dimple from both fore- and hindlimbs, and the suprasylvian sulcus from the forelimbs. (After P. B. C. Matthews, 1972.)

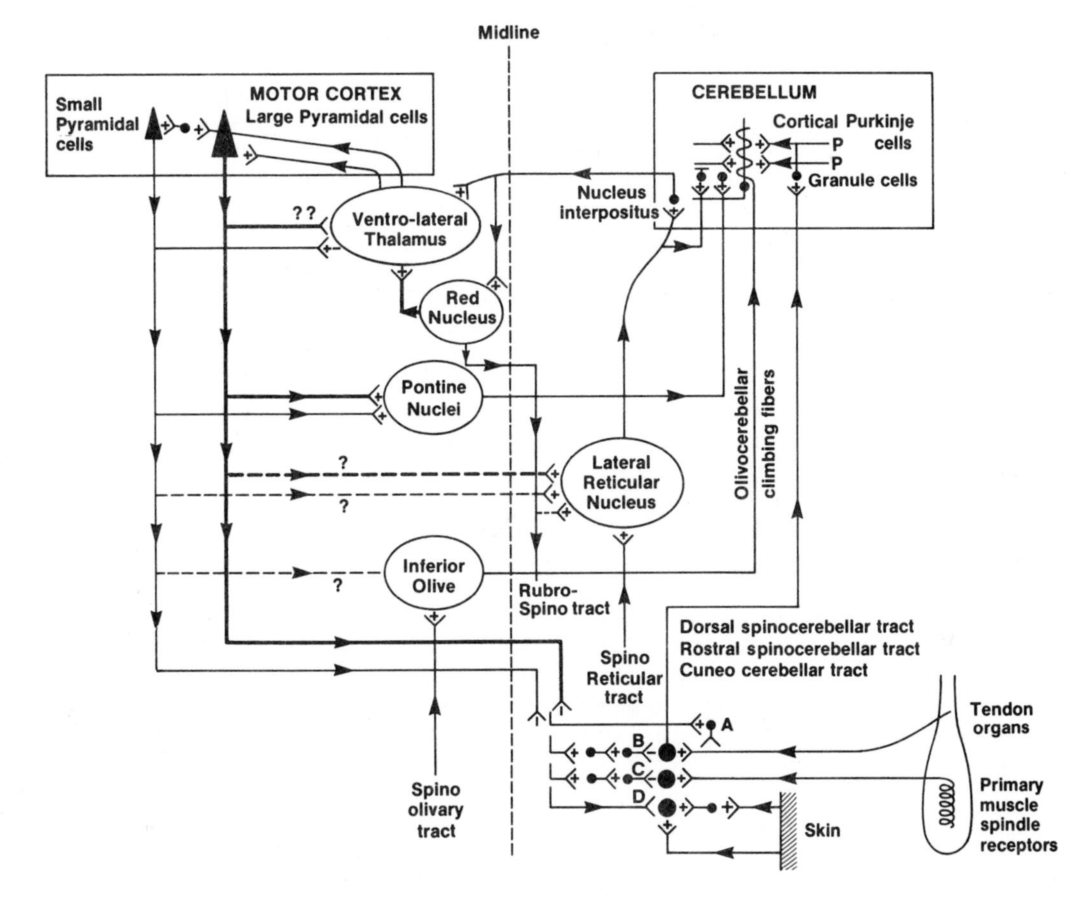

FIG. 7.15. Connections between peripheral receptors, motor cortex, and cerebellum. (Based in part on C. G. Phillips & Porter, 1977.) The fast pyramidal tract neurons send collaterals to the ventrolateral nucleus of the thalamus, the pontine nuclei, and possibly the lateral reticular nucleus. The slow pyramidal tract neurons send collaterals to the pontine nuclei, and possibly to the lateral reticular nucleus and inferior olive. The rubrospinal tract possibly sends collaterals to the lateral reticular nucleus. The input to the spinocerebellar tracts receives presynaptic (A) and postsynaptic (B,C) inhibition, and monosynaptic facilitation (D). Within the cerebellum, cortical Purkinje cells are excited by granular cells via parallel fibers (P) and by olivo-cerebellar climbing fibers. The cells inhibit cerebellar nuclei such as the nucleus interpositus. For details of the cerebellum, see Fig. 7.28. The nucleus interpositus excites the thalamus both directly and via the red nucleus. The thalamus excites the large pyramidal cells monosynaptically and the small pyramidal cells disynaptically.

The ventral spinocerebellar tract transmits impulses almost exclusively from group Ib fibers. Sensitivity is much less than for the dorsal spinocerebellar tract, contraction of several muscles being needed to activate a single ventral spinocerebellar neuron. The ventral pathway may serve to signal the stage a movement has reached (Oscarsson, 1967), although its functional role is still far from clear (P. B. C. Matthews, 1972). There are plainly many possibilities for modifying the activity of the corresponding neurons, including facilitatory and inhibitory signals transmitted from the skin and descending pathways from the brain.

Impact of Brain upon Receptors

Many regions of the central nervous system can modify the rate of discharge of the γ motoneurons, producing both excitatory and inhibitory effects (Granit and Kaada, 1952). Some descending pathways cause simultaneous excitation ("coactivation") of α and γ motoneurons (Granit, 1955; Eklund et al., 1964), but other pathways modify the γ-efferent discharge independently of any effect on the α motoneurons (Granit et al., 1955; Koeze, 1968).

Certain parts of the brain, such as the red nucleus, give rise to fibers that stimulate mainly the dynamic γ efferents (Appelberg, 1962). Other regions produce mainly an activation of the static γ efferents (Vedel and Paillard, 1965).

Servo Control of Movement

There seem several uses that the central nervous system could make of afferent information from the muscle receptors: (1) maintenance of central excitability and muscle tone, (2) reflex regulation of muscle length (Rossi, 1927), (3) servo control of muscle contraction (Kuffler and Hunt, 1952), (4) learning of complex movement patterns, and (5) development of appropriate discharge patterns from the motor cortex.

The stretch reflex can be regarded as a length-sensitive servo system tending to keep the length of the muscle fibers constant in the face of an external disturbance. The equilibrium setting is modified by activity of the γ-efferent fibers. Shortening of the spindles increases the discharge in group Ia afferents, and this induces an α motor discharge until equilibrium has been restored by a shortening of the muscle. The higher centers can thus produce a muscle contraction in two ways: directly, by a facilitation of the α motor discharge, and indirectly, by activation of the γ motor system (Merton, 1953; Granit, 1955; Hammond et al., 1956; D. Burke et al., 1978; Fig. 7.16). In the latter case, the system probably functions as a follow-up length servo; shortening continues until the desired length has been reached, independently of such factors as local fatigue and external loading.

By registering discrepancies between intended and actual movements, the body can make progressive improvements to the motor programs stored in the central nervous system (Figs. 7.17 and 7.29; also see below). Central transmission of proprioceptive information also allows some modification of the discharge from the motor cortex in the light of immediate loadings of the active body parts.

Scientists interested in the design of servo mechanisms have engaged in much speculation concerning other advantages of the complex pattern of receptors we have described, with its capacity to report both static tension and the rate of change of tension (M. C. Brown and Matthews, 1966; R. A. Schmidt, 1976). They have noted that the relative proportions of static and dynamic information can be varied by appropriate activity of the γ-efferent fibers. Engineers use such a system of "phase advance" to minimize oscillations in a mechanical servo system (J. C. West, 1953), and it is likely that a phase advance of the primary endings is one approach used by

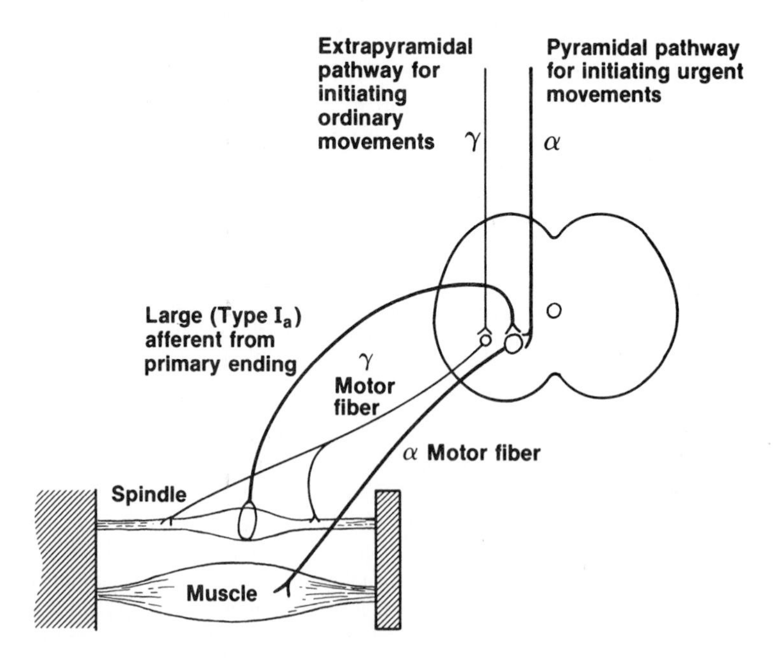

FIG. 7.16. Two possible pathways for initiation of muscle movement (after Merton, 1953). The effort demanded by the higher centers gives two inputs, α and γ, to a summing junction. The feedback loop passes through a multiplier. The critical motoneurons can be activated independently or coincident with input to the multiplier. Multiplication may occur by making more of the cortical motoneurons sensitive to feedback. Input from the skin (particularly on the thumb) contributes to the function of the multiplier.

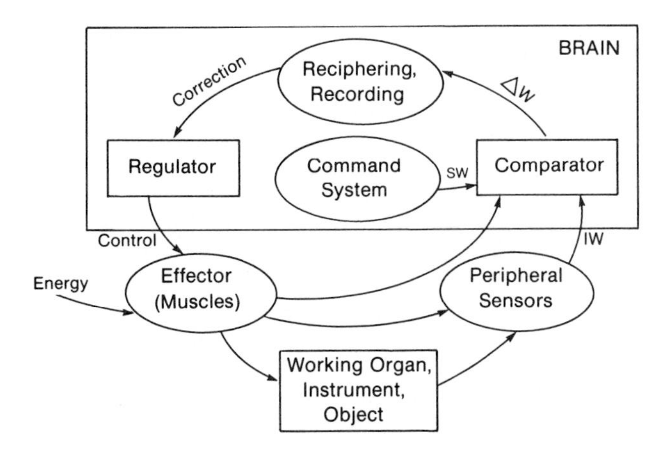

FIG. 7.17. Simplest possible diagram of an apparatus for the control of movements. (After N. Bernstein, 1967.)

the body to counteract the instability that stems from long neural loops and relatively slowly shortening muscles (P. B. C. Matthews, 1972; J. C. Joyce, Rack, and Ross, 1974). Further phase advance, and thus additional damping, occurs as impulses are transmitted through the spinal cord (Westbury, 1970). Such phase advance has an important advantage over one alternative approach to stability (an increase of viscous damping) in that it does not reduce the power of the system. Some viscous damping is provided, however, by the properties of the contracting muscles, since they develop more than the standard isometric force while they are being stretched and less than the standard isometric force while they are shortening (Fig. 1.9). The need for damping increases if a limb is moving a heavy mass. It is possible that under such circumstances the dynamic fusimotor fibers increase the extent of phase advance by making the primary spindle endings more sensitive to muscle velocity. Certainly, the loop would become less stable if the static sensitivity of the spindles was increased without at least a matching gain of dynamic sensitivity.

Engineers usually design a servo system with a high gain so that it can respond to a given command despite strong external resistance. The gain of the muscle servo system is more modest (Vallbo, 1973a,b), although it can still induce a useful increase of force in the face of an unexpected obstacle.

The Command Signal

The static γ efferents are largely responsible for injecting the command signal into the servo loop since they alone can make the spindle organ fire faster during shortening of the extrafusal muscle fibers. The function of the dynamic fibers is probably to increase the sensitivity of the primary endings (Fig. 7.18; Lennerstrand and Thoden, 1968).

Many authors still query how far movement can be *initiated* by γ-efferent activity (M. C. Brown and Matthews, 1966). Certainly, rapid movements depend on the α pathway, and most other movements are characterized by simultaneous activation ("coactivation") of α and γ pathways (Kuffler and Hunt, 1952; Sears, 1964; C. Von Euler, 1966b; Vallbo, 1973a,b; Stein, 1974; J. L. Smith, 1976).

Direct α activation of the ventral horn cells speeds responses by about 50 msec (Merton, 1953), while the parallel activity of the γ fibers provides a form of servo assistance to the contraction initiated via the pyramidal pathway (Eldred et al., 1953; Granit, 1975). Further servo assistance may be provided through complex intracerebral loops, to be discussed in following sections (Koeze et al., 1968; C. G. Phillips, 1969). The γ system also plays an important role in reducing the sensitivity of spindle receptors in antagonistic muscles (Clough et al., 1971); without an appropriate reduction of γ discharge, stiffening of the antagonists would occur as they underwent a passive lengthening.

Practical Implications

The servo mechanism is independent of conscious voluntary control, yet it allows very sensitive proprioception. In respiratory experiments (E. J. M. Campbell, 1966; Pengelly, Rebuck, and Campbell, 1974), a

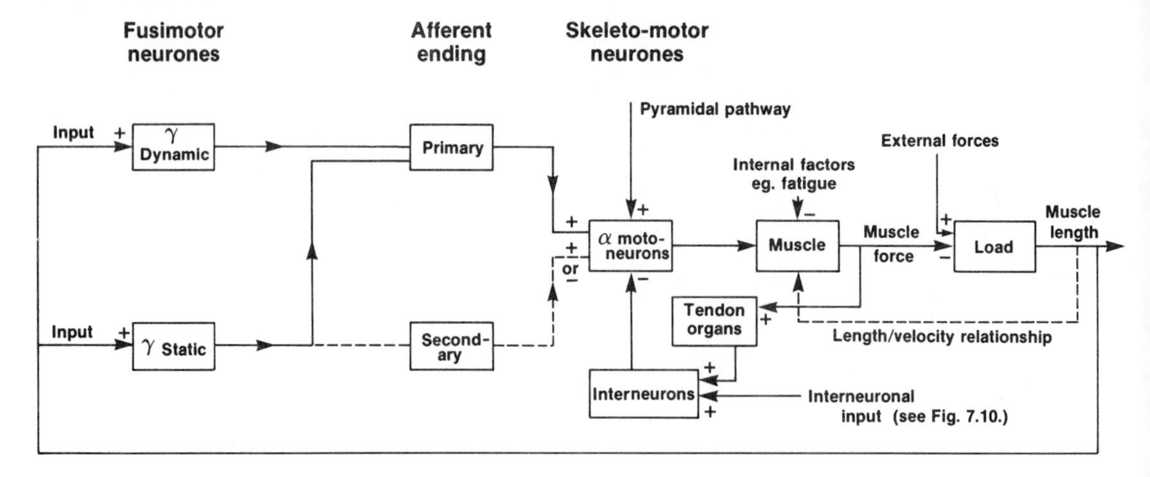

FIG. 7.18. Block diagram of possible mode of γ efferent muscle activation. The dynamic fibers control the sensitivity of the primary endings. The static fibers have the potential for injection of a "command" into the servo-loop, although according to some authors this potential is rarely exercised unless there is a direct coactivation of the α motoneurons via pyramidal pathways. (Based in part on M. C. Brown & Matthews, 1966.)

subject can detect an added elastic load equivalent to a water pressure of 10–15 mm (2.5 cm $H_2O \cdot l^{-1}$). The threshold for the detection of viscous respiratory resistance is even smaller (0.6 cm $H_2O \cdot l^{-1} \cdot sec^{-1}$, equivalent to a water pressure of ~3 mm during normal quiet breathing.

Since the spindle receptors can detect rate of change of tension, they are particularly likely to be stimulated by bouncing and jerky movements. Such forms of activity should be avoided in warming up and in exercises designed to improve flexibility. Gains of flexibility are more likely to result if the stretch is sustained and of sufficient intensity to stimulate the Golgi receptors, thereby relaxing the corresponding muscle group.

Special Senses

Space permits no more than a very brief comment upon the special senses; further discussion of their contribution to human performance will be found in Poulton (1970) and Shephard (1974b).

Vision

Certain aspects of visual acuity and the perception of movement are noted earlier in this chapter, while the contribution of vision to maintenance of posture is discussed later in the chapter.

Normal visual acuity allows perception of two parallel lines that subtend an angle of 1 min at the retina, but well-trained subjects can distinguish angles as small as 25 sec, corresponding to a separation of ~2 μm at the retinal surface. This is hard to interpret, since individual cone cells have a diameter of ~3 μm. Possibly, the explanation lies in a combination of direct stimulation of receptors and inhibition of adjacent pathways. Further, some cells or group of cells seem activated by stimuli falling in the central part of their field but are inhibited by peripheral stimuli, while other cells or groups of cells have the reverse pattern of activation (Kuffler, 1953). In daylight, visual acuity depends upon the cone cells of the retina, but under conditions of poor illumination (twilight or swimming underwater) visual function is taken over by the monochromatic rod cells. The spatial distribution of the two types of receptor is such that visual acuity in daylight is greatest when looking straight ahead, but when illumination is restricted perception is facilitated by looking 15–20° to one side of an object (Fig. 7.19). In the foveal region, where the cones are numerous, there is a 1:1 relationship between individual receptor cells and transmission lines to the brain. However, in the peripheral parts of the retina, up to 5000 transducers converge on a single ganglion cell; this arrangement is well suited to dim lighting, increasing the probability of firing in a given ganglion cell through a process of summation.

The retinal cells convert electromagnetic energy to nerve impulses through the breakdown of a pigment (visual purple, rhodopsin) into its components retinene and a protein opsin. The opsin may stimulate the receptors by binding certain ions. In dim light, there is a spontaneous recombination of retinene and opsin, but in daylight much of the reti-

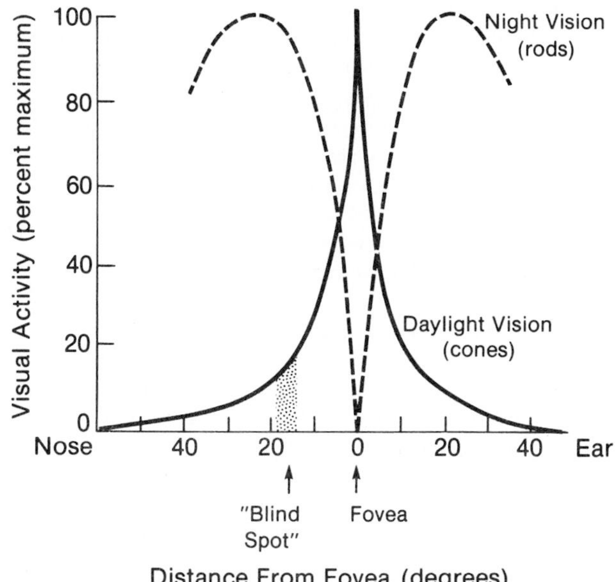

FIG. 7.19. Visual acuity at varying angular distances from the fovea. Note that the sensitivity of the cones is much greater than that of the rods.

nene is reduced by NADH to form vitamin A. The latter reacts only slowly with opsin.

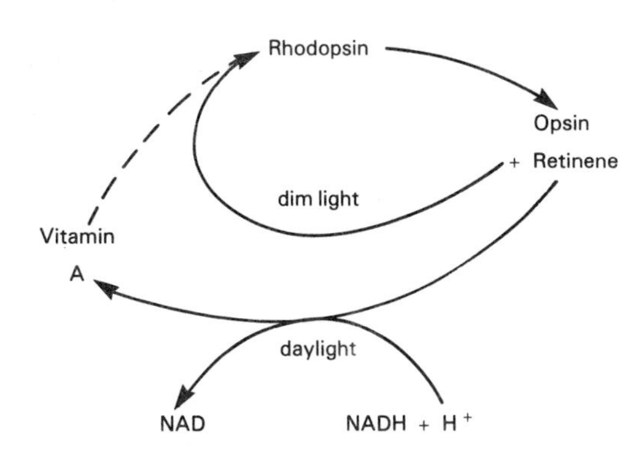

There is thus a progressive loss of night vision on emerging into daylight. Recovery of night vision follows a two-component curve (Fig. 7.20). The first component represents the recovery of the rod cells. Red goggles may be worn to speed the resynthesis of visual purple in preparation for underwater exploration.

The normal technique of viewing an object is a saccade. The eyes focus on the target for an instant, and as the receptors adapt they jerk rapidly to the next fixation point. A moving target can be tracked at angular velocities of up to $30° \cdot sec^{-1}$, discrimination of detail decreasing with angular velocity (Ludvigh, 1955). Cricket and tennis balls may move at a faster velocity than this. Some help can be obtained by rotating the head as well as the eyes, but at speeds above $50° \cdot sec^{-1}$ there is a marked deterioration of performance. The eyes tend to lag behind the target and then catch up with a jerk. If light is good, the time required for perception in the peripheral part of the visual field is a little over 100 msec; ~60 msec is occupied in rotating the foveal receptors to receive the critical information, and a further 50-msec lag is imposed by the receptor mechanism (Strughold, 1951).

Appropriate performance requires not only stimulation of the visual receptor organs, but also the correct interpretation of signals by the visual areas of the cerebral cortex. The retinal receptors pass infor-

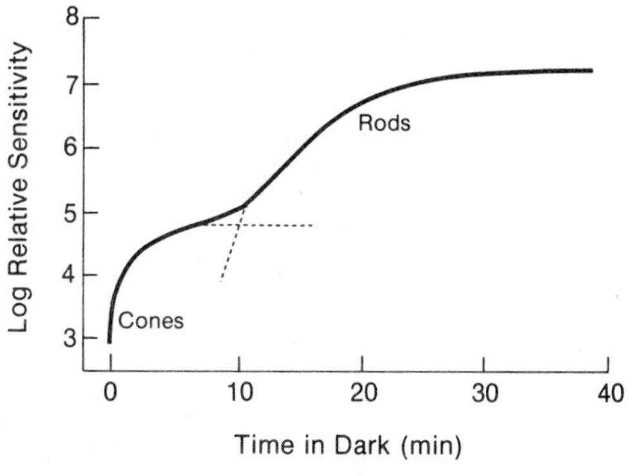

FIG. 7.20. The normal course of dark adaptation. (After Wald, 1954.)

mation via the lateral geniculate nuclei to the "simple" cells of the visual cortex. Each of these cells is fed by a number of retinal receptors, and responds to a line of adequate length, provided it falls on an appropriate part of the retina with a correct orientation toward the receptors. A number of simple cells, all responding to lines with a similar orientation, feed to a "complex" cell; this responds to a stimulus of appropriate orientation in any part of the retina. Lastly, impulses pass from a number of complex cells to a "hypercomplex" cell with an ability to recognize shapes (Hubel and Wiesel, 1965; Fig. 7.21).

The duration of a stimulus determines whether it is perceived vaguely, clearly, or in detail. Brighter illumination, learning, and vigilance all decrease the time needed for correct recognition of a given visual pattern.

Hearing

The auditory areas of the cortex have a similar capacity for storing and recognizing sound patterns that can be exploited in various types of sport. Differences in the timing and intensity of sound between the two ears may make some contribution to the judgment of external movements.

There is often difficulty in communication underwater. When sound hits an air/water interface, less than 0.1% of the energy is transmitted. Speech underwater is further complicated by alterations in the density of respired gases (Sergeant, 1969) and resonance in breathing equipment.

Posture and Motion

The labyrinth comprises the otolith organ (utricle and saccule) and the semicircular canals. The utricle and saccule each contain a projecting ridge (the macula) which is covered by long, hairlike receptors embedded in a gelatinous material and containing many chalky particles (the otoliths). The otoliths facilitate bending and therefore stimulation of the receptors in response to body tilting and linear acceleration; some authors maintain that the saccule is more sensitive to vibration, while the utricle responds to tilting and acceleration.

The semicircular canals each originate in a swelling (ampulla), which contains a ridge of tissue bearing a column of matted hairs (the cupula). Movement of fluid (endolymph) in the canals displaces the hairs, thus signaling acceleration; once the body has attained a constant velocity, the countermovement of the endolymph ceases and the hairs return to their resting position. A given receptor is stimulated by angular acceleration in one direction and inhibited by angular acceleration in the opposite sense. The corresponding semicircular canal on the opposite side of the body has the reverse pattern of response. All angular accelerations influence the discharge pattern of at least two semicircular canals, and many movements modify the output of all six receptors, the resultant information being synthesized by the cerebellum and postural areas of the cerebral cortex. Linear accelerations do not seem to influence the semicircular canals.

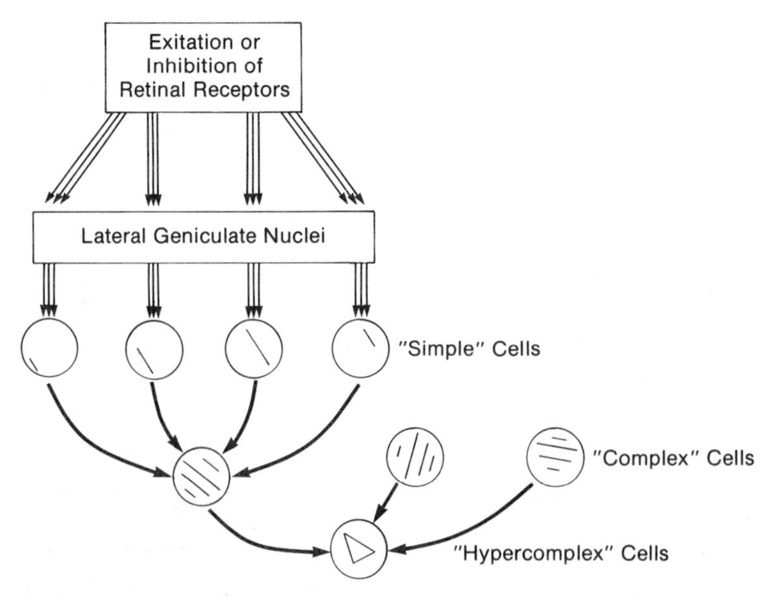

FIG. 7.21. A possible basis of pattern recognition in the visual cortex.

CENTRAL CONTROL MECHANISMS

Sensory Cortex and Processing of Sensory Data

Much more is known about the receptor organs (see above) than the central treatment of sensory information. We have already noted several possible pathways as the nerve fibers enter the dorsal root of the spinal cord:

- Direct synapse with a motoneuron (Fig. 7.9)
- Ascent or descent to an adjacent segment of the spinal cord (Fig. 7.9)
- Synapse with interneurons (Fig. 7.9)
- Synapse with nerve cells whose axons pass to the thalamus and cerebral cortex via the dorsal columns and spinothalamic tracts (Fig. 7.15)

The various synapses make important contributions to the processing of sensory information through both convergence and inhibition (see above). For example, afferent fibers from the cutaneous touch receptors over a wide range of the body surface may converge on a single dorsal column cell. Further, pre- or postsynaptic inhibition of transmission in adjacent cells adds greatly to the precision of localization of a stimulus.

All sensory pathways converge on the thalamus (Fig. 7.22). The spinothalamic, dorsal column, and trigeminal thalamic tracts pass via the medial lemniscus to the ventrobasal complex. The optic tract fibers pass to the lateral geniculate nucleus (see above), while auditory information is transmitted to the corresponding part of the sensory cortex (somatic sensory, visual, or auditory). The main projection of the ventrobasal complex is to the area immediately posterior to the central sulcus of the cortex. Here are represented in an inverted order (the lower limb at the upper end of the sulcus) the individual segments from the opposite side of the body, with a spatial representation of information about mechanical stimulation of the skin and joint movements. Some sensory fibers also reach the motor areas of the brain (zones 4 and 6; Fig. 7.23). Behind the classical sensory area lies the association areas, concerned with learning and complex interpretations of stimuli (Mountcastle, 1975); the association areas receive a sensory input from both sides of the body.

In interpreting sensory input, the brain makes a number of judgments. These include

- Type of stimulus (gauged from which of the sensory receptors are active)
- Intensity of stimulus (gauged from the frequency of afferent impulses and the number of cells that are active)
- Rate of stimulation (gauged from the behavior of rapidly adapting receptors)
- Duration of the stimulus (gauged from the behavior of slowly adapting receptors)
- Extent of the stimulus (gauged from spatial patterns of excitation and inhibition)

Motor Cortex

Patterns of Organization

The motor cortex lies in that part of the brain just anterior to the central sulcus. In most people, the left side of the brain is dominant, so that movements are performed most readily by the right hand. The left cerebral hemisphere is then concerned with various aspects of consciousness, analysis, and sequential

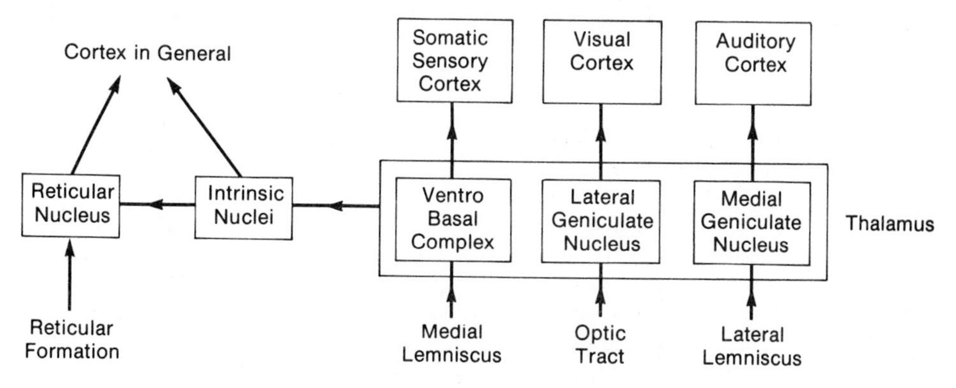

FIG. 7.22. General arrangement for processing sensory information.

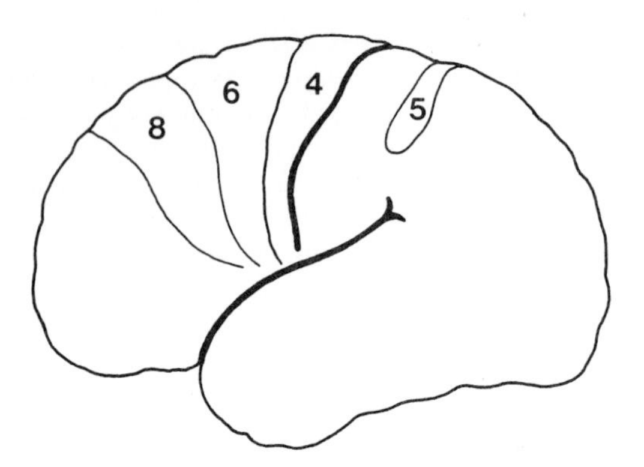

FIG. 7.23. Areas of the brain concerned in motor activity.

events, while the right side of the brain is nonverbal, being responsible for synthetic and musical ability (Eccles, 1973).

Microscopic examination of the motor cortex shows a large number of motoneurons (pyramidal cells) packed perpendicular to the cortical surface. Clinical neurophysiologists such as Hughlings Jackson (1932) held the view that the neural connections of these cells led to a representation of complex movement patterns within the motor cortex. Impulses from an appropriate group of "upper motoneurons" traveled via the pyramidal nerve fibers to a corresponding group of "lower motoneurons" in the ventral horn of the spinal cord, thus initiating a well-defined movement. Hughlings Jackson envisaged a hierarchy of movement patterns, relatively automatic activities ("middle-level") being stored in the sensory and motor cortex, and the least automatic activities being initiated from higher centers, possibly in the prefrontal cortex.

Unfortunately, the technique of electrical stimulation used in many topographic studies maps structure rather than function, and where a number of muscles have responded to a given stimulus, it is difficult to rule out the possibility that afferent fibers were activated or that there was a spread of the electrical disturbance to adjacent regions of the cortex. Certainly, more precise and localized electrical stimulation suggests that the upper motoneurons each control a small and highly specific "field" of muscle fibers (Asanuma et al., 1968; Fetz and Finocchio, 1972; Fetz, Finocchio, and Baker, 1973), although there are difficulties in explaining how the available store of DNA can "code" the necessary neural connections (Wyman, 1973; R. A. Schmidt, 1976).

Assuming that there are indeed many small units, this creates the possibility of initiating a vast repertory of movements by appropriate combinations of cortical activity. Possibly, the body can call upon both generalized programs that produce a number of similar movements and additional instructions that determine details of how the movement is to be carried out (Pew, 1974; Summers, 1975; Shapiro, 1976).

In species such as humans, where vision is a dominant sensation, the cortical motoneurons project largely to the opposite (heterolateral) side of the body. They reach not only the ventral horn cells of the spinal cord, but also the dorsal horn, where it is possible that they modify the sensory input to spinal reflex arcs and the spinocerebellar tracts (C. G. Phillips and Porter, 1977; Figs. 7.10, 7.23, and 7.24). The cortical motoneurons also give rise to collateral branches that pass within the cortex, to the striatum, the large-celled part of the red nucleus, and the midline reticular formation (D. M. Armstrong, 1965; Endo et al., 1973; Tsukahara et al., 1968; Kemp and Powell, 1971); these send a corollary discharge to parallel cortical efferent pathways (C. G. Phillips and Porter, 1977; Figs. 7.24 and 7.25). The network creates a potential for recurrent facilitation and inhibition of both pyramidal tract and corticorubral neurons, improving the information content of impulses directed to lower levels of the nervous system (Tsukahara et al., 1968). Positive feedback builds up a strong excitation in the column of pyramidal cells innervating a particular muscle group, while negative feedback to adjacent columns ensures inhibition of antagonistic muscles.

The cortical representation of the gross muscle groups is in an inverse order, motoneurons that project to the foot being found in the uppermost part of the motor cortex (Brodman's area 4; Fig. 7.23). The number of muscle fibers controlled by a single upper motoneuron is least in the case of muscles concerned with fine movements. In consequence, a large part of area 4 is concerned with muscles of the tongue, lips, larynx, thumbs, and fingers, and much smaller areas are allocated to the control of the leg and trunk muscles. Along the posterior border of the central sulcus, afferent information from peripheral touch receptors shows a similar topographic distribution (Asanuma and Rosén, 1972; Lemon and Porter, 1976).

Traditional Control Theory

According to traditional views, as a task was learned and became automatic, control was transferred from the motor cortex (area 4; Fig. 7.23) to the premotor cortex (area 6, and to a lesser extent areas 5 and 8) and/or the supplemental motor area (a zone buried deep within the medial longitudinal fissure of the

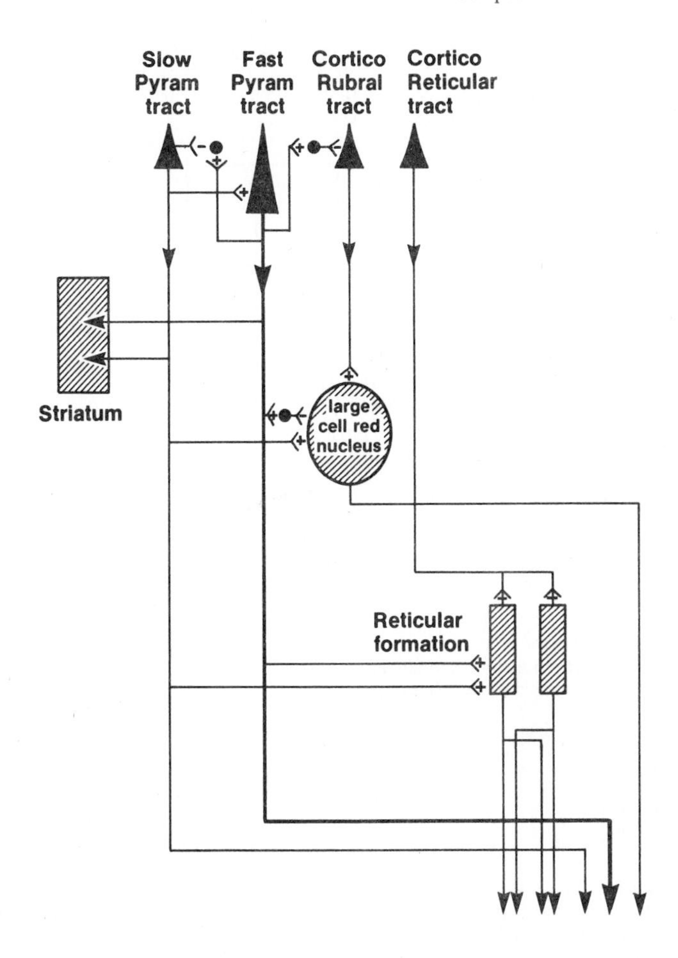

FIG. 7.24. Collateral connections of the pyramidal axons (see Fig. 7.25). (After C. G. Phillips & Porter, 1977.)

brain) (Wiesendanger et al., 1973). Efferent impulses then passed via extrapyramidal pathways; instead of traveling directly to the ventral horn cells, they were relayed in the reticular formation of the brainstem, and traveling thence modified settings of the γ loop (Figs. 7.24 and 7.25). The cerebellum played a most important role in this schema. While the movement was programmed and initiated by the premotor cortex, the strength, duration, and range of the resultant movement was modulated by impulses from the cerebellum. This received a continuous input of information from the eyes, the labyrinth, and the proprioceptors, compared this information with previously stored γ-loop settings, and adjusted the discharge of the γ-efferent fibers accordingly.

Such a control system presupposes a minimal program unit of ~250–300 msec duration, this limit being set by the visual reaction time. Events within the 300-msec span cannot be modified by external feedback (Welford, 1974). In contrast, signals generated internally from muscle stretch receptors can initiate activity of pyramidal tract neurons with a latency of 10–20 msec, and a similar adjustment of response can develop within 30–40 msec if an active movement is in progress, particularly if the necessary corrective action has been learned (Evarts, 1973b); possibly, there is a cortical loop from muscle receptors to pyramidal tract neurons concerned with the initiation and control of muscle contraction.

Current Control Theory

Current views (J. M. Kemp and Powell, 1971; C. G. Phillips and Porter, 1977) set the motor cortex at the center of a complex neural network (Fig. 7.25). One group of fibers originates in almost every part of the cerebral cortex, converging on the classical motor areas via the basal ganglia and thalamus (De Long, 1971; J. M. Kemp and Powell, 1971; Asanuma et al., 1974; Massion and Paillard, 1974; Kievit and Kuypers, 1975; Strick, 1975). The second pathway also has its origin in most parts of the motor cortex, but it loops back to the motor area via the pontine

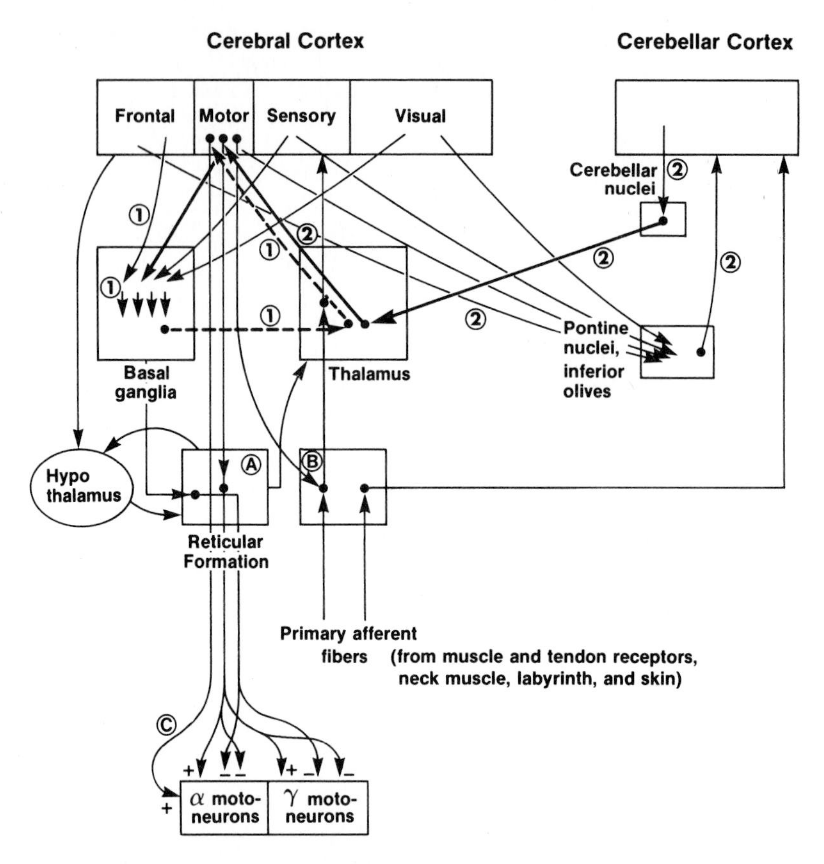

FIG. 7.25. Interconnections of motor pathways. There are two main loops for activation of the motor cortex: (1) from other areas of the cortex, through the basal ganglia and thalamus, and (2) from other areas of the cortex to the pontine nuclei, inferior olives, cerebellar cortex, cerebellar nucleus, and thalamus. Efferents from the motor cortex follow three main routes: (A) to the terminations of the primary sensory inputs in the dorsal horn of the spinal cord, the spinal trigeminal nucleus, the dorsal column nuclei, etc., (B) to the reticular formation, red nucleus, and striopallidum, whence they are relayed to the spinal reflex arc, with a potential for both facilitation and inhibition of α and γ motoneurons, and (C) corticobulbar and corticospinal axons that pass directly to the motoneurons of the appropriate cranial or spinal segment. (In part after J. M. Kemp & Powell, 1971.)

nuclei, inferior olives, cerebellar cortex, cerebellar nuclei, red nucleus, and thalamus (Eccles, Ito, and Szentágothai, 1967; Thach, 1970).

According to this theory, it also participates in a "downstream" loop that passes from the motor cortex via the pons and olive to the cerebellum and back to the motor cortex via the nucleus interpositus; this loop plays an important role in governing automatic movements (G. I. Allen and Tsukahara, 1974). Some signals are triggered by internal programs, related to environmental situations although remote from any direct connection with peripheral receptors. Other signals are generated by feedback (externally, from the moving parts, and internally, from cortico-cerebellar-thalamocortical loops). Nevertheless, conscious muscle sensations typically depend more upon the central command than upon a peripheral feedback of muscle force (Cafarelli and Bigland-Ritchie, 1978). As can be gauged from Figs. 7.15 and 7.25, looping is so extensive that it is difficult to mimic the behavior of a freely moving conscious animal in well-controlled laboratory experiments (V. B. Brooks and Stoney, 1971).

Descending Projections

The descending projections of the motor cortex form three functional groups. The first group (Fig. 7.10) terminates in relation to the endings of afferent sensory axons, in such sites as the dorsal horn of the spinal cord, the spinal trigeminal nucleus, and the dorsal column nuclei. Here, there is the possibility of controlling the input from muscles, tendons, joints, and the skin surface. The second group reaches the spinal reflex arc via relay stations in the brainstem: the reticular formation, red nucleus, and striopallidum (Fig. 7.15). The third group passes directly from the motor cortex to the motoneurons of the appropriate cranial or spinal segment; their origin is typically in the large pyramidal cells, and their action on target neurons is facilitated when several impulses are repeated at short intervals (Fig. 7.26).

The phenomenon illustrated in Fig. 7.26 seems an example of summation at the segmental synapse (Chap. 4). It is presumably brought about by a persistent alteration of ionic permeability in the target cells (Kernell and Sjöholm, 1973), and it has practical importance as a mode of varying the power of motor activity (Kernell, 1973; Muir and Porter, 1973). Apparently, some pyramidal cells code the force needed to produce a given movement in terms of the frequency of their discharge (Evarts, 1968; M. McD. Lewis and Porter, 1974; Fetz, German, and Cheney, 1975). Other pyramidal cells give a discharge proportional to the first derivative of the torque that is to be generated (Humphrey, 1972). The force developed can be further graded by a varying recruitment of pyramidal cells, and their order of discharge apparently determines the timing of muscle contraction sequences, so important to the skill and precision of a movement (Woolsey et al., 1972).

Timing of Electrical Activity

Activity of the pyramidal cells sometimes antedates movement by 100 msec or more (Deecke et al., 1969; Evarts, 1972a,b; Fig. 7.27). Some authors have suggested early pyramidal impulses have a "priming" function, allowing a later discharge to activate the ventral horn motoneurons. However, it seems more likely that the early electrical activity reflects passage of the impulse around some of the complex cerebral feedback loops described above (R. Porter and Lewis, 1975; Fig. 7.25).

The minimum time needed to alter the characteristics of a programmed movement by verbal instructions (Hammond, 1956), proprioception (Marsden, Merton et al., 1976a,b,1977), or other forms of feedback (Megaw, 1972; B. Jones, 1974) seems ~50–70 msec.

Cerebellum

The cerebellar cortex contains ~30 million Purkinje cells that receive afferent information via the spinocerebellar tracts, and less directly via the cerebrum and thalamus (Fig. 7.25). There are two types of afferent fiber within the cerebellum: climbing and mossy. Each climbing fiber innervates one Purkinje cell, making hundreds of excitatory synapses with its dendritic tree. The mossy fibers branch profusely, each synapsing with ~450 "granule cells." The latter give rise to parallel fibers which again form excitatory synapses with the Purkinje cells. Responses initiated via the mossy fibers occur ~10 msec faster than those occurring via the climbing fiber system.

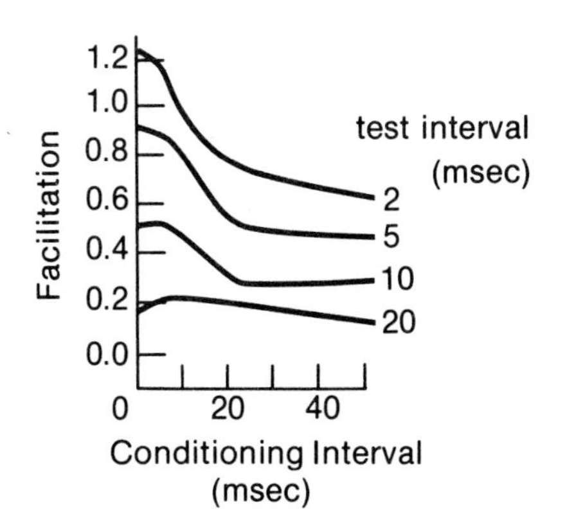

FIG. 7.26. Facilitation of target neuron by triplet of impulses in pyramidal fiber plotted in relation to conditioning interval and test interval. Facilitation is calculated in terms of excitatory postsynaptic potentials (EPSP): $(V_2 - V_0)/V_0$. (After Muir & Porter, 1973.)

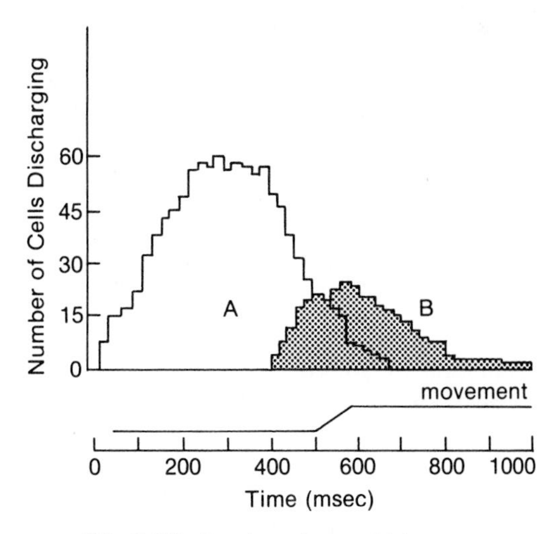

FIG. 7.27. Number of pyramidal tract neurons firing at varying times in relation to initiation of movement. A, neurons that commence firing >100 msec before movement; B, neurons that commence firing <100 msec before movement. (After R. Porter & Lewis, 1975.)

The cerebellar cortex also contains inhibitory cells (Fig. 7.28; basket, stellate, and Golgi cells). These are activated by both parallel and mossy fibers. The Golgi cells branch extensively and can inhibit up to 10,000 granule cells. One important function of such circuits is to inhibit Purkinje cells on either side of an active segment of cerebellum. They also "switch off" active groups of Purkinje cells after they have been firing for ~10 msec. As in a computer, the cerebellar cortex is then cleared ready for its next operation (Eccles, 1973).

The efferent fibers originate in the Purkinje cells. A group of ~200 cells sends axons that converge on a single neuron in the intracerebellar nuclei. Here an inhibitory action competes with direct excitation from the climbing fibers (Fig. 7.28), thus modulating the transmission of information to the anterior horn cells of the spinal cord, via both thalamocerebral pathways and, more directly, rubrospinal and vestibulospinal tracts (Grillner et al., 1969; Fig. 7.25). The convergent design of cerebellar pathways gives "signal averaging" (as in ECG and EEG analysis), improving the reliability of signals and overcoming problems resulting from "noise."

Role of Nuclei

The extrapyramidal tracts pass to the motoneurons via the basal ganglia, red nucleus, reticular formation, vestibular nucleus, and cerebellum.

The basal ganglia are relay stations on the pathway connecting the cerebral cortex and thalamus with the brainstem nuclei (Fig. 7.25). Electrical activity can be detected in the basal ganglia in advance of muscular contraction (Evarts, 1973a).

The red nucleus is a relay station for transmission of the cerebellar output down the rubrospinal tract (Fig. 7.15). It seems specifically concerned with static fusimotor behavior (Koeze et al., 1974).

The reticular formation is a diffuse collection of nuclei in the midbrain, pons, and tegmentum of

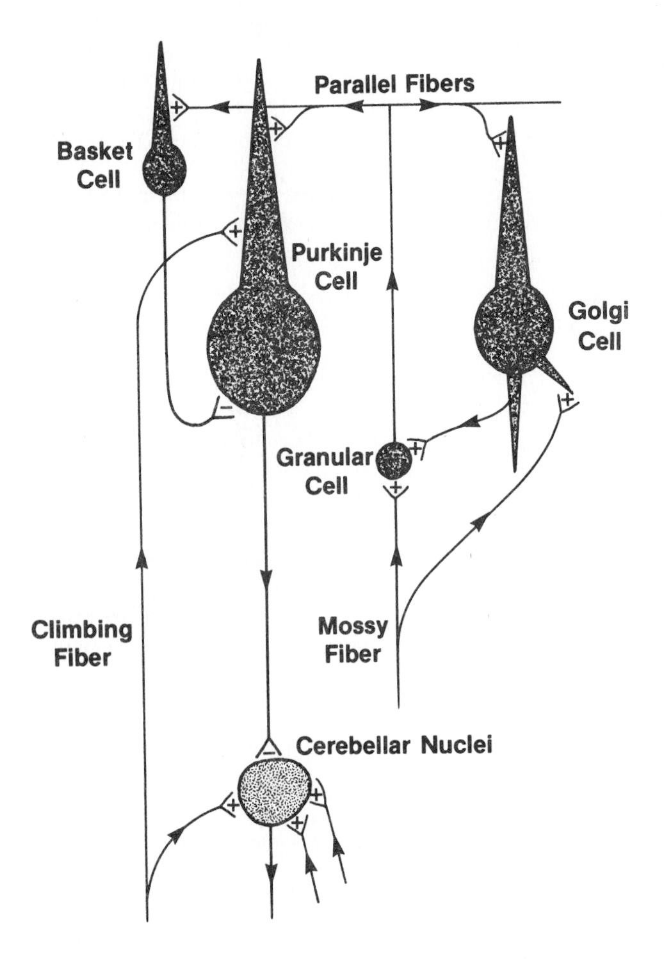

FIG. 7.28. Some of the synaptic connections in the cerebellar cortex. (After Eccles, 1973.)

the medulla. It is concerned with arousal, and sends excitatory and facilitatory reticulospinal fibers to both α and γ motoneurons in the spinal cord (Vedel and Mouillac-Bandevin, 1969; Appelberg and Jeneskog, 1972; Grillner, 1975).

The thalamus is an important sensory relay station (see above) and is also concerned in the transmission of information between the cortex and the cerebellum (Fig. 7.25).

Learning of Movements

Corticocerebellar Shift

While an analysis of movement patterns (see above) may be helpful in correcting specific faults of performance, the body tends to think of a movement as a whole rather than as a series of isolated muscle contractions. As a task is learned and becomes automatic, the main site of control is shifted from the cortex to the cerebellum (Chap. 3) and there is reliance upon proprioceptive rather than visual feedback (M. Robb, 1968). The sole role of the cortex is then to call the movement into play in response to an appropriate external stimulus. A shift of attention from the elements of the task to the initiating signal is the essence of motor learning.

Nature of Memory

Memory comprises both short- and long-term components (Stelmach, 1974; Gentile and Nacson, 1976); the short-term store has a relatively limited capacity, while the potential for long-term storage of rehearsed, reinforced, and meaningful information is exceedingly large. As in the sensory domain (Poulton, 1970), there is some evidence that many rapidly presented movements can overload the processing capacity of the short-term store (P. M. Fitts, 1954; Salmella, 1974); however, limitations of "channel capacity" are much less obvious than for the assimilation of verbal information (Kantowitz, 1972; S. W. Keele and Ellis, 1972). Information that does not pass beyond the short-term store is quite rapidly forgotten; presumably there is decay of some "memory trace" (Broadbent, 1958) unless it is rapidly reinforced (Stelmach, 1974).

Motor learning seems highly task-specific, with little evidence of general motor educability (Marteniuk, 1974). Intraindividual variations in the skill of performance are affected little by learning. Intraindividual variations change early in learning, but remain more constant in the middle and late stages of practice.

The nature of coding is still discussed. Unlike a manmade computer, the memory does not store unique solutions to a problem (Gentile and Nacson, 1976); the response evoked is situational and subject to continuous change in the light of various forms of feedback. The information stored includes not only γ-loop settings, but also visual and auditory signals associated with a well-performed movement. In some instances, such impressions may override kinesthetic cues (J. A. Adams, 1971). Gentile and Nacson (1976) argue that evaluation covers both the observed change in the environment relative to the expected environmental outcome and the observed movement pattern relative to the intended movement pattern. Concordance with expectation on both criteria gives strong reinforcement, while concordance on one of the two criteria gives a partial reinforcement.

Various techniques are adopted to exploit these concepts of coding. If a task calls for a limb to be moved at a steady pace, the subject may use verbal cues, such as counting, to regulate his performance. Less consciously, use may also be made of the output of the cortical motoneurons (Teuber, 1964).

Anatomical Basis of Learning

The anatomical substratum of learning is plainly the establishment of new synaptic connections within the central nervous system; although there is no gross change of structure, the brain retains a substantial plasticity of function (Eccles, 1973; P. Gilbert, 1975; Buser, 1976). Frequent use of specific synapses seems to stimulate protein synthesis, with branching of the spines of dendrites, hypertrophy of synaptic knobs, and formation of additional synapses. Eccles (1973) and P. Gilbert (1975) speculated that one important function of the cerebellar climbing fibers may be to increase the strength of parallel fiber synapses on the Purkinje cell spines (Fig. 7.28). The resulting motor output is monitored by a deeper lying nucleus (nucleus locus caeruleus), and if the outcome of the activity is appropriate, the nucleus sends signals to the Purkinje cell facilitating a more permanent strengthening of the synapse; in essence, the climbing fibers develop short-term memory, and axons from the nucleus locus caeruleus transform useful γ-loop settings into a long-term memory (Fig. 7.29).

The chemical basis of such memory is a special form of ribonucleic acid formed in the nucleus of the nerve cells (Hydén and Egyhazi, 1962). In animals, it has proved possible to facilitate learning by injecting brain extracts from trained donors (Byrne et al., 1966), thus raising the specter of new and horrendous techniques for the doping of athletes.

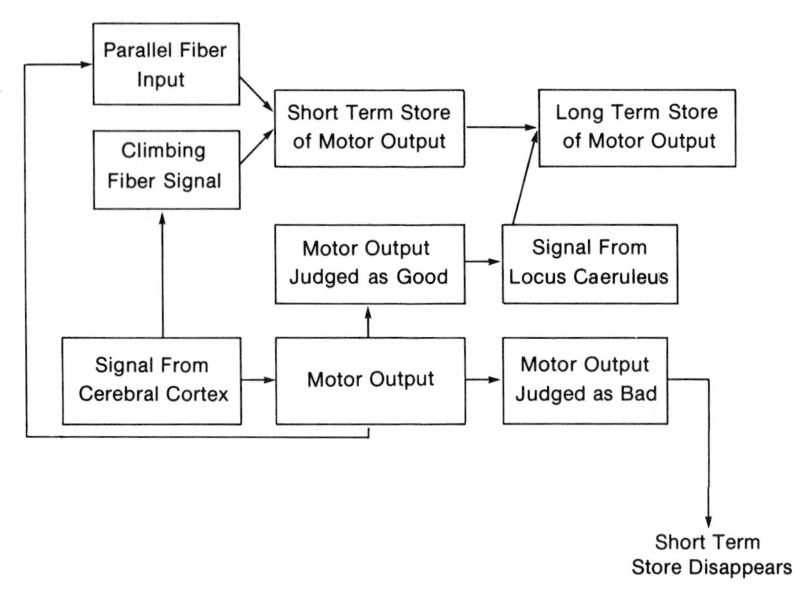

FIG. 7.29. Hypothetical basis of short- and long-term learning of movement paterns by the cerebellum. (After P. Gilbert, 1975.)

Practical Applications

In general, there is little "transfer" of learning; skill at one task has relatively little influence upon the ability to perform some other activity. Indeed, if the tasks are closely related but dissimilar (for example, the playing of tennis and squash), "negative transfer" may occur, practice of one sport leading to a worsening of performance in the other, with difficulty in the recall of appropriate skills. "Proactive interference" is a problem arising from a previously learned skill, while "retroactive interference" is a problem attributable to activity undertaken subsequent to the criterion learning (McGeoch, 1932). If a task is learned by the dominant limb, there can be some "cross-education" of the opposite limb (Hellebrandt and Waterland, 1962). This is shown by an increase not only in skill, but also in strength and endurance. In contrast, training of the nondominant limb has little effect on the performance of the dominant limb.

The physiological basis of cross-education has yet to be explained. Training involves overloading of the active limb, and in such circumstances an involuntary tensing of muscles on the opposite side of the body might be anticipated; however, this is not seen during either simple, nonresistive exercises or isometric contraction (R. A. Gregg, Mastellone, and Gersten, 1957; Samilson and Morris, 1964). One possible anatomical factor contributing to cross-education is the fact that 15 to 30% of the fibers from the motor cortex reach ventral horn cells on the same rather than the opposite side of the body. The augmentation of force in a weak or tired muscle by contraction of the homologous muscle in the opposite limb ("cocontraction") has been explained on a similar anatomical basis (Cratty, 1968).

If the muscles normally responsible for a specific movement are weakened or paralyzed by a disease such as anterior poliomyelitis, the body learns by a process of trial and error techniques for the use of alternate muscles. Thus, if the leg muscles are affected, walking is achieved by the trick of swinging the hip bones, using the quadratus lumborum muscle. Once these abnormal patterns of activity are established and the corresponding γ-loop settings are stored in the brain, the new forms of movement in turn become automatic, and are difficult to eradicate if there is a later recovery of normal muscle function. It is thus necessary to reeducate the patient, restoring normal movement patterns as soon as possible. If this is not done, further wasting of the unused muscles may occur, possibly associated with a shortening of tendons and the development of permanent deformities.

MUSCLE TONE AND POSTURE

Muscle Tone

General Considerations

Although there is some allocation of function between red (postural) and white (voluntary) muscles, all movements are superimposed on a background of postural tone, and most movements involve a disturbance of normal equilibria with an increase of postural work.

The extent of postural work varies with body mass, the skill of the individual, and the location of his center of gravity (the point within the body at which the total mass may be thought as acting without altering responses to gravitational acceleration; Fig. 7.30). Interindividual differences in postural work contribute to differences of mechanical efficiency (Chap. 1) from one person to another. Learning of a task generally leads to a reduction of postural activity, and thus a decrease of oxygen cost. One of the tasks of the physical educator and of the ergonomist is to teach patterns of movement that minimize postural work.

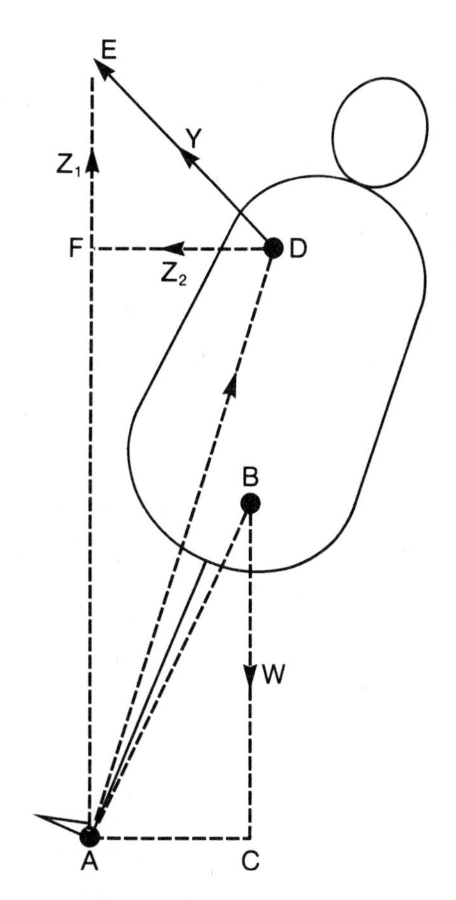

FIG. 7.30. The principle of center of gravity. A subject leans backwards, holding his legs as a rigid pillar and pivoting about his heels (A). The body mass W behaves as if concentrated at point B, and exerts a torque W • AC, tending to cause the body to fall. In this example, the arms resist falling by exerting a counterforce Y in the direction DE; this may be resolved into components Z_1 and Z_2 acting respectively in the axes EF and DF. For equilibrium, Z_1 • DF must equal W • AC, and Z_2 must be opposed by an appropriate friction between the heels and the ground.

Reflex Basis of Muscle Tone

The fundamental basis of muscle tone is the stretch reflex, already discussed. Some authors have thus insisted (Kelton and Wright 1949; Clemmesen, 1951; Ralston, 1957; Basmajian, 1967,1973; Vallbo, 1973b) that the muscles of a well-relaxed subject show quite long phases of electrical silence (2 min or more). Nevertheless, in most situations sensitive recording equipment will reveal occasional action potentials, suggesting that resting tone depends upon not only "passive elasticity or turgor of muscular and fibrous tissues" (Basmajian, 1967,1973) but also a "rotation of duty" among the motor units that make up an anatomical muscle (Jacobsen, 1943; Herman, 1970; de Vries, 1974).

The intensity of such activity depends on phasic variations in the sensitivity of individual spindle receptors, controlled through the γ-loop mechanism. The γ motoneurons are activated by the cells in the reticular formation of the brain (see above and Figs. 7.15 and 7.24). The activity of the reticular cells in turn reflects an individual's wakefulness or "arousal." Hence, both muscle tone and the sensitivity of the stretch reflex are increased when a person is aroused (for example, an athlete prior to a contest, or an examination candidate who is "scared stiff"). As we have seen from the length/tension diagram of skeletal muscle (Fig. 1.9), there is an optimum fiber length for maximum performance. The happy, alert individual tends to operate at optimum tone. Fear gives an excess of tension, while sadness and depression cause a loss of tone.

The "facilitatory" effect of increased reticular activity (Figs. 7.15 and 7.25) is normally counteracted by impulse from the basal ganglia of the brain. However, degeneration of the basal ganglia sometimes occurs in middle age, and a gross increase of muscle viscosity then develops (Parkinson's disease; Gottlieb and Agarwal, 1973). Other parts of the reticular formation also have an inhibitory influence on the γ loop. The input to these areas is derived from the cerebral cortex and cerebellum, and their output plays an important role in ensuring an appropriate relaxation of antagonistic muscles during the performance of complex movements.

Assessment of Muscle Tone

The assessment of muscle tone is a common clinical procedure. The subject is asked to relax, and the observer then assesses the firmness of the muscle to palpation, the resistance to passive movement about a given joint, and the briskness of stretch reflexes such as the response to a tap on the patellar tendon.

De Vries (1974) devised an apparatus for quantitative palpation of a muscle. It resembles the tonometer used by ophthalmologists. The main difficulty in practical use of the instrument is that the force recorded depends as much on the thickness of subcutaneous fat as on the inherent turgor or elasticity of the muscle belly.

Electromyography permits assessment of electrical activity; if surface electrodes are used, the overall status of a muscle is indicated. However, there are difficulties in recording the small voltages associated with resting tonus because of attenuation by overlying tissues (fascia, fat, and skin), and the signal is sometimes difficult to distinguish from noise.

Other dynamic approaches to the assessment of muscle tone include the recording of grip pressure from an idle hand, the recording of the pressure on a writing stylus, and measurements of blood lactate level.

The Art of Relaxation

Since resting muscle tone is influenced by anxiety, the tension measured by the integrated EMG diminishes as a subject becomes familiar with the laboratory and the recording equipment; in other words, a form of habituation occurs. The recording of muscle tone is sometimes useful in teaching the art of relaxation. One approach suggested by Jacobsen (1938) is to train the individual to recognize progressively lower levels of voluntarily induced tension. When his perceptive powers have been improved in this way, he becomes able to recognize and correct involuntary tension (G. Paul, 1966; Goldfried, 1971; Pinel and Schults, 1978; T. J. O'Hara and Orlick, 1978).

The practice of yoga is a second method of reducing involuntary tension. The expert practitioner of shavasana, the yogic technique of relaxation, is said to be able to reduce his oxygen consumption by as much as 50% with parallel reductions of heart rate, respiratory rate, blood cholesterol, and blood sugar (Mookerjee, 1974).

Most athletes are able to reduce the integrated EMG more readily than sedentary subjects. This is partly a question of practice, since sedentary people trained in the art of relaxation can often accomplish a greater reduction of muscle tension than athletes. However, there is also some evidence that vigorous exercise reduces muscle tension, particularly in those individuals in whom muscle tone is initially high. Athletes can also learn to activate individual motor units (Carlsöö and Edfeldt, 1963; Basmajian, 1967).

Posture

Postural Oscillations

At many body joints, the upright posture is maintained through slight oscillations about the equilibrium position. It is impossible for the line of gravity to pass through the center of every joint, and for this reason the upright position cannot be sustained without muscular effort (Steindler, 1955). As equilibrium becomes unstable (Fig. 7.30), the muscles on the opposite side of the joint ("antigravity muscles") are stretched, and through the γ loop a compensatory contraction is initiated. If compensation is excessive, overshoot occurs, leading to oscillation of the moving parts (Basmajian, 1967; Joseph, 1969). The frequency of such oscillation is determined by the length of the equivalent compound pendulum (see above).

Damping of the oscillation normally occurs through a process of phase advance (Gottlieb and Agarwal, 1973). However, if the sensitivity of the γ loop is increased (by fear or disease), a frank oscillation (tremor) can develop. Shivering results from a brisk and simultaneous contraction of agonists and antagonists; again there is oscillation at the natural frequency of the part (V. B. Brooks, Cooke, and Thomas, 1973).

Maintenance of Equilibrium

A body remains in a state of stable equilibrium as long as its center of gravity (Fig. 7.30) is held within the vertical projection of the area bounded by its supports. This is likely if the area of support is large and the center of gravity is low. Unfortunately, in humans the area bounded by the feet is small, and when standing the center of gravity is high. Thus, even a minor displacement of the trunk moves the center of gravity outside the area of support, creating an unstable equilibrium.

The center of gravity can be located by lying a subject between two or more scales and noting their respective readings (D. K. Mathews, 1963; Rasch and Burke, 1971; Niinimaa, Wright et al., 1977). In a typical adult man, the center of gravity lies at ~55% of standing height (Webb, 1964; Sen et al., 1978; Table 7.5). Tall individuals have a high center of gravity in both relative and absolute terms, putting them at a disadvantage in terms of maintaining their balance.

The tendency to fall is resisted by a number of skeletomuscular mechanisms. Where possible, the

TABLE 7.5
Location of Centers of Gravity for Body as a Whole and for Individual Segments

Segment	% of Body Mass	Vertical Axis (% of Stature)	Anteroposterior[a] (% of Stature)	Lateral[b] (% of Stature)
Whole body	100.0	55.3	0	0
Head	6.9	93.5	0	0
Trunk and neck	46.1	71.1	0	0
Upper arm	6.6	71.7	0	10.7
Lower arm	4.2	55.3	0	10.7
Hand	1.7	43.1	0	10.7
Upper leg	21.5	42.5	0	5.0
Lower leg	9.6	18.2	0	5.0
Foot	3.4	1.8	3.9	6.2

[a]Relative to ankle joint.
[b]Relative to midline.
Note: The overall center of gravity is set 1–2% lower in women and is somewhat higher in children and adolescents.
Source: Based on data of P. Webb, 1964.

joints are positioned to form rigid supports. Thus, when standing, thee knee joint is "locked" by an outward rotatory movement. If mechanical locking is not possible, the joint must be braced by continuous muscular activity, either unilateral contraction against the gravitational field or a simultaneous contraction of agonists and antagonists. In the latter case, there is a slight oscillation about the position of equilibrium (see above).

A given static posture must sometimes be maintained for a long period, and the muscle responsible must then be resistant to fatigue. The high myoglobin content, long contraction period, and other peculiarities of type 1 muscle fibers (Chap. 4) are important in this regard. Fatigue reflects largely an impairment of blood flow and may be anticipated when contractions are sustained at more than 15% of maximum voluntary force (Chap. 3). Postural requirements normally do not reach this level. However, they can do so (1) if an awkward posture is maintained, with bad alignment of the center of gravity of the various body parts, (2) if the supporting muscles are weakened through disease or lack of physical training, and (3) if the subject is supporting a heavy load or is himself obese.

Control of Posture

SPINAL MECHANISMS. The primary control of static posture is in the spinal cord, through the stretch and other segmental reflexes discussed above. Thus, a cat whose spinal cord has been deliberately severed in the lower lumbar region can support the weight of its hind quarters for several minutes. However, if stance is to be maintained for a longer period, more coordination is necessary than can be achieved at a segmental level.

COORDINATION. Coordinating centers are located mainly in the cerebellum. Here, impulses from the eyes, the inner ear, the neck, and the soles of the feet are weighed against the state of activity in the segmental stretch reflexes (Magnus, 1926; Fig. 7.10). The segmental response is then suitably modified by facilitatory or inhibitory impulses from the reticular system. The importance of the cerebellum in both modulating and damping spinal reflexes is shown by the effects of disease and injury (Gilman, 1973; Gottlieb and Agarwal, 1973). An individual with cerebellar damage shows *ataxia* (clumsy, slow, and incomplete movements), *asthenia* (weakness), and *atonia* (loss of tone). The muscles also fail to act in a coordinated sequence (*asynergia*), and there is a coarse tremor because the mechanism for damping body oscillations has been lost. There is some evidence that the cerebellum learns and stores movement patterns that are particularly effective in restoring balance after equilibrium has been disturbed. A footballer who slips on a muddy field or a boxer who receives a staggering blow both recover their equilibrium more quickly than would an untrained person.

VISION. A healthy person has a fair idea of his body position even if the eyes are closed (C. W. Mann et al., 1949). However, *visual impressions* are important to the maintenance of posture (Witkin and Wapner, 1950), particularly if other receptors are no longer functional. The eyes play an added role in the control of rapid movements. Visual stimuli from the retina are supplemented by information from the stretch receptors in the ocular muscles and the attachments of the lens, as convergence and accommodation occur (Bach and Rita, 1971; Bizzi et al., 1971). Paralysis of accommodation leads to a surprising loss of postural control. One common example of this is the incoordination of a person treated with an atropine mydriatic prior to ophthalmic examination of his eyes.

INNER EAR. The *labyrinth* of the inner ear contains organs that sense both dynamic changes of posture (the semicircular canals; G. M. Jones and Milsum, 1971; Fernandez and Goldberg, 1971; Nashner, 1972) and the static orientation of the head (the otolith organs, saccule, and utricle; L. R. Young and Meiry, 1968). The discharge from the receptors of the semicircular canals is proportional to the acceleration of the head, and if a steady rotation is maintained the neural activity ceases within about 30 sec. The *otolith organs* are stimulated by tilting the head through 2.5° or more. The *saccule* detects a lateral tilt, and the *utricle* a fore and aft tilt (Fig. 7.31). The intensity of discharge is proportional to the angle of tilt, and the receptors show little "adaptation," continuing their discharge for as long as the head remains tilted. Individuals with a good sense of balance are thought to have a greater sensitivity of their labyrinthine receptors. There is some evidence that sensitivity is increased by frequent stimulation and that it is reduced or lost in the absence of normal gravitational acceleration (as in space voyages). Overstimulation of the receptors (by whirling or tumbling movement, and by various forms of travel) leads to loss of speed and accuracy in reaching movements, particularly in the nondominant limb (Canfield et al., 1953). If the stimulus is maintained, dizziness and impairment of balance are supplemented by nausea and vomiting, the familiar picture of "motion sickness" (D. E. Goldman and Von Gierke, 1960; Guignard, 1965; Whiteside, 1965).

The problem is essentially related to head rather than body movement, and if the head is kept still, sickness can be avoided (W. H. Johnson et al., 1951). In travel, a firm shoulder harness is helpful, while in dancing the experienced performer learns to watch a distant point until the neck can no longer be turned comfortably, finally moving the head quickly to "fix" the eyes on a second distant point. Training reduces the liability to dizziness partly through the technique of visual fixation and partly through suppression of stimuli. A champion figure skater often performs maneuvers that would cause an incapacitating labyrinthine reaction in a novice (McCabe, 1960; W. E. Collins, 1965). One of the techniques during a fast pirouette is to suppress the visual input most of us use in balancing.

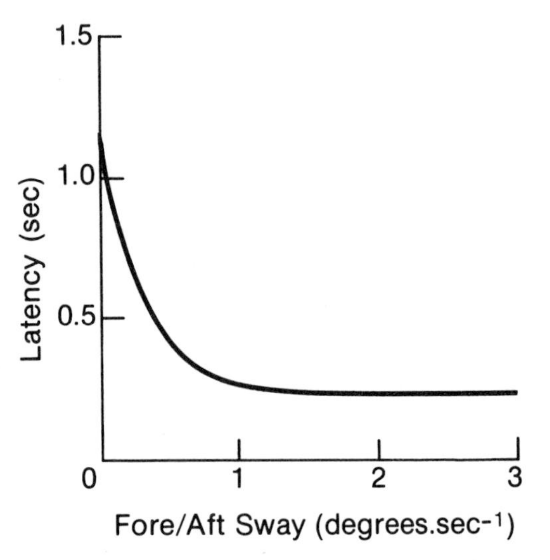

FIG. 7.31. The relationship between the rate of fore/aft sway and latent period for its detection. (After Nashner, 1973.)

NECK REFLEX. Stimuli received by the eyes and labyrinth tend to maintain an appropriate orientation of the head during tumbling or falling. This increases the tension in the neck muscles and the attitude of the body is adjusted by the neck reflex (V. C. Abrahams, 1972a); spindle receptors in the neck then excite or inhibit the vestibulocervical output (Gernandt and Proler, 1965) and inhibit the vestibulolumbosacral output (V. C. Abrahams, 1972b). In springboard diving and gymnastic tumbling, a slight error in the positioning of the head greatly reduces the precision of body movement.

CUTANEOUS PRESSURE RECEPTORS. Much information on posture is derived from the deep-pressure endings in the soles of the feet (Mori et al., 1970). The cerebellum compares the intensity of discharge from different parts of the feet, sensing a loss of balance from the increased discharge on the side toward which tilting has occurred. The relevant sensory impulses travel up the spinal cord in the dorsal columns (Fig. 7.15). Degeneration of these columns may occur in the late stages of syphilis (the clinical condition of "tabes dorsalis"). Affected patients are able to maintain their balance while the eyes are open (Dubrovsky et al., 1971; Beck, 1973), but once the eyes are closed, compensation is no longer possible. A common complaint is of tumbling into the basin when washing.

Assessment of Posture

It is difficult to provide a general description of good or bad posture, because optimal standing, sitting, and working positions vary with body type (the lean ectomorph, the muscular mesomorph, or the fat endomorph) and the nature of the task to be undertaken. There are also marked interindividual differences in the structure of the vertebral column and in the center of gravity of the various body segments. In general, an ideal posture minimizes static work. Faults of posture may have an organic basis (injury, weakness or disease of muscle, inherited abnormalities of the vertebral column), but more frequently they are acquired from poor habits, the carriage of heavy loads, or the wearing of high-heeled shoes. A caricature of "bad" posture is shown in Fig. 7.32. The initial fault is thought to be a hyperextension of the knee joints, possibly due to overflexible joints, or a lack of tone in the hamstring muscles (M. Turner, 1965). The defect is compensated by a hollow back, an increase of the pelvic angle, and a protruding abdomen. Such changes lead in turn to a rounding of the shoulders and a thrusting forward of the head (poke chin).

Many defects of posture, such as those illustrated, can be detected by simple observation and the study of silhouettes (Brownell, 1928; C. H. Hubbard, 1935). In the anteroposterior view, account is taken of tilting of the head, differences of shoulder height, differences in prominence of the hip bones and of the two halves of the rib cage, alignment of the legs (knock knees, bow legs, inward rotation of the thighs), and deformities of the feet (pronation/supination, hallux valgus, and hammer toes; D. K. Mathews, 1963; M. Turner, 1965; Rasch and Burke, 1971). A *plumb line* is helpful in making a simple overall assessment of posture. If the stance is good, the lobe of the ear should be in line with the middle of the shoulder tip, the middle of the greater trochanter of the femur, the back of the patella, and the front of the fibular malleolus (Steindler, 1955).

More detailed examinations of static posture usually include a radiographic examination; a reliability of ~1% has been claimed for simple mechanical devices used in studying spinal curvature (the "*conformateur*" and the "*spinograph*"; T. K. Cureton et al., 1935), but in practice they do not add greatly to a subjective rating.

Some faults of dynamic posture can be detected by direct observation. The runner, for example, should maintain a forward lean, with a natural alignment of the head and shoulders. However, toward the end of a race a tired competitor may allow his head to be thrown back, straightening his trunk and shortening his stride (Dyson, 1970). More complete analysis of dynamic posture is based on careful frame-by-frame examination of cinematograph film. The dynamic efficiency can also be assessed in physiological terms, such as the oxygen cost of a given activity and the corresponding load imposed on the heart, ventilation, and postural muscles (quantitative electromyography). The cumulative effects of poor posture can be documented in terms of rising blood lactate levels, oxygen debt, and the delayed recovery of heart rate and respiratory minute volume following effort.

Consequences of Poor Posture

APPEARANCE. The main argument for improvement of static posture is undoubtedly cosmetic

FIG. 7.32. A caricature of bad posture. Note the forward thrusting ("poke") chin, rounded shoulders, hollow back, increase of pelvic angle, protruding abdomen, and hyperextension of the knee joint. (After M. Turner, 1965.)

(Rathbone, 1938). A better appearance can in turn have beneficial effects on body image (Glassow, 1932) and consequently the attitude to minor aches and pains that might induce a medical consultation. As long ago as 1743, Audrey called sitting upright a "good posture" and sitting in full flexion a "bad posture." In the latter position, the back was "crooked and round" and the body form was "ungraceful." A bad dynamic posture is esthetically displeasing, both in top-level competitors and in ordinary citizens.

WORKING EFFICIENCY. Rasch and Burke (1971) noted that a rigid military posture required 20% more energy than normal relaxed standing, and they suggested that top athletes learn to relax completely between competitive efforts. Nevertheless, there are many circumstances whereby an improvement of posture reduces static work. There is then less tendency to accumulate lactate, and if a given body position has to be maintained for a long period, there is less likelihood of fatigue. An awkward performance leads to loss of accuracy, and an excessive expenditure of energy. It may also contribute to organic injuries such as prolapse of an intervertebral disc.

Rasch suggested that forward displacement of the lumbar vertebrae and injury of the erector spinae muscles were both made more likely by a combination of pelvic tilting, hyperextension of the spine, and weak counterpressure from the abdominal muscles. However, Chaffin and Moulis (1968) found no relationship between the inclination of the sacrum and the incidence of low-back pain. Klausen (1965b) suggested that while the short muscles of the back controlled individual intervertebral joints, the stability of the spine as a whole depended upon the long muscles of the back and/or the abdominal muscles. If a load was carried high upon the back, the tendency was to lean forward, increasing the load on the back muscles (Carlsöo, 1964). However, when lifting from a stooping position, an increased proportion of the needed counterforce was developed by the anterior abdominal muscles (Asmussen and Poulsen, 1968).

VISCERAL DISPLACEMENT. At one time, various abdominal complaints, such as dropped kidney and visceroptosis, were attributed to bad posture. It is now recognized that surprisingly large displacements of the abdominal organs can occur without harmful effects, and such terms as visceroptosis have been largely discarded by the medical profession.

Posture and the Cardiorespiratory System

CARDIOVASCULAR PROBLEMS. Substantial cardiovascular adjustments are necessary on moving from the recumbent to the upright posture and vice versa (Chap. 6). When standing, pooling of blood in the leg veins reduces central blood volume, diastolic filling pressure, and (through Starling's law of the heart) stroke output and systemic blood pressure. The decrease in impulse traffic from the aortic and carotid sinus pressure receptors then initiates several compensatory mechanisms, including tachycardia, peripheral vasoconstriction, and constriction of the leg veins (Abboud et al., 1979). In some situations, these adjustments are inadequate and fainting can occur, particularly:

- After exercise, when the body is hot and the muscle vasculature is widely dilated
- In a hot environment, where the capacity vessels are further relaxed and blood volume is depleted by sweating
- In poorly trained subjects with a low blood volume and/or diminished reflex adjustments to gravitational stress ("vasoregulatory asthenia")
- In people exposed to weightlessness for long periods, so that the normal reflex adjustments to gravity have been "forgotten"

A number of older circulatory tests of fitness (Crampton, 1905; Schneider, 1920) are based on the briskness of the reflex response on moving from the supine to the vertical position.

During maximum, symptom-limited work on a bicycle ergometer, the heart rate is 9% higher, the systolic pressure 16% less, and the cardiac output 17% less sitting than lying (Bevegård, Holmgren, and Jonsson, 1963; Kubicek and Gaul, 1977). However, Thadani and Parker (1978) argue that the *change* in cardiovascular variables from rest to work is similar in the two positions. Pulmonary arterial pressures during exercise are slightly greater sitting than lying (Kubicek and Zwick, 1976).

Standing increases capillary pressure in the legs, favoring effusion of fluid. Prolonged standing thus leads to a tiring swelling of the limbs, especially on a hot day (when capillary permeability is increased). The peripheral pooling and diminution of central blood volume can compromise renal flow. This is reflected by a loss of protein in the urine (the so-called orthostatic proteinuria, generally regarded as a normal phenomenon; Schultze and Heremans, 1966; Poortmans, 1969).

The healthy heart finds little problem in adapting to the increased central blood volume on lying down. However, if the output of the left side of the heart is impaired, as in such pathological conditions as stenosis of the mitral valve, acute breathlessness may develop.

RESPIRATORY PROBLEMS. The importance of posture to ventilation of the lungs has sometimes been overemphasized. The upright position (standing, and especially sitting with a relaxed abdominal wall) is associated with a higher proportion of diaphragmatic breathing (Goldthwait et al., 1930). At one time, this was thought to improve ventilation of the lower part of the lungs. However, this is unlikely, as changes of pressure are distributed rapidly throughout the pleural cavity. The compliance of the lungs is increased by the reduction of central blood volume, and this—coupled with changes of intrapleural pressure—increases the various static lung volumes by up to 500 ml (Shephard, 1961c; Agostoni and Mead, 1964). Greater expansion of the lungs minimizes the tendency to collapse of dependent lung regions, but this benefit is largely offset by an increased gradient of pressure in the pleural fluid (Agostoni et al., 1970). Perfusion of the upper parts of the lung might be thought more difficult when standing, since pulmonary arterial pressures are small relative to the height of the average thorax (Chap. 6). However, this problem is helped somewhat by the increase of pulmonary arterial pressures in the sitting relative to the supine position (see above).

During exercise, the respiratory rate is higher in the sitting than in the supine position (McGregor et al., 1961).

Practical Applications

POSTURAL COSTS OF WORK. When lying in a resting (but not basal) condition, the energy expenditure of the average man is 5.0 kJ • min^{-1}. On standing, this rises to 6.7-8.4 kJ • min^{-1}, the additional 1.7-2.4 kJ • min^{-1} representing the cost of maintaining an upright posture (Durnin and Passmore, 1967; Chap. 14). Most forms of laboratory exercise include an element of postural work; a graph of external work against energy expenditure will thus show a zero intercept substantially greater than 5.0 kJ • min^{-1}.

Postural work, and thus the energy cost of industrial and domestic tasks (Bedale, 1924), varies with the level of the working surface. The optimal height varies with body build and the nature of the task to be performed, but is commonly ~90-92 cm. If the bench height is increased, the body weight can no longer be used to assist in performing a task. On the other hand, if the bench is too low, the worker must stoop continually, and a substantial postural effort is required. In one specific example (J. R. Brown, 1971), energy expenditures were as follows:

Height of Working Surface (cm)	Energy Expenditure (kJ • min^{-1})
69	17.1
93	11.7
162	13.0

In many commercial and domestic tasks, much reaching and bending are involved, and a large part of the required work is due to a lifting and lowering of the center of gravity of the body (Godin and Shephard, 1973b). This can be illustrated by data for housewives (Droese et al., 1949):

Sitting (e.g., sewing, cooking)	6.7 kJ • min^{-1}
Standing (e.g., dishwashing, ironing)	9.6 kJ • min^{-1}
Reaching up (e.g., dusting, polishing, window cleaning)	17.1 kJ • min^{-1}
Bending down (e.g., bed making)	23.4 kJ • min^{-1}

Walking with a stoop may increase energy expenditure by as much as 30-50% (Bedford and Warner, 1955). The effects of awkward posture are seen particularly when men must work in cramped quarters (for instance, miners cutting a very shallow seam of coal). Further costs are incurred if balance must be maintained on a moving vehicle (Vos, 1966).

SPECIFIC PROBLEMS OF LIFTING. Back injuries are a tremendous and apparently increasing hazard of modern employment (Table 7.6). In the Province of Ontario (population ~7 million), the total loss to the economy from back injuries was set at $6 million per annum (J. R. Brown, 1972), measured in 1967 dollars.

A proportion of the injuries are due to external trauma or psychoneurosis, but the most common

TABLE 7.6
Frequency of Pathological Changes (%) at Selected Joints in Relation to Type of Occupation (Manual Handling Versus Bank Employees)

	Frequency of Pathological Change	
Joint	Manual Workers (%)	Bank Staff (%)
Vertebral column	98	37
Elbow	35	3
Knee	32	13
Hip	28	6
Shoulder	12	5

Source: Based on data of Schroter, 1958.

TABLE 7.7
Strain on Fifth Lumbar Intervertebral Disc Imposed by Bent-Back Lifting

Inclination of Trunk	Load (kg)			
(Degrees)	0	50	100	150
0	50	100	150	200
30	150	350	600	850
60	250	650	1000	1350
90	300	700	1100	1500

Source: Based on data of International Occupational Safety and Health Information Center, 1962.

cause is an internal trauma, arising from the association of poor posture, obesity, lack of physical fitness, and faulty techniques in the lifting and carrying of heavy weights (CIS, 1962; Guthrie, 1963). Often there are contributory factors: an uneven or slippery floor surface, a twisting motion while lifting, fatigue, and organic abnormalities of the spine. But on many occasions a painful back injury causing prolonged disability has no other cause than the lifting of an excessive load. In Ontario, the commonest anatomical form of low back injury is a sacroiliac strain (82% of claims); these have a relatively short period of disability (average 28 days). A further 12% present as visceral hernias, and ~6% have a "slipped disc" (a forward herniation of the central spongy portion of the intervertebral cartilage). This last category is the most important, since disability averages 135 days, and ultimately it often becomes necessary to arrange alternative employment.

Injuries are perhaps most common in industries where lifting is not a regular occurrence (Shephard, 1974b). Presumably, in these circumstances the back muscles are less well developed and the task is less formalized. The worker concerned is unlikely to have learned "tricks" for minimizing the stress imposed on the vertebral column and will have received no formal instruction in lifting techniques.

D. Jones (1969) summarized some principles of safe lifting. The spine has a greater resistance to compression than to tension, shear, or torsion. Curvature of the vertebral column creates a bending moment, increasing the strain on the lumbar vertebrae (Table 7.7). According to traditional wisdom, the back is thus more stable when locked in a given position than when in process of changing its curvature. In considering the mass to be lifted, account must be taken not only of the external load, but also of body mass and momentum. The momentum of the load and/or the body can be used to move the system through positions whereby the muscles are acting at a poor mechanical advantage and/or the curvature of the spine is changing.

The classical technique of lifting involved extension of the flexed knees while the back was locked rigidly in a straight position. Because of unfavorable leverage, this method was impracticable if a heavy load had to be lifted from the floor (P. R. Davis et al., 1965). A dynamic lifting technique (D. Jones, 1969) is currently recommended. The momentum of the body and load assist in the lifting process, and the back is positioned to provide the required support and thrust at different stages of the operation. The spine is not necessarily straight, but the curvature is altered only when the instantaneously applied load will safely permit this. The dynamic lift in fact conforms closely to the intuitive actions of those who have had many years of experience in dealing with heavy loads, and an analysis of forces suggests that stresses on the vertebral column may be less with this technique than with classical straight-back lifting (D. Jones, 1969). Postural work is minimized if the load is held close to the body. For this reason, it is much more exhausting to carry a bulky package than a compact box of equal mass.

The mechanical efficiency of a task such as lifting 25-kg boxes onto a truck is quite low (2–4%; Shephard, 1974b). The reason is that a large proportion of the total energy expenditure is performed against body mass rather than the external load. Efficiency is greatest over the range 50–100 cm above the ground, and falls markedly if stooping or reaching are required (Table 7.8). In such activities, the efficiency of obese subjects is particularly poor.

TABLE 7.8
Energy Cost of Lifting Boxes Weighing 9.1, 18.2, and 27.3 kg

Load (kg)	Energy Cost Over Two Lifting Ranges ($kJ \cdot min^{-1}$) 0–51 cm	51–102 cm
9.1	23.4	11.7
18.2	26.8	15.9
27.3	34.3	24.2

Source: Based on experiments of Dr. H. Davies, Rochester.

The safe load for the sedentary worker is still disputed. The Swiss Accident Insurance specified in 1961 that a man could lift a load diminishing from 400 to 50 kg as the back was inclined from the vertical to a 90° forward stoop. Further downward adjustments were made for bulky loads, jerky or infrequently repeated movements, aging, and spinal deformity. Figures for bent-back lifting were only half as great. Women were assumed to have 60% of the physical capacity of men (Asmussen, 1964), and even lighter loads were prescribed for adolescents, partly because of the danger of damaging developing bone structures and partly because adolescents were thought to tire more quickly.

More recently, the International Labor Organization (ILO, 1964) proposed a limit of 40 kg for men working in jobs that involve some carriage of heavy weights. Asmussen and Poulsen (1968) suggested that the back muscles should not exert more than 40–55% of their maximum isometric force. On this basis, a maximum load of 32 kg was recommended for 35-year-old men lifting with a 45° forward stoop. A third alternative (Table 7.9) is to allow workers to vary loads until a maximum acceptable level is reached; limits thus defined are similar to the figures proposed on theoretical grounds.

While moderate loads can be lifted slowly and deliberately, the very large loads manipulated by professional weight lifters (P. R. Davis, 1959) require the simultaneous contraction of all motor units in the active muscles. Such intense activity can only be sustained for very short periods, hence the usual lifting technique of a "clean and jerk." The lifting process is usually accompanied by a dramatic rise of systemic blood pressure. This is partly a reaction to the Valsalva maneuver (expiration against a closed glottis, raising the intrathoracic and intraabdominal pressures; Chap. 6); this provides countersupport for the spine and fixation points for the active muscles of the limbs and shoulder girdle. Accumulation of anerobic metabolites in the working muscles (Chaps. 3 and 6) also contributes to the systemic hypertension.

TABLE 7.9
Relationship of the Range of Lift to Subjectively Acceptable Loadings (kg)

Range of Lift	Acceptable to 50% of Population	Acceptable to 90% of Population
Floor to knuckle	30	24
Knuckle to shoulder	28	23
Shoulder to arm reach	24	22

Source: Based on data of Snook and Irvine, 1967.

AUTONOMIC NERVOUS SYSTEM AND HYPOTHALAMUS

General Considerations

There has been relatively little study of the autonomic nervous system during physical activity (Ginzel, 1976). Nevertheless, it plays a vital role in preparing the body for the physiological demands of an increased release of energy.

Recent research has emphasized the close association of somatic and autonomic nervous function, to the point where Koizumi and Brooks (1972) spoke of "common centers which integrate responses of the two outflows" (Fig. 7.33).

Cortical Factors

We have noted in Chapter 6 that certain areas of the cerebral cortex can alter the balance of sympathetic and parasympathetic activity, calling into play the fight-or-flight reaction that precedes vigorous physical effort (Brod et al., 1959; Folkow et al., 1965; O. A. Smith, 1974; Flynn, 1977).

Activation of the autonomic centers was originally conceived as a diffuse process, but more recent study has suggested that the blood supply to individual muscles has a localized representation in the cortex (Ingram, 1960; Hoff et al., 1963; N. P. Clarke, Smith, and Shearn, 1968). There is also some evidence of a negative feedback loop from the viscera (heart, lungs, and great vessels), excessive stretch, pressure, or mechanical deformation of receptors inhibiting both α and γ motor activity. Ginzel and Eldred (1976) suggest this protects an individual

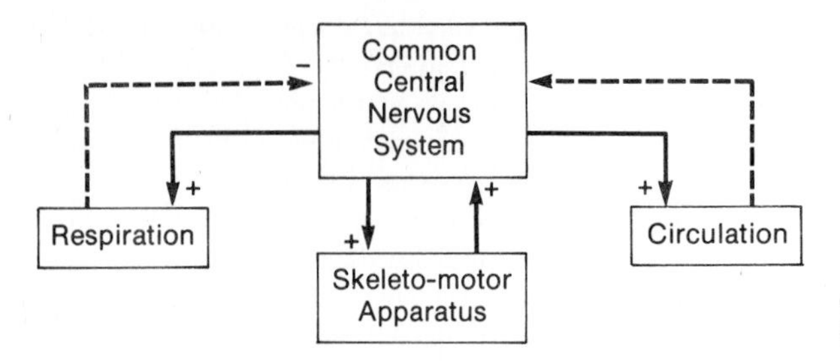

FIG. 7.33. Negative feedback concept of autonomic nervous system. (After Koizumi & Brooks, 1972, and Ginzel, 1976.)

against excessive exertion; a preferential inhibition of tonic and polysynaptic activity (Ginzel, 1975) encourages reduction of postural tone and thus the onset of sleep.

The responsiveness of the autonomic system, like that of the skeletomotor apparatus, is influenced by the level of arousal and thus the activity of the reticular formation (Magoun, 1963). With repetition of a stimulus, the response becomes less marked; habituation is said to have occurred (Glaser, 1966). Such an adjustment has practical importance, for example, in the interpretation of the heart rates attained during submaximum exercise. Habituation might be conceived as a lessening of arousal with a familiar stimulus, although Glaser and Griffin (1962) showed that habituation does not occur after damage to the prefrontal area of the cortex; in these circumstances, even mild stimuli give rise to an exaggerated response.

Hypothalamus

The hypothalamus plays an important coordinating role in initiating the fight-or-flight reaction. It also sets a very precise balance between the intake of food and the energy expended in habitual physical activity (Hervey, 1969), it controls body temperature (Edinger and Eisenman, 1970) and fluid intake, and it induces secretion of the pituitary hormones (Chap. 8).

Energy Balance

A considerable regulation of energy balance can be accomplished by adjustment of energy usage (an increase of physical activity and thermogenesis; D. S. Miller and Payne, 1962; Stirling and Stock, 1968; Apfelbaum et al., 1971). However, food intake is also closely regulated; eating is increased in response to cold (Brobeck, 1948), dilution of food (Adolph, 1947b), and food deprivation (Adolph, 1947b), but is decreased by experimental obesity (C. Cohn and Joseph, 1962).

The sensory input includes gustatory factors (taste and smell), short-term satiety stimuli (fullness of the stomach and entry of food into the duodenum with stimulation of osmo- and chemoreceptors; G. P. Smith et al., 1973), and a longer term error signal relating to body nutrient depletion/repletion. There is an innate set point (probably genetically programmed), with a circadian modulation. Hormones (insulin, glucagon, estrogens, and the adrenal hormones; Bray, 1973) and blood temperature (Edinger and Eisenman, 1970) influence food intake and energy expenditure either in their own right or through one of the above mechanisms (such as an alteration of set point). There may also be an interaction of short- and long-term mechanisms for the regulation of energy balance (for example, long-term satiety may increase sensitivity to short-term satiety signals; Panksepp, 1973).

Damage to the ventromedial part of the hypothalamus causes overeating and obesity (Hetherington and Ranson, 1942), while damage to the lateral area of the hypothalamus is associated with prolonged loss of appetite and reduction of body mass (Anand, 1967). Some authors have thus postulated distinct excitatory and inhibitory areas for behaviors such as eating and voluntary physical activity (Stellar, 1954; Morrison, 1968; Balagura et al., 1969). However, Panksepp (1973) and Van Atta and Sutin (1971) maintain that both excitation and inhibition have their origin in the ventromedial part of the hypothalamus (Fig. 7.34), while short-term satiety acts upon the lateral hypothalamus. Mayer (1955) proposed that the receptor for short-term satiety was a glucostat, located in the ventromedial hypothalamus. More recent investigations have disclosed two distinct types of glucose receptor; those in the ventromedial part of the hypothalamus are sensitive to insulin and the toxic action of gold thioglucose (Baile et al., 1970; C. J. V. Smith and Britt, 1971), while those in the lateral hypothalamus are stimulated by a competitive inhibitor of glucose (2-deoxy-D-glucose; Colin-Jones and Himsworth, 1970). The long-term regulation of food intake may also involve aminostats and lipostats. Appetite is influenced by blood levels of amino acids (Mellinkoff et al., 1956); the brain has systems with a high affinity for amino acids such as

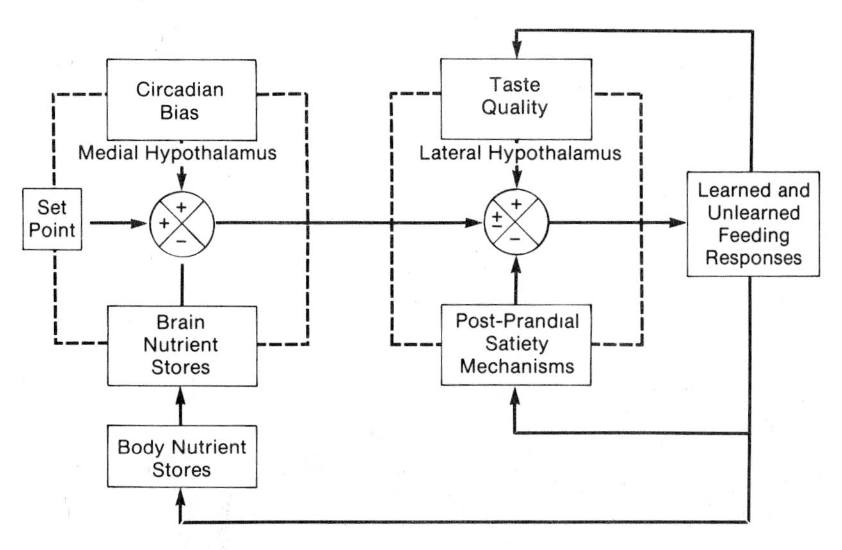

FIG. 7.34. Model for hypothalamic control of energy intake. (After Panksepp, 1973.)

glycine, which can hyperpolarize neurons (Logan and Snyder, 1972; Hösli and Haas, 1972). G. C. Kennedy (1953) proposed a long-term response of the ventromedial part of the hypothalamus to body fat levels. One objection to this concept is that the brain does not normally metabolize fat. Spitzer and Wolf (1971) demonstrated a limited catabolism of free fatty acids in the brain; however, it is likely that long-term energy balance is based upon brain stores of nutrients rather than the immediate blood-borne supply of fuel for a process of rapid catabolism (Panksepp, 1973).

The set point is also affected by the ionic balance within the ventromedial thalamus (Myers et al., 1972), eating being increased by the injection of cations such as calcium and potassium ions (Baile, 1973).

Hyperactivity

Children with brain injuries tend to be overactive. The cause may be insufficient inhibition of sensory input and motor outflow (Conners and Rothschild, 1973). Central stimulant drugs such as amphetamine are helpful in treatment and it is postulated that any improvements of behavior and learning result from increased arousal of the brain and thus increased inhibition (Sprague et al., 1970). While there undoubtedly is a clinical syndrome of hyperactivity, there is also much overdiagnosis. A survey in the United States found psychiatrists labeling 40% of their child patients as hyperactive. Further, ~70% of children attending psychiatric clinics were prescribed stimulant drugs (*British Medical Journal*, 1975b).

Thermal Regulation

Although humans are regarded as homiothermic, substantial deviations of body temperature can occur during exercise and exposure to adverse environmental conditions. Nevertheless, a reasonable thermal equilibrium is maintained in the face of widely varying rates of heat production.

Recent concepts of thermal regulation have been discussed by Bligh (1977). The search has been for an economical model that would explain the known behavior of the heat controller (Fig. 7.35). One scheme envisages the regulator-inhibiting pathways for both heat loss and heat production, a synaptic gate opening when the integrated sensory input exceeds a limit set by the inhibitory discharge (Fig. 7.34a). It is possible (and indeed the simplest explanation) that the gating inhibition is derived from collateral branches of the pathway protecting the body against the opposite extreme of temperature (Fig. 7.35b) rather than from an independent group of neurons that generate an inherent hypothalamic set point (Bligh, 1977). On this hypothesis, the magic of the system lies in the opposing heat and cold receptors, with the resultant potential for servo control of both heat production and heat loss. The substantial thermal inertia of the body mass and the deep-seated location of many of the receptors also contribute to the stability of body temperatures.

Exercise increases heat loss, but also leads to an apparent increase of set point, with a plateau temperature of perhaps 39°C rather than 37°C (Fig. 3.23). This could readily be explained by appropriate excitation and inhibition of the two gating synapses.

(a) Simplest model of heat loss

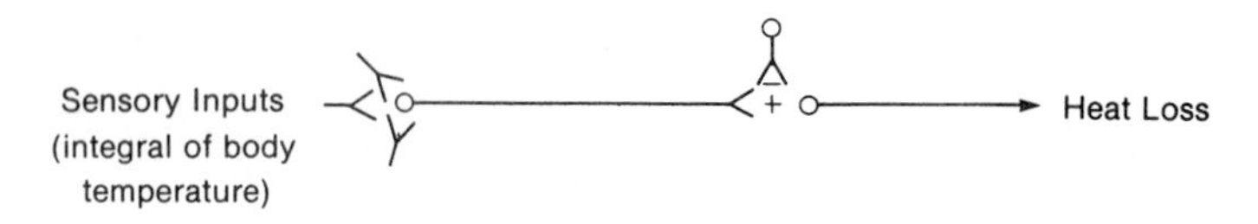

(b) Combined model of heat production and heat loss

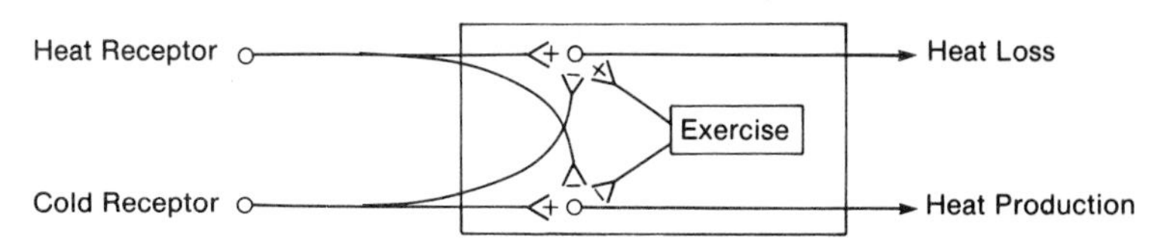

(c) More detailed model

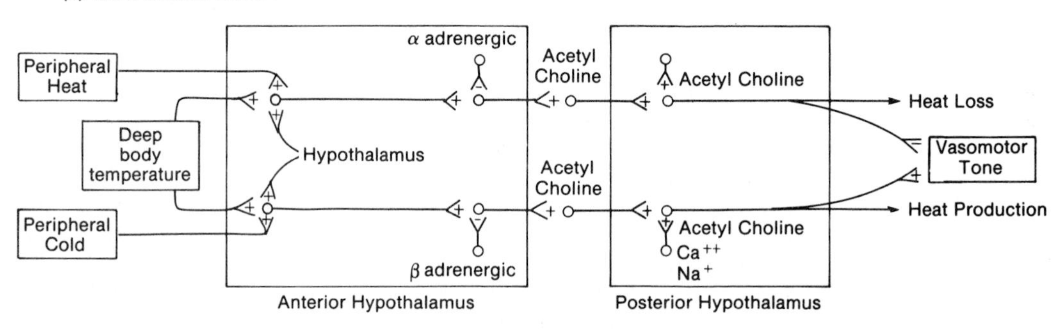

FIG. 7.35. Possible models of the hypothalamic thermal regulating centers. (After Bligh, 1977, and D. L. Jones et al., 1977.)

Acclimation to warm or cold climates also alters the pattern of heat loss and heat production in a given environment. Again, this could be explained by facilitation and/or inhibition of the gating synapses.

More detailed study has identified controlling centers in both the anterior and the posterior parts of the hypothalamus (Fig. 7.34c). The anterior hypothalamus integrates information reported from peripheral heat and cold receptors, deep-body temperature, and the local temperature of blood within the hypothalamus. The channel concerned with heat loss is inhibited by α-adrenergic fibers, while that concerned with heat production has β-adrenergic inhibition. Additional synapses within the posterior hypothalamus are excited by cholinergic fibers.

Fluid Balance

A few hours of exercise in a warm environment can generate a fluid deficit of several liters (Chap. 3). Nevertheless, long-term fluid regulation is quite precise, and indeed some calculations of body composition assume a constant 73% water content of lean tissue. The osmolar concentration of extracellular fluid is even more stable.

The volume of extracellular fluid is regulated by control of both sodium and water excretion, while osmolarity is determined largely by the excretion of water (H. W. Smith, 1957). The immediate bases of regulation are the control of water output by the antidiuretic hormone of the posterior pituitary gland (Chap. 8) and of sodium output by adrenal steroids (Chap. 8). However, the antidiuretic hormone is apparently synthesized in the hypothalamus along with an adrenoglomerulotropin, which stimulates the adrenal cortex to produce aldosterone. The hypothalamic region of the brain integrates various sources of information on blood volume. Different authors have set the receptors in the venous system (Viar et al.,

1951; Strauss, 1957), the atria (Gauer et al., 1961), the arterial side of the circulation (F. H. Epstein, 1956; Bartter and Gann, 1960), and the afferent renal arterioles (J. O. Davis, 1961). Since 60–80% of the total blood volume lies in the venous part of the circulation, it seems likely that receptors in the large veins or atria contribute a major part of the necessary sensory input. The system also responds to alterations in distribution of the available blood volume and coordinates this data with information on osmolarity in a way that is poorly understood.

During exercise, urine flow drops to almost zero. One of the main reasons is the reduction of renal blood flow (Chap. 6). Nevertheless, if there is a substantial reduction of central blood volume by sweating and peripheral sequestration of fluid, it is likely that there is also a stimulation of volume receptors and secretion of antidiuretic hormone; similar reactions follow prolonged standing and the application of venous tourniquets. Exercise may have more direct effects on the hypothalamic centers, since it is well recognized that the antidiuretic system can be activated by stimuli such as pain, fear, apprehension, and loud sounds (R. F. Pitts, 1963).

CHAPTER 8
Hormonal Control Systems

GENERAL CONSIDERATIONS

While the central nervous system is well adapted to regulating acute problems associated with an increased energy usage, the endocrine system makes an important contribution to long-term control. Substances known as hormones are secreted into the bloodstream by the endocrine glands, with profound and prolonged effects upon tissues in distant parts of the body (Table 8.1).

Such hormones have at least three possible modes of action: (1) they may alter the rate of synthesis of enzyme proteins, (2) they may alter the rate of synthesis of cyclic AMP and prostaglandins, thus altering either enzyme activity or membrane permeability, and (3) they may have a direct effect upon the permeability of cell membranes.

Most of the available human data refer to blood levels of the various hormones rather than rates of secretion. An increase of blood concentration during exercise may indicate an increased rate of hormone secretion by an endocrine gland, but it can also result from (1) the psychological stress of blood sampling (Copinschi et al., 1967; Helge et al., 1969) or physical activity, (2) diurnal rhythms (C. T. M. Davies and Few, 1973a,b), (3) a decreased degradation or excretion of the compound, due, for example, to a diminished hepatic or renal blood flow (Chap. 6), (4) a decreased uptake of hormone by the target tissue, or (5) a loss of fluid from the plasma (sustained exercise normally gives a 5–10% hemoconcentration; Terjung and Tipton, 1971). The effectiveness of a given concentration of hormone is also modified by (1) a change in blood flow to the target organ, (2) an alteration in the concentration or activity of binding substances in the bloodstream, and (3) a change in the permeability of cell membranes in the active tissues (Elgee et al., 1954; O. Stein and Gross, 1959).

HYPOTHALAMIC AND PITUITARY HORMONES

Hypothalamus

Certain functions of the hypothalamus were discussed in Chap. 7. Little is known about the influence of exercise upon the various "tropins" and "statins" that are secreted by this region of the brain. An increased output of a number of tropins might be inferred from the increased blood concentrations of materials secreted by the target glands (anterior pitu-

TABLE 8.1
Hormones Possibly Involved in the Response to Physical Activity

Site of Production	Hormone	Main Functions
Hypothalamus	Somatoliberin	Stimulates release of somatotropin
	Somatostatin	Inhibits release of somatotropin
	Thyroliberin	Stimulates release of thyrotropin
	Corticoliberin	Stimulates release of corticotropin
	Luliberin	Stimulates release of lutotropin
	Prolactoliberin	Stimulates release of prolactin
	Prolactostatin	Inhibits release of prolactin
	Antidiuretic hormone	Released from posterior pituitary, increases renal water retention
Anterior pituitary	Somatotropin (growth hormone)	Stimulates bone growth and fat mobilization
	Thyrotropin (thyroid stimulating hormone)	Stimulates production and release of thyroxine
	Corticotropin (adrenocorticotropic hormone)	Stimulates production and release of adrenal cortical hormones
	Lutropin (luteinizing hormone)	Stimulates production of testosterone by testes, promotes development of corpus luteum in females.
	Prolactin	Stimulates renal water retention, fat mobilization, and (in females) lactation
Thyroid gland	Thyroxine	Stimulates mitochondrial function, cell growth
	Calcitonin	Reduces blood calcium, phosphate levels
Adrenal cortex	Cortisol and others	Promotes fat utilization, conserves glucose, reduces inflammation
	Aldosterone and others	Promotes renal sodium and water retention
Adrenal medulla	Adrenaline, noradrenaline	Enhances cardiac output, vasoconstriction, glycogen breakdown, fat mobilization
Pancreas	Insulin	Promotes glucose uptake by cells, increases glycogen storage
	Glucagon	Promotes glucose release from liver, mobilizes fat, increases cardiac output
Parathyroid glands	Parathyroid hormone	Increases blood calcium, decreases blood phosphate
Testes	Testosterone	Increases muscle mass, decreases body fat, increases muscle glycogen and red cell production
Liver	Somatomedin	Activated by somatotropin, stimulates cartilage and bone growth
Various tissues	Prostaglandins	Function varies with specific chemistry of prostaglandin e.g., vasodilatation, bronchodilation or constriction, increase of cardiac output

Source: After D. R. Lamb, 1978.

itary growth hormone, adrenocortical and thyroid hormones). However, it remains possible that hemoconcentration and a decreased destruction of the hypothalamic tropins could provide a sufficient stimulus to increase the output of the target glands without any augmentation of hypothalamic secretion.

Vigorous exercise inhibits urine formation. While this is due in part to the decrease of renal blood flow, it probably reflects also an enhanced secretion of antidiuretic hormone from the posterior pituitary gland, in response to either exercise itself or the associated emotional stress. Dehydration is particularly likely under warm conditions (Chap. 3), and in such circumstances an increased secretion of antidiuretic hormone makes an important contribution to the restoration of fluid balance (Kozlowksi, Szczepanska, and Zielinski, 1967). Following exhausting exercise, there is no change in the hormone content of the primary secretory sites (the supraoptic and paraventricular nuclei of the hypothalamus), but stores of hormone in the posterior pituitary gland are depleted.

Anterior Pituitary

Growth Hormone

Under resting conditions, the most important function of growth hormone is to stimulate anabolism—growth of muscle, strengthening of ligaments and tendons, and an increase of bone thickness (Harper, 1969). It remains unclear as to whether the mode of action is an increase of RNA synthesis or a greater transport of amino acids to the active cells brought about by an increase of membrane permeability. Growth hormone also stimulates synthesis of an inactive adipolytic lipase (Fig. 8.1) through its action on RNA; the enzyme is then activated by cyclic AMP. In muscle, growth hormone counteracts the effects of insulin, although it is uncertain as to whether this is accomplished through an inhibition of membrane transport of glucose or an inhibition of glycolysis brought about by the increased plasma concentrations of glucose and free fatty acids. In liver, glycogen stores are increased, possibly through an increase of gluconeogenesis (Fig. 8.2).

Exercise leads to a substantial rise of blood growth hormone concentrations (20- to 40-fold; Sutton, Jones, and Toews, 1976; Sutton and Lazarus, 1976). The threshold load is ~40% of maximum oxygen intake in subjects who are unfit and 50% in those who are more active (Sutton, 1978; Metivier et al., 1978). Growth hormone concentrations usually increase after ~15–20 min of sustained activity; if work is prolonged, the hormone may appear as a series of bursts of secretion, with an overall tendency for plasma concentrations to return toward normal values (Fig. 8.3; Hunter et al., 1965; Hartley, Mason et al., 1972; Lassarre et al., 1974; Hartley, 1975; Metivier, 1975; Shephard and Sidney, 1975b; Sidney and Shephard, 1978a; Von Glutz et al., 1978). There are also occasional reports of increased blood levels after quite short bouts of submaximal work (Nilsson et al., 1975; Shephard and Sidney, 1975b).

The increase in the plasma concentration of growth hormone is sufficiently large that a true increase of secretion seems likely (Shephard and Sidney, 1975b). The half-time for hepatic degradation of growth hormone is 17–45 min (Fig. 8.3; Cameron et al., 1969), so that even if hepatic flow were completely arrested, this would cause no more than a doubling of plasma concentration in 20 min. In fact, exercise is unlikely to cause more than a 50% reduction of visceral flow, and in such circumstances the factor limiting hepatic breakdown of the hormone remains polypeptide metabolism rather than local blood flow.

Potential triggers to the release of growth hormone include psychological stress, a rise of core tem-

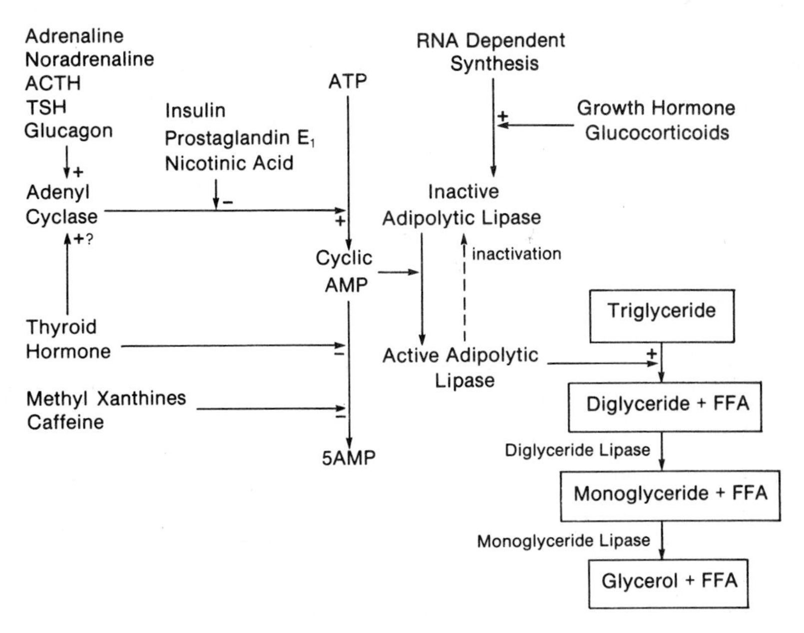

FIG. 8.1. The role of various hormones in the activation of adipolytic lipase. (After Harper, 1969.)

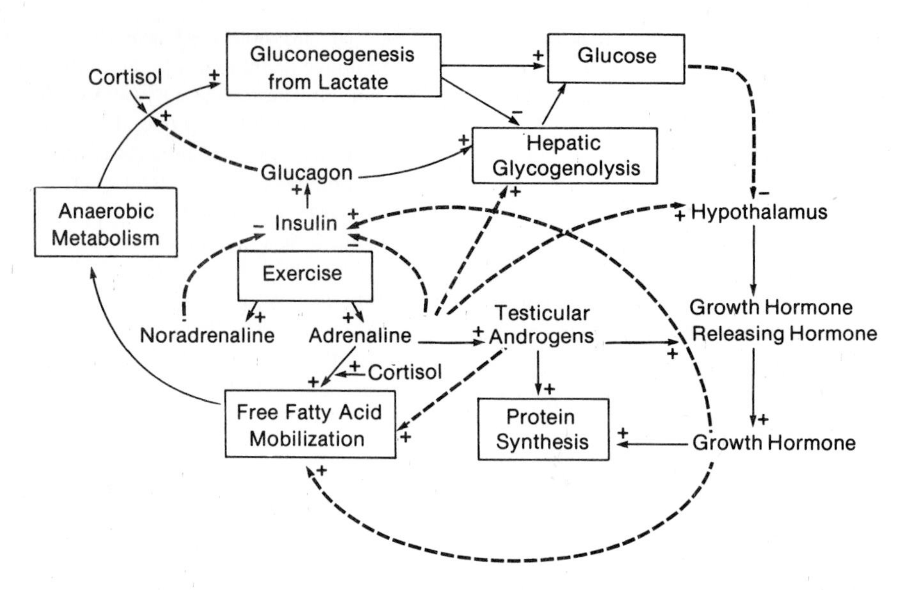

FIG. 8.2. Interrelationships of various hormones during exercise.

perature, the onset of anerobic effort (hydrogen ion concentration, blood lactate, or arterial oxygen saturation), and changes in plasma concentrations of amino acids and glucose.

The late onset of the rise in growth hormone concentration (Fig. 8.3) and the greater response in subjects who are unfit (Sutton, Young et al., 1969) might suggest that the cumulative stress of vigorous work is responsible. However, well-trained runners can exercise to the point of severe distress without any increase of plasma growth hormone concentrations (Sutton, Young et al., 1969), and other studies of

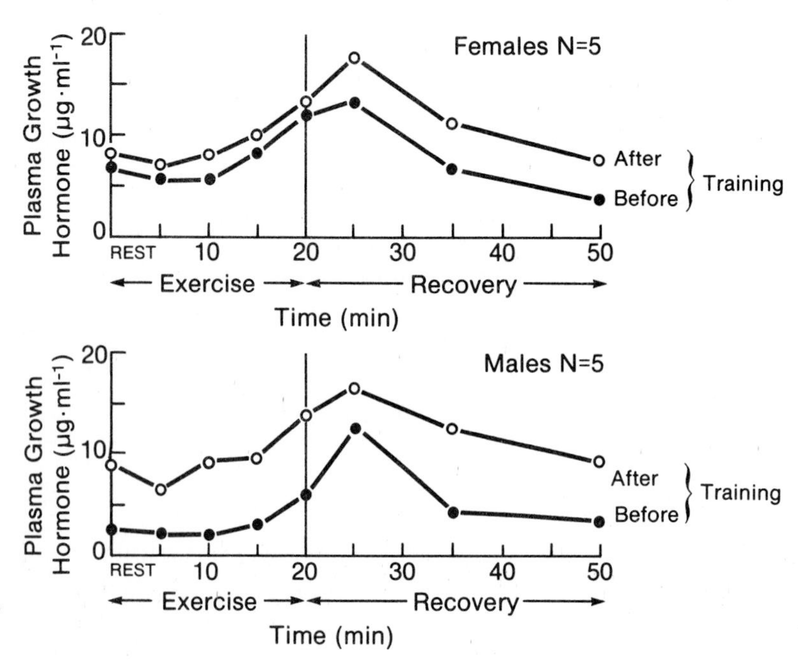

FIG. 8.3. Plasma growth hormone levels of elderly subjects during four 5-minute stages of progressive treadmill exercise, before and after training. Final loading was approximately 85% of maximum oxygen intake (Sidney & Shephard, 1968a).

less well-trained subjects have found a response at levels of exercise that were not stressful (Sutton, Coleman et al., 1973; A. P. Hansen, 1973; R. H. Johnson and Rennie, 1973; Sidney and Shephard, 1978a).

A rise of core temperature could act directly on hypothalamic centers, or it could increase hormone levels less directly by increasing the nonspecific stress of exercise. Buckler (1973) found some association between growth hormone levels and body temperature, but the low level of response seen in obese subjects (Schwarz et al., 1969; A. P. Hansen, 1973; but not Sutton, Young et al., 1969) is difficult to reconcile with a thermal hypothesis.

Hydrogen ion and lactate concentrations do not show any close parallel with increases in growth hormone concentrations (Sutton, Jones, and Toews, 1976). Some authors have described a relationship between growth hormone levels and the fall of arterial oxygen saturation (Hartog et al., 1967; Metivier et al., 1978), but the reported change in the latter variable (arterial saturation ~81% at 66% of maximum oxygen intake) is surprisingly large.

Amino acids are well recognized as a potent stimulus to the release of growth hormone under resting conditions (Roth et al., 1963; Buckler, 1969; Root and Oski, 1969), and it is possible that a reversible breakdown of mitochondrial protein occurs during effort. Carlsten et al. (1962) found an increase in only 1 of 14 amino acids (alanine) during effort; however, proponents of an amino acid trigger note that these authors failed to assay the most potent stimulant of growth hormone (arginine). If the release of growth hormone is increased, this could in turn facilitate the entry of amino acids into the active cells (Fig. 8.2), functioning as a linear amplifier of the anabolic process (Goldberg, 1967; Goldberg and Goodman, 1967). It has been suggested that growth hormone also increases the synthesis of ribosomes and messenger RNA, as well as facilitating attachment of ribosomes to the messenger RNA of muscle.

The rate of rise of plasma growth hormone concentration is rather slow to suggest that this mobilizes energy for effort except in sustained exercise (Pruett, 1970a,b), when reserves of blood sugar and muscle and liver glycogen are becoming depleted. If rapid lipolysis was required to sustain physical activity, this could be brought about more readily by the action of the nervous system on adenyl cyclase (Hartog et al., 1967; Fig. 8.1). However, the increased concentration of growth hormone is sufficiently persistent (Fig. 8.3) to have medium-term effects on protein, glucose, and fat metabolism. Insulin release may be stimulated (J. Campbell and Rastogi, 1969) but the direct effect of growth hormone is to inhibit the phosphorylation of glucose by hexokinase (Figs. 8.4 and 8.5). Muscle glucose uptake is thus decreased and blood glucose levels are stabilized. There is some association between growth hormone and blood glucose levels (Sutton, Young et al., 1969; R. H. Johnson and Rennie, 1973), and the exercise-induced rise of growth hormone levels can be inhibited by feeding large doses of glucose or sucrose (Hunter et al., 1965). Lastly, growth hormone may contribute to the late mobilization of fat, particularly if exercise is continued to the point where glycogen reserves are depleted (Raben and Hollenberg, 1959; A. P. Hansen, 1971).

In many studies, training leads to a less dramatic rise of growth hormone levels during exercise (Hartley, Mason et al., 1972; Metivier, 1975; Shephard and Sidney, 1975b; Sutton, 1978). This could be due to (1) a lessening of psychological stress and

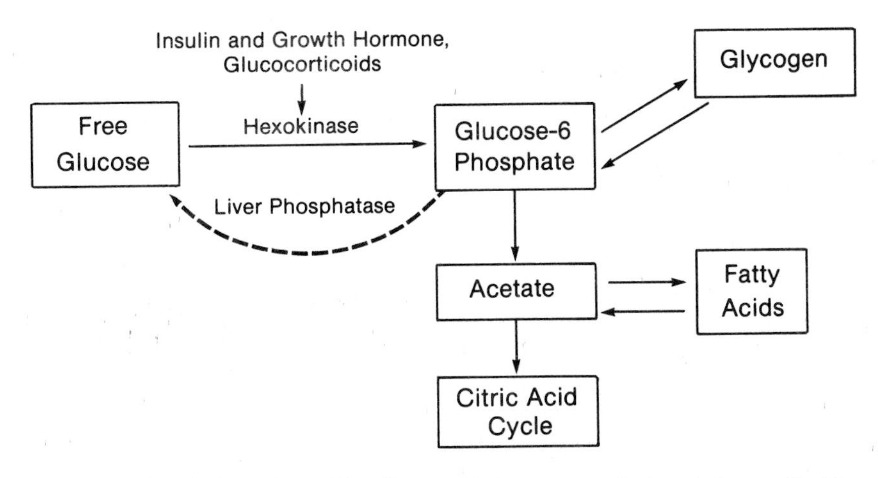

FIG. 8.4. Metabolic actions of insulin, growth hormone, and adrenal glucocorticoids.

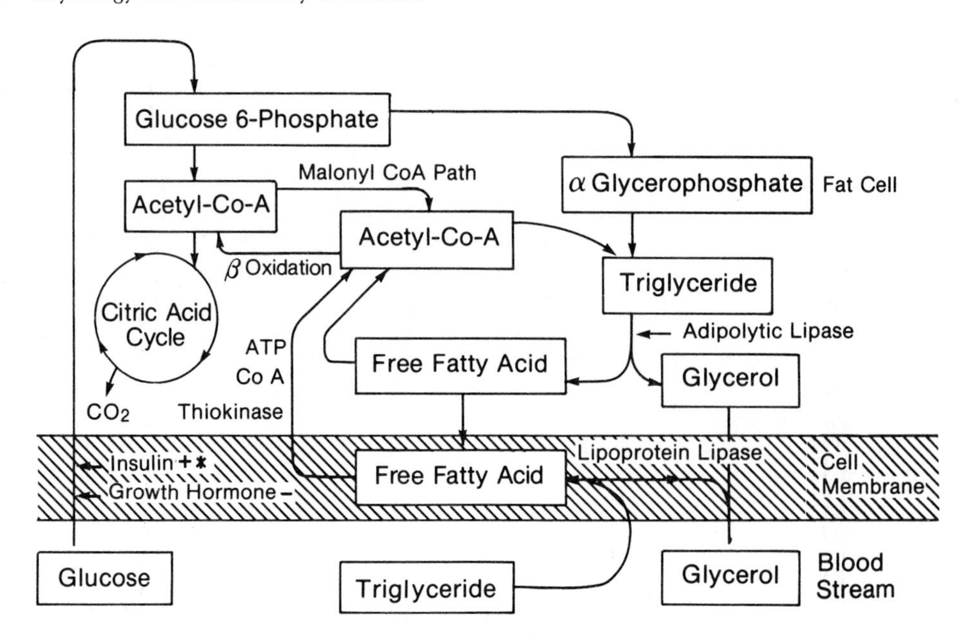

FIG. 8.5. Regulation of the relative use of glucose and fat at the membrane of the fat cell. Note that lipoprotein lipase favors uptake of fat, and adipolytic lipase its release.
*See Fig. 8.4 for details

(2) the development of alternative pathways for the mobilization of free fatty acids. However, in one group of elderly subjects, Sidney and Shephard (1978a) noted an increase of plasma growth hormone concentrations after training. They speculated that in this age group a lessening of androgen secretion weakened anabolic processes, creating a greater need for growth hormone during exercise.

Thyrotropin (Thyroid Stimulating Hormone)

Since thyroxine secretion is increased by exercise (see below), an increase in secretion of thyrotropin might also be expected. Current observations show an increase in blood levels of thyrotropin during anticipation of physical activity (J. W. Mason et al., 1973), but not while exercising (Terjung and Tipton, 1971). The half-life of the thyroid stimulating hormone is about 1 hr (Odell et al., 1967), so it is possible that stimulation of thyroxine secretion can be based on an "anticipatory" rise of the pituitary hormone rather than an increased output during effort.

Corticotropin (Adrenocorticotropic Hormone)

Stressful, exhausting exercise produces an increased output of cortisol (see below), and an increased output of corticotropin would thus be anticipated in such circumstances. Unfortunately, corticotropin is difficult to estimate and is rapidly destroyed (circulatory half-life 4–18 min); there is thus no conclusive proof that secretion is augmented during vigorous physical activity (D. R. Lamb, 1978).

Lutropin (Luteinizing Hormone)

The luteinizing hormone stimulates development of the corpus luteum in the female, while in the male it stimulates secretion of testosterone by the interstitial cells of the testes. Since androgen levels are boosted by exercise, an increased secretion of lutropin might be inferred. Surprisingly, long-term weight training does not increase resting plasma levels (Strömme et al., 1974), and a single bout of such vigorous activities as swimming and rowing has no influence upon plasma readings (Sutton, Coleman et al., 1978; Brisson et al., 1977).

Prolactin

Prolactin mobilizes fat and has an antidiuretic effect upon the kidneys. Malarkey (1976) reported increased blood levels following exercise. The circulatory half-time is 15–30 min.

THYROID HORMONES

Thyroxine

Thyroxine production and secretion is controlled by the thyroid stimulating hormone of the anterior pituitary. One important action of thyroxine is to "uncouple" carbohydrate mechanisms for the regeneration of ATP (Chap. 1), thereby increasing body heat production. The thyroid hormone can also mobilize fatty acids, promote cardiac hypertrophy, and inhibit the secretion of the thyroid stimulating hormone of the anterior pituitary. The circulatory half-life of thyroxine is prolonged (6–7 days; D. R. Lamb, 1978).

Exercise increases the free thyroxine level by ~35% (Terjung and Tipton, 1971). The liver uptake is enhanced, but the uptake by the muscles apparently remains unchanged (Terjung and Winder, 1975). The total plasma thyroxine (bound + free) may actually decline during exercise because increased secretion (Irvine, 1967) is more than matched by increased hepatic degradation (Irvine, 1968; Winder and Heniger, 1973; Fig. 8.3) and increased fecal excretion (Balsam and Leppo, 1974).

Thyroxine (T_4) is bound by T-4-binding globulin and, to a lesser extent, by T-4 binding prealbumin and albumin. The concentration of the various binding proteins apparently remains unchanged during exercise (Terjung and Tipton, 1971), but their thyroxine-binding ability is reduced (Terjung and Winder, 1975). The increase in the proportion of free hormone facilitates hepatic breakdown and seems largely responsible for the increased turnover of thyroxine during exercise.

Training decreases the resting concentrations of total thyroxine but increases the concentration of free thyroxine. Both secretion and degradation are accelerated, the half-time diminishing from 7 to 4 days (Irvine, 1968; Terjung and Winder, 1975; Fig. 8.6). However, these changes are reversed quite rapidly if athletes cease training. Despite alterations in plasma thyroxine, the consensus is that basal metabolism remains unchanged by training (Steinhaus, 1933; Terjung and Tipton, 1970; Terjung and Winder, 1975). Either more free T-4 is "needed" to accommodate the altered physiological function produced by chronic physical activity, or the excess T-4 is degraded by some metabolically inactive process (Terjung and Winder, 1975).

Many training responses such as bradycardia and the strengthening of ligaments can occur after thyroidectomy (Tipton, Terjung, and Barnard et al., 1968). However, the thyroid hormones seem necessary to reduce liver cholesterol (Barnard, Terjung, and Tipton, 1968) and increase fatty acid mobilization from adipose tissue (Paul, 1971). Furthermore, thyroid administration, like training, induces cardiac hypertrophy.

Calcitonin and Parathyroid Hormone

Both calcitonin and parathyroid hormone regulate blood calcium and phosphate levels. We have already noted the role of these ions in muscle contractility (Chap. 6) and the contractile process (Chap. 4). However, there have been no studies examining the effects of exercise upon calcitonin and parathyroid hormone.

ADRENAL CORTEX

The adrenal cortex secretes glucocorticoids (of which the most important is cortisol), mineralocorticoids (of which the principal component is aldosterone), and sex hormones.

Cortisol

Cortisol mobilizes fat and protein, thus conserving carbohydrate and tending to elevate blood glucose

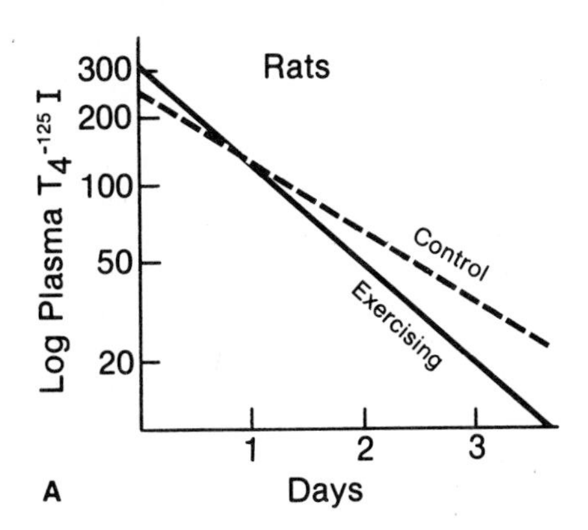

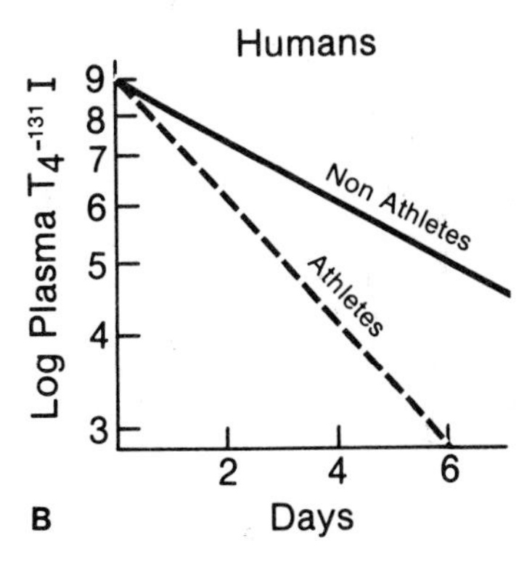

FIG. 8.6. The influence of exercise and athletic participation upon the rate of clearance of T_4 (thyroxine) from the blood stream. **A.** After Winder & Heniger (1973). **B.** After Irvine (1968).

(Tharp, 1975). Its effect is probably mediated by RNA synthesis, with an increase in the hepatic enzymes involved in amino acid metabolism (such as alanine α-ketoglutarate and tyrosine transaminases) and gluconeogenesis (such as pyruvate carboxylase and phosphoenol pyruvate carboxykinase). Catabolic effects are seen in muscle (increased usage of amino acids) and in bone (reduction of matrix and an increased calcium loss) after administration of cortisol. Within muscle, there is inhibition of glucose uptake and decreased usage of carbohydrate. Within adipose tissue, lipolysis is increased. The synthesis of adipolytic lipase is enhanced (Raben and Hollenberg, 1959). Cortisol may also facilitate epinephrine; this activates adenyl cyclase, which converts ATP to cyclic AMP, the activator of adipolytic lipase (Fig. 8.1). Cortisol initiates a general stress response, as shown by suppression of immune reactions, a reduction of eosinophil count, and involution of the thymus. Lastly, water balance may be modified through retention of sodium and chloride ions, with an increased excretion of potassium ions.

The half-life of circulating cortisol is ~4 hr (Harper, 1969). Light and moderate exercise have no consistent effect upon cortisol levels (Shephard and Sidney, 1975b; Sidney and Shephard, 1978a), but if exercise is heavy, prolonged, and exhausting, or involves the stress of serious competition, blood levels of cortisol are increased, the output of urinary metabolites (17-ketosteroids) is augmented, and there is a triphasic change in the number of circulating eosinophils (Hartley, Mason et al., 1972; C. T. M. Davies and Few, 1973a,b; Follenius and Bradenberger, 1975; Metivier, 1975; Shephard and Sidney, 1975b; Sundsfjord et al., 1975; Tharp, 1975; Newmark et al., 1976; LeClerq and Poortmans, 1978). These changes are thought to represent increased secretion and concentration dependent excretion rather than a dissociation of bloodstream complexes (transcortin/cortisol and albumin/cortisol) or a decrease in the rate of tissue uptake and degradation (Shephard and Sidney, 1975b; LeClerq and Poortmans, 1978).

Early studies based on the urinary excretion of adrenal hormones (Schonholzer, 1957) postulated that if effort was sufficiently prolonged, the adrenal cortex would become exhausted, with disastrous consequences for the regulation of metabolism and fluid and mineral balance. Some, but not all, blood studies have shown such a trend (Chin and Evonuk, 1971; Shephard and Sidney, 1975b; Von Glutz et al., 1978).

In animals, removal of the adrenal glands shortens the maximum possible working time, and cortisol is more effective than adrenaline or aldosterone as a means of restoring performance (Tharp, 1975). However, the hormone seems essential only during exhausting effort, and adrenalectomized rats have no problem adjusting to moderate work (Berdanier and Moser, 1972; Tipton, Struck et al., 1972).

In brief, exhausting effort (for example, 1 min at 140% of maximum oxygen intake), the appearance of cortisol is too rapid to postulate a triggering by adrenocorticotropic hormone. Other forms of stress also cause cortisol release and this may be the signal during rapidly exhausting exercise, locally secreted catecholamines serving as the chemical intermediary (S. Hill et al., 1956; Raymond et al., 1972; LeClerq and Poortmans, 1978). Passive heating causes an early fall of plasma cortisol (K. J. Collins et al., 1969), and it is unlikely that a rise of core temperature is involved in the exercise response (C. T. M. Davies and Few, 1973a,b).

In animals, training gives rise to an early increase of plasma glucocorticoids (Dieter et al., 1969), with a later return to normal values (Tharp, 1975), this being reminiscent of Hans Selye's general adaptation syndrome. The mass of the adrenal glands may also increase, but the release of glucocorticoids during exercise is decreased (Tharp and Buuck, 1974). Chronic exercise apparently has little consistent effect upon the adrenocortical responsiveness of human subjects (Hartley, Mason, et al., 1972; Shephard and Sidney, 1975b), although it may lengthen the time to exhaustion of the adrenal gland (Viru, 1973).

Aldosterone

Plasma aldosterone concentrations rise progressively with exercise of increasing intensity (Costill, Coté et al., 1975; Kirsch et al., 1975; Maher et al., 1975; Sundsfjord et al., 1975; Newmark et al., 1976), the response of female subjects being enhanced in the luteal phase of the menstrual cycle (Jurkowski et al., 1978). Hepatic clearance of aldosterone is reduced by exercise (Sundsfjord, et al., 1975), but peak exercise readings are so high (up to 6 times the resting concentration) that an increase of secretion must be postulated. Plasma concentrations remain high for 7–12 hr after exercise has ceased (Costill, Coté et al., 1975; Costill, Coté, and Fink, 1976). The trigger to aldosterone release is a reduction of renal blood flow, perhaps initiated by a falling central venous pressure (Kirsch et al., 1975). The decrease of renal perfusion stimulates a release of renin into the bloodstream

(Zambraski et al., 1977), which in turn increases angiotensin production. Finally, the angiotensin causes the adrenal cortex to secrete aldosterone. Other factors that may regulate aldosterone secretion include plasma levels of adrenocorticotropic hormone, and sodium and potassium ions (Von Glutz et al., 1978). Training does not curtail the exercise-induced rise of plasma renin activity (A. S. Leon, Pettinger, and Saviano, 1973).

SYMPATHOADRENAL HORMONES

The sympathetic nerve endings secrete noradrenaline, plus small amounts of adrenaline, while the catecholamine output of the adrenal medulla is 75% adrenaline and 25% noradrenaline. The normal trigger to activation of the sympathetic nervous system during exercise may be a fall of central blood pressure or an irradiation of impulses from the motor cortex. The stimuli to release of adrenaline from the adrenal medulla include emotional stress and/or a decrease of blood glucose below 50–70 mg • dl^{-1} (U. S. Von Euler and Luft, 1952; Crone, 1963).

We have already noted that much of the noradrenaline liberated at the nerve endings is broken down locally by the enzyme monoamine oxidase. The circulatory half-life of the hormone is only a few min. A proportion of noradrenaline and adrenaline is excreted unchanged in the urine, but much of both compounds is metabolized (Fig. 8.7). Brain catecholamines are excreted largely as 3-methoxy-4-hydroxyphenylglycol (MHPG), while catecholamines from other sources appear as a mixture of MHPG and vanillyl mandelic acid (VMA) (Schanberg et al., 1968; Mass and Landis, 1968; R. J. Goode et al., 1973).

Recent observations have shown that sustained exercise at 40% of maximum oxygen intake induces increases of both plasma MHPG and VMA, with an increased urinary output of MHPG and VMA. The plasma MHPG/VMA ratio decreases during exercise, but rises during recovery (Tang et al., 1979). These changes may be interpreted as a peripheral secretion of catecholamines, with a more ready excretion of VMA than of MHPG.

During exercise, the functional role of the sympathetic nerves in maintaining blood pressure, redistributing blood flow, and increasing cardiac output is self-evident (Chap. 6). After administration of sympathetic ganglion-blocking agents such as hexamethonium (Levine, 1960), the viscera are engorged and working capacity is diminished. Adrenaline serves to activate adipolytic lipase (Fig. 8.1) and raises blood glucose by activating the hepatic enzymes involved in glycogen breakdown (Hartley, 1975; Fig. 1.2). Receptors for both of these actions are of the β type, being blocked by drugs such as propranolol. The adrenal medulla itself does not seem essential to the exercise response, at least in rats, since glycogen breakdown still occurs in liver and muscle, and free fatty acids are still mobilized in plasma and adipose tissue after experimental extirpation of the adrenal medulla (Gollnick and Ianuzzo, 1975).

With combined adrenal medullectomy and administration of a drug to destroy the sympathetic nerve endings (6-hydroxydopamine), glycogen utilization is still unaltered, although there is less mobilization of fatty acids than in a normal animal (Gollnick and Ianuzzo, 1975). It is perhaps not surprising that the body can meet the metabolic needs of exercise without catecholamines, since there are several alternative mechanisms of regulation available. In muscle, inactive phosphorylase b can be converted to active phosphorylase a by calcium ions even in the absence of cyclic AMP (Bostrom et al., 1971). In the liver, the breakdown of glycogen may be stimulated by hypoxia (Bernelli-Zazzera and Gaja, 1964) and by secretion of glucagon (Kibler et al., 1964). Likewise, fat mobilization may be affected by pituitary hormones or testosterone (Fig. 8.2); certainly, plasma free fatty acids show their greatest increase during light work, whereas catecholamine secretion is typical of heavy work.

Many early studies were based on the urinary elimination of adrenaline and noradrenaline (U. S. Von Euler and Lishajko, 1961; Vaage, 1974). This integrates the nonmetabolized output over a fairly long period, but nevertheless can be used to demonstrate an increased secretion of catecholamines in response to sustained exercise (U. S. Von Euler and Hellner, 1952), cold (Lamke et al., 1972), and occupational stress (I. Åstrand et al., 1973b; Holmbøe et al., 1975). During exercise, noradrenaline output is greater than that of adrenaline; if work is of short duration, adrenaline excretion does not increase until subjects reach an oxygen intake of 2 l • min^{-1} (U. S. Von Euler and Hellner, 1952), unless there is associated emotional stress (as in an ice hockey competition; Elmadjian et al., 1958; Metivier, 1975; Howley, 1976).

Plasma noradrenaline and adrenaline levels both remain unchanged unless (1) there is psychological stress (E. S. Williams et al., 1978), (2) the intensity of exercise exceeds ~70% percent of maximum

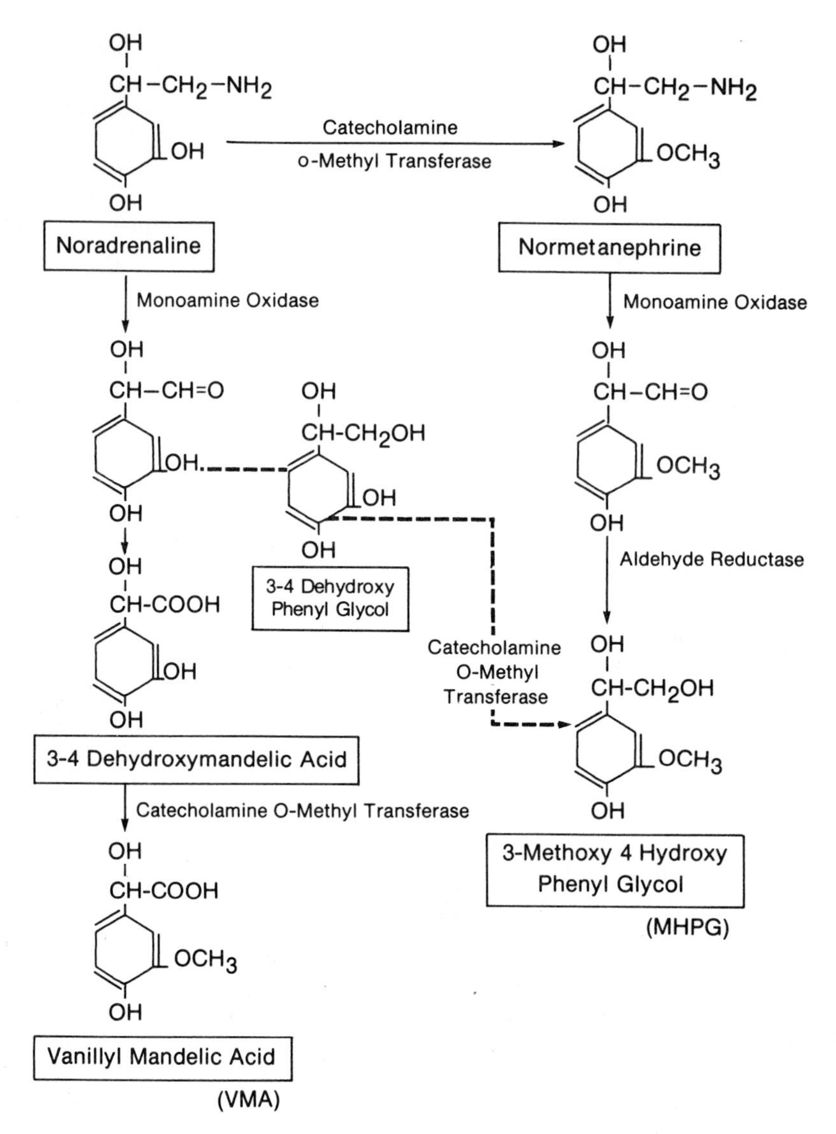

FIG. 8.7. Pathways of catecholamine metabolism.

oxygen intake (E. W. Banister and Griffiths, 1972; U. S. Von Euler, 1974; C. T. M. Davies et al., 1974; Galbo et al., 1975; Hartley, 1975), or (3) activity is prolonged (Peronet et al., 1978). However, the apparent absence of response in light work may reflect the sensitivity of the methods currently available, since increases of catecholamine metabolites can be demonstrated with quite mild exercise (Tang et al., 1979; Fig. 8.8). With moderate work, more noradrenaline appears than adrenaline (Fig. 8.9). This suggests that the sympathetic nerve terminals are the main site of formation of the catecholamine. Exhausting work may eventually lead to a decrease of plasma adrenaline levels, implying that the adrenal medulla has been exhausted.

Hyperbaric oxygen (Fagraeus, Haggendal, and Linnarson, 1973) reduces the exercise-induced catecholamine output, probably by reducing the relative work load. Surprisingly, habituation to the stress of a particular physical task does not diminish adrenaline excretion (Kärki, 1956). Training usually (Hartley, Mason et al., 1972; R. H. Johnson, Park et al., 1974; Bloom et al., 1976; Hickson et al., 1978; Winder et al., 1978) but not always (Brundin and Cernigliaro, 1975) reduces the output of catecholamine for a given work load. However, the response

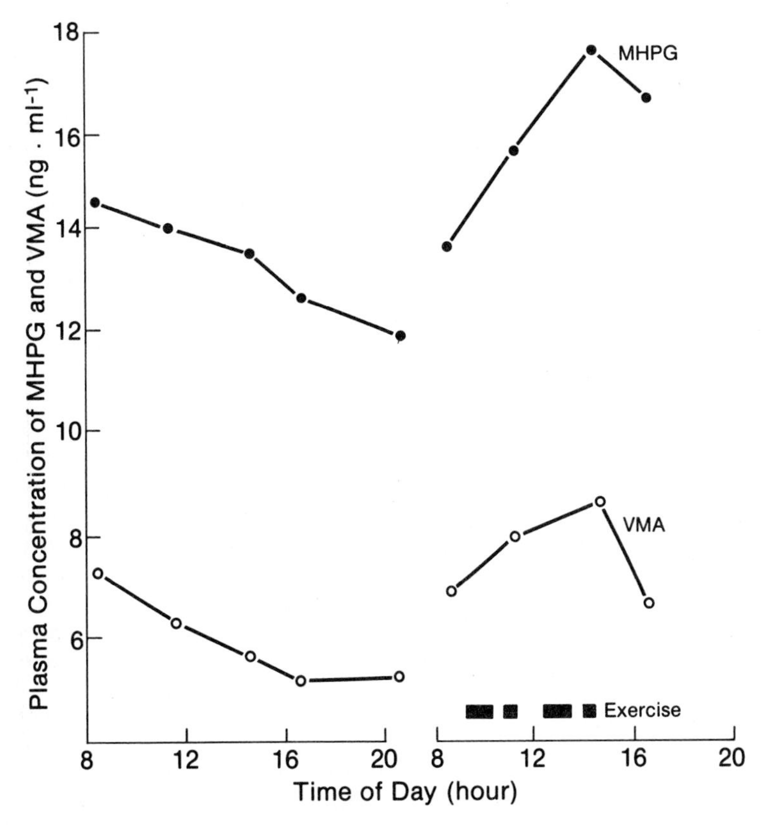

FIG. 8.8. Increases of plasma VMA and MHPG with exercise at 40 percent of maximum oxygen intake. (Based on experiments of Tang et al., 1979.)

at a given percentage of maximum oxygen intake remains unchanged (Hartley, 1975). Training also brings about an increase in the resting adrenaline content of the adrenal glands (Östman and Sjöstrand, 1971).

PANCREATIC HORMONES

Insulin

Insulin plays a vital role in regulating the transfer of glucose from the blood stream into tissues such as skeletal muscle. It also promotes muscle glycogen storage and is essential for the appearance of glycogen "supercompensation" in the active fibers (Chap. 2; Ivy, 1977). The circulatory half-life of insulin is ~ 40 min.

Plasma insulin levels may drop to ~ 50% of resting values during or immediately after exercise (Fig. 2.18; Pruett, 1970a,b; Hartley, Mason et al., 1972; Galbo et al., 1975; Metivier, 1975; Vranic et al., 1975). The effect is particularly marked with

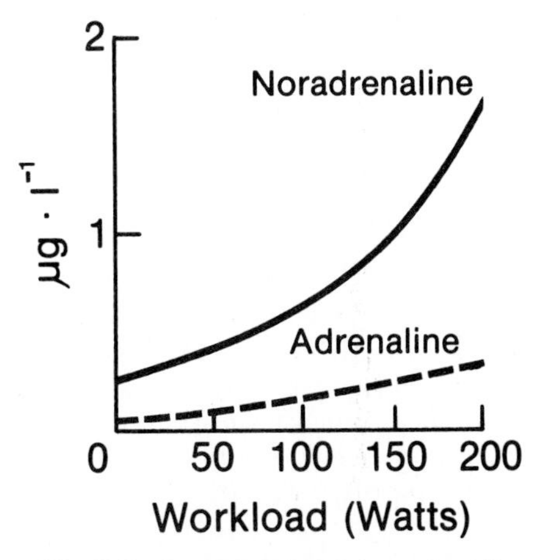

FIG. 8.9. Concentrations of noradrenaline and adrenaline in venous plasma with varying intensities of physical activity (after Vendsalu, 1960). It has been argued that arm vein noradrenaline levels underestimate noradrenaline secretion in leg work (Von Euler, 1974). Peak noradrenaline levels of up to 10 mg • l^{-1} have been observed by Haggendal et al. (1970) and R. H. Johnson, Park et al. (1974).

prolonged and intensive exercise (Hartley, Mason et al., 1972; Galbo et al., 1975), despite some release of bound insulin by the heart and inactive muscles (Park et al., 1975). Normal resting concentrations can be restored quite quickly (2–10 min) after effort is halted (Wahren, Felig, and Hendler, 1973), and if blood sampling is delayed the fall of insulin concentration may pass undetected (Von Glutz et al., 1978). There is both a decrease of pancreatic secretion and an increased uptake of insulin by the skeletal muscles (Vranic et al., 1975). The decreased levels of insulin do not hamper glucose uptake by the active muscles, since these have an increased blood flow. However, the combination of a low plasma insulin and a reduced visceral blood flow favors the release of glucose from hepatic glycogen stores. There is little fall of blood glucose with most types of exercise, so this can hardly be the trigger for reduced insulin secretion. Some authors have attributed the change to increased plasma levels of adrenaline and noradrenaline (Porte and Williams, 1966; Pruett, 1970b; Lundquist, 1971; Hartley, Mason et al., 1972; Metivier, 1975), although the time course of catecholamine secretion is very different.

Training usually (Hartley, Mason et al., 1972; Bloom et al., 1976) but not invariably (J. A. White et al., 1968) reduces the depression of plasma insulin during exercise. This has been attributed to a lesser secretion of catecholamines. In diabetics, an increase of habitual physical activity reduces the need for insulin. Several mechanisms are involved, including the direct burning of sugar following a carbohydrate meal, an increased avidity of the muscles for glucose, and a sparing of functionally weak pancreatic islet cells following the ingestion of sugar.

Glucagon

Glucagon has the opposite type of action to insulin, raising blood glucose by activating the enzymes that break down liver glycogen. It also increases cardiac contractility and releases fatty acids into the blood stream by an action upon adenyl cyclase (Fig. 8.1). The circulatory half-life is 5–10 min.

Short periods of exercise to exhaustion give a two-fold increase of hormone concentration (Bottger et al., 1972; Felig, Wahren et al., 1972; Galbo et al., 1975; Nilsson et al., 1975), which persists for up to 30 min after exercise. The trigger may be an enhanced sympathetic nerve activity in the islet cells. Doubt remains as to the practical importance of pancreatic secretion of glucagon during exercise, since a pancreatectomized dog can mobilize liver glycogen much as a normal animal (Vranic et al., 1975). Apparently, a substance immunologically similar to or identical with pancreatic glucagon is released from some extrapancreatic source during exercise.

At a given percentage of maximum oxygen intake, the increase of glucagon is less in highly trained than in untrained subjects (Bloom et al., 1976).

TESTOSTERONE

The principal natural masculinizing hormones are testosterone (secretion 5–10 mg • day^{-1} in men, 0.1 mg • day^{-1} in women) and androstenedione (secretion 1–2 mg • day^{-1} in men, 2–4 mg • day^{-1} in women). The main sites of secretion are the interstitial cells of the testes in men and the ovaries in women, although androstenedione is also produced by the adrenal glands. In some tissues, testosterone is "activated" by conversion to dehydrotestosterone, but this does not seem to be the case in skeletal muscle.

The administration of testosterone to experimental animals produces effects somewhat akin to repetitive muscular work. At the cellular level, there is a stimulation of amino acid uptake, RNA synthesis, and DNA synthesis, with growth of individual muscle fibers and facilitation of glycogen storage. Gross changes include an increase of muscle mass (O'Shea, 1971; L. C. Johnson, Fisher et al., 1972; Ward, 1973), a decrease of body fat content, red cell formation (Hait et al., 1973; Shephard, Killinger, and Fried, 1977), an increase of bone thickness, and aggressive behavior (D. R. Lamb, 1975).

Some 2% of testosterone is normally in a free form; 57% is bound to β-globulin, 40% to albumin, and small amounts to corticosteroid-binding globulin and red cells. The liver converts androgens to inactive 17-ketosteroids. The main vehicle of excretion is the urine, although small amounts are also lost in sweat and feces. There are three metabolic pools, with respective half-lives of 6.6 min, 33 min, and 3.4 hr (Baulieu and Robel, 1970); the last of the three pools is the largest and most important.

Male sex hormones do not seem to play a vital role in fat mobilization during exercise, since castrated rats show a normal increase of free fatty acids with endurance running (A. W. Taylor, Secord et al., 1973). However, testosterone may contribute to the normal development of muscular tissue and the enhanced growth that follows physical training. Adolescent boys show a rapid development of strength coincident with puberty (Chap. 4), and this has been attributed to an increased secretion of testosterone by the interstitial cells of the testes. Again, in middle-aged and older men the administration of testosterone enhances the response of the muscles to training, particularly following injury or immobilization of a body part. On the other hand, the response of a young man to a physical training regimen seems uninfluenced by testosterone (M. H. Williams, 1974; Shephard, 1977a; American College of Sports Medicine 1977), at least if this is administered in double-blind fashion. Presumably, the circulating androgens of a young man are already at an optimum level for muscular development and there is little rationale for the dangerous practice (Shephard, Killinger, and Fried, 1977) of administering such compounds in an attempt to enhance athletic performance (Chap. 14).

In Olympic-caliber athletes and weight lifters, blood levels of testosterone are increased by strenuous work (Sutton, Coleman, et al., 1973; Fahey, Rolph et al., 1976), but in untrained subjects there does not seem to be any change (D. R. Lamb, 1975; Fahey, Rolph et al., 1976).

The level of the pituitary regulator lutropin (Fig. 8.10) does not rise during exercise (Sutton, Coleman et al., 1973,1978), and the increase of testosterone is unchecked by drugs that suppress adrenocorticotropic hormone (Sutton, Coleman, and Casey, 1974). The response thus cannot be attributed to an increased secretion of either interstitial cells or adrenal cortex (Sutton, Coleman, and Casey, 1978). The change (40–50% increase) is too large to attribute simply to hemoconcentration, and there is no evidence of altered binding of testosterone to protein (Sutton, Coleman et al., 1973). The most likely explanation of the rise is a diminished hepatic degradation (secondary to a diminished hepatic blood flow).

There is disagreement as to whether training increases resting levels of plasma testosterone (Strömme, Meen et al., 1974; D. R. Lamb, 1975; R. J. Young et al., 1976; Fahey, Rolphe et al., 1976). However, plasma concentrations of lutropin are unchanged by 8 weeks of weight training (Strömme, Meen et al., 1974).

OTHER HORMONES

Cycling to exhaustion causes an increase in the plasma concentration of some prostaglandins (Kochan and Lamb, 1976). There is no information as to whether somatomedin concentration is increased by training, although this seems likely in view of the stimulation of cartilage formation and bone growth with chronic activity.

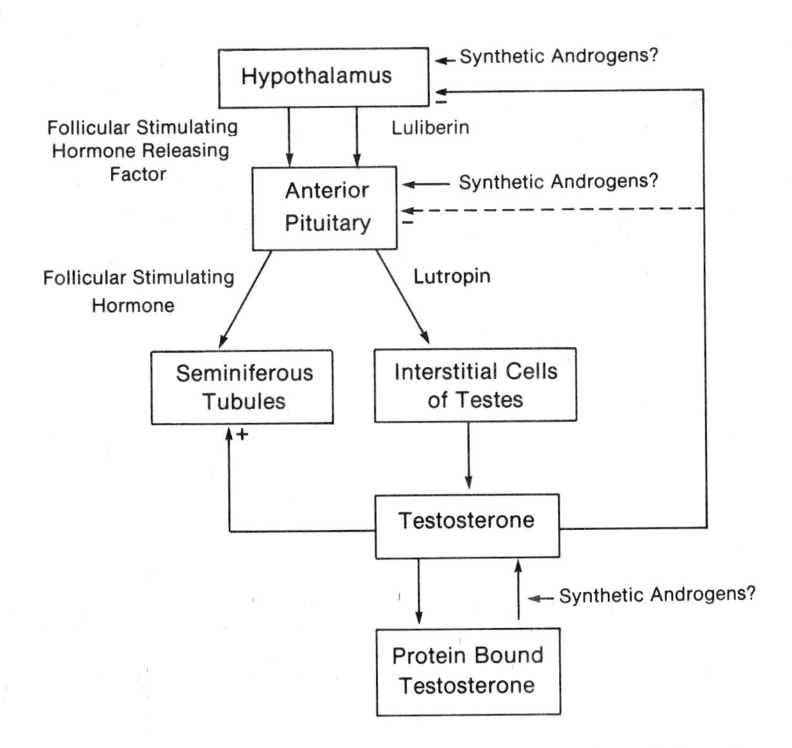

FIG. 8.10. The interrelationship of the testes, pituitary, and synthetic androgens.

There have been numerous attempts to document changes of performance over the female menstrual cycle (Ryan and Allman, 1974; Dalton and Williams, 1976). The effects cited have been quite limited, and it has been difficult to dissociate a direct hormonal response from associated problems of personal hygiene (Chap. 10). In general, skilled performance has deteriorated during the phase of premenstrual tension, and some items have been performed slightly better than normal during the phase of menstrual flow. Petrofsky et al. (1976) found that isometric strength and the associated heart rate and blood pressure response remained constant over the menstrual cycle, but isometric endurance was maximal at the midpoint of the ovulatory phase. Further research may clarify the basis for such findings. But in any event, a deterioration of function with menstruation has become rather an academic question, since the timing of the menstrual cycle can now be adjusted quite readily by use of hormone preparations.

CHAPTER 9
Other Body Systems

Blood and Body Fluids
Gastrointestinal Tract and Viscera

While there is extensive documentation of the behavior of the cardiac, respiratory, and musculoskeletal systems during exercise, relatively little is known of responses in other parts of the body. In general, function is thought to be depressed, although certain changes in the blood contribute to the transport of oxygen, while alterations of renal function contribute to the maintenance of fluid balance.

BLOOD AND BODY FLUIDS

Blood

Erythrocytes

The acute effect of vigorous exercise is commonly a 5–10% increase of both hemoglobin level and red cell count. It is thus important that any necessary samples of blood are collected before exercise is performed. In some animals, such as the dog, large changes of red cell count can arise from an exercise-induced contraction of the splenic capsule (J. Barcroft and Stephens, 1927). In humans, the capsule contains little muscle, although red cells trapped in the splenic sinusoids may be released into the general circulation during physical activity (Cruickshank, 1926). If protracted exercise is performed, hemoconcentration usually occurs (Dill et al., 1930; Keys and Taylor, 1935). Nevertheless, changes of hematocrit, plasma volume, and plasma proteins are not always proportionate (Kaltreider and Meneely, 1940; P. O. Åstrand and Saltin, 1964; Saltin, 1964a; Pugh, 1969a; Van Beaumont, Greenleaf, and Julios, 1972).

Fluid lost in sweat and expired air may exceed 30 ml • min^{-1} (see below). There is also a substantial exudation of fluid into the active tissues (M. H. Williams and Ward, 1977; Chap. 6). Vigorous exercise increases the fluid content of the active muscles by as much as 20%. Thus, if 20 kg of muscular tissue is active, exudation could theoretically reach the improbable total of 5 l. During bicycle ergometry, accumulations of ~1100 ml are seen (Lundvall, 1972; Lundvall et al., 1972); plasma losses are smaller (~600 ml), but nevertheless exceed sweat losses for all except protracted periods of exercise.

There is disagreement as to whether habitual exercise increases hemoglobin level and red cell count (Schneider and Havens, 1915; J. E. Davis and Brewer, 1935; K. L. Andersen, Heusner, and Pohndorf, 1955; Holmgren et al., 1964), but it is accepted that training increases the total body hemoglobin. Certainly, the hemoglobin is increased by other demands upon the oxygen transport system, such as exposure to carbon monoxide (in a cigarette smoker; Rode, Ross, and Shephard, 1972) and high altitude (in a mountaineer; Pugh, 1962), both circumstances leading to an increased red cell formation (erythropoiesis).

Whether the red cell count is increased by hemoconcentration or enhanced production (erythropoiesis), there is an increase of blood viscosity (Fig. 9.1). Over the normal range of change associated with acute and chronic exercise (an increase of hematocrit from 45 to 55%), viscosity rises by ~25%, but the resultant increase in the viscous component of

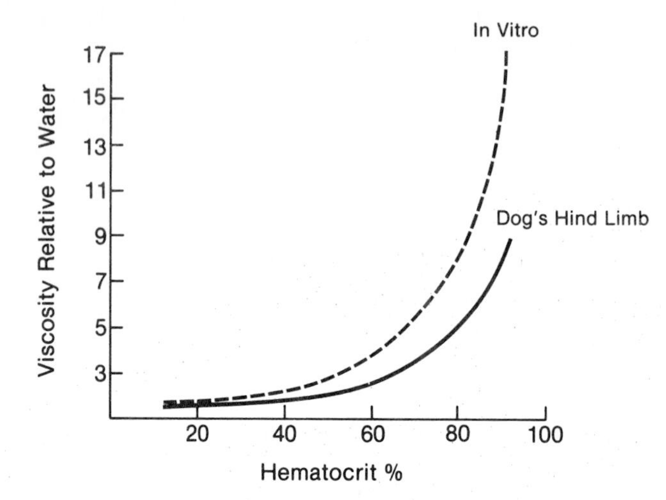

FIG. 9.1. Apparent plasma viscosity and hematocrit level. (Based on data of Whittaker & Winton, 1933.)

cardiac work is insufficient to reduce the maximum cardiac output (Froese and Gledhill, 1979); indeed, cooling of the extremities to near freezing has a 10 times larger effect upon blood viscosity (A. C. Burton, 1965b). The increased hemoglobin concentration thus augments the maximum oxygen carriage, giving a possibility of boosting performance by "blood doping" (Chap. 6). In the large vessels, plasma "skimming" reduces effective viscosity (A. C. Burton, 1965a,b). Under physiological conditions, axial accumulation of the erythrocytes is complete, and the effective viscosity of the blood is thus independent of normal variations in blood flow rates. In the smaller arterioles and capillaries it is even more difficult to quote an appropriate figure for blood viscosity; in such vessels, the plasma moves relatively independently of the red cells, and the effective resistance during vigorous effort is much less than would be predicted from *in vitro* measurements of viscosity (Fahraeus-Lindquist effect).

Some forms of physical activity, such as marching and running, expose the red cells to mechanical trauma (Buckle, 1965). This causes hemolysis (Kiiskinen, Kemppinen, and Hasen, 1975), a reduction in the serum proteins that bind hemoglobin (haptoglobins), and the appearance of hemoglobin in the urine (Bichler et al., 1972; Poortmans, 1975; M. H. Williams and Ward, 1977). The breakdown of the erythrocytes may be one factor contributing to anemia in the endurance athlete (de Wijn et al., 1971).

Leucocytes

Exercise generally causes a substantial increase in the overall white cell count, from perhaps 5000–7000/mm^2 to 25,000–30,000/mm^2. This cannot be explained simply in terms of hemoconcentration. One additional factor is probably the "washing out" of white corpuscles from such storage sites as bone marrow, spleen, liver, and lungs. With short but intense exercise, a lymphocytic response predominates, but with protracted exercise the main increase may be in the neutrophils (Egoroff, 1924). If the activity causes significant "stress," the eosinophil count shows an early increase and a subsequent decrease associated with a rapid outpouring of hormones from the adrenal cortex (Chap. 7). In the event that the adrenal gland becomes exhausted, there is a final increase of eosinophil count. However, it is rare for exercise to be pushed to this point, except in an excessively hot climate (Sulman et al., 1977). Training apparently has no effect on white cell counts (C. Hawkins, 1937).

Platelets

The response of the blood platelet count varies with the intensity of effort. Light or moderate work produces an acute decrease, followed by a transient increase in the number of circulating platelets. With heavier work, the increase is larger and more persistent, reaching a peak in ~30 min (Schneider and Havens, 1915; A. A. Dawson and Ogston, 1969; G. Lee, Amsterdam et al., 1977; Pelliccia, 1978). The mechanism again seems a flushing out of the cells from storage sites such as the pulmonary vessels, and the net effect is usually an acute increase in the coagulability of the blood (Poortmans et al., 1971; Korsan-Bengtsen et al., 1973). Some authors have held that the long-term effect of habitual activity is a decrease in the coagulability of the blood, basing this belief on comparisons of exercise with caged cockerels and of booking clerks with physically active railway "switchmen" (Montoye, 1960). On the other hand, Korsan-Bengtsen et al. (1973) noted a continuing enhancement of clotting ability in men who were physically active at work relative to their supposedly more sedentary counterparts.

The thrombocytosis of exercise can be mimicked by the administration of adrenaline, but it is still debated as to how far β-adrenergic receptors contribute to the exercise-induced increase of blood clotting (A. A. Dawson and Ogston, 1969). Certainly, there is not the once postulated splenic contraction, and an increase of thrombocyte count occurs with exercise even after administration of β-blocking agents such as propranolol.

There is also disagreement as to exercise-induced changes in the inherent "stickiness" of the platelets (G. Lee, Amsterdam et al., 1977); probably, if allowance is made for the increase in thrombocyte count, the stickiness is normal or even decreased. However, changes in the plasma contribute to the enhancement of coagulation. Detailed tests show a shortened partial thromboplastin time (Finkel and Cumming, 1965) but no change of prothrombin time (G. Lee, Amsterdam et al., 1977). This reflects an augmented plasma concentration of clotting factor VIII (Iatradis and Ferguson, 1963; Egeberg, 1963; Pelliccia, 1978) and possibly factor XII (Pelliccia, 1978). Factor XII (Hageman factor) is the first step in the long clotting sequence and is normally activated by physical contact. Through two intermediary stages (XI and IX) it activates factor VIII, the antihemophilic globulin. In the presence of calcium ions, factor VIII accelerates the breakdown of platelets that have been in contact with damaged tissue, and through an action upon factor X (Stuart) converts proaccelerin (factor V) to an active form (Fig. 9.2).

Although clotting is accelerated during exercise, the "bleeding time" may be increased. When making a clinical measurement of bleeding time, a small stab wound is wiped clear of blood at 15-sec intervals. Under such circumstances, the duration of bleeding is independent of clotting mechanisms and reflects instead the tendency of damaged capillary vessels to retract. During exercise, there is an increase of capillary perfusion pressure (Chap. 6). This holds the small blood vessels open, thereby lengthening the bleeding time.

If these were the only effects of exercise upon the coaguability of the blood, physical activity might be thought a disadvantage for the coronary-prone individual. However, G. A. McDonald and Fullerton (1958) suggested that exercise also curtails the increase of coagulability associated with a fatty meal, while E. Ferguson and Guest (1974) noted that training reduced the acute effect of exercise upon clotting time. More important, fibrinolysis is increased immediately after exercise (Cash et al., 1969; Berkada et al., 1971; McAlpine et al., 1971; Poortmans et al., 1971; Åstrup, 1973). This seems due to the release of various activators, including factor XII and thrombin (Karp and Bell, 1974); these convert plasminogen to the active, clot-destroying enzyme plasmin. If exercise is unduly prolonged, there may be a secondary decrease of fibrinolysis as stores of the plasminogen activator are exhausted. The "dose" of exercise needed by a postcoronary patient in order to bring about a reduction of intravascular clotting is thus rather critical. Physical training has little effect upon either the acute, exercise-induced fibrinolysis or resting fibrinolytic activity (Moxley et al., 1970; Ferguson and Guest, 1974).

Plasma Protein

Prolonged exercise leads to an increase in the prealbumin, albumin, α_2-macroglobulin, and IgG content of the plasma, with no change in other protein constituents (Poortmans, 1975; Table 9.1). There is a 12% increase of albumin over the first 30 min of activity; the effect is identical for running and swimming, but is not seen with repeated weight lifting. Some (Joye and Poortmans, 1970; Poortmans, 1970), but not all, authors (Van Beaumont, Greenleaf, and Julios,

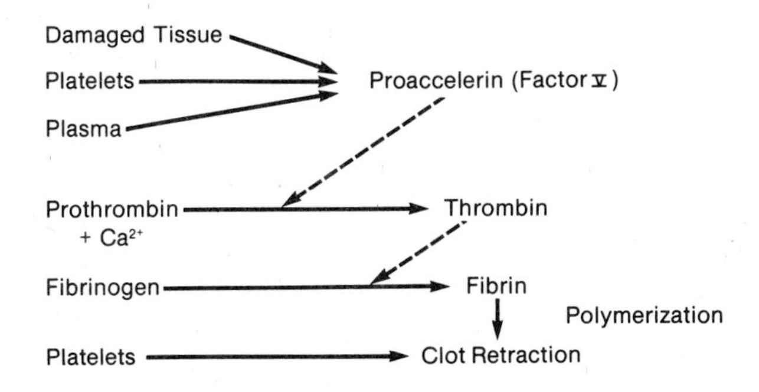

FIG. 9.2. Simplified schema of blood-clotting mechanisms.

TABLE 9.1
Changes in the Plasma Protein Components with 60 min Bicycling

	Plasma Concentration (mg • dl^{-1})		
Variable	Resting	Immediate Post-exercise	30 min Post-exercise
Prealbumin	40	45	42
Albumin	4250	4650	4610
Haptoglobin	162	159	152
α_2-macroglobulin	198	219	204
Transferrin	202	211	218
IgG	1003	1040	1019

Source: After Poortmans, 1975.

1972; Van Beaumont, Strand et al., 1973) have suggested that during exercise the proteins are mobilized from extravascular sources. Perhaps because of changes in protein content, physical activity is sometimes (Black and Karpovich, 1945) but not always (Hannisdahl, 1940) followed by an increase of erythrocyte sedimentation rate.

Body Fluids

Information on changes in body fluid volumes has already been presented at several points in this book. However, a brief coordinating review may be helpful.

Blood Volume

The blood volume of the average man is ~5 l, ~45% of this being attributable to suspended red cells. Estimates of blood volume are usually based on a combination of hematocrit readings and dilution of an indicator that is freely miscible with the plasma but escapes relatively slowly from the circulation. Candidates include dyes (indocyanine green, T-1824) and ^{131}I-labeled plasma protein. Alternatively, red cells may be tagged by a radioactive tracer (^{51}Cr, ^{55}Fe, ^{59}Fe, ^{32}P) or by carbon monoxide (Sjöstrand, 1962; Lawson 1962). During exercise, relative changes of plasma volume may be inferred from hemoglobin levels or hematocrit readings, the latter requiring adjustment for any change of erythrocyte volume (Van Beaumont, 1973; Costill and Fink, 1974). High molecular weight proteins such as IgM are also sometimes used as markers (Kavanagh and Shephard, 1978). Changes in the optical density of the plasma with exercise can lead to errors in the interpretation of dye concentrations (Ebert and Stead, 1941).

Brief bouts of exercise may produce a 10% decrease of plasma volume, with resultant hemoconcentration (Kaltreider and Meneely, 1940; Cassels and Morse, 1942; J. E. Wilkerson et al., 1977). With sustained effort, substantial amounts of fluid are lost from sweat, muscle exudate, and expired gas, but there is still controversy as to which body compartment is depleted the most. Available information suggests that during endurance activity plasma volume is well preserved until dehydration is severe (P. O. Åstrand and Saltin, 1964; Saltin, 1978; Kavanagh and Shephard, 1978); however, if the major stress is heat rather than vigorous exercise, the loss of plasma fluid may rise to 2.5% per 1% of dehydration (Costill, 1978a,b). Exercise also alters the distribution of the available blood volume; leg work increases central blood volume (R. J. Marshall and Shepherd, 1961; Levinson et al., 1966) but arm work diminishes it (Asmussen and Christensen, 1939b; Asmussen and Nielsen, 1955). As a consequence of these various factors, the tolerance of prolonged work is substantially reduced in the heat (Saltin, 1964a,b). The oxygen transport during brief (3–5 min) bouts of bicycle ergometer work is apparently unchanged by dehydration (Saltin, 1964a,b) but the treadmill maximum oxygen intake is significantly reduced if the plasma volume is reduced by use of diuretics (Costill, 1978a,b). Possibly, these discordant results are one more indication of the peripheral limitation of bicycle ergometer work.

Deliberate dehydration (as practiced by some wrestlers) leads to a decrease of cardiac stroke volume (T. E. Allen et al., 1977) and bicycle ergometer physical working capacity at a heart rate of 170 beats per min (Herbert and Ribisl, 1972).

Scandinavian authors have suggested that physical conditioning increases the circulating blood volume (Kjellberg et al., 1949a; Sjöstrand, 1949; Holmgren, 1967b; Saltin, Blomqvist et al., 1968). The majority of studies have used carbon monoxide as the marker substance, and results could thus have been influenced by an increase of muscle myoglobin concentrations. Some North American studies (perhaps using a less intense training regimen) have shown no change of T-1824 space after conditioning (D. E. Bass et al., 1958), but others have confirmed expansion of the plasma volume by the [^{131}I] albumin technique (Oscai, Williams, and Hertig, 1968). One possible mechanism for such a change would be a reduction in the sensitivity of the atrial stretch receptors (Chap. 7). One may presume that repeated exercise leads to an adaptation of the receptors in association with an increase of venous return, a displacement of the Starling curve to the left (Chap. 6, Fig. 6.8), and a more complete diastolic emptying.

Extracellular Fluid

The extracellular fluid accounts for 15–20% of body mass. It includes the plasma, cerebrospinal fluid, gastrointestinal juice, and interstitial fluid. The volume is estimated from the dilution of appropriately distributed marker substances. The value obtained depends upon the completeness with which the various extracellular fluid compartments are penetrated by the marker, and the extent to which it crosses cell membranes to enter the intracellular fluid. The highest results are obtained with markers of low molecular weight (sodium or chloride ions, radioactive sulfate ions, thiosulfate, and mannitol), and lesser volumes are reported where sucrose, raffinose, or inulin have been used as tracer substances. The "transcellular component" contained within the various hollow organs, such as the stomach, is particularly liable to remain undetected by commonly used techniques (Wolf, 1958).

Factors that increase capillary permeability and thus the formation of interstitial fluid include tissue oxygen lack, a rise of body temperature, and an increase of capillary hydrostatic pressure (Kjellmer, 1964). All of these factors operate in exercise (Chap. 6), together with an increase of capillary surface, escape of plasma protein, and a local accumulation of metabolites that increases the osmotic pressure of the extracellular fluid (Cullumbine and Koch, 1949; Kronfeld et al., 1958; Mellander, Johansson et al., 1967). We have noted above that a substantial swelling of the muscles accompanies physical activity. The rising hydrostatic pressure within the tissues and the rising osmotic pressure of the plasma make the process self-limiting. At the same time, the muscle contractions serve as a pumping mechanism that induces a several-fold increase in the return of fluid to the circulation via lymphatic channels (Elkins et al., 1953).

Training apparently has no effect upon the extracellular fluid volume (Pascale et al., 1955; Boileau, 1969).

Intracellular Fluid

The volume of intracellular fluid cannot be measured directly. The standard approach is to measure the total body water and to subtract from this figure the volume of extracellular fluid, determined by the methods described above. A chloride method has been described for the estimation of muscle water and electrolytes in biopsy specimens (Bergström, 1962), but unfortunately measurements during physical activity are upset by changes of membrane potential and associated ionic shifts.

The total body water is determined by dilution of a marker substance such as urea, antipyrine, deuterium oxide (D_2O), or tritium oxide (^3H_2O). The last two compounds are perhaps the most reliable sources of information, although with all methods there is difficulty in ensuring inclusion of water sequestered in sites such as bone and cartilage. Typical body water estimates range from 45 to 70% of body mass, averaging ~62% in the male and 51% in the female. Most lean tissues contain a relatively fixed percentage of water (~73%). It is thus possible to proceed from calculations of body water to an estimation of lean body mass and percentage body fat.

The increase of metabolism during exercise increases the osmotic pressure of the cellular contents. Other factors being equal, there is then an increase of free intracellular water in an attempt to restore normal osmotic relationships. Muscle contains a substantial volume of water (~1.6 l) bound to glycogen, and this is released if the glycogen is broken down by prolonged exercise. Available techniques are not sufficiently precise to determine whether the proportion of cell water is changed by physical conditioning. Although electrochemical considerations make it unlikely that there is much change of free water (McCance, 1956; F. D. Moore, Olesen et al., 1963), bound water presumably rises with any increase in glycogen stores.

Water Balance

The daily water requirement varies widely from 1.5 to 7.0 l, demand being increased by exercise and exposure to heat or cold. Water is derived from the ingestion of food (~1000 ml • day^{-1}), the oxidation of food products (~100 ml for each 5 MJ of energy expenditure), the water of hydration of glycogen (a "once-only" reserve of ~1600 ml), and the deliberate drinking of fluids. Drinking habits vary greatly, depending on both the intensity of effort undertaken and the thermal comfort of the environment. During exercise or heat exposure, thirst is not a sufficient indicator of fluid needs (Adolph, 1947a). On the other hand, during periods of rest, habit, and/or social demands may cause a greater fluid intake than is necessary to maintain the water balance of the body.

A minimum "obligatory" urine secretion of ~150 ml • day^{-1} is set by the need to excrete waste products through renal tubules that have a finite concentrating power (see next section). However, the enthusiastic and bibulous celebration of success in a football match can easily boost urinary output to 1 liter • hr^{-1}. Quite mild exercise greatly reduces urine secretion, and during exhausting effort output may decrease to almost zero.

Losses through the skin include "insensible" perspiration (relatively constant at 700 ml • day^{-1}) and sweating (slight or nonexistent when seated in a comfortable, thermally neutral room, but rising to ~2 l • hr^{-1} with heavy exercise in a hot climate).

Losses of water vapor in expired air depend upon the absolute water content of the atmosphere and the respiratory minute volume (Kerslake, 1972; Chap. 3). In the summer, the inspired gas may be almost fully saturated with water vapor at 30°C, giving it a water content of 30 mg • l^{-1}. Expired gas is fully saturated at 32°C, giving it a water content of ~33 mg • l^{-1}. Under such conditions, a respiratory minute volume of 80 l • min^{-1} would cause a respiratory water loss <15 ml • hr^{-1}. During winter, the situation is very different. The inspired air is now perhaps 50% saturated at 0°C (water content 2 mg • l^{-1}) while the water content of expired gas remains unchanged at 33 mg • l^{-1}. The respiratory loss is now 15 ml • hr^{-1} under resting conditions (ventilation 8 l • min^{-1} BTPS), and rises to 150 ml • hr^{-1} with an exercise ventilation of 80 l • min^{-1} BTPS. It is easy to appreciate from these figures how a combination of dry mountain air, vigorous exercise, and oxygen hunger can lead to progressive dehydration (Pugh, 1962). Fecal water loss is normally no more than 100 ml • day^{-1}. However, it can become quite large if there is diarrhea. Fluid losses secondary to a gastrointestinal disturbance have sometimes led to poor performance when athletes have competed in unfamiliar parts of the world (Goddard, 1967). The problem arises not from any well-recognized bacterial pathogen, such as the typhoid or paratyphoid bacillus, but from a host of minor bacteria that have little effect upon the normal resident of the country in question.

Gross Water Depletion

Problems of water deprivation in the marathon runner and in the wrestler endeavoring to achieve a specific weight category have been noted elsewhere (Chap. 3). Here, we shall focus upon severe water depletion as encountered by a wanderer lost in the desert or adrift in a small boat.

In such a situation, neither drinking water nor water-containing foods are available, but nevertheless, a certain minimum loss of water from the body is unavoidable. Urine production drops to the *volume obligatoire*. The required volume depends on the availability of foodstuffs. Whereas 850 ml of urine must be excreted when eating a normal mixed diet, 550 ml • day^{-1} suffices for a fasting man, and if energy requirements are met from sugar as little as 150 ml of urine can be passed per day. As dehydration develops, water is drawn from the extracellular fluid, and osmotic pressure rises in this body compartment. This in turn draws water from the intracellular compartment. The renal excretion of electrolytes is increased in an attempt to restore the normal ionic composition of extracellular fluid, and the mineral elements lost in this way must be restored during subsequent treatment.

The blood volume may change relatively little during the early stages of water deprivation. The decrease of capillary and venous pressures and the increase of plasma osmotic pressure both draw water into the bloodstream at the expense of the extracellular space, and if reliance is placed upon estimations of hemoglobin level or plasma volume, there is a danger that the severity of the patient's condition may be underestimated. (See Table 9.2.)

If no water is available, the body initially loses about 1 kg of its mass per day. The affected person is conscious of thirst and weakness, the skin becomes dry, and the eyes are sunken. When ~4 kg of mass has been lost, both the kidneys and the circulation show signs of failure. Death usually occurs if the water loss exceeds 15 kg.

Can survival of a water-deprived mariner be extended by drinking seawater? Much depends on the salt content of the ocean in question. Typically, seawater contains about 3 times as much salt as the plasma and it is difficult for the ailing kidney to excrete the excess sodium ions. Thus, the drinking of seawater leads to a further increase in the osmotic pressure of both plasma and extracellular fluid. This

TABLE 9.2
Body Water Loss During Moderate Exercise (60–70% of Maximum Oxygen Intake) in a Hot Environment (39°C, 25% relative humidity)

Total Dehydration (%)	Contribution of Compartment (% of Total Loss)		
	Plasma	Interstitial Fluid	Intracellular Fluid
2.2	10	60	30
4.1	10	38	52
5.8	11	39	50

Source: Based on data of Costill, 1978a.

in turn increases intracellular dehydration, and the survival time is therefore shortened rather than lengthened.

GASTROINTESTINAL TRACT AND VISCERA

Mouth

Personal experience illustrates the decreased volume and the increased viscosity of salivary gland secretions in anxiety-provoking situations. In competitive exercise, drying of the mouth may be further exacerbated by inhaling large volumes of dry air.

Stomach

Physical activity generally retards emptying of the stomach. During hard physical work, the combined effects of fluid absorption and gastric emptying permit a fluid intake of ~600 ml • hr^{-1} (Costill, 1978 a,b; Kavanagh and Shephard, 1978; Saltin, 1978; Chap. 3). Light activity, such as walking, may increase gastric motility (Hellebrandt and Tepper, 1934). On the other hand, vigorous exercise slows the gastric emptying of both a gruel test meal (J. M. H. Campbell et al., 1928) and a farina-barium-sulfate meal (Hellebrandt and Tepper, 1934) by at least 20%, and it also reduces the contractions recorded by an intragastric balloon (Hellebrandt and Dimmitt, 1934). The exercise-induced inhibition of gastric contractions seems transient; indeed, emptying is accelerated during the recovery period, so that the overall rate of gastric clearance for exercise plus recovery may be normal (Stickney and Van Liere, 1974).

Whether the stomach is fasting or is stimulated by food or histamine, the rate of gastric secretion is enhanced by moderate exercise. However, it is inhibited by more severe exertion (Crandall, 1928). There is sometimes a "rebound" of increased secretion in the immediate recovery period, but the long-term effects of regular physical activity are a decrease of histamine-induced secretion (Frenkl et al., 1964) and an increased production of mucin. The acidity of the secreted fluid is reduced by heavy work, (J. M. H. Campbell et al., 1928; Hellebrandt and Miles, 1932), although such inhibition is lessened as subjects become habituated to a given task (Hellebrandt et al., 1935). Enzyme secretion is apparently reduced by physical activity.

There is considerable discussion as to whether regular physical activity reduces the risk of gastric ulceration. In animals, ulcers have several pathological bases. Starvation apparently causes ulceration by an increased secretion of gastric hydrochloric acid, while deliberate restraint of an animal leads to loss of RNA, with a reduced production of protective mucoproteins (Ludwig and Lipkin, 1969). Hydrocortisone administration is a further possible method of ulcer production, the lesion then being associated with a depressed synthesis of connective tissue mucopolysaccharides (Ezer and Szporny, 1970). Lilehei and Wagensteen found that the incidence of histamine-induced gastric ulcers was increased if animals were exercised immediately following the administration of histamine. On the other hand, several weeks of swim training enhanced mucopolysaccharide synthesis, presumably with a protective effect against ulceration. Further recent studies (Frenkl et al., 1964; T. H. Johnson and Tharp, 1974) noted that regular exercise affords some protection against reserpine-induced ulceration; T. H. Johnson and Tharp attributed this to a reduced secretion of acid or an increased production of mucin.

The physiological basis of these various responses is uncertain. The marked influence of emotions such as fear and anger upon the character of the gastric secretions is known from classical observations made on a technician "Tom," whose gastric lining was visible through an old stomach wound (Beaumont, 1833). Other possible ways in which exercise could affect the stomach include an alteration in the balance of sympathetic and parasympathetic activity (Chap. 7), a restriction of gastric blood flow (Rowell, 1971,1974), and the liberation of some circulating hormone. Support for this last concept has been obtained in cross-circulation experiments (Hammer and Obrink, 1953).

Intestines

Exercise apparently has no acute effect on the movements of the small intestine (Stickney et al., 1956), although there is a very obvious blanching of the vasculature. The chronic response to physical conditioning is a small increase of motility (Van Liere, Hess, and Edwards, 1954), but it is less clear as to how much this is due to an alteration of eating habits and how much it reflects changes in the balance of sympathetic and parasympathetic nervous activity.

In animals, at least, movements of the large intestines are stimulated by the acute phase of physical activity (Barcroft and Florey, 1929; de Young et al., 1931) but following exercise there is a period of subnormal bowel movements. The effects of exercise upon intestinal digestion and absorption of food are unknown. However, since splanchnic blood flow is reduced, we may suspect that the acute response is depression of absorption.

Under resting conditions, the splanchnic region is substantially overperfused, receiving as much as a quarter of the total cardiac output. If maximum exercise is performed in a hot environment, regional flow can be reduced by as much as 80% (Rowell, 1971,1974; Chap. 6), thereby contributing 1.0–1.5 $l \bullet min^{-1}$ toward perfusion of the active tissues.

The composition of meals and their timing prior to an athletic contest are discussed in Chap. 3. Some authors have given athletes a light meal as little as half an hour before running and swimming events, apparently without adverse effects upon performance. However, a full stomach is uncomfortable during vigorous effort, and it presents an obvious danger of vomiting to swimmers and those participating in contact sports.

Liver and Gall Bladder

The liver participates in the general reduction of splanchnic blood flow that accompanies physical activity. Depression of liver function is shown by a decreased rate of elimination of substances such as the dye bromsulphthalein blue (Rowell, 1971). Tissue oxygen lack may reach the point where hepatic enzymes are liberated into the bloodstream (L. I. Rose, Bousser, and Cooper, 1970; Magazanik et al., 1974; M. E. Houston, Waugh, et al., 1978; Pate et al., 1978). Enzyme release parallels a decrease of intracellular ATP levels. It has therefore been suggested that the enzymes are bound to structural proteins but that the tightness of this binding is influenced by changes of ionic strength and thus the activity of ATP-dependent membrane pumps (Fig. 4.12; Pate et al., 1978). Not all of the serum enzymes come from the liver. Creatine phosphokinase is derived from muscle and glutamine oxaloacetic transminase escapes from a variety of tissues. However, glutamic pyruvate transaminase is derived largely from hepatic tissue (M. E. Houston, Waugh et al., 1978). The extent of oxygen lack and the associated reversible cellular damage depends largely upon the intensity of effort (Y. Shapiro et al., 1973). The magnitude of the enzyme response at a given work load is thus reduced by physical training. However, if the test exercise is kept to a fixed percentage of aerobic power, it is less certain that tissue injury is reduced by conditioning. Certainly, interindividual differences in response are much reduced if data on either blood flow reduction or functional impairment are compared at the same relative work load (Rowell, 1971).

The reduction of hepatic blood flow persists and may even develop over an hour or more of sustained exercise. Nevertheless, there is little evidence that permanent harm is caused to a healthy individual. During exercise, the oxygen consumption of the liver is increased by at least the amount that would be predicted from the general rise of body temperature, and transient nausea and gastrointestinal disturbances offer the only suggestion of an adverse effect carried over into the postexercise period.

Long-distance runners tend to have chronically high serum enzyme and bilirubin levels, but plasma protein concentrations are well sustained (R. P. Martin et al., 1977). The apparent "abnormalities" thus reflect an acute leakage of enzymes and a hemolysis of red cells rather than an impairment of liver function. In animals, 4 hr of exercise per day has no adverse effect upon hepatic regeneration (Mellinkoff and Machella, 1950), and indeed there are reports that physical training augments liver mitochondria and ribosomal content. However, the safety of intense and prolonged exercise is less certain in patients with a pathological limitation of maximal cardiac output. In such subjects, repeated hepatic oxygen lack may contribute to the development of centrilobular necrosis of the liver.

The liver normally contains ~100 g of glycogen (270 mmol of glucose units per kg of tissue). This can be boosted to 500 $mmol \bullet kg^{-1}$ by a high-carbohydrate diet but is reduced to less than 40 $mmol \bullet kg^{-1}$ by a single day of carbohydrate depletion (Hultman, 1978). Synthesis is enhanced by vagal stimulation but is depressed by sympathetic nerve activity. Exhausting exercise causes a rapid breakdown of liver glycogen, the maximum release of glucose (1–2 $g \bullet min^{-1}$; Rowell, 1971) being seen when initial stores are large. With a well-stocked liver, gluconeogenesis accounts for no more than 3–9% of hepatic glucose output, but in subjects who have been following a carbohydrate-poor diet, 40–70% of the exercise-induced output of glucose may be attributable to gluconeogenesis (Hultman, 1978).

Possible signals inducing hepatic glycogen breakdown during physical activity include an altered insulin/glucagon ratio (Fig. 2.18) and an increase of circulating adrenaline (Chap. 7). Certainly, glucose output runs parallel with an increase in the hepatic venous concentration of cyclic AMP (Broadus et al., 1970; Hultman, 1978). Irrespective of mechanisms, the "purpose" of the increased glycolysis is probably to maintain the glucose supply to the brain.

Intense exercise reduces the output of bile from the gall bladder through a reduction in the vagal drive to this organ. The probable consequence is a slow-down in the emulsification, digestion, and absorption of intestinal fat.

Kidneys

General Considerations

The kidneys play a vital role in homeostasis, maintaining the blood composition within closely defined limits and permitting an even more precise regulation of the immediate milieu of the tissues. Other less well-recognized functions may also be important in exercise—the secretion of a substance erythropoietin which stimulates erythrocyte formation, and the formation of glucose from amino acids and lactate (gluconeogenesis). The quantitative significance of the latter is uncertain, but the process seems accelerated by both exercise and a low-carbohydrate diet.

The volume of fluid and associated solutes filtered by the glomeruli is very large, ~180 l • day^{-1}. However, 178–179 l of this total is usually reabsorbed during passage of the filtrate through the renal tubules. Part of the resorptive process is active in nature, involving usage of ATP; reabsorption of sodium and potassium ions, glucose, and amino acids all require the expenditure of significant amounts of chemical energy. A limit to such resorption is thus set by the power of the transport mechanism and the energy available to it. For example, the maximum reabsorption of glucose is ~375 mg • min^{-1}, and if a sharp peak of blood glucose concentration is created by the ingestion of a sweet drink, glucose may escape into the urine (glycosuria).

The reabsorption of other materials proceeds passively—urea in response to a concentration gradient (as water is reabsorbed), chloride ions in association with positively charged ions such as sodium, and water in association with osmotically active ions such as sodium. Reabsorption is facilitated by the secretion of aldosterone and antidiuretic hormone. Aldosterone acts primarily by speeding the active reabsorption of sodium ions, while the antidiuretic hormone apparently enlarges the pores of the distal convoluted tubules, facilitating passive reabsorption of water.

Ammonia, hydrogen ions, and potassium ions are actively secreted into the distal convoluted tubules.

Urinary Volume

Changes of posture, central blood volume, and physical activity all modify the volume of urinary secretion against a background of diurnal phasic change. Urine secretion is less during sleep than when awake, and during the period of wakefulness the output of urine is increased either by assumption of a horizontal body position or by an increase of blood volume (such as may follow the ingestion of fluids).

Older reports described a dramatic decrease of urine output with vigorous exercise (MacKeith et al., 1923; Eggleton, 1943; Patel, 1964). However, the response undoubtedly depends upon the initial hydration of the subject, the duration and severity of activity, the environmental temperature, and the fitness of the subject (Kachadorian, 1972; Wesson, 1974). In many studies, the urinary flow has been augmented artificially by administration of fluids and even diuretics such as tea. Given moderate activity, a cool environment, and a normal initial urinary output of ~1 ml • min^{-1}, the renal clearance of water may be unchanged or even increased by exercise (Castenfors, 1967,1977; Aurell et al., 1967; Kachadorian and Johnson, 1970).

When urine output is suppressed, this probably reflects an increased secretion of antidiuretic hormone (Chap. 7). Triggers to release of the hormone could include not only exercise and associated changes of central blood volume and osmolarity, but also emotional reactions to physical activity. The secretion of antidiuretic hormone is sometimes depressed by "negative conditioning," the output of the hormone falling as a subject becomes habituated to a given exercise by its frequent repetition.

Renal Blood Flow

During supine rest, the renal blood flow is ~1.2 l • min^{-1}, a fifth of the cardiac output. The renal plasma flow is thus ~700 ml • min^{-1}. Some 10–15% of renal plasma flow is filtered by the glomeruli. A marked dimunition of renal blood flow occurs during intense exercise (Chap. 6; C. B. Chapman, Henschel et al., 1948; Radigan and Robinson, 1949; Wade and Bishop, 1962; Grimby, 1965; Rowell, 1971, 1974).

It is not yet established whether this change is due to increased sympathetic nerve activity, increased circulating levels of lactate and catecholamines (J. H. Smith et al., 1952; Wexler and Kao, 1970), or an autoregulatory response to local chemical changes in the kidneys (Millard et al., 1972). Perhaps because renal and other visceral circulatory adjustments are helping to meet the ever increasing needs of heat dissipation, renal blood flow continues to drop over at least an hour of sustained exercise, and the return to a normal rate of perfusion is equally slow after effort has ceased. Nevertheless, it is unlikely that circulatory changes contribute to the reduced formation of urine. Autoregulation tends to preserve glomerular filtration, so that this is less markedly decreased than renal flow (Castenfors, 1977). It has been suggested that the efferent arterioles of the glomeruli constrict more than the afferent arterioles and that the resulting rise of capillary pressure increases the filtration fraction (Kachadorian, 1972; Wesson, 1974) to 25 or even 30% of renal plasma flow.

Renal function deteriorates in proportion to the intensity of the combined exercise/environmental stress (Radigan and Robinson, 1949), its duration (Castenfors, 1967; Kachadorian, 1972), and the extent of dehydration that is provoked (Kachadorian, 1972). Glomerular filtration can be estimated from the clearance of substances such as endogenous creatinine,* inulin, mannitol, and diodone. Moderate work does not decrease inulin or creatinine clearance (Barath, 1953; Starlinger and Berghoff, 1962; Grimby, 1965; Kachadorian, 1972), but function is depressed by more severe and sustained activity such as running or cycling (Barclay et al., 1947; H. L. White and Rolf, 1948; Raisz et al., 1959; Patel, 1964; Refsum and Strømme 1975). In an 85-km ski event, renal clearance may decrease by 30% (Castenfors et al., 1967; Castenfors, 1967). With a combination of exercise and heat stress, urea clearance can drop to 50% of normal (Covian and Rehberg, 1936; L. E. Duncan, 1955); this reflects not only reduced glomerular filtration, but also an increased reabsorption of urea from the concentrated urine in the distal tubules of the kidneys (Wesson, 1954).

The decrease of renal function at any given work load is inversely related to the individual's cardiorespiratory fitness. Exercise-induced reductions of renal flow and filtration are thus lessened by physical training (A. B. Light and Warren, 1936; Cantone and Cerretelli, 1960; Grimby, 1965; Castenfors, 1967; Schrier et al., 1970).

Proteinuria

Exercise is associated with the appearance of substantial amounts (>20 mg • dl^{-1}) of protein, casts, and even red cells in the urine (Siltanen and Kekki, 1959; Coye and Rosandich, 1960). This phenomenon is sometimes dignified by the name "athletic pseudonephritis" (K. D. Gardner, 1956; Kachadorian et al., 1970). In the past, it was attributed to renal trauma (Castenfors, 1977). However, it can be demonstrated following nontraumatic sports (Alyea et al., 1958) and is more likely to reflect the combined effects of tissue oxygen lack, a rising blood acidity, and an increase of body temperature upon either permeability of the glomerular intercellular "cement" or subsequent tubular reabsorption of protein (Wesson, 1974). Normally, the urine is almost free of protein. There has been disagreement as to whether this indicates a very limited glomerular filtration of large molecules or a moderate filtration with subsequent very efficient tubular reabsorption. Some authors have described a steady filtration and resorption of up to 360 g of protein per day (Poortmans, 1969).

When proteinuria develops, the excreted material is often of low molecular weight (Table 9.3). Nevertheless, this characteristic is less obvious after exercise than after prolonged standing (Poortmans and Jeanloz, 1968). In particular, exercise does not change the excretion of very low molecular weight proteins such as ribonuclease, which are conserved by tubular reabsorption (Castenfors, 1977). This is strong evidence that exercise-induced proteinuria is not due to either failure or saturation of tubular mechanisms for protein reabsorption.

It thus seems that the extent of exercise-induced proteinuria in any given individual depends upon the size of the "pores" in the glomerular membrane (influenced by both inheritance and disease) and the intensity of effort relative to the aerobic power of the individual (A. Taylor, 1960b; Poortmans, 1975).

TABLE 9.3
Urinary Protein Excretion at Rest and 30 Min After a 60-Min Bout of Exercise at 67% of $\dot{V}_{O_2}$(max)

	Excretion Rate (mg • min^{-1})	
Constituent	Rest	30 Min Postexercise
Albumin	9.27	11.15
α_1-acid glycoprotein	0.52	2.02
α_1-antitrypsin	1.05	4.32
α_2-HS-glycoprotein	0.28	1.38
Zn α_2-glycoprotein	0.94	4.60
Transferrin	0.37	1.35
IgG	0.37	9.83

Source: Based on data of Poortmans, 1975.

*There is some active secretion of creatinine; nevertheless, the proportion is sufficiently small for creatinine clearance to provide a useful measure of glomerular filtration. The creatinine is produced from muscle creatine and is thus proportional to muscle mass (Zambraski et al., 1974). Excretion is increased acutely by exercise, making it unwise to express the excretion of other urinary constituents as a ratio to creatinine output, a tactic adopted in some clinical laboratories.

Training decreases the proteinuria at a fixed work load (A. Taylor, 1960b) but has little influence upon protein loss at a given percentage of maximum oxygen intake.

Exercise proteinuria is a transient phenomenon. It is first noted ~30 min after activity and for the next 15-60 min excretion occurs at 10-20 times the control rate. However, unless activity is repeated, the urine is protein-free by the following day (Wesson, 1974). It is thus unlikely that physical exercise causes permanent damage to the renal system. The fact that protein often cannot be detected in the urine until exercise has ceased suggests the possibility that flow to anoxic glomeruli is completely interrupted during vigorous work.

The appearance of casts seems to run in parallel with urinary concentrations of mucoprotein (Patel, 1964). Immunological studies suggest that hyaline casts are composed largely of Tamms-Horsfall mucoprotein (McQueen, 1966). This substance is normally secreted by the renal cells but is apparently precipitated when high concentrations of plasma protein enter the urine.

Ionic Composition of Urine

Exercise induces changes in many urinary constituents. Commonly, the urinary concentrating mechanism is impaired, so that the ratio of urine/plasma osmotic pressures is less than under resting conditions (Raisz et al., 1959). Local oxygen lack apparently impairs active tubular reabsorption of sodium and thus of water. The problem may be compounded by a low blood flow through the associated vasa recta, with a local buildup of resorbed sodium ions. Pugh, Corbett, and Johnson (1967) observed urinary specific gravities ranging from 1.005 to 1.030 after participation in a marathon race.

Ammonium ion excretion is increased during and following vigorous activity (MacKeith et al., 1923; Eggleton, 1943; Wesson, 1974; Poortmans, 1975). Part of the ammonia may be derived from adenosine monophosphate and amino acids within the exercising muscle and part may be derived from the liver (where ammonia is normally converted to urea). Although urea production is enhanced by sustained exercise (Shephard and Kavanagh, 1975a), the combination of increased protein catabolism and decreased hepatic blood flow may still cause an overloading of pathways for urea formation. A third important source of ammonia is active secretion by the tubular cells. These cells contain the enzyme glutaminase, which liberates ammonia from glutamic acid amide. The gas diffuses into the tubules, where it can combine with hydrogen ions, as discussed below.

Light activity has no consistent effect upon the urinary concentrations of sodium ions (Aurell et al., 1967), but concentrations are decreased within 15 min of commencing vigorous work (Werkö, Varnauskas et al., 1954; Castenfors, 1967; Refsum and Strømme, 1975). Part of this change is due to a release of aldosterone (Chap. 7). The decrease of renal blood flow stimulates certain specialized cells (the juxtaglomerular apparatus) to secrete renin into the bloodstream; this catalyzes the conversion of a plasma protein angiotensinogen into angiotensin I, with further conversion to angiotensin II. The last compound serves as a messenger, stimulating release of aldosterone by the adrenal cortex. However, the changes of sodium ion excretion occur too rapidly to be attributed entirely to the aldosterone mechanism. The magnitude of the effect is also too large to explain simply in terms of reduced glomerular filtration, and a completely satisfactory theory has yet to be developed.

Some authors have described an increase of potassium ion excretion during sustained activity (Castenfors, 1977), but others report a very variable effect (rise, fall, or no change; Kachadorian, 1972; Haralambie, 1975). Factors influencing the response include an increase of serum potassium (possibly offset by the reduced renal blood flow), increased aldosterone secretion, hematuria, and the increased excretion of hydrogen ions.

Chloride ion excretion is decreased during exercise (Eggleton, 1943; Barclay and Nutt, 1944), this being a "passive" response to changes in sodium ion excretion.

Phosphate excretion is usually diminished during exercise, with a rebound during the recovery period (Wesson, 1974). The decrease is probably due to the reduced glomerular filtration rate, while the subsequent rise reflects both increased glomerular filtration and reduced reabsorption by the kidney tubules.

Acid/base Regulation

During moderate aerobic exercise, there is no consistent change of urinary pH (Passmore and Johnson, 1960; Kachadorian, 1972), but if the anerobic threshold is suppressed there is usually a substantial increase in the urinary output of acid, persisting for 30 min into the recovery period (Wesson, 1974). This reflects both increased blood levels of lactic acid and an increased excretion of acid by the tubular cells. These cells have a substantial carbonic anhydrase activity and can thus accept carbon dioxide from the bloodstream, converting it to carbonic acid (H_2CO_3).

Ionization to H^+ and HCO_3^- then occurs. Castenfors (1967) observed a drop of urinary pH from 6.2 to 5.6 units after 45 min of exercise, and after a marathon race Pugh, Corbett, and Johnson (1967) found a pH of 5.0 in 27 of 35 runners. Nevertheless, the total contribution of the kidneys to stabilization of body pH during heavy work is small relative to that of the respiratory system.

Two mechanisms serve to neutralize hydrogen ions within the urine. Some hydrogen ions combine with ammonia secreted by the tubular cells, and since the tubular epithelium is permeable to ammonia gas but not to ammonium ions, the latter are trapped in the tubular lumen and excreted from the body. Buffering is also provided by disodium phosphate, already present in the tubules:

$$Na_2HPO_4 + H^+ \rightarrow NaH_2PO_4 + Na^+ \quad (9.1)$$

Hydrogen ions leaving the tubular cells are replaced by sodium ions, and since chloride resorption passively follows the movement of sodium ions, the chloride content of the urine is inevitably reduced.

Hemoglobinuria

We have noted the excretion of hemoglobin following prolonged exercise such as marching (Stahl, 1957; see also above). Hemoglobin, like albumin, is a protein of relatively low molecular weight (68,000), and it can pass into the urine if the kidney is suffering from oxygen lack. Normally, the small quantities of hemoglobin liberated into the bloodstream combine with haptoglobins to form complexes of high molecular weight. A substantial breakdown of red cells must thus occur before there is frank hemoglobinuria. If hemoglobinuria does develop, it is seen in the first few hours after exercise, and must be distinguished carefully from myoglobinuria. Myoglobin is sometimes excreted for 1–2 days after very strenuous muscular activity (Stahl, 1957). Having a molecular weight of only 17,000, it is filtered relatively easily by the renal glomeruli, and an increased urinary excretion is not a sign of renal damage.

CHAPTER 10
Effects of Age and Sex Upon Energy Exchange

Children
Women
Aging

CHILDREN

Cellular Hypertrophy and Hyperplasia

The cellular changes induced by growth in children have their parallel in the adaptive response of older subjects to repeated exercise. It is thus useful to examine these changes in some detail. Hypertrophy is an increase in the size of an existing cell or organelle, without an increase in the number of formed elements; in the context of added exercise during childhood, it implies an increase beyond what would be anticipated from normal growth. Hyperplasia, in contrast, is an increase in the number of cells or associated organelles such as nuclei (Goss, 1966).

The two processes differ in their basic control mechanisms. Hypertrophy involves the synthesis of cell constituents using existing DNA templates, while hyperplasia implies a preliminary replication of DNA. Growth is associated with increases in the cell content of RNA, ribosomes, enzymic and nonenzymic proteins, and mitochondria. In some tissues, including muscle, there may also be an increase in the number of nuclei, with a rise of DNA content. Acute bouts of exercise activate existing enzymes and stimulate existing ribosomes; the controlling signal is likely a metabolically linked reaction such as the ADP/ATP ratio (Chap. 2). Frequent repetition of the exercise stimulus induces changes of the type associated with growth; endurance training increases the number of glycogen-synthesizing enzymes and the number of mitochondria, while isometric activity alters the balance of anabolism and catabolism within active muscles, so that the mass of both contractile protein and protective connective tissue is increased.

Age apparently plays a key role in determining whether hypertrophy or hyperplasia will occur. The number of myocytes in the heart (Korecky and Rakusan, 1978) and of fibers in skeletal muscle (Goldspink, 1972) becomes relatively fixed after the first few months of extrauterine life (Zak, 1974). Contrary reports can be explained on the basis of fiber splitting and an increase in the complement of nuclei through (1) the migration of satellite nuclei from the sarcolemma (Cheek, 1968; Chap. 4) and (2) the proliferation of connective tissue elements (Grove et al., 1969). The fat cells (adipocytes) are also most prone to hyperplasia in the neonatal period. At this stage of development, overfeeding leads to a rapid increase of cell number (Hager et al., 1977). However, regular exercise continues to have a restraining influence upon the development of fat cells until their ability to divide is lost in adolescence.

General Growth Curve

Growth Spurts

If the course of development is assessed in terms of a general size criterion, such as standing height or body mass, the resulting curve (Figs. 10.1, 10.2) shows rapid progress over the first 2 years, a period of steady development for the following 6 or more years, an acceleration before and during puberty, and a final deceleration as adult size is reached (LaBaere, 1954; Tanner, 1962; G. A. Harrison et al., 1964). From the biological point of view the period of growth is longer than in many species, increasing the likelihood of interaction between environment and genetic potential (LaBaere, 1954; Malina, 1978).

In early childhood, girls are marginally smaller than boys. Because they reach puberty earlier, the girls show an earlier "growth spurt," becoming heavier (and sometimes taller) than boys between the ages of 12 and 14 years. If averaged data for an entire population is examined, the growth spurt is not a very impressive phenomenon (Tanner, 1962). This is because the spurt occurs at different ages in different children. When the average "velocity" is calculated, a deceleration of growth in early developers masks the spurt in those who mature more slowly.

Height

Some 50% of adult height is reached by 2 years. Between the ages of 5 and 10, stature increases by ~5 cm • year^{-1}, but the velocity of growth increases to 7–8 cm • year^{-1} immediately prior to puberty (peaking at ~12 years of age in the girls and 14 years in the boys). Thereafter, there is an exponential decline of growth rate. Adult height is generally reached at ~18 years in the boys and 16½ years in the girls, although a small increase in length of the vertebral column may continue for a few more years (Tanner, 1962; G. A. Harrison et al., 1964).

Body Mass

Body mass shows greater variability than stature (Fig. 10.2). By 10 years of age, the average boy or girl has reached about a half of the adult value. Girls show a rapid gain of body mass at 12–13 years of age; this reflects an increase in the percentage of body fat associated with development of the breasts and other secondary sexual characteristics (Tanner, 1962; Pařízková, 1978b; Fig. 10.3). The pubertal spurt of the boys occurs later and lasts longer; in their case, it reflects partly an increase of stature and partly an increase of muscle mass.

Athletic Selection

In view of the influence of body build upon athletic performance (Shephard, 1978a), there have been attempts to select optimally endowed children at an early age by referring to their percentile status on growth curves such as those of Figs. 10.1 and 10.2. However, predictions of adult size must be made with caution, since the growth percentile at any given age may differ quite widely from the adult figure (Fig. 10.4). Discrepancies are particularly marked as puberty is approached; early maturers at first seem large for their age, but end their growth with less than average stature. There are considerable regional differences of stature at any given age (Demirjian

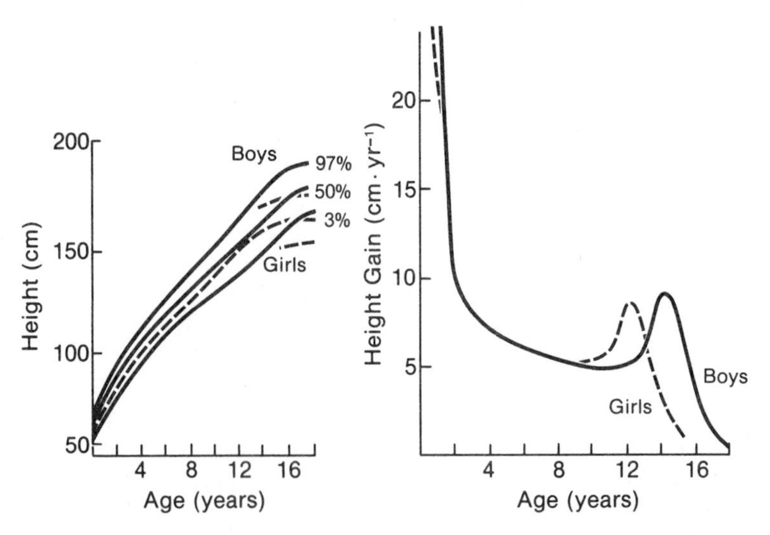

FIG. 10.1. Cross-sectional data for standing height in relation to age. Based on values of Tanner et al. (1962) for British subjects, with schematic velocity curve for individual child of each sex. (After Edington & Edgerton, 1976.)

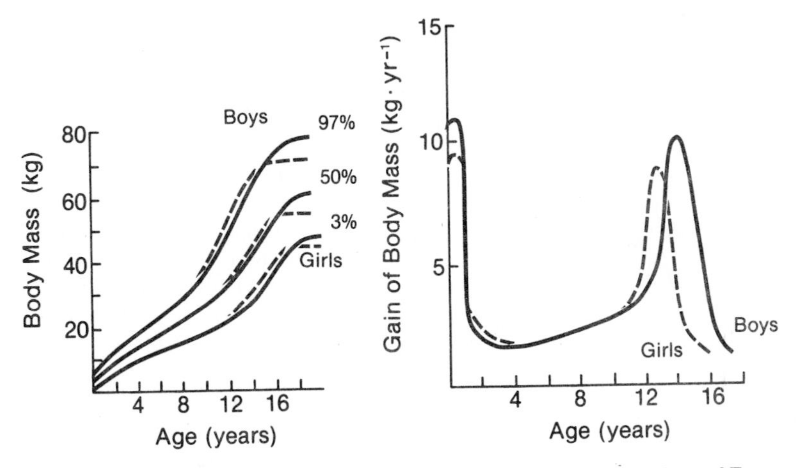

FIG. 10.2. Cross-sectional data for body mass in relation to age. Based on values of Tanner et al. (1962) for British subjects, with schematic velocity curve for individual child of each sex. (After Edington & Edgerton, 1976.)

et al., 1972; Rajic et al., 1978; Shephard, 1978a), this variation being linked more to nutrition and socioeconomic conditions than to genetic factors (Fig. 10.5; also see below).

Secular Trends

Public Health records show that there has been an increase of stature and an advance of puberty in many nations over the past century. The pattern of change in Western Europe is well illustrated by data for Norwegian adults, which go back to 1741. There was apparently little change of stature between 1741 and 1830, but a gain of 0.3 cm per decade from 1830 to 1875 and 0.6 cm per decade from 1875 to the present day (G. A. Harrison et al., 1964). Over the immediate past century, the standing height of British prepubertal boys has increased by ~1 cm per decade for those with professional parents and 1.5 cm per decade for those with parents of the laboring class. Adult stature in Britain has shown a smaller increase, the rate of change amounting to 0.5 and 0.8 cm per decade for the highest and the lowest social classes, respectively (G. A. Harrison et al., 1964). Advances in the age of puberty are best documented for girls (Fig. 10.6), where the age of menarche has decreased by about 0.35 years per decade (Tanner, 1962). The most recent statistics suggest that the increase of size and the advancement of maturation are coming to an

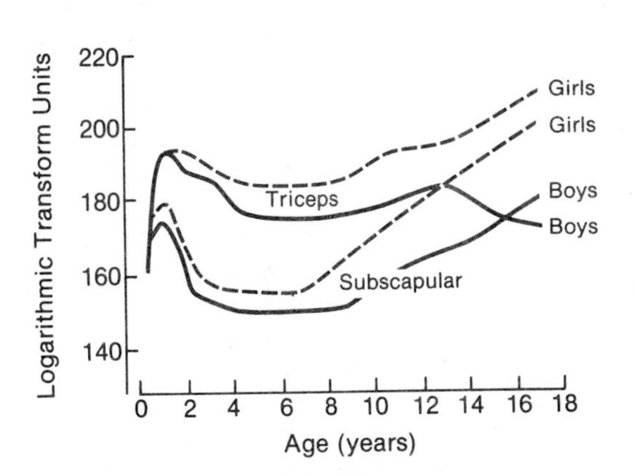

FIG. 10.3. Influence of age upon thickness of triceps and subscapular skinfolds. (Logarithmic transform units, schematic, after data accumulated by Tanner, 1962.)

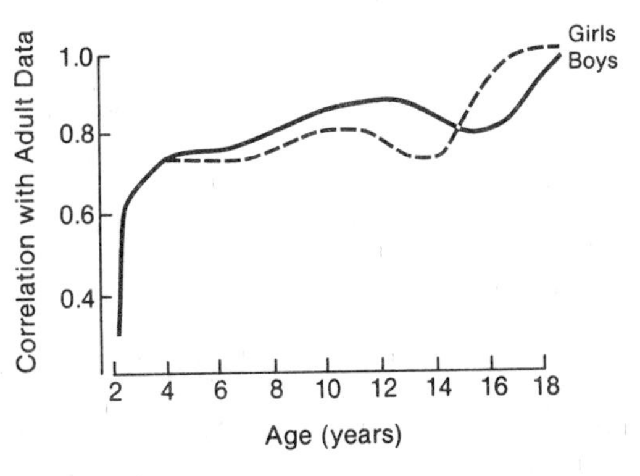

FIG. 10.4. Correlation between height at specified age and adult height. Note that although the correlation is statistically very significant, individual subjects can show wide temporary deviations from an idealized growth curve, particularly at the time of the pubertal growth spurt. (Schematic, based on data accumulated by Tanner, 1962.)

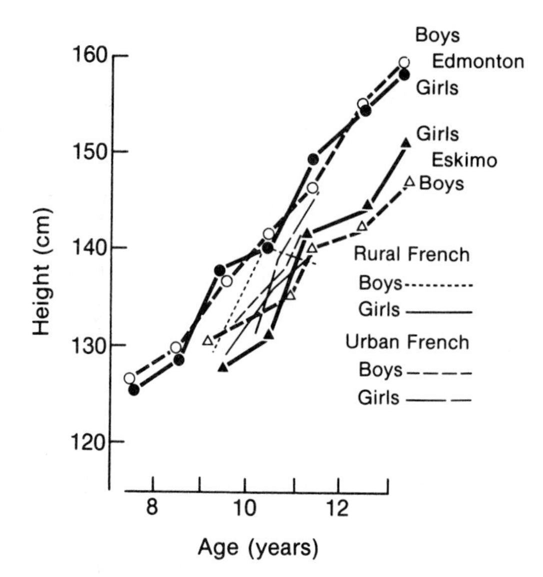

FIG. 10.5. Regional differences in growth of standing height. Cross-sectional data for Canadian Anglophones (Edmonton ○ and •; Eskimos ▲ and △). (From Shephard, 1978a, by permission of Cambridge University Press.)

end both in North America (Malina, 1978) and in Western Europe (Brundtland and Walloe, 1973; Tanner, 1973; D. F. Roberts, 1977; Salzler, 1977). Nevertheless, in many countries maturation still occurs earliest in those with professional-class parents who live in large cities.

Among previously small populations such as Eskimos, Lapps, and Japanese, the rate of increase of stature has been particularly rapid (Skrobak-Kaczynski, and Lewin, 1976; Shephard, 1978a).

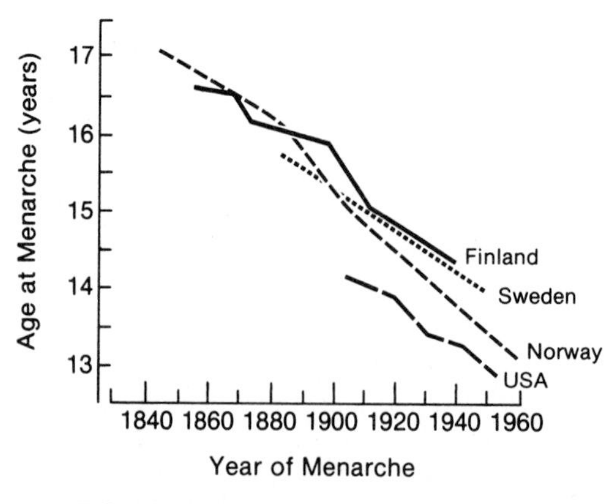

FIG. 10.6. Secular trend in age at menarche in selected nations. (Adapted from Tanner, 1962.)

The main basis for the more rapid growth and greater ultimate size of the world population seems an improvement of nutrition (Suzuki, 1970). This probably involves both an increased intake of essential amino acids and an increased consumption of refined carbohydrates. Among the poorer social classes, a reduction in the incidence of disease may also have played a role. A further factor affecting all classes of society has been a change in marriage patterns. Two hundred years ago it was uncommon for a man to marry outside his immediate village. However, the development of transportation has allowed an ever increasing choice of marriage partners, and the resulting genetic hybridization has almost certainly contributed to greater growth (Jamison, 1970; Wolanski, 1977).

Athletic records have developed progressively over the past century (Jokl and Jokl, 1968a; Shephard, 1978c). For example, Olympic and world record times for the 5000-m track event have decreased by ~2% per decade, and despite a smaller residual potential for improvement the progression of records has accelerated since 1950. In many events, including running and jumping, the increase of average stature has made a large contribution to the improvement of scores. Other possible factors influencing international records include the increased world population, a more complete searching of this population for potential athletes, improvements in training techniques, and, in some sports, improvements in equipment and facilities.

Effects of Physical Activity

Relatively little is known about interactions between regular physical activity and growth (Malina, 1969). Rats that are allowed spontaneous exercise apparently grow faster than similar animals that are kept to their cages (Ring et al., 1970) or are forced to swim every day for 100 days; the swimmers show not only a reduction in the mass of organs such as the kidneys, liver, adrenals, and spleen, but also associated reductions in cell numbers (E. E. Gordon et al., 1967a,b; Bloor et al., 1970). However, it is difficult to equate exercise stress in animals and in children. Animal experiments have sometimes required the equivalent of a daily 8-km run for 15 years, with attendant difficulties in maintaining adequate nutrition. Further, although the cardiac mass is typically lower in exercised animals than in control litter mates, it sometimes still accounts for a larger proportion of total body mass.

Early reports suggested that the growth of boys could be slowed by a vigorous program of interscholastic sports (F. A. Rowe, 1933; Schuck, 1962). In contrast, P. O. Åstrand, Engström et al. (1963) found very normal growth patterns in 30 top Swedish girl swimmers. Unfortunately, it is difficult to draw useful conclusions from examinations of athletes, since such groups are selected initially for atypical patterns of growth and maturation. In the Trois Rivières regional experiment, children were allocated arbitrarily to either a normal school program (40 min of physical education per week) or a special program requiring an hour of additional physical education per day. The added activity had no effect on the growth of either stature or body mass between the ages of 6 and 12 years (Shephard and Lavallée, in preparation). Furthermore, radiographs showed no influence of the added activity upon dental maturation, although there was some slowing of ossification at the wrist (Suzuki, 1970; Shephard, Lavallée, et al., 1978a). Cerny (1969) found no differences in the skeletal maturity of older (11–15 years) boys attending special sports schools.

Regional Growth

General Considerations

Specific tissues grow most rapidly at different stages in the course of an individual's overall development. There are also differences in the growth rate of any given tissue at different locations within the body. Thus, the hands develop earlier than the forearms, and the forearms in turn are consistently closer to their final adult size than are the upper arms. A similar pattern of growth (feet > lower legs > upper legs) is seen in the lower limbs (G. A. Harrison et al., 1964).

Head and Central Nervous System

Some 90% of growth in the head and brain occurs in the first 5 years of life, development of this region being virtually complete at 10 years. The rapid maturation is related in part to the course of cranial ossification. Up to the age of 18 months, the individual bony plates of the skull are separated at the fontanelles, and expansion of the brain cavity can occur relatively easily. Between 2 and 10 years of age cranial growth is still possible at the "sutures" between individual bony plates, but thereafter expansion can occur only if bone is eaten away from the inner surface of the skull and is deposited on its exterior.

The neural tissue is largely formed at birth, although it is not fully developed. Over the first few months of life, there is a rapid proliferation of axons and dendrites, with coincident improvements of coordination (Tables 10.1, 10.2). Structures controlling the upper extremities are further developed at a given age than are those controlling the lower limbs (Eckert, 1973; Espenschade and Eckert, 1974). The extent to which the neural potential for coordination is realized depends upon cultural factors (Malina, 1978) and the related sensory input (H. G. Williams, 1978). During infancy, there are no obvious sexual differences of motor development, but girls soon excel in jumping, hopping, rhythmic locomotion, and balance, while boys perform better in throwing, catching, and tasks requiring strength and speed (D. C. Sinclair, 1969).

Sensory and motor axons in general grow at a rate determined by the size of the body. In the developmental stage there is an oversupply of axons, but those that fail to make synaptic contact degenerate. The neurons of the adult continue to synthesize protein, and this material apparently is transported along the nerve tubules with the assistance of an actomyosin-like chemical (neurostenin; see below).

The cerebellum undergoes at least a sevenfold increase in mass after birth; there is a correspondingly extensive synthesis of DNA, giving the potential for large increases in the number and interconnections of neurons. Little is known about the influence of regular physical activity upon this developmental process. However, it is recognized that the ultimate form of a neural structure is modified by its use or disuse. For example, if mice reared in the dark are compared with normal litter mates, the former have fewer synaptic spines on the dendrites of their visual

TABLE 10.1
Some Measures of Neural Coordination in the Typical Growing Child

Age (Months)	Characteristic
0–4	Decrease of reflex and rhythmic movements
4–8	Voluntary movements develop in upper half of body, head raising progressing to sitting
8–14	Voluntary movements develop in lower part of body progressing to walking
14–20	Development of symbolic and conditional movements, especially speech.
28	Vertical jump, both feet together
29	Standing on one foot
37	Jump from 26-cm step
49	Hopping (less than 2 m)
60	Throwing a ball in a basket at 1.5 m

Note: While delay in motor development is often associated with mental deficiency, a fair range of sitting and walking ages is compatible with normal mental health.

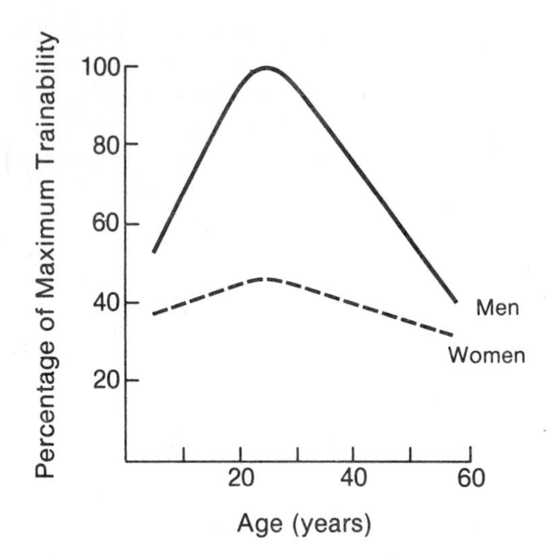

FIG. 10.7. The influence of growth and aging upon the ability of men and women to be trained. All data expressed relative to trainability of the young adult male. (Schematic, after Hettinger, 1961.)

cortex and less myelin in their optic nerves. It has also been shown that axonal sprouting is stimulated by lipids similar to those found in myelin.

Muscle

The total muscle mass of a male increases from 25% of body weight at birth to at least 40% in the adult. Much of the relative increase in mass occurs at puberty. Female subjects do not show a comparable gain at puberty, and the muscle force of the mature female averages only 60% of that for a mature male (Asmussen, 1964).

There is little increase in the number of muscle fibers after the first few months of postnatal life, but the diameter of existing fibers increases two- to threefold before their adult size is reached. Myosin, actin, and tropomyosin are synthesized on large polyribosome aggregates that lie between existing myofibrils. As new filaments are formed, these become attached to the Z discs (Chap. 4; Heywood et al., 1967). There is a progressive change in the characteristics of the myosin that is synthesized—particularly its Ca^{2+}-activated ATPase activity (Perry, 1970). Muscle fiber size attains a maximum in the fourth decade of life, with a subsequent decline. Muscle length increases in proportion with bone length. New sarcomeres are incorporated at the musculotendinous junctions (MacKay and Harrop, 1969) and individual sarcomeres are lengthened somewhat by a reduction in overlap of the myofilaments (Rowe and Goldspink, 1968). Additional nuclei are derived from the satellite cells (Chap. 4). These undergo mitotic division, one of the resultant nuclei migrating into the elongated muscle (Moss and Leblond, 1971). However, it is less certain that the additional DNA thus formed contributes to regulation of protein synthesis within the growing fibers.

Regular physical activity modifies muscle growth. Endurance-type training causes only a moderate hypertrophy of individual fibers, but there is nevertheless an increase in the number and size of mitochondria (Howald, 1975) and an increase in the concentration of enzymes concerned with the oxidation of carbohydrate, oxidation of fat, and production of ATP (Holloszy, 1973; Holloszy and Booth, 1976). High resistance or isometric exercise provides a much greater stimulus to synthesis of the contractile elements. The myofibrils increase in size and eventually divide, increasing their number per muscle fiber (E. E. Gordon et al., 1967b). Perhaps because the intensity of exercise has been insufficient, some training studies have not shown clear-cut increases in the enzymes associated with anerobic metabolism (Gollnick and Hermansen, 1973). However, Eriksson (1972) noted a 30% increase in the "aerobic" enzyme succinic dehydrogenase and an 83% increase of the "anerobic" enzyme phosphofructokinase when a small group of 11-year-old boys was trained for 6 weeks. Interestingly, the maximum oxygen intake of this group increased by only 8%, suggesting that tissue enzyme concentrations were not limiting oxygen conductance.

According to Hettinger (1961), the trainability of muscle strength reaches a peak in early adult life, with women less trainable than men (Fig. 10.7). However, Rohmert (1968b) found that when the training response was plotted as a percentage of the final value, the rate of increase was similar in children and adults, while Ikai (1967) noted that it was easier to increase muscular endurance in 12- to 15-year-old children than in adults. The Japanese investigators

TABLE 10.2
Percentage of Children with Proficiency in Specific Skills at 5 and 7 Years of Age

Skill	Percent of Sample Proficient	
	5 Years	7 Years
Jumping	58	84
Hopping	33	84
Skipping	14	91
Galloping	43	92
Throwing	20	74

Source: Based upon data of Gutteridge (1939) for United States.

also reported that the local blood flow of adolescents could be increased more readily than that of adults. One possible explanation for the poor training response of female subjects is that there are fewer nuclei per unit of muscle mass than in the male. However, such a hypothesis would imply that existing nuclear DNA was synthesizing muscle protein at its maximum potential rate and that there was an insufficient reserve of satellite cells available for integration into the muscle proper; neither of these points is yet established.

Bone and Connective Tissue

In the short bones, growth of cartilage precedes ossification, with the process of ossification eventually overtaking cartilage formation (chondrogenesis). The long bones first show ossification at the periphery of their shafts. Growth requires an eating away of the marrow surface (osteoclastic activity), with a simultaneous deposition of new bone at the outer surface (osteoblastic activity). Ossification centers later develop at the ends of the long bones; these are separated from the ossified part of the shaft by a cartilaginous epiphyseal plate. Further growth in length can occur until the dividing plate is ossified (at ~16 to 18 years of age in women and 18 to 21 years of age in men). In the arms, growth is greater at the wrists and the shoulders than at the elbows, but in the legs the knee ends of the tibia and femur grow more than the epiphyses of the hip and ankle (Sinclair, 1969; Larson, 1973).

Patterns of bone growth are sometimes exploited in assessing an individual's maturity ("radiographic age"; see below). Immature bones are also vulnerable to injury (Larson, 1973). The partially ossified bone may develop an incomplete ("greenstick") fracture. The majority heal quite quickly, but a proportion involve the epiphyseal plate (Krogman, 1955; Table 10.3). Vulnerable sites include the distal epiphysis of the tibia, the phalanges, and the distal radial epiphysis. Separation typically occurs in the zone of hypertrophied cells without damaging the growth plate or its blood supply. However, ~10% of epiphyseal injuries lead to premature closure or other disturbances of growth.

Vigorous sport (for example, baseball pitching or jumping) can lead to inflammation of an epiphyseal region (epiphysitis), with permanent damage to the growing bone that is involved (J. E. Adams, 1965).

It is generally accepted that in humans physical activity encourages bone growth. Thus (Steinhaus, 1933) the bones of the dominant limb are longer and wider than those of their counterpart

TABLE 10.3
Injuries Experienced by American Children Under 15 Years of Age, with Frequency of Epiphyseal Involvement

Activity	Total Number of Injuries	Percent Involving Epiphyseal Region
American football	188	11.7
Physical education	82	7.3
Track	78	11.5
Skiing	71	11.2
Basketball	70	7.1
Baseball	48	14.5
Miscellaneous	136	21.3
Total	673	12.8

Source: Based on data of R. L. Larson, 1973.

(Jokl, 1978), although it is conceivable that arm length could have determined dominance rather than the reverse being the case. Prolonged bed rest, space exploration, and reduced activity due to arthritis lead to calcium loss from bones (Kazarian and Von Gerke, 1969; Rummel et al., 1975). This is reversed by the normal tensile and compression forces of muscle contraction and load bearing. It is thought that muscle contraction generates electrical potentials within the bone structure, and that these potentials stimulate bone growth. There is probably an optimum amount of activity, since rats trained by very heavy programs of chronic running have shorter and thinner bones than corresponding control animals.

Tendons show little change of collagen content from birth to old age, although the elasticity of the collagen may deteriorate with aging (Shephard, 1978b; see below). Endurance training causes an increase in the cross-sectional area of tendons (Booth and Gould, 1975; Tipton, Matthes et al., 1977), with associated increases in the number of nuclei and the content of complex sugars (mucopolysaccharides). In contrast, immobilization of a joint leads to loss of cartilage from the joint surface, infiltration of the capsule with fatty and fibrous tissue, and considerable loss of mobility.

Body Fat

Body fat shows a rather interesting growth curve (Pařizková, 1977). There is a substantial (10–20%) increase of subcutaneous fat between the time of birth and ~9 months of age. Then, as the child becomes more mobile, the percentage of body fat decreases, and a minimum is reached between 6 and 8 years of age. Girls are on average slightly fatter than boys even in early childhood, and the difference between the sexes becomes more marked after the age of 8 (Fig. 10.3). From menarche to 18 years of age, both early- and late-maturing girls gain ~4.5 kg of fat (Frisch, 1976), so that variables expressed per unit of body mass (aerobic power or muscle force, for example) inevitably deteriorate by ~8% at this stage of growth.

As children become older, there is a progressive skewing in the distribution curve of percentage body fat, implying that a large proportion of the population have accumulated an excessive amount of body fat (Fig. 10.8). Observations on rats suggest that the adipose tissue of young animals is more cellular in nature and contains more DNA than tissue from a mature animal (Tenorová and Hruza, 1962). Young adipose tissue is therefore metabolically more active, and free fatty acids are more readily mobilized from it during physical activity (Altschuler et al., 1962; Pǎřizková, 1966).

Reproductive Tissues

The reproductive tissues remain relatively dormant until just before puberty. Very rapid growth then occurs, the adult size of the reproductive organs being reached in the space of 4 or 5 years (Tanner, 1962; G. A. Harrison et al., 1964).

In boys, the first sign of puberty is an accelerated growth of the testes and scrotum. This occurs between the 10th and the 13th years. About a year later, accelerated growth of the penis begins, and this is accompanied by the height spurt (Fig. 10.1). Pubic hair appears between the 10th and the 15th years, and grows more vigorously with enlargement of the penis. Axillary and facial hair first develop some 2 years after pubic hair. Enlargement of the larynx and breaking of the voice occur when growth of the penis is almost complete; at this stage there is a marked growth of lean mass (Fig. 1.2) and strength increases in relation to standing height.

In girls, the breasts begin to develop between 8 and 13 years. Growth of pubic hair, the development of the uterus and vagina, and the height spurt run roughly parallel with breast growth. The first menstrual period (menarche) occurs between the 10th and the 16th years, toward the end of the other pubertal changes. Girls do not show a strength spurt; on the contrary, the rate of increase of strength slows at puberty. There are some constitutional grounds for this (such as the low nuclear/cytoplasmic ratio, discussed above), but social pressures also exert a strong influence upon the attained strength, greater development being seen in communities where the life style demands a continued vigorous use of mus-

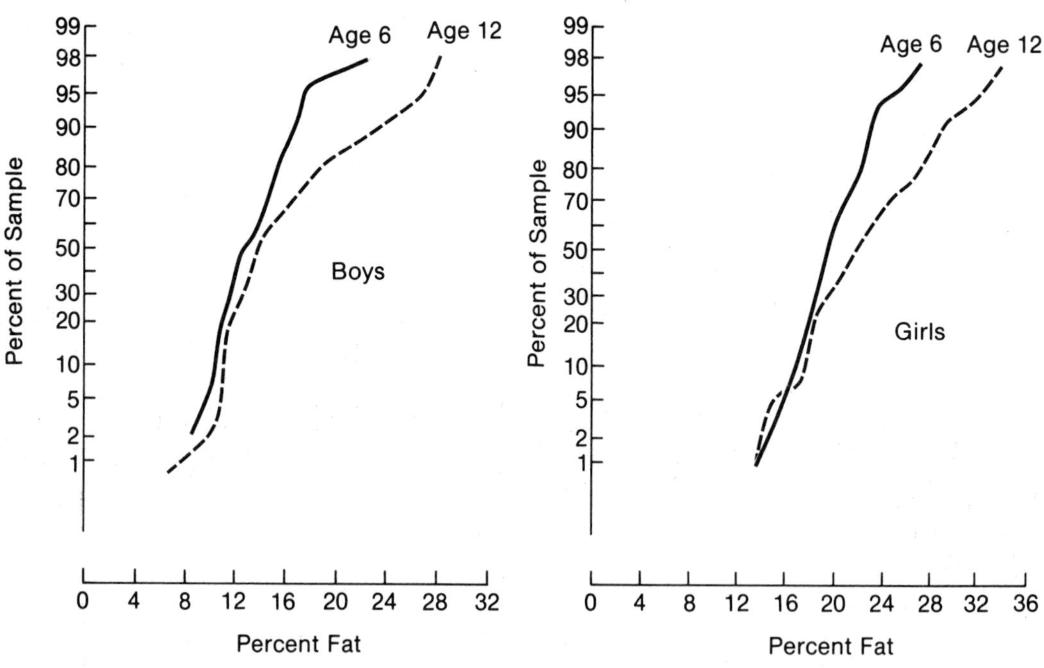

FIG. 10.8. Distribution of percentage body fat in a population of French Canadian schoolchildren at ages 6 and 12. (After Shephard, 1978k.)

cles throughout adolescence (Rode and Shephard, 1973b; Fig. 10.9).

Assessment of Maturity

To evaluate physiological tests or performance measurements in an adolescent child, it is helpful to determine "biological age" and, in particular, to decide whether the child in question has reached or passed the period of accelerated growth. Scales of biological age may be based upon a quantitation of secondary sex characteristics, such as pubic hair and the size of the testes (Tanner, 1962), but the most reliable indices of overall maturity are bone growth as assessed from radiographs, and eruption of the teeth (Tables 10.4 and 10.5).

In radiographs of the limbs, attention is directed to the size of centers of ossification and the extent to which the epiphyses have fused with the shafts of the long bones, appearances in a given child being compared with a standard "atlas" (Greulich and Pyle, 1959; Tanner, Whitehouse et al., 1975). The accuracy of the biological age thus calculated varies with the calendar age of the child, results being most precise around the period of puberty. Individual observers can rate films with a test/retest variation of 6 months or less, but there are also systematic differences of 6 months and more between observers (Shephard, Lavallée et al., 1979c). Radiographic estimates of dental maturity have a similar precision. A simple inspection of the mouth can describe the number of teeth that are visible and the completeness of their eruption (Shephard, Allen et al., 1968c), while more detailed analysis of eruption, root closure, and calcification of the crowns is based upon radiographs of the jaw (Garn, Rohmann, and Silverman, 1967).

TABLE 10.4
Assessment of Skeletal Age from Radiographs of the Wrist and Long Bones

Skeletal Age (Years)	Carpal Ossification	Other Bones Showing Centers of Ossification
1	Os magnum (capitate)	Head of femur
2	Unciform (hamate)	Lower epiphysis of radius, tibia, and fibula
3	Cuneiform (triquetral)	Patella, head of humerus
4	–	Lower epiphysis of ulna, upper epiphysis of fibula, and greater trochanter
5	Trapezoid, semilunar	Upper epiphysis of radius
6	Scaphoid	Centers in head of humerus coalesce
10	–	Upper epiphysis of ulna, tuberosity of os calcis
12	Pisiform	–

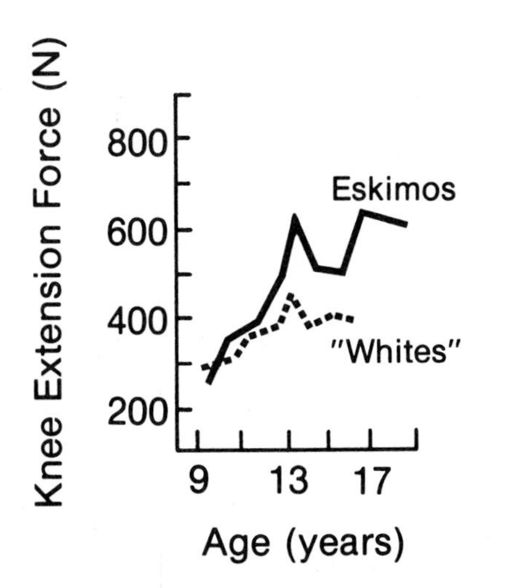

FIG. 10.9. A comparison of the growth of knee extension force in Eskimo and "white" girls. (Based on cross-sectional data of Rode & Shephard, 1973b, and M. L. Howell et al., 1965.)

TABLE 10.5
Eruption of Teeth in Relation to Developmental Age

Dental Age (Years)	Teeth Erupted
	Deciduous ("milk") Teeth
$^{6}/_{12}$–$^{9}/_{12}$	Lower central incisors
$^{8}/_{12}$–$^{10}/_{12}$	Upper incisors
$1^{3}/_{12}$–$1^{9}/_{12}$	Lower lateral incisors and first molars
$1^{4}/_{12}$–$1^{8}/_{12}$	Canines
$1^{8}/_{12}$–2	Second molars
	Permanent Teeth
6	First molars
7	Two central incisors
8	Two lateral incisors
9	First premolars
10	Second premolars
11–12	Canines
12–13	Second molars
17–25	Third molars

Girls are generally more advanced in skeletal and dental development than boys. If children of either sex are classified in terms of biological age, those whose bone development is in advance of their calendar years usually show intellectual and emotional maturity and early puberty. The radiographic age is particularly advanced among groups enjoying favorable socioeconomic conditions (Shephard, Lavallée et al., 1979c). We have noted above that physical activity apparently has a local mechanical effect; exercises involving the wrists (such as gymnastics) retard development of this region, while leaving dental maturation unaltered (Shephard, Lavallée et al., 1978a).

The value of biological age in interpreting physiological data and in ensuring that competition occurs between children at a similar stage of development is controversial. The concept probably has some merit in the pubertal years (G. R. Cumming, Garand, and Borysyk, 1972b), but at all ages radiographic age shows a fairly close correlation with calendar age, so that the former adds little information that could not have been obtained from simpler estimates of maturity such as measurements of height and body mass (Shephard, Lavallée et al., 1978a). In view of the undesirability of exposing children to X-irradiation, determinations of biological age in normal students should be restricted to an inspection of the mouth and an assessment of secondary sexual characteristics.

Standardization of Data

Many resting measurements in children—basal heat production, cardiac output, and respiratory minute volume—have traditionally been standardized for differences of body size by expressing results per unit of body surface area, this being estimated by the nomogram of E. F. DuBois (1927; Fig. 1.6). Since (1) resting heat loss depends upon the available surface for evaporation, radiation, and convection of heat and (2) a large part of resting energy expenditure is allocated to maintaining body heat, there is some logic to this approach. However, it has less obvious relevance to the standardization of physiological data collected during vigorous physical activity.

The simplest and the most widely used method of standardizing data such as maximum oxygen intake is to express results per unit of body mass. There are many theoretical objections to such a procedure, including changes in the relative proportions of fat and lean tissue as a child grows older. In practice, mass standardization seems to work quite well; when expressed per kg, maximum oxygen intake, working capacity, muscle force, and other variables remain very stable over a wide range of body sizes.

A theoretical basis for the standardizing of biological data has been developed by Scandinavian investigators (Von Döbeln 1966; Asmussen and Christensen, 1967; P. O. Åstrand and Rodahl, 1977). Their concepts can be traced to Archimedes' laws for geometrically similar bodies and two observations of Borelli (geometrically similar animals jump to the same height, while the maximum velocity of running is independent of body size). According to the Scandinavian dimensional theory, various parts of the body grow at rather similar rates, so that overall stature H can be used as a marker (Fig. 10.10). Variables such as leverage and stride length should then be proportional to standing height (H), muscle force (a function of the cross-section of the active fibers) should be proportional to H^2, and volumetric variables such as body mass, total body hemoglobin heart volume, and lung volumes should be proportional to H^3.

Unfortunately, many of the variables of interest to the exercise physiologist are measurements of power rather than size. In an attempt to circumvent this problem, Von Döbeln (1966) made the ingenious but debatable assumption that time (t) could be represented by an equivalent length L (and thus height). His reasoning may be summarized as follows:

$$\text{Acceleration} = \frac{\text{force}}{\text{mass}} = \frac{L^2}{L^3} \quad (10.1)$$

$$\text{Acceleration} = \frac{L}{t^2} \quad (10.2)$$

Thus

$$\frac{L}{t^2} \propto \frac{L^2}{L^3} \text{ and } t \propto L \quad (10.3)$$

On this basis, measurements of human power should vary as the square of standing height, or as (body mass)$^{2/3}$:

$$\text{Work} = \text{force} \times \text{distance} = L^2L = L^3 \quad (10.4)$$

$$\text{Power} = \frac{\text{work}}{\text{time}} = \frac{L^3}{L} = L^2 \quad (10.5)$$

A similar conclusion was reached by Von Hoesslin in 1888. How good is the agreement be-

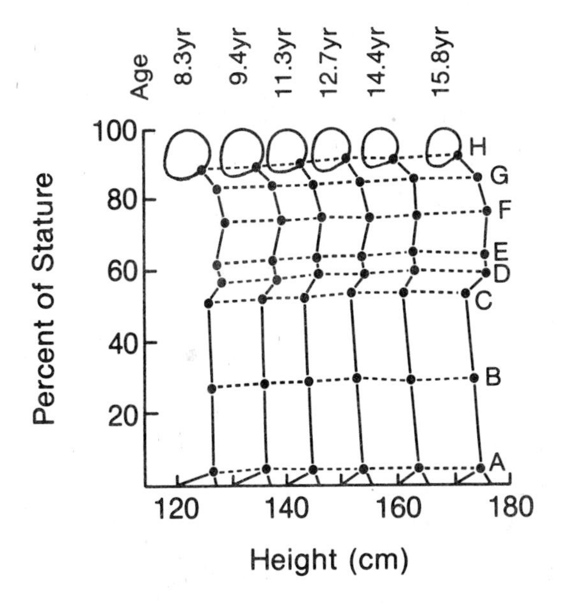

FIG. 10.10. Relative size of various body segments in boys aged 8 to 16 years. Subjects grouped by 10 cm intervals of stature. A, lateral malleolus; B, lateral femoral epicondyle; C, greater trochanter of femur; D, posterio-superior iliac spine; E, deepest point of lumbar lordosis; F, highest point of dorsal kyphosis; G, vertebra prominens; H, external auditory meatus. (Data of Heebøll-Nielsen, as reported by Asmussen, 1973a.)

tween the dimensional theory (Table 10.6) and practice? Asmussen took P. O. Åstrand's (1952) cross-sectional vital capacity data, calculating exponents of $H^{3.1}$ for boys and $H^{3.0}$ for girls. However, it could be argued that the concord between the data and theoretical expectation was fortuitous, since the children concerned had grown through a period of restricted nutrition (during World War II) and a subsequent "catch-up" phase. In contrast, De Muth et al. (1965) demonstrated exponents of $H^{2.81}$ and $H^{2.82}$ for vital capacity measurements made on a large sample of U.S. boys and girls; we may speculate that in the U.S sample, the potential for development was not realized due to progressive inactivity during adolescence. In Canadian Eskimo boys, Rode and Shephard (1973b,c) found that vital capacity varied as $H^{3.28}$, while the 1-sec forced expiratory volume was proportional to $H^{3.31}$; corresponding values for Eskimo girls were $H^{3.01}$ and $H^{2.50}$.

Discrepancies between theory and practice have been wider for other variables. In analyses of data for boys, Asmussen (1973a) found average exponents of $H^{2.89}$ for muscle force and $H^{2.90}$ for aerobic power, log/log plots showing a steep increase of slope at age 13; in girls, the slopes were generally a little less, with no change of slope at puberty. W. D. Ross, Bailey et al. (1977) reported an exponent of $H^{2.50}$ for the aerobic power of Canadian boys. Other data are summarized in Tables 10.7 and 10.8.

Von Döbeln and Eriksson (1973) suggested that departures from the theoretical exponents might reflect a lack of normality in the children concerned due to insufficient habitual activity. Certainly comparisons between populations generally indicate higher exponents in the more active groups (see, for example, the girls of Table 10.7). Asmussen (1973a) also stressed that improved coordination could yield unexpectedly large gains of score, especially in performance measurements.

From the viewpoint of standardizing data, the choice of a height exponent remains insecure, and it is wrong to accuse young children of a lack of aerobic power and muscle force (P. O. Åstrand and Rodahl, 1977) merely because they fail to conform with what may prove an erroneous theory! The practically minded investigator will be content to note (1) that H, H^2, and H^3 are closely correlated with each other and (2) that for many physiological measurements body mass or lean mass remain as effective methods of reducing the variance of data as use of any height exponent (Gilliam et al., 1978; Shephard, Lavallée et al., 1979b; Table 10.9).

Physiological Responses

Anerobic Power and Capacity

Stores of the energy-rich phosphagens (ATP and CP; Chap. 2) are much as in an adult (B. Eriksson,

TABLE 10.6
Relationship of Physiological Variables to Body Length L

L^{-1}	L^0	L	L^2	L^3
Frequency	Density	Length	Surface	Mass
Acceleration	Velocity	Time	Flow	Volume
	Pressure		Force	Energy
	Temperature		Power	

Source: Based on concepts of Von Döbeln, 1966.

TABLE 10.7
Power Exponents Relating Selected Physiological Variables to Standing Height

Variable	Boys		Girls	
	Active	Sedentary	Active	Sedentary
Maximum oxygen intake ($l \cdot min^{-1}$)	2.60	3.21	2.76	2.66
PWC_{170} ($kg \cdot m \cdot min^{-1}$)	3.58	3.37	3.60	2.94
Vital capacity (l, BTPS)	2.61	2.71	2.76	2.68
R. hand grip force (kN)	3.12	3.29	3.34	3.16
Back extension force (kN)	2.62	2.69	2.91	2.76
Leg extension force (kN)	2.40	2.80	3.42	2.96
Sit-ups (total)	2.28	2.81	2.06	1.05
Flexed arm hang (sec)	1.78	1.12	1.42	−0.25
Standing broad jump (m)	1.19	1.39	1.16	1.18
Shuttle run (sec)	−0.58	−0.62	−0.52	−0.41
50-yard run (sec)	−0.76	−1.04	−0.84	−0.87
300-yard run (sec)	−0.76	−0.95	−0.75	−0.84

Source: Based on data for children aged 6–12 years (Shephard, Lavallée, et al., 1979b).

1978). However, children show low muscle lactate concentration in both submaximum and maximum exercise, with a correspondingly small oxygen deficit (Fig. 10.11). This reflects a low concentration of the key, rate-limiting enzyme of glycolysis, phosphofructokinase (B. O. Eriksson, Gollnick, and Saltin, 1973).

Aerobic Power

The maximum oxygen intake of children has been studied from the ages of 4 through to adult life. In the youngest children, exercise has sometimes been performed on pedal cars or a step test has been carried out in the company of a technician (Pářizková, 1978a), but the usual test method has been progressive exercise to exhaustion (Shephard, Allen et al., 1968c; G. R. Cumming, 1974; Klimt et al., 1974; Mrzena and Máček, 1978) using a treadmill or a bicycle ergometer. Some authors have noted difficulty in reaching a plateau of maximum oxygen intake (P. O. Åstrand, 1952; G. R. Cumming and Friesen, 1967). If a plateau is attained, the maximum oxygen intake measurements are quite reliable (test/retest r = 0.92), but in the absence of a plateau reproducibility is poor (r = 0.27; D. A. Cunningham et al., 1977). One suggested alternative to the standard progressive increase of work load is to require a brief (2–3 min) period of supramaximal exercise (G. R. Cumming and Friesen, 1967); this is said to yield quite realistic maximum oxygen intake data.

Boys show a steady growth of absolute maximum oxygen intake. Thus, if data are expressed per unit of body mass, results remain relatively constant at ~50 $ml \cdot kg^{-1} \cdot min^{-1}$ STPD (Shephard, 1971c, 1978a). In some North American samples, scores have declined over the teenage years, presumably due to insufficient physical activity (D. A. Bailey, 1974); in contrast, results for Eskimos (Rode and Shephard, 1973b) and what was possibly a biased sample of Swedish children (P. O. Åstrand, 1952) rose to a peak of ~60 $ml \cdot kg^{-1} \cdot min^{-1}$ in late adolescence. In young girls, the maximum oxygen intake is almost as large as in boys, but a deterioration sets in at ~10 years of age; this is possibly more marked in North America than in Scandinavia, teenage girls in the United States and Canada having as low an aerobic power as adult women (~40 $ml \cdot kg^{-1} \cdot min^{-1}$; Shephard, 1971c,1978a; Nagle, Hagberg, and Kamei, 1977).

P. O. Åstrand (1952) found a very high maximum heart rate (210–215 beats per min) in his sample of boys. However, few other laboratories have reported maximum readings higher than 195–200 beats per min (Shephard, 1978a), despite apparently comparable intensities of effort.

The average blood lactate at exhaustion (~9 $mmol \cdot l^{-1}$) is somewhat lower in a child than in

TABLE 10.8
Some Height Exponents Calculated by Asmussen (1973) for Performance Tests Involving Use of the Leg Muscles

Test	Exponent
Horizontal running	2.00*
Isometric strength	3.01*
Vertical jump	3.59*
Start to sprint	4.54†

Note: Results in girls* and boys† aged 7–17 years.

TABLE 10.9

Relative Effectiveness of H, H^2, H^3, Body Surface Area, Body Mass, and Lean Mass in the Standardization of Selected Variables, Boys and Girls Aged 6–12 Years*

Variable	H	H^2	H^3	Surface Area	Total Mass	Lean Mass
Maximum oxygen intake ($l \cdot min^{-1}$)	0.73	0.73	0.73	0.78	0.76	0.81
PWC_{170} ($kg \cdot m \cdot min^{-1}$)	0.65	0.64	0.63	0.65	0.61	0.72
Vital capacity (l, BTPS)	0.83	0.83	0.83	0.83	0.77	0.83
R. hand grip force (kN)	0.79	0.79	0.79	0.79	0.74	0.82
Back extension force (kN)	0.68	0.68	0.68	0.69	0.64	0.72
Leg extension force (kN)	0.58	0.60	0.60	0.61	0.56	0.58
Sit-ups (total)	0.32	0.32	0.31	0.25	0.18	0.26
Flexed arm hang (sec)	0.13	0.12	0.12	0.04	−0.03	0.08
Standing broad jump (m)	0.57	0.57	0.56	0.48	0.37	0.48
Shuttle run (sec)	−0.42	−0.41	−0.40	−0.33	−0.25	−0.34
50-yard run (sec)	−0.56	−0.55	−0.53	−0.47	−0.38	−0.47
300-yard run (sec)	−0.51	−0.50	−0.49	−0.41	−0.30	−0.41

*All data presented as coefficients of correlation between the variable and the index of body size. The largest correlation coefficients are underlined.

Source: After Shephard, Lavallée et al., 1979b.

an adult. This may be related in part to differences of muscle enzyme activities, as noted above, but the main factor is probably difficulty in persuading children to undertake maximal effort. The IBP working group found an average blood lactate of almost 12 mmol • l^{-1} in 11-year-old children judged to have made a good maximal effort (Shephard, Allen et al., 1968c; Table 10.10).

Submaximum Exercise

The aerobic power can be predicted from submaximum exercise tests, as in an adult. At the first visit to the laboratory, anxiety may increase the exercise heart rate, leading to a 10% underprediction of maximum oxygen intake (Shephard, 1971c). If oxygen consumption is estimated from the work performed, note must also be taken of possible departures from the mechanical efficiency found in adults.

The energy cost of running per unit of body mass is higher in a child than in an adult (Shephard, Lavallée et al., 1974), probably because of a shorter stride. Problems of leg length also cause a small child to perform a step test inefficiently. Most observers (Shephard, Allen et al., 1968c; Seliger, 1970; K. L. Anderson and Ghesquière, 1972; Stewart and Gutin, 1975; Shephard, 1978a; but not Wilmore and Siegerseth, 1967) even find a reduction in the mechanical efficiency of cycling in the child, despite the reputed "constancy" of efficiency on the bicycle ergometer. The calculated efficiency of cycling in children rises with workload. This is probably due to the onset of anerobic work; efficiencies of up to 25% can thus be calculated from the slope of the oxygen-consumption/work-performance line (Rutenfranz and Mocellin, 1967; K. L. Andersen, Shephard et al., 1971).

Interpolated measurements of leg power output such as the PWC_{170} (physical working capacity at a heart rate of 170 beats per min) provide an alternative basis of assessment. The PWC scores show a regular development with age, normal values averaging ~140 $N \cdot m \cdot kg^{-1} \cdot min^{-1}$ in boys and 110 $N \cdot m \cdot kg^{-1} \cdot min^{-1}$ in girls.

The systemic blood pressure rises progressively with work rate, but available figures suggest that the increase is relatively less than in an adult (Thorén, 1968). The main handicap of the growing child is seen when exercise is prolonged for an hour or more. Fatigue appears more readily than in the adult, partly because the muscles are weaker (leading to problems in perfusion of the vigorously contracting fibers; Chap. 6) and partly because glycogen stores are smaller (B. O. Eriksson, 1972).

Oxygen Conductance

Individual links in the oxygen transport chain develop roughly in proportion to body size. Alveolar ventilation accounts for a somewhat larger proportion of total ventilation in children (80–85%) than in adults (75–80%; Shephard and Bar-Or, 1970). The maximum cardiac output of the preadolescent is a little smaller than would be predicted from oxygen consumption (Bar-Or, Shephard, and Allen, 1971), a "hypokinetic" circulation reflecting the greater relative surface area available for heat dissipation, less subcutaneous fat, a smaller visceral blood flow, and greater peripheral oxygen extraction than in the adult (Fig. 10.12). However, Raven et al. (1972)

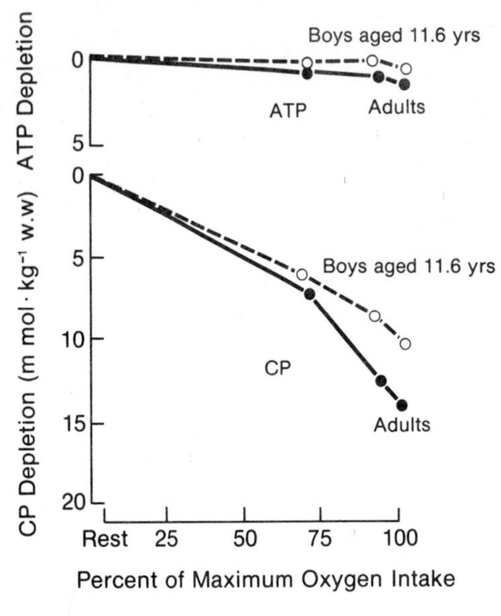

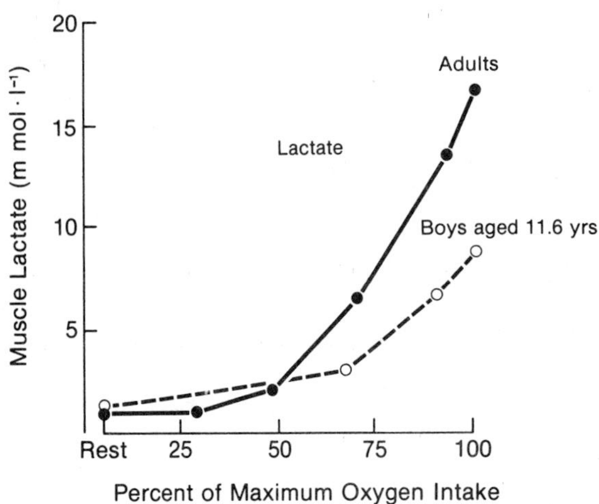

FIG. 10.11. Depletion of phosphagen and build-up of muscle lactate during bicycle ergometer work at selected percentages of maximum oxygen intake. (Based on data of Karlsson, 1971b.)

found no evidence that this characteristic persisted in well-trained 13- to 15-year-old girls; they concluded the hypokinetic response could be reversed by a combination of growth and training.

The hemoglobin concentration of a young child is less than in an adult, ranging from ~12 g • dl^{-1} at 1 year to 14 g • dl^{-1} at puberty (Shephard, 1956e). Children show at least a corresponding deficit of blood pigment if the total hemoglobin is calculated per unit of body mass (Fig. 10.13). In the girls, there is no postpubertal increase of hemoglobin concentration or red cell count, but in the boys a peak hemoglobin of more than 16 g • dl^{-1} is reached at the age of 18 years, the red cell count increasing from its adolescent value of 4.5×10^6 cells per mm^3 to as much as 5.4×10^6 cells per mm^3.

Strength

People commonly talk of children outgrowing their strength. There is some physiological support for this concept, since the boys' height spurt (associated with a lengthening of the long bones) precedes the strength spurt (associated with a rapid increase of body mass; Stolz and Stolz, 1951; Tanner, 1962). In the girls, also, height continues to increase beyond the age of 12 or 13 years, although at least in "civilized" Western society there is no further development of muscular strength after puberty.

TABLE 10.10
Maximum Oxygen Intake (Direct Treadmill Measurement) and Related Variables in 10- to 12-year-old Children Making a Good Maximum Effort

Variable	Boys	Girls
Maximum oxygen intake		
($l \cdot min^{-1}$)	1.79 ± 0.32	1.43 ± 0.26
($ml \cdot kg^{-1} \cdot min^{-1}$)	47.9 ± 6.3	38.6 ± 3.9
Error of Åstrand prediction		
($l \cdot min^{-1}$)	0.00 ± 0.32	−0.05 ± 0.12
Maximum heart rate		
($beats \cdot min^{-1}$)	195 ± 7	199 ± 7
Respiratory gas-exchange ratio	1.09 ± 0.16	1.28 ± 0.22
Arterial lactate concentration		
($mmol \cdot l^{-1}$)	11.7 ± 2.0	12.2 ± 3.5

Source: After Shephard, Allen et al., 1968c.

The independent growth of stature and strength are a normal feature of adolescence and should not cause alarm or restriction of physical activity.

Control of Growth

Environmental Factors

The relative importance of environmental and constitutional factors in determining the rate of maturation and ultimate body size is still debated. One possible approach is to compare identical and nonidentical twins. The age of menarche differs by an average of 2 months between identical twins and by 10 months between twins who are not identical (Tanner, 1962). This suggests that genetic factors influence the situation strongly, but the comparison is unfair to environment, since identical twins normally share rather similar living conditions and a certain variance of environment is necessary to demonstrate the full possible environmental effect.

A second approach is to compare growth patterns between children of differing social class (G. A. Harrison et al., 1964). It has been recognized for many years that social status is positively associated with growth and maturation. Indeed, differences were more obvious in the last century than at the present time, when traditional social boundaries are breaking down. In the Great Britain of the 1870s, a growing boy with laboring class parents was 10 cm shorter than a boy whose father had a white-collar job, and he was 12 cm shorter than a student of comparable age who was attending the very privileged public school system. Differences were smaller, but remained quite obvious (3 and 5 cm, respectively) when data for early manhood were considered. In the 1950s, the children of parents falling into British social classes I and II (professional and semiprofessional groups) were still 2–3 cm taller than the national average, but in West Germany more recent statistics showed no influence of parental income upon the stature of the child (Walter, 1977). Factors that once contributed to the greater growth of those at the upper end of the social scale probably included better nutrition, a larger home, and thus opportunity

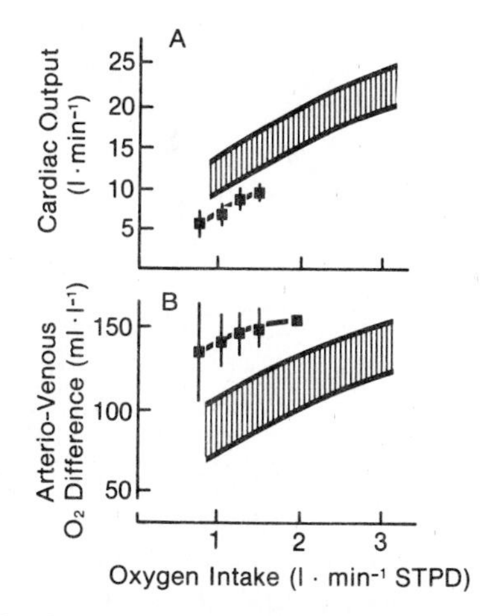

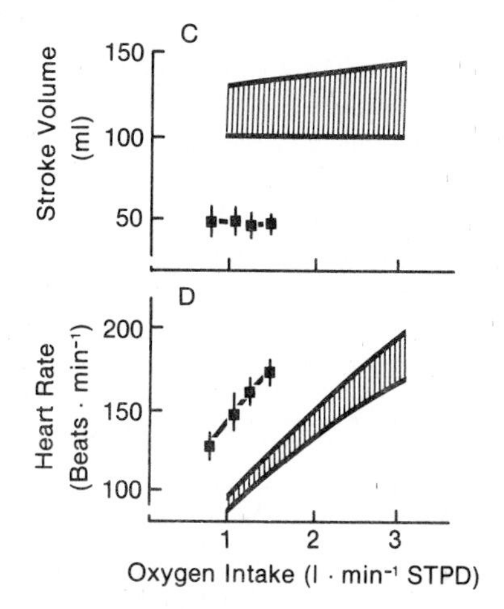

FIG. 10.12. A comparison between the circulatory function of 10- to 13-year-old boys and 115 sedentary adults. Adult range illustrated by shading. (Bar-Or, Shephard, & Allen, 1971.)

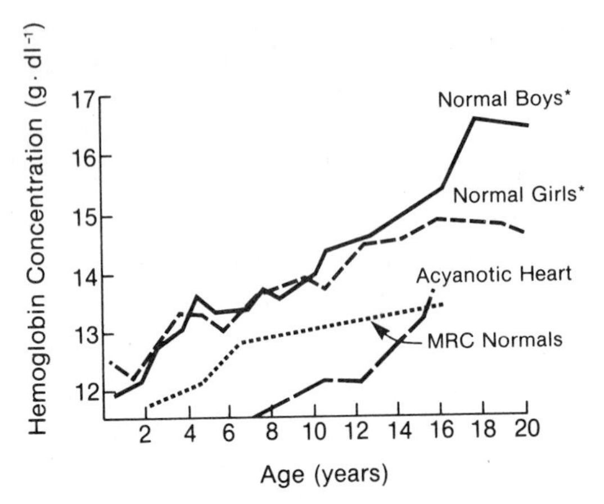

FIG. 10.13. Influence of age upon hemoglobin level. (Based on data of Mugrage & Andresen, 1938. (*) Medical Research Council, 1945, and Shephard, 1956e.)

for more adequate sleep, a smaller average family size, a better organized pattern of life, and possibly (at least for students attending the British public schools) more deliberate exercise (Tanner, 1962; J. W. B. Douglas and Simpson, 1964).

The specific retardation of growth due to nutritional deficiencies is seen in data from Germany and Japan (Ashizawa et al., 1978). The average height of German children increased by ~ 10 cm between 1920 and 1940, but decreased by 3 cm during the subsequent war period (Howe and Schiller, 1952). Animal experiments also demonstrate the need of first class protein if growth is to be maintained. The body is unable to synthesize certain amino acids (Table 10.11), and if these are lacking from the diet, animals fail to grow at the speed of their control litter mates. One important criticism of many animal-training experiments is that the exercised group fails to show a normal increase of body mass, presumably because food intake has failed to satisfy the combined demands of exercise and growth. Where malnutrition is temporary, its effect is to delay rather than to restrict development, and if observations are extended for a long enough period the affected men or animals reach the same size as well-fed controls (Tanner, 1962). In underdeveloped countries where protein/energy malnutrition is a more permanent way of life, permanent effects upon working efficiency and therefore productivity may occur (Viteri, 1971; Barac-Nieto et al., 1974). Areskog et al. (1969) noted a low physical working capacity (PWC_{170}) in poorly nourished Ethiopian children, although the effect largely disappeared if power output was expressed per unit of body mass. C. T. M. Davies (1974) also observed a low maximum oxygen intake in malnourished, anemic East African children; this was related to their low hemoglobin and reduced lean tissue mass, normal values being restored by appropriate treatment. Chronic illness commonly leads to a retardation of growth (Fig. 10.14; Hauspie et al., 1977; Shephard, 1978a). It is less certain as to whether this is a specific effect of disease, a secondary consequence of the accompanying reduction in food intake, or an interaction between the two adverse factors (Scrimshaw et al., 1968).

Psychological factors also influence growth. This was well illustrated in a study of two German orphanages (Widdowson, 1951). In one institution, growth was much slower than in the second. The difference was traced to a stern and rather unpleasant matron. When she was transferred from the first to the second orphanage, the children at the first

TABLE 10.11
Amino Acids Essential to Growth

Threonine*	Lysine
Valine*	Methionine*
Leucine	Phenylalanine*
Isoleucine*	Tryptophan*
Arginine	Histidine

*Essential also for maintenance of body weight.

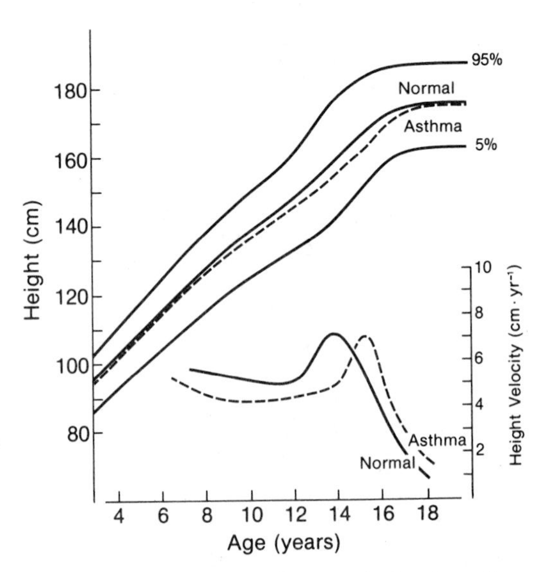

FIG. 10.14. The influence of chronic asthma upon the height and height velocity curves of 442 asthmatic boys. (Based on data of Hauspie et al., 1977.)

establishment gained rapidly in both height and body mass, while the development of those at the second institution was drastically slowed (Fig. 10.15).

At one time, it was held that a tropical environment favored a linear pattern of growth (tall, lean subjects), while a cold climate was associated with a stocky growth pattern (Malina, 1978). However, it was always difficult to dissociate the relative importance of constitution, nutrition, and disease in producing such differences of adult physique, and advances of civilization now seem to be abolishing differences of physique formerly attributed to climate.

Hormonal Factors

The prenatal regulation of growth is still somewhat obscure, although it is known that the Y chromosome stimulates the development of the testes from the seventh week of intrauterine life and that from the twelfth week the testicular cells produce androgens that lead to a differentiation of secondary sexual characteristics. An adequate secretion from the fetal thyroid gland is also essential to normal development, particularly growth of the brain.

After birth, one important regulator of development is the polypeptide growth hormone secreted by the pituitary gland (Chap. 8). This hormone is known to promote the incorporation of amino acids into tissue protein, favoring the formation of muscle and bone rather than fat (J. Campbell and Rastogi, 1969). At the molecular level, it increases synthesis of ribosomes and messenger RNA, facilitating the attachment of ribosomes to messenger RNA (R. E. Martin and Wool, 1968; O'Malley, 1969). Despite its major anabolic role, growth hormone is not essential to exercise-induced hypertrophy, since a relatively normal training response can occur in hypophysectomized animals (Goldberg and Goodman, 1969a,b). At best, it may be considered a "linear amplifier" of stimuli received from other sources.

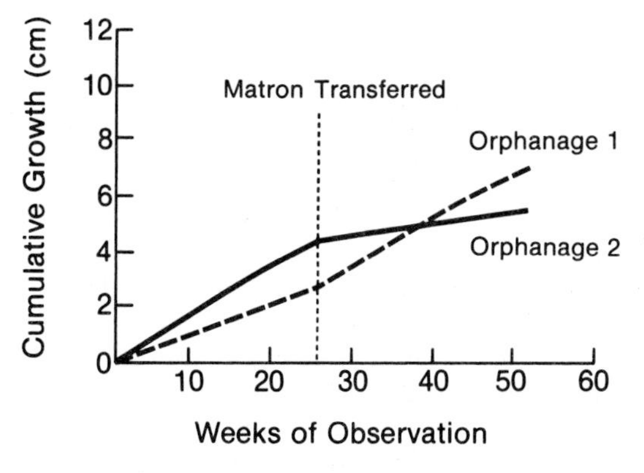

FIG. 10.15. The influence of psychological environment upon growth. A stern and unpleasant matron was transferred from Orphanage 1 to Orphanage 2 at the point indicated by the broken line. (Based on observations of Widdowson, 1951.)

The continued presence of thyroid hormone is important to growth. If thyroid secretion is deficient in the postnatal period, development is delayed. Thyroxine has a general stimulating effect upon body metabolism and has an activating effect upon many key enzyme systems.

The adolescent growth spurt seems a response to androgens, secreted by the adrenal cortex in both sexes, and by the testes in the male. Testosterone is probably responsible for the increase of muscle strength and red cell count that occurs in boys. The secretion of androgens is triggered by pituitary gonadotrophins (Chap. 8), release of the latter being held in check by the hypothalamus until an appropriate stage of development has been reached. Among other molecular effects, androgens stimulate RNA polymerase and thus nucleolar RNA synthesis plus DNA polymerase and thus DNA synthesis (O'Malley, 1969). There is therefore an increased incorporation of precursors into muscle nucleic acids and proteins (Sobel and Kaufman, 1970; Fig. 10.16). Appetite is improved, nitrogen is retained, and body mass is increased.

Other hormones also contribute to the growth process. Insulin causes a reaggregation of existing ribosomes and increased protein synthetic activity (R. E. Martin and Wool, 1968), although exercise-induced hypertrophy can occur without insulin (Goldberg, 1968). Glucocorticoids act on RNA polymerase, making DNA more available for transcription (Elson et al., 1965; Mansour and Nass, 1970). The resultant alteration of protein metabolism is closely related to cellular energy requirements; although protein synthesis can be promoted, cortisone often causes muscle atrophy. There is a tendency for selective loss of protein from slow-twitch fibers, resisted by a program of endurance exercise. Estrogen has a limited androgenic effect, although it is much less potent than testosterone, while progesterone has a weak antiandrogenic effect.

Neural Factors

The course of muscle growth and related enzyme development is markedly influenced by the local nerve supply, as has been shown experimentally by cross-inervation of selected muscle groups (Chap. 4; Buller

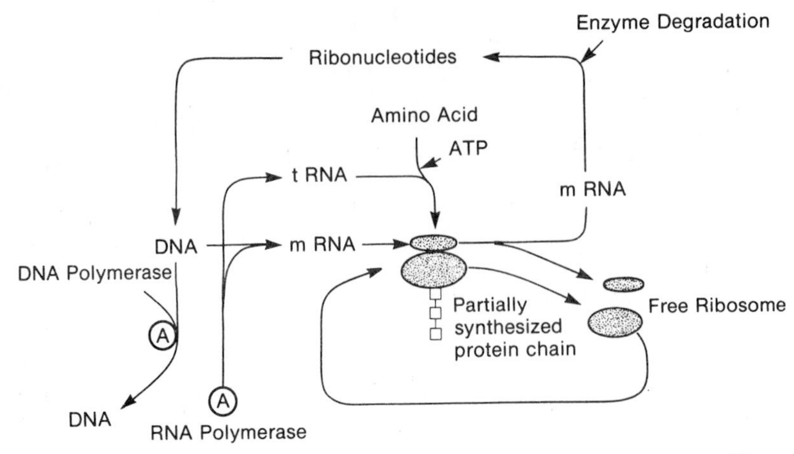

FIG. 10.16. Influence of androgens upon protein synthesis. Sites of action indicated Ⓐ. (After Edgerton, 1973.)

et al., 1960; Robbins et al., 1969). Both the small motoneurons and the associated slow-twitch fibers have a faster rate of incorporation of amino acids per unit volume than large motoneurons and fast-twitch fibers (R. P. Peterson, 1966; Dreyfus, 1967), although there is evidence of a rapid protein turnover rate in the fast-twitch fibers. It is still not clear as to how far control of development is exercised by chemicals flowing down the nerve fiber and how far the frequency of electrical impulses is responsible. Possibly, the passage of electrical impulses down the nerve speeds an axial flow of regulatory chemicals (Kow et al., 1967; Lentz, 1971).

Genetic Control

Methods of estimating the relative influence of environment and genes upon growth have been reviewed by Bouchard (1978). At least a half and perhaps two-thirds of interindividual differences in aerobic power are inherited (Shephard, 1978a), although traditional calculations (Klissouras, 1971) have overestimated the magnitude of the genetic effect (Chap. 11); by training one of a pair of monozygous twins, Howald (1976) developed a 15% intersubject difference of aerobic power, a 19% difference in the surface area of the inner mitochondrial membranes, and a 28% difference in succinate dehydrogenase activity.

Critical Age for Training

Comment has already been made on age and the training of muscle strength (see above). Several authors have postulated that there is a critical age for the development of cardiorespiratory endurance, the adolescent period usually being regarded as most favorable to a training response (B. Ekblom, 1969a,b; Sprynarová, 1974). We have already noted the problem of standardizing data in the growing child (see above). If a change in the growth of maximum oxygen intake and related variables is observed in response to regular exercise, it becomes very difficult to determine whether this is a true training effect or merely the consequence of accelerated maturation (Shephard, 1978a). In some groups of children, normal daily activity is sufficiently great that no training response can be elicited (G. R. Cumming, Goodwin et al., 1967; Weber et al., 1976; K. J. Stewart and Gutin, 1976). However, average students show a substantial response to well-designed physical education programs in both primary schools (Shephard, Lavallée et al., 1977d; Fig. 10.17) and secondary schools (R. C. Goode, Virgin et al., 1976).

Limitations of Performance for Children

At one time, it was held that excessive effort was dangerous for children. In physiological terms, there is little evidence to support this contention. The large heart of an active child is a healthy manifestation of training, as is the development of a firm and adequate musculature. Furthermore, various authors have shown that fears of athlete's heart, of becoming "muscle-bound" or (in the female) of masculinization (Erdelyi, 1962), and damage to the reproductive organs (Erdelyi, 1962; P. O. Åstrand, Engström et al., 1963) are unwarranted.

One possible exception to this generalization is the repetitive lifting of heavy loads, either at work or

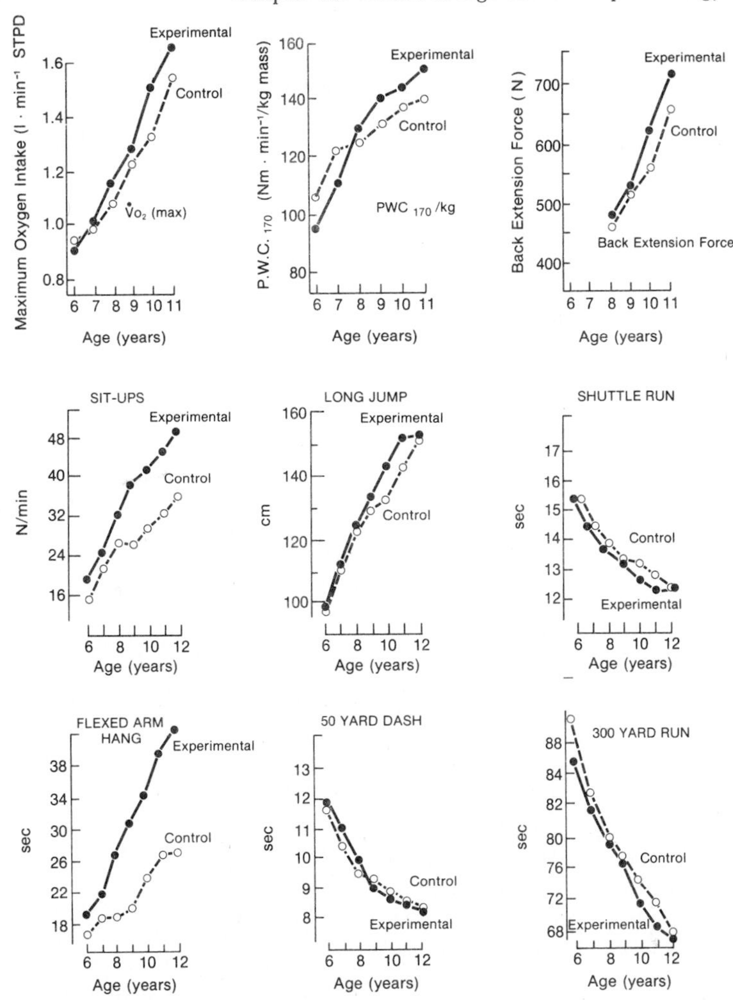

FIG. 10.17. Influence of added physical education upon performance of children. Between ages 6 and 12, experimental children had 5 hours of required physical education per week, whereas control children had standard 40-minute period. (After Shephard, Lavallée et al., 1977.)

in the gymnasium. If an excessive mass is carried before closure of the epiphyses, this can lead to deformation of the long bones and the pelvic girdle. The International Labour Organization (1964) has recommended that boys aged 16 to 18 years should lift not more than 20 kg and that the limiting load for girls at this age should be 15 kg.

The safety of normal exercise does not imply a wholesale endorsement of competitive athletics for young children. Major competitions expose youngsters to undesirable psychological and social pressures (American Academy of Pediatrics, 1966; Rarick, 1973). Although the immediate side effects of moderate competition such as participation in a hockey or baseball tournament are slight (D. L. Hanson, 1967; Rivard et al., 1977), the hours of training required for international sport can easily stunt the social and psychological development of a child, interfering with the broad educational goals of a school or university (Shephard, 1975c).

WOMEN

For various reasons, women have only occasionally participated in physiological experiments that require exhausting exercise. However, the clamor of the women's movement for equality of opportunity in employment has created a sudden need for information on the working capacity and power of the female. The present section reviews available data on sex-related differences of effort tolerance, examining possible changes of performance induced by the menstrual cycle. Brief comment is also made on nutri-

TABLE 10.12
Predicted Maximum Oxygen Intake in a Telephone Sample of Men and Women at Various Ages*

Age (Years)	Men: Step ($ml \cdot kg^{-1} \cdot min^{-1}$)	Men: Bicycle Ergometer ($ml \cdot kg^{-1} \cdot min^{-1}$)	Women: Step ($ml \cdot kg^{-1} \cdot min^{-1}$)	Women: Bicycle Ergometer ($ml \cdot kg^{-1} \cdot min^{-1}$)
15–19	47.2	42.5	39.2	33.7
20–29	40.1	36.3	37.3	30.6
30–39	37.1	32.4	34.6	28.1
40–49	33.8	27.0	31.8	24.4
50–59	32.3	25.7	31.2	21.9
60–69	30.7	22.5	30.2	18.9

*Comparison of step test and bicycle ergometer data.
Source: Based on data of D. A. Bailey, Shephard et al., 1974.

tional needs and factors related to exercise and pregnancy.

Effort Tolerance

Aerobic Power

During the prepubertal period, differences of aerobic power between boys and girls are quite small (see above). However, if comparisons are drawn between sedentary men and women, the men have a substantial (15–20%) advantage of aerobic power in all except the oldest age categories (Table 10.12). If scores are expressed in traditional units ($ml \cdot kg^{-1} \cdot min^{-1}$), a part of this difference can be explained on the basis of the extra mass of fat carried by the female; since the fat is metabolically inert, this inevitably limits the maximum oxygen intake of a woman by 6–7% (Von Döbeln, 1956; C. T. M. Davies, 1972). It might be argued that much of the residual difference is cultural, a reflection of the lesser physical activity of women. A comparison of international competitors reveals larger (20–25%) sex differences of performance in most sports (Table 10.13), but this is not conclusive proof of an inherent physiological limitation of female competitors, since the number of women entrants is still much smaller than that for men (Govaerts, 1978). From the viewpoint of employment, individual assessment is needed; although average values are lower in women than in men, many females achieve higher scores than their male counterparts (Drinkwater, 1973).

Performance of the maximum oxygen intake test follows a similar plan for men and women. Women reach slightly lower blood lactate levels than men for a given intensity of effort (Table 10.14), as far as can be judged from heart rate and plateauing of the oxygen-consumption/work-load relationship; in young women, this probably reflects a smaller active muscle mass relative to their blood volume. The maximal heart rate is usually a little higher than in men (Table 10.14), although in one series of Swedish women tested by I. Åstrand (1960) it was relatively low (187 beats per min).

If exercise is performed on a peripherally limited instrument such as the bicycle ergometer, the maximum oxygen intake attained depends upon the lean tissue mass of the thigh and calf (C. T. M. Davies, 1971). Presumably, many women are at a disadvantage in this type of exercise because their leg muscles are small; contraction must be made at a high percentage of maximal force, and there is difficulty in perfusing the active muscles (Chap. 6). Differences of maximum oxygen intake between men and women are much smaller if scores are expressed per unit mass of lean tissue, but this is probably an unfair basis of comparison (Buskirk and Taylor, 1957) since most tasks involve the displacement of body mass (lean and adipose tissue).

TABLE 10.13
Comparison of Maximum Oxygen Intake between Female and Male Athletes

Sport	Oxygen Transport ($ml \cdot kg^{-1} \cdot min^{-1}$ STPD): Female	Male	Percent Difference in Female
Cross-country skiing	63	82	−30.1
Alpine skiing	61	68	−11.5
Orienteering	59	77	−30.5
Swimming	57	67	−17.5
Running (400–800 m)	56	68	−21.4
Speed skating	53	78	−47.2
Fencing	43	59	−37.2
Table tennis	43	58	−34.9

Source: Based on data collected by P. O. Åstrand upon Swedish National Champions.

TABLE 10.14
Comparison of Maximum Oxygen Intake Measurements in Men and Women

Variable	Women		Men	
	Age 20–29	Age 60–69	Age 20–29	Age 60–69
Maximum oxygen intake				
$l \cdot min^{-1}$	1.95 ± 0.37	1.64 ± 0.21	3.81 ± 0.76	2.35 ± 0.33
$ml \cdot kg^{-1} \cdot min^{-1}$	35.7	26.8	49.4	31.4
Arterial lactate, $mmol \cdot l^{-1}$	11.0 ± 2.5	9.1 ± 2.0	13.6 ± 2.3	11.1 ± 3.3
Heart rate	195 ± 8*		190 ± 5*	172 ± 12

*P. O. Åstrand (1952) previously reported maxima of 195 beats per min for men and 198 beats per min for women aged 20–25.
Source: Based on data of Shephard, Allen et al., 1968a; and Sidney and Shephard, 1977a.

If data are expressed per kg of total body mass, female subjects reach a peak of aerobic power just before puberty, and scores decline as the adult complement of fat is laid down. In some sports, such as swimming, performance is best in the early teen years (Shephard, 1978c), but in other forms of competition peak achievements are seen after maturity has been reached; indeed, in those athletes who can restrict the development of body fat, aerobic power may continue to improve into the adult years.

One early study (I. Åstrand, 1960) gave rise to the opinion that Scandinavian women had a higher aerobic power than North Americans (Shephard, 1966e); however, the Swedish sample was associated with the Central Gymnastic Institute in Stockholm, and other data from both Norway (K. L. Andersen, 1964) and Sweden (Kilbom, 1971) have not confirmed the supposed outstanding endowment of Scandinavian women.

Submaximal Exercise

At any given submaximal work load, the typical female must work at a larger fraction of her aerobic power than a male. In consequence, the heart rate and respiratory minute volume of the women are larger, and there is a greater production of lactate (P. O. Åstrand, 1956; Hermansen and Andersen, 1965; Cotes, Davies et al., 1969; Máček and Vávra, 1971) with development of a larger oxygen debt (Nöcker and Bohlau, 1955). Women are also more likely to show an exercise-induced ST segmental depression than men (G. R. Cumming, 1978a); there is no evidence that middle-aged females suffer more coronary vascular narrowing than men of similar age, but the force that they exert per unit cross-section of heart muscle may be greater than in men.

Considerations of both anatomy and culture lead to differences of mechanical efficiency between men and women. This is well documented for walking. Women produce more rotation of the hips than men, with less vertical displacement of their center of gravity. The energy cost of movement per unit of body mass is thus less in a woman (Booyens and Keatinge, 1957). During swimming, women also have some advantages. The high percentage of body fat affords protection against cold and gives a general buoyancy. Since the thighs contain more fat and less bone than in a man, they can be held more horizontally in the water, reducing the impedance to forward movement (Rennie, Prendergast et al., 1973; Chap. 14). Nevertheless, because of their lesser muscle mass, women cannot swim as fast as men (Table 10.15).

The greater subcutaneous fat of a woman hampers the elimination of body heat during sustained effort in a warm environment (Hertig et al., 1963; T. Morimoto et al., 1967; Nunneley, 1978) and this may be one reason for a slower recovery of the resting heart rate in women than in men (Sidney and Shephard, 1973). Women apparently have a larger number of active sweat glands (Kawahata, 1960), but when working in the heat they often produce less sweat than men (Brouha, Smith et al., 1960; Wyndham, Morrison, and Williams, 1965; R. H. Fox et al., 1969). Wyndham and his associates have commented that man is "a prolific sweat waster, whereas the female adjusts her sweat rate better to the required heat loss." Nevertheless, some authors have found a poor heat tolerance in the female. Thus Wyndham, Morrison, and Williams (1965) noted that 92% of women could not complete a standard heat exposure tolerated by 50% of men. In contrast, Weinman et al. (1967) found lower rectal temperatures and lower heart rates in their female subjects; however, their experiment was complicated because the women tested were fitter than the men.

Oxygen Conductance

The various links in the oxygen transport chain are all somewhat less well developed in women than in men. Static lung volumes are 20–25% smaller in the female (Shephard, 1971b), with corresponding de-

TABLE 10.15
Difference in 1975 World Records for Female and Male Competitors

Event	Female	Male	Percent Poorer Performance in Women
Swimming (free style)			
100 m	57.5 sec	51.2 sec	12.3
200	123.6	112.8	9.6
400	257.4	238.2	8.1
1500	1003.0	932.0	7.6
Track			
100 yards	10.0 sec	9.0 sec	11.1
220	22.6	20.0	13.0
440	52.2	44.5	17.3
880	121.0	104.6	15.7
1 mile	269.5	231.1	16.6
Field			
High jump	1.94 m	2.30 m	18.3
Long jump	6.84	8.90	30.1
Javelin throw	66.1	94.1	42.4
Speed skating			
500 m	41.8 sec	38.0 sec	10.0
1000	86.4	77.6	11.3
1500	134.0	118.7	11.3
3000	286.5	248.3	11.5
5000	541.6	429.8	26.0

creases of maximum voluntary ventilation and maximum exercise ventilation. The cardiac volume (M. Burger, 1955), stroke volume, and maximum cardiac output are also less than in a man, while lower hemoglobin concentrations (average 13.8 $g \cdot dl^{-1}$, compared with 15.6 $g \cdot dl^{-1}$ in male subjects) lead to a lesser maximum oxygen carriage per unit of blood flow (M. Burger, 1955; Cotes, Dobbs, et al., 1969). Finally, as noted above, weaker muscles cause some difficulty in perfusing active tissues.

Muscle Force

Girls do not show an adolescent spurt of muscle growth (see above). While boys show a 14-fold increase in the number of muscle cells from the age of 2 months to 16 years, in girls there is only a 10-fold increase (Cheek, 1968). The maximum muscle force of the average adult woman is ~60% of that in a man. Asmussen (1964) argued that on dimensional theory a 20% difference would be expected from the shorter stature of the female. He assumed that muscle force was related to the square of height, although with the observed curve ($H^{2.9}$; see above) a difference of up to 30% could arise from considerations of size alone.

Judging from athletic records, women are at the greatest disadvantage in activities such as jumping, where a tall stature and forceful muscular contraction are required. However, they perform well in swimming and short-distance running events (Table 10.15). According to Hettinger (1961), the muscles of women are less easily trained than those of men.

Skeletal Structure

On average, bone strength and density are less in a woman than in a man (Bradbury, 1949). This leads to a high rate of sports injuries, with a particular risk of conditions involving overstrain (contractures, inflammations of tendons and tendon sheaths, bursae, foot problems, and periostitis; Klaus, 1964). The female anatomy is characterized by a greater pelvic tilt, and the femoral axes are more oblique than in a man (Caffey et al., 1956). The arms also have a greater carrying angle, with more tendency to outward rotation (cubitus valgus) at the elbow (W. B. Atkinson and Elftman, 1945). These various characteristics are a disadvantage in many sports; nevertheless, women show a greater potential range of motion than men at most joints (Hupprich and Sigerseth, 1950), leading to excellence in activities such as gymnastics.

Coordination

The center of gravity of the body is lower in a woman than in a man (Chap. 7). Partly for this reason, and partly because of cultural factors, women generally perform better than men at tasks requiring balance, dexterity, and fine coordination (Klaus and Noack, 1971). True reaction times show no sex difference, but movement times and thus observed response times are faster in men (Pierson and Lockhart, 1964; G. R. Wright and Shephard, 1978a; Fig 10.18).

Cardiorespiratory Training Response

If due allowance is made for the initial fitness of subjects, the response of women to vigorous cardiorespiratory training seems much as in men (Kilbom, 1971; Sidney and Shephard, 1978b; Lesmes, Fox et al., 1978; Pedersen and Jørgensen, 1978). Because the initial working capacity of a woman is often very low (J. R. Brown and Shephard, 1967; D. A. Bailey et al., 1974), many who seek relatively strenuous jobs need physical training to bring the proposed daily task below 50% of maximal aerobic power, the level at which fatigue is usually incurred.

For preadolescent girls, C. H. Brown et al. (1972) noted that 12 weeks of competitive cross-country running produced a 26% increase of maximal oxygen intake. P. O. Åstrand, Engström et al. (1963) made a longitudinal study of female swimming champions. As a group, they were taller than untrained girls, with a larger functional capacity (heart volume, total hemoglobin, and lung volume). Further, differences between athletes and nonathletes were correlated with the amount of training undertaken, suggesting that these differences were causally related to the training program. Follow-up studies of this group suggested that final attitudes toward physical activity were not particularly favorable (B. O. Eriksson, Engström et al., 1971). By early middle age, the majority of the women had ceased their swim training, and their maximum oxygen intake had regressed to 37 ml • kg^{-1} • min^{-1}, less than average figures for a Swedish housewife. Nevertheless, heart and lung volumes remained large.

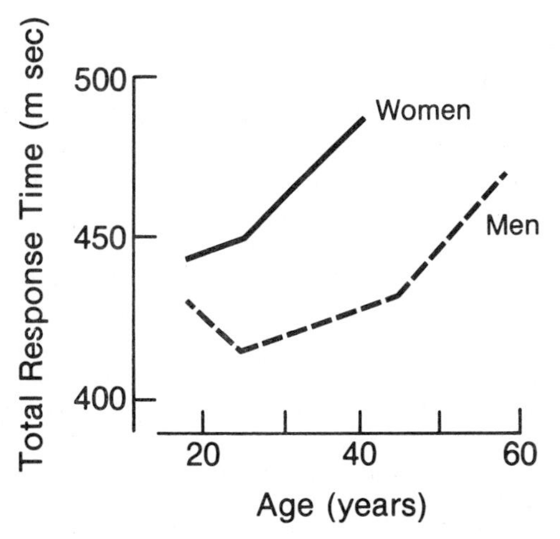

FIG. 10.18. Influence of age and sex upon response time. Subjects operating a foot brake in response to a red signal light (G. R. Wright & Shephard, 1978a.)

Menstrual Cycle

Effects of the menstrual cycle upon selected variables are listed in Table 10.16. The impact of such changes upon exercise physiology remains largely unexplained. M. Phillips (1968) concluded that neither pulse rate nor blood pressure responses showed any menstrual variations. Others (A. W. Sloan, 1961; Garlick and Bernauer, 1968) found no cyclic alterations of cardiorespiratory performance. Kawahata (1960) reported that estradiol had an inhibitory effect upon sweating. He found an earlier onset of sweating during menstrual flow than at other times. Sargent and Weinman (1966) and Haslag and Hertzman (1965) were unable to duplicate these observations, but D. Chapman and Horvath (1973) also found earlier sweating during menstruation. Both D. Chapman and Horvath (1973) and Haslag and Hertzman (1965) noted that higher core tempera-

TABLE 10.16
Changes in Some Physiological Variables over the Menstrual Cycle

Variable	Nature of Change
Body temperature	Decreased at ovulation, sharp postovulatory rise
Systemic blood pressure	Lower at mid-cycle
Respiratory minute volume	Increased in luteal phase
Carbohydrate metabolism	Increased fasting blood sugar during menstruation
Blood lactate	Increased at ovulation
Body mass	Premenstrual increase
Daily activity	Greatest in postmenstrual phase
Mental efficiency	Decreased in premenstrual phase

Source: After Southam and Gonzaga, 1965.

tures were reached during the luteal phase of the menstrual cycle.

Body mass is increased by water retention in the premenstrual phase; in itself, this has only a minor effect upon performance, but related sensations may lead to a deterioration of all-out effort (Dalton and Williams, 1976), while a temporary rise of intraocular pressure may impair activities requiring a high level of visual acuity (Dalton and Williams, 1976). During the time of menstrual flow, poor physical performance (Wearing et al., 1972) may arise from practical problems of personal hygiene.

Psychological tension during the premenstrual period may lead to an increase of maximal muscle force, but there is generally a decrease of hand steadiness in this phase of the cycle (Zimmerman and Parlee, 1973). Simple reaction time is unaltered (Pierson and Lockhart, 1963; Loucks and Thompson, 1968), while there is a deterioration of skilled mental and physical performance, with an increased risk of accidents (Noack, 1954; Dalton, 1960,1968; Erdelyi, 1962). Many athletes use estrogen and progesterone preparations to adjust the timing of their menstrual cycle prior to major competitions (Dalton and Williams, 1976). Severe competitive stress may in itself modify both the length of the cycle and the duration of menstrual flow (Dalton, 1968); certain authors (Klaus and Noack, 1971; Erdelyi, 1962) regard such irregularities as a sign of overtraining. However, painful menstruation (dysmenorrhea) seems to be less frequent in athletes (T. W. Anderson, 1965) than in sedentary girls.

It has sometimes been suggested that heavy training leads to a masculinization of female competitors. Certainly, body fat is reduced, and the secretion of androgenic hormones may be increased by vigorous activity (Chap. 8). However, the main basis of any difference in secondary sex characteristics between athletes and other women is an initial selection rather than a response to subsequent activity (Erdelyi, 1962; Klaus and Noack, 1971).

Nutrition

The general principles of nutrition are the same for a woman as for a man, but due allowance must be made for the menstrual loss of protein and minerals. The need for iron is particularly critical, since a normal person only absorbs 5–10% of dietary intake (10–15 mg • day^{-1}), yet menstrual losses are 0.5–1.0 mg • day^{-1}. Women with heavy menstrual bleeding often suffer from minor degrees of iron deficiency anemia and they should have an iron intake of 3–4 times the male standard (Wintrobe, 1967). The influence of hemoglobin level upon performance is discussed in Chap. 6. Iron turnover is increased by vigorous training. Kilbom (1971) noted a 25% decrease of serum iron levels as a result of physical conditioning.

During lactation and pregnancy, the intake of nutrients must be further augmented to cover the secretion of milk, growth of the fetus, and the energy costs of displacing a heavier body.

Pregnancy

Exercise During Pregnancy

A substantial fraction of the cardiac output is directed to the placenta during pregnancy. Venous return is impaired (Ueland et al., 1969), and the reserve capacity of the liver and kidneys is severely taxed. In view of the known effect of exercise in restricting visceral blood flow (Chap. 6), maximal or near-maximal effort is undesirable during late pregnancy. Vigorous diaphragmatic movements are also impeded by fetal growth, making heavy breathing difficult and uncomfortable. The previously sedentary woman should thus restrict herself to moderate activity during pregnancy.

Excessive muscular effort should be avoided, as this may initiate abortion or premature labor (Javert, 1960). Nevertheless, some champion athletes have continued to compete until a few days before the onset of labor (Rumpf, 1952), apparently without harmful effects.

The energy cost of all movement is increased during pregnancy. This is due partly to an increase of body mass (ultimately at least 10 kg) and partly to a displacement of the center of gravity of the body (with awkward and unaccustomed patterns of movement).

Exercise and Labor

Although fears were once expressed that athletic participation by a girl might hamper childbirth during her adult years, Erdelyi (1962) noted that labor was shorter than average in 87% of Hungarian women athletes and that Caesarian sections were less frequent than in control subjects. Zaharieva (1972) also noted easier childbirth in former Olympic competitors.

Exercise Postpartum

There is now little argument that a graded return to vigorous activity is beneficial in the postpartum period. However, the energy requirements of lactation are such that serious conditioning should not be contemplated until after the infant has been weaned.

Noack (1954) reported that 10 of 15 German athletes returned to competition after pregnancy. Eight made objective improvements over their previous best performance, and the remaining two at least equaled previous records. Klaus and Noack (1971) suggested that the circulatory demands of pregnancy provided the equivalent of 9 months rigorous conditioning!

AGING

There is now an extensive literature on physical activity and aging (Shephard, 1978b). Limitations of space preclude more than a brief review of the normal aging process, with comments upon the training response of the elderly and possible interactions between physical activity and the rate of functional deterioration.

Normal Aging Process

Molecular Basis

Comfort (1973) suggested that aging is a question of information loss. The process occurs in all living organisms and is partly an inherited, programmed characteristic. There are associated chemical changes in the master template of protein synthesis (DNA) due to radiation, chemical damage by free radicals, viruses or autoantibodies, and failure of normal mechanisms for template repair. Other errors may develop in the chain of protein transcription (Fig. 10.16), affecting various constituents of the chromosome, RNA, and protein synthetases. In consequence, the rate of production of key enzymes and structural proteins becomes less than their rate of breakdown. Moreover, abnormal molecules are synthesized, accumulating within the cell as simple precipitates and complexes with vital constituents (Shephard, 1978b).

Many aging cells show increased glycogen storage, fatty infiltration or degeneration, and an accumulation of pigments such as lipofuscin. Changes of membrane permeability and/or failure of the sodium pump (Chap. 4; Fig. 4.12) decrease cellular retention of potassium ions and the exclusion of sodium ions and water (Timiras, 1972). Cross-linkages between collagen molecules decrease the compliance of connective tissue (Viidik, 1967b); there is also a loss of elastic tissue, with infiltration by cellulose and "pseudoelastin" (an abnormal variant of collagen; D. A. Hall, 1973). The connective tissue cells form less polysaccharide ground substance, but its turnover rate is increased; this reduces the stability of materials such as cartilage (Sylven and Malmgren, 1952) and possibly makes the tissues less permeable to nutrients. Other problems of the elderly person who attempts to exercise—back and joint disorders, tendon rupture, and altered pressure/volume relationships for the lungs, heart, and great vessels—can be traced at least in part to these various changes in the molecular structure of connective tissue.

Studies of liver and red cells do not suggest that aging leads to a general reduction of tissue enzyme activity (Bertolini, 1966; Timiras, 1972). Nervous tissue shows a reduction of cholinesterase and choline acetylase activity per unit mass, but this could reflect an increase in the proportion of white matter rather than a more fundamental change of enzyme activity. There are reports of reduced ATPase and lactic dehydrogenase activity in specimens of muscle from elderly rats (Rockstein and Brandt, 1961; Schmukler and Barrows, 1966; Edington and Edgerton, 1976; Pokrovsky, 1978); it may thus be that tissues which have lost the power of cell division (postmitotic tissues such as muscle and brain) are more vulnerable to aging than are the frequently renewed tissues studied by earlier investigators.

The characteristics of cellular organelles change with aging (Bakerman, 1969). The nucleus becomes larger, the nucleoli increase in size and number, and there are alterations in the size, shape, cristal pattern, and matrix density of the mitochondria. However, the functional importance of these changes remains unclear.

Some cells retain an ability to divide even in old age. For example, the output of red cells from the bone marrow of an elderly person can still be increased by hypoxic stress (Das, 1969). However, there is a progressive death of cells in postmitotic tissue such as brain due to a combination of metabolic errors and arteriosclerotic hypoxia (Brody, 1955). Death or malfunction of key cells in turn impairs the overall response to any homeostatic challenge (including exercise). Ultimately, a minor threat to homeostasis, be it infection, exertion, or a change in environment, is sufficient to terminate life.

Gross Function

Aging is associated with a progressive deterioration of organ function as active cells die and are replaced by fibrous and fatty tissue. Further, loss of coordination between various systems contributes to impairment of homeostasis (Pokrovsky, 1978).

In male subjects, many physiological variables reach their peak between 20 and 25 years of age, with a progressive decline thereafter (Fig. 10.19). Often, there is a 25–30% loss of function by 60 years, with a more rapid subsequent deterioration. In the female, the same variables usually reach their peak at the time of menarche, but it is by no means clear as to how far the early adult loss of performance is dictated by physiology and how far it is a consequence of cultural factors such as a progressive restriction of physical activity.

The major portion of available data refers to the aging of male subjects, but even in men there are several obstacles to precise study. How should a "normal" population be defined? At the age of 20 years, relatively few persons suffer from chronic disease, but by the age of 70 years up to 50% of a population show some medical abnormality (Heikkinen, 1978); in many of these individuals there is a resultant deterioration of physical performance (J. R. Brown and Shephard, 1967; Shephard, 1979d). Those who volunteer for exercise testing are often the fit and healthy residue (Shephard, 1978a,1979d), both health-conscious and nonsmokers (Sidney and Shephard, 1976); unfit members of the cohort are eliminated both by the volunteer bias and by a selective high mortality rate.

There are difficulties in standardizing data for elderly subjects, since height is reduced by kyphosis and vertebral compression (Quetelet, 1835; Miall et al., 1967; Friedlander et al., 1977), while any given body mass contains more fat and less lean tissue than in a younger person (Shephard, 1979d). Lean body mass might be thought the ideal method of data standardization, but there are technical problems when measuring this variable in an older person. Whole-body counter estimates of lean tissue are complicated by fat shielding of ^{40}K emissions, changes in the potassium content of lean tissue, and a reduction in the proportion of potassium-rich muscle in the lean body compartment (Myrhe and Kessler, 1966).

Calculation of aging rates is often distorted by the fitting of linear regression equations when the loss of function conforms to a curved relationship (T. W. Anderson, Brown et al., 1968). Longitudinal studies are generally considered preferable to cross-sectional investigations, but in some instances they have indicated impossibly high rates of aging. The explanation seems that the true loss of function due to aging has

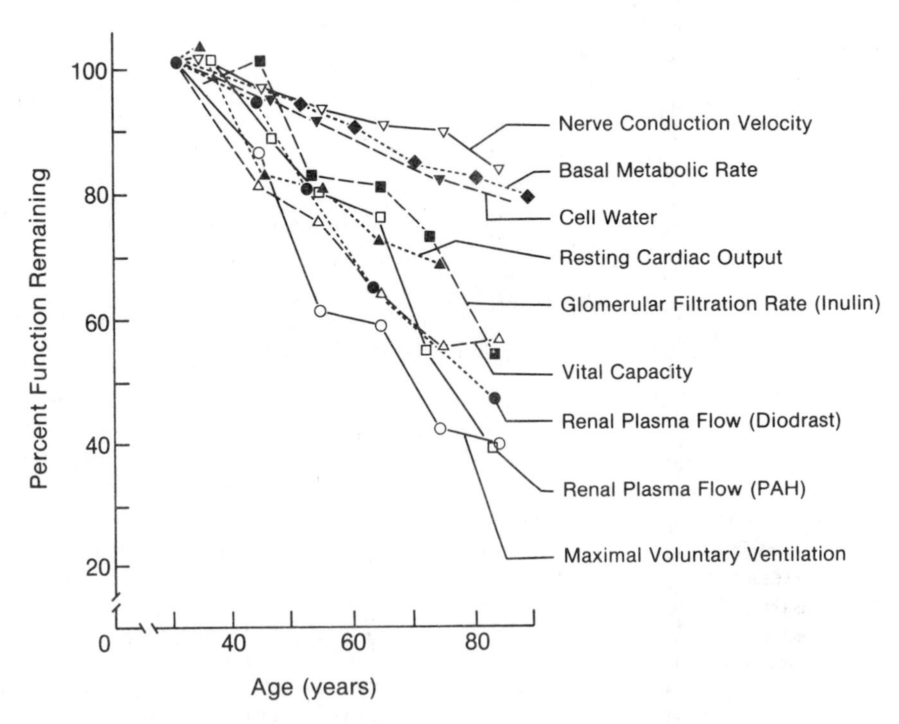

FIG. 10.19. Changes in selected physiological variables with increasing age. (Based on the data of Shock, 1967.)

been confounded with the effects of a decrease in physical activity as a person becomes older (McPherson, 1978; Fig. 10.20). In cross-sectional investigations, it is possible to group results at each age by activity level. The estimated rate of aging may then be more precise than values obtained from a longitudinal investigation (Shephard, 1979c), although in some parts of the world the sample becomes modified by a differential migration within the population (Illsley et al., 1963; Shephard, 1979c).

Anerobic Capacity and Power

There do not seem to have been any biopsy estimates of muscle phosphagen reserves in the elderly. An indirect assessment of anerobic power is possible from the timing of staircase sprints. The equivalent oxygen delivery decreases from 165 ml • kg^{-1} • min^{-1} in a young adult (Chap. 3) to ~90 ml • kg^{-1} • min^{-1} in a 65-year-old person. However, it seems fair comment that not all of the observed deterioration is physiological. Young people are easily persuaded to engage in such vigorous activity, but the elderly may limit their performance due to fear of stumbling, poor vision, instability of the knee joints, and lack of recent familiarity with the required exercise.

Elderly people show a marked creatinuria. D. A. Hall (1973) inferred from this that they have difficulty in resynthesizing creatine phosphate after vigorous physical activity.

The capacity of the lactate mechanism is often said to be reduced in the elderly (S. Robinson, 1938; I. Åstrand, 1960; W. C. Adams, McHenry, and Bernauer, 1972) on the grounds that arterial lactate concentrations (~7 mmol • l^{-1}), respiratory gas-exchange ratios (W. C. Adams, McHenry, and Bernauer, 1972), and oxygen debt (Fischer et al., 1965; Tlusty, 1969) are low following exhausting effort. In our experience (Sidney and Shephard, 1977a), restricted lactate accumulation is partly a question of motivation, since with sufficient urging readings of 10–12 mmol • l^{-1} can be attained. Any residual difference from a young person does not necessarily reflect a deficiency of cellular mechanisms for anerobic metabolism, as the lactate has been determined in blood rather than muscle. In an old person, the ratio of muscle mass to blood volume is smaller than in a young adult, and the lactate also escapes less readily from muscle to blood, so that peak intraarterial lactate concentrations are blunted.

Aerobic Power

The aerobic power of a sedentary man declines from 40–50 ml • kg^{-1} • min^{-1} at the age of 20 years to perhaps 25–30 ml • kg^{-1} • min^{-1} at the age of 60 years (Fig. 10.21). This is due partly to a reduction of absolute aerobic power and partly to an increase of body mass, the latter commonly occurring between 30 and 40 years of age.

There is an associated decrease of maximum heart rate. Some early Scandinavian reports suggested that the peak reading dropped to 155–160 beats per min in a 65-year-old adult, but more recent data from North America have shown average maxima of 170 beats per min and higher (Lester et al., 1968; Sidney and Shephard, 1977a; Table 10.17). The maximum heart rate is also reduced at high altitude (Chap. 12), and it is thus tempting to blame the age-related change upon myocardial oxygen lack.

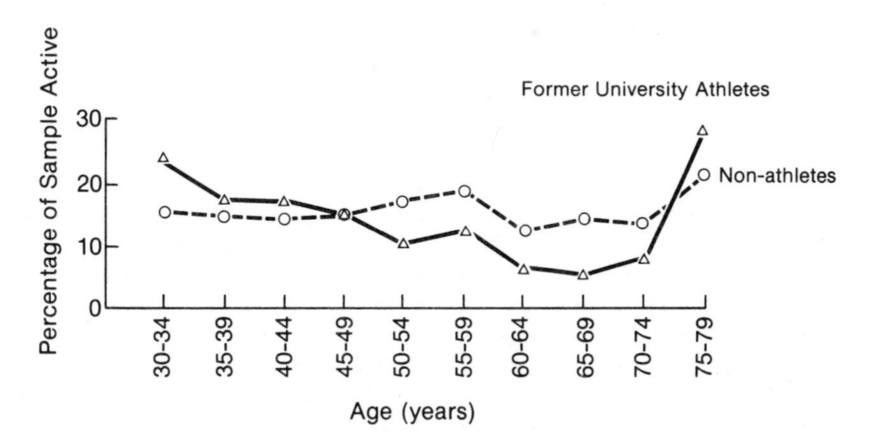

FIG. 10.20. Changes of habitual activity with age. Based on data of Montoye, Van Huss et al. (1957) for former athletes and nonathletes, both alumnae of a U.S. university. The data shown for ages 70–74 and 75–79 is based on a very small sample.

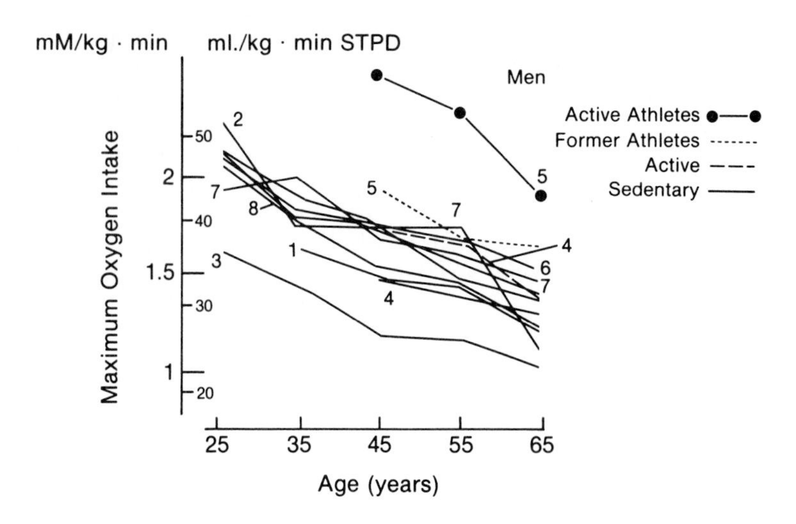

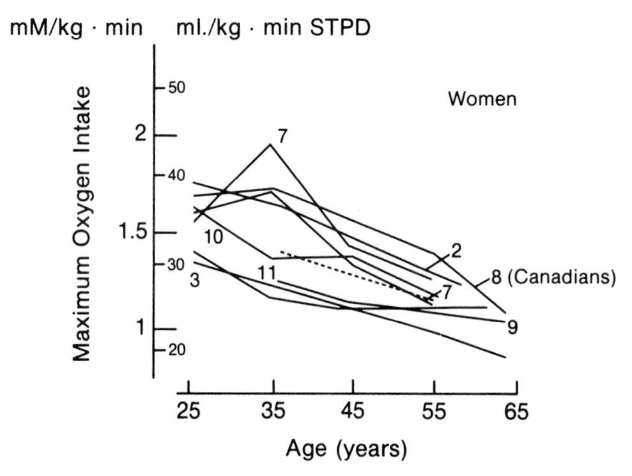

FIG. 10.21. The influence of age upon aerobic power. (Based largely on data collected by Sidney, see Shephard, 1978b for references.) 1. W. C. Adams et al. (1972); 2. I. Åstrand (1960); 3. Bailey et al. (1974); 4. Dehn & Bruce (1972); 5. Grimby & Saltin (1966); 6. Robinson (1938); 7. Shephard (1966e); 8. Shephard (1977a); 9. Cumming et al. (1973); 10. Kilbom (1971); 11. Profant et al. (1972).

Pathological impairment of blood flow to the cardiac pacemaker can indeed lead to a slowing of maximum heart rate, but such a phenomenon is by no means common even in patients with severe coronary vascular disease (Kavanagh and Shephard, 1976). Furthermore, the slow maximum heart rate of an old person cannot be corrected by administration of oxygen during physical activity. A reduced compliance of the heart wall is a more plausible explanation; this could increase the time required for filling of the ventricles and thus modify the feedback of information on venous filling to the cardioregulatory centers. Last, the sympathetic drive to the cardiac pacemaker may diminish with age.

Other physiological characteristics of "good" maximal effort in a 65-year-old subject are summarized in Table 10.17. In a sedentary older person, effort is often halted before such values are attained. Reasons include dyspnea (Von Döbeln, Åstrand, and Bergström, 1967; G. Grimby, Bjure et al., 1972), fear of overexertion (Julius et al., 1967; Lester et al., 1968), muscular weakness (I. Åstrand, Åstrand, and Rodahl, 1959; Barry, Daly et al., 1966; Von Döbeln, Åstrand, and Bergström, 1967; G.

TABLE 10.17
Characteristics of Maximum Effort in 65-year-old Subjects Making a Good Maximum Effort

Variable	Men (n = 19)	Women (n = 20)
$\dot{V}_{O_2}$(max)		
$l \cdot min^{-1}$	2.35 ± 0.35	1.64 ± 0.21
$ml \cdot kg^{-1} \cdot min^{-1}$	31.4 ± 4.4	26.8 ± 2.9
Plateau <2 $ml \cdot kg^{-1} \cdot min^{-1}$ (% of subjects)	78.9	75.0
Maximum heart rate (beats $\cdot min^{-1}$)	172 ± 12	161 ± 12
Respiratory gas-exchange ratio	1.11 ± 0.08	1.07 ± 0.09
Arterial blood lactate ($mmol \cdot l^{-1}$)	11.1 ± 3.3	9.1 ± 2.0
Systolic blood pressure (torr)	210 ± 31	192 ± 27

Source: Based on data of Sidney and Shephard, 1977a.

Grimby, Bjure et al., 1972), poor motivation (G. Grimby, Bjure et al., 1972), and the appearance of electrocardiographic abnormalities (Barry, Daly et al., 1966).

Submaximum Exercise

Since maximum aerobic power is low and the muscles are weaker than in a younger subject, performance of a given intensity of submaximal exercise by an old person is often characterized by an increase of heart rate and respiratory minute volume relative to data for a younger person (Shephard, 1978b; Shephard and Sidney, 1979); there is an earlier and more substantial increase of lactic acid (Robinson, 1938; Strandell, 1964b; Grimby and Saltin, 1966; C. T. M. Davies, 1972), and a higher blood pressure is reached (Sidney and Shephard, 1977a). The last finding may reflect in part an increase of resting blood pressure (de Vries, 1978). The cardiac work rate, as estimated by the rate-pressure product (Chap. 6) shows some increase, and this added load is generally borne by a weaker myocardium with an impaired vascular supply.

The rate of adaptation to exercise (the "on-transient") slows as a person becomes older (Robinson, 1938; E. A. Harris and Thomson, 1958; Shock, 1961; J. Skinner, 1970). The mechanical efficiency of most forms of exercise is a little less than in a younger person. A greater mass of fat must be supported and/or displaced, a high blood pressure and aging of the airways increase cardiac and respiratory workloads, while joint stiffness, poor motor coordination, and lack of recent familiarity with laboratory exercise all increase the oxygen cost of a given effort. Sidney and Shephard (1977a) found that elderly subjects performed bicycle ergometer exercise with a mechanical efficiency of 21.5% compared with 23% in a healthy young adult. Parallel changes in the efficiency of running and stepping are likely (Robinson, 1938; Molina and Giorgi, 1951; Ryhming, 1953; A. H. Norris et al., 1955; Durnin and Mikulic, 1956), while walking becomes characterized by short and inefficient steps (Steinberg, 1966; Ayalon and Van Gheluwe, 1975; Hebbelinck, 1978).

Attempts to predict aerobic power from submaximum exercise test data are less satisfactory in the elderly than in the young. Problems of reduced mechanical efficiency can be circumvented by direct measurements of oxygen consumption, but there is still uncertainty regarding the maximum heart rate to be used in extrapolations (see above). Even in subjects making a good maximum effort, we have found peak heart rates to vary by ±12 beats per min (Sidney and Shephard, 1977a). Further, failure to sustain stroke volume at high work loads leads to a disproportionate increase of heart rate with oxygen intake, in contrast to the linear relationship seen in younger subjects:

Subjects	Deviation from linearity (beats $\cdot min^{-1}$)
Young men	− 3.2 ± 7.3
Young women	− 7.2 ± 14.3
Old men	+ 8.1 ± 7.2
Old women	+11.5 ± 9.7

With both bicycle and treadmill procedures, predicted maxima for the elderly show a scatter of at least 15% relative to directly measured values, discrepancies being even larger for women exercising on the treadmill. Bicycle ergometer results also systematically underestimate the true maximum oxygen intake (D. A. Bailey, Shephard, et al., 1974; Sidney and Shephard, 1977a), reflecting the tachycardia associated with weak and heavily stressed quadriceps muscles in this form of exercise. If submaximum predictions are to be made on older subjects, it is preferable to use a treadmill or step test rather than a bicycle ergometer.

Interpolated tests such as the physical working capacity at a heart rate of 170 beats per min (PWC_{170}) avoid uncertainties regarding the maximum heart rate. However, there remain problems of low mechanical efficiency and quadriceps overloading associated with bicycle ergometer exercise. Furthermore, while the PWC_{170} is a relatively light test for a young person, it equals or even exceeds the maximum power output of someone who is older.

During sustained submaximal activity, an elderly person is liable to fatigue because of a substantial accumulation of lactate. The thick layer of subcutaneous fat also leads to problems of temperature regulation. There have been no in vivo measurements of muscle glycogen stores in the elderly, but postmortem reports frequently comment on the high glycogen content of liver and kidney cells (Timiras, 1972); this has been attributed to the high blood sugar readings of the elderly. The heart, kidney, and liver may also show fatty degeneration at postmortem; possible explanations include an increased transport of fat from the periphery, an increased capture of circulating chylomicrons, an increased synthesis of fat by the liver, and a decreased utilization of fat associated with respiratory impairment and impending death. It is also conceivable that the fat of an elderly subject is stored in a different form from that of a young person. An impairment of phospholipid production, for example, could transform dispersed "micellar" fat into discrete globules readily seen under the light microscope.

Finally, the recovery period following effort is prolonged in the elderly. This is due to a combination of a greater relative work rate, an increased proportion of anerobic metabolism, a slower heat elimination, and a lower level of physical fitness (Robinson, 1938; A. H. Norris et al., 1955; Shock, 1961; Wessel et al., 1968; Montoye, Willis, and Cunningham, 1968; Tlusty, 1969).

Oxygen Conductances

Individual links in the oxygen transport chain all show some impairment with age. The rate of loss of vital capacity depends upon respiratory health and smoking habits (T. W. Anderson, Brown et al., 1968; Niinimaa and Shephard, 1978a); commonly, there is a 25% loss from age 25 to age 65 years. Studies of elderly swimmers show a parallel decrement of performance, possibly due in part to a loss of buoyancy (Rahe and Arthur, 1975). The reduction of static lung volumes, stiffening of the joints in the thoracic cage, and an increase of airway resistance (often exacerbated by chronic bronchitis and emphysema) reduces the maximum voluntary ventilation from 200 to 120 l • min^{-1} (Shephard, 1971b). During treadmill work, the number of liters of ventilation needed to transport 1 l of oxygen is much as in a younger person (25.2 l in men, 27.4 l in women), but in bicycle ergometry the cost is increased (to 34.0 l in men and 33.9 l in women), probably because of quadriceps weakness and lactate accumulation (Sidney and Shephard, 1977a). A given ventilation tends to be developed at a faster respiratory rate and a smaller tidal volume in the elderly (de Vries, 1978). However, the maximum exercise ventilation remains ~50% of the maximum voluntary ventilation, as in a younger individual. In seven studies of elderly men, the maximum exercise ventilation averaged 62.7 l • min^{-1} STPD, while in three investigations of old women, the average was 47.3 l • min^{-1} STPD. The age-related loss of respiratory function is substantially larger in smokers than in nonsmokers (Niinimaa and Shephard, 1978a).

Some deterioration of alveolar ventilation would be anticipated with aging, since (1) the resting dead space increases by ~1 ml per year of adult life and (2) there is a progressive destruction of both the alveolar surface and the capillary bed. In practice, the ventilatory equivalent remains much as in a young person, suggesting that any change of alveolar ventilation during vigorous activity must be small. The older person has one advantage in that exercise leads to a substantial increase of pulmonary arterial pressure. This improves the matching of ventilation and perfusion in the upper parts of the lungs.

Aging increases the work of breathing several fold, changes occurring in both the chest and the lungs. Movements of the rib cage are impeded by kyphosis, a flattening or a barrel-shaped deformity of the chest, and stiffening or ankylosis of the joints, so that the proportion of diaphragmatic breathing is increased. Within the lungs, the dimensions of the larger air passages are unaffected by aging, but the number and thickness of the radial elastic fibers that maintain patency of the small airways is reduced; there is thus a greater likelihood that the effort-independent portion of the flow/volume curve will be reached during exercise (Fig. 5.16). Inability to make a rapid expiration compounds the problem of a 20–25% reduction in vital capacity, so that dyspnea is likely at a ventilation of 60–65 l • min^{-1} in men and 50 l • min^{-1} in women. Because of loss of elastic tissue, a quarter of the vital capacity range is affected by airway closure at the age of 65 (P. LeBlanc et al., 1970).

The resting pulmonary diffusing capacity falls progressively with aging; this reflects poorer gas mixing, a lesser pulmonary blood flow, a decrease in the proportion of the alveolar surface covered by capillaries, and a decrease of the alveolar capillary blood volume (Niinimaa and Shephard, 1978a). However, perhaps because of the rise in pulmonary arterial pressure during exercise, the increase of diffusing capacity with oxygen intake is more steep than in a younger person; the estimated maximum diffusing capacity of a healthy 65-year-old person (~18.7 mmol • min^{-1} • kPa^{-1} in a man, 13.8 mmol • min^{-1} • kPa^{-1} in a woman) is therefore not much poorer than in a young adult (Shephard, 1978b).

The heart volume is well maintained. Indeed, in some cross-sectional studies values have increased with age; this is probably due to a greater selection of subjects in the older age groups (Strandell, 1964a; P. O. Åstrand, 1968; C. T. M. Davies, 1972). In light exercise, the stroke volume is much as in a younger individual, but as maximum effort is approached the stroke output diminishes (Granath, Johnson, and Strandell, 1964; Becklake, Frank et al., 1965; Grimby, Nilsson and Saltin, 1966; Hanson, Tabakin, and Levy, 1968; Niinimaa and Shephard 1978b); at exhaustion, values are thus 10–20% smaller than in a younger individual (Strandell, 1964a; Grimby and Saltin, 1966; Kilbom and Åstrand, 1971). Factors contributing to this aging effect include poorer myocardial perfusion, lesser cardiac compliance, and poorer cardiac contractility. There is a slowing of tension development in cardiac muscle due to reduced ATPase activity (Albert et al., 1967), and impaired coordination of the contractile process (Starr, 1964). After 40 years of age, an ever increasing proportion of the apparently healthy population shows abnormalities of the exercise electrocardiogram, particularly a horizontal or downward-sloping displacement of the ST segment (Fig. 10.22). In male subjects (but less obviously in fe-

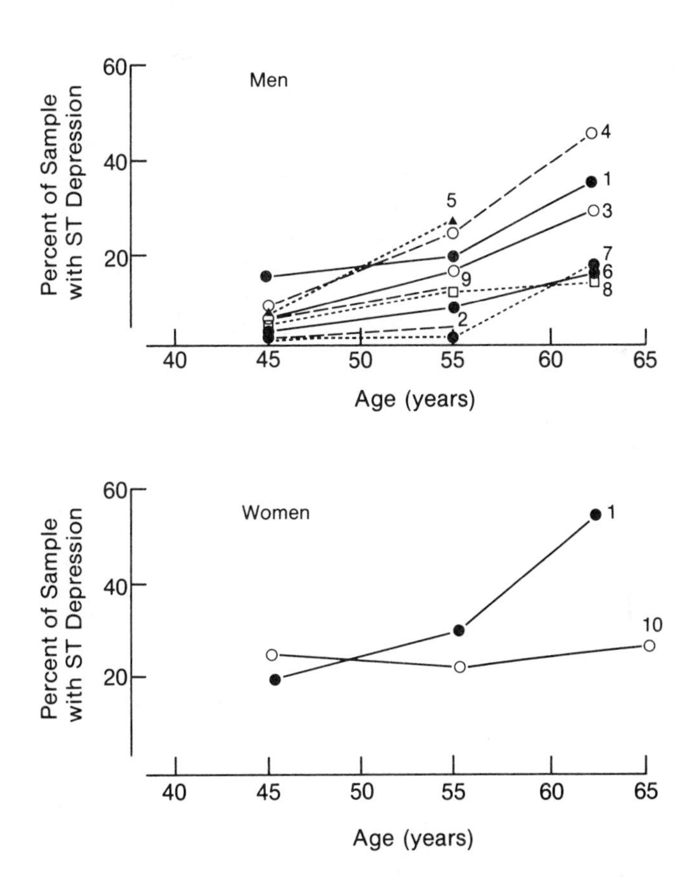

FIG. 10.22. The percentage of subjects showing depression of the ST segment of the electrocardiogram during or following maximum or near-maximum exercise. (For details of references, see Shephard, 1978b.) 1. I. Åstrand (1965); 2. Buzina et al. (1972); 3. Cumming et al. (1972); 4. Doan et al. (1965); 5. Goldbarg et al. (1970); 6. Holmdahl & Ingelmark (1949); 7. Lester et al. (1967); 8. Li et al. (1967); 9. Taylor et al. (1970); 10. Cumming et al. (1973). (See Shephard 1978b for nos. 4,5,6,8, and 9 references.)

males), this sign is associated with an increased risk of myocardial infarction and sudden death. Given a restricted stroke volume and a reduction of maximum heart rate as noted above, the maximum cardiac output of a 65-year-old person is 17–20 l • min^{-1}, some 20–30% smaller than in a young adult. Anemia is a common finding in elderly people admitted to institutions. This sometimes has a pathophysiological basis (gastric atrophy, chronic bleeding, or fatty degeneration of red bone marrow), but often the cause is poor nutrition due to a restricted income and/or loss of interest in cooking. Elwood (1971) stressed that anemia need not occur in a healthy senior citizen; he found average hemoglobin readings of 14.5 and 14.9 g • dl^{-1} for men and 13.3 and 12.9 g • dl^{-1} for women in two community surveys of subjects older than 65 years.

The maximum arteriovenous oxygen difference tends to diminish with age. Contributory factors include (1) a reduction of physical fitness levels, (2) a reduction of arterial oxygen saturation, (3) a lowering of hemoglobin concentration, (4) a poor peripheral distribution of blood flow, (5) a loss of activity in tissue enzyme systems (Kraus, 1971; Björntorp, Fahlén et al., 1971), and (6) a large relative blood flow to the skin and viscera. Nevertheless, some studies of healthy 65-year-old subjects have found arteriovenous oxygen differences as large as 140–150 ml • l^{-1} (Niinimaa and Shephard, 1978b).

There have been no specific studies of blood flow distribution in the elderly. However, skeletal muscle weakness and a poor myocardial contractility combine to increase the likelihood of peripheral flow limitation during forceful muscular contractions. Likewise, skin blood flow tends to be larger than in a young person because of a thicker layer of subcutaneous fat and reduced sweating. The number of capillaries per muscle fiber is reduced with aging (Pářizková et al., 1971). Because of associated fiber wasting, the number of vessels per mm^2 and thus the average diffusion path from the bloodstream to metabolic sites within the mitochondria may remain unaltered, but nevertheless diffusion of metabolites is hampered by changes in the characteristics of the mucopolysaccharides in the amorphous intercellular substance (D. A. Hall, 1973).

In view of the reduced maximum cardiac output and loss of myocardial contractility, one might anticipate a reduction of maximum systolic pressure in the elderly. However, in practice these factors are outweighed by the reduced compliance of the major vessels, and the peak pressure is often greater than in a younger individual (Granath, Johnson, and Strandell, 1964; I. Åstrand, 1965; Julius et al., 1967; Kasser and Bruce, 1969; Sheffield and Roitman, 1973). Using a standard sphygmomanometer cuff, our laboratory found average maxima of 217 ± 38 mmHg (28.9 ± 5.1 kPa) in elderly men and 206 ± 32 mmHg (27.5 ± 4.3 kPa) in elderly women, compared with 180 mmHg (24 kPa) in the young adult. Both isometric and prolonged rhythmic work can be limited by the exercise-induced rise of systemic blood pressure and the resultant heavy cardiac work load. Reflex adjustments of resting blood pressure such as those that occur on standing are less efficient in an older person, due in part to a general diminution of regulatory reflexes, in part to a poor responsiveness of varicose capacitance vessels (Carlsten, 1972), and in part to a lesser level of cardiorespiratory fitness than in a younger person (Fig. 4.21).

Averaging data over several published reports, the muscle strength of a 65-year-old man is ~80% while that of a 65-year-old woman is 90% of readings for a younger person (Shephard, 1978b). Loss of muscle bulk parallels the decrease of strength, as can be seen from formal determinations of lean tissue mass (E. C. Anderson and Langham, 1959; Forbes and Reina, 1970; Novak, 1972). If wasted muscle is not replaced by fat, the overall body mass should fall with age (Shephard, 1977a). Indeed, at age 65 a man should weigh 12 kg less than at age 25 and a woman 5 kg less (Forbes and Reina, 1970).

In practice, the recorded isometric force of an old person is sometimes limited by a central inhibition of muscular contraction. The apparent loss of strength in a very elderly subject can thus be due in part to diminished expectations and enhanced fears of overexertion.

Aging is associated with an increase in both absolute and relative amounts of body fat, as can be deduced from increasing skinfold thicknesses (Wessel et al., 1963; Shephard, 1977a) and decreases of body density (Friis-Hansen, 1965; Myrhe and Kessler, 1966). Men may accumulate fat faster than women, and in some studies an initial sex difference has been reduced with aging (Skerlj et al., 1953; C. M. Young, 1965). However, in our Toronto sample, the men gained an average of 4.0 mm from the third to the seventh decade of life, while women gained 7.1 mm over the same period (Shephard, Jones et al., 1969). In both sexes, fat tends to accumulate over the lower part of the trunk (Skerlj et al., 1953), while in the women the thigh and knee regions also show substantial thickening (Table 10.18).

Skeletal Structures

In addition to loss of lean tissue from muscle, the elderly are affected by a progressive loss of organic

TABLE 10.18
Percentage Increase in the Thickness of Selected Skinfolds from 25 to 65 Years

Skinfold	Increase of Skinfold 25–65 Years (%)	
	Male	Female
Chin	+39	+67
Subscapular	+31	+77
Triceps	+12	+26
Suprailiac	+8	+59
Waist	+62	+101
Suprapubic	+111	+68
Chest	+49	+106
Knee	+37	+90

Source: See Shephard, 1978b for data sources.

matter and minerals from bone (Carlson et al., 1976). This increases susceptibility to such problems as osteoporosis, vertebral compression, thoracic kyphosis, back pain, and fracture of the bones (Rowe and Sorbie, 1963; Rodstein, 1964; Nordin, 1971). There is a suggestion that vigorous activity may arrest (Sidney et al., 1977) if not reverse (E. L. Smith and Reddan, 1976) such changes.

Integrity of the synovial joints is impaired by the aging of collagen (Hass, 1943; Grahame, 1973). The thickness of the cartilage is decreased, and frank defects appear over some weight-bearing areas.

The elderly are very vulnerable to muscle and tendon ruptures (Kilbom, Hartley et al., 1969; G. V. Mann et al., 1969). Possible contributory factors include (1) muscle stiffness due to fatigue, (2) slow relaxation of antagonists, (3) loss of elastic tissue and alterations in the structure of collagen (D. A. Hall, 1973), (4) loss of flexibility in the joints (Allman, 1974), and (5) a decrease in the capillary blood supply to the tendons (Rothman and Parke, 1965).

Coordination of Effort

Detailed reviews of age-related change in psychomotor performance are available (Welford and Birren, 1965; Burr, 1971; Brocklehurst, 1973). We may note a reduction in the field of vision, lack of accommodation and loss of visual acuity (Leighton, 1973), problems of hearing (Fisch, 1973), and impairment of balance (Exton-Smith, 1977; Overstall et al., 1977).

There is a slow reaction to sensory cues (Botwinik, 1965) due to (1) a poor signal-to-noise ratio in areas of the brain concerned with sensory processing (Botwinik, 1965; Welford and Birren, 1965), (2) deterioration of sensory receptors, (3) slowing of nerve conduction, and (4) temporal dispersion of sensory information.

Problems of memory limit the ability to acquire new skills (Brinley, 1965) and there is difficulty in carrying out sequential tasks (Talland, 1961). Pattern identification is delayed (Wallace, 1956) and there is a deterioration of performance in all tasks demanding rapid movement (Jalavisto, 1965), particularly if the skill in question has been neglected (Larivière and Simonson, 1965). Reaction and movement times are substantially slowed (Hodgkins, 1962; Szafran, 1965; G. R. Wright and Shephard, 1978a; Fig. 10.18).

The heart rate needed to attain a given "rating of perceived exertion" (RPE) declines progressively with aging, but since there is a parallel decline of maximum oxygen intake, the RPE at a given percentage of maximum aerobic power remains relatively constant (Borg and Linderholm, 1967; Sidney and Shephard, 1977c).

Hormonal Reactions to Exercise

The reduced efficiency of homeostasis in the elderly (Palmer et al., 1978) has been linked to altered endocrine function (Timiras, 1972; Weg, 1975), including an impaired ability to secrete hormones, an altered metabolism or excretion of hormones by the liver and kidney, and changes in the responsiveness of the target cells (Roth and Adelman, 1974).

Basal levels of human growth hormone change surprisingly little from childhood to old age (Frantz and Rabkin, 1965; Cartlidge et al., 1970), and the increase during prolonged exercise occurs much as in a young person (Hunger et al., 1965; Sidney and Shephard, 1978a; Fig. 8.3).

Resting plasma cortisol concentrations are not greatly affected by aging (Serio et al., 1970; Grand et al., 1971). Exercise to 80–90% of maximum oxygen intake has no effect upon cortisol readings (Sidney and Shephard, 1978a), this result being much as in younger subjects.

Noradrenaline blood levels and excretion increase with age (M. R. P. Hall, 1973; Zeigler et al., 1976). Increases of heart rate, systemic blood pressure, and associated catecholamine secretion with stressful exercise or cold exposure are also greater in the elderly than in the younger person (Palmer et al., 1978).

The secretion of androgenic hormones decreases with advancing years. Simonson, Kearns et al. (1944) and Hettinger (1961) reported increases of strength and endurance when older subjects were given a daily course of methyl testosterone.

Insulin and thyroid secretions are reduced in the elderly, but the implications for physical activity have yet to be examined.

Industrial Implications

Opinions on the occupational potential of an older employee vary in almost direct relation to the age of the person making the judgment. Much depends upon the type of work required and the criteria of satisfactory performance.

In a desk task, age brings caution, experience, and cumulative skills in the handling of people. On the other hand, creativity and initiative are lost. In neurophysiological terms, one may envisage a progressive decrease in the pool of functioning neurons, but the establishment of an increased number of functional connections. This pattern of change gives both "experience" and "rigidity" of attitudes.

With respect to tasks requiring physical effort, a worker aged 50 may have much weaker muscles than an employee of 30, and aerobic power may also have deteriorated to the point where quite light work is a fatiguing 50% of maximum effort. Nevertheless, unless he is stricken by disease, a 50-year-old man is well able to continue in his employment (Shephard, 1974b). Skill is developed, so that heavy work can be performed more economically. The oxygen cost of the heaviest tasks is reduced by performing them more slowly. Moreover, population statistics tend to give the misleading impression that all men of 50 are physically weak. In fact, fitness is better preserved in those who perform heavy work regularly. Unfortunately, the heaviest tasks in a modern factory often fall to the "odd-job" man, and the older worker may be relegated to this role when advances in technology have outdated a traditional sedentary craft. In some instances, an increase in the physical demands of employment may be unavoidable, and the gain of aerobic power promised by a formal program of physical training can then do much to minimize industrial fatigue.

Training Response

Trainability

Some investigators have suggested that the response to cardiorespiratory training is reduced in older subjects (Hollmann, 1964; Roskamm, 1967; Saltin, Hartley et al., 1969; Wilmore, Royce et al., 1970; Kilbom, 1971), particularly if the individuals concerned have not previously undertaken much physical activity (Hollmann, 1964). Others find at least as great a response as in younger subjects (Tzankoff et al., 1972; Stamford, 1973; Eisenman and Golding, 1975; Sidney and Shephard, 1978b; de Vries, 1978). Divergent conclusions reflect differences in initial levels of fitness, training methods, and methods of expressing the training response (absolute or percentage changes).

The relationship between initial maximum oxygen intake and improvement of cardiorespiratory fitness is less obvious than in younger subjects (Saltin, Hartley et al., 1969; de Vries, 1970,1971; Kilbom, 1971; Sidney and Shephard, 1978b). This may reflect (1) homogenously low fitness levels in the elderly and (2) a poor motivation of the least fit subjects toward exercise.

In sedentary old people, the threshold intensity of effort needed to initiate training may be as low as 40% of the heart rate range (de Vries, 1971), so that vigorous walking (heart rate 100–120 beats per min) for 30–60 min 3 times per week is sufficient to commence training. However, a progressive increase of work load is needed if the training response is to continue (Sidney and Shephard, 1978b). Much of the ultimate 10–20% gain of oxygen transport is realized over the first 2–3 months of conditioning, but the loss of body fat, increase of lean mass, and other favorable changes in body composition continue over at least a year of increased activity (Sidney et al., 1977).

Physiological Responses

In general, physiological responses to training are similar in type but less marked than those occurring in a younger person. One investigator reported a small increase of vital capacity (de Vries, 1970), but most observers have found no change of static or dynamic lung volumes (Barry et al., 1966; Buccola and Stone, 1975; Niinimaa and Shephard, 1978a). Any decrease of resting heart rate is small (~3–5 beats per min; Shephard and Sidney, 1979). Improvements of cardiovascular function include (1) an increase of total blood volume and total hemoglobin, but not hemoglobin concentration (Benestad, 1965), (2) a decrease of left ventricular ejection time (G. M. Adams et al., 1977), and (3) a reduction of heart rate (Barry et al., 1966; Niinimaa and Shephard, 1978b), systolic blood pressure (Barry et al., 1966; Stamford, 1973), and lactate levels (Barry et al., 1966; Suominen et al., 1977a) during submaximal exercise. However, there is little change of stroke volume or diastolic pressure (Shephard and Sidney, 1979). Some authors have seen no change in the electrocardiogram with conditioning (Siegel et al., 1970). However, Sidney and Shephard (1977b) found that endurance training induced elevation of the ST segment at rest and during recovery, with a lessening of ST depression at a given heart rate during exercise. Such a response might reflect a lessening of the leakage of potassium ions from skeletal muscles or a reduction in the work load of the myocardium secondary to myocardial hypertrophy.

Many of the training responses seen in the elderly have a regulatory rather than a structural basis. Biopsy of skeletal muscles has indicated an increase of resting glycogen concentrations, with augmented activity of aerobic enzymes and possibly anerobic enzymes (Suominen et al., 1977a,b). The percentage of slow-twitch fibers remains unchanged. Nevertheless, measurements of total body potassium indicate a gain of lean tissue (Sidney et al., 1977), and the normal loss of calcium from the bones is arrested or reversed (E. L. Smith and Reddan, 1976). Over a year of consistent endurance activity, as much as three-quarters of the adult accumulation of subcutaneous fat is lost, equivalent to a 3- to 4-mm reduction in the thickness of an average skinfold and a 20% decrease of body fat (Sidney et al., 1977). Muscle force can be augmented by specific muscular training (E. A. Chapman, de Vries, and Swezey, 1972), although it increases relatively little with exercise programs that have a cardiovascular emphasis (de Vries, 1970; Sidney et al., 1977).

Following training, Sidney and Shephard (1978a) observed an increased growth hormone response to vigorous exercise. They suggested that this might contribute to either a mobilization of body fat or the conservation of lean tissue.

Other beneficial effects of training in the elderly include improvements of balance, body mechanics, and flexibility (Katsuki, 1972; Wessel, 1975), and a reduction of anxiety (de Vries, 1978).

The Elderly Athlete

Since most training studies of the elderly have covered a period of only a few months, there is some interest in the physiological characteristics of elderly athletes. It must be stressed that (1) the athlete is a highly selected representative of his community, (2) the proportion of athletes and thus the degree of selection increases with age, and (3) the vigor of participation in sport often diminishes with age. Former athletes may become less active than their supposedly sedentary counterparts by the time that they reach middle age (Montoye, Van Huss et al., 1957).

The average age of a successful athlete is influenced markedly by the relative requirements of a given event in terms of agility, physiological power, and experience. Activities such as roller skating, gymnastics, and speed events are performed best in the late teens. Endurance runners and cyclists, footballers, and tennis players reach their peak in the middle 20s, and a skillfull soccer player may remain a useful member of a professional team until his early 40s. Sports such as golf and bowling make limited demands on the physiology of the body, and they are performed best in the early 30s (Lehman, 1953; Shephard, 1978b; Hebbelinck, 1978).

While ex-athletes become heavier than their contemporaries (Montoye et al., 1957), those who continue with their sport avoid middle-age obesity (Saltin and Grimby, 1968; Pollock, 1974; Kavanagh and Shephard, 1977b; Asano et al., 1978). Some authors find lean tissue and a large heart volume are well preserved in elderly athletes (C. T. M. Davies, 1972; Pollock, 1974; Kavanagh and Shephard, 1977b), but others have described a gradual reduction of lean mass even in those who claim to be continuing with their sport (Saltin and Grimby, 1968).

The initial maximum oxygen intake is higher, and the age-related loss of oxygen transport (0.3–0.5 $ml \cdot kg^{-1} \cdot min^{-1} \cdot year^{-1}$) seems somewhat smaller in athletes than in the general population (Dill et al., 1967; Dehn and Bruce, 1972; Pollock, 1974; Kavanagh and Shephard, 1977b; Asano et al., 1978). However, this last finding probably reflects differences in the age-related decline of daily activity between the athletes and the general population rather than a true retardation of the aging process.

Evidence concerning the exercise electrocardiogram is conflicting. Holmgren and Strandell (1959) reported a high frequency of ST segmental abnormalities in former competitors, but other investigators found a normal or even a low incidence of abnormal records (Pyörälä, Karvonen et al., 1967; Saltin and Grimby, 1968; Kavanagh and Shephard, 1977b). Such discrepancies probably relate to differences in the training intensities sustained as athletes become older. With the possible exception of continuing cross-country skiers (Karvonen et al., 1974), there is little evidence that athletes live longer than the general population (Yamaji and Shephard, 1978).

Exercise and Rate of Aging

A number of longitudinal studies have compared athletes who persisted with their training against the general population or former athletes who reverted to a sedentary pattern of life. Some of these reports (for example, Hollmann, 1965; Dehn and Bruce, 1972) concluded that aging proceeds more slowly in the continuing athlete. However, the rates of deterioration of aerobic power cited by these authors (0.70 and

0.56 ml • kg^{-1} • min^{-1} • $year^{-1}$, respectively) were much as in other cross-sectional studies of the general population. We must thus conclude that the rate of loss was inflated in their supposed control series, perhaps by an increase of body mass and/or a decrease of habitual activity. Certainly, most data suggest that the loss of aerobic power proceeds at a similar rate in sedentary and athletic populations. Nevertheless, the continuing athlete has two practical consolations: (1) his effective rate of aging would become faster than that of a sedentary person if he compounded the normal biological deterioration with the effects of a cessation of regular activity and (2) while his physical condition is maintained, the margin of aerobic power for meeting emergencies remains larger than that of a sedentary contemporary.

B. O. Eriksson, Engström et al. (1971) suggested that former athletes maintain their large heart volume. This is probably true for the first few years after training has ceased. However, Saltin and Grimby (1968) noted that the heart volume of 45-year-old former orienteers had dropped to 835 ml (11.1 ml • kg^{-1}) compared with 1047 ml (15.0 ml • kg^{-1}) in those who had remained active.

CHAPTER 11
Training the Energy Transducers

Concepts of Fitness
Constitution and Training
Responses to Training
Types of Training
Detraining and Bed Rest

CONCEPTS OF FITNESS

General Considerations

Concepts of fitness are reviewed elsewhere (Shephard, 1977a). For our present purpose, we may note that fitness is highly task-specific. Energy transducers well fitted to one pattern of working are quite inappropriate for activity that requires some other pattern of energy release. Distinction must thus be drawn between tasks that require, respectively, a high degree of coordination and psychomotor skill, explosive force, muscular endurance, cardiorespiratory endurance, and so on. To develop fitness for any one category of task, specific training is required, and progress must be measured by an appropriate battery of tests (Table 3.12).

The type of fitness sought by an individual is influenced by age, sex, occupation, personality, and cultural factors. In childhood, fitness is valued mainly in the context of ability to perform complex motor acts requiring speed, coordination, and agility. A young man commonly seeks endurance for athletic feats, but an older person wishes to control obesity, avoid a heart attack, or even improve tolerance for the physical duties of daily life. Women show little correlation between perceived fitness and maximum oxygen intake (D. A. Bailey, Shephard et al., 1974); it has thus been suggested that they perceive fitness in terms of items such as a good posture, a good figure, and a good body carriage.

Many athletic events are completed within a few seconds. However the typical activities of an average citizen last from 1 to 60 min. Industrial shifts have a nominal length of 4 or 8 hr, but there are interruptions for items such as conversation, a cigarette, or repair of machinery (Shephard, 1974b), so that in practice few occupations demand more than an hour of continuous physical effort. Muscular strength and endurance, mechanical efficiency and body mass all have some influence on performance over the first hour of vigorous exercise, but the main factor limiting effort of this duration is the oxygen conductance (Shephard, 1971a) of the cardiorespiratory system. A training regimen designed to increase maximum oxygen intake and thus cardiorespiratory fitness is also optimal for the control of obesity and the possible prevention of cardiorespiratory disease (Shephard, 1977a).

A physically fit individual differs from someone who is unfit in at least three respects: (1) he can achieve a higher maximum rate of working, (2) physiological variables such as heart rate and respiratory minute volume show a smaller displacement from their resting values at any given rate of working, and (3) the same variables return more rapidly to the resting state after activity has ceased. Depending on the category of fitness that has been developed, there may also be molecular and structural adaptations leading to altered resting values for such variables as heart rate and stroke volume. Cardiorespiratory fitness has been defined (Shephard, 1977a) as the "ability of a man to maintain the various processes involved in metabolic exchange as close to the resting state as is mutually possible during the performance of a strenuous and fully learnt task for moderate time (1–60 minutes), with a capacity to reach a higher steady rate of working than the 'unfit,' and to restore promptly after exercise all equilibria which are disturbed."

Fitness for Brief Activity (0–1 Min)

Objective tests sampling individual determinants of performance during brief activity have been discussed in previous chapters.

Reaction Times

Reaction times, both simple and complex, are readily measured by standard devices capable of timing events to 1 msec or better.

Static Strengths

Static strengths are usually assessed by dynamometers and/or tensiometers (Tables 4.6, 4.7, 4.9). Where there is doubt concerning motivation (for example, in compensation cases), such information can be supplemented by electromyography (Fig. 3.8), soft-tissue radiography, and calculations of lean body mass per unit of stature. Isometric endurance is expressed as the time for which a specified fraction of maximum force can be sustained.

Dynamic Strength and Related Variables

Dynamic strength can be tested by a simple system of weights and pulleys, or by more precise isokinetic equipment that resists contraction while rotating at a predetermined speed (Lesmes et al., 1978). The explosive force can be assessed roughly from a jump-and-reach test (Table 4.8) but is better determined as the reaction recorded from a force platform (Chap. 3). Dynamic endurance is estimated from the period for which an isokinetic apparatus can be operated or the period of time for which a given set of weights can be raised and lowered. The endpoint of both isotonic and isometric effort is largely psychological, depending on the motivation of the subject and his reactions to increasing acidity within the active muscle fibers.

Flexibility can be assessed under both passive and dynamic conditions, using various types of goniometer. The correct *posture* is important in many brief activities (H. H. Clarke, 1979). *Speed* and *accuracy* are sampled by a wide range of psychomotor devices, such as pursuit meters. *Coordination* and *steadiness* are measured by tapping tests and various forms of tiltboard such as the stabilometer.

Motivation

Motivation plays a large role in peak performance, irrespective of the duration of activity. It is quite difficult to evaluate. One approach has been to incorporate an element of performance into the test itself, leaving speed or loading to the subject's choice. In an industrial situation calling for 8 hr of endurance-type activity, a rate of 40% of maximum oxygen intake is commonly selected (A. L. Hughes and Goldman, 1970), but a poorly motivated person will choose a slower and more variable pace. The rate of working adopted by the individual can be related to the appearance of effort, as gauged by an experienced observer (Moores, 1970), or note can be taken of physiological reactions such as the heart rate and blood lactate at the end of the working day. In the laboratory, an alternative approach is to require subjects to work at a fixed percentage of aerobic power, making subjective ratings of effort on a psychophysical scale such as that of Borg (1971; Fig. 11.1). The Borg rating of perceived exertion (RPE) has been widely used by physiologists over the past decade. The scores reported are related to the duration D of effort ($D^{<1.0}$), the load L ($L^{\sim 1.4}$), and (at constant power output) to the inverse of pedal speed (Cafarelli, 1977; Cafarelli et al., 1977; Löllgen et al., 1977; Moffatt and Stamford, 1978). However, if other factors are equal, the determinant of RPE becomes the relative power output (percentage of maximum oxygen intake; Sargeant and Davies, 1973; Sidney and Shephard, 1977c).

Oxygen Debt

Traditionally, the study of *anerobic "capacity"* and *power* has involved the measurement of oxygen consumption throughout 15–30 min of recovery from vigorous exercise. The cumulative excess oxygen consumption has then been calculated relative to initial resting readings and interpreted in terms of repayment of an oxygen debt (Fig. 2.11). By use of "curve-stripping" techniques, it is mathematically possible to divide a semilogarithmic plot of the repayment process into two linear components corresponding, respectively, with repayment of alactate and lactate debts (Chap. 2; di Prampero, 1971a). Unfortunately, the rates of repayment thus calculated do not always coincide

Points	Sensation
6	
7	very very light
8	
9	very light
10	
11	fairly light
12	
13	somewhat hard
14	
15	hard
16	
17	very hard
18	
19	very very hard
20	

FIG. 11.1. A psychophysical scale for use in rating perceived exertion (Borg, 1971).

with the rates at which debts can be incurred. An alternative approach to testing is thus to examine the power and capacity of the forward reactions. The power of the alactate component can be evaluated from a staircase sprint (Margaria, 1966; Fig. 2.3). The subject runs swiftly up a flight of stairs. Some 3 sec is allowed for acceleration (15 standard domestic steps or 30 horizontal m), and the rate of ascent over the next 2 sec (a further 10–15 steps) is timed by means of light beam relays. The physical power developed is readily determined from body mass and the speed of ascent, and, depending on assumptions about mechanical efficiency, an equivalent oxygen transport can also be calculated. The total stored energy ("alactate capacity") can be found by integrating power output over the 8-sec period needed to exhaust this resource. Estimates of both power and capacity per unit of body mass are ~30% larger in a well-trained athlete than in an average young man. The advantage of the athlete can probably be explained by (1) greater familiarity with sprinting, (2) greater flexibility, and (3) a greater muscle/fat ratio, without invoking a more specific increase in the alactate power and capacity of unit muscle mass.

The maximum rate of lactate production has been estimated by plotting the blood lactate readings immediately after exercise against the intensity and duration of the effort undertaken (di Prampero, 1971a). The critical assumption with this approach is that the metabolism of lactate is slow. Blood samples collected 3–5 min after exercise can then be considered as representative of the lactate produced and distributed throughout body water. According to the Milan laboratory, results obtained by this approach (a power equivalent to an oxygen transport of ~68 ml • kg^{-1} • min^{-1}, exhausted in 40–50 sec, and a capacity equivalent to an oxygen store of 45 ml • kg^{-1}) are comparable with those yielded by analysis of oxygen debt repayment (di Prampero, 1971a). Other methods of assessing anerobic capacity include estimation of the total work performed during a 30–60-sec bicycle ergometer sprint (Tuttle, 1950; V. Katch, Weltman et al., 1977; Bar-Or, 1978; Bar-Or and InBar, 1978; Table 11.1), and timing of an uphill treadmill run that exhausts the average subject in ~45 sec (Niinimaa, Wright et al., 1977). Capacity is related strongly to the maximum tolerated intramuscular lactate concentration. The well-trained sprint competitor thus gains some advantage from (1) a greater volume of active muscle, (2) a higher proportion of fast-twitch, glycolytic fibers, (3) greater

TABLE 11.1
Correlations Between 30-sec Power Output on Bicycle Ergometer and Other Tests of Anerobic Capacity

Alternative Test	Coefficient of Correlation
Maximum O_2 debt	0.86
300-m sprint	0.86
25-m swim	0.87–0.90
Margaria's staircase sprint	0.79
40-m sprint	0.86
Proportion of fast-twitch fibers	0.60–0.75

Source: After Bar-Or, 1978.

concentrations of rate-limiting glycolytic enzymes in the active fibers, and (4) greater motivation, leading to activation of a higher proportion of the total muscle mass.

If repeated bursts of anerobic activity are required (as in a soccer or a hockey game; H. J. Green et al., 1978a,b), glycolysis may be limited by intramuscular glycogen reserves. However, this is most unlikely in a single anerobic effort. Let us suppose that 20 kg of muscle is called into action by a treadmill sprint. Given an initial glycogen content of 1.5 g • dl^{-1} (Hultman, 1971), a total reserve of 300 g can be broken down to pyruvate, the anerobic process yielding ~390 kJ of energy, equivalent to an oxygen consumption of 18.6 l, or 248 ml • kg^{-1} in a 75-kg man. This is ~6 times the observed capacity of the glycolytic reaction. Thus, unless we assume an improbably small active muscle mass of 3–4 kg, exhaustion cannot be blamed upon a general depletion of glycogen stores. Nor is a selective depletion of glycogen particularly likely; fast-twitch fibers are recruited preferentially for sprint-type activities, and such fibers have an above-average content of glycogen (Chap. 4).

Fitness for Moderate-Duration Activity

Resting Measurements

HEART RATE. Endurance athletes have a much slower heart rate than sedentary individuals (Chap. 6). The measurement of resting pulse rate might thus seem a useful test of fitness and, indeed, an indirect assessment of cardiac stroke volume. Unfortunately, a slow heart rate can also be a sign of disease. Even if this possibility is ruled out by appropriate tests, the heart rate remains a somewhat labile measurement, being influenced by such factors as emotional disturbances, effective environmental temperature, posture, recent physical activity, consumption of cigarettes, beverages, and food. There is also a diurnal rhythm of resting heart rate that parallels the daily variation in body temperature (C. T. M. Davies and Sargeant, 1975a), the highest readings being observed in the late afternoon and evening, and the lowest values in the small hours of the morning (Chap. 14).

Many of these sources of variation can be minimized or avoided if the heart rate is measured during sleep. Values may be high for the first 2 or 3 hr after retiring (Shephard, 1967c), particularly if the subject has been physically active during the evening, but readings obtained in the early hours of the morning are quite consistent if brief periods of disturbing dreams can be avoided. Devices such as portable tape recorders (Holter, 1961) and electrochemical integrators of the electrocardiogram (H. Wolff, 1966) may thus find increasing use in the assessment of cardiorespiratory fitness.

RESTING PULSE PRESSURE. The magnitude of the systemic pulse pressure depends upon cardiac stroke volume, contractility, and the elasticity of the aortic vessels. Thus, if one assumes a value for aortic elasticity, a qualitative estimate of stroke volume can be derived from the simple measurement of pulse pressure (Wezler and Böger, 1937), while if the pulse wave data are combined with recordings of heart sounds and the electrocardiogram various indices of myocardial contractility can be calculated (Chap. 6).

Unfortunately, aortic elasticity is not a constant; indeed, it is influenced by many of the same variables that disturb the resting pulse rate, including anxiety, recent exercise, and arteriosclerotic disease. Information obtained from pulse wave recordings is thus of relatively limited value in assessing cardiorespiratory fitness (Chap. 6; T. K. Cureton, 1947; B. D. Franks, 1969; G. R. Cumming and Borysyk 1971; J. F. Joyce, 1977).

POSTURAL TESTS. Both the Crampton index and the Schneider test are based on the changes of heart rate and blood pressure that occur when a subject is moved rapidly from the supine to the vertical position (Crampton, 1905; Schneider, 1920). Unfit individuals show a marked increase of heart rate and a fall of blood pressure, while in a well-trained endurance athlete the systemic blood pressure rises slightly, with little change of heart rate. The differences of postural response found in the well-trained individual reflect partly a larger total blood volume (Holmgren, 1967b) and partly a more rapid reflex adjustment of the capacity vessels to gravitational stress (J. P. Henry, 1951).

Postural tests are rarely included in current fitness assessments, partly because pulse rate and blood pressure are subject to many extraneous influences and partly because individuals with a poor postural response also have a poor score on standard cardiorespiratory fitness tests. The syndrome of a poor postural response and a limited tolerance for physical work is sometimes dignified by the name neurasthenia (Holmgren, 1967a).

OTHER RESTING TESTS. *Breath holding* (Kallfelz, 1962) may be of interest in assessing fitness for specific activities such as swimming underwater;

however, it is markedly influenced by motivational factors and is no longer used in standard assessments. The same is true of the *Flack test* (the time for which a column of mercury can be blown to a height of 40 mm; Flack, 1920).

The *vital capacity* was once considered an important measure of fitness (Dreyer, 1920). More recent analysis has played down the role of ventilation in oxygen transport and it would be surprising if there were a strong relationship between endurance performance and lung volumes. On the other hand, a large vital capacity aids the competitive performance of a middle-distance swimmer by increasing his buoyancy (Shephard, Godin, and Campbell, 1973; Shephard, 1978a). Vital capacity may also exceed the predicted value if the individual under test has well-developed chest muscles (Sidney and Shephard, 1973), as in a rower or a white-water paddler (Chap. 14). Given strong thoracic muscles, the initial pulmonary blood volume can be reduced by a Valsalva-type maneuver (forced expiration against a closed glottis) and more force can also be exerted in compression of the thoracic cage as residual lung volume is approached.

Maximum Effort Tests

The least controversial tests of fitness for moderate-duration activity involve maximum effort to exhaustion. Nevertheless, such procedures have their own peculiar problems. Motivation for all-out effort can be difficult, particularly in young children and elderly subjects, and the risk of precipitating a cardiac emergency may be increased by the use of maximum effort. McDonough et al. (1969) estimated that ventricular fibrillation developed in 1 of 3000 maximum tests; however, about half of the population evaluated had known heart disease, and the residue included many coronary-prone individuals enrolled in the Seattle Heart-Watch program. Rochmis and Blackburn (1971) set the chance of provoking ventricular fibrillation at 1 in 10,000 tests, with no difference in the statistics for maximal and submaximal procedures. All of the Rochmis and Blackburn series who developed fibrillation had previously documented cardiac disease. More recent calculations (Shephard, 1975b) suggest that the risk of a cardiac emergency is too low to calculate accurately for a person without known cardiac disease.

ENDURANCE TIME. Measurements of endurance and maximum performance were popular before automated analyzers of oxygen consumption became widely available. Fitness might be assessed from the endurance time when running on a treadmill at a speed and slope normally exhausting in 5–15 min (Table 11.2) or the maximum amount of work performed on a bicycle ergometer over a specified time, such as 10 min.

Exhaustion in such tests is determined not only by maximum oxygen intake, but also by the ability of the circulation to perfuse active muscles, the tolerance of the tissues toward lactate accumulation, and the general reactions of the body to a variety of unpleasant sensations.

Very large improvements of endurance time may be brought about by an appropriate training program. However, this does not imply an equal gain of cardiorespiratory fitness. A small gain of maximum oxygen intake with some improvement of muscle strength greatly reduces lactate accumulation in near-maximum effort, while habituation increases tolerance of unpleasant sensations. A 1200% improvement of endurance time may thus coincide with only a 10–20% gain of oxygen transport.

HARVARD STEP TEST. The usual form of Harvard step test (Brouha, 1943) requires ascent of a 20-in. (50.8-cm) bench 30 times per min to exhaus-

TABLE 11.2
Treadmill Performance and Corresponding Oxygen Cost When Children and Adults Undertake Progressive Treadmill Exercise According to the Bruce Protocol

			Oxygen Cost ($ml \cdot kg^{-1} \cdot min^{-1}$)			
			Children Age 11–12		Adults	
Stage	Speed ($km \cdot hr^{-1}$)	Slope (%)	Boys	Girls	Men	Women
I	2.7	10	18.2	17.9	17.4	16
II	4.0	12	24.6	24.1	24.8	23
III	5.4	14	36.8	33.5	34.3	33
IV	6.7	16			43.8	45

Source: From G. R. Cumming, 1977,1978b.

tion (maximum test duration 300 sec). About a third of healthy young men complete the full test period. The pulse rate is counted 1–1½, 2–2½, and 4–4½ min after exercise, and an empirical fitness index is calculated according to the formula

$$\text{Fitness index} = \frac{\text{test duration (sec)} \times 100}{2 \times \text{pulse sum } (1\text{–}1\tfrac{1}{2}) + (2\text{–}2\tfrac{1}{2}) + (4\text{–}4\tfrac{1}{2})} \quad (11.1)$$

The timing of the three pulse readings is intended to give a good description of the pulse recovery curve. An index of less than 55 indicates poor fitness, 65 is average, 80–90 is good, and > 90 is excellent.

The rate of recovery of the pulse is clearly faster in a fit than in an unfit subject, but the form of the curve is influenced by many factors, including:

1. Intensity of exercise stress relative to the maximum oxygen intake of the individual
2. Age of the subject
3. Rise of body temperature during exercise and rate of cooling during recovery
4. Extent and rate of repayment of any oxygen debt
5. Depletion of blood volume during exercise, posture adopted after the test, and rate at which blood volume is restored
6. Extent to which exercise tachycardia is mediated by the cerebral cortex
7. Extent of metabolic rate elevation during recovery, reflecting such factors as the liberation of catecholamines, a continuing high body temperature, and increased work of breathing

The Harvard protocol is still used sometimes for the evaluation of Olympic athletes. Scores do not correlate well with the directly measured maximum oxygen intake (Miyamura et al., 1975), but the Harvard score may give a better prediction of endurance performance than does aerobic power (Ishiko, 1967). One reason is that the Harvard test also measures motivation. The gross metabolic cost of climbing the 20-in. step (45–50 ml • kg^{-1} • min^{-1} STPD) is high even for a young man, and many subjects cannot complete the required 5-min period. A lower step (17 in., 43.2 cm) is commonly used for testing girls and older men, while if subjects are superbly fit, effort can be increased 20–25% by wearing a 20-kg pack (*Harvard pack test*; Ladell and Kenney, 1955).

Maximum Oxygen Intake. Most authors are now agreed (Shephard, Allen et al., 1968a; Weiner and Lourie, 1969; Shephard, 1977a) that the best approach to the assessment of cardiorespiratory fitness is the direct measurement of maximum oxygen intake during exhausting work such as uphill treadmill running, stepping, or pedaling a bicycle ergometer (Chap. 3).

The internationally recommended technique commences with a warm-up at ~70% of the individual's presumed maximum effort, then jumps to 90–100% of maximum, with further 5% increases of work rate every 2 min until exhaustion is reached (Weiner and Lourie, 1969). If the subject is well motivated, the oxygen consumption either fails to show an increase or actually declines with further augmentation of loading (Fig. 11.2). In order to define a "plateau," the oxygen cost of three successive work loads must agree to within 2 ml • kg^{-1} • min^{-1}, or 0.15 l • min^{-1} STPD. Other subsidiary criteria of a good maximum effort include (1) a heart rate close to the age-related maximum (Chap. 6), (2) a respiratory gas-exchange ratio of 1.15 or more, and (3) a high arterial lactate level (11–16 mmol • l^{-1} in a young man, possibly reduced to 9 mmol • l^{-1} in a child and 7 mmol • l^{-1} in an old person; see Chap. 10). With young and athletic adults, there is little difficulty in reaching a plateau of oxygen consumption, and results show a day-to-day variation of only 4–5% (H. L. Taylor, Buskirk, and Henschel, 1955; Wyndham, Strydom et al., 1959; T. E. Graham and Andrew, 1973; G. R. Wright, Sidney, and Shephard, 1978). However, the body must be driven into a substantial oxygen debt before a plateau is demonstrated, and this may be difficult to realize in children or older subjects (Chap. 10).

The highest values of maximum oxygen intake are obtained by uphill treadmill running, with results 3% smaller during stepping, 7% smaller during bicycle ergometer exercise, and 20–30% smaller during arm ergometry (Table 3.2). The low results with arm exercise are due partly to a reduction of central blood volume (a consequence of pooling in the leg veins) and partly to problems of perfusing a relatively small mass of active muscles. In a typical uphill treadmill running experiment the subject's face becomes blue and then ashen grey as exhaustion is approached. He may complain of nausea and dizziness, and is seen to be acutely breathless, with impaired coordination and a confused response to questions. In contrast, a subject on a bicycle ergometer ceases exercise because of weakness, pain, and exhaustion localized to the active muscles (Shephard, Allen et al., 1968a).

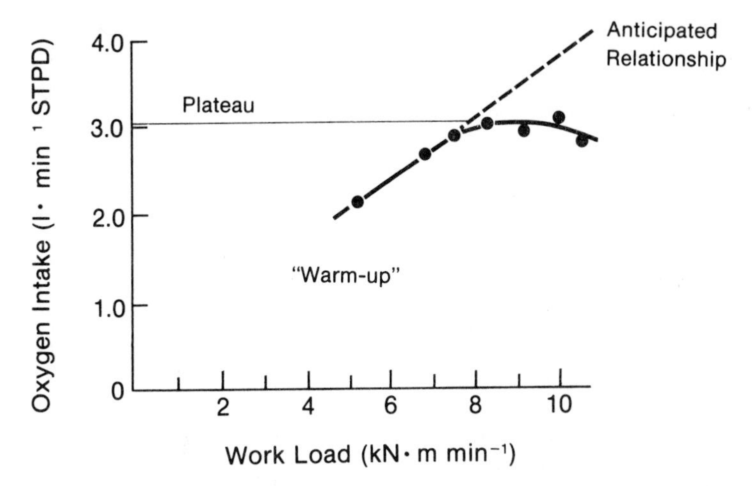

FIG. 11.2. The technique of measuring the maximum oxygen intake. After a "warm-up," the workload is increased progressively every two minutes until three successive increments change the oxygen consumption by a total of less than 2 ml • $kg^{-1}min^{-1}$ STPD or 0.15 1 • min^{-1} STPD.

Details of measurement techniques are given elsewhere (Weiner and Lourie, 1969; K. L. Andersen, Shephard et al., 1971; Shephard, 1977a,1978a). In general, results for treadmill exercise vary little with protocol (Binkhorst and Leeuwen, 1963; Falls and Humphrey, 1973; McArdle et al., 1973; Stamford, 1976; Shephard, 1978a), although D. H. Paterson and Cunningham (1978) noted that in children walking gives a lower heart rate, poorer reliability, and lower maximum oxygen intake than jogging or running. There is some evidence that the traditional warm up at 70% of maximum effort minimizes the risks of electrocardiographic abnormalities (Barnard, Gardner et al., 1973) and tendon injuries. Furthermore, the maximum oxygen intake attained is a little larger if a thorough warm-up is practiced (H. L. Taylor, Buskirk, and Henschel, 1955; Pirnay, Petit et al., 1966; B. J. Martin et al., 1975).

Normal values for male and female subjects are summarized in Figs. 3.19 and 3.20 (Shephard, 1966e). Data from many parts of the world show remarkably consistent figures of 48–50 ml • kg^{-1} • min^{-1} for nonathletic boys and young men, but by the age of 60 as much as 40–50% of the initial aerobic power is lost (Fig. 10.21). Some studies from Sweden (for example, P. O. Åstrand, 1952) have shown higher readings, particularly for young subjects, but this may be an artifact of sample selection (Shephard, 1978a). Women have lower scores than men except in old age, where values for both sexes are similar. Both male and female athletes have higher scores than sedentary individuals, differences being particularly marked for competitors in endurance sports (Shephard, 1978a; Table 3.5). The advantage of the sportsman reflects both training and selection (see below).

Submaximal Test Principles

A variety of submaximal procedures have been proposed for the assessment of cardiorespiratory fitness. The majority are based upon the supposed linear relationship of heart rate and oxygen consumption between 50 and 100% of maximum oxygen intake; in effect, the graph is extrapolated to the theoretical maximum heart rate of the individual, and the corresponding oxygen intake is then read from the abscissa. The postulated relationship presupposes either a constant stroke volume and arteriovenous oxygen difference, or equal and opposing changes in these two variables as power output is increased. A constant stroke volume is more likely in the young than in the elderly; the latter show a reduction of stroke volume and consequent alinearity of the heart-rate/oxygen-consumption line at high rates of working (Sidney and Shephard, 1977a; Niinimaa and Shephard, 1978b).

An important source of difficulty with most submaximum tests is that the line relating oxygen consumption and heart rate can be displaced to the left by extraneous influences such as anxiety (Shephard, 1969), a high environmental temperature (C. G. Williams, Bredell et al., 1962; Rowell, 1974), dehydration (Saltin, 1964a,b), a recent meal (N. L. Jones and Haddon, 1973), ingestion of alcohol (Blomqvist, Saltin et al., 1970), pyrogen-induced fever (Grimby and Nilsson, 1963), and prolonged heavy exercise (Saltin, 1964a,b). The maximum heart rate is also modified by age (Chap. 6), sex, altitude (Buskirk, Kollias et al., 1967a,b), disease (for example, the sick sinus syndrome), semistarvation (Keys, Brozek et al., 1950), and possibly fitness for sports similar to the test activity (for example, treadmill testing of a track competitor; Koeslag and Sloan, 1976; G. R. Wright, Sidney, and Shephard, 1978). Last, some authors have described an "oxygen consumption asymptote"—oxygen consumption rising by 0.2–0.5 l • min⁻¹ STPD with no matching increase of heart rate (Chap. 6; Maritz et al., 1961).

For these several reasons, predictions show a substantial scatter about the true maximum oxygen intake (Table 11.3). Some authors have reported systematic errors as large as 25% (Rowell, Taylor, and Wang, 1964; C. T. M. Davies, 1968), but this probably reflects a combination of anxiety (Rowell's subjects were undergoing cardiac catheterization) and lack of habituation to the test environment (Shephard, 1969). A more usual finding is a systematic discrepancy of 0–10% from the directly measured value, with a superimposed standard deviation of ~10% (Shephard, Allen et al., 1968b). A single prediction tells an individual relatively little about his cardiorespiratory fitness, since the 95% confidence limits of a 40 ml • kg⁻¹ • min⁻¹ estimate range from 32 to 48 ml • kg⁻¹ • min⁻¹. However, populations can be surveyed by submaximum tests, since random errors cancel out in a large sample. Further, predicted results for a given individual are fairly reliable (coefficient of variation 4–5%, test/retest correlation as high as 0.97). Discrepancies between submaximum and maximum data remain consistent in a given individual (Wright et al., 1978; Fig. 11.3), so that predicted results can be used to follow changes in an individual's physical condition.

ÅSTRAND NOMOGRAM. The nomogram of I. Åstrand (1960) is perhaps the best known of several submaximum prediction methods. It is based on single observations of heart rate and oxygen consumption obtained in the sixth min of bicycle ergometer or stepping exercise. The heart rate must lie within the range 125–170 beats per min. It is further assumed (P. O. Åstrand and Ryhming, 1954) that the heart rate at 50% of aerobic power is 128 in men and 135 in women, while the corresponding rates at 100% of aerobic power are 195 and 198. The formulas for calculation of maximum oxygen intake might thus seem

For men:

$$\dot{V}_{O_2(max)} = 2\dot{V}_{O_2(observed)} \times \frac{195 - 128}{f_h - 61} \tag{11.2}$$

where 61 is a theoretical heart rate at zero metabolism,

For women:

$$\dot{V}_{O_2(max)} = 2\dot{V}_{O_2(observed)} \times \frac{198 - 135}{f_h - 72} \tag{11.3}$$

Such formulas give similar answers to the published nomograms. However, the latter now incorporate certain "adjustments," and the formulas on which the current versions are based seem to be

For men:

$$\dot{V}_{O_2(max)} = 2\dot{V}_{O_2(observed)} \times \frac{193 - 128}{P - 63} \tag{11.4}$$

TABLE 11.3
Relative Accuracy of 3 Methods for the Prediction of Maximum Oxygen Intake*

Prediction Method	"Naive" Subjects: Step Test	Bicycle Ergometer	Treadmill	Subjects after 5 Days Habituation: Step Test	Bicycle Ergometer	Treadmill
Åstrand nomogram	−0.18	+0.30	−0.27	+0.16	+0.33	−0.04
	±0.46	±0.31	±0.37	±0.25	±0.35	±0.39
Margaria nomogram	−0.09	+0.18	−0.35	+0.12	+0.13	−0.39
	±0.39	±0.27	±0.67	±0.28	±0.35	±0.61
Maritz et al.	−0.03	+0.19	−0.17	+0.12	+0.18	−0.22
extrapolation	±0.36	±0.28	±0.57	±0.28	±0.34	±0.44

*Mean ± S.D. of difference from treadmill maximum oxygen intake, experiments on 24 young men with directly measured $\dot{V}_{O_2(max)}$ of 3.81 ± 0.76 l • min⁻¹ STPD.

Source: Based on the results of the I.B.P. Working Party (Shephard, Allen et al., 1968b).

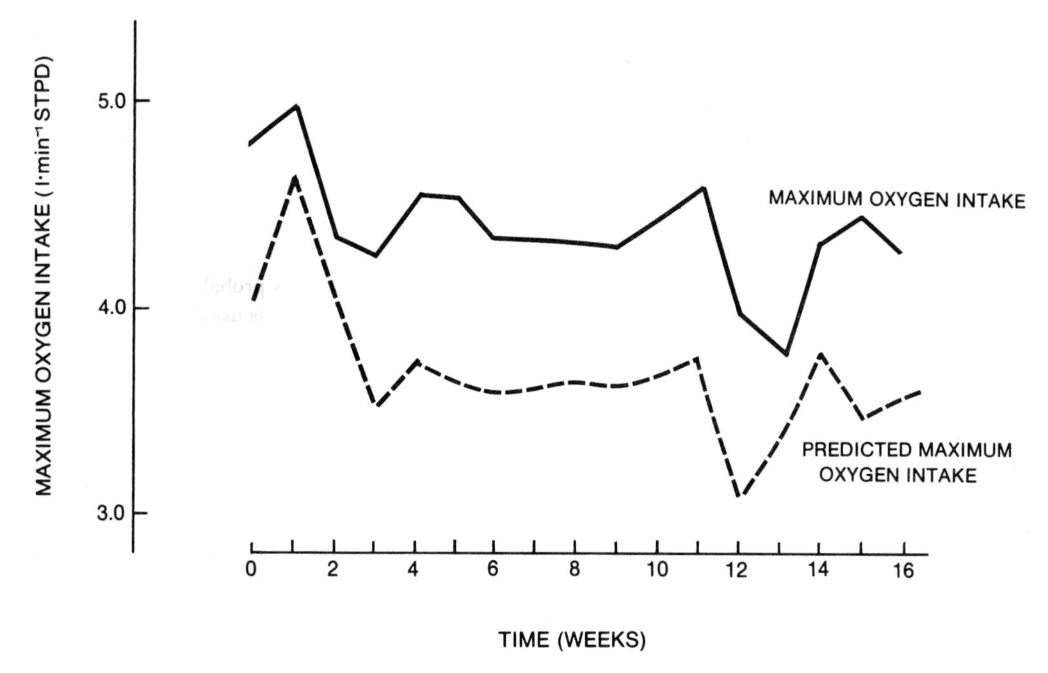

FIG. 11.3. Repeated comparisons of directly measured and predicted maximum oxygen intake in a subject with large systematic discrepancy. Note that the discrepancy remains relatively constant from week to week. (After G. R. Wright, Sidney & Shephard, 1978.)

For women:

$$\dot{V}_{O_{2(max)}} = 2\dot{V}_{O_{2(observed)}} \times \frac{203 - 138}{P - 73} \quad (11.5)$$

The original populations tested by the Åstrands were young and relatively athletic physical education students, so that the question of a decline in heart rate with age did not arise. Subsequently, the procedure was extended to older age groups, and it became necessary to introduce an empirical correction, ranging from 1.00 at age 25 years to 0.65 at 65 years (Table 11.4). A certain mystique developed around this correction factor, and it was claimed to yield more accurate predictions of maximum oxygen intake than use of a maximum heart rate appropriate to the subject's age (Von Döbeln, Åstrand, and Bergström, 1967).

Field workers still find it convenient to use the Åstrand nomogram as such, but in laboratory investigations the necessary formulas are better programmed for a small desk-top computer (Shephard, 1970b). If facilities are not available for the measurement of oxygen consumption, it is possible to substitute the equivalent rate of working, assuming a net mechanical efficiency of 16% for stepping and 23% for bicycle ergometry. However, the efficiency varies somewhat with experience of the test, age (Sidney and Shephard, 1977a), and (in the case of bicycle ergometry) the rate of pedaling (Gueli and Shephard, 1976).

Other Prediction Methods. Margaria, Aghemo, and Rovelli (1965) developed a procedure based on heart rates observed during the climbing of a 40-cm step at 15 and 25 ascents per min. One useful innovation in this test was the introduction of a nomogram with three readout scales corresponding to maximum heart rates of 200, 180, and 160 beats per min.

TABLE 11.4
Empirical Age Correction Factors To Be Applied to Maximum Oxygen Intake as Predicted by the Nomogram of I. Åstrand (1960)

Age (Years)	Correction Factor
25	1.00
35	0.87
45	0.78
55	0.71
65	0.65

Wyndham and his colleagues (Maritz et al., 1961) proposed extrapolation of a linear regression based on four rates of stepping and the corresponding heart rates. Partly because Wyndham's laboratory was in Johannesburg (altitude 1830 m), he recommended that extrapolations be made to the somewhat low maximum heart rate of 180 beats per min. However, most authors who use the Wyndham procedure now extrapolate to a maximum heart rate appropriate for the population they are studying.

At first inspection, one might anticipate that the relative accuracy of the three prediction methods would vary as $\sqrt{N}$, N being the number of work rates tested. In practice, gains of accuracy with an increase of N are slight (Table 11.3). The placing of data points provides a likely explanation for this anomaly. When using the Åstrand nomogram, a high work rate can be selected, little extrapolation is required, and the error of prediction is correspondingly small. On the other hand, Wyndham's procedure is based upon four heart rates measured at quite low intensities of exercise, so that substantial extrapolation is necessary. Further, in calculating the slope of the oxygen-consumption/heart-rate line, undue account is taken of the lowest of the four data points, where physiological responses are particularly vulnerable to influence by environmental factors (Shephard, 1975c).

Issekutz and his associates (Issekutz, Birkhead, and Rodahl, 1962) used measurements of the respiratory gas-exchange ratio during submaximum exercise as a means of extrapolating work rates to maximum power output. They assumed a steady increase of R from a resting value of 0.83 to ~ 1.15 in maximum effort, with a linear relationship between the work rate and log ΔR (ΔR being the observed ratio -0.75). The maximum aerobic work rate ($\dot{W}_{max}$) was then given by

$$\dot{W}_{max} = \dot{W}_{observed} \frac{(\log 0.40 - \log 0.08)}{(\log \Delta R - \log 0.08)} \tag{11.6}$$

In order to convert $\dot{W}_{max}$ to $\dot{V}_{O_2(max)}$, it was necessary to assume a figure for the mechanical efficiency of effort and to add the resting energy expenditure. The Issekutz prediction is fairly satisfactory if one is dealing with experienced young volunteers leading well-regulated lives (Shephard, 1967b), but in other groups it may be upset by anxiety hyperventilation, diets that give a resting respiratory quotient other than 0.83, and maximum values of R that differ from 1.15 (Shephard, 1975d). One possible application of the Issekutz procedure is in the evaluation of cardiac patients treated by β-blocking drugs such as propranolol. In subjects who are receiving this type of medication, there is little change of heart rate with exercise, and procedures like the Åstrand nomogram give very misleading answers.

Irrespective of the prediction technique that is to be used, a multistage test procedure can be adopted in order to reach the desired power output (K. L. Andersen, Shephard et al., 1971). A graded protocol confers an advantage of safety, since reactions are first assessed at light work rates. The test can be performed in either a progressive manner (where the rate of working is increased every 3 or 4 min) or as a series of discontinuous tests each of 5–6 min duration. Results are essentially similar with the two approaches (Shephard, Allen et al., 1968a). The discontinuous protocol is probably a little safer, but the progressive procedure is usually preferred since it conserves the time of both subject and investigator.

Some authors have advocated a "triangular" test profile, the rate of working being increased continuously or at 1-min intervals (Arstila, 1972). This technique slightly reduces oxygen consumption for a given work rate (Fernandez et al., 1974). It also gives a lower heart rate for a given oxygen consumption (and thus a somewhat greater predicted maximum oxygen intake; Shephard and Kavanagh, 1978a). Perhaps because time is not allowed for metabolites to induce coronary vasodilation, electrocardiographic ST segmental depression is somewhat greater than with a steady-state or near-steady-state procedure.

INTERPOLATED TESTS. The various submaximal procedures discussed so far have been based upon extrapolations of data. Critics of such an approach point out that the "information content" of a set of observations is reduced rather than increased by extrapolation. Is it not preferable to report interpolated or, at worst, briefly extrapolated results?

Perhaps the best known technique of interpolation is the PWC_{170}. This reports the rate of working ("physical working capacity") that can be sustained for 6 min with a final heart rate of 170 beats per min (Wahlund, 1948; Sjöstrand, 1960,1967). The main objection to the test is that the imposed stress varies with the age of the individual. In a 20-year-old subject, the effort required is quite moderate ($\sim 80\%$ of aerobic power), but by 60 years of age the final work rate may be exhausting in 1 or 2 min. For this reason, some authors also describe a PWC_{150} and even a PWC_{130} for the testing of older and hospitalized patients.

An analogous approach is to report the heart rate at an oxygen consumption of 1.0 or 1.5 l • min⁻¹ ($f_{h,1.0}$, $f_{h,1.5}$) (Cotes, 1966; Spiro et al., 1974). Again, there is the problem that the relative stress for a given level of oxygen consumption increases with aging.

In Germany, data is sometimes reported as the *Leistungspulsindex* (LPI, the slope of the heart-rate/work-output relationship) or the *Oxygenpuls* (the slope of the heart-rate/oxygen-consumption relationship). A triangular test profile is usually adopted (Müller, 1950) and, as with the two preceding tests, interpretation of the slope must change with the age of the subject. Ulmer et al. (1971) rejected the LPI on the grounds that there was little difference of score between top cyclists and untrained subjects.

The most recent approach to interpolation is the use of *target heart rates* and/or *relative work rates* (Tables 11.5 and 11.6; Shephard, 1971d,1978g). By use of an age-specific target heart rate, electrocardiograms from subjects of differing ages and differing fitness levels can be compared at the same relative intensity of effort. With the relative work rate approach, a subject is required to perform a task that is, for example, 70% of aerobic power in a typical sedentary North American of the same age. Cardiorespiratory fitness is then gauged from the time for which the task can be performed and/or the pulse rate in the first few sec of the recovery process (Table 11.7). If the pulse count is obtained in the first 10–15 sec of recovery, there is a good correlation between this count and the immediately preceding exercise heart rate (r = 0.96–0.98; Cotton and Dill, 1935; Shapiro et al., 1976), with no systematic difference between the two sets of data (D. A. Bailey, Shephard et al., 1974).

TABLE 11.5
Target Heart Rates Corresponding to ~75 and 85% of Aerobic Power

Age (Years)	Heart Rate (beats • min⁻¹) 75% $\dot{V}O_{2(max)}$	85% $\dot{V}O_{2(max)}$
20–30	160	170
30–40	150	160
40–50	140	150
50–60	130	140

Source: From Shephard, 1971d.

Performance Tests

Until recently, many physical educators assessed fitness by means of "performance test" batteries rather than by physiological measurements. Subjects were required to complete six or seven simple physical procedures (such as sit-ups, pull-ups, and sprint runs); these were chosen to sample ability in activities of brief and moderate duration (Kraus and Hirschland, 1954; American Alliance for Health, Physical Education and Recreation, 1965; Canadian Association for Health, Physical Education and Recreation, 1966).

It is now recognized that such performance test scores are strongly influenced by standing height (Shephard, Lavallée et al., 1979b), body mass (G. R. Cumming and Keynes, 1967), motivation, recent familiarity with a gymnasium (Drake et al., 1968), experience of the specific test activity (W. R. Campbell and Pohndorf, 1961), and environmental conditions (Strydom, 1978). Thus, for most purposes objective physiological tests are preferred.

Specific attempts to predict maximum oxygen intake from the speed of distance running (Margaria, Aghemo, and Limas, 1975) are summarized in Table 3.4. Balke (1954) measured the distance a man could run in 15 min, and more recently K. H. Cooper (1968a,b) popularized measurement of the distance covered in 12 min (Table 11.8). There is quite a close correlation between running distance and maximum oxygen intake in active young men, women, and boys (r = 0.9; K. H. Cooper, 1968a,b; Getchell et al., 1977).

TABLE 11.6
Pace for Ascent of Double 8-in. (20.3-cm) Step, Corresponding to ~70% of Aerobic Power in Sedentary North American

Age (Years)	Stepping Rhythm (beats • min⁻¹)[a] Men	Women
Spare band	156	132
15–19	156 (144)[b]	120
20–29	144	114
30–39	132	114
40–49	114	102
50–59	96 (102)[b]	84
60–69	78 (84)[b]	66 (84)[b]
Warm-up for oldest age group	60 (66)[b]	60 (66)[b]

[a]With a double step, each ascent and descent requires a total of six paces. A person normally commences to exercise at the rhythm set for a person 10 years older than himself, then exercises at the pace appropriate to his age, and if his condition permits he can conclude by exercising at the rhythm for a person 10 years younger than himself.

[b]To simplify recording, the cadences shown in parenthesis have been substituted for the theoretical stepping rhythm.

Source: Based on rhythm adopted for Canadian Home Fitness Test, Shephard, 1979c.

TABLE 11.7
Criteria for the Assessment of Cardiorespiratory Fitness from the Performance of the Constant Work Rate Test of Table 11.5 Based on Duration of Stepping and 10-sec Recovery Pulse Count*

Age (Years)	First 3 Min: Undesirable Fitness Level	Second 3 Min: Minimum Fitness Level	Second 3 Min: Recommended Fitness Level
15–19	≥30	≥27	≤26
20–29	≥29	≥26	≤25
30–39	≥28	≥25	≤24
40–49	≥26	≥24	≤23
50–59	≥25	≥23	≤22
60–69	≥24	≥23	≤22

*If the test is performed in the laboratory, it is possible to predict the maximum oxygen intake from the test data using the equation of Jetté, Campbell et al. (1976):

$$\dot{V}_{O_{2(max)}} = 42.5 + 16.6\,(E) - 0.12\,(M) - 0.12\,(f_h) - 0.24\,(A)$$

where E is the energy cost of stepping in $l \cdot min^{-1}$, M is the body mass in kg, f_h is the recovery heart rate, and A is the age in years.
Source: From Shephard, 1979c.

However, results depend strongly upon motivation, efficiency of running (Sidney and Shephard, 1977a; J. R. Morrow et al., 1978), and ability to choose an appropriate pace. One suggestion has been that an observer set the speed of running (Léger and Boucher, 1978). In young children, correlations between the 12-min distance and aerobic power are poor (K. J. Cureton et al., 1974) and, despite theoretical arguments for a run of at least 1.5 miles (2.4 km), better predictions of maximum oxygen intake can be obtained by running over the distances adopted in current performance test batteries (300–600 yards, 275–550 m; Lavallée et al., 1974). In older and poorly conditioned adults, there is some danger that 12 min of maximum exercise may provoke ventricular fibrillation (Shephard, 1974c).

The principles of a submaximum stepping test (D. A. Bailey et al., 1974) have been discussed above. If the heart rate during performance of this test is recorded by electrocardiogram, the accuracy of the resultant predicted maximum oxygen intake compares favorably with that of other submaximum test procedures, but some precision is inevitably lost if the subject is allowed to count his own pulse rate during the early recovery period (Shephard, 1979c).

In testing frail elderly subjects, Bassey et al. (1976) suggested that free walking be allowed at three different speeds; the heart rate at each speed could be recorded by a tape recorder and the data interpolated to give the heart rate at a constant pace of walking. The authors claimed that scores obtained by such a procedure correlated well with data obtained by submaximal bicycle ergometry.

Anerobic Threshold

Several recent authors have advocated that fitness should be gauged from the anerobic threshold, the intensity of effort at which lactate accumulation provokes a disproportionate hyperventilation. Some investigators have even claimed that the anerobic threshold gives a better indication of cardiorespiratory condition than a prediction of maximum oxygen intake (Donohue and Lavoie, 1978).

Whipp, Torres et al. (1977) suggested that continuous measurement of oxygen consumption should be made during performance of a bicycle ergometer test with a triangular ("ramp function") profile (steady increase in power output of 50 $W \cdot min^{-1}$). In theory, such a test allows the estimation of four variables: (1) anerobic threshold, (2) mechanical efficiency, (3) time constant of cardiorespiratory adjustment to effort, and (4) maximum oxygen intake.

In practice, the rate of accumulation of blood lactate depends very much upon the test profile and

TABLE 11.8
Estimation of the Maximum Oxygen Intake from the Distance Run in 12 min

Distance Covered in 12 Min	Maximum Oxygen Intake ($ml \cdot kg^{-1} \cdot min^{-1}$)
<1.0 miles (1.6 km)	<28.0
1.0–1.24 (1.6–2.0 km)	28.0–34.0
1.25–1.49 (2.0–2.4 km)	34.1–42.0
1.50–1.74 (2.4–2.8 km)	42.1–52.0
>1.75 (>2.8 km)	>52.0

Source: After K. H. Cooper, 1968a.

on the ratio of muscle volume to blood volume. There is thus no unique anerobic threshold at a fixed percentage of maximum oxygen intake (Shephard, 1977a; Fig. 2.6). J. A. Davis et al. (1976) set the anerobic threshold at 46.5% of maximum oxygen intake for operation of an arm crank but 63.8% of maximum oxygen intake for bicycle ergometry. Furthermore, C. G. Williams, Kok et al. (1966), Takaoka (1973), MacDougall (1977), and Withers (1977) all showed that the percentage of maximum oxygen intake at which lactate begins to accumulate is increased by training, and Costill (1976) noted that the threshold depends strongly on the proportion of slow-twitch fibers in the active muscles.

Fitness for Sustained Activity (>60 min)

Glycogen Stores

The importance of glycogen stores to the tolerance of sustained effort is noted in Chap. 2. Unfortunately, there is no well-established and simple method of determining muscle glycogen reserves. One promising suggestion is to time the number of rapid ascents of a staircase that can be completed while carrying a load equal to 10% of body mass (Manz et al., 1978).

Reasonably accurate but rather localized values can be obtained by muscle biopsy (Hultman, 1971; H. J. Green, Houston, and Thomson, 1978; H. J. Green, Daub et al., 1978). A small specimen of muscle is taken for chemical analysis using a broad-gauge needle (3–5 mm outer diameter). If a large muscle, such as the vastus lateralis, is sampled, the procedure is tolerated remarkably well. Many subjects have had 10–20 biopsies performed, both at rest and during exercise, with no complaints other than mild discomfort persisting for ~24 hr. Nevertheless, repeated muscle trauma of this order seems undesirable unless the procedure is clinically necessary.

Blood Volume

Fluid is lost from the circulation during exercise due to exudation into the active tissues and sweating. The initial blood volume is therefore an important determinant of tolerance for prolonged effort (Saltin, 1964 a,b). The blood volume is usually measured by dilution following injection of a known volume of "marker" substance (Chap. 9).

Heat Tolerance

A third factor limiting prolonged effort is a progressive rise of deep-body temperature (Chap. 3). A well-trained individual has a lower body temperature for a given cumulative stress, often coupled with ability to reach a higher core temperature than an untrained person.

Mouth temperatures cannot be used to study body temperature during vigorous exercise because the tongue is cooled by mouth breathing. If a man exercises hard in a cool environment, the oral temperature is reduced by as much as 0.5–1.0°C. At one time, the rectal temperature was a popular index of deep-body conditions (E. Shvartz et al., 1978). The results obtained from the rectum are reasonably representative of core temperature when a man is sitting at rest, but during exercise the rectal temperature is boosted above the core value by warm venous blood returning from the active muscles of the lower limbs. The recorded temperature also varies with the depth of rectal penetration (5–20 cm). Some subjects find repeated rectal examinations distasteful. One alternative is to swallow a radio-transmitting capsule that reports gastrointestinal temperatures for up to 3 days (Brox and Ackles, 1973). For shorter investigations, a thermocouple can be passed into the esophagus. This instrument gives quite an accurate indication of overall body temperature if it is positioned carefully, but it is a little unpleasant to retain during activity. Probably the most practical method of estimating core temperature during exercise is to pass a thermocouple into the auditory meatus until light contact is made with the tympanic membrane (C. T. M. Davies and Sargeant, 1975a); this technique carries the advantage that the probe lies in close proximity to the regulatory centers of the hypothalamus.

Heat dissipation during exercise depends mainly on sweating (Chap. 3). Well-trained individuals sweat earlier and in greater amounts than those who are unfit (Chap. 12). The rate of sweating is determined quite readily if a subject is weighed on a beam balance that is accurate to 5–10 g. Given a sweat rate of 2 l • hr^{-1}, weighings at 15-min intervals will have an accuracy of 1–2%. In calculating sweat loss over longer periods, account must be taken of fluid intake, urination, and (particularly if the air is dry) respiratory water losses (Chap. 3).

In the industrial setting, the simplest sign of poor adaptation to hard physical work and a high environmental temperature is a rise of pulse rate over the working day. If a progressive tachycardia is observed, the intensity of activity must be reduced and/or rest pauses must be lengthened, at least until the individual concerned has achieved a greater level of fitness. In this context, French investigators have advanced the concept of the *puissance maximale supportée* (Sadoul et al., 1966); this is the heaviest rate of working that can be sustained for 20 min without signs of physiological stress. Criteria of tolerance include an increase of respiratory minute volume <5%, a respiratory gas-exchange ratio <1.00, and a ventilatory equivalent <30 ml • l^{-1}.

Other Aspects of Fitness

Medical Fitness

A high level of fitness is unlikely to be achieved while a person has overt disease. This is obvious for some acute maladies, but is equally true of many chronic disorders (J. R. Brown and Shephard, 1967; Carlsten, 1972; Shephard, 1978a). A thorough medical examination is essential not only on grounds of safety, but also as an integral part of a fitness assessment.

Where disease is already present, exercise tests must be performed more cautiously and their interpretation must be more guarded. Assumptions of a linear heart-rate/oxygen-consumption relationship and a maximum heart rate coinciding with the population average may prove unwarranted. For example, in ischemic heart disease, the maximum heart rate may be limited by anginal pain or the sick sinus syndrome, while in chronic obstructive lung disease effort may be halted by shortness of breath before the anticipated limiting heart rate has been reached.

When assessing reasons for lack of fitness in a specific patient, it is helpful to make a detailed study of those links in the oxygen transport chain that are likely to be affected by the disease process.

TABLE 11.9
Average Values of "Ideal" Body Mass in Relation to Height and Sex for Subjects of Medium Frame

Height (cm) (No Shoes)	Ideal Body Mass (kg) (Indoor Clothing)	
	Male	Female
147.3	–	48.5
149.9	–	49.9
152.4	–	51.2
155.0	–	52.6
157.5	57.6	54.2
160.0	58.9	55.8
162.6	60.3	57.8
165.1	61.9	60.0
167.6	63.7	61.7
170.2	65.7	63.5
172.7	67.6	65.3
175.3	69.4	66.8
177.8	71.4	68.5
180.3	73.5	–
182.9	75.5	–
185.4	77.5	–
188.0	79.8	–
190.5	82.1	–
193.0	84.3	–

Notes: (1) A subject can have substantially less than the ideal mass and yet be in good health; many primitive peoples weigh 10–15 kg below the ideal (Shephard, 1978a). (2) A well-muscled subject may exceed the ideal mass; "excess weight" must thus be interpreted in conjunction with observations on body fat and skeletal dimensions. (3) A reduction of body mass may reflect loss of lean tissue rather than fat (Shephard, 1977a).

Source: Based on records of individually assured lives over the period 1935–54, from the 1959 Build and Blood Pressure Study of the Society of Actuaries.

Nutrition

In many areas of the world, fitness is still restricted by a lack of food. Body mass then provides a useful index of tolerance for prolonged work (Wyndham, 1966). Many recruits to the South African mines, for example, have a "weight" substantially less than would be predicted from their height. The deficit is corrected by provision of adequate food and participation in a strenuous training program. North America, also, has groups that show varying degrees of malnutrition. Examples include (1) ghetto children, (2) old people living on restricted incomes, (3) women who attempt "crash" dieting, (4) alcoholics (particularly those who have a low energy intake due to inactivity), and (5) other food faddists (including some athletes).

Perhaps the simplest measure of general nutritional status is the blood hemoglobin level. A low hemoglobin concentration inevitably impairs cardiorespiratory fitness (Chap. 6).

Overnutrition (Chap. 13) is reflected in an excess body mass relative to actuarial standards (Table 11.9). Many clinicians regard obesity as a discontinuous variable, first appearing when a person is 7 or 10 kg above "ideal" body mass. In advising an individual, there may be need of an arbitrary criterion whereby fat accumulation is regarded as dangerous to health and therefore in need of particularly aggressive treatment. However, a buildup of fat is more easily corrected before it becomes extreme. Furthermore, population diagrams suggest a continuous rather than a discontinuous distribution of "excess weight," the normal shading imperceptibly into the pathological (Shephard, 1978f; Fig. 10.8).

Simple weighing cannot distinguish between muscularity and an accumulation of body fat. Accordingly, determinations of body mass should be supplemented by measurements of skinfold thickness (Table 11.10) or estimation of body fat by a hydro-

TABLE 11.10
Average Thickness (mm) of Eight Skinfolds in Subjects Approximating the Actuarial "Ideal" Body Mass

Skinfold	Male (mm)	Female (mm)
Chin	5.8 ± 8.7	7.1 ± 2.8
Triceps	7.8 ± 4.1	15.6 ± 6.2
Chest	12.0 ± 7.9	8.6 ± 3.7
Subscapular	11.9 ± 5.1	11.3 ± 4.2
Suprailiac	12.7 ± 7.0	14.6 ± 8.0
Waist	14.3 ± 8.2	15.3 ± 7.5
Suprapubic	11.0 ± 6.4	20.5 ± 8.2
Knee (medial aspect)	8.6 ± 4.1	11.8 ± 4.2
Average, all folds	10.4 ± 4.9	13.9 ± 5.1

Source: From Shephard, 1977a.

static technique. Although it does not add to the information content of data, many authors convert skinfold readings to estimated percentages of body fat, typically through the intermediate step of calculating body density. As an example, we may cite the equations of Durnin and Womersley (1974) for density and Siri (1956) for body fat:

For men aged 17–72:

$$\text{body density} = 1.1704 - 0.0731 \log \Sigma s \qquad (11.7)$$

For women aged 16–68:

$$\text{body density} = 1.1327 - 0.0643 \log \Sigma s \qquad (11.8)$$

where Σs is the sum of triceps, subscapular, and suprailiac folds, in mm.

Then,

$$\text{body fat, percent} = \left(\frac{4.95}{\text{density}} - 4.50\right) \times 100 \qquad (11.9)$$

A desirable level of body fat may be established in terms of physical appearance, or an arbitrary criterion may be applied (for example, a ceiling of 14% for a young soldier; Haisman, 1974). Alternatively, one may make a probit plot of the distribution of percentage body fat within a given population, noting the percentage fat at which the graph shows skewing (Fig. 10.8), or one may calculate the percentage of body fat corresponding to the ideal skinfold readings of Table 11.10 (17.1% fat in men, 21.4% fat in women).

Many measurements of fitness, such as maximum oxygen intake and muscle strength, are commonly expressed relative to body mass. This imposes a heavy penalty on those who are obese, but in the context of fitness it is usually a fair penalty, since the muscles of a heavy person must perform more work and the oxygen cost of most physical activities is increased by obesity (J. R. Brown, 1966; Godin and Shephard, 1973b; Hanson, 1973). The added cost is due largely to the burden of displacing body mass, although there are also increased expenditures relating to maintenance of posture, friction of body parts, and the clumsiness of a habitually inactive person. J. R. Brown (1966) suggested that the oxygen cost of any given activity ($\dot{V}_{O_2}$) was given by an equation of the type

$$\dot{V}_{O_2} = a + b\,(M')^n \qquad (11.10)$$

where a and b were constants, M′ was body mass, and n an exponent ranging from 0.75 to 1.00. Putting n = 1.0, b ranged from insignificant in some tasks such as hand-press operation to 12 ml • kg^{-1} • min^{-1} in heavy or very heavy work. Godin and Shephard (1973b) pointed out that the total energy cost of any activity stemmed from two main components. One was related to basal energy expenditure and varied as the 0.75th power of body mass. The other corresponded to the energy cost of displacing body mass, having an exponent that ranged from near zero for a sedentary task such as cycling to 1.0 for tasks whereby the body was lifted repeatedly against gravity:

$$\dot{V}_{O_2} = a(M')^{0.75} + b(M')^{0-1.0} \qquad (11.11)$$

Assessments of nutrition often include determinations of blood lipids (Allain et al., 1974; Wahlefeld, 1974; Chap. 13). A serum cholesterol higher than 200 mg • dl^{-1} is undesirable, although a large proportion of older North Americans exceed this standard (Hewitt et al., 1977; Table 11.11); the risk of myocardial infarction is increased, particularly when high total cholesterol readings are associated with a decrease in the high density lipoprotein (HDL) fraction (Castelli et al., 1977; T. Gordon et al., 1977). The prognostic importance of serum triglyceride levels is less certain, although again it is probably an advantage to health if the serum concentration is less than 200 mg • dl^{-1}. An active life style is associated with a low serum cholesterol, a high HDL fraction, and a low serum triglyceride concentration (C. L. Allen et al., 1978; Shephard, Cox, and

West, 1980), although the causal nature of the relationship has yet to be proven.

Drugs and Fitness

Note should be taken of the consumption of alcohol and of cigarettes; recent consumption of cigarettes, in particular, can affect scores for many fitness tests (Shephard, 1977a; Chap. 14).

CONSTITUTION AND TRAINING

General Considerations

Argument continues over the relative importance of constitution and training as determinants of fitness. Many exercise physiologists have taken the view that constitution plays a dominant role (Klissouras, 1976). Thus P. O. Åstrand (1967) wrote, "I am convinced that anyone interested in winning Olympic Gold Medals must select his parents very carefully," while Garn, Clark et al. (1960) marshalled evidence for a "packaged" transfer of genes that contributed to success in competitive sports. We may note that while such success could stem from inheritance of a favorable body build, it could also reflect the inheritance of a personality that strove for high achievement in competition or accepted the rigors of prolonged training. Garn, Clark et al. (1960) were interested primarily in anthropometry, and they commented that a muscular mesomorph was much more likely to be the offspring of large-framed parents than of either large- and small-framed parents or two small-framed parents.

Twin Studies

The majority of investigators have adopted the traditional genetic approach of comparing variances between monozygous and dizygous twins. The contribution of heredity (H_{est}) to the total variance in any given aspect of fitness can then be estimated as

$$H_{est} = \frac{(s^2_{DZ} - s^2_{DZm}) - (s^2_{MZ} - s^2_{MZm})}{s^2_{DZ} - s^2_{DZm}} \times 100 \qquad (11.12)$$

where s^2_{DZ} is the variance in dizygous twins, s^2_{MZ} is the variance in monozygous twins, and s^2_{DZm} and s^2_{MZm} are the corresponding items of methodological variance. Some authors have simplified the calculation, ignoring the methodological item or making the debatable assumption that s^2_m was the same for monozygous and dizygous twins. A more serious fallacy in such calculations is the assumption that monozygous and dizygous twins encounter the same range of environmental influences as the general population. Perhaps for this reason, values of H_{est} are highly variable and usually overstate the genetic component (Bouchard, 1978).

There have been reports that the genotype influences body build (Vandenburg, 1962), muscular development (Hewitt, 1958; Garn, Clark et al., 1960; Garn, 1961a), fitness (Hewitt, 1958; Tanner and Israelsohn, 1963), early motor development (L. R. T. Williams and Hearfield, 1973), and skills such as balancing (R.S. Wilson and Harpring, 1972)

TABLE 11.11
Population Averages for Plasma Cholesterol and Triglycerides in Relation to Age, Sex, and Use of Hormonal Contraceptives

	Cholesterol (mg • dl⁻¹)			Triglycerides (mg • dl⁻¹)		
Age	Men	Women Using Hormones	Women Not Using Hormones	Men	Women Using Hormones	Women Not Using Hormones
20–24	168	180	165	109	115	78
25 29	184	187	172	128	119	77
30–34	198	188	177	145	127	76
35–39	2[illegible]2	205	189	150	118	85
40–44	207	197	198	167	134	98
45–49	212	212	202	166	145	103
50–54	211	224	221	158	121	112
55–59	210	218	231	143	106	110
60–64	206	224	226	129	103	111
65–69	210	229	221	136	83	128

Source: From Hewitt et al., 1977.

and running (Klissouras, 1973). Klissouras (1971) estimated that H_{est} was 93.4% for maximum oxygen intake per unit of body mass, 81.4% for maximum blood lactate, and 85.9% for maximum heart rate. Two years later, the same author was involved in tests on Finnish children (Komi et al., 1973); in stark contrast with the earlier report, H_{est} was found to be close to zero for most observations except the Margaria test of anerobic power (H_{est} = 99.2%), the patellar reflex time (H_{est} = 97.5%), and the reaction time (H_{est} = 85.7%). Despite some strong claims for inheritance, a comparison of monozygous twins disclosed a 47% difference of aerobic power and a 61% difference of maximum oxygen intake (Klissouras, 1971). Further, muscle biopsy has shown that monozygous twins can develop substantial differences in the surface area of their mitochondria and the activity of enzymes involved in oxidative metabolism (Howald, 1976; Chap.10).

Familial Correlations

A second possible approach to genetic analysis is based on the partitioning of parent-parent ($r_{p\text{-}p}$), parent-offspring ($r_{p\text{-}o}$), and full sib-sib ($r_{s\text{-}s}$) data correlations (Cavalli-Sforza and Bodmer, 1971; Cruz-Coke et al., 1973). The relevant equations can be summarized as follows:

$$\frac{s_A^2}{s_P^2} = 2r_{p\text{-}o} \quad (11.13)$$

$$\frac{s_D^2}{s_P^2} = 4(r_{s\text{-}s} - r_{p\text{-}o}) \quad (11.14)$$

$$H_{est} = \frac{s_A^2}{s_P^2} + \frac{s_D^2}{s_P^2} \quad (11.15)$$

Where s_A^2 is the additive component of variance, s_D^2 the dominant component, and s_P^2 the phenotypic component.

One important limitation of this type of analysis is that fetal development is influenced by both inheritance and the placental environment. Thus the correlation between the characteristics of a mother and her offspring is closer than that between a father and his children (Wolanksi, 1969). Montoye and Gayle (1978) found that the correlation of maximum oxygen intake between father and son was 0.66 for fathers under 40 years of age and 0.34 for all men; however, correlations of maximum oxygen intake between husband and wife and between siblings were not statistically significant.

Genetic Markers

Studies of elite athletes have shown an unusually low frequency of the "genetic marker" for phenylthiourea taste sensitivity. Top competitors thus have some unusual genes. However, other genetically linked characteristics (dermatoglyphics and the presence of various blood groups) occur at about the rates that would be anticipated in the general population (de Garay et al., 1974).

Empirical Estimates

A simpler approach to the partitioning of fitness between constitution and training is to estimate how far the characteristics of a superb endurance athlete can be attributed to selection of an individual with unusual genetic potential (Shephard, 1978a). The aerobic power of an average young man living in Toronto is 48 ± 8 ml • kg^{-1} • min^{-1} STPD (Shephard, 1977a). The best that could be discovered by exhaustive search of the national population would be an individual with a maximum oxygen intake 4 standard deviations above the population mean; this would give him an aerobic power of 80 ml • kg^{-1} • min^{-1} STPD. Nevertheless, occasional endurance athletes have values as large as 92 ml • kg^{-1} • min^{-1}. The difference between 80 and 92 ml • kg^{-1} • min^{-1} must represent a response to training. We might thus assign 27% of variance (92 − 80/92 − 48) to training and 73% (80 − 48/92 − 48) to constitutional factors. However, it is also known that at least 36% of the variance of aerobic power within a population is due to intersubject differences of habitual activity (Shephard and Callaway, 1966; D. A. Bailey, Shephard et al., 1974). Genetic influences unrelated to habitual activity thus account for no more than [(100 − 36)/100] × 73 = 47% of the variance in maximum oxygen intake. It of course remains arguable that at least a part of the intersubject difference of habitual activity is genetically determined.

RESPONSES TO TRAINING

Methodological Problems

Cross-sectional Comparisons

Many older reports, supposedly discussing training responses, were content to make cross-sectional comparisons between athletes and sedentary individuals. There are several limitations to this approach: (1) observed differences may result from selection of the athlete rather than from training (see above), (2) the athlete may not be well trained for the variable under review (for example, a champion weight lifter usually has a rather poorly developed cardiorespiratory fitness), (3) the presumed sedentary subject may follow an active lifestyle despite lack of participation in formal athletics, and (4) the quality of test performance may differ between the two groups. (Many athletes are "high achievers" and habituated to intense effort. In some tests, they thus push themselves closer to a true physiological limit than would a sedentary person. On the other hand, if an important competition is approaching, an athlete may be less willing to undertake maximum effort than would an average person).

TABLE 11.12
Dropout Rate for Participants in an Exercise Training Program for Postcoronary Patients (Rechnitzer et al., 1975)[a]

Period of Training (Months)	Dropout Rate (Percent of Sample Defecting Over 6-month Period) High-intensity Exercise	Low-intensity Exercise
0–6	8.4	13.1
6–12	13.1	15.4
12–18	28.0	11.1
18–24	10.4	6.6
24–30	7.5	9.9
30–36	9.9	3.7
36–42	19.4	9.0
42–48	(30.0)[b]	(3.4)[b]

[a]All data expressed as six monthly losses from residual sample.
[b]Small sample. Estimated cumulative loss 75, 53% for two regimens.
Source: After Shephard, 1979e.

Longitudinal Surveys

In longitudinal training experiments, a test group of subjects follows a specified training program week by week, and their subsequent physiological status is compared with that of a control group, whose members agree not to alter their level of habitual activity.

Such an experiment is open to less objection than the cross-sectional comparison described above. Nevertheless, there remain a number of pitfalls. The type of person who volunteers for a longitudinal exercise study is usually health-conscious (Shephard, 1978a,1979d) and may be concerned about his heart or some other facet of health. Psychological tests often indicate a "neurotic" or a "hysterical" type of personality. The very existence of a group tends to alter the lifestyle of its members with respect to diet, cigarette consumption, and the like, and, depending upon experimental design (Kavanagh and Shephard, 1973; Rechnitzer et al., 1975), the controls may or may not show parallel changes of lifestyle. Longitudinal studies of exercise are plagued by a high percentage of dropouts, usually for psychosocial rather than for medical reasons (Massie and Shephard, 1971; Oldridge, 1979; Table 11.12), and the residual sample inevitably becomes even less typical of the general population than the initial group who volunteered for study.

The subjects for the usual experiment are supposedly sedentary members of a community, but unfortunately there is no simple criterion to establish just how sedentary such individuals are at the commencement of an investigation. If a volunteer has a poor initial showing on the tests detailed in the first section, this may mean that he is very sedentary in his habits, but it is also conceivable that he is moderately active, with a poor genetic endowment. It is thus necessary to attempt a quantification of the initial level of activity, using retrospective questioning, a diary, or 24-hr heart rate counts (Sidney and Shephard, 1977d; Shephard, 1978a; K. L. Andersen, Masironi et al., 1978).

Some investigators simply followed the response to voluntarily selected activities; they have observed the progress of hockey, football, or swimming teams as they prepared for a competitive season or the gains of condition in older adults who joined fitness programs at YMCAs and similar institutions. This approach has the advantage that the natural enthusiasm of the participants is given full rein, but it is exceedingly difficult to quantitate the intensity, frequency, and duration of activity that has been undertaken. The more intelligent subjects may be persuaded to keep an exercise log, and this can later be translated into units of activity using the points scheme of K. H. Cooper (1968a) or a modification thereof (Massie, Rode et al., 1970). However, the process of translation is time consuming and needs either a skilled clerk or a sophisticated computer program. Further, the log necessarily describes not only the intensity but also the frequency and duration of activity, so that some decision on the relative importance of the three training variables is an essential preliminary to interpretation of the data. Finally, it is difficult to persuade subjects to accept random allocation to sport and sedentary programs, yet without randomization a study is beset by many of the problems of self-selection encountered in a cross-sectional comparison.

The intensity and duration of training can be controlled more rigidly if the exercise is performed in the laboratory, using a device such as a treadmill or a bicycle ergometer (Shephard, 1968c). Unfortunately, the required regimen may then become boring for the subject, and in a large city much time may be lost in driving to and from the exercise machine. Furthermore, if an adequate sample of subjects is to be tested, a large part of the laboratory day becomes occupied with the provision of facilities for routine exercise.

Types of Training Response

Two main types of adaptation to repeated exercise are distinguished. The first type has been called a "regulatory" response (Holmgren, 1967c). It involves the establishment of new functional connections within the central nervous system, and it has such manifestations as a shift from sympathetic to parasympathetic activity, an increase of central blood volume, a redistribution of blood flow from other parts of the body to the active muscles, and an initiation of sweating at a lower core temperature. The time course of this process is quite rapid, new patterns of regulation being fully developed within a few weeks of commencing a given program of enhanced activity.

The second type of response involves structural change—hypertrophy and even hyperplasia. The mass of muscle, cardiac tissue, ligaments, and bones is increased, usually with at least parallel development of the capillary blood supply. This type of response continues for months and even years, although eventually a limit is reached, possibly because a maximum useful cell size has been reached in the active tissues. The extent of any structural response depends upon the adequacy of diet relative to the required energy expenditure. In many animal-training experiments, there is a nutritional problem. For example, Oscai, Molé, and Holloszy (1971) found that male rats on an endurance training program gained "weight" more slowly than their sedentary litter mates, and as an apparent consequence of energy imbalance heart sizes were not increased in the active animals relative to the sedentary controls.

Some authors have suggested that both regulatory and structural changes are produced more readily in children than in older adults (Chap. 10). However, much depends on the allowances that are made for initial fitness (Sidney and Shephard, 1978b) and the method used to express the training response (absolute gain or percent change). Taking subjects of comparable initial status, the gains seen in a senior citizen are at least as great as those in a growing child (Shephard, 1978b).

Gross Physiological Changes

Overall Response

Repeated vigorous physical activity produces substantial adaptive changes in almost all body systems. A training plan for the ordinary individual will pay due regard to the individual's perception of fitness but will nevertheless seek to improve oxygen transport, strength, joint mobility, health of the articular cartilage, and coordination. If the subject is fat, it will also increase the overall energy consumption, creating a small energy deficit.

MAXIMUM OXYGEN INTAKE. Various conditioning programs have succeeded in increasing aerobic power by 5–30% (Shephard, 1965,1977a; Pollock, 1973). The magnitude of the training response has depended upon the initial status of those tested, the largest effects being seen in the most sedentary members of a group (Shephard, 1968c,1975e; Pollock, 1973). Where subjects have undergone preliminary deconditioning by bed rest, subsequent gains have been as large as 100% (Saltin, Blomqvist et al., 1968; Kavanagh, Shephard, and Kennedy, 1977).

In many training experiments, the period of observation has been quite short, and the impression has been formed that the adaptive process is complete within a few weeks (Saltin, Blomqvist et al., 1968)—a concept contrary to coaching experience. Where training has extended for 1 or 2 years, the end results have sometimes been similar to responses observed after a few weeks (for example, Sidney and Shephard, 1978b). However, the individuals concerned have been interested in light recreation rather than in serious athletics. In contrast, a group of postcoronary patients who decided to prepare themselves for marathon running showed continuing gains of maximum oxygen intake for 3–4 years after commencement of training (Kavanagh, Shephard, and Kennedy, 1977; Fig. 11.4).

LACTATE TOLERANCE. Training—particularly training of the interval type—may build up a tolerance for lactate (P. O. Åstrand, 1956; Gollnick and Hermansen, 1973); this may reflect an increase in the alkaline reserve of the tissues and/or a rise of pain

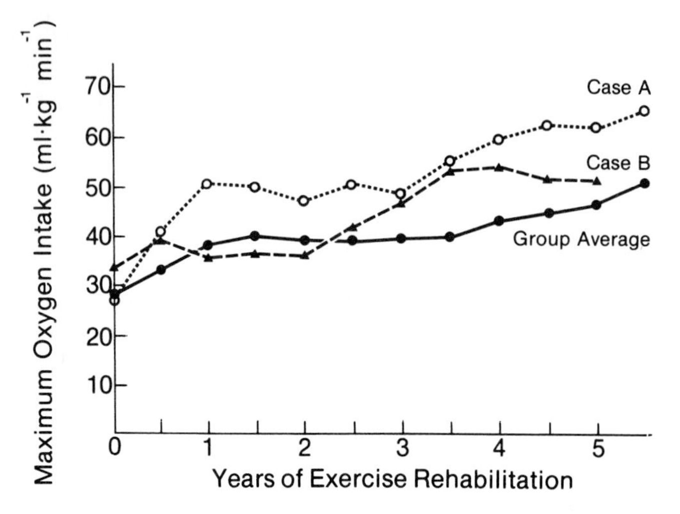

FIG. 11.4. Time course of increase in maximum oxygen intake with training of postcoronary patients for marathon participation. Average data for 13 subjects, with individual curves for 2 subjects. (Based on unpublished data of Kavanagh & Shephard.)

threshold (see above). The terminal muscle lactate concentration is usually higher in the trained than in the untrained individual, but even if this is not the case, a greater muscle mass, a greater blood volume, and a larger alactate capacity give the trained person a greater anerobic capacity and power than his sedentary counterpart (G. R. Wright, Bompa, and Shephard, 1976). The proportion of anerobic work performed at any given fraction of aerobic power is also reduced by training, mainly because the muscles are strengthened and perfusion of the active fibers is less readily impeded during contraction (Shephard, Allen et al., 1968b; Fig. 2.6).

ENDURANCE TIME. Endurance (see above) often shows a dramatic improvement with training. If performance is measured at a constant work rate, this is readily explained; what had been a maximal or near-maximal response of either a muscle group or of the entire body is reduced to perhaps 80% of maximum effort. There is much less reliance upon anerobic work and less chance of reaching an intramuscular pH that will limit the function of enzymes such as phosphorylase and phosphofructokinase (Chap. 2). However, even if effort is held to a constant percentage of the individual's current maximum potential, the endurance of strenuous effort is still extended by training. The respective contributions from psychological gains (improved body image, reduced perception of exertion) and physiological changes (improved local perfusion of muscle, increased tissue enzyme concentrations) have yet to be clarified.

Cardiovascular System

Exercise of the main muscles of the body leads to a training of the central circulation (factors such as heart rate, stroke volume, and cardiac output). Specific training of muscles required for a particular sport or activity causes an improvement of peripheral circulation (an increase of local blood flow).

HEART RATE. Perhaps the most obvious characteristic of the trained person is a very slow heart rate, both at rest and during submaximum effort (Scheuer and Tipton, 1977). The mean pulse rate of 260 athletes participating in the Amsterdam Olympic games was 50 beats per min (Hoogerwerf, 1929). Eighteen reports on sedentary subjects collected by Pollock (1973) showed an average decrease of resting readings from 70.4 to 63.8 beats per min in response to endurance training. Some (C. T. M. Davies, 1967b; Ekblom, 1969a,b; Saltin, Hartley et al., 1969), but not all (Maksud, Coutts et al., 1972; J. S. Skinner, 1973), authors have found a small decrease of maximum heart rate with training. In 19 reports collected by Pollock (1973), the average was 183.7 beats per min before and 179.8 beats per min after conditioning. A decrease of maximum heart rate seems more likely if the test and training exercise are similar (G. R. Wright, Sidney, and Shephard, 1978) and the subject is relatively young (Wilmore, Royce et al., 1970).

Since cardiac oxygen consumption is strongly related to heart rate, bradycardia decreases myocardial oxygen usage in rest and submaximal exercise, a particular advantage in patients with ischemic heart disease. The reason for the slow heart rate is still debated. The sinuatrial node has an increased sensitivity to β-receptor stimulants such as isoproterenol after training (Dowell and Tipton, 1970). However, it seems unlikely that the activity of atrial stretch receptors is increased by hypertrophy or dilation since a rapid rather than a slow heart rate is observed when the atria are distended by cardiac failure. The main cause of the slow resting heart rate is probably a centrally mediated increase of parasympathetic nerve activity (Raab and Krzywanek, 1966). Bradycardia can still develop if the sympathetic innervation of the heart is destroyed immunologically (Tipton, 1965), and indeed the administration of β-blocking drugs such as propranolol shows that there is little sympathetic nerve drive to the human heart at rest (Maksud, Coutts et al., 1972; Bodem et al., 1973). The acetylcholine content of the atria is greater in trained than in untrained animals, despite an unchanged cholinesterase activity (Herrlich et al., 1960). Further, the dose of atropine needed to block vagal activity is increased by training (Tipton, 1969). There is thus good evidence of increased parasympathetic drive. There may also be local, nonneural factors leading to increased acetylcholine production, since excised strips of atrial muscle from trained animals have a low intrinsic rate of contraction. Some authors (De Schryver and Mertens-Strjhagen, 1972; Brundin and Cernigliaro, 1975; but not Östman et al., 1972) described a decrease of cardiac catecholamine concentrations with chronic exercise, but this change is unrelated to the intensity of conditioning and does not have the same time course as the training bradycardia. Possibly the density of adrenaline-binding sites is reduced by cardiac hypertrophy (Salzman et al., 1970; Brundin and Cernigliaro, 1975).

During submaximum exercise, responses to β-blocking drugs such as propranolol are reduced by training (Frick, 1967; Ekblom, Kilbom, and Soltysiak 1973). There is also a decrease in myocardial turnover and excretion of catecholamines (Östman et al., 1972). It thus seems that exercise causes less sympathetic activation in a trained person.

The trained state is also characterized by a more rapid recovery of heart rate following exercise (see above).

Stroke Volume. The stroke volume is increased by training, both at rest and in all levels of activity (Wolf and Cunningham, 1978), the largest gains being seen when the subject is standing erect. This reflects a general increase of blood volume (Holmgren, 1967b), and an increase in tone of the leg veins with a consequent rise of central blood volume. Studies of isolated muscle strips suggest that there may also be an increase of myocardial contractility, with more complete emptying of the ventricular chamber (Molé and Rabb, 1973; Molé, 1978).

Some authors found no change in the amount of myofibrillar protein per g of heart tissue (Bhan and Scheuer, 1972). Others found an enlargement of individual myofibrils and an addition of new myofibrils (Richter and Kellner, 1963), without change in the dimensions of individual myosin filaments (Carney and Brown, 1964). In experimental animals, both the maximum contractile force and the rate of development of tension are increased by training; this suggests an increased myosin ATPase activity (Wilkerson and Evonuk, 1971; Bhan and Scheuer, 1972; Scheuer, 1973) and/or an increased availability of calcium ions facilitating activation of cross-bridges (Chap. 4). In humans, the evidence is conflicting. Some reports provide indirect evidence of increased myocardial contractility, for example, a shortening of the cardiac preejection period (Winters et al., 1973), but others note a lengthening of the isovolumetric phase of contraction (B. D. Franks and Cureton, 1969; Wiley, 1971). Under resting conditions, cardiac catheterization shows no difference of myocardial contractility between trained and untrained subjects, but during maximum effort the rate of pressure rise is slower in athletes, presumably due to a lesser sympathetic activation (Roskamm, 1973a).

Cardiac Output. Training augments the maximum output of the heart in direct proportion to any gain of maximum oxygen intake (Ekblom, Åstrand et al., 1968; Simmons and Shephard, 1971b; Rowell, 1974); thus if there is no increase of aerobic power, the maximum cardiac output also shows little change (Douglas and Becklake, 1968). During submaximum work, the cardiac output tends to be less in a trained than in an untrained subject; the muscle perfusion at a given level of oxygen consumption is usually reduced with conditioning, but this may be counteracted by an increased perfusion of the viscera (Rowell, 1974).

Arteriovenous Difference. The maximum arteriovenous oxygen difference is usually widened by training (Saltin, Hartley et al., 1969; Simmons and Shephard, 1971b; Roskamm, 1973a). There are several reasons for this. During maximal effort, a larger fraction of the cardiac output can be redistributed away from the skin (Simmons and Shephard, 1971b), since earlier sweating and the loss of subcutaneous fat facilitate elimination of body heat. It is less certain that the reduction of visceral blood flow at maximal rates of working is enhanced by training (Rowell, 1974), but an increase of enzyme activity within the working muscles may make some contribution to the widening of arteriovenous oxygen differences. During submaximum exercise, the blood flow to the active musculature is reduced by training

(Treumann and Schroeder, 1968; Douglas and Becklake, 1968; Clausen, Klausen et al., 1973); in contrast, muscle perfusion during maximum effort is increased (Rochelle et al., 1971; de Marées and Barbey, 1973).

SYSTEMIC BLOOD PRESSURE. There have been reports that training leads to a small fall of resting blood pressure (Jokl, Jokl-Ball et al., 1970), particularly in older subjects and patients suffering from hypertension (Kilbom et al., 1969; de Vries, 1970; Choquette and Ferguson, 1973). Possible artifacts include habituation of the subjects to the test environment, alterations of peripheral pulse wave reflection (Chap. 6), and an improved fit of the sphygmomanometer cuff secondary to loss of subcutaneous fat. Fifteen reports collected by Pollock (1973) showed systemic blood pressures of 128.6/78.8 mmHg (17.1/10.5 kPa) before and 121.0/75.1 mmHg (16.1/10.0 kPa) after training. Further investigation may prove that this is a physiological response rather than a technical artifact, but the change is so small as to have little therapeutic significance (Frick, Konttinen, and Sarajas, 1963; Tabakin et al., 1965; Ekblom, Åstrand et al., 1968).

A widening of pulse pressure might be anticipated with training due to the increase of cardiac stroke volume. Some experiments show such an effect but (perhaps because of reduced anxiety) the average of data collected by Pollock (1973) reveals a narrowing of pulse pressure from 49.8 to 45.9 mmHg (6.6 to 6.1 kPa).

Improvements of myocardial contractility with training may allow a subject to reach a higher maximum systemic pressure during exercise; in contrast, a diminution of the blood pressure response to a given work rate often provides early notice of worsening ischemic heart disease (Shephard, 1978j,1979b).

HEART VOLUME. The average endurance athlete has a somewhat larger heart than an ordinary individual or a contestant in a "muscular" sport (Herxheimer, 1929; Blümchen et al., 1966; Roskamm and Reindell, 1972; Table 11.13); however, it is less certain as to whether this is due to training or selection. Studies on sedentary subjects have examined changes in the heart shadow at radiography; since training may bring about an increase of contractility with greater emptying of the ventricles, a constant heart volume can mask an increase of ventricular mass. Roskamm and Reindell (1972) found no change of cardiac dimensions in young men who for 2½ years exercised 1 hr • day^{-1} at 70–80% of maximum oxygen intake. Others who have carried out similar or more vigorous training experiments report an increase in the cardiac shadow (Frick, Konttinen, and Sarajas, 1963; Ekblom, 1969a,b; Cermak, 1973; but not Perrault et al., 1978). Animal experiments generally show an increase of cardiac mass with training (Thörner, 1949–50; Van Liere and Northrop, 1957; Wyatt and Mitchell, 1974), although a response can be masked by an inadequate diet (see above).

TABLE 11.13
Radiographic Estimates of Heart Volume for Normal Men and Male Participants in Various German National Teams

Subjects	Heart Volume (ml • kg^{-1})
Normal subject	11.7
Weight lifters	10.8
Gymnasts	11.7
Wrestlers	12.2
Handball players	12.4
Skaters	12.4
Boxers	12.7
Pentathlon	12.8
Cross-country skiers	13.2
Cyclists (amateur)	14.4
Cyclists (professional)	14.8

Source: Based on data of Roskamm and Reindell, 1972.

The essential stimulus to cardiac hypertrophy seems an overload of the ventricle, sustained for a substantial part of each day. Permanent pressure loading of a ventricle caused by stenosis of its outlet valve leads to massive hypertrophy. One difference between physiological and pathological hypertrophy is that the latter fails to induce a matching improvement of myocardial blood flow. The exercise electrocardiogram then shows ischemic changes, and contractility is depressed rather than increased (Gunning et al., 1973).

Once induced, hypertrophy persists for some years; thus B. O. Eriksson, Engström et al. (1971) found persistent large heart shadows in former swimming champions (Chap. 10). However, the maximum oxygen intake of the subjects was quite low at this stage in their careers, so that the normal relationship between radiographic heart volume and maximum oxygen intake was invalidated. Some of the girls underwent retraining after 10 years of sedentary life. This group realized a 14% gain of aerobic power, with no change in the size of their already large heart shadows (B. O. Eriksson, 1978). Pathological hypertrophy of the heart (as in congenital pulmonary stenosis) also persists after surgical relief of the obstructed valve. However, Saltin and Grimby (1968) noted that the heart volumes of former orienteers gradually reverted to sedentary values after the sport had been abandoned.

BLOOD SUPPLY. Cardiac hypertrophy is associated with an increase in the cross-section of the major coronary vessels (Tepperman and Pearlman, 1961; Stevenson, 1967), an increase of capillary density (Thorner, 1935; Petrén et al., 1936; Tomanek, 1970; Bell and Rasmussen, 1972; Rasmussen et al., 1978), and an increase of myocardial perfusion at a given heart rate after load and perfusion pressure (Penpargkul and Scheuer, 1970; Scheuer and Stezoski, 1972). Some animal studies have further shown an increase of the collateral blood supply (Chap. 6; Eckstein, 1957; Bloor and Leon, 1970; but not Burt and Jackson, 1965; Kaplinsky et al., 1968; T. M. Sanders, White et al., 1978). Positive results were noted, particularly if training followed rather than preceded occlusion of the main vascular supply. In humans, there is as yet little evidence that exercise increases collateral perfusion in postcoronary patients (Kattus and Grollman, 1972; R. J. Ferguson, Choquette et al., 1973; R. J. Ferguson, Petitclerc et al., 1974). Possible problems are exercise at too low an intensity and advanced disease with irreversible changes in the coronary vessels.

OTHER CIRCULATORY CHANGES. Training usually augments the total blood volume (Deitrick et al., 1948; H. L. Taylor, Henschel, et al., 1949; Hollmann and Venrath, 1963; Sjöstrand, 1967; Holmgren, 1967b). There may also be an increase in the hemoglobin content per unit volume of blood, although this is a less consistent finding; indeed, some forms of athletic training are associated with anemia (Chap. 6). The blood sometimes shows an increase of alkaline reserve, particularly if training has emphasized interval work, although often there is no change in this variable (P. O. Åstrand, 1956).

Volume work consumes less oxygen than an equivalent amount of pressure work (Chap. 6). One might thus anticipate a decrease in myocardial oxygen consumption at a given cardiac output due to the training-induced increase of cardiac stroke volume. Frick, Konttinen, and Sarajas (1963) commented on a decrease in tension/time index at a given cardiac output. Further factors that reduce the cardiac oxygen usage of a conditioned subject are (1) a decreased output of catecholamines and (2) a decreased sensitivity to catecholamines (Crews and Aldinger, 1967). However, the intrinsic mechanical efficiency of the muscle remains unaltered (Penpargkul and Scheuer, 1970).

Respiratory System

Substantial coefficients of correlation can be demonstrated linking vital capacity with performance data such as times for a 5000-m race. Such correlations persist after standardization for body mass (Ishiko, 1967). Especially large vital capacities are encountered in athletes who rely upon either well-developed thoracic muscles (for example, rowers, G. R. Wright, Bompa, and Shephard, 1976; and white-water paddlers, Sidney and Shephard, 1973) or the buoyancy obtained from large thoracic lung volumes (middle-distance swimmers; Bloomfield and Sigerseth, 1965; Shephard, Godin, and Campbell, 1973). However, these findings probably arise more by selection than by conditioning.

Athletic training does not normally augment vital capacity (T. K. Cureton, 1936; Kollias, Boileau et al., 1972), mean expiratory flow rate, or maximum voluntary ventilation (G. R. Cumming, 1971). However, gains can be produced by a specific training of the respiratory muscles (Delhez et al., 1967–68; Leith and Bradley, 1976), and in adolescent girl swimmers increases of vital capacity are larger than would be predicted simply from an increase of stature (Engström et al., 1971). Conceivably, a strengthening of the chest muscles increases lung volumes by forcible expulsion of blood from the pulmonary vessels or compression of the thoracic cage during the final stages of expiration. In some clinical conditions, respiratory failure is associated with a reduction of ATP and CP stores in the intercostal muscles, and this abnormality disappears as respiration again meets the demands of oxygen transport (Gertz et al., 1977). In healthy subjects isometric training of the thoracic muscles increases peak respiratory pressures by up to 55%, while isotonic training increases the maximum 15-min ventilation from 81 to 96% of the 15-sec maximum voluntary ventilation (Leith and Bradley, 1976).

The most obvious respiratory consequence of training is the adoption of a slower and deeper pattern of breathing (Shephard, 1979f), both at rest and during moderate exercise. This feature is particularly marked in swimmers and others who learn to combine exercise with periods of breath holding. There is an associated increase in the extraction of oxygen from unit volume of respired gas (that is, the ventilatory equivalent for oxygen decreases from 30–35 $l \cdot l^{-1}$ to around 25 $l \cdot l^{-1}$).

At any given level of submaximal effort, there is a speeding of the ventilatory on-transient (Beaver and Wasserman, 1970; Chap. 5). The steady-state respiratory minute volume is less in trained than in untrained subjects. This reflects three main adaptations: (1) a general increase of mechanical efficiency and thus a lowering of oxygen cost for a given work output, (2) a centrally mediated decrease of ventilatory drive in moderate exercise, (3) a possible reduction in the sensitivity of the carotid chemoreceptors (Chap. 5), and (4) a lesser production of lactate in severe effort. The increase of anerobic threshold (Whipp et al., 1977) is due to (1) a decrease in the relative rate of working associated with augmentation of maximum oxygen intake, (2) a faster cardiovascular on-transient at the beginning of exercise (Chap. 6), and (3) an improved perfusion of the active muscles associated with an increase of their maximum voluntary force (Chap. 6).

The respiratory minute volume developed during maximal work increases roughly in proportion to the gain of aerobic power brought about by training (Ekblom, Åstrand et al., 1968). In a series of 18 reports collected by Pollock (1973) the average change was +9.8%.

Some authors have suggested that athletes have a less acute perception of dyspnea than sedentary individuals (Comroe, 1956). Endurance competitors develop respiratory minute volumes averaging ~160 $l \cdot min^{-1}$ (Saltin and Åstrand, 1967), values that would be acutely uncomfortable for a sedentary individual; since there is little increase of either static lung volumes or the maximum respiratory rate, the athletes necessarily use a larger than normal fraction of their vital capacities. Specific psychophysical studies of perceived ventilatory effort have yet to be carried out; training leads to a decrease in the overall RPE (Borg, 1971; Fig. 11.1) at a fixed work rate, but scores show little alteration if expressed at a constant percentage of maximum oxygen intake (Sidney and Shephard, 1977c).

It would be impractical for a sedentary person to sustain a ventilation of 160 $l \cdot min^{-1}$ because the oxygen cost of breathing would surpass the "crossover" point (Shephard, 1966a) where it exceeded the oxygen delivery accomplished by the added ventilation (Fig. 5.25). Training has little effect upon the mechanical work of breathing, but nevertheless the crossover point is displaced to the right, mainly because an increase of maximum cardiac output augments the oxygen delivered per unit of ventilation.

Several studies noted a large pulmonary diffusing capacity in athletes (Bates et al., 1955; Bannister, Cotes et al., 1960; Newman et al., 1962; Mostyn et al., 1963), both at rest and during physical activity. This is partly due to a large central blood volume and pulmonary blood flow, but there are also artifacts arising from the slow breathing pattern (T. W. Anderson and Shephard, 1968a; Chap. 5). If care is taken to avoid such artifacts, then the increase of maximum diffusing capacity with training is no greater than would be anticipated from gains of aerobic power, pulmonary blood flow, and diffusing surface (T. W. Anderson and Shephard, 1968b; Hanson, 1969; K. L. Andersen and Magel, 1970).

Body Composition

MUSCLE. Most forms of vigorous training induce some muscle hypertrophy (Hollmann and Hettinger, 1976; Jansson and Kaijser 1977). An increased proportion of the muscle becomes occupied by myofibrils as opposed to fat, connective tissue, and supporting proteins, although paradoxically fibroblasts also become more evident (Jablecki et al., 1973) and collagen content may increase—particularly if training is pursued to the point of causing local injury (Turto et al., 1974). Isometric or heavily loaded isotonic programs (for example, Chapman et al., 1972; Edström and Ekblom, 1972; Thorstensson, 1976) produce quite large gains of muscle bulk (see below), with specific increase of fast-twitch (type II) fibers. In contrast, activities designed to increase cardiorespiratory fitness produce relatively slight changes of gross muscle dimensions (de Vries, 1970; Sidney et al., 1977) and fiber areas (Jansson and Kaijser, 1977; Nygaard, Bentzen et al., 1977).

When muscle dimensions are increased, this is generally held to reflect an expansion of existing fibers rather than an increase in the total number of muscle cells (Chap. 10). Nevertheless, recent animal experiments suggest that excessive training can cause fiber splitting and possibly even new fiber formation (Gonyea et al., 1977). Opinions are divided as to whether regular physical activity increases the ratio of capillaries to muscle fibers (Carrow et al., 1967; Tomanek, 1970; Hermansen and Wachtlová, 1971; Rakusan et al., 1971; Wachtlová and Pařízková, 1972; Brodal et al., 1977; Chap. 4). If indeed there is an increase of capillarity, this occurs relatively uniformly among the different fiber types (P. Andersen and Henriksson, 1977) and does little more than maintain a constant diffusion distance between the capillary and the center of the enlarged muscle fiber (Fig. 11.5). However, the collateral blood supply of skeletal muscle is typically increased by training, and this can improve the exercise tolerance of individuals

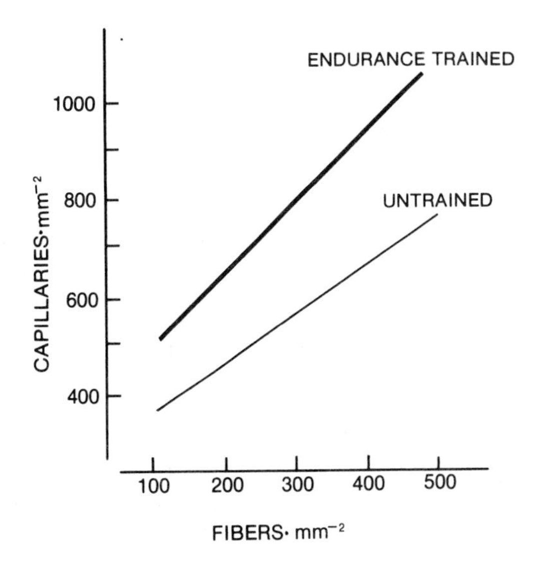

FIG. 11.5. Relation between capillary density and fiber area. Note that if fiber size is increased by hypertrophy, fiber density and the related capillary density fall. Thus, a well-trained subject with large fibers can have a lower capillary density per unit area than an untrained subject, although the capillary fiber ratio is normal or even increased. (Based on data of Brodal et al., 1977.)

with peripheral vascular disease. Within individual fibers (see below), there are increases of several key constituents, including mitochondria, the enzymes of aerobic metabolism (Holloszy et al., 1973,1975), adenosine triphosphate and creatine phosphate (Palladin, 1945; Yakovlev, 1958; Karlsson, Diamant, and Saltin, 1971; but not Hearn and Gollnick, 1961), myoglobin (Holloszy, 1975; Meldon, 1976), glycogen (A. W. Taylor, 1975), and potassium (Nöcker et al., 1958). Particularly during recovery from injury, gains of muscle force may exceed gains of muscle cross-section (Fried and Shephard, 1970; Ikai and Fukunaga, 1968; D. H. Clarke, 1973; Hollman and Hettinger, 1976). Presumably, training lessens voluntary inhibition of muscle contraction, and it may also allow a more effective coordination of action between various motor units capable of contributing to a given isometric effort.

FAT. At the same time that the muscles are developing, there is a depletion of fat deposits within the subcutaneous tissues and elsewhere. Training brings about a reduction in both the size and the fat content of individual adipose cells, although the cell number apparently remains unchanged (Hirsch and Han, 1969; Salans et al., 1971). The sensitivity of the β-receptors is also increased, so that the adipocytes liberate more free fatty acids in response to a given catecholamine stimulus (Pářizková and Stankova, 1964; Gollnick, 1971). Intramuscular lipid stores are increased by a factor of 2.0–2.5 (T. E. Morgan et al., 1969; Askew et al., 1973; Howald, 1975; Orlander and Kiessling, 1978).

Studies summarized by Pollock (1973) show that the average training program reduces the overall percentage of body fat 0–3.8%. Changes are largest when substantial amounts of energy are expended for some months, and perhaps for this reason long periods of vigorous walking do more to reduce body fat than brief sessions of jogging (Pollock, Dimmick et al., 1975). Carter and Phillips (1969) noted that fat loss continued over at least 1 year of increased physical activity. Nevertheless, a few months of vigorous training can correct most of the fat accumulation that occurs over the life span of an average person (Sidney et al., 1977; Table 11.14). Expressing changes as a percent of initial skinfold readings, losses vary from a modest 3% in the suprapubic region to 70% in the triceps fold.

O'Hara, Allen et al. (1979) demonstrated that the fat loss for a given energy expenditure is greater in a cold than in a warm environment (Fig. 11.6). Possible explanations of this phenomenon are discussed in Chap. 12. Irrespective of environmental temperature, severe training induces a reduction of

TABLE 11.14
Changes in Skinfold Thickness with 14 Weeks of Endurance Training for 65-year-old Subjects

	Loss with 14 Weeks Training (Men and Women) (%)	
Skinfold	**Absolute Change (%)**	**Relative to Gain Age 25–65***
Chin	−10	19
Subscapular	−17	31
Triceps	−14	70
Suprailiac	−13	39
Waist	−11	7
Suprapubic	− 3	3
Chest	−19	25
Thigh	−11	
Calf	−18	
Knee	−10	16
Average, all folds	−12	32

*For gain age 25–65 years, see Table 10.18.
Source: Based on data of Shephard, 1978b

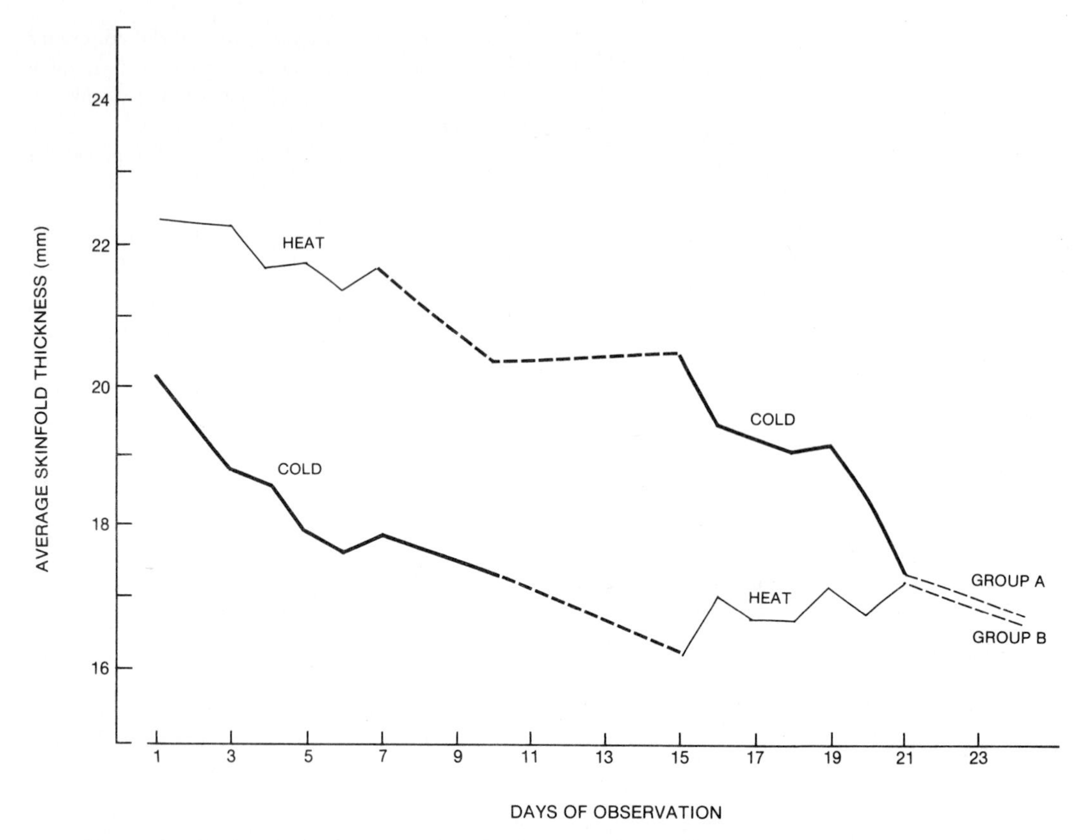

FIG. 11.6. Fat loss with vigorous activity in (A) an arctic and (B) a warm environment (marked "heat"). (Based on data of W. J. O'Hara, Allen et al., 1979, from a "cross-over" experiment on young men.)

serum triglycerides and low density lipoprotein cholesterol; sometimes there is also a decrease of liver (L. A. Carlson, 1967a,b; Barnard et al., 1968) and serum cholesterol (Altekruse and Wilmore, 1973; Lopez et al., 1974; but not R. C. Goode, Firstbrook, and Shephard, 1966; Björntorp, 1970; or Pyörälä, Kärävä et al., 1971). The net effect of fat loss is an increase of body density (J. S. Skinner, Holloszy, and Cureton, 1964; Pařizková, 1977; O'Hara, Allen et al., 1979). Often, there is a parallel synthesis of lean tissue (shown by analysis of the carcass in animals, Table 11.15, and in man by an increase of $^{40}K^{+}$ readings), so that body mass changes relatively little (Sidney et al., 1977). Plainly, the clinical emphasis on "weight loss" is fallacious. Indeed, encouragement of lean tissue synthesis helps fat loss, since substantial amounts of energy are required for the anabolic process.

SKELETAL SYSTEM. Bones involved in physical activity are often strengthened (Saville and Whyte, 1969; Kiiskinen and Heikkinen, 1975; Booth

TABLE 11.15
Effects of Exercise and Dieting upon Body Composition

Group	Carcass Mass (g)	Protein (g)	Fat (g)	Ash (g)	Water (g)
Exercise	496	84	105	12	295
Diet	498	73	135	12	278
Control	676	93	245	14	325

Source: Based on cross-sectional study of rats by Oscai and Holloszy, 1969.

and Gould, 1975). Possible changes include an increased content of minerals and hydroxyproline (Chvapil, 1967; E. L. Smith and Babcock, 1973; Kiiskinen and Heikkinen, 1975), an increased density (King and Pengelly, 1973), and a strengthening of architecture through the development of new trabeculae (Kohlrausch, 1924; J. A. Ross, 1950). On the other hand, excessive activity can inhibit bone growth and give rise to stress fractures (Tipton, Matthes, and Maynard, 1972; Kiiskinen and Heikkinen, 1975; Chap. 10).

Regular exercise increases the strength of tendons and ligaments in male (Viidik, 1967a; Tipton, Martin et al., 1975; Tipton et al., 1977) but not female animals (A. Adams, 1966). Possibly male sex hormones contribute to the anabolic process. The thickness of a given tendon is increased and its attachment to bone is strengthened (Tipton et al., 1977). The biosynthesis of collagen is boosted (Suominen and Heikkinen, 1975a) and individual fibrils become thicker. The hydroxyproline content is increased (Kiiskinen and Heikkinen, 1975), and prolyl hydroxylase activity is augmented (Suominen and Heikkinen, 1975b). Collagen turnover also rises, thereby reducing cross-linkages between individual collagen molecules and improving the overall elasticity of the tendons (Viidik, 1973).

Training thickens articular cartilage, making it more resistant to compression (Holmdahl and Ingelmark, 1948). There is also an increase in the flexibility of the joint capsule (E. A. Chapman et al., 1972).

Central Nervous System

Regular repetition of exercise brings about a number of alterations in the functional responses of the central nervous system. Most authors regard such changes as "regulatory" rather than "structural," although recent evidence suggests that some synthesis of neural protein is associated with the learning process (Chap. 7; Fig. 7.29).

Motor learning leads to a storage of information concerning γ-loop settings in the premotor cortex and/or the cerebellum; practical consequences are a brisker reaction time, improved coordination (Albinson and Andrew, 1976; Stelmach, 1976), and (usually) performance of a given task at a higher mechanical efficiency. While the ordinary person is undoubtedly pleased to achieve an economy of effort (Shephard, 1969) an athlete may prefer to improve his speed or extend his range of performance even if efficiency is worsened thereby (Lauru, 1957).

Training is also associated with the storage of information that permits a more rapid and complete adjustment of the circulation to such stresses as physical activity (Chap. 6) and adoption of a vertical posture (Chap. 7). An impaired orthostatic tolerance is a well-recognized complication of bed rest (Fried and Shephard, 1969), weightlessness (Müller, 1963; P. C. Johnson et al., 1973; Rummel et al., 1973, 1975), and lack of cardiorespiratory fitness (A. Holmgren, 1967a). Unfortunately, the space traveler cannot overcome these problems by daily endurance-type activity. If orthostatic reflexes are to be preserved, there must also be a simulation of gravity by applying negative counterpressure to the legs.

Other changes within the central nervous system fall under the rubric of *habituation*, or negative conditioning (Glaser, 1966; Shephard, 1969). They reflect the overall adjustment of the subject to the circumstances under which a given type of exercise must be performed (Table 11.16). In animal experiments, habituation does not occur if the prefrontal cortex is destroyed (Glaser and Griffin, 1962); parallel observations have yet to be made on patients who have sustained an accidental or a surgical prefrontal leucotomy. About a quarter of the normal population shows an excessive heart rate on the first occasion that they attempt exercise in an unfamiliar situation. The tachycardia is most obvious if the test appears dangerous (for example, running on a high-speed treadmill). Several repetitions of the required exercise progressively reduce such an emotional response. Subconscious inhibition of maximum effort is also lessened by familiarity with the investigator and the required task.

Hormonal Response

Physical activity modifies the secretion of several hormones in a way that could stimulate anabolic processes. The acute response to exercise is an increase

TABLE 11.16
Habituation of Young Adult Males to Successive Days of Standard Submaximal Exercise on a Step Test, Bicycle Ergometer, and Treadmill

Mode of Exercise	Heart Rate (beats • min^{-1}) Day 1	Day 2	Day 3	Day 4	Day 5
Treadmill	127	122	124	120	117
Bicycle ergometer	117	118	118	118	115
Step test	126	121	118	117	119

Source: After Shephard, Allen et al., 1968.

in blood levels of growth hormone and androgens, with a decrease of insulin (Chap. 8). After training, insulin levels at a given work rate are increased, but glucagon levels are decreased (Hartley, Mason et al., 1972; Bloom et al., 1976). Thyroxine turnover may be increased (Terjung and Winder, 1975), while catecholamine output is diminished (Hartley, 1975). Changes in growth hormone response are more variable (Shephard and Sidney, 1975a,b).

The amount of any hormone secreted in general remains proportional to relative stress. Thus, if maximum oxygen intake is increased, it might seem inevitable that the hormone output at a given absolute work rate would be diminished. However, in some cases the initial hormonal response of an untrained person is sufficiently great to exhaust the capacity of an endocrine gland (for example, cortisol secretion from the adrenal cortex); training then permits a better sustained response to prolonged exercise (Shephard and Sidney, 1975b). In old people, an increased output of growth hormone could conceivably have a specific protein sparing (anabolic) action during vigorous exercise (Sidney and Shephard, 1978a).

Cellular Changes

General Considerations

There has been considerable controversy concerning cellular adaptations to repeated exercise. Much of the confusion seems attributable to the units in which data have been expressed. Concentrations have sometimes been stated per unit of muscle mass. Such values inevitably increase if the proportion of fat or connective tissue in the "muscle" diminishes. Other data have been related to muscle protein or to mitochondrial protein. Thus, if the mitochondrial protein increases (K. H. Kiessling, Pilstrom et al., 1973), there may appear to be a decrease of extramitochondrial enzymes, and even increases of mitochondrial enzymes may be overlooked (Holloszy, 1973).

A further problem concerns the exercising of small animals. It is difficult to quantitate swimming or treadmill running for rats. Sometimes the activity selected has lacked the necessary vigor to induce training, and on other occasions it has been so stressful as to reduce overall body mass and restrict growth (see above). Swimming has further exposed animals to the combined stress of exercise plus hypothermia.

Response of Organelles

Cross-sectional comparisons of muscle biopsy specimens from athletes and sedentary subjects (Howald, 1975) demonstrate that the former group have (1) more mitochondria per myofibril, (2) larger mitochondria, and (3) mitochondria with an enlarged inner surface (the inner surface being where the respiratory enzymes are carried). Longitudinal training experiments in rats show a 60% increase of mitochondrial protein (Holloszy, 1967), with increases in the size and number of mitochondria (Gollnick and King, 1969). In humans, training induces increases of mitochondrial volume in young but not in middle-aged subjects (K. H. Kiessling, Pilstrom et al., 1973,1975). If mitochondrial protein is indeed augmented by ~60% (Holloszy, 1967,1973), this is less than the increase of activity reported for certain enzyme systems; the implication is that the activity of other enzyme systems is increased little if at all.

The transverse and longitudinal tubular systems of muscle (Chap. 4) are unaffected by training, but there is a 2- to 2½-fold increase in the lipid content of the muscle fibers (T. E. Morgan et al., 1969; Askew et al., 1973; Howald, 1975; Orlander and Kiessling, 1978). It was once suggested that endurance training converted white to red muscle fibers. The current consensus is that there may be some increase in the myoglobin content of fast-twitch fibers (conversion of type IIb to IIa) but that there is no significant interchange of fast- and slow-twitch characteristics (Essén, Jansson et al., 1975; Kaijser and Jansson, 1977). Indeed, all three types of fiber share roughly equally in the increase of aerobic power (K. M. Baldwin et al., 1972). Much of the confusion can be traced to crude histochemical techniques of fiber typing. A "white" fiber initially has little aerobic enzyme activity, and increase of staining thus makes it look like a "red" fiber. On the other hand, an increase of staining in a red muscle that is already darkly stained before training may pass unnoticed.

Most studies of heart muscle have described mitochondrial enlargement with training (Wollenberger and Schulze, 1961; Laguens et al., 1966; Aldinger and Rajindar, 1970). However, Cosmas and Edington (1975) reported a shift toward smaller mitochondria, arguing that the increase of surface/volume ratio was a useful adaptation to the increased energy needs of functional overload.

Anerobic Metabolism

Reports discussed in this and the following section concern increases in the activity of key, rate-limiting enzymes in response to training. It should be stressed at the outset (Poortmans, 1975) that while such changes could arise from an increased synthesis of the enzyme concerned (changes in RNA and DNA), other possibilities include alterations in the activity of repressor or inducer proteins and more favorable conditions for the function of existing enzymes (for example, hexokinase activity might be increased by a lesser availability of glucose-6-phosphate; see Chap. 2).

Early reports did not reveal any increase in the enzymes of the glycolytic pathway in response to training (Gollnick and Hermansen, 1973). This may have reflected low intensities of exercise, such as mild running; in some reports, activity was even insufficient to induce training of aerobic enzyme systems. Holloszy et al. (1971) exercised their rats vigorously enough to augment the oxidative power of skeletal muscle, but nevertheless they found no increase in such key glycolytic enzymes as phosphorylase and phosphofructokinase.

K. M. Baldwin et al. (1972) commented that in their experiments the response varied with fiber type; training reduced the glycolytic potential of fast-twitch red muscle, but it increased the glycolytic activity of slow-twitch intermediate fibers.

Gollnick et al. (1972) found no consistent difference of phosphofructokinase (PFK) activity in a cross-sectional comparison between athletes and nonathletes. However, B. O. Eriksson, Gollnick, and Saltin (1973) were able to produce a 40% increase in the PFK activity of 10- to 11-year-old boys by 2 months of endurance training. Gollnick et al. (1973a) also found an increase of PFK activity when men underwent a 5-month training program; the latter authors deduced from changes of α-glycerophosphate dehydrogenase activity that the increase of glycolytic capacity was confined to the fast-twitch fibers. A. W. Taylor, Stothart et al. (1974), A. W. Taylor (1975), and A. W. Taylor, Lavoie et al. (1978) described increases of phosphorylase, synthetase I and D, and glycogen-branching and debranching enzymes; the most recent experiments thus suggest a facilitation of both glycogen synthesis and breakdown (Holloszy and Booth, 1976).

Measurable increases of ATP and creatine phosphate concentrations are associated with the increase of mitochondrial density in trained muscle (Karlsson, Diamant, and Saltin, 1971). Nevertheless, the quantities involved are too small to have any practical effect upon anerobic capacity (Chap. 2) or power (Chap. 3).

There have been few studies of changes in the anerobic capacity of heart muscle with training. Available reports indicate little change in the activity of such enzymes as lactate dehydrogenase, aldolase, adenylate kinase, or creatine kinase (Hearn, 1965; Gollnick, Struck, and Bogyo, 1967; Walpurger and Anger, 1970).

Aerobic Metabolism

Myoglobin levels are greater in active than in sedentary animals (Whipple, 1926; Shenk et al., 1934). Endurance training increases concentrations within active muscles by up to 80% over the course of 12 weeks (Pattengale and Holloszy, 1967), thus facilitating intracellular transport of oxygen (Chap. 3).

Some early studies (Hearn and Wainio, 1956; M. K. Gould and Rawlinson, 1959) found no response of aerobic enzyme systems to training, but it is possible that either the intensity or the duration of exercise was insufficient. Most recent experiments have demonstrated at least twofold increases in the activity of various enzymes in the Krebs cycle and the mitochondrial respiratory chain (for example, succinic dehydrogenase, cytochrome c reductase, cytochrome c, and cytochrome oxidase), with a tight "coupling" of oxidative reactions to the resynthesis of ATP (Holloszy, 1967,1973; Molé, Oscai, and Holloszy, 1971; Suominen and Heikkinen, 1975b). Such changes have functional significance, at least in vitro, since specimens from trained muscle show a substantially increased ability to oxidize pyruvate (Barnard, Edgerton, and Peter, 1970; T. E. Morgan et al., 1971). Furthermore, since there are more mitochondrial cristae per g in trained than in untrained muscle, a given oxygen consumption can be attained with a smaller reduction of ATP and CP; lower levels of AMP, ADP, and phosphate; and possibly less formation of ammonia (Tornheim and Lowenstein, 1972). All of these changes encourage utilization of fat in preference to glycogen as the energy source for the Krebs cycle.

Utilization of fat rather than carbohydrate during moderate exercise is a well-recognized response to training. One important consequence is that muscle stores of glycogen (Chap. 2) are spared for bursts of anerobic activity (Christensen and Hansen, 1939a,b; Issekutz, Miller, and Rodahl, 1966; Hermansen, Hultman, and Saltin, 1967; P. Paul and Holmes, 1975). The likelihood of hypoglycemic fatigue is also reduced thereby (Christensen and Hansen, 1939b; Pruett, 1970a,b). The ability to oxidize fat is approximately doubled (Molé, Oscai, and Holloszy, 1971). There are increases in the activity of such enzymes as palmityl-CoA synthetase, carnitine palmityltransferase, and palmityl-CoA dehydrogenase (Molé, Oscai, and Holloszy, 1971; Baldwin, Klinkerfuss et al., 1972). However, such enzyme changes probably contribute little to the increased usage of fat, since there is an almost equal increase in the activity of enzymes concerned with carbohydrate metabolism. The main rate-limiting factor is rather the mobilization of fat (Bremer, 1967). Nevertheless, at any given blood level of fatty acids, tissues with a high level of fat-burning enzymes (for example, heart and red muscle) consume more fat than poorly endowed tissues (such as white muscle). The regulatory influence of phosphagen ratios has been noted above. A further basis of control is that increased fat oxidation inhibits both glycolysis and pyruvate oxidation (Newsholme and Randle, 1964; Paul, Issekutz, and Miller, 1966); citrate accumulation inhibits phosphofructokinase (Parmeggiani and Bowman, 1963), while acetyl-CoA inhibits pyruvate dehydrogenase (Randle et al., 1966).

Some authors have attributed the training-induced increase of maximum oxygen intake to an increased activity of aerobic enzymes. Although there is a widening of the arteriovenous oxygen difference with training (Chap. 6), this probably reflects largely a diversion of blood flow from the skin to the active muscles (Chap. 3). Even before training, oxygen extraction in the working muscles is relatively complete, and little additional widening of arteriovenous difference could be anticipated in the muscle capillaries. Further, studies by the ^{133}Xe clearance method suggest that local muscle flow increases at least in proportion to the added mass of muscle (Grimby, Häggendal, and Saltin, 1967) and, to the extent that gains of maximum oxygen intake parallel gains of lean mass, there is no need to postulate additional oxygen extraction within the active muscles. Finally, Henriksson and Reitman (1977) demonstrated an asynchrony between changes of local enzyme activity and changes of aerobic power. Eight to ten weeks of endurance training gave a 19% increase of maximal aerobic power but a 32–35% increase of succinate dehydrogenase. With two weeks of detraining, enzyme activities had returned to control values, but maximum oxygen intake remained 16% above its initial level.

The activity of certain mitochondrial enzymes (for example, creatine phosphokinase and adenylate kinase) is unchanged by endurance training (Oscai and Holloszy, 1971).

Occasional studies reported some increase in the respiratory power of cardiac muscle with training (Arcos et al., 1968; Kraus and Kirsten, 1970), but others failed to confirm these findings (Oscai, Molé, and Holloszy, 1971; Dohm et al., 1972; Gollnick and Ianuzzo, 1972). Absence of a training response may be related to the fact that the heart muscle has a very high oxidative power even in a sedentary individual.

Training Stimulus

Hypothetical causes of cardiac and skeletal muscle hypertrophy have included a local deficit of nutrients, an increased secretion of key hormones, an increased rate of external working, a lengthening of the fibers, and an increased local rate of energy expenditure (Badeer, 1964). Oxygen lack, whether secondary to functional overload or general hypoxia (Burdette and Ashford, 1965; Sulkin and Sulkin, 1965), has also been considered seriously. However, there are obvious situations, such as high-altitude experiments and patients with ischemic heart disease, wherein chronic oxygen lack fails to provoke hypertrophy.

The most satisfactory unifying hypothesis is that training is a response to functional overload (Pelosi and Agliati, 1968). An increase of muscular tension increase metabolic rate per unit mass of muscle and this stimulates protein synthesis (Badeer, 1964; Schreiber et al., 1970,1975; Hjalmarson and Isaksson, 1972; Zimmer et al., 1972; Goldberg, Erlinger et al., 1975). Hormones such as pituitary growth hormone, testosterone, thyroid hormone, and insulin are not essential to this process but may serve as "linear amplifiers" of the exercise-induced change (Gollnick, 1971). There is both an increase of anabolic activity and a slowing of protein catabolism. Responses have been studied most frequently in the heart, where hypertrophy follows afterloading (a rise of ventricular pressure) but not preloading (an increase of diastolic filling). For the first few hours, afterloading slows protein synthesis (Swartman et al., 1978), but supranormal rates develop within 24 hr. It is suggested that an increase in the force developed per unit cross-section of the muscle fibers activates the nucleic acid DNA-RNA system (Fig. 10.16), stimulating additional protein synthesis (Meerson, 1962; Badeer, 1964; Hamosh et al., 1967).

There is an increase in the RNA content of microsomal (Hamosh et al., 1967; Rogozkin, 1976) and ribosomal (Rogozkin, 1976) fractions of the sarcoplasm, although specific activity (in terms of amino acid incorporation per unit of mass) remains unchanged (Hamosh et al., 1967). Nuclear synthesis of RNA is also increased within 24 hr of inducing muscle overload (Sobel and Kaufman, 1969), and the permeability of the sarcolemma rises, thereby allowing an increased influx of amino acids (Goldberg and Goodman, 1969a). The immediate stimulus to protein synthesis may be an increased cytoplasmic level of amino acids. The acute effect of overloading is to cause protein degradation (Hatt et al., 1965; Bozner and Meessen, 1969), and this could generate triggering amino acid levels. However, stretching of the cell membrane also facilitates entry of amino acids into active fibers (Poortmans, 1978). A transient reduction of intracellular ATP probably activates the enzyme ornithine decarboxylase, which increases the rate of synthesis of polyamines (Caldarera et al., 1971; Gibson and Harris, 1974). At the same time, catecholamines activate a protein kinase via cyclic AMP. The protein kinase in turn could remove inhibition of RNA polymerase, allowing aggregation of the ribosomes into polyribosomes, with increased protein synthesis (Poortmans, 1978).

Studies on rat hindlimbs have shown that if muscles are immobilized under tension their growth is enhanced, but if they are immobilized in a relaxed position there is atrophy due to both a decrease of protein synthesis and an increase of protein catabolism (D. F. Goldspink, 1977a,b).

The training process becomes self-limiting unless the overload is augmented regularly. In the heart, for example, hypertrophied fibers develop a given ventricular pressure with less force per unit of cross-section. Tension falls, and the stimulus to increased protein formation is lost (Hood et al., 1968; Grossman et al., 1975).

TYPES OF TRAINING

Continuous Endurance Exercise

Laboratory training is usually based upon sustained periods of endurance exercise, 15–30 min of running on a treadmill or a comparable period of exercise on a bicycle ergometer. If the rate of working is kept constant from one visit to the next, a plateau of training is reached within a few weeks. To continue the conditioning process, the intensity of activity must be increased progressively. Adoption of a constant heart rate provides a simple guide to exercise prescription, although if the subject is initially in poor physical condition it may prove possible to augment the required heart rate once some training has occurred (Shephard, 1978j).

Continuous endurance exercise leads to some hypertrophy of both heart and the active skeletal muscle, with an increase of lean mass (O'Hara, Allen et al., 1979). The efficiency of movement improves (Paez et al., 1967; Shephard, Allen et al., 1968a), although much of this gain is specific to the activity that has been practiced. There is also a decrease of heart rate at a given power output as the subject becomes habituated to the laboratory and the required type of exercise (Glaser, 1966; Shephard, Allen et al., 1968b; Shephard, 1969). There may be changes of resting heart rate, blood pressure, and responses to submaximal work, with gains of maximal performance and endurance. However, the most consistent and best documented physiological response to endurance conditioning is a substantial rise of aerobic power (Shephard, 1965,1977a; Pollock, 1973); this is seen particularly in subjects with a low initial level of fitness (Müller, 1962; Shephard, 1968c; Sharkey, 1970; Gledhill and Eynon, 1972; Wenger and MacNab, 1975; but not Nordesjö, 1974; Table 11.17).

In a widely quoted study, Karvonen, Kentala, and Mustala (1957) found that a certain minimum intensity of exercise was needed to induce a training response. The subjects were young medical students. No response was seen at a heart rate of 135 beats per min, but gains of condition were found at 153 beats per min. He thus concluded that the training thresh-

TABLE 11.17
Gains (%) of Maximum Oxygen Intake (% age-related normal) with Endurance Training in Relation to Age and Initial Fitness (latter expressed relative to data of Shephard, 1977a)

Age 10–19		Age 20–29		Age 30–39		Age 40–49		Age 50+	
Initial	Gain	Initial	Gain	Initial	Gain	Initial	Gain	Initial	Gain
76	28	94	12	66	93	84	18	85	19
108	10	97	5	73	43	88	22	90	24
108	10	103	4	89	17	88	19	122	4
121	0	110	9	90	35	88	30	126	8
				96	17	94	15	150	0
				100	21	101	14		
				100	7	104	14		
						108	14		
						108	10		
						124	24		

Note: In percentage terms, the gain of maximum oxygen intake is similar in young and older adults, but because of lower initial values the old person shows a smaller absolute gain. Each entry shows the mean initial fitness and training response cited by an individual author.
Source: Based on data for male subjects collected by Pollock, 1973.

old was ~140 beats per min, or 60% of the way from the resting heart rate to the value anticipated in maximal effort. Others reported training with prolonged exercise at lower heart rates. For instance, Hollmann, Herkenrath et al. (1966) set the threshold at 130 beats per min, Durnin, Brockway, and Whitcher (1960) saw a response in soldiers who had marched 10 km at a heart rate of ~120 beats per min, and the subjects of Huibregtse et al. (1973) also responded to walks of ~10 km.

A laboratory investigation (Shephard, 1968c) exercised subjects 1, 3, or 5 times per week for 5-, 10-, or 20-min sessions at 39, 75, or 96% of maximum oxygen intake. Multiple-regression analysis confirmed that the main variables influencing the training response were initial fitness and the intensity of exercise undertaken, with the frequency and duration of sessions playing a less important role. Perhaps because subjects were initially sedentary, small (5–10%) gains were registered even in groups exercising at 39% of maximum oxygen intake. A heart rate of 120 beats per min is a relatively uncommon finding in the sedentary city dweller of North America (Shephard, 1967c; Table 11.18). Accordingly, it is not surprising that it provides an effective initial training stimulus. Indeed, a previously bedridden person can profitably exercise at a heart rate of 100–110 beats per min (American College of Sports Medicine, 1975b; Kavanagh, 1976). On the other hand, if an athlete is already running marathon distances at average heart rates of 170–180 beats per min, very intense effort will be needed to bring about a further improvement of his cardiorespiratory condition. There have been recent suggestions that the anerobic threshold provides a better guide to training intensity than does heart rate (McLellan and Skinner, 1978).

There remains a need for closer definition of the minimum stimulus to development and maintenance of aerobic power (Pollock, 1973; Shephard, 1975e). In particular, there is difficulty in equating the training significance of brief intense effort and moderate but prolonged activity. If training is continued for some months, subjects who undertake less intense or less frequent activity eventually approach the gains shown by those with a higher weekly work output (Sidney and Shephard, 1978b; Fig. 11.7). In an experiment of fixed and short duration, the critical variable may thus prove the total work performed (Sharkey, 1970). However, if all subjects are willing to accept a continuing increase in the intensity of their training program, those undertaking frequent and fairly long sessions are successful in maintaining their advantage (Pollock, 1973; Gestman et al., 1977).

Many businessmen are interested to know the minimum time required for conditioning (see below). K. H. Cooper (1968b) implied that if the intensity of exercise is adequate, 5 min • day^{-1} may bring subjects to a minimum standard of fitness, and his view is supported by the findings of Hollmann, Herkenrath et al. (1966), Shephard (1968c), and Bouchard (1975). However, the training effect can be augmented by longer sessions (Wilmore, Royce et al., 1970; Yeager and Brynteson, 1970). Thus others (for example, P. O. Åstrand, 1967) advocate five 30-min sessions per week to improve physical condition and three 30-min sessions per week to maintain the status quo.

If intense rhythmic effort is sustained for as long as 30 min without remission, there may be a substantial rise of systemic blood pressure (Chap. 6). This is particularly likely if exercise surpasses the anerobic threshold. Exercise-induced hypertension can be an important point when planning a conditioning program for a middle-aged, coronary-prone individual. If angina develops during exercise, an in-

TABLE 11.18
Periods of Day at Specified Heart Rate[a]

Period of Day (Min)	Heart Rate (beats • min^{-1})			
	<100	100–119	120–139	140–170
Mean period of day (min)	1329	91	16[b]	2[c]
Range (min)	1155–1404	4–269	0–54	0–10

[a]Data obtained by tape recordings of electrocardiogram on eight relatively sedentary men living in Toronto.
[b]Two of the eight subjects had no readings greater than 120 beats per min.
[c]Six of the eight subjects had no readings greater than 140 beats per min.
Source: Based on data of Shephard, 1967C.

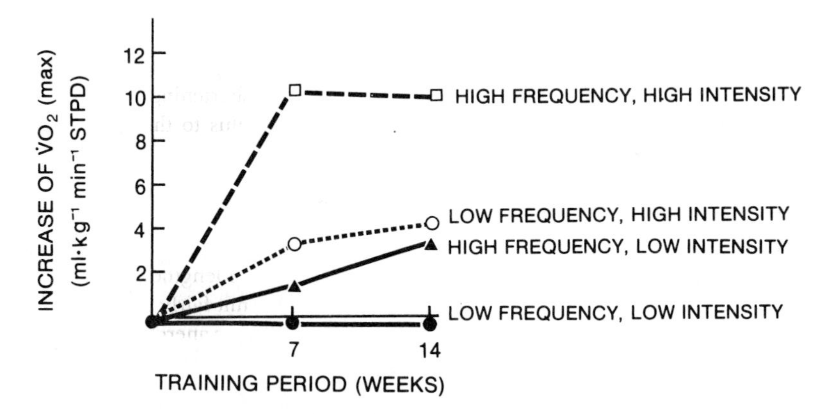

FIG. 11.7. Change of predicted maximum oxygen intake with varying frequency and intensity of training. (Data of Sidney & Shephard, 1978b, for 65-year-old subjects.)

terval training plan can provide a much more effective basis of conditioning than continuous activity (Kavanagh and Shephard, 1975b).

In the context of athletics, continuous exercise is commonly recommended during the early phases of preparation, particularly for endurance competitors such as distance runners or cyclists. The selected daily training period typically lasts for 3 to 5 times the duration of a race. Speed is adjusted to produce a steady pulse rate of 150–160 beats per min, and for the first few weeks the objective is a general conditioning of the cardiovascular system. Once this has been accomplished, the competitor progresses to faster speeds, producing pulse rates of up to 180 beats per min and covering distances of about twice the intended range. At each training session, several repetitions are alternated with perhaps 5 min of walking and jogging. Finally, as competition approaches, activity is undertaken over the intended distance; preferably, the actual running track is used, so that the correct pace is learned, with mechanical efficiency and habituation of the contestant being brought to a maximum.

An athlete naturally trains by repetition of his sport. For the ordinary citizen, there is little to choose between such activities as running, cycling, and walking, provided that a comparable intensity, duration, and frequency of activity is attained (Pollock, 1973). However, it is important that a substantial fraction of the body musculature be activated.

The specificity of endurance training (Clausen, 1977) has been discussed briefly (Chap. 3). Gains of maximum oxygen intake that have been developed on the treadmill can apparently improve subsequent performance of bicycle ergometer exercise, but bicycle ergometer training has less influence upon the subsequent tolerance of treadmill running (Pechar, McArdle et al., 1974). Perhaps because of specific habituation, training-induced decreases of heart rate during one form of submaximal activity are not transferred to the other (J. Roberts and Alspaugh, 1972).

Gains of maximum oxygen intake induced by leg exercise improve the maximum power output of the arms, but arm training has a much smaller impact upon leg performance (Clausen, 1973; Magel, Foglia et al., 1975). The explanation seems that when doing short-crank arm work, effort is limited substantially by local muscle strength. Training leads to a strengthening of the active muscles. Fewer motor units are then activated in submaximum work (giving a lesser tachycardia), and perfusion is more readily sustained (giving a higher maximum cardiac output, a higher maximum oxygen intake, and a lesser accumulation of lactate). Several authors have compared the response to training of one and two legs. When a subject is required to operate a bicycle ergometer using only one leg, there is again some peripheral restriction of effort, and the training resulting from repetition of such activity is not fully generalized to normal two-legged operation of the ergometer. Thus C. T. M. Davies and Sargeant (1975b) reported that a program of one-legged training increased the one-leg maximum oxygen intake by 14%, but the two-leg maximum oxygen intake by only 5%. Saltin, Nazar et al. (1976) trained one leg by sprint exercise (150% of maximal oxygen intake) and the other leg by endurance activity (75% of maximal); respective gains of one-legged maximum intake were 0.32 and 0.54 l • min^{-1}, while the two-legged maximum oxygen intake was increased by only 0.30 l • min^{-1}. Last, Henriksson (1977) found gains of 11 and 4% of maximum oxygen intake in trained and untrained legs, the trained leg showing a 27% greater activity of the aerobic enzyme succinate dehydrogenase and a corresponding increase in the percent usage of fat during exercise.

Interval Training

Interval training is widely used as a method of preparation for competitive running (E. L. Fox, Bartels et al., 1975). It also has its advocates in the conditioning of sedentary middle-aged men and seems particularly suitable after a coronary attack (Kavanagh and Shephard, 1975b; Kavanagh, 1976). If equal amounts of work are performed, interval training gives approximately the same gain of maximum oxygen intake as a continuous training program (Eddy et al., 1977). A typical interval training plan involves alternation of a fast activity (such as running) with a slower activity (such as walking or jogging). Depending upon the type of any intended competition and the preference of the individual coach, there is substantial variation in the distance and the speed of the fast runs, the duration and type of recovery activity, and the total number of repetitions per training session (Wilt, 1968).

Slow-interval training is used in the early stages of a competititve season to develop cardiorespiratory fitness. A typical regimen might require three repetitions of a 7-unit sequence in which three 400-m runs (pulse 180 beats per min) were alternated with three 400-m jogs, followed by a 400-m walk.

A middle-distance runner later progresses to *fast-interval training*. This is intended to develop anerobic endurance. Cellular gains include an increase of glycogen stores, alkaline reserve, ATP, myoglobin, and probably some increase of glycolytic enzymes (see above). A competitor in a 1500-m track event might prepare himself by 7-unit sequences as above, running each 400-m interval at 1–2% above his average competitive speed. Physiological data (Table 11.19) suggest that a lengthening of the active phase and/or a shortening of the recovery interval increase the stimulus to the development of anerobic endurance (I. Åstrand, Åstrand et al., 1960b; Christensen, Hedman, and Saltin, 1960; R. H. T. Edwards, Ekelund et al., 1968).

Repetition running is a variant of interval training popular with contestants in long-distance events. Active phases are lengthened, and intervals are also extended to permit fairly complete recovery. Whether aerobic power or anerobic capacity is developed by this type of training depends largely upon the speed attained during the running phases.

Fartlek training is an informal type of interval training, carried out under cross-country conditions. Unless carefully monitored, it may fail to make the necessary physiological demands on the body.

I. Åstrand, Åstrand, et al. (1960b), Christensen, Hedman, and Saltin, (1960) and R. H. T. Edwards, Ekelund et al. (1968) provided useful physiological comparisons of several possible interval regimens (Tables 11.19 and 11.20). With very brief intermittent activity (< ½ min), the muscles perform a full complement of work, but the cardiorespiratory system shows no real approach to a steady-state. The oxygen consumption during the active phase is only 63% of maximum, the blood lactate remains low, and the stimulus to the cardiorespiratory system is

TABLE 11.19
Comparison of Responses to 1 hr of Intermittent or Continuous Exercise

Power Output	Work (min)	Rest (min)	O_2 Intake (l • min^{-1} STPD)	Resp. Min. Volume (l • min^{-1})	Heart Rate (beats • min^{-1})	Blood Lactate (mmol • l^{-1})
Intermittent (21.2 kN • m • min^{-1} during active phase)	½	½	2.9	63	150	2.2
	1	1	2.9	65	167	5.0
	2	2	4.4	95	178	10.6
	3	3	4.6	107	188	13.3
Continuous						
(10.6 kN • m • min^{-1})	60	—	2.4	49	134	1.3
(21.2 kN • m • min^{-1})	9	—	4.6	124	190	16.7

Source: Based on data of I. Åstrand, Åstrand et al., 1960a,b.

TABLE 11.20

Influence of Varying Patterns of Interval Training on Oxygen Intake, Respiratory Minute Volume, and Blood Lactate*

Work Phase (sec)	Rest Phase (sec)	Oxygen Intake ($l \cdot min^{-1}$ STPD) Active Phase	Recovery Interval	Respiratory Minute Volume ($l \cdot min^{-1}$) Active Phase	Recovery Interval	Blood Lactate ($mmol \cdot l^{-1}$)
(Continuous, max. 2400)	—	5.6	—	158	—	16.7
5	5	4.3	4.5	101	101	2.6
5	10	3.4	3.0	81	77	1.8
10	5	5.1	4.9	142	140	4.9
10	10	4.4	3.8	104	95	2.2
15	10	5.0	4.5	139	144	5.7
15	15	4.6	3.8	90	95	2.3
15	30	3.6	2.8	79	64	1.8

*Treadmill speed 20 $km \cdot hr^{-1}$, running time 30 min.
Source: Based on data of Christensen, Hedman, and Saltin, 1960.

relatively mild. However, if expressed per unit of work rate, the heart rate, ventilation, and oxygen cost of exercise are all greater during intermittent than continuous activity; this presumably reflects the cost of lactate and phosphagen resynthesis (R. H. Edwards, Ekelund et al., 1968). With brief-interval work, we may speculate that much of the oxygen need is shelved temporarily by development of an alactate oxygen debt (Christensen and Höberg, 1950b; Christensen, Hedman, and Saltin, 1960). Plainly, such a debt could be repaid over the half-minute recovery intervals. Brief interval exercise thus seems likely to develop muscle strength, muscle endurance, and alactic power.

With rather longer intervals (~1 min), a fair load is imposed upon the cardiovascular system, but there is still only a limited buildup of anerobic metabolites (Table 11.17). We have seen (Chap. 2) that light activity aids lactate elimination during the recovery intervals (Hermansen and Stensvold, 1972; Belcastro and Bonen, 1975; Weltman et al., 1977). A 1-min jog/walk interval program might thus seem a good pattern of training to suggest for the post-coronary patient with a tendency to angina of effort (Kavanagh and Shephard, 1975b). Depending on the age of the subject (Chap. 6), a maximum stroke volume is reached at a power output corresponding to maximum oxygen intake or less, and there is little reason to tax anerobic mechanisms by extended interval work if the objective is to produce an increase of aerobic power (Karlsson, Åstrand, and Ekblom, 1967).

G. R. Cumming (personal communication) argued that stroke volume is largest in the early phases of recovery, so that an interval program (with many recovery phases) may provide more effective cardiovascular conditioning than continuous exercise. His data were obtained in the supine position. P. O. Åstrand and Rodahl (1977) maintain that with upright bicycle ergometer exercise, the maximum stroke volume is seen during rather than following activity; however, recent data from this laboratory suggest that Cumming's thesis is also correct for upright exercise.

If the active phase is continued beyond 1 min, there is a greater accumulation of lactate, and with 3 min of exercise, the terminal lactate concentration approaches the limit of endurance. Such training is much less pleasant than either short-interval or aerobic work, and it should be reserved for later in the training season. It is doubtful if interval activity with a 3-min/3-min timing and an intensity of 21.2 $kN \cdot m \cdot min^{-1}$ could be sustained much beyond the 1-hr period of I. Åstrand's study. Nevertheless, the total work performed in this fashion (30 × 21.2 kN-m) is in marked contrast with the limiting achievement for continuous work (9 × 21.2 kN-m). Presumably, the recovery intervals permit general dispersal of lactate throughout body fluids (Chap. 2). To the extent that there is a potential for development of anerobic capacity, long-interval work with a somewhat reduced power output might seem the optimal method of accomplishing this objective (E. L. Fox, Bartels et al., 1977). However, an equally large buildup of lactate and presumably a comparable stimulus to anerobic systems can be realized through much shorter bursts of activity, provided that the intensity of effort is high and the recovery period is foreshortened (as during fast-interval work, G. M. Davis, Eynon, and Cunningham, 1977).

Roskamm (1967) compared physiological responses to two interval programs and one form of continuous training. A 2½-min exercise/2½-min-recovery format yielded the largest increase of maximum performance (measured as Watts per pulse, and reflecting the ability to carry out a combination of aerobic and anerobic work). On the other hand, the tolerance of submaximal work (as indicated by the heart rate at 70% of aerobic power) was greatest in the group who had undertaken continuous endurance training. Knuttgen et al. (1973) compared a ½-min/½-min and a 3-min/3-min program; they concluded that the latter was more effective in developing maximum oxygen intake. Such data are in accord with the suggestions of Pollock (1973) that the pattern of training should be dictated by its purpose. Continuous training develops mainly aerobic power, but by suitable adjustment of active and recovery phases interval work can build strength, anerobic power, anerobic capacity, or aerobic power.

Development of anerobic power and capacity is relatively specific to the muscle fibers engaged in a training program (see above). The wise athlete thus concentrates his preparation upon the type of activity that he intends to follow in competition. By appropriate adjustment of the intensity, he can also regulate the fiber type that is trained (Terjung, 1976; Chap. 4). With low intensities of effort, the slow-twitch (type I) fibers are first recruited, although if the activity is continued for a sufficient time, glycogen depletion forces a secondary recruitment of fast-twitch (type II) fibers. Gains from distance training include an increase of capillary density, a rise in the myoglobin content of the active fibers, and a development of mechanisms for fat metabolism. With intensive forms of interval training, both type I and type II fibers are recruited from the outset of activity and both types of fiber share in the resultant development. A final training variable is the extent of glycogen depletion and its rate of replenishment (Chap. 2). Up to 48 hr are needed to replace glycogen reserves (Piehl, 1974), a time at variance with the daily training schedule of many athletes; the possible effects of exercising for long periods on alternate days has yet to be studied rigorously. The current competitor spends many hours of training per day. This probably confers significant gains of skill and possibly helps to augment the usage of fat relative to carbohydrate. However, available data (Table 3.5) suggest that the aerobic power per unit of body mass is much the same as in earlier groups of endurance athletes who lacked the government funding to follow such demanding schedules.

Sprint Training

The sprinter requires a brisk reaction time, explosive strength, experience in moving at the maximum attainable speed (~ 40 km • hr^{-1}), and a large anerobic power. However, competition makes little demand upon his anerobic capacity. The duration of training sprints should thus be sufficient to permit acceleration to maximum speed (~ 6 sec), but little advantage is gained if the effort is continued for more than 10 sec per bout.

The heart rates of the sprinter may exceed the anticipated steady-state "maximum" value. However, the duration of such activity is too short to allow a full adaptation of the peripheral circulation, and the stroke volume may actually be less than the resting value, particularly if the breath is held during the sprint. For these reasons, sprinting normally has little effect upon aerobic power.

If an athlete wishes to improve his cardiorespiratory fitness while practicing the technique of sprinting, he may alternate 50-m sprints with jogging or walking an equal distance ("*interval sprinting*"). The physiological situation is then analogous to brief but intense intermittent work (Table 11.19). A second variant, designed to improve explosive strength, is "*acceleration sprinting.*" Here, a typical sequence involves jogging, striding, sprinting, and walking over equal distances. The objectives are twofold: to warm up gradually to maximal effort, thereby avoiding muscle and tendon injuries, and to develop maximum speed at each sprint without cumulative fatigue. An extended recovery period is thus essential. "*Hollow sprints*" include two maximum efforts, separated by a period of jogging and followed by a period of walking. This technique extends the period of intense activity, while preserving its sprint characteristics; more time is permitted for glycolysis, with a consequent stimulus to the development of anerobic capacity.

Strength Training

Isotonic Training

Muscle force and endurance can be increased by the isotonic use of weights, isokinetic and eccentric muscle contraction, and isometric exercises.

Regular isotonic weight lifting increases isotonic strength but has much smaller effects on isometric strength (Berger, 1976; Table 11.21). Isotonic gains can be demonstrated over a wide range of joint angles (D. H. Clarke, 1973). As might be expected from the nature of the training, local blood flow is improved, and there is an associated improvement of endurance (particularly endurance of isotonic activity). If the exercise is repeated over many months, there may also be some development of muscle bulk (DeLorme, 1945; McMorris and Elkins, 1954). However, there is commonly a discrepancy between gains of muscle force and the increase of muscle bulk as assessed either clinically (muscle girth) or by use of soft-tissue radiographs (Fried and Shephard, 1970). It is thus probable that a part of the observed increase in both strength and endurance reflects a lesser inhibition of the motor neuron pool (Ikai and Steinhaus, 1961).

The largest and most rapid gains of strength occur when a muscle performs rhythmic exercise under a condition of overload. A muscle is said to be overloaded when the initial fiber length is greater than the optimum for subsequent development of tension (Fig. 1.9). A typical daily training routine comprises several sets of contractions, each involving 2–10 repetitions of an overloaded lifting effort.

If 10 repetitions are used, the first set may involve the repeated lifting of a load 50% of the maximum tolerated for 10 efforts; in the second set, the load is increased to 75% and in the final set it is further increased to 100% of the 10-repetitions maximum (DeLorme and Watkins, 1948; O'Shea, 1976). One disadvantage of this approach is that progressive fatigue causes pain and a restricted range of motion during the final set. Zinovieff (1951) thus proposed reducing the load as fatigue developed. According to DeLorme (1945), the use of heavy loads with few repetitions develops power, while a program involving frequent repetitions with light loads develops endurance. Nevertheless, loads as light as 60% of the one repetition maximum produce gains of strength if practiced conscientiously for periods of 6 weeks or longer (Bonde-Petersen, 1960; Hansen, 1967). The response seems to be similar with slow (2/min) or fast (15/min) repetitions (Salter, 1955; Chui, 1964).

Strictly speaking, exercise in the weight room is dynamic or rhythmic rather than isotonic. The subject must first develop an isometric force equal to the mass that is to be displaced, and muscle tension is modified further to accommodate changes in leverage as the movement progresses. The initial isometric effort is of sufficient intensity to occlude or partially occlude the local circulation, and for this reason the systemic blood pressure tends to rise, increasing the cardiac work load (Chap. 6). Many people hold their breath with the glottis closed while weight lifting, a form of Valsalva maneuver. If the effort is sustained (as in an exhibition weight lifter), the initial increase of systemic blood pressure is followed by a dramatic fall (due to obstruction of venous return), and there is a second increase of pressure as the expiration is released and blood again enters the thorax. The Valsalva effect is unpleasant even for a healthy person and can be dangerous to an individual with cardiac disease. Although fixation of the chest is essential at certain points in the lifting of a heavy mass, the weight lifter should endeavor to breathe as naturally as possible throughout his performance.

Despite the potential risks of weight lifting,

TABLE 11.21
Influence of Various Types of Muscular Training upon Maximal Strength and Endurance

Type of Training	Gain of Maximal Strength (%)		Gain of Endurance at 60% Maximum (%)		Author
	Isometric	Isotonic	Isometric	Isotonic	
Isotonic, 60%, 150 contractions per day for 30 days	+6	+29	—	+5040	Bonde-Petersen et al.(1961)
Isotonic, 60%, 100 contractions per hr for 30 days	—	+13	−2	+630	Hansen (1967)
Isometric, 60%, 150 contractions per day for 30 days	+4	+ 6	+1060	+41	Hansen (1961)

Source: Based on reports collected by P. O. Åstrand and Rodahl, 1977.

practical experience is not particularly alarming. Karpovich found 494 accidents in a group of 31,702 individuals performing various types of weight training. The majority of problems were due to injuries of the back, shoulders, and fingers, and there were only five instances of hernia and one of sudden death.

Isotonic exercise does little to improve cardiorespiratory fitness. The endurance time for a treadmill run or all-out cycle ride may be extended due to better perfusion of the strengthened muscles (Howell, Kimoto, and Morford, 1962), but there is little gain of aerobic power. The lack of change in maximum oxygen intake can be explained in terms of the low oxygen cost of weight lifting. If a mass of 50 kg is lifted 30 times through a total distance of 2 m, the work performed is just under 30 kN • m; spread over 30 min of gymnasium time, this requires an average rate of working of only 1 kN • m • min^{-1}, compared with 12 kN • m • min^{-1} during a hard bicycle ergometer ride. Many athletes appreciate this point, undertaking supplementary exercise to develop their cardiorespiratory endurance, but the occasional exerciser may believe he has met the requirements of health through the use of weights. Such an individual could build an impressive shoulder musculature and be obliged to carry this increased body mass with an unimproved and inadequate heart and circulation.

Isometric Training

In isometric training, no appreciable external shortening of the muscle occurs. Contractions may be made against a rigid external device such as a dynamometer, or the muscles may be tensed against their natural antagonists. This offers the possibility of a program to maintain muscle strength while a limb is immobilized within a plaster cast. If the patient is unwilling to contract his muscles voluntarily, activity may be induced electrically (by faradic stimulation). Although no external work is performed, appreciable amounts of energy may be expended (D. H. Clarke, 1960a).

Much of the interest in isometric training was engendered by Hettinger and Müller (1953) and Hettinger (1961), who maintained that increases in muscle strength of up to 5% per week could be brought about by as little as one isometric contraction per day. If the contraction was of maximum strength, a 1- to 2-sec effort provided an adequate training stimulus, while at two-thirds of maximum strength 4–6 sec of activity was needed. The maximum potential response could still be attained with more sustained contractions at 40–50% of maximum voluntary force, but if the intensity was reduced to 20% of maximum effort a gradual waning of strength occurred. The current recommendation of German investigators is thus a contraction at 50–70% of maximum isometric force, held for 3–6 sec and repeated 5 times per day.

Several authors have duplicated Müller's findings (Rarick and Larsen, 1959; Morehouse, 1967; Cotten, 1967; Röhmert, 1968b; Hollman and Hettinger, 1976), but other investigators have been less successful (Mayberry, 1959; Bonde-Petersen, 1960; Doré et al., 1977). Lack of response could reflect the use of subjects with a low training potential (Müller and Röhmert, 1963). By analogy with cardiorespiratory training, it is reasonable to suppose that possible gains of maximum voluntary force are influenced greatly by the initial condition of the muscles concerned. However, there is also a dangerous fallacy in Müller's approach of using a similar device to both produce and measure training. Part of the apparent increase in strength that he has reported could be no more than a learning of technique. In support of this criticism, gains of strength are very specific to a particular movement executed at a particular angulation of the joint (Meyers, 1967; Belka, 1968). Development of isometric strength and endurance in this position are not associated with more general gains of dynamic strength and endurance. Indeed, if there is an appreciable fiber hypertrophy, the resultant decrease of capillary density may lessen endurance (Fig. 11.5; Table 11.21).

Some human studies have found little change of muscle dimensions with isometric training (D. L. Rose et al., 1957; Hislop, 1963; Ikai and Fukunaga, 1968; Van Uytvanck and Vrijens, 1971), although Meyers (1967) demonstrated a significant increase of relaxed arm girth. Histological studies suggest that any hypertrophy occurs in type II (fast-twitch) fibers; unfortunately, these are particularly vulnerable to fatigue from a diminution in their capillary density (Chap. 4; Thorstensson, 1976). Strength training has no influence upon reaction time (D. H. Clarke and Henry, 1961) and does little to increase the speed of movement (D. H. Clarke and Henry, 1961; D. H. Clarke, 1973).

Distance runners have no greater isometric strength than the average citizen, but their tolerance of repeated isotonic contractions is many times that of a sedentary subject. The athletes with the greatest isometric strength are sprinters and throwers. However, sprinters have a normal isotonic endurance, and if loading is expressed as a percentage of maximum isometric strength, the isotonic endurance of the throwers is less than average.

Physiological mechanisms contributing to gains of strength and dynamic endurance include improved coordination (J. W. Hansen, 1967) and a decrease of central inhibition (Ikai and Steinhaus, 1961). The intense anerobic work involved in maximum or near-maximum static contractions may also stimulate protein anabolism, with an increase of muscle bulk and a resultant development of alactate power (as in repetitive sprints). However, it would be surprising if the potential for improvement of alactate power could be realized with one intense contraction per day, as a literal acceptance of Müller's work might suggest.

Eccentric Training

Isotonic training typically involves both concentric and eccentric contractions (Chap. 4), but most authors have not distinguished the relative contributions of weight lifting and weight lowering to the development of muscular strength. Seliger, Dolejš et al. (1968) and B. L. Johnson, Adamczyk et al. (1976) both concluded that the two possible modes of training yielded approximately equal gains of isometric strength, although B. L. Johnson, Adamczyk et al. (1976) also noted that subjects found the eccentric training easier. Others have confirmed the effectiveness of eccentric muscle-building programs (Singh and Karpovich, 1967; Laycoe and Marteniuk, 1971), a response occurring not only in the agonist muscles, but also in their antagonists.

Cardiorespiratory training can also be performed in an eccentric manner; for example, subjects may be required to run downhill on a treadmill.

Isokinetic Exercise

One of the more recent training fads is isokinetic exercise, whereby a torque generator allows the limb to contract at a predetermined speed against an infinite resistance (Hislop and Perrine, 1967; Chu and Smith, 1976). Since maximum force can be sustained throughout the possible range of joint motion, this technique of training combines elements of isotonic and isometric effort. Critics of the method point out that an isokinetic movement has few natural counterparts, with the possible exception of swimming. Nevertheless, comparisons with isotonic and isometric training suggest that greater gains of strength can be realized by the isokinetic procedure (Thistle et al., 1967; Moffroid et al., 1969).

Strength and Performance

Strength has little influence upon the maximum speed of limb movement (D. H. Clarke, 1960b; F. M. Henry and Whitley, 1960; L. E. Smith, 1961; Chap. 4). It is thus not surprising that weight training does little to improve speed (D. H. Clarke, 1973). On the other hand, there is no evidence to support the fear that muscle development slows movement and causes an individual to become "muscle-bound" (Zorbas and Karpovich, 1951).

Some authors have found gains of sprint performance from isotonic exercise, particularly when this is combined with flexibility training (Dintiman, 1964). Others have found direct practice of the sport a more effective method of preparation (A. L. Thompson and Stull, 1959; Schultz, 1967).

Tasks requiring explosive force, such as a vertical jump, are helped by isotonic training, and most comparisons have favored this method over isometric preparation (Chui, 1950; R. L. Campbell, 1962; Berger, 1963). Isotonic training also improves performance of tasks requiring muscular endurance, such as chin-ups and sit-ups (Capen, 1950; Bonde-Petersen, 1960).

Cross-Transfer

Many authors have described a cross-transfer of strength, exercise of one limb producing gains in the maximum voluntary force of its counterpart (Scripture et al., 1894). This has variously been ascribed to "indirect practice" (Scripture et al., 1894), "cross-education" (W. W. Davis, 1898), a diffusion of motor impulses to the contralateral arm (Wissler and Richardson, 1900), associated activity to maintain balance and counteract shifts in the center of gravity (Hellebrandt et al., 1947), a transfer of coordination or a development of tolerance to fatigue (Slater-Hammel, 1950). Many of these mechanisms are regulatory rather than structural in type; however, there is also some histological evidence of muscle fiber development in the contralateral limb (Reitsma, 1969).

Other Methods of Training

Sports Participation

Sport has little place in the training of a professional athlete unless the activity is the one in which competition is envisaged (Painter and Green, 1978). Time is occupied that could have been devoted to more effective methods of preparation; there may be little gain in performance of the chosen activity (Daub et al., 1978) or, even worse, the skills that have been learned may interfere with the techniques needed in competition ("negative transfer"; Cratty, 1971). Further, local muscle hypertrophy may add to body mass while hampering the desired performance (Fried and Shephard, 1971).

Recreational sport might seem a good basis for improving the fitness of the general population. However, there are several practical problems with such an approach. The energy cost of a given sport depends very much upon the personality of the player and the skill of his fellow team members and/or opponents. A postcoronary patient with a type A personality is in danger of exercising too hard (Shephard, 1978j), while many sports that are popular among the middle classes—golf, sailing, bowling, and the like—fail to reach the threshold intensity of effort needed for endurance training (Durnin and Passmore, 1967; Shephard, 1977a). Further, many sports pursued at school are not continued into middle and older age. Finally, mass participation in many games would create a prohibitive demand for land and facilities (Shephard, 1977a).

Some indication of the training obtained from different sports can be inferred from estimates of their energy cost (Durnin and Passmore, 1967) and from surveys of aerobic power and muscular strength in successful contestants (Tables 3.5, 11.22; Shephard, 1978a,b). Although golf does little to develop cardiorespiratory fitness, the prolonged periods of moderate energy expenditure may be quite effective in burning excess body fat.

The aerobics point system (K. H. Cooper, 1968a,b) is a practical attempt to equate various recreational activities in terms of their respective abilities to improve cardiorespiratory condition. Limitations of current point values include (1) discrepancies between the points awarded and the energy used in different sports and (2) uncertainties regarding the equivalence of training stimulation by short, intense activity and moderate but prolonged effort (Massie, Rode et al., 1970). Nevertheless, Cooper has made an important contribution by showing how a person can combine training stimuli developed through participation in a number of sports.

Circuit Training

The concept of circuit training was developed by R. E. Morgan and Adamson (1965). Eight to twelve "stations" are arranged around a gymnasium and different forms of calisthenics are performed at each. Morgan and Adamson suggested a good circuit should include exercises for the arms, shoulders, back, abdomen, legs, and combinations of these several areas.

The individual to be trained determines for himself how many repetitions he can make at each station. He then moves three complete times around the circuit, performing half the maximum possible number of repetitions at each station. On subsequent

TABLE 11.22
Muscle Strength and Endurance in Normal Subjects and Athletes

Subjects	Isometric Force (N)		Maximum Number of Isotonic Contractions at 1/3 Maximum Force	
	Leg Ext.	Arm Flexion	Leg Ext.	Arm Flexion
Women				
Average-normal	—	88	—	70
Middle- and long-distance runners	520	98	68	68
Hurdlers	471	98	67	67
Jumpers	549	118	57	58
Sprinters	471	118	71	71
Throwers	667	167	43	43
Men				
Average-normal	540	167	48	75
Middle- and long-distance runners	540	186	399	48
Hurdlers	598	186	67	46
Jumpers	667	206	49	45
Sprinters	697	186	52	65
Throwers	863	255	38	51

Note: (1) Subjects with a large isometric force tend to have a low level of isotonic endurance (and vice versa). (2) Development in athletes may be limb-specific (see especially the leg endurance of the middle-distance male runners).
Source: Based on data of Ikai, 1966.

days, he attempts to move faster around the circuit, and gradually increases the number of repetitions.

The type of physiological response induced by circuit training depends very much upon the content of the individual circuits (Banister, 1964). A substantial development of aerobic power is possible, particularly if items such as stepping, skipping, and running are included. However, in the more usual circuit, the emphasis is upon muscle building. Comparison with interval and continuous training shows that the latter methods give a greater development of aerobic power but lesser gains of muscular strength (Roskamm, 1967).

Calisthenics

The energy expenditure in calisthenics as commonly performed is quite low. Weiss and Karpovich (1947) found that the oxygen cost of trunk bending was only 20% above the resting value, and the most vigorous of 40 common exercises yielded an oxygen consumption of only 0.81 • min^{-1}. However, much depends upon the cadence, and with a sufficiently forceful pattern of movement calisthenics can induce gains of strength, body composition, and cardiorespiratory performance, along with improved flexibility (G. M. Andrew et al., 1974; M. Cox and Shephard, 1979).

Membership of a group with an enthusiastic leader provides useful motivation, particularly for the gregarious extravert (Massie and Shephard, 1971), but the repetitive nature of calisthenics can become boring, and unless classes are carefully graded, the intensity of effort demanded of the individual participant will be poorly related to his potential power output.

The Canadian 5BX and 10BX plans are personal schemes of progressive exercise that require no formal apparatus. Each day's program is divided between 5 min of normal calisthenics and 6 min of stationary running or longer periods of outdoor running and walking (Orban, 1962). The required activities are detailed in six charts, the starting point being graded by age, sex, and initial fitness level. The energy cost of the program (averaged over 11 min), increases from 33 to 67–71 kJ • min^{-1} as a subject moves from chart 1 to chart 6. The main stimulus to development of cardiovascular condition is the period of running. A person who covers 1.6 km in 7.5 min (chart 4c) is developing a final oxygen consumption of 46 ml • kg^{-1} • min^{-1}. This intensity of activity provides a good cardiovascular stimulus (Kappagoda et al., 1977), although its duration is rather short. In contrast with normal continuous training, there is also some development of strength (Malhotra, Sen Gupta, and Joseph, 1973), although less than would be obtained from an isotonic program. There seem two main drawbacks to the 5BX training plan: (1) boredom and (2) the twisting exercises of charts 4–6 may precipitate injuries of the intervertebral discs.

Strategy of Training

The Athlete

Examination of the characteristics of successful competitors (Table 3.5) shows that the performance of an endurance athlete will be helped by a selective development of his oxygen transport system. Depending on the type of contest and the initial physique, strengthening of key muscles may be helpful. However, if muscular development is disproportionate to the improvement of aerobic power, the added body mass can set the athlete at a disadvantage, particularly in events that call for the lifting of body mass against gravity (Fried and Shephard, 1971). In sports in which a substantial fraction of the total effort is sustained by a relatively small muscle group, performance may be helped by specific development of the muscles concerned—presumably improvements of local blood flow (Chap. 6) make a substantial contribution to observed gains. Other events call for enhancement of explosive force, isometric strength, or muscular endurance (Table 3.12). All forms of activity are helped by improvements of flexibility. The strengthening of articular cartilage, ligaments, and bone is of particular significance in contact sports, where it helps to minimize the risks of injury.

A suitable training emphasis for different classes of runner is illustrated in Fig. 11.8. Similar graphs can be created for contestants in other sports. Repetitive and acceleration sprints develop speed, suitable forms of fast-interval work develop anerobic capacity, and continuous running or moderate-length interval work builds up aerobic power (Wilt, 1968).

Let us suppose an athlete competes for 4 months of each year. Following the season, he is allowed 1 month "holiday" of active relaxation. During this period, he participates in regular endurance-type activities such as vigorous swimming; his twin objectives are (1) to prevent a gross deterioration of physical condition and (2) to avoid sports that will interfere with his specific acquired skills. The first 3 months of renewed training are devoted to the development of aerobic power and strength. Depending on the type of event in which the athlete competes, there is subsequently an increasing emphasis upon

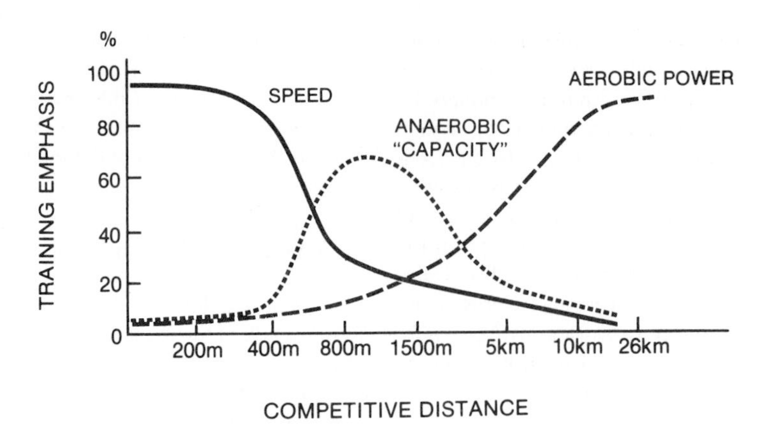

FIG. 11.8. The strategy of training for track events. The relative training emphasis is suggested in relation to the intended distance of competition. Similar curves can be developed for other types of competition.

speed and anerobic capacity. Over the competitive season, the quantity of training is adjusted to maintain—and, if possible, to improve—the physiological status of the athlete (J. S. Hanson, 1975) while allowing sufficient rest for optimum performance on days of competition. Precompetition rest is important to the psychological preparation of the athlete (Vanek, 1979) and also allows time for restoration of fluid balance, replenishment of glycogen stores, and recovery from minor (subclinical) injuries.

The daily duration of training depends upon the persistence of the individual, but in general is increased as physical condition improves. Except at times of competition, a distance competitor may cover several times his competitive distance, and a speed athlete may make 10–20 repetitions of his chosen event. There is no strong evidence that physiological rewards result from the prolonged hours of preparation of the current international competitor.

It remains uncertain as to whether there is a physiological basis for the alleged deterioration of performance with overtraining. Conceivable problems include a depletion of body mineral reserves (Chap. 2) and glycogen stores (Chap. 2), a disturbance of fluid balance (decrease of blood volume and/or intracellular fluid), and buildup of subclinical injuries (Stanesçu, 1971).

Muscle soreness was once held to be a reflection of spasm, but it shows no obvious relationship to the electromyogram; since it seems to be correlated with the hydroxyproline/creatine ratio, it probably reflects a disruption of connective tissue elements in muscle and/or its attachments (W. H. Abraham, 1977). Certain investigators (for example, Laguens and Gomez Dunn, 1967; E. W. Banister, 1971) observed changes of subcellular structures, including mitochondrial swelling and destruction of cristae after exhaustive exercise A high-protein diet apparently averts such changes (Cvorkov et al., 1974). It has been claimed that the abnormal appearances are a fixation artifact, but this leaves unexplained why exercise alters the fixation process (Terjung, Klinkerfuss et al., 1973; Bowers et al., 1974).

The perceived demands of international competition are now such that activity is often continued to the detriment of rest and recreation, and this can become burdensome to the athlete (Shephard, 1975c,1978c). The diminishing physiological returns for a given investment of time and effort have a negative psychological impact upon competitive performance. The increase of muscle mass at the expense of cardiorespiratory conditioning can be a further hazard for the distance athlete (Fried and Shephard, 1971).

The Sedentary Adult

The ordinary citizen seeks from a training program the aspect of fitness that he values (see above), be it physical condition for sports participation, control of body "weight," protection against a heart attack, or "energy" to meet the physical demands of daily living.

A high proportion of recruits have reached middle age (40–50 years) and are somewhat obese. Their main needs are then a substantial increase of energy expenditure (to help in reducing body fat) and progression to exercise of sufficient intensity to improve cardiorespiratory condition. Within this framework, activities such as walking, running, and cycling are equally effective modes of conditioning (Pollock, Dimmick et al., 1975).

Some items can be built into the daily routine.

Thus aerobic power can be increased and body fat can be reduced by climbing 25 flights of stairs per day (Fardy and Ilmarinen, 1975). A gradual increase in the intensity of the formal program greatly reduces the likelihood of minor orthopedic problems that otherwise have a negative influence upon the motivation of an exercise "class" (G. V. Mann, Garrett et al., 1969; Kilbom, Hartley et al., 1969). Pollock, Dimmick et al. (1975) and Pollock, Miller et al. (1975) noted that the frequency of injuries rose from 0 to 39% as the frequency of training was increased from 1 to 5 sessions per week. Injuries also rose from 22 to 54% as individual sessions were extended from 15 to 45 min. Pollock, Dimmick et al. concluded that 45 min of exercise 5 days • week^{-1} was too taxing an initial program for previously sedentary subjects. A suitable graded progression of effort is also important in reducing the risks of provoking cardiac arrest during exercise (Shephard, 1974c,1979b).

The development of cardiorespiratory fitness calls for some form of physical activity that will increase the oxygen consumption to at least 60% of aerobic power. Particularly if outdoor activity is contemplated during the winter months, a warm-up reduces both injuries and cardiac dysrhythmias (Chap. 3). A mask that will warm and humidify the inspired gas (Kavanagh, 1970) helps to avoid both exercise bronchospasm (Shephard, 1978j) and exercise-induced angina. Some development of muscles in the legs, abdomen, and back is plainly in order, although the necessary changes are often brought about by rhythmic activity, without specific recourse to weight lifting or isometric exercises. Development of the shoulder and arm muscles is probably unnecessary unless the nature of the subject's work calls for the frequent lifting of heavy objects (Shephard, 1978j).

Postcoronary Patient

The responses of the postcoronary patient to training are much as in a normal sedentary individual. However, the initial level of fitness is low, so that there is a greater than average opportunity for improvement of condition (Kavanagh, Shephard, and Kennedy, 1977). The heart muscle is also more irritable than normal, and there is a danger that exercise may provoke a sudden electrical death of the heart (Shephard, 1974c).

Precautions include (1) setting a pulse ceiling less than that needed to evoke significant dysrhythmia or exercise-induced ST segmental depression (Chap. 6), (2) training of the patient to recognize dysrhythmias and angina, (3) careful attention to mineral balance (Kavanagh and Shephard, 1977a, 1978), (4) avoidance of factors likely to increase systemic blood pressure and thus cardiac workload (prolonged exercise, support of the body by the arms, isometric activity), and (5) reduction of the exercise prescription in any unfavorable circumstances (domestic or business stress, hot weather, increase of symptoms; Shephard, 1979g), (6) avoidance of solitary exercise, (7) regular monitoring of progress by a physician, and (8) specific attention to details of the warm-down process (including the potential hazard of ventricular fibrillation in shower areas; McDonough and Bruce, 1969; Shephard, 1979b).

If there is extensive vascular disease, effort may be limited by angina. In such cases, a combination of brief interval training (Kavanagh and Shephard, 1975b) with the hypotensive action of small doses of nitroglycerine may allow the patient to reach a sufficient intensity of daily activity to commence the training process (Kavanagh, 1976). Extensive scarring of the myocardium may give a tendency to left ventricular failure, and this is also provoked by severe myocardial ischemia during physical activity; exercise training is probably contraindicated for individuals with progressive angina of effort (Parker, diGiorgi, and West, 1966; Shephard and Kavanagh, 1978b). Further details of postcoronary programs are discussed in Chap. 13.

DETRAINING AND BED REST

Detraining, and the more marked deterioration of physical condition associated with bed rest (H. L. Taylor et al., 1949; Saltin, Blomqvist et al., 1968; Fried and Shephard, 1969,1970; Greenleaf, Bernauer et al., 1977) have long been recognized, but the twin problems of enforced inactivity and the removal of normal gravitational stimuli have assumed new urgency because of the loss of physical condition that occurs in the cramped and "weightless" quarters of a space capsule (Müller, 1963; P. C. Johnson, Leach, and Rambant, 1973; Rummel et al., 1973, 1975).

The physiological manifestations of detraining are the reverse of those associated with training. A loss of working capacity is detectable within 2 weeks of stopping an activity program (Roskamm, 1967), and most of the gains of cardiovascular condition are lost within 12 weeks (Kendrick et al., 1971; Drinkwater and Horvath, 1972). If complete bed rest is required, aerobic power diminishes by 20–30% over the course of 2–3 weeks. The extent of deterioration is similar in athletes and sedentary subjects, but once the functional loss has occurred, it is more readily reversed in sedentary than in athletic individuals.

The basis for the loss of aerobic power (Table 11.23) seems a diminution of cardiac stroke volume (Saltin, Blomqvist et al., 1968; Buderer et al., 1976). This reflects (1) a reduction of total blood volume, (2) an increase of the peripheral at the expense of the central blood volume, and (3) a decrease of myocardial contractility (S. A. Bergman, Hoffler et al., 1976). There is no change of maximum heart rate, and the heart rate at a given percentage of aerobic power also remains unchanged, so that the usual methods of predicting maximum oxygen intake continue to function satisfactorily. The arteriovenous oxygen difference is usually increased while performing submaximum work, but it remains at approximately the prebed rest value during maximum effort. The cardiac shadow is reduced in size, but this is due more to a change of chamber volume than to loss of cardiac tissue (Nicogossian et al., 1976).

Losses of muscle strength also occur with bed rest (Fried and Shephard, 1970). Some authors have suggested a selective response, for example, a loss of cytochrome oxidase from red but not white fibers (Booth and Kelso, 1973) or a conversion of aerobic white muscle fibers into the anerobic, red variety (Edgerton, 1976). Others find similar reductions in the dimensions of type I and type II fibers. The time course of muscular detraining is apparently slower than that for aerobic power. No reductions of muscle strength or endurance can be seen 4–10 weeks after periods of isotonic training (Houtz et al., 1946; Berger, 1965; D. H. Clarke et al., 1973), and a substantial fraction of initial gains may be retained over 12–18 months of more sedentary living (McMorris and Elkins, 1954; Egolinskii, 1961).

TABLE 11.23
Effects of Bed Rest and Subsequent Training on the Cardiovascular Function of Young Men

Variable	Initial Value	After 20 Days Bed Rest	After 60 Days Retraining
Maximum oxygen intake (ml • kg^{-1} • min^{-1} STPD)	43.0	31.8	51.1
Maximum exercise ventilation (l • min^{-1} BTPS)	128.7	98.6	156.4
Maximum heart rate (beats • min^{-1})	192.8	196.6	190.8
Maximum stroke volume (ml)	104.0	74.2	119.8
Cardiac output (l • min^{-1})	20.0	14.8	22.8
Maximal arteriovenous oxygen difference (ml • l^{-1})	162	165	171
Heart volume (ml)	860	770	895

Source: Based on data of Saltin, Blomqvist et al., 1968.

The usual experimental study is rather short, and there are then few significant structural changes. Longer periods of detraining or bed rest give rise to a negative nitrogen balance, a decrease of lean body mass, a decrease of heart volume, and an increase in both the percentage and the absolute amount of body fat (Saltin and Grimby, 1968; Fried and Shephard, 1969,1970; P. C. Johnson, Leach, and Rambant, 1973).

The urinary elimination of calcium and hydroxyproline is increased, with a tendency to decalcification of the long bones and an increased risk of "stone" formation within the kidneys (Claus-Walker et al., 1975; Rambant et al., 1975; Vogel and Whittle, 1976). There is also a debilitating loss of balancing skill (Haines, 1974). This last change is corrected by 2–3 days of normal activity and we must thus presume that it reflects a loss of functional connections ("coding") within the brain (Fig. 7.29) rather than muscle weakness.

If the sensations of weightlessness are simulated by complete submersion of the body, the deterioration of cardiovascular regulatory responses is rapid, and an impaired tilt-table response can be demonstrated in as little as 6 hr.

The performance of physical work while in bed or traveling in a space capsule minimizes the loss of aerobic power (Sawin et al., 1975). However, it fails to avert either decalcification of bones or impairment of tolerance to gravitational stimuli. It is necessary to stand for at least 3 hr • day^{-1} to avoid calcium loss. There seems some association between the support of body mass and the maintenance of calcium balance. The regular application of negative counterpressure (suction) to the legs and abdomen prevents the development of orthostatic intolerance in either a bedridden or a weightless subject (R L. Johnson, Nicogossian et al., 1976). Unfortunately, it fails to alleviate the calcium loss.

Much mental effort has been expended by physiologists in their attempts to discover the minimum weekly program of physical activity that is necessary to sustain fitness. The quest is ephemeral, since the required amount undoubtedly depends upon the level of fitness it is hoped to maintain. For the ordinary middle-aged adult, two or three 30-min sessions per week at 60% of maximum oxygen intake probably provides a reasonable dose of activity to maintain aerobic power (P. O. Åstrand, 1967), while there have been reports that muscle strength is sustained by training as infrequently as once every 2 (Morehouse, 1967) or even 4 (D. L. Rose et al., 1957) weeks. Naturally, the continued training requirements of an international class athlete will be much greater than this.

If subjects undertake further training after a period of deconditioning, there is no evidence that retraining occurs any faster than initial conditioning (B. O. Eriksson, Lundin, and Saltin, 1975).

CHAPTER 12
Challenges to Energy Balance

Hot Environments
Cold Environments
Underwater Activity
High Altitudes

HOT ENVIRONMENTS

Environmental Comfort and Homiothermy

Certain aspects of adaptation to a hot environment have already been discussed (Chap. 3). Reptiles, insects, and various "lower" forms of animal life are *poikilothermic*—their body temperature varies almost directly with that of the external environment. In contrast humans, like other mammals, are *homiothermic*, maintaining a relatively constant body temperature in the face of wide variations in both energy expenditure and thermal environment.

The main advantage of the homiothermic condition is that the rate of biochemical reactions is predictable. Since the speed of most biochemical processes is doubled by a 10°C rise of tissue temperature, it is difficult for a reptile to gauge the force of its body movements. Humans fare much better in this respect, although limb temperatures still vary substantially with environmental conditions. The main disadvantage of homiothermy is that if body temperature passes outside rather narrow limits, death will occur. Individual cells tolerate a temperature range −1°C to 45°C (Chap. 3), but a homiothermic animal cannot long tolerate core temperatures outside the range 32–41°C. Protection against a cold environment is fairly simple given adequate clothing, but in a hot environment the only possible adaptive tactic is to reduce energy expenditure (Hardy, 1967).

The average mouth temperature in humans is ~37.0°C, while the rectal temperature is ~0.3°C higher. Both oral and rectal temperatures show a diurnal variation of ~0.5°C (Chap. 14), and in women there is a rise of ~0.5°C coincident with ovulation. Much larger changes of temperature can be induced by exercise, fever, and heat exposure. The highest reading compatible with a full recovery of cerebral function is ~42°C, although terminal values of ~43.5°C have been recorded in patients with fatal hyperpyrexia.

Despite the limited range of permissible body temperatures, humans have a very variable heat production, from ~5 kJ • min^{-1} at rest to ~100 kJ • min^{-1} during maximum effort. A delicate matching of physical activity and environmental heat exchange is thus necessary even under temperate conditions.

An environment feels comfortable when a person is neither gaining nor losing heat. The corresponding temperature depends upon the rate of working, the amount of clothing that is worn, and the extent of any heat or cold acclimatization. A sedentary office worker commonly prefers a room temperature of 21–25°C, while a range of 18–21°C is more comfortable if moderate activity must be performed. Europeans accept lower room temperatures than North Americans, partly because they wear heavier clothing and partly because they have become acclimated to a colder domestic environment (A. C. Burton and Edholm, 1969).

In assessing the comfort of a given environment, the most important variable is skin temperature (Edholm, Fox, and Wolf, 1973); signals from thermoreceptors in the skin are apparently interpreted in relation to the level of physical activity, skin blood flow, and sweating or shivering (Hardy, Stolwijk, and Gagge, 1971). Subjects are influenced more by limb than by trunk temperatures, so that an appropriate choice of footwear may increase comfort more than the wearing of an additional sweater. The preferred skin temperature lies between 32 and 35.5°C (Precht et al., 1973).

The rate of heat loss is much greater in water than in air. Comfortable temperatures are 35–37°C for inactive bathers and 20–30°C for active swimmers (Chap. 3). A compromise between comfort and hygiene is sometimes necessary because high water temperatures promote bacterial growth.

Heat Exchange

Heat exchange occurs by conduction, convection, radiation, and evaporation of sweat (Chap. 3).

Conduction

Conduction implies the transfer of thermal energy to and from the skin by direct contact, without the interposition of air. Normally, it accounts for only a small fraction of total heat transfer. Appreciable conduction occurs (1) when walking barefoot on hot sand, (2) when sitting on a hot car seat, wearing only light clothing, and (3) when immersed in cold water.

Convection

Convection implies a forced transfer of heat induced by the movement of gas or fluid. Humans have three main barriers to convection: (1) subcutaneous fat, (2) clothing, and (3) a thin film of stationary air or water in immediate contact with skin and/or clothing. The rate of energy transfer across each of these impedances depends on the temperature gradient, the thermal conductivity of the convected matter, and the convective flux.

In the case of subcutaneous fat, the thickness of the barrier and the thermal conductivity of the blood are relatively fixed quantities. Heat transfer therefore varies with the temperature gradient from the core of the body to the skin surface and with the convective flux (skin blood flow). Heat loss at the surface of skin or clothing is increased by wind or water movement; the "stationary" film of air or water is disturbed, and convective flux across the film is augmented. Heat loss is also increased if normal ambient air is replaced by a gas with a high thermal conductivity (such as helium). At high altitude, the number of gas molecules per unit volume of the stationary film is less and the impedance to heat transfer is correspondingly increased. Conversely, at depth gas is compressed and impedance is reduced; compounding the problem of maintaining body temperature in a diver, water has a combined conductive/convective coefficient ~20 times greater than that of normal ambient air.

Radiation

Radiation implies the transfer of heat as a wave motion. All objects radiate energy at a rate dependent upon the nature of their surface and the absolute temperature.

Low-temperature radiation (293–323° K,* 20–50°C) has a relatively long (infrared) wavelength. Most materials (including clothing and skin) emit and absorb such radiation much as a black surface would (that is, with little reflection). Whether heat is gained from or lost to surrounding surfaces thus depends simply on the gradient of absolute temperatures and geometric relationships.

Solar radiation originates at a much higher temperature and is of short wavelength; outward radiation from humans to objects in space can therefore be neglected. However, appreciable reflection occurs from skin and light-colored clothing.

The radiant heat load is measured by a radiometer or a thermometer enclosed in a "black" globe. Such a device overestimates the stress encountered in sunlight, since it absorbs nearly all of the incident radiation and presents a larger surface to the sun than does a standing person. In order to adjust for the individual's skin reflectance and postural effects, the globe can be painted a suitable shade of gray. A globe modified in such a fashion still functions as a "black" surface with respect to lower frequency radiations emanating from terrestrial objects.

Evaporation

Evaporation of 1 g of water dissipates ~2.43 kJ of energy (the latent heat of vaporization). An exercising subject can evaporate substantial quantities of both sweat and externally applied water. Light clothing facilitates evaporation, since it provides an ex-

*Radiant temperatures are usually expressed on the Kelvin scale; 0° K = −273°C.

tended wicklike surface from which evaporation can occur and conserves sweat that would otherwise have accumulated as a useless pool on the floor. However, as the clothing becomes saturated, the efficiency of cooling drops, up to 50% of the heat of vaporization being drawn from ambient air rather than the body (F. N. Craig and Moffit, 1974).

The rate of evaporation of fluid from the skin normally depends on (1) skin temperature, (2) the gradient of water vapor pressure across the film of stationary air surrounding the skin, and (3) the rate of air movement and thus the thickness of the stationary film. An increase of atmospheric pressure impedes the movement of water molecules, while the converse is true of a low ambient pressure.

The evaporative loss from the lungs depends on the BTPS ventilation, the dryness of the atmosphere, and the barometric pressure. Men who are working hard in the dry air of high mountains have a respiratory loss of up to 3 l of water per day, imposing a substantial demand for the replacement of both fluid and thermal energy.

Energy Balance

It follows from the basic laws of thermodynamics (Chap. 1) that during any period of sustained activity, the algebraic sum of energy exchange due to radiation (E_r'), convection (E_c'), conduction (E_k'), and evaporation of sweat (E_s') must equal the total energy consumption (E′) less any external work (W′) that has been performed (Winslow et al., 1939; Kerslake, 1965,1972; Figs. 12.1 and 12.2):

$$E_s' \pm E_r' \pm E_c' \pm E_k' = E' \pm W' \qquad (12.1)$$

On a short-term basis, this equation of equilibrium may be disturbed by a decrease of body mass, the drinking of fluids, the passing of urine, or a storage of heat (Chap. 3). Physical activity also modifies the left-hand side of the equation. A subject may be forced to emerge from the shade into the radiant heat of the sun as he competes on a hot track. If the environmental temperature is above skin temperature, body movement increases convective heat gain, while if the air is cooler than the body, movement facilitates both convective heat loss and the evaporation of sweat. Given a skin temperature of 35°C, an air temperature of 20°C, and a track speed of 5 m • sec^{-1}, the convective heat loss of an adult male amounts to ~32 kJ • min^{-1}, about half the heat production of a marathon runner (Wyndham and Strydom, 1972).

Body Temperature

The initial effect of exposure to a hot environment is to increase the core temperature for a given rate of working. This does not necessarily provoke discomfort, since thermal sensations depend more upon skin than upon deep-body temperatures (Edholm, Fox, and Wolf, 1973). However, the temperature gradient between the body and its surrounding is increased, and this automatically augments heat elimination.

In any given individual, there is a linear relationship between the steady-state deep-body temperature and rate of working; this response is maintained to at least 75% of maximum oxygen intake and is relatively independent of environmental temperature (Stolwijk et al., 1968; B. Nielsen, 1969). Comparisons of "positive" and "negative" exercise (B. Nielsen, 1969) suggest that changes of core temperature are related to energy output rather than the heat production of the body. At the highest rates of working, the relationship becomes curvilinear (C. T. M. Davies, Brotherhood, and Zeidifard, 1976). Interindividual differences of temperature response are smallest if the rate of working is expressed as a percentage of maximum oxygen intake (Chapter 3). The

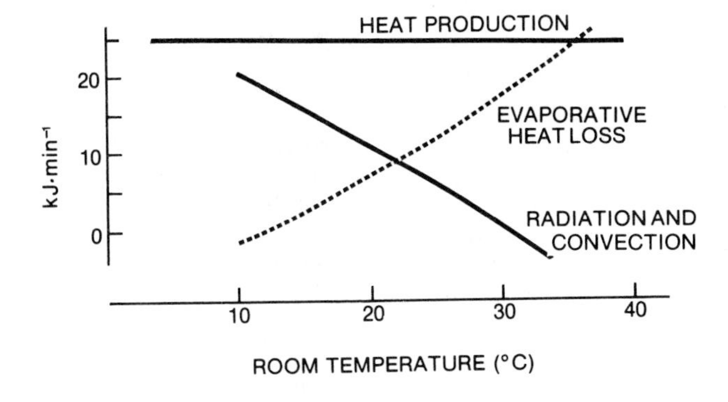

FIG. 12.1. The influence of room temperature upon the relative proportions of total heat loss due to evaporation, radiation, and convection. Based on data of M. Nielsen (1938) for nude subject exercising on a bicycle ergometer at a power output of 150 Watts.

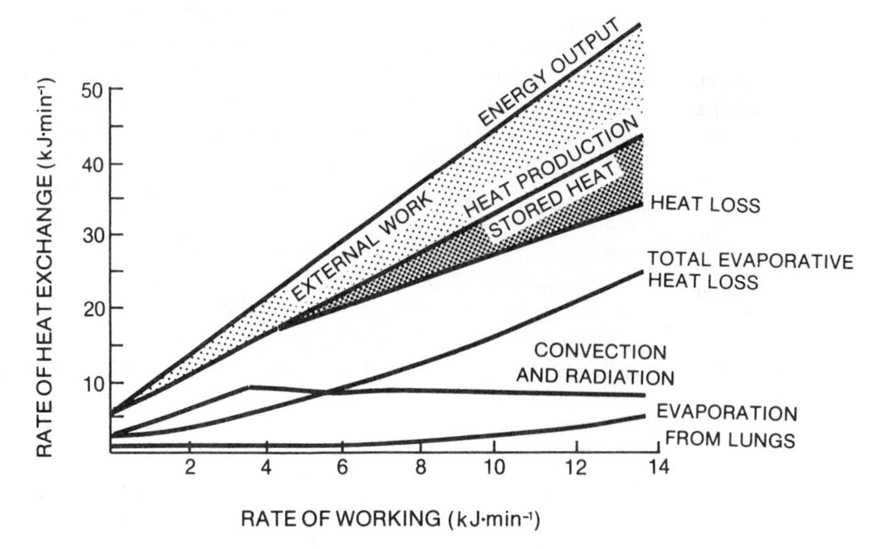

FIG. 12.2. The several routes of heat transfer during bicycle ergometer exercise in a temperate climate. Note that the proportion of convective loss may rise if the body is displaced, as in running, or is exposed to air movement. (Based on data of M. Nielsen, 1938.)

rise of body temperature might suggest that exercise elevates the "set point" of the hypothalamic temperature-regulating center (Chap. 9); on the other hand, an increase of sweat production can occur within a few sec of starting exercise, and this would be difficult to explain other than by a lowering of set point (Meyer et al., 1962; Van Beaumont and Bullard, 1963). Possibly, there is an initial stimulation of heat loss mechanisms, but this is not sustained as exercise continues.

While the core temperature depends mainly upon work rate, the skin temperature is affected mainly by ambient conditions.

Cardiovascular Effects

A hot environment provokes a substantial increase of skin blood flow, thus facilitating conduction of heat from the core to the skin surface per unit of temperature gradient. The extent of the cardiovascular reaction varies with the intensity and duration of physical activity. If the subject is resting, or engaged in light exercise, the cutaneous flow can reach 9–10 l • min^{-1} (Koroxenidis et al., 1961; Folkow, Heymans, and Neil, 1965; Rowell, Brengelmann et al., 1969a; Detry et al., 1972). During a brief bout of vigorous activity, the response of the skin vessels is changed but little by exposure to a hot environment (Rowell, 1974; Drinkwater et al., 1976a). In more sustained effort, the demands of the working muscles restrict the heat-induced increase of skin blood flow, even though the exercise-induced reduction of visceral flow is exaggerated by a hot environment (Chap. 3; Rowell, 1971,1974; J. M. Johnson et al., 1974; W. C. Adams, Fox et al., 1975b).

The increase of skin blood flow in the heat is mediated by (1) a release of vasoconstrictor tone in the sympathetically innervated vessels of the hands and feet, (2) an active dilation of vessels in the trunk and proximal parts of the limbs, (3) a response to bradykinin liberated from the sweat glands, and (4) local effects of temperature on the vessels, including cooling by evaporation of sweat (A. C. Burton, 1965b). The neural component of the vascular response is regulated by the hypothalamic centers (Chap. 7; Bligh, 1977). At high altitudes, the normal vasodilator reaction may be overridden by a vasoconstricter drive associated with carbon dioxide loss (Raynaud et al, 1973).

The rising core temperature normally brings about an opening of arteriovenous anastomoses, particularly in the extremities. This allows a much greater circulation of blood through the limbs than could be accommodated in the normal capillary bed (Chap. 6). There is an associated rerouting of venous blood (P. G. Wright, 1977). In a cold environment, most of the venous return occurs via venae comitantes, deeply sited paired veins that lie immediately alongside the major limb arteries. When this pathway is operational, heat from the arteries is passed directly to returning venous blood and the limbs remain relatively cool. In contrast, under hot conditions, most of the blood is returned to the central circulation via the superficial veins (Rowell, 1974). The arterial flow to the limbs is no longer cooled by the venae comitantes, and in consequence heat is dissipated from the warm superficial capacity vessels (Bazett et al., 1948; Schmidt-Nielsen, 1963).

The veins share in the vasodilator response to a rise of core temperature, their pressure/volume characteristics being modified to increase the extent of peripheral pooling for a given distending pressure (J. P. Henry, 1951; Bevegard and Shepherd, 1966a; Webb-Peploe, 1969; Fig. 6.12).

Central blood volume is thus reduced. The trend is exaggerated by a rapid exudation of fluid into the tissues due to an increase of (1) capillary pressure, (2) capillary permeability, and (3) extravascular leakage of protein (Chap. 8). The decrease of plasma volume in the heat is relatively larger in women than in men (Senay and Fortney, 1975).

Stroke volume is less well maintained when exercising in the heat than with a comparable duration of physical activity in a cool environment (Saltin, 1964a,b; Oddershede et al., 1977). Sustained exercise also provokes an unusually large reduction of visceral blood flow (Chap. 3); this can lead to (1) an escape of hepatic enzymes into the circulation, (2) temporary reduction of hepatic and renal function, and (3) in extreme environments to a failure of skin blood flow (fatal hyperpyrexia), renal blood flow (renal failure), and suprarenal flow (adrenal failure).

Little oxygen is extracted from blood perfusing the cutaneous vessels. The arteriovenous oxygen differences at a given power output is thus reduced in the heat. Peripheral vascular resistance is lower than when exercising in a temperate environment, and there is also a lesser rise of systemic blood pressure. The working muscles are perfused less adequately than under normal ambient conditions. Triglyceride utilization is reduced, glycogen utilization is increased, and lactate accumulation in submaximal exercise may be doubled (Fink et al., 1975). On the other hand, the accumulation of blood lactate in maximal effort is reduced (Saltin, 1964a,b).

The increase of skin flow, decline of stroke volume, loss of plasma volume, and increase of core temperature all contribute to the observed tachycardia at a given power output (Pirnay, Deroanne, and Petit, 1970; Claremont et al., 1976). Mostardi et al. (1974) noted a rise of 17–18 beats per min per 1°C rise of rectal temperature at an oxygen intake of 22 ml • kg^{-1} • min^{-1}, with a lesser increase of 7–8 beats per min^{-1} • $°C^{-1}$ at an oxygen intake of 32 ml • kg^{-1} • min^{-1}. Miller and Walters (1974) expressed test data for submaximal exercise relative to environmental temperature; for each degree Celsius increase of core temperature, the heart rate was augmented by 1.4% and the respiratory minute volume rose by 0.4%.

With acute heat exposure, most authors have found little change of maximum oxygen intake (Chap. 3; C. G. Williams et al., 1962; Rowell, Blackmon et al., 1965; Saltin, Gagge et al., 1972), but if either heat exposure or exercise is of sufficient duration to cause a peripheral sequestration of fluid, maximum oxygen intake (F. N. Craig and Cummings, 1966; Klausen, Dill et al., 1967; Pirnay, Deroanne, and Petit, 1970) and endurance (MacDougall, Reddan et al., 1974) are both impaired. There is also a reduction in the voluntarily selected rate of working (Snook and Ciriello, 1974).

Sweat Production

Emotion is not normally a major stimulus to sweat production (Abram et al., 1973; Allen et al., 1973); the prime signals are cutaneous and core temperature (Fig. 12.3; Nielsen, 1969). Integration is normally in the hypothalamus (Fig. 7.35; Hammel, 1965; Hardy, 1967; Smiles et al., 1976; Bligh, 1977), although in some circumstances both vasodilation and sweating can be mediated at a spinal level (Walther et al., 1971; Cabanac, 1975).

Other possible inputs to the temperature-regulating centers arise from venous thermoreceptors (Hensel, 1974), neuromuscular reflexes (Gisolfi and Robinson, 1970), and the osmotic pressure of the plasma (M. H. Harrison, 1976). Perhaps because of such additional thermoregulatory input, perhaps because of nonlinearity in the work rate response, and perhaps because of local effects of limb movement, some authors find that heat loss is less in interval than in continuous work (Ekblom, Greenleaf et al., 1971; but not B. Nielsen, 1968).

The immediate effect of physical activity is to bias the control setting of the heat-regulating centers in a downward direction (Tam et al., 1978). At any given work rate, women sweat less and store more heat than men, especially if they are in the postovulatory phase of the estrous cycle (C. Wells and Horvath, 1974; Bittel and Henane, 1975). With very prolonged exercise (for example, a 9-hr walk), muscle damage and the resultant release of pyrogens elevate the set point (Haight and Keatinge, 1973a).

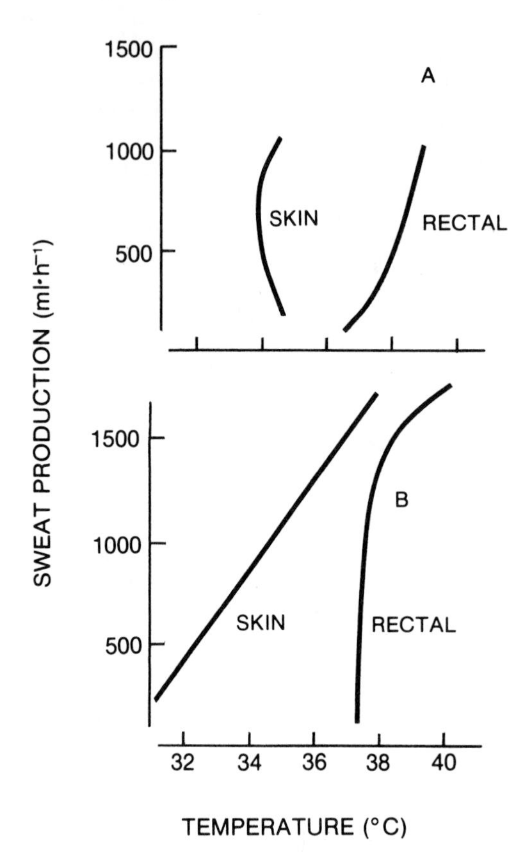

FIG. 12.3. The influence of skin and core temperatures upon the rate of sweating. **A.** Varying rate of working. **B.** Constant rate of working but varying environmental conditions. (Based on material of S. Robinson, 1957.)

The maximum rate of sweat production (up to 2 l • hr^{-1} in a man; Pugh, Corbett, and Johnson, 1967) is sufficiently large that activity is usually limited by the ability to evaporate sweat rather than by the ability to produce it (Table 12.1). Provided that the added burden of radiant heating is avoided, the hot dry climate of a desert is tolerated much better than the hot wet climate of a jungle region. The rate of evaporation of sweat normally depends upon (1) skin temperature, (2) water vapor pressure, and (3) air movement. However, until the skin is completely wetted, the area of the sweat film is the controlling factor (Kerslake, 1965,1972). Tolerance of warm water is less than for an apparently comparable air environment, since in water sweat production depletes fluid reserves without cooling the body (Hori et al., 1975).

If the requirements of energy balance are not met by evaporation [equation (12.1)], both skin and core temperatures rise. This increases the rate of evaporation but also increases sweat production. Unfortunately, much of the added sweat drips from the body without evaporation (Pugh, Corbett, and Johnson, 1967). Kerslake (1972) calculated that evaporation becomes incomplete once the overall sweat production has reached a third of the maximum evaporative capacity. Sweating is particularly inefficient if parts of the body are excessively clothed. The skin temperature rises in these regions, leading to a local overactivity of the sweat glands.

Individuals vary greatly in their ability to produce sweat. If a standard exercise/heat-stress is applied, then a person acclimatized to a hot environment sweats earlier and in greater quantities than a person accustomed to more temperate conditions. Trained individuals sweat less than the untrained at a given absolute rate of working. On the other hand, they produce as much sweat as an unfit person at a lower rectal temperature (Hénane et al., 1974; Baum et al., 1976; Shvartz et al., 1976; Senay and Kok, 1977; M. F. Roberts, Wenger et al., 1977) and probably sweat more if pushed to maximum effort (Davies, Barnes, and Sargeant, 1971). There seems to be considerable interaction between physical training and heat acclimatization (Piwonka et al., 1965; Gisolfi, 1973).

If a given session of exercise or heat exposure is prolonged, sweat production gradually declines even if body fluid volumes are well maintained. This is thought to reflect a blockage of the sweat ducts, secondary to maceration of the skin (Wyndham and Strydom, 1972). It occurs most readily if excessive clothing is worn or conditions are unduly humid.

Measures of Thermal Stress

The thermal stress imposed by a given combination of environment, clothing, and physical activity is

TABLE 12.1
Typical Decrease of Body Mass Due to Sweating in a 90-kg Athlete Participating in an 80-min Football Game under Specified Environmental Conditions

Temperature (°C)	Decrease of Body Mass (kg) with Relative Humidity			
	<40%	40–60%	60–80%	80–100%
38	2.7	3.1	3.3	3.5
32	2.3	2.6	2.8	3.1
27	1.8	2.2	2.4	2.6
22	1.4	1.7	1.9	2.2
16	0.3	0.5	0.7	0.9

often summarized as a predicted 4-hr sweat rate (Kerslake, 1965). This is read from a rather complex nomogram (Fig. 12.4).

The basic 4-hr sweat rate is first predicted by joining the dry bulb to an appropriate wet bulb scale, noting the intercept on the corresponding B-4 SR scale. The point of entry to the wet bulb scale of the nomogram is augmented by a factor Δ_t under the following circumstances:

1. If the globe temperature (T_G) exceeds the dry bulb temperature (T_D):

$$\Delta_t = 0.4(T_G - T_D) \qquad (12.2)$$

2. If the metabolic rate exceeds 3.8 kJ • m^{-2} • min^{-1}, Δ_t is as shown in the insert to Fig. 12.4.
3. If the clothing provides more insulation than shorts (for example, Δ_t for overalls = 1°C).

The predicted 4-hr sweat rate is equal to the B-4 SR if men sit while wearing shorts. If they wear overalls, the 4-hr sweat volume must be increased by 0.25 l, and physical activity modifies the reading further according to the equation:

$$P4SR = B4SR + 0.2(E - 3.8) \qquad (12.3)$$

where E is the energy expended by the individual, measured in kJ • m^{-2} • min^{-1}. The nomogram can be used to estimate unevaporated sweat. For this purpose, we enter the chart at the lowest possible wet bulb reading and repeat the calculations; the difference between the figure thus obtained and the actual P4SR represents unevaporated sweat.

An alternative basis for assessing the severity of a given environment is to calculate an "effective temperature" based on relative humidity and air velocity. The value thus obtained expresses the equivalent comfort of still air saturated with water vapor. Allowance can also be made for radiant heating if the globe temperature is measured in place of the dry bulb temperature (Fig. 12.5). The original scale was

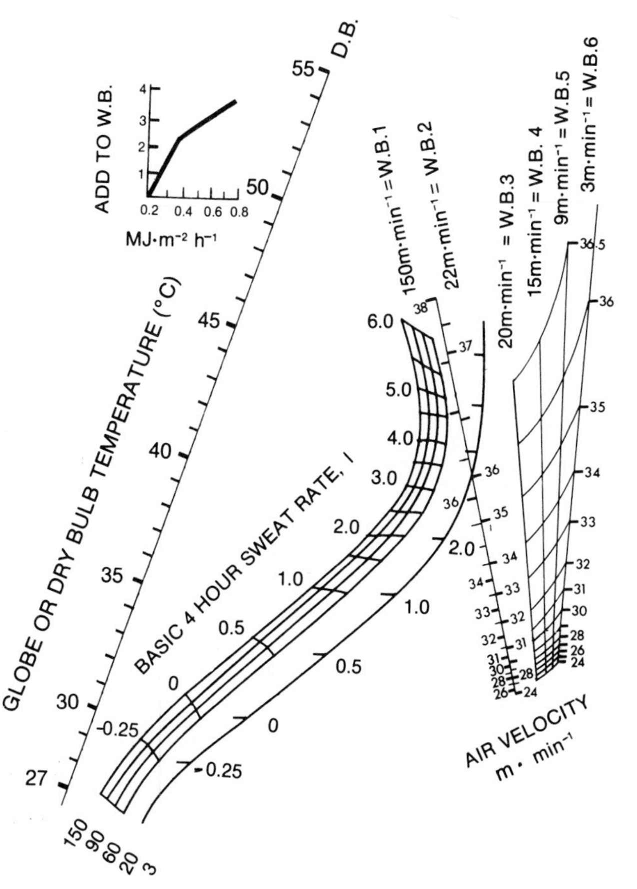

FIG. 12.4. A nomogram for prediction of the four-hour sweat rate. (Modified from Kerslake, 1965.)

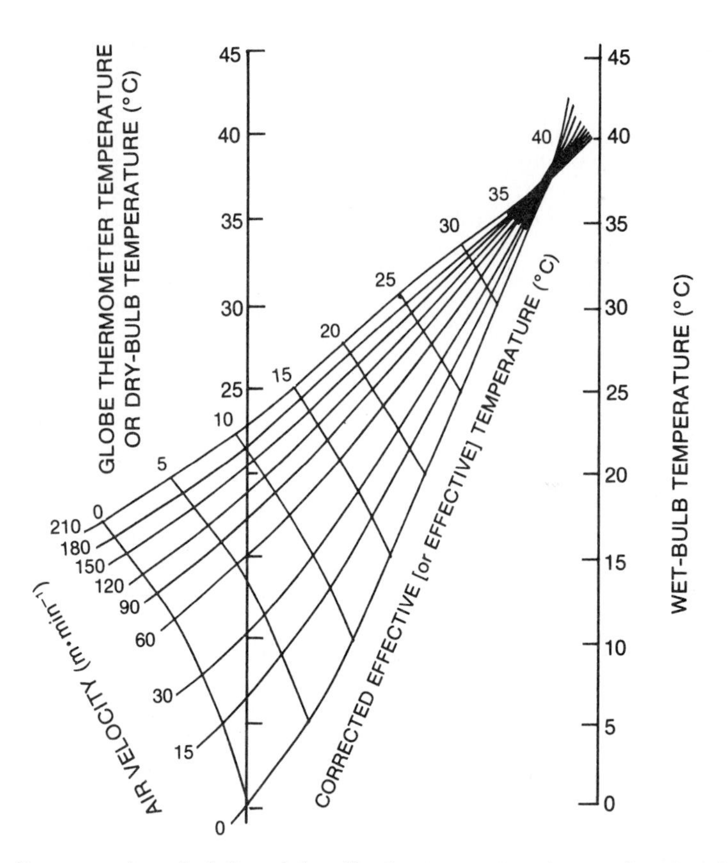

FIG. 12.5. Nomogram for calculation of the effective temperature (men stripped to the waist and performing moderate activity). Based on a chart published by the American Society of Heating, Refrigerating and Air Conditioning Engineers (ASHRAE Handbook of Fundamentals, 1967).

based on men stripped to the waist and engaged in moderate activity. Some modification of readings is necessary if vigorous exercise is to be performed.

A third method of data synthesis is due to Minard (1961). He developed a wet bulb globe temperature index (WBGT) as follows:

For men outdoors:

$$WBGT = 0.7(T_W) + 0.2(T_G) + 0.1(T_D) \quad (12.4)$$

For men indoors:

$$WBGT = 0.7(T_W) + 0.3(T_D) \quad (12.5)$$

where T_W is the wet bulb temperature, T_G is the globe temperature, and T_D is the dry bulb temperature. Minard developed the index for personnel of the U.S. Marine Corps, carrying out normal duties while wearing combat uniforms. He recommended caution if the index surpassed 28°C, a restriction of activity for unacclimatized men at 29.5°C, a limitation of effort for all except fully acclimatized individuals at 31°C, and a replacement of outdoor activity by lectures and demonstrations at 32°C.

The U.S. National Road Running Club has advised cancelling contests under the following conditions: (1) $T_D > 35°C$; (2) $T_D > 29.4°C$, $T_W > 24.3°C$; (3) $T_D > 26.7°C$, $T_W > 23.3°C$ (Shephard, 1976b); however, one more recent report suggested that when running at a speed of 176 $m \cdot min^{-1}$, the upper limit for a thermal steady-state is substantially higher (T_D 38.5°C, T_W 28.5°C; J. R. Duncan et al., 1978).

If intermittent activity is required (for example, the running of many heats by a sprinter) it is difficult to set safe limits based on environmental temperature readings, and it is better to be guided by physiological evidence of thermal stress (such as an increase of heart rate and body temperature; Horvath and Colwell, 1973).

Fluid and Mineral Loss

There has been much discussion as to the relative fluid loss from the various body compartments when subjects exercise in the heat. It is probable that divergent results reflect varying proportions of thermal and exercise stress (Chap. 3). Heat stress tends to deplete the plasma volume, while exercise decreases the intracellular fluid (Adolph, 1947a; Pugh, Corbett, and Johnson, 1967; Kozlowski and Saltin, 1973; Shephard, Kavanagh, and Moore, 1978; but not Costill and Fink, 1974; or Costill and Saltin, 1975). Long and repeated periods of exercise in a hot climate cause cumulative disturbances of fluid and mineral balance, with errors in the estimation of body composition by underwater weighing. A 1-l fluid loss brings about an 0.7% error in the estimated fat content of the body (Girandola, Wiswell, and Romero, 1977).

Chronic depletion of sodium reserves leads to muscle cramps (Leithead and Lind, 1964; Wyndham and Strydom, 1972), while the associated reduction of body water causes chronic irritability, weakness, and fatigue (heat neurasthenia; Wyndham and Strydom, 1972). Loss of other minerals contributes to anemia, and (particularly in soft-water areas) to a loss of force in skeletal and cardiac muscle (altered calcium/magnesium balance; Chap. 4) with an increased vulnerability to sudden death through electrical failure of the heart.

Psychomotor Performance

The threshold temperature for a decrement of psychomotor performance depends upon the acclimatization of the subject and the sensitivity of the test used. A well-acclimatized man tolerates a temperature of 30°C, whereas someone who is unaccustomed to heat may show a loss of psychomotor skills at 25°C. The deterioration of function is less obvious in well-motivated than in uncooperative subjects. The observed changes may be largely a reaction to the minor discomforts encountered in the heat. However, if physical work is performed, a reduction of cerebral blood flow probably contributes to the terminal decrement in score.

Acclimatization

When an individual is repeatedly exposed to heat (or other forms of environmental stress), acclimatization tends to occur (Fig. 12.6; Leithead and Lind, 1964; Ladell, 1964; Rowell, 1974). A fully acclimatized person can function normally despite an adverse environment that would incapacitate an unacclimatized individual. The extent and rate of acclimatization is influenced more by the intensity than by the frequency or duration of the applied stress. On removal from the adverse environment, acclimatization is gradually lost (Adolph, 1964).

The primary change during heat acclimatization is an earlier and a greater production of sweat for a given rise of core temperature (Rowell, Kraning et al., 1967; Cleland et al., 1969; Shvartz et al., 1976; Baum et al., 1976; Senay and Kok, 1977). Sweat production may be augmented as much as 100% (Leithead and Lind, 1964). Runners also show a depression of the shivering threshold, as though the set point of their thermoregulatory center is displaced downward.

The greater sweat secretion is not always de-

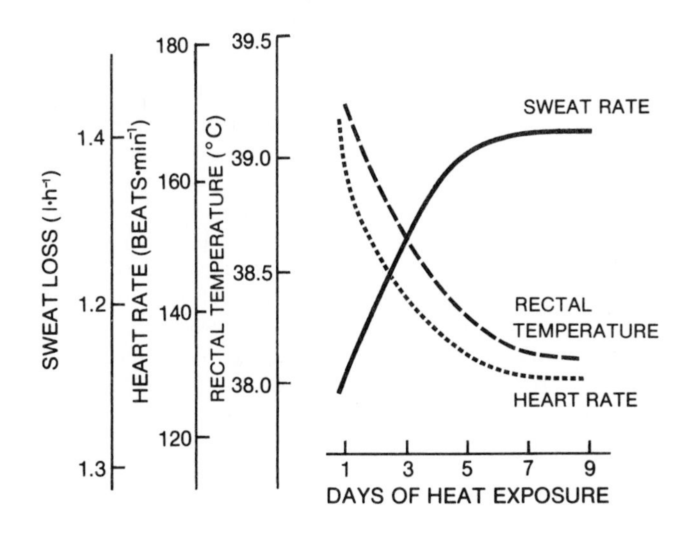

FIG. 12.6. Effect of heat acclimatization upon rectal temperature, heart rate, and sweat rate. Young men performing moderate work (1.2 MJ • hr^{-1}) in a hot dry climate (49°C dry bulb, 27°C wet bulb). (After Lind & Bass, 1963.)

tected by the subject, partly because the increased output is largely evaporated and partly because distribution is altered, more sweat being secreted over the trunk and proximal parts of the limb. The increase of sweat secretion lowers core temperature and disturbs the linearity of the relationship between tympanic temperature and oxygen consumption (C. T. M. Davies, Barnes, and Sargeant, 1971). The temperature of the skin surface is also reduced, and in consequence a greater quantity of heat is carried from the core to the skin per unit of blood flow. If heat dissipation was inadequate prior to acclimatization, then adaptation to a given combination of exercise and thermal stress may be associated with a greater blood flow to the skin, particularly over the upper half of the body (Glaser and Shephard, 1963). On the other hand, if equilibrium was possible prior to acclimatization, the steady-state condition will be sustained with a smaller cutaneous blood flow, thereby permitting better perfusion of both the viscera and the active muscles (D. E. Bass, 1963; Rowell, 1974). These changes, together with some increase in the tone of the capacity vessels and possibly an increase of plasma volume (Senay, Mitchell, and Wyndham, 1976), account for the lower heart rate and increased stroke volume of the acclimatized individual at any given rate of working (Rowell, Kraning et al., 1967). Cardiac output often shows little change (Rowell, 1974; Wyndham, Rogers et al., 1976).

The excretion of salt in sweat and urine diminishes with acclimatization. This probably reflects an increased secretion of the hormone aldosterone, with retention of sodium ions and an increased loss of potassium ions (Ladell and Shephard, 1961).

Long-term adaptations to heat have been described. In some species (but perhaps not in humans), the secretion of thyroid hormone is reduced, with a consequent closer coupling of carbohydrate metabolism to phosphorylation (Chap. 1). Behavior is modified. Unnecessary and exuberant activity is avoided. The subject learns to keep away from sources of intense heat; for instance, he walks in the shade and avoids hot tarmac. Finally, there is some habituation to the sensations encountered in a hot environment (Glaser and Shephard, 1963; Glaser, 1966). A given disturbance of physiological function is thus endured with a minimum psychological reaction.

There are substantial interindividual differences in heat tolerance. However, black people from tropical regions have no systematic advantage over acclimatized whites (Ladell, 1964).

Cardiorespiratory training and heat acclimatization have many common features. Adaptation to both stresses is temporarily impaired by overindulgence in alcohol or lack of sleep (D. E. Bass, 1963). Individuals who are in good physical condition adapt well to a warm climate, while the elderly and obese respond more poorly. Cardiorespiratory training partially prepares a man for the heat (Piwonka, Robinson, et al., 1965; Gisolfi and Robinson, 1969; Gisolfi, 1973; Inbar et al., 1978), while detraining has the converse effect. Nevertheless, if hard physical work is to be undertaken in the heat, maximum acclimatization is attained only through a combination of physical activity and heat exposure (Strydom and Williams, 1969; Wyndham, Strydom et al., 1973a,b).

Acclimatization occurs in an exponential fashion. Much of the total adaptation develops over the first 3 days of exposure, and changes are largely completed within 2 weeks (Glaser and Shephard, 1963; Wyndham and Strydom, 1972). On leaving the tropics, adaptations persist for only a few weeks (Cleland et al., 1969), but they can apparently be restored at an accelerated pace if the individual returns to a hot environment (Glaser, 1966).

In terms of preparing either servicemen or athletes for a hot climate, it is unfortunate that the full adaptive potential cannot be realized through a series of heat exposures in an environmental chamber (Edholm and Bacharach, 1965; Strydom and Williams, 1969). Features of the tropics such as an unfamiliar pattern of life and a range of new microorganisms can be countered only when they are met by the traveler.

Pathological Reactions to Heat

Heat Syncope

Perhaps the commonest reaction to an excessive combination of effort and thermal stress is heat syncope. The venous return to the heart is inadequate and the blood pressure falls. This triggers a vasovagal attack, with a slow pulse rate and muscular vasodilation (Wyndham, 1951). The problem is commonly initiated by a decrease of central blood volume due to (1) fluid loss in the sweat (water exhaustion), (2) excessive pooling of blood in the relaxed peripheral veins, and (3) increased exudation of fluid (heat edema of the dependent parts). A competition between the circulatory needs of muscle and skin contributes to the fall of blood pressure. The situation is rapidly restored by (1) lying down with the legs elevated, (2) tepid sponging, and (3) oral administration of fluids.

Mild Heat Exhaustion

Mild heat exhaustion is a lesser degree of circulatory inadequacy that develops over a day of physical activity in the heat. Symptoms include fatigue, irritability, a deterioration of mechanical efficiency, and an increase of accidents (Chap. 3; Strydom, Wyndham, et al., 1966b).

If ignored, exhaustion may progress to a chronic neurotic reaction ("heat neurasthenia"), with loss of energy, initiative, and interest (Wyndham and Strydom, 1972), plus complaints of dizziness and blackouts. The picture is not dissimilar to the effort neurasthenia seen in more temperature climates, and recovery is usually rapid on moving from the hot environment. Before dismissing complaints as heat neurasthenia, one must exclude more important causes of exhaustion, such as chronic salt or fluid depletion.

Heat Stroke

Heat stroke is a dangerous and potentially irreversible failure of heat-regulating mechanisms (Buskirk, 1968; Spickard, 1968; MacFarlane, 1978). Although it is relatively uncommon, the mortality is high (Chap. 3). The likely victim is an obese, unfit, and unacclimatized person who is forced to exercise at a heavy relative work load in a hot environment. A rectal temperature higher than 40°C is commonly regarded as a warning of impending heat stroke, although one recent report found that half of the runners in a marathon race had temperatures of ~42°C without signs of heat illness (Maron et al., 1977). In subjects that develop heat stroke, a progressive rise of body temperature is compounded by dehydration, a diminishing secretion of sweat (anhidrotic heat exhaustion), and a diminished blood flow to the skin. The combination of rising body temperature and an inadequate blood flow to the brain may cause cerebral irritability, hallucinations, coma, and irreversible cerebral damage. A combination of visceral vasoconstriction and a failing circulation may also damage the kidneys (heat anuria) and the adrenal cortex (heat shock). A 2-fold increase in plasma enzyme levels (serum glutamic oxaloacetic transaminase, serum glutamic pyruvic transaminase, and lactic dehydrogenase, LDH) is a normal response to marathon running (Wyndham and Strydom, 1972; Shephard and Kavanagh, 1975a), but if there is heat stroke and visceral damage, 10- to 20-fold increments of serum enzyme concentrations are seen, together with massive hematuria, proteinuria, and a rapidly rising blood urea. Diagnosis is helped by a fractionation of serum LDH; normal runners show increments of the LDH-3, LDH-4, and LDH-5 fractions, while patients with cardiac and renal damage show increases of LDH-1, and LDH-2 (L. I. Rose, Bousser, and Cooper, 1970).

A number of international competitors have died of heat stroke (Chap. 3). Contributory factors include the wearing of clothing that is impermeable to sweat, restriction of skin blood flow through the illegal administration of amphetamines, and a deliberate limitation of water intake (sometimes forced upon an athlete by the rules of competition).

Heat Cramps

A man who is sweating hard loses up to 2 g of salt per hr, and this progressively depletes the body pool of ~175 g (Chap. 2). The usual physiological manifestation of salt deficiency is the onset of painful muscular cramps ("Stoker's cramps"). The symptom is common soon after arrival in the tropics, partly because adaptive changes in the salt content of the sweat and urine are incomplete (Ladell and Shephard, 1961; Ladell, 1964) and partly because the newcomer has not acquired the habit of deliberately supplementing his salt intake.

Salt depletion exacerbates dehydration, since the body attempts to restore the osmotic pressure of the blood by excreting water. *Salt deficiency exhaustion* may present as a general chronic weakness or, if more acute, as a loss of body mass with constipation, a scanty urine, and headache (Wyndham and Strydom, 1972). If dehydration is advanced, there may be nausea, vomiting, sunken eyes, an inelastic skin, and circulatory failure (Chap. 9). Once vomiting is established, chloride loss is increased, and the condition becomes fatal if a normal plasma composition is not soon restored.

Heat Rash

There are several forms of heat rash (Shephard, 1976b). The primary pathology is a blockage of the sweat ducts due to excessive sweating and maceration of the skin. In long-standing cases, the situation is complicated by secondary infection. Other disorders of the skin, particularly fungal infections, flourish under hot and moist conditions.

COLD ENVIRONMENTS

Several aspects of energy balance in a cold environment have been discussed (Chap. 3). We here refer specifically to clothing, thermoregulation, fat loss, acclimatization, and certain pathological reactions to cold.

Clothing

Adequate clothing can maintain thermal balance both in air and in water. Insulation depends mainly upon an increase in the film of stationary air in contact with the skin. Such air tends to be displaced by body or air movement unless the outer layer of clothing is windproof. If ankle and wrist closures are incomplete, the stationary film is also displaced by limb movements. Finally, wetting of the clothing by mist, rain, spray, water immersion, or sweating leads to a rapid and almost complete loss of insulation (Pugh, 1972). For this reason, it is vital to avoid sweating when work is performed in the cold. The design of clothing must allow an appropriate adjustment of insulation as the intensity of activity is increased, and effort must be held below the level at which a significant accumulation of sweat would be expected to occur.

The insulation offered by various clothing assemblies was originally rated in arbitrary "clo" units. One clo provides sufficient insulation for the continuing comfort of a man sitting in a room heated to 21°C, with an air movement of 6.1 m • min^{-1} and a relative humidity of less than 50%. At the time of its description, this particular environment corresponded quite closely with normal British domestic life, from which we may deduce that the heavy indoor garments popular in Britain provide ~1 clo of insulation. The corresponding thickness of clothing is ~0.6 cm.

More recently, the performance of garments has been rated in physical terms. Given a heat flow of 10 kJ • hr^{-1} per m^2 of clothed surface, 1 clo of insulation creates a temperature gradient of 0.37°C. In general, insulation is independent of the nature of the fiber. However, a high wind penetrates loosely woven material, displacing trapped air and reducing insulation. Clothing for arctic conditions should therefore include an outer windproof layer (Pugh, 1972; Renbourn, 1972). Some types of clothing, such as closely woven nylon, offer an almost insuperable barrier to sweat. This can create a dangerous heat load. The ideal fiber for a hot environment "wets" easily; it conserves fluid that would otherwise have dripped from the body, and functions as a wick, transferring water vapor from the skin surface to the atmosphere.

Thermoregulation

Most physiological reactions to the cold are the opposite of those described for the heat. There is an immediate and vigorous general constriction of the cutaneous blood vessels, with increases of peripheral resistance, systematic blood pressure, and afterloading of the heart. Nevertheless, a certain minimum blood flow is needed to meet the metabolic requirements of subcutaneous tissues, and this can cause an excessive flow of heat from the core to the periphery. The likelihood of such a problem is reduced through a diversion of venous return from the superficial vessels to the venae comitantes (see above).

An excessively cold environment may provoke a "paradoxical" dilation of the skin vessels. Dilation alternates with periods of intense vasoconstriction, creating a "hunting" reaction. The phenomenon is blamed upon a paralysis of the arteriovenous anastomoses by extreme cold, and it can cause a severe heat loss. In less extreme climates, vasoconstriction minimizes heat losses by radiation, convection, and conduction, but substantial evaporative heat losses may be unavoidable (Chap. 3).

Over the ambient range 5–30°C, air temperature has a relatively slight effect upon core temperature (Candas et al., 1973). If body temperature falls despite vasoconstriction, metabolic heat production is increased through shivering (an increase in muscle tone, or a frank simultaneous contraction of agonists and their antagonists; Chap. 3) plus an increase of voluntary physical activity (Pugh, Edholm et al., 1960; Hart and Jansky, 1963; C. A. Dawson et al., 1970; Nadel, Holmér et al., 1974). There may also be an increased secretion of thyroid hormone, adrenaline, and adrenocorticoids. The stimulus to shivering is disputed; hypotheses include the stimulation of cutaneous cold receptors (Benzinger, 1963) or a less direct response to the total heat content of the periphery (A. Craig and Dvorak, 1968). Partly because body movement increases convective heat loss and partly because an active person has less extensive cutaneous vasoconstriction, shivering is often a more effective method of increasing core temperature than an equivalent amount of physical exercise (Keatinge, 1961; A. Craig, and Dvorak, 1968).

Tolerance of body cooling is quite limited. Discomfort and a deterioration of psychomotor performance are seen with a heat loss of ~170 kJ • m^{-2}, and the maximum safe heat loss is ~340 kJ • m^{-2}. Nerve conduction velocity decreases as local temperature falls, and nerve block is complete at 8–10°C (Chap. 3); this causes progressive difficulty in the performance of skilled tasks (Glaser and Shephard, 1963; Vangaard, 1975; J. LeBlanc, 1975). Frostbite occurs if the blood supply to an area of skin is sufficiently reduced to permit tissue destruction by freezing. Owing to the substantial osmotic pressure of tissue fluid, freezing occurs at −1°C rather than 0°C.

Fat Loss

Adipose tissue has a low metabolic rate and thus requires a limited blood flow. Individuals with a substantial layer of subcutaneous fat are better insulated than thinner subjects. Nevertheless, circumpolar peoples who have not yet adopted a sedentary "Western" life-style have a relatively thin layer of subcutaneous fat (Shephard, Hatcher, and Rode, 1973). Presumably, it is better to obtain a variable insulation from clothing than more permanent insulation from a thick layer of body fat (which in any event may disappear during a period of starvation).

A combination of cold exposure and prolonged vigorous activity leads to a loss of fat greater than that encountered with equal activity under warm conditions (W. J. O'Hara, Allen et al., 1979; Chap. 11). Possible contributing factors include mobilization of fat to sustain body temperature, incomplete combustion of fat through a "ketone shunt" (cold exposure usually causes a massive ketosis), uncoupling of oxidative reactions due to an increased secretion of thyroxine, and added energy expenditure (due to the cost of walking through snow and the "hobbling" effect of heavy clothing). The cold stress inducing such an effect is not extreme (W. J. O'Hara, Allen et al., 1977a–c), suggesting that a vigorous winter holiday (for example, several weeks of cross-country skiing) might be of value in treating middle-aged obesity (W. J. O'Hara, Allen et al., 1978b).

Acclimatization

It is hard to demonstrate physiological adaptations to a cold environment. While the general level of skin blood flow is reduced in an attempt to conserve heat, there may be an increase in perfusion of the exposed extremities, restoring the dexterity of the hands (Glaser and Shephard, 1963). Nevertheless, it is difficult to measure the extent of such changes because blood flow remains small and the precision of plethysmography is restricted by shivering.

Some animal species show an increase of basal metabolic rate with prolonged exposure to cold (Itoh, 1974). This has been attributed to nonshivering thermogenesis, a phenomenon linked to the metabolism of brown fat (Hart and Jansky, 1963; C. A. Dawson et al., 1970). It is less certain that humans respond in this fashion. Some early reports described a high resting metabolism in circumpolar peoples (Shephard, 1978a), but this may have reflected anxiety or a "specific dynamic action" of the traditional fat diet. More recent data for arctic populations eating Western foods show no increase of metabolism (Rodahl, 1952; Rodahl and Bang, 1957; Rodahl, 1963; Shephard and Godin, 1973). Repeated exposure to severe cold is associated with nitrogen loss (Issekutz, Rodahl, and Birkhead, 1962); while this may be no more than a catabolic reaction to the energy demands of sustained shivering, some authors consider it as evidence of a hormonal adaptation to cold.

Acclimated individuals feel more comfortable and shiver less while exposed to cold (Glaser and Shephard, 1963; J. LeBlanc, 1975; Golden 1979). However, body temperatures often drop to a lower level than those seen in unacclimatized men exposed to the same stress. Much of the apparent adaptation may thus be no more than habituation to a variety of unpleasant sensations (Glaser, 1966).

With prolonged residence in the cold, various tricks in the organization of house, clothing, and outdoor movement are adopted. These minimize the cold stress to which the individual is exposed; indeed, adjustment of personal microclimate may become so efficient that there is no longer scope for physiological acclimatization (A. F. Rogers and Sutherland, 1978).

Extreme cooling of the hands is a particular problem among fishermen. With repeated exposure, the extremities undergo a considerable degree of local acclimatization (J. Nelms and Soper, 1962; Strömme et al., 1963; Glaser and Shephard, 1963; Hellström, 1965; J. LeBlanc, 1975). Vasoconstriction is suppressed, and the hand remains nimble enough to perform precise work while in cold water (Glaser and Shephard, 1963; J. LeBlanc, 1975). Raynaud's phenomenon is the converse situation: an excessive sensitivity of the digital blood vessels to cold.

Chin et al. (1973) have noted that physical training reduces catecholamine release when rats are exposed to severe cold. However, there is little interaction between training and cold acclimatization in humans (K. L. Andersen, 1967; Bryan, 1967).

Cold Pathology

If local cooling is sufficient to freeze intracellular

fluid, the cells involved may suffer irreversible damage ("frostbite").

Exposure to cold air or cold water causes a cutaneous vasoconstriction and hypertensive reaction (Renson and Van Gerven, 1969; J. LeBlanc, 1975) that increases both pre- and afterloading of the heart. During moderate work, an increase of muscle tone and/or frank shivering also augments the oxygen cost of activity (Chap. 3; Nadel, Holmér et al., 1974). The heart-rate/oxygen-consumption relationship is shifted to the right, and there is a related increase of cardiac stroke volume (Fig. 12.7; A. Craig and Dvorak, 1968; McArdle, Magel et al., 1976). These various factors increase cardiac work load, making angina of effort particularly likely when exercising out of doors in cold weather. Impingement of cold, dry air on bronchial receptors may also provoke an intense bronchospasm; the likelihood of both bronchial spasm and angina can be reduced by use of a jogging mask that preheats the respired air (Kavanagh, 1970).

General body cooling by hill walking or cold-water immersion can lead to ketonuria, cold exhaustion, postural hypotension, collapse, and hypothermia (Pugh, 1969b; Haight and Keatinge, 1973b; Shephard, 1976b; Chap. 3). Treatment following cold exposure is quite difficult, and patients frequently die. Complications include respiratory depression, metabolic acidosis, circulatory failure, and renal failure. Cardiac fibrillation may also develop. The usual plan of treatment includes fairly rapid rewarming, assistance to ventilation including the administration of oxygen, and injection of hydrocortisone to counteract circulatory failure (Pugh, 1972; de Villota et al., 1973).

A sudden plunge into cold water causes simultaneous and intense stimulation of cutaneous cold receptors, with a reflex hyperventilation, breathlessness, and an inability to control respiration (Canabac et al., 1964; R. C. Goode, Duffin et al., 1975). Occasionally, cardiac arrest or fibrillation may be induced by the shock of immersion. Slowing of muscular contraction, malfunction of sensory receptors, and a progressive deterioration of cerebral function all compound the difficulty of the swimmer. The effort required for propulsion is also increased due to an appreciable rise in the viscosity of water (1.8 centipoise at 0°C, compared with 1.0 centipoise at 20°C).

UNDERWATER ACTIVITY

The challenge of exercise at increased ambient pressure is being encountered by an increasingly large proportion of the total population in such pursuits as recreational diving, tunneling through swampy land, building the foundations of bridges, escaping from submarines, and exploiting the ocean bed. The hazards that are met depend upon the type of equipment that is used and the experience of the user.

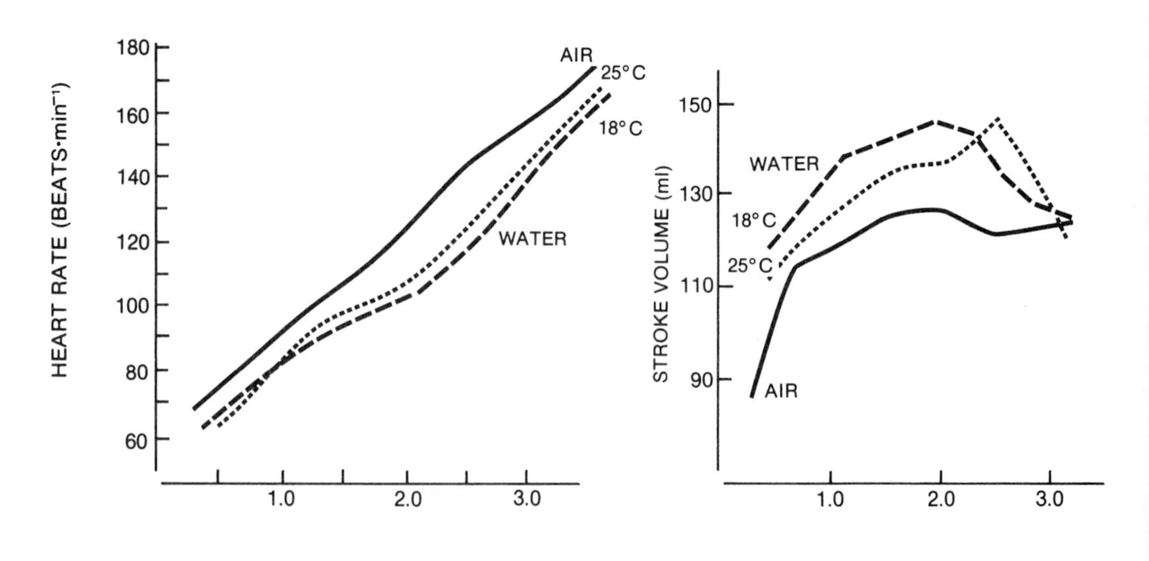

FIG. 12.7. Influence of immersion in cold water upon cardiac response to bicycle ergometer exercise. (After McArdle, Magel et al., 1976.)

Small increments of ambient pressure are relatively harmless in tunneling operations, but there is no "safe" minimum depth for a diver. Drowning can occur in a few cm of water, and lung rupture is theoretically possible following submersion to 1½-2 m.

Types of Equipment

Breath-hold Diving

An increase of ambient pressure extends the possible period of breath holding (Hesser, 1965). Nevertheless, a diver can only spend a limited period underwater using the breath-holding technique (Lanphier and Rahn, 1963). In 1968, Croft made a 148-sec dive, reaching a depth of 73 m. His only pieces of equipment were contact lenses and saline-filled goggles.

The Traditional Diver

The traditional diver (Haldane and Priestley, 1935) wears a rigid helmet, breast-plate, and weighted suit. The heavy nature of the equipment and use of a surface line greatly restrict mobility. Air is supplied from a compressor or gas tanks. The necessary flow is determined by the rate of working and the depth. Activity demanding an air flow of 50 l • min^{-1} BTPS at the surface requires 100 l • min^{-1} at a depth of 10 m and 200 l • min^{-1} at 30 m. At very great depths, the nitrogen content of the air is replaced by helium to avoid problems of nitrogen narcosis (see below). A partial recirculation of gas through a CO_2 absorbent canister economizes on the use of expensive helium mixtures (S. Williams, 1969; Table 12.2).

A Surface Demand System

A surface demand system provides a practical alternative for the underwater worker. A mouthpiece is connected by a neutrally buoyant hosepipe to a demand valve at the surface. Because the buoyancy of the subject is less than in a traditional diving suit, the weighted boots can be replaced by fins, with a considerable improvement of mobility. On the other hand, the light surface line is vulnerable to kinking and fracture, and it is thus a wise precaution to carry a small self-contained underwater breathing apparatus for emergency use.

Open Circuit

Self-contained systems are widely used for recreational purposes (S. Williams, 1969). Gas is supplied via a demand valve from compressed air tanks carried on the back. The limited endurance of the cylinders can create problems if a lengthy period of decompression is needed, and from the military viewpoint the telltale trail of bubbles provides a potential adversary with a clue to the presence of a swimmer. It is difficult to read cylinder gauges while swimming, and accordingly exhaustion of the system is signaled by a valve that imposes a noticeable resistance to breathing when three-quarters of the air supply have been used. Economy of gas usage may be achieved if the diver breathes through a "snorkel" tube when he is close to the water surface.

TABLE 12.2
Relative Efficiency of Various Underwater Breathing Systems

System	Efficiency at 4 Atmospheres (%)*
Free flow	0.2–1.8
Open-circuit demand	0.8–1.7
Semiclosed (constant mass of gas delivered)	3.2–40
Semiclosed (constant ratio of delivery to ventilation)	8.4–15.5
Closed circuit	100

*Percentage of delivered oxygen relative to that consumed by the body.
Source: After S. Williams, 1969.

Closed-Circuit Systems

Closed-circuit systems were in disfavor for a time due to problems from oxygen toxicity (Miles, 1966; MacKay, 1976). The popularity of this type of equipment has returned with the development of small cylinders of liquid oxygen. During normal shallow-water operations, nitrogen is first flushed from the subject's lungs and blood. Oxygen is then supplied to a bag of ~7-l capacity at a rate of ~1 l • min^{-1}. Excess gas escapes from a blow-off valve adjusted to a little above water pressure (10–30 cm H_2O).

There are two main types of apparatus: pendulum and recirculating systems (Fig. 12.8). The pendulum arrangement has the advantage of simplicity, with only a blow-off valve to maintain, and because of the bidirectional air flow, the efficiency of CO_2 absorption is high. On the other hand, the pendulum equipment inevitably has a substantial dead space relative to that of the recirculating system. In both types of device, the canister holds ~1 kg of soda-lime granules, sufficient to absorb ~200 l of carbon dioxide.

If oxygen is breathed, there is a danger of

oxygen poisoning at depths greater than 8–10 m. However, the operating range can be extended by use of nitrogen/oxygen mixtures. At 43 m, 40% oxygen is provided at a flow of ~8 l • min^{-1}, and at 55 m, 32.5% oxygen at a flow of ~13 l • min^{-1}. Decompression schedules (see below) are calculated as for an equivalent dive breathing air.

Diving Suits

Diving suits are designed to provide thermal insulation (Chap. 3) and protection against sharp objects such as coral and wreckage. A rubberized suit tends to trap air, and this can allow a subcutaneous "squeeze" (see below). At depth, compression of the air pockets largely destroys insulation (Miles, 1966). Possible methods of overcoming this problem include use of an incompressible plastic or, more simply, periodic inflation of the suit (preferably by gas with a low heat capacity, such as a Freon/CO_2 mixture; Rawlins and Tauber, 1971).

A "wet" suit is an alternative diving outfit (R. F. Goldman et al., 1966). Since this is permeable to water, problems of a subcutaneous squeeze are avoided. Water is trapped beneath the suit, and this is warmed by the body, providing some thermal insulation. The main problems of a wet suit are loss of buoyancy and discomfort when out of the water. Also, when working hard the enclosed water becomes hot, although the swimmer can then open his suit and let in some cold water (Beckman, 1963; A. B. Craig, 1971).

Caissons and Tunnel Workings

Caissons and tunnel workings exclude water by use of compressed air (H. E. Lewis and Paton, 1957; Haxton and Whyte, 1969; Shephard, 1972d). Decompression schedules are relatively easy to calculate for such environments since pressure remains constant over the working day.

Submarines

Submarines normally operate at sea level pressures, but in the event of damage, the escape compartment must be flooded, bringing air and any environmental contaminants to the local hydrostatic pressure. There is a danger that with compression partial pressures of carbon dioxide could reach narcotic levels, and for this reason modern craft have a separate escape chamber with an independent CO_2-free supply of oxygen and nitrogen. Ascent from the escape chamber is at a rate of ~1 m • sec^{-1}, air being exhaled steadily. The aim is to keep the lungs comfortably inflated, yet to complete the ascent before hypoxia becomes too severe (Fig. 12.9). The partial pressure of CO_2 usually remains relatively constant over the ascent, but oxygen pressures drop from ~40 kPa at 4 atmospheres to ~4 kPa in the final 2–3 m of a 30-m ascent (Schaefer, 1969).

Underwater Habitats

Underwater habitats are used in exploration of the continental shelf (R. W. Hamilton and Schreiner, 1968; Chouteau, 1969; Schaefer et al., 1971; MacInnis, 1971; Summitt et al., 1971.) The aquanaut spends several weeks saturated with gas at a pressure of 10–12 atmospheres and makes periodic excursions to much greater depths. In this way, he avoids loss of working time through adherence to lengthy decompression schedules.

Gills

Gills can be designed that allow a man to sustain his resting energy expenditure while underwater; however, they are bulky, dangerously fragile, and must be exposed to a brisk stream of moving water to allow an adequate oxygen intake (Kylstra, 1969); further, because CO_2 has a high water solubility, there is a danger of hypocapnia.

Problems of Increased Ambient Pressure

As during descent from high altitude, problems may arise from failure to equalize the pressure of gas pockets in the middle ear (barootitis; H. G. Armstrong and Heim, 1937; J. D. Harris, 1971), teeth (aerodontalgia), and sinuses (aerosinusitis). One of the Canadian contestants at the 1970 Commonwealth games ruptured an eardrum while diving, and in her case it was thought that the initial collapse of the eustachian tube had developed on the flight from Canada. Problems are less likely when breathing helium/oxygen mixtures, since such gases pass more rapidly through the Eustachian tube (J. D. Harris, 1971). Prophylactic measures include (1) avoidance of deep diving when entrances to the sinuses and Eustachian tubes are narrowed by an acute respiratory infection and (2) regular dental examinations to exclude gas pockets in teeth.

Both submersion (Rennie, di Prampero, and Cerretelli, 1974) and diving lead to a reflex slowing of the heart rate, but this response is attenuated by simultaneous isometric muscle contraction (S. A. Bergman, Campbell, and Wildenthal, 1972). There is a large rise of blood lactate after a dive, suggesting

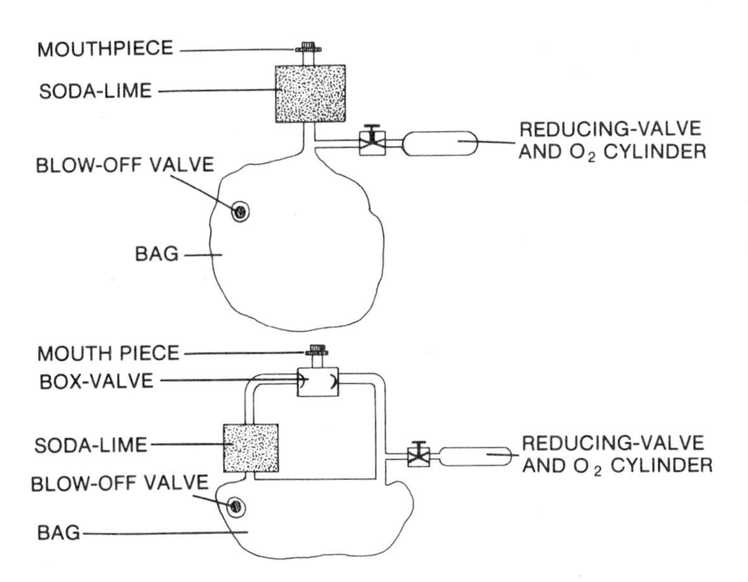

FIG. 12.8. Self-contained underwater breathing systems. **Top.** Pendulum type. **Bottom.** Recirculating system.

a restriction of muscle blood flow while submerged (Scholander et al., 1962).

If breathing is checked during descent (either deliberately, as in breath-hold diving, or through inexperience, as in the use of self-contained underwater breathing equipment), the intrathoracic gas volume is compressed toward residual volume. Depending on the initial gas volume in the lungs, the increase of external pressure, the respiratory quotient, and the retention of carbon dioxide, the chest cage may be brought to its minimum dimensions without abolishing all of the pressure differential between the water and lung gas. The intrathoracic vessels are normally protected from negative pressures by collapse of the large veins where they enter the thorax (Hong, Ting, and Rahn, 1960). However, when submerged, capacity vessels external to the rib cage are compressed, and as much as a liter of blood is driven into the pulmonary circulation (Burki, 1976). The intravascular pressure therefore rises with respect to alveolar gas pressure, and the pulmonary vessels tend to rupture into the lungs, with hemorrhage and edema formation (A. B. DuBois, 1965). Collapse of the lung may also occur, with thoracic displacement of the abdomen and diaphragm (Agostoni, 1965). The total syndrome is described as a "thoracic squeeze" (Chap. 5). One prophylactic trick adopted by pearl divers is to whistle prior to diving. This drives blood out of the thorax, increasing total lung capacity.

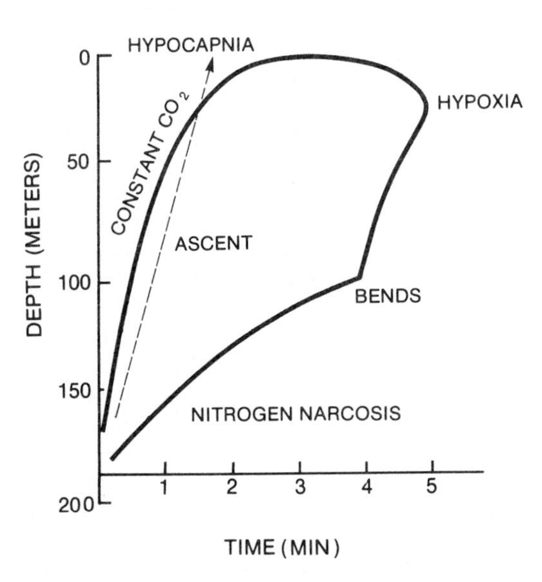

FIG. 12.9. Typical controlled ascent in submarine escape, showing "safe" zone. (After A. B. DuBois et al., 1963.)

A pressure differential may also develop behind rigid goggles, causing conjunctival hemorrhage, with bleeding into the eyelids and the fat pocket behind the eye (retroorbital fossa). Air pockets trapped behind a waterproof immersion suit allow hemorrhage into folds and ridges of skin, while if the ears are covered, blood blisters may develop in the external auditory meatus and on the tympanic membrane.

Problems of Maintained Pressure

Oxygen Toxicity

Pure oxygen produces harmful effects if it is breathed for 6–12 hr (Bean, 1945; J. D. Wood, 1969). Problems include lung collapse (Fig. 12.10), pulmonary congestion, depressed function of the surfactant-forming alveolar macrophages (Giammona et al., 1965), oxidation of the unsaturated fatty acid used in surfactant production, and absorption of oxygen in occluded airways (Penrod, 1956). There is also some retention of CO_2, pulmonary vasoconstriction, and an increase in pulmonary vascular permeability.

Initial symptoms of oxygen poisoning include substernal distress, cough, nausea, and pallor, with an increase of pulse rate and blood pressure. Respiratory changes lead to a progressive reduction of vital capacity, while cerebral toxicity is indicated by constriction of the retinal vessels and a circumscription of the visual field (Behnke, Forbes, and Motley, 1936; K. W. Donald, 1947; Nichols and Lambertsen, 1971). There may also be incoordination, light-headedness, dizziness, nausea, euphoria, confusion, and convulsions (Stadie et al., 1944; Gillen, 1966; J. M. Young, 1971; Criscuoli and Albano, 1971).

At rest, 2 atmospheres of oxygen are tolerated for 6 hr, 3 atmospheres for 3 hr, and 4 atmospheres for ~30 min (Behnke, 1942a,b; Donald, 1947). During exercise, tolerance decreases to 30 min at 3 atmospheres. The tolerance limit also seems less for diving than for deliberate chamber exposure, so that a limit of 1 hr at a depth of 8 m is normally advised for a swimmer using oxygen (Table 12.3). Oxygen toxicity can arise when breathing compressed air if the ambient pressure exceeds 3–4 atmospheres.

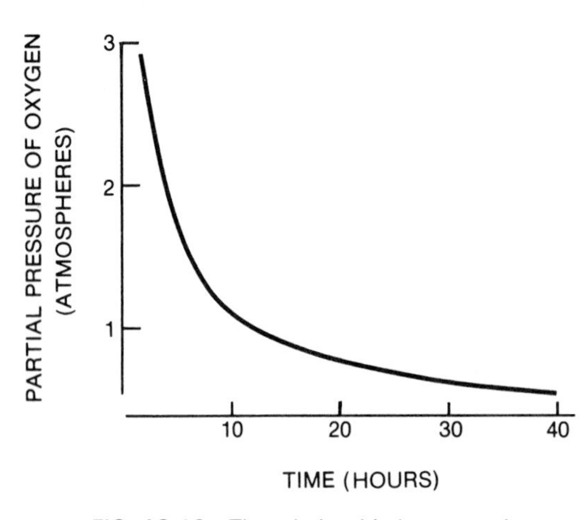

FIG. 12.10. The relationship between the partial pressure of oxygen and the time to onset of respiratory symptoms. (After J. D. Wood, 1969.)

The respiratory problems are related to gas absorption. If a segment of lung becomes closed off from the breathing apparatus, the large oxygen pressure gradient between the affected alveoli and the pulmonary capillaries leads to a rapid and complete absorption of the trapped gas; there is a loss of the normal "respiratory strut" provided by equal pressures of nitrogen in the alveoli and bloodstream.

Cerebral changes may reflect an inhibition of vital enzymes by the oxygen itself or associated free radicals (H. C. Davies and Davies, 1965; Haugaard, 1968). Possibly, essential sulfhydryl groups are oxidized (A. P. Sanders and Currie, 1971). Vulnerable enzymes and metabolites include cytochrome c reductase, glutamic acid decarboxylase, γ-aminobutyric acid (GABA), and diphenyl nitrohydrazine (NADH). GABA plays an important role in normal cerebral function, and the seizures of oxygen poisoning are associated with low GABA levels. High-pressure oxygen could reduce brain GABA by (1) inhibition of the GABA-forming enzyme glutamic acid decarboxylase, (2) activation of the GABA-destroying enzyme (mitochondrial GABA α-oxoglutarate transaminase), or (3) alteration of membrane characteristics so that GABA diffuses more readily from the synaptic vesicles to the mitochondria (J. D. Wood, 1971). GABA therapy protects against both lung damage and convulsions. Other possible sources of disturbed function are oxidation of structural lipids in the cell membranes (Zirkle et al., 1965) and an accumulation of carbon dioxide (Lambertsen et al., 1953; I. G. Walker, 1961).

Inert Gas Narcosis

Historical aspects of inert gas narcosis and decompression sickness are reviewed by D. I. Fryer (1968). As early as 1835, Junod noted that when breathing compressed air "the functions of the brain are activated, imagination is lively, thoughts have peculiar charm, and in some persons symptoms of intoxication are present." The highest functions of the brain are inhibited, and there is a cheerful but dangerous euphoria, often with an exaggeration of normal personality. At depths of more than 100 m, there may be a sense of impending blackout, manic or depressive states, and changes of sensory perception. If these symptoms are ignored, neuromuscular incoordination and loss of consciousness may occur (Criscuoli and Albano, 1971). Recovery is rapid on decompression.

Loss of reasoning ability, reaction speed, and dexterity can be detected within 12 min at a pressure

TABLE 12.3
Limiting Times for Dives Breathing Oxygen

Depth (m)	Time (min)
3.0	240
4.5	150
6.1	110
7.6	75
9.1	45
10.7	25
12.2	10

Source: After Duffner and Lanphier, 1974.

of 4 atmospheres (Kiessling and Maag, 1960; Frankenhauser et al., 1963). More prolonged exposure gives difficulty with simple mental arithmetic and a decrease in the frequency at which a flickering light appears as a continuous source (flicker-fusion threshold) (J. G. Dickson et al., 1971; P. B. Bennett, 1971).

Symptoms are related largely to the narcotic action of the increasing partial pressure of nitrogen (P. B. Bennett and Blenkarn, 1974), although carbon dioxide buildup can have a potentiating effect (P. B. Bennett, 1963), whether the latter is due to hypoventilation (Bühlman, 1961) or to inhalation of oxygen (Bean, 1950; P. B. Bennett, 1965). Other inert gases have an even greater effect, and it seems possible to draw a parallel between the lipid solubility of a given gas and its narcotic potency (P. B. Bennett, 1969; E. B. Smith, 1969; Table 12.4). Possibly there is a solution of gas in the lipid phase of the cell membrane (Schreiner, 1968); this could displace oxygen, leading to histotoxic hypoxia and a depression of neuronal transmission.

TABLE 12.4
Relationship Between Lipid Solubility of Various Gases and Their Narcotic Potency

Gas	Lipid Solubility	Inverse of Narcotic Potency
Helium	0.015	4.26
Neon	0.019	3.58
Hydrogen	0.036	1.83
Nitrogen	0.067	1.00
Argon	0.140	0.43
Krypton	0.430	0.14
Xenon	1.700	0.04

Source: After P. B. Bennett, 1969.

Poulton et al. (1963) described a deterioration of cerebral function with exposure to an ambient pressure of 2 atmospheres. This is not yet generally accepted (P. B. Bennett, 1966), although one recent report found that air at normal atmospheric pressure had some anesthetic effect relative to helium/oxygen mixtures (P. M. Winter et al., 1975).

From the technical viewpoint, it is quite possible to carry out diving operations to depths of ~90 m while breathing compressed air, but unless a team is very experienced it is more wise to set a limit of 30 m. At greater depths, oxygen can be mixed with a relatively nontoxic gas such as helium. This facilitates the work of breathing, but has the disadvantage of increasing heat loss in cold water (H. A. Leon and Cook, 1960).

Work of Breathing

During submaximal exercise, the ventilation of a diver is decreased partly by the high work of breathing (Lanphier, 1969) and partly by the high partial pressure of oxygen. The normal sea level relationship between alveolar CO_2 pressure and respiratory minute volume is distorted (L. D. H. Wood and Bryan, 1971; Fig. 12.11), and in an attempt to reduce the cost of respiration there is an increase of tidal volume and a decrease of breathing frequency (Hesser and Holmgren, 1959; M. E. Bradley et al., 1971). There is also a decrease of heart rate, due partly to an increase of parasympathetic tone (the diving reflex, as discussed above) and partly to the high partial pressure of oxygen (Jarrett, 1966; Fagraeus, 1974). At moderate depths, the maximum oxygen intake is increased (Fagraeus, 1974), but at greater depths ventilatory work limits performance unless nitrogen is replaced by helium. There is a tendency for CO_2 accumulation, but this does not seem to influence the limiting blood and muscle concentrations of lactate during anerobic work.

Among the many factors that conspire to increase the work of breathing at depth, the most important single consideration is an increase in the density of respired gas. The turbulent component of airway resistance increases in proportion to $\sqrt{D}$, where D is the density of the respired gas mixture. There is also an increase in the force required for convective acceleration of gas as it passes from the lung spaces to the airways of narrower total cross-section. The pressure drop along the airways is more rapid than at sea level, and the "equal pressure point" (Fig. 5.15) is displaced peripherally, beyond the bronchi with cartilaginous supports. The diver thus has a low maximum expiratory flow rate while he is at depth, and attempts at more vigorous expiration merely lead to

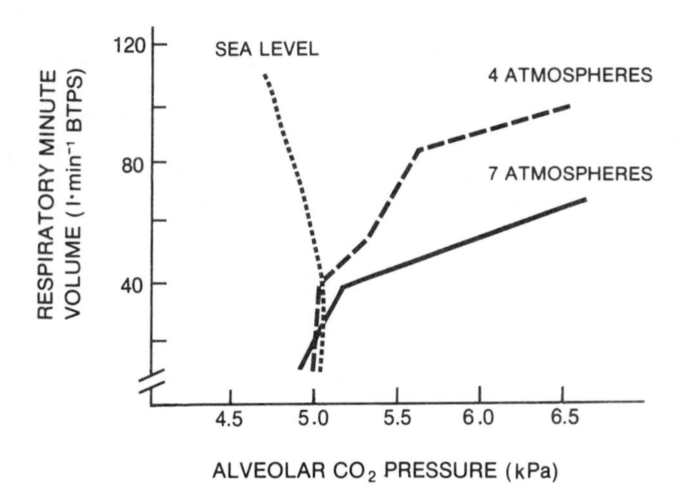

FIG. 12.11. Relationship between respiratory minute volume and alveolar CO_2 pressure at rest and at increasing rates of working. (After L. D. H. Wood & Bryan, 1971.)

collapse of the airway and "wasted" ventilatory effort (Lord et al., 1966; Maio and Farhi, 1967; Varène et al., 1967). The remedies are twofold: to lengthen expiration at the expense of inspiration, thereby decreasing the expiratory pressure gradient, and to increase the mean chest volume, thereby augmenting the elastic forces resisting collapse.

Some ventilatory resistance is imposed by the viscosity of the water itself. Compression of the thoracic cage (for example, in snorkel diving) forces a swimmer to breathe over an unfavorable segment of his chest compliance curve (Fig. 5.18); it also narrows the air passages, increasing both viscous resistance and the likelihood of expiratory airway collapse. In other situations, air pressure may exceed water pressure. The chest is then forced toward an inspiratory position. A small positive intrapulmonary pressure eases the work of breathing, but a greater excess pressure induces an unnatural pattern of breathing—a forced expiration, with relaxation during inspiration; the body then responds by developing a ventilaton that is excessive relative to the required rate of working.

Respiratory equipment used at high ambient pressures should have no more than a moderate air flow resistance. Although a high-resistance breathing system is tolerated at sea level, it becomes an unbearable burden when the gas density is increased eight- or tenfold. The effect of a more moderate external resistance depends on the locus of the "equal pressure" point. If a diver has already reached the maximum possible expiratory flow rate for a given gas density and is wasting respiratory work in fruitless attempts to produce a greater air flow, then a moderate external load may be accepted with little cost to the work of breathing.

Carbon Dioxide Accumulation

Even if a subject is not hampered by breathing equipment, the increase in the work of breathing at depth is sufficient to depress maximum voluntary ventilation (MVV).

At 30 m depth (4 atmospheres), the MVV is reduced to 50% and at 120–150 m (13–16 atmospheres) to only 25% of the sea level figure (Fig. 12.12: Sevsing and Drube, 1960; Maio and Farhi, 1967; Lanphier, 1969). It is further unlikely that an exercising diver who is breathing through valves, tubing, and a soda-lime canister can sustain more than 50% of his maximum voluntary ventilation over a 15-min period. Even if a diver has a sea level MVV of 200 $l \cdot min^{-1}$, he is thus restricted to an MVV of 50 $l \cdot min^{-1}$ and an external ventilation of 25 $l \cdot min^{-1}$ BTPS when swimming at a depth of 120 m.

The increase of gas density tends to worsen alveolar ventilation, both by slowing the diffusional exchange of alveolar and bronchial gas, and by enhancing inequality of ventilation. But even if alveolar ventilation were to remain the same fraction of external ventilation as at sea level, it would amount to no more than 20 $l \cdot min^{-1}$ BTPS at 120 m. Carbon dioxide accumulation is thus very likely in the deep-sea diver (Behnke and Willmon, 1939; Schaefer, Bond et al., 1968; Schaefer, 1969; L. D. H. Wood and Bryan, 1971).

The alveolar carbon dioxide pressure P_{A,CO_2} (kPa) is given by the equation

$$P_{A,CO_2} = \frac{115\ \dot{V}_{CO_2}\ (l \cdot min^{-1}\ STPD)}{\dot{V}_A (l \cdot min^{-1}\ BTPS)} \qquad (12.6)$$

where $\dot{V}_{CO_2}$ is the carbon dioxide output and $\dot{V}_A$ is alveolar ventilation. Given a CO_2 production of 2 l • min^{-1} and an alveolar ventilation of 20 l • min^{-1} BTPS, there is inevitably a CO_2 partial pressure gradient of 11.5 kPa between inspired and alveolar gas. Let us further suppose that dead space and/or inefficiencies in the CO_2 scrubbers raise the inspired CO_2 concentration to 0.25%; at 10 atmospheres, this adds a further burden of 2.5 kPa, bringing the total alveolar CO_2 pressure to the dangerously toxic level of 13.9 kPa.

Lanphier (1969) pointed out the very large ventilation needed to avoid an accumulation of CO_2 in equipment such as diving helmets; in one example with a CO_2 production of 1 l • min^{-1}, a ventilation of 420 l • min^{-1} was needed to keep the helmet CO_2 pressure to 1.3 kPa at a breathing pressure of 4 atmospheres; at sea level, the same pressure was seen with a ventilaton of 130 l • min^{-1}.

Other factors exacerbating the tendency to CO_2 accumulation while underwater are (1) inhibition of blood CO_2 transport by the high pressure of oxygen, (2) a restriction of cerebral blood flow, again due to the high oxygen pressure, (3) a depression of peripheral chemoreceptors (by oxygen) and the medulla (by inert gas narcosis), and (4) a long-term reduction in sensitivity of the respiratory centers (Goff and Bartlett, 1957). It is hardly surprising that it is difficult to sustain a vigorous work rate at extreme depths or that divers develop a tolerance to high pressures of carbon dioxide (Lanphier, 1964; Pingree, 1977; but not Froeb, 1961).

The likelihood of CO_2 retention is greatly reduced if all of the nitrogen and part of the oxygen in the respired mixture are replaced by helium (Anthonisen et al., 1971; M. E. Bradley et al., 1971; J. N. Miller et al., 1971; Fig. 12.12). The density of the inhaled gas is then about one-fifth of normal, and

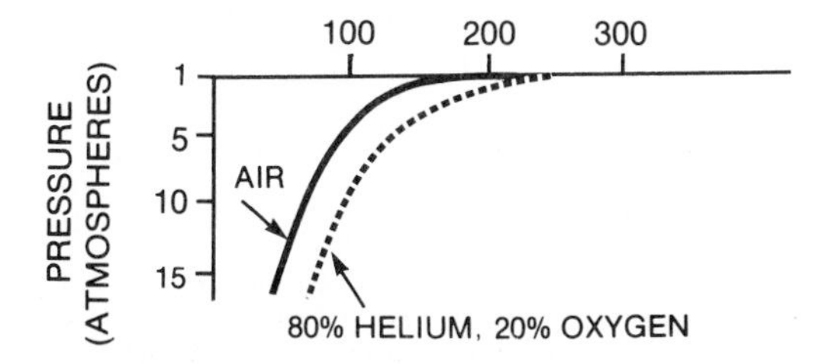

FIG. 12.12. Maximum voluntary ventilation at increasing ambient pressures, breathing air and at an 80/20 helium/oxygen mixture. (After J. N. Miller et al., 1971.)

turbulent resistance is reduced by at least $\sqrt{5}$. Convective acceleration also demands less pressure and the maximum expiratory flow rate is greatly improved. The main disadvantages of the helium mixture are (1) a small increase in the viscous resistance to breathing and (2) an increase of body heat loss and thus of resting energy expenditures (H. A. Leon and Cook, 1960; L. W. Raymond et al., 1968).

Other Problems

There are many other problems when working underwater. Vision is distorted (Kinney and Luria, 1971). Images are altered in size and distance, colors are changed, brightness is dimmed, and outlines become blurred. Communication is hampered by distortion of speech (due to increased gas pressures, abnormal gas mixtures, and resonant equipment; Sergeant, 1969; Fant et al., 1971). The intelligibility of speech while breathing helium gas mixtures is 90% at sea level but only 47% at 7 atmospheres pressure. Most subjects compensate by talking more slowly, and this reduces alveolar oxygen pressure. The body is also cooled rapidly, the combination of increased central blood volume and cold provoking a marked diuresis (McCally, 1965).

A variety of forms of animal and plant life can injure the swimmer, as may collision with rocks, wreckage, and overhead obstacles.

Particularly in warm weather, swimming pools, shallow rivers, lakes, and associated changing rooms can all transmit infection to a swimmer.

The physiological processes associated with drowning have received detailed study (D. G. Greene, 1965; Segerra and Redding, 1974). Initially, inhaled water irritates the bronchial tree, giving rise to spasm of both larynx and bronchi (Lougheed et al., 1939) and breath holding. However, with continued submersion, breathing recommences, and the lungs become inundated with substantial quantities of water, vomit, and other materials (D. G. Greene, 1965). This leads in turn to hypoxia, hypercapnia, and acidosis, with the usual chain of events seen in impaired gas exchange (a rise of blood pressure, a slowing of the heart, the development of dysrhythmias, a fall of blood pressure, hyperpnea, and terminal gasping).

In fresh water, the osmotic pressure of the inhaled fluid is less than that of the blood. Water is thus drawn into the blood, lowering plasma sodium ion concentration. There is also an increase of plasma potassium ion concentration, attributed to tissue hypoxia rather than rupture of red cells. These changes predispose to the onset of ventricular fibrillation (Swann, 1956). Saltwater drowning, in contrast, is associated with the passage of fluid from the blood into the lungs; plasma sodium ion concentration rises, and there is general hemoconcentration with a risk of pulmonary edema.

After removal to a place of safety, the first step in the treatment of a drowned person is to check the pulse, applying cardiac massage if necessary (Chap. 6). Obvious debris is removed from the mouth and throat, and artificial respiration is then instituted by an accepted technique. Because of bronchospasm and a decrease of lung compliance, quite high ventilatory pressures may be necessary.

Hospital treatment includes correction of acid/base imbalance, with administration of bronchodilator drugs, antibiotics to control infection, and oxygen if pulmonary edema has developed as a reaction to the saltwater or irritant chemicals in swimming pool water.

Problems of Pressure Reduction

Decompression Sickness

Decompression sickness arises from a supersaturation of the tissues with inert gas (D. H. Elliott, 1969; Kidd and Elliott, 1969). When a certain excess pressure is developed relative to ambient air (a ratio of 1.75–2.25:1.00), bubbles form intracellularly or extracellularly (Ruff and Müller, 1966; Buckles, 1968; Van Liew and Hlastala, 1969); the process may be abetted by overexpansion of the alveoli, as part of the thoracic squeeze syndrome (Hartmann and Müller, 1962). Pain and injury arise from (1) distortion and disruption of tissues, blockage of the circulation by intravascular bubbles (Walder, 1963; Gillis et al., 1968) and lipid emboli (Cockett et al., 1971), (2) arterial spasm induced by perivascular bubbles (Lambertsen, 1968), and (3) the after-effects of circulatory arrest (Saumarez et al., 1973), including changes in the morphology and carbonic anhydrase content of the red cells (Carlyle et al., 1979). Evidence for these statements includes (1) visualization of bubbles by X rays and ultrasonic techniques (Gillis et al., 1968; MacKay and Rubissow, 1971), although the bubbles do not always coincide in size or site with the severity of symptoms, (2) the preventive effect of a change in the composition of the respired gas, (3) the prompt resolution of most symptoms by recompression, and (4) the preventive value of computers based on the theory of bubble formation (Stubbs and Kidd, 1967; Hennessy, 1973; Fig. 12.13).

The most common clinical complaint is of the "bends"—a discomfort or pain felt in or near one the large limb joints (Kooperstein and Schuman, 1957). This causes clumsiness and weakness of the affected limb, and it sometimes progresses to an unbearable pain with secondary collapse. Symptoms are quickly corrected by recompression, but if a second dive is made within a few hours, pain may recur at the site previously affected.

Other forms of decompression sickness (Shephard, 1972d) include skin lesions (itching, erythema, a capillary rash, or a patchy swelling of the subcutaneous tissue), respiratory symptoms (the "chokes," shortness of breath, a dry cough, and a feeling of suffocation; El Ghawabi et al., 1971), neurological symptoms (migraine-like visual field defects; G. L. Engel et al., 1944; epileptiform attacks or spastic paralysis), vestibular dysfunction (Rubenstein and Summitt, 1971), and collapse (Romano et al., 1943; Cannon and Gould, 1964—sometimes secondary to

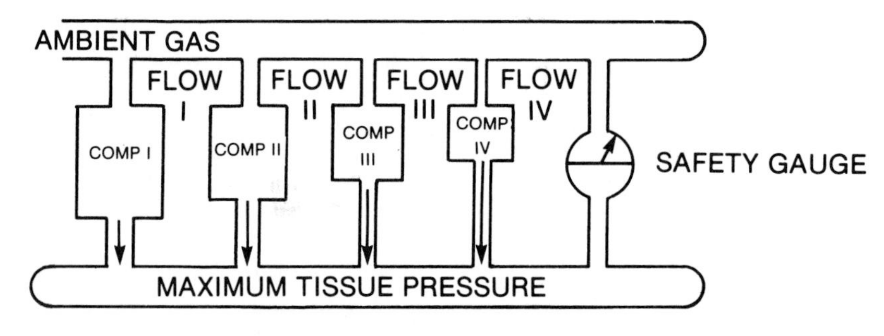

FIG. 12.13. Schematic diagram showing the arrangement of a pneumatic analogue used in monitoring the safety of diving operations. The four rigid compartments (Comp I to Comp IV) correspond to the volume/solubility products of different tissues, and the apertures are scaled to represent the corresponding regional blood flows (Flow I–IV). The gauge indicates the maximum pressure differential between ambient gas and any one of the four compartments. If this exceeds 1.75:1, the diver halts his ascent until a safe pressure differential is indicated.

the bends, but in its more serious and potentially fatal form appearing after the chokes, neurological symptoms, or as an unheralded primary manifestation).

Bone necrosis can develop some months after exposure. The areas particularly affected are the head and lower diaphysis of the long bones—the femur, humerus, and tibia (Coley and Moore, 1940; Rendlick and Harrington, 1940; Medical Research Council, 1971; El Ghawabi et al., 1971; D. H. Elliott and Harrison, 1971).

Both the saturation of the tissues with gas during a dive and its subsequent release from the body follow an exponential course (C. D. Stevens et al., 1947; Kety, 1951; Lundin, 1960; Schreiner, 1967). The exponent for any given tissue depends upon the ratio of local blood flow to the product of tissue volume and gas solubility (Schreiner and Kelley, 1971; Fig. 12.14). The half-time of both saturation and desaturation is quite long. A worker fails to reach equilibration over the usual caisson shift of 4 hr, and elimination of dissolved nitrogen occupies a similar period. Exercise speeds decompression. Because it has a lower fat solubility, helium is eliminated from slowly decompressed tissues more than twice as rapidly as nitrogen (Bühlmann, Frei, and Keller, 1967; Lever et al., 1971; Fig. 12.15).

Exercise, heat, cold, oxygen lack, injury, infection, overindulgence in alcohol, recent previous exposure to inert gas, a poor circulation (including aging or lack of fitness), and obesity all increase susceptibility to decompression sickness (Behnke, 1942a,b; E. B. Ferris et al., 1944; MacKenzie and Riesen, 1944; Van Liew, 1971). However, aging and obesity present more of a problem to the aviator (who has become saturated over a lifetime) than to the diver who is only exposed to high gas pressures for a few hours (Gribble, 1960). Repeated exposure to high pressures apparently lessens the likelihood of decompression sickness (Walder, 1969).

Until recently, divers have relied upon tables (Table 12.5), indicating the permissible rate of ascent in terms of depth and duration of exposure (Boycott et al., 1908; U.S. Navy, 1963; Workman, 1965; Hempelman, 1969). This approach is relevant to fixed-shift tasks, such as tunneling, but is difficult to apply to diving (where neither time nor depth are constant). Even breath-hold dives, if repeated at frequent intervals, can give rise to decompression sickness (Paulev, 1965), but tables to avert such a contingency would necessarily be overrestrictive. The possible frequency, duration, and depth of diving have been greatly extended by (1) use of analog computers (Göransson et al., 1963; B. A. Hills, 1967,1969; Kidd and Stubbs, 1969; Fig. 12.13) and saturation diving (where the worker eats and sleeps for several weeks at an intermediate but relatively high pressure). In an underwater habitat, the diver may change his breathing gas periodically, so that no foreign gas accumulates to a dangerous pressure (Keller and Bühlmann, 1965; Bühlmann, 1971).

Most forms of decompression sickness respond well to immediate recompression and a subsequent more gradual release of pressure (Behnke and Shaw, 1937; Bornmann, 1968). Recompression does not lead to immediate solution of the offending bubbles, but it reduces their volume drastically, lessening the chances of permanent injury. Oxygen administration is also recommended (Workman, 1968), although its supposed beneficial effect upon nitrogen elimination is sometimes offset by an associated slowing of the circulation. Medical treatment may be needed for circulatory collapse (Cockett et al., 1965).

Once bubble formation has occurred, decom-

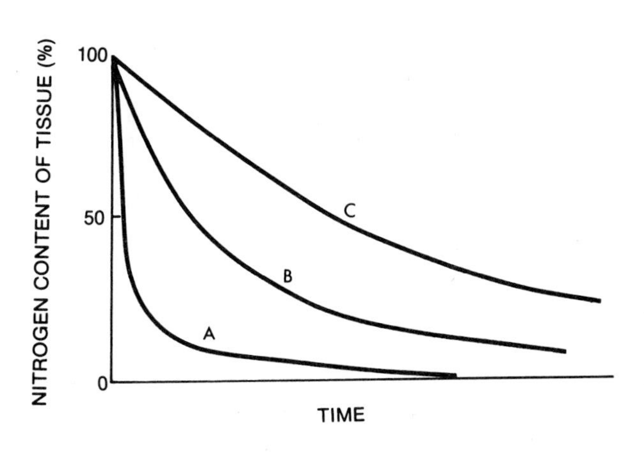

FIG. 12.14. Schema of nitrogen elimination for a subject breathing pure oxygen. In tissue A, the ratio of blood flow to the product (tissue volume × N_2 solubility) is large, and elimination is rapid. In tissues B and C, the ratio is smaller, and elimination is consequently much slower.

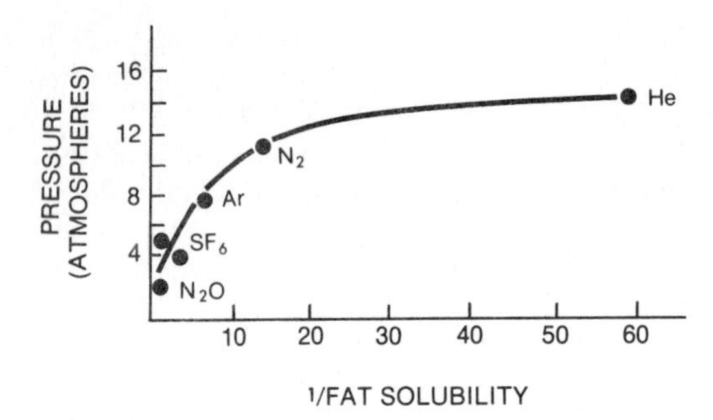

FIG. 12.15. Relationship between fat solubility of gas and initial pressure, causing decompression sickness in 50% of animals when reduced to pressure of 1 atmosphere. (After Lever et al., 1971.)

pression sickness is liable to recur in the affected region for several days. This is difficult to explain on normal curves of gas elimination. It has thus been postulated that a bubble/tissue complex is formed with a very long half-life.

Gas Expansion

Trapped gas expands during decompression, and this can give rise to structural injuries, particularly in the lungs. No difficulty is likely unless the breath is held during ascent. However, if the glottis is kept closed (for example, due to anxiety in a submarine escape), the chest expands progressively until inspiratory capacity is reached. A differential pressure then develops between lung gas on the one hand and water (and thus the circulatory system) on the other hand.

TABLE 12.5
Depth/Time Limits for Air Dives Not Requiring Decompression during Ascent

Depth (m)	Time (min)*
10.7	310
12.2	200
15.2	100
18.3	60
21.3	50
24.4	40
27.4	30
30.5	25
33.5	20
36.6	15
42.7	10
57.9	5

*Time is total interval from leaving water surface. Rate of ascent should not exceed 20 m • min^{-1}.
Source: After Duffner and Lanphier, 1974.

Experiments on anesthetized dogs suggest that the limiting pressure is 10–11 kPa, although in a conscious human subject counterpressure from clothing and a voluntary tensing of the chest and abdominal muscles may allow higher differentials without injury (Malhotra and Wright, 1960,1961).

If the lung ruptures into the mediastinum, surgical emphysema develops, with possible problems from displacement of the heart. Rupture into the pleura causes a pneumothorax, with collapse of the affected lung and again some mediastinal displacement. The most dangerous possibility is rupture of the lung into a pulmonary vein, with air embolism. Depending on posture, and thus the intravascular distribution of the gas bubbles, death from cerebral vascular occlusion is then very likely.

Other Problems of Ascent

The breath-hold diver may show a loss of consciousness during ascent (Chap. 5; A. B. Craig, 1961).

HIGH ALTITUDES

Human responses to high altitudes were studied originally in the context of balloon exploration, aviation, and mountaineering (Bert, 1878; J. Barcroft, 1914; Haldane and Priestley, 1935). More recently, the needs of mining companies and of athletes have focused attention upon acute and chronic responses to exercise at more moderate altitudes, ranging from Mexico City (2240 m) to certain Andean mines (up to 6000 m).

Acute Responses

Overview

The essential physical change in the atmosphere is a decrease of total ambient pressure (Table 12.6). This occurs in a logarithmic fashion, so that at 5500 m the pressure is approximately halved. It follows from Dalton's law of partial pressures that there is a corresponding reduction in the partial pressure of inspired oxygen. In alveolar gas, the change is more serious; since there is little alteration in the alveolar partial pressure of either water vapor or carbon dioxide, the alveolar oxygen pressure suffers a more drastic reduction. Nevertheless, there is surprisingly little change of arterial oxygen content at moderate altitudes. This stems from the shape of the oxygen dissociation curve of hemoglobin (C. S. Houston and Riley, 1947; Ernsting and Shephard, 1951). The alveolar oxygen pressure can be reduced from 13.3 to

TABLE 12.6
Alterations in Environment with Altitude

Altitude (m)	Temperature (°C)	Barometric Pressure (kPa)	Oxygen Partial Pressure (kPa)	Air Density ($g \cdot l^{-1}$)
0	15.0	101	19.8	1.23
610	11.0	94	18.4	1.15
1220	7.1	87	16.9	1.09
1829	3.1	81	15.7	1.02
2438	−0.8	75	14.4	0.96
3048	−4.8	70	13.3	0.90
3658	−8.8	64	12.1	0.85
4267	−12.7	59	11.2	0.80

9.3 kPa (100–70 mmHg) with little reduction of arterial oxygen saturation. Since the venous oxygen content also changes little, it is as though the slope of the oxygen dissociation curve becomes steeper (Fig. 12.16).

Expressed in terms of oxygen conductance theory (Shephard, 1971a;1972c; Chap. 5), it is as if more oxygen is transported per unit of cardiac output because the effective solubility of oxygen in the blood has been increased (Shephard, 1971a;1972c). This inherent property of hemoglobin plays a large role in the adaptation of humans to altitudes of ~2000 m, and it explains the rapid appearance of adverse reactions with a further small increase of altitude. Local hypoxia may also alter the relative proportions of 1,3-and 2,3-diphosphoglycerate in the red cell, leading to a rightward shift of the oxygen dissociation curve (Fig. 12.17); this facilitates oxygen extraction in the tissue capillaries (Lenfant et al., 1968).

Respiratory System

Mexico City has an altitude of ~2240 m. On first arriving at this altitude, the alveolar oxygen pressure drops to ~9.7 kPa (73 mmHg), with a corresponding drop in arterial oxygen tension. This in itself causes a 7–8% decrease of arterial oxygen saturation. A further 4–5% decrease may later arise from loss of buffers and a decrease of pH (Doll, Keul et al., 1967). The discharge of the carotid chemoreceptors is increased, this effect becoming marked at altitudes that bring the arterial oxygen pressure below 8 kPa (~60 mmHg). The medullary centers controlling respiration and circulation (Chaps. 5 and 6) are depressed by oxygen lack. Nevertheless, this central effect seems outweighed by peripheral stimulation of the chemoreceptors. The immediate overall response to altitude is thus an increase of heart rate, both at rest and during brief or more prolonged submaximum exercise (Fig. 12.18; Consolazio, Nelson

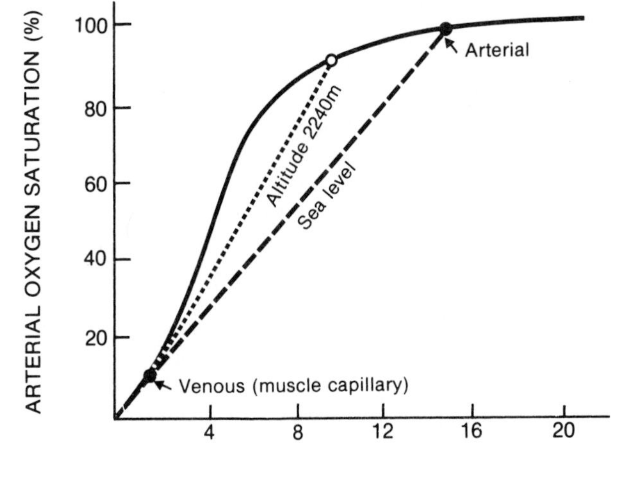

FIG. 12.16. The effective increase in slope of the oxygen dissociation curve of hemoglobin when exercising at an altitude of 2240 meters.

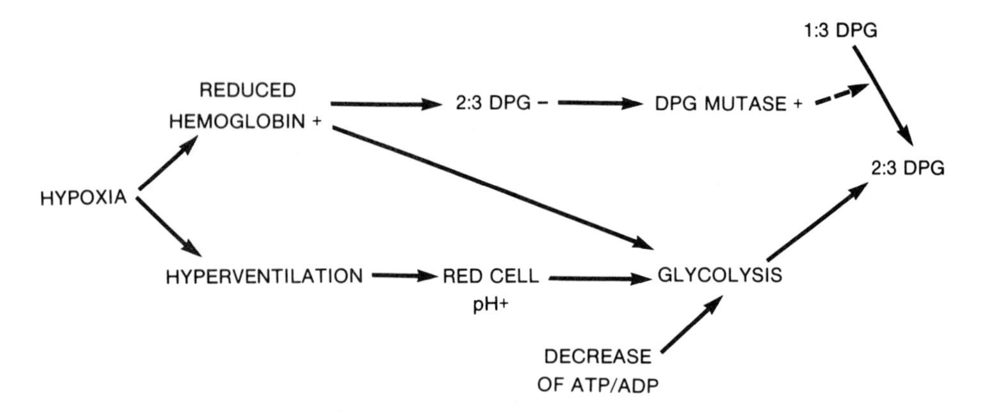

FIG. 12.17. Possible chain of events leading to an increase of red cell 2–3 diphosphoglycerate under hypoxic conditions.

et al., 1966; Shephard, 1967d; S'Jongers, Hebbelinck et al., 1967; McManus et al., 1974). Likewise, there is an increase in the rate (and to a lesser extent the depth) of breathing.

According to Bhattacharyya et al. (1970), the characteristic laboratory response to hypoxia (an increase in slope of the ventilation/P_{A,CO_2} line) is reduced or abolished by simultaneous exercise. On the other hand, a recent analysis of variance shows altitude and work rate as providing independent and additive stimuli to the respiratory and cardiac centers (Sato and Sakate, 1974). The respiratory minute volume is increased in BTPS terms, but it maintains a normal relationship to relative work load (percent of maximum oxygen intake) or heart rate (Grover and Reeves, 1967). The STPD volume of respired gas may be less than normal while at altitude (Consolazio et al., 1966). If it also is increased, then an excessive elimination of carbon dioxide occurs. The resultant depression of both the chemoreceptors and the respiratory centers diminishes the steady ventilatory response to a given reduction of arterial oxygen tension.

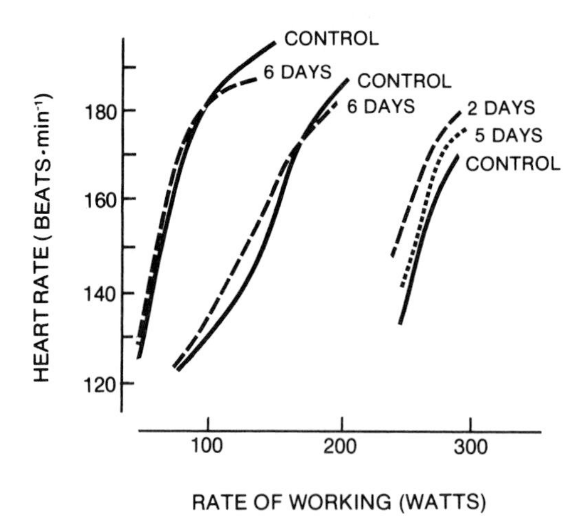

FIG. 12.18. Influence of acclimatization on performance of progressive exercise test by three subjects of differing fitness. Control observations in Toronto (altitude ~230 m). Repeat data in Mexico City (altitude 2240 m). (Based on report of Shephard, 1967d.)

Breathing becomes intermittent because of the phase lag between respiratory drive (now coming from the carotid bodies, in intimate contact with arterial blood) and the depressant effect of CO_2 washout upon the activity of the medullary control centers. Intermittent breathing is most noticeable at night, when CO_2 production is low, and the reticular activating system is depressed. It is a disturbing sensation, and can rob the newcomer to altitude of much sleep (Lange and Hecht, 1962; Hecht, 1967).

A decrease of atmospheric density raises the critical velocity at which air flow becomes turbulent as it lowers the energy cost of turbulent air movement. The cost of a given BTPS ventilation is thus less than at sea level, and the required STPD ventilation can be sustained by a normal (Jacquemin and Varène, 1967) or slightly increased (Thoden et al., 1969) oxygen usage. One factor that augments respiratory work is an increase in compressibility of the air. Oxygen lack may also induce an STPD hyperventilation, with a corresponding increase in the oxygen consumed by the respiratory muscles.

The diffusing capacity of the lungs ($\dot{D}_L$) has more influence upon maximum oxygen transport than it does at sea level (Fig. 12.19). We have seen (Chap. 5) that the impedance term describing the in-

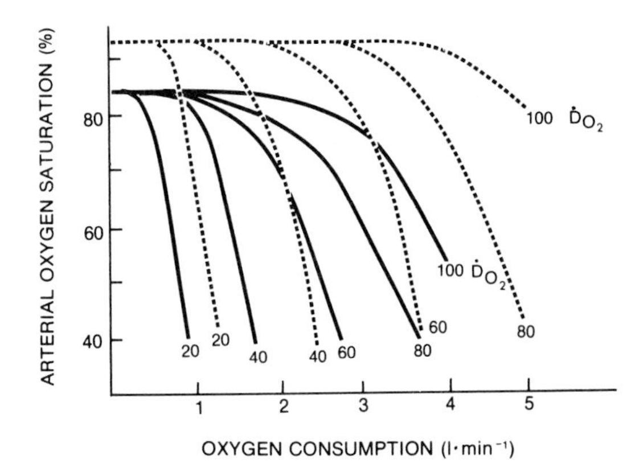

FIG. 12.19. Relationship between pulmonary diffusing capacity (ml • $min^{-1}torr^{-1}$), oxygen consumption (1 • min^{-1}), and arterial oxygen saturation at alveolar oxygen pressures of 10.7 kPa (80 mm Hg) and 6.7 kPa (50 mm Hg). (Based on J. B. West, 1967.)

teraction between pulmonary diffusion and blood transport includes the exponent $e^{-\int \frac{\dot{D}_L}{\lambda \dot{Q}}}$. The cardiac output $\dot{Q}$ may be changed relatively little at altitude, but the oxygen solubility factor λ is increased since gas exchange now occurs along the steep part of the oxygen dissociation curve (where a small change of oxygen pressure causes a large change in blood oxygen content). Further, if D_L is expressed in units of gas conductance (ml STPD • min^{-1} per ml • l^{-1} concentration gradient) rather than in traditional clinical units (ml • min^{-1} per unit gradient of partial pressure), $\dot{D}_L$ decreases in proportion to the drop in ambient pressure (Doll et al., 1967; Fig. 12.16). Equilibration between gas and blood remains fairly complete in the average alveolus, but in parts of the lung where the $D_L/\lambda\dot{Q}$ ratio is low or the transit time through the pulmonary capillaries is less than average, saturation of pulmonary venous blood becomes incomplete. The diffusion component of the alveolar-arterial (A-a) oxygen tension gradient is widened, particularly if the subject is exercising hard, and in this situation an individual with a large maximum diffusing capacity gains some advantage (Blomqvist, Johnson, and Saltin, 1969). J. B. West (1962b) found that at an altitude of 5790 m, maximum exercise reduced the arterial oxygen saturation of acclimatized mountaineers from 67 to 56%, and during short bursts of supramaximal activity, even lower oxygen saturations were recorded.

Two other components of the A-a gradient well recognized at sea level (inequalities of ventilation/perfusion ratio and shunting of venous blood past the lungs) become less important when exercising at altitude. Increased ventilation leads to a more uniform distribution of inspired gas, increased pulmonary arterial pressures lead to more uniform perfusion of the lungs, and the lower pulmonary venous oxygen tension reduces the effect of a given venous-arterial shunt on the oxygen pressure of mixed arterial blood. Nevertheless, the reduction of alveolar-arterial oxygen pressure gradient at altitude is sometimes less than would be anticipated from these changes (Kreuzer et al., 1964; Haab et al., 1967).

Cardiovascular System

There are interindividual differences in the exercise heart rate response to altitude, and inevitably this increases the scatter of maximum oxygen intake predictions (Flandrois and LaCour, 1971). The cardiac output tends to be greater than at sea level because of the high heart rate. The stroke volume may remain at the normally anticipated value during submaximum effort, although often there is a diminution of both resting and exercise stroke volume over the first few days at altitude (Alexander et al., 1967).

Contributory factors include (1) disturbances of fluid balance related to alterations of acid/base status (Fig. 12.25), (2) hyperventilation in cold, dry mountain air (Nevison, 1967; Bühlmann, Spiegel, and Straub, 1969), and (3) an increased exudation of fluid into the tissues. Plasma volume falls (Surks et al., 1966) and the hemoglobin content of unit volume of blood rises at the expense of some increase in blood viscosity (Cerretelli and Debijadji, 1964). Restoration of the normal, sea level plasma volume by dextran infusion does not always restore cardiac output. One suggestion has been that myocardial function is depressed by hypoxia secondary to lowered coronary arterial oxygen tension or coronary blood flow (Alexander et al., 1967). However, this is not supported by electrocardiographic evidence such as right axis deviation or myocardial ischemia (Saltin, Grover et al., 1968).

Maximum cardiac output falls. This reflects not only the decrease of stroke volume, but also a decline of maximum heart rate. Typical figures for a young man are ~195 beats per min at sea level, but only ~140 beats per min at an altitude of 7300 m. This relative bradycardia has been attributed to myocardial oxygen lack, since in some subjects it can be reversed by the administration of oxygen (MacDougall et al., 1976). However, a central reflex slowing of heart rate is a more likely explanation, since the effect can also be reversed by administration of atropine (Hartley, Vogel, and Cruz, 1974).

A decrease in the CO_2 tension of arterial blood reduces cerebral blood flow, and for this reason excessive hyperventilation may worsen psychomotor performance (Otis, Rahn, et al., 1951). There is an optimum respiratory minute volume for each combination of activity and altitude, but in general little advantage is gained if alveolar CO_2 tension is reduced below 3.3–4.0 kPa (25–30 mmHg).

The cutaneous blood flow is also reduced by hypocapnia (Durand and Martineaud, 1971; Raynaud et al., 1973), and visceral flow decreases more than would be anticipated at a comparable intensity of activity at sea level (Tucker and Horvath, 1974).

Aerobic Power

The overall aerobic power changes little between sea level and 1800 m (Buskirk, Kollias et al., 1967a,b; Fig. 12.20), but at higher altitudes there is a progressive reduction. This amounts to 7–8% in Mexico City, with a further loss of 3.2% for each additional 300 m of altitude, giving a cumulative decrement of ~40% at 5000 m. In the newcomer to altitude, such losses are augmented by effects due to altered climate, loss of sleep, and gastrointestinal infections.

Below 1800 m, it is relatively easy for the body to increase chest movements to the point where the STPD respiratory minute volume is restored. At higher altitudes, such methods of compensation are no longer fully effective. Chest movements continue to increase until a large part of the maximum oxygen intake is diverted to the muscles of respiration. However, the sea level STPD ventilation can no longer be maintained. The effects of the diminishing $\dot{D}_L$ (see above) also become more serious, and because the subject is already operating on the steepest part of his oxygen dissociation curve, there is no mechanism for a further increase in the effective blood solubility coefficient λ.

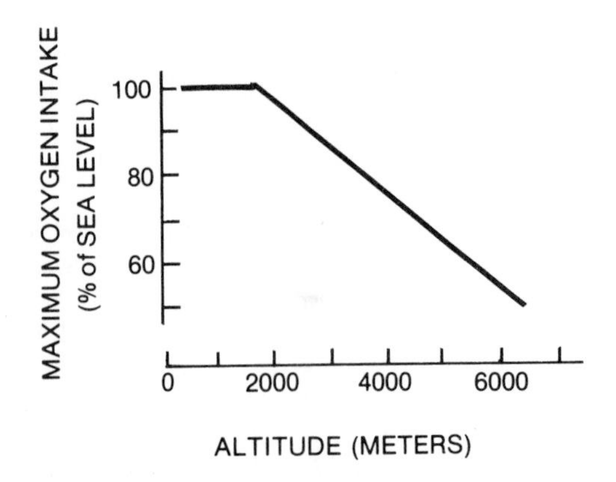

FIG. 12.20. Influence of altitude on maximum oxygen intake. (Based on data in world literature, collected and analysed by Buskirk, Kollias et al., 1967a,b).

In sedentary subjects, a reduction of body mass due to loss of both fat and fluid may partially compensate for the reduction of oxygen transport (Krzywicki et al., 1969; Hannon, Chinn, and Shields, 1969; Hannon, Shields, and Harris, 1969a,b).

Anerobic Capacity and Power

Anerobic capacity and power show little immediate change at actual or simulated altitude. Hollmann, Venrath et al., (1966) found no alteration in the 12-or 25-sec power output on a bicycle ergometer when breathing 12 or 15% oxygen. On the other hand, there was some impairment of the 50-sec all-out cycling score. A reduced partial pressure of oxygen leads to more lactate accumulation in submaximal work (Naimark, Jones, and Lal, 1965; Keul, Keppler, and Doll, 1967; Hollmann, Grünewald et al., 1968; Bühlmann, Spiegel, and Straub, 1969; Wyndham, 1972), but if loads are expressed in relative units (percentage of maximum oxygen intake), responses at sea level and at altitude become closely comparable (Hermansen and Saltin, 1967). A given rate of working decreases the standard bicarbonate to a greater extent at altitude than at sea level (Weidemann et al., 1968), but the limiting blood and muscle concentrations of lactate remain surprisingly normal even after plasma and tissue bicarbonate have been reduced as an adaptation to the new environment (Cerretelli, 1967; J. E. Hanson, Stelter, and Vogel, 1967; Durand, Paunier et al., 1967).

Effects on Performance

Physiologists have studied athletic performance with particular reference to athletic competitions held in Mexico City (Pugh, 1965; B. B. Lloyd, 1967; Jokl and Jokl, 1968b; A. B. Craig, 1969; Faulkner, 1971). Because of diminished air resistance, performance is improved in short-distance running and in throwing events (discus and javelin) (A. B. Craig, 1969; Faulkner, 1971; Pugh, 1976). In contrast, problems of oxygen transport limit endurance competition (Fig. 12.21) and slow the recovery from shorter events (Shephard, 1967d). Nevertheless, the observed deterioration of distance performance in the Mexico City Olympic Games was less than the theoretical 7–8% (Shephard, 1974f,1976b) due to (1) the progressive development of athletic records (Jokl and Jokl, 1968a,b), (2) competition by high-altitude natives, and (3) substantial periods of acclimatization adopted by some sea level competitors.

Mountaineers

Mountaineers maintain a surprisingly similar pace of climbing between sea level and 6000 m. Oxygen consumption is typically 25–35 ml • kg^{-1} • min^{-1} (Pugh, 1964; Eigelsreiter et al., 1968; Fig. 12.22). At sea level, this is ~50% of aerobic power, but at 6000 m it amounts to 80–90% of maximum oxygen intake. The possible duration of individual climbing bouts thus becomes very short at high altitudes. On the upper slopes of Everest, Pugh (1964) noted that climbers stopped every 12 paces.

Industry

As with mountaineers, workers in the mines of the Chilean Andes often find themselves exercising at more than 50% of maximum oxygen intake. Conditions have been simulated by the use of low-oxygen mixtures in the laboratory. The endurance time for exhausting exercise is reduced to 55% of control when breathing 16% oxygen and to 22% of control with 12% oxygen (Gleser and Vogel, 1973). The individual who is performing heavy work at altitude thus has a much greater need of rest pauses than a person who is doing equivalent work at sea level.

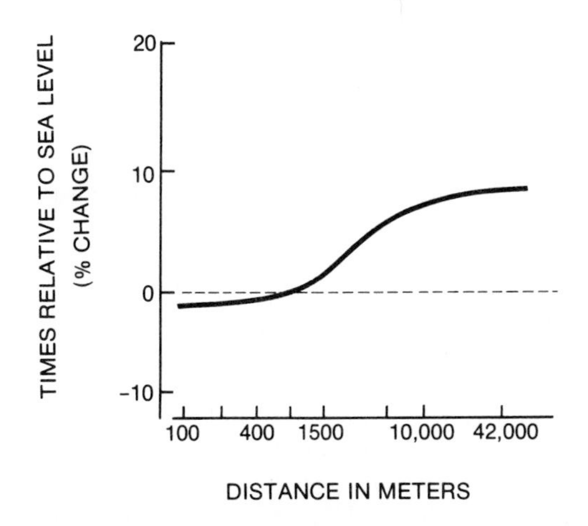

FIG. 12.21. Relationship between winning times (Olympic Games, Mexico, 1968) and existing world records. (After Pugh, 1969c.)

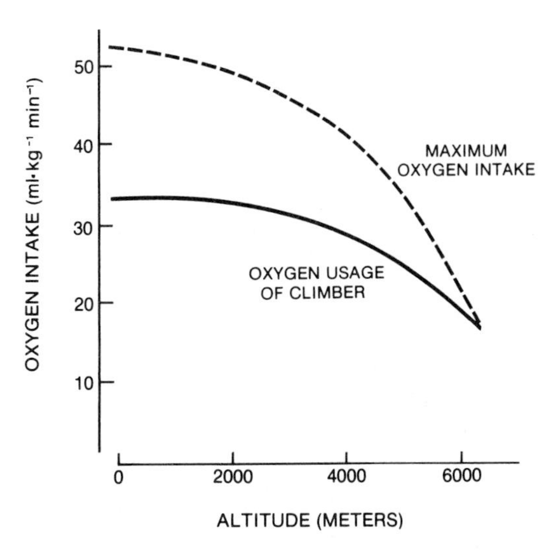

FIG. 12.22. Relationship of altitude with maximum oxygen intake and oxygen usage of climber. (After Pugh, 1964.)

Pathology of Altitude

Mountain Sickness

The clinical syndrome of mountain sickness is normally first encountered at an altitude of ~3300 m and is well marked at 5000–5500 m (Hultgren and Lundberg, 1968). However, participation in frequent "heats" or very vigorous training programs may produce symptoms at lower altitudes (Shephard, 1967c; Jokl, 1968).

Complaints include loss of appetite, nausea, vomiting, muscular weakness and incoordination, lassitude or fatigue, headache, irritability, and loss of sleep (Favour, 1966; Carson et al., 1969). The respiratory minute volume is increased, but both arterial oxygen and CO_2 tensions are low. The condition develops after a few hr at altitude. It reaches a peak in ~48 hr and regresses as the individual becomes acclimatized. It does not bear a simple relationship to oxygen lack, since symptoms are often worst when the arterial oxygen tension is recovering.

While the main features reflect the combined impact of low intravascular oxygen and CO_2 pressures upon the tissue oxygen supply, there are added effects from an altered fluid balance due to either acid/base shifts or anoxic failure of the sodium pump (Chap. 4). Gastrointestinal symptoms are associated with alterations in the rate of secretion and composition of the pancreatic juice; however, gastric secretion and motility remain unchanged (Hartiala, 1967). Most symptoms are reduced by the administration of drugs such as phenformin (a stimulant of anerobic glycolysis) and acetazolamide, although it is not clear as to whether the latter functions as a carbonic anhydrase inhibitor or as a diuretic. Further, the long-term value of administering such drugs is debatable. They slow the adaptive diminution of tissue bicarbonate, exacerbate the decrease of plasma volume, and do not necessarily correct the deterioration of psychomotor performance (Carson et al., 1969).

Exceptional individuals such as the Sherpa Tensing have conquered Mount Everest without oxygen, but an altitude of 5500 m seems the limit of permanent adaptation for the average person. There are sulfur mines as high as 5800 m in Chile, but the workers prefer to live at 5300 m, complaining of loss of sleep, appetite, and enjoyment of food when quartered nearer to the mines.

High-altitude Deterioration

Even in Mexico City, there may be a progressive decrease of cardiac stroke volume over the first few weeks of residence. With life at very high altitudes there is a more marked worsening of condition. The physical working capacity declines, the appetite becomes poor, sleep is lost; body mass diminishes, and the affected individual becomes increasingly lethargic.

Many factors contribute to such a deterioration of physical condition. The syndrome typically develops over several weeks of arduous climbing. A large volume of water is lost from the body due to changes of acid/base balance, vigorous ventilation in the dry mountain air, and sweating induced by the solar radiation that is reflected from snow-covered glaciers. A daily water intake of ~3 l of fluid may be needed to avoid dehydration (Chap. 3). The diet of the mountaineer also tends to be poor and unpalatable, and an energy expenditure of 16–20 $MJ \cdot day^{-1}$ may be offset by a food intake of no more than 6 $MJ \cdot day^{-1}$. These problems are often compounded by illness, intense mental and physical stress, biting cold, and lack of sleep. The air within small tents may become heavily contaminated by carbon monoxide or carbon dioxide (D. W. Rennie, 1977). Hyperventilation is so intense that a large part of the oxygen intake is diverted to the respiratory muscles; even a bowel movement creates an excessive ventilatory demand. Finally, the heat loss involved in saturation of expired gas with water vapor may exceed the potential metabolic heat production.

Under such severe conditions, a progressive deterioration in the condition of a mountaineer is hardly surprising. Preventive measures include provision of a high-energy diet and insistence on adequate replacement of fluids. Once the syndrome is established, the only practical remedy is rapid evacuation to a lower altitude.

Some authors distinguish several more chronic forms of mountain sickness (Hecht, 1967; Hultgren and Lundberg, 1968; Arias-Stella, 1971; Peñaloza et al., 1971). These are associated with chronic alveolar hypoventilation and the development of cor pulmonale.

High-altitude Edema

An intense pulmonary edema may develop from 9 to 36 hr after reaching high altitude (Hecht, 1967; Singh et al., 1968). In normal industrial and recreational activity, attacks do not occur below 3000 m, and indeed they are rare below 3500 m (C. S. Houston, 1960; Singh et al., 1965; Hultgren and Lundberg, 1968). Fears were expressed that the very intense efforts of the Mexico Olympic Games might induce edema in some of the endurance contestants; fortunately, these anxieties proved groundless, although there are rumors that edema has developed in some participants at regional competitions in Bolivia.

Clinical features of the edema syndrome include an acute shortness of breath (dyspnea), with a blood-stained and watery phlegm (hemoptysis), chest discomfort and cough, nausea and vomiting. Confirmatory evidence includes the usual clinical signs of alveolar exudate, the electrocardiographic picture of right heart strain (prominent R waves in leads V_1 and V_2, often with other abnormal features of these leads, including right bundle-branch block, depression of the ST segment, and T-wave inversion), with an intense pulmonary vascular congestion on the chest radiograph.

Attacks of high-altitude edema are most common in those returning to vigorous activity following a period of relative leisure nearer to sea level. Teenagers are frequently affected, and there is often a history of recent respiratory infection.

Suggested causal factors include (1) an increase of central blood volume secondary to CO_2 washout, peripheral vasoconstriction, and a disturbed acid/base balance, (2) pulmonary venous constriction secondary to a reduced alveolar oxygen tension, (3) an increase of total blood volume from previous acclimatization to high altitude, (4) an increase of cardiac output induced by a combination of vigorous physical activity and oxygen lack, (5) a reduction of left ventricular contractility and an increase of diastolic pressure secondary to myocardial oxygen lack, and (6) an increase in the permeability of the pulmonary vessels due to infection or oxygen lack. The net results of these various changes are an increased transudation of fluid from the pulmonary vessels plus a rapid and potentially fatal flooding of the lungs with edema fluid (Severinghaus, 1971).

Attacks are generally avoided if effort is preceded by appropriate acclimatization. Established cases should be treated by bed rest, oxygen, and antibiotics to avoid secondary infection.

Exposure to altitudes of 5000 m and above may also provoke episodes of acute cerebral edema accompanied by retinal hemorrhages (C. Clarke and Duff, 1976).

Acclimatization

Continued residence at altitude leads to a progressive adaptation of the various mechanisms concerned with oxygen transport. Some studies suggest that acclimatization occurs more readily in children than in adults (Frisancho et al., 1973).

Respiratory Minute Volume

The "hunting" pattern of intermittent respiration is gradually lost over the first 3–6 weeks at moderate altitude, so that eventually the sea level ventilation can be maintained quite comfortably at least while sitting at rest.

The earliest biochemical change is a decrease in the buffering capacity of the cerebrospinal fluid (Fig. 12.23; Severinghaus and Mitchell, 1964). The bicarbonate content of this fluid is actively regulated by the choroid plexus (Loeschcke and Mitchell, 1963), so that within a few hr of reaching altitude, the medullary CO_2 receptors are again bathed by a fluid of the accustomed pH. Adaptation can be speeded artificially by deliberate hyperventilation prior to ascent (Severinghaus and Mitchell, 1964). The buffering capacity of arterial blood also decreases, but this is a slower process, occurring over the course of several weeks.

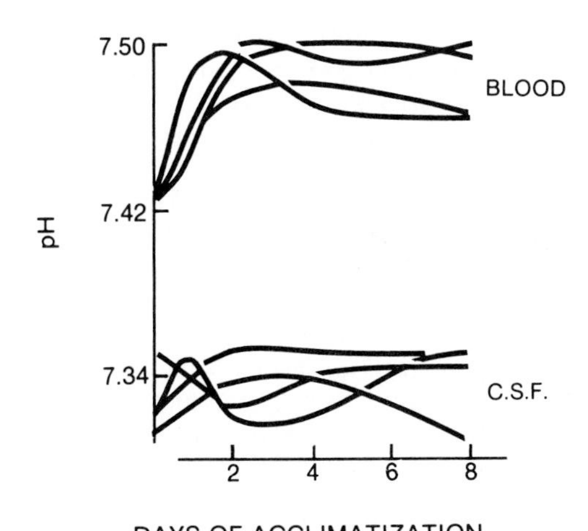

FIG. 12.23. Time course of adjustments in blood and C.S.F. pH with exposure to 3800 m altitude. (After Severinghaus & Mitchell, 1964.)

On first arrival at altitude, the drive to the respiratory centers is derived largely from the carotid chemoreceptors. During the next few weeks, excretion of bicarbonate and diminished secretion of gastric acid (Kellogg, 1963; Kreuzer, 1967) restore the dominance of the medually chemoreceptors (Fig. 12.24). Ultimately, the altitude-adapted subject has an enhanced responsiveness to CO_2 at a given oxygen tension. However, the increased discharge rate of the chemoreceptors does not disappear with prolonged residence at altitude (Klausen, Dill, and Horvath, 1970), and indeed at very high altitudes ventilation continues to be regulated by oxygen lack.

An increase of respiratory minute volume is an effective method of adaptation at altitudes between 2500 and 4500 m. Acclimatization increases the alveolar oxygen pressure by 1.0–1.2 kPa, equivalent to a 1000-m decrease of altitude (Pugh, 1964) and a 20% gain of oxygen transport (Dempsey, 1971).

Although the respiratory adjustments facilitate oxygen delivery, the diminution of alkaline reserve hampers removal of CO_2 from the active tissues. Fortunately, the loss of bicarbonate buffer is largely offset by an increase of hemoglobin, since hemoglobin also is a very important blood buffer.

Pulmonary Diffusion

The properties of the alveolar membrane are unaltered by brief periods of altitude acclimatization. Nevertheless, there is a small increase of pulmonary diffusing capacity (J. West, 1962b) due to (1) an increase in the hemoglobin content of the blood and (2) a speeding of the reaction rate θ at low partial pressures of oxygen (Chap. 5). Further, if growing animals are kept at altitude for long periods, they apparently develop some increase of both alveoli and pulmonary capillaries (Burri and Weibel, 1971).

Blood Transport

A small increase of hemoglobin is seen within a few hr of reaching altitude (Shephard, 1967d; Reynafarje, 1967). This is due mainly to the early decrease

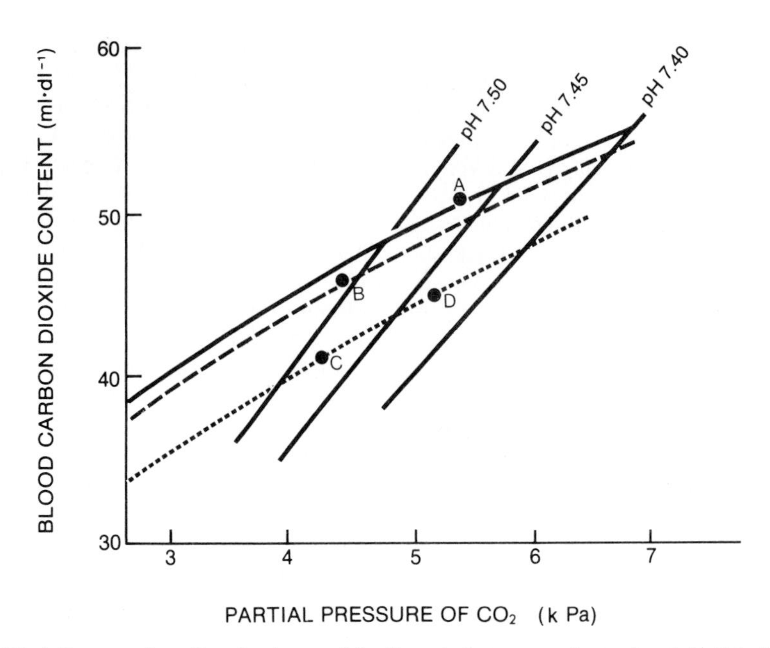

FIG. 12.24. Influence of acclimatization on CO_2 dissociation curve. A, sea level; B, 5th day at 3500 m; C, 6th–40th day at 3500 m; D, 2nd to 5th day back at sea level. (After Luft, 1941.)

of plasma volume (Fig. 12.25; Hannon, Chinn, and Shields, 1969; Hannon, Shields, and Harris, 1969a,b). If the subject continues at altitude, the plasma volume remains subnormal, but the total blood volume is gradually restored by an increase of red cell mass (Samek et al., 1968). An increased formation of red cells can be demonstrated within 2 hr of arrival at altitude, although new production raises the count by only 1–2% per day. The massive 50% adjustment seen at altitudes of 5500–6000 m requires at least a month to achieve in men (Pugh, 1958;Reynafarje, 1964,1967; Samek et al., 1968) and may occur even more slowly in women (Hannon, Shields, and Harris, 1969b). There have thus been suggestions of accelerating adaptation by red cell transfusions (Rewald, 1970).

On initial arrival at altitude, rather immature red cells may be released into the circulation. These are larger than normal, with a poor hemoglobin content. However, when adaptation is complete, the cells have a normal hemoglobin content and a normal life span. The increase of hemoglobin increases the solubility factor λ and thus the potential transport of oxygen per unit volume of blood (Shephard, 1971a,1972c). Unfortunately, this advantage is offset by an increase of blood viscosity (Chap. 9; Fig. 9.1), which becomes of practical significance at hemoglobin levels above 20 g • dl^{-1}; the resulting impairment of capillary flow within the muscles may explain the increased venous oxygen content of altitude-acclimatized subjects (D. Rennie, 1977). A small rightward shift of the oxygen dissociation curve facilitates oxygen delivery to the tissues (Lenfant et al., 1968).

At rest and during moderate activity, the heart rate, cardiac output, and regional blood flow revert toward sea level values as the respiratory minute volume and hemoglobin level increase (Shephard, 1967d; Tucker and Horvath, 1974; Fig. 12.18). Nevertheless, the stroke volume and cardiac output at any given rate of working remain subnormal even among permanent residents at high altitude (Klausen, 1966; Buskirk, Kollias, et al., 1967a,b; Hartley, Alexander et al., 1967; Vogel and Hansen, 1967). This suggests that hypoxia may have a relatively permanent effect upon myocardial contractility (Poupa, 1967). Animal studies (Bischoff et al., 1969) described mitochondrial swelling, expansion of cisternae in the sarcoplasmic reticulum, an increase of lipid droplets, and an increased separation of the myofilaments with residence at 4300 m. In humans, the maximum cardiac output is partially restored toward its sea level value as a normal blood volume is regained, but the maximum heart rate remains less than at sea level even after prolonged periods at high altitudes.

Residents at altitudes above 3000 m have a high pulmonary arterial pressure (pulmonary hy-

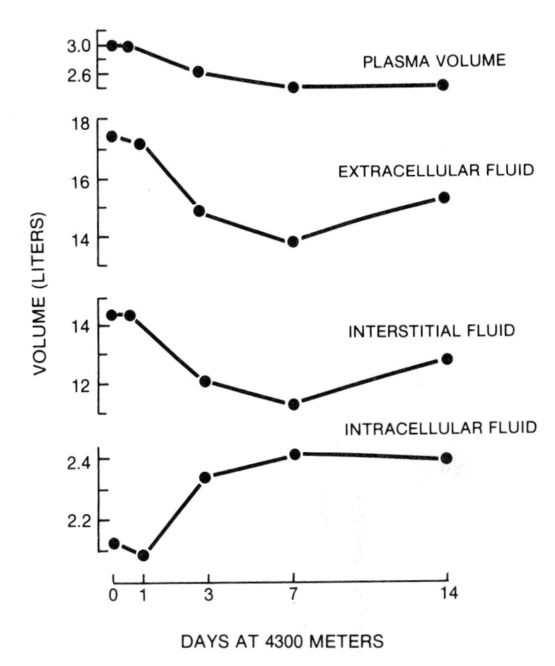

FIG. 12.25. Changes in volume of body fluid compartments with residence at 4300 meters. (After Hannon, Chinn et al., 1969.)

pertension) and there are associated anatomical changes, particularly an increase in the muscularity of the pulmonary arterioles (Grover et al., 1968). The hypertension is "useful" in the sense that it increases blood flow through the poorly perfused apical regions of the lungs (Dawson and Grover, 1974). However, it does not seem to induce any cardiac hypertrophy (Vogel, Genovese et al., 1969).

Tissue Adaptations

There are reports of mitochondrial enlargement, increased activity of respiratory enzyme systems, and increases in the myoglobin content of muscle fibers in response to life at altitude (Reynafarje, 1962; Hurtado, 1971). The heart muscle also shows an increase of oxidative enzymes (for example, succinic dehydrogenase) but not anerobic enzymes (for example, lactic dehydrogenase; P. Harris, 1971). High-altitude natives incur a smaller oxygen debt than partially acclimatized athletes when working at the same relative intensity of submaximal effort. This suggests that muscle blood flow or capillarity is greater in the permanent resident at high altitude (Valdivia, 1956).

Optimum Recommendation

It is generally agreed that acclimatization increases endurance for submaximal exercise and partially restores the maximum oxygen intake at any given altitude (Faulkner, 1967; Buskirk, Kollias et al., 1967a,b; Shephard, 1974f; Maher, Jones, and Hartley, 1974). However, the ideal period of acclimatization for the person who must compete at altitude is not yet decided. Some authors favor several weeks of adaptation (Asahina et al., 1967; Jungmann, 1967; Weihe, 1967; Roskamm, Weidemann et al., 1968), but this carries several risks: (1) a slackening of training schedules in the adverse environment, (2) a progressive loss of fluid and decline of stroke volume, and (3) adverse medical and psychological reactions to an unfamiliar environment. There is therefore much to commend a shorter period, particularly at altitudes over 3000 m (where the decline of stroke volume is substantial). Noble and Maresh (1978) found no impairment of $\dot{V}_{O_{2(max)}}$ or anerobic power in basketball players over the first 24 hr at altitude, although the optimum recommendation is probably 3 days adaptation (Shephard, 1974f;1978c,l). This allows time for adjustment of acid/base balance; there is some increase of hemoglobin level, recovery from mountain sickness, and an opportunity to learn the pace appropriate to the new environment without courting the problems of a longer sojourn.

High-altitude Natives

Certain advantages of the high-altitude native, such as a high hemoglobin level (Hurtado, 1964) and an increased cardiac output (Vogel, Hartley, and Cruz, 1974), mirror the pattern of response seen with long-term acclimatization of sea level residents to the same environment. The relative contribution of genes and environment to the development of anatomical adaptations (increased static lung volumes, greater alveolar size, increased pulmonary capillary bed, greater heart volume, and larger bone marrow; J. Barcroft, 1914; Burri and Weibel, 1971; Bharadwaj et al., 1973) is more controversial. The carotid bodies of the high-altitude native are enlarged (C. W. Edwards, 1971), and the chemoreceptive response is marked by a reduced sensitivity to both CO_2 and O_2 lack; again, the basis of such findings (genetic or environmental) has yet to be resolved (Lahiri and Milledge, 1968; H. V. Forster et al., 1969; J. Dempsey, 1971; J. Dempsey, Reddan et al., 1971; Lahiri, Milledge, and Sørensen, 1972; Byrne-Quinn et al., 1972; Bisgard et al., 1974; Lahiri, Brody et al., 1978).

Loss of Acclimatization

Kenyan endurance athletes have had an excellent

performance in recent international competitions, raising the possibility that the endurance of other sportsmen might be improved by a period of training at an "altitude camp" (Shephard, 1974f, 1978l).

On return to sea level, there is an immediate resting hyperventilation, reflecting a deficiency of buffers within the cerebrospinal fluid. This could well cause an athlete to overbreathe during a competition, diverting an excessive fraction of the total oxygen supply to the respiratory muscles. However, medullary and cerebrospinal bicarbonate concentrations return to their normal sea level values within a few hr.

The other main adaptation to altitude (the increased red cell count) is a little more persistent, but nevertheless two-thirds of the increase developed at 6000 m is lost within 17 days of return to sea level. A decrease in production and an increase in destruction of red cells both contribute to this change.

In theory, a small bonus remains for the first few days after return from a mountain camp. However, in practice, this is generally offset by loss of training and the learning of an incorrect pace while at altitude. Mountain training camps apparently increase maximum oxygen intake only in individuals who are initially below peak condition (Buskirk, Kollias et al., 1967b; Faulkner, Daniels, and Balke, 1967; Consolazio, 1967; Shephard, 1974f; W. C. Adams et al., 1975a), and they are no longer favored by most endurance competitors (R. G. Richardson, 1974). This conclusion is now endorsed by the laboratory experiments of C. T. M. Davies and Sargeant (1974c). They compared the effects of training one leg at normal ambient pressures of oxygen and the other at hypoxic pressures. There was no evidence that the training process was potentiated in the leg exposed to oxygen lack.

—CHAPTER 13

Energy Balance and Health

In this chapter, we review briefly the topics of dietary regulation, obesity, and its correction. Lipid disorders are discussed with specific reference to exercise and heart disease, and comment is made concerning the possible impact of increased physical activity upon productivity, cardiac disease, and other forms of illness. A final section evaluates the risks associated with unaccustomed physical activity.

ENERGY BALANCE

Regulation of Food Intake

The energy stores of the body are limited (Chap. 2). Long-term homeostasis thus depends upon a careful matching of food intake with energy expenditures. There is a narrow prescriptive zone, whereby a greater food intake is compensated by greater physical activity (or vice versa), but outside of such limits lie, on the one hand, an accumulation of fat and, on the other, protein catabolism and tissue wasting (Gasnier and Mayer, 1939; Ingle and Nezamis, 1947; Mayer et al., 1954).

Details of the coordinating centers are discussed in Chap. 7. While other areas may also regulate feeding (S. D. Morrison, 1977), the main control resides in the hypothalamus (Debons et al., 1977). Energy balance is surprisingly precise; indeed, the 10-kg accumulation of fat seen in a typical middle-aged adult has probably arisen from a 1% reduction of daily energy expenditure, with no matching decrease of food intake (see below).

Energy Expenditure

Sources of expenditure and the energy yields of protein, fat, carbohydrate, and alcohol are described in Chap. 1. The costs of specific tasks are detailed in Tables 13.1–13.3. Notice that rather different ratings of intensity are made by nutritionists (who think of a 24-hr day), industrial physicians (who anticipate 6–8 hr of activity), and sports physicians (who are often interested in very brief periods of intense effort).

Many of the values for industrial and domestic activities are hallowed by a certain antiquity. It is hard to find either a blacksmith or a woman who scrubs floors by hand, except in a reconstructed pioneer village! Energy requirements for other tasks have also fallen progressively with the introduction of improved tools and packaging, and the tuition of more economical working techniques. Even traditional "heavy" work, such as mining, farming, and lumbering, often uses less than 20.9 kJ • min^{-1}, and there are few remaining occupations that call for an expenditure of more than 3 times the resting metabolism (Montoye, 1975; Menotti et al., 1978). The in-

TABLE 13.1
Differences in the Categorization of Activity Between Nutritionists, Industrial Physicians, and Sports Physicians

Rating of Activity	Level of Energy Expenditure (kJ • min^{-1})		
	Nutritionists	Industrial Physicians	Sports Physicians
Sedentary	<5.2	8.4	8.4
Light	–	8.4–13.8	8.4–20.9
Moderate	5.2–10.5	13.8–22.6	20.9–41.8
Heavy	10.5–20.9	22.6–37.7	41.8–62.8
Very heavy	>20.9	>37.7	>62.8

tensity of recreational activity has varied less dramatically than the cost of industrial work. Nevertheless, the passage of time has seen alterations in rules, the development of better but heavier protective equipment, and the introduction of new techniques of play. All of these changes challenge the current accuracy of published data. Perhaps the most extreme example of reduced energy costs is the use of an air-conditioned motor-driven cart when crossing a golf course. Cycling is a second area of change. Most published information refers to the use of narrow, 66-cm-wheel racing machines, whereas many current recreational users often purchase smaller wheeled cycles with broad tires. In many sports, the energy cost also varies widely with the skill of the player and the vigor and enthusiasm of his partners and or opponents.

Daily Energy Needs

When world food requirements were calculated, an energy intake of 13.4 MJ • day^{-1} was set for the "reference man" and 9.6 MJ • day^{-1} for the "reference woman" (Food and Agriculture Organisation, 1968; Table 13.4). United States nutritionists have commonly used somewhat lower figures (12.1 MJ for men, 8.8 MJ for women). Although the lower standards are realistic for most people in the United States, it is arguable that the average level of activity in the United States has now fallen below the minimum needed for complete health; on this basis, nutritionists should press for a combination of more energy intake and more physical activity.

The total food requirements of any given population naturally depend upon the age profile of

TABLE 13.2
Approximate Gross Energy Cost of Selected Industrial and Domestic Activities[a]

Category	Level of Energy Expenditure (kJ • min^{-1})			
	≤5.2	5.2 – 10.5	10.5 – 20.9	≥20.9
General[b]	Sleeping Sitting	Standing Walking Washing	Fast walking Playing with children Dressing	Running Stair climbing Playing with children
Male occupations	Office worker Tailor Printer	Shop assistant Painter Carpenter Light-metal worker Shoemaker	Rivetter Painter Carpenter Sheet metal worker Blacksmith	Miner Lumberjack Construction laborer Steel worker Machine fitter Mason Small farmer Postman
Female occupations	Writer Typist Sewing Knitting Ironing Dishwashing	Sweeping Dusting Washing	Polishing Scrubbing Laundering	

[a]The figures cited refer mainly to small groups of male subjects, and take no account of rest pauses.
[b]Values are slightly higher for men than for women.
Source: Based largely on data collected by Durnin and Passmore, and presented in "Energy, Work and Leisure," Heinemann, 1967.

TABLE 13.3
Gross Energy Cost of Various Sports and Leisure Activities*

Sport	Cost (kJ • min^{-1})
Archery	13–25
Athletes	Up to 84
Badminton	25
Basketball	38
Billiards	12.5
Bowls (lawn)	17
Boxing	38–59
Canoeing	13–29
Climbing (mountain)	29–42
Cross-country running	42–46
Cycling	17–84
Dancing	17–34
Field hockey	38
Gardening	13–21
Golf (no cart)	21
Gymnastics	13–50
Horse-riding	13–42
Rowing	17–53
Skiing	
(cross-country)	42–79
(downhill)	Up to 42
Squash racquets	42–75
Swimming	21–63
Table tennis	17–21
Tennis	25–38
Volleyball	13–17

*The figures refer to small numbers of male subjects, and no account has been taken of rest pauses.

Source: Based large on data collected by Durnin and Passmore, and presented in "Energy, Work and Leisure," Heinemann, 1967.

the community, since both voluntary physical activity and the energy needed for tissue maintenance decrease with aging. However, energy usage cannot drop below about 10.5 MJ for men and 8.8 MJ for women, even if daily work is entirely sedentary. If the food supply is inadequate, body tissue is sacrificed to meet energy requirements and voluntary activity is reduced. The output of manual workers with a limited diet (for example, men living in prison camps or underdeveloped countries) is held in close relationship to the available food supplies.

The daily energy needs of a competitive athlete are much larger than for a sedentary person, 25 MJ • day^{-1} for an active skier or football player, and up to 50 MJ • day^{-1} in some marathon cycling contestants.

Energy Resources

Since the various foodstuffs are largely interconvertible (Fig. 1.4), energy can be obtained from protein, fat, or carbohydrate (Table 13.5). However, if the anerobic threshold (Chaps. 5 and 11) is surpassed, there is an immediate need for glycogen; for example, in the 85-km Vasa ski race, the performance of competitors is impaired by an exhaustion of glycogen reserves in their arm muscles (Bergström et al., 1973b). The endurance athlete may find immediate advantage in boosting glycogen stores by taking a high-carbohydrate diet for some days before a major competition (Chap. 2), but it is in his long-term interest to accustom the body to consumption of fat, conserving glycogen reserves for occasional bursts of anerobic activity.

Foods such as butter have a high fat content and a relatively low water content; they thus have a high "caloric density" relative to carbohydrate-rich foods like vegetables. Over a protracted event such as a 100-km foot race, as much as 25% of the energy

TABLE 13.4
Energy Expenditure of a Reference Man (aged 25 years, weight 65 kg) and a Reference Woman (aged 25 years, weight 55 kg) at a Mean Environmental Temperature of 10°C

	Reference Man			Reference Woman		
Activity	Daily Duration (hr)	Cost (kJ • min^{-1})	Total per Day (kJ)	Daily Duration (hr)	Cost (kJ • min^{-1})	Total per Day (kJ)
Work	8	10.5	5,021	8	7.7	3,682
Washing and dressing	1	12.6	753	1	10.5	628
Walking	1½	22.2	2,008	1	15.1	920
Sitting	4	6.4	1,548	5	5.9	1,757
Recreation and domestic work	1½	21.8	1,966	1	14.6	879
Bed rest	8	4.4	2,092	8	3.7	1,757
Total	24	9.3	13,388	24	6.7	9,623

Source: Based on data of Food and Agricultural Organization of the United Nations, F.A.O. Nutritional Studies No. 15, 1968.

TABLE 13.5
Approximate Energy Yield of Common Foodstuffs, Cooked Where Appropriate (kJ)

Foods	Energy Yield (kJ)
1 apple	314
Serving apple pie	1381
2 slices bacon	397
1 banana	376
1 cup lima beans	628
1 cup pork beans	1234
85 g beef	1163
1 cup beets	293
1 tea biscuit	544
1 slice bread	272
1 cup broccoli	188
1 tablespoon butter	418
1 cup cabbage	167
1 slice angel food cake	460
1 slice chocolate cake	1548
1 cup carrots	188
1 cup cauliflower	126
1 cup celery	84
28 g cheddar cheese	481
28 g cottage cheese	105
1 cup chicken soup	314
1 cup chocolate	586
1 cup cocoa	983
28 g codfish	439
Clear coffee	0
1 cup cornflakes	397
1 ear corn	356
1 tablespoon whipped cream	209
1 doughnut	565
1 egg	314
¼ cup flour	418
1 tablespoon French dressing	251
½ grapefruit	167
1 fillet haddock	669
½ cup ice cream	690
1 tablespoon jam	230
85 g lamb	962
1 tablespoon lard	523
2 large lettuce leaves	21
1 serving lemon meringue	1255
57 g liver	502
1 cup macaroni/cheese	1883
1 tablespoon margarine	418
1 cup whole milk	690
1 cup skimmed milk	356
1 cup mushrooms	126
1 cup noodles	439
1 cup oatmeal	628
1 tablespoon salad oil	523
1 orange	293
1 cup orange juice	460
1 pancake	251
1 cup parsnips	397
1 peach	188
1 tinned peach	331
1 pear	397
1 cup tinned peas	586
1 cup tinned pineapple	397
1 plum	126
85 g pork	1192
10 potato chips	460
1 medium potato	502
Serving pumpkin pie	1109
1 cup rhubarb + sugar	1611
1 cup rice (converted)	858
85 g salmon	502
85 g sardines	753
114 g sausage (pork)	1423
85 g shrimp	460
1 cup spaghetti	920
1 cup spinach	188
1 cup summer squash	146
1 cup strawberries	230
1 tablespoon sugar	209
1 cup tomato juice	209
1 tomato	188
85 g tuna	711
1 cup turnips	188
85 g veal	744
½ slice watermelon	188
1 cup white sauce	1799
1 cup wheat flour	1674

needs may be supplied from intracellular fat reserves (Howald, 1977; Dagenais et al., 1978). However, if an undue proportion of energy is obtained from the combustion of either dietary or body fat, incomplete breakdown leads to an accumulation of toxic ketones in the body fluids (ketosis). In contrast to the athlete, a sedentary individual curtails the delivery of fat to skeletal and cardiac muscle, boosts delivery to adipose tissue, and reduces the turnover of palmitate (Pařízková and Poledne, 1974).

The body has a minimum need for certain amino acids that it cannot synthesize (Chap. 2). Nevertheless, an excessive intake of protein is expensive, and the associated burden of animal fat predisposes to atherosclerosis.

The nature of food intake can be assessed by inventories and retrospective questionnaires. However, the information thus obtained has very limited accuracy. The current approach of a dietician is to live with a family for a period and to observe the food consumption of individual householders. The food cooked may be assessed by weighing or (a little less accurately) in terms of portions. After due allowance for wastage, the results may be converted to kJ using a schema such as that in Table 13.5. The data thus obtained are accurate to ~15%.

In the affluent metropolitan centers of North America, a typical middle-class citizen derives 12–15% of his food energy from protein, 35–40% from fat, and 45–55% from carbohydrate; some individuals also meet 15–30% of their energy from alcohol. Ethnic, geographic, and economic factors bring about wide variations of dietary composition.

Some groups, such as the urban poor, have a high carbohydrate intake while others, such as traditional Eskimos, have an unusually large intake of fat.

Meals for Athletes

The immediate recommendation for the athlete is relatively simple. Meals should be frequent but of moderate size (Hutchinson, 1952; Mayer and Bullen, 1960). A heavy meal should be avoided for some hours prior to competition, but a light and not too sugary snack is permissible 2–3 hr before an event. If exercise is prolonged, the ability to maintain a high power output is further enhanced by regular (for example, hourly) doses of a dilute sucrose or glucose solution (Chaps. 2 and 3). This tactic also increases the ratio of carbohydrate to lipid utilization during exercise (Brodan and Kuhn, 1971).

Some athletes, particularly cross-country skiers, prefer quite high concentrations of glucose (up to 40%), but such solutions are unphysiological and delay gastric emptying (Chap. 2). One possible tactic is the use of "glucose syrups"–partially hydrolyzed preparations of starch with a relatively low sweetness and osmotic pressure (Ford et al., 1968). Glucose preparations are claimed to reduce the accident rate, improve team work, extend the tolerance of fatiguing work, and speed early recovery following glycogen depletion (H. J. Green and Bagley, 1972; Muckle, 1973; J. D. Brooke and Green, 1974; J. D. Brooke, 1978b). However, since the first step in the metabolism of glucose is phosphorylation (Chap. 2), a limit to usage of exogenous glucose may be set by hexokinase activity at the membrane of the muscle cells (D. G. Walker, 1966).

Water, sodium, and potassium losses should be made good, particularly in a hot climate (Chaps. 2 and 12).

OBESITY

Fat Stores

If food is eaten in excess of energy requirements, storage occurs. Liver and muscle can accumulate small quantities of glycogen (Chap. 2), but the main site of energy storage is the adipose tissue. This consists of 86% fat, 12% water, and 2% protein.

Lipoprotein-bound triglycerides (see below) are liberated near the surface of the capillary endothelium through the action of the enzyme lipoprotein lipase. This enables the fat to penetrate into the adipocyte. Glucose plays a facilitative role, allowing the formation of glycerophosphate, but the direct conversion of glucose to fat within the adipose tissue cell is of less practical importance (Angel, 1974). Lack of activity slows the outflow of fat from the adipose cells (Table 13.6; Poledne and Pařízková, 1975; Pařízková, 1978b).

A positive energy balance brings about a marked increase of adipocyte fat content not only in the young child (where there is a hyperplasia; Salans et al., 1973; Brook et al., 1975) but in the adult (where there is hypertrophy; Stern and Greenwood, 1974; Angel, 1974). Hyperplasia, in particular, leads to difficulty in subsequent dieting (Rognum and Kindt, 1973; Bjöntorp, de Jounge et al., 1973; Björntorp, Carlgren et al., 1975). Conversely, activity in early life (for example, rats forced to swim for their first 28 weeks of life) seems to protect against subsequent obesity (Oscai and Holloszy, 1969). Studies of rats have shown a positive correlation between the blood flow per fat cell and the size of the fat pads (DiGirolamo and Esposito, 1975). Regular vigorous exercise reduces the size, but not the number, of fat cells (A. W. Taylor, Garrod et al., 1976).

Assessment of Obesity

Excess Body Mass

Obesity is essentially an excessive accumulation of fat or adipose tissue. However, assessment is commonly

TABLE 13.6
Influence of Activity Level upon Fat Metabolism of 90-day-old Rats

Variable	Activity Classification		
	Exercised	Control	Hypokinetic
Body mass (g)	342	381	345
Body fat (%)	5.9	12.7	12.1
Plasma free fatty acid (μmol • ml^{-1})	0.49	0.51	0.29
Free fatty acid pool (μmol)	6.6	6.1	3.4
Outflow rate (μmol • min^{-1})	8.3	7.7	4.4
Inflow rate to adipose tissue (mmol • min^{-1} • g^{-1})	2.0	4.2	1.9

Source: After Pařízková, 1978b.

TABLE 13.7

Changes of Body Composition Associated with Several Forms of Weight Gain

Cause of Increase in Body Mass	Change in Body Composition (%)		
	Fat	Cells	Extracellular Fluid
Gluttony	+66	+20	+14
Gluttony plus indolence	+109	−20	+11
Physical activity	−38	+120	+18

Source: After Keys, Brozek et al., 1950.

made in terms of the excess mass relative to actuarial standards (Chap. 11). This is satisfactory in most population studies but provides limited information on the status of the individual. Total body mass shows daily fluctuations of 0.5–1.0 kg due to variations of eating, hydration, and physical activity (Edholm, Adam, and Best, 1974). Further, there is no basis for deciding the reason for an increase of body mass in a single subject. Keys et al (1950) illustrated three possible patterns of "weight gain" (Table 13.7) associated, respectively, with gluttony, gluttony plus indolence, and enhanced physical activity; depending on the cause, the added mass may reflect either lean tissue or fat.

Appearance provides a simple subjective method of distinguishing these possibilities, but inevitably decisions made on this basis differ markedly from one examiner to another. Anthropometric measures used in somatotyping also provide some guidance; for instance, muscle leads to an increase in breadth of the chest and the proximal parts of the limbs, while fat tends to increase abdominal dimensions.

Body Fat Determination

A more satisfactory assessment of obesity is based on body fat determinations—measurements of the thickness of subcutaneous fat, using skinfold calipers (Weiner and Lourie, 1969; Durnin and Womersley, 1974), or estimates of body density and thus total body fat by underwater weighing (Von Döbeln, 1959; Brožek, 1963,1965; Beeston, 1965; Garrow, 1974; Pollock, Hickman et al., 1976; Chap. 11). Studies by Mendez and Kollias (1977) suggest that the density of the two principal body components (fat and fat-free mass) remain relatively constant during large changes of body mass (for example, starvation); however, changes can arise with dehydration.

Adults do not normally object to the ritual of a hydrostatic weighing, but infants and young children are less willing to submit to complete submersion while strapped to a chair (Zuti and Corbin, 1978). In pediatric studies, the external volume of the body is sometimes estimated from the displacement of air within a closed chamber. Alternatively, the fat content of the body can be estimated from its capacity to absorb a tracer substance selectively soluble in adipose tissue, from soft-tissue radiographs (Garn, 1961b; Maresh, 1963; Comstock and Livesay, 1963; Tanner, 1965) or from the use of ultrasound (Haymes et al., 1976).

Lean Mass Measurements

It is also possible to estimate lean body mass and to calculate the mass of fat by difference (fat mass = total body mass minus lean body mass). Some authors use a whole-body counter to determine the body content of the naturally occurring isotope ^{40}K (McNeill and Green, 1959). Since potassium is located primarily in muscle, the lean body mass can be estimated from this figure and an assumed potassium ion content of lean tissue. Potassium concentrations of 65 mmol • kg^{-1} for men and 56 mmol • kg^{-1} for women apply over a considerable range of body builds but are unsatisfactory in malnourished individuals (Edmonds et al., 1975). Problems also arise after exercise because of a shift of potassium-containing blood to superficial vessels (Lane et al., 1977).

Other investigators have studied the fluid compartments of the body. By the use of markers such as deuterium or tritium (J. A. Johnson et al., 1951; Langham et al., 1956; Luft et al., 1963; Friis-Hansen, 1965) the intra- and extracellular components of the total body water can be calculated, and after marking certain assumptions about uniform distribution of the isotopes (Halliday and Hopkinson, 1977) and the water content of lean tissue (72 or 73%; Pace and Rathbun, 1946; Olesen, 1965), the lean body mass can be predicted. Problems again arise with extreme body types. Thus Pierson et al. (1976) found a ratio of extracellular to intracellular water of 0.42 in normal subjects, but 0.74 in the obese; it was suggested (Wang and Pierson, 1976) that new equations were needed for body water, as follows:

$$\text{Total body water} = 0.14\ (F) + 0.80\ (L) \tag{13.1}$$

$$\text{Extracellular water} = 0.11\,(F) + 0.24\,(L) \quad (13.2)$$

where F is the fat and L is the lean mass of the body.

Obesity and Health

Possible tactics in treating obesity are to require an actuarial "ideal weight," the corresponding percentage of fat, or a more empirical anthropometric criterion (ideal skinfold or ideal girth).

On average, the actuarial ideal weight corresponds to ~17.1% fat in men and 21.4% fat in women (Chap. 11). Corresponding skinfold readings have been documented (Table 11.10). A very simple anthropometric index of excess body mass can be based on the adjustment of abdominal girth (ΔG) needed to make this measurement 12 cm less than the average chest circumference (Jetté and Mongeon, 1978):

$$\text{Excess body mass} = 0.45 + 0.758\,\Delta G \quad (13.3)$$

Active young men and women have, respectively, ~10 and 20% body fat. Many endurance athletes and primitive peoples fall far below the percentages found in individuals of "ideal" body mass (Shephard, 1978a). On the other hand, figures of 25% fat in men and 45% fat in women are by no means uncommon for North Americans.

Often, moderate obesity has no obvious effect upon the health of an individual, but in a statistical sense there is an increased risk of death from a number of diseases, particularly cardiac and cerebral vascular conditions, diabetes, and abnormalities of the digestive system (Angel, 1974; Table 13.8). If a man of 45 years is of average body mass for his age, he may expect to live 1½ years less than if he is of the ideal mass, and if he is 20% above average body mass his life expectancy is shortened by a further 2½ years. Further, if the obesity is corrected, the risk of death returns toward the ideal value. Unfortunately, most Western men tend to exceed the ideal body mass as they age. However, the adverse trend is not inevitable, and indeed is rare among "primitive" peoples who follow a traditional active life-style.

Cause of Obesity

Error of Energy Balance

The excess energy intake of an obese person is often small; indeed, he may complain that he eats less than his slimmer colleagues. Let us suppose that 10 kg of fat accumulates at a rate of 1 kg • $year^{-1}$. Since the energy equivalent of fatty tissue is ~29 kJ • g^{-1}, a simple energy balance calculation would suggest an excess intake of 29 MJ • $year^{-1}$, an overeating (or underactivity!) of less than 1%. In practice, heroic overeating experiments (Simms et al., 1973) have shown that the tolerance of an excess energy intake is somewhat greater than this, partly because costs are incurred in depositing fat and partly because body heat production is increased by ingestion of fat.

TABLE 13.8
Mortality of Grossly Obese Men and Women in Relation to Excess Body Mass[a]

Disease	Men[b] +24 kg	Men[b] +33 kg	Men[b] +42 kg	Women[b] +28 kg	Women[b] +37 kg	Women[b] +46 kg
Diabetes	179	385	629	270	242	250
Digestive diseases	147	197	298	140	200	225
Renal diseases	146	230	298	93	122	—
Vascular diseases of brain	136	183	215	143	142	210
Heart and circulation	131	155	185	175	178	217
Pneumonia and influenza	128	103	242	148	110	—
Accidents and homicides	109	126	120	85	98	—
Suicides	71	104	142	47	—	—
All causes	123	145	168	130	138	178

[a]Death rates for subjects aged 15–69, expressed as a percentage of standard values for subjects of same sex.
[b]The excess mass varies slightly with stature.
Source: Based on data of Build and Blood Pressure Study, Society of Actuaries, 1959.

Constitutional Factors

In some animal species, there is an inherited component to energy balance regulation; a recessive gene favors an increase in the number and size of adipocytes and an increased hepatic synthesis of very low density lipoproteins, with a consequent tendency to obesity (Bray, 1977). In humans, also, obesity often runs in families, the babies of obese mothers being fatter than those of normal mothers even after controlling data for smoking habits (Whitelaw, 1976).

Comparisons of monozygotic and dizygotic twins (C. G. D. Brook et al., 1975) suggest that 84–98% of the variation in skinfold readings after the age of 10 years is inherited. Nevertheless, there are fallacies to twin studies (Chap. 11) and undoubtedly much familial obesity reflects poor eating or activity patterns rather than genetic factors.

Potential constitutional defects favoring obesity include (Galton et al., 1973) a malfunction of the adrenergic receptor on the plasma membrane of the fat cell, a deficiency of the kinase that activates plasma lipase, or a deficiency of plasma lipase.

Psychological Factors

Psychological factors also play an important role in the failure of regulation. A mesomorphic child finds rewards from sports participation, but for the endomorph physical activity is a source of negative reinforcement; positive rewards are then sought in sedentary activities and in eating (Hendry, 1978).

Crisp and McGuinness (1975) found a relationship between obesity and depression (in men only), but the obese of both sexes had low scores for anxiety. Obese individuals also tend to be "habit" eaters, consuming all that is readily available (Nisbett, 1968). Lean subjects, on the other hand, are readily influenced by their level of satiety and by past and anticipated activity. Men allow substantial degrees of personal obesity to pass unnoticed, but women generally have a correct perception of their ideal body mass (*British Medical Journal*, 1974).

Retention of Food

Other factors influencing the body's control over its energy resources are the extent of absorption and excretion. Not all ingested food is retained within the body. A variable proportion of complex organic molecules is lost in the feces, and further energy is excreted in the urine, sweat, and breath (for example, alcohol, urea, and ketone bodies). On a high-fat diet, ~82% of ingested energy is retained in the body, but on the high-carbohydrate diet enjoyed by many obese persons, retention increases to ~94%. Gastrointestinal hurry and secretion of hormones favoring fat metabolism are thus factors that help to keep an individual slim.

Physical Activity

Once an individual has become obese, he typically restricts physical exercise (Bruch, 1940; Stefanik et al., 1959; Chirico and Stunkard, 1960; Durnin, 1967; Mayer, 1972), ofter operating in the zone of extreme inactivity where normal regulating mechanisms are no longer effective. Furthermore, heat loss from the body is reduced by the thick layer of subcutaneous fat. Thus, the daily energy needs of an obese individual are often less than those of a slimmer person. However, if exercise is undertaken, the energy expenditure is greater than in a slim person, the added cost being in almost direct proportion to the increase of body mass (Chap. 11).

Miscellaneous Factors

The replacement of a cigarette habit by a food habit is a common finding among patients attending smoking withdrawal centers (Rode, Ross, and Shephard, 1972). A 5- to 10-kg increase of body mass often occurs in the first year that smoking is stopped—a specific illustration of the importance of the psyche in regulating food intake.

In women, obesity is more common in the lower social classes (*British Medical Journal*, 1974). There is an association between the use of oral contraceptives and obesity (A. J. Lewis et al., 1978) and those using oral contraceptives also have a high serum cholesterol level (Hewitt et al., 1977; Shephard, Cox, and West, 1980d,e; Shephard, Youldon et al., 1979).

ENERGY DEFICIT

Moderate Fat Loss

A negative energy balance may arise from a temporary shortage of food or from deliberate dieting. The change of body composition that occurs depends upon the initial status of the individual and the level of activity that is maintained (Table 13.9).

In the early stages of dieting, there is a substantial water loss. The body stores of glycogen (~500 g) are mobilized, and since ~3 g of water is stored with each of glycogen, the body mass inevitably drops by ~1.5 kg (Runcie and Hilditch, 1974). This fact is vigorously exploited by those selling "cures" for obesity. The early water loss is accentuated by a ketogenic diet (Van Itallie and Yang, 1977), but is made good as glycogen stores are replenished and the body mass stabilizes.

Even a small reduction of energy intake leads to protein catabolism in a sedentary individual. However, the problem of tissue wasting is reduced or overcome if a firm diet is combined with increased exercise (Keys et al., 1950; Pařízková, 1964; Larsson, 1967; Babirak et al., 1974; Zuti and Golding, 1976). Not only is energy consumption increased, but activity also helps to quell the pangs of hunger (Edholm, Fletcher et al., 1955; B. M. Thomas and Miller, 1958; Mayer, 1960; Stevenson, 1967; Karvonen, Saarela, and Votila, 1978), possibly through an altered setting of the hypothalamic glucose receptors and an increase of body temperature. Further, a reduction in the size of the fat cells increases their effective sensitivity to lipase (McGarr et al., 1976).

The patient who wishes to lose a few kg of fat should arrange regular and evenly spaced meals (Allred and Raehrig, 1973). Gorging leads to greater hunger and greater protein catabolism than steady nibbling (Leveille and Romsos, 1974; Litvinova et al., 1976; Durrant et al., 1978). The overall food intake should be reduced by ~ 500 kJ • day^{-1}, with a corresponding increase of physical activity. The intended increase of energy expenditure is more likely to be realized by sustained moderate exercise than by short bursts of intense effort. A brisk 3-km walk uses ~ 400 kJ of additional energy, whereas the jogging of 1.6 km in 8 min adds only 230 kJ to the daily energy expenditure. Furthermore, a moderate intensity of activity is most likely to increase fat loss (Girandola, 1967), since fat is the preferred fuel for light exercise (Chap. 2).

TABLE 13.9
Changes of Body Composition Associated with a Decrease of Body Mass

Method of Reducing Body Mass	Contribution to Loss of Mass (%)		
	Fat	Cell	Extracellular Fluid
Obese			
diet alone	75	10	15
Obese			
diet + exercise	98	−10	12
Thin			
starvation	50	50	0
late starvation	30	90	−20
terminal starvation ("wet")	10	40	−150
terminal starvation ("dry")	2	18	80

Source: After Keys, Brozek et al., 1950.

The suggested regimen of 500 kJ dietary restriction and 500 kJ of new energy expenditure causes a minimum fat loss of 0.2 kg • week^{-1} (more if new protein is being synthesized in the working muscles; W. J. O'Hara, Allen et al., 1977a–c,1979). If lean mass is increasing, body fat can be reduced without change of total body mass (W. J. O'Hara, Allen et al., 1977a–c; Oja et al., 1978). The middle-aged fat accumulation of ~ 10 kg is easily corrected over the space of 1 year (Sidney et al., 1977; Table 11.14). A slow but steady loss of fat is the optimum mode of treatment for most purposes, since this allows time for the modification of attitudes toward both food and activity. If a more rapid loss is needed (for example, to facilitate impending surgery), more vigorous dieting may be added to create an energy deficit of 2000–4000 kJ • day^{-1}.

In animals, fat loss is similar with a carbohydrate or protein/fat diet of equal energy content (Tsuji et al., 1972). However, man often finds it easier to achieve an energy deficit by the use of skimmed milk, lean meat, less butter, and reduced servings of most carbohydrates, together with increased exercise (Birchwood, 1975). Drugs of the amphetamine class were once used for slimming but are now banned for this purpose; they probably acted partly by altering the setting of appetite-regulating centers in the hypothalamus and partly by elevating the mood of the dieter. Methyl cellulose preparations act like green vegetables, giving a sensation of fullness by virtue of their large unabsorbed bulk. They are open to the same objection as a natural cellulose diet; there is an increased rate of gastric emptying, so that early satiety is followed by a renewed feeling of hunger.

Drastic Reductions of Body Mass

Very drastic restrictions of food and fluid intake are sometimes adopted by boxers and wrestlers who wish to achieve a specific weight category (Chap. 3; Table 3.6). Such practices are roundly condemned by sports scientists (American College of Sports Medicine, 1976). They are hazardous to the individual who decreases his body mass (sodium loss in the sweat and potassium leakage from muscle predispose to ventricular fibrillation). Furthermore, athletic performance remains impaired even if rehydration is attempted before competition, and the opponent is faced by an unfair contest against a person who is larger than himself.

Drastic starvation is sometimes adopted in the treatment of obesity, but it is not usually effective. Innes et al. (1974) found only 16 long-term successes (with 8 relapses) among 75 grossly obese patients.

Anxious young women may find themselves far below their anticipated body mass. If serious organic disorders can be excluded, a diagnosis of *anorexia nervosa* is likely. The treatment is to advise a very palatable diet with a high caloric density. Often, appetite remains poor and prolonged medical treatment is needed to ensure recovery.

Accidental starvation may follow a catastrophe in a remote place (climbing, flying, or sailing). The potential for survival depends upon the initial size of the fat depots and on the amount of energy expended in trying to reach a place of safety. Reserves of carbohydrate (mainly as glycogen) are exhausted within a day, and thereafter energy requirements are met from fat and protein. Protein catabolism is essential to meet the blood sugar requirements of the central nervous system. At first, protein usage amounts to no more than 15% of total energy ($\sim$60 g • day^{-1}); this is taken from a labile reserve of $\sim$300 g protein, distributed throughout the body tissues, but particularly the liver. Once this resource has been exhausted, tissue wasting is inevitable. At the cellular level, the rate of production of DNA is decreased, as is the number of DNA units per cell (Millward and Waterlow, 1978). The effects of protein catabolism are most obvious in the skeletal muscles. As starvation progresses, the percentage of energy derived from protein rises. Loss of plasma protein leads to tissue edema and water retention. After $\sim$40 days, both fat and protein stores have been reduced to the minimum compatible with survival.

The intestines participate in the tissue atrophy, and great care must be taken in rehabilitating starved patients. The first meals provided should always be small and readily digested.

EXERCISE AND BLOOD LIPIDS

Dietary Fat and Ischemic Heart Disease

Dietary Imprudence

Since fatty plaques are a prominent feature of ischemic heart disease, it has been tempting to postulate that the underlying problem is the high intake of animal (saturated) fat characteristic of the current Western diet (Fig. 13.1; Keys, 1970; Wen and Gershoff, 1973).

Unsaturated fats (Table 13.10) include in their molecule fatty acids with one or more polyvalent carbon linkages. These linkages have an affinity for iodine, and the extent of unsaturation of dietary fat is thus reported as an "iodine value"; transition from an animal to a vegetable fat diet raises the iodine value from perhaps 50 to 100 or more.

Is the current North American diet imprudent? The average North American consumes $\sim$50% more food than a "primitive" and unacculured "hunter-gatherer" or pastoralist. Part of this difference is due to a larger body size and wastage of food, either in the kitchen or on the table. Nevertheless, the obesity of a typical middle-aged Western man reflects an excessive food intake. He also eats twice as much protein, and 4 or 5 times as much animal fat as many primitive peoples, together with large quantities of sugar. "Civilized" communities have shown a particularly dramatic increase in the per capita consumption of sugar over the last two centuries, from 2.3 kg per annum in 1770 to 11.4 kg per annum in 1870 and over 55 kg per annum at the present time.

The case against animal fat is not particularly strong. Some primitive people—traditional Eskimos, Mongols, and East African tribes such as the Samburu and Masai—have a high-fat diet yet remain relatively free of ischemic heart disease (G. V. Mann, Teel et al., 1955; G. V. Mann, Shaffer et al., 1964; H. H. Draper, 1976; Sayed et al., 1976). Furthermore, the consumption of animal fat by the Western world was already high in the early 1900s, and it is difficult to blame the subsequent "epidemic" of ischemic heart disease upon the relatively small increment of fat consumption since the turn of the present century. Again, if individual patients from the Western world are matched in other respects, the association between fat intake and the occurrence of overt ischemic heart disease is weak.

On the other hand, a high serum cholesterol level is associated with a greater than average risk of the various manifestations of ischemic heart disease (C. W. Frank, Weinblatt, and Shapiro, 1970). Various authors set the border line of "abnormality" at 250–270 mg • dl^{-1}, although the North American population average of $\sim$200 mg • dl^{-1} (Hewitt et al., 1977; Table 11.11) may already be above a desirable level.

Reduction of Serum Cholesterol

The immediate effect of a low-cholesterol diet (Table 13.11; Fig. 13.2) is a dramatic reduction of serum

Hypothesis A

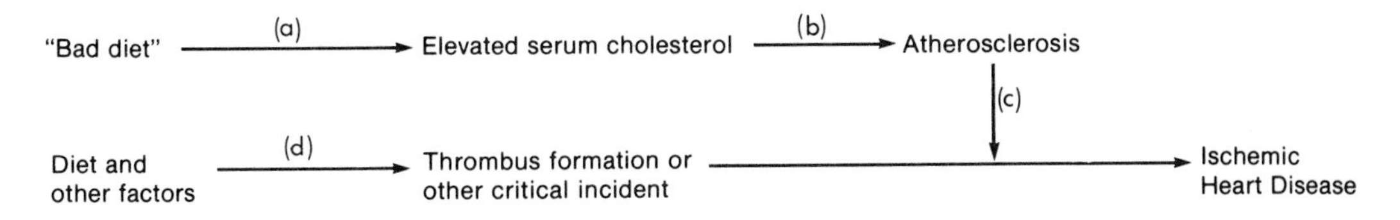

Hypothesis B

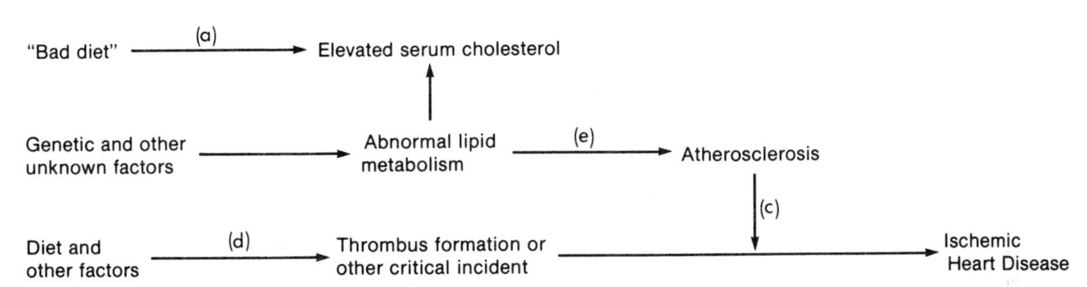

FIG. 13.1. Two possible hypotheses concerning the relationship between diet and ischemic heart disease (after G. Beaton). Proof of steps **a** and **c** is well-established. Proof of step **b** rests largely upon the feeding of high cholesterol diets to caged animals (Montoye, 1960). Step **d** probably encompasses a number of factors such as alteration of serum electrolytes and an increase of cardiac workload by exercise or anxiety. The role of diet at stage **d** is suggested by the association between hyperlipidemia and an increased clotting tendency of the blood. **Hypothesis A** is in keeping with epidemiological data, but so is **Hypothesis B** if it is assumed that a number of unknown factors contributing to abnormal metabolism are by-products of our technically oriented Western society. If **Hypothesis A** is correct, then a change of diet should have a direct effect on serum cholesterol and thus the extent of atherosclerosis. If **Hypothesis B** is correct, however, little advantage would be gained from a change of dietary patterns.

cholesterol. In one study from our laboratory (R. C. Goode, Firstbrook, and Shephard, 1966), a small group of healthy young men followed a rigid low-cholesterol diet for 8 weeks. The serum cholesterol initially dropped from 170–230 mg • dl^{-1} to ~90 mg • dl^{-1}. However, when the subjects were also required to run for 25 min each day, they were unable to sustain body mass without an increased intake of sugar, and there was then an associated return of serum cholesterol toward baseline values. It seemed that hepatic cholesterol synthesis had increased.

An experiment by the U.S. National Heart Institute covered a span of 8 years and involved 800 veterans (Dayton et al., 1969). Half of the group were fed a vegetable oil diet, lowering their cholesterol intake from 653 to 365 mg • day^{-1}. At the end of the investigation, the average serum cholesterol reading for the experimental group remained 18% below its initial value. They also had a marginal advantage over the controls in terms of overt ischemic heart disease. Unfortunately, the study was beset by a number of the problems that plague long-term prospective investigations: the subjects were relatively elderly (average age 65 years), adherence to the required dietary regimen was less than 60%, and faithful members of both test and control groups became more health-conscious as the experiment progressed. Thus, the control group also experienced a substantial (13%) decrease of serum cholesterol, and it may well be that nonspecific factors rather than the unsaturated fat diet were responsible for the improvement of prognosis in the test group.

High-cholesterol Diets and Natural Synthesis

Plaques resembling (but not identical with) human atherosclerosis can be produced by feeding massive doses of cholesterol or saturated fat to certain animal species (Montoye, 1960; Vesselinovitch and Wissler, 1977; H. Y. C. Wong and Johnson, 1977). However, it is necessary to precondition the animals by prolonged caging and/or alterations of hormonal

TABLE 13.10
Proportions of Saturated and Unsaturated Fat in Various Foodstuffs

Food	Percentage of Fatty Acids in Selected Fats and Oils		
	Saturated	Monounsaturated	Polyunsaturated
Bacon	39	51	10
Butter	70	27	3
Chicken	30	48	22
Coconut	97	1	2
Corn	12	33	55*
Cottonseed	26	22	52
Herring	21	0	79*
Lard	40	48	12
Linseed	11	64	25
Olive	9	86	5
Palm fruit	49	42	9
Palm kernel	88	11	1
Peanut	19	50	31
Safflower	14	35	58*
Sesame	14	43	43
Soya bean	18	18	64*
Sunflower	12	35	53*
Whale	81	0	19

*Foods marked with an asterisk have a particularly high ratio of polyunsaturated to saturated fatty acids.
Source: Based mainly on data of Joliffe et al., 1959.

TABLE 13.11
Cholesterol Content of Some Common Foodstuffs (mg per average serving)

Foodstuff	Cholesterol Content (mg per serving)
Beef (raw)	84
Butter (1 pat)	38
Cheese (Cheddar)	30
Cheese (skim milk)	0
Chicken (raw)	60
Crab	190
Egg (1)	250
Fish	70
Heart (raw)	150
Ice cream	35
Kidney (raw)	375
Lamb (raw)	80
Lard (1 tablespoon)	13
Liver (raw)	330
Margarine (vegetable)	0
Milk (1 cup)	26
(skim)	7
Oysters (6)	1200
Pork	84
Shrimps (12)	150
Veal	118

Note: Vegetables contain no cholesterol, unless butter, cheese, or other animal product is added during cooking.
Source: Based mainly on data of Joliffe et al., 1958.

secretion before lesions develop, and the combination of such unusual circumstances with a highly abnormal food pattern reduces the relevance of such investigations to normal human life.

The human cholesterol balance sheet is such that a high intake of animal fat or cholesterol-rich foods does not necessarily lead to hypercholesterolemia (Oshima and Suzuki, 1975; Ruys and Hickie, 1976). Furthermore, the influence of cholesterol feeding upon serum cholesterol is independent of dietary fat (J. T. Anderson et al., 1976). The usual foods of Western man provide 500–600 mg • day^{-1} of cholesterol. This figure is overshadowed by synthesis in the liver and intestines (Ho et al., 1970), normally 1000 mg • day^{-1}, but rising still higher if there is an excessive intake of energy, either as fat or carbohydrate. The daily circulation (Fig. 13.3) and excretion of cholesterol (~1000 mg) is also large, so that blood levels soon reflect any changes in synthesis or secretion.

Other Dietary Factors

In their search for a nutritional explanation of ischemic heart disease, some investigators have focused on possible alternative evils of our Western diet—a low fiber (Trowell, 1976) and a high sugar intake (Yudkin, 1957; McGandy et al., 1966; Grande,

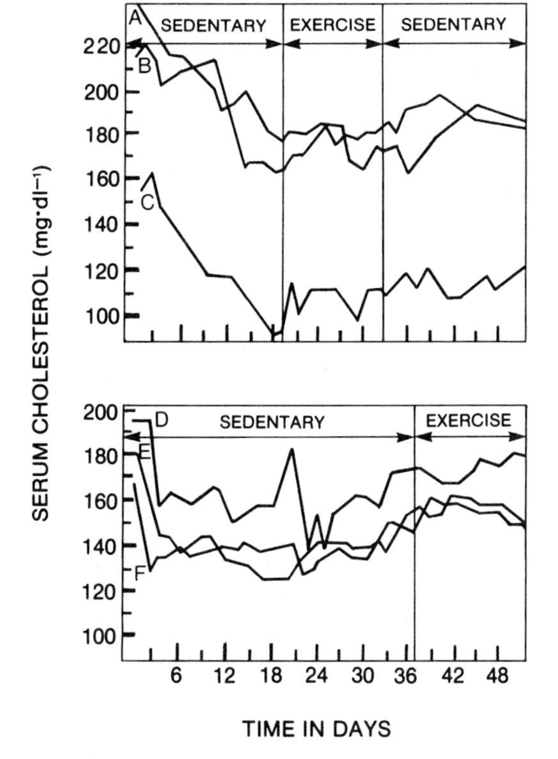

FIG. 13.2. The influence of a cholesterol-free diet and regular daily exercise upon serum cholesterol levels. (Based on data of R. C. Goode, Firstbrook & Shephard, 1966 for six subjects; exercise consisted of 25-minute treadmill running per day.)

1967; Bjørkerud, 1970). The latter would link up with the known association between diabetes and heart disease, and a possible relationship between heart disease and the frequent consumption of sweetened tea or coffee (Jick et al., 1973). Sugar may be injurious partly because it encourages overeating. It also has adverse effects upon the insulin-secreting islets of Langerhans. The immediate response of animals to the feeding of large quantities of sugar is an increased output of insulin, but if the sugar diet is maintained the islet cells become exhausted and show irreversible damage.

Overall Energy Balance

In any dietetic experiment, habitual activity and the total energy balance are important variables. This is particularly true with regard to the regulation of blood lipids. It has been argued that primitive peoples are able to withstand a wide range of diets because they have a high daily energy expenditure (G. V. Mann, Teel et al., 1955; G. V. Mann, Shaffer et al., 1964; H. H. Draper, 1976; Sayed et al., 1976), and it is probable that an excessive intake of food is necessary before the harmful effects of either animal fat or sugar are revealed. Exercise curtails the rise of blood sugar following a carbohydrate meal, thereby diminishing the insulin demanded of the pancreatic islets. An active person can thus tolerate a larger intake of sugar before the islet cells become exhausted.

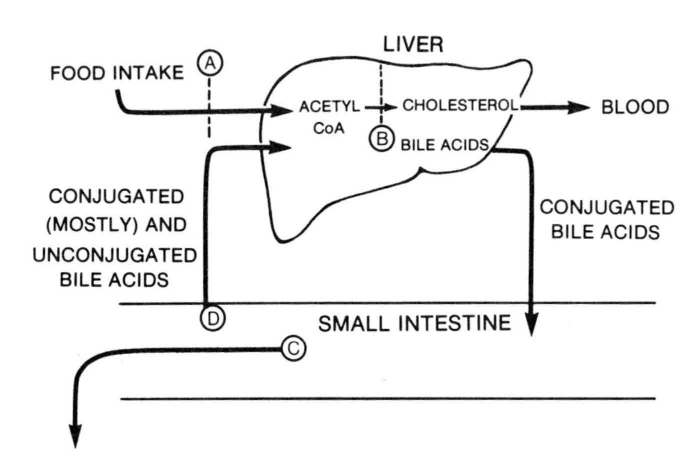

FIG. 13.3. Possible chemical mechanisms for reducing serum cholesterol. **A.** Energy intake may be reduced. **B.** Hepatic synthesis of cholesterol may be reduced by clofibrate. **C.** Fecal excretion can be increased by a bile-sequestering resin such as cholestyramine. **D.** Uptake can be slowed by competing sterols, for example, β sitosterol, found in certain plants. Note that the body pool of cholesterol is 3 to 5 g. The daily circulation between the small intestine and the liver is 20 to 30 g, but because of the complex nature of the cholesterol molecule, breakdown is slow and the half-life long (40 to 60 days).

Timing of Intervention

Irrespective of mechanisms, it is unlikely that any short-term program of exercise and/or dietary modification will have much influence on the course of atherosclerosis in an adult. Deposition of fat in the lining of the blood vessels commences during infancy (Jaffé and Manning, 1971) and is often quite advanced in young adults (Enos et al., 1955). A preventive program should commence immediately after cessation of breast feeding (Hanefeld et al., 1977).

Once the atherosclerotic process has progressed to a fibrinoid or calcified state, no dietary regimen can be of great avail. A suitable diet (see below) may help but cannot correct any genetically induced problem in the transport or metabolism of fat. Control of obesity reduces the load placed upon the heart by a sudden burst of physical activity, but it may not check the trigger factors that initiate ventricular fibrillation or myocardial ischemia once atherosclerosis is present. Conversely, some populations tolerate quite extensive atherosclerosis without a high incidence of ischemic heart disease. Possibly, they lack trigger factors that would convert a silent lesion to an overt coronary attack.

Classification of Hyperlipidemias

There is general agreement that the risk of ischemic heart disease is increased in those individuals who have a high serum lipid level (hyperlipidemia). As many as a third of all coronary victims are grossly hyperlipemic, and five out of six have a serum cholesterol greater than 210 mg • dl^{-1}.

Serum Lipid Fractions

In recent years, interest has attached to fractionation of the serum lipids by such techniques as electrophoresis (Fig. 13.4), ultracentrifugation, and selective precipitation. The four main constituents are as follows (V.Stein and Stein, 1970; Shore and Shore, 1970; Gotto and Jackson, 1977):

α (high-density) fraction. The specific gravity ranges from 1.06 to 1.21 and the molecular weight from 1.8 to 3.6 × 10^5. The major apoproteins are Apo A-I and Apo A-II, and the major lipids are phospholipids and cholesterol esters.

β (low-density) fraction. The specific gravity ranges from 1.01 to 1.06 and the molecular weight from 2.7 to 4.8 × 10^6. The major apoprotein is Apo B, and the major lipids are cholesteryl esters and phospholipids. This fraction is concerned particularly with the transfer of fat from the liver to the plasma.

Pre-β (very low density) fraction. The specific gravity ranges from 0.95 to 1.01 and the molecular weight from 5 to 10 × 10^6. A variety of apoproteins are found in these particles (Apo B, Apo C-I, Apo C-II, Apo C-III, and Apo E). A large fraction (50–80%) of each particle is triglyceride; this is largely endogenous, originating in the liver.

Chylomicrons. The specific gravity is <0.95 and the molecular weight is very high ($>0.4 \times 10^9$). Apoproteins are Apo B, Apo C-I, Apo C-II, and Apo C-III, but 80–95% of each particle is triglyceride; the chylomicrons serve to transport triglycerides via the lymphatics and plasma to the tissue stores.

The α fraction apparently has some protective effect against atherogenesis (Glueck et al., 1976; T. Gordon et al., 1977; P. J. Jenkins et al., 1978), although the exact mechanism is not yet decided. Possibly, the high density lipoprotein (HDL) cholesterol occupies sites on the cell membrane, preventing entry of the irritant low density lipoprotein (LDL) fraction, or possibly the α-lipoprotein acts as a scavenger, carrying cholesterol back to the liver before damage is caused to the vasculature.

Hyperlipidemia

Hyperlipidemia is described in terms of a high total cholesterol, a low HDL/LDL ratio, and an accumulation of β and pre-β fractions. Five patient groups

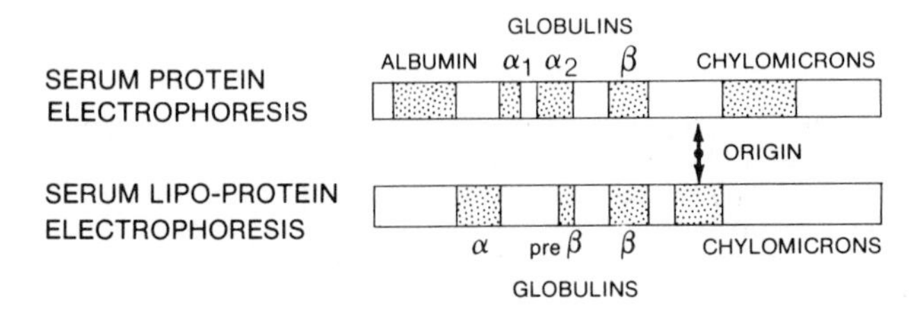

FIG. 13.4. Electrophoretic study of serum lipoproteins. After completion of electrophoresis, the serum protein film is stained with brom-phenol blue, and a fat stain such as Oil Red O is used to identify the different lipoprotein fractions.

are described (Fredrickson, 1974); types II and IV are the most common.

Type I patients have an increase of low density lipoproteins, normal or reduced levels of very low density lipoproteins, and yellow xanthomatous deposits of carotene and lipochrome pigments around the tendons and in the eyelids. The problem in this class of patient may be a deficiency of the enzyme that breaks down chylomicrons. The fat levels remain unimproved by diets low in animal fat and sugar, and there is little response to drugs such as clofibrate (Steinberg, 1970; Paoletti et al., 1977), an inhibitor of cholesterol synthesis, and cholestyramine (Miettinen, 1970), a bile-sequestering resin which increases the fecal excretion of cholesterol (Fig. 13.3). However, large doses of thyroid hormone are sometimes helpful to type I patients, and there may also be value in administering medium-chain triglycerides that can be transported directly to the liver without chylomicron formation.

Type II patients have a similar lipoprotein pattern to type I, but xanthomata are absent and concentrations of serum bile acids are low (Angelin, 1977). Type II hyperlipoproteinemia is an inherited disorder (Slack and Mills, 1970; M. D. Morris and Greer, 1970; Gotto, 1977). Normal concentrations of lipoprotein can be restored by (1) administration of clofibrate or cholestyramine and (2) increasing the dietary ratio of polyunsaturated to saturated fat.

Type III patients have an increase of very low density lipoproteins and normal concentrations of low density lipoproteins and xanthomata. They respond particularly well to clofibrate, dietary treatment, and reduction of obesity.

Type IV patients have an increase of very low density lipoproteins but a reduced level of low density lipoproteins. Synthesis of cholesterol is increased (Angelin, 1977). This group responds well to dietary restriction and reduction of obesity. Clofibrate reduces the level of very low density lipoproteins but it may at the same time increase the level of low density lipoproteins.

Type V patients have elevated concentrations of both high and low density lipoprotein. Patients of this type respond best to combined thyroxine and clofibrate treatment, with reduction of obesity and a high-protein diet.

Where equipment is not available to determine a complete lipid profile, some information can be derived simply from total cholesterol and triglyceride readings as follows:

	Cholesterol	**Triglycerides**
Type I	+	+ +
Type II	+	±
Type III	+ +	+
Type IV	±	+ +
Type V	+	+ +

Serum triglyceride readings are usually less than 150 mg • dl^{-1} in slim subjects, and readings greater than 200 mg • dl^{-1} may be regarded as substantially elevated.

Lipid Fractions and Heart Disease

Part of the apparent association between the various lipid fractions and ischemic heart disease stems from a mutual association with other cardiac risk factors such as body fat and physical inactivity (Shephard, Youldon et al., 1979; Shephard, Cox, and West, 1980). Nevertheless, multivariate analysis suggests that both total cholesterol and a low HDL/LDL ratio make independent contributions to the risk of ischemic heart disease. It is easier to change the very low density than the low density fraction of the serum lipids, and this is particularly unfortunate since the LDL (β) fraction seems the main culprit in atherogenesis.

Exercise and Lipids Fractions

The acute effect of sustained exercise, such as a marathon race, is a substantial decrease of serum triglycerides (Liesen et al., 1975); further, the proportion of oleate is increased relative to stearate and palmitate, since an increased proportion of the circulating triglyceride is derived from lipolysis as opposed to hepatic synthesis.

Some early investigators reported that the chronic effect of exercise was to decrease serum cholesterol (D. E. Campbell, 1965; G. V. Mann, Garrett et al., 1969). Such findings were typically associated with a negative energy balance and a diminution of body mass. Even if habitual activity is unchanged, a loss of body fat usually leads to a fall of serum cholesterol. Conversely, serum cholesterol rises during periods when body mass is rising.

R. C. Goode, Firstbrook, and Shephard (1966) studied the effect of 25 min of vigorous treadmill running per day under conditions whereby food intake was sufficient to conserve body mass. There was no change of fasting serum cholesterol, but a small decrease of fasting serum triglycerides was observed. Others obtained similar data (Grimby, Wilhelmsen et al., 1971; McTaggart and Ribas-Cardus, 1978). To the limited extent that high triglyceride readings are associated with an increased risk of ischemic heart disease (Cohn, Gabbay, and Weglicki, 1976), the decrease in this variable should be beneficial to prognosis.

More recently, interest has centered upon changes in the lipid profile induced by exercise. There is little alteration in response to the standard businessman's calisthenics program (Shephard, Youldon et al., 1979; Shephard, Cox, and West, 1980), but more vigorous and long-lasting activity brings about at least a 15–18% increase of HDL cholesterol (L. A. Carlson, 1967a; Lopez, 1974; A. S. Leon, Conrad et al., 1977; Table 13.12). Further analysis shows this is due mainly to an increase in the HDL-2 fraction (Fig. 13.5). This is normally more obvious in women than in men, but in both sexes is increased by running (P. D. S. Wood, Haskell et al., 1978; Table 13.12). There is an associated increase of apolipoprotein A-I but not apolipoprotein A-II. In contrast, cigarette smoking is negatively correlated with HDL cholesterol (Garrison et al., 1978; Shephard, Youldon et al., 1979; Shephard, Cox, and West, 1980).

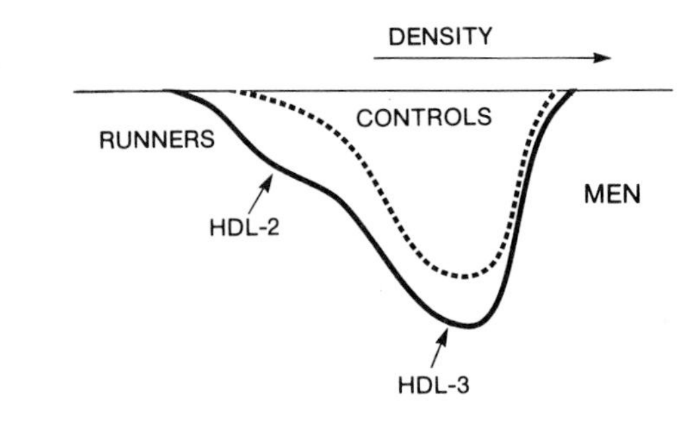

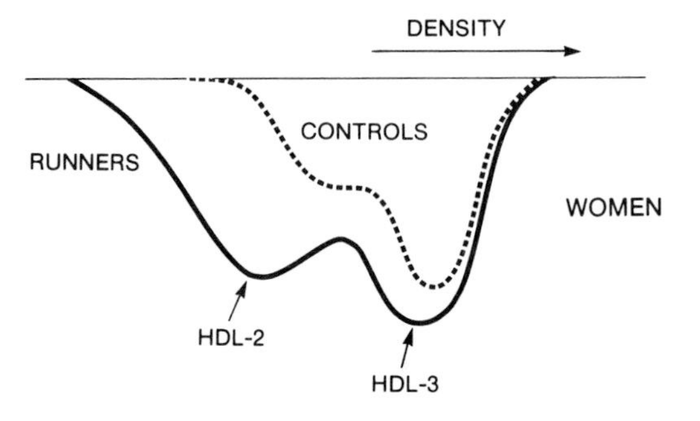

FIG. 13.5. A comparison of lipid profile between runners and controls. (After P. D. Wood, 1979.)

Problems of Dietary Modification

Given the association between overt ischemic heart disease and certain lipid constituents, there is substantial current pressure for nationwide adoption of diets containing less animal and more vegetable fat (Stamler, 1971; Blackburn, 1974a). Unfortunately, there is little solid evidence to recommend such a policy.

The problem of the timing of intervention has been noted (see above). At best, the adult population would succeed in the rather unpalatable task of eating a high percentage of vegetable fat and there might then be a marginal reduction in both serum cholesterol and the incidence of ischemic heart disease (Van Itallie and Hashim, 1965; Rinzler, 1968; Turpeinen et al., 1968; Dayton et al., 1969; Medalie, 1970). However, it is likely that many people would replace fat by sugar, and this could be more harmful than the original diet of animal fat.

TABLE 13.12
Comparison of Lipid Profiles for Runners and Controls

Variable	Runners		Controls	
	Men	Women	Men	Women
Triglycerides ($mg \cdot dl^{-1}$)	70	56	146	122
Total cholesterol ($mg \cdot dl^{-1}$)	200	193	210	210
VLDL cholesterol	11	5	28	27
LDL cholesterol	125	113	139	128
HDL cholesterol	64	75	43	55
Ratio HDL/LDL	0.51	0.66	0.31	0.43

Source: After P. D. S. Wood et al., 1978.

While a low-fat/high-carbohydrate diet many depress serum cholesterol, it can also lead to an increase of serum triglycerides. There is also the possible harmful effect of the added sugar upon the pancreatic islets (see above). Moreover, it is by no means certain that vegetable oils are themselves devoid of risk (A. N. Howard, 1970). Polyunsaturated fats have a tendency to polymerize to a compound called ceroid and this may provide a rather permanent basis of clot formation; while clots of animal fat can be broken down by the enzymes of the body, the enzymes are unable to attack the waxy ceroid. The hydrogenation process used in the production of synthetic fats converts as much as 40–60% of the unsaturated fatty acids from the cis to the trans form. This leads to a straightening and a lengthening of the fatty acid molecule (Enig, 1979), and the revised configuration can lead to distortion of the normal bipolar layer of the cell membrane (Figs. 4.12 and 13.6). Finally, there have been reports that rape seed oil (with a high erucic acid content) causes myocardial lipidosis, with subsequent focal necrosis of the heart muscle.

Despite the several objections to dietary modification for the general population, there is some advantage in modifying the diet of patients with a high risk of coronary disease (H. B. Brown, 1970). The inevitable disruption of shopping and cooking routines is particularly worthwhile in patients suffering from group IV hyperlipidemia (see above).

PHYSICAL ACTIVITY AND HEALTH

Physical Activity and Productivity

Potential industrial production may be lost for a variety of reasons, including physical and mental fatigue, absenteeism, minor uncertified illness, major illness, accidents, and strikes.

Physical fatigue is relatively uncommon in modern industry. Nevertheless, an old person who has a heavy physical task may find his performance limited through use of an excessive fraction of his maximum oxygen intake, with an accumulation of anerobic metabolites. In such a case, physical well-being and work output may be helped (Kilbom, 1971) if maximum oxygen intake is increased by (1) endurance training (potential gain ~20%; Chap. 11), (2) correction of any anemia, (3) correction of obesity (and thus increase of oxygen transport per unit of body mass), and (4) cessation of smoking (potential gain 5–10%; Chap. 14).

Mental fatigue is most likely in repetitive, boring work. Exercise may help to relieve this boredom and restore arousal to an optimal level (Chap. 3). Several studies (Geissler, 1960; Manguroff et al., 1966; Galevskaya, 1970) suggest that during light or sedentary work, fatigue is relieved more effectively by gymnastics than by a rest pause. On the other hand, there are some groups (for example, air traffic controllers) who tend to remain aroused throughout a day's work, and in such groups the additional arousal induced by physical activity could have an adverse influence upon productivity.

Uncertified illness and minor absenteeism is one of the major problems of modern companies. This can sometimes be related to a physiological problem (Lindén, 1969; Buzina, 1972)—the worker may have an aerobic power that is inadequate to meet the demands of both home and an 8-hr day. But more frequently, absenteeism has a psychological basis, arising as a reaction to the drudgery of modern production line tasks. Exercise helps to relieve this boredom and provides an outlet whereby personal achievement and success can be recognized. This probably accounts for a large part of the reductions of absenteeism that have been described (Table 13.13; Lindén, 1969).

Major illness is discussed below. Fatigue is a potent source of accidents, and certain categories of injury, such as back strains, have been associated with unaccustomed effort by workers who are lacking in fitness and are physically weak (Alston et al., 1966; Magora and Taustein, 1969; J. R. Brown, 1972).

Strikes are not usually related to physical activity or its lack. Nevertheless, it is conceivable that participation in a sports program could provide a sense of status and achievement that would otherwise be sought in the organization of disputes with management (Shephard, 1974b).

Physical Activity and Ischemic Heart Disease

Current Epidemic

During the first half of the present century, the population of the Western world experienced a rapid in-

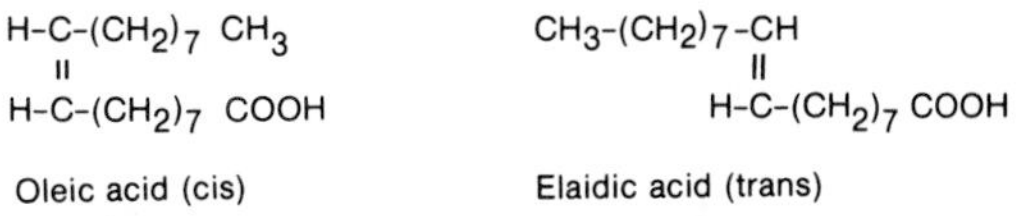

Oleic acid (cis)	Elaidic acid (trans)
$H-C-(CH_2)_7\ CH_3$	$CH_3-(CH_2)_7-CH$
‖	‖
$H-C-(CH_2)_7\ COOH$	$H-C-(CH_2)_7\ COOH$

FIG. 13.6. A comparison of cis and trans forms of fatty acid. The natural fatty acids exist in the cis form, with the molecule angulated at each double bond. The trans form, in contrast, is straight chained, like a saturated fatty acid.

crease in various manifestations of ischemic heart disease (sudden death, myocardial infarction, the onset of anginal pain, and related electrocardiographic abnormalities) (T. W. Anderson and LeRiche, 1970). The epidemic apparently peaked in the late 1960s, and there has subsequently been a drop of more than 20% in the annual death rate from ischemic heart disease.

It is tempting for exercise enthusiasts to blame the initial epidemic on physical inactivity and to attribute its subsequent decline to a return of interest in personal fitness. However, other factors have equal or greater importance (the increase and subsequent decline of cigarette smoking, changes in animal fat consumption, control of hypertension, and improvements in the treatment of patients who have suffered a heart attack).

Role of Occupation

Retrospective, prevalence, and prospective surveys suggest that those who work hard have a less than average chance of subsequent ischemic heart disease (Tables 13.14–13.16; Pedley, 1942; Ryle and Russel, 1949; Stamler, Lindberg et al., 1960; H. L. Taylor, Klepetar et al., 1962; Chapman and Massey, 1964; McDonough et al., 1965; Paffenbarger and Hale, 1975; Paffenbarger, Hale et al., 1977; Stamler, 1978). However, several trends have blurred this evidence: (1) higher socioeconomic groups take more leisure activity and have better health habits than lower grades of worker, (2) continuing cigarette smokers are found primarily among blue-collar workers, (3) access to medical advice and treatment is greater for white-collar workers, (4) occupational categories are increasingly linked with milieu and ethnic groups, and (5) few occupations now require what has been thought a protective dose of physical activity (Table 13.15). Furthermore, myocardial scarring is almost as common in active as in inactive subjects, and active workers are more liable to angina than are their inactive counterparts (J. N. Morris, Heady et al., 1953; J. N. Morris and Crawford, 1958; G. A. Rose, Prineas, and Mitchell, 1967; Kannel, Sorlie, and McNamara, 1971). If there is indeed benefit from an active occupation, then at best it confers an ability to live with a certain amount of atheroma and fibrous scarring of the myocardium. A strong objection to almost all occupational surveys is the element of self-selection. People choose their employment, and there are thus initial differences of health, activity, and cardiac "risk factors" between recruits to sedentary and physically active jobs (J. N. Morris, Heady, and Raffle, 1956; Oliver, 1967; Hickey et al., 1975; Paffenbarger, 1977). Unfortunately, this selective process is often exaggerated by a transfer of diseased subjects from active to inactive job categories (Kahn, 1963; Paffenbarger, Hale et al., 1977).

Studies of Athletes

Comparisons of the incidence of ischemic heart disease and of longevity between "athletic" and "nonathletic" populations (Yamaji and Shephard, 1978) have not demonstrated any striking differences. Unfortunately, the supposed athletes have often ceased to participate in their sport long before the age of vulnerability to ischemic heart disease, while sometimes the nonathletes chosen as controls have en-

TABLE 13.13
Absenteeism Before and After Introduction of an Employee Fitness Program

Group	Absenteeism (per 1000 employees per month)			
	Before Program	During Program	△	% Change
Program adherents (n = 171)	603	351	−252	−41.8
Remainder of experimental company (n = 1110)	711	572	−139	−19.5
Control company (n = 577)	844	661	−183	−21.7

Note: All three groups showed a reduction of absenteeism during the program; this reflects both the well-known Hawthorne effect and an epidemic of influenza during the preceding period. Nevertheless, the reduction of absenteeism was greater and the final value lower in program adherents.
Source: Data of Shephard and Cox, unpublished.

TABLE 13.14
Influence of Occupational Activity upon Various Indices of Ischemic Heart Disease

Index	Mean	Range	Number of Studies
Coronary heart disease attack rate	0.60	0.17–1.03	16
Coronary heart disease mortality	0.66	0.28–1.22	21
Myocardial infarction	0.56	0.33–0.98	9
Angina pectoris	1.36	0.65–1.98	7
Vascular pathology	0.76	0.51–1.00	7
Myocardial pain	0.48	0.21–0.68	8

Source: Based on data accumulated by S. M. Fox and Haskell, 1968.

gaged in noncompetitive physical activities while at university. Montoye, Van Huss et al (1957) further demonstrated poor health habits of middle-aged former university sportsmen. They were less active than their contemporaries, had experienced a larger increase of body mass with aging, and were more likely to be smokers and drinkers of beverage alcohol than their nonathletic contemporaries.

One group with a favorable prognosis is Finnish cross-country ski champions. As a group, they live several years longer than their inactive counterparts (Karvonen, Klemola et al., 1974). The skiers continue their sport into old age, but most of them

TABLE 13.15
Estimates of Excess Daily Occupational Energy Expenditure Needed to Protect against Ischemic Heart Disease

Author	Daily Excess Energy Expenditure (kJ)
J. S. Skinner, Benson et al. (1966)	16.8–21.0
H. L. Taylor, Blackburn et al. (1969)	25.1–50.3
S. M. Fox and Skinner (1966)	16.8–21.0 3–5 times per week
Paffenbarger, Laughlin et al. (1970)	38.8

Source: After S. M. Fox and Skinner, 1966.

TABLE 13.16
Potential Reduction in Rates of Fatal Heart Attack in California Longshoremen with Selected Changes of Life-style

Life-style Change	Initial Risk (Attacks per 10,000 Man-Years)	Potential Reduction (%)
1. Low- → high-energy output	69.7	48.8
2. Heavy cigarette smoking → cessation	94.3	27.9
3. High systemic blood pressure → control of hypertension	89.1	28.8
Combination		
1 + 2	95.7	64.7
1 + 3	91.5	73.5
2 + 3	161.6	50.3
1 + 2 + 3	151.9	88.2

Source: After Paffenbarger, 1977.

are also nonsmokers and this could account for much of their favorable prognosis. Further, we have no proof that the skiers did not commence their careers with an advantage of health status or body build relative to inactive Finns.

Influence of Leisure Activity

Population surveys of leisure activity (for example, Gyntelburg, 1977) are hampered by the poor reliability of diaries and questionnaires (Zukel et al., 1959), correlations between vigorous physical activity and social class (Elmfeldt et al., 1976), and denial of disability by those who have already suffered infarction (Froehlicher and Oberman, 1972).

Recent prospective studies have focussed on narrow segments of society. J. N. Morris, Adams et al. (1973) found that the incidence of ischemic heart disease among executive-grade British civil servants was reduced among those who engaged in either 5 min • day^{-1} of near-maximal activity (active recreation, "keep-fit" classes, or "vigorous getting about") or 30 min • day^{-1} of heavy work ($\geq$31 kJ • min^{-1}). Paffenbarger (1977) surveyed a large group of Harvard graduates. Substantial protection against ischemic heart disease was seen with an additional weekly energy expenditure of 8000 kJ, a figure reminiscent of the occupational surveys (Table 13.15). The favorable prognosis extended to those who had become active since leaving university but did not include former athletes who were currently inactive.

While both of these studies are strongly suggestive of a benefit from regular exercise, they do not overcome the problem of self-selection, and it could further be argued that the variable studied (deliberate vigorous physical activity) is merely serving as an indicator of positive health attitudes.

Problems of Direct Experimentation

The problems of obtaining more direct experimental proof of the exercise hypothesis are well recognized. Animal and bird models are unsatisfactory since restrictive caging and a high-cholesterol diet are needed to produce ischemic heart disease (Warnock et al., 1957; Myasnikov, 1958; McAllister et al., 1959; Montoye, 1960; Vesselinovich and Wissler, 1977; H. Y. C. Wong and Johnson, 1977). Coronary events in initially healthy men are sufficiently rare that an experiment with random allocation of subjects to exercise and control groups requires an unmanageably large sample size. A pilot trial indicated that even if attention was concentrated upon "high-risk" individuals, the likely cost of a definitive experiment would be $31 million (measured in U.S. 1967 currency; H. L. Taylor, Parlin et al., 1967). Furthermore, it would be difficult to draw generalizable conclusions since the likely dropout rate would be 50% over a 6-month period of exercise (Ilmarinen and Fardy, 1977).

Prospective assessments of exercise therapy for patients who have already sustained an infarction (Kentala, 1972; Wilhelmsen, Sanne et al., 1975; Rechnitzer et al., 1975; Naughton, 1978) have likewise been inconclusive (see below).

It is unlikely that categoric proof of the benefits of exercise will be obtained in the near future. However, the potential remedy is both agreeable and non-addictive, with few serious complications, and on this basis it may reasonably be commended to high-risk patients and healthy individuals alike (Blackburn, 1974b).

Protection against Ischemic Heart Disease

If there were no obvious mechanism whereby exercise could protect the myocardium against ischemic heart disease, the suggestive evidence from the epidemiologists would carry little weight. However, several authors have made quite plausible suggestions as to how physical activity can correct certain risk factors, thereby protecting the individual against a future episode of ischemic heart disease. We first examine the risk factors and then consider how they are modified by physical activity.

The Risk Factors

Prospective studies have identified specific risk factors that increase an individual's likelihood of developing an overt clinical manifestation of ischemic heart disease (Dawber, Moore, and Mann, 1957; Keys, 1970; Borhani, 1977; Karvonen, 1977).

The value of individual items as screening tools is assessed in terms of *sensitivity* and *specificity* (G. D. Friedman, 1974; Mausner and Bahn, 1974; Linderholm, 1977). Consider, for instance, serum cholesterol. In a community study at Framingham, Massachusetts, 156 of every 1000 men aged 40–59 years had a *high serum cholesterol* (>260 mg • dl^{-1}). Over a 10-year period of observation, 32 of these men developed clinically recognized ischemic heart disease. The total number of new episodes of ischemic heart disease over the same period was 120 per 1000 men. The sensitivity of the test was thus 32/120 (27%) and its predictive value was 32/156 (21%). A total of 844 men had normal serum cholesterol readings, but 880 remained free of clinical ischemic heart disease over the 10-year period. The specificity of the test was thus 844 − (120 − 32)/880, or 86% and the predictive value of a negative result was (120 − 32)/844 = 10%. The test procedure yielded (156 − 32) false positive and (120 − 32) false negative results. Test optimization involves selection of a criterion of "normality" that maximizes sensitivity without an excessive loss of specificity. Note that the chosen criterion is influenced by the period of observation; with a long study, some false positive results convert to true positives, while there is also an increase in the number of false negative results.

The overall usefulness of any given risk indicator may be summarized by the risk ratio. This compares the predictive value of positive and negative test results. For serum cholesterol, it amounts to 2.1 (21%/10%).

A second factor predisposing to ischemic heart disease is a high *systemic blood pressure* (systolic > 21.3 kPa, 160 mmHg; diastolic > 12.6 kPa, 95 mmHg). In the 10-year study of Kannel, this factor had a sensitivity of 18% and a specificity of 86%, with a predictive value of 15% for positive tests and 12% for negative tests (risk ratio 1.3). Combining the information from serum cholesterol and blood pressure measurements, the risk ratio rose to 2.5, and if either abnormality was taken as a positive finding, a risk ratio of 2.9 was attained.

Many other factors, when considered singly, discriminate significantly between the coronary-prone and the coronary-resistant individual. However, few of these factors, with the possible exception of cigarette smoking, contribute much additional information after patients have been classified in terms of serum cholesterol and systemic blood pressure (Wilhelmsen, Tibblin et al., 1976).

Ischemic heart disease is more common in men than in women. This might be due largely to differences of life-style (pressures of work, frequency and intensity of physical activity, cigarette consumption, and so on). In support of this view, the sex difference is diminishing as women adopt a more masculine life-style. On the other hand, the sex difference diminishes after the female menopause, suggesting that hormonal factors are also implicated. Several recent investigators have stressed that the use of oral contraceptives by young women has a synergistic action, multiplying the danger from other risk factors (J. I. Mann et al., 1976).

The victims of overt ischemic heart disease often give an adverse *family history* of coronary attacks and sudden death. In some cases, a familial hypertension or hyperlipoproteinemia may be implicated (Lundman, 1977; Nikkila and Rissanen, 1977; see above), but there is often a common life-style (Sackett et al., 1975) and a shared environment which may include adverse attitudes to smoking, diet, and physical activity. It is also likely that a heart attack leads to a more careful searching of family records, and thus the discovery of more affected relations.

Cigarette smoking has a strong influence upon risk, although its mode of action is still discussed (McGill, 1977). The carbon monoxide, cyanide, and nitrogen oxides of the gas phase may all induce atherosclerosis (Åstrup, 1977). Nicotine itself could act in many ways: disturbing carbohydrate and fat metabolism, raising blood pressure, reducing coronary perfusion, and increasing the irritability of the heart (thus provoking arrhythmias); however, nicotine effects are probably minimal (Schievelbein, 1977). There is an increase in hematocrit and red cell count (Eisen and Hammond, 1956; Isager and Hagerup, 1971; Billimoria et al., 1975), and the life span of the platelets is decreased, increasing the tendency to clotting of the blood (Mustard and Murphy, 1963). Finally, there is an association between cigarette smoking and hyperlipoproteinemia (Shephard, Cox, and West, 1980).

The risks of *physical inactivity* have been discussed (see above). A time-conscious *type A personality* is sometimes (M. Friedman and Rosenman, 1974), but not always (E. H. Friedman and Hellerstein, 1973), an adverse finding. Business stress is not a major factor except for those with an upwardly mobile career (Hinckle, 1972). However, a series of social readjustments has an impact on the likelihood of disease (Holmes and Rahe, 1967).

An *excess body mass* exerts little effect on prognosis after allowance for the associated high cholesterol and blood pressure readings (Keys, Aravanis et al., 1972). Nevertheless, obesity is of some practical importance because if it is corrected there is a good chance that the hypertension and the hyperlipoproteinemia will also be resolved (F. W. Ashley and Kannel, 1974). Other adverse metabolic data include a *poor glucose tolerance*, a *high serum uric acid*, and possibly a *high serum triglyceride reading*. More detailed analysis of the serum lipid profile is discussed above.

Adverse characteristics of a laboratory *exercise test* include a low aerobic power, a horizontal or downward sloping depression of the ST segment of the electrocardiogram, and polyfocal ventricular premature contractions (Chap. 6).

If information from all possible sources is combined through appropriate multivariate statistical analysis, as many as 50% of potential coronary patients can be identified. This is an interesting statistical exercise and has some value in selecting high-risk groups for prospective experiments. On the other hand, the value of such knowledge to the individual patient is debatable (J. M. G. Wilson and Jungner, 1968; Lauzon, 1978).

Some authors have predicted substantial prognostic gains from risk factor modification (for instance, Table 13.16; Paffenbarger, 1977; Karvonen, 1977; Wilhelmsen, 1977). Nevertheless, many of the patients tested are inevitably wrongly classified, and since there is no specific cure for ischemic heart disease, the management of the individual is largely unchanged. Gross coronary vascular disease will call for an angiogram and possible bypass surgery. Gross diabetes or hypertension will also require appropriate therapy. With these specific exceptions, the same advice is tendered to high- and to low-risk patients: avoid overeating, reduce body weight, take more exercise, stop smoking, and learn to relax at home and at work. However, knowledge of the poor prognosis may improve compliance with the required regimen (Shephard, Rode, and Ross, 1972).

Multiphasic screening brings to light patients with subclinical atheromatous vascular disease. The individual concerned is inevitably aware of his disease for a longer period than if he had not been tested, and this apparently increases his period of survival. Nevertheless, the age at death remains unchanged by test participation.

Burning of Excess Energy

Since exercise both consumes energy and suppresses appetite, it has a potential to correct such problems as obesity and hyperlipoproteinemia. In animals, it has been demonstrated that enforced activity decreases the extent of atheromatous plaques, particularly if a diet rich in saturated fat is supplied at the same time (Montoye, 1960). Unfortunately, it becomes difficult to demonstrate such an effect if the animal remains on a normal diet, undertaking the more moderate levels of activity practical for a presently sedentary man. Nevertheless, studies of primitive populations and of soldiers engaged on long forced marches provides some evidence that man can eat large quantities of animal fat and/or sugar without raising serum lipids, provided that sufficient physical activity is also undertaken (G. V. Mann, Teel et al., 1955; G. V. Mann, Shaffer et al., 1964). We have noted further the role of exercise in the control of obesity and blood sugar (see above).

Changes in Myocardial Oxygen Supply

Some (but not all) animal experiments have suggested that exercise stimulates the development of collateral vessels within the myocardium (Chap. 6). In the event of coronary occlusion, the heart muscle then has an alternative route of oxygen supply. Observations on postcoronary patients have to date failed to show a comparable response (Kattus and Grollman, 1972; Moccetti and Lichtlen, 1973; Barmeyer, 1976). It is possible that in the middle-aged coronary victim the vessels are already too sclerosed to respond, although exercise is effective in developing a collateral blood supply to the legs even when atherosclerotic disease of this region is quite advanced (Ericsson et al., 1970). Alternatively, it may be that the level of activity attained in some postcoronary exercise classes is insufficient to stimulate collateral development.

Nevertheless, there is sometimes evidence of improved myocardial oxygenation, to the point where ST segmental depression may be completely corrected (Kavanagh, Shephard et al., 1973). This reflects a diminution of cardiac work load (Frick, 1968,1976; Kellerman et al., 1977). Body mass may be reduced by correction of obesity, while training increases blood volume and stroke volume, allowing a lower heart rate and thus ventricular work rate (D. H. Paterson, Shephard, et al., 1979; Chap. 6). An increase in hemoglobin level further reduces the cardiac output needed for a given oxygen transport, as does an increase of peripheral oxygen extraction (Detry, Rousseau et al., 1971). Strengthening of the skeletal muscles facilitates perfusion of the active tissues, with a smaller rise of systemic blood pressure for a given power output. Finally, changes of cardiac dimensions and wall thickness (hypertrophy) reduce the intramural tension associated with a given systemic blood pressure.

Changes in Blood Coagulability

Exercise modifies the coagulability of the blood (Chap. 9). The acute response is usually an enhancement of clotting, but the more long-term reaction is an increase of fibrinolysis (Åstrup, 1973).

General Effects

Regular activity habituates the body to the sensations of vigorous exercise. If effort is demanded in an emergency, this is less of a sympathoadrenal shock to a well-trained than to a poorly-trained individual. Catecholamine output is diminished by training (Chap. 8) but cortisol output is better sustained; an increased output of growth hormone helps in the mobilization of fat while an increased secretion of thyroxine burns excess energy.

Last, an exercise class may give the coronary-prone individual a pleasant and possibly therapeutic (Eliot et al., 1976) escape from the pressures of business and domestic life, with opportunities to discover a new camaraderie and *joie de vivre* (Heinzelmann and Bagley, 1970; S. M. Fox, Naughton, and Gorman, 1972; Heinzelmann 1973).

Required Intensity of Exercise

We may note that the several potential preventive mechanisms discussed above require very different intensities of exercise. The benefits of pleasurable relaxation, camaraderie, and *joie de vivre* might well be realized through very mild exercise or, indeed, through a club devoted to some sedentary pursuit. Control of energy balance and blood lipids requires sustained, but not necessarily very vigorous, effort; again, a diet would provide a practical if less satisfactory alternative treatment. The reduction of heart rate and cardiac work load in contrast requires regular exercise to the training threshold (60–70% of aerobic power; Chap. 11). Finally, if we seek to reduce the stress of sudden unanticipated vigorous effort or to dilate the coronary collaterals, it will be necessary to exercise to the rather dangerous level where myocardial oxygen lack is induced.

Exercise for the Postcoronary Patient

Benefits of Secondary Prophylaxis

As with primary prophylaxis, there is little hard evidence of the value of exercise to the postcoronary patient. It can be shown that fatal and nonfatal recurrences of infarction are 50–70% less frequent among an exercise class than in the average patient who does not receive such treatment (Shephard, 1979e,h), but attempts to carry out formal controlled trials (Kentala, 1972; Rechnitzer et al., 1975; Wilhelmsen, Sanne et al., 1975; Naughton, 1978) have been inconclusive due to (1) an insufficient period of observation, (2) a high dropout rate, (3) some contamination of the control group with an interest in exercise, (4) a less than expected rate of reinfarction, and (5) simultaneous changes in other health habits.

It remains difficult to avoid the suspicion that a fair part of the apparent benefit in exercised patients is due to (1) selection of individuals with a better than average prognosis and (2) indirect benefits of the exercise class, such as advice on smoking and diet and the value of a more disciplined approach to life. The main justification of the postcoronary exercise class at present lies not in a possible increase of longevity among its members, but rather in the enjoyment of an improved quality of life. Given the choice, most coronary victims would undoubtedly prefer 4 years of full activity to 5 years of a very restricted life.

Immediate Treatment

At one time it was feared that premature activity might lead to a rupture of the heart at the site of infarction. It is now recognized that such occurrences are rare. The main hazards are cardiac failure and the development of an arrhythmia, and if the first 24 hr is passed safely, the prognosis improved rapidly.

Unless the patient shows signs of heart failure, shock, intractable pain, or uncontrolled arrhythmia, exercise may be commenced within 24 hr of infarction (Acker, 1973; Brock, 1973; N. K. Wenger, 1973; Zohman, 1973). Over the next week, effort is restricted to a maximum of ~ 10.5 kJ • min^{-1}. This initially includes such items as sitting at 45° in bed, feeding oneself, and carrying out light physiotherapeutic exercises to individual muscle groups. By the end of the week, the patient is sitting in a chair for three 1-hr periods per day.

In the second and third weeks, activity is gradually increased to occasional peaks of 16 kJ • min^{-1}. The patient undertakes such tasks as washing, cleaning his teeth, and light craft work while preparing himself for home care. Walking is commenced in the second week, and by the third week amounts to 30 m per trip. Each new stage in the activity program is tried before rather than after a meal, and the electrocardiographic response is carefully watched for signs of an adverse reaction.

In the fourth week, the patient may be transferred to a convalescent facility (Brusis, 1977), and at this stage he should be walking 0.8–1.0 km at a stretch. Within 8 weeks, he should reach his preinfarction level of activity and be prepared to resume most types of work. However, much further progress can be made through specific outpatient exercise rehabilitation classes.

Subsequent Treatment

Even 8 weeks after infarction, there is still some risk that exercise may precipitate ventricular fibrillation or infarction. This risk is minimized by regular laboratory evaluation. Such tests enable the supervising physician to monitor accurately the rate of improvement in cardiorespiratory fitness and to define a level of physical activity that can be sustained without dysrhythmias or excessive ST segmental depression (Shephard, 1979g). A patient of average intelligence can be quoted a "safe" walking distance and speed and he can be trained to count his immediate postexercise heart rate accurately (Pollock, 1973). He may also learn to recognize premature ventricular contractions and angina and can be counselled to halt exercise briefly if such symptoms develop (Chap. 11).

A physician should be present when the safe limit of exercise is defined or revised. For the first year, it is helpful for the patient to attend at least one physician-supervised exercise class per week. This allows observation of any symptoms and gives the individual a clear picture of his capacity. Thereafter, progress can be monitored by diary sheets with a reinforcing group session once in 4 or 8 weeks. The younger postcoronary patients can eventually return to very active lives, to the point where their exercise response does not differ from that of a person with a normal myocardium. A substantial number have indeed participated in marathon runs, some with times of a little over 3 hr (Kavanagh, Shephard, and Kennedy, 1977). On the other hand, older patients with more extensive disease of the myocardium and/or greater occlusion of the coronary vessels sometimes show a very disappointing response despite several years of conscientious training (Kavanagh, Shephard et al., 1973).

Other Forms of Disease

We have seen (Chap. 11) that prolonged bed rest or

inactivity following bone injury is accompanied by a loss of aerobic power, muscle wasting, decalcification of bones, and an increase of subcutaneous fat. All of these defects can be corrected by a suitable progressive training regimen during the period of convalescence.

Exercise has its advocates in the treatment of other medical conditions. It is an effective method for the treatment of obesity (see above) and is also helpful to the diabetic (Layani, 1978). If the middle-aged and overweight diabetic can be persuaded to undertake regular physical activity, then he will require a smaller dosage of insulin (Devlin, 1963; Perley and Kipuis, 1965; Grodsky and Benoit, 1967; Soll et al., 1975). Exercise may help to dispel the anxieties of neurotic patients (Folkins et al., 1973; Sidney and Shephard, 1977e). Some authors have claimed that training leads to a dramatic improvement in patients suffering from chronic obstructive lung disease (W. F. Miller and Taylor, 1962; Pierce, Paez, and Miller, 1965; Paez et al., 1967; J. S. Alpert et al., 1974; H. Bass, 1974). Certainly, the duration of slow treadmill walking is greatly extended (Woolf and Suero, 1969; H. Bass et al., 1970; Nicholas et al., 1970), but it is less clear as to whether this is due to an improvement of cardiac status or whether it merely reflects improved posture, more efficient walking and breathing, and a breaking of the vicious cycle of inactivity, breathlessness, and further inactivity (Christie, 1968; Shephard, 1976c; Mertens et al., 1978). Many patients with chronic chest disease are in a detrained state, and it is likely that a proportion of such patients would respond favorably to an increase of habitual physical activity. However, the right heart is already under considerable strain in many cases of chronic obstructive lung disease, and it is then more surprising that an increase of physical activity is beneficial (H. Bass et al., 1970; Brundin, 1975).

TABLE 13.17
Nature of Activity Being Undertaken at Time of Myocardial Infarction

Activity	Percent of Sample	Anticipated Percentage*
Sleeping	17.3	33.4
Working	11.2	29.2
Relaxing at home	21.4	16.6–20.8
Driving	8.2	5.4
Dining	7.1	5.4
Personal toilet	3.1	4.2
Odd jobs	10.2	2.0
Sport and vigorous activity	9.2	2.0
Walking	8.2	1.2
Other	4.1	0.4

*Based on normal pattern of daily living.

Source: Based on information obtained by retrospective questioning of survivors, Kavanagh and Shephard, 1973.

Risks of Physical Activity

Myocardial Infarction and Cardiac Arrest

Cardiac problems are a rare occurrence when a normal adult undertakes vigorous physical activity. Nevertheless, reports from certain very long distance races suggest a small (three- to fourfold) increase of cardiac emergencies relative to anticipated statistics for the resting state (Shephard, 1975b). This view is supported by retrospective questioning of the survivors of myocardial infarction (Kavanagh and Shephard, 1973; Table 13.17). Haskell (1978) studied the experience of 30 rehabilitation programs at 103 locations. His survey covered 13,570 participants and 1,629,634 patient-hours; he encountered 50 cardiac arrests, with 42 resuscitations, 7 myocardial infarctions, 2 of which were fatal, and 4 fatalities due to acute pulmonary disorders. The occurrence rate was thus 1 nonfatal episode per 34,673 patient-hours and 1 fatal episode per 116,402 patient-hours. Haskell commented that the incidence of emergencies was lower in situations where a physician was monitoring the ECG and was also lower if information prior to 1970 was discarded. Plainly, any large center that is arranging exercise programs for the coronary-prone middle-aged adult must anticipate a cardiac emergency every few years, and staff should be trained to deal with such a situation.

Factors that increase the vulnerability of the myocardium during exercise include (1) secretion of catecholamines, (2) increase of serum potassium, and (3) increase of cardiac work load. The majority of episodes are sudden electrical emergencies (ventricular fibrillation or asystole), although it is also possible to precipitate myocardial infarction during exercise through (1) hemorrhage into an atherosclerotic plaque, (2) rupture of a plaque, with lodging of the resultant embolus in a distal part of the coronary arterial tree, and (3) relative hypoxia of the myocardium beyond a point of narrowing in a coronary vessel. A surprisingly high proportion of problems arise during the recovery period (for instance, when showering; McDonough and Bruce, 1969; Shephard, 1974c). This is partly because of a phase lag in the factors increasing myocardial irritability. There may also be the sudden shock of a shower that is too hot or too cold. Diastolic pressure and thus coronary blood flow falls due to a cessation of the venous pump (Chap. 6; relaxation of the muscles) and wide dilation of the capacity vessels in a humid changing area. Finally, a reduction of heart rate gives more opportunity for the appearance of abnormal cardiac rhythms.

Among postcoronary patients, the most vulnerable individuals are those with angina (Parker et al., 1966) and an exercise-induced depression of the ST segment of the electrocardiogram (Shephard, 1979i).

Other possible cardiovascular problems include heart failure (secondary to myocardial oxygen lack or valvular disease) and rupture of an aneurysmal swelling of the heart, great vessels, or vessels at the base of the brain.

TABLE 13.18
Number of Days Participation in Sport Before an Injury is Sustained

Sport	Number of Days
Wrestling	50
Canadian football	51
Rugby football	54
Basketball	107
Association football	140
Hockey	190
Squash	253
Boxing	375
Swimming	1399
Rowing	7865

*All injuries were of sufficient severity to require treatment by the attendant surgeon.

Source: Based on data of MacIntosh et al., 1972.

Other Adverse Effects of Exercise

Musculoskeletal injuries can be a problem in exercise programs for the middle-aged coronary-prone or post-coronary patient (Kilbom, Hartley et al., 1969; G. V. Mann, Garrett et al., 1969). The incidence of such episodes is reduced by (1) an adequate warm-up and warm-down (Chap. 3), (2) a modest initial level of training, and (3) a gradual increase of training.

Among younger participants, the average number of days to injury depends upon the type of activity that is undertaken (Table 13.18), the skill of the competitor, the level of competition, fatigue, and a host of environmental factors (MacIntosh et al., 1972); however, in many sports the average injury rate is undesirably high.

Other adverse consequences of physical activity include occasional deaths from drowning, cold exposure, and heat stress. Serious infectious diseases such as poliomyelitis and typhoid fever may be contracted while swimming in infected water, and there is an increased risk of certain minor infections of the skin, external ear, and conjunctiva. Exercise may also exacerbate a preexisting infection. For example, increased respiration can activate a quiescent tuberculous lesion of the lungs, while if the polio virus is circulating in the bloodstream, it may become localized in the active ventral horn cells. Most of these problems can be avoided if appropriate safety and hygienic measures are adopted (Shephard, 1972d).

CHAPTER 14
Miscellaneous Topics

DIURNAL RHYTHMS

Many physiological variables show a regular 24-hr (diurnal or circadian) rhythm (Bunning, 1964; Guberan et al., 1969; J. N. Mills, 1973a,b). Some of these rhythms are inherent, endogenous characteristics that persist (1) in the absence of external time cues, (2) in the face of a phase shift in external cues, (3) in the presence of false time cues, and (4) in constant environmental conditions (for example, no fluctuations of light or temperature). Other rhythms are secondary to external environmental change or internal disturbances brought about by such factors as a diurnal variation of physical activity.

The heart rate shows a circadian variation during both rest (Kleitman and Kleitman, 1953; P. R. Lewis and Lobban, 1957; Howitt et al., 1966) and submaximal exercise (Voigt et al., 1967; Klein et al., 1968; Crockford and Davies, 1969; C. T. M. Davies and Sargeant, 1975a). However, the maximum heart rate remains constant. These findings probably reflect diurnal variations in (1) physical activity, (2) arousal, and (3) core temperature. There are consequent changes of up to 25% in both the physical working capacity at a heart rate of 170 beats per min and the predicted maximum oxygen intake. The minimum exercise heart rate (and thus maximum PWC_{170}) occurs between 2 and 4 a.m., while the maximum exercise heart rate (and minimum PWC_{170}) is seen between 4 and 6 p.m. There is much less variation in the directly measured oxygen intake than in the PWC_{170}, although Wojtezak-Jaroszowa and Banaskiewicz (1974) noted a 5% larger maximum oxygen intake in the morning (9 a.m.-1 p.m.) than at night (1-5 a.m.).

The core temperature shows a diurnal variation of ~0.5°C (Kleitman and Kleitman, 1953; Glaser and Shephard, 1963; Cabanac et al., 1976). Peak values are reached soon after noon. This inevitably reduces the tolerance of prolonged work, particularly in a hot environment. Thresholds for sweating and vasodilation are also higher by day than at night (C. B. Wenger et al., 1976). The diurnal rhythm of muscle strength has already been discussed.

At the cellular level, many tissue constituents, such as glycogen, show a 24-hr rhythm (Conlee et al., 1976). However, this is probably secondary to diurnal variations of physical activity.

From the practical point of view, most circadian changes are relatively small. Swimming performance is apparently improved by the rise of core temperature over the day (A. Rodahl et al., 1976), but performance of most activities is influenced more by diurnal changes of environmental temperature than by changes of a physiological nature. Nevertheless, the careful investigator ensures that his experimental design takes account of diurnal rhythms.

FATIGUE

Fatigue is one of the commonest complaints of both worker and athlete. It may have a physiological basis. More commonly it is of psychological origin, and occasionally it may also have a medical component.

Physiological Fatigue

Physiological fatigue can be demonstrated as a diminution of power output, for example, a decrease in the maximum speed of a runner or a falling work output on a bicycle ergometer. Over longer periods, such as an industrial day, it is more conveniently shown as a failure of homeostasis—a rising heart rate, respiratory minute volume, gas-exchange ratio, core temperature, and blood lactate (see, for example, the *puissance maximale supportée*; Sadoul et al., 1966; Chap. 11). However, there is a substantial individual variation in the limiting values of variables such as blood lactate, disparities between objective evidence of fatigue, and subjective feelings of tiredness (Pierson, 1963), emphasizing the wide margin between the physiologically possible and the psychologically acceptable.

Physiological problems relate to a failure of general and local circulatory homeostasis, exhaustion of the various stores discussed in Chap. 2, and a breakdown of central regulatory mechanisms.

Circulatory Homeostasis

We have noted a decline of stroke volume and a rising heart rate during sustained work (Chap. 6). This is due to such factors as (1) a dilation of the peripheral capacitance vessels, (2) a reduction of blood volume by sweating and exudation of fluid into the active tissues, and (3) possible exhaustion of the sympathetic regulatory mechanisms. Under hot environmental conditions, the body may be unable to sustain an adequate blood flow to the brain and other vital organs; this leads to mental confusion, incoordination, and eventual loss of consciousness.

More commonly, failure of circulatory homeostasis is a local phenomenon. Perfusion of the actively contracting muscles is insufficient to prevent a local accumulation of lactic acid (C. Kay and Shephard, 1969).

When the local pH drops to ~7.0, glycolysis is inhibited (Danforth, 1965; Trivedi and Danforth, 1966; Hofer and Pette, 1968) and fatigue sets in (Chap. 3). Nevertheless, fatigue cannot always be blamed upon lactate accumulation. For instance, it still occurs in McArdle's syndrome, a rare congenital anomaly whereby a lack of phosphorylase prevents glycogen breakdown. Further, slowly contracting muscles, such as the soleus, can produce substantial quantities of lactate without signs of contracture or fatigue, whereas muscles with a high percentage of fast-twitch fibers (such as the gastrocnemius and the triceps; Saltin, Henriksson et al., 1977) are exhausted if they generate an equal quantity of lactate (Brust, 1971).

An accumulation of anerobic metabolites is particularly likely if local blood flow is interrupted by isometric muscular contraction. The circulation is restricted if the muscle develops more than 15% of its maximum voluntary force, and vascular occlusion becomes complete at 70–80% of maximum voluntary force (Lind and McNicol, 1967; Chap. 6; Fig. 14.1). Local circulatory arrest also occurs during rhythmic activity if an equivalent muscular force is developed; depending on the mode of exercise, anerobic metabolites accumulate with expenditures of more than 50–80% of maximum oxygen intake (Shephard, Allen et al., 1968b; Fig. 2.6).

Interruption of local blood flow is more likely in a subject who has weak muscles than in a person who is strong. It is also more likely when a task is carried out by small muscles than when a large muscle is active. The intrinsic and extrinsic muscles of the eyes are particularly vulnerable to fatigue (J. R. Brown,

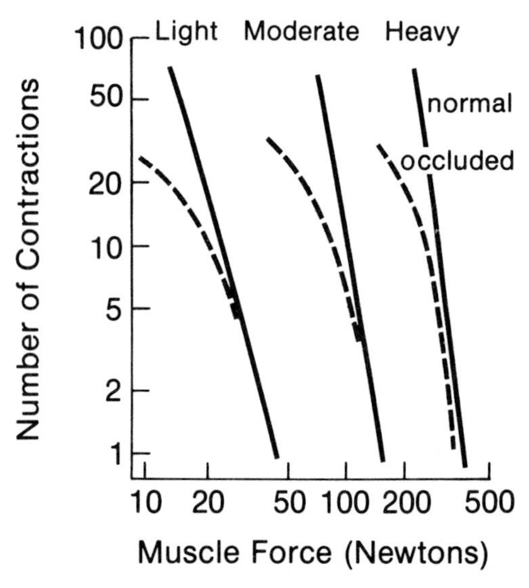

FIG. 14.1. Influence of deliberate blood flow occlusion upon force developed with light, moderate, and heavy repeated muscle contraction. (After J. C. Stevens & Krimsley, 1975.)

1964). Writer's cramp (Edholm, 1967) is another example of fatigue in small muscles; it can often be traced to application of an excessive grip force to the pen or pencil. In other instances, a heavy lever or control pedal must be held in position when it would be better to arrange a locking device or servo mechanism to assist the muscles (McCormick, 1957; Shephard, 1974b). Often there is associated postural strain; isometric effort is spent supporting the body mass or an external load (Sato, 1966; K. Jorgensen, 1970) when it would be better to eliminate the need for such effort by providing a chair or altering the height of a workbench (I. Åstrand, 1971; Chaffin, 1973; Shephard, 1974b).

The maximum oxygen intake is commonly used as a measure of the circulatory reserve available for rhythmic work. While an effort demanding 100% of maximum oxygen intake can be sustained for a few minutes, most subjects find difficulty in operating at more than 75% of maximum oxygen intake for an extended period such as a marathon contest (Shephard and Kavanagh, 1975a; Costill, 1976); indeed, one unusual feature of the successful long-distance runner is an ability to operate for a long period at a high percentage of maximum aerobic power. In industry, the objective is to avoid excessive fatigue, and quite conservative limits of aerobic activity are proposed. Bonjer (1968) recommended calculating the allowable power output as

$$\text{Allowable power output } (\% \text{ aerobic power}) = 32.3(\log 5700 - \log t) \quad (14.1)$$

where t is the duration of the activity in min. This corresponds to 64% of aerobic power for 1 hr, 54% for 2 hr, 45% for 4 hr, and 35% for 8 hr. Others (Burger, 1964; I. Åstrand, 1967b; A. L. Hughes and Goldman, 1970) have suggested a somewhat higher maximum (40–50% of aerobic power for an 8-hr day). Such recommendations translate into a heart rate ceiling of 120–130 beats per min for a young adult and 110 beats per min for an older worker.

The appropriate limit is inevitably situational. A lower ceiling must be imposed if working conditions are unfavorable (for example, a hot climate, an awkward posture, overhead work, or use of small muscle groups). Nevertheless, the 40–50% figure is realistic for most types of employment. It is the pace adopted when work is self-regulated (A. L. Hughes and Goldman, 1970), and a greater rate of working is often observed in the mines of South Africa. In the latter situation, men on 12-month contracts spend 8–9 hr • day^{-1} underground. Conditions are very hot and they must use their arms to provide a substantial fraction of the required energy.

Exhaustion of Energy Reserves

Neither the immediate glucose content of the blood (~6 g) nor the potential delivery of glucose from the liver (1–2 g • min^{-1}) can satisfy the sustained energy demands of heavy work. The body thus depends heavily upon local reserves of glycogen and fat within the active muscle fibers (Chap. 2).

If near-maximal activity is performed, glycogen reserves are exhausted in 90–100 min (Hultman, 1971; Chap. 2). Subsequent transport of fat to the active cells can sustain ~50% of maximum aerobic power, but unfortunately does not provide fuel for anerobic activity. A sensation of intense weakness and fatigue thus develops whenever postural or isometric activity is needed. If anerobic activity cannot be avoided, it is undertaken at the expense of an already low blood sugar. Despite some hepatic gluconeogenesis, blood glucose readings of 50 mg • dl^{-1} and less are possible with severe, protracted effort. The brain depends upon carbohydrate for its metabolism, so that a falling blood sugar may add the problem of central nervous system fatigue to peripheral weakness of the muscle fibers (J. D. Brooke, 1978a,b).

Measurements of oxygen cost and integrated electromyograms show that movements change their characteristics with the onset of fatigue (J. G. Wells, 1963; Chaffin, 1973). The low-frequency component of the electromyogram is increased (Fig. 14.2); this may reflect increased facilitation in the spinal cord, with an attempt to recruit less fatigued muscle groups (Chaffin, 1973). Possible factors include an alteration in the synchrony of motor unit firing (Person and Libkind, 1970), recruitment of slowly firing high-threshold fibers (Denny-Brown, 1949; Scherrer et al., 1960), and alterations in the electrophysio-

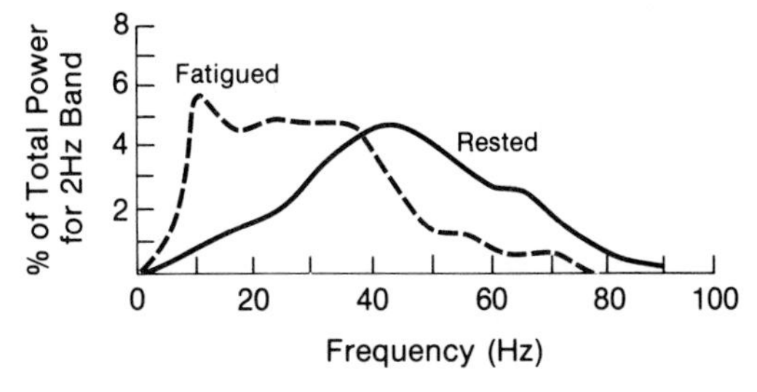

FIG. 14.2. Shift in EMG frequency spectrum with exhausting contraction. (After Chaffin, 1973.)

logical characteristics of the conducting membranes (Mortimer et al., 1970).

Facilitation of the spindle organs leads to misinterpretation of proprioceptive cues, and this exacerbates clumsiness attributable to performance of the task by unusual muscle groups. There may also be a failure of central coordination, evidenced by a speeding of motor reactions when small doses of sugar or glucose are administered.

Occasionally, problems may be traced to a lack of a key mineral ion, failure of the calcium pump (Chap. 4), or chronic loss of sodium ions and associated fluid depletion (Chap. 12). It may be significant that fatigue-relieving drugs such as caffeine increase the flux of calcium into skeletal muscle (Eberstein and Sandow, 1963).

Central Regulatory Mechanisms

Simonson (1971) reviewed other factors that can contribute to fatigue. Almost every step from the genesis of the nerve impulse to the final restoration of the status quo can be implicated.

The disease of myasthenia gravis is associated with marked weakness and fatigue. Here, the problem seems a failure of transmission at the neuromuscular junction. Excessive stimulation of a given nerve causes some depletion of transmitter substances, but it is unlikely that this contributes significantly to the fatigue of a healthy person. Animal studies show a comparable time to fatigue, whether contractions are produced via the motor nerve or by direct stimulation of the muscle (Wacholder, 1931) and the magnitude of action potentials usually remains unchanged as fatigue develops (Merton, 1954).

A second fatigue-causing pathology is Addison's disease, a deficiency in the secretions of the adrenal cortex (Chap. 8). Cortisol is necessary to maintenance of blood sugar and liver glycogen stores, while aldosterone promotes sodium retention and potassium excretion. It is thus tempting to hypothesize that prolonged work leads to exhaustion of the adrenal cortex, with consequent weakness and fatigue (Derevenco et al., 1967; Selye, 1974). In support of this view, we may cite (1) reports of diminishing cortisol secretion with very prolonged exercise, (2) correction of this trend with training (Shephard and Sidney, 1975a,b; Chap. 8), and (3) more obvious exhaustion of the adrenal cortex where men have been forced to work hard under very hot conditions (Chap. 11). In animals, administration of pitressin and aldosterone increases transcellular gradients of potassium and sodium and reduces fatigue (S. M. Friedman et al., 1963).

Influence of Training

Several of the usual responses to physical training lessen the likelihood of physiological fatigue. If there is a 20% increase of maximum oxygen intake (Chap. 11), a given task can be performed at a lower percentage of aerobic power, and this may bring the operation below the anerobic threshold. Likewise, a local increase of muscle strength reduces the percentage of maximum voluntary force that must be exerted, lessening problems of perfusing the active fibers (C. Kay and Shephard, 1969); on the other hand, more general muscular development has the opposite effect, increasing body mass and adding to postural work (Godin and Shephard, 1973b).

Bad posture often reflects a poor body image (Chap. 7). One useful by-product of increased personal fitness is thus an enhanced body image; this may bring about improvements of posture that can be reinforced by appropriate instruction (M. Turner, 1965).

Regular physical activity increases muscle glycogen reserves (Hultman, 1971). At the same time, fat provides a higher proportion of the fuel for vigorous activity (Chap. 11). It may thus become possible to complete a day's work before carbohydrate reserves are exhausted.

Several improvements of thermoregulation help to lessen fatigue, including (1) a reduction of skin blood flow (Piwonka and Robinson, 1967), (2) an increase in tone of the peripheral capacitance vessels (Holmgren, 1967b,c), and (3) a reduction in the salt content of the sweat (Ladell and Shephard, 1961).

Last, training may induce hormonal adjustments, including a more sustained secretion of cortisol during sustained activity (Shephard and Sidney, 1975a,b).

Psychological Fatigue

Psychological fatigue or mental tiredness is a very real complaint if a man is engaged in repetitive, production line work (G. C. E. Burger, 1964). It also presents as "staleness" in an athlete who has followed an exhausting training schedule (Counsilman, 1955).

While such tiredness can arise acutely, it is more often chronic. Typically, it has an emotional or situational rather than a physiological basis (Ikai and Steinhaus, 1961). Factor analyses (Kogi et al., 1970; Yoshitake, 1971; Kinsman et al., 1973) distinguish three elements: *projected fatigue*, sensed as leg weakness, shaking and aching muscles, a pounding heart, shortness of breath, and a dry mouth; *task aversion*, perceived as sweating, discomfort, and a wish to do something else; and *poor motivation*, encompassing feelings of reduced drive, vigor, and determination. The first factor has a substantial physiological component, while the second and third factors are related more simply to psychological fatigue.

Attempts have been made to quantify psychological fatigue through such statistics as a diminution in the quality or quantity of industrial output, an increased loss of usable time, absenteeism, and a high labor turnover (Lindén, 1969). Others (Bartley and Chute, 1947; Pierson, 1963) maintain that fatigue and impaired performance are distinct entities.

Occasionally, there may be evidence that a fatigued employee is overworked, but more usually the complaint is that his capacity exceeds the demands of the task he is called to perform.

The basic situation is of underarousal, with loss of vigilance (Poulton, 1970; Chap. 3). In contrast with physical tiredness, the worker demands a change rather than a rest. Although there may be some changes in the electromyogram, these are much less than can be induced by a physically exhausting task (Chaffin, 1973).

An unpleasant environment exacerbates the situation, not by any direct physiological mechanism, but rather by increasing aversion to work. In contrast, productivity is augmented by procedures that increase arousal while preserving a favorable mood (Chap. 3).

In the athlete, psychological tiredness has a similar basis, reflecting (1) boredom with repeated training sessions, (2) an aversive reaction to the associated loss of social life, and (3) an appreciation that as training continues the rewards of improving performance become progressively smaller (Shephard, 1978c).

Highly trained athletes undoubtedly habituate themselves to the discomforts of their event (L. R. T. Williams et al., 1976) and tolerate intensities of effort that would not be acceptable to an average person. On the other hand, short-term training programs do little to change perception of effort (C. Kay and Shephard, 1969; Sidney and Shephard, 1977c).

Medical Problems

Exhausting effort may induce slowly reversible tissue injuries, with liberation of pain-producing chemicals such as histamine and kinins in the connective tissue (Asmussen, 1973b). Disregarding occasional gross pathologies of the middle-aged exerciser, such as tendon ruptures and fatigue fractures of bones (Kilbom, Hartley et al., 1969; G. V. Mann, Garrett et al., 1969), there remain a number of subcellular changes. Evidence of altered membrane permeability can be found in modifications of the ionic balance of the plasma; the escape of various intracellular enzymes into the bloodstream; the appearance of proteins in the urine; ultrastructural changes in muscle, heart, and nerve (E. W. Banister, 1971; Chap. 11); and possible exhaustion of the adrenal cortex (Derevenco et al., 1967). It has been suggested that such changes are a necessary concomitant of an adaptive response to heavy exercise. Nevertheless, the dividing line that separates such phenomena from irreversible tissue injury is fine, and it would be surprising if extensive ultrastructural changes did not contribute to the fatigue syndrome.

In medical practice, fatigue is a common sequel to bed rest (Fried and Shephard, 1969,1970). It may also be a manifestation of organic disease such as anemia or carcinoma. However, at least 50% of patients who complain to their doctors of fatigue have a psychological rather than an organic problem. The commonest diagnosis is an anxiety state (Spaulding, 1964).

ERGOGENIC AIDS

Ergogenic aid is a euphemism for any procedure or drug used to augment human performance. A plethora of approaches have been tried by some competitive athletes, usually against the dictates of common sense and the rules of the sport, commonly against the rules of the state, and almost always contrary to medical advice (W. P. Morgan, 1972; M. H. Williams, 1974; Shephard, 1977a).

Modern techniques for the detection of drugs have reduced the use of stimulants, for example, the percentage of positive urine samples in Belgian cyclists decreased from 26% in 1965 to 3.5% in 1975 (Dirix, 1978). We will thus content ourselves with a brief discussion of certain procedures that are reputedly current among athletes.

Blood Doping

The concept of blood doping is simple, A sample of 250–500 ml of blood is withdrawn from an athlete and stored for a period of about a month. During this time, the blood volume and the total hemoglobin of the competitor is restored by natural processes. The blood sample is then reinfused.

Performance could theoretically be improved by either an increase of hemoglobin (the use of packed red cells) or an expansion of blood volume and total hemoglobin (Chap. 6). On the other hand, the increased red cell count reduces maximum cardiac output by its effect upon blood viscosity, while any increase of blood volume is quickly corrected by mechanisms controlling fluid balance.

To date, some experiments have shown an improvement of performance (Keroes et al., 1969; Kirschner, 1972; Ekblom, Goldbarg, and Gullbring, 1972b; Ekblom, Wilson, and Åstrand, 1976; Buick et al., 1978; Froese and Gledhill, 1979), but others have not seen any effect (B. F. Robinson et al., 1966; M. Roberts, Pirnay et al., 1972; M. H. Williams, 1974). Problems of published experiments include less than optimal techniques of blood storage (Gledhill, Buick et al., 1978; Froese and Gledhill, 1979), infusion of relatively small volumes of blood, and failure to use appropriate control procedures (M. H. Williams, 1974).

The competitive gains from blood doping have yet to be demonstrated unequivocally, and the procedure carries inevitable risks, such as infection associated with the collection, storage, and reinfusion of blood. The only available method of control is to estimate red cell metabolism; this is appreciably less than normal in the reinfused cells.

Anabolic Steroids

One objective of steroid administration is to supplement the supposed role of endogenous testosterone in the synthesis of lean tissue (W. M. Fowler et al., 1965; L. C. Johnson and O'Shea, 1969; Fahey and Brown, 1973; Ward, 1973; Stamford and Moffatt, 1974; Ariel, 1974; L. C. Johnson, Roundy et al., 1975; Fahey, Rolph et al., 1976; Chap. 8). Androgens are also reputed to augment appetite, improve general well-being, and increase competitiveness (R. V. Brooks, 1978). Massive doses of synthetic steroids (up to 300 mg; Oseid, 1976) have thus been taken by some athletes seeking an increase of muscular strength, endurance, or bulk (M. H. Williams, 1974; Shephard, 1977a,1977b,1978c).

The anabolic role of the androgens is far from clear (Chaps. 8 and 10; Fig. 10.16). The prime stimulus to protein synthesis is probably a development of tension within the active muscle fibers. Essential requirements are an adequate supply of first class protein and an appropriate training regimen. If these needs are met, it has yet to be demonstrated that the normal serum testosterone of a young man is insufficient for muscle hypertrophy to occur (Table 14.1; Ianuzzo and Chen, 1978). Indeed, because of the feedback control of pituitary gonadotrophin output by blood levels of androgen, it is improbable that a long-term increase in blood androgen levels can be induced except by massive doses of synthetic hormones. Such treatment produces a major inhibition of the interstitial cells of the testes (Kilshaw et al., 1975), dangerous toxic effect upon the liver (F. L. Johnson, 1975; Shephard, Killinger, and Fried, 1977), and (in young competitors) premature closure of epiphyses (Kochakian, 1976; Table 14.2).

Experimental administration of androgens has rarely produced more muscle development than a comparable control regimen. Occasionally, increases of mass have been observed relative to control subjects, but it has not been demonstrated clearly that the extra mass is lean tissue rather than fluid (American College of Sports Medicine, 1977).

Measures to prevent doping with anabolic steroids include (1) records of body mass and (2) unheralded inspection of urine samples for synthetic androgens. If such compounds have value to the athlete, effects are likely to persist for a long period. It is thus essential that spot checks of urine be made away from the competition site for some months prior to an event. Unfortunately, administration of testosterone itself is impossible to detect. It is a potentially carcinogenic drug and causes severe masculinization of female competitors.

TABLE 14.1
Influence of Varying Doses of Steroid (Methandrostenolone) on Development of Levator Ani Muscle Overloaded by Ipsilateral Gastrocnemius Myectomy

Steroid Dosage	Muscle Mass (Wet) (mg)		Total DNA Content (μg per muscle)		Protein Concentration (mg per g wet mass)	
(mg • kg^{-1} • day^{-1})	Control	Hypertrophied	Control	Hypertrophied	Control	Hypertrophied
0	213	334	240	672	171	139
3	219	300	176	660	165	132
6	216	291	268	568	165	156
9	219	312	232	544	170	164
18	215	324	236	620	168	153

Source: Based on data of Ianuzzo and Chen, 1978.

TABLE 14.2

Clinical Findings in Six Subjects Who Admitted Self-Administration of Danabol Intermittently in Doses of up to 20 mg• day^{-1}

Variable	Laboratory Normal	Four Subjects Not on Current Treatment				Two Subjects on Current Treatment	
Blood Composition							
Hemoglobin (g • dl^{-1})	14–16	15.5	14.9	15.4	15.5	14.4	16.8*
Hematocrit (%)	43–47	44	42*	44	45	44	49*
Liver Function							
SGOT (U)	5–19	10	35*	14	12	17	27*
Alkaline phosphatase (U)	25–90	106*	77	88	81	55	90*
Endocrine Function							
Plasma testosterone (μg • dl^{-1})	400–1000	439	362*	399*	726	104*	189*
Luteinizing hormone (μg • dl^{-1})	2–10	<0.5*	0.5*	<0.5*	<0.5*	3.4	3.1
Follicle stimulating hormone (μg • dl^{-1})	10–35	48.7*	19.6	8.1*	11.3	9.4*	8.5*
Serum thyroxine (μg • dl^{-1})	4–11	8.5	4.2	8.2	11.4*	4.6	4.1
T_3 resin uptake	25–35	31	42*	32	30	39*	43*
General Metabolism							
Blood sugar (mg • dl^{-1})	<110	122*	76	56	–	58	–
Serum cholesterol (mg • dl^{-1})	<220	228*	248*	161	206	194	188
Serum triglycerides (mg • dl^{-1})	50–145	267*	88	111	88	76	82
Serum uric acid (mg • dl^{-1})	3–7.6	8.5	3.2	5.6	–	4.5	–

*Abnormal findings.
Source: Based on data of Shephard, Killinger, and Fried, 1977.

Nerve Doping

There are reports that athletes have attempted to modify the inherent characteristics of their muscle fibers by use of selected patterns of neural stimulation. There is at present no evidence that such manipulation does as much for the muscles as a normal training program (A. W. Taylor, Kots, and Lavoie, 1978); while there is some possibility of converting type IIa to type IIb fibers by selective stimulation patterns, type I and type II fibers can only be interconverted by the drastic procedure of neural transplantation.

Hypnosis

Since the ultimate limitation of energy output is often a centrally mediated inhibition of muscular contraction, there is a possibility of augmenting performance by modification of arousal or consciousness (Ikai and Steinhaus, 1961; Mookerjee et al., 1978). Athletes have tried such techniques as hypnosis and yoga. However, there is no proof that the resultant release of inhibition is greater than that normally experienced when performing before a large crowd.

Oxygen

There is a small possibility of augmenting body stores by breathing pure oxygen before a contest. Any added oxygen content of the lungs is eliminated within a few breaths of return to ambient air breathing. However, stores in blood and tissue fluid persist for 1–2 min. Over this time, the added oxygen reserve of ~140 ml may augment brief performances (L. Hill and Flack, 1910; Karpovich, 1934; Shephard, 1977a).

The breathing of pure oxygen apparently has little effect upon recovery from vigorous exercise

(A. T. Miller, 1952; Elbel et al., 1961; Bjorgum and Sharkey, 1966; Hagerman, Bowers et al., 1968; B. A. Wilson, Hermiston et al., 1978). There are several possible explanations: (1) an increase of arterial oxygen tension reduces muscle blood flow, (2) the limiting factor in recovery from exhausting work is diffusion of lactic acid from the muscle into the bloodstream (Chap. 2), (3) the breathing of pure oxygen has a relatively small influence on arterial oxygen content, and (4) once effort has ceased, the somewhat smaller arterial oxygen content of an air-breathing subject can be made good by a greater peripheral extraction of oxygen.

SOCIAL ABUSED DRUGS

Tobacco

Endurance competitors rarely smoke, since in theory, at least, tobacco abuse has several adverse effects upon endurance performance. One survey of British athletes found that 16% were smokers, but these were mainly skill rather than endurance competitors. Participation in a recreational jogging program has much less influence upon smoking habits, but cigarette smokers who take up long-distance running generally stop smoking (P. Morgan et al., 1976).

Smoking increases the oxygen cost of breathing by causing acute bronchospasm along with more permanent changes of airway resistance (Table 14.3; Rode and Shephard, 1971; Da Silva and Hamosh, 1973; Backhause, 1975; McCarthy et al., 1976) and histamine sensitivity (Higenbottam et al., 1978). Oxygen transport per unit volume of blood is reduced by an increased burden of carboxyhemoglobin (Rode, Ross, and Shephard, 1972; Ekblom and Huot, 1972; Wald et al., 1975; L. H. Hawkins et al., 1976) and a leftward shift of the oxygen dissociation curve (Åstrup et al., 1966; Minty and Guz, 1978). The heart rate is increased both at rest and during exercise, and there is an increase of both cardiac output and systemic blood pressure (C. B. Thomas et al., 1956; Blackburn, Brozek, and Taylor, 1960; Regan, Frank et al., 1961; Irving and Yamamoto, 1963; Chevalier et al., 1963; Rode, Ross, and Shephard, 1972; Cellina et al., 1975). The work of the heart is increased, but the increase of coronary flow does not always match the added demand; this may reflect an increased secretion of vasopressin, a posterior pituitary hormone that constricts the coronary vessels (Regan, Frank et al., 1961). If the coronary vessels are already narrowed by atherosclerosis, smoking may induce ST segmental depression and anginal pain (Aronow, Cassidy et al., 1974). The combination of local hypoxia with increased concentrations of noradrenaline (Burn, 1960) and nicotine (Aronow, 1976) would be expected to lower the threshold for various dysrhythmias (including ventricular fibrillation), although McHenry, Faris et al. (1977) saw no difference in the frequency of exercise-induced premature ventricular contractions between smokers and nonsmokers. Recent experiments on beagle dogs further suggest that smoking can cause a long-term impairment of myocardial function (Ahmed et al., 1976).

Blood flow to the skin is usually diminished by smoking (Shepherd, 1963), but flow to the muscles is increased (Rottenstein et al., 1960).

TABLE 14.3
Oxygen Cost of Breathing: Influence of Cigarette Smoking

	Oxygen Cost ($ml \cdot l^{-1}$)[a]			
	Smoking Days		Nonsmoking Days	
Subject	Voluntary Hyperventilation	Dead Space Hyperventilation	Voluntary Hyperventilation	Dead Space Hyperventilation
A	15.1	12.7	11.4	6.5
B	10.3	5.4	11.3	4.8
C	7.9	7.0	7.2	5.7
D[b]	5.0	5.6	3.8	−1.6[c]
E[b]	13.5	12.4	6.1	5.7
F[b]	8.5	1.1	3.4	3.0
Mean	10.1	7.4	7.2	4.0

[a]Measured over ventilation range from 90 to 130 $l \cdot min^{-1}$ while exercising at 80% of aerobic power.
[b]Light smokers.
[c]The implication of the negative cost is that the added dead space reduced the cost of spontaneous ventilation.
Source: Based on data of Rode and Shephard, 1971.

In practice, it is not always possible to demonstrate the anticipated deterioration of endurance in either cross-sectional or longitudinal comparisons of smokers with nonsmokers. Women who smoke tend to be of a masculine, "liberated" type (Rode, Ross, and Shephard, 1972), and for this reason they may show an above-average maximum oxygen intake relative to their nonsmoking counterparts. In contrast, male smokers are typically unsure of their masculinity, with low-average values for maximum oxygen intake and endurance time (Shephard and Pimm, 1975; McHenry, Faris et al., 1977; Table 14.4). In both sexes, the problem is further clouded by a relationship between abstention from cigarettes and health consciousness, nonsmokers adopting a cluster of favorable life-styles.

In longitudinal studies, smokers develop some tolerance of chronic carbon monoxide poisoning through a small increase of blood hemoglobin concentration (Styka and Penney, 1978). Withdrawal of cigarettes commonly leads to a 4- to 5-kg accumulation of body fat, and this reduces aerobic power per unit of body mass (Rode, Ross, and Shephard, 1972; Table 14.5).

Alcohol

Alcohol has acute effects upon central nervous function and the peripheral vasculature. Arousal is reduced, and in a subject who was previously overaroused, a small amount of alcohol could conceivably improve skilled performance. There have been instances where contestants such as pistol shooters have taken alcohol to improve their competitive ability (M. H. Williams, 1974; Shephard, 1978c). However, alcohol is a depressant drug, and more usually it causes a loss of judgment, reaction time, and coordination, with a deterioration of mechanical efficiency and performance (Hebbelinck, 1959,1963). Inhibitions are lessened, and in theory this could increase muscular force, depressing normal sensations of fatigue. In practice, gains of this type are rarely realized (M. H. Williams, 1974); for example, Asmussen and Bøje (1948) found that a blood alcohol of 100 mg • dl^{-1} had no effect upon bicycle ergometer power output during either a 15-sec sprint or a 5-min all-out ride.

Cutaneous vasodilation develops with quite small doses of alcohol, but this has little effect upon ergometer performance (Schuerch et al., 1978). Blomqvist, Saltin et al. (1970; Table 14.6) studied subjects with blood alcohol readings of 90–200 mg • dl^{-1}. When submaximal exercise was performed under such conditions, there was an increase of heart rate and cardiac output, but a reduction of peripheral resistance.

Heart rate, stroke volume, arteriovenous oxygen difference, and oxygen intake were all unchanged during a few min of maximum activity, although we may suspect that the increase of peripheral pooling would have lowered the circulatory tolerance of prolonged work.

Acute overuse of alcohol impairs myocardial contractility (L. Gould, 1970), with loss of myoglobin from the skeletal muscles. Chronic alcoholism is asso-

TABLE 14.4
Influence of Smoking Behavior upon the Fitness Levels of Physical Education Students

Feature	Males: Nonsmokers	Males: Smokers	Females: Nonsmokers	Females: Smokers
Excess body mass (kg)	5.5 ± 5.8	9.9 ± 7.6	−1.1 ± 5.9	1.8 ± 5.2
Skinfold thickness (average of 3 folds, mm)	12.2 ± 4.2	19.5 ± 13.6	13.3 ± 4.0	15.3 ± 8.1
Body fat (%)	18.0 ± 4.5	21.9 ± 3.9	28.5 ± 4.0	27.7 ± 3.0
Lean mass (kg • cm^{-1})	0.34 ± 0.03	0.35 ± 0.03	0.25 ± 0.03	0.25 ± 0.02
Vital capacity (l BTPS)	5.59 ± 0.74	5.71 ± 0.86	3.94 ± 0.53	4.22 ± 0.38
One-second forced expiratory volume (l BTPS)	4.62 ± 0.63	4.46 ± 0.74	3.37 ± 0.47	3.41 ± 0.29
Maximum oxygen intake				
l • min^{-1}	4.04 ± 0.44	4.01 ± 0.53	2.34 ± 0.36	2.80 ± 0.49
ml • kg^{-1} • min^{-1}	54.7 ± 5.8	53.6 ± 4.9	39.1 ± 4.6	46.3 ± 4.8

Source: Based on data of Shephard and Pimm, 1975.

TABLE 14.5
Influence of Smoking Cessation Program on Fitness Levels*

Variable	Men: Continuing Smokers	Men: Ex-smokers	Women: Continuing Smokers	Women: Ex-smokers
Absolute aerobic power (l • min^{-1} STPD)				
Visit 1	2.58	2.61	2.24	2.16
Visit 2	2.58	2.66	2.11	2.35
Relative aerobic power (ml • kg^{-1} • min^{-1} STPD)				
Visit 1	34.7	35.0	37.7	36.4
Visit 2	34.0	33.9	34.3	35.1

*Retest data obtained 1 year after cessation of smoking.
Source: Based on data of Rode, Ross, and Shephard, 1972.

ciated with vitamin deficiencies and chronic degeneration of the heart muscle (*British Medical Journal*, 1978).

Other Drugs

Other socially abused drugs can cause a deterioration of physical performance; for a full review, see Shephard (1972d).

ENERGY COST OF SELECTED ACTIVITIES

The energy costs of various industrial and athletic pursuits are summarized in Tables 13.2 and 13.3. Such data are helpful when (1) preparing an exercise prescription and (2) calculating daily energy expenditures and thus dietary allowances. In this section, a more detailed examination is made of the costs of specific activities.

Standing

On average, a man uses ~4.8 kJ • min^{-1} of energy while lying and ~9% more when standing (Benedict and Murschhauser, 1915), the added expenditure reflecting maintenance of the upright posture. The cost of standing is lowest if the knee joints are securely locked and a large part of the body weight is transmitted directly through the long bones to the ground. Added effort is required if the center of gravity either oscillates about a neutral position (as with an excess of muscular tension) or is permanently displaced to one side of equilibrium (as with a spinal deformity or a slouching posture). We have noted (Chap. 7) that improvement of posture reduces the energy cost of many industrial, domestic, and athletic activities.

Walking

No net physical work is performed during level walking (Chap. 1). Nevertheless, energy is used in lifting and lowering the center of gravity of the body and in increasing the forward speed of individual body parts (Cavagna, Thys, and Zamboni, 1976). At moderate velocities, walking is more efficient than running (Fellingham et al., 1978). According to Margaria (1971,1976), a proportion of the potential

TABLE 14.6
Effects of Moderate Dose of Alcohol upon Hemodynamic Responses to Maximum Effort

Condition	Heart Rate	Stroke Volume	Cardiac Output	Arteriovenous Oxygen Difference	Peripheral Resistance
Seated rest	+	0	+	–	–
Submaximal exercise	+	0	+	–	–
Maximal exercise	0	0	0	0	0

Source: Based on data of Blomqvist, Saltin et al., 1970.

energy developed in raising the center of gravity of the body is reconverted to kinetic energy as the body descends (Fig. 14.3; Chap. 1).

Depending upon the hardness of the heel, the nature of the ground, and the degree of muscular control exerted during descent, ground impact transmits substantial peak accelerations (2–8 g) to the tibia (Light and McLellan, 1977).

Mechanical efficiency is at least 25% for the performance of positive work (the lifting of body mass) and about −120% for negative work (the lowering of body mass; Margaria, 1938; Fig. 14.4). The latter figure implies that the energy cost of descent is 20% of that for ascent, or 80% of the potential energy that has to be dissipated (Fig. 14.3; Chap. 1). Some of the remaining 20% of stored energy can be applied to the performance of forward movement; thus, the cost for downhill walking is lower than that for moving on the level. However, if the slope is very steep, this no longer holds, and costs become higher than for progression on the level.

If allowance is made for work performed on individual body segments, the mechanical efficiency of level walking averages 20.7%. The most economical pace is that naturally selected (Zarrugh and Radcliffe, 1978), ~4 km • hr^{-1}. At this speed, approximately equal amounts of work are performed in lifting the center of gravity of the body and in increasing the forward movement of body parts (Cavagna, Thys, and Zamboni, 1976).

Over the speed range from 3.2 to 6.4 km • hr^{-1}, the energy expenditure E of a man with a body mass of 70 kg is given by

$$E = 5.36V + 2.09 \text{ kJ} \cdot \text{min}^{-1} \qquad (14.2)$$

where V is his speed in km • hr^{-1}. As with other tasks (Godin and Shephard, 1973b), the energy expenditure is also a function of body mass. At a pace of 4.8 km • hr^{-1}:

$$E = 0.197M + 4.27 \text{ kJ} \cdot \text{min}^{-1} \qquad (14.3)$$

where M is the body mass in kg (Mahadeva et al., 1953). A lower cost per unit of body mass has sometimes (Booyens and Keatinge, 1957) but not always (Durnin and Passmore, 1967; Ralston, 1960) been found in women. This is partly because of differences in the anatomy of the hips and partly because women move less exuberantly.

The usually quoted energy costs apply to walking over ideal surfaces. Expenditures are much increased by a rough field (as in cross-country running) or by a heavy snowfall (Givoni and Goldman, 1971). There is a fivefold increase of energy costs when walking on ground covered by 45 cm of soft snow (Pandolf et al., 1976). The rate of working is also increased by head winds (Pugh, 1971,1976) and by weather conditions that require the wearing of heavy clothes. Clothing has a “hobbling” effect and also increases the mass that must be carried.

The oxygen cost usually changes but little with repetition of a laboratory walking task (Erickson et al., 1946). However, there are exceptions to this generalization. During rapid walking, a subject may learn to move his arms less vigorously. Maintenance of a fixed position on an inclined treadmill is also an acquired skill, particularly if the belt is supported by rollers (Knehr et al., 1942).

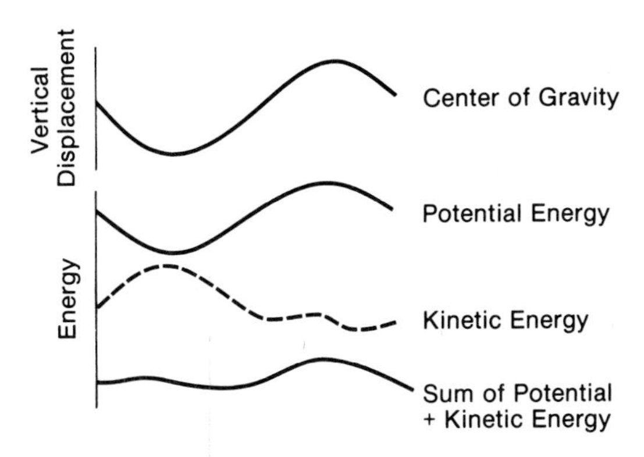

FIG. 14.3. Changes in energy content of man over stepping cycle, speed 4 km • hr^{-1}. (After Cavagna, Saibene, & Margaria, 1963.)

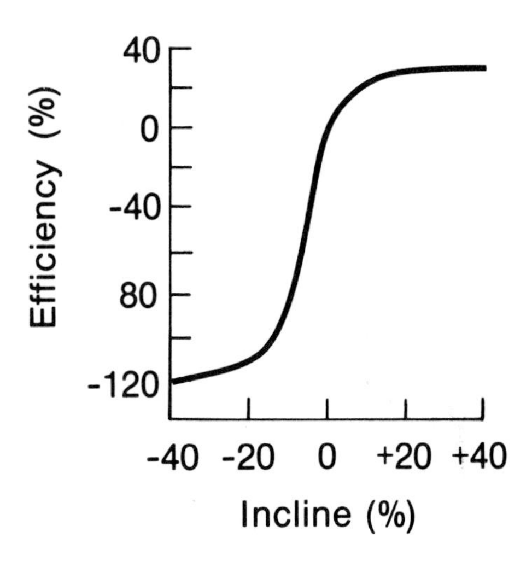

FIG. 14.4. The efficiency of walking in relation to slope. (After Margaria, 1938.)

The cost of walking rises steeply at speeds above 6 km • hr^{-1}, and a champion walker who moves at 13 km • hr^{-1} has an oxygen consumption as great as or greater than that of a man running at the same speed (Fig. 14.5). Ralston (1960) suggested a quadratic formula for estimating the energy expenditure during fast walking:

$$E = 8.49 + 0.433 (V^2) \text{ kJ} \cdot \text{min}^{-1} \quad (14.4)$$

where V is the speed in km • hr^{-1}. At 12.8 km • hr^{-1}, the calculated energy expenditure is 79.4 kJ • min^{-1}; this corresponds quite well with the oxygen consumption of an Olympic walker (4 l • min^{-1}, ~84 kJ • min^{-1}).

The best method of carrying a load depends on the terrain. On firm ground, the energy expenditure needed to carry a 20-kg pack can be used to pull a 100-kg cart (Haisman and Goldman, 1974); however, in soft sand or heavy brush the use of a back pack is the more efficient of the two approaches. A load placed on the center of the back increases oxygen cost slightly less than an equivalent increase of body mass (Zuntz and Schumberg, 1901; Brezina and Kolmer, 1912). A load applied elsewhere (for instance, a pair of heavy boots) is less well tolerated (J. Draper et al., 1953). Efficiency also deteriorates if the load exceeds 30% of body mass (Cathcart et al., 1923).

Stair Climbing

Stair climbing is of interest as one of the few current activities in most North American homes. It is the probable basis of cardiorespiratory health in the London bus conductor (Chap. 13). Prospective studies of the general population have also found an association between the daily ascent of 25 or more flights of stairs, increased fitness, and a low incidence of heart attacks (Fardy and Ilmarinen, 1975; J. N. Morris, 1975).

The overall efficiency of the usual laboratory step test is ~16% (Ryhming, 1953; Shephard, 1967b; Shephard, Allen et al., 1968b; Chap. 1). Work is performed to control descent of the body parts (for example, the triceps surae muscle contracts to prevent the foot hitting the ground too hard; W. Freedman et al., 1976). If due allowance is made for energy expended during descent (up to a third of the cost of ascent; Nagle, Balke, and Naughton, 1965), the efficiency becomes similar to that of many other large-muscle tasks (~22%).

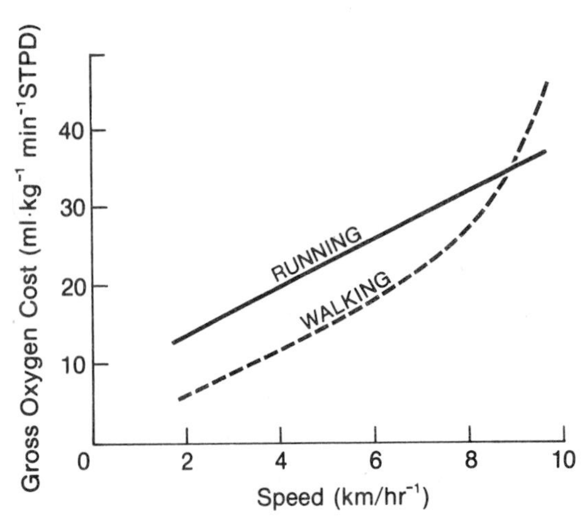

FIG. 14.5. A comparison of the oxygen cost of walking and running on a level treadmill.

Running

There are problems in calculating both the cost to the body and the work performed during a short horizontal run (Norman et al., 1976). As with walking, level running has a mechanical efficiency of zero in strictly physical terms (Chap. 1). On the other hand, if the rate of working is calculated from vertical oscillations of the trunk and changes of limb speed as assessed by cinematography, an impossibly high efficiency of 40–45% can be calculated (Margaria, 1971,1976); the explanation of such values is that kinetic energy is absorbed by the contracting leg muscles, to be released during the next stride (Cavagna, Saibene, and Margaria, 1964b; Thys et al., 1972; Komi and Bosco, 1978).

With downhill running, the efficiency rises to about −120% (see above) at a 15% slope, and is then fairly constant to at least a 35% slope (Margaria 1971,1976; Gregor and Costill, 1973). At a very steep slope (45%), an efficiency of −141% has been observed (C. T. M. Davies et al., 1974).

Friction between the foot and the ground limits running speed. The force developed per stride can be resolved into a vertical component proportional to body mass and a horizontal component proportional to speed (Fig. 14.6). Slipping of the foot is likely if the resultant of the two forces makes an angle of less than 45° with the track (Margaria, 1971). If the coefficient of friction is reduced by a dusty surface, then speed must be reduced. Conversely, if friction is increased by wearing spiked shoes, greater speeds are possible.

In moon walking, the running speed is severe-

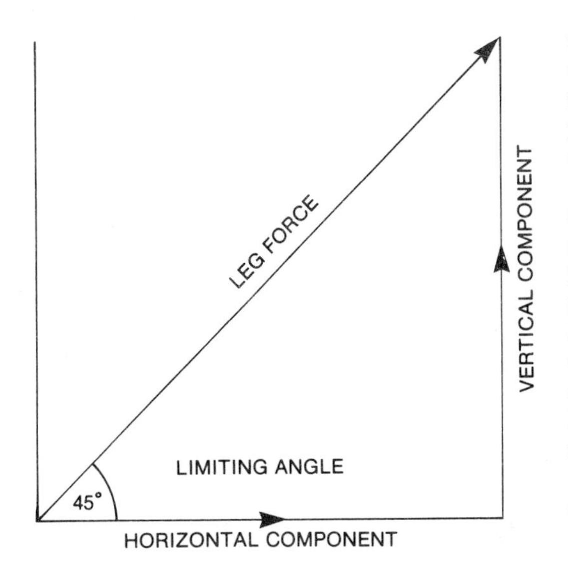

FIG. 14.6. Resolution of forces developed by the leg during running. There is a vertical component (proportional to body mass) and a horizontal component (proportional to running speed). Slipping occurs if the resultant force operates at an angle of less than 45° to the track.

ly restricted by a decrease of gravitation and thus of vertical force; slipping occurs at quite low speeds, and the most effective mode of progression becomes a form of jumping (Margaria, 1976).

It is often claimed that the energy cost of traveling a given distance is independent of speed. This concept holds for men walking or running on a treadmill at moderate speeds (Margaria, 1971,1976), but it does not apply to an athlete moving at high velocity on a track. Under the conditions of a race, wind resistance is a nonlinear function of velocity, and the energy cost of running varies as the 2.8th power of speed (Pugh, 1976). By running behind another competitor, a runner covering 1500 m in 4 min can reduce his energy expenditure by ~6%.

The oxygen cost of treadmill running is related almost linearly to speed and slope (Fig. 14.5) and is influenced little by athletic ability (Margaria, Aghemo, and Limas, 1975). However, costs are greater in children (due to dimensional factors or poor technique; P. O. Åstrand, 1952; Shephard, Lavallée et al., 1974) and in older adults (possibly due to loss of flexibility; Sidney and Shephard, 1977a).

Many subjects are unfamiliar with treadmill running. The optimum stride is shorter and faster on the treadmill than on the track, and the foot spends a larger fraction of the running cycle in contact with the "ground" (B. C. Elliott and Blanksby, 1976). In consequence, energy expenditures show a small (5–10%) decrease if the technique is learned by repeating the run on successive days (Shephard, Allen et al., 1968b; Bransford and Howley, 1977).

At moderate speeds on a level or a positive slope, the velocity is increased by a lengthening of stride. At higher speeds, there is also an increase of pace frequency (Ballreich, 1976) with a shortening of ground contact time per pace (Cavagna, Thys, and Zamboni, 1976). The optimum stride length increases with speed (Knuttgen, 1961) from 79–81 cm at 8 km • hr^{-1} to as much as 200–230 cm at 29–32 km • hr^{-1}. The maximum stride length in a 100-m race is reached when 50–60 m has been covered (M. Singh, Irwin, and Gutoski, 1978). Cavanagh et al., (1977) noted that the stride of top runners is typically 165% of leg length; hip flexion and knee flexion are also modified to reduce work against gravity. Stride frequency is limited not only by the natural frequency of the part, but also by the ability of the central nervous system to make rapidly alternating movements (P. F. Radford and Upton, 1976). Typical values increase from 170 paces per min at moderate speeds to 230 paces per min in maximum effort. While the experienced distance runner chooses a running pattern close to his optimum, the sprinter may decide to sacrifice the economy of ballistic movement to speed. The natural limb frequency is then exceeded, forced movements being executed throughout the range of leg motion. If a long-distance runner were to adopt similar tactics, he would be outclassed by competitors with equal aerobic power but a more economical technique of running.

The standing height and leg length of women average ~90% of male values. Relative to their height, women take longer strides than men (Nelson, Brooks, and Pike, 1977). In 1945, the speed of female runners averaged only 78% of that for male competitors, but by 1975 the average had increased to 91% of the male value.

Van der Walt and Wyndham (1973) suggested that the oxygen cost of running can be calculated from pace length (P, meters) and leg length (L, meters) as follows:

$$\dot{V}_{O_2} = 0.36 + 0.042\,(M) + 0.000135\,(MV^2) + 1.219\,(P) - 2.086\,(L) \qquad (14.5)$$

where M is the body mass in kg and V is the speed in km • hr^{-1}.

A well-trained distance runner maintains a very uniform speed and thus a uniform oxygen consumption over his event (Caldwell and Zauner, 1978). However, there is probably some advantage to a slight tapering of speed; this serves to maintain a constant physiological stress in the face of diminishing aerobic power (Ariyoshi, Tanaka et al., 1979; Ariyoshi, Yamaji et al., 1979.)

Bicycling

Most data refer to the use of a laboratory cycle ergometer. In recreational cycling, energy is expended against a combination of frictional resistance in the tires and drive mechanism, wind resistance, and any opposing wind force. There are also small internal losses from the acceleration and deceleration of body parts (Whitt, 1971). The average man can sustain an external power output of ~12 kN • m • min^{-1} for a few min, and perhaps 6 kN • m • min^{-1} over an 8-hr day. Let us suppose a speed of 20 km • hr^{-1}; the distance covered per min is then 0.33 km, and the force exerted is 18–36 N, depending on the duration of effort.

The frictional force varies with the mass of the cycle plus rider, the nature of the road surface, and the tire width. Estimates for recreational bicycles have been in the range 0.05–0.30 N force per kg mass (Faria and Cavanagh, 1978). Assuming a mass of 90 kg and a coefficient of 0.2 N • kg^{-1}, the frictional force would amount to 18 N. A recent study of bicycles used in Olympic events (di Prampero, Cortili et al., 1977) found a much lower rolling force (~3.3 N).

Over the first few minutes of effort, the recreational rider would have a margin of 18 N to meet wind resistance and force, but thereafter he would need to slow down, particularly if he was riding into the wind. Wind resistance is a cubic function of road speed and in a racing cyclist accounts for up to 90% of the total force exerted. The use of "drop" handlebars is thus valuable, even though it places an abnormal stress on the back muscles (Rowland and Rice, 1973). As in speed skating (see above), a large person has some competitive advantage over a smaller individual (diPrampero, Cortili et al., 1977).

The possible linkage between breathing rate and speed of pedaling is discussed in Chap. 5. The most efficient pedal speed for the recreational cyclist is ~50 rpm (Gueli and Shephard, 1976). Competitive cyclists prefer a much faster rate. Their perceived effort falls as pedal speed is increased (Karpovich and Pestrecov, 1941), and the faster movements may facilitate perfusion of the active muscles. Nevertheless, efficiency remains less than when using a slower rate of pedaling (Grosse-Lordemann and Müller, 1936; Ulmer, 1969; Löllgen, Ulmer et al., 1975).

The optimum design of machine for the sprint cyclist is discussed by Kyle and Mastropaolo (1978). Power output over a distance of 100 m is maximal with a gear of 180–200 cm. There have been suggestions that efficiency is increased by use of an elliptical chain wheel. The optimal saddle height is 103–104% of leg length (Hamley and Thomas, 1978; Shennum and de Vries, 1976). Most subjects show some asymmetry in the force developed by the two legs, although differences are not consistent from one day to the next (Daly and Cavanagh, 1976).

In Canada, bicycles can only be operated 6 months of the year, and there is thus some interest in a "snow bicycle," equivalent to the petrol-driven snowmobile. One prototype has been driven 0.5 km at a speed of 8 km • hr^{-1} (Shephard, unpublished data). However, the power requirement (240 W) is high for a single operator.

Swimming

The energy cost of various swimming strokes is illustrated in Fig. 3.34. These curves refer to skilled swimmers, and much higher costs are encountered if an inexperienced person attempts to move through the water at comparable speeds.

Depending on the relationship of lean mass to lung volumes and the percentage of body fat (Von Döbeln and Holmér, 1974), a relatively small power output is needed to stay afloat. However, on attempting to move rapidly through the water, energy is expended against viscous and turbulent water resistance, with additional effects from hydroplaning and the creation of a bow wave (Clarys, 1978).

At racing speeds, drag increases as the square of water velocity (Faulkner, 1968), with a tendency to hydroplaning between 2.1 and 7.2 km • hr^{-1} and the appearance of a bow wave at 7.2 km • hr^{-1}. The drag may be doubled by the fitting of scuba equipment; on the other hand, it is reduced if the legs can be held in a horizontal position (Pendergast et al., 1978). A lighter bone structure and greater thigh fat give women a substantial advantage in this respect, and there is also evidence that successful male swimmers have narrower leg bones than the average person (Shephard, Godin, and Campbell, 1973). The rate of working of the swimmer has traditionally been estimated roughly from the force required to pull him through the water at a steady pace (Karpovich and Pestrecov, 1939; Kent and Atha, 1973). This approach ignores such causes of energy loss as the increased drag of moving limbs (diPrampero, Pendergast et al., 1974) and inertial changes associated with the acceleration and deceleration of body parts. The resultant estimate of efficiency is ~2%, compared with 18–20% for other forms of arm work (Shephard and Simmons, unpublished data). Higher values are obtained when a loaded subject swims in a tank or flume (diPrampero, Pendergast et al., 1974; Kemper et al., 1976; Miyashita, 1978). Observations on top contestants have suggested an efficiency of 5.6–6.6% for the arm movements of free-style swimming and 3.8–5.7% for breaststroke (Holmér, 1974a,b), corresponding figures for leg movements being 1.3–2.4 and 2.4–3.8%.

In cold water, efficiency is lowered by shivering. Nadel, Holmér et al. (1974) set convective heat transfer at 230 $W \cdot m^{-2} \cdot {}^{\circ}C^{-1}$ in still water, 460 $W \cdot m^{-2}$ in moving water, and 580 $W \cdot m^{-2}$ when swimming, values for the swimmer being influenced very little by skin temperature or water speed.

Experienced swimmers develop almost 100% of their maximum oxygen intake when working with their arms (Holmér, Lundin, and Eriksson, 1974; Charbonnier et al., 1975). In submaximal work, stroke volume and heart rate for a given oxygen consumption are similar in swimming and running, but both maximum ventilation and maximum cardiac output are usually less when swimming (Dixon and Faulkner, 1971; Holmér, Stein et al., 1974; Kipke, 1978). Speed in events of 400–1500 m distance can be predicted from the formula

$$\text{Speed} = k\left[\dot{V}_{O_2}\,\text{arm} + \frac{\dot{V}_{O_2}\,\text{leg} - \dot{V}_{O_2}\,\text{arm}}{6}\right] \quad (14.6)$$

where k is a constant, and $\dot{V}_{O_2}$ arm and $\dot{V}_{O_2}$ leg are the maximum values of oxygen consumption that can be attained by use of the arms and legs, respectively.

A well-performed crawl is the most economical racing stroke. The "butterfly" causes rapid fatigue of the back and shoulder muscles, and it is an economical mode of progression only at high speeds, where water turbulence becomes an increasing problem. Differences in the energy expenditure of recreational swimmers reflect the experience of the individual more than the type of stroke he is using (Chap. 3). A poor swimmer may spend 5 times as much energy as a champion performer who is moving at the same speed (Karpovich, 1933). In particular, the experienced swimmer learns to avoid sudden accelerations and decelerations of both the arms and the body (Belokovsky and Kuznetsov, 1976; Chap. 3).

The energy cost of moving at moderate speeds is reduced somewhat if fins are worn (Goff, Brubach, and Specht, 1957).

Other Water Sports

Sailing

Sailing is often considered undemanding from the physiological point of view. Nevertheless, modern dinghy sailing demands long periods of isometric activity in the abdominal and quadriceps muscles. Competitors are favored by strength and endurance in these muscles, anerobic capacity and aerobic power, along with a high center of gravity, a well-developed sense of balance, and a resistance to mental fatigue (Niinimaa, Wright et al., 1977).

Rowing

A rower works against friction in the water and the moving parts of the boat; this component of his total energy expenditure is proportional to the third power of speed (Secher, 1973). He also encounters wind resistance, and sometimes a head wind. The speed of the boat varies about its mean value during each rowing stroke; from the viewpoint of external work, it is thus an advantage to adopt a rapid stroke rate and minimize variation in the speed of the vessel.

Rowing ergometers (Niu et al., 1966) may have some role in the winter training of oarsmen (G. R. Wright, Bompa, and Shephard, 1976). However, the forces that are encountered differ from those experienced during normal rowing (di Prampero, Cortili et al., 1971); there may also be a peripheral limitation of effort not seen in normal practice of the sport (Schwartz, 1973; G. R. Wright, Bompa, and Shephard, 1976).

The current competitive rower is a tall individual with well-developed muscles (Yamakawa and Ishiko, 1966; De Pauw and Vrijens, 1971; Hagerman, Addington, and Gaensler, 1972; G. R. Wright, Bompa, and Shephard 1976) and a large aerobic power (Klavora, 1973; Hagerman, McKirnan, and Pompei, 1975; G. R. Wright, Bompa, and Shephard, 1976; Hagerman, Connors et al., 1978). Height gives an advantage of leverage on the oars. Body mass and aerobic power are both greater in a large person. Unfortunately, a large body mass increases frictional work, but this effect is discounted by (1) the mass of the boat itself and (2) frictional losses in moving parts such as the seat and rowlocks.

The tall rower thus has a substantial advantage over his shorter counterpart. The force exerted on the oars is ~800 N (Ishiko, 1971), but as would be expected from the need to optimize force/velocity relationships and local perfusion, this amounts to no more than 40% of the maximum voluntary force for the muscle groups concerned (Secher, 1975).

Canoeing

Canoeing events include simple sprint and distance races, and white-water competitions whereby contestants must also maneuver their vessels through narrow gates of swiftly running water. The latter type of event has several points of physiological interest, including a heavy reliance upon arm work and a demand for combinations of aerobic and anerobic activity (Seliger, Pachlopniková et al., 1969). Top competitors are able to develop a high percentage of their maximum oxygen intake by using their arms (Tesch, Piehl et al., 1974; Ridge et al., 1976; dal Monte and Leonardi, 1976), and they are also able to reach high levels of blood lactate in sprint and white-water events (Tesch, Piehl et al., 1976).

An arbitrary score based upon cumulative experience and selected physiological variables (height, muscle force, vital capacity, aerobic power, anerobic capacity, and lean mass) gives a fair prediction of a white-water paddler's performance in friendly competitions, but under the stress of an international contest, unmeasured psychological variables assume much greater significance than physiological characteristics (Sidney and Shephard, 1973).

Aquabics

The performance of gymnastic exercises in water is becoming quite popular, particularly with middle-aged subjects. Because body mass is largely supported, joint injuries are less likely to occur than with comparable activity on land. The energy cost of walking and jogging at a given speed is 2–3 times as great in water as in air (B. W. Evans et al., 1978).

Winter Sports

Cross-Country Skiing

Cross-country skiing is a popular endurance sport both in Scandinavia (Vuori, 1975) and in North America. It has attracted epidemiological interest in that cross-country ski champions live several years longer than the general population (Chap. 13; Karvonen et al., 1974).

Energy is spent in overcoming wind resistance and friction at the snow surface, with additional demands for potential energy accumulated during the ascent of hills (Niinimaa, Shephard, and Dyon, 1979).

Wind resistance is a power function of speed, and depends also upon the body profile. Advantage is gained from the forward leaning posture adopted by a competitive skier. At a speed of 15 km • hr^{-1}, the drag force is typically ~8 N (Pugh, 1976). The coefficient of kinetic friction varies greatly with wax and snow conditions. Typical values range from 0.07 to 0.08 (Penniman and Jerard, 1969; Outwater, 1970). The frictional force depends also upon the mass of the body and any additional load that must be carried. Mechanical efficiency has been estimated at ~21% while skiing on a level course (Niinimaa, Shephard, and Dyon, 1979). Ascent of hills is more costly, and the oxygen consumption when moving rapidly over a cross-country course is very large (Christensen and Högberg, 1950a; Fig. 14.7).

Because of needs for a large aerobic power (Bergh, 1974; J. S. Hanson, 1975) and a low friction, success in competitive skiing (S) is predicted by the formula

$$S = 0.52\,(\dot{V}_{O_2}) + 2.94\,(Y) - 0.98\,(F) \qquad (14.7)$$

where $\dot{V}_{O_2}$ is the maximum oxygen intake in ml • kg^{-1} • min^{-1}, Y is the competitive experience in years, and F is the percentage of body fat. Sinning et al. (1977) found an average of only 7.2% body fat in male competitors.

A novice finds difficulty in developing more than 90% of his maximum oxygen intake while skiing, but an experienced racer can develop an equal or

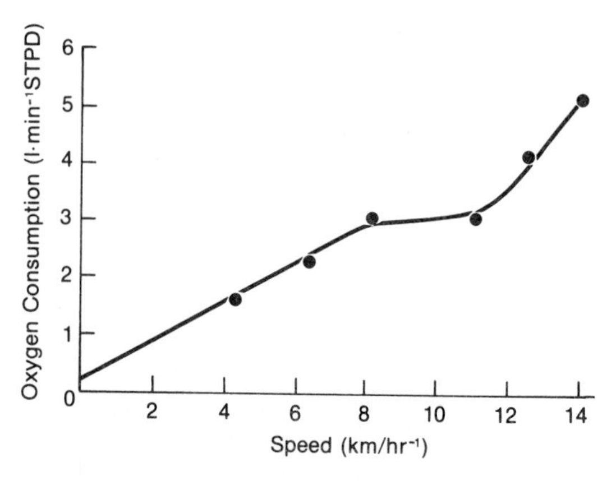

FIG. 14.7. The influence of speed upon the energy cost of cross-country skiing. (Based on data of Christensen & Högberg, 1950a.)

larger oxygen consumption on skis or roller skis than on the treadmill ((P. O. Åstrand and Saltin, 1961a; K. L. Anderson, Elsner et al., 1962; Meen et al., 1972; Bergh, 1974; Niinimaa, Shephard, and Dyon, 1979). At moderate speeds ($\sim$5 km • hr^{-1}), a laden soldier traversing level ground covered by firm snow has a similar oxygen consumption whether walking or skiing; however, perhaps because the skiing task is distributed over more muscle, it feels easier.

At higher speeds and in soft snow, skiing is substantially more economical than walking (Rönningen, 1976), at least for the experienced skier (Harkins, 1978). The current record for a 50-km ski race is a speed of $\sim$5.9 m • sec^{-1}, compared with the marathon pace of $\sim$5.5 m • sec^{-1} for a 42-km distance unencumbered by snow.

On snow that is soft and deep, snow shoes also provide a more economical method of movement than walking (Christensen and Högberg, 1950a).

Downhill Skiing

The main source of energy for the downhill skier is the potential energy developed during the chair-lift ascent. This is dissipated against aerodynamic drag and friction between the skis and the snow as in cross-country skiing, although high speeds greatly increase the drag component and kinetic friction ($\sim$0.14) is also greater for the unwaxed skis. Wind tunnel experiments by Watanabe and Ohtsuki (1977) estimated that the drag force increases from 12 N at a speed of 10 m • sec^{-1} to 256 N at 30 m • sec^{-1}. Aerodynamic forces are minimized by adoption of the "egg" posture, but drag is greatly increased by any lateral extension of the arms (Watanabe, 1978).

While the chair-lift mechanism provides most of the energy used by the downhill skier, a fast course still throws a substantial isometric load upon the legs, and champions often have very strong extensor muscles. Top competitors develop oxygen consumptions of 3–4 l • min^{-1} while skiing, although most recreational skiers have a much lower rate of metabolism (Agnevik et al., 1969). The blood lactate levels of top competitors often rise to 15 mmol • l^{-1} and figures as high as 24 mmol • l^{-1} have been recorded (E. Eriksson, Nygaard, and Saltin, 1977). There is also considerable psychic stress during slalom events; the coupling of anxiety with a substantial oxygen consumption and intense isometric contractions of the leg muscles yields briefly sustained supramaximal heart rates (Chap. 6; Agnevik et al., 1969).

Muscle biopsies show a progressive depletion of glycogen reserves over a day of skiing. It is mainly slow-twitch (type I) fibers that are emptied (E.Eriksson, Nygaard, and Saltin, 1977). This suggests that there is a substantial aerobic component to downhill skiing and implies a need to encourage aerobic training for serious competitors.

Skating

The main energy costs of skating arise from friction at the skate blades and aerodynamic drag; smaller components relate to gravitational work (raising and lowering of the body mass) and inertia (acceleration and deceleration of individual body parts).

As with walking and running, the oxygen cost of recreational skating is linearly related to the speed (Fig. 14.8). This is because the main energy loss is frictional while moving at modest velocities. Friction is particularly great if the ice surface is soft. The required force is proportional to body mass.

At the velocities encountered in speed skating, air resistance accounts for most of the energy cost (>70%). This component of work is performed with a mechanical efficiency of $\sim$11% (di Prampero, Cortili et al., 1976). The relationship between velocity and oxygen consumption now becomes curvilinear. The ideal skating pattern comprises a high stride rate, a forward lean, a low take-off angle, and placement of the recovery foot directly under the body (Marino and Dillman, 1978). Over a long-distance race, it is also an advantage to maintain as constant a speed as possible (Kuhlow, 1976).

Since the drag profile depends on body surface area (proportional to height^2) but aerobic power is a cubic function of height (Chap. 10), the champion speed skater tends to be a large person (di Prampero, Cortili et al., 1976). International competitors develop 90–95% of their treadmill maximum oxygen intake while skating (Ekblom, Hermansen, and Saltin, 1967).

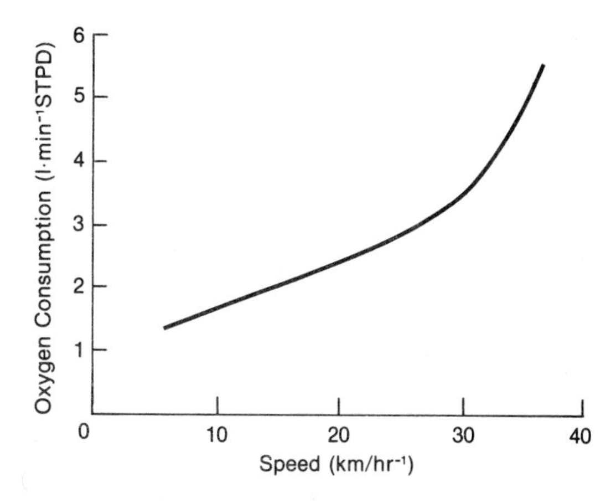

FIG. 14.8. The energy cost of skating. (Based on the data of Ekblom, Hermansen, & Saltin, 1967.)

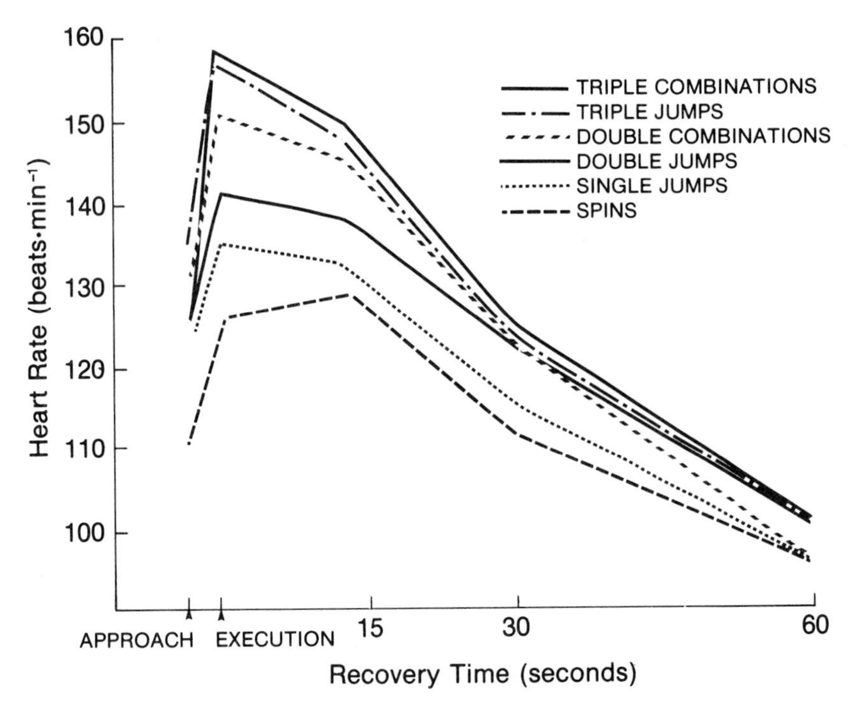

FIG. 14.9. Heart rates of champion figure skaters during approach, execution, and recovery from maneuvers of increasing difficulty (Woch et al., 1979).

In figure skating, there is a considerable emotional tachycardia immediately before a jump, the increase of heart rate reflecting the anticipated difficulty of the maneuver (Woch et al., 1979; Fig. 14.9). The aerobic demand is not particularly large (75–80% of maximum oxygen intake; Niinimaa, Woch, and Shephard, 1978). Competitors are small and light, with a relatively low aerobic power. Their one striking endowment is a well-developed lateral head of the quadriceps femoris muscle (Ross, Brown et al., 1977; Niinimaa, Woch, and Shephard, 1978).

Ball Sports

Most team sports such as hockey, association football, and North American football call for frequently repeated brief bursts of anerobic activity (Seliger, Navara, and Pachlopniková, 1970). Physiological studies of top players show only moderate levels of aerobic power (Damoiseau et al., 1966; M. Houston and Green, 1976; Withers et al., 1977) but well-developed anaerobic capacity and power (Stanescu, 1970; Seliger, Navara, and Pachlopniková, 1970; H. J. Green and Houston, 1975; Withers et al., 1977), with a tendency for depletion of glycogen reserves over the course of a game (H. J. Green, Daub et al., 1978). The traditional pattern of play in association football calls for a light and agile player (Shephard, 1975a) with a good sense of teamwork. In other sports with a strong emphasis on body checking (such as North American football), the heavy person is at an advantage (Kollias, Buskirk et al., 1972).

Other Sports and Leisure Activities

Details of the energy demands of other sports are given by Durnin and Passmore (1967). The physical and physiological characteristics of outstanding participants are discussed by Shephard (1978a,c).

References

Abbott, B. E., Aubert, X. M., and Hill, A. V. The absorption of work by a muscle stretched during a single twitch or short tetanus. Proc. Roy. Soc. Lond. B 139: 86–104, **1951**.

Abboud, F. M., Eckberg, D. L., Johannsen, U. J., and Mark, A. L. Carotid and cardiopulmonary baroreceptor control of splanchnic and forearm vascular resistance during venous pooling in man. J. Physiol. (Lond.) 286: 173–184, **1979**.

Abraham, W. H. Factors in delayed muscle soreness. Med. Sci. Sport. 9: 11–20, **1977**.

Abrahams, V. C. Neck muscle proprioceptors and a role of the cerebral cortex in postural reflexes in subprimates. Rev. Can. Biol. 31 (Suppl.): 115–130, **1972a**.

——. Neck muscle proprioceptors and vestibulospinal outflow at lumbosacral levels. Can. J. Physiol. Pharmacol. 50: 17–21, **1972b**.

Abrahams, V. C., Hilton, S. M., and Zbrozyna, A. Active muscle vasodilatation produced by stimulation of the brain stem: Its significance in the defense reaction. J. Physiol. (Lond.) 154: 491–513, **1960**.

Abram, W. P., Allen, J. A., and Roddie, I. C. The effect of pain on human sweating. J. Physiol. (Lond.) 235: 741–748, **1973**.

Acker, J. Early activity after myocardial infarction, pp. 311–314. In: Exercise testing and exercise training in coronary heart disease. Eds.: J. P. Naughton, H. K. Hellerstein, and I. C. Mohler. New York: Academic Press, **1973**.

Adal, M. N. The fine structure of the sensory region of cat muscle spindles. J. Ultrastruct. Res. 26: 332–354, **1969**.

Adal, M. N. and Barker, D. Intramuscular diameters of afferent nerve fibers in the rectus femoris muscle of the cat, pp. 249–256. In: Symposium on muscle receptors. Ed.: D. Barker. Hong Kong: Hong Kong University Press, **1962**.

——. Intramuscular branching of fusimotor fibres. J. Physiol. (Lond.) 177: 288–299, **1965**.

Adams, A. Effect of exercise upon ligament strength. Res. Quart. 37: 163–167, **1966**.

Adams, G. E., Bonner, E. A., Ribisl, P. M., and Miller, H. S. Blood pressure during heavy work on the treadmill and bicycle ergometer. Med. Sci. Sport. 10: 50, **1978**.

Adams, G. M., de Vries, H. A., Girandola, R. M., and Burren, J. S. The effect of exercise training on systolic time intervals in elderly men. Med. Sci. Sport. 9: 68, **1977**.

Adams, I. P., Imms, F. J., Prestidge, S. P., and Weff, R. Cardiovascular responses to isometric muscle contractions in patients recovering from lower limb fractures. J. Physiol. (Lond.) 284: 56p–57p, **1978**.

Adams, J. A. A closed-loop theory of motor learning. J. Motor Behav. 3: 111–150, **1971**.

Adams, J. E. Injury to the throwing arm. A study of traumatic changes in the elbow joints of boy baseball players. Calif. Med. 102: 127–132, **1965**.

Adams, W. C. Influence of exercise mode and selected ambient conditions on skin temperature. Ann. N.Y. Acad. Sci. 301: 110–127, **1977**.

Adams, W. C., Bernauer, E. M., Dill, D. B., and Bomar, J. B. Effects of equivalent sea-level and altitude training on $\dot{V}_{O_2\,max}$ and running performance. J. Appl. Physiol. 39: 262–266, **1975a**.

Adams, W. C., Fox, R. H., Fry, A. J., and MacDonald, I. C. Thermoregulation during marathon running in cool, moderate, and hot environments. J. Appl. Physiol. 38: 1030–1037, **1975b**.

Adams, W. C., McHenry, M. M., and Bernauer, E. M. Multistage treadmill walking performance and associated cardiorespiratory responses of middle-aged men. Clin. Sci. 42: 355–370, **1972**.

Adamson, G. T. and Cotes, J. E. Static and explosive muscle force: Relationship to other variables J. Physiol. (Lond.) 189: 76–77p, **1967**.

Adee, D. The effect of environmental temperature on heart rate, deep body temperature, and performance in swimming. Ph.D. thesis. Minneapolis: University of Minnesota, **1953**.

Adolph, E. F. Physiology of Man in the Desert. New York: Interscience. **1947a**.

——. Urges to eat and drink in rats. Amer. J. Physiol. 151: 110–125, **1947b**.

——. Perspectives of adaptation. Some general properties. In: Handbook of Physiology, Section 4. Adaptation to the Environment. Ed.: D. B. Dill. Washington, D.C.: Amer. Physiol. Soc., **1964**.

Adolph, E. F. and Molnar, G. W. Exchanges of heat and tolerance to cold in men exposed to outdoor weather. Amer. J. Physiol. 146: 507–537, **1946**.

Adrian, M. Cinematographic, electromyographic, and electrogoniometric techniques for analyzing movements. Exercise Sport. Sci. Rev. 1: 339–363, **1973**.

Agnevik, G., Karlsson, J., Diamant, B., and Saltin, B. Oxygen debt, lactate in blood and muscle tissue during maximal exercise in man. In: Biochemistry of Sport, pp. 62–65. Ed.: J. R. Poortmans. Baltimore, Md.: University Park Press, **1969**.

Agnevik, G., Wallström, B., and Saltin, B. A physiological analysis of alpine skiing. Cited by P. O. Åstrand and Rodahl, **1970**.

Agostoni, E. Limitations to depths of diving: Mechanics of chest wall, pp. 139–145. In: Physiology of Breath-hold diving of the Ama of Japan. Eds.: H. Rahn and T. Yokoyama. Washington, D.C.: Nat. Acad. Sci. Nat. Res. Council Publ. 1341, **1965**.

Agostoni, E. and D'Angelo, E. The effect of limb movements on the regulation of depth and rate of breathing. Resp. Physiol. 27: 33–52, **1976**.

Agostoni, E., D'Angelo, E., and Bonanni, M. V. Topography of pleural surface pressure above resting volume in relaxed animals. J. Appl. Physiol. 29: 297–306, **1970**.

Agostoni, E., Gurtner, G., Torri, G., and Rahn, H. Respiratory mechanics during submersion and negative-pressure breathing. J. Appl. Physiol. 21: 251–258, **1966**.

Agostoni, E. and Mead, J. Statics of the respiratory system, pp. 387–409. In: Handbook of Physiology: Section 3, Respiration. Vol. 1. Eds.: W. O. Fenn and H. Rahn. Washington, D.C.: Amer. Physiol. Soc., **1964**.

Agostoni, E. and Rahn, H. Abdominal and thoracic pressures at different lung volumes. J. Appl. Physiol. 15: 1087–1092, **1960**.

Agrifoglio, G., Thorburn, G. D., and Edwards, E. S. Measurement of blood flow in human lower extremity by indicator dilution method. Surg. Gynaecol. Obstetr. 113: 641–645, **1961**.

Ahlborg, B., Bergström, J., Ekelund, L. G., Guarnieri, G., Harris, R. G., Hultman, E., and Nordesjø, L. O. Muscle metabolism during isometric contraction performed at constant force. J. Appl. Physiol. 33: 224–228, **1972**.

Ahlborg, G. and Felig, P. Influence of glucose ingestion on fuel-hormone response during prolonged exercise. J. Appl. Physiol. 41: 683–688, **1976**.

Ahlborg, G., Felig, P., Hagenfeldt, R., Hendler, R., and Wahren, J. Substrate turnover during prolonged exercise in man: Splanchnic and leg metabolism of glucose, free fatty acids and amino acids. J. Clin. Invest. 53: 1080–1090, **1974**.

Ahlborg, G. and Wahren, J. Brain substrate utilization during prolonged exercise. Scand. J. Clin. Lab. Invest. 29: 397–402, **1972**.

Ahlman, K. L., Eränko, C., Karvonen, M. J., and Lepänen, V. Effects of hard competitive muscular work on electrolyte content of thermal sweat. J. Appl. Physiol. 4: 911–915, **1952**.

Ahlquist, R. P. Study of adrenotropic receptors. Amer. J. Physiol. 153: 586–600, **1948**.

Ahmed, S. S., Moschos, C. B., Lyons, M. M., Oldewurtel, H. A., Coumbis, R. J., and Regan, T. J. Cardiovascular effects of long-term cigarette smoking and nicotine administration. Amer. J. Cardiol. 37: 33–40, **1976**.

Ainsworth, M. and Eveleigh, G. W. A method of estimating lung airway resistance in humans.

Porton Technical Paper 320, **1952**.

Ainsworth, M. and Shephard, R. J. The intrabronchial distribution of soluble vapours at selected rates of gas flow, pp. 233–248. In: Inhaled particles and vapours. Ed.: C. N. Davies. Oxford: Pergamon Press, **1961**.

Aitken, R. S. and Clark-Kennedy, A. E. On the fluctuation in the composition of the alveolar air during the respiratory cycle in muscular exercise. J. Physiol. (Lond.) 65: 389–411, **1928**.

Alam, M. and Smirk, F. H. Observations in man on a pulse accelerating reflex from the voluntary muscles of the leg. J. Physiol. (Lond.) 92: 167–177, **1938**.

Albert, N. R., Gale, H. H., and Taylor, N. The effect of age on contractile protein ATPase activity and the velocity of shortening. In: Factors influencing myocardial contractility. Eds.: R. D. Tanz, F. Kavaler, and J. Roberts. New York: Academic Press, **1967**.

Albinson, J. G. and Andrew, G. M. Child in sport and physical activity. Baltimore: University Park Press, **1976**.

Albuquerque, E. X., Warnicke, J. E., Tasse, J. R., and Sansone, F. M. Effects of vinblastine and colchicine on neural regulation of the fast and slow skeletal muscle of the rat. Exp. Neurol. 37: 607–634, **1972**.

Aldinger, E. E. and Rajindar, S. S. Effects of digitoxin on the ultrastructural myocardial changes in the rat subjected to chronic exercise. Amer. J. Cardiol. 26: 369–374, **1970**.

Al-Dulymi, R. and Hainsworth, R. A new open-circuit method for estimating carbon dioxide tension in mixed venous blood. Clin. Sci. Mol. Med. 52: 377–382, **1977**.

Alexander, J. K., Hartley, L. H., Modelski, M., and Grover, R. F. Reduction of stroke volume during exercise in man following ascent to 3,100 m altitude. J. Appl. Physiol. 23: 849–858, **1967**.

Alexander, L. Treatment of shock from prolonged exposure to cold, especially in water. Washington, D.C.: Dept. of Commerce Publication Board Report 250, **1946**.

Allain, C. C., Poon, L. S., Chan, C. S., et al. Enzymatic determination of total serum cholesterol. Clin. Chem. 20: 470–475, **1974**.

Allan, E. V., Barker, N. W., and Hines, E. A. Peripheral vascular diseases. Philadelphia: Saunders, **1962**.

Allen, C. L., Brown, T. E., and O'Hara, W. J. Aerobic fitness and coronary risk factors. In: Proceedings of the first RSG4 Physical Fitness Symposium with special reference to military forces. Ed.: C. Allen. Toronto, Ont.: Defence and Civil Institute of Environmental Medicine, **1978**.

Allen, G. I. and Tsukahara, N. Cerebrocerebellar communication systems. Physiol. Rev. 54(4): 957–1006, **1974**.

Allen, J. A., Armstrong, J. E., and Roddie, I. C. The regional distribution of emotional sweating in man. J. Physiol. (Lond.) 235: 749–760, **1973**.

Allen, T. E., Smith, D. P., and Miller, D. K. Hemodynamic response to submaximal exercise after dehydration and rehydration in high school wrestlers. Med. Sci. Sport. 9: 159–163, **1977**.

Alles, G. A. and Feigen, G. A. The influence of benzedrine on work-decrement and patellar reflex. Amer. J. Physiol. 136: 392–400, **1942**.

Allman, F. L. Conditioning for Sports. In: Sports Medicine. Eds.: A. J. Ryan and F. L. Allman. New York: Academic Press, **1974**.

Allred, J. B. and Raehrig, K. R. Metabolic oscillations and food intake. Fed. Proc. 32: 1727–1730, **1973**.

Alpert, J. S., Bass, H., Szues, M. M., Banas, J. S., Dalen, J. E., and Dexter, L. Effects of physical training on hemodynamics and pulmonary function at rest and during exercise in patients with chronic obstructive pulmonary disease. Chest 66: 647–651, **1974**.

Alpert, N. R. and Root, W. S. Relationship between excess respiratory metabolism and utilization of intravenously infused sodium racemic lactate and sodium L(−) lactate. Amer. J. Physiol. 177: 455–462, **1954**.

Alston, W., Carlson, K. E., Feldman, D. J., Grimm, Z., and Gerontinos, L. A quantitative study of muscle factors in the chronic low back syndrome. J. Amer. Geriatr. Soc. 14: 1041–1047, **1966**.

Altekruse, E. B. and Wilmore, J. H. Changes in blood chemistries following a controlled exercise program. J. Occup. Med. 15: 110–113, **1973**.

Altschuler, H., Lieberson, M., and Spitzer, J. J. Effect of body weight on free fatty acid release by adipose tissue in vitro. Experientia 18: 91–92, **1962**.

Alyea, E. P., Parish, M. H., and Durham, N. C. Renal response to exercise—urinary findings. J. Amer. Med. Assoc. 167: 807–813, **1958**.

Amar, J. The human motor, pp. 129–131. New

York: Dutton, **1920**.

American Academy of Pediatrics. Report of Committee on School Health, pp. 71–75. Evanston, Ill.: Amer. Acad. Pediatrics, **1966**.

American Alliance for Health, Physical Education, and Recreation. Youth Fitness Test Manual (revised ed.). Washington, D.C.: National Education Association, **1965**.

American College of Sports Medicine. Position statement on prevention of heat injuries during distance running. Med. Sci. Sport. 7: vii–viii, **1975a**.

——. Guidelines for graded exercise testing and prescription. Philadelphia: Lea & Febiger, **1975b**.

——. Position Stand on Weight Loss in Wrestlers. Med. Sci. Sport. 8(2): xi, **1976**.

——. Position statement on the use and abuse of anabolic-androgenic steroids in sports. Med. Sci. Sport. 9: xi–xiii, **1977**.

American Heart Association. Report of Committee on Electrocardiography (Chairman C. E. Kossmann). Recommendations for standardization of leads and of specifications for instruments in electrocardiography and vectorcardiography. Circulation 35: 583–602, **1967**.

American Medical Association. Weight loss in amateur wrestling. Statement, House of Delegates, Annual Clinical Meeting, Philadelphia, **1976**.

Amery, A., Billiet, L., Conway, J., and Reybrouck, T. Comparison of cardiac output determined by a CO_2 rebreathing method at rest and during graded submaximal exercise. J. Physiol. (Lond.) 267: 34p–35p, **1977**.

Aminoff, M. J. and Sears, T. A. Spinal integration of segmental, cortical and breathing inputs to thoracic motoneurones. J. Physiol. (Lond.) 215: 577–575, **1971**.

Amphlett, G. W., Perry, S. V., Syska, H., Brown, M. D., and Vrbova, G. Cross-innervation and the regulatory protein system of rabbit soleus muscle. Nature 257: 602–604, **1975**.

Amsterdam, E. A. and Mason, D. T. Prognostic importance of exercise tests, pp. 183–193. In: Coronary Heart Disease, Exercise Testing and Cardiac Rehabilitation. Eds.: W. E. James and E. A. Amsterdam. Miami: Symposia Specialists, **1977**.

Anand, B. K. Central chemosensitive mechanisms related to feeding. Chapter 19, p. 249. In: Handbook of Physiology, Section 6: Alimentary Canal. Vol. 1. Eds.: C. F. Code and W. Heidel. Washington, D.C.: Amer. Physiol. Soc., **1967**.

Andersen, F. Registration of the pressure power (the force) of the body on the floor during movements, especially vertical jumps, pp. 87–89. Biomechanics 1, 1st Int. Seminar, Zurich, **1967**.

Andersen, H. Körner, L., Landgren, S., and Silfvenius, H. Fibre components and cortical projections of the elbow joint nerve in the cat. Acta Physiol. Scand. 69: 373–382, **1967**.

Andersen, K. L. Physical fitness: studies of healthy men and women in Norway, pp. 489–500. In: International Research in Sport and Physical Education. Eds.: E. Jokl and E. Simon. Springfield, Ill.: C. C. Thomas, **1964**.

——. The effect of physical training with and without cold exposure upon physiological indices of fitness for work. Can. Med. Assoc. J. 96: 801–803, **1967**.

Andersen, K. L., Elsner, R. E., Saltin, B., and Hermansen, L. Physical fitness in terms of maximal oxygen intake of nomadic Lapps. Fort Wainwright, Alaska: USAF Arctic Medical Laboratory Tech Rept AALTDR 61–53 pp. 1–32, **1962**.

Andersen, K. L. and Ghesquière, J. Sex differences in maximal oxygen uptake, heart rate and oxygen pulse at 10 and 14 years in Norwegian children. Human Biol. 44: 413–432, **1972**.

Andersen, K. L., Heusner, W. W., and Pohndorf, R. H. The progressive effects of athletic training on the red and white blood cells and the total plasma protein. Int. Z. Angew. Physiol. 16: 120–128, **1955**.

Andersen, K. L. and Magel, J. R. Physiological adaptation to a high level of habitual physical activity during adolescence. Int. Z. Angew. Physiol. 28: 209–227, **1970**.

Andersen, K. L., Masironi, R., Rutenfranz, J., and Seliger, V. Habitual physical activity and health. Copenhagen: World Health Organization Regional Office for Europe, **1978**.

Andersen, K. L., Shephard, R. J., Denolin, H., Varnauskas, E., and Masironi, R. Fundamentals of exercise testing. Geneva: World Health Organization, **1971**.

Andersen, P. Capillary density in skeletal muscle of man. Acta Physiol. Scand. 95: 203–205, **1975**.

Andersen, P. and Henriksson, J. Capillary supply

of the quadriceps femoris muscle of man: adaptive response to exercise. J. Physiol. (Lond.) 270: 677–690, **1977**.

Anderson, E. C. and Langham, W. H. Average potassium concentration of the human body as a function of age. Science 130: 713, **1959**.

Anderson, J. T., Grande, F., and Keys, A. Independence of the effects of cholesterol and degree of saturation of the fat in the diet on serum cholesterol in man. Amer. J. Clin. Nutr. 29: 1184–1189, **1976**.

Anderson, P. and Phillips, C. E. P. Unpublished observations cited in Phillips, C. G., Powell, T. P. S. and Wiesedanger, M. J. Physiol. (Lond.) 217: 419–446, **1971**.

Anderson, S. D. Physiological aspects of exercise: Induced bronchoconstriction. London University: Ph.D. thesis, **1972**.

Anderson, T. W. Swimming and exercise during menstruation. J. Health Phys. Educ. Recr. 36: 66–68, **1965**.

Anderson, T. W., Brown, J. R., Hall, J. W., and Shephard, R. J. The limitations of linear regressions for the prediction of vital capacity and forced expiratory volume. Respiration 25: 140–158, **1968**.

Anderson, T. W., and LeRiche, W. H. Ischaemic heart disease and sudden death, 1901–1961. Brit. J. Prev. Soc. Med. 24: 1–9, **1970**.

Anderson, T. W. and Shephard, R. J. A theoretical study of some errors in the measurement of pulmonary diffusing capacity. Respiration 26: 102–115, **1968a**.

——. Physical training and exercise diffusing capacity. Int. Z. Angew. Physiol. 25: 198–209, **1968b**.

——. The effects of hyperventilation and exercise upon the pulmonary diffusing capacity. Respiration 25: 465–484, **1968c**.

Anderson, T. W., Suranyi, G., and Beaton, G. H. The effect on winter illness of large doses of vitamin C. Can. Med. Assoc. J. 111: 31–36, **1974**.

Andrew, B. L. and Dodt, E. The deployment of sensory nerve endings at the knee joint of the cat. Acta Physiol. Scand. 28: 287–296, **1953**.

Andrew, G. M. and Baines, L. Relationship of pulmonary diffusing capacity ($\dot{D}_L$) and cardiac output ($\dot{Q}_c$) in exercise. Europ. J. Appl. Physiol. 33: 127–137, **1974**.

Andrew, G. M., Brooker, B., and Brawley, L. Effects of adult fitness classes on heart and lung functions at rest and exercise. J. Can. Assoc. Health Phys. Ed. Recr. 41: 33–37, **1974**.

Angel, A. Pathophysiology of obesity. Can. Med. Assoc. J. 110: 540–548, **1974**.

Angelin, B. Cholesterol and bile acid metabolism in normo and hyperlipoproteinaemia. Acta Med. Scand. Suppl. 610: 1–40, **1977**.

Angers, D. Modèle mécanique de fuseau neuromusculaire déefferenté; terminaisons primaires et secondaires. C. R. Séanc. Sci., Paris, 261: 2255–2258, **1965**.

Angus, G. E. and Thurlbeck, W. M. Number of alveoli in the human lung. J. Appl. Physiol. 32: 483–485, **1972**.

Anthonisen, N. R., Bradley, M. E., Vorosmarti, J., and Lineweaver, P. G. Mechanics of breathing with helium-oxygen and neon-oxygen mixtures in deep saturation diving, pp. 339–345. In: Underwater physiology. Ed.: C. J. Lambertsen. New York: Academic Press, **1971**.

Antonini, E. Structure and function of haemoglobin and myoglobin, pp. 121–137. In: Oxygen in the animal organism. Eds.: F. Dickens and E. Neil. New York: Macmillan, **1964**.

Apfelbaum, M., Bostsarron, J., and Lacatis, D. Effect of caloric restriction and excessive caloric intake on energy expenditure. Amer. J. Clin. Nutr. 24: 1405–1409, **1971**.

Appelberg, B. The effect of electrical stimulation in nucleus ruber on the response to stretch in primary and secondary muscle spindle afferents. Acta Physiol. Scand. 56: 140–151, **1962**.

Appelberg, B., Besson, P., and LaPorte, Y. Action of static and dynamic fusimotor fibres on secondary endings of cat's spindles. J. Physiol. (Lond.) 185: 160–171, **1966**.

Appelberg, B. and Jeneskog, T. Mesencephalic fusimotor control. Exp. Brain Res. 15: 97–112, **1972**.

Appelgren, L. Perfusion and diffusion in shock. A study of disturbed tissue-blood exchange in low flow states in canine skeletal muscle by a local clearance technique. Acta Physiol. Scand. (Suppl. 378): 1–72, **1972**.

Arcos, J. C., Sohal, R. S., Sun, S. C., Argus, M. F., and Burch, G. E. Changes in ultrastructure and respiratory control in mitochondria of rat heart hypertrophied by exercise. Exp. Mol. Pathol. 8: 49–65, **1968**.

Arenander, E. Hemodynamic effects of varicose veins and results of radical surgery. Acta Chir. Scand. Suppl. 260: 1–76, **1960**.

Areskog, N. H., Selinus, R., and Vahlquist, B.

Physical work capacity and nutritional status in Ethiopian male children and young adults. Amer. J. Clin. Nutr. 22: 471–479, **1969**.

Arias-Stella, J. Chronic mountain sickness: pathology and definition, pp. 61–78. In: High altitude physiology: cardiac and respiratory aspects. Eds.: R. Porter and J. Knight. Edinburgh: Churchill-Livingstone, **1971**.

Ariel, G. Prolonged effects of anabolic steroid upon muscular contractile force. Med. Sci. Sport. 6: 62–64, **1974**.

Ariyoshi, M., Tanaka, H., Kanamori, K., Obara, S., Yoshitake, H., Yamaji, K., and Shephard, R. J. Influence of running pace upon performance; effects upon oxygen intake, blood lactate and rating of perceived exertion. Can. J. Appl. Sport. Sci. 4: 210–213, **1979**.

Ariyoshi, M., Yamaji, K., and Shephard, R. J. Influence of running pace upon performance; effects upon treadmill endurance time and oxygen cost. Europ. J. Appl. Physiol. 41: 83–91, **1979**.

Armstrong, B. W., Hurt, H. H., Blide, R. W., and Workman, J. M. The humoral regulation of breathing. Science 133: 1897–1906, **1961**.

Armstrong, D. M. Synaptic excitation and inhibition of Betz cells by antidromic pyramidal volleys. J. Physiol. (Lond.) 178: 37p–38p, **1965**.

Armstrong, H. G. and Heim, J. W. Effect of flight on middle ear. J. Amer. Med. Assoc. 109: 417–421, **1937**.

Armstrong, R. B., Saubert, C. W., Sembrowich, W. L., Shepherd, R. E., and Gollnick, P. D. Glycogen depletion in rat skeletal muscle fibers at different intensities and durations of exercise. Pflügers Arch. 352: 243–256, **1974**.

Aronchick, J. and Burke, E. J. Psycho-physical effects of varied rest intervals following warm-up. Res. Quart. 48: 260–264, **1977**.

Aronow, W. S. Effect of cigarette smoking and of carbon monoxide on coronary heart disease. Chest 70: 514–518, **1976**.

Aronow, W. S. and Cassidy, J. Five years follow-up of double Master's test, maximal treadmill stress test and resting and post-exercise apexcardiogram in asymptomatic persons. Circulation 52: 616–618, **1975**.

Aronow, W. S., Cassidy, J., Vangrow, J. S., March, H., Kern, J. C., Goldsmith, J. R., Khemka, M., Pagano, J., and Vawter, M. Effect of cigarette smoking and breathing carbon monoxide on cardiovascular hemodynamics in anginal patients. Circulation 50: 340–347, **1974**.

Arstila, M. Pulse-conducted triangular exercise: ecg test. Acta Med. Scand. Suppl. 529: **1972**.

Asahina, K., Ikai, M., Ogawa, S., and Kuroda, Y. A study on acclimatization to altitude in Japanese athletes. Rev. Suisse Med. Sportive 14: 240–245, **1967**.

Asano, K., Ogawa, S., and Furuta, Y. Aerobic work capacity in middle- and old-aged runners, pp. 465–471. In: Exercise physiology. Eds.: F. Landry and W. R. Orban. Miami, Fla.: Symposia Specialists, **1978**.

Asanuma, H., Fernandez, J., Scheibel, M. E., and Scheibel, A. B. Characteristics of projections from the nucleus ventralis lateralis to the motor cortex in cats: an anatomical and physiological study. Exp. Brain Res. 20: 315–330, **1974**.

Asanuma, H. and Rosén, I. Topographical organization of cortical efferent zones projecting to distal forelimb muscles in the monkey. Expl. Brain Res. 14: 243–256, **1972**.

Asanuma, H., Stoney, S. D., and Abzug, C. Relationship between afferent input and motor outflow in cat motorsensory cortex. J. Neurophysiol. 31: 670–681, **1968**.

Ascoop, C. A. What is an abnormal ischemic ECG response to exercise? pp. 143–154. In: Coronary heart disease, exercise testing and cardiac rehabilitation. Eds.: W. E. James and E. A. Amsterdam. Miami: Symposia Specialists, **1977**.

Ashizawa, K., Takahashi, C., and Yanagisawa, S. Stature and body weight growth during adolescence based on longitudinal data of Japanese children born during World War II. J. Human Ergol. 7: 3–14, **1978**.

Ashkar, E. Circulatory changes during exercise in denervated dogs with intact splanchnic nerves. Acta Physiol. Latino-Americana 15: 351–356, **1965a**.

——. Circulatory response to exercise in vagotomized dogs with partial sympathectomy. Acta Physiol. Latino-Americana 15: 344–350, **1965b**.

Ashley, C. C., and Ridgeway, E. B. Simultaneous recording of membrane potential, calcium transient and tension in single muscle fibres. Nature (Lond.) 219: 1168–1169, **1968**.

Ashley, F. W. and Kannel, W. B. Relation of weight change to changes in atherogenic traits:

the Framingham study. J. Chron. Dis., 27: 103-114, **1974**.

Ashton, H. Critical closure in human limbs. Brit. Med. Bull. 19: 149-154, **1963**.

Askew, E. W., Huston, R. L., and Dohm, G. L. Effect of physical training on esterification of glycerol-3-phosphate by homogenates of liver, skeletal muscle, heart and adipose tissue of rats. Metabolism 22: 473-480, **1973**.

Asmussen, E. The cardiac output in rest and work in humid heat. Amer. J. Physiol. 131: 54-59, **1940**.

——. Positive and negative muscular work. Acta Physiol. Scand. 28: 364-382, **1953**.

——. Muscular Exercise. In: Handbook of Physiology, Section 3: Respiration. Part 2. pp. 939-978. Eds.: W. Fenn and H. Rahn. Washington, D.C. Amer. Physiol. Soc., **1964**.

——. Exercise and regulation of ventilation. Circ. Res. 20: 1-132, **1967**.

——. Growth in muscular strength and power. In: Physical activity, human growth and development, pp. 60-79. Ed.: G. L. Rarick. New York: Academic Press, **1973a**.

——. Fatigue and physiological capacity for work. Work-environm.-health, 10: 1-8, **1973b**.

Asmussen, E. and Bøje, O. Body temperature and capacity for work. Acta Physiol. Scand. 10: 1-22, **1945**.

——. The effects of alcohol and some drugs on the capacity for work. Acta Physiol. Scand. 15: 109-118, **1948**.

Asmussen, E. and Bonde-Petersen, F. Storage of elastic energy in skeletal muscles in man. Acta Physiol. Scand. 91: 385-392, **1974a**.

——. Apparent efficiency and storage of elastic energy in human muscles during exercise. Acta Physiol. Scand. 92: 537-545, **1974b**.

Asmussen, E. and Christensen, E. H. Einfluss der Blutverteilung auf den Kreislauf bei Körperliche Arbeit. Skand. Arch. Physiol. 82: 185-192, **1939b**.

——. Kompendium: Legemsövelsernes Specielle Teori. Copenhagen: Kobenhavns Universities Fond til Tilvejebringelse af Läremidler, **1967**.

Asmussen, E., Christensen, E. H., and Nielsen, M. Kreislaufgrösse und cortikalmotorische Innervation. Skand. Arch. Physiol. 83: 181-187, **1940**.

Asmussen, E. and Christensen, H. Die Mittelkapazität der Lungen bei erhöhtem O_2—Bedarf. Skand. Arch. Physiol. 82: 201-211, **1939a**.

Asmussen, E., Hansen, O., and Lammert, O. The relation between isometric and dynamic muscle strength in man. Comm. Test. Obs. Inst., Hellerup, Denmark 20: 3-11, **1965**.

Asmussen, E. and Heebøll-Nielsen, K. A dimensional analysis of physical performance and growth in boys. J. Appl. Physiol. 7: 593-603, **1955**.

——. Isometric muscle strength of adult men and women. Comm. Test. Obs. Inst. 11: 3-44, Hellerup, Denmark, **1961**.

Asmussen, E., Heebøll-Nielsen, K., and Molbech, S. V. Methods for evaluation of muscle strength. Comm. Test. Obs. Inst. 5: 3-13, Hellerup, Denmark. and supplement pp. 1-60, **1959**.

Asmussen, E. and Mathiasen, P. Some physiologic functions in physical education students reinvestigated after twenty five years. J. Amer. Geriatr. Soc. 10: 379-387, **1962**.

Asmussen, E. and Molbech, S. V. Methods and standards for evaluation of the physiological working capacity of patients. Comm. Test. Obs. Inst. 4 Hellerup, Denmark, **1959**.

Asmussen, E. and Nielsen, M. Studies on the regulation of respiration in heavy work. Acta Physiol. Scand. 12: 171-188, **1946**.

——. Studies on the initial changes in respiration at the transition from rest to work and from work to rest. Acta Physiol. Scand. 16: 270-285, **1948**.

——. The cardiac output in rest and work determined simultaneously by the acetylene and dye injection methods. Acta Physiol. Scand. 27: 217-250, **1952**.

——. Cardiac output during muscular work and its regulation. Physiol. Rev. 35: 778-800, **1955**.

——. Physiological dead space and alveolar gas pressures at rest and during exercise. Acta Physiol. Scand. 38: 1-21, **1956**.

Asmussen, E. and Poulsen, E. On the role of intra-abdominal pressure in relieving the back muscles while holding weights in a forward inclined position. Comm. Test. Obs. Inst., Hellerup 28, **1968**.

Asmussen, E. and Sorensen, N. The "wind-up" movement in athletics. Travail Humain 34: 147-156, **1971**.

Asmussen, E., Von Döbeln, W., and Nielsen, M. Blood lactate and oxygen debt after exhaustive work at different oxygen tensions. Acta Physiol. Scand. 15: 57-62, **1948**.

Åstrand, I. Aerobic work capacity in men and

women with special reference to age. Acta Physiol. Scand. 49 (Suppl. 169): 1-92, **1960**.

——. Blood pressure during physical work in a group of 221 women and men 48-63 years old. Acta Med. Scand. 178: 41-46, **1965**.

Åstrand, I., Chairwoman. The Scandinavian committee on ECG classification. The "Minnesota Code" for ECG classification. Adaptation to CR leads and modification of the code for ECG[s] recorded during and after exercise. Acta Med. Scand. Suppl. 481, **1967a**.

Åstrand, I. Degree of strain during building work as related to individual aerobic work capacity. Ergonomics 10: 293-303, **1967b**.

——. Circulatory responses to arm exercise in different work positions. Scand. J. Clin. Lab. Invest. 27: 293-297, **1971**.

——. ST depression, heart rate and blood pressure during arm and leg work. Scand. J. Clin. Lab. Invest. 30: 411-414, **1972**.

Åstrand, I., Åstrand, P. O., and Rodahl, K. Maximal heart rate during work in older men. J. Appl. Physiol. 14: 562-566, **1959**.

Åstrand, I., Åstrand, P. O., Christensen, E. H., and Hedman, R. Myohemoglobin as an oxygen store in man. Acta Physiol. Scand. 48: 454-460, **1960a**.

Åstrand, I., Åstrand, P. O., Christensen, E. H., and Hedman, R. Intermittent muscular work. Acta Physiol. Scand. 48: 448-453, **1960b**.

Åstrand, I., Åstrand, P. O., Hallbäck, I., and Kilböm, A. Reduction in maximal oxygen intake with age. J. Appl. Physiol. 35: 649-654, **1973a**.

Åstrand, I., Fugelli, P., Karlsson, C. G., Rodahl, K., and Vokac, Z. Energy output and work stress in coastal fishing. Scand. J. Clin. Lab. Invest. 31: 105-113, **1973b**.

Åstrand, I., Gahary, A., and Wahren, J. Circulatory responses to arm exercise with different arm position. J. Appl. Physiol. 25: 528-532, **1968**.

Åstrand, P. O. Experimental studies of physical working capacity in relation to age and sex. Copenhagen: Munksgaard, **1952**.

——. Human physical fitness with special reference to sex and age. Physiol. Rev. 36: 307-335, **1956**.

——. Breath holding during and after muscular exercise. J. Appl. Physiol. 15: 220-224, **1960**.

——. Concluding remarks. In: Proc. Int. Symp. Phys. Activ. Cardiovasc. Health. Can. Med. Assoc. J. 96: 907-911, **1967**.

——. Physical performance as a function of age. J. Amer. Med. Assoc. 205: 729-733, **1968**.

Åstrand, P. O., Cuddy, T. E., Saltin, B., and Stenberg, J. Cardiac output during submaximal and maximal work. J. Appl. Physiol., 19: 268-274, **1964**.

Åstrand, P. O., Ekblom, B., Messin, R., Saltin, B., and Stenberg, J. Intra-arterial blood pressure during exercise with different muscle groups. J. Appl. Physiol. 20: 253-256, **1965**.

Åstrand, P. O., Ekblom, B., and Goldberg, A. N. Effects of blocking the autonomic nervous system during exercise. Acta Physiol. Scand., 82: 18A, **1971**.

Åstrand, P. O., Engström, L., Eriksson, B., Karlberg, P., Nylander, I., Saltin, B., and Thorén, C. Girl swimmers. With special reference to respiratory and circulatory adaptation and gynaecological and psychiatric aspects. Acta Paediatr. (Suppl.) 147: 1-75, **1963**.

Åstrand, P. O., Hällback, I., Hedman, R., and Saltin, B. Blood lactates after prolonged severe exercise. J. Appl. Physiol. 18: 619-622, **1963**.

Åstrand, P. O. and Rodahl, K. Textbook of work physiology (2nd ed.). New York: McGraw Hill, **1977**.

Åstrand, P. O. and Ryhming, I. A nomogram for calculation of aerobic capacity (physical fitness) from pulse rate during sub-maximal work. J. Appl. Physiol. 7: 218-221, **1954**.

Åstrand, P. O. and Saltin, B. Maximal oxygen uptake and heart rate in various types of muscular activity. J. Appl. Physiol. 16: 977-981, **1961a**.

——. Oxygen uptake during the first minutes of heavy muscular exercise. J. Appl. Physiol. 16: 971-976, **1961b**.

——. Plasma and red cell volume after prolonged severe exercise. J. Appl. Physiol. 19: 829-832, **1964**.

Åstrup, P. Atherogenic compounds of tobacco smoke, pp. 156-161. In: Atherosclerosis IV. Eds.: G. Schettler, Y. Goto, Y. Hata, and G. Klose. Berlin: Springer Verlag, **1977**.

Åstrup, P., Hellung-Larsen, P., Kjeldsen, K., and Mellemgaard, K. The effect of tobacco smoking on the dissociation curve of oxyhaemoglobin. Scand. J. Clin. Lab. Invest. 18: 450-457, **1966**.

Åstrup, T. The effects of physical activity on blood

coagulation and fibrinolysis, pp. 169-192. In: Exercise testing and exercise training in coronary heart disease. Eds.: J. P. Naughton, H. K. Hellerstein, and I. C. Mohler. New York: Academic Press, **1973**.

Atkinson, P. J., Weatherell, J. A., and Weidmann, S. M. Changes in density of the human femoral cortex with age. J. Bone Joint Surg. 44: 496-502, **1962**.

Atkinson, W. B. and Elftman, H. Carrying angle of the human arm as a secondary sex characteristic. Anat. Rec. 91: 49-52, **1945**.

Atwater, A. E. Cinematographic analyses of human movement. Exercise Sport. Sci. Rev. 1: 217-258, **1973**.

Atwater, W. O. and Benedict, F. G. Metabolism of matter and energy in the human body. U.S. Dept. of Agric. Bull. 69 (cited by Durnin and Passmore, 1967), **1899**.

Aubert, X. Le couplage énergétique de la contraction musculaire. Ed. Arscia, Brussels, pp. 1-315, **1956**.

Aurell, M., Carlson, M., Grimby, G., and Hood, B. Plasma concentration and urinary excretion of certain electrolytes during supine work. J. Appl. Physiol. 22: 633-638, **1967**.

Aviado, D. M. and Schmidt, C. F. Reflexes from stretch receptors in blood vessels, heart and lungs. Physiol. Rev. 35: 247-300, **1955**.

Axelrod, J. Neurotransmitters. Sci. Amer. 230: 59-71, **1974**.

Ayalon, A. and Van Gheluwe, B. A comparison study of some mechanical variables from daily activities in elderly and young people, pp. 209-220. In: Physical Exercise and Activity for the Aging. Natanya, Israel: Wingate Institute, **1975**.

Ayotte, B., Seymour, J., and McIlroy, M. A new method for measurement of cardiac output with nitrous oxide. J. Appl. Physiol. 28: 863-866, **1970**.

Azémar, G. Sport et latéralité. Paris: Editions Universitaires, **1970**.

Babirak, S. P., Dowell, R. T., and Oscai, L. B. Total fasting and total fasting plus exercise: effects on body composition of the rat. J. Nutr. 104: 452-457, **1974**.

Bach, P. and Rita, Y. Neurophysiology of eye movements. In: The control of eye movements, pp. 7-45. Eds.: P. Bach, Y. Rita, and C. C. Collins. New York: Academic Press, **1971**.

Bachorik, P. S., Wood, P. D., Albers, J. J., Steiner, P., Dempsey, M., Kuba, K., Warnick, R., and Karlsson, L. Plasma high-density lipoprotein cholesterol concentration determined after removal of other lipoprotein by heparin/manganese precipitation or by ultracentrifugation. Clin. Chem. 22: 1828-1834, **1976**.

Backhause, C. I. Peak expiratory flow in youths with varying cigarette smoking habits. Brit. Med. J. (1): 360-362, **1975**.

Bäcklund, L. and Nordgren, L. A new method for testing isometric muscular strength under standardized conditions. Scand. J. Clin. Lab. Invest. 21: 33-41, **1968**.

Badeer, H. S. Effect of heart size on myocardial oxygen consumption. Amer. Heart J. 60: 948-954, **1960**.

——. The stimulus to hypertrophy of the myocardium. Circulation 30: 128-136, **1964**.

——. Metabolic basis of cardiac hypertrophy. Prog. Cardiovasc. Dis. 11: 53-63, **1972**.

Baertschi, A. J. and Gann, D. S. Responses of atrial mechanoreceptors to pulsations of atrial volume. J. Physiol. (Lond.) 273: 1-21, **1977**.

Baile, C. A. Putative neurotransmitters in the hypothalamus and feeding. Fed. Proc. 33: 1166-1175, **1973**.

Baile, C. A., Mayer, J., Baumgardt, B. R., and Peterson, A. Comparative gold thioglucose effects on goats, sheep, dogs, rats and mice. J. Dairy Sci. 53: 801-807, **1970**.

Bailey, D. Exercise, fitness and physical education for the growing child, pp. 13-22. In: Proceedings of the National Conference on Fitness and Health. Ed.: W. A. R. Orban. Ottawa: Information Canada, **1974**.

Bailey, D., Holmlund, B., and Orban, W. Paper presented to American College of Sports Medicine, Los Angeles, **1964**.

Bailey, D. A., Carron, A. V., Teece, R. G., and Wehner, H. Effect of vitamin C supplementation upon the physiological response to exercise in trained and untrained subjects. Int. Z. Vitaminforsch. 40: 435-441, **1970a**.

Bailey, D. A., Carron, A. V., Teece, R. G., and Wehner, H. J. Vitamin C supplementation related to physiological response to exercise in smoking and non-smoking subjects. Amer. J. Clin. Nutr. 23: 905-912, **1970b**.

Bailey, D. A., Shephard, R. J., and Mirwald, R. L. Validation of a self-administered home-test of cardio-respiratory fitness. Can. J. Appl. Sport. Sci. 1: 67-78, **1976**.

Bailey, D. A., Shephard, R. J., Mirwald, R. L., and McBride, G. A. Current levels of Canadian cardio-respiratory fitness. Can. Med. Assoc. J. 111: 25-30, **1974**.

Bailey, J. J., Itscoitz, S. B., Hirshfield, J. W., Grauer, L. E., and Horton, M. R. A method for evaluating computer programs for electrocardiographic interpretation. I. Application to the experimental IBM program of 1971. Circulation 50: 73-79, **1974a**.

Bailey, J. J., Itscoitz, S. B., Grauer, L. E., Hirshfield, J. W., and Horton, M. R. A method for evaluating computer programs for electrocardiographic interpretation. II. Application to version D of the PHS program and the Mayo clinic program of 1968. Circulation 50: 80-87, **1974b**.

Bailey, J. J., Horton, M., and Itscoitz, S. B. A method for evaluating computer programs for electrocardiographic interpretation. III. Reproducibility testing and the sources of program errors. Circulation 50: 88-93, **1974c**.

Bailin, S. and Stewart, K. J. Cardiorespiratory mobilization and endurance performance. J. Human Ergol. 5: 71-78, **1976**.

Bakan, P., Belton, J. A., and Toth, J. C. Extraversion-introversion and decrement in an auditory vigilance task. In: Vigilance: A symposium. Eds.: Buckner and McGrath. New York: McGraw Hill, **1960**.

Bake, B. Role of mechanical factors in ventilation distribution, pp. 225-229. In: Muscular exercise and the lungs. Eds.: J. A. Dempsey and C. E. Reed. Madison, Wisc: University of Wisconsin, **1977**.

Bake, B., Bjure, J., and Widimsky, J. The effect of sitting and graded exercise on the distribution of pulmonary blood flow in healthy subjects studied with the ^{133}Xenon technique. Scand. J. Clin. Lab. Invest. 22: 99-106, **1968**.

Bakerman, S. Aging life processes, p. 10, Springfield, Ill.: C. C. Thomas, **1969**.

Bakers, J. H. and Tenney, S. M. The perception of some sensations associated with breathing. Resp. Physiol. 10: 85-92, **1970**.

Balagot, R. C. and Bandelin, V. R. Comparative evaluation of some DC cardiac defibrillators. Amer. Heart J. 77: 489-497, **1969**.

Balagura, S. Wilcox, R. H., and Coscina, D. B. The effect of diencephalic lesions on food intake and motor activity. Physiol. Behav. 4: 629-633, **1969**.

Baldwin, E. Dynamic aspects of biochemistry (5th ed.). New York: Cambridge University Press, **1967**.

Baldwin, K. M., Klinkerfuss, G. H., Terjung, R. L., Molé, P. A., and Holloszy, J. O. Respiratory capacity of white, red, and intermediate muscle: Adaptive response to exercise. Amer. J. Physiol. 222: 373-378, **1972**.

Balke, B. Optimale korperliche Leistungsfahigkeit, ihre Messung und Veränderung infolge Arbeitsmüdung. Int. Z. Angew. Physiol. 15: 311-323, **1954**.

Ball, R. M., Bache, R. J., Coff, F. R., and Greenfield, J. C. Regional myocardial blood flow during graded treadmill exercise in the dog. J. Clin. Invest. 55: 43-49, **1975**.

Ballreich, R. Model for estimating the influence of stride length and stride frequency on the time in sprinting events, pp. 208-212. In: Biomechanics V. Ed.: P. V. Komi. Baltimore: University Park Press, **1976**.

Balsam, A. and Leppo, L. E. Stimulation of the peripheral metabolism of 1-thyroxine and 3,5,3',-1-triiodothyronine in the physically trained rat. Endocrinology 95: 299-302, **1974**.

Band, D. M., Cameron, I. R., and Semple, S. J. G. Effect of different methods of CO_2 administration on oscillations of arterial pH in the cat. J. Appl. Physiol. 26: 268-273, **1969**.

Bang, O. The lactate content of the blood during and after muscular exercise in man. Skand. Arch. Physiol. 74 (Suppl. 10): 49-82, **1936**.

Banister, E. W. A comparison of fitness training methods in a school program. Res. Quart. 36: 387-392, **1964**.

Banister, E. W. Energetics of muscular contraction, pp. 5-36. In: Frontiers of Fitness. Ed.: R. J. Shephard. Springfield, Ill.: C. C. Thomas, **1971**.

Banister, E. W., Brown, S. R., Loewen, H. R., and Nordan, H. C. The Royal Canadian Air Force 5BX Program. A metabolic evaluation. Med. Serv. J. Can. 23: 1237-1244, **1967**.

Banister, E. W. and Griffiths, J. Blood levels of adrenergic amines during exercise. J. Appl. Physiol. 33: 674-676, **1972**.

Banister, E. W. and Jackson, R. C. The effect of speed and load changes on oxygen intake for equivalent power outputs during bicycle ergometry. Int. Z. Angew. Physiol. 24: 284-290, **1967**.

Banister, J. and Torrance, R. W. The effects of the tracheal pressure upon flow: pressure relations in the vascular bed of isolated lungs. Quart. J. Exp. Physiol. 45: 352-367, **1960**.

Banks, R. W., Barker, D., Harker, D. W., and

Stacey, M. J. Correlation between ultrastructure and histochemistry of mammalian intrafusal muscle fibers. J. Physiol. (Lond.) 252: 16p-17p, **1975**.

Bannister, R. G., Cotes, J. E., Jones, R. S., and Meade, F. Pulmonary diffusing capacity on exercise in athletic and non-athletic subjects. J. Physiol. (Lond.) 152: 66p-67p, **1960**.

Bannister, R. G. and Cunningham, D. J. C. The effects on the respiration and performance during exercise on adding oxygen to the inspired air. J. Physiol. (Lond.) 125: 118-137, **1954**.

Barac-Nieto, M., Spurr, G. B., Lotero, H., and Maksud, M. G. Physical work capacity, endurance and undernutrition in Colombian rural dwellers. Fed. Proc. 33: 678 (abstract), **1974**.

Bárány, M. ATPase activity of myosin correlated with the speed of muscle shortening. J. Gen. Physiol. 50: 197-216, **1967**.

Bárány, M., Bárány, K., Reckard, T., and Volpe, A. Myosin of fast and slow muscles of the rabbit. Arch. Biochem. 109: 185-191, **1965**.

Bárány, M. and Close, R. I. The transformation of myosin in cross-innervated rat muscles. J. Physiol. (Lond.) 213: 455-474, **1971**.

Barath, G. Untersuchungen über die Nieren Funktion nach sportlichen Übungen und nach Einwirkung von trockener Wärme. Arbeitsphysiologie 15: 383-387, **1953**.

Barclay, J. A., Cooke, W. T., Kenney, R. A., and Nutt, M. E. Effects of water diuresis and exercise on volume and composition of urine. Amer. J. Physiol. 148: 327-337, **1947**.

Barclay, J. A. and Nutt, M. E. The effect of exercise on the composition of the urine. J. Physiol. (Lond.) 103: 21p, **1944**.

Barcroft, H. Sympathetic control of vessels in the hand and forearm skin. Physiol. Rev. 40 (Suppl.) 4: 81-92, **1960**.

——. Circulation in skeletal muscle. In: Handbook of Physiology, Section 2: Circulation, Vol. 2. Ed.: W. F. Hamilton. Washington, D.C.: Amer. Physiol. Soc., **1963**.

Barcroft, H. and Dornhorst, A. C. Blood flow through the human calf during rhythmic exercise. J. Physiol. (Lond.) 109: 402-411, **1949**.

Barcroft, H. and Swan, H. J. C. Sympathetic control of human blood vessels. London: Edward Arnold, **1953**.

Barcroft, J. The respiratory function of the blood. London: Cambridge University Press, **1914**.

Barcroft, J. and Florey, H. The effects of exercise on the vascular conditions in the spleen and the colon. J. Physiol. (Lond.) 68: 181-189, **1929**.

Barcroft, J. and Stephens, J. G. Observations upon the size of the spleen. J. Physiol. (Lond.) 64: 1-22, **1927**.

Bard, C. and Fleury, M. Manipulation de l'information visuelle et complexité de la prise de décision, pp. 63-68. In: Motor learning, sport, psychology, pedagogy and didactics of physical activity. Eds.: F. Landry and W. R. Orban. Miami: Symposia Specialists, **1978**.

Barker, D. The structure and distribution of muscle receptors, pp. 227-240. In: Symposium on Muscle Receptors. Ed.: D. Barker. Hong Kong: Hong Kong University Press, **1962**.

——. The innervation of mammalian skeletal muscle, pp. 3-15. In: Myotatic, Kinesthetic and Vestibular Mechanisms. Eds.: A. V. S. de Reuck and J. Knight. London: Churchill, **1967**.

——. Morphology of muscle receptors: In: Muscle receptors: handbook of sensory physiology. Vol. 3, Pt. 2. Ed: C. C. Hunt. New York: Springer Verlag, **1974**.

Barker, D. and Ip, M. C. The motor innervation of cat and rabbit muscle spindles. J. Physiol. (Lond.) 177: 278, **1965**.

Barker, D., Stacey, M. J., and Adal, M. N. Fusimotor innervation in the cat. Phil. Trans. Roy. Soc. B, 258: 315-346, **1970**.

Barmeyer, J. Physical activity and coronary collateral development. Adv. Cardiol. 18: 104-112, **1976**.

Barnard, R. J., Edgerton, V. R., Furukawa, T., and Peter, J. B. Histochemical, biochemical and contractile properties of red, white and intermediate fibers. Amer. J. Physiol. 220: 410-414, **1971**.

Barnard, R. J., Edgerton, V. R., and Peter, J. B. Effect of exercise on skeletal muscle. I. Biochemical and histochemical properties. J. Appl. Physiol. 28: 762-766, **1970**.

Barnard, R. J., Gardner, G. W., Diaco, N. V., MacAlpin, R. N., and Kattus, A. A. Cardiovascular responses to sudden strenuous exercise: Heart rate, blood pressure and ECG. J. Appl. Physiol. 34: 833-837, **1973**.

Barnard, R. J., Terjung, R. L., and Tipton, C. M. Hormonal involvement in the reduction of cholesterol associated with chronic exercise. Int. Z. Angew Physiol. 25: 303-309, **1968**.

Barnard, R. J. and Thorstensson, A. T. Effect of exhaustive exercise on the rat heart, pp. 431-

435. In: Metabolic adaptation to prolonged physical exercise. Eds.: H. Howald and J. R. Poortmans. Basel: Birkhauser Verlag, **1975**.

Barnes, C. D. and Pompeiano, O. Inhibition of monosynaptic extensor reflex attributable to presynaptic depolarisation of the group Ia afferent fibres produced by vibration of a flexor muscle. Arch. Ital. Biol. 108: 233–258, **1970**.

Bar-Or, O. A new anaerobic capacity test: characteristics and applications. Paper presented to 21st World Congress of Sports Medicine, Brasilia, **1978**.

Bar-Or, O. and Inbar, O. Relationships among anaerobic capacity, sprint and middle-distance running of schoolchildren, pp. 142–147. In: Physical Fitness Assessment: Principles, Practice and Application. Eds.: R. J. Shephard and H. Lavallée. Springfield, Ill.: C. C. Thomas, **1978**.

Bar-Or, O. and Shephard, R. J. Cardiac output determination in exercising children: Methodology and feasibility. Acta Paed. Scand. Suppl. 217: 49–52, **1971**.

Bar-Or, O., Shephard, R. J., and Allen, C. L. Cardiac output of 10–13 year old boys and girls during sub-maximal exercise. J. Appl. Physiol. 30: 219–223, **1971**.

Barry, A. J., Daly, J. W., Pruett, E. D. R., Steinmetz, J. R., Page, H. F., Birkhead, N. C., and Rodahl, K. The effects of physical conditioning on older individuals. I. Work capacity, circulatory-respiratory function, and electrocardiogram. J. Gerontol. 21: 192–199, **1966**.

Bartelstone, H. J. Role of the veins in venous return. Circulation Res. 8: 1059–1076, **1960**.

Bartle, S. H. and Sanmarco, M. E. Comparison of angiocardiographic and thermal washout technics for left ventricular volume measurement. Amer. J. Cardiol. 18: 235–252, **1966**.

Bartlett, D. Effects of Valsalva and Mueller maneuvers on breath-holding time. J. Appl. Physiol. 42: 717–721, **1977**.

Bartlett, R. G., Brubach, H. F., and Specht, H. Oxygen cost of breathing. J. Appl. Physiol. 12: 413–424, **1958**.

Bartley, S. H. and Chute, E. Fatigue and impairment in man. New York: McGraw Hill, **1947**.

Bartter, F. C. and Gann, D. S. On the hemodynamic regulation of the secretion of aldosterone. Circulation 21: 1016–1023, **1960**.

Basmajian, J. V. Muscles Alive. Their Functions Revealed by Electromyography. 2nd ed., pp. 1–421. Baltimore: Williams & Wilkins, **1967**.

Basmajian, J. V. Electromyographic analyses of basic movement patterns. Exercise Sport. Sci. Rev. 1: 259–284, **1973**.

Bason, R., Billings, C. E., Fox, E. L., and Gerke, R. Oxygen kinetics for constant work loads at various altitudes. J. Appl. Physiol. 35: 497–500, **1973**.

Bass, D. E. Thermoregulatory and circulatory adjustments during acclimatization to heat in man. In: Temperature: Its Measurement and Control in Science and Industry. Ed.: J. D. Hardy. New York: Reinhold, **1963**.

Bass, D. E., Buskirk, E. R., Iampietro, P. F., and Mager, H. Comparison of blood volume during physical conditioning, heat acclimatization, and sedentary living. J. Appl. Physiol. 12: 186–188, **1958**.

Bass, H. Exercise and respiratory disease. In: Sports Medicine. Eds.: A. J. Ryan and F. L. Allman. San Francisco: Academic Press, **1974**.

Bass, H., Whitcomb, J. F., and Forman, R. Exercise training: Therapy for patients with chronic obstructive pulmonary disease. Dis. Chest 57: 116–121, **1970**.

Bassenge, E., Holtz, J., and Kolin, A. Autonomic control of local venous capacity and total vascular compliance in the conscious dog. J. Physiol. (Lond.) 284: 105p–106p, **1978**.

Bassett, C. A. L. Biophysical principles affecting bone. In: The Biochemistry and Physiology of Bone. Ed.: G. H. Bourne. Vol. 3, pp. 1–68. New York: Academic Press, **1972**.

Bassey, E. J., Fentem, P. H., MacDonald, I. C., and Scriven, P. M. Self-paced walking as a method for exercise testing in elderly and young men. Clin. Sci. 51: 609–612, **1976**.

Batelli, F. and Stern, L. Oxydation des p-phenylendiamins durch die Tiergewebe. Biochem. Z. 46: 317–366, **1912**.

Bates, D. V., Boucot, G., and Dormer, A. E. The pulmonary diffusing capacity in normal subjects. J. Physiol. (Lond.) 129: 237–252, **1955**.

Baulieu, E. E. and Robel, P. Catabolism of testosterone and androstenedione. In: The androgens of the testis. Ed.: K. B. Eik-Nes. New York: Marcel Dekker, **1970**.

Baum, E., Bruch, K., and Schwennicke, P. Adaptive modifications in the thermoregulatory system of long-distance runners. J. Appl. Physiol. 40: 404–410, **1976**.

Baxter, B. A study of reaction time using factorial design. J. Exp. Psychol. 31: 430–437, **1942**.

Bazett, H. C., Love, L., Newton, M., Eisenberg,

L., Day, R., and Forster, R. Temperature changes in blood flowing in arteries and veins in man. J. Appl. Physiol. 1: 3-19, **1948**.

Bean, J. W. Effects of oxygen at increased pressure. Physiol. Rev. 25: 1-147, **1945**.

——. Tensional changes of alveolar gas in reactions to rapid compression and decompression and question of nitrogen narcosis. Amer. J. Physiol. 161: 417-425, **1950**.

Beard, E. F. and Owen, C. A. Cardiac arrhythmias during exercise testing in healthy men. Aerospace Med. 44: 286-289, **1973**.

Beaty, O. and Donald, D. O. Role of potassium in the transient reduction in vasoconstrictive responses of muscle resistance vessels during rhythmic exercise in dogs. Circ. Res. 41: 452-460, **1977**.

Beaumont, W. Observations on the gastric juice and the physiology of digestion. Plattsburg, N.Y.: Personal Publication, **1833**.

Beaver, W. L. and Wasserman, K. Transients in ventilation at start and end of exercise. J. Appl. Physiol. 25: 390-399, **1968**.

——. Tidal volume and respiratory rate changes at start and end of exercise. J. Appl. Physiol. 29: 872-876, **1970**.

Beaver, W. L., Wasserman, K., and Whipp, B. J. On-line computer analysis and breath-by-breath graphical display of exercise function tests. J. Appl. Physiol. 34: 128-132, **1973**.

Bechbache, R. R. and Duffin, J. The entrainment of breathing frequency by exercise rhythm. J. Physiol. (Lond.) 272: 553-562, **1977**.

Beck, C. H. M. Compensation of postural control by squirrel monkeys following dorsal column lesions, pp. 421-424. In: Control of Posture and Locomotion. Eds.: R. B. Stein, K. G. Pearson, R. S. Smith, and J. B. Redford. New York: Plenum Press, **1973**.

Becklake, M. R., Frank, H., Dagenais, G. R., Ostiguy, G. L., and Guzman, G. A. Influence of age and sex on exercise cardiac output. J. Appl. Physiol. 20: 938-947, **1965**.

Becklake, M. R., Varvis, C. J., Pengelly, L. D., Kenning, S., McGregor, M., and Bates, D. V. Measurement of pulmonary blood flow during exercise using nitrous oxide. J. Appl. Physiol. 17: 579-586, **1962**.

Beckman, E. L. Thermal protection during immersion in cold water, pp. 247-266. In: Proceedings of Second Symposium on Underwater Physiology. Eds.: C. J. Lambertsen and L. J. Greenbaum. Publ. 1181. Washington, D.C.: Nat. Acad. Sci. Nat. Res. Council, **1963**.

Beckner, G. L. and Winsor, T. Cardiovascular adaptations to prolonged physical exercise. Circulation 9: 835-846, **1954**.

Bedale, E. M. Industrial fatigue. U. K. Research Board Rept. 29, **1924**.

Bedford, T. and Warner, C. G. The energy expended while walking in stooping postures. Brit. J. Industr. Med. 12: 290-295, **1955**.

Beeckmans, J. and Shephard, R. J. Computer calculations of exercise dead space: the role of laminar flow and development of a clinical prediction formula. Respiration 28: 232-252, **1971**.

Beeston, J. W. U. Determination of specific gravity of live sheep and its correlation with fat percentage, pp. 49-55. In: Human Body Composition. Approaches and Applications. Ed.: J. Brozek. Oxford: Pergamon Press, **1965**.

Behnke, A. R. Investigations concerned with problems of high altitude flying and deep diving; application of certain findings pertaining to physical fitness to the general military service. Military Surg. 90: 9-28, **1942a**.

——. Symposium on industrial medicine; effects of high pressures; prevention and treatment of compressed air illness. Med. Clin. N. Amer. 26: 1213-1237, **1942b**.

Behnke, A. R., Forbes, H. S., and Motley, E. P. Circulatory and visual effects of oxygen at 3 atmospheres pressure. Amer. J. Physiol. 114: 436-442, **1936**.

Behnke, A. R. and Shaw, L. A. Use of oxygen in treatment of compressed-air illness. U.S. Naval Med. Bull. 35: 61-73, **1937**.

Behnke, A. R. and Willmon, T. L. U.S.S. Squalus; medical aspects of rescue and salvage operations and use of oxygen in deep-sea diving. U.S. Naval Med. Bull. 37: 629-640, **1939**.

Belcastro, A. N. and Bonen, A. Lactic acid removal rates during controlled and uncontrolled recovery exercise. J. Appl. Physiol. 39: 932-936, **1975**.

Belka, D. E. Comparison of dynamic, static and combination training on dominant wrist flexor muscles. Res. Quart. 39: 244-250, **1968**.

Bell, R. D. and Rasmussen, R. L. Exercise and the myocardial fiber capillary ratio. In: Training. Scientific Basis and Application. Ed.: A. W. Taylor. Springfield, Ill.: C. C. Thomas, **1972**.

Belokovsky, V. V. and Kuznetsov, V. V. Analysis of dynamic forces in crawlstroke swimming. In: Biomechanics V. Ed.: P. V. Komi. Baltimore: University Park Press, pp. 235-242,

1976.

Benade, A. J. S. and Heisler, N. Comparison of efflux rates of hydrogen and lactate ions from isolated muscles in vitro. Resp. Physiol. 32: 369-380, **1978**.

Benchimol, A., Li, Y. B., and Dimond, E. G. Cardiovascular dynamics in complete heart block at various heart rates. Circulation 30: 542-553, **1964**.

Benedict, F. G. and Murschhauser, H. Energy transformations during horizontal walking. Washington: Carnegie Inst. Publ. 231, **1915**.

Benestad, A. M. Trainability of old men. Acta Med. Scand. 178: 321-327, **1965**.

Bennett, D., Goldstein, R., and Leach, G. Echocardiographic left ventricular dimensions in the supine and upright positions and on upright exercise in normal subjects. J. Physiol. (Lond.) 275: 77p, **1977**.

Bennett, D. H. and Evans, D. W. Correlation of left ventricular mass determined by echocardiography with vector-cardiographic and electrocardiographic voltage measurements. Brit. Heart. J. 36: 981-987, **1974**.

Bennett, P. B. The cause and prevention of depth intoxication. J. Roy. Navy Sub. Serv. 18: 5-14, **1963**.

——. Cortical CO_2 and O_2 at high pressures of argon, nitrogen, helium, and oxygen. J. Appl. Physiol. 20: 1249-1252, **1965**.

——. The aetiology of compressed air intoxication and inert gas narcosis. Oxford: Pergamon Press, **1966**.

——. Inert gas narcosis, pp. 155-182. In: Physiology and Medicine of Diving. Eds.: P. B. Bennett and D. H. Elliott. London: Baillière, Tindall & Cassell, **1969**.

——. Psychological, physiological and biophysical studies of narcosis, pp. 457-469. In: Underwater Physiology. Ed.: C. J. Lambertsen. New York: Academic Press, **1971**.

Bennett, P. B. and Blenkarn, G. D. Arterial blood gases in man during inert gas narcosis. J. Appl. Physiol. 36: 45-48, **1974**.

Benzinger, T. H. Peripheral cold and central warm reception, main origins of thermal discomfort. Proc. Natl. Acad. Sci. U.S., 49: 832-839, **1963**.

Berdanier, C. and Moser, P. Metabolic responses of adrenalectomized rats to exercise. Proc. Soc. Exp. Biol. Med. 141: 490-493, **1972**.

Berger, R. A. Comparison of static and dynamic strength increase. Res. Quart. 33: 329-333, **1962**.

——. Effects of dynamic and static training on vertical jumping ability. Res. Quart. 34: 419-424, **1963**.

——. Comparison of the effect of various weight training loads on strength. Res. Quart. 36: 141-146, **1965**.

——. Effects of knowledge of isometric strength during performance on recorded strength. Res. Quart. 38: 507, **1967**.

——. Weight training hints for conditioning athletes, p. 28. In: An Analysis and Evaluation of Modern Trends Related to the Physiological Factors for Optimal Sports Performance. Ed.: E. J. Burke. Ithaca, N.Y.: Ithaca State College, **1976**.

Bergh, U. Längdlöpning. Idrottsfysiologi Report II, Trygg-Hansa, Stockholm, **1974**.

Bergman, N.A. Effects of varying waveforms on gas exchange. Anesthesiology 28: 390-395, **1967**.

Bergman, S. A., Campbell, J. K., and Wildenthal, K. "Diving reflex" in man: Its relation to isometric and dynamic exercise. J. Appl. Physiol. 33: 27-31, **1972**.

Bergman, S. A., Hoffler, G. W., Johnson, R. L., and Wolthuis, R. A. Pre- and post-flight systolic time intervals during LBNP: The second manned sky lab mission. Aviat. Space Env. Med. 47: 359-362, **1976**.

Bergström, J. Muscle electrolytes in man. Determined by neutron activation analysis on needle biopsy specimens. A study on normal subjects, kidney patients and patients with chronic diarrhoea. Scand. J. Clin. Lab. Invest. 14 (Suppl.) 68: 1-110, **1962**.

——. Local changes of ATP and phosphoryl-creatine in human muscle tissue in connection with exercise. Circ. Res. (Suppl. to Vols. 20-21) I: 91-96, **1967**.

Bergström J., Guarnieri, G., and Hultman, E. Carbohydrate metabolism and electrolyte changes in human muscle tissue during heavy work. J. Appl. Physiol. 30: 122-125, **1971**.

Bergström, J., Guarnieri, G., and Hultman, E. Changes in muscle water and electrolytes during exercise, pp. 173-178. In: Limiting Factors of Physical Performance. Ed.: J. Keul. Stuttgart: Thieme, **1973a**.

Bergström, J., Harris, R. C., Hultman, E., and Nordesjö, L. O. Energy rich phosphagens in dynamic and static work, pp. 341-355. In: Muscle Metabolism during Exercise. Eds.: B. Pernow and B. Saltin. New York: Plenum

Press, **1971**.

Bergström, J., Hermansen, L., Hultman, E., and Saltin, B. Diet, muscle glycogen and physical performance. Acta Physiol. Scand. 71: 140–150, **1967**.

Bergström, J., Hultman, E., and Saltin, B. Muscle glycogen consumption during cross-country skiing (the Vasa ski race). Int. Z. Angew. Physiol. 31: 71–73, **1973b**.

Berkada, B., Akokan, G., and Derman, V. Fibrinolytic response to physical exercise in males. Atherosclerosis 13: 85–91, **1971**.

Bernard, C. Leçons sur les Phenomènes de la Vie. Paris: Baillière, **1878**.

Bernelli-Zazzera, A., and Gaja, G. Some aspects of glycogen metabolism following reversible and irreversible liver ischemia. Exp. Mol. Pathol. 3: 361–368, **1964**.

Bernstein, L., D'Silva, J. L., and Mendel, D. The effect of the rate of breathing on the maximum breathing capacity determined with a new spirometer. Thorax 7: 255–262, **1952**.

Bernstein, L., and Mendel, D. The accuracy of spirographic recording at high respiratory rates. Thorax 6: 297–309, **1951**.

Bernstein, N. The coordination and regulation of movements. Oxford: Pergamon Press, **1967**.

Berson, A. S. and Pipberger, H. V. The low-frequency response of electrocardiographs, a frequent source of recording errors. Amer. Heart J. 71: 779–789, **1966**.

Bert, P. La pression barométrique. Paris: Masson, **1878**.

Bertolini, A. M. Aging in red cells. In: Perspectives in Experimental Gerontology. Ed.: N. W. Shock. Springfield, Ill.: C. C. Thomas, **1966**.

Berzelius. In: Lehmann, C. G. Lehrbuch der physiologischen Chemie 1, 103 Leipzig (1850). Cited by R. E. Davies. Energy Rich Phosphagens. In: Muscle metabolism during exercise. Eds.: B. Pernow and B. Saltin. New York: Plenum Press, **1971**.

Bessou, P., Dejours, P., and Laporte, Y. Effects ventilatoires réflexes de la stimulation de fibres afférentes de grand diamètre d'origine musculaire chez le chat. C. R. Soc. Biol. 153: 477–481, **1959**.

Bessou, P., Emonet-Dénand, F., and Laporte, Y. Motor fibers innervating extrafusal and intrafusal muscle fibers in the cat. J. Physiol. (Lond.) 180: 649–672, **1965**.

Bessou, P., Laporte, Y., and Pagès, B. Frequencygrams of spindle primary endings elicited by stimulation of static and dynamic fusimotor fibres. J. Physiol. (Lond.) 196: 47–63, **1968**.

Beswick, F. W. and Jordan, R. C. Cardiological observations of the 6th British Empire and Commonwealth Games. Brit. Heart J. 23: 113–130, **1964**.

Bevegärd, S., Freyschuss, H., and Strandell, T. Circulatory adaptations to arm and leg exercise in supine and sitting positions. J. Appl. Physiol. 21: 37–46, **1966**.

Bevegärd, S., Holmgren, A., and Jonsson, B. The effect of body position of the circulation at rest and during exercise with special reference to the influence on the stroke volume. Acta Physiol. Scand. 49: 279–298, **1960**.

Bevegärd, S., Holmgren, A., and Jonsson, B. Circulatory studies in well-trained athletes at rest and during heavy exercise, with special reference to stroke volume and the influence of body position. Acta Physiol. Scand. 57: 26–50, **1963**.

Bevegärd, B. S. and Shepherd, J. T. Changes in tone of limb veins during supine exercise. J. Appl. Physiol. 20: 1–8, **1965**.

——. Reaction in man of resistance and capacity vessels in forearm and hand to leg exercise. J. Appl. Physiol. 21: 123–132, **1966a**.

——. Circulatory effects of stimulating the carotid arterial stretch receptors in man at rest and during exercise. J. Clin. Invest. 45: 132–142, **1966b**.

——. Regulation of the circulation during exercise in man. Physiol. Rev. 47: 178–213, **1967**.

Bhan, A. K. and Scheuer, J. Effects of physical training on cardiac actomyosin adenosine triphosphatase activity. Amer. J. Physiol. 223: 1486–1490, **1972**.

Bharadwaj, H., Singh, A. P., and Malhotra, M. S. Body composition of the high-altitude natives of Ladakh. A comparison with sea-level residents. Human Biol. 45: 423–434, **1973**.

Bhattacharyya, N. K., Cunningham, D. J. C., Goode, R. C., Howson, M. G., and Lloyd, B. B. Hypoxia, Ventilation, P_{CO_2} and exercise. Resp. Physiol. 9: 329–347, **1970**.

Bianchi, A. L. Localisation et étude des neurones respiratoires bulbaires. Mise en jeu antidromique par stimulation spinale ou vagale. J. Physiol. (Paris) 63: 5–40, **1971**.

Bianco, S., Griffin, J. P., Kalamburoff, P. L., and Prince, F. J. Prevention of exercise induced asthma by indoramin. Brit. Med. J.

4: 18-20, **1974**.

Bichler, K. H., Lachmann, E., and Porzsolt, F. Untersuchungen zur mechanischen Hämolyse bei langstrecken Läufern. Sportarzt Sportmedizin 23: 9-14, **1972**.

Bierman, C. W., Pierson, W. E., and Shapiro, G. G. Effect of drugs on exercise-induced bronchospasm, pp. 289-300. In: Muscular Exercise and the Lung. Eds.: J. A. Dempsey and C. E. Reed. Madison, Wisc.: University of Wisconsin Press, **1977**.

Bigland, B. and Lippold, O. C. J. The relation between force, velocity and integrated electrical activity in human muscles. J. Physiol. (Lond.) 123: 214-224, **1954a**.

——. Motor unit activity in voluntary contraction of human muscle. J. Physiol. (Lond.) 125: 322-335, **1954b**.

Bigland-Ritchie, B., and Woods, J. J. Integrated EMG and oxygen uptake during dynamic contractions of human muscles. J. Appl. Physiol. 26: 475-479, **1974**.

Bigland-Ritchie, B. and Woods, J. J. Integrated electromyogram and oxygen uptake during positive and negative work. J. Physiol. (Lond.) 260: 267-277, **1976**.

Bigland-Ritchie, B. Jones, D. A., Hosking, G. P., and Edwards, R. H. T. Central and peripheral fatigue in sustained maximum voluntary contractions of human quadriceps muscle. Clin. Sci. Mol. Med. 54: 609-614, **1978**.

Billimoria, J. D., Pozner, H., Metselaar, B., Best, F. W., and James, D. C. O. Effect of cigarette smoking on lipids, lipoproteins, blood coagulation, fibrinolysis and cellular components of human blood. Atherosclerosis 21: 61-76, **1975**.

Binder, M. D., Kroin, J. S., Moore, G. P., Stauffer, E. K., and Stuart, D. G. Correlation analysis of muscle spindle responses to single motor unit contractions. J. Physiol. (Lond.) 257: 325-336, **1976**.

Bing, R. J., Hammond, M. M., Handelsman, J. C., Powers, S. R., Spencer, F. C., Eckenhoff, J. E., Goodale, W. T., Hafkenshiel, J. H., and Kety, S. S. The measurement of coronary blood flow, oxygen consumption and efficiency of the left ventricle in man. Amer. Heart J. 38: 1-24, **1949**.

Binkhorst, R. A. and Leeuwen, P. Van. A rapid method for the determination of aerobic capacity. Int. Z. Angew. Physiol. 19: 459-467, **1963**.

Binkhorst, R. A. and van't Hof, M. A. Force-velocity relationship and contraction time of the rat fast plantaris muscle due to compensatory hypertrophy. Pflüg. Archiv. 342: 145-158, **1973**.

Birchwood, B. What's wrong with carbohydrate? Can. Fam. Phys. 21: 69-72, **1975**.

Birge, S. J. and Whedon, G. D. Bone, pp. 213-235. In: Hypodynamics and Hypogravics. Ed.: M. McCally. New York: Academic Press, **1968**.

Birnbaum, G. L. and Thompson, S. A. The mechanism of asphyxial resuscitation. Resuscitation with inert-asphyxiating-gas in advanced asphyxia. Surg. Gynecol. Obstetr. 75: 79-86, **1942**.

Birren, J. E. Sensation, perception and modification of behaviour in relation to the process of aging. In: The Process of Aging in the Nervous System. Eds.: J. E. Birren, H. A. Imus, and W. F. Windle. Springfield, Ill.: C. C. Thomas, **1959**.

Bischoff, M. B., Dean, W. D., Bucci, T. J., and Frics, L. A. Ultrastructural changes in myocardium of animals after five months residence at 14,110 feet. Fed. Proc. 28: 1268-1273, **1969**.

Biscoe, T. J. Carotid body: structure and function. Physiol. Rev. 51: 437-495, 1971.

Biscoe, T. J. and Purves, M. J. Carotid chemoreceptor and cervical sympathetic activity during passive third limb exercise in the anaesthetized cat. J. Physiol. (Lond.) 178: 43p, **1965**.

Bisgard, H. E., Will, J. A., Tyson, I. B., Dayton, L. M., Henderson, R. R., and Grover, R. F. Distribution of regional lung function during mild exercise in residents of 3,100 m. Resp. Physiol. 22: 369-379, **1974**.

Bishop, J. M. and Segel, N. The circulatory effects of intravenous pronethalol in man at rest and during exercise in the supine and upright positions. J. Physiol. (Lond.) 169: 122p, **1963**.

Bishop, V. and Horwitz, L. D. Quantitative assessment of cardiac pump performance. J. Physiol. (Lond.) 269: 355-370, **1977**.

Bittel, J. and Henane, R. Comparison of thermal exchanges in men and women under neutral and hot conditions. J. Physiol. (Lond.) 250: 475-489, **1975**.

Bizzi, E., Kalil, R. E., and Tagliasco, V. Eye-head coordination in monkeys: Evidence for centrally patterned organization. Science 173: 452-454, **1971**.

Bjorgum, R. K. and Sharkey, B. J. Inhalation of oxygen as an aid to recovery after exertion. Res. Quart. 37: 462–467, **1966**.

Bjørkerud, S. Synthesis of lipids from glucose and incorporation of palmitate in atherosclerotic and normal rabbit aorta, pp. 126–129. In: Atherosclerosis. Proc. 2nd Int. Symp. Ed.: R. J. Jones. Berlin: Springer Verlag, **1970**.

Björksten, M. and Jonsson, B. Endurance limit of force in long term intermittent static contractions. Scand. J. Work Environ. Health 3: 23–27, **1977**.

Björntorp, P. Metabolism in patients with ischemic heart disease and obesity after training, pp. 493–504. In: Muscle Metabolism During Exercise. Eds.: B. Pernow and B. Saltin. New York: Plenum Press, **1970**.

Björntorp, P., Carlgren, G., Isaksson, B., Krotkiewski, M., Larsson, B., and Sjöström, L. Effect of an energy-reduced dietary regimen in relation to adipose tissue cellularity in obese women. Amer. J. Clin. Nutr. 28: 445–452, **1975**.

Björntorp, P., deJounge, K., Krotkiewski, M., Sullivan, L., Sjöström, L., and Steinberg, J. Physical training in human obesity. 3. Effects of long term physical training on body composition. Metabolism 22: 1467–1475, **1973**.

Björntorp, P., Fahlén, M., Holm, J., Scherstén, T., and Steinberg, J. Changes in the activity of skeletal muscle succinic oxidase after training, pp. 138–142. In: Coronary heart disease and physical fitness. Eds.: O. A. Larsen and R. O. Malmborg. Copenhagen: Munksgaard, **1971**.

Black, A. M. S., McCloskey, D. I., and Torrance, R. W. The responses of carotid body chemoreceptors in the cat to sudden changes of hypercapnic and hypoxic stimuli. Resp. Physiol. 13: 36–49, **1971**.

Black, A. M. S. and Torrance, R. W. Chemoreceptor effects in the respiratory cycle. J. Physiol. (Lond.), 189: 59–61, **1967**.

———. Respiratory oscillations in chemoreceptor discharge in control of breathing. Resp. Physiol. 13: 221–237, **1971**.

Black, J. E. Blood flow requirements of the human calf after walking and running. Clin. Sci. 18: 89–93, **1959**.

Black, W. A. and Karpovich, P. V. Effect of exercise upon the erythrocyte sedimentation rate. Amer. J. Physiol. 144: 224–226, **1945**.

Blackburn, H. The exercise electrocardiogram. Technological, procedural and conceptual developments, pp. 220–258. In: Measurement in Exercise Electrocardiography. The Ernst Simonson Conference. Ed.: H. Blackburn. Springfield, Ill.: C. C. Thomas, **1969**.

———. Progress in the epidemiology and prevention of coronary heart diseases. In: Progress in Cardiology. Eds.: P. N. Yu and J. F. Goodwin. Philadelphia: Lea & Febiger, **1974a**.

———. Disadvantages of intensive exercise therapy after myocardial infarction, p. 162. In: Controversy in Internal Medicine. Ed.: F. Ingelfinger. Philadelphia: Saunders, **1974b**.

Blackburn, H., Blomqvist, G., Freiman, A., Friesinger, G. C. et al. The exercise electrocardiogram: differences in interpretation. Amer. J. Cardiol. 21: 871–880, **1968**.

Blackburn, H., Brozek, J., and Taylor, H. L. Common circulatory measurements in smokers and non-smokers. Circulation 22: 1112–1124, **1960**.

Blackburn, H. and Katizbak, R. What electrocardiographic leads to take after exercise. Amer. Heart. J. 67: 184–185, **1964**.

Blackburn, H., Taylor, H. L., Hamrell, B., Buskirk, E. R., Nicholas, W. C., and Thorsen, R. D. Premature ventricular complexes induced by stress testing. Amer. J. Cardiol. 31: 441–449, **1973**.

Blackley, J., Knochel, J. P., and Long, J. Impaired muscle glycogen synthesis and prevention of muscle glycogen supercompensation by potassium deficiency Clin. Res. 22: 517A, **1974**.

Blair, D. A., Glover, W. E., Greenfield, A. D. M., and Roddie, I. C. Excitation of cholinergic vasodilator nerves to human skeletal muscles during emotional stress. J. Physiol. (Lond.) 148: 633–647, **1959a**.

———. The increase in tone in forearm resistance blood vessels exposed to increased transmural pressure. J. Physiol. (Lond.) 149: 614–625, **1959b**.

Blair, S. N. The effect of stimulus and movement complexity upon reaction time and movement time. In: Contemporary Psychology of Sport. Eds.: G. S. Kenyon and T. M. Grogg. Chicago: Athletic Institute, **1970**.

Blalock, A. and Duncan, G. W. Traumatic shock—consideration of several types of injuries. Surg. Gynecol. Obstetr. 75: 401–409, **1942**.

Blick, E. F. and Stein, P. D. Work of the heart: A

general thermodynamic analysis. New York: Pergamon Press, **1977**.

Bligh, J. Thermoregulation: What is regulated and how? Proc. Int. Union Physiol. Sci., Vol. 12: 22-23, Paris, **1977**.

Blinks, J. R. and Koch-Weser, J. Physical factors in the analysis of the action of drugs on myocardial contractility. Pharm. Rev. 15: 601-652, **1963**.

Blockley, W. V. and Taylor, C. L. Human tolerance limits for extreme heat. In: Heat Stress and Heat Disorders. Eds.: C. S. Leithead and A. R. Lind (1964). London: Cassell. Heat-Pip. Air Cond. 21: 111, **1949**.

Blomqvist, G. The Frank lead exercise electrocardiogram: A quantitative study based on averaging technic and digital computer analysis. Acta Med. Scand. 178, (Suppl.) 440: 1-98, **1965**.

——. Exercise physiology related to diagnosis of coronary artery disease, pp. 2-1 to 2-26. In: Coronary Heart Disease: Prevention, Detection, Rehabilitation with Emphasis on Exercise Testing. Ed.: S. M. Fox. Denver, Colorado: International Medical Corporation, **1974**.

Blomqvist, G., Johnson, R. L., and Saltin, B. Pulmonary diffusing capacity limiting human performance at altitude. Acta Physiol. Scand. 76: 284-287, **1969**.

Blomqvist, G., Saltin, B., Dean, W. F., and Mitchell, J. H. Acute effect of ethanol ingestion on the response to submaximal and maximal exercise in man. Circulation 42: 463-470, **1970**.

Bloom, D. S. and Vecht, R. H. Circulatory changes during isometric exercise measured by transcutaneous aortovelography. J. Physiol. (Lond.) 281: 21p-22p, **1978**.

Bloom, S. R., Johnson, R. H., Park, D. M., Rennie, M. J., and Sulaiman, W. R. Differences in the metabolic and hormonal response to exercise between racing cyclists and untrained individuals. J. Physiol. (Lond.) 258: 1-18, **1976**.

Bloomfield, J. and Sigerseth, P. Anatomical and physiological differences between sprint and middle-distance competitors at the university level. J. Sport. Med. Phys. Fitness 5: 76-81, **1965**.

Bloor, C. M. and Leon, A. S. Interaction of age and exercise on the heart and its blood supply. Lab. Invest. 22: 160-164, **1970**.

Bloor, M., Pasyk, S., and Leon, A. S. Interaction of age and exercise on organ and cellular development. Amer. J. Pathol. 58: 185-189, **1970**.

Blum, H. F. The physiological effects of sunlight on man. Physiol. Rev. 25: 483-530, **1945**.

Blumberger, K. J. and Sigisbert, M. Studies of cardiac dynamics, pp. 4-372 to 4-377. In: Cardiology, and encyclopaedia of the cardiovascular system. Ed.: A. A. Luisada. New York: McGraw Hill, **1959**.

Blumchen, G., Roskamm, H., and Reindell, H. Herzvolumen und korperliche Leistungsfähigkeit. Kreislaufforschung 55: 1012-1016, **1966**.

Bock, A. V., Vancaulert, C., Dill, D. B., Fölling, A., and Hurxthal, L. Studies in muscular activity, part 4: "Steady State" and the respiratory quotient during work. J. Physiol. (Lond.) 66: 162-174, **1928**.

Bodem, G., Brammell, H. L., Weil, J. V., and Chidsey, C. A. Pharmacodynamic studies of beta adrenergic antagonism induced in man by propranolol and practolol. J. Clin. Invest. 52: 747-754, **1973**.

Boerth, R. C., Covell, J. W., Pool, P. E., and Ross, J. Increased myocardial oxygen consumption and contractile state associated with increased heart rate in dogs. Circ. Res. 24: 725-734, **1969**.

Boileau, R. A. Effects of a nine week physical conditioning program upon body composition of obese and lean individuals. Penn. State University: Ph.D. dissertation, **1969**.

Bonde-Petersen, F. Muscle training by static, concentric and eccentric contractions. Acta Physiol. Scand. 48: 406-416, **1960**.

Bonde-Petersen, F., Henrickson, J., and Lundin, B. Blood flow in thigh muscle during bicycling exercise at varying work rates. Europ. J. Appl. Physiol. 34: 191-197, **1975a**.

Bonde-Petersen, F., Mørk, A. L., and Nielsen, E. Local muscle blood flow and sustained contractions of human arm and back muscles. Europ. J. Appl. Physiol. 34: 43-50, **1975b**.

Bonen, A. and Belcastro, A. N. Comparison of self-selected recovery methods on lactic acid removal rates. Med. Sci. Sport. 8: 176-178, **1976**.

Bonen, A., Campbell, C. J., Kirby, R. L., and Belcastro, A. N. Relationship between slow-twitch muscle fibres and lactic acid removal. Can. J. Appl. Sport. Sci. 3: 160-162, **1978**.

Bonjer, F. H. Relationship between working time, physical working capacity and allowable caloric expenditure. In: Muskelarbeit und Muskeltraining. Darmstadt: Gentner Verlag, **1968**.

Bookwalter, K. W. Grip strength norms for males. Res. Quart. 21: 249–273, **1950**.

Booth, F. W. and Gould, E. W. Effects of training and disuse on connective tissue. In: Exercise and Sports Sciences Reviews. Vol. 3, 83–112. Eds.: J. H. Wilmore and J. F. Keogh. New York: Academic Press, **1975**.

Booth, F. W. and Kelso, J. R. Cytochrome oxidase of skeletal muscle-adaptive response to chronic disuse. Can. J. Physiol. 51: 679–681, **1973**.

Booyens, J. and Keatinge, W. R. The expenditure of energy by men and women walking. J. Physiol. (Lond.) 138: 165–171, **1957**.

Borg, G. The perception of physical performance. In: Frontiers of Fitness. Ed.: R. J. Shephard. Springfield, Ill.: C. C. Thomas, **1971**.

Borg, G. and Linderholm, H. Perceived exertion and pulse rate during exercise in various age groups. Acta Med. Scand. Suppl. 472: 194–206, **1967**.

Borg, G. A. V. and Noble, B. J. Perceived exertion. Exercise Sport. Sci. Rev. 2: 131–154, **1974**.

Borhani, N. O. Epidemiology of coronary heart disease, pp. 1–12. In: Exercise in cardiovascular health and disease. Eds.: E. A. Amsterdam, J. H. Wilmore, and A. N. DeMaria. New York: Yorke Books, **1977**.

Bornmann, R. C. Limitations in the treatment of diving and aviation bends by increased ambient pressure. Aerospace Med. 39: 1070–1076, **1968**.

Bosco, J. S., Terjung, R. L., and Greenleaf, J. E. Effects of progressive hypohydration on maximal isometric muscular strength. J. Sport. Med. 8: 81, **1968**.

Bosley, M. A., O'Donovan, M. J., Stephens, J. A., Taylor, A., and Usherwood, T. P. Unit studies in normal human gastrocnemius by microstimulation, glycogen depletion and needle biopsy. J. Physiol. (Lond.) 260: 11p–12p, **1976**.

Bostrom, C. O., Hunkeler, F. L., and Krebs, E. G. The regulation of skeletal muscle phosphorylase kinase by Ca^{2+}. J. Biol. Chem. 246: 1961–1967, **1971**.

Bottger, I., Schlein, E. M., Faloona, G. R., Knochel, J. P., and Unger, R. H. The effects of exercise on glucagon secretion. J. Clin. Endocrinol. 35: 117–125, **1972**.

Botwinik, J. Theories of antecedent conditions of speed of response, pp. 67–87. In: Behaviour, Aging and the Nervous System. Springfield, Ill.: C. C. Thomas, **1965**.

Bouchard, C. Le development du système de transport de l'oxygène chez les jeunes adultes. Quebec City: Editions du Pélican, **1975**.

——. Genetics, growth and physical activity, pp. 29–45. In: Physical activity and human well-being. Eds.: F. Landry and W. R. Orban. Miami: Symposia Specialists, **1978**.

Boughner, D. R. and Roach, M. R. Effect of low frequency vibration on the arterial wall. Circ. Res. 29: 136–144, **1971**.

Bouhuys, A. Lung volumes and breathing patterns in wind-instrument players. J. Appl. Physiol. 19: 967–975, **1964**.

Bouhuys, A., Proctor, D. F., and Mead, J. Kinetic aspects of singing. J. Appl. Physiol. 21: 483–496, **1966**.

Bouisset, S. EMG and muscle force in normal motor activity. In: New Developments in Electromyography and Clinical Neurophysiology, Vol. 1, pp. 547–583. Ed.: J. E. Desmedt. Basel: Karger, **1973**.

Bouisset, S. and Goubel, F. Integrated electromyographical activity and muscle work. J. Appl. Physiol. 35: 695–702, **1973**.

Bounos, G., Sutherland, N. G., McArdle, A. H., and Gard, F. N. The prophylactic use of an "elemental" diet in experimental hemorrhagic shock and intestinal ischemia. Ann. Surg. 166: 312–343, **1967**.

Bourne, G. H. Vitamins and muscular exercise. Brit. J. Nutr. 2: 261–263, **1948**.

Bowers, W. D., Hubbard, R. W., Smoake, J. A., Daum, R. C., and Nilson, E. Effects of exercise on the ultrastructure of skeletal muscle. Amer. J. Physiol. 227: 313–316, **1974**.

Bowie, W. and Cumming, G. R. Sustained handgrip reproducibility. Effects of hypoxia. Med. Sci. Sport. 3: 24–31, **1971**.

Boycott, A. E., Damant, G. C. C., and Haldane, J. S. The prevention of compressed air illness. J. Hyg. (Camb.) 8: 343–443, **1908**.

Boyd, I. A. The structure and innervation of the nuclear bag muscle fibre system and the nuclear chain muscle fibre system in mammalian muscle spindles. Phil. Trans. Roy. Soc. B 245: 81–136, **1962**.

——. The mammalian muscle spindle: An advanced study (film). J. Physiol. (Lond.) 214: 1p, **1971**.

Bozner, A. and Meessen, H. Die Feinstruktur des Herzmuskels der Ratte nach einmaligem und nach wiederholtem Schwimmtraining. Virchows Arch. B, 3: 248-269, **1969**.

Bradbury, C. E. Anatomy and construction of the human figure. New York: McGraw Hill, **1949**.

Bradley, C. A., Harris, E. A., Seelye, E. R., and Whitlock, R. M. L. Gas exchange during exercise in healthy people. I. The physiological dead space volume. Clin. Sci. Mol. Med. 51: 323-333, **1976**.

Bradley, G. W. Von Euler, C., Marttila, I., and Roos, B. A. model of the central and reflex inhibition of inspiration in the cat. Biol. Cybernetics 19: 105-116, **1975**.

Bradley, K., Easton, D. M., and Eccles, J. C. An investigation of primary or direct inhibition. J. Physiol. (Lond.) 122: 474-488, **1953**.

Bradley, K. and Eccles, J. C. Analysis of fast afferent impulses from thigh muscles. J. Physiol. (Lond.) 122: 462-473, **1953**.

Bradley, M. E., Anthonisen, N. R., Vorosmarti, J., and Linaweaver, P. G. Respiratory and cardiac responses to exercise in subjects breathing helium-oxygen mixtures at pressures from sea level to 19.2 atmospheres, pp. 325-337. In: Underwater Physiology. Ed.: C. J. Lambertsen. New York: Academic Press, **1971**.

Bradley, R. D. and Semple, S. J. G. The comparison of blood and cerebrospinal fluid changes in man in various acute and chronic acid-base disturbances with special reference to the chemical control of ventilation, pp. 235-242. In: The Regulation of Human Respiration. Eds.: D. J. C. Cunningham and B. B. Lloyd. Oxford: Blackwell Scientific, **1963**.

Bransford, D. R. and Howley, E. T. Oxygen cost of running in trained and untrained men and women. Med. Sci. Sport. 9: 41-44, **1977**.

Braunwald, E. and Kelly, E. R. The effect of exercise on central blood volume in man. J. Clin. Invest. 39: 413-419, **1960**.

Braunwald, E., Ross, J., and Sonnenblick, E. H. Mechanism of contraction of the normal and failing heart. Boston: Little, Brown, **1968**.

Braunwald, E., Sarnoff, S. J., Case, R. B., Stainsby, W. N., and Welch, G. H. Hemodynamic determinants of coronary flow: Effect of changes in aortic pressure and cardiac output on the relationship between myocardial oxygen consumption and coronary flow. Amer. J. Physiol. 192: 157-163, **1958**.

Braunwald, E., Sonnenblick, E. H., Ross, J., Glick, G., and Epstein, S. E. An analysis of the cardiac response to exercise. Circ. Res. 20 and 21 (Suppl.): I-44 to 1-58, **1967**.

Bray, G. A. Endocrine factors in the control of food intake. Fed. Proc. 33: 1140-1145, **1973**.

Bray, G. A. The Zucker fatty rat. Fed. Proc. 36: 148-153, **1977**.

Brebbia, D. R., Goldman, R. F., and Buskirk, E. R. Water vapor loss from the respiratory tract during outdoor exercise in the cold. Tech. Rept. EP-57. Natick: Headquarters, U.S. Quartermaster Research and Development Command, **1957**.

Bremer, J. Factors influencing the carnitine-dependent oxidation of fatty acids, pp. 65-88. In: Cellular Compartmentalization and Control of Fatty Acid Metabolism. Eds.: J. Bremer and F. C. Gran. New York: Academic Press, **1967**.

Brener, J. and Kleinman, R. A. Learned control of decreases in systolic blood pressure. Nature 226: 1063-1064, **1970**.

Brezina, E. and Kolmer, W. Ueber den Energieverbrauch der Geharbeit unter dem Einfluss verschiedener Geschwindigkeiten und verschiedener Belastung. Biochem. Z. 38: 129-153, **1912**.

Bridgman, C. F. The structure of tendon organs in the cat: A proposed mechanism for responding to muscle tension. Anat. Rec. 162: 209-220, **1968**.

Brigden, W., Howarth, S., and Sharpey-Schafer, E. P. Postural changes in the peripheral blood flow of normal subjects with observations on vasovagal fainting reactions as a result of tilting, the lordotic posture, frequency and spinal anaesthesia. Clin. Sci. 9: 79-91, **1950**.

Briggs, R. S., Brown, P. M., Crabb, M. E., Cox, T. J., Ead, H. W., Hawkes, R. A., Jequier, P. W., Southall, D. P., Grainger, R., Williams, J. H., and Chamberlain, D. A. The Brighton resuscitation ambulances: a continuing experiment in prehospital care by ambulance staff. Brit. Med. J. 2: 1161-1165, **1976**.

Brinley, J. F. Cognitive sets, speed and accuracy of performance in the elderly, pp. 114-149. In: Behaviour, Aging and the Nervous System. Springfield, Ill.: C. C. Thomas, **1965**.

Brisson, G. R., Volle, M. A., Desharnais, M., Dion, M., and Tanaka, M. Pituitary gonadal axis in exercising men. Med. Sci. Sport. 9: 47, **1977**.

Bristow, J. D., Brown, E. B., Cunningham,

D. J. C., Howson, M. G., Strange-Petersen, E., Pickering, T., and Sleight, P. Effect of bicycling on the baroreflex regulation of pulse interval. Circ. Res. 28: 582-592, **1971**.

——. Chronic effects of alcohol. Brit. Med. J. 2: 381-382, **1978**.

British Medical Journal (Ed.). Significance of ectopic beats. Brit. Med. J. 2: 191-192, **1973**.

——. Men, women, and obesity. Brit. Med. J. 4: 249-250, **1974**.

——. Measuring blood pressure. Brit. Med. J. 4: 366-367, **1975a**.

——. Hyperactivity in children. Brit. Med. J. 4: 123-124, **1975b**.

Britman, N. A. and Levine, H. J. Contractile element work: a major determinant of myocardial oxygen consumption. J. Clin. Invest. 43: 1397-1408, **1964**.

Broadbent, D. E. Perception and Communication. Oxford: Pergamon Press, **1958**.

Broadus, A. E., Kaminsky, N. I., Northcutt, R. C. et al. Effects of glucagon on adenosine 3-5-monophosphate and guanosine 3-5-monophosphate in human plasma and urine. J. Clin. Invest. 49: 2237-2245, **1970**.

Brobeck, J. R. Food intake as a mechanism of temperature regulation. Yale J. Biol. Med. 20: 545-552, **1948**.

Brock, L. Early reconditioning for post-myocardial infarction patients: Spalding Rehabilitation Center, pp. 315-323. In: Exercise Testing and Exercise Training in Coronary Heart Disease. Eds.: J. P. Naughton, H. K. Hellerstein, and I. C. Mohler. New York: Academic Press, **1973**.

Brocklehurst, J. C. Textbook of Geriatric Medicine and Gerontology. Edinburgh: Churchill-Livingstone, **1973**.

Brod, J., Fencl, V., Hejl, Z., and Jirka, J. Circulatory changes underlying blood pressure elevation during acute emotional stress (mental arithmetic) in normo-tensive and hypertensive subjects. Clin. Sci. 18: 269-279, **1959**.

Brodal, P. Leddinnervasjon—et forsömt kapittel. Fysioterapeuten 39: 65, **1972**.

Brodal, P., Ingjer, F., and Hermansen, L. Capillary supply of skeletal muscle fibers in untrained and endurance-trained men. Amer. J. Physiol. 232: H705-H712, **1977**.

Brodan, V. and Kuhn, E. Influence of intravenous glucose administration on metabolism during physical exercise. Nutr. Metab. 13: 54-65, **1971**.

Brody, H. Organization of the cerebral cortex. III. A study of aging in the human cerebral cortex. J. Comp. Neurol. 102: 511-556, **1955**.

Broman, S. and Wigertz, O. Transient dynamics of ventilation and heart rate with changes in work load from different load levels. Acta Physiol. Scand. 81: 54-74, **1971**.

Brook, C. G. D., Huntley, R. M. C., and Slack, J. Influence of heredity and environment in determination of skinfold thickness in children. Brit. Med. J. 2: 719-721, **1975**.

Brooke, J. D. Effect of initial exercise on blood glucose. Can. J. Appl. Sport. Sci. 3: 181, **1978a**.

——. Effects on fractionated reaction time of sustained movement to deplete carbohydrate, pp. 171-180. In: Exercise Physiology. Eds.: F. Landry and W. A. R. Orban. Miami, Fla.: Symposia Specialists, **1978b**.

Brooke, J. D. and Green, L. F. The effect of a high carbohydrate diet on human recovery following prolonged work to exhaustion. Ergonomics 17: 489-497, **1974**.

Brooke, M. H. and Kaiser, K. K. Three "myosin ATPase" systems: the nature of their pH liability and sulphydryl dependence. J. Histochem. Cytochemistry. 18: 670-672, **1970**.

Brooks, G. A., Brauner, K. E., and Cassens, R. G. Glycogen synthesis and metabolism of lactic acid after exercise. Amer. J. Physiol. 224: 1162-1166, **1973**.

Brooks, G. A., Hittelman, K. J., Faulkner, J. A., and Beyer, R. E. Temperature, liver mitochondrial respiratory functions and oxygen debt. Med. Sci. Sport. 3: 72-74, **1971a**.

——. Temperature, skeletal muscle, mitochondrial functions and oxygen debt. Amer. J. Physiol. 220: 1053-1059, **1971b**.

Brooks, R. V. Some aspects of the action of anabolic steroids on athletes, pp. 219-229. In: Physical Activity and Human Well-Being. Eds.: F. Landry and W. R. Orban. Miami, Fla.: Symposia Specialists, **1978**.

Brooks, V. B. Some examples of programmed limb movements. Brain Res. 71: 299-308, **1974**.

Brooks, V. B., Cooke, J. D., and Thomas, J. S. The continuity of movements, pp. 257-272. In: Control of Posture and Locomotion. Eds.: R. B. Stein, K. G. Pearson, R. S. Smith, and J. B. Redford. New York: Plenum Press, **1973**.

Brooks, V. B. and Stoney, S. D. Motor mechanisms: The role of the pyramidal system in motor control. Ann. Rev. Physiol. 33: 337-

392, **1971**.

Brostrom, C. O., Hunkeler, F. L., and Krebs, E. G. The regulation of skeletal muscle phosphorylase kinase by Ca^{2+}. J. Biol. Chem. 246: 1961-1967, **1971**.

Brouha, L. The step test: a simple method of measuring physical fitness for muscular work in young men. Res. Quart. 14: 31-36, **1943**.

Brouha, L. and Radford, E. P. The cardiovascular system in muscular activity. In: Science and Medicine of Exercise and Sports (1st ed.). Ed.: W. R. Johnson. New York: Harper, **1960**.

Brouha, L., Smith, P. E., Delanne, R., and Maxfield, M. E. Physiological reactions of men and women during muscular activity and recovery in various environments. J. Appl. Physiol. 16: 133-140, **1960**.

Brown, B. H., Pryce, W. I. J., Baumber, D., and Clarke, R. G. Impedance plethysmography: Can it measure changes in limb blood flow? Med. Biol. Eng.: 674-682, **1975**.

Brown, C. H., Harrower, J. R., and Deeter, M. F. The effects of cross country running on pre-adolescent girls. Med. Sci. Sports. 4: 1-5, **1972**.

Brown, H. B. Diets that lower blood cholesterol in man, pp. 426-435. In: Atherosclerosis. Proceedings 2nd Int. Symp. Ed.: R. J. Jones. Berlin: Springer Verlag, **1970**.

Brown, J. R. Industrial fatigue. Med. Serv. J. Can. 20: 221-231, **1964**.

——. The metabolic cost of industrial activity in relation to weight. Med. Serv. J. Can. 22: 262-272, **1966**.

——. Lifting as an Industrial Hazard. Labour Safety Council of Ontario, Toronto, Ont., **1971**.

——. Manual Lifting and Related Fields. An Annotated Bibliography. Toronto: Labour Safety Council, Ont. Dept. of Labour, **1972**.

Brown, J. R. and Shephard, R. J. Some measurements of fitness in older female employees of a Toronto department store. Can. Med. Assoc. J. 97: 1208-1213, **1967**.

Brown, J. S., Knauft, E. B., and Rosenbaum, G. The accuracy of positioning reactions as a function of their direction and extent. Amer. J. Psychol. 61: 167-182, **1948**.

Brown, M., Cotter, M., Hudlická, O., Smith, M., and Vrbová, G. Metabolic changes in long-term stimulated fast muscles, pp. 471-475. In: Metabolic Adaptation to Prolonged Physical Exercise. Eds.: H. Howald and J. R. Poortmans. Basel: Birkhauser-Verlag, **1975**.

Brown, M. C., Crowe, A., and Matthews, P. B. C. Observations on the fusimotor fibres of the tibialis posterior muscle of the cat. J. Physiol. (Lond.) 177: 140-159, **1965**.

Brown, M. C., Engberg, I., and Matthews, P. B. C. The relative sensitivity to vibration of muscle receptors of the cat. J. Physiol. (Lond.) 192: 773-800, **1967**.

Brown, M. C. and Matthews, P. B. C. On the subdivision of the efferent fibres to muscle spindles into static and dynamic fusimotor fibres, pp. 18-31. In: Control and Innervation of Skeletal Muscle. Ed.: B. L. Andrew. Dundee: Thomson, **1966**.

Brownell, C. L. A Scale for Measuring Posture of Ninth Grade Boys. New York: Teachers College, Columbia University, **1928**.

Brox, W. T. and Ackles, K. N. SDL-1 Physiological Diver-Monitoring System. Progress Report 1. Downsview, Ont.: D.C.I.E.M. Operational Report 73-OR-989, **1973**.

Brožek, J. Body composition. Ann. N. Y. Acad. Sci. 110: 1-1018, **1963**.

Brožek, J. Research on body composition and its relevance for human biology, pp. 85-119. In: Human Body Composition: Approaches and Applications. Ed.: J. Brožek. Oxford: Pergamon Press, **1965**.

Bruce, R. A., Alexander, E. R., Li, Y. B., Chiang, B. N., Ting, N., and Hornsten, T. R. Electrocardiographic responses to maximal exercise in American and Chinese population samples, pp. 413-444. In: Measurement in Exercise Electrocardiography. Ed.: H. Blackburn. Springfield, Ill.: C. C. Thomas, **1969**.

Bruce, R. A., Blackmon, J. R., Jones, J. W., and Strait, G. Exercise testing in adult normal subjects and cardiac patients. Pediatrics 32 (Suppl.): 742-756, **1963**.

Bruce, R. A., Kusumi, F., Culver, B. H., and Butler, J. Cardiac limitation to maximal oxygen transport and changes in components after jogging across the U.S. J. Appl. Physiol. 39: 958-964, **1975**.

Bruch, H. Obesity in childhood. Energy expenditure of obese children. Amer. J. Dis. Child. 60: 1082-1109, **1940**.

Bruderman, I., Somers, K., Hamilton, W. K., Tooley, W. H., and Butler, J. Effect of surface tension on circulation in the excised lungs of dogs. J. Appl. Physiol. 19: 707-712, **1964**.

Bruley, D. F., Reneau, D. D., Richer, H. I., and Kniseley, M. H. Modeling cerebral tissue oxygenation with autoregulation. In: Oxygen Supply: Theoretical and Practical Aspects of Oxygen Supply and Microcirculation of Tissues. Eds.: M. Kessler, D. F. Bruley, L. C. Clark, D. W. Lubbers, I. A. Silver, and J. Strauss. Baltimore: University Park Press, **1973**.

Brundin, A. Physical training in severe chronic obstructive lung disease. 1. Clinical course, physical working capacity and ventilation. 2. Observations on gas exchange. Scand. J. Resp. Dis. 55: 25-36, **1975**.

Brundin, T. and Cernigliaro, C. The effect of physical training on the sympathoadrenal response to exercise. Scand. J. Clin. Lab. Invest. 35: 525-530, **1975**.

Brundtland, G. H. and Walloe, L. Menarcheal age in Norway: Halt in the trend towards earlier maturation. Nature 241: 478-479, **1973**.

Brunton, T. L. On the action of nitrite of amyl on the circulation. J. Anat. Physiol. 5: 92-101, **1871**.

Brusis, O. A. Guidelines for supervised and nonsupervised cardiac rehabilitation programs, pp. 233-245. In: Coronary Heart Disease, Exercise Testing and Cardiac Rehabilitation. Eds.: W. E. James and E. A. Amsterdam. Miami, Fla.: Symposia Specialists, **1977**.

Brust, M. Effects of temperature and twitch rate on fatigue in fast and slow muscles. Proc. Int. Union Physiol. Sci. 9: 87, **1971**.

Bryan, A. C. Commentary, Can. Med. Assoc. J. 96: 804, **1967**.

Bryan, A. C., Bentivoglio, L. G., Beerel, F., MacLeish, H., Zidulka, A., and Bates, D. V. Factors affecting regional distribution of ventilation and perfusion in the lung. J. Appl. Physiol. 19: 395-402, **1964**.

Bubenheimer, P., Samek, L., Schmeisser, H. J., and Roskamm, H. Echocardiographic evaluation of left ventricular function during exercise in untrained young men and athletes. In: Sports Cardiology. Eds.: T. Lubich and A. Venerando. Bologna: Aulo Gaggi, **1979**.

Buccola, V. A., and Stone, W. J. Effects of jogging and cycling programs on physiological and personality variables in aged men. Res. Quart. 46: 134-139, **1975**.

Bücher, T. N. and Rüssmann, W. Gleichgewicht und Ungleichgewicht in System der Glykolyse. Angew. Chem. 75: 881-893, **1963**.

Buchthal, F. and Schmalbruch, H. Contraction times of twitches evoked by H-reflexes. Acta Physiol. Scand. 80: 378-382, **1970**.

Buckle, R. M. Exertional (March) hemoglobinuria. Reduction of hemolytic episodes by use of Sorbo-Rubber insoles in shoes. Lancet 1: 1136-1138, **1965**.

Buckler, J. M. The effect of age, sex and exercise on the secretion of growth hormone, Clin. Sci. 37: 765-774, **1969**.

——. The relationship between changes in plasma growth hormone levels and body temperature occurring with exercise in man. Biomedicine 19: 193-197, **1973**.

Buckles, R. G. The physics of bubble formation and growth. Aerospace Med. 39: 1062-1069, **1968**.

Buckley, J. M. and Souhrada, J. F. A comparison of pulmonary function tests in detecting exercise-induced broncho-constriction. Pediatrics 56 (Suppl.): 883-889, **1975**.

Buderer, M. C., Rummel, J. A., Michel, E. L., Mauldin, D. G., and Sawin, C. F. Exercise cardiac output following Sky Lab missions: the second manned Sky Lab mission. Aviat. Space Environ. Med. 47: 365-372, **1976**.

Bühlmann, A. Atemsphysiologische Aspekte des Tauchens. In: Der Weg in die Tiefe. Bulletin 3: Documenta Geigy. Basel: Geigy, **1961**.

Bühlmann, A. A. Decompression in saturation diving, pp. 221-227. In: Underwater Physiology. Ed.: C. J. Lambertsen. New York: Academic Press, **1971**.

Bühlmann, A. A., Frei, P., and Keller, H. Saturation and desaturation with N_2 and He_2 at 4 atm. J. Appl. Physiol. 23: 458-462, **1967**.

Bühlmann, A. A., Spiegel, M., and Straub, P. W. Hyperventilation und Hypovolämie bei Leistungssport in mittleren Höhen. Schweiz. Med. Wschr. 99: 1886-1894, **1969**.

Buick, F. J., Gledhill, N., Froese, A. B., Spriet, L., and Meyers, E. C. Double blind study of blood boosting in highly trained runners. Med. Sci. Sport. 10: 49, **1978**.

Bullard, R. W. and Rapp, G. M. Problems of body heat loss in water immersion. Aerospace Med. 41: 1269-1277, **1970**.

Buller, A. J. The muscle spindle and the control of movement, pp. 11-17. In: Breathlessness. Eds. J. B. L. Howell and E. J. M. Campbell. Oxford: Blackwell Scientific, **1966**.

Buller, A. J., Eccles, J. C., and Eccles, R. M. Interactions between motoneurons and muscles in respect of the characteristic speeds of their responses. J. Physiol. (Lond.) 150: 417-439,

1960.

Buller, N. P., Garnett, R., and Stephens, J. A. The use of skin stimulation to produce reversal of motor unit recruitment order during voluntary contraction in man. J. Physiol. (Lond.) 277: 1p--2p, **1978**.

Bunning, E. The Physiological Clock. Berlin: Springer Verlag, **1964**.

Burch, G. E., Ray, C. T., and Cronwick, J. A. Certain mechanical peculiarities of human cardiac pump in normal and diseased states. Circulation 5: 504–513, **1952**.

Burcher, E. and Garlick, D. Effects of exercise metabolites on adrenergic vaso-constriction in gracilis muscle of the dog. J. Pharmacol. Exp. Ther. 192: 149–156, **1975**.

Burdette, W. J. and Ashford, T. P. Structural changes in human myocardium following hypoxia. J. Thor. Cardiovasc. Surg. 50: 210–220, **1965**.

Burger, G. C. E. Permissible load and optimal adaptation. Ergonomics 7: 397–417, **1964**.

Burger, M. Zur Pathophysiologie der Geschlecter. Munch. Med. Wschr. 97: 981–988, **1955**.

Burgess, P. R. and Clark, J. F. Characteristics of knee joint receptors in the cat. J. Physiol. (Lond.) 203: 317–335, **1969**.

Burgess, P. R. and Perl, E. R. Cutaneous mechanoreceptors and nociceptors. In: Handbook of sensory physiology, Vol. 2. Ed.: A. Iggo. New York: Springer Verlag, **1973**.

Burke, D., Hagbarth, K. E., and Skuse, N. E. Recruitment order of human spindle endings in isometric voluntary contractions. J. Physiol. (Lond.) 285: 101–112, **1978**.

Burke, R. E. On the central nervous system control of fast and slow twitch motor units. In: New Developments in Electromyography and Clinical Neurophysiology. Ed.: J. E. Desmedt. Vol. III, pp. 69–94. Basel: Karger, **1973**.

Burke, R. E. and Edgerton, V. R. Motor unit properties and selective involvement in movement. In: Exercise and Sport Sciences Reviews. Ed.: J. H. Wilmore, Vol. 3, pp. 31–81. New York: Academic Press, **1975**.

Burke, R. E., Rudomin, P., and Zajac, F. E. Catch property in single mammalian motor units. Science 168: 122–124, **1970**.

Burke, R. E., Rymer, W. Z., and Walsh, J. V. Functional specialization in the motor unit population of cat medial gastrocnemius muscle, pp. 29–44. In: Control of Posture and Locomotion. Eds.: R. B. Stein, K. G. Pearson, R. S. Smith, and J. B. Redford. New York: Plenum Press, **1973**.

Burke, R. E. and Tsairis, P. Anatomy and innervation ratios in motor units of cat gastrocnemius. J. Physiol. (Lond.) 234: 749–765, **1973**.

——. The correlation of physiological properties with histochemical characteristics in single muscle units. Ann. N.Y. Acad. Sci. 228: 145–159, **1974**.

Burke, R. K. Relationship between physical performance and warm-up procedures of varying intensity and duration. Ph.D. thesis. Los Angeles: University of Southern California, **1957**.

Burki, N. K. Effects of immersion in water and changes in intrathoracic blood volume on lung function in man. Clin. Sci. Mol. Med. 51: 303–311, **1976**.

Burn, J. H. The action of nicotine on the peripheral circulation. Ann. N. Y. Acad. Sci. 90: 81–84, **1960**.

Burns, J. W. and Covell, J. W. Myocardial oxygen consumption during isotonic and isovolumic contractions in the intact heart. Amer. J. Physiol. 223: 1491–1497, **1972**.

Burr, H. T. Psychological Functioning of Older People. Springfield, Ill.: C. C. Thomas, **1971**.

Burri, P. H. and Weibel, E. R. Morphometric evaluation of changes in lung structure due to high altitude. In: High Altitude Physiology; Cardiac and Respiratory Aspects. Eds.: R. Porter and J. Knight. Edinburgh: Churchill-Livingstone, **1971**.

Burt, J. J. and Jackson, R. The effects of physical exercise on the coronary collateral circulation of dogs. J. Sport. Med. Phys. Fitness 5: 203–206, **1965**.

Burton, A. C. Human calorimetry. Ill. The average temperature of the tissues of the body. J. Nutr. 9: 261–280, **1935**.

——. Relation of structure to function of the tissues of the wall of blood vessels. Physiol. Rev. 34: 619–642, **1954**.

——. The importance of the shape and size of the heart. Amer. Heart J. 54: 801–810, **1957**.

——. Hemodynamics and the physics of the circulation, pp. 523–542. In: Physiology and Biophysics. Eds.: T. C. Ruch and H. D. Patton. Philadelphia: W. B. Saunders, **1965a**.

——. Physiology and Biophysics of the Circulation, pp. 1–217. Chicago: Year Book Publishers, **1965b**.

Burton, A. C. and Edholm, O. G. Man in a Cold

Environment. New York: Hafner, **1969**.

Burton, H. W., Taylor, P. B., and Hermiston, R. T. A perfused rat hindquarter system for examining the relationship between oxygen consumption and muscle temperature. Can. J. Appl. Sport. Sci. 3: 193, **1978**.

Buser, P. Higher functions of the nervous system. Ann. Rev. Physiol. 38: 217–245, **1976**.

Buskirk, E. R. Problems related to conduct of athletes in hot environments. In: Physiological Aspects of Sports and Physical Fitness. Chicago: Athletic Institute, **1968**.

Buskirk, E. R. and Beetham, W. Dehydration and body temperature as a result of marathon running. Medicina Sportiva 14: 493, **1960**.

Buskirk, E. R., Iampietro, P. F., and Bass, D. E. Work performance after dehydration: Effects of physical conditioning and heat acclimatization. J. Appl. Physiol. 12: 189–194, **1958**.

Buskirk, E. R., Kollias, J., Reatigui, E. P. Akers, R., Prokop, E., and Baker, P. Physiology and performance of track athletes at various altitudes in the United States and Peru, pp. 65–71. In: International Symposium on the Effects of Altitude on Physical Performance. Ed.: R. Goddard. Chicago: Athletic Institute, **1967a**.

Buskirk, E. R., Kollias, J., Akers, R. F., Prokop, E. K., and Reategui, E. P. Maximal performance at altitude and on return from altitude in conditioned runners. J. Appl. Physiol. 23: 259–266, **1967b**.

Buskirk, E. R. and Mendez, J. Nutrition, environment and work performance with special reference to altitude. Fed. Proc. 26: 1760–1767, **1967**.

Buskirk, E. R. and Taylor, H. L. Maximal oxygen intake and its relation to body composition, with special reference to chronic physical activity and obesity. J. Appl. Physiol. 11: 72–78, **1957**.

Buzina, R. Nutrition status, working capacity and absenteeism in industrial workers. Proceedings, 1st Internat. Symposium on "Alimentation et Travail." Paris: Masson, **1972**.

Byrne, W. L., Samuel, D., Bennett, E. L. et al. Memory transfer. Science 153: 658–659, **1966**.

Byrne-Quinn, E., Sodal, I. E., and Weil, J. V. Hypoxic and hypercapnic ventilatory drives in children native to high altitude. J. Appl. Physiol. 32: 44–46, **1972**.

Byrne-Quinn, E., Weil, J. W., Sodal, I. E., Filley, G. F., and Grover, R. F. Ventilatory control in the athlete. J. Appl. Physiol. 30: 91–98, **1971**.

Byrnes, W. C. and Kearney, J. T. Measures of Max $\dot{V}_{O_2}$ and their relationship to running performance among three subject groups. Amer. Corr. Ther. J. Sept/Oct. 1974: 145–150, **1974**.

Cabanac, M. Temperature regulation. Ann. Rev. Physiol. 37: 415–439, **1975**.

Cabanac, M., Hildebrandt, G., Massonnet, B., and Strempel, H. A study of the nycthemeral cycle of behavioural temperature regulation in man. J. Physiol. (Lond.) 257: 275–291, **1976**.

Caceres, C. A. and Hochberg, H. M. Performance of the computer and the physician in the analysis of the electrocardiogram. Amer. Heart J. 79: 439–443, **1970**.

Cafarelli, E. Peripheral and central inputs to the effort sense during cycling exercise. Europ. J. Appl. Physiol. 37: 181–189, **1977**.

Cafarelli, E. and Bigland-Ritchie, B. Sensation of force in muscles of different lengths. Med. Sci. Sport. 10: 66, **1978**.

Cafarelli, E., Cain, W. S., and Stevens, J. S. Effort of dynamic exercise: Influence of load, duration and task. Ergonomics 20: 147–158, **1977**.

Caffey, J., Ames, R., Silverman, W. A., Ryder, C. T., and Hough, G. Contradiction of congenital dysphasia: Predislocation hypothesis of congenital dislocation of the hip through study of normal variation in acetabular angles at successive periods in infancy. Paediatrics 17: 632–640, **1956**.

Cain, D. F., Infante, A. A., and Davies, R. E. Chemistry of muscle contraction. Adenosine triphosphate and phosphoryl creatine as energy supplies for single contractions of working muscle. Nature (Lond.) 196: 214–217, **1962**.

Caldarera, C. M., Casti, A., Rossoni, C., and Visioli, O. Polyamines and noradrenaline following myocardial hypertrophy. J. Mol. Cell Cardiol. 3: 121–126, **1971**.

Caldwell, M. D. and Zauner, C. W. Oxygen uptake in well-trained athletes at various points in a distance run. J. Sport. Med. Phys. Fitness 18: 19–23, **1978**.

Callaway, D. H. Nitrogen balance of men with marginal intakes of protein and energy. J. Nutr. 105: 914–923, **1975**.

Cameron, D. P., Burger, H. G., Catt, K. J. et al.

Metabolic clearance rate of radioiodinated human growth hormone in man. J. Clin. Invest. 48: 1600–1608, **1969**.

Campbell, D. E. Influence of several physical activities on serum cholesterol concentrations in young men. J. Lipid Res. 6: 478–480, **1965**.

Campbell, E. J. M. Motor pathways. In: Handbook of Physiology. Section 3. Respiration, Vol. 1. Eds.: W. O. Fenn and H. Rahn. Washington, D. C.: American Physiological Society, **1964**.

——. The relationship of the sensation of breathlessness to the act of breathing. In: Breathlessness. Eds.: J. B. L. Howell and E. J. M. Campbell. Oxford: Blackwell, **1966**.

Campbell, E. J. M. et al. Respiratory Muscles: Mechanics and Neural Control (2nd ed.). Philadelphia: Saunders, **1970**.

Campbell, J. and Rastogi, K. S. Action of growth hormone: Enhancement of insulin utilization with inhibition of insulin effect on blood glucose in dogs. Metab. Clin. Exp. 18: 930–944, **1969**.

Campbell, J. M. H., Mitchell, M. B., and Powell, A. T. W. The influence of exercise on digestion. Guy's Hosp. Rept. 78: 279–293, **1928**.

Campbell, R. L. Effects of supplemental weight training on the physical fitness of athletic squads. Res. Quart. 33: 343–348, **1962**.

Campbell, W. R. and Pohndorf, R. H. Physical fitness of British and United States children. In: Health and Fitness in the Modern World. Ed.: L. A. Larson. Chicago: Athletic Institute, **1961**.

Campion, D. S. Resting membrane potential and ionic distribution in fast and slow twitch mammalian muscle. J. Clin. Invest. 54: 514–518, **1974**.

Canabac, M., Lacaisse, A., Pasquis, P., and Dejours, P. Caractères et méchanisme des réactions ventilatoires au frisson thermique chez l'homme. C. R. Soc. Biol., Paris, 158: 80–84, **1964**.

Canadian Association for Health, Physical Education and Recreation. Fitness performance test manual. Toronto: Canad. Assoc. for Health, Physical Education and Recreation, **1966**.

Candas, V., Vogt, J. J., and Libert, J. P. Évolution des températures rectales et cutanées lors de l'exercise musculaire effectué dans des températures d'air comprises entre 10 et 30°C. Arch. Sci. Physiol. 27: A239–A246, **1973**.

Cander, L. and Forster, R. E. Determination of pulmonary parenchymal tissue volume and pulmonary capillary blood flow in man. J. Appl. Physiol. 14: 541–551, **1959**.

Canfield, A. A., Comrey, A. L., and Wilson, R. C. The influence of increased positive G on reaching movements. J. Appl. Psychol. 37: 230–235, **1953**.

Cannon, P. and Gould, T. R. Treatment of severe decompression sickness in aviators. Brit. Med. J. 1: 278–282, **1964**.

Cantone, A. and Cerretelli, P. Effect of training on proteinuria following muscular exercise. Int. Z. Angew. Physiol. 18: 324–329, **1960**.

Capel, L. H. and Smart, J. The forced expiratory volume after exercise, forced inspiration, and the Valsalva and Müller manoeuvres. Thorax 14: 161–165, **1959**.

Capen, E. K. The effect of systematic weight training on power, strength, and endurance. Res. Quart. 21: 83–93, **1950**.

Carlile, F. Effect of preliminary passive warming-up on swimming performance. Res. Quart. 27: 143–151, **1956**.

Carlson, D. S., Armelagos, G. J., and Van Gerven, D. P. Patterns of age-related cortical bone loss (osteoporosis) within the femoral diaphysis. Human Biol. 48: 295–314, **1976**.

Carlson, F. D. and Wilkie, D. R. Muscle Physiology. Englewood Cliffs, N.J.: Prentice Hall, **1974**.

Carlson, L. A. Plasma lipids and lipoproteins and tissue lipids during exercise. In: Nutrition and Physical Activity, p. 16. Ed.: G. Blix. Uppsala: Almqvist & Wiksell, **1967a**.

——. Lipid metabolism and muscular work. Fed. Proc. 26: 1755–1759, **1967b**.

Carlson, L. A. and Lindstedt, S. The Stockholm prospective study. I. The initial values for plasma lipids. Acta Med. Scand. Suppl. 493: 1–135, **1968**.

Carlsöo, S. Influence of frontal and dorsal loads on muscle activity and on the weight distribution in the feet. Acta Orthop. Scand. 34: 299–309, **1964**.

Carlsöo, S. and Edfeldt, A. W. Attempts at muscle control with visual and auditory impulses as auxiliary stimuli. Scand. J. Psychol. 4: 231–235, **1963**.

Carlsöo, S. and Johansson, O. Stabilization of and load on the elbow joint in some protective movements. Acta Anat. 48: 224–231, **1962**.

Carlsten, A. Influence of leg varicosities on the physcal work performance, pp. 207–214. In: Environmental Effects on Work Performance.

Eds.: G. R. Cumming, D. Snidal, and A. W. Taylor. Ottawa: Can. Assoc. Sports Sciences, **1972**.

Carlsten, A., Hallgren, B., Jagenburg, R., Svanborg, A., and Werko, L. Arterial concentration of free fatty acids and free amino acids in healthy human individuals at rest and at different work loads. Scand. J. Clin. Lab. Invest. 14: 185–191, **1962**.

Carlyle, R. F., Nichols, G., Paciorek, J. A., Rowles, P., and Spencer, N. Changes in morphology and carbonic anhydrase content of red blood cells from men subjected to simulated dives. J. Physiol. (Lond.) 288: 46p–47p, **1979**.

Carney, J. A. and Brown, A. L. Myofilament diameter in the normal and hypertrophic rat myocardium. Amer. J. Pathol. 44: 521–529, **1964**.

Carrière, L. Les effets de la competition des résponses et du contexte sur la prise de décision dans des situations problèmes, pp. 77–84. In: Motor Learning, Sport, Psychology, Pedagogy and Didactics of Physical Activity. Eds.: F. Landry and W. R. Orban. Miami: Symposia Specialists, **1978**.

Carron, A. V. Motor performance under stress. Res. Quart. 39: 463–469. **1968**.

Carrow, R. E., Brown, R. E., and Van Huss, W. D. Fiber sizes and capillary to fiber ratios in skeletal muscle of exercised rats. Anat. Rec. 159: 33–39, **1967**.

Carruthers, B. Ponte, J., and Purves, M. J. Observations on the partial pressure of CO_2 in carotid arterial blood in cats. J. Physiol. (Lond.) 284: 166p–167p, July, **1978**.

Carson, R. P., Evans, W. O., Shields, J. L., and Hannon, J. P. Symptomatology, pathophysiology, and treatment of acute mountain sickness. Fed. Proc. 28: 1085–1091, **1969**.

Carter, J. E. and Phillips, W. H. Structural changes in exercising middle-aged males during a 2-year period. J. Appl. Physiol. 27: 787–794, **1969**.

Cartlidge, N. E. F., Black, M. M., Hall, M. R. P., and Hall, R. Pituitary function in the elderly. Gerontologica Clinica 12: 65–70, **1970**.

Casaburi, R., Whipp, B. J., Wasserman, K., and Koyal, S. Ventilatory and gas exchange responses to cycling with sinusoidally varying pedal rate. J. Appl. Physiol. Resp. Env. Ex. Physiol. 44: 97–103, **1978**.

Cash, J. D. and McGill, R. C. Fibrinolytic response to moderate exercise in young male diabetics and non-diabetics. J. Clin. Pathol. 22: 32–35, **1969**.

Cassels, D. E. and Morse, M. Blood volume and exercise. J. Pediatr. 20: 352–364, **1942**.

Castelli, W. P., Doyle, J. T., Gordon, T., Hames, C. G., Hjortland, M. C., Hulley, S. B., Kagan, A., and Zuke, W. J. HDL cholesterol and other lipids in coronary heart disease: The cooperative lipoprotein phenotyping study. Circulation 55: 767–772, **1977**.

Castenfors, J. Renal function during exercise. With special reference to exercise proteinuria and the release of renin. Acta Physiol. Scand. 70 (Suppl. 293): 1–44, **1967**.

——. Renal function during prolonged exercise. In: The marathon: Physiological, medical, epidemiological, and psychological studies. Ann. N.Y. Acad. Sci. 301: 151–159, **1977**.

Castenfors, J., Mossfeldt, F., and Piscator, M. Effect of prolonged heavy exercise on renal function and urinary protein excretion. Acta Physiol. Scand. 70: 194–206, **1967**.

Cater, D. B. The measurement of P_{O_2} in tissues. In: Oxygen in the Animal Organism, pp. 239–246. Eds.: F. Dickens and E. Neil. New York: Macmillan, **1964**.

Cathcart, E. P. and Burnett, W. A. Influence of muscle work on metabolism in varying conditions of diet. Proc. Roy. Soc. Lond. (B) 99: 405–426, **1926–27**.

Cathcart, E. P., Richardson, D. T., and Campbell, W. On the maximum load to be carried by the soldier. J. Roy. Army Med. Corps. Lond. 40: 35; 41: 12, 87: 161; **1923**.

Cavagna, G. A. Elastic bounce of the body. J. Appl. Physiol. 29: 279–282, **1970**.

Cavagna, G. A., Citterio, G., and Jacini, P. The additional mechanical energy delivered by the contractile component of the previously stretched muscle. J. Physiol. (Lond.) 251: 65p–66p, **1975**.

Cavagna, G. A., Dusman, B., and Margaria, R. Positive work done by a previously stretched muscle. J. Appl. Physiol. 24: 21–32, **1968**.

Cavagna, G. A., Komarek, L., and Mazzoleni, S. The mechanics of sprint running. J. Physiol. (Lond.) 217: 709–721, **1971**.

Cavagna, G. A., Margaria, R., and Arcelli, R. A high speed motion picture analysis of the work performed in sprint running. Res. Film 5: 309–319, **1965**.

Cavagna, G. A., Saibene, F. P., and Margaria, R. External work in walking. J. Appl. Physiol.

18: 1–9, **1963**.

——. Effect of negative work on the amount of positive work performed by soladid muscle. J. Appl. Physiol. 20: 157–158, **1964a**.

——. Mechanical work in running. J. Appl. Physiol. 19: 249–256, **1964b**.

Cavagna, G. A., Thys, H. A., and Zamboni, A. The sources of external work in level walking and running. J. Physiol. (Lond.) 262: 639–657, **1976**.

Cavalli-Sforza, L. L., and Bodmer, W. C. The Genetics of Human Populations. San Francisco: Freeman, **1971**.

Cavanagh, P. R. Pollock, M. L., and Landa, J. A biomechanical comparison of elite and good distance runners. Ann. N. Y. Acad. Sci. 301: 328–345, **1977**.

Celejowa, I. Energy balance and nutrient requirements of Polish athletes, pp. 185–196. In: Nutrition, Dietetics and Sport. Eds.: G. Ricci and A. Venerando. Torino: Minerva Medica, **1978**.

Celejowa, I. and Homa, M. Food intake, nitrogen and energy balance in Polish weight-lifters during a training camp. Nutr. Metab. 12: 259–274, **1970**.

Cellina, G. U., Honour, A. J., and Littler, W. A. Direct arterial pressure, heart rate and electrocardiogram during cigarette smoking in unrestricted patients. Amer. Heart J. 89: 18–25, **1975**.

Cermak, J. Changes of the heart volume and of the basic somatometric indices in 12–15 years old boys with an intense exercise regime. A long term study. Brit. J. Sport. Med. 7: 241–244, **1973**.

Cerny, F. Protein metabolism during two hour ergometer exercise. In: Metabolic Adaptation to Prolonged Physical Exercise. Eds.: H. Howald and J. R. Poortmans. Basel: Birkhauser Verlag, **1975**.

Cerny, L. The results of an evaluation of skeletal age of boys 11–15 years old with different regime of physical activity. In: Physical Fitness Assessment. Ed.: V. Seliger. Prague: Charles University, **1969**.

Cerretelli, P. Lactacid O_2 debt in acute and chronic hypoxia, pp. 58–64. In: Exercise at Altitude. Ed: R. Margaria. Dordrecht: Excerpta Medica Fdn, **1967**.

Cerretelli, P., Ambrosoli, G., and Fumagalli M. Anaerobic recovery in man. Europ. J. Appl. Physiol. 34: 141–148, **1975**.

Cerretelli, P. and Debijadji, R. Work capacity and its limiting factors at high altitude, pp. 233–247. In: The Physiological Effects of High Altitude. Ed.: W. H. Weihe. London: Pergamon Press, **1964**.

Cerretelli, P., Sikand, R., and Farhi, L. E. Readjustments in cardiac output and gas exchange during onset of exercise and recovery. J. Appl. Physiol. 21: 1345–1350, **1966**.

Chaffin, D. B. Localized muscle fatigue: Definition and measurement. J. Occup. Med. 15: 346–354, **1973**.

——. Ergonomics guides. Ergonomic guide for the assessment of human static strength. Amer. Industr. Hyg. J. 36: 505–511, **1975**.

Chaffin, D. B. and Monlis, E. J. An empirical investigation of low back strains and vertebral geometry. Unpublished rept. Western Electric Company USA, **1968**.

Chamberlain, D. A. and Clark, A. N. G. Atrial fibrillation complicating Wolff-Parkinson White syndrome treated with amiodarone. Brit. Med. J. 2: 1519–1520, **1977**.

Chance, B. Cellular oxygen requirements. Fed. Proc. 16: 671–680, **1957**.

——. Molecular basis of O_2 affinity for cytochrome oxidase. In: Oxygen and physiological function. Ed.: F. F. Jöbsis. Dallas, Texas: Professional Information Library, **1977**.

Chance, B. and Pring, M. Logic in the design of the respiratory chain. In: Biochimie des Sauerstoffs, pp. 120–130. Eds.: B. Hess and H. Standinger. Berlin: Springer Verlag, **1968**.

Chance, B., Schoener, B., and Schindler, F. The intracellular oxidation-reduction state. In: Oxygen in the Animal Organism, pp. 367–392. Eds.: F. Dickens and E. Neil. New York: Macmillan, **1964**.

Chandler, B. M., Sonnenblick, E. H., and Pool, P. E. The mechanochemistry of cardiac muscle. III. The effects of norepinephrine on the utilization of high energy phosphates. Circ. Res. 22: 729–735, **1968**.

Chapman, C. B., Henschel, A., Minckler, J., Forsgren, A., and Keys. A. The effect of exercise on renal plasma flow in normal male subjects. J. Clin. Invest. 27: 639–644, **1948**.

Chapman, C. B., Taylor, H. L., Baden, C., Evert, R. V., Keys, A., and Carlson, W. S. Simultaneous determination of the resting arterio-venous oxygen difference by acetylene and direct Fick methods. J. Clin. Invest. 29: 651–659, **1950**.

Chapman, D. and Horvath, S. M. Unpublished report, cited by Drinkwater, **1973**.

Chapman, E. A., deVries, H. A., and Swezey, R. Joint stiffness: Effect of exercise on young and old men. J. Gerontol. 27: 218–221, **1972**.

Chapman, J. M. and Massey, F. J. The inter-relationship of serum cholesterol, hypertension, body weight and risk of coronary disease. Results of the first ten years' follow up in the Los Angeles Heart Study. J. Chron. Dis. 17: 933–949, **1964**.

Charbonnier, J. P., LaCour, J. R., Riffat, J., and Flandrois, R. Experimental study of the performance of competition swimmers. Europ. J. Appl. Physiol. 34: 157–167, **1975**.

Chauveau, A. Source et Nature du Potential Directement Utilisé dans le travail musculaire d'après les échanges respiratoires, chez l'homme en état d'abstinence. C. R. Acad. Sci. (Paris) 122: 1163–1986, **1896**.

Cheek, D. B. Human Growth. Philadelphia: Lea & Febiger, **1968**.

Cheng, T. O., Godfrey, M. P., and Shepard, R. H. Pulmonary resistance and state of inflation of lungs in normal subjects and in patients with airway obstruction. J. Appl. Physiol. 14: 727–732, **1959**.

Cherniack, N. S. and Longobardo, G. S. Oxygen and carbon dioxide gas stores in the body. Physiol. Rev. 50: 196–243, **1970**.

Chevalier, R. B., Bowers, J. A., Bondurant, S., and Ross, J. C. Circulatory and ventilatory effects of exercise in smokers and nonsmokers. J. Appl. Physiol. 18: 357–360, **1963**.

Chiang, B. N., Perlman, L. V., Ostrander, L. D., and Epstein, F. H. Relationship of premature systoles to coronary heart disease and sudden death in the Tecumseh epidemiologic study. Ann. Int. Med. 70: 1159–1166, **1969**.

Chiang, S. T., Steigbigel, N. H., and Lyons, H. A. Pulmonary compliance and non-elastic resistance during treadmill exercise. J. Appl. Physiol. 20: 1194–1198, **1965**.

Chin, A. and Evonuk, E. Changes in plasma catecholamines and corticosterone levels after muscular exercise. J. Appl. Physiol. 30: 205–207, **1971**.

Chin, A. K., Seaman, R. and Kapileshwarker, M. Plasma catecholamine response to exercise and cold adaptation. J. Appl. Physiol. 34: 409–412, **1973**.

Chinard, F. P. The permeability characteristics of the pulmonary blood-gas barrier, pp. 106–147. In: Advances in Respiratory Physiology. Ed.: C. G. Caro. London: Arnold, **1966**.

Chirico, A. M. and Stunkard, A. J. Physical activity and human obesity. N. Engl. J. Med. 263: 935–940, **1960**.

Choquette, G. and Ferguson, R. J. Blood pressure reduction in "borderline" hypertensives following physical training. Can. Med. Assoc. J. 108: 699–703, **1973**.

Chouteau, J. Saturation diving: The Con shelf experiments. In: The Physiology and Medicine of Diving and Compressed Air Work, pp. 491–504. Eds.: P. B. Bennett and D. H. Elliott. London: Baillière, **1969**.

Christensen, E. H. and Hansen, O. Zur Methodik der Respiratorischen Quotient: Bestimmungen in Ruhe und Arbeit. Scand. Arch. Physiol. 81: 137–151, **1939a**.

——. Arbeitsfähigkeit und Ernährung. Scand. Arch. Physiol. 81: 150–171, **1939b**.

Christensen, E. H., Hedman, R., and Saltin, B. Intermittent and continuous running. Acta Physiol. Scand. 50: 269–286, **1960**.

Christensen, E. H. and Högberg, P. Physiology of skiing. Int. Z. Angew. Physiol. 14: 292–303, **1950a**.

——. The efficiency of anaerobical work. Arbeitsphysiologie 14: 249–250, **1950b**.

Christie, D. Physical training in chronic obstructive lung disease. Brit. Med. J. 1: 150–151, **1968**.

Chu, D. A. and Smith, G. S. Isokinetic exercise: controlled speed and accommodating resistance p. 29–31. In: Analysis and Evaluation of Modern Trends Related to the Physiological Factors for Optimal Sports Performance. Ed.: E. J. Burke. Ithaca, N. Y. Ithaca State College, **1976**.

Chui, E. The effect of systematic weight training on athletic power. Res. Quart. 21: 188–194, **1950**.

Chui, E. F. Effects of isometric and dynamic weight training exercises upon strength and speed of movement. Res. Quart. 35: 246–257, **1964**.

Chvapil, M. In: Physiology of Connective Tissue, p. 333. Ed.: M. Chvapil. London: Butterworth, **1967**.

C. I. S. International Occupational Safety and Health Information Centre, ILO. Information Sheet, 3. Geneva: I. L. O., **1962**.

Clamann, H. P. Activity of single motor units during isometric tension. Neurology 20: 254–260, **1970**.

Clancy, L. J., Critchley, J. A. J. H., and Leitch, A. G. Circulating noradrenaline in the potentiation by hypoxia of the hyperventilation in exercise. J. Physiol. (Lond.) 246: 77p–78p, **1975**.

Claremont, A. D., Costill, D. L. Fink, W., and Van Handel, P. Heat tolerance following diuretic induced dehydration. Med. Sci. Sport. 8: 239–243, **1976**.

Claremont, A. D., Nagle, F., Reddan, W. D., and Brooks, G. A. Comparison of metabolic temperature, heart rate and ventilatory responses to exercise at extreme ambient temperatures (0°C and 35°C). Med. Sci. Sport. 7: 150–154, **1975**.

Clark, F. J. and Von Euler, C. On the regulation of depth and rate of breathing. J. Physiol. (Lond.) 222: 267–295, **1972**.

Clarke, C. and Duff, J. Mountain sickness, retinal haemorrhages, and acclimatisation on Mount Everest in 1975. Brit. Med. J. 2: 495–497, **1976**.

Clarke, D. H. Energy cost of isometric exercise. Res. Quart. 31: 3–6, **1960a**.

——. Correlation between the strength/mass ratio and the speed of an arm movement. Res. Quart. 31: 570–574, **1960b**.

——. Adaptations in strength and muscular endurance resulting from exercise. Ex. Sport. Sci. Rev. 1: 73–102, **1973**.

Clarke, D. H. and Henry, F. M. Neuromotor specificity and increased speed from strength development. Res. Quart. 32: 315–325, **1961**.

Clarke, H. H. Muscular Strength and Endurance in Man, 5th ed. Englewood Cliffs, N.J.: Prentice Hall, **1966**.

——. Posture. Phys. Fitness Res. Dig. 9: 1–23, **1979**.

Clarke, H. H. and Glines, D. Relationship of reaction, movement and completion times to motor strength, anthropometric and maturity measures of 13-year-old boys. Res. Quart. 33: 194–201, **1962**.

Clarke, K. S. Predicting certified weight of young wrestlers: a field study of the Tcheng-Tipton method. Med. Sci. Sports. 6: 52–57, **1974**.

Clarke, N. P., Smith, O. A., and Shearn, D. W. Topographical representation of vascular smooth muscle of limbs in primate motor cortex. Amer. J. Physiol. 214: 122–129, **1968**.

Clarke, R. S. J., Hellon, R. F., and Lind, A. R. The duration of sustained contractions of the human forearm at different muscle temperatures. J. Physiol. (Lond.) 143: 454–473, **1958**.

Clarkson, B., Thompson, D., Horwith, M., and Luckley, E. H. Cyclical edema and shock due to increased capillary permeability. Amer. J. Med. 29: 193–216, **1960**.

Clarys, J. P. An experimental investigation of the application of fundamental hydrodynamics to the human body. In: Swimming Medicine IV. Eds.: B. Eriksson and B. Fürberg. Baltimore: University Park, **1978**.

Clausen, J. P. Muscle blood flow during exercise and its significance for maximal performance, pp. 253–266. In: Limiting Factors of Human Performance. Ed.: J. Keul. Stuttgart: Thieme, **1973**.

——. Effects of physical training on cardiovascular adjustments to exercise in man. Physiol. Rev. 57: 779–815, **1977**.

Clausen, J. P. Klausen, K., Rasmussen, B., and Trap-Jensen, J. Central and peripheral circulatory changes after training of the arms and legs. Amer. J. Physiol. 225: 675–682, **1973**.

Clausen, J. P. and Lassen, N. A. Muscle blood flow during exercise in normal man studied by the ^{133}Xenon clearance method. Cardiovasc. Res. 5: 245–254, **1971**.

Clausen, J. P., Trap Jensen, J., and Lassen, N. A. The effects of training on the heart rate during arm and leg exercise. Scand. J. Clin. Lab. Invest. 26: 295–301, **1970**.

Claus-Walker, J., Spencer, W. A., Carter, R. E., Halstead, L. S., Meier, R. H., and Campos, R. J. Bone metabolism in quadriplegia: Dissociation between calciuria and hydroxyprolinuria. Arch. Phys. Med. Rehab. 56: 327–332, **1975**.

Cleland, T. S., Horvath, S. M., and Phillips, M. Acclimatization of women to heat after training. Int. Z. Angew. Physiol. 27: 15–24, **1969**.

Clement, D. B., Asmundson, R. C., and Medhurst, C. W. 3: Hemoglobin values: Comparative survey of the 1976 Canadian Olympic Team. Can. Med. Assoc. J. 117: 614–616, **1977**.

Clemmesen, S. Some studies of muscle tone. Proc. Roy. Soc. Med. 44: 637–646, **1951**.

Clode, M. and Campbell, E. J. M. The relationship between gas exchange and changes in blood lactate concentration during exercise. Clin. Sci. 37: 263–272, **1969**.

Close, R. O. Dynamic properties of mammalian skeletal muscles. Physiol. Rev. 52: 129–197,

1972.

Clough, J. F. M., Kernell, D., and Phillips, C. G. The distribution of monosynaptic excitation from the pyramidal tract and from primary spindle afferents to motoneurones of the baboon's hand and forearm. J. Physiol. (Lond.) 198: 145-166, **1968**.

Clough, J. F. M., Phillips, C. G., and Sheridan, J. D. The short latency projection from the baboon's motor cortex to fusimotor neurones of the forearm and hand. J. Physiol. (Lond.) 216: 257-279, **1971**.

Cobb, L., Baum, R. S., Alvarez, H. et al. Resuscitation from out-of hospital ventricular fibrillation: 4 years follow-up. Circulation 52 (Suppl. 3): 223-235, **1975**.

Coburn, R. F. and Mayers, L. B. Myoglobin O_2 tension determined from measurements of carboxy-myoglobin in skeletal muscle. Amer. J. Physiol. 220: 66-74, **1971**.

Cockett, A. T. K., Nakamura, R. M., and Kado, R. T. Physiological factors in decompression sickness. Arch. Environ. Health 11: 760-764, **1965**.

Cockett, A. T. K., Pauley, S. M., Saunders, J. C., and Hirose, F. M. Coexistence of lipid and gas emboli in experimental decompression sickness, pp. 245-250. In: Underwater Physiology. Ed.: C. J. Lambertsen. New York: Academic Press, **1971**.

Cohen, C. Architecture of the α class of fibrous proteins. In: Molecular Architecture in Cell Physiology, pp. 169-190. Eds: T. Hayashi and A. G. Szent-Györgyi. Englewood Cliffs, N. J.: Prentice Hall, **1966**.

Cohen, I. and Zimmerman, A. L. Changes in serum electrolyte levels during marathon running. S. Afr. Med. J. 53: 449-453, **1978**.

Cohen, R. D. and Iles, R. A. Lactic acidosis: Some physiological and clinical considerations. Clin. Sci. Mol. Med. 52: 405-410, **1977**.

Cohen, R. D. and Woods, H. F. Clinical and biochemical aspects of lactic acidosis. Oxford: Blackwell Scientific, **1976**.

Cohn, C. and Joseph, D. Influence of body weight and body fat on appetite of "normal" lean and obese rats. Yale J. Biol. Med. 34: 598-607, **1962**.

Cohn, P. F., Gabbay, S. I., and Weglicki, W. B. Serum lipid levels in angiographically-defined coronary artery disease. Ann. Int. Med. 84: 241-245, **1976**.

Cole, J. P. and Ellestad, M. H. Significance of chest pain during treadmill exercise: correlation with coronary events. Amer. J. Cardiol. 41: 227-232, **1978**.

Coleman, A. E. Validity of distance runs with elementary school children. Paper presented at 1974 Southern District Convention of AAHPER (Norfolk, Va). Cited by Disch, **1975**.

Coleridge, J. C. G. and Kidd, C. Electrophysiological evidence of baroreceptors in the pulmonary artery of the dog. J. Physiol. (Lond.) 150: 319-331, **1960**.

——. Reflex effects of stimulating baroreceptors in the pulmonary artery. J. Physiol. (Lond.) 166: 197-210, **1963**.

Coley, B. L. and Moore, M. Caisson disease with special reference to bones and joints; report of 2 cases. Ann. Surg. 111: 1065-1075, **1940**.

Colin-Jones, D. J. and Himsworth, R. L. The location of the osmoreceptor controlling gastric acid secretion during hypoglycaemia. J. Physiol. (Lond.) 206: 397-409, **1970**.

Collins, K. J., Few, J. D., Forward, T. J., and Giec, L. A. Stimulation of adrenal glucocorticoid secretion in man by raising the body temperature J. Physiol. (Lond.) 202: 645-660, **1969**.

Collins, W. E. Adaptation to vestibular disorientation. II. Nystagmus and vertigo following high velocity angular accelerations. Oklahoma City: Federal Aviation Agency, **1965**.

Comfort, A. Theories of aging. In: Textbook of Geriatric Medicine and Gerontology. Ed.: J. C. Brocklehurst. Edinburgh: Churchill-Livingstone, **1973**.

Comroe, J. H. Dyspnea. Mod. Conc. Cardiovasc. Dis. 25: 347-349, **1956**.

——. Physiology of Respiration, pp. 1-245. Chicago: Year Book Publishers, **1943**.

Comroe, J. H., Forster, R. E., DuBois, A. B., Briscoe, W. A., and Cournand, A. The Lung: Clinical Physiology and Pulmonary Function Tests. Chicago: Year Book Publishers, **1955**.

Comroe, J. H. and Schmidt, C. F. Reflexes from the limbs as a factor in the hyperpnea of muscular exercise. Amer. J. Physiol. 138: 536-547, **1943**.

Comstock, G. W. and Livesay, V. T. Subcutaneous fat determinations from a community-wide chest X-ray survey in Muscogee County, Georgia. Ann. N.Y. Acad. Sci. 110: 475-491, **1963**.

Conlee, R. K., Rennie, M. J., and Winder, W. W.

Skeletal muscle glycogen: diurnal variation and effects of fasting. Amer. J. Physiol. 231: 614-618, **1976**.

Conners, C. K. and Rothschild, G. H. The effect of dextroamphetamine on habituation of peripheral vascular response in children. Abn. Child Psychol. 1: 16-25, **1973**.

Conradi, S. Ultrastructure of dorsal root boutons on lumbosacral motoneurons of the adult cat, as revealed by dorsal root section. Acta Physiol. Scand. 78 (Suppl. 332): 85-105, **1970**.

Consolazio, C. F. Submaximal and maximal performance at altitude, pp. 91-96. In: The International Symposium on the Effects of Altitude on Physical Performance. Ed.: R. Goddard. Chicago: Athletic Institute, **1967**.

Consolazio, C. F., Johnson, H. L., Nelson, R. Q., Dramise, J. G., and Skala, J. H. Protein metabolism of intensive physical training in the young adult. Amer. J. Clin. Nutr. 28: 29-35, **1975**.

Consolazio, C. F., Nelson, R. A., Matoush, L. R. O., and Hansen, J. E. Energy metabolism at high altitude (3,475 m). J. Appl. Physiol. 21: 1732-1740, **1966**.

Conway, E. J. Critical energy barriers in the excretion of sodium. Nature 187: 394-396, **1960**.

Coon, R. L. and Kampine, J. P. Role of CO_2 and O_2 tension, pp. 185-196. In: Muscular Exercise and the Lung. Eds.: J. A. Dempsey and C. E. Reed. Madison, Wisc.: University of Wisconsin Press, **1977**.

Cooper, E. A. The work of ventilating respirators, pp. 67-75. In: Design and Use of Respirators. Ed.: C. N. Davies. New York: Macmillan, **1961**.

Cooper, J. M. Biomechanics. Chicago: Athletic Institute, **1971**.

Cooper, K. E., Hunter, A. R., and Keatinge, W. R. Accidental hypothermia. Int. Anesthesiol. Clin. 2: 999-1013, **1964**.

Cooper, K. H. A means of assessing oxygen intake. J. Amer. Med. Assoc. 203: 201-204, **1968a**.

——. Aerobics. New York: Evans, **1968b**.

Coote, J. H., Hilton, S. M., and Perez-Gonzalez, J. F. The reflex nature of the pressor response to muscular exercise. J. Physiol. (Lond.) 215: 789-804, **1971**.

Copinschi, G., Hartog, M., Earll, J. M., and Havel, R. J. Effect of various blood sampling procedures on serum levels of immunoreactive human growth hormone. Metab. Clin. Exp. 16: 402-409, **1967**.

Coppin, C. M. L., Jack, J. J. B., and MacLennan, C. R. A method for the selective activation of tendon organ afferent fibres from the cat soleus muscle. J. Physiol. (Lond.) 219: 18p-20p, **1970**.

Corcondilas, A., Koroxenidis, G. T., and Shepherd, J. T. Effect of a brief contraction of forearm muscles on forearm blood flow. J. Appl. Physiol. 19: 142-146, **1964**.

Corcoran, D. W. J. Personality and the inverted-U relation. Brit. J. Psychol. 52: 267-273, **1965**.

Corliss, R. J. Cardiac catheterization: Varying views and different angles. In: Heart Disease and Rehabilitation: State of the Art. Eds.: M. Pollock and D. Schultze. Boston: Houghton Mifflin, **1979**.

Corvaja, N. Muscle spindles in the lumbrical muscle of the adult cat. Electron microscopic observations and functional considerations. Arch. Ital. Biol. 107: 365-543, **1969**.

Cosmas, A. C. and Edington, D. W. Mitochondrial distributions in hearts of male rats as a function of long-term physical training. In: Metabolic Adaptations to Prolonged Physical Exercise. Eds.: H. Howald and J. R. Poortmans. Basel: Birkhauser Verlag, **1975**.

Costa, G. and Gaffuri, E. Studies of perceived exertion rate on bicycle ergometer in conditions reproducing some aspects of industrial work (shift work; noise) pp. 297-305. In: Physical Work and Effort. Ed.: G. Borg. Oxford: Pergamon Press, **1975**.

Costill, D. L. Physiology of marathon running. J. Amer. Med. Assoc. 221: 1024-1029, **1972a**.

——. Fluid replacement during and following exercise. In: Physiology of Fitness and Exercise. Eds.: J. F. Alexander, R. C. Serfass, and C. M. Tipton. Chicago: Athletic Institute, **1972b**.

——. Championship material, pp. 8-9. In: An Analysis and Evaluation of Modern Trends Related to the Physiological Factors for Optimal Sports Performance. Ed.: E. J. Burke. New York: Ithaca College, **1976**.

——. Fluids for athletic performance: Why and what you should drink during prolonged exercise. In: Towards an Understanding of Human Performance. Ed.: E. J. Burke. Ithaca, N.Y.: Mouvement Publications, **1977a**.

——. Sweating: Its composition and effects on body fluids. Ann. N.Y. Acad. Sci. 301: 160-174, **1977b**.

——. Electrolytes and water: changes in body fluid

compartments during exercise and dehydration. In: 3rd International Symposium on Biochemistry of Exercise. Eds.: F. Landry and W. R. Orban. Miami: Symposia Specialists, **1978a**.

——. Muscle water and electrolytes during acute and repeated bouts of dehydration, pp. 98-116. In: Nutrition, Physical Fitness and Health. Eds.: J. Pařízková and V. A. Rogozkin. Baltimore: University Park Press, **1978b**.

Costill, D. L., Branam, G., Fink, W., and Nelson, R. Exercise-induced sodium conservation-changes in plasma renin and aldosterone. Med. Sci. Sport. 8: 209-213, **1976**.

Costill, D. L., Cahill, P. J., and Eddy, D. Metabolic responses to submaximal exercise in three water temperatures. J. Appl. Physiol. 22: 628-632, **1967**.

Costill, D. L., Coté, R., and Fink, W. Muscle water and electrolytes following varied levels of dehydration in man. J. Appl. Physiol. 40: 6-11, **1976**.

Costill, D. L., Coté, R. Miller, E., Miller, T., and Wynder, S. Water and electrolyte replacement during repeated days of work in the heat. Aviat. Space Environ. Med. 46: 795-800, **1975**.

Costill, D. L., Daniels, J., Evans, W., Fink, W., Krahenbuhl, G., and Saltin, B. Skeletal muscle enzymes and fibre composition in male and female track athletes. J. Appl. Physiol. 90: 149-154, **1976**.

Costill, D. L. and Fink, W. J. Plasma volume changes following exercise and thermal dehydration. J. Appl. Physiol. 37: 521-525, **1974**.

Costill, D. L., Fink, W. J., and Pollock, M. L. Muscle fiber composition and enzyme activities of elite distance runners. Med. Sci. Sport. 8: 96-100, **1976**.

Costill, D. L., Jansson, E. D., Gollnick, P. D., and Saltin, B. Glycogen utilization in leg muscles of men during level and uphill running. Acta Physiol. Scand. 91: 475-481, **1974**.

Costill, D. L., Kammer, W. F., and Fisher, A. Fluid ingestion during distance running. Arch. Env. Health 21: 520-525, **1970**.

Costill, D. L. and Saltin, B. Factors limiting gastric emptying during rest and exercise. J. Appl. Physiol. 37: 679-683, **1974**.

——. Muscle glycogen and electrolytes following exercise and thermal dehydration. In: Metabolic Adaptations to Prolonged Physical Exercise. Eds.: H. Howald and J. Poortmans. Basel: Birkhauser Verlag, **1975**.

Costill, D. L., Sparks, K., Gregor, R., and Turner, C. Muscle glycogen utilization during exhaustive running. J. Appl. Physiol. 31: 353-356, **1971**.

Costill, D. L., and Sparks, K. E. Rapid fluid replacement following thermal dehydration. J. Appl. Physiol. 34: 299-303, **1973**.

Cotes, J. Occupational Safety and Health Series Rept. 6. Geneva: I.L.O., **1966**.

Cotes, J. E. Physiological aspects of respirator design, pp. 32-47. In: Design and Use of Respirators. Ed.: C. N. Davies. New York: Macmillan, **1961**.

Cotes, J. E., Davies, C. T. M., Edholm, O. G., Healy, M. J. R., and Tanner, J. M. Factors relating to the aerobic capacity of 46 healthy British males and females, ages 18 to 28 years. Proc. Roy. Soc. Ser. B, 174: 91-114, **1969**.

Cotes, J. E., Dobbs, J. M., Elwood, P. C., Hall, A. M., McDonald, A., and Saunders, M. J. The response to submaximal exercise in adult females: Relationship to haemoglobin concentration. J. Physiol. (Lond.) 203: 79p, **1969**.

Cotten, D. Relationship of the duration of sustained voluntary isometric contraction to changes in endurance and strength. Res. Quart. 38: 366-374, **1967**.

Cotton, F. Ș. and Dill, D. B. On the relationship between the heart rate during exercise and that of the immediate post-exercise period. Amer. J. Physiol. 111: 554-558, **1935**.

Coulson, R. L. Energetics of isovolumic contractions of the isolated rabbit heart. J. Physiol. (Lond.) 260: 45-53, **1976**.

Counsilman, J. E. Fatigue and staleness. Athlet. J. 35: 1-8, **1955**.

Cournand, A. and Richards, D. W. Pulmonary insufficiency: 1. Discussion of physiological classification and presentation of clinical tests. Amer. Rev. Tuberc. 44: 26-41, **1941**.

Covell, J. W., Braunwald, E., Ross, J., and Sonnenblick, E. H. Studies on digitalis. XVI. Effects on myocardial oxygen consumption. J. Clin. Invest. 45: 1535-1542, **1966**.

Covian, F. G. and Rehberg, P. B. Über die Nierenfunktion während schwerer Muskelarbeit. Skand. Arch. Physiol. 75: 21-37, **1936**.

Cox, M. and Shephard, R. J. Some observations on an industrial fitness programme. Ergonomics: in press, **1981**.

Coye, R. D. and Rosandich, R. R. Proteinuria during the 24 hour period following exercise. J.

Appl. Physiol. 15: 592-594, **1960**.

Coyle, E. F., Costill, D. L., Fink, W. J., and Hoopes, D. G. Gastric emptying rates for selected athletic drinks. Res. Quart. 49: 110-124, **1978**.

Craig, A. B. Causes of loss of consciousness during underwater swimming. J. Appl. Physiol. 16: 583-586, **1961**.

——. Olympics 1968. A post-mortem. Med. Sci. Sport. 1: 177-180, **1969**.

——. Heat exchange: man and the water environment, pp. 425-433. In: Underwater Physiology. Ed.: C. J. Lambertsen. New York: Academic Press, **1971**.

Craig, A. B. and Dvorak, M. Thermal regulation of man exercising during water immersion. J. Appl. Physiol. 25: 28-35, **1968**.

——. Comparison of exercise in air and in water of different temperatures. Med. Sci. Sport. 1: 124-130, **1969**.

Craig, F. N., Cummings, E. G., and Blevins, W. Regulation of breathing at the beginning of exercise. J. Appl. Physiol. 18: 1183-1187, **1963**.

Craig, F. N. and Cummings, E. G. Dehydration and muscular work. J. Appl. Physiol. 21: 670-674, **1966**.

Craig, F. N. and Froehlich, H. L. Endurance of pre-heated men in exhausting work. J. Appl. Physiol. 24: 636-639, **1968**.

Craig, F. N. and Moffitt, J. T. Efficiency of evaporative cooling from wet clothing. J. Appl. Physiol. 36: 313-316, **1974**.

Craig, M. M. and Raftery, E. B. The response to exercise in hypertensive subjects. Clin. Sci. Mol. Biol. 54: 18p-19p, **1978**.

Crampton, C. W. A test of condition. Med. News 87: 529-535, **1905**.

Crandall, L. A. The effect of physical exercise on the gastric secretion. Amer. J. Physiol. 84: 48-55, **1928**.

Cranefield, P. F., Wit, A. L., and Hoffman, B. F. Genesis of cardiac arrhythmias. Circulation 47: 190-204, **1973**.

Cratty, B. J. Psychology and physical activity. Englewood Cliffs, N.J.: Prentice-Hall, **1968**.

——. Activation and athletic endeavour. In: International Symposium on the Art and Science of Coaching. Ed.: L. Percival. Toronto: Fitness Institute, **1971**.

Crews, J. and Aldinger, E. E. Effect of chronic exercise on myocardial function. Amer. Heart J. 74: 536-542, **1967**.

Criscuoli, P. M. and Albano, G. Neuropsychological effects of exposure to compressed air, pp. 471-478. In: Underwater Physiology. Ed.: C. J. Lambertsen. New York: Academic Press, **1971**.

Crisp, A. H. and McGuinness, B. Jolly fat: Relation between obesity and psychoneurosis in general population. Brit. Med. J. 1: 7-9, **1975**.

Crittenden, R. H. Physiological Economy in Nutrition. New York: Frederick Stokes, **1904**.

Crockett, J. L. and Edgerton, V. R. Muscle and end-plate choline-esterase in three fiber types of guinea pig muscle. Cited by R. E. Burke and V. R. Edgerton, 1975. Amer. J. Physiol., **1975**.

Crockford, G. W. and Davies, C. T. M. Circadian variations in responses to submaximal exercise on a bicycle ergometer. J. Physiol. (Lond.) 201: 94p-95p, **1969**.

Crone, C. Which part of the central nervous system activates the adrenal medulla during hypoglycemia? Acta Physiol. Scand. 59 (Suppl. 213): 29 (abstract), **1963**.

Cropp, G. J. A. Grading, time course, and incidence of exercise-induced airway obstruction and hyperinflation in asthmatic children. Pediatrics 56 (Suppl.): 868-879, **1975**.

Cropp, G. J. A. and Comroe, J. H. Role of mixed venous blood P_{CO_2} in respiratory control. J. Appl. Physiol. 16: 1029-1033, **1961**.

Crowell, J. W. Oxygen transport in the hypotensive state. Fed. Proc. 29: 1848—1853, **1970**.

Cruickshank, E. W. H. On the output of hemoglobin and blood by the spleen. J. Physiol. (Lond.) 61: 455-464, **1926**.

Cruz-Coke, R., Donoso, H., and Barrera, R. Genetic ecology of hypertension. Clin. Sci. Mol. Med. 45: 55s-65s, **1973**.

Cullumbine, H. and Koch, A. C. E. Changes in plasma and tissue fluid volume following exercise. Quart. J. Exp. Physiol. 35: 39-46, **1949**.

Cumming, G. Alveolar Ventilation: recent model analysis, pp. 139-166. In: M.T.P. International Review of Science. Respiratory Physiology. Ed.: J. G. Widdicombe. London: Butterworths, **1974**.

Cumming, G., Horsfield, K., and Preston, S. B. Diffusion equilibrium in the lungs examined by nodal analysis. Resp. Physiol. 12: 329-345, **1971**.

Cumming, G. R. Correlation of athletic performance with pulmonary function in 13 to 17 year old boys and girls. Med. Sci. Sport. 1: 140-143, **1969**.

——. Correlation of physical performance with laboratory measures of fitness, pp. 265–279. In: Frontiers of Fitness. Ed.: R. J. Shephard. Springfield, Ill.: C. C. Thomas, **1971**.

——. Yield of ischaemic electrocardiograms in relation to exercise intensity in a normal population. Brit. Heart J. 34: 919–923, **1972**.

——. Attempts at maximal exercise testing in 3–6 year old children. Proc. Can. Assoc. Sports Sciences, Edmonton, **1974**.

——. Exercise studies in clinical pediatric cardiology, pp. 17–45. In: Frontiers of Activity and Child Health. Eds.: H. Lavallée and R. J. Shephard. Quebec City: Editions du Pélican, **1977**.

——. Resolve that the exercise electrocardiogram is of limited diagnostic value, pp. 13–29. In: Sports Medicine. Eds.: F. Landry and W. A. R. Orban. Miami: Symposia Specialist, **1978a**.

——. Body size and the assessment of physical performance, p. 18–31. In: Physical Fitness Assessment: Principles, Practice and Application. Eds.: R. J. Shephard and H. Lavallée. Springfield, Ill.: C. C. Thomas, **1978b**.

Cumming, G. R. and Alexander, W. D. The calibration of bicycle ergometers. Can. J. Physiol. Pharm. 46: 917–919, **1968**.

Cumming, G. R. and Borysyk, L. Lack of correlation of heartometer measurements with maximum oxygen uptake. Med. Sci. Sport. 3: 166–168, **1971**.

Cumming, G. R., Borysyk, L., and Dufresne, C. The maximal exercise e.c.g. in asymptomatic men. Can. Med. Assoc. J. 106: 649–653, **1972**.

Cumming, G. R. and Carr, W. Hemodynamic response to exercise after propranolol in normal subjects. Can. J. Physiol. Pharmacol. 44: 465–474, **1966**.

——. Hemodynamic response to exercise after beta-adrenergic and parasympathetic blockade. Can. J. Physiol. Pharmacol. 45: 813–819, **1967**.

Cumming, G. R., Dufresne, C., and Samm, J. Exercise e.c.g. changes in normal women. Can. Med. Assoc. J. 109: 108–111, **1973**.

Cumming, G. R. and Edwards, A. H. Indirect measurement of left ventricular function during exercise. Can. Med. Assoc. J. 89: 219–221, **1963**.

Cumming, G. R. and Friesen, W. Bicycle ergometer measurement of maximal oxygen uptake in children. Can. J. Physiol. Pharm. 45: 937–946, **1967**.

Cumming, G. R., Garand, T., and Borysyk, L. Correlation of performance in track and field events with bone age. J. Pediatr. 80: 970–973, **1972b**.

Cumming, G. R., Goodwin, A., Baggley, G., and Antel, A. Repeated measurements of aerobic capacity during a week of intensive training at a youth's track camp. Can. J. Physiol. Pharm. 45: 805–811, **1967**.

Cumming, G. R. and Keynes, R. A fitness performance test for school children and its correlation with physical working capacity and maximal oxygen uptake. Can. Med. Assoc. J. 96: 1262–1269, **1967**.

Cundiff, D. E. and Corbun, C. B. Some observations on selected electrophysical components of the resting cardiac cycles of world championship wrestlers, pp. 107–116. In: Exercise and Fitness 1969. Ed.: B. D. Franks. Chicago: Athletic Institute, **1969**.

Cunningham, D. A., Van Waterschoot, B. M., Paterson, D. H., Lefcoe, N., and Sangal, S. P. Reliability and reproducibility of maximal oxygen uptake measurement in children. Med. Sci. Sport. 9: 104–108, **1977**.

Cunningham, D. J. C. Integrative aspects of the regulation of breathing: a personal view, pp. 303–369. In: MTP International Reviews of Science. Physiology, Series One, Respiratory Physiology. Ed.: J. G. Widdicombe. London: Butterworths, **1974**.

Cunningham, D. J. C., Elliott, D. H., Lloyd, B. B., Miller, J. B., and Young, J. M. A comparison of the effects of oscillatory and steady alveolar partial pressure of oxygen and carbon dioxide on pulmonary ventilation. J. Physiol. (Lond.) 179: 498–508, **1965**.

Cunningham, D. J. C., Howson, M. G., Strange-Petersen, E., Pickering, T. G., and Sleight, P. Changes in the sensitivity of the baroreflex in muscular exercise. Acta Physiol. Scand. 79: 16A–17A, **1970**.

Cunningham, D. J. C., Pearson, S. B., Marsh, R. H. K., and Kellogg, R. H. The effects of various time patterns of alveolar CO_2 and O_2 on breathing in man. In: Int. Symp. Neural Control of Breathing. Eds.: W. Karczewski and J. G. Widdicombe. Warsaw: Acta Neurobiol. Exp. 33: 123, **1973**.

Cunningham, D. J. C., Strange-Petersen, E., Peto, R., Pickering, T., and Sleight, P. Comparison of the effect of different types of exercise on the baroreflex regulation of heart rate. Acta Physiol. Scand. 86: 444–455, **1972a**.

Cunningham, D. J. C., Strange-Petersen, E., Pickering, T. G., and Sleight, P. The effects of hypoxia, hypercapnia, and asphyxia on the baroreceptor-cardiac reflex at rest and during exercise in man. Acta Physiol. Scand. 86; 456–465, **1972b**.

Cunningham, D. M. and Brown, G. W. Two devices for measuring the forces acting on the human body during walking. Proc. Soc. Exp. Stress Anal. 9: 75–90, **1952**.

Cureton, K. J., Boileau, R. A., Lohman, T. G., and Misner, J. E. Determinants of distance running performance in children: Analysis of a path model. Res. Quart. 48: 270–279, **1974**.

Cureton, K. J., Sparling, P. B., Evans, B. W., Johnson, S. M., Kong, U. D., and Purvis, J. W. Effect of experimental alterations in excess weight on aerobic capacity and distance running performance. Med. Sci. Sport. 10: 194–199, **1978**.

Cureton, T. K. Analysis of vital capacity as a test of condition for high school boys. Res. Quart. 7: 80–92, **1936**.

——. Physical Fitness Appraisal and Guidance. St. Louis: C.V. Mosby, **1947**.

——. Physical Fitness of Champion Athletes. Urbana: University of Illinois Press, **1951**.

Cureton, T. K., Wickens, J. S., and Elder, H. P. Reliability and objectivity of Springfield postural measurements. Res. Quart. 6(2). Supplement, **1935**.

Cvorkov, N., Banister, E. W., and Liskop, K. S. Effect of high protein diet on rat heart mitochondria after exhaustive exercise. Amer. J. Physiol. 226: 996–1000, **1974**.

Dagenais, G. R., Tancredi, R. G., and Zierler, K. L. Participation d'un réservoir liquidique intramusculaire au maintien de l'oxydation des acides gras libres dans le muscle squelettique humain au repos, pp. 337–344. In: 3rd International Symposium on Biochemistry of Effort. Eds.: F. Landry and W. A. R. Orban. Miami: Symposia Specialists, **1978**.

dal Monte, A. and Leonardi, L. M. Functional evaluation of kayak paddlers from biomechanical and physiological viewpoints. In: Biomechanics V. Ed.: P. V. Komi. Baltimore: University Park Press, **1976**.

Dalton, K. Menstruation and accidents. Brit. Med. J. 2: 1425–1426, **1960**.

——. Menstruation and examinations. Lancet 2: 1386–1388, **1968**.

Dalton, K. and Williams, J. G. P. Women in sport, pp. 200–225. In: Sports Medicine, 2nd ed. Eds.: J. G. P. Williams and P. N. Sperryn. London: Edward Arnold, **1976**.

Daly, D. J. and Cavanagh, P. Asymmetry in bicycle ergometer pedalling. Med. Sci. Sport. 8: 204–208, **1976**.

Daly, M. de B. Reflex circulatory and respiratory responses to hypoxia, pp. 267–275. In: Oxygen in the Animal Organism. Eds.: F. Dickens and E. Neil. Oxford: Pergamon Press, **1964**.

Daly, M. de B. and Scott, M. J. The effect of stimulation of the carotid body chemoreceptors on heart rate in the dog. J. Physiol. (Lond.), 144: 148–166, **1958**.

Damato, A. N., Galante, J. G., and Smith, W. M. Hemodynamic response to treadmill exercise in normal subjects. J. Appl. Physiol. 21: 959–966, **1966**.

Damoiseau, J., Bottan, R., Petit, J. M., and Legros, R. Intéret des mesures de consommation maximum d'O_2 chez des jouers de football. Théorie de l'Ed. Phys. 4: 1–10, **1966**.

Dancaster, C. P., Duckworth, W. C., and Roper, C. J. Nephropathy in marathon runners. S. Afr. Med. J. 43: 758–760, **1969**.

Dancaster, C. P. and Whereat, S. J. Fluid and electrolyte balance during the Comrades marathon. S. Afr. J. Med. Sci. 36: 147–149, **1971**.

Danforth, W. H. Activation of glycolytic pathway in muscle, p. 287. In: Control of Energy Metabolism. Eds.: B. Chance, R. W. Estabrook, and J. R. Williamson. New York: Academic Press, **1965**.

Danforth, W. H., Helmreich, E., and Cori, C. F. The effect of contraction and of epinephrine on the phosphorylase activity of frog sartorius muscle. Proc. Nat. Acad. Sci., Wash. 48: 1191–1199. **1962**.

D'Angelo, E. and Torelli, G. Neural stimuli increasing respiration during different types of exercise. J. Appl. Physiol. 30: 116–121, **1971**.

Danielli, J. F. Surface chemistry of cell membranes. In: Surface Phenomena in Chemistry & Biology. Eds.: J. F. Danielli, K. G. A. Pankhurst, and A. C. Riddleford, pp. 246–365. New York: Pergamon Press, **1958**.

Danzer, C. S. The cardiothoracic ratio: An index of cardiac enlargement. Amer. J. Med. Sci. 157: 513–521, **1919**.

Darling, R. C., Johnson, R. E., Pitts, G. C., Con-

solazio, F. C., and Robinson, P. F. Effects of variations in dietary protein on the physical well-being of men doing manual work. J. Nutr. 28: 273–281, **1944**.

Das, B. C. An examination of variability of blood chemistry, hematology and proteins in relation to age. Gerontologia 15: 275–287, **1969**.

da Silva, A. M. T. and Hamosh, P. Effect of smoking a single cigarette on the "small airways." J. Appl. Physiol. 34: 361–5, **1973**.

da Silva, A. R. The role of the spectator in the soccer player's dynamics. In: Contemporary Psychology of Sport. Eds.: G. S. Kenyon and T. M. Grogg. Chicago: Athletic Institute, **1970**.

Daub, B., Green, H., Houston, M., and Thomson, J. Alternate forms of leg training: Cross-adaptive effects. Can. J. Appl. Sport. Sci. 3: 180, **1978**.

Dautrebande, L. Reactions of lung receptors to pharmacological and inert air-borne particulates, pp. 121–147. In: The Regulation of Human Respiration. Eds.: D. J. C. Cunningham and B. B. Lloyd. Oxford: Blackwell Scientific, **1963**.

——. Experimental observations on the participation of alveolar spaces in airway dynamics, pp. 153–168. In: Airway Dynamics: Physiology and Pharmacology. Ed.: A. Bouhuys. Springfield, Ill.: C. C. Thomas, **1970**.

Davidson, S., Passmore, R., and Brock, J. F. Human Nutrition and Dietetics. Edinburgh: Churchill-Livingstone, **1972**.

Davies, A. The effect of irritant receptor activity on expiratory time. J. Physiol. (Lond.) 275: 39p–40p, **1978**.

Davies, A. and Kohl, J. Patterns of accelerated breathing provoked by lung irritant receptor activity. J. Physiol. (Lond.) 284: 59p–60p, July, **1978**.

Davies, C. N. A formalized anatomy of the human respiratory tract, pp. 82–91. In: Inhaled Particles and Vapours. Ed.: C. N. Davies. Oxford: Pergamon, **1961**.

Davies, C. T. M. Commentary. In: International Symposium on Physical Activity and Cardiovascular Disease. Can. Med. Assoc. 96: 781, **1967a**.

——. Commentary. In: International Symposium on Physical Activity and Cardiovascular Disease. Can. Med. Assoc. J. 96: 743–744, **1967b**.

——. Limitations to the prediction of maximum oxygen intake from cardiac frequency measurements. J. Appl. Physiol. 24: 700–706, **1968**.

——. Human power output in exercise of short duration in relation to body size and composition. Ergonomics 14: 245–256, **1971a**.

——. Body composition in children: a reference standard for maximum aerobic power output on a stationary bicycle ergometer. Acta Paediatr. Scand. Suppl. 217: 136–137, **1971b**.

——. The oxygen transporting system in relation to age. Clin. Sci. 42: 1–13, **1972**.

——. The physiological effects of iron deficiency anaemia and malnutrition on exercise performance in East African school children. Acta Paediatr. Belg. 28 (Suppl.): 253–256, **1974**.

Davies, C. T. M. and Barnes, C. Plasma FFA in relation to maximum power output in man. Int. Z. Angew. Physiol. 30: 247–257, **1972**.

Davies, C. T. M., Barnes, C., and Sargeant, A. J. Body temperature in exercise. Effects of acclimatization to heat and habituation to work. Int. Z. Angew. Physiol. 30: 10–19, **1971**.

Davies, C. T. M., Brotherhood, J. R., and Zeidifard, E. Temperature regulation during severe exercise with some observations on effects of skin wetting. J. Appl. Physiol. 41: 772–776, **1976**.

Davies, C. T. M., di Prampero, P. E., and Cerretelli, P. Kinetics of cardiac output and respiratory gas exchange during exercise and recovery. J. Appl. Physiol. 32: 618–625, **1972**.

Davies, C. T. M., Ekblom, B., Bergh, U., and Kanstrup-Jensen, I. L. The effects of hypothermia on sub-maximal and maximal work performance. Acta Physiol. Scand. 95: 201–202, **1975**.

Davies, C. T. M. and Few, J. Adreno-cortical function during prolonged exercise. Proc. 2nd Int. Symp. Exercise Biochemistry, Magglingen, Switzerland, **1973a**.

Davies, C. T. M. and Few, J. D. Effects of exercise on adreno-cortical function. J. Appl. Physiol. 35: 887–891, **1973b**.

Davies, C. T. M., Few, J., Foster, K. G., and Sargeant, A. J. Plasma catecholamine concentration during dynamic exercise involving different muscle groups. Europ. J. Appl. Physiol. 32: 195–206, **1974a**.

Davies, C. T. M. and Sargeant, A. J. Physiological responses to one- and two-leg exercise breathing air and 45% oxygen. J. Appl. Physiol. 36: 142–148, **1974b**.

——. Effects of hypoxic training on normoxic maximal aerobic power output. Europ. J. Appl. Physiol. 33: 227-236, **1974c**.

——. Circadian variation in physiological responses to exercise on a stationary bicycle ergometer. Brit. J. Industr. Med. 32: 110-114, **1975a**.

——. Effects of training on the physiological responses to one- and two-leg work. J. Appl. Physiol. 38: 375-385, **1975b**.

Davies, C. T. M., Sargeant, A. J., and Smith, B. The physiological responses to running downhill. Europ. J. Appl. Physiol. 32: 187-194, **1974**.

Davies, C. T. M., Tuxworth, W., and Young, J. M. Physiological effects of repeated exercise. Clin. Sci. 39: 247-258, **1970**.

Davies, H. C. and Davies, R. E. Biochemical aspects of oxygen poisoning. In: Handbook of Physiology, Section 3, Vol II. Respiration, pp. 1047-1058. Eds.: W. O. Fenn and H. Rahn. Baltimore: Williams & Wilkins, **1965**.

Davies, R. E. A molecular theory of muscle contraction: Calcium dependent contractions with hydrogen bond formation plus ATP-dependent extensions of part of the myosin-actin crossbridges. Nature 199: 1068-1074, **1963**.

——. The dynamics of the energy-rich phosphates, pp. 56-62. In: Limiting Factors of Physical Performance. Ed.: J. Keul. Stuttgart: Thieme, **1973**.

Davies, W. R. and Shephard, R. J. Physiological characteristics of manikins used for the teaching of expired-air resuscitation. J. Physiol. (Lond.) 168: 3p-5p, **1963**.

Davis, G. M., Eynon, R. B., and Cunningham, D. A. Effect of varied rest periods during interval training upon aerobic and anaerobic fitness. Med. Sci. Sport. 9: 58, **1977**.

Davis, J. A., Vodak, P., Wilmore, J. H., Vodak, J., and Kurtz, P. Anaerobic threshold and maximal anaerobic power for three modes of exercise. J. Appl. Physiol. 41: 544-550, **1976**.

Davis, J. E. and Brewer, N. Effect of physical training on blood volume, hemoglobin, alkali reserve, and osmotic resistance of erythrocytes. Proc. Soc. Exp. Biol. Med. 32: 1276-1277, **1935**.

Davis, J. H. Fatal underwater breath holding in trained swimmers. J. Forensic Sci. 6: 301-306, **1961**.

Davis, J. O. A critical evaluation of the role of receptors in the control of aldosterone secretion and sodium excretion. Progr. Cardiovasc. Dis. 4: 27-46, **1961**.

Davis, P. R. Posture of the trunk during the lifting of weights. Brit. Med. J. 1: 87-89, **1959**.

Davis, P. R., Troup, J. D. G., and Burnard, J. H. Movements of the thoracic and lumbar spine when lifting: a chrono-cyclo-photographic study. J. Anat. (Lond.) 99: 13-26, **1965**.

Davis, W. W. Research in cross-education. Stud. Yale Psychol. Lab. 6: 6-50, **1898**.

Davson, H. A. and Danielli, J. F. The permeability of natural membranes (2nd ed.). London: Cambridge University Press, **1952**.

Dawber, T. R., Kannel, W. B., and Friedman, G. D. Vital capacity, physical activity and coronary heart disease, pp. 254-265. In: Prevention of Ischemic Heart Disease. Ed.: W. Raab. Springfield, Ill.: C. C. Thomas, **1966**.

Dawber, T. R., Moore, F. E., and Mann, G. V. II. Coronary heart disease in the Framingham study. Amer. J. Publ. Health 47: 4-24, **1957**.

Dawson, A. Regional pulmonary blood flow in sitting and supine man during and after acute hypoxia. J. Clin. Invest. 48: 301-310, **1969**.

Dawson, A. and Grover, R. F. Regional lung function in natives and long-term residents of 3,000 m altitude. J. Appl. Physiol. 36: 294-298, **1974**.

Dawson, A. A. and Ogston, D. Exercise-induced thrombocytosis. Acta Haematol. 42: 241-246, **1969**.

Dawson, C. A., Roemer, R. B., and Horvath, S. M. Body temperature and oxygen uptake in warm- and cold-adapted rats during swimming. J. Appl. Physiol. 29: 150-154, **1970**.

Dawson, P. M. Cardiac fatigue, pp. 576-583. In: Physiology of Physical Education for Physical Educators and Their Pupils. Ed.: P. M. Dawson. Baltimore: Williams & Wilkins, **1935**.

Dayton, S., Hashimoto, S. D., Dixon, W. J., and Tomiyasu, W. A controlled clinical trial of a diet high in unsaturated fat in preventing complications of atherosclerosis. Circulation 40 (Suppl. II): 1-63, **1969**.

de Andrade, B. J. F. and de Rose, E. H. Study of the rest ecg of marathon runners. In: Sports Cardiology. Eds.: T. Lubich and A. Venerando. Bologna: Aulo Gaggi, **1979**.

Debons, A. F., Krimsky, I., Maayan, M. L., Fani, K., and Jimenez, F. A. Gold thioglucose obesity syndrome. Fed. Proc. 36: 143-147, **1977**.

de Coster, A. Present concepts of the relationship between lactate and oxygen debt. In: Frontiers of Fitness. Ed.: R. J. Shephard. Springfield, Ill.:

C. C. Thomas, **1971**.

Deecke, L., Scheid, P., and Kornhuber, H. H. Distribution of readiness potential, premotion positivity, and motor potential of the human cerebral cortex preceding voluntary finger movements. Exp. Brain Res. 7: 158-168, **1969**.

Defares, J. G. Determination of $P_{\bar{V},CO_2}$ from the exponential CO_2 rise during rebreathing. J. Appl. Physiol. 13: 159-164, **1958**.

——. On the use of mathematical models in the analysis of the respiratory control system. In: The Regulation of Human Respiration. Ed.: B. B. Lloyd. Oxford: Blackwell Scientific, **1963**.

de Garay, A. L., Levine, L., and Carter, J. E. Genetic and Anthropological Studies of Olympic Athletes. New York: Academic Press, **1974**.

Degré, S., deCoster, A., Messin, R., and Denolin, H. Normal pressure-flow relationship during exercise in the sitting position. Int. Z. Angew. Physiol. 31: 53-59, **1972**.

De Haan, E. J., Groot, G. S. P., Scholte, H. R., Tager, J. M., and Wit-Peeters, E. M. Biochemistry of muscle mitochondria, pp. 417-469. In: The Structure and Function of Muscle, 2nd ed., Vol. 3. Physiology and Biochemistry. Ed.: G. H. Bourne. New York: Academic Press, **1973**.

Dehn, M. M. and Bruce, R. A. Longitudinal variations in maximal oxygen intake with age and activity. J. Appl. Physiol. 33: 805-807, **1972**.

Deitrick, J. E., Whedon, G. D., and Shorr, E. Effects of immobilization upon various metabolic and physiologic functions of normal men. Amer. J. Med. 4: 3-36, **1948**.

Dejours, P. La régulation de la ventilation au cours de l'exercice musculaire chez l'homme. J. Physiol. (Paris) 51: 163-261, **1959**.

——. Chemoreflexes in breathing. Physiol. Rev. 42: 335-358, **1962**.

——. The regulation of breathing during muscular exercise in man. A neuro-hormonal theory. In: The Regulation of Human Respiration. Eds.: D. J. C. Cunningham and B. B. Lloyd. Oxford: Blackwell Scientific, **1963**.

——. Control of respiration in muscular exercise. In: Handbook of Physiology. Section 3. Respiration, Vol. 1. Eds.: W. O. Fenn and H. Rahn. Washington, D. C.: American Physiological Society, **1964**.

——. Neurogenic factors in the control of ventilation during exercise. Circ. Res. 20: 146-153, **1967**.

——. Principles of comparative respiratory physiology. New York: American Elsevier, **1975**.

Dejours, P., Bechtel-Labrousse, Y., Lefrançois, R., and Raynaud, J. Étude de la mécanique ventilatoire au cours de l'exercice musculaire. J. Physiol. (Paris) 53: 318-319, **1961**.

Dejours, P., Flandrois, R., Lefrançois, R., and Teillac, A. Étude de la régulation de la ventilation au cours de l'exercise musculaire chez l'homme. J. Physiol. (Paris) 53: 321-322, **1961**.

Dejours, P., Raynaud, J., Cuenod, C. L., and Labrousse, Y. Modifications instantanées de la ventilation au début et à l'arrêt de l'exercice musculaire Interprétation. J. Physiol. (Paris) 47: 155-159, **1953**.

del Castillo, J. and Katz, B. Biophysical aspects of mutual muscular transmission. Progr. Biophys. 6: 121-170, **1956**.

Delgado, J. M. R. Circulatory effects of cortical stimulation. Physiol. Rev. 40, Suppl. 4: 146-171, **1960**.

Delhez, L., Bottin-Thonon, A., and Petit, J. M. Influence de l'entraînement sur la force maximum des muscles respiratoires. Trav. Soc. Méd. Belg. d'Ed. Phys. et de Sports 20: 52-63, **1967-68**.

DeLong, M. R. Activity of pallidal neurons during movement. J. Neurophysiol. 34: 414-427, **1971**.

De Lorme, T. L. Restoration of muscle power by heavy-resistance exercises. J. Bone Joint Surg. 27: 645-667, **1945**.

De Lorme, T. L. and Watkins, A. L. Techniques of progressive resistance exercises. Arch. Phys. Med. 29: 263-273, **1948**.

de Marées, H. and Barbey, K. Änderung der peripheren Durchblutung durch Ausdauertraining. Z. F. Kardiol. 62: 653-663, **1973**.

Demedts, M. and Anthonisen, R. Effects of increased external airway resistance during steady-state exercise. J. Appl. Physiol. 35: 361-366, **1973**.

Demirjian, A., Jenicek, M., and Dubuc, M. B. Les normes staturo-ponderales de l'enfant urbain canadien-français d'âge scolaire. Can. J. Publ. Health 63: 14-30, **1972**.

Dempsey, J. A. Quantity and significance of pulmonary adaptations to work in adolescent altitude residents. Acta Paediatr. Scand. Suppl. 217: 99-104, **1971**.

Dempsey, J., Hanson, P., and Masterbrook, M.

Imperfections in the response of the healthy pulmonary system to exercise. Med. Sci. Sport. 10: 43, **1978**.

Dempsey, J. A., Reddan, W. G., Birnbaum, M. L., Forster, H. V., Thoden, J. S., Grover, R. F., and Rankin, J. Effects of acute through life-long hypoxic exposure on exercise pulmonary gas exchange. Resp. Physiol. 13: 62–89, **1971**.

Dempsey, Y. The relationship between the strength of the elbow flexors and muscle size of the upper arm in children. University of Wisconsin, Madison: M.Sc. thesis, **1955**.

De Muth, R. G., Howatt, F. W., and Hill, M. B. The growth of lung function. Paediatrics, Suppl. 35(1): 161–218, **1965**.

Denison, D., Edwards, R. H. T., Jones, G., and Pope, H. Direct and rebreathing estimates of the O_2 and CO_2 pressures in mixed venous blood. Resp. Physiol. 7: 326–334, **1969**.

——. Estimates of the CO_2 pressures in systemic arterial blood during rebreathing on exercise. Resp. Physiol. 11: 186–196, **1971**.

Denniston, J. C., Maher, J. J., Reeves, J. T., Cruz, J. C., Cymerman, A., and Grover, R. F. Measurement of cardiac output by electrical impedance at rest and during exercise. J. Appl. Physiol. 40: 91–95, **1976**.

Denny-Brown, D. The interpretation of the electromyogram. Arch. Neurol. Psychiatr. 61: 99–128, **1949**.

De Pauw, D. and Vrijens, J. Untersuchungen bei Elite-Ruderern in Belgien. Sportarzt und Sportmedizin 8: 176–179, **1971**.

Depue, R. H. and Rice, R. V. F-actin is a right-handed helix. J. Mol. Biol. 12: 302–303, **1965**.

Derevenco, P., Florea, E., Derevenco, V., Anghel, I., and Simu, Z. Einige physiologische Aspekte des Übertrainings. Sportarzt und Sportmedizin 4: 151–156, **1967**.

Dervichian, D. G. The physical chemistry of phospholipids. Progr. Biophy. Mol. Biol. 14: 265–342, **1964**.

De Schryver, C. and Mertens-Strythagen, J. Intensity of exercise and heart tissue catecholamine content. Pflügers Arch. 336: 345–354, **1972**.

Desmedt, J. E. and Godaux, E. Fast motor units are not preferentially activated in rapid voluntary contractions in man. Nature (Lond.) 267: 717–719, **1977**.

——. Ballistic contractions in fast or slow human muscles: discharge patterns of single motor units. J. Physiol. (Lond.) 285: 185–196, **1978**.

Detmer, R. A., Raush, J., Fletcher, E., and Gordon, A. S. Ideal wave form and characteristics for direct current defibrillators. Surg. Forum 15: 249–251, **1964**.

Detry, J. M., Brengelmann, G. L., Rowell, L. B., and Wyss, C. Skin and muscle components of forearm flow in directly heated resting man. J. Appl. Physiol. 32: 506–511, **1972**.

Detry, J. R., Rousseau, M., Vandenbronche, G., Kusumi, F., and Brasseur, I. A. Increased arteriovenous oxygen difference after physical training in coronary heart disease. Circulation 44: 109–118, **1971**.

de Villota, E. D., Barat, G., Peral, P., Juffe, A., de Miguel, J. M. F., and Avello, F. Recovery from profound hypothermia and cardiac arrest after immersion. Brit. Med. J. 4: 394, **1973**.

Devlin, J. The effect of training and acute physical exercise on plasma insulin-like activity. Irish J. Med. Sci. 6: 423–425, **1963**.

de Vries, H. A. The looseness factor in speed and O_2 consumption of an anaerobic 100 yard dash. Res. Quart. 34: 305–313, **1963**.

——. Physiological effects of an exercise training regimen upon men aged 52 to 88. J. Gerontol. 25: 325–336, **1970**.

——. Exercise intensity threshold for improvement of cardiovascular-respiratory function in older men. Geriatrics 26: 94–101, **1971**.

——. Physiology of Exercise for Physical Education and Athletics. Dubuque, Iowa: W. C. Brown, **1974**.

——. Physiology of exercise and aging, pp. 79–94. In: Physical Activity and Human Well-Being. Eds.: F. Landry and W. A. R. Orban. Miami: Symposia Specialists, **1978**.

de Wijn, J. F., de Jongste, J. L., Mosterd, W., and Willebrand, D. Haemoglobin, packed cell volume, serum iron-binding capacity of selected athletes during training. J. Sport. Med. Phys. Fitness 11: 42–51, **1971**.

de Young, V. R., Rice, H. A., and Steinhaus, A. H. Studies in the physiology of exercise. 7. The modification of colonic motility induced by exercise and some indications for a nervous mechanism. Amer. J. Physiol. 99: 52–63, **1931**.

Diamant, B., Karlsson, J., and Saltin, B. Muscle tissue lactate after maximal exercise in man. Acta Physiol. Scand. 72: 383–384, **1968**.

Dichgans, J. and Brandt, T. Visual-vestibular interaction and motion perception. Bibl. Ophthalmol. 82: 327–338, **1972**.

Dickson, J. G., Lambertsen, C. J., and Cassils,

J. G. Quantitation of performance decrements in narcotized man, pp. 449–455. In: Underwater Physiology. Ed.: C. J. Lambertsen. New York: Academic Press, **1971**.

Dieter, M., Altland, P., and Highman, B. Exercise tolerance of cold-acclimated rats: serum and liver enzymes and histological changes. Can. J. Physiol. Pharmacol. 47: 1025–1031, **1969**.

Diete-Spiff, K. Tension development by isolated muscle spindles of the cat. J. Physiol. (Lond.) 193: 31–43, **1967**.

DiGirolamo, M., and Esposito, J. Adipose tissue blood flow and cellularity in the growing rat. Amer. J. Physiol. 229: 107–112, **1975**.

Dill, D. B., Edwards, H. T., Bauer, P. S., and Levenson, E. J. Physical performance in relation to external temperature. Arbeitsphysiol. 4: 508–518, **1931**.

Dill, D. B., Edwards, H. T., and Talbott, J. H. Studies in muscular activity. J. Physiol. (Lond.) 69: 267–305, **1930**.

Dill, D. B., Robinson, S., and Ross, J. C. A longitudinal study of 16 champion runners. J. Sport. Med. 7: 4–32, **1967**.

Dillman, C. J. Kinematic analyses of running. Ex. Sport. Sci. Rev. 3: 193–218, **1975**.

Dintiman, G. B. Effects of various training programs on running speed. Res. Quart. 35: 456–463, **1964**.

di Prampero, P. E. Anaerobic capacity and power. In: Frontiers of Fitness. Ed.: R. J. Shephard. Springfield, Ill.: C. C. Thomas, **1971a**.

——. The alactic oxygen debt: Its power, capacity and efficiency, pp. 371–382. In: Muscle Metabolism During Exercise. Eds.: B. Pernow and B. Saltin. New York: Plenum Press, **1971b**.

di Prampero, P. E., Cerretelli, P., and Piiper, J. Capacita di diffusione del polmone per l'O_2 nel cane a riposo e durante il lavoro. Boll. Soc. Ital. Biol. Sper. 45: 375–377, **1969**.

di Prampero, P. E., Cortili, G., Celentano, F., and Cerretelli, P. Physiological aspects of rowing. J. Appl. Physiol. 31: 853–857, **1971**.

di Prampero, P. E., Cortili, G., Mognoni, P., and Saibene, F. Energy cost of speed skating and efficiency of work against resistance. J. Appl. Physiol. 40: 584–591, **1976**.

——. Effects of body size and altitude on energy expenditures and speed in cycling. Proc. Int. Union Physiol. Sci. 13: 188, **1977**.

di Prampero, P. E., Pendergast, D. R., Wilson, D. W., and Rennie, D. W. Energetics of swimming in man. J. Appl. Physiol. 37: 1–5, **1974**.

Dirix, A. Examens anti-doping effectués in 1975 par la ligue vélocipédique Belge, pp. 301–305. In: Sports Medicine. Eds.: F. Landry and W. A. R. Orban. Miami: Symposia Specialists, **1978**.

Disch, J., Frankiewicz, R., and Jackson, A. Construct validation of distance run tests. Res. Quart. 46: 169–176, **1975**.

Dixon, R. W. and Faulkner, J. A. Cardiac outputs during maximum effort running and swimming. J. Appl. Physiol. 30: 653–656, **1971**.

Dlužniewska, K., Obtulowicz, A., and Sieñko, S. Badania nad gospodarka witamina C u robotnickŏw pracujacych w goracych dzialach przemyslu. (Studies on Vitamin C metabolism in workmen employed in hot departments in industry.) Folia Medica Cracov 7: 209–219, **1965**.

Dodge, H. T. and Sandler, H. Clinical applications of angiocardiography, pp. 171–201. In: Cardiac Mechanics: Physiological, Clinical and Mathematical Considerations. Eds.: I. Mirsky, D. Ghista and H. Sandler. New York: Wiley, **1974**.

Dohm, G. L., Huston, R. L., Askew, H. N., and Weiser, P. C. Effects of exercise on activity of heart and muscle mitochondria. Amer. J. Physiol. 223: 783–787, **1972**.

Dole, V. P. Fat as an energy source. In: Fat as a tissue. Eds.: K. Rodahl and B. Issekutz. New York: McGraw Hill, **1964**.

Doll, E., Keul, J., Brechtel, A., Limon-Lason, R., and Reindell, H. 1. Der Einfluss körperlicher Arbeit auf die arteriellen Blutgase in Freiburg und Mexico City. Sportarzt und Sportmedizin, 8: 317–326, **1967**.

Doll, E., Keul, J., Brechtel, A., and Reindell, H. Die arteriellen Blutgase bei Verminderung der Sauerstoff-konzentration in der Inspirationsluft während körperlicher Arbeit. Int. Z. Angew. Physiol. 25: 46–59, **1968**.

Doll, E., Keul, J., and Maiwald, C. Oxygen tension and acid-base equilibria in venous blood of working muscle. Amer. J. Physiol. 215: 23–29, **1968**.

Dollery, C. T., Dyson, N. A., and Sinclair, J. D. Regional variations in uptake of radioactive CO in the normal lung. J. Appl. Physiol. 15: 411–417, **1960**.

Donald, D. E., Milburn, S. E., and Shepherd, J. T. Effect of cardiac denervation on the maximal capacity for exercise in the racing greyhound. J. Appl. Physiol. 19: 849-852, **1964a**.

Donald, D. E., Rowlands, D. J., and Ferguson, D. A. Similarity of blood flow in the normal and sympathectomized dog hind limb during graded exercise. Circ. Res. 26: 185-199, **1970**.

Donald, D. E. and Samueloff, S. L. Exercise tachycardia not due to blood-borne agents in canine cardiac denervation. Amer. J. Physiol. 211: 703-711, **1966**.

Donald, D. E., and Shepherd, J. T. Response to exercise in dogs with cardiac denervation. Amer. J. Physiol. 205: 393-400, **1963**.

——. Initial cardiovascular adjustment to exercise in dogs with chronic cardiac denervation. Amer. J. Physiol. 207: 1325-1329, **1964b**.

——. Sustained capacity for exercise in dogs after complete cardiac denervation. Amer. J. Cardiol. 14: 853-859, **1964c**.

Donald, K. W. Oxygen poisoning in man. Brit. Med. J. 1: 667-672 and 712-717, **1947**.

Donohue, S. and Lavoie, N. A simple method for measuring maximum aerobic capacity and its relation to state of training. Can. J. Appl. Sports. Sci. 3: 188, **1978**.

Doolittle, T. L. and Bigbee, R. The twelve minute walk run: test of cardiorespiratory fitness of adolescent boys. Res. Quart. 37: 192-201, **1968**.

Doré, C., Hackett, A. J., Imms, F. J., and Prestidge, S. P. Measurement of the strength of extensor and flexor muscles of the knee in patients recovering from menisectomy and from fractures of the lower limb. Proc. Physiol. Soc. U.K., January, **1977**.

Dosman, J., Bode, F., Urbanetti, J., Antic, R., Martin, R., and Macklem, P. T. Role of inertia in the measurement of dynamic compliance. J. Appl. Physiol. 38: 64-69, **1975**.

Doss, W. S. and Karpovich, P. V. A comparison of concentric, eccentric, and isometric strength of elbow flexors. J. Appl. Physiol. 20: 351-353, **1965**.

Douglas, F. G. V. and Becklake, M. R. Effects of seasonal training on maximal cardiac output. J. Appl. Physiol. 25: 600-605, **1968**.

Douglas, J. W. B. and Simpson, H. R. Height in relation to puberty, family size and social class. Millbank Memorial Fund Quart. 40: 20-35, **1964**.

Douglas, W. W. and Ritchie, J. M. A technique for recording functional activity in specific groups of medullated and non-medullated fibres in whole nerve trunks. J. Physiol. (Lond.) 138: 19-30, **1957**.

Dowell, R. T. and Tipton, C. M. Influence of training on heart rate responses of rats to isoproterenol and propranolol. Physiologist 13: 182, **1970**.

Doyle, F., Brown, J., and Lachance, C. Relation between bone mass and muscle weight. Lancet 1: 391-393, **1970**.

Doyle, J. T. and Kinch, S. H. The prognosis of an abnormal electrocardiographic stress test. Circulation 41: 545-553, **1970**.

Drake, A. J. Preferential uptake of lactate by normal myocardium J. Physiol. (Lond.) 289: 89p, **1978**.

Drake, A. J., Noble, M. I. M., and Van den Bus, G. C. The effect of changes in arterial P_{CO_2} on myocardial blood flow, oxygen consumption and mechanical performance. J. Physiol. (Lond.) 266: 45p-46p, **1977**.

Drake, V., Jones, G., Brown, J. R., and Shephard, R. J. Fitness performance tests and their relationship to maximum oxygen intake. Can. Med. Assoc. J. 99: 844-848, **1968**.

Draper, H. H. Nutritional research in circumpolar populations. In: Circumpolar Health. Eds.: R. J. Shephard and S. Itoh. Toronto: University of Toronto Press, **1976**.

Draper, J., Edwards, R., and Hardy, R. Method of estimating the respiratory cost of task by use of minute-volume determinations. J. Appl. Physiol. 6: 297-303, **1953**.

Dreyer, G. The assessment of physical fitness. London: Cassell, **1920**.

Dreyfus, J. C. Sur les différences métaboliques entre les muscles rouges et blancs chez le lapin normal. Rev. Fr. Etud. Clin. Biol. 12: 343-348, **1967**.

Drinkwater, B. L. Physiological responses of women to exercise, pp. 125-153. In: Exercise and Sport Sciences Reviews. Ed.: J. H. Wilmore. New York: Academic Press, **1973**.

Drinkwater, B. L., Denton, J. E., Raven, P. B., and Horvath, S. M. Thermoregulatory response of women to intermittent work in the heat. J. Appl. Physiol. 41: 57-61, **1976a**.

Drinkwater, B. L., Denton, J. E., Kupprat, I. C., Talag, T. S., and Horvath, S. M. Aerobic power as a factor in women's response to work in hot environments. J. Appl. Physiol. 41: 815-821, **1976b**.

Drinkwater, B. L. and Horvath, S. M. Detraining effects on young women. Med. Sci. Sports 4: 91-95, **1972**.

Droese, W., Kofranyi, E., Kraut, H., and Wilde-

mann, L. Energetische Untersuchung der Hausfrauarbeit. Arbeitsphysiol. 14: 63-81, **1949**.

Drury, D. R. and Wick, A. N. Chemistry and metabolism of L(+) and D(−) lactic acids. Ann. N.Y. Acad. Sci. 119: 1061-1069, **1965**.

Drysdale, D. B. and Whipp, B. J. Interaction of peripheral and intracranial respiratory drive in man. J. Physiol. (Lond.) 263: 147p, **1976**.

DuBois, A. B. Alveolar gas exchange during 20-, 30- and 40-meter dives: theory, pp. 159-164. In: Physiology of Breath-Hold Diving and the Ama of Japan. Eds.: H. Rahn and T. Yokoyama. Washington, D.C. Natl. Acad. Sci. Natl. Res. Council Publ. 1341, **1965**.

DuBois, A. B., Bond, G. F., and Schaefer, K. E. Alveolar gas exchange during submarine escape. J. Appl. Physiol. 18: 509-512, **1963**.

DuBois, E. F. Basal Metabolism in Health and Disease. Philadelphia: Lea & Febiger, **1927**.

Dubrovsky, G., Davelaar, E., and Garcia-Rill, E. The role of dorsal columns in serial order acts. Exp. Neurol. 33: 93-102, **1971**.

Duffner, G. J. and Lanphier, E. H. Medicine and science in sport diving. In: Science and Medicine of Exercise and Sports. Ed.: W. R. Johnson and E. R. Buskirk. New York: Harper, **1974**.

Duffy, E. The psychological significance of the concept of arousal or activation. Psychol. Rev. 64: 265-275, **1957**.

Duling, B. R. and Pittman, R. N. Oxygen tension: Dependent or independent variable in local control of blood flow. Fed. Proc. 34: 2012-2019, **1975**.

Duncan, J. R., Squires, W. G., Gregory, W. B., and Jessup, G. T. Thermoregulatory constraints to distance running in the heat. Med. Sci. Sport. 10: 40, **1978**.

Duncan, L. E. The effect of exercise on the excretion of water by patients with congestive failure. Circulation 12: 90-95, **1955**.

Dunnett, J. and Nayler, W. G. Effect of pH on the uptake and efflux of calcium from cardiac sarcoplasmic reticulum vesicles. J. Physiol. (Lond.) 281: 16p-17p, **1978**.

Durand, J. and Martineaud, J. P. Resistance and capacitance vessels of the skin in permanent and temporary residents at high altitude. In: High Altitude Physiology: Cardiac and Respiratory Aspects. Eds.: R. Porter and J. Knight. Edinburgh: Churchill-Livingstone, **1971**.

Durand, J., Paunier, C. I., de Lattre, J., Martineaud, J. P., and Verpillat, J. M. The cost of the oxygen debt at altitude. In: Exercise at Altitude. Ed.: R. Margaria. Dordrecht: Excerpta Medica Fdn., **1967**.

Durnin, J. V. G. A. Activity patterns in the community. In: Proc. Int. Symp. Physical Activity and Cardiovascular Health. Can. Med. Assoc. J. 96: 882-886, **1967**.

——. Protein requirements and physical activity, pp. 53-60. In: Nutrition, Physical Fitness, and Health. Eds.: J. Pařízková and V. A. Rogozkin. Baltimore: University Park Press, **1978**.

Durnin, J. V. G. A., Brockway, J. M., and Whitcher, H. W. Effects of a short period of training of varying severity on some measurements of physical fitness. J. Appl. Physiol. 15: 161-165, **1960**.

Durnin, J. V. G. A. and Mikulic, V. Influence of graded exercises on the oxygen consumption, pulmonary ventilation and heart rate of young and elderly men. Quart. J. Exp. Physiol. 41: 442- 452, **1956**.

Durnin, J. V. G. A. and Passmore, R. Energy, Work and Leisure. London: Heinemann Educational Books Ltd., **1967**.

Durnin, J. V. G. A. and Womersley, J. Body fat assessed from total body density and its estimation from skinfold thickness: Measurements on 481 men and women aged from 16 to 72 years. Brit. J. Nutr. 32: 77-79, **1974**.

Durrant, M. L., Stalley, S. F., Warwick, P. M., and Garrow, J. S. The effects of meal frequency on body composition and hunger during weight reduction. Clin. Sci. 54: 48, **1978**.

Dydynska, M. and Wilkie, D. R. The chemical and energetic properties of muscles poisoned with fluorodinitrobenzene. J. Physiol. (Lond.) 184: 751-769, **1966**.

Dyson, G. H. G. The Mechanics of Athletics, 5th ed. London: University of London Press, **1970**.

Ebashi, S. Structural proteins and their interaction. Symp. Biol. Hung. 8: 77-87, **1968**.

Ebashi, S., Ebashi, F., and Maruyama, K. A new protein factor promoting contraction of actinomysin. Nature 203: 645-646, **1964**.

Ebashi, S., Endo, M., and Otsuki, I. Control of muscle contraction. Quart. Rev. Biophys. 2: 351-384. **1969**.

Ebashi, S. and Kodama, A. A new protein factor promoting aggregation of tropomyosin. J. Biochem. Tokyo, 58: 107-108, **1965**.

Ebashi, S. and Nonomura, Y. Proteins of the myofibril, pp. 285-362. In: The Structure and Function of Muscle, 2nd ed., Vol. 3. Physiology and Biochemistry. Ed.: G. H. Bourne. New York: Academic Press, **1973**.

Eberstein, A. and Sandow, A. Fatigue mechanisms in muscle fibres, p. 515. In: Effect of Use and Disuse. Eds.: E. Gutman and P. Hnik. Prague: Czech Academy of Science, **1963**.

Ebert, R. V. and Stead, E. A. Error in measuring changes in plasma volume after exercise. Proc. Soc. Exp. Biol. Med. 46: 139-141, **1941**.

Eccles, J. C. The synapse. Sci. Amer. 212: 56-66, **1965**.

——. The Inhibitory Pathways of the Central Nervous System, pp. 1-135. Liverpool: Liverpool University Press, **1969**.

——. The Understanding of the Brain. New York: McGraw Hill, **1973**.

Eccles, J. C., Eccles, R. M., and Lundberg, A. Synaptic actions on motoneurons caused by impulses in Golgi tendon organ afferents. J. Physiol. (Lond.) 138: 227-252, **1957a**.

Eccles, J. C., Eccles, R. M., and Lundberg, A. The convergence of monosynaptic excitatory afferents on to many different species of alpha mononeurones. J. Physiol. (Lond.) 137: 22-50, **1957b**.

Eccles, J. C., Eccles, R. M., and Magni, F. Central inhibitory action attributable to presynaptic depolarization produced by muscle afferent volleys. J. Physiol. (Lond.) 159: 147-166, **1961**.

Eccles, J. C., Fatt, P., and Langdren, S. The central pathway for the direct inhibitory action of impulses in the largest afferent nerve fibres to muscle. J. Neurophysiol. 19: 75-98, **1956**.

Eccles, J. C., Ito, M., and Szentágothai, J. The Cerebellum as a Neuronal Machine. New York: Springer Verlag, **1967**.

Eccles, J. C. and Sherrington, C. S. Numbers and contraction values of individual motor units examined in some muscles of the limb. Proc. Roy. Soc. B, 106: 326-357, **1930**.

Eckberg, D. L. Temporal response patterns of the human sinus node to brief carotid baroreceptor stimuli. J. Physiol. (Lond.) 258: 769-782, **1976**.

Eckert, H. M. Age changes in motor skills, pp. 155-175. In: Physical Activity. Human Growth and Development. Ed.: G. L. Rarick. New York: Academic Press, **1973**.

Eckstein, R. W. Effect of exercise and coronary artery narrowing on coronary collateral circulation. Circ. Res. 5: 230-235, **1957**.

Eddy, D. O., Sparks, K. L., and Adelizi, D. A. The effects of continuous and interval training in women and men. Europ. J. Appl. Physiol. 37: 83-92, **1977**.

Edgerton, V. R. Morphology and histochemistry of the soleus muscle from normal and exercised rats. Amer. J. Anat. 127: 81-87, **1970**.

——. Exercise and growth of Muscle Tissue, pp. 1-31. In: Physical Activity: Human Growth and Development. Ed.: G. L. Rarick. New York: Academic Press, **1973**.

——. Neuromuscular adaptation to power and endurance work. Can. J. Appl. Sport. Sci. 1: 49-58, **1976**.

Edgerton, V. R., Essén, B., Saltin, B., and Simpson, D. R. Glycogen depletion in specific types of human skeletal muscle fibers in intermittent and continuous exercise, pp. 402-415. In: Metabolic Adaptation to Prolonged Physical Exercise. Eds.: H. Howald and J. R. Poortmans. Basel: Birkhauser Verlag, **1975**.

Edgerton, V. R., Smith, J. L., and Simpson, D. R. Muscle fiber type populations of human leg muscles. Histo-Chem. J. 7(3): 259-266, **1975**.

Edgren, B. and Borg, G. The cycling strength test (CST) as a measure of dynamic muscular leg strength. Stockholm: Reports, Institute of Applied Psychol. 64: 1-7, **1975**.

Edholm, O. The Biology of Work. New York: McGraw Hill, **1967**.

Edholm, O. G., Adam, J. M., and Best, T. W. Day to day weight changes in young men. Ann. Hum. Biol. 1: 3-12, **1974**.

Edholm, O. G. and Bacharach, A. L. The physiology of human survival. London: Academic Press, **1965**.

Edholm, O. G., Fletcher, J. G., Widdowson, E. M., and McCance, R. A. Energy expenditure and food intake of individual men. Brit. J. Nutr. 9: 286-300, **1955**.

Edholm, O. G., Fox, R. H., and Wolf, H. S. Body temperature during exercise and rest in cold and hot climates. Archiv. Sci. Physiol. 27: A339-A355, **1973**.

Edinger, H. M., and Eisenman, J. S. Thermosensitive neurons in tuberal and posterior hypothalamus of cats. Amer. J. Physiol. 219: 1098-1103, **1970**.

Edington, D. and Edgerton, V. R. The Biology of Physical Activity. Boston: Houghton Mifflin, **1976**.

Edler, E. I. The diagnostic use of ultrasound in heart disease. In: Ultrasonic Energy. Ed.: E. Kelly. Urbana, Ill.: Univ. of Illinois Press, **1965**.

Edman, K. A. P. Maximum velocity of shortening

in relation to sarcomere length and degree of activation of frog muscle fibres. J. Physiol. (Lond.) 278: 9p-10p, **1978**.

Edmonds, C. J., Jasani, B. M., and Smith, T. Total body potassium and body fat estimation in relationship to height, sex, age, malnutrition and obesity. Clin. Sci. 48: 431-440, **1975**.

Edström, L. and Ekblom, B. Differences in sizes of red and white muscle fibres in vastus lateralis of musculus Quadriceps Femoris of normal individuals and athletes. Relation to physical performance. Scand. J. Clin. Lab. Invest. 30: 175-181, **1972**.

Edström, L. and Ingelberg, E. Histochemical composition, distribution of fibres and fatiguability of single motor units. Anterior tibial muscle of the rat. J. Neurol. Neurosurg. Psychiatr. 31: 424-433, **1968**.

Edwards, C. W. The carotid body in animals at high altitude. In: High altitude physiology; cardiac and respiratory aspects. Eds.: R. Porter and J. Knight. Edinburgh: Churchill-Livingstone, **1971**.

Edwards, H. T. Lactic acid in rest and work at high altitudes. Amer. J. Physiol. 116: 367-375, **1936**.

Edwards, R. H. T. Methods of studying the unsteady state at the start of exercise. J. Physiol. (Lond.) 202: 12p, **1969**.

Edwards, R. H. T., Ekelund, L. G., Harris, R. C., Hesser, C. M., Hultman, E., Melcher, A., and Wigertz, O. Cardio-respiratory and metabolic costs of continuous and intermittent exercise in man. J. Physiol. (Lond.) 234: 481-497, **1968**.

Edwards, R. H. T., Hill, D. K., and Jones, D. A. Metabolic changes associated with the slowing of relaxation in fatigued mouse muscle. J. Physiol. (Lond.) 251: 287-301, **1975a**.

——. Heat production and chemical changes during isometric contractions of the human quadriceps muscle. J. Physiol. (Lond.) 251: 303-315, **1975b**.

Edwards, R. H. T., Hill, D. K., Jones, D. A., and Merton, P. A. Fatigue of long duration in human skeletal muscle after exercise. J. Physiol. (Lond.) 272: 769-778, **1977**.

Edwards, R. H. T., Nørdesjo, L. O., Koh, D., Harris, R. C., and Hultman, E. Isometric exercise-factors influencing endurance and fatigue. In: Muscle Metabolism During Exercise. Eds.: B. Pernow and B. Saltin. New York: Plenum Press, **1971**.

Egeberg, O. The effect of exercise on the blood clotting system. Scand. J. Clin. Lab. Invest. 15: 8-13, **1963**.

Eggleton, M. G. The effect of exercise on chloride excretion in man during water diuresis and during tea diuresis. J. Physiol. (Lond.) 102: 140-154, **1943**.

Egolinskii, I. A. Certain data on experimental endurance training in men. Fiziol. Zh. SSSR im I.M. Sechenova 47: 38-45, **1961**.

Egoroff, A. Die Myogene Leukocytose. Z. f. Klin. Med. Berlin 100: 485-497, **1924**.

Eide, E., Fedina, L., Jansen, J. Lundberg, A., and Vyklicky, L. Unitary components in the activation of Clarke's column neurones. Acta Physiol. Scand. 77: 145—158, **1969**.

Eigelsreiter, H., Weimann, J., and Zeravik, J. Der Energieumsatz beim Bergsteigen in grossen Höhen. Int. Z. Angew. Physiol. 25: 373-376, **1968**.

Einthoven, W. Die Registrierung der menschlichen Herztöne mittels des Saitengalvanometers. Pflüg. Arch. ges Physiol. 117: 461-472, **1907**.

Eisele, J. H., Ritchie, B. C., and Severinghaus, J. W. Effect of stellate ganglion blockade upon the hyperpnea of exercise. J. Appl. Physiol. 22: 966-969, **1967**.

Eisele, R., Kaczmarczyk, G., Mohnhaupt, R., Reinhardt, H. W., and Schimmrich, B. Left atrial pressure and sodium excretion. A film on experiments in chronically instrumented conscious dogs. J. Physiol. (Lond.) 284: 34p, **1978**.

Eisen, M. E. and Hammond, E. C. The effect of smoking on packed cell volume, red blood cell counts, haemoglobin and platelet counts. Can. Med. Assoc. J. 75: 520-523, **1956**.

Eisenberg, E. and Gordan, G. S. Skeletal dynamics in man measured by nonradioactive strontium. J. Clin. Invest. 40: 1809-1825, **1961**.

Eisenman, P. A. and Golding, L. A. Comparison of effects of training on $V_{O_{2(max)}}$ in girls and young women. Med. Sci. Sport. 2: 136-138, **1975**.

Ekblom, B. Effect of physical training on oxygen transport system in man. Acta Physiol. Scand. Suppl. 328: 9-45, **1969a**.

——. Effect of physical training on adolescent boys. J. Appl. Physiol. 27: 350-355, **1969b**.

——. Effect of training on circulation during prolonged severe exercise. Acta Physiol. Scand. 78: 145-150, **1970**.

Ekblom, B., Åstrand, P. O., Saltin, B., Stenberg, J., and Wallstrom, B. Effect of training on circulatory response to exercise. J. Appl. Physiol. 24: 518-528, **1968**.

Ekblom, B., Goldbarg, A. N. Kilbom, Å., and

Åstrand, P. O. Effects of atropine and propranolol on the oxygen transport system during exercise in man. Scand. J. Clin. Lab. Invest. 30: 35-42, **1972a**.

Ekblom, B., Goldbarg, A. N., and Gullbring, B. Response to exercise after blood loss and reinfusion. J. Appl. Physiol. 33: 175-180, **1972b**.

Ekblom, B., Greenleaf, C. J., Greenleaf, J. E., and Hermansen, L. Temperature regulation during exercise dehydration in man. Acta Physiol. Scand. 79: 475-483, **1970**.

——. Temperature regulation during continuous and intermittent exercise in man. Acta Physiol. Scand. 81: 1-10, **1971**.

Ekblom, B. and Hermansen, L. Cardiac output in athletes. J. Appl. Physiol. 25: 619-625, **1968**.

Ekblom, B., Hermansen, L., and Saltin, B. Hastighetsåkning på Skridsko. Idrottsfysiologi Report 5. Framtiden, Stockholm, **1967**.

Ekblom, B. and Huot, R. Responses to sub-maximal and maximal exercise at different levels of carboxyhemoglobin. Acta Physiol. Scand. 86: 474-482, **1972**.

Ekblom, B., Kilbom, A., and Soltysiak, J. Physical training, bradycardia and autonomic nervous system. Scand. J. Clin. Lab. Invest. 32: 251-256, **1973**.

Ekblom, B., Wilson, G., and Åstrand, P. O. Central circulation during exercise after venesection and reinfusion of red blood cells. J. Appl. Physiol. 40: 379-383, **1976**.

Ekelund, L. G. and Holmgren, A. Circulatory and respiratory adaptation during long term non-steady state exercise in the sitting position. Acta Physiol. Scand. 62: 240-255, **1964**.

Eklund, B. and Kaijser, L. Effect of regional α and β adrenergic blockade of blood flow in the resting forearm during contralateral isometric handgrip. J. Physiol. (Lond.) 262: 39-50, **1976**.

Eklund, G., Von Euler, C., and Rutkowski, S. Spontaneous and reflex activity of intercostal gamma motoneurons. J. Physiol. (Lond.) 171: 139-163, **1964**.

Elam, J. O. Respiratory and circulatory resuscitation, pp. 1265-1312. In: Handbook of Physiology, Section 3, Respiration, Vol. 2. Eds.: W. Fenn and H. Rahn. Baltimore: Williams & Wilkins, **1965**.

Elbel, E. R., Ormond, D., and Close, D. Some effects of breathing oxygen before and after exercise. J. Appl. Physiol. 16: 48-52, **1961**.

Elder, G. C. B. The heterogeneity of fibre-type populations in human muscle (abstr.) Med. Sci. Sport. 9: 64, **1977**.

Eldred, E., Granit, R., and Merton, P. A. Supraspinal control of the muscle spindles and its significance. J. Physiol. (Lond.) 122: 498-523, **1953**.

Eldred, E., Yellin, H., Gadbois, L., and Sweeney, S. Bibliography on muscle receptors; their morphology, pathology, and physiology. Exp. Neurol. Suppl. 3: 1-154, **1967**.

Eldridge, F. L. Neural drive mechanisms of central origin, pp. 149-158. In: Muscular Exercise and the Lung. Eds.: J. A. Dempsey and C. E. Reed. Madison: University of Wisconsin, **1977**.

Elftman, H. The measurement of the external forces in walking. Science 88: 152-153, **1938**.

Elgee, N. J., Williams, R. H., and Lee, M. D. Distribution and degradation studies with ^{131}I insulin. J. Clin. Invest. 33: 1252-1260, **1954**.

El Ghawabi, S. H., Mansour, M. B., Youssef, F. L., El Ghawabi, M., and Abd El Latif, M. Decompression sickness in caisson workers. Brit. J. Industr. Med. 28: 323-329, **1971**.

Elgrishi, I., Ducimetière, P., and Richard, J. L. Reproducibility of analysis of the electrocardiogram in epidemiology using the "Minneosta Code." Brit. J. Prev. Soc. Med. 24: 197-200, **1970**.

Elias, H. M. and Pauly, J. E. Human Micro-anatomy. Chicago: Da Vinci, **1960**.

Eliot, R. S., Forker, A. D., and Robertson, R. J. Aerobic exercise as a therapeutic modality in the relief of stress. Adv. Cardiol. 18: 231-242, **1976**.

Elkins, E. C., Herrick, J. F., Grindlay, J. H., Mann, F. C., and de Forest, R. E. Effect of various procedures on flow of lymph. Arch. Phys. Med. Rehab. 34: 31-39, **1953**.

Ellaway, P. H. Recurrent inhibition of fusimotor neurones exhibiting background discharges in the decerebrate and the spinal cat. J. Physiol. (Lond.) 216: 419-440, **1971**.

Ellestad, M. H. Stress Testing: Principles and Practice. Philadelphia: F. A. Davis, **1975**.

Elliott, B. C. and Blanksby, B. A. A cinematographic analysis of overground and treadmill running by males and females. Med. Sci. Sport. 8: 84-87, **1976**.

Elliott, D. H. The pathological processes of decompression sickness, pp. 414-436. In: Physiology and Medicine of Diving. Eds.: P. B. Bennett and D. H. Elliott. London: Baillière, Tindall & Cassell, **1969**.

Elliott, D. H. and Harrison, J. A. B. Aseptic bone

necrosis in Royal Navy Divers, pp. 251–262. In: Underwater Physiology. Ed.: C. J. Lambertsen. New York: Academic Press, **1971**.

Elliott, G. F. Variations of the contractile apparatus in smooth and striated muscles. X-ray diffraction studies at rest and in contraction. J. Gen. Physiol. 50: 171–124, **1967**.

Elliott, G. F., Lowy, J., and Worthington, C. An X-ray and light diffraction study of the filament lattice of striated muscle in the living state and in rigor. J. Mol. Biol. 6: 295–305, **1963**.

Ellis. F. P., Ferres, H. M., and Lind, A. R. Effects of water intake on sweat production in hot environments. J. Physiol. (Lond.) 125: 61p–62p, **1954**.

Ellis, F. P., Nelson, F., and Pincus, L. Mortality during heat waves in New York City, July, 1972 and August and September, 1973. Environ. Res. 10: 1–13, **1975**.

Ellis, G. Units, Symbols and Abbreviations. A Guide for Biological and Medical Editors and Authors. London: Royal Society of Medicine, **1971**.

Ellis, M. J. Coping with the stresses of intensive competition, pp. 233–241. In: Motor Learning: Sport, Psychology, Pedagogy, and Didactics of Physical Activity. Eds.: F. Landry and W. A. R. Orban. Miami: Symposia Specialists, **1978**.

Ellis, N. R. and Pryer, R. S. Quantification of gross bodily activity in children with severe neuropathology. Amer. J. Ment. Defic. 63: 1034–1037, **1959**.

Elmadjian, F., Hope, J. M., and Lamson, E. T. Excretion of epinephrine and norepinephrine under stress. Rec. Progr. Hormone Res. 14: 513–553, **1958**.

Elmfeldt, D., Wilhelmsson, C., Vedin, A., Tibblin, G., and Wilhelmsen, L. Characteristics of representative male survivors of myocardial infarction compared with representative population samples. Acta. Med. Scand. 199: 387–398, **1976**.

Elson, D. Metabolism of nucleic acids (macromolecular DNA and RNA). Ann. Rev. Biochem. 34: 449–486, **1965**.

Elwood, P. C. Epidemiological aspects of iron deficiency in the elderly. Geront. Clin. 13: 2–11, **1971**.

Emonet-Dénand, F., and La Porte, Y. Frequencygrams of rabbit spindle primary endings elicited by stimulation of fusimotor fibres. J. Physiol. (Lond.) 201: 673–684, **1969**.

Endo, K., Araki, T., and Yagi, N. The distribution and pattern of axon branching of pyramidal tract cells. Brain Res. 57: 484–491, **1973**.

Engel, G. L., Weff, J. P., Ferris, E. B., Ramano, J., Ryder, H., and Blankenhorn, M. A. A migraine-like syndrome complicating decompression sickness. War Med. 5: 304–314, **1944**.

Engel, W. K. The essentiality of histo- and cytochemical studies of skeletal muscle in the investigation of neuromuscular disease. Neurology 12: 778–784, **1962**.

——. Fiber-type nomenclature of human skeletal muscle for histochemical purposes. Neurology 25: 344–348, **1974**.

Engelhardt, W. A. and Ljubimova, M. N. Myosin and adenosine triphosphate. Nature (Lond.) 144: 668–669, **1939**.

Engström, I., Eriksson, B. O., Karlberg, P., Saltin, B., and Thorén, C. Preliminary report on the development of lung volumes in young girl swimmers. Acta Paediatr. Scand. Suppl. 217: 73–76, **1971**.

Enig, M. G. The problem of trans-fatty acids in modern diet. In: Topics in Ischaemic Heart Disease: An International Symposium. Toronto: Toronto Rehabilitation Centre, **1979**.

Enos, W. F., Bayer, J. C., and Holmes, R. H. Pathogenesis of coronary disease in American soldiers killed in Korea. J. Amer. Med. Assoc. 158: 912–914, **1955**.

Epstein, F. H. Renal excretion of sodium and the concept of a volume receptor. Yale J. Biol. Med. 29: 282–298, **1956**.

Epstein, S. E., Robinson, B. F., Kahler, R. L., and Braunwald, E. Effects of beta-adrenergic blockade on the cardiac response to maximal and sub-maximal exercise in man. J. Clin. Invest. 44: 1745–1753, **1965**.

Epstein, S. E., Stampfer, M., and Beiser, G. D. Role of the capacitance and resistance vessels in vaso-vagal syncope. Circulation 37: 524–533, **1968**.

Epuran, M., Horghidan, V., and Muresanu, I. Variations of psychical tension during the mental preparation of sportsmen for contest, pp. 241–245. In: Contemporary Psychology of Sport. Eds.: G. S. Kenyon and T. M. Grogg. Chicago: Athletic Institute, **1970**.

Erbel, R., Kreuzer, H., Neuhaus, L., and Spiller, P. Die determinanten des Intramyokardialen Drückes. Basic Res. Cardiol. 70: 647–660, **1975**.

Erdelyi, G. J. Gynecological survey of female athletes. J. Sport. Med. Phys. Fitness 2: 174–179, **1962**.

Erickson, L., Simonson, E., Taylor, H. L., Alexander, H., and Keys, A. The energy cost of horizontal and grade walking on the motor-driven treadmill. Amer. J. Physiol. 145: 391–401, **1946**.

Ericsson, B., Haeger, K., and Lindell, S. E. Effect of physical training on intermittent claudication. Angiology 21: 188–192, **1970**.

Eriksson, B. O. Physical training, oxygen supply and muscle metabolism in 11–13 year old boys. Acta Physiol. Scand. Suppl. 384: 1–48, **1972**.

——. Physical activity from childhood to maturity: medical and pediatric considerations, pp. 47–55. In: Physical Activity and Human Well-Being. Eds.: F. Landry and W. R. Orban. Miami: Symposia Specialists, **1978**.

Eriksson, B. O., Engström, I., Karlberg, P., Saltin, B., and Thorén, C. A physiological analysis of former girl swimmers. Acta Paediatr. Scand. Suppl. 217: 68–72, **1971**.

Eriksson, B. O., Gollnick, P. D., and Saltin, B. Muscle metabolism and enzyme activities after training in boys 11–13 years old. Acta Physiol. Scand. 87: 485–497, **1973**.

Eriksson, B. O., Lundin, H., and Saltin, B. Cardiopulmonary function in former girl swimmers and the effects of physical training. Scand. J. Clin. Lab. Invest. 35: 135–145, **1975**.

Eriksson, B. and Saltin, B. Muscle metabolism and anaerobic metabolism in prepubertal boys before and after physical training, pp. 173–182. In: Pediatric Work Physiology. Proceedings of 4th International Symposium. Ed.: O. Bar-Or. Natanya, Israel: Wingate Institute, **1973**.

Eriksson, E. and Myrhage, R. Microvascular dimensions and blood flow in skeletal muscle. Acta Physiol. Scand. 86: 211–222, **1972**.

Eriksson, E., Nygaard, E., and Saltin, B. Physiological demands in downhill skiing. Physician Sport. Med. 5(12): **1977**.

Ernsting, J. The physiology of pressure breathing. In: A Textbook of Aviation Physiology. Ed.: J. A. Gillies. Oxford: Pergamon Press, **1965**.

Ernsting, J. and Parry, D. J. Some observations on the effects of stimulating the stretch receptors in the carotid artery in man. J. Physiol. (Lond.) 137: 45p, **1957**.

Ernsting, J., and Shephard, R. J. Respiratory adaptations in congenital heart disease. J. Physiol. (Lond.) 112: 332–343, **1951**.

Espenschade, A. and Eckert, H. Motor development, pp. 322–333. In: Science and Medicine of Exercise and Sport. Eds.: W. R. Johnson and E. R. Buskirk. New York: Harper & Row, **1974**.

Espenschade, A. S. and Meleney, H. E. Motor performances of adolescent boys and girls of today in comparison with those of 24 years ago. Res. Quart. 32: 186–189, **1961**.

Essén, B. Intramuscular substrate utilization during prolonged exercise. Ann. N.Y. Acad. Sci. 301: 30–44, **1977**.

Essén, B., Hagenfeldt, L., and Kaijser, L. Utilization of blood-borne and intramuscular substrates during continuous and intermittent exercise in man. J. Physiol. (Lond.) 265: 489–506, **1977**.

Essén, B., Jansson, E., Henriksson, J., Taylor, A. W., and Saltin, B. Metabolic characteristics of fibre types in human skeletal muscles. Acta Physiol. Scand. 95: 153–165, **1975**.

Essén, B., Pernow, B., Gollnick, P. D., and Saltin, B. Muscle glycogen content and lactate uptake in exercising muscles, pp. 130–134. In: Metabolic Adaptation to Prolonged Physical Exercise. Eds.: H. Howald and J. R. Poortmans. Basel: Birkhauser Verlag, **1975**.

Evans, B. W., Cureton, K. J., and Purvis, J. W. Metabolic and circulatory responses to walking and jogging in water. Res. Quart. 49: 442–449, **1978**.

Evans, C. L. Recent Advances in Physiology. London: Churchill, **1930**.

Evarts, E. V. Relation of pyramidal tract activity to force exerted during voluntary movement. J. Neurophysiol. 31: 14–27, **1968**.

——. Pre- and postcentral neuronal discharge in relation to learned movement from movement and unitary activity in sensorimotor cortex, pp. 449–458. In: Corticothalamic Projections and Sensorimotor Activities. Eds.: T. Frigyesi, E. Rinvik, and M. D. Yahr. New York: Raven Press, **1972a**.

——. Contrasts between activity of precentral and postcentral neurons of cerebral cortex during movement in the monkey. Brain Res. 40: 25–31, **1972b**.

——. Motor cortex reflexes associated with learned movement. Science 179: 501–503, **1973a**.

——. Report to conference on the control of movement and posture. Eds.: R. Granit and R. E. Burke, Brain Res. 53: 1–28, **1973b**.

Exton-Smith, A. N. Care of the elderly: Meeting the challenge of dependency. Eds.: A. N. Exton-Smith and J. G. Evans. London: Academic Press, **1977**.

Ezer, E. and Szporny, L. Prevention of experimental gastric ulcer in rats by a substance which increases biosynthesis of acid mucopolysaccharides. J. Pharm. Pharmacol. 22: 143-145, **1970**.

Fabian, J., Stolz, I., Janota, M. et al. Reproducibility of exercise tests in patients with symptomatic ischaemic heart disease. Brit. Heart J. 37: 785-789, **1975**.

Fagraeus, L. Cardiorespiratory and metabolic functions during exercise in the hyperbaric environment. Acta Physiol. Scand. Suppl. 414: 1-40, **1974**.

Fagraeus, L., Haggendal, J., and Linnarson, D. Heart rate, arterial blood pressure and noradrenaline levels during exercise with hyperbaric oxygen and nitrogen. Swed. J. Defence Med. 9: 265-270, **1973**.

Fagraeus, L. and Linnarson, D. Autonomic origin of heart rate fluctuations at the onset of exercise. Report of Lab. Aviat. Naval Med., Karolinska Inst., pp. 1-8, **1974**.

Fahey, J. D. and Brown, C. H. The effects of an anabolic steroid on the strength, body composition and endurance of college males when accompanied by a weight training program. Med. Sci. Sport. 5: 272-276, **1973**.

Fahey, T. D., Rolph, R., Moungmee, P., Nagel, J., and Mortara, S. Serum testosterone, body composition, and strength of young adults. Med. Sci. Sport. 8: 31-34, **1976**.

Fahrenbach, W. H. Sarcoplasmic reticulum: Ultrastructure of the triadic junction. Science 147: 1308-1310, **1965**.

Falls, H. and Humphrey, L. D. A comparison of methods for eliciting maximum oxygen uptake from college women during treadmill walking. Med. Sci. Sport. 5: 239-241, **1973**.

Falls, H. B., Ismail, A. H., and MacLeod, D. F. Estimation of maximum oxygen uptake in adults from American Association of Health, Physical Education and Recreation Youth Fitness Test Items. Res. Quart. 37: 192-201, **1966**.

Fant, G. M., Lindqvist, J., Sonesson, B., and Hollien, H. Speech distortion at high pressures, pp. 293-299. In: Underwater Physiology. Ed.: C. J. Lambertsen. New York: Academic Press, **1971**.

F.A.O./W.H.O. Requirements of Vitamin A, Thiamine, Riboflavin and Niacin. WHO Tech. Rept. Ser. 362: 1-86, **1967**.

Fardy, P. S. and Hellerstein, H. K. A comparison of continuous and intermittent progressive multi-stage exercise testing. Med. Sci. Sport. 10: 7-12, **1978**.

Fardy, P. S. and Ilmarinen, J. Evaluating the effects and feasability of an at work stair-climbing intervention program for men. Med. Sci. Sport. 7: 91-93, **1975**.

Farhi, A., Burko, H., Newman, E., Klatte, E. C., Carwell, G., and Arnold, T. The effect of exercise on the heart and pulmonary vasculature. Radiology 91: 488-492, **1968**.

Farhi, L. E. Ventilation-perfusion relationship and its role in alveolar gas exchange, pp. 148-197. In: Advances in Respiratory Physiology. Ed.: C. G. Caro. London: Edward Arnold, **1966**.

Faria, I. E. and Cavanagh, P. R. The Physiology and Biomechanics of Cycling. New York: Wiley, **1978**.

Fatt, P. and Katz, B. An analysis of the end-plate potential recorded with an intracellular electrode. J. Physiol. (Lond.) 115: 320-370, **1951**.

Faulkner, J. A. Physiology of swimming and diving, pp. 415-446. In: Exercise Physiology. Ed.: H. B. Falls. New York: Academic Press, **1968**.

———. Maximum exercise at medium altitude, pp. 360-375. In: Frontiers of Fitness. Ed.: R. J. Shephard. Springfield, Ill.: C. C. Thomas, **1971**.

Faulkner, J. A., Daniels, J. T., and Balke, B. Effects of training at moderate altitude on physical performance capacity. J. Appl. Physiol. 23: 85-89, **1967**.

Faulkner, J. A., Roberts, D. E., Elk, R. L., and Conway, J. Cardiovascular responses to submaximum and maximum effort cycling and running. J. Appl. Physiol. 30: 457-461, **1971**.

Favour, C. A. Mountain sickness in fit subjects acutely exposed to high altitude. U.S. Army Contract DA-49-193-MX 3059, **1966**.

Feigenbaum, H. Echocardiography. Philadelphia: Lea & Febiger, **1972**.

———. Use of echocardiography to evaluate cardiac performance, pp. 203-231. In: Cardiac Mechanics: Physiological, Clinical and Mathematical Considerations. Eds.: I. Mirsky, D. Ghista, and H. Sandler. New York: Wiley, **1974**.

Feil, H. and Siegel, M. L. Electrocardiographic

changes during attacks of angina pectoris. Amer. J. Med. Sci. 175: 255-260, **1928**.

Felig, P. Amino acid metabolism in exercise. Ann. N.Y. Acad. Sci. 301: 56-63, **1977**.

Felig, P. and Wahren, J. Interrelationship between amino acid and carbohydrate metabolism during exercise: the glucose alanine-cycle. In: Muscle Metabolism During Exercise. Eds.: B. Pernow and B. Saltin. New York: Plenum Press, **1971**.

Felig, P., Wahren, J., Hendler, R., and Ahlborg, G. Plasma glucagon levels in exercising man. N. Engl. J. Med. 287: 184-185, **1972**.

Fellingham, G. W., Roundy, E. S., Fisher, A. G., and Bryce, G. R. Caloric cost of walking and running. Med. Sci. Sport. 10: 132-136, **1978**.

Fenn, W. D., Otis, A. B., Rahn, H., Chadwick, L. E., and Hegnauer, A. H. Displacement of blood from the lungs by pressure breathing. Amer. J. Physiol. 151: 258-269, **1947**.

Fenn, W. O. A quantitative comparison between the energy liberated and the work performed by the isolated sartorius of the frog. J. Physiol. (Lond.) 58: 373-395, **1923**.

Fenner, A., Jansson, E. H., and Avery, M. E. Enhancement of the ventilatory response to carbon dioxide tube breathing. Resp. Physiol. 4: 91-100, **1968**.

Ferguson, E., and Guest, M. M. Exercise, physical conditioning, blood coagulation and fibrinolysis. Thromb. Diath. Haemorrh. 31: 63-71, **1974**.

Ferguson, R. J., Choquette, G., Chaniotis, L., Petitclerc, R., Huot, R., Gauthier, P., and Campeau, L. Coronary arteriography and treadmill exercise capacity before and after 13 months physical training. Med. Sci. Sport. 5: 67, **1973**.

Ferguson, R. J., Gauthier, P., Coté, P., and Bourassa, M. G. Coronary hemodynamics during upright exercise in patients with angina pectoris. Circulation 52 (Supl. II): 115 (Abstract), **1975**.

Ferguson, R. J., Petitclerc, R., Choquette, G. et al. Effect of physical training on treadmill exercise capacity, collateral circulation and progression of coronary disease. Amer. J. Cardiol. 34: 764-769, **1974**.

Fernandez, C. and Goldberg, J. M. Physiology of peripheral neurons innervating semicircular canals of the squirrel monkey. II. Response to sinusoidal stimulation and dynamics of peripheral vestibular system. J. Neurophysiol. 34: 661-675, **1971**.

Fernandez, E. A., Mohler, J. G., and Butler, J. P. Comparison of oxygen consumption measured at steady state and progressive rates of work. J. Appl. Physiol. 37: 982-987, **1974**.

Fernández-Morán, H. Cell membrane ultrastructure: low-temperature electronmicroscropy and x-ray diffraction studies of lipoprotein components in lamellar systems. Circulation 26: 1039-1065, **1962**.

Ferrer, M. I. The sick sinus syndrome. Circulation 47: 635-641, **1973**.

Ferrero, C. and Doret, J. P. Interpretation hemodynamique de l'onde U del'electrocardiogramme. Cardiologia 25: 112-116, **1954**.

Ferris, B. J. and Pollard, D. S. Effect of deep and quiet breathing on pulmonary compliance in man. J. Clin. Invest. 39: 143-149, **1960**.

Ferris, E. B., Webb, J. P., Engel, G., and Brown, E. W. A comparative study of decompression sickness under varied conditions of exercise, altitude and denitrogenation. Division of Medical Sciences, U.S. National Research Council, **1944**.

Fetz, E. E. and Finocchio, D. V. Operant conditioning of isolated activity in specific muscles and precentral cells. Brain Res. 40: 19-24, **1972**.

Fetz, E. E., Finocchio, D. V., and Baker, M. A. Motor fields of precentral cells elicited by operant reinforcement of unit activity, pp. 187-190. In: Control of Posture and Locomotion. Eds.: R. B. Stein, K. G. Pearson, R. S. Smith, and J. B. Redford. New York: Plenum Press, **1973**.

Fetz, E. E., German, D. C., and Cheney, P. D. Connections between motor cortex cells and motoneurons revealed by correlation techniques in awake monkeys. Neuroscience Abstr. 1: 163, **1975**.

Field, R. A. Glycogen depletion diseases, pp. 141-177. In: The Metabolic Basis of Inherited Diseases. New York: McGraw Hill, **1966**.

Fillenz, M. and Widdicombe, J. G. Receptors of the lungs and airways. In: Handbook of Sensory Physiology, Vol. 3, p. 81. Ed.: E. Neil. Heidelberg: Springer Verlag, **1971**.

Fine, J. Shock and peripheral circulatory insufficiency, pp. 2037-2069. In: Handbook of Physiology, Section 2, Circulation, Vol. 3. Ed.: W. F. Hamilton. Baltimore: Williams & Wilkins, **1965**.

Fink, W. J., Costill, D. L., and Van Handel, P. J.

Leg muscle metabolism during exercise in the heat and cold. Europ. J. Appl. Physiol. 34: 183–190, **1975**.

Finkel, A., and Cumming, G. R. Effect of exercise in the cold on blood clotting and platelets. J. Appl. Physiol. 20: 423–424, **1965**.

Fisch, L. Special senses: the aging auditory system, pp. 265–279. In: Textbook of Geriatric Medicine. Ed.: J. C. Brocklehurst. Edinburgh: Churchill-Livingstone, **1973**.

Fischer, A., Pařízková, J., and Roth, Z. The effect of systematic physical activity on maximal performance and functional capacity in senescent men. Int. Z. Angew. Physiol. 21: 269–304, **1965**.

Fisher, F. D. and Tyroler, H. A. Relationship between ventricular premature contractions in routine electrocardiograms and subsequent death from coronary heart disease. Circulation 47: 712–719, **1963**.

Fisher, M. B. and Birren, J. E. Age and strength. J. Appl. Psychol. 31, 490–497, **1948**.

Fishman, A. P. Dynamics of the pulmonary circulation. In: Handbook of Physiology. Section 2: Circulation. Vol II. Ed.: W. F. Hamilton. Baltimore: Williams & Wilkins, **1963**.

Fitch, K. D. Comparative aspects of available exercise systems. Pediatrics 56 Suppl: 904–907, **1975a**.

———. Exercise-induced asthma and competitive athletics. Pediatrics 56 Suppl: 942–943, **1975b**.

Fitts, P. M. The information capacity of the human motor system in controlling the amplitude of movement. J. Exp. Psychol. 47: 381–391, **1954**.

Fitts, R. H., Nagle, F. J., and Cassens, R. G. Characteristics of skeletal muscle fiber types in the miniature pig and the effect of training. Can. J. Physiol. 51: 825–831, **1973**.

Flack, M. The medical and surgical aspects of aviation. Oxford: Oxford University Press, **1920**.

Flandrois, R., Favier, R., and Pequignot, J. M. Role of adrenaline in gas exchanges and respiratory control in the dog at rest and exercise. Resp. Physiol. 30: 291–303, **1977**.

Flandrois, R. and LaCour, J. R. The prediction of maximal oxygen intake in acute moderate hypoxia. Int. Z. Angew. Physiol. 29: 306–313, **1971**.

Flandrois, R., LaCour, J. R., Maroquin, J. I., and Charlot, J. Essai de mise en évidence d'un stimulus neurogénique articulaire de la ventilation lors de l'exercise musculaire chez le chien. J. Physiol. (Paris) 58: 222–223, **1966**.

Flandrois, R., LaCour, J. R., and Osman, H. Control of breathing in the exercising dog. Resp. Physiol. 13: 361–371, **1971**.

Fleischmann, P. and Kellermann, J. J. Persistent irregular tachycardia in a successful athlete without impairment of performance. Isr. J. Med. Sci. 5: 950–952, **1969**.

Fletcher, J. G. and Lewis, H. E. Photographic methods for estimating external lifting work in man. Ergonomics 2: 114–115, **1959**.

———. Human power output: the mechanics of pole vaulting. Ergonomics 3: 30–34, **1960**.

Floyd, W. W. and Neil, E. The influence of the sympathetic innervation of the carotid bifurcation on chemoreceptor and baroreceptor activity in the cat. Arch. Int. Pharmacodyn. 91: 230–239, **1952**.

Flynn, J. P. Attack behavior in cats. Proc. Int. Union Physiol. Sci. 12: 38, **1977**.

Folinsbee, L., Shephard, R. J., and Silverman, F. Decrease of maximum work performance following ozone exposure. J. Appl. Physiol. 42: 531–536, **1977**.

Folkins, C. H., Lynch, S., and Gardner, M. M. Psychological fitness as a function of physical fitness. Arch. Phys. Med. Rehab. 53: 503–508, **1973**.

Folkow, B. Range of control of the cardiovascular system by the central nervous system. Physiol. Rev. 40 (Suppl. 4): 93–101, **1960**.

Folkow, B., Gaskell, P., and Waaler, B. A. Blood flow through limb muscles during heavy rhythmic exercise. Acta Physiol. Scand. 80: 61–72, **1970**.

Folkow, B., Heymans, C., and Neil, E. Integrated aspects of cardiovascular regulation pp. 1787–1823. In: Handbook of Physiology. Section 2, Vol. 3. Ed.: W. F. Hamilton. Washington, D.C.: Amer. Physiol. Soc. **1965**.

Folkow, B. and Neil, E. Circulation. London: Oxford University Press, **1971**.

Follenius, M. and Bradenberger, G. Effect of muscular exercise on day-time variations of plasma cortisol and glucose, pp. 322–325. In: Metabolic Adaptation to Prolonged Physical Exercise. Eds.: H. Howald and J. R. Poortmans. Basel: Birkhauser Verlag, **1975**.

Food and Agricultural Organisation. Calorie Requirements. Geneva: Food and Agric. Org. of U.N., **1968**.

Forbes, G. B. Towards a new dimension in human growth. Pediatrics 36: 825–835, **1965**.

Forbes, G. B. and Reina, J. C. Adult lean body mass decline with age: Some longitudinal observations. Metabolism 19: 653-663, **1970**.

Ford, M. A., Brooke, J. D., and Thomas, V. Some effects of a high carbohydrate supplement containing glucose syrup on physical work, pp. 211-216. In: Nutrition, Dietetics and Sport. Eds.: G. Ricci and A. Venerando. Torino: Edizioni Minerva Medica, **1978**.

Fordtran, J. S. and Saltin, B. Gastric emptying and intestinal absorption during prolonged exercise. J. Appl. Physiol. 23: 331-335, **1967**.

Forrester, T. and Hamilton, I. J. D. Functional hyperaemia in soleus muscle of the cat. J. Physiol. (Lond.) 249: 20p-21p, **1975**.

Forster, H. V., Dempsey, J. A., Birnbaum, M. L., Reddan, W. G., Thoden, J. S., Grover, R. F., and Rankin, J. Comparison of ventilatory responses to hypoxic and hypercapnic stimuli in altitude-sojourning lowlanders, lowlanders residing at altitude, and native altitude residents. Fed. Proc. 28: 1274-1279, **1969**.

Forster, R. E. Factors affecting the rate of exchange of O_2 between blood and tissues, pp. 393-407. In: Oxygen in the Animal Organism. Eds.: F. Dickens and E. Neil. New York: Macmillan, **1964**.

Fowler, K. T. The vertical gradient of perfusion in the erect human lung. J. Appl. Physiol. 20: 1163-1172, **1965**.

Fowler, W. M., Gardner, G. W., and Egstrom, G. H. Effect of an anabolic steroid on physical performance of young men. J. Appl. Physiol. 20: 1038-1040, **1965**.

Fowler, W. S. Breaking point of breath-holding. J. Appl. Physiol. 6: 539-545, **1954**.

Fox, E. L., Bartels, R. L., Billings, C. E., O'Brien, R., Bason, R., and Mathews, D. K. Frequency and duration of interval training programs and changes in aerobic power. J. Appl. Physiol. 38: 481-484, **1975**.

Fox, E. L., Bartels, R. L., Klinzing, J., and Ragg, K. Metabolic responses to interval training programs of high and low power output. Med. Sci. Sport. 9: 191-196, **1977**.

Fox, R. H. Heat stress and athletics. Ergonomics 3: 307-313, **1960**.

Fox, R. H., Lofstedt, B. E., Woodward, P. M., Eriksson, E., and Werkstrom, B. Comparison of thermoregulatory function in men and women. J. Appl. Physiol. 26: 444-453, **1969**.

Fox, S. M. Exercise and stress testing workshop report. J.S. Carol Med. Assoc. Suppl. 1: 77, **1969**.

Fox, S. M. and Haskell, W. L. Physical activity and the prevention of coronary heart disease. Bull. N.Y. Acad. Med. 44: 950-965, **1968**.

Fox, S. M., Naughton, J. P., and Haskell, W. L. Physical activity and the prevention of coronary heart disease. Ann. Clin. Res. 3: 404-432, **1971**.

Fox, S. M., Naughton, J. P., and Gorman, P. A. Physical activity and cardiovascular health. Mod. Concepts Cardiovasc. Dis. 41: 17-30, **1972**.

Fox, S. M. and Skinner, J. S. Some planning in the United States for further studies to define the relationship between physical activity and coronary disease. In: Physical Activity and the Heart. Eds.: K. Evang. and K. L. Andersen. Baltimore: Williams & Wilkins, **1966**.

Fozzard, H. A. and Das Gupta, D. S. Electrophysiology and the electrocardiogram. Mod. Concepts Cardiovasc. Dis. 44: 29-34, **1975**.

Frank, C. W., Weinblatt, E., and Shapiro, S. Prognostic implications of serum cholesterol in coronary heart disease, pp. 390-395. In: Atherosclerosis. Proceedings of 2nd International Symposium. Ed.: R. J. Jones. Berlin: Springer, **1970**.

Frank, D. Physical warm-up, pp. 20-23. In: An analysis and evaluation of modern trends related to the physiological factors for optimal sports performance. Ed.: E. J. Burke. Ithaca, N.Y.: Ithaca College, **1976**.

Frank, E. An accurate, clinically practical system for spatial vectorcardiography. Circulation 13: 737-749, **1956**.

Frank, M. N. and Kinlaw, W. B. Indirect measurement of isovolumetric contraction time and tension period in normal subjects. Amer. J. Cardiol. 10: 800-806, **1962**.

Frank, N. R. Influence of acute pulmonary vascular congestion on recoiling force of excised cat's lung. J. Appl. Physiol. 14: 905-908, **1959**.

Frank, O. Die grundform des Arteriellen Pulses. Die Berechnung der Herzarbeit. Sekt E. Z. Biol 19: 483-526, **1898**.

Frankenhauser, M., Graff-Lonnevig, V., and Hesser, C. M. Effects on psychomotor functions of different nitrogen-oxygen gas mixtures at increased ambient pressures. Acta Physiol. Scand. 59: 400-409, **1963**.

Franks, B. D. Effects of different types and amounts

of training on selected fitness measures, pp. 139-160. In: Exercise and Fitness, 1969. Ed.: B. D. Franks. Chicago: Athletic Institute, **1969**.

Franks, B. D. and Cureton, T. K. Effects of training on time components of the left ventricle. J. Sport Med. Phys. Fitness 9: 80-88, **1969**.

Franks, W. R. Report C-2829. Committee on Aviation Medicine, National Research Council, Canada, **1940**.

Frantz, A. G. and Rabkin, M. T. Effects of estrogen and sex differences on secretion of human growth hormone. J. Clin. Endocrinol. Metab. 25: 1470-1480, **1965**.

Franzini-Armstrong, C. and Porter, K. R. Sarcolemmal invaginations constituting the T-system in fish muscle fibers. J. Cell Biol. 22: 675-696, **1964**.

Frech, W. E., Schultehinrichs, D., Vogel, H. R., and Thews, G. Modelluntersuchungen zum Austausch der Atemgase. 1. Die O_2-Aufnahmezeiten des Erythrocyten unter den Bedingungen des Lungen capillarblutes. Pflüg. Archiv. 301: 292-301, **1968**.

Fredholm, B. B. Inhibition of fatty acid release from adipose tissue by high arterial lactate concentrations. Acta Physiol. Scand. 77 (Suppl. 330): 77, **1969**.

Fredrickson, D. S. Function and structure of plasma lipoprotein. In: Lipid Metabolism, Obesity and Diabetes Mellitus: Impact upon Atherosclerosis. Eds.: H. Greten, R. Levine, E. F. Pfeiffer, and A. E. Renold. Stuttgart, Thieme, **1974**.

Freedman, S. Sustained maximum voluntary ventilation. Resp. Physiol. 8: 230-244, **1970**.

——. The effects of added loads in man: Conscious and anaesthetized, pp. 22-25. In: Loaded Breathing. Eds.: L. D. Pengelly, A. S. Rebuck, and E. J. M. Campbell. Don Mills, Ont.: Longman Canada, **1974**.

Freedman, W., Wannstedt, G., and Herman, R. EMG patterns and forces developed during step down. Amer. J. Phys. Med. 55: 275-290, **1976**.

Freeman, M. A. R. and Wyke, B. The innvervation of the knee joint. An anatomical and histological study in the cat. J. Anat. 101: 505-532, **1967**.

Frenkl, R., Csalay, L., Makra, G., and Sonifal, Z. The effect of regular muscle activity on the histamine sensitivity of the rat. Acta Physiol. Acad. Sci. Hung. 25: 199-202, **1964**.

Freund, H. J., Büdingen, H. J., and Dietz, V. Activity of single motor units from human forearm muscles during voluntary isometric contractions. J. Neurophysiol. 38: 933-946, **1975**.

Frey, R. and Nolte, H. Wiederbelebung am Unfallort und auf dem Transport. Anaesthetist 17: 113-116, **1968**.

Freyschuss, U. Cardiovascular adjustment to somatomotor activation. Acta Physiol. Scand. Suppl. 342: **1970**.

Frick, M. H. Significance of bradycardia in relation to physical training, pp. 33-41. In: Physical Activity and the Heart. Eds.: M. J. Karvonen, and A. J. Barry. Springfield, Ill.: C. C. Thomas, **1967**.

——. Coronary implications of hemodynamic changes caused by physical training. Amer. J. Cardiol. 22: 417-425, **1968**.

——. Long-term excess physical activity and central haemodynamics in man. Adv. Cardiol. 18: 136-143, **1976**.

Frick, M. H., Konttinen, A., and Sarajas, H. S. S. Effects of physical training on circulation at rest and during exercise. Amer. J. Cardiol. 12: 142-147, **1963**.

Fried, T. and Shephard, R. J. Deterioration and restoration of physical fitness after training. Can. Med. Assoc. J. 100: 831-837, **1969**.

——. Assessment of a lower extremity training programme. Can. Med. Assoc. J. 103: 260-266, **1970**.

——. A team approach to sports medicine. J. Amer. Med. Assoc. 216: 1777-1778, 1971.

Friedlander, J. S., Costa, P. T., Bosse, R., Ellis, E., Rhoads, J. E., and Stoudt, H. W. Longitudinal physique changes among healthy white veterans at Boston. Human Biol. 49: 541-558, **1977**.

Friedman, E. H. and Hellerstein, H. K. Influence of psychosocial factors on coronary risk and adaptation to a physical fitness evaluation program, pp. 225-251. In: Exercise Testing and Exercise Training in Coronary Heart Disease. Eds.: J. P. Naughton, H. K. Hellerstein, and I. C. Mohler. New York: Academic Press, **1973**.

Friedman, G. D. Primer of Epidemiology. New York: McGraw Hill, **1974**.

Friedman, M. and Rosenman, R. H. Type A behavior and your heart. Greenwich, Conn: Fawcett, **1974**.

Friedman, S. M., Sréter, F. A., and Friedman,

C. L. The effect of vasopressin and aldosterone on the distribution of water, sodium and potassium and on work performance in old rats. Gerontologia 7: 65–76, **1963**.

Friis-Hansen, B. Hydrometry of growth and aging, pp. 191–209. In: Human Body Composition: Approaches and Applications. Symposia of the society for the study of human biology, Vol. 7. Ed.: J. Brozek. Oxford: Pergamon Press, **1965**.

Frisancho, A. R., Martinez, C., Velasquez, T., Sanchez, J., and Montoye, H. J. Influence of developmental adaptation on aerobic capacity at high altitude. J. Appl. Physiol. 34: 176–180, **1973**.

Frisch, R. E. Fatness of girls from menarche to age 18 years, with a nomogram. Human Biol. 48: 353–359, **1976**.

Fröberg, S. O., Carlson, L. A., and Ekelund, L. G. Local lipid stores and exercise, pp. 307–313. In: Muscle Metabolism During Exercise. Eds.: B. Pernow and B. Saltin. New York: Plenum Press, **1971**.

Froeb, H. F. Ventilatory response of SCUBA divers to CO_2 inhalations. J. Appl. Physiol. 16: 8–10, **1961**.

Froehlicher, V. F. Animal studies of the effect of chronic exercise on the heart and atherosclerosis. A review. Amer. Heart J. 84: 496–506, **1972**.

Froehlicher, V. F. and Oberman, A. Analysis of epidemiologic studies of physical inactivity as risk factor for coronary artery disease. Progr. Cardiovasc. Dis. 15: 41–65, **1972**.

Froehlicher, V. F., Yanowitz, F., Thompson, A. J., and Lancaster, M. C. Physiological responses in aircrewmen and the detection of latent coronary artery disease. Neuilly-sur-Seine, France: NATO Agardograph, 210: 1–60, **1975**.

Froese, A. B. and Gledhill, N. 1979. See **Spriet, L. L., Gledhill, N., Froese, A. B., Wilkes, D. L., and Meyers, E. C.** The effect of induced erythrocythemia on central circulation and oxygen transport during maximal exercise. Med. Sci. Sports 12: 122, **1980**.

Fromm, C. and Noth, J. Vibration-induced autogenic inhibition of gamma motoneurons. Brain Res. 83: 495–497, **1975**.

Fryer, D. I. Evolution of concepts in the etiology of bends. Aerospace Med. 39: 1058–1061, **1968**.

Fuchs, F., Reddy, Y., and Briggs, F. N. The interaction of cations with the calcium binding site of troponin. Biochem. Biophys. Acta. 221: 407–409, **1970**.

Fujihara, Y., Hildebrandt, J. R., and Hildebrandt, J. Cardiorespiratory transients in exercising man. I. Tests of superposition. J. Appl. Physiol. 35: 58–67, **1973a**.

Fujihara, Y., Hildebrandt, J., and Hildebrandt, J. R. Cardiorespiratory transients in exercising man. II. Linear Models. J. Appl. Physiol. 35: 68–76, **1973b**.

Fujita, A. A study of the training effect upon selective body reaction time, pp. 127–138. In: Motor Learning, Sport, Psychology, Pedagogy and Didactics of Physical Activity. Eds.: F. Landry and W. R. Orban. Miami: Symposia Specialists, **1978**.

Fukuchi, Y., Roussos, C. S., Macklem, P. T., and Engel, L. A. Convection, diffusion and cardiogenic mixing of inspired gas in the lung: An experimental approach, Resp. Physiol. 26: 77–90, **1976**,

Fulton, J. F. and Pi-Suner, J. A note concerning the probable function of various afferent end-organs in skeletal muscle. Amer. J. Physiol. 83: 554–562, **1928**.

Funderburk, C. F., Hipskind, S. G., Welton, R. C., and Lind, A. R. Development of and recovery from fatigue induced by static effort at various tensions. J. Appl. Physiol. 37: 392–396, **1974**.

Fung, Y. C. and Sobin, S. S. Pulmonary alveolar blood flow. Circ. Res. 30: 470–490, **1969**.

Furchgott, R. F. Pharmacological characterization of receptors: Its relation to radioligand-binding studies. Fed. Proc. 37: 115–120, **1978**.

Furness, P. and Jessop, J. Prolonged changes in physiological tremor following a brief voluntary contraction of human muscle. J. Physiol. (Lond.) 258: 72p–73p, **1976**.

Gabbato, F. and Media. A. Analysis of the factors that may influence the duration of isotonic systole in normal conditions. Cardiologia, 29: 114–131, **1956**.

Gaensler, E. A. Personal communication.

Gaesser, G. A. and Brooks, G. A. Muscular efficiency during steady-rate exercise: Effects of speed and work rate. J. Appl. Physiol. 38: 1132–1139, **1975**.

Gagge, A. P. Partitional calorimetry in the desert, pp. 23–51. In: Physiological Adaptations, Desert and Mountain. Eds.: M. K. Yousef, S. M. Horvath, and R. W. Bullard. New York: Academic Press, **1972**.

Gaisl, G. and Harnoncourt, K. The behaviour of

stress acidosis during ergometric examinations, pp. 70–72. In: Metabolic adaptation to prolonged physical exercise. Eds.: H. Howald and J. R. Poortmans. Basel: Birkhauser Verlag, **1975**.

Galbo, H., Holst, J. J., and Christensen, N. J. Glucagon and plasma catecholamine responses to graded and prolonged exercise in man. J. Appl. Physiol. 38: 70–76, **1975**.

Galevskaya, E. N. On the system and timing of gymnastic exercises for skilled industrial labourers at work. Theory and practice of physical culture (Moscow) I. 52–54. Cited by N. Schneidman. Soviet studies in the fitness of the aged. Can. Fam. Phys. (October 1972): 53–56, **1970**.

Galichia, J., Daniel, J., Jorgensen, C., and Gobel, F. The effect of age on central to peripheral amplification of the arterial pulse during upright exercise. Amer. J. Cardiol. 37: 137 (Abstract), **1976**.

Galton, D. J., Gilbert, C., Reckless, J. P. D., and Kaye, J. Triglyceride storage disease—a group of inborn errors of triglyceride metabolism. Proc. Med. Res. Soc., U.K., June 29th, **1973**.

Gandevia, S. C. and McCloskey, D. I. Effects of related sensory inputs on motor performances in man studied through changes in perceived heaviness. J. Physiol. (Lond.) 272: 653–672, **1977a**.

——. Changes in motor commands, as shown by changes in perceived heaviness, during partial curarization and peripheral anaesthesia in man. J. Physiol. (Lond.) 272: 673–689, **1977b**.

Gandy, A., Green, H., Houston, M., Thomson, J., and Williams, I. Reproducibility of repeated sampling using the muscle biopsy technique. Can. J. Appl. Sports Sci. 3: 184, **1978**.

Garcia-Austt, E., Bogacz, J., and Vanzulli, A. Effects of attention and inattention upon visual evoked response. Electroenceph. Clin. Neurophysiol. 17: 136–143, **1964**.

Gardner, E. and Haddad, B. Pathways to the cerebral cortex for afferent fibres from the hind-leg of the cat. Amer. J. Physiol. 172: 475–482, **1953**.

Gardner, K. D. Athletic pseudonephritis: Alteration of urinary sediment by athletic competition. J. Amer. Med. Assoc. 161: 1613–1617, **1956**.

Garlick, M. A. and Bernauer, E. M. Exercise during the menstrual cycle: Variations in physiological baselines. Res. Quart. 39: 533–542, **1968**.

Garn, S. M. The genetics of human growth, pp. 415–434. In: De Genetica Medica II. Ed.: L. Gedda. Rome: Edizioni Istituto Mendel, **1961a**.

——. Radiographic analysis of body composition, pp. 36–58. In: Techniques for Measuring Body Composition. Eds.: J. Brožek and A. Henschel. Washington, D.C. Natl. Acad. Sci., Natl. Res. Council, **1961b**.

Garn, S. M., Clark, A., Landkof, L., and Newell, L. Parental body build and developmental progress in the offspring. Science 132: 1555–1556, **1960**.

Garn, S. M., Rohmann, C. G., and Silverman, F. N. Radiographic standards for post-natal ossification and tooth calcification. Med. Radiograph Photogr. 43(2): 45–66, **1967**.

Garnett, R., O'Donovan, M. J., Stephans, J. A., and Taylor, A. Evidence for the existence of three motor unit types of normal human gastrocnemius. J. Physiol. (Lond.) 280: 65p, **1978**.

Garrison, R. J., Kannel, W. B., Feinleif, M., Castelli, W. P., McNamara, P. M., and Padgett, S. J. Cigarette smoking and HDL cholesterol. The Framingham Offspring Study. Atherosclerosis 30: 17–25, **1978**.

Garrow, J. S. Energy balance and obesity in man. Amsterdam: North Holland, **1974**.

Gasnier, A. and Mayer, A. Recherches sur la régulation de la nutrition. Ann. Physiol. 15: 145–214, **1939**.

Gauer, O. H., Henry, J. P., and Sieker, H. O. Cardiac receptors and fluid volume control. Progr. Cardiovasc. Dis. 4: 1–26, **1961**.

Gauthier, G. F. On the relationship of ultrastructural and cytochemical features to color in mammalian skeletal muscle. Z. Zellforsch. Mikrosk. Anat. 95: 462–482, **1969**.

Gautier, H., Lacaisse, A., Pasquis, P., and Dejours, P. Réaction ventilatoire à la stimulation des fuseaux neuro-musculaires par la succinylcholine chez le chat. J. Physiol. (Paris) 56: 560–561, **1964**.

Gebber, G. L. and Snyder, D. W. Hypothalamic control of baroreceptor reflexes. Amer. J. Physiol. 218: 124–131, **1970**.

Geddes, L. A. and Baker, L. E. The relationship between input impedance and electrode area in recording the ECG. Med. Biol. Eng. 4: 439–450, **1966**.

Geddes, L. A., Tacker, W. A., McFarlane, J., and Bourland, J. Strength-duration curves for ventricular defibrillation in dogs. Circ. Res. 27: 551-560, **1970**.

Geiser, M. and Trueta, J. Muscle action, bone rarefaction and bone formation. J. Bone Joint Surg. 40: 282-311, **1958**.

Geissler, H. J. Zu einigen Untersuchungsergebnissen auf dem Gebiete der Ausgleichs-gymnastik wahrend der Arbeitszeit. Wiss. Zschr. DHFK. Leipzig 3: 229-242, **1960**.

Gelerntner, H. L. and Swihart, J. C. A mathematical-physical model of the genesis of the electrocardiogram. Biophys. J. 4: 285-301, **1964**.

Gelfand, S. The relationship between experimental pain tolerance and pain threshold. Can. J. Psychol. 18: 36-42, **1964**.

Gellhorn, E. The physiology of the supraspinal mechanisms. In: Science and Medicine of Exercise and Sports. Ed.: W. R. Johnson. New York: Harper & Row, **1960**.

Geltes, L. S. Electrophysiologic basis of arrhythmias and acute myocardial ischemia. In: Modern Trends in Cardiology, Vol. 3, pp. 219-246. Ed.: M. F. Oliver. London: Butterworths, **1975**.

Genaud, P. E. M., Legendre, R., and Saury, A. External cardiac massage and drowning. Lancet 1: 1383-1384, **1965**.

Genov, F. The nature of the mobilization readiness of the sportsman and the influence of different factors upon its formation. In: Contemporary Psychology of Sport. Eds.: G. S. Kenyon and T. M. Grogg. Chicago: Athletic Institute, **1970a**.

——. Peculiarity of the maximum motor speed of sportsmen when in mobilized readiness, pp. 233-240. In: Contemporary Psychology of Sport. Eds.: G. S. Kenyon and T. M. Grogg. Chicago: Athletic Institute, **1970b**.

Gentile, A. M. and Nacson, J. Organizational processes in motor control. Exercise Sport Sci. Rev. 4: 1-33, **1976**.

George, P. and Rutman, R. J. The "high energy phosphate bond" concept. Progr. Biophys. 10: 2-53, **1960**.

Gergely, J. The regulatory proteins of myofibrils and their role in the mechanism of contraction and relaxation, pp. 6-12. In: Limiting Factors of Physical Performance. Ed.: J. Keul. Stuttgart: Thieme, **1973**.

Gerking, S. D. and Robinson, S. Decline in the rates of sweating of men working in severe heat. Amer. J. Physiol. 147: 370-378, **1946**.

Gernandt, B. E. and Proler, M. L. Medullary and spinal accessory nerve responses to vestibular stimulation. Exp. Neurol. 11: 27-37, **1965**.

Gero, J. and Gerova, M. Significance of the individual parameters of pulsating pressure in stimulation of baroreceptors, p. 17. In: Baroreceptors and Hypertension. Ed.: P. Kezdi. Oxford: Pergamon Press, **1965**.

Gerstenblith, G., Lakatta, E. G., and Weisfeldt, M. L. Age changes in myocardial function and exercise response. Progr. Cardiovas. Dis. 19: 1-21, **1976**.

Gertler, M. M. and Leetma, H. E. Biochemical adaptation in the heart secondary to physical effort, pp. 199-209. In: Exercise Testing and Exercise Training in Coronary Heart Disease. Eds.: J. P. Naughton, H. K. Hellerstein, and I. C. Mohler. New York: Academic Press, **1973**.

Gertz, I., Hedenstierna, G., Hellers, G., and Wahren, J. Muscle metabolism in patients with chronic obstructive lung disease and acute respiratory failure. Clin. Sci. Mol. Med. 52: 395-403, **1977**.

Gestman, L. R., Pollock, M. L., Durstine, J. L., Ward, A., Ayres, J., and Linnerud, A. C. Physiological responses of men to 1, 3, and 5 day per week training programs. Res. Quart. 47: 638-646, **1977**.

Getchell, L. H., Kirkendall, D., and Robins, G. Prediction of maximum oxygen uptake in young adult women joggers. Res. Quart. 48: 61-67, **1977**.

Gheron, E. Mechanisms of voluntary effort, pp. 187-196. In: Contemporary Psychology of Sport. Eds.: G. S. Kenyon and T. M. Grogg. Chicago: Athletic Institute, **1970**.

Ghista, D. N. and Sandler, H. Oxygen utilization of the human left ventricle: An indirect method for its evaluation and clinical considerations, pp. 463-482. In: Cardiac Mechanics: Physiological, Clinical, and Mathematical Considerations. Eds.: I. Mirsky, D. N. Ghista, and H. Sandler. New York: Interscience, **1971**.

Giammona, S. T., Kerner, D., and Bondurant, S. Effect of oxygen breathing at atmospheric pressure on pulmonary surfactant. J. Appl. Physiol. 20: 855-858, **1965**.

Gibson, K. and Harris, P. The in-vitro and in-vivo effects of polyamines on cardiac protein biosynthesis. Cardiovasc. Res. 8: 668-673, **1974**.

Gilbert, C., Kretzschmar, K. M., Wilkie, D. R.,

and Woledge, R. C. Chemical change and energy output during muscular contraction. J. Physiol. (Lond.) 218: 163–193, **1971**.

Gilbert, C. A. and Stevens, P. A. Forearm vascular responses to lower body negative pressure and orthostasis. J. Appl. Physiol. 21: 1265–1272, **1966**.

Gilbert, P. How the cerebellum could memorize movements. Nature 254: 688–689, **1975**.

Gilbert, R. and Auchincloss, J. H. Mechanics of breathing in normal subjects during brief, severe exercise, J. Lab. Clin. Med. 73: 439–450, **1969**.

——. Comparison of cardiovascular responses to steady- and unsteady-state exercise. J. Appl. Physiol. 30: 388–393, **1971**.

Gilbert, R., Baule, G. H., and Auchincloss, J. H. Theoretical aspects of oxygen transfer during early exercise. J. Appl. Physiol. 21: 803–809, **1966**.

Gillen, H. W. Oxygen convulsions in man, pp. 217–223. In: Proceedings of the Third National Conference on Hyperbaric Medicine. Eds.: I. W. Brown and B. G. Cox. Publ. 1404. Washington, D.C.: Natl. Acad. Sci. Natl. Res. Council, **1966**.

Gillespie, J. A. The nature of bone changes associated with nerve injuries and disuse. J. Bone Joint Surg. 36: 464–473, **1954**.

Gilliam, T. B., Sady, S., Thorland, W. G., and Weltman, A. L. Comparison of peak performance measures in children ages 6 to 8, 9 to 10 and 11 to 13 years. Res. Quart. 48: 695–702, **1978**.

Gilligan, D. R., Altschule, M. D., and Katersky, E. M. Physiological intravascular hemolysis of exercise hemoglobinemia and hemoglobinuria following cross-country runs. J. Clin. Invest. 22: 859–869, **1943**.

Gillis, M. F., Peterson, P. L., and Karagianes, M. T. In vivo detection of circulating gas emboli associated with decompression sickness using the Doppler flowmeter. Nature (Lond.) 217: 965–967, **1968**.

Gilman, S. A cerebello-thalamocortical pathway controlling fusimotor activity. In: Control of Posture and Locomotion. Eds.: R. B. Stein, K. G. Pearson, R. S. Smith, and J. B. Redford. New York: Plenum Press, **1973**.

Ginzel, K. H. The importance of sensory nerve endings as sites of drug action. Naunyn-Schmiedeberg's Arch. Pharm. 228: 29–56, **1975**.

——. Interaction of somatic and autonomic functions in muscular exercise. Exercise Sport Sci. Rev. 4: 35–86, **1976**.

Ginzel, K. H. and Eldred, E. Reflex depression of somatic motor activity from heart, lungs and carotid sinus. In: Respiratory Adaptations, Capillary Exchange, and Reflex Mechanisms. Eds.: A. S. Paintal and P. Gill-Kumar. Delhi: Vallabhbhai Patel Chest Institute, **1976**.

Girandola, R. N. Body composition changes in women: Effects of high and low intensity exercise. Arch. Phys. Med. Rehab. 57: 297–300, **1967**.

Girandola, R. N. and Katch, F. I. Effects of physical training on ventilatory equivalent and respiratory exchange ratio during weight-supported, steady-state exercise. Europ. J. Appl. Physiol. 35: 119–125, **1976**.

Girandola, R. N., Wiswell, R. A., and Romero, G. Body composition changes resulting from fluid ingestion and dehydration. Res. Quart 48: 299–303, **1977**.

Gisolfi, C. V. Work-heat tolerance derived from interval training. J. Appl. Physiol. 35: 349–354, **1973**.

Gisolfi, C. V. and Copping, J. R. Thermal effects of prolonged treadmill exercise in the heat. Med. Sci. Sport. 6: 108–113, **1974**.

Gisolfi, C. and Robinson, S. Relations between physical training, acclimatization, and heat tolerance. J. Appl. Physiol. 26: 530–534, **1969**.

——. Central and peripheral stimuli regulating sweating during intermittent work in men. J. Appl. Physiol. 29: 761–768, **1970**.

Giuntini, C., Mariani, M., Maseri, A., and Donato, C. Il controllo del volume di sangue polmonare nell'uomo e le sue implicazioni cliniche. Atti 72 Congr. S.I.M.I., Montecatini, **1971**.

Givoni, B. and Goldman, R. F. Predicting metabolic energy cost. J. Appl. Physiol. 30: 429–433, **1971**.

Glanville, A. D. and Kreezer, G. The maximum amplitude and velocity of joint movements in normal male adults. Human. Biol. 9: 197–211, **1937**.

Glaser, E. M. The Physiological Basis of Habituation, pp. 1–102. London: Oxford University Press, **1966**.

Glaser, E. M. and Griffin, J. P. Influence of the cerebral cortex on habituation. J. Physiol. (Lond.) 160: 429–445, **1962**.

Glaser, E. M. and Shephard, R. J. Simultaneous

experimental acclimatization to heat and cold in man. J. Physiol. (Lond.) 169: 592-602, **1963**.

Glassow, R. Fundamentals in physical education. Philadelphia: Lea & Febiger, **1932**.

Glazier, J. B., Hughes, J. M. B., Maloney, J. E., and West, J. B. Measurements of capillary dimensions and blood volume in rapidly frozen lungs. J. Appl. Physiol. 26: 65-76, **1969**.

Gledhill, N. and Eynon, R. B. The intensity of training. In: Training: Scientific Basis and Application. Ed.: A. W. Taylor. Springfield, Ill.: C. C. Thomas, **1972**.

Gledhill, N., Buick, F. J., Froese, A. B., Spriet, L., and Meyers, E. C. An optimal method of storing blood for blood boosting. Med. Sci. Sport. 10: 40, **1978**.

Gledhill, N., Froese, A. B., and Dempsey, J. A. Ventilation to perfusion distribution during exercise in health, pp. 325-342. In: Muscular Exercise and the Lung. Eds.: J. A. Dempsey and C. E. Reed. Madison: University of Wisconsin, **1977**.

Gleser, M. A., Horstman, D. H., and Mello, R. P. The effect on $\dot{V}_{O_{2(max)}}$ of adding arm work to maximal leg work. Med. Sci. Sport. 6: 104-107, **1974**.

Gleser, M. A. and Vogel, J. A. Effects of acute alterations of $\dot{V}_{O_{2(max)}}$ on endurance capacity of men. J. Appl. Physiol. 34: 443-447, **1973**.

Glueck, C. J., Gartside, P., Fallat, R. W., Sielski, J., and Steiner, P. M. Longevity syndromes: Familial hypo-beta and familial hyper-alpha lipoproteinemia. J. Lab. Clin. Med. 88: 941-957, **1976**.

Glynn, I. M. and Karlish, S. J. D. The sodium pump. Ann. Rev. Physiol. 37: 13-56, **1975**.

Goddard, R. (Ed.) The effects of altitude on athletic performance. Chicago: Athletic Institute, **1967**.

Godfrey, S. Circulatory and respiratory responses to exercise of children with heart or lung disease. In: Pediatric Work Physiology. Proceedings of the 4th International Symposium, pp. 295-300. Ed.: O. Bar-Or. Natanya, Israel: Wingate Institute, **1973**.

——. Problems peculiar to the diagnosis and management of childhood asthma. Brit. Tuberc. Thor. Assoc. Rev. 4: 1, **1974**.

——. Clinical variables of exercise-induced bronchospasm, pp. 247-261. In: Muscular Exercise and the Lung. Eds.: J. A. Dempsey and C. E. Reed. Madison: University of Wisconsin Press, **1977**.

Godfrey, S. and Campbell, E. J. M. The role of afferent impulses from the lung and chest wall in respiratory control and sensation. In: Breathing: Hering-Breuer Centenary Symposium. Ed.: R. Porter. London: Churchill, **1970**.

Godfrey, S. Edwards, R. H. T., Copland, G. M., and Gross, D. L. Chemosensitivity in normal subjects, athletes, and patients with chronic airways obstruction. J. Appl. Physiol. 30: 193-199, **1971**.

Godfrey, S., Silverman, M., and Anderson, S. D. Problems of interpreting exercise-induced asthma. J. Allergy 52: 199-209, **1973**.

Godfrey, S. and Wolf, E. An evaluation of rebreathing methods for measuring mixed venous P_{CO_2} during exercise. Clin. Sci. 42: 345-353, **1972**.

Godin, G. and Shephard, R. J. Activity patterns of the Canadian Eskimo. In: Human Polar Biology. Eds.: O. G. Edholm and E. K. E. Gunderson. London: Heinemann, **1973a**.

——. Body weight and the energy cost of activity. Arch. Environ. Health 27: 289-293, **1973b**.

Godin, G., Wright, G., and Shephard, R. J. Urban exposure to carbon monoxide. AMA Arch. Environ. Health, 25: 305-313, **1972**.

Goff, L. G. and Bartlett, R. G. Elevated end-tidal CO_2 in trained underwater swimmers. J. Appl. Physiol. 10: 203-206, **1957**.

Goff, L. G., Brubach, H. F., and Specht, H. Measurements of respiratory responses and work efficiency of underwater swimmers utilizing improved instrumentation. J. Appl. Physiol. 10: 197-202, **1957**.

Gold, W. M. and Nadel, J. A. Dyspnoea and hyperventilation associated with unilateral disease of the chest wall relieved by blocking the intercostal nerves. In: Breathlessness. Eds.: J. B. L. Howell, E. J. M. Campbell. Oxford: Blackwell Scientific, **1966**.

Goldberg, A. L. Work-induced growth of skeletal muscles in normal and hypophysectomized rats. Amer. J. Physiol. 213: 1193-1198, **1967**.

——. Role of insulin in work-induced growth of skeletal muscle. Endocrinology 83: 1071-1073, **1968**.

Goldberg, A. L., Etlinger, J. D., Goldspink, D. F., and Jeblecki, C. Mechanisms of work-induced hypertrophy of skeletal muscle. Med. Sci. Sport. 7: 185-198, **1975**.

Goldberg, A. L. and Goodman, H. M. Relationship between growth hormone and muscular work in determining muscle size. J. Physiol. (Lond.) 200: 655-666, **1967**.

——. Amino acid transport during work-induced

growth of skeletal muscle. Amer. J. Physiol. 216: 1111-1115, **1969a**.

——. Relationship between growth hormone and muscular work in determining muscle size. J. Physiol. (Lond.) 200: 655-666, **1969b**.

Goldberg, D. I. and Shephard, R. J. Stroke volume during recovery from upright bicycle exercise. J. Appl. Physiol. Respirat. Exercise Physiol. 48: 833-837, **1980**.

Golden, St. C. F., Hampton, I. F. G., and Smith D. Cold tolerance in long-distance swimmers. J. Physiol. (Lond.) 290: 48p, **1979**.

Goldfried, M. R. Systematic desensitization as training in self-control. J. Consult. Clin. Psychol. 37: 228-234, **1971**.

Goldhammer, S. and Scherf, D. Elektrokardiographische untersuchunger bei Kranken mit Angina Pectoris ("ambulatorischer" Typus). Zeit. Klin. Med. 122: 134-151, **1932**.

Goldman, D. E. and Von Gierke, H. E. The effects of shock and vibration on man. Bethesda, Md.: U.S. Naval Medical Research Inst. Lecture and Rev. Series 60: 3, **1960**.

Goldman, M. D. Mechanical coupling of the diaphragm and rib cage, pp. 50-63. In: Loaded Breathing. Eds.: L. D. Pengelly, A. S. Rebuck, and E. J. M. Campbell. Don Mills, Ont.: Longman Canada, **1974**.

——. Respiratory muscles: The generator. In: Muscular Exercise and the Lungs, pp. 7-15. Eds.: J. A. Dempsey and C. E. Reed. Madison: University of Wisconsin Press, **1977**.

Goldman, R. F., Breckenridge, J. R., Reeves, E., and Beckman, E. L. "Wet" versus "dry" suit approaches to water immersion protective clothing. Aerospace Med. 37: 485-487, **1966**.

Goldschlager, N., Cake, D., and Cohn, K. Exercise induced ventricular arrhythmias in patients with coronary artery disease. Their relation to angiographic findings. Amer. J. Cardiol. 31: 434-440, **1973**.

Goldspink, D. F. The influence of immobilization and stretch on protein turnover of rat skeletal muscle. J. Physiol. (Lond.) 264: 267-282, **1977a**.

——. The influence of activity on muscle size and protein turnover. J. Physiol. (Lond.) 264: 283-296, **1977b**.

Goldspink, G. Fixation of muscle. Nature 192: 1305-1306, **1961**.

——. Postembryonic growth and differentiation of striated muscle, pp. 179-236. In: The Structure and Function of Muscle, Vol. 1, Structure, Part 1. Ed. G. H. Bourne. New York: Academic Press, **1972**.

——. The proliferation of myofibrils during muscle fiber growth. J. Cell Sci. 6: 593-603, **1970**.

Goldspink, G., Larson, R. E., and Davies, R. E. The immediate energy supply and the cost of maintenance of isometric tension for different muscles in the hamster. Z. Vergl. Physiol. 66: 389-397, **1970**.

Goldstein, I., Goldstein, S., Urbanetti, J. A. and Anthonisen, N. R. Effects of expiratory threshold loading during steady-state exercise. J. Appl. Physiol. 39: 697-701, **1975**.

Goldthwait, J. E., Brown, L. T., Swaim, L. T., and Kuhus, J. G. Body Mechanics in the Study and Treatment of Disease. Philadelphia: Lea & Febiger, **1930**.

Golenhofen, K. and Felix R. Local heat clearance probes with alternative heating and their application in the measuring of human muscle blood flow. Pflüg. Archiv. 331: 145-152, **1972**.

Golenhofen, K., and Hildebrandt, G. Psychische Einflüsse auf die Muskeldurchblutung. Pflüg. Archiv. 263: 637-646, **1957**.

Gollnick, P., Karlsson, J., Piehl, K. and Saltin, B. Selective glycogen depletion in skeletal muscle fibers of man following sustained contractions. J. Physiol. (Lond.) 241: 59-68, **1974**.

Gollnick, P. D. Free fatty acid turnover and the availability of substrates as a limiting factor in prolonged exercise. Ann. N.Y. Acad. Sci. 301: 64-71, **1977**.

Gollnick, P. D., Armstrong, R. B., Saubert, C. W., Piehl, K., and Saltin, B. Enzyme activity and fiber composition in skeletal muscle of untrained and trained men. J. Appl. Physiol. 33: 312-319, **1972**.

Gollnick, P. D., Armstrong, R. B., Saubert, C. W., Sembrowich, W. L., Shepherd, R. E., and Saltin, B. Glycogen depletion patterns in human skeletal muscle fibres during prolonged work. Pflüg. Archiv. 334: 1-12, **1973a**.

Gollnick, P. D., Armstrong, R. B., Saltin, B., Saubert, C. W., Sembrowich, W. L., and Shepherd, R. E. Effect of training on enzyme activity and fiber composition of human skeletal muscle. J. Appl. Physiol. 34: 107-111, **1973b**.

Gollnick, P. D. and Hermansen, L. Biochemical adaptations to exercise: Anaerobic metabolism. Exercise Sport Sci. Rev. 1: 1-43, **1973**.

Gollnick, P. D. and Ianuzzo, C. D. Hormonal deficiencies and the metabolic adaptations of rats to training. Amer. J. Physiol. 223: 278-282,

1972.

——. Acute and chronic adaptations to exercise in hormone deficient rats. Med. Sci. Sport. 7: 12-19, **1975**.

Gollnick, P. D. and King, D. W. Effect of exercise and training on mitochondria of rat skeletal muscle. Amer J. Physiol 216: 1502-1509, **1969**.

Gollnick, P. D., Piehl, K., and Saltin, B. Selective glycogen depletion pattern in human muscle fibres after exercise of varying intensity and at varying pedalling rates. J. Physiol. (Lond.) 241: 45-57, **1974**.

Gollnick, P. D., Piehl, K., Karlsson, J., and Saltin, B. Glycogen depletion patterns in human skeletal muscle fibers after varying types and intensities of exercise, pp. 416-421. In: Metabolic Adaptation to Prolonged Physical Exercise. Eds.: H. Howald and J.R. Poortmans. Basel: Birkhauser Verlag, **1975**.

Gollnick, P. D., Struck, P. J., and Bogyo, T. P. Lactic dehydrogenase activities of rat heart and skeletal muscle after exercise and training. J. Appl. Physiol. 22: 623-627, **1967**.

Gollnick, P.W. Cellular adaptations to exercise, pp. 112-126. In: Frontiers of Fitness. Ed.: R. J. Shephard. Springfield, Ill.: C. C. Thomas, **1971**.

Gomez, O. A. and Hamilton, W. F. Functional cardiac deterioration during development of hemorrhagic circulatory deficiency. Circ. Res. 14: 327-336, **1964**.

Gonyea, W., Ericson, G. C., and Bonde-Petersen, F. Skeletal muscle fiber splitting induced by weight-lifting exercise in cats. Acta Physiol. Scand. 99: 105-109, **1977**.

Goode, R. C., Brown, E. B., Howson, M. G., and Cunningham, D. J. C. Respiratory effects of breathing down a tube. Resp. Physiol. 6: 343-359, **1969**.

Goode, R. C., Duffin, J., Miller, R., Romet, T. T., Chout, W., and Ackles, K. Sudden cold water immersion. Resp. Physiol. 23: 301-310, **1975**.

Goode, R. C., Firstbrook, J. B., and Shephard, R. J. Effects of exercise and a cholesterol-free diet on human serum lipids. Can. J. Physiol. Pharm. 44: 575-580, **1966**.

Goode, R. C., Virgin, A., Romet, T. T., Crawford, P., Duffin, J., Pallandi, T., and Woch, Z. Effects of a short period of physical activity in adolescent boys and girls. Can. J. Appl. Sport. Sci. 1: 241-250, **1976**.

Goode, R. J., Dekirmenjian, H., Meltzer, H. Y., and Maas, J. W. Relation of exercise to MHPG excretion in normal subjects. Arch. Gen. Psychiatr. 29: 391-396, **1973**.

Goodman, R. F. The effects of football equipment on heat transfer. In: Physiological Aspects of Sports and Physical Fitness. Chicago: Athletic Institute, **1968**.

Goodwin, G. M. The sense of limb position and movement. Exercise Sport Sci. Rev. 1: 87-124, **1976**.

Goodwin, G. M., McCloskey, D. I., and Matthews, P. B. C. A systematic distortion of position sense produced by muscle vibration. J. Physiol. (Lond.) 221: 8p-9p, **1972**.

Goodwin, G. M., McCloskey, D. I., and Mitchell, J. H. Cardiovascular and respiratory responses to changes in central command during isometric exercise at a constant muscle tension. J. Physiol. (Lond.) 266: 173-190, **1972**.

Goransson, A., Lundgren, C., and Lundin, G. A theoretical model for the computation of decompression tables for divers. Nature 199: 384-385, **1963**.

Gordon, A. M., Huxley, A. F., and Julian F. J. The variation in isometric tension with sarcomere length in vertebrate muscle fibres. J. Physiol. (Lond.) 184: 170-192, **1966**.

Gordon, E. E., Kowalski, K., and Fritts, M. Protein changes in quadriceps muscle of rat with repetitive exercises. Arch. Phys. Med. Rehabil. 48: 296-303, **1967a**.

——. Changes in rat muscle fiber with forceful exercises. Arch. Phys. Med. Rehabil. 48: 577-582, **1967b**.

Gordon, T., Castelli, W. P., Hjortland, M. C., Kannel, W. B., and Dawber, T. R. High density lipoprotein as a protective factor against coronary heart disease: The Framingham study. Amer. J. Med. 62: 707-714, **1977**.

Goss, C. M. In: Anatomy of the Human Body, 29th ed., pp. 271, 274, and 287. Cited by Booth and Gould. Philadelphia: Lea & Febiger, **1973**.

Goss, R. J. Hypertrophy versus hyperplasia. Science 153: 1615-1620, **1966**.

Gott, P. H., Rosalie, H. A., and Crampton, R. S. The athletic heart syndrome. Arch. Int. Med. 122: 340-344, **1968**.

Gottlieb, G. L. and Agarwal, G. C. Postural adaptation. The nature of adaptive mechanisms in the human motor system, pp. 197-210. In:

Control of Posture and Locomotion. Eds.: R. B. Stein, K. G. Pearson, R. S. Smith, and J. B. Redford. New York: Plenum Press, **1973**.

Gottlieb, G. L. Agarwal, G. C., and Stark, L. Studies in Postural Control Systems. Part III. A Muscle Spindle Model. I.E.E.E. Trans. on systems science and cybernetics. Vol. SSC-6, pp. 127-132. New York: Institute of Electrical Electronic Eng., **1970**.

Gotto, A. M. Current concepts of hyperlipoproteinemia, pp. 209-219. In: Atherosclerosis IV. Eds.: G. Schettler, Y. Goto, Y. Hata, and G. Klose. Berlin: Springer Verlag, **1977**.

Gotto, A. M. and Jackson, R. L. Structure of the plasma lipoproteins: A review, pp. 177-188. In: Atherosclerosis IV. Eds.: G. Schettler, Y. Goto, Y. Hata, and G. Klose. Berlin: Springer Verlag, **1977**.

Gould, L. Cardiac effects of alcohol. Amer. Heart J. 79: 422-425, **1970**.

Gould, M. K. and Rawlinson, W. A. Effect of swimming on the levels of lactic dehydrogenase, malic dehydrogenase and phosphorylase in muscles of 8-, 11- and 15- week-old rats. Biochem. J. 73: 41-44, **1959**.

Govaerts, F. Sports, emancipation des femmes et femininité, pp. 445-463. In: Physical Activity and Human Well-Being. Eds.: F. Landry and W. R. Orban. Miami: Symposia Specialists, **1978**.

Gower, D. and Kretzschmar, K. M. Heat production and chemical change during isometric contraction of rat soleus muscle. J. Physiol. (Lond.) 258: 659-671, **1976**.

Graham, D. T., Kabler, J. D., and Lunsford, L. Vaso-vagal fainting: A diphasic response. Psychosom. Med. 23: 493-507, **1961**.

Graham, T. E. and Andrew G. The variability of repeated measurements of oxygen debt in man following a maximal treadmill exercise. Med. Sci. Sport. 5: 73-78, **1973**.

Graham, T. P., Covell, J. W., Sonnenblick, E. H., Ross, J., and Braunwald, E. Control of myocardial oxygen consumption: Relative influence of contractile state and tension development. J. Clin. Invest. 47: 375-385, **1968**.

Grahame, R. Diseases of the joints. In: Textbook of Geriatric Medicine and Gerontology. Ed.: J. C. Brocklehurst. Edinburgh: Churchill — Livingstone, **1973**.

Gramiak, R. and Shah, P. M. Cardiac ultrasonography: A review of current applications. Radiol. Clin. N. Amer. 9: 469-490, **1971**.

Granath, A., Horie, E., and Linderholm, H. Compliance and resistance of the lungs in sitting and supine positions at rest and during work. Scand. J. Clin. Lab. Invest. 11: 226-234, **1959**.

Granath, A., Johnson, B. and Strandell, T. Circulation in healthy old men studied by right heart catheterization at rest and during exercise in supine and sitting position. Acta Med. Scand. 176: 425-446, **1964**.

Grand, B., Rosenberg, G. M., Liberman, H., Trachtenberg, J., and Kral, V. A. Diurnal variation of the serum cortisol level of geriatric subjects. J. Gerontol. 26: 351-357, **1971**.

Grande, F. Dietary carbohydrates and serum cholesterol. Amer. J. Clin. Nutr. 20: 176-184, **1967**.

Grandjean, E. Fitting the Task to the Man: An Ergonomic Approach. London: Taylor & Francis, **1971**.

Granger, H. J., Goodman, A. H., and Cook, B. H. Metabolic models of microcirculatory regulation. Fed. Proc. 34: 2025-2030, **1975**.

Granit, R. Receptors and Sensory Perception, pp. 1-369. New Haven: Yale University Press, **1955**.

——. The Basis of Motor Control. London: Academic Press, **1970**.

——. Constant errors in the execution and appreciation of movement. Brain 95: 451-460, **1972**.

——. Demand and accomplishment in voluntary movement, pp. 3-18. In: Control of Posture and Locomotion. Eds.: R. B. Stein, K. G. Pearson, R. S. Smith, and J. B. Redford. New York: Plenum Press, **1973**.

——. The functional role of the muscle spindles: Facts and hypotheses. Brain 98 (Part IV): 531-536, **1975**.

Granit, R., Holmgren, B., and Merton, P. A. Two routes for excitation of muscle and their subservience to the cerebellum. J. Physiol. (Lond.) 130: 213-224, **1955**.

Granit, R. and Kaada, B. R. The influence of stimulation of central nervous structures in muscle spindles in the cat. Acta Physiol. Scand. 27: 130-160, **1952**.

Gray, J. S. Pulmonary Ventilation and Its Physiological Regulation. Springfield, Ill.: C. C. Thomas, **1950**.

Grayson, J. Internal calorimetry in the determination of thermal conductivity and blood flow. J.

Physiol. (Lond.) 118: 54–72, **1952**.

Green, D. E. and Goldberger, R. F. Molecular Insights into the Living Process. New York: Academic Press, **1967**.

Green, H. J., Bishop, P., Houston, M., McKillop, R., Norman, R. and Stothart, P. Time-motion and physiological assessment of ice hockey performance. J. Appl. Physiol. 40: 159–163, **1976**.

Green, H. J., Daub, B. D, Painter, D. C., and Thomson, J. A. Glycogen depletion patterns during ice-hockey performance. Med. Sci. Sport. 10: 289–293, **1978**.

Green, H. J. and Houston, M. E. Effect of a season of ice-hockey on energy capacities and associated functions. Med. Sci. Sport. 7: 299–303, **1975**.

Green, H. J., Houston, M. E., and Thomson, J. A. Inter- and intragame alterations in selected blood parameters during ice-hockey performance, pp. 37–46. In: Ice Hockey. Eds.: F. Landry and W. A. R. Orban. Miami: Symposia Specialists, **1978**.

Green, L. F. and Bagley, R. Ingestion of a glucose syrup drink during long-distance canoeing. Brit. J. Sport. Med. 6: 125–128, **1972**.

Green, M., Mead, J., and Sears, T. A. Effects of loading on respiratory muscle control in man, pp. 73–80. In: Loaded Breathing. Eds.: L. D. Pengelly, A. S. Rebuck, and E. J. M. Campbell. Don Mills, Ont.: Longman Canada, **1974**.

Greene, D. G. Drowning. In: Handbook of Physiology: Section 3. Respiration, Vol. 2, pp. 1195–1204. Eds.: W. O. Fenn and H. Rahn. Washington, D. C.: Amer. Physiol. Soc. **1965**.

Greene, J. H. and Morris, W. H. M. The force platform: An industrial engineering tool. J. Industr. Eng. 9: 128–132, **1958**.

Greene, M. A., Boltax, A. J., and Ulberg, R. J. Cardiovascular dynamics of vasovagal reactions in man. Circ. Res. 9: 12–17, **1961**.

Greenfield, A. D. M. An emotional faint. Lancet 1: 1302–1303, **1951**.

Greenleaf, J. E., Bernauer, E. M., Adams, W. C., and Juhos, L. Fluid-electrolyte shift, and $\dot{V}_{O_2(max)}$ in man at simulated altitude (2,287 m). J. Appl. Physiol. Resp. Environ. Exercise Physiol. 44: 652–658, **1978**.

Greenleaf, J. E., Bernauer, E. M., Juhos, L. T., Young, H. L., Morse, J. T., and Staley, R. W. Effects of exercise on fluid exchange and body composition in man during 14-day bed rest. J. Appl. Physiol. 43: 126–132, **1977**.

Greenleaf, J. E. and Castle, B. I. Exercise temperature regulation in man during hypohydration and hyperhydration. J. Appl. Physiol. 30: 847–853, **1971**.

Greenleaf, J. E., Greenleaf, C. J., Card, D. H., and Saltin, B. Exercise-temperature regulation in man during acute exposure to simulated altitude. J. Appl. Physiol. 26: 290–296, **1969**.

Greenwood, P. V., Hainsworth, R., Karim, F., Morrison, G. W., and Sofola, O. A. The effect of stimulation of carotid chemoreceptors on the inotropic state of the left ventricle. J. Physiol. (Lond.) 266: 47p–48p, **1977**.

Gregg, D. E. In: Cardiovascular Functions. Ed.: A. A. Luisada. New York: McGraw Hill, **1962**.

——. The natural history of coronary collateral development. Circ. Res. 35: 335–344, **1974**.

Gregg, D. E. and Fisher, L. C. Blood supply to the heart. In: Handbook of Physiology, Section 2. Circulation, Vol. 1. Ed.: W. F. Hamilton. Washington, D.C.: Amer. Physiol. Soc., **1964**.

Gregg, R. A., Mastellone, A. F. and Gersten, J. W. Cross-exercise: A review of the literature and study utilizing electromyographic techniques. Amer. J. Phys. Med. 36: 269–280, **1957**.

Gregor, R. J. and Costill, D. L. A comparison of the energy expenditure during positive and negative grade running. J. Sport. Med. Phys. Fitness 13: 248–252, **1973**.

Greulich, W. W. and Pyle, S. I. Radiographic atlas of skeletal development of the hand and wrist. Oxford: Oxford University Press, **1959**.

Gribble, M. de G. A comparison of the "high-altitude" and "high pressure" syndromes of decompression sickness. Brit. J. Industr. Med. 17: 181–186, **1960**.

Grillner, S. A consideration of stretch and vibration data in relation to the tonic stretch reflex, pp. 397–405. In: Control of Posture and Locomotion. Eds.: R. B. Stein, K. G. Pearson, R. S. Smith, and J. B. Redford. New York: Plenum Press, **1973**.

——. Locomotion in vertebrates: Central mechanisms and reflex interaction. Physiol. Rev. 55: 247–304, **1975**.

Grillner, S., Hongo, T., and Lund, S. Descending monosynaptic and reflex control of γ

motoneurones. Acta Physiol. Scand. 75: 592-613, **1969**.

Grimby, G. Renal clearances during prolonged supine exercise at different loads. J. Appl. Physiol. 20: 1294-1298, **1965**.

——. Pulmonary mechanics: the load. In: Muscular Exercise and the Lung, pp. 17-24. Eds.: J. A. Dempsey and C. E. Reed. Madison: Univ. of Wisconsin Press, **1977**.

Grimby, G., Bjure, J., Aurell, M., Ekstrom-Jodal, B., Tibblin, G., and Wilhelmsen, L. Work capacity and physiologic responses to work. Men born in 1913. Amer. J. Cardiol. 30: 37-42, **1972**.

Grimby, G., Häggendal, E., and Saltin, B. Local Xenon 133 clearance from the quadriceps muscle during exercise in man. J. Appl. Physiol. 22: 305-310, **1967**.

Grimby, G. and Nilsson, N. J. Cardiac output during exercise in pyrogen-induced fever. Scand. J. Clin. Lab. Invest. 15, Suppl. 69: 44-61, **1963**.

Grimby, G., Nilsson, N. J., and Saltin, B. Cardiac output during sub-maximal and maximal exercise in active middle-aged athletes. J. Appl. Physiol. 21: 1150-1156, **1966**.

Grimby, G. and Saltin, B. A physiological analysis of physically well-trained middle-aged and old athletes. Acta Med. Scand. 179: 513-526, **1966**.

Grimby, G., Wilhelmsen, L., Björntorp, P., Saltin, B., and Tibblin, G. Habitual physical activity: aerobic power and blood lipids, pp. 469-481. In: Muscle Metabolism During Exercise. Eds.: B. Pernow and B. Saltin. New York: Plenum Press, **1971**.

Grimby, L. and Hannerz, J. Disturbances in voluntary recruitment order of low and high frequency motor units on blockade of proprioceptive afferent activity. Acta Physiol. Scand. 29: 207-216, **1976**.

Grodins, F. and Yamashiro, S. M. Modeling the respiratory system: essentials, pp. 25-38. In: Muscular Exercise and the Lung. Eds.: J. A. Dempsey and C.E. Reed. Madison: University of Wisconsin Press, **1977**.

Grodsky, G. M. and Benoit, F. Effect of massive weight reduction on insulin secretion in obese subjects. International Diabetes Symposium, Stockholm, **1967**.

Grollman, A. Zur Bestimmung des Minuten-Volumens mit der Azetylenmethode bei normaler Arbeit. Naunyn-Schmiedebergs Arch. Exp. Path. Pharmak. 162: 463-771, **1931**.

Grombach, J. V. The gravity factor in world athletics. Amateur Athlete 31: 24-25, **1960**.

Grose, J. E. Depression of muscle fatigue curves by heat and cold. Res. Quart. 29: 19-31, **1958**.

Grosse-Lordemann, H. and Müller, E. A. Der Einfluss der Leistung und der Arbeitsgeschwindigkeit auf das Arbeitsmaximum und Wirkungsgrad beim Radfahren. Arbeitsphysiol. 9: 454-475, **1936**.

Grossman, W., Jones, D., and McLaurin, L. P. Wall stress and patterns of hypertrophy in the left ventricle. J. Clin. Invest. 56: 56-64, **1975**.

Grove, D., Nair, K. G., and Zak, R. Biochemical correlates of cardiac hypertrophy. 3. Changes in DNA content; the relative contributions of polyploidy and mitotic activity. Circ. Res. 25: 463-471, **1969**.

Grover, R. F. and Reeves, J. T. Pulmonary ventilation during exercise at altitude, pp. 33-39. In: Exercise at Altitude. Ed.: R. Margaria. Amsterdam: Excerpta Medica Foundation, **1967**.

Grover, R. F., Vogel, J. H. K., Voigt, G. C., and Blount, S. G. Reversal of high altitude pulmonary hypertension, pp. 155-164. In: Exercise and Altitude. Eds.: E. Jokl and P. Jokl. Basel: Karger, **1968**.

Grunewald, W. The influence of the three dimensional pattern on the intercapillary oxygen diffusion: A new composed model for comparison of calculated and measured oxygen distribution, pp. 5-17. In: Oxygen Supply: Theoretical and Practical Aspects of Oxygen Supply and Microcirculation of Tissue. Eds.: M. Kessler, D. F. Bruley, L. C. Clark, D. W. Lubbers, I. A. Silver, and J. Strauss. Baltimore: University Park Press, **1973**.

Gualtière, W. S. Transition from exercise stress test to physical training prescription, pp. 320-342. In: Changing Concepts in Cardiovascular Disease. Eds.: H. I. Russek and B. L. Zohman. Baltimore: Williams & Wilkins, **1972**.

Guberan, E., Williams, M. K., Walford, J. and Smith, M. M. Circadian variations of FEV in shift workers. Brit. J. Industr. Med. 26: 121-125, **1969**.

Gueli, D. and Shephard, R. J. Pedal frequency in bicycle ergometry. Can. J. Appl. Sport. Sci. 1: 137-142, **1976**.

Guignard, J. C. Noise. In: Textbook of Aviation Physiology. Ed.: J. A. Gillies. Oxford: Pergamon Press, **1965**.

Gunning, J. F., Cooper, G., Harrison, C. E., and Coleman, H. N. Myocardial oxygen consumption in experimental hypertrophy and congestive heart failure due to pressure overload. Amer. J. Cardiol. 32: 427–436, **1973**.

Gurewich, V., Sasahara, A. A., Quinn, J. S., Peffer, C. J., and Littmann, D. Aortic pressures during closed-chest cardiac massage. Circulation 23: 593–595, **1961**.

Gurtner, G. H. and Forster, R. E. Can alveolar PCO_2 exceed pulmonary end-capillary CO_2? J. Appl. Physiol. 42: 323–328, **1977**.

Gurvich, N. L. and Yuniev, G. S. Restoration of heart rhythm during fibrillation by a condenser discharge. Amer. Rev. Sov. Med. 4: 252–256, **1947**.

Guth, L. "Trophic" influences of nerve on muscle. Physiol. Rev. 48: 645–687, **1968**.

Guth, L. and Watson, P. K. The influence of innervation on the soluble proteins of slow and fast muscles of the rat. Exp. Neurol. 17: 107–117, **1967**.

Guthrie, D. I. A new approach to handling in industry. A rational approach to the prevention of low back pain. S. Afr. Med. J. 37: 651–656, **1963**.

Gutteridge, M. V. A study of motor achievements of young children. Arch. Psychol. (N.Y.) 244: 1–178, **1939**.

Guy, A. J. and Patrick, J. M. Some properties of the non-chemical drive to breathe at the breaking point of breath-holding. J. Physiol. (Lond.) 272: 77p–78p, **1977**.

Gyntelburg, F. Coronary heart disease and physical activity. Nordic Council Arctic Med. Res. Rep. 19: 64–69, **1977**.

Haab, P. E., Held, D. R., and Farhi, L. E. Readjustments of ventilation/perfusion relationships in the lung at altitude, pp. 108–111. In: Exercise at Altitude. Ed.: R. Margaria. Dordrecht: Excerpta Medica Fdn., **1967**.

Haas, E. Über die art der tätigkeit unserer muskeln beim halten verschieden schwerer gewichte. Pflüg. Arch. 212: 651–656, **1926**.

Hackel, D. B. and Breitenecker, R. Time factor in reversibility of myocardial metabolic changes in hemorrhagic shock. Proc. Soc. Exp. Biol. Med. 113: 534–537, **1963**.

Haddy, F. J. and Scott, J. B. Metabolically linked vasoactive chemicals in local regulation of blood flow. Physiol. Rev. 48: 688–707, **1968**.

——. Metabolic factors in peripheral circulatory regulation. Fed. Proc. 34: 2006–2011, **1975**.

Hagbarth, K. E., Hongell, A., and Wallin, G. Parkinson's disease: Afferent muscle activity in rigid patients. A preliminary report. Acta Med. Uppsala 75: 76, **1970**.

Hager, A., Sjöstrom, L., Arvidsson, B., Björntorp, P., and Smith, U. Body fat and adipose tissue cellularity in infants: A longitudinal study. Metabolism 26: 607–614, **1977**.

Hagerman, F. C., Addington, W. C., and Gaensler, E. A. A comparison of selected physiological variables among outstanding competitive oarsmen. J. Sport. Med. Phys. Fitness 12: 12–22, **1972**.

Hagerman, F. C., Bowers, R. W., Fox, E. L., and Ersing, W. W. The effects of breathing 100 percent oxygen during rest, heavy work and recovery. Res. Quart. 39: 965–974, **1968**.

Hagerman, F. C., Connors, M. C., Gault, J. A., Hagerman, G. R., and Polinski, W. J. Energy expenditure during simulated rowing. J. Appl. Physiol. Resp. Env. Ex. Physiol. 45: 87–93, **1978**.

Hagerman, F. C., McKirnan, M. D., and Pompei, J. A. Maximal oxygen consumption of conditioned and unconditioned sportsmen. J. Sport. Med. Phys. Fitness. 15: 43–48, **1975**.

Haggendal, J., Hartley, L. H., and Saltin B. Arterial noradenaline concentration during exercise in relation to the relative work levels. Scand. J. Clin. Lab. Invest. 26: 337–342, **1970**.

Haight, J. S. J. and Keatinge, W. R. Elevation in set point body temperature regulation after prolonged exercise. J. Physiol. (Lond.) 229: 77–86, **1973a**.

——. Failure of thermoregulation in the cold during hypoglycaemia induced by exercise and ethanol. J. Physiol. (Lond.) 229: 87–97, **1973b**.

Haike, H. J., Heymann, P., and Wagner, K. Experimentelle Untersuchungen über den Einfluss der immobilisation auf die Knochenfestigkeit und Knochenelastizität sowie über die Regenerationsfähigkeit derselben bei der Remobilisation. Z. Orthop. Grenzgeb. 102: 200–208, **1967**.

Haines, R. F. Effect of bed rest and exercise on body balance. J. Appl. Physiol. 36: 323–327, **1974**.

Haisman, M. F. Research cited by Amor, A. F.: Assessment of obesity from measurements of skinfold thickness. NATO sub-committee on nutritional aspects of military feeding, October 1–3, **1974**.

Haisman, M. F. and Goldman, R. F. Effect of terrain on the energy cost of walking with backloads and handcart loads. J. Appl. Physiol. 36: 545–548, **1974**.

Hait, W. N., Gorshein, D., Bess, E. C. et al. The effect of steroid metabolites on the hematopoietic stem cell pool. J. Pharmacol. Exp. Ther. 186: 656–661, **1973**.

Haldane, J. S. and Priestley, J. G. Respiration, pp. 1–493. Oxford: Clarendon Press, **1935**.

Hall, D. A. Metabolic and structural aspects of aging. In: Textbook of Geriatric Medicine and Gerontology. Ed.: J. C. Brocklehurst. Edinburgh: Churchill-Livingstone, **1973**.

Hall, M. R. P. Hypophyso-adrenal axis, pp. 431–432. In: Textbook of Geriatric Medicine and Gerontology. Ed.: J. C. Brocklehurst. Edinburgh: Churchill-Livingstone, **1973**.

Hall-Craggs, E. C. B. and Lawrence, A. Longitudinal fibre division in skeletal muscle: A light and electron microscopic study. Z. Zellforsch. Mikrosk. Anat. 109: 481–494, **1970**.

Hallett, M., Shahani, B. T. and Young, R. R. E. M. G. analysis of stereotyped voluntary movements in man. J. Neurol. Neurosurg. Psychiatr. 38: 1154–1162, **1975**.

Halliday, D. and Hopkinson, W. I. High precision measurement of total body water in man. J. Physiol. (Lond.) 287: 16p–17p, **1977**.

Halttunen, P. K. The voluntary control in human breathing. Acta Physiol. Scand. Suppl. 419: 1–47, **1974**.

Hamer, J. and Sowton, E. Cardiac output after beta-adrenergic blockade in ischaemic heart disease. Brit. Heart J. 27: 892–895, **1965**.

Hamilton, R. W. and Schreiner, H. K. Putting and keeping man in the sea. Chem. Eng. 75 (13): 263–270, **1968**.

Hamilton, W. F. Measurement of the cardiac output, pp. 551–584. In: Handbook of Physiology. Section 2. Circulation, Vol. I. Ed.: W. F. Hamilton. Washington, D.C.: American Physiological Society, **1962**.

Hamley, E. J. and Thomas V. Physiological and postural factors in the calibration of the bicycle ergometer. J. Physiol. (Lond.) 191: 55p–57p, **1978**.

Hammel, H. T. Neurons and temperature regulation. In: Physiological Controls and Regulations, p. 71. Eds.: W. S. Yamamoto and J. R. Brobeck. Philadelphia: Saunders, **1965**.

Hammer, S., and Obrink, K. J. The inhibitory effect of muscular exercise on gastric secretion. Acta Physiol. Scand. 28: 152–161, **1953**.

Hammond, P. H. The influence of prior instruction to the subject on an apparently involuntary neuro-muscular response. J. Physiol. (Lond.) 132: 17p, **1956**.

Hammond, P. H., Merton, P. A., and Sutton, G. G. Nervous gradation of muscular contraction. Brit. Med. Bull. 12: 214–218, **1956**.

Hamosh, M., Lesch, M., Baron, J., and Kaufman, S. Enhanced protein synthesis in a cell-free system from hypertrophied skeletal muscle. Science 157: 935–937, **1967**.

Hampton, J. R., Dowling, M., and Nicholas, C. Comparison of results from a cardiac ambulance manned by medical or non-medical personnel. Lancet 1: 526–529, **1977**.

Hanefeld, M., Leonhardt, W., and Haller, H. Coronary risk factors in adults: the influence of nutrition in early life, pp. 104–108. In: Atherosclerosis IV. Eds.: G. Schettler, Y. Goto, Y. Hata, and G. Klose. Berlin: Springer Verlag, **1977**.

Hanke, D., Schlepper, M., Westermann, K., and Witzleb, E. Venentonus, Haut-und Muskeldurchblutung an Unterarm und Hand bei Beinarbeit. Pflüg. Archiv. 309: 115–127, **1969**.

Hanne-Paparo, N., Drory, Y., Schoenfeld, Y., Shapiro, Y., and Kellermann, J. J. Common E.C.G. changes in athletes. Cardiology 61: 267–278, **1976**.

Hannerz, J. Discharge properties of motor units in relation to recruitment order in voluntary contraction. Acta Physiol. Scand. 91: 374–385, **1974**.

Hannisdahl, B. Der Einfluss von Muskelarbeit auf die Blutsenkung. Arbeitsphysiol. 11: 165–174, **1940**.

Hannon, J. P., Chinn, K. S. K., and Shields, J. L. Effects of acute high-altitude exposure on body fluids. Fed. Proc. 28: 1178–1184, **1969**.

Hannon, J. P., Shields, J. L., and Harris, C. W. Effects of altitude acclimatization on blood composition of women. J. Appl. Physiol. 26: 540–547, **1969a**.

——. Anthropometric changes associated with high altitude acclimatization in females. Amer. J. Phys. Anthrop. 31: 77–84, **1969b**.

Hansen, A. P. The effect of adrenergic receptor blockade on the exercise-induced serum growth hormone rise in normals and juvenile diabetics. J. Clin. Endocrinol. Metab. 33: 807–812, **1971**.

——. Abnormal serum growth hormone response to exercise in maturity-onset diabetes. Diabetes, 22: 619-628, **1973**.

Hansen, J. E., Stelter, G. P. and Vogel, J. A. Arterial pyruvate, lactate, pH and pCO_2 during work at sea level and high altitude. J. Appl. Physiol. 23: 523-530, **1967**.

Hansen, J. W. The training effect of repeated isometric muscle contractions. Int. Z. Angew. Physiol. 18: 474-477, **1961**.

——. Effect of dynamic training on the isometric endurance of the elbow flexors. Int. Z. Angew. Physiol. 23: 367-370, **1967**.

Hansen, T. Osmotic pressure effect of the red cells: possible physiological significance. Nature 190: 504-508, **1961**.

Hanson, D. L. Cardiac response to participation in little league baseball as determined by telemetry. Res. Quart. 38: 384-388, **1967**.

Hanson, J. and Huxley, H. E. The structural basis of the cross-striations in muscle. Nature (Lond.) 172: 530-532, **1953**.

——. The structural basis of contraction in striated muscle. Symp. Soc. Exp. Biol. 9: 228-264, **1955**.

Hanson, J. and Lowy, J. The structure of F-actin and of actin filaments isolated from muscle. J. Mol. Biol. 6: 46-60, **1963**.

——. The structure of actin filaments and the origin of the axial periodicity in the I substance of vertebrate striated muscle. Proc. Roy. Soc. B, 160: 449-458, **1964**.

Hanson, J. S. Physical training and the pulmonary diffusing capacity. Dis. Chest 56: 488-493, **1969**.

——. Exercise responses following production of experimental obesity. J. Appl. Physiol. 35: 587-591, **1973**.

——. Maximal exercise performance in members of the US Nordic ski team. J. Appl. Physiol. 35: 592-595, **1974.**

——. Decline of physiologic training during the competitive season in members of the U.S. Nordic ski team. Med. Sci. Sport. 7: 213-216, **1975**.

Hanson, J. S., Tabakin, B. S., and Levy, A. M. Appendix by D. B. Hill. Comparative exercise-cardiorespiratory performance of normal men in the third, fourth and fifth decades of life. Circulation 37: 345-360, **1968**.

Hanson M. A., Nye, P. C. G., Rao, P. S., and Torrance, R. W. Effects of acetazolamide and benzolamide on the response of the carotid chemoreceptors to CO_2. J. Physiol. (Lond.) 284: 165p-166p, **1978**.

Haralambie, G. Excitabilité neuromusculaire et magnésiémie chez les sportifs. Int. Z. Angew. Physiol. 25: 181-189, **1968**.

——. Changes in electrolytes and trace elements during long-lasting exercise. In: H. Howald and J. R. Poortmans, pp. 340-351. Basel: Birkhauser Verlag, **1975**.

Haralambie, G. and Keul, J. Der Einfluss von Muskelarbeit auf den Magnesiumspiegel und neuromuskuläre Erregbarkeit beim Menschen. Med. Klin. 65: 1445-1448, **1970a**.

——. Das Verhalten von Serum-Coeruloplasmin und-Kupfer bei langdauernder Körperbelastung. Ärztl. Forsch 24: 112-115, **1970b**.

Hardy, J. D. Central and peripheral factors in physiological temperature regulation. In: Les Concepts de Claude Bernard sur le milieu intérieur. Colloque International Claude Bernard. 29th June-2nd July 1965. Paris: Masson, **1967**.

Hardy, J. D. and DuBois, E. F. The technique of measuring radiation and convection. J. Nutr. 15: 461-497, **1938**.

Hardy, J. D., Gagge, A. P., and Stolwijk, A. J. Physiological and Behavioral Temperature Regulation. Springfield, Ill.: C. C. Thomas, **1970**.

Hardy, J. D., Stolwijk, J. A. J., and Gagge, A. P. Man. In: Comparative Physiology of Thermoregulation. Vol. II, p. 327. Ed.: G. C. Whittow. New York: Academic Press, **1971**.

Harf, A., Pratt, T., and Hughes, J. M. B. Regional distribution of Va/Q in man at rest and with exercise measured with Krypton-81. J. Appl. Physiol. Resp. Environ. Exercise Physiol. 44: 115-123, **1978.**

Harkins, K. J. Metabolic cost comparison of cross-country skiing between elite and non-elite skiers. Can. J. Appl. Sport. Sci, 3: 186, **1978**.

Harkness, R. D. Mechanical properties of collagenous tissues. In: Treatise on Collagen. Ed.: B. S. Gould. Vol. 2, Part A, pp. 247-310. New York: Academic Press, **1968**.

Harper, H. A. Review of physiological chemistry. Los Altos, Calif.: Lange, **1969**.

Harris, E. A. and Thomson, J. G. The pulmonary ventilation and heart rate during exercise in healthy old age. Clin. Sci. 17: 349-359, **1958**.

Harris, J. D. Hearing loss in decompression, pp. 277-286. In: Underwater Physiology. Ed.: C. J. Lambertsen. New York: Academic Press, **1971**.

Harris P. Some observations on the biochemistry of

the myocardium at high altitude. In: High Altitude Physiology: Cardiac and Respiratory Aspects. Eds.: R. Porter and J. Knight. Edinburgh: Churchill-Livingstone, **1971**.

Harris, P., Bateman, M., Bayley, T. J., Gloster, J., and Whitehead, J. Observations on the course of the metabolic events accompanying mild exercise. Quart. J. Exp. Physiol. 53: 43-64, **1968**.

Harris, R. C., Hultman, E., Kaijser, L., and Nordesjø, L. O. The effect of circulatory occlusion on isometric exercise capacity and energy metabolism of the quadriceps muscle in man. Scand. J. Clin. Lab. Invest. 35: 87-95, **1975**.

Harris, W. S. Systolic time intervals in the noninvasive assessment of left ventricular performance in man, pp. 233-292. In: Cardiac Mechanics: Physiological, Clinical and Mathematical Considerations. Eds.: I. Mirsky, D. Ghista and H. Sandler. New York: Wiley, **1974**.

Harrison, G. A., Weiner, J. S., Tanner, J. M., and Barnicot, N. A. Human Growth. An Introduction to Human Evolution, Variation and Growth. Oxford: Clarendon Press, **1964**.

Harrison, M. H. Intravascular volume and electrolyte changes with acclimatization to heat in man. J. Physiol. (Lond.) 258: 30p-31p, **1976**.

Harrison, T. R., Dixon, K., Russell, R. O., Bidwai, P. S., and Coleman, H. N. The relation of age to the duration of contraction, ejection, and relaxation of the normal human heart. Amer. Heart J. 67: 189-199, **1964**.

Hart, J. S. Commentary. Can. Med. Assoc. J. 96: 803-804, **1967**.

Hart, J. S. and Jansky, L. Thermogenesis due to exercise in warm and cold acclimated rats. Can. J. Biochem. Physiol. 41: 629-634, **1963**.

Hartiala, K. Digestive functions in altitude conditions. In: International Symposium on the Effects of Altitude on Physical Performance. Ed.: R. F. Goddard. Chicago: Athletic Institute, **1967**.

Hartley, L. H. Growth hormone and catecholamine response to exercise in relation to physical training. Med. Sci. Sport. 7: 34-36, **1975**.

——. Central circulatory function during prolonged exercise. Ann. N.Y. Acad. Sci. 301: 189-194, **1977**.

Hartley, L. H., Alexander, J. K., Modelski, M., and Grover, R. F. Subnormal cardiac output at rest and during exercise in residents at 3,100 m altitude. J. Appl. Physiol. 23: 839-848, **1967**.

Hartley, L. H., Mason, J. W., Hogan, R. P., Jones, L. G., Kotchen, T. A. Mougey, E. H., Wherry, F. E., Pennington, L. L., and Ricketts, P. T. Multiple hormonal responses to graded exercise in relation to physical training. J. Appl. Physiol. 33: 602-606, **1972**.

Hartley, L. H., Pernow, B., Häggendahl, J., Lacour, J., de Lattre, J., and Saltin, B. Central circulation during sub-maximal work preceeded by heavy exercise. J. Appl. Physiol. 29: 818-823, **1970**.

Hartley, L. H. and Saltin, B. Blood gas tensions and pH in brachial artery, femoral vein and brachial vein during maximal exercise. Med. Sci. Sport. 3: 66-72, **1969**.

Hartley, L. H., Vogel, J. A., and Cruz, J. C. Reduction of maximal exercise heart rate at altitude and its reversal with atropine. J. Appl. Physiol. 36: 362-365, **1974**.

Hartmann, H. and Müller, K. G. Mechanische Belastung der Lunge beim Druckfall. Z. Flugwiss. 10: 203-216, **1962**.

Hartmann, H., Weiner, K. H., Fust, H. D., and Seifert, R. Über einen Dauerversuch von 100 Stunden bei einem Druck von 11 ata. Int. Z. Angew. Physiol. 22: 30-44, **1966**.

Hartog, M., Havel, R. J., Copinschi, G. et al. The relationship between changes in serum levels of growth hormone and mobilization of fat during exercise in man. Quart. J. Exp. Physiol. 52: 86-96, **1967**.

Harwood-Nash, D. C. F. Thumping of the praecordium in ventricular fibrillation. S. Afr. Med. J. 36: 280-281, **1962**.

Hashimoto, I., Sembrowich, W. L., and Gollnick, P. D. Calcium uptake by isolated sarcoplasmic reticulum and homogenates in different fiber types following exhaustive exercise. Med. Sci. Sport. 10: 42, **1978**.

Haskell, W. L. Cardiovascular complications during exercise training of cardiac patients. Circulation 57: 920-924, **1978**.

Haslag, W. M. and Hertzman, A. B. Temperature regulation in young women. J. Appl. Physiol. 20: 1283-1288, **1965**.

Hass, G. M. Studies of cartilage (iv). A morphological and clinical analysis of aging human costal cartilage. Arch. Pathol. 35: 275-284, **1943**.

Hatch, T. and Cook, K. M. Partitional Respirometry. AMA Arch. Industr. Health 11: 142-

158, **1955**.

Hatch, T. and Kindsvatter, V. H. Lung retention of quartz dust smaller than ½ micron. J. Industr. Hyg. Toxicol. 29: 342–346, **1947**.

Hatfield, H. S. and Pugh, L. G. C. E. Thermal conductivity of human fat and muscle. Nature (Lond). 168: 918–919, **1951**.

Hatt, P. Y., Ledoux, C., Bonvalet, J. P., and Guillemat, H. Lyse et synthise des protéines myocardiaques au cours de l'insuffisance cardiaque experimentale. (Etude au microscope electronique.) Arch. Mal. Coeur Vaiss. 58: 1703–1721, **1965**.

Hattner, R. S. and McMillan, D. E. Influence of weightlessness upon the skeleton: a review. Aerospace Med. 39: 849–855, **1968**.

Haugaard, N. Cellular mechanisms of oxygen toxicity. Physiol. Rev. 48: 311–373, **1968**.

Hauspie, R., Susanne, C., and Alexander, F. Growth and maturation in children with chronic asthma, pp. 203–210. In: Growth and Development. Physique. Ed.: O. G. Eiben. Budapest: Akadémia Kiadó, **1977**.

Hawkins, C. The effects of conditioning and training upon the differential white cell count. Ph.D. dissertation, New York University, **1937**.

Hawkins, L. H., Cole, P. V., and Harris, J. R. W. Smoking habits and blood carbon monoxide levels. Environ. Res. 11: 310–318, **1976**.

Haxton, A. F. and Whyte, H. E. The compressed air environment, pp. 1–16. In: The Physiology and Medicine of Diving. Eds.: P. B. Bennett and D. H. Elliott. London: Baillière, Tindall & Cassell, **1969**.

Hayden, F. J. and Yuhasz, M. The CAHPER Fitness–Performance Test Manual for Boys and Girls 7 to 17 years of age. Toronto: Canadian Association for Health, Physical Education and Recreation, **1966**.

Haymes, E. M., Lundegren, H. M., Loomis, J. L., and Buskirk, E. R. Validity of the ultrasonic technique as a method of measuring subcutaneous adipose tissue. Ann. Human Biol. 3: 245–251, **1976**.

Hayward, J. Man in cold water, physiological basis for survival techniques. Can. Physiol. 6: 89–90, **1975**.

Heaney, R. P. Radiocalcium metabolism in disuse osteoporosis in man. Amer. J. Med. 33: 188–200, **1962**.

Hearn, G. R. The effects of terminating and detraining on enzyme activities of heart and skeletal muscle of trained rats. Int. Z. Angew. Physiol. 21: 190–194, **1965**.

Hearn, G. R. and Gollnick, P. D. Effects of exercise on the adenosine triphosphatase activity in skeletal and heart muscle of rats. Int. Z. Angew. Physiol. 19: 23–26, **1961**.

Hearn, G. R. and Wainio, W. W. Succinic dehydrogenase activity of the heart and skeletal muscle of exercised rats. Amer. J. Physiol. 185: 348–350, **1956**.

Hebbelinck, M. The effects of a moderate dose of alcohol on a series of functions of physical performance in man. Arch. Int. Pharm. Ther. 120: 402–405, **1959**.

———. The effects of a small dose of alcohol on certain basic components of human physical performance. II. The effect on neuromuscular performance. Arch. Int. Pharmacodyn. 143: 247–257, **1963**.

———. Kinanthropometry and Aging: Morphological, structural, body mechanics and motor fitness aspects of aging, pp. 95–110. In: Physical Activity and Human Well-Being. Eds.: F. Landry and W. A. R. Orban. Miami: Symposia Specialists, **1978**.

Hecht, H. H. Certain vascular adjustments and maladjustments at altitudes, pp. 189–200. In: Exercise at Altitude. Ed.: R. Margaria. Dordrecht: Excerpta Medica Fdn., **1967**.

Hedberg, G. and Jansson, E. Skelettmuskelfiberkomposition. Kapacitet och intresse för olika fysiska aktiviteter bland elever i gymnasieskolan. Umeå: Pedagogical Institute. Report 54, **1976**.

Hedman, R. The available glycogen in man and the connection between the rate of oxygen intake and carbohydrate usage. Acta Physiol. Scand. 40: 305–321, **1957**.

Heebøll-Nielsen, K. R. Muscle asymmetry in normal young men. Comm. Test. Obs. Inst. 18: 3–9. Hellerup, Denmark, **1964**.

Hegg, J. J. Achievement motivation: Origin and development. Introduction, pp. 394–396. In: Sport in the Modern World: Chances and Problems. Eds.: O. Grupe, D. Kurz, and J. M. Teipel. Berlin: Springer Verlag, **1972**.

Heigenhauser, G. J. F. and Faulkner, J. A. Estimation of cardiac output by the CO_2 rebreathing method during tethered swimming. J. Appl. Physiol. Resp. Env. Ex. Physiol. 44: 821–824, **1978**.

Heikkinen, E. Studies on aging, physical fitness and health, pp. 331–337. In: Sports Medicine, Eds.: F. Landry and W. A. R. Orban. Miami:

Symposia Specialists, **1978**.

Heikkinen, E., Vihersaari, T., and Penttinen, R. Effect of previous exercise on fracture healing: A biochemical study with mice. Acta Orthop. Scand. 45: 481–489, **1974**.

Heintzen, P. H., Moldenhauer, K., and Lange, P. E. Three-dimensional computerized contraction pattern analysis: Description of methodology and its validation. Europ. J. Cardiol. 1: 229–239, 1974.

Heinzelmann, F. Social and psychological factors that influence the effectiveness of exercise programs, pp. 275–287. In: Exercise Testing and Exercise Training in Coronary Heart Disease. Eds.: J. P. Naughton, H. K. Hellerstein, and I.C. Mohler. New York: Academic Press, **1973**.

Heinzelmann, F. and Bagley, R. W. Response to physical activity programs and their effects on health behaviour. Public Health Rep. 85: 905–911, **1970**.

Helge, H., Weber, B., and Quabbe, H. J. Growth hormone release and venipuncture. Lancet. 1: 204, **1969**.

Hellebrandt, F. A., Brogdon, E., and Hoopes, S. L. The disappearance of digestive inhibition with the repetition of exercise. Amer. J. Physiol. 112: 442–450, **1935**.

Hellebrandt, F. A. and Dimmitt, L. L. Studies in the influence of exercise on the digestive work of the stomach. 3. Its effect on the relation between secretory and motor function. Amer. J. Physiol. 107: 364–369, **1934**.

Hellebrandt, F. A. and Miles, M. M. The effect of muscular work and competition on gastric acidity. Amer. J. Physiol. 102: 258–266, **1932**.

Hellebrandt, F. A., Parrish, A. M., and Houtz, S. J. Cross education; influence of unilateral exercise on contralateral limb. Arch. Phys. Med. 28: 76–85, **1947**.

Hellebrandt, F. A. and Tepper, R. H. Studies on the influence of exercise on the digestive work of the stomach. 2. Its effect on emptying time. Amer. J. Physiol. 197: 355–363, **1934**.

Hellebrandt, F. A. and Waterland, J. C. Indirect learning: the influence of unimanual exercise on related muscle groups of the same and opposite side. Amer. J. Phys. Med. 35: 144–159, **1962**.

Hellerstein, H. K. A misguided goal or unrealized objective? Introduction, pp. 125–135. In: Critical Evaluation of Cardiac Rehabilitation. Ed.: J. J. Kellermann and H. Denolin. Basel: Karger, **1977**.

Hellerstein, H. K., Hirsch, E. L., Ader, R., Greenblot, N., and Siegel, M. Principles of exercise testing for normals and cardiac subjects, pp. 129–168. In: Exercise Testing and Exercise Training in Coronary Heart Disease. Eds.: J. P. Naughton, H. K. Hellerstein, and I. C. Mohler. New York: Academic Press, **1973**.

Hellström, B. Local Effects of Acclimatization to Cold in Man. Oslo, Norway: Universitetsforlag, **1965**.

Helson, H. Design of equipment and optimal human operation. Amer. J. Psychol. 62: 473–479, **1949**.

Hempleman, H. V. British decompression theory and practice, pp. 291–318. In: Physiology and Medicine of Diving. Eds.: P. B. Bennett and D. H. Elliott. London: Baillière, Tindall & Cassell, **1969**.

——. Investigation into the decompression tables: A new theoretical basis. U.K. Royal Navy Rep. RNP 52/708 UPS 131, **1977**.

Hénane, R., Flandrois, R., Charbonnier, J. P., and Bittell, J. Comparaison des réponses thermorégulatrices à la chaleur chez des athlètes et des sujets non-entrainés. J. Physiol. (Paris) 69: 257a (Abstr.), **1974**.

Hendry, L. B. The stigmatized body and social identity (physique, physical activities, diet; a study of high and low ponderal index adolescents), pp. 91–98. In: Nutrition, Dietetics and Sport. Eds.: G. Ricci and A. Venerando. Torino: Edizioni Minerva Medica, **1978**.

Henneman, E., Somjen, G. G., and Carpenter, D. O. Functional significance of cell size in spinal motoneurons. J. Neurophysiol. 28: 599–620, **1965**.

Hennessy, T. R. The equivalent bulk-diffusion model of the pneumatic decompression computer. Med. Biol. Eng. 11: 135–137, **1973**.

Henquell, L., Odoroff, C. L., and Honig, C. R. Coronary intercapillary distance during growth: relation to Pto_2 and aerobic capacity. Amer. J. Physiol. 231: 1852–1859, **1976**.

Henriksson, J. Training induced adaptation of skeletal muscle and metabolism during submaximal exercise. J. Physiol. (Lond.) 270: 661–675, **1977**.

Henriksson, J. and Reitman, J. S. Time course of changes in human skeletal muscle succinate dehydrogenase and cytochrome oxidase activities and maximal oxygen uptake with physical activity and inactivity. Acta Physiol. Scand. 99: 91–97, **1977**.

Henry, F. M. Influence of motor and sensory sets on reaction latency and speed of discrete movements. Res. Quart 31: 459-468, **1960**.

——. Stimulus complexity, movement complexity, age and sex in relation to reaction latency and speed in limb movements. Res. Quart. 32: 353-366, **1961**.

Henry, F. M. and Rogers, D. E. Increased response latency for complicated movements and "memory drum" theory of neuromotor reaction. Res. Quart. 31: 448-458, **1960**.

Henry, F. M. and Whitley, J. D. Relationships between individual differences in strength, speed, and mass in an arm movement. Res. Quart. 31: 24-33, **1960**.

Henry, J. G. and Bainton, C. R. Human core temperature increase as a stimulus to breathing during moderate exercise. Resp. Physiol. 21: 183-191, **1974**.

Henry, J. P. The significance of the loss of blood volume into the limbs during pressure breathing. J. Aviat. Med. 22: 31-38, **1951**.

Henschel, A. Water balance—A problem in occupational health. Occup. Health Rev. 17: 11-13, **1965**.

Hensel, H. Thermoreceptors. Ann. Rev. Physiol. 36: 233-249, **1974**.

Hensel, H., Ruef, J., and Golenhofen, K. Fortlaufende Registrieurung der Muskeldurchblutung am Menschen mit einer calorimetersonde. Pflüg. Archiv. 259: 267-280, **1954**.

Henson, P. L., Cooper, J., and Wilkerson, J. Pace and grade related to the oxygen and energy requirements and the mechanics of treadmill running. Med. Sci. Sport. 9: 61, **1977**.

Herbert, W. G. and Ribisl, P. M. Effects of dehydration upon physical working capacity of wrestlers under competitive conditions. Res. Quart. 43: 416-422, 1972.

Herman, R. The myotatic reflex. Brain 93: 273-312, **1970**.

Hermansen, L. Lactate production during exercise. In: Muscle Metabolism During Exercise, pp. 401-408. Eds.: B. Pernow, and B. Saltin. New York: Plenum Press, **1971**.

Hermansen, L. and Andersen, K. L. Aerobic work capacity in young Norwegian men and women. J. Appl. Physiol. 20: 425-431, **1965**.

Hermansen, L., Ekblom, B., and Saltin, B. Cardiac output during submaximal and maximal treadmill and bicycle exercise. J. Appl. Physiol. 29: 82-86, **1970**.

Hermansen, L., Hultman, E., and Saltin, B. Muscle glycogen during prolonged severe exercise. Acta Physiol. Scand. 71: 129-139, **1967**.

Hermansen, L., Maehlum, S., Pruett, E. D. R., Vaage, O., Waldum, H., and Wessel-Aas, T. Lactate removal at rest and during exercise. In: Metabolic Adaptation to Prolonged Physical Exercise. Eds.: H. Howald and J. R. Poortmans. Basel: Birkhauser Verlag, **1975**.

Hermansen, L. and Osnes, J. B. Blood and muscle pH after maximal exercise in man. J. Appl. Physiol. 32: 304-308, **1972**.

Hermansen, L. and Saltin, B. Blood lactate concentration during exercise at acute exposure to altitude, pp. 48-57. In: Exercise at Altitude. Ed.: R. Margaria. Dordrecht: Excerpta Medica Fdn., **1967**.

Hermansen, L. and Stensvold, I. Production and removal of lactate during exercise in man. Acta Physiol. Scand. 86: 191-201, **1972**.

Hermansen, L. and Wachtlová, M. Capillary density of skeletal muscle in well-trained and untrained man. J. Appl. Physiol. 30: 860-863, **1971**.

Herrlich, H. C., Raab, W., and Gigee, W. Influence of muscular training and of catecholamines on cardiac acetylcholine and cholinesterase. Arch. Int. Pharmacodyn. Ther. 129: 201-215, **1960**.

Hertig, B. A., Belding, H. S., Kraning, K. K., Batterton, D. L. Smith, C. R., and Sargent, F. Artificial acclimatization of women to heat. J. Appl. Physiol. 18: 383-386, **1963**.

Hervey, G. R. Regulation of energy balance. Nature 223: 629-631, **1969**.

Hervey, G. R. Hutchinson, I., and Knibbs, A. V. Effects of methandrostenolone on body composition in male students undergoing athletic training. J. Endocrinol. 65: 49p, **1975**.

Herxheimer, H. Untersuchungen über die Änderung der Herzgrösse unter dem Einfluss bestimmter Sportarten. Z. Klin. Med. 3: 376-393, **1929**.

Hess, W. R. Das Zwischenhirn. Basel: Schwabe, **1949**.

——. Das Zwischenhirn. 2. Auflage. Basel: Schwabe, **1954**.

Hesser, C. M. Breath-holding under high pressure, pp. 165-181. In: Physiology of Breath-Hold Diving and the Ama of Japan. Eds.: H. Rahn and T. Yokoyama. Washington, D.C. Natl. Acad. Sci. Natl. Res. Council Publ. 1341, **1965**.

Hesser, C. M. and Holmgren, B. Effects of raised barometric pressures on respiration in man.

Acta Physiol. Scand. 47: 28-43, **1959**.

Hesser, C. M., Linnarson, D., and Bjurstedt, H. Cardiorespiratory and metabolic responses to positive, negative and minimum load dynamic exercise. Resp. Physiol. 30: 51-67, **1977**.

Hetherington, A. W. and Ranson, S. W. The spontaneous activity and food intake of rats with hypothalamic lesions. Amer. J. Physiol. 136: 609-617, **1942**.

Hetland, Ø, Brubak, E. A., Refsum, H. E., and Strømme, S. B. Serum and erythrocyte zinc concentrations after prolonged heavy exercise, pp. 367-370. In: Metabolic Adaptation to Prolonged Physical Exercise. Eds.: H. Howald and J. R. Poortmans. Basel: Birkhauser Verlag, **1975**.

Hettinger, T. L. Physiology of Strength. Springfield, Ill.: C. C. Thomas, **1961**.

Hettinger, T. L. and Hollmann, W. Dynamometrische Messungen und Muskeln. Sportarzt und Sportmedizin 1(18), **1969**.

Hettinger, T. L. and Müller, E. A. Muskelleistung und Muskeltraining, arbeitsphysiologie. Arbeitsphysiol. 15: 111-126, **1953**.

Hewitt, D. Sib resemblance in bone, muscle and fat measurements of the human calf. Ann. Human Genet. 22: 213-221, **1958**.

Hewitt, D., Jones, G. J. L., Godin, G. J., McComb, K., Breckenridge, W. C. Little, J. A., Steiner, G., Mishkel, M. A., Baillie, J. H., Martin, R. H., Gibson, E. S., Prendergast, W. F., and Parliament, W. J. Normative standards of plasma cholesterol and triglyceride concentrations in Canadians of working age. Can. Med. Assoc. J. 117: 1020-1024, **1977**.

Heymans, C. Introduction to the Regulation of Blood Pressure and Heart Rate. Springfield, Ill.: C. C. Thomas, **1950**.

Heymans, C. and Neil, E. Reflexogenic Areas of the Cardiovascular System. London: Churchill, **1958**.

Heyward, V. and McCreary, L. Comparison of the relative endurance and critical occluding tension levels of men and women. Res. Quart. 49: 301-307, **1978**.

Heywood, S. M., Dowben, R. M., and Rich, A. The identification of polyribosomes synthesizing myosin. Proc. Natl. Acad. Sci. U.S. 57: 1002-1009, **1967**.

Hickam, J. B., Cargill, W. H., and Golden A. Cardiovascular reactions to emotional stimuli. Effect on the cardiac output, arterio-venous oxygen difference, arterial pressure, and peripheral resistance. J. Clin. Invest. 27: 290-298, **1948**.

Hickey, N., Mulcahy, R., Bourke, G. J., Graham, I., and Wilson-Davis, K. Study of coronary risk factors related to physical activity in 15,171 men. Brit. Med. J. 3: 507-509, **1975**.

Hickson, R. C., Hagberg, J. M., Coulee, R. K., Jones, D. A., Ehsani, A. A., and Winder, W. W. Effect of training on hormonal responses to exercise in competitive swimmers. Med. Sci. Sport. 10: 41, **1978**.

Higenbottam, T. W., Hamilton, D., and Clark, T. J. H. Changes in airway size and bronchial response to inhaled histamine in smokers and non-smokers. Clin. Sci. 54: 11p, **1978**.

Hill, A. V. The physiological basis of athletic records. Sci. Month. 21: 409-428, **1925**.

——. The heat of shortening and the dynamic constants of muscle. Proc. Roy. Soc. B. 126: 136-195, **1938**.

——. On the time required for diffusion and its relation to processes in muscle. Proc. Roy. Soc. B. 135: 446-453, **1948**.

——. Chemical change and mechanical response in stimulated muscle. Proc. Roy. Soc. B. 141: 314-320, **1953**.

——. The influence of the external medium on the internal pH of muscle. Proc. Roy. Soc. Ser. B. 144: 1-22, **1955-1956**.

——. Production and absorption of work by muscle. Science 131: 897-903, **1960**.

Hill, A. V., Long, C. N., and Lupton, H. Muscular exercise, lactic acid and the supply and utilization of oxygen. Pts. IV-VI. Proc. Roy. Soc. Lond. 97: 84-138, **1924-25**.

Hill, D. K. Tension due to interaction between the sliding filaments in resting striated muscle. The effects of stimulation. J. Physiol. (Lond.) 199: 637-684, **1968**.

Hill, E. P., Power, G. G., and Longo, L. D. Mathematical simulation of pulmonary O_2 and CO_2 exchange. Amer. J. Physiol. 224: 904-917, **1973**.

Hill, L. and Flack, M. The influence of oxygen inhalations on muscular work. J. Physiol. (Lond.) 40: 347-372, **1910**.

Hill, S., Goetz, F., Fox, H., Murawski, B., Krakauer, L., Reifenstein, R., Gray, S., Reddy, W., Hedberg, S., St. Marc J., and Thorn, G. Studies on adrenocortical and psychological responses to stress in man. Arch. Int. Med. 97: 269-298, **1956**.

Hills, B. A. A pneumatic analogue for predicting the occurrence of decompression sickness, Med. Biol. Eng. 5: 421-432, **1967**.

Hills, B. A. Thermodynamic decompression: An approach based upon the concept of phase equilibration in tissue, pp. 319-356. In: Physiology and Medicine of Diving. Eds.: P. B. Bennett and D. H. Elliott. London: Baillière, Tindall & Cassell, **1969**.

Hinckle, L. E. An estimate of the effects of "stress" on the incidence and prevalence of coronary heart disease in a large industrial population in the United States. Thromb. Diath. Haemorrh. Suppl. 51: 15-65, **1972**.

Hinckle, L. E., Carver, S. T., and Stevens, M. The frequency of asymptomatic disturbances of cardiac rhythm and conduction in middle-aged men. Amer. J. Cardiol. 24: 629-650, **1969**.

Hirsch, J. and Han, P. W. Cellularity of rat adipose tissue: effects of growth, starvation, and obesity. J. Lipid Res. 10: 77-82, **1969**.

Hislop, H. J. Quantitative changes in human muscular strength during isometric exercise. J. Amer. Phys. Ther. Assoc. 43: 21-38, **1963**.

Hislop, H. J. and Perrine, J. J. The isokinetic concept of exercise. Phys. Ther. 47: 114-117, **1967**.

Hjalmarson, A. and Isaksson, O. In vitro work load and rat heart metabolism. 1. Effect on protein synthesis. Acta Physiol. Scand. 86: 126-144, **1972**.

Ho, K. J., Taylor, C. B., and Biss, K. Overall control of sterol synthesis in animals and man. In: Atherosclerosis. Proc. 2nd Int. Symp. Ed.: R. J. Jones. New York: Springer Verlag, **1970**.

Hodgkin, A. L. The Conduction of the Nervous Impulse. Liverpool: Liverpool University Press, **1964**.

Hodgkins, J. Influence of age on the speed of reaction and movement in females. J. Gerontol. 17: 385-391, **1962**.

Hodgkins, J. Reaction time and speed of movements in males and females of various ages. Res. Quart. 34: 335-343, **1963**.

Hodgson, H. J. F., Marsden, C. D., and Meadows, J. C. The effect of adrenaline on the response to muscle vibration in man. J. Physiol. (Lond.) 202: 98p, **1969**.

Hodgson, H. J. F. and Matthews, P. B. C. The ineffectiveness of excitation on the primary endings of the muscle spindle by vibration as a respiratory stimulant in the decerebrate cat. J. Physiol. (Lond.) 194: 555-563, **1968**.

Hoebel, B. G. and Teitelbaum, P. Weight regulation in normal and hypothalamic hyperphagic rats. J. Comp. Physiol. Psychol. 2: 189-193, **1966**.

Hoes, M., Binkhorst, R. A., Smeekes-Kuyl, A., and Vissurs, A. C. Measurement of forces exerted on a pedal crank during work on the bicycle ergometer at different loads. Int. Z. Angew. Physiol. 26: 33-42, **1968**.

Hofer, H. W. and Pette, D. Wirkungen und Wechselwirkungen von substraten und Effektoren an der Phosphofructokinase des Kaninchen Skeletmuskeln. Z. Physiol. Chem. 349: 1378-1392, **1968**.

Hoff, E. C., Kell, J. F., and Carroll, M. N. Effects of cortical stimulation and lesions on cardiovascular function. Physiol. Rev. 43: 68-114, **1963**.

Högberg, P. and Ljunggren, O. Uppvärmningens inverkan pa löpprestationerna. Cited by Åstrand & Rodahl, 1977. Svensk. Idrott, 40, **1947**.

Hogg, W., Brunton, J., Kryger, M., Brown, R., and Macklem, P. Gas diffusion across collateral channels. J. Appl. Physiol. 33: 568-575, **1972**.

Hohorst, H. J., Reim, M., and Bartels, H. Studies on the creatinekinase equilibrium in muscle and the significance of ATP and ADP levels. Biochem. Biophys. Res. Comm. 7: 142-146, **1962**.

Holland, J., Milic-Emili, J. Macklem, P., and Bates, D. V. Regional distribution of pulmonary ventilation and perfusion in elderly subjects. J. Clin. Invest. 47: 81-92, **1968**.

Hollander, A. P. and Bouman, L. N. Cardiac acceleration in man elicited by a muscle-heart reflex. J. Appl. Physiol. 38: 272-278, **1975**.

Hollmann, W. Changes in the capacity for maximal and continuous effort in relation to age, pp. 369-371. In: International Research in Sport and Physical Education. Eds.: E. Jokl and E. Simon. Springfield, Ill.: C. C. Thomas, **1964**.

——. Körperliches Training als Prävention von Herz-Krieslauf Krankheiten. Stüttgart: Hippokrates Verlag, **1965**.

Hollmann, W., Grünewald, B., Chirdel, K., and Kastner, K. Körperliche Leistung und Hypoxie. Mat. Med. Nordmark 20: 481-493, **1968**.

Hollmann, W., Herkenrath, G., Grunewald, B., Budinger, H., Jonath, U., Russmann, H., and Hain, D. Untersuchungen über Möglichkeiten zur Steigerung des Körperlichen Leistungsverinögens von Rekruten. Sportarzt und Sportmedizin 12: 582–592, **1966**.

Hollmann, W. and Hettinger, T. H. Sportmedizin: Arbeits und Trainingsgrundlagen. Stüttgart: Schattauer, 1976.

Hollmann, W. and Venrath, H. Die Beeinflussung von Herzgrösse, maximaler O_2 Aufnahme und Ausdauergranze durch ein Ausdauertraining mittlerer und hoher Intensität. Sportarzt. 14: 189, **1963**.

Hollmann, W., Venrath, H., Herkenrath, G., and Barwisch, B. Der Einfluss unterschiedlicher O_2- Konzentrationen in der Inspirationsluft auf das Kardio-pulmonale. Verhalten bei 12- bis 50 sekundigen Maximalbelastungen. Sportarzt und Sportmedizin 4: 137–143, **1966**.

Holloszy, J. O. Biochemical adaptations in muscle-effects of exercise on mitochondrial oxygen uptake and respiratory enzyme activity in skeletal muscle. J. Biol. Chem. 242: 2278–2282, **1967**.

——. Biochemical adaptations to exercise: aerobic metabolism. Exercise Sports Sci. Rev. 1: 45–71, **1973**.

——. Discussion. Ann. N.Y. Acad. Sci. 301: 453, **1977**.

Holloszy, J. O. and Booth, F. W. Biochemical adaptations to endurance exercise in muscle. Ann. Rev. Physiol. 38: 273–291, **1976**.

Holloszy, J. O. Booth, F. W., Winder, W. W., and Fitts, R. H. Biochemical adaptation of skeletal muscle to prolonged physical exercise, pp. 438–447. In: Metabolic Adaptation to Prolonged Physical Exercise. Eds.: H. Howald and J.R. Poortmans. Basel: Birkhauser Verlag, **1975**.

Holloszy, J. O., Oscai, L. B., Molé, P. A., and Don, I. J. Biochemical adaptations to endurance exercise in skeletal muscle, pp. 51–61. In: Muscle Metabolism During Exercise. Eds.: B. Pernow and B. Saltin. New York: Plenum Press, **1971**.

Holmberg, S., Serzysko, W., and Varnauskas, E. Coronary circulation during heavy exercise in control subjects and patients with coronary heart disease. Acta Med. Scand. 190: 465–480, **1971**.

Holmbøe, J., Bell, H., and Norman, N. Urinary excretion of catecholamines and steroids in military cadets exposed to prolonged stress. Försvarsmedizin 11: 183, **1975**.

Holmdahl, D. E. and Ingelmark, B. E. Der Bau des Gelenkknorpels unter verschiedenen funktionellen Verhaltnissen. Acta Anat. 6: 309–375, **1948**.

——. The contact between the articular cartilage and the medullary cavities of the bones. Acta Anat. 12: 341–349, **1951**.

Holmér, I. Physiology of swimming man. Acta Physiol. Scand. Suppl. 407: 1–55, **1947a**.

——. Propulsive efficiency of breast-stroke and free-style swimming. Europ. J. Appl. Physiol. 33: 95–103, **1947b**.

Holmér, I. and Bergh, U. Metabolic and thermal response to swimming in water at varying temperatures. J. Appl. Physiol. 37: 702–705, **1974**.

Holmér I., Lundin, A., and Eriksson, B. O. Maximum oxygen uptake during swimming and running by elite swimmers. J. Appl. Physiol. 36: 711–714, **1974**.

Holmér, I., Stein, E. M., Saltin, B., Ekblom, B., and Åstrand, P. O. Hemodynamic and respiratory responses compared in swimming and running. J. Appl. Physiol. 37: 49–54, **1974**.

Holmes, T. H. and Rahe, R. H. The social readjustment rating scale. J. Psychosom. Res. 11: 213–218, **1967**.

Holmgren, A. Circulatory changes during muscular work in man. Scand. J. Clin. Lab. Invest. 8 (Suppl. 24): 1–97, **1956**.

——. On the variation of $D_{L, CO}$ with increasing oxygen uptake during exercise in healthy trained young men and women. Acta Physiol. Scand. 65: 207–220, **1965**.

——. Vasoregulatory asthenia. In: Proc. Int. Symp. Physical Activity and Cardiovascular Health. Can. Med. Assoc. J. 96: 853, **1967a**.

——. Cardio-respiratory determinants of cardiovascular fitness. In: Proc. Int. Symp. Physical Activity and Cardiovascular Health. Can. Med. Assoc. J. 96: 697–702, **1967b**.

——. Commentary. In: Proc. Int. Symp. Physical Activity and Cardiovascular Health. Ed.: R. J. Shephard. Can. Med. Assoc. J. 96: 794, **1967c**.

——. Vasoregulatory asthenia, pp. 34–37. In: Coronary Heart Disease and Physical Fitness. Eds.: O. A. Larsen and R. O. Malmborg. Baltimore: University Park Press, **1971**.

Holmgren, A. and McIlroy, M. B. Effect of temperature on arterial blood gas tensions and pH during exercise. J. Appl. Physiol. 19: 243–245, **1964**.

Holmgren, A., Mossfeldt, F., Sjöstrand, T., and Ström, G. Effect of training on work capacity, total haemoglobin, blood volume, heart volume and pulse rate in recumbent and upright positions. In: International Research in Sport and Physical Education. Eds.: E. Jokl and E. Simon. Springfield, Ill.: C. C. Thomas, **1964**.

Holmgren, A. and Ovenfors, C. O. Heart volume at rest and during muscular work in the supine and in the sitting position. Acta Med. Scand. 167: 267-277, **1960**.

Holmgren, A. and Pernow, B. Spectrophotometric measurement of oxygen saturation of blood in the determination of cardiac output. A comparison with the Van Slyke method. Scand. J. Clin. Lab. Invest. 11: 143-149, **1959**.

Holmgren, A. and Strandell, T. Relationship between heart volume, total hemoglobin and physical work capacity in former athletes. Acta Med. Scand. 163: 146-160, **1959**.

Holmgren, E. Untersuchungen über die morphologisch nachweisbaren stofflichen Umsetzungen der quergestreiften Muskelfasern. Arch. f. Mikr. Anat. (Bonn) 75: 240-336, **1910**.

Holmqvist, B. and Lundberg, A. Differential supraspinal control of synaptic actions evoked by volleys in the flexion reflex afferents in alpha motoneurones. Acta Physiol. Scand. 54 (Suppl. 186): 1-51, **1961**.

Holter, N.J. New method for heart studies. Science 134: 1214-1220, **1961**.

Holzer, H. Regulation of enzymes by enzyme-catalyzed chemical modification. Adv. Enzymol. 32: 297-326, **1969**.

Høncke, P. Investigations on the structure and functions of living, isolated, cross-striated muscle fibres of mammals. Acta Physiol. Scand. 15 (Suppl 48): **1947**.

Hong, S. K., Cerretelli, P., Cruz, J. C., and Rahn, M. Mechanics of respiration during submersion in water, pp. 29-35. In: Studies in Pulmonary Physiology: Mechanics, Chemistry and Circulation of the Lung, Vol. III. U.S. Air Force, Brooks A.F.B. Texas, **1970**.

Hong, S. K., Ting, E. Y., and Rahn, H. Lung volumes at different depths of submersion. J. Appl. Physiol. 15: 550-553, **1960**.

Hongo, T., Jankowska, E., and Lundberg, A. The rubrospinal tract. II. Facilitation of interneuronal transmission in reflex paths to motoneurones. Exp. Brain Res. 7: 365-391, **1969**.

Honig, C. R., Feldstein, M. L., and Frierson, J. L. Capillary lengths, anastomoses, and estimated capillary transit times in skeletal muscle. Amer. J. Physiol. 233: H122-H129, **1977**.

Honig, C. R., Frierson, J. L., and Patterson, J. L. Comparison of neural controls of resistance and capillary density in resting muscle. Amer. J. Physiol. 218: 937-942, **1970**.

Hood, W. P., Rackley, C. E., and Rolett, E. L. Wall stress in the normal and hypertrophied human left ventricle. Amer. J. Cardiol. 22: 550-558, **1968**.

Hoogerwerf, A., and Hoitink, A. W. J. H. The influence of vitamin C administration on the mechanical efficiency of the human organism. Int. Z. Angew. Physiol. 20: 164-172, **1963**.

Hoogerwerf, S. Elektrokardiographische Untersuchungen der Amsterdamer Olympiadekampfer. Arbeitsphysiol. 2: 61-75, **1929**.

Höök, O. and Tornvall, G. Apparatus and method for determination of isometric muscle strength in man. Scand. J. Rehab. Med. 1: 139-142, **1969**.

Horan, L. G., Flowers, N. C., and Brody, D. A. Body surface potential distribution. Comparison of naturally and artificially produced signals as analysed by digital computer. Circ. Res. 13: 373-387, **1963**.

Hori, S., Ihzuka, H., and Inoye, A. Physiological responses to whole body and hot air exposure with special reference to assessment of heat tolerance. Jap. J. Physiol. 25: 563-573, **1975**.

Hornbein, T. F. and Roos, A. Specificity of H ion concentration as a carotid chemoreceptor stimulus. J. Appl. Physiol. 18: 580-584, **1963**.

Hornbein, T. F., Sørensen, S. C., and Parks, C. R. Role of muscle spindles in lower extremities in breathing during bicycle exercise. J. Appl. Physiol. 27: 476-479, **1969**.

Horsfield, K., Davies, A., and Cumming, G. Role of conducting airways in partial separation of inhaled gas mixtures. J. Appl. Physiol. Resp. Ex. Environ. Physiol. 43: 391-396, **1977**.

Horstman, D. H. Nutrition, pp. 343-365. In: Ergogenic Aids and Muscular Performance. Ed.: W. P. Morgan, New York: Academic Press, **1972**.

——. Exercise performance at 5°C. Med. Sci. Sport. 9: 52, **1977**.

Horvath, S. M. and Colwell, M. O. Heat stress and the new standards. J. Occup. Med. 15: 524-528, **1973**.

Horwitz, L. D., Atkins, J. M., and Leshin, S. J. Role of the Frank-Starling Mechanism in Exercise. Circ. Res. 31: 868-875, **1972**.

Hösli, L. and Haas, H. L. The hyperpolarization of neurones of the medulla oblongata by glycine. Experientia 28(a): 1057–1058, **1972**.

Houk, J. and Henneman, E. Responses of Golgi tendon organs to active contractions of the soleus muscle of the cat. J. Neurophysiol. 30: 466–481, **1967**.

Houk, J. C., Harris, D. A., and Hasan, Z. Nonlinear behaviour of spindle receptors, pp. 147–163. In: Control of Posture and Locomotion. Eds.: R. B. Stein, K. G. Pearson, R. S. Smith, and J. B. Redford. New York: Plenum Press, **1973**.

Houston, C. S. Acute pulmonary edema of high altitude. N. Engl. J. Med. 263: 478–480, **1960**.

Houston, C. S. and Riley, R. L. Respiratory and circulatory changes during acclimatization to high altitude. Amer. J. Physiol. 149: 565–588, **1947**.

Houston, M. E. and Green, H. J. Physiological and anthropometric characteristics of elite Canadian ice-hockey players. J. Sport. Med. 16: 123–128, **1976**.

Houston, M. E., Waugh, M. R., Green, H. J., and Noble, E. G. Response of serum enzymes to intermittent work, pp. 187–192. In: 3rd International Symposium on Biochemistry of Exercise. Eds.: F. Landry and W. R. Orban. Miami: Symposia Specialists, **1978**.

Houtz, S. J., Parrish, A. M., and Hellebrandt, F. A. Influence of heavy resistance exercise on strength. Physiother. Rev. 26: 299–304, **1946**.

Howald, H. Ultrastructural adaptation of skeletal muscle to prolonged physical exercise, pp. 372–383. In: Metabolic adaptation to prolonged physical exercise. Eds.: H. Howald and J. R. Poortmans. Basel: Birkhauser Verlag, **1975**.

——. Ultrastructure and biochemical function of skeletal muscle in twins. Ann. Human Biol. 3: 455–462, **1976**.

——. Cellular changes in human skeletal muscle during submaximal exercise. Int. Union Physiol. Sci. (Paris) 12: 723, **1977**.

Howard, A. N. Recent advances in nutrition and atherosclerosis, pp. 408–413. In: Atherosclerosis. Proc. 2nd Int. Symp. Ed.: R. J. Jones. Berlin: Springer Verlag, **1970**.

Howard, P. The physiology of positive acceleration, pp. 551–687. In: A Textbook of Aviation Physiology. Ed.: J. A. Gillies. Oxford: Pergamon Press, **1965**.

Howard, P., Paul, D., and Shephard, R. J. The decrease of radioactive sodium during pressure breathing. Farnborough, Hants: Royal Air Force FPRC 938, **1955**.

Howe, P. E. and Schiller, M. Growth responses of the school child to changes in diet and environmental factors. J. Appl. Physiol. 5: 51–61, **1952**.

Howell, J. B. L. and Campbell, E. J. M. Breathlessness. Oxford: Blackwell Scientific, **1966**.

Howell, M. L., Kimoto, R., and Morford, W. R. Effect of isometric and isotonic exercise programs upon muscular endurance. Res. Quart. 33: 536–540, **1962**.

Howell, M. L., Loiselle, D. S., and Lucas, W. G. Strength of Edmonton School Children. Edmonton: Fitness Research Unit, University of Alberta. Unpublished report, **1965**.

Howells, K. F. and Jordan, T. C. The myofibril content of histochemically characterized rat muscle fibre types. J. Physiol. (Lond.) 284: 35p, July, **1978**.

Howitt, J. S., Balkwill, J. S., Whiteside, T. C. D., and Whittingham, P. D. G. A preliminary study of flight deck work loads in civil air transport aircraft. U.K. Ministry of Defence FPRC 1240, **1966**.

Howley, E. T. The effect of different intensities of exercise on the excretion of epinephrine and norepinephrine. Med. Sci. Sport. 8: 219–222, **1976**.

Hubbard, A. W. Homokinetics: Muscular function in human movement, pp. 5–23. In: Science and Medicine of Exercise and Sports. Ed.: W. R. Johnson. New York: Harper & Row, **1960**.

Hubbard, C. H. Advantages of a new shadow-silhouetteograph over the original. Res. Quart. 6(1), Suppl., **1935**.

Hubel, D. H. and Wiesel, T. N. Receptive fields and functional architecture in two non-striate visual areas (18 and 19) of the cat. J. Neurophysiol. 28: 229–289, **1965**.

Huckabee, W. E. Relationships of pyruvate and lactate during anaerobic metabolism. II. Exercise and formation of O_2-debt. J. Clin. Invest. 37: 255–271, **1958a**.

——. The role of anaerobic metabolism in the performance of mild muscular work. II. The effect of asymptomatic heart disease. J. Clin. Invest. 37: 1593–1602, **1958b**.

Hudlická, O. and Cotter, D. The role of capillary density on the improvement of muscle performance. In: Proceedings of the International Union of Physiological Sciences, Paris. Volume 13, 337, **1977**.

Hughes, A. L. and Goldman, R. F. Energy cost of hard work. J. Appl. Physiol. 29: 570–572, **1970**.

Hughes, J. M. B., Glazier, J. B., Maloney, J. E., and West, J. B. Effect of lung volume on the distribution of pulmonary blood flow in man. Resp. Physiol. 4: 58-72, **1968**.

Hughes, R. L., Clode, M., Edwards, R. H. T., Goodwin, T. J., and Jones, N. L. Effect of inspired O_2 on cardiopulmonary and metabolic responses to exercise in man. J. Appl. Physiol. 24: 336-347, **1968**.

Huibregtse, W. H., Hartley, L. H., Jones, L. G., Doolittle, W. H., and Criblez, T. L. Improvement of aerobic work capacity following non-strenuous exercise. Arch. Env. Health 27: 12-15, **1973**.

Hukuhara, T. Neuronal organisation of the central respiratory mechanisms in the brain stem of the cat. In: International Symposium on Neural Control of Breathing. Eds.: W. A. Karczewski and J. G. Widdicombe. Warsaw: Acta Neurobiol. Exp. 33: 219-244, **1978**.

Hultgren, H. N. and Lundberg, E. Medical problems of high altitude, pp. 110-121. In: Exercise and Altitude. Eds.: E. Jokl and P. Jokl. Basel: Karger, **1968**.

Hultman, E. Muscle glycogen stores and prolonged exercise. In: Frontiers of Fitness. Ed.: R. J. Shephard. Springfield, Ill.: C. C. Thomas, **1971**.

——. Regulation of carbohydrate metabolism in the liver during rest and exercise with special reference to diet, pp. 99-126. In: 3rd International Symposium on Biochemistry of Exercise. Eds.: F. Landry and W. R. Orban. Miami: Symposia Specialists, **1978**.

Hultman, E., Bergstrom, J., and McLennan Anderson, N. Break-down and resynthesis of phosphorylcreatine and adenosine-triphosphate in connection with muscular work in man. Scand. J. Lab. Invest. 19: 56-66, **1967**.

Humphrey, D. R. Relating motor cortex spike trains to measures of motor performance. Brain Res. 40: 7-18, **1972**.

Hunsicker, P. A. and Donelly, R. J. Instruments to measure strength. Res. Quart. 26 (Suppl. 408), **1955**.

Hunt, E. A. E.C.G. study of 20 champion swimmers before and after 110 yard sprint swimming competition. Can. Med. Assoc. J. 88: 1251-1253, **1963**.

Hunt, J. N. The osmotic control of gastric emptying. Gastroenterology 41: 49-51, **1961**.

Hunt, J. N. and Pathak, J. D. The osmotic effects of some simple molecules and ions on gastric emptying. J. Physiol. (Lond.) 154: 254-269, **1960**.

Hunt, J. N. and Stubbs, D. F. The volume and energy content of meals as determinants of gastric emptying. J. Physiol. 245: 209-225, **1975**.

Hunter, W. M., Fonseka, C. C., and Passmore, R. The role of growth hormone in the mobilization of fuel for muscular exercise. Quart. J. Exp. Physiol. 50: 406-416, **1965**.

Hunyor, S. N., Flynn, J. M., and Cochineas, C. Comparison of performance of various sphygmomanometers with intra-arterial blood pressure readings. Brit. Med. J. 2: 159-162, **1978**.

Hupprich, F. L. and Sigerseth, P. O. The specificity of flexibility in girls. Res. Quart. 21: 25-33, **1950**.

Hurtado, A. Acclimatization to high altitudes. In: The Physiological Effects of High Altitude. Ed.: W. H. Weihe. New York: Macmillan, **1964**.

——. The influence of high altitude on physiology. In: High Altitude Physiology: Cardiac and Respiratory Aspects. Eds.: R. Porter and J. Knight. Edinburgh: Churchill-Livingstone, **1971**.

Hussain, R. and Patwardham, V. N. Iron content of thermal sweat in iron-deficiency anaemia. Lancet 1: 1073-1074, **1959**.

Hutchinson, R. C. Meal habits and their effects on performance. Nutr. Abstr. Rev. 22: 283-297, **1952**.

Hutt, C., Hutt, S. J., and Ounsted, C. A method for the study of children's behaviour. Dev. Med. Child Neurol. 5: 233-245, **1963**.

Hutten, H., Thews, G., and Vaupel, P. Some special problems concerning the oxygen supply to tissue as studied by an analogue computer, pp. 25-31. In: Oxygen Supply: Theoretical and Practical Aspects of Oxygen Supply and Microcirculation of Tissues. Eds.: M. Kessler, D. F. Bruley, L. C. Clark, D. W. Lübbers, I. A. Silver, and J. Strauss. Baltimore: University Park Press, **1973**.

Huxley, A. F. Muscle structure and theories of contraction. Progr. Biophys. 7: 255-318, **1957**.

——. Skeletal muscle. In: Muscle as a tissue, p. 3. Ed.: K. Rodahl and S. M. Horvath. New York: McGraw Hill, **1962**.

Huxley, A. F. and Niedergerke, R. Structural changes in muscle during contraction. Interference microscopy of living muscle fibres. Nature (Lond.) 173: 971-973, **1954**.

Huxley, A. F. and Taylor, R. E. Local activation of striated muscle fibres. J. Physiol. (Lond.) 44: 426-441, **1958**.

Huxley, H. E. Muscle cells. In: The Cell, Vol. 4, pp. 365-481. Eds.: J. Brachet and A. E. Mirsky. New York: Academic Press, **1960**.

——. The mechanism of muscular contraction. Science 164: 1356-1366, **1969**.

——. Structural changes in the actin- and myosin-containing filaments during contraction. Cold Spring Harbor Symp. Quart. Biol. 37: 361-376, **1972**.

Huxley, H. E. and Brown, W. The low-angle X-Ray diagram of vertebrate striated muscle and its behaviour during contraction and rigor. J. Mol. Biol. 30: 383-434, **1967**.

Hvorslev, C. M. Studien über die Bewegungen der Schulter. Skand. Arch. Physiol. 53: 1-136, **1928**.

Hyatt, R. Reaction to Dr. Shephard's paper Ventilatory mechanics during exercise in health and disease, pp. 19-30. In: Fitness and Exercise. Eds.: J. F. Alexander, R. F. Serfass, and C. M. Tipton. Chicago: Athletic Institute, **1972**.

Hyden, H. and Egyhazi, E. Nuclear RNA changes of nerve cells during a learning experiment in rats. Proc. Natl. Acad. Sci. U.S. 48: 1366-1373, **1962**.

Ianuzzo, C. D. and Chen, V. Effects of methandrostenolone on the acute compensatory growth of rat skeletal muscle, pp. 373-380. In: 3rd International Symposium on Biochemistry of Exercise. Eds.: F. Landry and W. A. R. Orban. Miami: Symposia Specialists, **1978**.

Iatradis, S. G. and Ferguson, J. H. Effect of physical exercise on blood clotting and fibrinolysis. J. Appl. Physiol. 18: 337-344, **1963**.

Iggo, A. Non-myelinated afferent fibres from mammalian skeletal muscle. J. Physiol. (Lond.) 155: 52p-53p, **1960**.

Ikai, M. The effects of training on muscular endurance, pp. 145-157. In: Proceedings of International Congress of Sports Sciences, 1964. Ed.: K. Kato. Tokyo: Japanese Union of Sports Sciences, **1966**.

——. Report to 10th Int. Congress of ICHPER, pp. 29-35. Cited by Asmussen, 1973. Vancouver, B.C., **1967**.

Ikai, M. and Fukunaga, T. Calculation of muscle strength per unit cross-sectional area of human muscle by means of ultrasonic measurement. Int. Z. Angew. Physiol. 26: 26-32, **1968**.

Ikai, M. and Steinhaus, A. H. Some factors modifying the expression of human strength. J. Appl. Physiol. 16: 157-163, **1961**.

Ikai, M., Yabe, K., and Ishii, K. Muskelkraft und Muskuläre Ermüdung bei willkurlicher Anspannung und electrischer Reizung des Muskels. Sportarzt und Sportmedizin 5: 197, **1967**.

Iliev, I. B. and Velvev, V. V. Physical performance and functional systolic cardiovascular murmurs in young athletes 11-14 years of age, pp. 357-362. In: Frontiers of Activity and Child Health. Eds.: H. Lavallée and R. J. Shephard. Springfield, Ill.: C. C. Thomas, **1977**.

Illsley, R. A., Finlayson, A., and Thompson, B. The motivation and characteristics of internal migrants. Millbank Memorial Quarterly Fund 41: 217-248, **1963**.

Ilmarinen, J. and Fardy, P. S. Physical activity intervention for males with high risk of coronary heart disease: A three year follow-up. Prev. Med. 6: 416-425, **1977**.

Inbar, O. and Bar-Or, O. Relationships of anaerobic and aerobic arm and leg capacities to swimming performance of 8-12 year old children, pp. 283-292. In: Frontiers of Activity and Child Health. Eds.: H. Lavallée and R. J. Shephard. Quebec City: Editions du Pélican, **1977**.

Inbar, O., Gutin, B., Dotan, R., and Bar-Or, O. Conditioning vs heat exposures as methods for acclimatizing 8-10 year old boys to dry heat. Med. Sci. Sport. 10: 62, **1978**.

Infante, A. A., Klanpiks, D., and Davies, R. D. Length, tension, and metabolism during short isometric contractions of frog sartorius muscles. Biochim. Biophys. Acta 88: 215-217, **1964**.

Ingelmark, B. E. Der Bau der Sehnen wahrend verschiedener Altersperioden unter wechselnden funktionellen Bedingungen. Acta Anat. 6: 113-140, **1948**.

Ingelmark, B. E. and Ekholm, R. A study on variations in the thickness of articular cartilage in association with rest and periodical load. Uppsala Läkareförenings Forhandlingar 53: 61, **1948**.

Ingham, A. G. and Smith, M. D. Social implications of the interactions between spectators and athletes. Exercise Sport Sci. Rev. 2: 189-224, **1974**.

Ingjer, F. and Brodal, P. Capillary supply of skeletal muscle fibers in untrained and endurance-trained women. Europ. J. Appl. Physiol. 38: 291-299, **1978**.

Ingle, D. J. and Nezamis, J. E. The effect of insulin on the tolerance of normal male rats to the overfeeding of a high carbohydrate diet. Endocrinology 40: 353-357, **1947**.

Ingram, W. R. Central autonomic mechanisms. In: Handbook of Physiology: Neurophysiology Vol. 2. Eds.: J. Field, H. W. Magoun, and V. E. Hall. Washington, D.C.: Amer. Physiol. Soc., **1960**.

Innes, J. A., Campbell, I. W., Campbell, C. J., Needle, A. L., and Munroe, J. F. Long-term follow-up of therapeutic starvation. Brit. Med. J. 2: 357-359, **1974**.

Ino, T. A new method for the measurement of venous return and its application. Regulation of the venous return and its regional distribution, a study of chief emphasis on the carotid sinus reflex. Jap. Circ. J. 24: 1297-1314, **1960**.

International Labour Organization. Maximum permissible weight to be carried by one worker. I.L.O. Report. Occupational Safety & Health Series 5, Geneva, **1964**.

Irvine, C. H. G. Thyroxine secretion rate in the horse in various physiological states. J. Endocrinol. 39: 313-320, **1967**.

——. Effect of exercise on thyroxine degradation in athletes and non-athletes. J. Clin. Endocrinol. 28: 942-948, **1968**.

Irving, D. W. and Yamamoto, T. Cigarette smoking and cardiac output. Brit. Heart J. 25: 126-132, **1963**.

Irving, L. Bradycardia in human divers. J. Appl. Physiol. 18: 489-491, **1963**.

——. Adaptations to cold. Sci. Amer. 214: 94-101, **1966**.

Isager, H. and Hagerup. L. Relationship between cigarette smoking and high packed-cell volume and haemoglobin levels. Scand. J. Haematol. 8: 241-244, **1971**.

Ishiko, T. Aerobic capacity and external criteria of performance. Can. Med. Assoc. J. 96: 746-749, **1967**.

——. Biomechanics of rowing, pp. 249-252. In: Biomechanics II. Eds.: J. Vredenbregt and J. Wartenweiler. Basel: Karger, **1971**.

Ismail-Beigi, F. and Edelman, I. S. Mechanism of thyroid calorigenesis: role of active sodium transport. Proc. Natl. Acad. Sci. U.S. 67: 1071-1078, **1970**.

Issekutz, B. Effect of exercise on the metabolism of plasma free fatty acids. In: Fat as a Tissue, Chapter 11. Eds.: K. Rodahl and B. Issekutz. New York: McGraw Hill, **1964**.

Issekutz, B., Birkhead, N. C., and Rodahl, K. Use of respiratory quotients in assessment of aerobic work capacity. J. Appl. Physiol. 17: 47-50, **1962**.

Issekutz, B. and Miller, H. Plasma free fatty acids during exercise and the effect of lactic acid. Proc. Soc. Exp. Biol. Med. 110: 237-239, **1962**.

Issekutz, B., Miller, H. I., and Rodahl, K. Lipid and carbohydrate metabolism during exercise. Fed. Proc. 25: 1415-1420, **1966**.

Issekutz, B. and Rodahl, K. Respiratory quotient during exercise. J. Appl. Physiol. 16: 606-610, **1961**.

Issekutz, B., Rodahl, K., and Birkhead, N. C. Effect of severe cold stress on the nitrogen balance of men under different dietary conditions. J. Nutr. 78: 189-197, **1962**.

Issekutz, B., Shaw, W. A. S., and Issekutz, T. B. Effect of lactate on FFA and glycerol turnover in resting and exercising dogs. J. Appl. Physiol. 39: 349-353, **1975**.

Issekutz, B., Shaw, W. A. S., and Issekutz, A. C. Lactate metabolism in resting and exercising dogs. J. Appl. Physiol. 40: 312-319, **1976**.

Itoh, S. Physiology of cold-adapted man. Medical Library Series 7, pp. 1-173. Sapporo, Japan: Hokkaido University School of Medicine, **1974**.

Ivanitsky, M. F. Cited by R. L. Larson. Growth of Bone and Joint Structures, pp. 33-59. In: Physical Activity. Human Growth and Development. Ed.: G. L. Rarick. New York: Academic Press, **1973**.

Ivy, J. L. The role of insulin during a glycogen loading process and its effect on adenosine 3-5 monophosphate levels of striated muscle. Med. Sci. Sport. 9: 49, **1977**.

Jablecki, C. K., Henser, J. E., and Kaufman, S. Autoradiographic localization of new RNA synthesis in hypertrophying skeletal muscle. J. Cell. Biol. 57: 743-759, **1973**.

Jack, J. J. and MacLennan, C. R. The lack of an electrical threshold discrimination between group 1a and group 1b fibres in the nerve to the cat peroneus longus muscle. J. Physiol. (Lond.) 212: 35p-36p, **1971**.

Jackson, H. Selected writings of John Hughlings Jackson. Vol. II. Ed.: J. Taylor. London: Hodder & Stoughton, **1932**.

Jackson, L. K., Simmons, R., Leinbach, R. C., Rosner, S. W., Presto, A. J., Weihrer, A. L., and Caceres, C. A. Noise reduction and representative complex selection in the computer analyzed exercise electrocardiogram, pp. 73-107. In: Measurement in Exercise Electrocardiography. The Ernst Simonson Conference. Ed.: H. Blackburn. Springfield, Ill.: C. C. Thomas, **1969.**

Jacobsen, E. Progressive Relaxation. Chicago:

University of Chicago Press, **1938**.

——. Innervation and tonus of striated muscle in man. J. Nerv. Ment. Dis. 97: 197–203, **1943**.

Jacquemin, C. H. and Varène, P. Etude de la bioénergétique ventilatoire au cours de l'exercice musculaire en altitude. Int. Z. Angew. Physiol. 24: 164–180, **1967**.

Jaeger, M. J. and Mattys, H. The pressure flow characteristics of the human airway. In: Airway Dynamics. Physiology and Pharmacology. Ed.: A. Bouhuys. Springfield, Ill.: C. C. Thomas, **1970**.

Jaffé, D. and Manning, M. Coronary arteries in early life. Proc. 13th Ann. Cong. Pediatrics, Vienna, **1971**.

Jalavisto, E. The role of simple tests measuring speed of performance in the assessment of biological vigour: A factorial study in elderly women, pp. 353–365. In: Behaviour, Aging and the Nervous System. Eds.: A. T. Welford and J. E. Birren. Springfield, Ill.: C. C. Thomas, **1965**.

James, W. E. and Patnoi, C. M. Instrumentation Review, pp. 7-1 to 7-46. In: Coronary Heart Disease. Prevention, Detection, Rehabilitation with Emphasis on Exercise Testing. Ed.: S. M. Fox. Denver, Colorado: International Medical Corporation, **1974**.

Jamieson, J. and Talbot, D. An examination of the effect of Vitamin E on the performance of highly trained swimmers. Can. J. Appl. Sport. Sci. 2: 67–70, **1977**.

Jamison, P. L. Growth of Wainwright Eskimos: Stature and weight. Arctic Anthropol. 7: 86–89, **1970**.

Jansen, J. K. S. and Matthews, P. B. C. The central control of the dynamic response of muscle spindle receptors. J. Physiol. (Lond.) 161: 357–378, **1962**.

Jansen, J. K. S. and Rudjord, T. On the silent period and Golgi tendon organs of the soleus muscle of the cat. Acta Physiol. Scand. 62: 364–379, **1964**.

Jansen, M. On Bone Formation: Its Relation to Tension and Pressure. New York: Longmans, Green, **1920**.

Jansson, E. and Kaijser, L. Muscle adaptation to extreme endurance training in man. Acta Physiol. Scand. 100: 315–324, **1977**.

Jansson, L., Johansson, K., Jonson, B., Olsson, L. G., Werner, O., and Westling, H. Computer assistance in the ecg laboratory: A new look. Scand. J. Clin. Lab. Invest. 36 (Suppl. 145): 1–43, **1976**.

Jarrett, A. S. Alveolar carbon dioxide tension at increased ambient pressures. J. Appl. Physiol. 21: 158–162, **1966**.

Javert, G. Role of the patient's activities in the occurrence of spontaneous abortion. Fert. Steril. 11: 550–558, **1960**.

Jean, C. F., Streeter, D. D., and Reichenbach, D. D. Fiber orientation in the normal and hypertensive cadaver left ventricle. Circulation 46 (Suppl.), 44 (Abstract), **1972**.

Jebavy, P. and Widimski, J. Lung-transfer factor at maximal effort in healthy men. Respiration 30: 297–310, **1973**.

Jelliffe, D. B. The Assessment of the Nutritional Status of the Community. Geneva: W.H.O. **1966**.

Jenkins, E. A. and Weightman, B. On the mechanism of fatigue in articular cartilage. J. Physiol. (Lond.) 278: 34p, **1978**.

Jenkins, P. J., Harper, R. W., and Nestel, P. J. Severity of coronary atherosclerosis related to lipoprotein concentration. Brit. Med. J. 2: 388–391, **1978**.

Jenkins, W. L., Maas, L. O., and Olson, M. W. Influence of inertia in making settings on a linear scale. J. Appl. Psychol. 36: 208–213, **1951**.

Jennings, D. B. and Macklin, R. D. The effects of O_2 and CO_2 and of ambient temperature on ventilatory patterns of dogs. Resp. Physiol. 16: 79–91, **1972**.

Jensen, C. Pertinent facts about warm-up, pp. 58–60. In: Towards an Understanding of Human Performance. Ed.: E. J. Burke. Ithaca, N.Y.: Mouvement Publications, **1977**.

Jernerus, R., Lundin, G., and Pugh, L. G. C. E. Solubility of acetylene in lung tissue as an error in cardiac output determinations with the acetylene method. Acta Physiol. Scand. 59: 1–6, **1963**.

Jervell, O. Investigation of the concentration of lactic acid in blood and urine. Acta Med. Scand. Suppl. 24: 1–135, **1928**.

Jessop, J. and Lippold, O. C. J. Altered synchronization of motor unit firing as a mechanism for long lasting increases in the tremor of human hand muscles following brief strong effort. J. Physiol. (Lond.) 269: 29p–30p, **1977**.

Jetté, M., Campbell, J., Mongeon, J., and Routhier, R. The Canadian Home Fitness Test as a Predictor of Aerobic Capacity. Can. Med. Assoc. J. 114: 680–682, **1976**.

Jetté, M. and Mongeon, J. Further evaluation of a field procedure for estimating ideal body weight in males. Can. J. Appl. Sport. Sci. 3: 186–187, **1978**.

Jick, H., Miettinen, O. S., Neff, R. K., Shapiro,

S., **Heinonen, O. P., and Slone, D.** Coffee and myocardial infarction. N. Engl. J. Med. 289: 63-67, **1973**.

Jirka, Z. The nutrition in top sport and vitamins. International Symposium on Athletes' Nutrition, Leningrad, **1975**.

Jöbsis, F. F. and O'Connor, M. J. Calcium release and reabsorption in the sartorius muscle of the toad. Biochem. Biophys. Res. Comm. 25: 246-252, **1966**.

Jöbsis, F. F. and Stainsby, W. N. Oxidation of N.A.D.H. during contractions of circulated mammalian skeletal muscle. Resp. Physiol. 4: 292-300, **1968**.

Joels, N. and White, H. The contribution of the arterial chemoreceptors to the stimulation of respiration by adrenaline and noradrenaline in the cat. J. Physiol. (Lond.) 197: 1-23, **1968**.

Johns, R. J. and Wright, V. Relative importance of various tissues in joint stiffness. J. Appl. Physiol. 17: 824-828, **1962**.

Johnson, B. L., Adamczyk, J. W., Tennøe, K. O., and Strømme, S. B. A comparison of concentric and eccentric muscle training. Med. Sci. Sport. 8: 35-38, **1976**.

Johnson, B. L. and Nelson, J. K. Effect of different motivational techniques during training and in testing upon strength performance. Res. Quart. 38: 630-636, **1967**.

Johnson, F. L. The association of androgenic-anabolic steroids and life-threatening disease. Med. Sci. Sport. 7: 284-286, **1975**.

Johnson, J. A., Cavert, H. M., Lifson, N., and Vissher, M. B. Permeability of the bladder to water studied by means of isotopes. Amer. J. Physiol. 165: 87-92, **1951**.

Johnson, J. M. Regulation of skin circulation during prolonged exercise. Ann. N.Y. Acad. Sci. 301: 195-212, **1977**.

Johnson, J. M., Rowell, L. B., and Brengelmann, G. L. Modification of the skin blood flow—body temperature relationship by upright exercise. J. Appl. Physiol. 37: 880-886, **1974**.

Johnson, L. C., Fisher, G., Silvester, L. J., and Hofhems, C. Anabolic steroid effects on strength, body weight, oxygen uptake and spermatogenesis upon mature males. Med. Sci. Sport. 4: 43-45, **1972**.

Johnson, L. C. and O'Shea, J. P. Anabolic steroid: Effects on strength development. Science 164: 957-959, **1969**.

Johnson, L. C., Roundy, E. S., Allsen, P. E., Fisher, A. G., and Silvester, L. G. Effect of anabolic steroid treatment on endurance. Med. Sci. Sport. 7: 287-289, **1975**.

Johnson, P. C. and Henrick, H. A. Metabolic and myogenic factors in local regulation of the microcirculation. Fed. Proc. 34: 2020-2024, **1975**.

Johnson, P. C., Leach, C. S., and Rambant, P. C. Estimates of fluid and energy balance of Apollo 17. Aerospace Med. 44: 1227-1230, **1973**.

Johnson, R. E., Darling, R. C., Sargent, F., and Robinson, P. Effects of variations in dietary Vitamin C on the physical well being of manual workers. J. Nutr. 29: 155-165, **1945**.

Johnson, R. E. and Edwards, H. T. Lactate and pyruvate in blood and urine after exercise. J. Biol. Chem. 118: 427-432, **1937**.

Johnson, R. H., Park, D. M., Rennie, M. J., and Sulaiman, W. R. Hormonal responses to exercise in racing cyclists. J. Physiol. (Lond.) 241: 23p-25p, **1974**.

Johnson, R. H., and Rennie, M. J. Changes in fat and carbohydrate metabolism caused by moderate exercise in patients with acromegaly. Clin. Sci. 44: 63-71, **1973**.

Johnson, R. L., Taylor, H. F., and Lawson, W. H. Maximal diffusing capacity of the lung for carbon monoxide. J. Clin. Invest. 44: 349-355, **1965**.

Johnson, R. L., Nicogossian, A. E., Bergman, S. A., and Hoffler, G. W. Lower body negative pressure: the second manned Skylab Mission. Aviation Space Environ. Med. 47: 347-353, **1976**.

Johnson, S. A., Roff, R. A., Greenleaf, J. F., Ritman, E. L., Lee, S. L., Herman, G. T., Sturm, R. E., and Wood, E. H. The problem of accurate measurement of left ventricular shape and dimensions from multiplane roentgenographic data. Europ. J. Cardiol. 1: 241-258, **1974**.

Johnson, T. H. and Tharp, G. D. The effect of chronic exercise on reserpine-induced gastric ulceration in rats. Med. Sci. Sport. 6: 188-190, **1974**.

Johnson, W. H., Stubbs, R. A., Kelk, G. F., and Franks, W. R. Stimulus required to produce motion sickness. Restriction of head movements as a preventative of airsickness-field studies in airborne troops. J. Aviation Med. 22: 365-374, **1951**.

Johnson, W. R. Hypnosis and muscular performance. J. Sports Med. Phys. Fitness 1: 71-79, **1961**.

Jokl, E. The clinical physiology of physical fitness and rehabilitation. Springfield, Ill.: C. C. Thomas, **1958**.

——. Indisposition after running. In: Exercise

and Altitude. Eds.: E. Jokl and P. Jokl. Basel: Karger, **1968**.

——. Presentation to World Congress of Sports Medicine, Brasilia, **1978**.

Jokl, E. and Jokl, P. The Physiological Basis of Athletic Records. Springfield, Ill.: C. C. Thomas, **1968a**.

——. The effect of altitude on athletic performance. In: Exercise and Altitude. Eds.: E. Jokl and P. Jokl. Basel: Karger, **1968b**.

Jokl, E. and Jokl, P., Eds. Exercise and Altitude. Basel: Karger, **1968c**.

Jokl, E., Jokl-Ball, M., Jokl, P., and Frankel, L. Notation of exercise, pp. 2–18. In: Medicine and Sport, Vol. 4. Physical Activity and Aging. Eds.: D. Brunner and E. Jokl. Basel: Karger, **1970**.

Joliffe, N., Rinzler, S. N., and Archer, M. The anti-coronary club, including a discussion of the effects of a prudent diet on the serum cholesterol of middle-aged men. Amer. J. Clin. Nutr. 7: 451–462, **1959**.

Jolly, W. A. On the time relations of the knee jerk and simple reflexes. Quart. J. Exp. Physiol. 4: 67–87, **1911**.

Jones, B. Is proprioception important for skilled performance? J. Motor Behav. 6: 33–45, **1974**.

Jones, D. Safe Lifting. Toronto: Canadian Occupational Safety Association, **1969**.

Jones, D. L., Veale, W. L., and Cooper, K. E. Thermoregulation in the conscious rabbit at various ambient temperatures: Effects of central administration of sympathomimetics, p. 359. Proceedings of the International Union of Physiological Sciences. Vol. 13, **1977**.

Jones, D. S., Beargie, R. J., and Pauly, J. E. An electromyographic study of some muscles of costal respiration in man. Anat. Rec. 117: 17–24, **1953**.

Jones, G. M. and Milsum, J. H. Frequency-response analysis of central vestibular unit activity resulting from rotational stimulation of the semicircular canals. J. Physiol. (Lond.) 219: 191–216, **1971**.

Jones, G. M. and Watt, D. G. D. Observations on the control of stepping and hopping movements in man. J. Physiol. (Lond.) 219: 709–737, **1971**.

Jones, H. E. Motor Performance and Growth. A Developmental Study of Static Dynamometric Strength. Berkeley, Calif.: Univ. of Calif. Press, **1949**.

Jones, N. L., Campbell, E. J. M., Edwards, R. H. T., and Wilkoff, G. Alveolar to blood PCO_2 difference during rebreathing in exercise. J. Appl. Physiol. 27: 356-360, **1969**.

Jones, N. L., Campbell, E. J. M., Edwards, R. H. T., and Robertson, D. G. Clinical Exercise Testing. Philadelphia: W. B. Saunders, **1975**.

Jones, N. L., Campbell, E. J. M., McHardy, G. J. R., Higgs, B. E., and Clode, M. The estimation of carbon dioxide pressure of mixed venous blood during exercise. Clin. Sci. 32: 311-327, **1967**.

Jones, N. L. and Haddon, R. W. T. Effect of a meal on cardiopulmonary and metabolic changes during exercise. Can. J. Physiol. Pharm. 51: 445–450, **1973**.

Jones, N. L., McHardy, G. J. R., Naimark, A., and Campbell, E. J. M. Physiological dead space and alveolar-arterial gas pressure differences during exercise. Clin. Sci. 31: 19–29, **1966**.

Jones, W. B., Finchum, R. N., Russell, R. O., and Reeves, T. J. Transient cardiac output responses to multiple levels of supine exercise. J. Appl. Physiol. 28: 183–189, **1970**.

Jordan, D. B., Zoneraich, S., Rhee, J., and Zoneraich, O. Systolic time intervals of endurance athletes. Med. Sci. Sport. 10: 49, **1978**.

Jorfeldt, L. Turnover of ^{14}C-L(+)-lactate in human skeletal muscle during exercise, pp. 409–417. In: Muscle Metabolism During Exercise. Eds.: B. Pernow and B. Saltin. New York: Plenum Press, **1971**.

Jorfeldt, L., Juhlin-Dannfelt, A., and Karlsson, J. Lactate release in relation to tissue lactate in human skeletal muscle during exercise. J. Appl. Physiol. Resp. Environ. Exer. Physiol. 44: 350-352, **1978**.

Jorfeldt, L. and Wahren, J. Human forearm muscle metabolism during exercise. V. Quantitative aspects of glucose uptake and lactate production during prolonged exercise. Clin. Sci. 41: 459–473, **1970**.

Jorgensen, C. R. Coronary blood flow and myocardial oxygen consumption in man, pp. 39–50. In: Physiology of Fitness and Exercise. Eds.: J. F. Alexander, R. C. Serfass, and C. M. Tipton. Chicago: Athletic Institute, **1972**.

Jorgensen, C. R., Gobel, F. L., Taylor, H. L., and Wang, Y. Myocardial blood flow and oxygen consumption during exercise. Ann. N.Y. Acad. Sci. 301: 213–223, **1977**.

Jorgensen, K. Back muscle strength and body weight as limiting factors for work in the standing slightly-stooped position. Hellerup, Denmark: Comm. Danish. Natl. Assoc. for Infantile Paralysis 30: 1–9, **1970**.

Josenhans, W. T. Commentary. In: Proc. Int. Symp. on Physical Activity and Cardiovascular Health. Can. Med. Assoc. J. 96: 713, **1967a**.

——. Muscle testing by accelerometry. A preliminary study. Int. Z. Angew. Physiol. 24: 121–128, **1967b**.

Joseph, J. Man's Posture. Electromyographic Studies. Springfield, Ill.: C. C. Thomas, **1969**.

Joyce, J. C. and Rack, P. M. H. The effects of load and force on tremor at the normal human elbow joint. J. Physiol. (Lond.) 240: 375–396, **1974**.

Joyce, J. C., Rack, P. M. H., and Ross, H. F. The forces generated at the human elbow joint in response to imposed sinusoidal movements of the forearm. J. Physiol. (Lond.) 240: 351–374, **1974**.

Joyce, J. F. The relationship of selected heartometer measurements to oxygen pulse. J. Sport. Med. Phys. Fitness, 17: 131–138, **1977**.

Joye, H. and Poortmans, J. Hematocrit and serum proteins during arm exercise. Med. Sci. Sport. 2: 187–190. **1970**.

Jude, J. R., Kouwenhaven, W. B., and Knickerbocker, G. G. External cardiac resuscitation. In: Monographs in the Surgical Sciences, 1: 59–117. Baltimore: Williams & Wilkins, **1964**.

Julius, S., Amery, A., Whitlock, L. S., and Conway, J. Influence of age on the hemodynamic response to exercise. Circulation 36: 222–230, **1967**.

Jungmann, H. Studies on the course and duration of acclimatization to an altitude of 2,000 m (6,562 ft.). In: The International Symposium on the Effects of Altitude on Physical Performance. Ed.: R. Goddard. Chicago: Athletic Institute, **1967**.

Junod, V. T. Récherches physiologiques et thérapeutiques sur les éffects de la compression et de la raréfaction de l'air, tant sur le corps que sur les membres isolés. Paris: Libraire de Devill-Cavellin, **1834**.

Jurkowski, J. E., Sutton, J. R., Kearn, P. M., and Viol, G. W. Plasma renin activity and plasma aldosterone during exercise in relation to the menstrual cycle. Med. Sci. Sport. 10: 41, **1978**.

Kachadorian, W. A. The effects of activity on renal function, pp. 97–116. In: Fitness and Exercise. Eds.: J. F. Alexander, R. C. Serfass, and C. M. Tipton. Chicago: Athletic Institute, **1972**.

Kachadorian, W. A. and Johnson, R. E. Renal responses to various rates of exercise. J. Appl. Physiol. 28: 748–752, **1970**.

Kachadorian, W. A., Johnson, R. E., Buffington, R. E., Lawler, L., Serbin, J. J., and Woodall, T. The regularity of "athletic pseudonephritis" after heavy exercise. Med. Sci. Sport. 2: 142–145, **1970**.

Kagaya, A. and Ikai, M. Res. J. Phys. Ed. 14: 127, **1970**. Cited by Asmussen (1973).

Kahn, H. The relationship of reported coronary heart disease mortality to physical activity of work. Amer. J. Pub. Health 53: 1058–1067, **1963**.

Kaijser, L. Limiting factors for aerobic muscle performance. Acta Physiol. Scand. Suppl. 346: 1–96, **1970**.

Kaijser, L. and Jansson, E. Effect of extreme endurance training on muscle fibre characteristics. Proc. Int. Union Physiol. Sci. II, 723 (Abstr.), **1977**.

Kaijser, L., Lassers, B. W., Wahlqvist, M. L., and Carlson, L. A. Myocardial lipid and carbohydrate metabolism in fasting men during prolonged exercise. J. Appl. Physiol. 32: 847–858, **1972**.

Kallfelz, I. Die apnoische Pause. Untersuchungen über ihre Beziehungen zum Sauerstoffmangel-Belastungstest. Zentrale für Wissenschaftliches Berichtowesen der Deutschen Versuchanstalt für Luft und Raumfahrt e.v. Porz-Wahn, Rheinland, W. Germany, **1962**.

Kaltreider, N. L. and Meneely, C. R. The effect of exercise on the volume of the blood. J. Clin. Invest. 19: 627–634, **1940**.

Kamon, E. and Pandolf, K. B. Maximal aerobic power during ladder-mill climbing, uphill running and cycling. J. Appl. Physiol. 32: 467–473, **1972**.

Kanda, K., Burke, R. E., and Walmsley, B. Differential control of fast and slow twitch motor units in the decerebrate cat. Exp. Brain Res. 29: 57–74, **1977**.

Kanitz, M. The effects of a seventeen month's training program on the strength and flexibility of adult males, pp. 31–44. In: Physical Exercise and Activity for the Aging. Natanya, Israel: Wingate Institute, **1975**.

Kannel, W. B., Sorlie, P., and McNamara, P. The relationship of physical activity to risk of coronary heart disease: The Framingham study. In: Coronary Heart Disease and Physical Fitness. Eds.: O. A. Larsen and R. O. Malmborg. Baltimore: University Park Press, **1971**.

Kantowitz, B. H. Interference in short-term memory: Interpolated task difficulty. J. Exp. Psychol. 95: 264–274, **1972**.

Kao, F. F. An experimental study of the pathways involved in exercise hyperpnoea employing cross-circulation techniques, pp. 461. In: The Regulation of Human Respiration. Eds.: D. J. C. Cunningham and B. B. Lloyd. Oxford: Blackwell Scientific, **1963**.

——. An Introduction to Respiratory Physiology. Amsterdam: Excerpta Medica, **1972**.

——. The peripheral neurogenic drive: An experimental study, pp. 71–88. In: Muscular Exercise and the Lung Eds.: J. A. Dempsey and C. E. Reed. Madison: University of Wisconsin Press, **1977**.

Kaplinsky, E., Hood, W. B., McCarthy, G., McCombs, H. L., and Lown, B. Effects of physical training in dogs with coronary artery ligation. Circulation 37: 556–565, **1968**.

Kappagoda, C. T., Linden, R. J., and Newell, J. P. Validation of a submaximal exercise test. J. Physiol. (Lond.) 268: 19p–20p, **1977**.

Karczewski, W. A. Organization of the brain stem respiratory complex, pp. 197–219. In: Organisation of the Brain Stem Respiratory Complex. Ed.: J. G. Widdicombe. London: Butterworths, **1974**.

Kärki, N. T. The urinary excretion of noradrenaline and adrenaline in different age groups, its diurnal variation and the effect of muscular work on it. Acta Physiol. Scand. 39 (Suppl. 132): 1–96, **1956**.

Karlson, P. In: Kurzes Lehrbuch der Biochemie für Mediziner und Naturwissenschaftler. Stuttgart: Thieme, **1966**.

Karlsson, J. Lactate and phosphagen concentrations in working muscles of man with special reference to oxygen deficit at the onset of work. Acta Physiol. Scand. Suppl. 358, **1971a**.

——. Muscle ATP, CP and lactate in submaximal and maximal exercise, pp. 383–393. In: Muscle Metabolism During Exercise. Eds.: B. Pernow and B. Saltin. New York: Plenum Press, **1971b**.

Karlsson, J., Åstrand, P. O., and Ekblom, B. Training of the oxygen transport system in man. J. Appl. Physiol. 22: 1061–1065, **1967**.

Karlsson, J., Diamant, B., and Saltin, B. Lactate dehydrogenase activity in muscle after prolonged severe exercise in man. J. Appl. Physiol. 25: 88–91, **1968**.

——. Muscle metabolites during submaximal and maximal exercise in man. Scand. J. Clin. Lab. Invest. 26: 385–394, **1971**.

Karlsson, J., Funderburk, C.F. Essén, B., and Lind, A. R. Constituents of human muscle in isometric fatigue. J. Appl. Physiol. 38: 208–211, **1975**.

Karnovsky, M. J. Morphology of capillaries with special reference to muscle capillaries, pp. 341–350. Alfred Benzon Symposium on Capillary Permeability, Copenhagen, **1969**.

Karp, J. E. and Bell, W. R. Fibrinogen-fibrin degradation products and fibrinolysis following exercise in humans. Amer. J. Physiol. 227: 1212–1215, **1974**.

Karpovich, P. V. Water resistance in swimming. Res. Quart.4: 21–28, **1933**.

——. The effect of oxygen inhalation on swimming performance. Res. Quart. 5: 24–30, **1934**.

——. Warm-up, pp. 166–171. In: Encyclopedia of Sports Sciences and Medicine. New York: Macmillan, **1971**.

Karpovich, P. V. and Millman, N. Energy expenditure in swimming. Amer. J. Physiol. 142: 140–144, **1944**.

Karpovich, P. V. and Pestrecov, K. Mechanical work and efficiency in swimming. Arbeitsphysiol. 10: 504–514, **1939**.

——. Effect of gelatin upon muscular work in man. Amer. J. Physiol. 134: 300–309, **1941**.

Karrasch, K. and Müller, E. A. Das Verhalten der Pulsfrequenz in der Erholungsperiode nach korperliche Arbeit. Arbeitsphysiol. 14: 369–382, **1951**.

Karvonen, M. J. Epidemiology of atherosclerosis and risk factor identification in the asymptomatic population, pp. 11–22. In: Coronary Heart Disease, Exercise Testing and Cardiac Rehabilitation. Eds.: W. E. James and E. A. Amsterdam. Miami: Symposia Specialists, **1977**.

Karvonen, M. J., Kentala, E., and Mustala, O. The effects of training on heart rate. A "longitudinal" study. Ann. Med. Exp. Fenn. 35: 307–315, **1957**.

Karvonen, M. J., Klemola, H., Virkajarvi, J., and Kekkonen, A. Longevity of endurance skiers. Med. Sci. Sport. 6: 49–51, **1974**.

Karvonen, M. J., Saarela, J., and Votila, E. The effect of repeated exercise on appetite in a 24-hour relay race, pp. 139–143. In: Nutrition, Dietetics and Sport. Eds.: G. Ricci and A. Venerando. Torino: Minerva Medica, **1978**.

Kasser, I. S. and Bruce, R. A. Comparative effects

of aging and coronary heart disease on submaximal and maximal exercise. Circulation 39: 759–774, **1969**.

Katch, F. I., McArdle, W. D., Czula, R., and Pechar, G. S. Maximal oxygen intake, endurance running performance and body composition in college women. Res. Quart. 44: 301–312, **1973**.

Katch, F. I. and Michael, E. D. Body composition of high school wrestlers according to age and wrestling weight category. Med. Sci. Sport. 3: 190–194, **1971**.

Katch, V., Weltman, A., Martin, R., and Gray, L. Optimal test characteristics for maximal anaerobic work on the bicycle ergometer. Res. Quart. 48: 319–327, **1977**.

Katch, V. L. The role of maximum oxygen intake in endurance performance. Paper presented at 1970 AAHPER Convention, Seattle, Wash., **1970**.

Katch, V. L. and Michael, E. D. The relationship between various segmental leg measurements, leg strength and relative endurance performance of college females. Human Biol. 45: 371–383, **1973**.

Katsuki, S. Some considerations on the effect of physical conditioning in middle and older age. Bull. Phys. Fitness Res. Inst. Tokyo, 23: 1–11, **1972**.

Kattus, A. and Grollman, J. Patterns of coronary collateral circulation in angina pectoris: relation to exercise training. In: Changing Concepts in Cardiovascular Disease. Eds.: H. I. Russek and B. L. Zohman. Baltimore: Williams & Wilkins, **1972**.

Katz, A. M. Contractile protein of the heart. Physiol. Rev. 50: 63–158, **1970**.

——. Effects of ischemia on the contractile process of heart muscle. Amer. J. Cardiol. 32: 456–460, **1973**.

——. Biochemical basis for cardiac contraction, pp. 67–86. In: Cardiac Mechanics: Physiological, Clinical and Mathematical Considerations. Eds.: I. Mirsky, D. N. Ghista, and H. Sandler. New York: Wiley, **1974**.

——. Congestive heart failure: Role of altered myocardial cellular control. N. Engl. J. Med. 293: 1184–1191, **1975**.

——. Physiology of the Heart, pp. 1–433. New York: Raven Press, **1977**.

Katz, A. M., Tada, M., and Kirchberger, M. A. Control of calcium transport in the myocardium by the cyclic AMP-protein kinase system. Adv. Cyclic Nucleotide Res. 5: 453–472, **1975**.

Katz, B. and Miledi, R. The statistical nature of the acetylcholine potential and its molecular components. J. Physiol. (Lond.) 224: 665–699, **1972**.

Katz, B. A. Accidental hypothermia in man. Can. Fam. Phys. 20: 56–58, **1974**.

Kaufman, D. A. and Ware, W. B. Effect of warm-up and recovery techniques on repeated running endurance. Res. Quart. 48: 328–332, **1977**.

Kaufman, W. C. and Marbarger, J. P. Pressure breathing: Functional circulatory changes in the dog. J. Appl. Physiol. 9: 33–37, **1956**.

Kavanagh, T. A cold-weather "jogging mask" for angina patients. Can. Med. Assoc. J. 103: 1290–1291, **1970**.

——. Heart Attack? Counter-attack! Toronto: Van Nostrand, **1976**.

Kavanagh, T. and Shephard, R. J. The immediate antecedents of myocardial infarction in active men. Can. Med. Assoc. J. 109: 19–22, **1973**.

——. Maintenance of hydration in post-coronary marathon runners. Brit. J. Sport. Med. 9: 130–135, **1975a**.

——. Conditioning of post-coronary patients: Comparison of continuous and interval training. Arch. Phys. Med. Rehabil. 56: 72–76, **1975b**.

——. Maximum exercise tests on post-coronary patients. J. Appl. Physiol. 40: 611–618, **1976**.

——. On the choice of fluid for the hydration of middle-aged marathon runners. Brit. J. Sport. Med. 11: 26–35, **1977a**.

——. Fluid and mineral needs of post-coronary distance runners, pp. 143–151. In: Sports Medicine. Eds.: F. Landry and W. A. R. Orban. Miami: Symposia Specialists, **1978**.

——. The effects of continued training on the aging process. Ann. N.Y. Acad. Sci. 301: 656–670, **1977b**.

——. **Doney, H., and Pandit, V.** Intensive exercise in coronary rehabilitation. Med. Sci. Sport. 5: 34–39, **1973**.

Kavanagh, T., Shephard, R. J., and Kennedy, J. Characteristics of postcoronary marathon runners. Ann. N.Y. Acad. Sci. 301: 455–465, **1977**.

Kavanagh, T. Shephard, R. J., and Pandit, V. Marathon running after myocardial infarction. J. Amer. Med. Assoc. 229: 1602–1605, **1974**.

Kawahata, A. Sex differences in sweating, pp. 169–184. In: Essential Problems in Climatic Physiology. Eds.: H. Yoshimura, K. Ogata, and S. Itoh. Kyoto, Japan: Nakado, **1960**.

Kawashiro, T., Piiper, J., and Scheid, P. Dependence of O_2 uptake on surface P_{O_2} in intact, excised skeletal muscle of the rat: Validity of the Warburg model. J. Physiol. (Lond.) 284: 45p-46p, **1978**.

Kawashiro, T., Sikand, R. S., Adaro, F., Takahashi, H., and Piiper, J. Study of intrapulmonary gas mixing in man by simultaneous wash-out of helium and sulfur hexafluoride. Resp. Physiol. 28: 261-275, **1976**.

Kay, C. and Shephard, R. J. On muscle strength and the threshold of anaerobic work. Int. Z. Angew. Physiol. 27: 311-328, **1969**.

Kay, J. D., Petersen, E. S., and Christensen, H. V. Breathing pattern in man during bicycle exercise at different pedalling frequencies. J. Physiol. (Lond.) 241: 123p-124p, **1974**.

Kazarian, L. E. and Von Gerke, H. E. Bone loss as a result of immobilization and chelation. Clin. Orthoped. Rel. Research 65: 67-75, **1969**.

Kearney, J. T. and Byrnes, W. C. Relationship between running performance and predicted maximum oxygen uptake among divergent ability groups. Res. Quart. 45: 9-15, **1974**.

Keatinge, W. R. The effect of work and clothing on the maintenance of body temperature in water. Quart. J. Exp. Physiol. 46: 69-82, **1961**.

Keele, C. A. and Neil, E. Samson Wright's Applied Physiology, 12th ed. Oxford Univ. Press, **1971**.

Keele, S. W. and Ellis, J. B. Memory characteristics of kinesthetic information. J. Motor Behav. 4: 127-134, **1972**.

Keller, A. D. Ablation and stimulation of the hypothalamus: Circulatory effects. Physiol. Rev. 40 (Suppl. 4): 116-135, **1960**.

Keller, H. and Bühlmann, A. A. Deep diving and short decompression by breathing mixed gases. J. Appl. Physiol. 20: 1267-1270, **1965**.

Kellerman, J. J., Ben-Ari, E., Chayet, M., Lapidot, C., Drory, Y., and Fisman, E. Cardiocirculatory response to different types of training in patients with angina pectoris. Cardiology 62: 218-231, **1977**.

Kellerth, J. O. Aspects on the relative significance of pre- and post- synaptic inhibition in the spinal cord, pp. 197-212. In: Structure and Function of Inhibitory Synapses. Eds.: C. Von Euler, S. Skoglund, and U. Söderberg. Oxford: Pergamon Press, **1968**.

Kellogg, R. H. The role of CO_2 in altitude acclimatization, pp. 379-395. In: The Regulation of Human Respiration. Eds.: D. J. C. Cunningham and B. B. Lloyd. Oxford: Blackwell Scientific, **1963**.

Kelton, I. W. and Wright, R. D. The mechanism of easy standing by man. Austr. J. Exp. Biol. Med. Sci. 27: 505-515, **1949**.

Kemp, J. M. and Powell, T. P. S. The connexions of the striatum and globus pallidus: Synthesis and speculation. Phil. Trans. Roy. Soc. B 262: 441-457, **1971**.

Kemp, R. G. Inhibition of muscle pyruvate kinase by creatine phosphate. J. Biol. Chem. 248: 3963-3967, **1973**.

Kemper, H. C. G., Verschuur, R., Clarys, J. P., Jiskoot, J., and Rijken, H. Efficiency in swimming the front crawl, pp. 243-249. In: Biomechanics V. Ed.: P. V. Komi. Baltimore: University Park Press, **1976**.

Kendrick, Z. V., Pollock, M. L., Hickman, T. N., and Miller, H. S. Effects of training and detraining on cardiovascular efficiency. Amer. Corr. Ther. J. 25: 79-83, **1971**.

Kennedy, G. C. The role of depot fat in the hypothalamic control of food intake in the rat. Proc. Roy. Soc. Lond. B 140: 578-596, **1953**.

Kennedy, J. L. and Travis, R. C. Prediction of speed at performance by muscle action potentials. Science 105: 410-411, **1947**.

Kennedy, J. W., Baxley, W. A., Figley, M. M., Dodge, H. T., and Blackmon, J. R. Quantitative angiography. 1. The normal left ventricle in man. Circulation 34: 272-278, **1966**.

Kent, R. M. and Atha, J. Intracycle retarding force fluctuations in breast-stroke. J. Sport. Med. 13: 274-281, **1973**.

Kentala, E. Physical fitness and feasibility of physical rehabilitation after myocardial infarction in men of working age. Ann. Clin. Res. 4 (Suppl. 9): 1-84, **1972**.

Kernell, D. Motoneurone properties and motor control, pp., 19-28. In: Control of Posture and Locomotion. Eds.: R. B. Stein, K. G. Pearson, R. S. Smith, and J. B. Redford.New York: Plenum Press, **1973**.

Kernell, D. and Sjöholm, H. Repetitive impulse firing: Comparisons between neurone models based on "voltage clamp equations" and spinal motoneurones. Acta Physiol. Scand. 87: 40-56, **1973**.

Keroes, J., Ecker, R. R., and Rapaport, E. Ventricular function curves in the exercising dog: effects of rapid intravenous infusions and of propranolol. Circ. Res. 25: 557-567, **1969**.

Kerslake, D. Mc K. Errors arising from the use of mean heat exchange coefficients in calculation of the heat exchanges of a cylindrical body in a transverse wind, p. 183. In: Temperature: Its Measurement and Control in Science and Industry. Vol. 3, Part 3. Ed.: J. D. Hardy. New York: Reinhold, **1963**.

——. The effects of thermal stress on the human body. In: A Textbook of Aviation Physiology. Ed.: J. A. Gillies. Oxford: Pergamon Press, **1965**.

——. Monographs of the Physiological Society: The Stress of Hot Environment. London: Cambridge University Press, **1972**.

Kety, S. S. Theory and application of exchange of inert gas at lungs and tissues. Pharmacol. Rev. 3: 1-41, **1951**.

Keul, J. Muscle metabolism during long lasting exercise, pp. 31-42. In: Metabolic Adaptation to Prolonged Physical Exercise. Eds.: H. Howald and J. R. Poortmans. Basel: Birkhauser Verlag, **1975**.

Keul, J. and Doll, E. Limiting Factors of Physical Performance. Stuttgart: Thieme, **1973**.

Keul, J., Doll, E., and Keppler, D. Energy Metabolism of Human Muscle (translation by J. Skinner). Baltimore: University Park Press, **1972**.

Keul, J., Doll, E., Limon-Lason, R., Merz, P., and Reindell, H. II. Der Einfluss körperlicher Arbeit auf die arteriellen Glukose-, Lactat-und Pyruvatspiegel in Freiburg und in Mexico City. Sportarzt und Sportmedizin 8: 327-334, **1967**.

Keul, J., Keppler, D., and Doll, E. Lactate-pyruvate ratio and its relation to oxygen pressure in arterial, coronary venous and femoral venous blood. Arch. Int. Physiol. Biochem. 75: 573-578, **1967**.

Keul, J., Doll, E., Steim, H., Fleer, U., and Reindell, H. Über den Stoffwechsel des Herzens bei Hochleistungs—sportlern. III. Der oxydative Stoffwechsel des trainierten menschlichen Herzens unter verschiedenen Arbeitsbedingungen. Z. Kreislaufforsch. 55: 477-488, **1966**.

Kew, M. C., Abrahams, C., Levin, N. W., Seftel, H. C., Rubenstein, A. H., and Bersohn, I. The effects of heat stroke on the function and structure of the kidney. Quart. J. Med. 36: 277-300, **1967**.

Kew, M. C., Abrahams, C., and Seftel, H. C. Chronic interstitial nephritis as a consequence of heat stroke. Quart. J. Med. 39: 189-199, **1970**.

Kew, M. C., Bersohn, I., Seftel, H. C., and Kent, K. G. Liver damage in heat stroke. Amer. J. Med. 49: 192-202, **1970**.

Kew, M. C., Tucker, R. B. K., Bersohn, L., and Seftel, H. C. The heart in heat stroke. Amer. Heart J. 77: 324-335, **1969**.

Keys, A. Physical performance in relation to diet. Fed. Proc. 2: 164-187, **1943**.

——. Coronary heart disease in seven countries. Amer. Heart Assoc. Monograph 29: 162-183, **1970**.

Keys, A., Aravanis, C., Blackburn, H., Van Buchem, F. S. P., Buzina, R., Djordjevic, B. S., Fidanza, F., Karvonen, M. J., Menotti, A., Puddu, V., and Taylor, H. L. Coronary heart disease: Overweight and obesity. Ann. Int. Med. 77: 15-27, **1972**.

Keys, A., Brozek, J., Henschel, A., Michelsen, O., and Taylor, H. L. The Biology of Human Starvation. Minneapolis: Univ. of Minnesota Press, **1950**.

Keys, A. and Henschel, A. F. Vitamin supplementation of U.S. Army rations in relation to fatigue and the ability to do muscular work. J. Nutr. 23: 259-269, **1942**.

Keys, A., Henschel, A. F., Mickelson, O., and Brozek, J. M. The performance of normal young men on controlled thiamine intakes. J. Nutr. 26: 399-415, **1943**.

Keys, A., Henschel, A. F., Mickelson, O., Brozek, J. M., and Crawford, J. H. Physiological and biochemical functions in normal young men on a diet restricted in riboflavin. J. Nutr. 27: 165-178, **1944**.

Keys, A., Henschel, A. F., Taylor, H. L., Mickelsen, O., and Brozek, J. M. Experimental studies on man with a restricted intake of the B vitamins. Amer. J. Physiol. 144: 5-42, **1945**.

Keys, A. and Taylor, H. The behavior of the plasma colloids in recovery from brief severe work and the question as to the permeability of the capillaries to proteins. J. Biol. Chem. 109: 55-67, **1935**.

Kibler, R. F., Taylor, W. J., and Meyers, J. D. The effect of glucagon on net splanchnic balance of glucose, amino acid, nitrogen, urea, ketones and oxygen in man. J. Clin. Invest. 43: 904-915, **1964**.

Kidd, D. J. and Elliott, D. H. Clinical manifestations and treatment of decompression sickness in divers, pp. 464-490. In: Physiology and Medicine of Diving. Eds.: P. B. Bennett and D. H. Elliott. London: Baillière, Tindall & Cassell, **1969**.

Kidd, D. J. and Stubbs, R. A. The use of the pneumatic analogue computer for divers, pp. 386-413. In: Physiology and Medicine of Diving. Eds.: P. B. Bennett and D. H. Elliott.

London: Baillière, Tindall & Cassell, **1969**.

Kiessling, K. H., Pilström, L., Karlsson, J., and Piehl, K. Mitochondrial volume in skeletal muscle from young and old physically untrained and trained healthy men and from alcoholics. Clin. Sci. 44: 547–554, **1973**.

Kiessling, K. H., Pilström, L., Bylund, A. C. H., Saltin, B., and Piehl, K. Morphometry and enzyme activities in skeletal muscle from middle-aged men after training and from alcoholics, pp. 384–389. In: Metabolic Adaptations to Prolonged Exercise. Eds.: H. Howald and J. R. Poortmans. Basel: Birkhauser Verlag, **1975**.

Kiessling, R. J. and Maag, C. H. Performance impairment as a function of nitrogen narcosis. U.S. Navy Experimental Diving Unit Research Rep. 3-60: 1–19, **1960**.

Kievit, J. and Knypers, H. G. J. M. Sub-cortical afferents to the frontol lobe in the rhesus monkey studied by means of retrograde horseradish peroxidase transport. Brain Res. 85: 261–266, **1975**.

Kiiskinen, A. and Heikkinen, E. Effects of physical training on development and strength of tendons and bones in growing mice. Scand. J. Clin. Lab. Invest. 29 (Suppl. 123): 20, **1973**.

——. Effect of prolonged physical training on the development of connective tissues in growing mice, pp. 253–261. In: Metabolic Adaptation to Prolonged Physical Exercise. Eds.: H. Howald and J. R. Poortmans. Basel: Birkhauser Verlag, **1975**.

Kiiskinen, A., Kemppinen, L., and Hasen, J. Is the anemia following repeated sessions of heavy physical exercise caused by a decreased rate of production of red cells? In: Metabolic Adaptation to Prolonged Physical Exercise. Eds.: H. Howald and J. R. Poortmans. Basel: Birkhauser Verlag, **1975**.

Kilbom, Å. Physical training in women. Scand. J. Clin. Lab. Invest. 28 (Suppl. 119): 1–34, **1971**.

——. Circulatory adaptations during static muscular exercise. Scand. J. Work Environ. Health 2: 1–13, **1976**.

Kilbom, Å. and Åstrand, I. Physical training with submaximal intensities in women. II. Effect on cardiac output. Scand. J. Clin. Lab. Invest. 28: 163–175, **1971**.

Kilbom, Å. and Brundin, T. Circulatory effects of isometric muscle contractions, performed separately and in combination with dynamic exercise. Europ. J. Appl. Physiol. 36: 7–17, **1976**.

Kilbom, Å., Hartley, L. H., Saltin, B., Bjure, J., Grimby, G., and Åstrand, I. Physical training in sedentary middle-aged and older men. 1. Medical evaluation. Scand. J. Clin. Lab. Invest. 24: 315–328, **1969**.

Kilburn, K. H. and Sieker, H. O. Hemodynamic effects of continuous positive and negative pressure breathing in normal man. Circ. Res. 8: 660–669, **1960**.

Kilshaw, B. H., Harkness, R. A., Hobson, B. M., and Smith, A. W. M. The effects of large doses of the anabolic steroid, methandrostenolone, on an athlete. Clin. Endocrinol. 4(5): 537–541, **1975**.

Kindermann, W. and Keul, J. Anaerobe Energiebereitstellung im Hochleistungssport, pp. 1–118. Schorndorf, W. Germany: Karl Hoffmann, **1977**.

Kindermann, W., Keul, J., and Reindell, H. Cardiac function in sports, pp. 47–58. In: Basic Book of Sports Medicine. Ed.: G. LaCava. Rome: International Olympic Committee, **1978**.

Kindig, N. B. and Hazlett, D. R. The effects of breathing pattern in the estimation of pulmonary diffusing capacity. Quart. J. Exp. Physiol. 59: 311–329, **1974**.

King, D. W. and Gollnick, P. Ultrastructure of rat heart and liver after exhaustive exercise. Amer. J. Physiol. 218: 1150–1155, **1970**.

King, D. W. and Pengelly, R. G. Effect of running on the density of rat tibias. Med. Sci. Sport. 5: 68–69, **1973**.

Kinney, J. A. S. and Luria, S. M. Vision and visibility, pp. 271–276. In: Underwater Physiology. Ed.: C. J. Lambertsen. New York: Academic Press, **1971**.

Kinsman, R. A., Weiser, P. C., and Stamper, D. A. Multidimensional analysis of subjective symptomatology during prolonged strenuous exercise. Ergonomics 16: 211–226, **1973**.

Kipke, L. Dynamics of oxygen intake during step-by-step loading in a swimming flume. In: Swimming Medicine IV. Eds.: B. Ericksson and B. Fürberg. Baltimore: University Park Press, **1978**.

Kirsch, K., Risch, W. D., Mund, U., Rocker, L., and Stoboy, H. Low pressure system and blood volume regulating hormones after prolonged exercise, pp. 315–321. In: Metabolic Adaptation to Prolonged Physical Exercise. Eds.: H. Howald and J. R. Poortmans. Basel: Birkhauser Verlag, **1975**.

Kirschner, H. Physical capacity after moderate blood loss in dogs. Acta Physiol. Pol. 23: 597–608, **1972**.

Kirsten, E., Kirsten, R., Arese, P., Kraus, H., and Snigula, E. A study on the interdependence of contractile tone and metabolite levels in vivo in rat skeletal muscle. Biochem. Z. 344: 233-237, **1966**.

Kissling, G. and Jacob, R. Limitation of the stroke volume during increased myocardial performance, pp. 218-224. In: Limiting Factors of Physical Performance. Ed.: J. Keul. Stuttgart: Thieme, **1973**.

Kitamura, K., Jorgensen, C. R., Gobel, F. L., Taylor, H. L., and Yang, Y. Hemodynamic correlates of myocardial oxygen consumption during upright exercise. J. Appl. Physiol. 32: 516-522, **1972**.

Kjellberg, S. R., Rudhe, U., and Sjöstrand, T. Increase of the amount of hemoglobin and blood volume in connection with physical training. Acta Physiol. Scand. 19: 146-151, **1949a**.

——. The amount of hemoglobin (blood volume) in relation to the pulse rate and heart volume during work. Acta Physiol. Scand. 19: 152-169, **1949b**.

Kjellmer, I. The effect of exercise on the vascular bed of skeletal muscle. Acta Physiol. Scand. 62: 18-30, **1964**.

——. The potassium ion as a vasodilator during muscular exercise. Acta Physiol. Scand. 63: 460-468, **1965a**.

——. Studies on exercise hyperemia. Acta Physiol. Scand. 64 (Suppl. 244): 1-27, **1965b**.

Klassen, G. A., Broadhurst, C., Peretz, D. I., and Johnson, A. L. Cardiac resuscitation in 126 medical patients. Using external cardiac massage. Lancet 1: 1290-1292, **1963**.

Klaus, E. J. The athletic status of women. In: International Research in Sport and Physical Education. Eds.: E. Jokl and E. Simon. Springfield, Ill.: C. C. Thomas, **1964**.

Klaus, E. J. and Noack, H. Frau und Sport. Stuttgart: Thieme, **1971**.

Klausen, K. Comparison of CO_2 rebreathing and acetylene methods for cardiac output. J. Appl. Physiol. 20: 763-766, **1965a**.

——. The form and function of the loaded human spine. Acta Physiol. Scand. 65: 176-190, **1965b**.

——. Cardiac output in man in rest and work during and after acclimatization to 3,800 m. J. Appl. Physiol. 21: 609-616, **1966**.

Klausen, K., Dill, D. B., and Horvath, S. M. Exercise at ambient and high oxygen pressure at high altitude and at sea level. J. Appl. Physiol. 29: 456-463, **1970**.

Klausen, K., Dill, D. B., Phillips, E. E., and McGregor, D. Metabolic reactions to work in the desert. J. Appl. Physiol. 22: 292-296, **1967**.

Klausen, K., Piehl, K., and Saltin, B. Muscle glycogen stores and capacity for anaerobic work, pp. 127-129. In: Metabolic Adaptation to Prolonged Physical Exercise. Eds.: H. Howald and J. R. Poortmans. Basel: Birkhauser Verlag, **1975**.

Klavora, P. Effect of training on working capacity of rowers and their aerobic responses to training. First Canadian Congress of Sport and Physical Activity, Montreal, **1973**.

Klein, K. E., Wegmann, H. M., and Bruner, H. Circadian rhythms in indices of human performance, physical fitness and stress resistance. Aerospace Med. 39: 512-518, **1968**.

Kleine, von T. O. Enzymmuster gesunder und pathologisch veränderter Muskeln des Menschen. Z. Klin. Chem. 5: 244-247, **1967**.

Kleinhauss, G. and Franke, W. Zum Aussagewert indirekter Blutdruckbestimmungen in Rühe und bei Kreislaufbelastung durch Ergometerarbeit. Z. Kreislauff 60: 588-599, **1971**.

Kleitman, N. Sleep and Wakefulness. Chicago: University of Chicago Press, **1963**.

Kleitman, N. and Kleitman, E. Effect of non-24 hour routines of living on oral temperature and heart rate. J. Appl. Physiol. 6: 283-291, **1953**.

Klevans, L. R. and Gebber, G. L. Facilitatory forebrain influence on cardiac component of baroreceptor reflexes. Amer. J. Physiol. 219: 1235-1241, **1970**.

Klimt, F., Pannier, R., and Paufler, D. Ausdauerbelastungen bei Vorschulkindern. Schweiz Z. Sportmedizin 201: 7-23, **1974**.

Klissouras, V. Heritability of adaptive variation. J. Appl. Physiol. 31: 85-94, **1971**.

——. Genetic limit of functional adaptability. Int. Z. Angew. Physiol. 30: 85-94, **1972**.

——. Prediction of potential performance with special reference to heredity. J. Sport. Med. Phys. Fitness. 13: 100-107, **1973**.

——. Prediction of athletic performance: Genetic considerations. Can. J. Appl. Sport. Sci. 1: 195-200, **1976**.

Knappeis, G. G. and Carlsen, F. The ultra-structure of the Z-disc in skeletal muscle. J. Cell. Biol. 13: 323-335, **1962**.

Knehr, C. A., Dill, D. B., and Neufeld, W. Training and its effects on man at rest and at work. Amer. J. Physiol. 136: 148-156, **1942**.

Knochel, J. P. Potassium deficiency during training in the heat. Ann. N.Y. Acad. Sci. 301: 175–182, **1977**.

Knochel, J. P., Dotin, L. N., and Hamburger, R. J. Pathophysiology of intense physical conditioning in a hot climate. 1. Mechanisms of potassium depletion. J. Clin. Invest. 51: 242–255, **1972**.

Knuttgen, H. G. Oxygen uptake and pulse rate while running with undetermined and determined stride lengths at different speeds. Acta Physiol. Scand. 52: 366–371, **1961**.

——. Oxygen debt, lactate, pyruvate and excess lactate after muscular work. J. Appl. Physiol. 17: 639–644, **1962**.

——. Oxygen debt after submaximal physical exercise. J. Appl. Physiol. 29: 651–657, **1970**.

Knuttgen, H. G., Nordesjø, L., Ollander, B., and Saltin, B. Physical conditioning through interval training with young adults. Med. Sci. Sports. 5: 220–226, **1973**.

Knuttgen, H. G. and Saltin, B. Muscle metabolites and oxygen uptake in short-term submaximal exercise in man. J. Appl. Physiol. 32: 690–694, **1972**.

Kobayashi, Y. Effect of vitamin E on aerobic work performance in man during acute exposure to hypoxic hypoxia. Ph.D. thesis, University of New Mexico, Albuquerque, N.M., **1974**.

Koch, G. Muscle blood flow after ischemic work and during bicycle ergometer work in boys aged 12 years. Acta Paediatr. Belg. Suppl.: 29–39, **1974**.

Kochakian, C. D. Anabolic-Androgenic Steroids, pp. 1–725. Berlin: Springer Verlag, **1976**.

Kochan, R. G. and Lamb, D. R. Prostaglandin B equivalents in plasma of exercised men, pp. 878–879. In: Advances in Prostaglandin and Thromboxane Research. Eds.: B. Samuelsson and R. Paseletti. New York: Raven Press, **1976**.

Koeslag, J. H. and Sloan, A. W. Maximal heart rate and maximal oxygen consumption of long-distance runners and other athletes. J. Sport. Med. Phys. Fitness 16: 17–21, **1976**.

Koeze, T. H. The independence of corticomotoneuronal and fusimotor pathways in the production of muscle contraction by motor cortex stimulation. J. Physiol. (Lond.) 197: 87–105, **1968**.

Koeze, T. H., Afshar, F., and Watkins, E. S. Fusimotor activation effects of stimulation of the primate red nucleus. Confinea Neurologia 36: 341–346, **1974**.

Koeze, T. H., Phillips, C. G., and Sheridan, J. D. Thresholds of cortical activation of muscle spindles and α motoneurones of the baboon's hand. J. Physiol. (Lond.) 195: 419–449, **1968**.

Kogi, K. and Hakamada, T. Slowing of surface electromyogram and muscle strength in muscle fatigue. Rep. Inst. Sci. Labor (Tokyo) 60: 27, **1962**. Cited P. O. Åstrand and K. Rodahl (1977).

Kogi, K., Saito, Y., and Matsuhashi, T. Validity of three components of subjective fatigue feelings. J. Sci. Labour 46: 251–270, **1970**.

Kohlrausch, W. Ueber den Einfluss funktioneller Beanspruchung auf das Längenwachstum von Knochen. Munch. Med. Wschr. 71: 513–514, **1924**.

Kohn, R. M., Ibrahim, M. A., and Feldman, J. G. Premature ventricular beats and coronary heart disease risk factors. Amer. J. Epidemiol. 94: 556–563, **1971**.

Koizumi, K. and Brooks, C. Mc C. The integration of autonomic system reactions. Rev. Physiol. 67: 1–68, **1972**.

Koizumi, K. and Sato, A. Influence of sympathetic innervation on carotid sinus baroreceptor activity. Amer. J. Physiol. 216: 321–329, **1969**.

Kok, R., Wyndham, C. H., Strydom, N. B., and Rogers, G. G. A comparison of certain physiological and anthropometric characteristics of heat intolerant and heat tolerant Bantu from climatic room acclimatization. Johannesburg: Chamber of Mines Res. Rep. 14: **1972**.

Kollias, J., Boileau, R. A., Bartlett, H. L., and Buskirk, E. T. Pulmonary function and physical condition in lean and obese subjects. Amer. Med. Assoc. Arch. Environ. Health 25: 146–150, **1972**.

Kollias, J. and Buskirk, E. R. Exercise and altitude. In: Science and Medicine of Exercise and Sport, 2nd ed. Eds.: W. R. Johnson and E. R. Buskirk. New York: Harper & Row, **1974**.

Kollias, J., Buskirk, E. R., Howley, E. T., and Loomis, S. L. Cardiorespiratory and body composition measurements of high school football players. Res. Quart. 43: 472–478, **1972**.

Komarevtsev, L. N. Kinetic methods of evaluation of athletes provision with vitamins, p. 159. International Symposium on Athletes' Nutrition, Leningrad, **1975**.

Komi, P. V. Relationship between muscle tension, EMG, and velocity of contraction under concentric and eccentric work. In: New Developments in Electromyography and Clinical

Neurophysiology, Vol. 1, p. 596. Ed.: J. E. Desmedt. Basel: Karger, **1973**.

Komi, P. V. and Bosco, C. Utilization of stored elastic energy in leg extensor muscles by men and women. Med. Sci. Sport. 10: 261-265, **1978**.

Komi, P. V., Klissouras, V., and Karvinen, E. Genetic variation in neuromuscular performance. Int. Z. Angew. Physiol. 31: 289-304, **1973**.

Kontos, H. A., Shapiro, W., and Patterson, J. L. Observations on dyspnea induced by combinations of respiratory stimuli. Amer. J. Med. 37: 374-385, **1964**.

Kooperstein, S. I. and Schuman, B. J. Acute decompression illness—a report of forty-four cases. Industr. Med. Surg. 26: 492-496, **1957**.

Korecky, B. and Rakusan, K. Normal and hypertrophic growth of the rat heart: Changes in cell dimensions and number. Amer. J. Physiol. 234: H123-H128, **1978**.

Körge, P. and Seene, T. Zur Bestimmung des extrarenalen Natrium and Kaliumverlustes bei Sportlern. Med. Sport. 13: 204-206, **1973**.

Korner, P. I. Integrative neural cardiovascular control. Physiol. Rev. 51: 312-367, **1971**.

Korotkoff, N. C. On the question of methods of determining the blood pressure. Reports of Imp. Military Acad. (St. Petersburg) 11: 365, **1905**.

Koroxenidis, G. T., Shepherd, J. T., and Marshall, R. J. Cardiovascular response to acute heat stress. J. Appl. Physiol. 16: 896-872, **1961**.

Korsan-Bengtsen, K., Wilhelmsen, L., and Tibblin, G. Blood coagulation and fibrinolysis in relation to degree of physical activity during work and leisure time. Acta Med. Scand. 193: 73-77, **1973**.

Kouwenhoven, W. B., Jude, J. R., and Knickerbocker, G. G. Closed chest cardiac massage. J. Amer. Med. Assoc. 173: 1064-1067, **1960**.

Kow, I. M., Wilkinson, P. N., and Chornock, F. W. Axonal delivery of neuroplasmic components to muscle cells. Science 155: 342-345, **1967**.

Kozlovskaya, I., Uno, M., and Brooks, V. B. Performance of a step-tracking task by monkeys. Comm. Behav. Biol. 5: 153-156, **1970**.

Kozlowski, S. Physical performance and maximum oxygen uptake in man in exercise dehydration. Bull. Acad. Pol. Sci. (Biol.) 14: 513-519, **1966**.

Kozlowski, S., Brzezinska, Z., Nazar, K., Kowalski, W., and Franczyk, M. Plasma catecholamines during sustained isometric exercise. Clin. Sci. 45: 723-731, **1973**.

Kozlowski, S., Rasmussen, B., and Wilkoff, W. G. The effect of high oxygen tensions on ventilation during severe exercise. Acta Physiol. Scand. 81: 385-395, **1971**.

Kozlowski, S. and Saltin, B. Effects of dehydration on muscle metabolism during exercise. Physiologist 16: 289, **1973**.

Kozlowski, S., Szczepanska, E., and Zielinski, A. The hypothalamo-hypophyseal antidiuretic system in physical exercise. Arch. Int. Physiol. Biochim. 75: 218-228, **1967**.

Krahenbuhl, G. S. and Martin, S. L. Adolescent body size and flexibility. Res. Quart. 48: 797-799, **1977**.

Kral, J., Zenisek, A., and Hais, I. Sweat and exercise. J. Sport. Med. 3: 105-112, **1963**.

Krantz, J. C., Lu, G. G., Bell, F. K., and Cascorbi, H. F. Nitrites. XIX. Studies on the mechanism of action of glyceryl trinitrate. Biochem. Pharmacol. 11: 1095-1099, **1962**.

Krasney, J. A., Levitsky, M. G., and Koehler, R. C. Sinoaortic contribution to the adjustment of systemic resistance in exercising dogs. J. Appl. Physiol. 36: 679-685, **1974**.

Kraus, H. Effects of training of skeletal muscle, pp. 134-137. In: Coronary Heart Disease and Physical Fitness. Eds.: O. A. Larsen and R. O. Malmborg. Copenhagen: Munksgaard, **1971**.

Kraus, H. and Hirschland, R. P. Minimum muscular fitness tests in school children. Res. Quart. 25: 178-188, **1954**.

Kraus, H. and Kirsten, R. Die Wirkung von Körperlichem Training auf die mitochondriale Energieproduktion im Herzmuskel und in der Leber. Pflüg. Archiv. 320: 334-347, **1970**.

Krebs, H. A. and Woodford, M. Fructose-1,6-diphosphatase in striated muscle. Biochem. J. 94: 436-445, **1965**.

Kreuzer, F. Transport of O_2 and CO_2 at altitude, pp. 149-158. In: Exercise at Altitude. Ed.: R. Margaria. Dordrecht: Excerpta Medica Fdn., **1967**.

——. Facilitated diffusion of oxygen and its possible significance: A review. Resp. Physiol. 9: 1-30, **1970**.

Kreuzer, F., Tenney, S. M., Mithoefer, J. C., and Remmers, J. Alveolar-arterial oxygen gradient in Andean natures at high altitude. J. Appl. Physiol. 19: 13-16, **1964**.

Kroemer, K. H. E. and Howard, J. M. Towards standardization of muscle strength testing. Med. Sci. Sport. 2: 224–230, **1970**.

Krogman, W. M. Child growth and football. J. Health Phys. Ed. Recreat. 26(6); 12, 77–79, **1955**.

Krogh, A. The supply of oxygen to the tissues and the regulation of the capillary circulation. J. Physiol. (Lond.) 52: 457–474, **1918–1919**.

——. Monographien aus dem Gesamtgebiet der Physiologie der Pflanzen und der Tiere. Berlin: Springer Verlag, **1929a**.

——. The Anatomy and Physiology of Capillaries. New Haven, Conn.: Yale University, **1929b**.

Krogh, A. and Lindhard, J. Regulation of respiration and circulation during the initial stages of muscular work. J. Physiol. (Lond.) 47: 112–136, **1913**.

——. Relative value of fat and carbohydrate as source of muscular energy. Biochem. J. (Cambr.) 14: 290–363, **1920**.

Kronfeld, D. S., MacFarlane, W. V., Harvey, N., Howard, B., and Robinson, K. W. Strenuous exercise in a hot environment. J. Appl. Physiol. 13: 425–429, **1958**.

Krzanowski, J. and Matschinsky, F. M. Regulation of phosphofructokinase by phosphocreatine and phosphorylated glycolytic intermediates. Biochem. Biophys. Res. Comm. 34: 816–823, **1969**.

Krzywicki, H. J., Consolazio, C. F., Matoush, L. O., Johnson, H. L., and Barnhart, R. A. Body composition changes during exposure to altitude. Fed. Proc. 28: 1190–1194, **1969**.

Kubicek, F. and Gaul, G. Comparison of supine and sitting body position during a triangular exercise test. 1. Experiences in healthy subjects. Europ. J. Appl. Physiol. 36: 275–283, **1977**.

Kubicek, F. and Zwick, H. Zum Normalverhalten des Pulmonalarteriendruckes unter ansteigender Ergometerbelastung. Med. Klin. 71: 409–413, **1976**.

Kubler, W., Bretschneider, H. J., Voss, W., Gehl, H., Wenthe, F., and Colas, J. L. Über die Milchsäure und Brentztraubensäurepermeation aus dem hypothermen Myokard. Pflüg. Archiv. 287: 203–223, **1965**.

Kuby, S. A. and Noltmann, E. A. ATP-Creatine transphosphorylase. In: The Enzymes, 2nd ed., pp. 515–603. Eds.: P. D. Boyer, H. Lardy, and K. Myrbäck. New York: Academic Press, **1962**.

Kuffler, S. W. Discharge patterns and functional organization of the mammalian retina. J. Neurophysiol. 16: 37–68, **1953**.

Kuffler, S. W. and Hunt, C. C. The mammalian small-nerve fibres; a system for efferent nervous regulation of muscle spindle discharge. Res. Publ. Assoc. Res. Nerv. Ment. Dis. 30: 24–37, **1952**.

Kugelberg, E. Histochemical composition, contraction speed and fatiguability of rat soleus motor units. J. Neurol. Sci. 20: 177–198, **1973**.

Kuhlow, A. Running economy in long-distance speed skating, pp. 291–298. In: Biomechanics V. Ed.: P. V. Komi. Baltimore: University Park Press, **1976**.

Kunze, Z. Das Sauerstoffdruckfeld in normalen und pathologisch veränderten Muskel. Berlin: Springer Verlag, **1969**.

Kurita, A., Chaitman, B. R., and Bourassa, M. G. Significance of exercise-induced junctional ST depression in evaluation of coronary artery disease. Amer. J. Cardiol. 40: 492–497, **1977**.

Kuroda, E., Klissouras, V., and Milsum, J. H. Electrical and metabolic activities and fatigue in human isometric contractions. J. Appl. Physiol. 29: 358–367, **1970**.

Kushmerick, M. J. and Davies, R. E. The chemical energetics of muscle contraction. 2. The chemistry, efficiency and power of maximally working sartorius muscles. Proc. Roy. Soc. B 174: 315–353, **1969**.

Kyle, C. R. and Mastropaolo, J. Predicting racing bicyclist performance using the unbraked flywheel method of bicycle ergometry. In: Biomechanics of Sports and Kinanthropometry. Eds.: F. Landry and W. A. R. Orban. Miami: Symposia Specialists, **1978**.

Kylstra, J. A. The feasibility of liquid-breathing and artificial gills, pp. 195–212. In: The Physiology and Medicine of Diving and Compressed Air Work. Eds.: P. B. Bennett and D. H. Elliott. Baltimore: Williams & Wilkins, **1969**.

Lab, M. J. Depolarization produced by mechanical changes in normal and abnormal myocardium. J. Physiol. (Lond.) 284: 143p–144p, **1978**.

LaBaere, W. The human animal. Chicago: University of Chicago Press, **1954**.

Ladell, W. S. S. Effects of water and salt intake upon performance of men working in hot and humid environments. J. Physiol. (Lond.) 127: 11–46, **1955**.

——. Terrestrial animals in humid heat: Man, pp. 625–660. In: Adaptation to the Environment. Ed.: D. B. Dill. Washington, D.C.: American Physiological Society, **1964**.

——. The Physiology of Human Survival. New York: Academic Press, **1965**.

Ladell, W. S. S. and Kenney, R. A. Some laboratory and field observations on the Harvard Pack test. Quart. J. Exp. Physiol. 40: 283-296, **1955**.

Ladell, W. S. S. and Shephard, R. J. Aldosterone inhibition and acclimatization to heat. J. Physiol. (Lond.) 160: 19p-20p, **1961**.

La Force, R. C. and Lewis, B. M. Diffusional transport in the human lung. J. Appl. Physiol. 28: 291-298, **1970**.

Laguens, R. P. and Gómez-Dunn, C. L. Fine structure of myocardial mitochondria in rats after exercise for one half to two hours. Circ. Res. 21: 271-279, **1967**.

Laguens, R. P., Lozada, B. B., Gómez-Dunn, C. L., and Beramendi, A. R. Effect of acute and exhaustive exercise upon the fine structure of heart mitochondria. Experientia 19: 244-246, **1966**.

Lahiri, S., Brody, J. S., Motoyama, E. K., and Velasquez, T. M. Regulation of breathing in newborns at high altitude. J. Appl. Physiol. Resp. Environ. Exer. Physiol. 44: 673-678, **1978**.

Lahiri, S. and Milledge, J. S. Relative respiratory insensitivity of Himalayan Sherpa altitude residents to hypoxia at 4,880 m and at sea level, pp. 387-392. In: Arterial Chemoreceptors. Ed.: R. W. Torrance. Oxford: Blackwell Scientific, **1968**.

Lahiri, S., Milledge, J. S., and Sørensen, S. C. Ventilation in man during exercise at high altitude. J. Appl. Physiol. 32: 766-769, **1972**.

Lakie, M. and Tsementzis, S. A. Torque-resonance relationship in the wrist: Sex difference. J. Physiol. (Lond.) 289: 86p, **1978**.

Lamb, D. R. Androgens and exercise. Med. Sci. Sport. 7: 1-5, **1975**.

——. Physiology of Exercise. Responses and Adaptations, pp. 1-438. New York: Macmillan, **1978**.

Lamb, L. E., Kelly, R. J., Smith, W. L., Le Blanc, A. D., and Johnson, P. C. Limiting factors in the capacity to achieve maximum cardiac work. Aerospace Med. 40: 1291-1296, **1969**.

Lamb, T. W. Ventilatory responses to intravenous and inspired carbon dioxide in anesthetized cats. Resp. Physiol. 2: 99-104, **1966**.

Lambert, O. The relationship between maximum isometric strength and maximum concentric strength at different speeds. Intern. Fed. Phys. Educ. Bull. 35: 13, **1965**.

Lambert, R. K. and Wilson, T. A. A model for the elastic properties of the lung and their effect on expiratory flow. J. Appl. Physiol. 34: 34-48, **1973**.

Lambertsen, C. J. Factors in the stimulation of respiration by carbon dioxide, pp. 257-276. In: The Regulation of Human Respiration. Eds.: D. J. C. Cunningham and B. B. Lloyd. Oxford: Blackwell Scientific, **1963**.

——. Concepts for advances in the therapy of bends in undersea and aerospace activity. Aerospace Med. 39: 1086-1093, **1968**.

Lambertsen, C. J., Gelfand, R., and Kemp, R. A. Dynamic response characteristics of several CO_2 reactive components of the respiratory control system. In: Cerebrospinal Fluid and the Regulation of Breathing. Eds.: Mc C. Brooks, F. F. Kao, and B. B. Lloyd. Oxford: Blackwell Scientific, **1965**.

Lambertsen, C. J., Kough, R. H., Cooper, D. Y., Emmel, G. L., Loeschcke, H. H., and Schmidt, C. F. Oxygen toxicity. Effects in man of oxygen inhalation at 1 and 3.5 atmospheres upon blood gas transport, cerebral circulation and cerebral metabolism. J. Appl. Physiol. 5: 471-486, **1953**.

Lamke, L. O., Lennquist, S., Liljedahl, S. O., and Wedin, B. The influence of cold stress on catecholamine excretion and oxygen uptake of normal persons. Scand. J. Clin. Lab. Invest. 30: 57-62, **1972**.

Lammert, O. Forholdet mellem maximal isometrisk Kraft og dynamisk Kraft. Tidsskrift for Legemsøvelser 4: 1-12, **1963**.

Landgren, S. and Silfvenius, H. Projection to the cerebral cortex of group I muscle afferents from the cat's hind limb. J. Physiol. (Lond.) 200: 353-372, **1969**.

Landis, E. M. and Pappenheimer, J. P. Exchange of substrates through the capillary walls. In: Handbook of Physiology, Section 2. Circulation, Vol. 2. Ed.: W. F. Hamilton. Washington, D.C.: Amer. Physiol. Soc., **1963**.

Lane, H. W., Roessler, G., Nelson, E. W., and Cerda, J. J. Whole body counter measurements of total body potassium before and after exercise. Fed. Proc. 36: 433, **1977**.

Lane, H. W., Roessler, G. S., Nelson, E. W., and Cerda, J. J. Effect of physical activity on human potassium metabolism in a hot and humid environment. Amer. J. Clin. Nutr. 31: 838-843, **1978**.

Lange, R. L. and Hecht, H. H. The mechanism of Cheyne-Stokes respiration. J. Clin. Invest. 41: 42-52, **1962**.

Langer, G. A. Effects of digitalis on myocardial ionic exchange. Circulation 46: 180-187, **1972**.

——. The structure and function of the myocardial cell surface. Amer. J. Physiol. 235: H461-H468, **1978**.

Langham, W. H., Eversole, W. J., Hayes, F. N., and Trujillo, T. T. Assay of tritium on activity in body fluids with use of a liquid scintillation system. J. Lab. Clin. Med. 47: 819-825, **1956**.

Langley, J. N. The nerve fibre constitution of peripheral nerves and of nerve roots. J. Physiol. (Lond.) 56: 382-396, **1922**.

Langner, P. H., Okada, R. H., Moore, S. R., and Fies, H. L. Comparison of four orthogonal systems of vectocardiography. Circulation 16: 46-54, **1958**.

Lanier, R. R. The effects of exercise of the knee joints of inbred mice. Anat. Rec. 94: 311-321, **1946**.

Lanphier, E. H. Man in high pressures, pp. 893-909. In: Handbook of Physiology, Section 4. Adaptation to the Environment. Eds.: D. B. Dill, E. F. Adolph, and C. G. Wilber. Washington: Amer. Physiol. Soc., **1964**.

——. Pulmonary function, pp. 58-112. In: The Physiology and Medicine of Diving and Compressed Air Work. Eds.: P. B. Bennett and D. H. Elliott. London: Baillière, Tindall & Cassell, **1969**.

Lanphier, E. H. and Rahn, H. Alveolar gas exchange during breath-hold diving. J. Appl. Physiol. 18: 471-477, **1963**.

La Porte, W. The influence of a gymnastic pause upon recovery following post-office work. Ergonomics 9: 501-506, **1966**.

Laporte, Y. and Emonet-Dénand, F. Evidence for common innervation of bag and chain muscle fibres in cat spindles, pp. 119-126. In: Control of Posture and Locomotion. Eds.: R. B. Stein, K. G. Pearson, R. S. Smith, and J. B. Redford. New York: Plenum Press, **1973**.

Laporte, Y. and Lloyd, D. C. Nature and significance of the reflex connections established by large afferent fibres of muscular origin. Amer. J. Physiol. 169: 609-621, **1952**.

Laritcheva, K. A., Yalovaya, N. I., Shubin, V. I., and Smirnov, P. V. Study of energy expenditure and protein needs of top weight lifters, pp. 155-163. In: Nutrition, Physical Fitness and Health. Eds.: J. Pařízková and V. A. Rogozkin. Baltimore: University Park Press, **1978**.

Larivière, G., Lavallée, H., and Shephard, R. J. Correlations between field tests of performance and laboratory measurements of fitness. Results in the ten years old school child. Acta Paediatr. Belg. 28: 19-28, **1974**.

Larivière, J. E. and Simonson, E. The effect of age and occupation on speed of writing. J. Gerontol. 20: 415-421, **1965**.

Laros, G. S., Tipton, C. M., and Cooper, R. R. Influence of physical activity on ligament insertions in the knees of dogs. J. Bone Joint Surg. Amer. 53: 275-286, **1971**.

Larson, R. L. Physical activity and the growth and development of bone and joint structures, pp. 33-59. In: Physical Activity. Human Growth and Development. Ed.: G. L. Rarick. New York: Academic Press, **1973**.

Larsson, S. Diet, exercise and body composition, p. 132. In: Nutrition and Physical Activity. Ed.: G. Blix. Uppsala: Almqvist & Wiksell, **1967**.

Lassarre, C., Girard, F., Durand, J., and Raynaud, J. Kinetics of human growth hormone during submaximal exercise. J. Appl. Physiol. 37: 826-830, **1974**.

Laszlo, G., Clark, T. J. H., Pope, H., and Campbell, E. J. M. Differences between alveolar and arterial pCO_2 during rebreathing experiments in resting human subjects. Resp. Physiol. 12: 36-52, **1971**.

Laubach, L. I. L. Comparative muscular strength of men and women: A review of the literature. Aviat. Space Environ. Med. 47: 534-542, **1976**.

Laubach, L. L. and McConville, J. T. The relationship of strength to body size and typology. Med. Sci. Sport. 1: 189-194, **1969**.

Laurenceau, J. L., Turcot, J., and Dumesnil, J. G. Echocardiographic study of Olympic athletes. In: Sports Cardiology. Eds.: T. Lubick and A. Venerando. Bologna: Aulo Gaggi, **1979**.

Lauru, L. Physiological study of motions. Adv. Management 22: 17-24, **1957**.

Lauzon, R. R. J. The Canadian Home Fitness Test in programs of community health. In:

Physical Fitness Assessment: Principles, Practice and Applications. Eds.: Roy J. Shephard and H. Lavallée. Springfield, Ill.: C. C. Thomas, **1978**.

Lavallée, H., Larivière, G., and Shephard, R. J. Correlations between field test of performance and laboratory measurements of fitness. Results in the ten year old school child. Acta Paediatr. Belg. 28 (Suppl.): 19–28, **1974**.

Lavenne, F. and Belayew, D. Exercise tolerance test at room temperature for the purpose of selecting rescue teams for training in a hot climate. Rev. Inst. Hyg. Mines 21: 48–58, **1966**.

Lawrie, R. A. The activity of the cytochrome system in muscle and its relation to myoglobin. Biochem. J. 55: 298–305, **1953**.

Lawson, H. C. The volume of blood—a critical examination of methods for its measurement, pp. 23–50. In: Handbook of Physiology, Section 2. Circulation, Vol. 1. Ed.: W. F. Hamilton. Washington, D.C. Amer. Physiol. Soc., **1962**.

Lawson, W. H. Rebreathing measurements of pulmonary diffusing capacity for CO during exercise. J. Appl. Physiol. 29: 896–900, **1970**.

Lawther, P. J. Human responses to air pollutants. Lecture at Gas Club, Toronto, **1967**.

Layani, D. Sport et Diabète, pp. 75–80. In: Nutrition, Dietetics and Sport. Eds.: G. Ricci and A. Venerando. Torino: Minerva Medica, **1978**.

Laycoe, R. R. and Marteniuk, R. G. Learning and tension as factors in static strength gains produced by static and eccentric training. Res. Quart. 42: 299–306, **1971**.

Layman, E. M. Psychological effects of physical activity. Exercise Sport. Sci. Rev. 2: 33–70, **1974**.

LeBlanc, J. Man in the Cold, pp. 1–195. Springfield, Ill.: C. C. Thomas, **1975**.

LeBlanc, P., Ruff, F., and Milic-Emili, J. Effects of age and body position on "airway closure" in man. J. Appl. Physiol. 28: 448–451, **1970**.

LeBlond, C. P. and Greulich, R. C. Autoradiographic studies of bone formation and growth. In: The Biochemistry and Physiology of Bone. Ed.: G. H. Bourne. New York: Academic Press, **1956**.

LeClerq, R. and Poortmans, J. R. Evolution of plasma cortisol during short-time exercise, pp. 203–207. In: 3rd International Symposium on Biochemistry of Exercise. Eds.: F. Landry and W. R. Orban. Miami: Symposia Specialists, **1978**.

Lee, D. H. K. Large mammals in the desert, pp. 109–125. In: Physiological Adaptations, Desert and Mountain. Eds.: M. K. Yousef, S. M. Horvath, and R. W. Bullard. New York: Academic Press, **1972**.

Lee, G., Amsterdam, E. A., de Maria, A. N., Davis, G., LaFave, T., and Mason, D. T. Effect of exercise on hemostatic mechanisms, pp. 122–136. In: Exercise in Cardiovascular Health and Disease. Eds.: E. A. Amsterdam, J. H. Wilmore, and A. N. DeMaria. New York: Yorke Medical Books, **1977**.

Lee, G. de J. and Dubois, A. B. Pulmonary capillary blood flow in man. J. Clin. Invest. 34: 1380–1390, **1955**.

Lee, K. D., Mayou, R. A., and Torrance, R. W. The effect of blood pressure upon chemoreceptor discharge to hypoxia and the modification of this effect by the sympathetic-adrenal system. Quart. J. Exp. Physiol. 49: 171–183, **1964**.

Leedy, H. E., Ismail, A. H., Kessler, W. V., and Christian, J. E. Relationships between physical performance items and body composition. Res. Quart. 36: 158–163, **1965**.

Lefcoe, N. M. The time course of maximum ventilatory performance during and after moderately heavy exercise. Clin. Sci. 36: 47–52, **1969**.

Lefcoe, N. M. and Yuhasz, M. S. The "second wind" phenomenon in constant load exercise. J. Sport. Med. Phys. Fitness 11: 135–138, **1971**.

Lefer, A. M. Role of myocardial depressant factor in the pathogenesis of circulatory shock. Fed. Proc. 29: 1836–1847, **1970**.

Léger, L. A. and Boucher, R. A continuous multistage field test. Can. J. Appl. Sport. Sci. 3: 186, **1978**.

Lehman, H. C. Age and Achievement. New York: Oxford University Press, **1953**.

Lehmann, G., Straub, H., and Szakáll, A. Pervitin als Leistungssteigerudesmitter. Arbeitsphysiol. 10: 680–691, **1939**.

Lehninger, A. L. The Mitochondrion: Molecular Basis of Structure and Function. Menlo Park, Calif.: W. A. Benjamin, **1964**.

——. Bioenergetics. Menlo Park, Calif.: W. A. Benjamin, **1973**.

Leighton, D. A. Special senses: Aging of the eye. In: Textbook of Geriatric Medicine and Gerontology. Ed.: J. C. Brocklehurst. Edinburgh: Churchill-Livingstone, **1973**.

Leighton, J. An instrument and technic for the measurement of range of joint motion. Arch.

Phys. Med. 36: 571–578, **1955**.

Leitch, A. G., Clancy, L., and Flenley, D. C. Maximal oxygen uptake, lung volume and ventilatory response to carbon dioxide and hypoxia in a pair of identical twin athletes. Clin. Sci. Mol. Med. 48: 235, **1975**.

Leith, D. E. and Bradley, M. Ventilatory muscle strength and endurance training. J. Appl. Physiol. 41: 508–516, **1976**.

Leithead, C. S., and Lind, A. R. Heat stress and heat disorders. London: Cassell, **1964**.

Leksell, L. The action potentials and excitatory effects of the small ventral root fibres to skeletal muscle. Acta Physiol. Scand. 10 (Suppl.) 31: 1–84, **1945**.

Lemon, R. N. and Porter, R. Afferent input to movement-related precentral neurones in conscious monkeys. Proc. Roy. Soc. B. 194: 313–339, **1976**.

Lenfant, C., Torrance, J., English, E., Finch, C. A., Reynafarje, C., Ramos, J., and Faura, J. Effect of altitude on oxygen binding by hemoglobin and on organic phosphate levels. J. Clin. Invest. 47: 2652–2656, **1968**.

Lennerstrand, G. and Thoden, U. Dynamic analysis of muscle spindle endings in the cat using length changes of different length-time relations. Acta Physiol. Scand. 73: 281–299, **1968**.

Lentz, T. L. Nerve trophic function: in vitro assay of effects of nerve tissue on muscle cholinesterase activity. Science 171: 187–189, **1971**.

Leon, A. S., Conrad, J., Hunninghake, D., Jacobs, D., and Serfass, R. Exercise effects on body composition, work capacity and carbohydrate and lipid metabolism of young obese men. Med. Sci. Sport. 9: 60, **1977**.

Leon, A. S., Pettinger, W. A., and Saviano, M. A. Enhancement of serum renin activity by exercise in the rat. Med. Sci. Sport. 5: 40–43, **1973**.

Leon, H. A. and Cook, S. F. A mechanism by which helium increases metabolism in small mammals. Amer. J. Physiol. 199: 243–245, **1960**.

Lepeschkin, E. Physiological factors influencing the electrocardiographic response to exercise, pp. 363–387. In: Measurement in Exercise Electrocardiography. The Ernst Simonson Conference. Ed.: H. Blackburn. Springfield, Ill.: C. C. Thomas, **1969**.

Lepley, D., Weisfeldt, M., Close, A., Schmidt, R., Bowler, J., Kory, R. C., and Ellison, E. H. Effect of low molecular weight dextran on hemorrhagic shock. Surgery 54: 93–103, **1963**.

Lesmes, G. R., Costill, D. L., Coyle, E. F., and Fink, W. J. Muscle strength and power changes during maximal isokinetic training. Med. Sci. Sport. 10: 266–269, **1978**.

Lesmes, G. R., Fox, E. L., Stevens, C., and Otto, R. Metabolic responses of females to high intensity interval training of different frequencies. Med. Sci. Sport. 10: 229–232, **1978**.

Lester, F. M., Sheffield, L. T., and Reeves, J. T. Electrocardiographic changes in clinically normal older men following near maximal and maximal exercise. Circulation 36: 5–14, **1967**.

Lester, F. M., Sheffield, L. T., Trammell, P., and Reeves, T. J. The effect of age and athletic training on the maximal heart rate during muscular exercise. Amer. Heart J. 76: 370–376, **1968**.

Leuba, J. H. The influence of the duration and of the rate of arm movements upon the judgement of their length. Amer. J. Psychol. 20: 374–385, **1909**.

Leveille, G. A. and Romsos, D. R. Meal eating and obesity. Nutr. Today, Nov./Dec., **1974**.

Lever, M. J., Paton, W. D. M., and Smith, E. B. Decompression characteristics of inert gases, pp. 123–136. In: Underwater Physiology. Ed.: C. J. Lambertsen. New York: Academic Press, **1971**.

Levine, R. The physiological disposition of hexamethonium and related compounds. J. Pharm. Exp. Ther. 129: 296–304, **1960**.

Levinson, G. E., Pacifico, A. D., and Frank, F. M. J. Studies of cardiopulmonary blood volume. Measurement of total cardiopulmonary blood volume in normal human subjects at rest and during exercise. Circulation 33: 347–356, **1966**.

Levison, H. and Cherniack, R. Ventilatory cost of exercise in chronic obstructive pulmonary disease. J. Appl. Physiol. 25: 21–27, **1968**.

Levy, L., Graichen, H., Stolwijk, J. A. J., and Galabresi, M. Evaluation of local tissue blood flow by continuous direct measurement of thermal conductivity. J. Appl. Physiol. 22: 1026–1029, **1967**.

Levy, M. N. and Zieske, H. Autonomic control of cardiac pacemaker activity and atrioventricular transmission. J. Appl. Physiol. 27: 465–470, **1969**.

Lewes, D. Electrode jelly in electrocardiography. Brit. Heart J. 27: 105–115, **1965**.

Lewis, A. J., Lawrence, L. V., Serfass, R., and Leon, A. Body composition, aerobic capacity and dietary intake of athletic and sedentary college women. Med. Sci. Sport. 10: 52, **1978**.

Lewis, H. E. and Paton, W. D. M. Decompression sickness during the sinking of a caisson. Brit. J. Industr. Med. 14: 5-12, **1957**.

Lewis, M. McD. and Porter, R. Pyramidal tract discharge in relation to movement performance in monkeys with partial anaesthesia of the moving hand. Brain Res. 71: 245-251, **1974**.

Lewis, P. R. and Lobban, M. C. Dissociation of diurnal rhythms in human subjects living on abnormal time routines. Quart. J. Exp. Physiol. 42: 371-386, **1957**.

Leyton, R. A. Cardiac ultrastructure and function in the normal and failing heart, pp. 11-65. In: Cardiac Mechanics: Physiological, Clinical and Mathematical Considerations. Eds.: I. Mirsky, D. N. Ghista, and H. Sandler. New York: Wiley, **1974**.

Liesen, H., Korsten, H., and Hollman, W. Effects of a marathon race on blood lipid constituents in younger and older athletes, pp. 194-200. In: Metabolic Adaptation to Prolonged Physical Exercise. Eds.: H. Howald and J. R. Poortmans. Basel: Birkhauser Verlag, **1975**.

Light, A. B. and Warren, C. R. Urea clearance and proteinuria during exercise. Amer. J. Physiol. 117: 658-661, **1936**.

Light, L. H. Implications of aortic blood velocity measurements in children. J. Physiol. (Lond.) 285: 17p-18p, **1978**.

Light, L. H. and McLellan, G. Skeletal transients associated with heel strike. J. Physiol. (Lond.) 272: 9p-10p, **1977**.

Liljestrand, G. Untersuchunger über die Atmungsarbeit. Skand. Arch. Physiol. 35: 199-293, **1918**.

Lin, Y. C., Lally, D. L., Moore, T. O., and Hong, S. K. Physiological and conventional breath-hold breaking points. J. Appl. Physiol. 37: 291-296, **1974**.

Lind, A. R. A physiological criterion for setting thermal environmental limits for everyday work. J. Appl. Physiol. 18: 51-56, **1963**.

Lind, A. R. and Bass, D. E. The optimal exposure time for the development of acclimatization to heat. Fed. Proc. 22: 704, **1963**.

Lind, A. R. and McNicol, G. W. Muscular factors which determine the cardiovascular responses to sustained and rhythmic exercise. Can. Med. Assoc. J. 96: 706-712, **1967**.

Lind, A. R., McNicol, G. W., Bruce, R. A., MacDonald, H. R., and Donald, K. W. The cardiovascular responses to sustained contractions of a patient with unilateral syringomyelia. Clin. Sci. 35: 45-53, **1968**.

Lind, A. R. and Williams, C. A. Changes in the forearm blood flow following brief isometric handgrip contractions at different tensions. J. Physiol. (Lond.) 272: 97p-98p, **1977**.

Linden V. Absence from work and physical fitness. Brit. J. Industr. Med. 26: 47-53, **1969**.

Linderholm, H. ECG in epidemiological studies of coronary heart disease. Nordic Council Arctic Med. Research Rep. 19: 50-57, **1977**.

Lindgren, P., Rosen, A., Strandberg, P., and Uvnas, B. The sympathetic vasodilator outflow: A cortico-spinal autonomic pathway. J. Comp. Neurol. 105: 95-109, **1956**.

Lindstrom, L., Magnusson, R., and Petersen, I. Muscular fatigue and action potential conduction velocity changes, studied with frequency analysis of EMG signals. Electromyography 10: 341-356, **1970**.

Linnarson, D. Dynamics of pulmonary gas exchange and heart rate changes at start and end of exercise. Acta Physiol. Scand. Suppl. 415: 1-68, **1974**.

Littman, D. Textbook of Electrocardiography. New York: Harper & Row, **1972**.

Litvinova, V. N., Morozov, V. Il., Pshendin, A. I., Fedorova, G. P., Feldkoren, B.I., Tshajkovski, V. C., and Rogozkin, V. A. Wlianie kratnosti pitania na obmen belkow w skeletnich mischzach; pecheni belich kris. Vopr. Pitan. 4: 36-41, **1976**.

Lloyd, A. J. Surface electromyography during sustained isometric contractions. J. Appl. Physiol. 30: 713-719, **1971**.

Lloyd, A. J., Voor, J. H., and Thieman, T. J. Subjective and electromyographic assessment of isometric muscle contractions. Ergonomics 13: 685-691, **1970**.

Lloyd, B. B. Presidential address, Section 1 (Physiology and Biochemistry). British Association. In: Advancement of Sciences, pp. 515-530, **1966**.

——. Theoretical effects of altitude on the equation of motion of a runner, pp. 65-72. In: Exercise at Altitude. Ed.: R. Margaria. Dordrecht: Excerpta Medica Fdn., **1967**.

Lloyd, B. B. and Cunningham, D. J. C. A quantitative approach to the regulation of human respiration, pp. 331-349. In: the Regulation of Human Respiration. Eds.: D. J. C. Cunningham and B. B. Lloyd. Oxford: Blackwell Scientific, **1963**.

Lloyd, D. P. C. Neuron patterns controlling transmission of ipsilateral hind limb reflexes in cat. J. Neurophysiol. 6: 293-315, **1943a**.

——. Reflex action in relation to pattern and peripheral source of afferent stimulation. J. Neurophysiol. 6: 111-120, **1943b**.

——. Facilitation and inhibition of spinal motoneurons. J. Neurophysiol. 9: 421–438, **1946**.

Lloyd, T. C. Cardiopulmonary baroreflexes: effects of staircase, ramp and square-wave stimulation. Amer. J. Physiol. 228: 470–476, **1975**.

Lloyd, T. C. and Wright, G. W. Pulmonary vascular resistance and vascular transmural gradient. J. Appl. Physiol. 15: 241–245, **1960**.

Loeschke, H. H. Central nervous chemoreceptors, pp. 168–196. In: MTP International Review of Science Physiology Series. 1. Respiratory Physiology. London: Butterworths, **1974**.

Loeschke, H. H. and Mitchell, R. A. Properties and localization of intracranial chemosensitivity, pp. 243–256. In: The Regulation of Human Respiration. Eds.: D. J. C. Cunningham and B. B. Lloyd. Oxford: Blackwell Scientific, **1963**.

Logan, W. J. and Snyder, S. H. High affinity uptake systems for glycine, glutamic and aspartic acids in synaptosomes of rat central nervous tissue. Brain Res. 42: 413–431, **1972**.

Lohmann, K. Über die enzymatische Aufspaltung der Kreatinphosphorsäure; zugleich ein Beitrag zum Chemismus der Muskelkontraktion. Biochem. Z. 271: 264–277, **1934**.

——. Chemische Vorgänge bei der Muskelkontraktion. Angew. Chem. 50: 97–100, **1937**.

Löllgen, H., Ulmer, H. V., Gross, R., Wilbert, G., and V. Nieding, G. Methodological aspects of perceived exertion rating and its relation to pedalling rate and rotating mass. Europ. J. Appl. Physiol. 34: 205–215, **1975**.

Löllgen, H., Ulmer, H. V., and Nieding, G. Heart rate and perceptual response to exercise with different pedalling speed in normal subjects and patients. Europ. J. Appl. Physiol. 37: 297–304, **1977**.

Lombardo, T. A., Rose, L., Taeschler, M., Tuluy, S., and Burg, R. J. The effect of exercise on coronary blood flow, myocardial oxygen consumption and cardiac efficiency in man. Circulation 7: 71–78, **1953**.

Longmuir, I. S. The oxygen electrode. Oxygen in the Animal Organism, pp. 219–237. Eds.: F. Dickens and E. Neil. New York: Macmillan, **1964**.

Lonne, E., Lonne, C. H., Fahrenberg, J., and Roskamm, H. Pulsfrequenz-messungen und EKG Registrierung bei Autorenne. Sportarzt und Sportmedizin 19: 103–111, **1968**.

Lopez, A., Vial, R., Balart, L., and Arroyave, G. Effect of exercise and physical fitness on serum lipids and lipoproteins. Atherosclerosis 20: 1–9, **1974**.

Lord, G. P., Bond, G. F., and Schaefer, K. E. Breathing under high ambient pressure. J. Appl. Physiol. 21: 1833–1838, **1966**.

Lotter, W. S. Interrelationships among reaction times and speeds of movement in different limbs. Res. Quart. 31: 147–155, **1960**.

Loucks, J. and Thompson, H. Effect of menstruation on reaction time. Res. Quart. 39: 407–408, **1968**.

Lougheed, D. W., Janes, J. M., and Hall, G. F. Physiological studies in experimental asphyxia and drowning. Can. Med. Assoc. J. 40: 423–428, **1939**.

Lowey, S., Slayter, H. S., Weeds, A. G., and Baker, H. Substructure of the myosin molecule. I. Subfragments of myosin by enzymic degradation. J. Mol. Biol. 42: 1–29, **1969**.

Lown, B., Newman, J., Amarasingham, R., and Berkovits, B. V. Comparison of alternating current with direct current electroshock across the closed chest. Amer. J. Cardiol. 10: 223–233, **1962**.

Lübbers, D. W. Local tissue PO_2: Its measurement and meaning, pp. 151–155. In: Oxygen Supply: Theoretical and Practical Aspects of Oxygen Supply and Microcirculation of Tissue. Eds.: M. Kessler, D. F. Bruley, L. C. Clark, D. W. Lübbers, I. A. Silver, and J. Strauss. Baltimore: University Park Press, **1973**.

——. Measuring methods for the analysis of tissue oxygen supply, pp. 62–71. In: Oxygen and Physiological Function. Ed.: F. F. Jöbsis. Dallas, Texas: Professional Information Library, **1977a**.

——. Quantitative measurement and description of oxygen supply to the tissue. In: Oxygen and Physiological Function. Ed.: F. F. Jöbsis. Dallas, Texas: Professional Information Library, **1977b**.

Ludbrook, J. The musculovenous pumps of the human lower limb. Amer. Heart J. 71: 635–641, **1966**.

Ludbrook, J., Faris, I. B., Iannos, J., Jamieson, G. G., and Russell, W. J. Lack of effect of isometric handgrip exercise on the responses of the carotid sinus baroreceptor reflex in man. Clin. Sci. Mol. Med. 55: 189–194, **1978**.

Ludvigh, E. Direction sense of the eye. Amer. J. Ophthalmol. 36: 139–142, **1955**.

Ludwig, W. M. and Lipkin, M. Biochemical and cytological alterations in gastric mucosa of guinea pigs under restraint stress. Gastroenterology 56: 895–902, **1969**.

Lueft, R. J. The effect of experimental design upon

expectancy and speed of response, pp.771-775. In: Contemporary Psychology of Sport. Eds.: G. S. Kenyon and T. M. Grogg. Chicago: Athletic Institute, **1970**.

Luff, A. R. Dynamic properties of fast and slow skeletal muscles in the cat and rat following cross-innervation. J. Physiol. (Lond.) 248: 83-96, **1975**.

Luff, A. R. and Atwood, H. L. Membrane properties and contraction of single muscle fibers in the mouse. Amer. J. Physiol. 222: 1435-1440, **1972**.

Luft, U. C. Die Hohenrpassung. Ergebn. Physiol. 44: 256-314, **1941**.

Luft, U. C., Cardus, D., Lim, T. P. K., Anderson, E. C., and Howarth, J. L. Physical performance in relation to body size and composition. Ann. N.Y. Acad. Sci. 110: 795-808, **1963**.

Lukin, L. and Ralston, H. J. Oxygen deficit and repayment in exercise. Int. Z. Angew. Physiol. 19: 183-193, **1962**.

Lumsden, T. Observations on the respiratory centres. J. Physiol. (Lond.) 57: 354-367, **1923**.

——. Chelonian respiration (tortoise). J. Physiol. (Lond.) 58: 259-266, **1924**.

Lundberg, A. The excitatory control of the Ia inhibitory pathway, pp. 333-340. In: Excitatory Synaptic Mechanisms. Eds.: P. Andersen and J. K. S. Jansen. Oslo: Universitets Forlaget, **1970**.

Lundberg, A. and Winsbury, G. Selective adequate activation of large afferents from muscle spindle and Golgi tendon organs. Acta Physiol. Scand. 49: 155-164, **1960**.

Lundgren, N. The physiological effects of time schedule work on lumber workers. Acta Physiol. Scand. 13 (Suppl. 41): 1-137, **1946**.

Lundgren, O. and Jodal, M. Regional blood flow. Ann. Rev. Physiol. 37: 395-414, **1975**.

Lundin, G. Nitrogen elimination from the tissues during oxygen breathing and its relationship to the fat: muscle ratio and the localization of bends. J. Physiol. (Lond.) 152: 167-175, **1960**.

Lundman, T. Genetic aspects of ischaemic heart disease. Experiences from twin studies. Nordic Council Arctic Med. Res. Rep. 19: 22-28, **1977**.

Lundquist, I. Insulin secretion: its regulation by monoamines and acid amylglucosidase. Acta Physiol. Scand. Suppl. 372: 1-47, **1971**.

Lundvall, J. Tissue hyperosmolarity as a mediator of vasodilatation and transcapillary fluid flux in exercising skeletal muscle. Acta Physiol. Scand. Suppl. 379: 1-142, **1972**.

Lundvall, J., Mellander, S., Westling, H., and White, T. Fluid transfer between blood and tissues during exercise. Acta Physiol. Scand. 85: 258-269, **1972**.

Lymm, R. W. and Huxley, H. E. X-ray diagrams from skeletal muscle in the presence of ATP analogs. Cold Spring Harbor Symp. Quart. Biol. 37: 449-453, **1972**.

MacAlpin, R. N., Abbasi, A. S., Grollman, J. H., and Eber, L. Human coronary artery size during life: a cinearteriographic study. Radiology 108: 567-576, **1973**.

MacAraeg, P. V. J. The importance of fluid and electrolyte in athletes. J. Sport. Med. Phys. Fitness 14: 213-217, **1974**.

——. A study of serum, sweat and urine electrolytes in athletes, pp. 429-435. In: Proceedings, 20th World Congress of Sports Medicine. Ed.: A. H. Toyne. Melbourne, Australian Sports Medicine Federation, **1975**.

MacCallum, J. B. On the histogenesis of the striated muscle fiber, and the growth of the human sartorius muscle. Bull. Johns Hopkins Hosp. 9: 208-215, **1898**.

MacDougall, J. D. The anaerobic threshold: Its significance for the endurance athlete. Can. J. Appl. Sport. Sci. 2: 137-140, **1977**.

MacDougall, J. D., Reddan, W. G., Dempsey, J. A., and Forster, H. Acute alterations in stroke volume during exercise at 3,100 m altitude. J. Human Ergol. 5: 103-111, **1976**.

MacDougall, J. D., Reddan, W. G., Layton, C. R., and Dempsey, J. A. Effects of metabolic hyperthermia on performance during heavy prolonged exercise. J. Appl. Physiol. 36: 538-544, **1974**.

Máček, M. and Vávra, J. Cardiopulmonary and metabolic changes during exercise in children 6-14 years old. J. Appl. Physiol. 30: 200-204, **1971**.

MacFarlane, A. Daily mortality and environment in English conurbations. II. Deaths during summer spells in greater London. Environ. Res. 15: 332-341, **1978**.

MacInnis, J. B. Performance aspects of an open-sea saturation exposure at 615 feet, pp. 513-518. In: Underwater Physiology. Ed.: C. J. Lambertsen. New York: Academic Press, **1971**.

MacIntosh, D. L., Skrien, T., and Shephard, R. J. Physical activity and injury. A study of sports injuries at the University of Toronto, 1951-1968. J. Sport. Med. Phys. Fitness, 12: 224-237, **1972**.

MacKay, B. and Harrop, T. J. An experimental study of the longitudinal growth of skeletal muscle in the rat. Acta Anat. 72: 38–49, **1969**.

MacKay, D. E. Diving. In: Sports Medicine. Eds.: J. G. P. Williams and P. N. Sperryn. London: Arnold, **1976**.

McKay, R. S. and Rubissow, G. Detection of bubbles in tissues and blood, pp. 151–160. In: Underwater Physiology. Ed.: C. J. Lambertsen. New York: Academic Press, **1971**.

MacKeith, N. W., Pembrey, M. S., Spurrell, W. R., Warner, E. C., and Westlake, H. J. W. J. Observations on the adjustment of the human body to muscular work. Proc. Roy. Soc. B. 95: 413–439, **1923**.

MacKenzie, C. G. and Riesen, A. H. The production by moderate exercise of a high incidence of bends. J. Amer. Med. Assoc. 124: 499–501, **1944**.

MacKenzie, G. J., Taylor, S. H., McDonald, A. H., and Donald, K. W. Haemodynamic effects of external cardiac compression. Lancet 1: 1342–1345, **1964**.

MacKenzie, J. The Study of the Pulse, Arterial, Venous and Hepatic and of the Movements of the Heart. Edinburgh: Y. J. Pentland, **1902**.

Macklem, P. T. Partitioning of the pressure drop in the airways, pp. 85–97. In: Airway Dynamics: Physiology and Pharmacology. Ed.: A Bouhuys. Springfield, Ill.: C. C. Thomas, **1970**.

Macklem, P. T. and Mead. J. The physiological basis of common pulmonary function tests. Arch. Environ. Health, 14: 5–9, **1967**.

Maehlum, S. Muscle glycogen synthesis after a glucose infusion during post-exercise recovery in diabetic and non-diabetic subjects. Scand. J. Clin. Lab. Invest. 38: 349–354, **1978**.

Magazanik, A., Shapiro, Y., Meytes, D., and Meytes, J. Enzyme blood levels and water balance during a marathon race. J. Appl. Physiol. 36: 214–217, **1974**.

Magel, J. R., Foglia, G. F., McArdle, W. D., Gutin, B., Pechar, G. S., and Katch, F. I. Specificity of swim training on maximum oxygen uptake. J. Appl. Physiol. 38: 151–155, **1975**.

Magnus, R. Some results of studies in the physiology of posture. Lancet 1: 531–536; 1: 585–588, **1926**.

Magora, A. and Taustein, I. An investigation of the problem of sick-leave in the patient suffering from low back pain. Industr. Med. 38: 80–90, **1969**.

Magoun, H. W. The Waking Brain, 2nd ed. Springfield, Ill.: C. C. Thomas, **1963**.

Mahadeva, K., Passmore, R., and Woolf, B. Individual variations in metabolic cost of standardized exercises: Effects of food, age, sex and race. J. Physiol. (Lond.) 121: 225–231, **1953**.

Maher, J. T., Goodman, A., Francesconi, R., Bowers, W., Hartley, L., and Anglakos, E. Responses of rat myocardium to exhaustive exercise. Amer. J. Physiol. 222: 207–212, **1972**.

Maher, J. T., Jones, L. G., and Hartley, L. H. Effects of high-altitude exposure on submaximal endurance capacity of men. J. Appl. Physiol. 37: 895–898, **1974**.

Maher, J. T., Jones, L. G., Hartley, L. H., Williams, G. H., and Rose, L. I. Aldosterone dynamics during graded exercise at sea level and high altitude. J. Appl. Physiol. 39: 18–22, **1975**.

Mai, J. V., Edgerton, V. R., and Barnard, R. J. Capillarity of red, white and intermediate muscle fibers in trained and untrained guinea-pigs. Experientia 26: 1222–1223, **1970**.

Maio, D. A. and Farhi, L. E. Effect of gas density on mechanics of breathing. J. Appl. Physiol. 23: 687–693, **1967**.

Maksud, M. G. and Coutts, K. D. Application of of the Cooper twelve minutes run-walk test to young males. Res. Quart. 42: 54–59, **1971**.

Maksud, M. G., Coutts, K. D., Tristani, F. E., Dorchak, J. R., Barvoriak, J. J., and Hamilton, L. H. The effects of physical conditioning and propranolol on physical work capacity. Med. Sci. Sport. 4: 225–229, **1972**.

Malarkey, W. B. Recently discovered hypothalamic-pituitary hormones. Clin. Chem. 22: 5–15, **1976**.

Malhotra, M. S., Sen Gupta, J., and Joseph, N. T. Comparative evaluation of different training programmes on physical fitness. Ind. J. Physiol. Pharm. 17: 356–364, **1973**.

Malhotra, M. S., Sridharan, K., and Venkataswamy, Y. Potassium losses in sweat heat stress. Aviat. Space Environ. Med. 47: 503–504, **1976**.

Malhotra, M. S. and Wright, H. C. Arterial air embolism during decompression and its prevention. Proc. Roy. Soc. (Lond.) B 154: 418–427, **1960**.

———. The effects of a raised intrapulmonary pressure on the lungs of fresh unchilled cadavers. J. Path. Bacteriol. 82: 198–202, **1961**.

Malina, R. M. Growth, maturation and performance of Philadelphia Negro and white elementary schoolchildren. Doctoral dissertation, Univ. of Pennsylvania, Philadelphia, **1968**.

———. Exercise as an influence upon growth. Clin.

Pediatr. 8: 16–26, **1969**.

——. Anthropometric correlates of performance, pp. 249–274. In: Exercise and Sports Sciences Reviews, Vol. 3. Eds.: J. H. Wilmore and J. R. Keogh. New York: Academic Press, **1975**.

——. Growth, physical activity and performance in an anthropological perspective, pp. 3–28. In: Physical Activity and Human Well Being. Eds.: F. Landry and W. R. Orban. Miami: Symposia Specialists, **1978**.

Malina, R. M. and Rarick, G. L. A service for assessing the role of information feed-back in speed and accuracy of throwing performance. Res. Quart. 39: 220–223, **1968**.

Malmo, R. B. Anxiety and behavioural arousal. Psychol. Rev. 64: 276–287, **1957**.

Maloney, J. E. Some studies in pulmonary biophysics. Ph.D. dissertation, University of Sydney, **1965**.

Mancia, G., Iannos, J., Jamieson, G. G., Lawrence, R. H., Sharman, P. R., and Ludbrook, J. The effect of isometric handgrip exercise on the carotid sinus baroreceptor reflex in man. Clin. Sci. Mol. Med. 54: 33–37, **1978**.

Mandel, L. J., Riddle, T. G., and LaManna, J. C. A rapid scanning spectrophotometer and fluorometer for in vivo monitoring of steady-state and kinetic optical properties of respiratory enzymes, pp. 79–89. In: Oxygen Supply and Physiological Function. Ed.: F. F. Jöbsis. Dallas, Texas: Professional Information Library, **1977**.

Manguroff, J., Channe, N., and Georgieff, N. Tempo, Dosierung, Anzahl und Charakter der Ubungen für Berufsgymnastik. Vuprosi na Fiz. Kult. Sofia 5: 161. Quoted by Laporte, **1966**.

Mann, C. W., Berthelot-Berry, N. H., and Dauterive, H. J. The perception of the vertical. I. Visual and non-labyrinthine cues. J. Exp. Psychol. 39: 538–547, **1949**.

Mann, G. V. and Garrett, H. L. Lactate tolerance, diet, and physical fitness, pp. 31–41. In: Nutrition, Physical Fitness and Health. Eds.: J. Pařízková and V. A. Rogozkin. Baltimore: University Park Press, **1968**.

Mann, G. V., Garrett, L. H., Farlie, A., Murray, H., and Billings, F. T. Exercise to prevent coronary heart disease. Amer. J. Med. 46: 12–27, **1969**.

Mann, G. B., Shaffner, R. D., Anderson, R. S., and Sandstead, H. H. Cardiovascular disease in the Masai. J. Atheroscler. Res. 4: 289–312, **1964**.

Mann, G. V., Teel, K., Haynes, O., McNally, A., and Bruno, D. Exercise in the disposition of dietary calories. N. Engl. J. Med. 253: 349–355, **1955**.

Mann, J. I., Doll, R., Thorogood, M., Vessey, M. P., and Waters, W. E. Risk factors for myocardial infarction in young women. Brit. J. Prev. Soc. Med. 30: 94–100, **1976**.

Mann, R. H. and Burchell, H. B. Premature ventricular contractions and exercise. Proc. Staff Meet. Mayo Clin. 27: 383–389, **1952**.

Mansour, A. M. and Nass, M. M. K. In vivo cortisol action on RNA synthesis in rat liver nuclei and mitochondria. Nature (Lond.). 228: 665–667, **1970**.

Manz, R. L., Reid, D. C., and Wilkinson, J. G. A modified stair run as an indicator of anaerobic capacity as shown by selective glycogen depletion. Med. Sci. Sport. 10: 56, **1978**.

Maréchal, R., Pirnay, F., and Petit, J. M. Débit circulatoire pendant le contraction isométrique. Arch. Int. Physiol. Biochem. 81: 273–281, **1973**.

Maresh, M. Tissue changes in the individual during growth from X-rays of the extremities. Ann. N.Y. Acad. Sci. 110: 465–474, **1963**.

Margaria, R. Sulla fisiologia e specialmente sul consumo energetico della marcia e della corsa a varie velocità ed inclinazioni del terreno. Atti Accad. Naz. Lincei Memorie, Serie VI 7: 299–368, **1938**.

——. An outline for setting significant tests of muscular performance. In: Human Adaptability and Its Methodology. Eds.: H. Yoshimura and J. S. Weiner. Tokyo: Society for the Promotion of Sciences, **1966**.

——. Current concepts of walking and running. In: Frontiers of Fitness. Ed.: R. J. Shephard. Springfield, Ill.: C. C. Thomas, **1971**.

——. Biomechanics and energetics of muscular exercise. Oxford: Clarendon Press, **1976**.

Margaria, R., Aghemo, P., and Limas, F. P. A simple relation between performance in running and maximal aerobic power. J. Appl. Physiol. 38: 351–352, **1975**.

Margaria, R., Aghemo, P., and Rovelli, E. Indirect determination of maximal O_2 consumption in man. J. Appl. Physiol 20: 1070–1073, **1965**.

Margaria, R., Cerretelli, P., and Mangili, F. Balance and kinetics of anaerobic energy release during strenuous exercise in man. J. Appl. Physiol. 19: 623–628, **1964**.

Margaria, R., Edwards, H. T., and Dill, D. B. The possible mechanism of contracting and paying the oxygen debt and the role of lactic acid in muscular contraction. Amer. J. Physiol. 106: 689–715, **1933**.

Marino, G. W. and Dillman, C. J. Multiple regression models of the mechanics of the acceleration phase of ice skating, pp. 193–201. In: Biomechanics of Sports and Kinanthropometry. Eds.: F. Landry and W. A. R. Orban. Miami: Symposia Specialists, **1978**.

Maritz, J. S., Morrison, J. F., Peter, J., Strydom, N. B., and Wyndham, C. H. A practical method of estimating an individual's maximum oxygen intake. Ergonomics 4: 97–122, **1961**.

Maron, B. J. Cardiac causes of sudden death in athletes and considerations for screening athletic populations. In: Sports Cardiology. Eds.: T. Lubick and A. Venerando. Bologna: Aulo Gaggi, **1979**.

Maron, M. B., Wagner, J. A., and Horvath, S. M. Thermoregulatory responses during competitive marathon running. J. Appl. Physiol. 42: 909–914, **1977**.

Marsden, C. D., Merton, P. A., and Morton, H. B. Servo-action in the human thumb. J. Physiol. (Lond.) 257: 1–44, **1976**a.

Marsden, C. D., Merton, P. A., and Morton, H. B. Servo-action in human posture. J. Physiol. (Lond.) 263: 187p–188p, **1976b**.

Marsden, C. D., Merton, P. A., Morton, H. B., Hallett, M., Adam, J., and Rushton, D. N. Disorders of movement in cerebellar disease in man, pp. 179–199. In: The Physiological Aspect of Clinical Neurology. Ed.: F. C. Rose. Oxford: Blackwell Scientific, **1977**.

Marsh, M. E. and Murlin, J. R. Muscular efficiency on high carbohydrate and high fat diets. J. Nutr. 1: 105–137, **1928**.

Marshall, J. M. The cardiovascular response to stimulation of carotid chemoreceptors. J. Physiol. (Lond.) 226: 48p–49p, **1977**.

Marshall, R. J. Relationships between stimulus and work of breathing at different lung volumes. J. Appl. Physiol. 17: 917–921, **1962**.

Marshall, R. J., and Shepherd, J. T. Interpretation of changes in "central" blood volume and slope volume during exercise in man. J. Clin. Invest. 40: 375–385, **1961**.

Marshall, R. J., Wang, Y., Semler, H. J., and Shepherd, J. T. Flow, pressure, and volume relationships in the pulmonary circulation during exercise in normal dogs and dogs with divided left pulmonary artery. Circ. Res. 9: 53–59, **1961**.

Marston, S. B. and Tregear, R. T. Evidence for a complex between myosin and ADP in relaxed muscle fibers. Nature New Biol. 235: 23–24, **1972**.

Marteniuk, R. G. Individual differences in motor performance. Exer. Sport. Sci. Rev. 2: 103–130, **1974**.

Marteniuk, R. G. and Sullivan, S. J. Utilization of information in learning and controlling slow and fast movements. In: Motor Learning, Sport, Psychology, Pedagogy and Didactics of Physical Activity. Eds.: F. Landry and W. R. Orban. Miami: Symposia Specialists, **1978**.

Martens, R. Arousal and motor performance. Exer. Sport. Sci. Rev. 2: 155–188, **1974**.

Martin, B. J., Robinson, S., Wiegman, D. L., and Aulick, L. H. Effect of warm-up on metabolic responses to strenuous exercise. Med. Sci. Sport. 7: 146–149, **1975**.

Martin, R. E. and Wool, I. G. Formation of active hybrids from subunits of muscle ribosomes from normal and diabetic rats. Proc. Natl. Acad. Sci. U.S. 60: 569–574, **1968**.

Martin, R. P., Haskell, W. L., and Wood, P. D. Blood chemistry and lipid profiles of elite distance runners. Ann. N.Y. Acad. Sci. 301: 346–360, **1977**.

Martonosi, A., Gouvea, M. A., and Gergely, J. Studies on actin. III. G-F transformation of actin and muscular contraction (experiments in vivo). J. Biol. Chem. 235: 1707–1710. **1960**.

Maruyama, K. Some physico-chemical properties of β-actinin. Actin factor isolated from striated muscle. Biochim. Biophys. Acta 102: 542–548, **1965**.

Marx, H. J., Rowell, L. B., Conn, R. D., Bruce, R. A., and Kusumi, F. Maintenance of aortic pressure and total peripheral resistance during exercise in heat. J. Appl. Physiol. 22: 519–525, **1967**.

Mason, D. T., Spann, J. F., and Zelis, R. Quantification of the contractile state of the intact human heart. Maximal velocity of the contractile element shortening determined by the instantaneous relation between the rate of pressure rise and pressure in the left ventricle during isovolumic systole. Amer. J. Cardiol. 26: 248–257, **1970**.

Mason, J. W., Hartley, L. H., Kotchen, T. A., Wherry, F. E., Pennington, L. L., and Jones, L. G. Plasma thyroid-stimulating hormone response in anticipation of muscular exercise in the human. J. Clin. Endocrinol. Metab. 37: 403–406, **1973**.

Mason, R. E., Likar, I., Biern, R. O., and Ross,

R. S. Correlation of graded exercise electrocardiographic response with clinical and coronary cinearteriographic findings, pp. 445–455. In: Measurement in Exercise Electrocardiography. The Ernst Simonson Conference. Ed.: H. Blackburn. Springfield, Ill.: C. C. Thomas, **1969**.

Mass, J. W. and Landis, J. H. In vivo studies of the metabolism of norepinephrine in the central nervous system. J. Pharmacol. Exp. Ther. 163: 147–162, **1968**.

Massie, J., Rode, A., Skrien, T., and Shephard, R. J. A critical review of the "aerobics" points system. Med. Sci. Sport. 2: 1–6, **1970**.

Massie, J. and Shephard, R. J. Physiological and psychological effects of training. Med. Sci. Sport. 3: 110–117, **1971**.

Massion, J. and Paillard, J. Motor aspects of behaviour and programmed nervous activities. Coll. Int. CNRS 226. Brain Res. 71: 189–575, **1974**.

Master, A. M. The electrocardiogram after exercise. A standardized heart function test. U.S. Nav. Med. Bull. 40: 346–351, **1942**.

——. The Master Two-Step test. Some historical highlights and current concepts. J. S. Carol Med. Assoc. Suppl. (Dec.): 12–17, **1969**.

Master, A. M. and Jaffe, H. L. Electrocardiographic changes after exercise in angina pectoris. J. Mt. Sinai Hosp. 7: 629–632, **1941**.

Master, A. M., Van Liere, E. J., Lindsay, H. A., and Hartroft, W. S. Arterial blood pressure. In: Biology Data Book. Eds.: P. L. Altman and D. S. Dittmer. Washington, D.C.: Fed. Amer. Soc. Exp. Biol., **1964**.

Masterson, J. P., Lewis, H. E., and Widdowson, E. M. Nutrition and energy expenditure during a polar expedition. Adv. Sci. 13: 414–416, **1956**.

Mastropaolo, J. A., Stamler, J., Berkson, D. M. et al. Validity of phonoarteriographic blood pressures during rest and exercise. J. Appl. Physiol. 19: 1219–1233, **1964**.

Masuda, M., Shibayama, H., and Ebashi, H. Changes in arterial blood pressure during running and walking determined by a kind of indirect method. Bull. Phys. Fitness Res. Inst. (Japan) 11: 1–16, **1967**.

Matell, G. Time-courses of changes in ventilation and arterial gas tensions in man induced by moderate exercise. Acta Physiol. Scand. 58 (Suppl. 206): 1–53, **1963**.

Mathews, D. K. Measurement in Physical Education. Philadelphia: W. B. Saunders, **1963**.

Mathews, R. E. and Douglas, G. J. Sulphur-35 measurements of functional and total extracellular fluid in dogs with hemorrhagic shock. Surg. Forum 20: 3–5, **1969**.

Matthews, B. H. C. Nerve endings in mammalian muscle. J. Physiol. (Lond.) 78: 1–53, **1933**.

Matthews, P. B. C. Mammalian Muscle Receptors and Their Central Actions, pp. 1–630. London: Arnold, **1972**.

Matthews, P. B. C. and Stein, R. B. The sensitivity of muscle spindle afferents to small sinusoidal changes of length. J. Physiol. (Lond.) 200: 723–743, **1969**.

Mattingly, T. W. The post-exercise electrocardiogram. Its value in the diagnosis and prognosis of coronary arterial disease. Amer. J. Cardiol. 9: 395–409, **1962**.

Mattsson, S. The reversibility of disuse osteoporosis. Acta Orthop. Scand. Suppl. 144, **1972**.

Mauro, A. and Adams, W. R. The structure of the sarcolemma of the frog skeletal muscle fiber. J. Biophys. Biochem. Cytol. 10: 177–183, **1961**.

Mausner, J. S. and Bahn, A. K. Epidemiology: An Introductory Text. Philadelphia: W. B. Saunders, **1974**.

Maxwell, L. C. Oxidative capacity and capillarization of skeletal muscle. Med. Sci. Sport. 10: 58, **1978**.

Maxwell, L. C., Barclay, J. K., Mohrman, D. E., and Faulkner, J. A. Physiological characteristics of skeletal muscles of dogs and cats. Amer. J. Physiol. 233: C14–C18, **1977**.

Mayberry, R. P. Cited by Clarke, D. H. (1973). Proc. Coll. Phys. Educ. Assoc. 62: 155–158, **1959**.

Mayer, J. Regulation of energy intake and body weight: the glucostatic theory and the lipostatic hypothesis. Ann. N.Y. Acad. Sci. 63: 15–43, **1955**.

——. Exercise and weight control. In: Science and Medicine of Exercise and Sports, 1st ed. Ed.: W. E. Johnson. New York: Harper, **1960**.

——. Human Nutrition. Its Physiological, Medical and Social Aspects. A Series of Eighty-Two Essays. Springfield, Ill.: C. C. Thomas, **1972**.

Mayer, J. and Bullen, B. Nutrition and athletic performance. Physiol. Rev. 40: 369–397, **1960**.

Mayer, J., Marshall, N. B., Vitale, J. J., Christensen, J. H., Mashayekhi, M. B., and Stare, F. J. Exercise, food intake and body weight in normal rats and genetically obese adult mice. Amer. J. Physiol. 177: 544–548, **1954**.

Mazzone, R. W., Modell, H. I., and Farhi, L. E. Interaction of convection and diffusion in pulmonary gas transport. Resp. Physiol. 28: 217-225, **1976**.

McAllen, R. M. and Spyer, K. M. The baroreceptor input to cardiac vagal motoneurones. J. Physiol. (Lond.) 282: 365-374, **1978**.

McAllister, F. F., Bartsch, R., Jacobson, J., and D'Allesio, G. The accelerating effect of muscular exercise on experimental atherosclerosis. Arch. Surg. 80: 54-60, **1959**.

McAlpine, V., Milojevic, S., and Monkhouse, F. C. Changes in the electrophoretic patterns of euglobulin fractions following activation of the fibrinolytic system by exercise. Can. J. Physiol. Pharmacol. 49: 672-677, **1971**.

McArdle, J. J. and Albuquerque, E. X. A study of the reinnervation of fast and slow mammalian muscles. J. Gen. Physiol. 61: 1-23, **1973**.

McArdle, W. D., Katch, F. I., and Pechar, G. S. Comparison of continuous and discontinuous treadmill and bicycle tests for Max VO_2. Med. Sci. Sport. 5: 156-160, **1973**.

McArdle, W. D., Magel, J. R., Lesmes, G. R., and Pechar, G. S. Metabolic and cardiovascular adjustment to work in air and water at 18, 25, and 33 C. J. Appl. Physiol. 40: 85-90, **1976**.

McCabe, B. F. Vestibular suppression in figure skaters. Trans. Amer. Acad. Ophthal. Otolaryng. 64: 264-268, **1960**.

McCafferty, W. B. and Edington, D. W. The effects of prolonged direct muscle stimulation and recovery on biochemicals associated with glycolysis in rat skeletal muscle, pp. 135-138. In: Metabolic Adaptation to Prolonged Physical Exercise. Eds.: H. Howald and J. R. Poortmans. Basel: Birkhauser Verlag, **1975**.

McCally, M. Body fluid volumes and the renal response to immersion, pp. 253-270. In: Physiology of Breath-Hold Diving and the Ama of Japan. Eds.: H. Rahn and T. Yokoyama. Washington, D.C.: Natl. Acad. Sci. Natl. Res. Council Publ. 1341, **1965**.

McCance, R. A. Unconsidered mechanisms responsible for maintaining the stability of the internal environment. Tisdall Oration. Can. Med. Assoc. J. 75: 791-798, **1956**.

McCance, R. A. and Widdowson, E. M. The composition of foods. Spec. Rep. Series, Med. Res. Council (Lond.) 297, **1960**.

McCarthy, D. S., Craig, D. B., and Cherniak, R. M. The effect of acute intensive cigarette smoking on maximal expiratory flows and the single breath nitrogen washout. Amer. Rev. Resp. Dis. 113: 301-304, **1976**.

McComas, A. J., Sica, R. E. P., Upton, A. R. M., Longmire, D., and Caccia, M. R. Physiological estimates of the numbers and sizes of motor units in man, pp. 55-72. In: Control of Posture and Locomotion. Eds.: R. B. Stein, K. G. Pearson, R. S. Smith, and J. B. Redford. New York: Plenum Press, **1973**.

McCormick, E. J. Human Engineering. New York: McGraw-Hill, **1957**.

McDonald, G. A. and Fullerton, H. W. Effect of physical activity on increased coagulability of blood after ingestion of high-fat meal. Lancet 2: 600-601, **1958**.

McDonald, I. G. and Feigenbaum, H. Analysis of left ventricular wall motion by reflected ultrasound—application to assessment of myocardial function. Circulation 46: 14-25, **1972**.

McDonald, R. H., Taylor, R. R., and Cingolani, H. E. Measurement of myocardial developed tension and its relation to oxygen consumption. Amer. J. Physiol. 211: 667-673, **1966**.

McDonough, J. R. and Bruce, R. A. Maximal exercise testing in assessing cardiovascular function. In: Proc. National Conf. on Exercise in the Prevention, in the Evaluation and in the Treatment of Heart Disease. J. S. Carol. Med. Assoc. 65 (Suppl. 1): 26-33, **1969**.

McDonough, J. R., Haines, C., Stulb, S., and Garrison, G. Coronary heart disease among Negroes and whites in Evans County, Georgia. J. Chron. Dis. 18: 443-468, **1965**.

McEvoy, J. D. S. and Jones, N. L. Arterialized capillary blood gases in exercise studies. Med. Sci. Sport. 7: 312-315, **1975**.

McFadden, E. R., Haynes, R. L., and Ingram, R. H. Effect of exercise on airway mechanics in asthma and observations on mechanisms, pp. 279-286. In: Muscular Exercise and the Lung. Eds.: J. A. Dempsey and C. E. Reed. Madison: University of Wisconsin Press, **1977**.

McGandy, R. B., Hegsted, D. M., Myers, M. L., and Stare, F. J. Dietary carbohydrate and serum cholesterol levels in man. Amer. J. Clin. Nutr. 18: 237-242, **1966**.

McGarr, J. A., Oscai, L. B., and Borensztajan, J. Effects of exercise on hormone-sensitive lipase activity in rat adipocytes. Amer. J. Physiol. 230: 385-388, **1976**.

McGeoch, J. A. Forgetting and the law of disuse. Psychol. Rev. 39: 352-370, **1932**.

McGill, H. C. Abnormalities potentially mediating the effect of cigarette smoking on atherosclerosis, pp. 161-166. In: Atherosclerosis IV. Eds.:

G. Schettler, Y. Goto, Y. Hata, and G. Klose. Berlin: Springer Verlag, **1977**.

McGilvery, R. W. The use of fuels for muscular work. In: Metabolic Adaptation to Prolonged Physical Exercise. Eds.: H. Howald and J. R. Poortmans. Basel: Birkhauser Verlag, **1975**.

McGregor, M., Adam, W., and Sekelj, P. Influence of posture on cardiac output and minute ventilation during exercise. Circ. Res. 9: 1089-1092, **1961**.

McGuire, T. F., Talbott, G. D., Rosenbaum, D. A., Webber, J. M., and White, S. C. Cardiovascular effects of breathing against unbalanced atmospheric pressures. J. Amer. Med. Assoc. 163: 1209-1213, **1957**.

McHenry, P. L., Faris, J. V., Jordan, J. W., and Morris, S. N. Comparative study of cardiovascular function and ventricular premature complexes in smokers and nonsmokers during maximal treadmill exercise. Amer. J. Cardiol. 39: 493-498, **1977**.

McHenry, P. L., Fisch, C., Jordan, J. W., and Corya, B. R. Cardiac arrhythmias observed during maximal treadmill exercise testing in clinically normal men. Amer. J. Cardiol. 29: 331-336, **1972**.

McHenry, P. L., Stowe, D. E., and Lancaster, M. C. Computer quantitation of the ST segment response during maximal treadmill exercise. Clinical correlation. Circulation 38: 691-701, **1968**.

McKechnie, J., Leary, W., and Joubert, S. Some electrocardiographic and biochemical changes recorded in marathon runners. S. Afr. Med. J. 41: 722-725, **1967**.

McKeever, W. P., Gregg, D. E., and Canney, P. C. Oxygen uptake of the non-working left ventricle. Circ. Res. 6: 612-623, **1958**.

McKenzie, R. T. Treatment of convalescent soldiers by physical means. Proc. Roy. Soc. Med. (Surg. Sect.) 9: 31-70, **1916**.

McLellan, J. and Skinner, J. S. Use of anaerobic threshold as a basis for training. Can. J. Appl. Sport. Sci. 3: 180, **1978**.

McManus, B. M., Horvath, S. M., Bolduan, N., and Miller, J. C. Metabolic and cardiorespiratory responses to long-term work under hypoxic conditions. J. Appl. Physiol. 36: 177-182, **1974**.

McMichael, J. and Sharpey-Schafer, E. P. Cardiac output in man by a direct Fick method: effects of posture, venous pressure change, atropine and adrenaline. Brit. Heart J. 6: 33-40, **1944**.

McMorris, R. O. and Elkins, E. C. Study of production and evaluation of muscular hypertrophy. Arch. Phys. Med. Rehabil. 35: 420-426, **1954**.

McNamara, H. I., Sikorski, J. M., and Clavin, H. The effects of lower body negative pressure on hand blood flow. Cardiovasc. Res. 3: 284-291, **1969**.

McNeill, K. G. and Green, R. M. Measurements with a whole body counter. Can. J. Physics 37: 683-689, **1959**.

McPherson, B. D. Aging and involvement in physical activity: a sociological perspective, pp. 111-125. In: Physical Activity and Human Well-Being. Eds.: F. Landry and W. A. R. Orban. Miami: Symposia Specialists, **1978**.

McQueen, E. C. Composition of urinary casts. Lancet 1: 397-398, **1966**.

McRitchie, R. J., Vatner, S. F., Boettcher, D., Hendrick, G. R., Patrick, T. A., and Braunwald, E. Role of arterial baroreceptors in mediating cardiovascular response to exercise. Amer. J. Physiol. 230: 85-89, **1976**.

McTaggart, W. G. and Ribas-Cardus, F. Relationships between physical exercise and concentration of plasma lipids, pp. 149-160. In: Exercise Physiology. Eds.: F. Landry and W. A. R. Orban. Miami: Symposia Specialists, **1978**.

Mead, J., Lindgren, I., and Gaensler, E. A. The mechanical properties of the lungs in emphysema. J. Clin. Invest. 34: 1005-1016, **1955**.

Mead, J., Takishima, T., and Leith, D. Stress distribution in lungs: a model of pulmonary elasticity. J. Appl. Physiol. 28: 596-608, **1970**.

Mead, J., Turner, J. M., Macklem, P. T., and Little, J. B. Significance of the relationship between lung recoil and maximum expiratory flow. J. Appl. Physiol. 22: 95-108, **1967**.

Mead, J. and Whittenberger, J. L. Lung inflation and hemodynamics, pp. 477-486. In: Handbook of Physiology. Vol. 3. Respiration. Eds.: W. O. Fenn and H. Rahn. Washington, D.C.: Amer. Physiol. Soc., **1964**.

Meakins, J. and Long, C. N. H. Oxygen consumption, oxygen debt and lactic acid in circulatory failure. J. Clin. Invest. 4: 273-293, **1927**.

Medalie, J. H. Current developments in the epidemiology of atherosclerosis in Israel. In: Atherosclerosis. Proc. 2nd Int. Symp. Ed.: R. J. Jones. New York: Springer Verlag, **1970**.

Medical Research Council. Haemoglobin levels in

Great Britain in 1943 (with observations upon serum protein level). Committee on Haemoglobin Survey (Drury A. N., Chairman). Special Report Series 252: 1-128, **1945**.

——. Decompression sickness and aseptic necrosis of bone. Brit. J. Industr. Med. 28: 1-21, **1971**.

Meen, H. D., Gullestad, R., and Strømme, S. B. En sammenligning av maksimalt oksygenopptak under skisprint i motbakke og löp på tredemölle. Kroppsöving 7: 134, **1972**.

Meerson, F. Z. Compensatory hyperfunction of the heart. Circ. Res. 10: 250-258, **1962**.

Megaw, E. D. Directional errors and their detection in a discrete tracking task. Ergonomics 15: 633-643, **1972**.

Meldon, J. H. The theoretical role of myoglobin in steady-state oxygen transport to tissue and its impact upon cardiac output requirements. Acta Physiol. Scand. Suppl. 440: 93 (Abstr.), **1976**.

Mellander, S. and Johansson, B. Control of resistance, exchange and capacitance functions in the peripheral circulation. Pharm. Rev. 20: 117-196, **1968**.

Mellander, S., Johansson, B., Gray, S., et al. The effects of hyperosmolarity on intact and isolated vascular smooth muscle. Possible role in exercise hyperemia. Angiologica 4: 310-322, **1967**.

Mellander, S. and Lewis, D. H. Effect of hemorrhagic shock on the reactivity of resistance and capacitance vessels and on capillary filtration transfer in cat skeletal muscle. Circ. Res. 13: 105-118, **1963**.

Mellerowicz, H. Ergometrie. Munich: Urban & Schwarzenburg, **1962**.

Mellinkoff, S. M., Frankland, M., Boyle, D., and Greipel, M. Relationship between serum amino acid concentration and fluctuations in appetite. J. Appl. Physiol. 8: 535-538, **1956**.

Mellinkoff, S. M. and Machella, T. E. Effect of exercise upon liver following partial hepatectomy in albino rats. Proc. Soc. Exp. Biol. Med. 74: 484-486, **1950**.

Mendell, L. M. and Henneman, E. Terminals of single Ia fibers, location, density and distribution within a pool of 300 homonymous motoneurones. J. Neurophysiol. 34: 171-187, **1971**.

Mendez, J. and Kollias, J. Diet and starvation on the composition and calculated density of fat-free body mass. J. Appl. Physiol. 42: 731-734, **1977**.

Menotti, A., Ricci, G., Urbinati, G. C., Borgogelli, C., Conte, R., and Tigli, R. The habitual physical activity in the Rome project of coronary heart disease prevention, pp. 49-53. In: Nutrition, Dietetics and Sport. Eds.: G. Ricci and A. Venerando. Torino: Minerva Medica, **1978**.

Merlino, L. Influence of massage on jumping performance. Res. Quart. 30: 66-74, **1959**.

Merrilees, N. C. R. Some observations on the fine structure of a Golgi tendon organ of a rat, pp. 199-205. In: Symposium on Muscle Receptors. Ed.: D. Barker. Hong Kong: Hong Kong University Press, **1962**.

Merrill, E. G. The lateral respiratory neurones in the medulla: their associations with nucleus ambiguus, nucleus retroambigualis, the spinal accessory nucleus and the spinal cord. Brain Res. 24: 11-28, **1970**.

Merritt, F. L. and Weissler, A. M. Reflex venomotor alterations during exercise and hyperventilation. Amer. Heart J. 58: 382-387, **1959**.

Mertens, D. J., Shephard, R. J., and Kavanagh, T. Long-term exercise for chronic obstructive lung disease. Respiration 35: 96-107, **1978**.

Merton, P. A. Speculations on the servo-control of movement, pp. 247-255. In: The Spinal Cord. Ed.: G. E. W. Wolstenholme. London: Churchill, **1953**.

——. Voluntary strength and fatigue. J. Physiol. (Lond.). 123: 553-564, **1954**.

Messer, J. V., Wagman, R. J., Levine, H. J., Neill, W. A., Krausow, N., and Gorlin, R. Patterns of human myocardial oxygen extraction during rest and exercise. J. Clin. Invest. 41: 725-742, **1962**.

Metias, E. F., Cunningham, D. J. C., Howson, M. G., Strange Petersen, E., and Wolf, C. B. The reflex effects on the human respiratory pattern of alternating the time profile of inspiratory PCO_2 during steady hypoxia. Pflüg. Arch. 373: R39 (Abstr.), **1978**.

Metivier, G. The effects of long-lasting physical exercise and training on hormonal regulation, pp. 276-292. In: Metabolic Adaptation to Prolonged Physical Exercise. Basel: Birkhauser Verlag, **1975**.

Metivier, G., Poortmans, J. R., and Vanroux, R. Metabolic controls of human growth hormone (HGH) in trained athletes performing various workloads, pp. 209-218. In: 3rd International Symposium on Biochemistry of Effort. Eds.: F. Landry and W. A. R. Orban. Miami: Symposia Specialists, **1978**.

Metz, K. F. and Alexander, J. F. An investigation of the relationship between maximal aerobic capacity and physical fitness in twelve to fifteen year-old boys. Res. Quart. 41: 75–82, **1970**.

Metzger, H. PO_2 histograms of three dimensional systems with homogenous and inhomogenous microcirculation. A digital computer study, pp. 18–24. In: Oxygen Supply: Theoretical and Practical Aspects of Oxygen Supply and Microcirculation of Tissue. Eds.: M. Kessler, D. F. Bruley, L. C. Clark, D. W. Lübbers, I. A. Silver, and J. Strauss. Baltimore: University Park Press, **1973**.

Meyer, F. R., Robinson, S., Newton, J. L., Ts'ao, C. H., and Holgersen, L. O. The regulation of the sweating response to work in man. Physiologist 5: 182 (Abstr.), **1962**.

Meyerhof, O. Die Energieumwandlungen im Muskel. I. Ueber die Beziehungen der Milchsaure zur Warmebildung und Arbeitleistung des Muskels in der Anaerobiose. Pflüg. Archiv. 182: 232–283, **1920**.

Meyerhof, O. and Lohmann, K. Über die naturlichen Guanidinophosphorsäuren (Phosphagene) in der quergestreiften Muskulatur. I. Das physiologische Verhalten der Phosphagene. Biochem. Z. 196: 22–48, **1928**.

——. Über energetische Wechselbeziehungen zwischen dem Umsatz der Phosphorsäureester im Muskelextrakt. Biochem. Z. 253: 431–461, **1932**.

Meyers, C. R. Effects of two isometric routines on strength, size, and endurance in exercised and non-exercised arms. Res. Quart. 38: 430–440, **1967**.

Miall, W. E., Ashcroft, M. T., Lovell, H. G., and Moore, F. A longitudinal study of the decline in adult height with age in two Welsh communities. Human Biol. 39: 445–454, **1967**.

Michaelis, H. and Muller, E. A. Die Bedeutung des alveolaren CO_2-Druckes für die Bestimmung des auf die Atmung entfallenden Energieverbrauches. Arbeitsphysiol. 12: 85–91, **1942**.

Michel, C. C. The transport of oxygen and carbon dioxide by the blood, pp. 67–104. In: M.T.P. International Review of Science. Respiratory Physiology. Ed.: J. G. Widdicombe. London: Butterworths, **1974**.

Middleman, S. Transport Phenomena in the Cardiovascular System. New York: Interscience, **1972**.

Middleton, E. Airway smooth muscle in exercise-induced bronchospasm: Some speculations. Pediatrics 56 (Suppl.): 944–947, **1975**.

Miettinen, T. A. Effect of cholestyramine on fecal steroid excretion and cholesterol synthesis in patients with hypercholesterolemia, pp. 558–562. In: Atherosclerosis. Proceedings of 2nd International Symposium. Ed.: R. J. Jones. Berlin: Springer Verlag, **1970**.

Miles, S. Underwater Medicine. London: Staples Press, **1966**.

Milhorn, H. T. and Guyton, A. C. An analog computer analysis of Cheyne-Stokes breathing. J. Appl. Physiol. 20: 328–333, **1965**.

Milic-Emili, J. Pulmonary statics, pp. 105–137. In: MTP International Review of Science. Respiratory Physiology. Ed.: J. G. Widdicombe. London: Butterworths, **1974**.

Milic-Emili, G., Cerretelli, P., Petit, J. M., and Falconi, C. La consommation d'oxygène en fonction de l'intensité de l'exercise musculaire. Arch. Int. Physiol. Biochim. 67: 10–14, **1959**.

Milic-Emili, J., Henderson, J. A. M., Dolovich, M. B., Trop, D., and Kaneko, K. Regional distribution of inspired gas in the lung. J. Appl. Physiol. 21: 749–759, **1966**.

Milic-Emili, G. and Petit, J. M. Il lavoro meccanico della respirazione a varia frequenza respiratoria. Arch. Sci. Biol. 43: 326–330, **1959**.

——. Mechanical efficiency of breathing. J. Appl. Physiol. 15: 359–362, **1960**.

Milic-Emili, G., Petit, J. M., and Deroanne, R. The effects of respiratory rate on the mechanical work of breathing during muscular exercise. Int. Z. Angew. Physiol. 18: 330–340, **1960**.

Millard, R. W., Higgins, C. B., Franklin, D., and Vatner, S. F. Regulation of the renal circulation during severe exercise in normal dogs and dogs with experimental heart failure. Circ. Res. 31: 881–888, **1972**.

Miller, A. T. The influence ot oxygen administration on cardiovascular function during exercise and recovery. J. Appl. Physiol. 5: 165–168, **1952**.

Miller, D. S. and Payne, P. R. Weight maintenance and food intake. J. Nutr. 78: 255–262, **1962**.

Miller, J. N., Wangensteen, O. D., and Lanphier, E. D. Ventilatory limitations on exertion at depth, pp. 317–323. In: Underwater Physiology. Ed.: C. J. Lambertsen. New York: Academic Press, **1971**.

Miller, N. C. and Walters, R. F. Interactive modelling as a forcing function for research in the physiology of human performance. Simulation 21: 1-13, **1974**.

Miller, W. F. and Taylor, H. F. Exercise training in the rehabilitation of patients with severe respiratory insufficiency due to pulmonary emphysema. The role of oxygen breathing. South. Med. J. 55: 1216-1221, **1962**.

Millikan, G. A. The kinetics of muscle haemoglobin. Proc. Roy. Soc. Ser. B. 120: 366-388, **1936**.

——. Experiments on muscle haemoblogin *in vivo*; the instantaneous measurement of muscle metabolism. Proc. Roy. Soc. (Biol.). 123: 218-241, **1937**.

Mills, J. E., Sellick, H., and Widdicombe, J. G. Epithelial irritant receptors in the lungs, p. 77. In: Breathing: Hering-Breuer Centenary Symposium. Ed.: R. Porter. London: Churchill, **1970**.

Mills, J. N. Transmission processes between clock and manifestations, pp. 28-84. In: Biological Aspects of Circadian Rhythms. Ed.: J. N. Mills. London: Plenum Press, 1973a.

——. Biological Aspects of Circadian Rhythms. London: Plenum Press, **1973b**.

Millward, D. J. Protein turnover in skeletal muscle. II. The effect of starvation and a protein free diet on the synthesis and catabolism of skeletal muscle proteins in comparison to liver. Clin. Sci. 39: 591-603, **1970**.

Millward, D. J. and Waterlow, J. C. Effect of nutrition on protein turnover in skeletal muscle. Fed. Proc. 37: 2283-2290, **1978**.

Milner-Brown, H. S., Stein, R. B., and Yemm, R. The orderly recruitment of human motor units during voluntary isometric contractions. J. Physiol. (Lond.) 230: 359-370, **1973**.

Milnor, W. R., Knickerbocker, G. G., and Kouwenhoven, W. B. Cardiac responses to transthoracic capacitor discharges in the dog. Circ. Res. 6: 60-65, **1958**.

Minaire, Y. and Forichion, J. Lactate metabolism and glucose lactate conversion in prolonged exercise, pp. 106-112. In: Metabolic Adaptation to Prolonged Physical Exercise. Eds.: H. Howald and J. R. Poortmans. Basel: Birkhauser Verlag, **1975**.

Minard, D. Prevention of heat casualties in Marine Corps recruits. Period of 1955-60, with comparative incidence rates and climatic heat stresses in other training categories. Milit. Med. 126: 261-272, **1961**.

Minard, D. and Copman, L. Elevation of body temperature in disease. In: Temperature: Its Measurement and Control in Science and Industry, Vol. 3(3): 253. Ed.: J. D. Hardy. New York: Reinhold, **1963**.

Minty, K. B. and Guz, A. The lower end of the human carboxyhaemoglobin (COHb) dissociation curve. Clin. Sci. Mol. Biol. 54: 10p, **1978**.

Mirsky, I. Left ventricular stresses in the intact human heart. Biophysical J. 9: 189-208, **1969**.

——. Review of various theories for the evaluation of left ventricular wall stresses, pp. 381-409. In: Cardiac Mechanics: Physiological, Clinical and Mathematical Considerations. New York: Wiley, **1974**.

Mirsky, I. and Parmley, W. W. Force-velocity studies in isolated and intact heart muscle, pp. 87-112. In: Cardiac Mechanics: Physiological, Clinical and Mathematical Considerations. Eds.: I. Mirsky, D. Ghista, and H. Sandler. New York: Wiley, **1974**.

Mirsky, I., Pasternac, A., Ellison, R. C., and Hugenholtz, P. G. Clinical applications of force-velocity parameters and the concept of a "normalized velocity," pp. 293-329. In: Cardiac Mechanics: Physiological, Clinical and Mathematical Considerations. Eds.: I. Mirsky, D. Ghista, and H. Sandler. New York: Wiley, **1974**.

Mitchell, J. H., Sproule, B. J., and Chapman, C. B. Factors influencing respiration during heavy exercise. J. Clin. Invest. 37: 1693-1701, **1958**.

Mitchell, J. W., Nadel, E. R., and Stolwijk, A. J. Respiratory weight losses during exercise. J. Appl. Physiol. 32: 474-476, **1972**.

Mitchell, R. A. Cerebrospinal fluid and the regulation of respiration, pp. 1-47. In: Advances in Respiratory Physiology. Ed.: C. G. Caro. London: Arnold, **1966**.

Miyamoto, Y. and Moll, W. Measurement of dimension and pathway of red cells in rapidly frozen lungs in situ. Resp. Physiol. 12: 141-156, **1971**.

Miyamura, M., Kuroda, H., Hirata, K., and Honda, Y. Evaluations of the step test scores based on the measurement of maximal aerobic powers. J. Sport. Med. Phys. Fitness, 15: 316-322, **1975**.

Miyamura, M., Yamashina, T. and Honda, Y. Ventilatory responses to CO_2 rebreathing at rest and during exercise in untrained subjects and athletes. Jap. J. Physiol. 26: 245-254, **1976**.

Miyashita, M. Method of calculating overall efficiency in swimming crawl stroke, pp. 135–142. In: Biomechanics of Sports and Kinanthropometry. Eds.: F. Landry and W. A. R. Orban. Miami: Symposia Specialists, **1978**.

Moccetti, T. and Lichtlen, P. Prognostic aspects of coronary angiography based on a one to six year control, pp. 142–148. In: Coronary Heart Disease. Eds.: M. Kaltenbach, P. Lichtlen, and G. C. Friesinger. Stuttgart: Thieme, **1973**.

Mocellin, R. and Sebening, W. Investigations on cardiac output and oxygen uptake in 8–14 year old boys. Acta Paediatr. Belg. 28 (Suppl.): 113–120, **1974**.

Moe, G. K., Harris, A. S., and Wiggers, C. J. Analysis of the initiation of fibrillation by electrocardiographic studies. Amer. J. Physiol. 134: 473–492, **1941**.

Moesch, H. and Howald, H. Hexokinase (HK), Glyceraldehyde-3P-dehydrogenase(GAPDH), succinate dehydrogenase (SDH), and 3-hydroxyacyl-CoA-dehydrogenase (HAD) in skeletal muscle of trained and untrained men, pp. 463–465. In: Metabolic Adaptation to Prolonged Physical Exercise. Eds.: H. Howald and J. R. Poortmans. Basel: Birkhauser Verlag, **1975**.

Moffatt, R. J. and Stamford, B. A. Effects of pedalling rate changes on maximal oxygen uptake and perceived effort during bicycle ergometer work. Med. Sci. Sport. 10: 27–31, **1978**.

Moffroid, M., Whipple, R., Hofkosh, J., Lowman, E., and Thistle, H. A study of isokinetic exercise. Phys. Ther. 49: 735–747, **1969**.

Mogensen, L. Exercise testing and continuous long-term e.c.g. recording in the detection of arrhythmics and ST-T changes, pp. 131–141. In: Coronary Heart Disease, Exercise Testing and Cardiac Rehabilitation. Eds.: W. E. James and E. A. Amsterdam. Miami: Symposia Specialists, **1977**.

Mohme-Lundholm, E., Svedmyr, N., and Vamos, N. Enzymatic micro-method for determining the lactic acid content of finger-tip blood. Scand. J. Clin. Lab. Invest. 17: 501–502, **1965**.

Molbech, S. V. Average percentage force at repeated maximal isometric muscle contractions at different frequencies. Comm. Test. Obs. Inst. 16: 3–12, **1963**.

Molé, P. A. Increased contractile potential of papillary muscles from exercise-trained rat hearts. Amer. J. Physiol. 234: H421–H425, **1978**.

Molé, P. A., Oscai, L. B., and Holloszy, J. O. Adaptation of muscle to exercise. Increase in levels of palmityl CoA synthetase, carnitine palmityl transferase and palmityl CoA dehydrogenase, and in the capacity to oxidize fatty acids. J. Clin. Invest. 50: 2323–2330, **1971**.

Molé, P. A. and Rabb, C. Force-velocity relations in exercise-induced hypertrophied rat heart muscle. Med. Sci. Sport. 5: 69, **1973**.

Molina, C. and Giorgi, E. Il metabolismo respiratorio dei soggetti anziani durante l'esercizio muscolare. Med. Lavoro 42: 315–325, **1951**.

Molnar, G. W., Towbin, E. J., Gosselin, R. E., Brown, A. H., and Adolph, E. F. A comparative study of water, salt and heat exchanges of men in tropical and desert environments. Cited by C. Wyndham. Amer. J. Hyg. 44: 411–433, **1946**.

Monod, H. and Scherrer, J. Capacité de travail statique d'un groupe musculaire synergique chez l'homme. C. R. Soc. Biol. Paris 151: 1358–1362, **1957**.

Monroe, R. G. and French, G. N. Left ventricular pressure-volume relationships and myocardial oxygen consumption in the isolated heart. Circ. Res. 9: 362–374, **1961**.

Montoye, H. J. Summary of research on the relationship of exercise to heart disease. J. Sport. Med. Fitness 2: 35–43, **1960**.

——. Physical Activity and Health: An epidemiologic Study of an Entire Community. Englewood Cliffs, N.J.: Prentice-Hall, **1975**.

Montoye, H. J. and Gayle, R. Familial relationships in maximal oxygen uptake. Human Biol. 50: 241–250, **1978**.

Montoye, H. J., Van Huss, W. D., Olson, H., Pierson, W. R., and Hudec, A. J. The longevity and morbidity of college athletes. Michigan State University, Phi Epsilon Kappa Fraternity, **1957**.

Montoye, H. J., Willis, P. W., and Cunningham, D. A. Heart rate response to submaximal exercise: Relation to age and sex. J. Gerontol. 23: 127–133, **1968**.

Mookerjee, S. M. Impact of yoga training on some physiological norms. Proceedings, Vol. 11, 26th International Congress, International Union of Physiological Sciences, **1974**.

Mookerjee, S., Chahal, K. S., and Giri, C. Impact of yogic exercises on the Indian field hockey team: Winners of the third world cup, 1975, pp. 389–396. In: Exercise Physiology. Eds.: F. Landry and W. A. R. Orban.

Miami: Symposia Specialists, **1978**.

Moore, F. D. Relevance of experimental shock studies to clinical shock problems. Fed. Proc. 20 (Suppl. 9): 227-232, **1961**.

Moore, F. D., Olesen, K. H., McMurrey, J. D., Parker, H. V., Ball, M. R., and Boyden, C. M. The Body Cell Mass and Its Supporting Environment. Philadelphia: W. B. Saunders, **1963**.

Moore, M. J., Rebeiz, J. J., Holden, M., and Adams, R. D. Biometric analyses of normal skeletal muscle. Acta Neuropathol. (Berl.) 19: 51-69, **1971**.

Moores, B. A. A comparison of work-load using physiological and time-study assessments. Ergonomics 13: 769-776, **1970**.

Morales, M. F., Botts, J., Blum, J., and Hill, T. L. Elementary process in muscle action: an examination of current concepts. Physiol. Rev. 35: 475-505, **1955**.

Morehouse, C. A. Development and maintenance of isometric strength of subjects with diverse initial strengths. Res. Quart. 38: 449-456, **1967**.

Morgan, H. E., Regen, D. M., and Park, C. R. Identification of a mobile carrier-mediated sugar transport system in muscle. J. Biol. Chem. 239: 369-374, **1964**.

Morgan, P., Gildiner, M., and Wright, G. R. Smoking reduction in adults who take up exercise: A survey of a running club for adults. CAHPER J. 42: 39-43, **1976**.

Morgan, R. E. and Adamson, G. T. Circuit Training. London: Bell, **1965**.

Morgan, T. E., Coff, L. A., Short, F. A., Ross, R., and Gunn, D. R. Effects of long-term exercise on human muscle mitochondria. In: Muscle Metabolism During Exercise, pp. 87-95. Eds.: B. Pernow and B. Saltin. New York: Plenum Press, **1971**.

Morgan, T. E., Short, F. A., and Coff, L. A. Alterations in human skeletal muscle lipid composition and metabolism induced by physical conditioning, pp. 116-121. In: Biochemistry of Exercise. Ed.: J. R. Poortmans. Basel: Karger, **1969**.

Morgan, W. P. Ergogenic Aids and Muscular Performance. New York: Academic Press, **1972**.

Morganroth, J. and Maron, B. J. The athlete's heart syndrome: A new perspective. Ann. N.Y. Acad. Sci. 301: 931-939, **1977**.

Mori, S., Reynolds, P. J., and Brookhart, J. M. Contribution of pedal afferents to postural control in the dog. Amer. J. Physiol. 218: 726-734, **1970**.

Morimoto, K. and Harrington, W. F. Isolation and physical chemical properties of an M-line protein from skeletal muscle. J. Biol. Chem. 247: 3052-3061, **1972**.

Morimoto, T., Slabochova, Z., Sargent, F., and Naman, R. K. Sex differences in physiological reactions to thermal stress. J. Appl. Physiol. 22: 526-532, **1967**.

Moritz, A. R., Henriques, F. C., and McLean, R. The effect of inhaled heat on the air passages and lungs: an experimental investigation. Amer. J. Pathol. 21: 311-331, **1945**.

Moritz, A. R. and Weisiger, J. R. Effects of cold air on the air passages and lungs. Arch. Int. Med. 75: 233-240, **1945**.

Moroff, S. V. and Bass, D. E. Effects of overhydration on man's physiological responses to work in the heat. J. Appl. Physiol. 20: 267-270, **1965**.

Morris, C. B. The measurement of the strength of muscle relative to the cross-section. Res. Quart. 19: 295-303, **1948**.

Morris, J. N. The epidemiology of coronary artery disease. In: International symposium on exercise and coronary artery disease. Ed.: T. Kavanagh. Toronto: Toronto Rehabilitation Centre, **1975**.

Morris, J. N., Adams, C., Chave, S. P. N., Sirey, C., Epstein, L., and Sheehan, D. J. Vigorous exercise in leisure time and the incidence of coronary heart disease. Lancet 1: 333-339, **1973**.

Morris, J. N. and Crawford, M. D. Coronary heart disease and physical activity of work. Brit. Med. J. 2: 1485-1496, **1958**.

Morris, J. N., Heady, J. A., and Raffle, P. A. B. Physique of London busmen, epidemiology of uniforms. Lancet 2: 569-570, **1956**.

Morris, J. N., Heady, J., Raffle, P., Roberts, C., and Parks, J. Coronary heart disease and physical activity of work. Lancet 2: 1053-1057, 1111-1120, **1953**.

Morris, J. N., Kagan, A., Pattison, D. C., Gardner, M. J., and Raffle, P. A. B. Incidence and prediction of ischaemic heart disease in London busmen. Lancet 2: 553-559, **1966**.

Morris, M. D. and Greer, W. E. Familial hyperbetalipoproteinemia (Type II hyperlipoproteinemia) in the Rhesus monkey, pp. 192-196. In: Atherosclerosis. Proc. 2nd Int. Symp. Ed.: R. J. Jones. Berlin: Springer Verlag, **1970**.

Morrison, S. D. The relationship of energy expenditure and spontaneous activity to aphagia of rats with lesions in the lateral hypothalamus.

J. Physiol. (Lond.) 197: 325-343, **1968**.
——. The hypothalamic syndrome in rats. Fed. Proc. 36: 139-142, **1977**.

Morrow, J. R., Jackson, A. S., and Bell, J. A. The function of age, sex, and body mass on distance running. Res. Quart. 49: 491-497, **1978**.

Morrow, P. E. Dynamics of dust removal from the lower airways: measurements and interpretations based upon radioactive aerosols, pp. 299-312. In: Airway Dynamics: Physiology and Pharmacology. Ed.: A. Bouhuys. Springfield, Ill.: C. C. Thomas, **1970**.

Morrow, P. E., Bates, D. V., Fish, B. R., Hatch, T. F., and Mercer, T. T. Deposition and retention models for internal dosimetry of the human respiratory tract. Health Physics 12: 173-207, **1966**.

Mortimer, J. T., Magnusson, R., and Petersen, I. Conduction velocity in ischemic muscle. Effect on EMG frequency spectrum. Amer. J. Physiol. 219: 1324-1329, **1970**.

Morton, A. R. and Fitch, K. D. Sodium cromoglycate BP in the prevention of exercise-induced asthma. Med. J. Austr. 2: 158-162, **1974**.

Morton, D. J. Human Locomotion and Body Form, pp. 1-285, chapters 31 and 33. Baltimore: Williams & Wilkins, **1952**.

Moss, F. P. and Leblond, C. P. Satellite cells as the source of nuclei in muscles of growing rats. Anat. Rec. 170: 421-436, **1971**.

Most, A. S., Brachfeld, N., Gorlin, R., and Wahren, J. Free fatty acid metabolism of the human heart at rest. J. Clin. Invest. 48: 1177-1188, **1969**.

Mostardi, R., Kubica, R., Veicsteinas, A., and Margaria, R. The effect of increased body temperature due to exercise on the heart rate and on the maximal aerobic power. Europ. J. Appl. Physiol. 33: 237-245, **1974**.

Mostyn, E. M., Helle, S., Gee, J. B. L., Bentivoglio, L. G., and Bates, D. V. Pulmonary diffusing capacity of athletes. J. Appl. Physiol. 18: 687-695, **1963**.

Motsay, G. J., Alho, A., Jaeger, T., Dietzman, R. H., and Lillehei, R. C. Effects of corticosteroids on the circulation in shock: experimental and clinical results. Fed. Proc. 29: 1861-1873, **1970**.

Mottram, R. F. Metabolism of exercising muscle. In: Frontiers of Fitness, pp. 61-78. Ed.: R. J. Shephard. Springfield, Ill.: C. C. Thomas, **1971**.

Mountcastle, V. B. The view from within: Pathways to the study of perception. Johns Hopkins Med. J. 136: 109-131, **1975**.

Moxley, R. T., Brakman, P., and Åstrup, T. Resting levels of fibrinolysis in blood in inactive and exercising men. J. Appl. Physiol. 28: 549-552, **1970**.

Mrzena, B. and Máček, M. Use of treadmill and working capacity assessment in preschool children, pp. 29-31. In: Pediatric Work Physiology. Eds.: J. Borms and M. Hebbelinck. Basel: Karger, **1978**.

Muckle, D. S. Glucose syrup ingestion and team performance in soccer. Brit. J. Sport. Med. 7: 340-343, **1973**.

Mugrage, E. R. and Andresen, M. I. Red blood cell values in adolescence. Amer. J. Dis. Child. 56: 997-1003, **1938**.

Muido, L. The influence of body temperature on performance in swimming. Acta Physiol. Scand. 12 (Suppl.): 36-39, 102-109, **1946**.

Muir, R. B. and Porter, R. The effect of a preceeding stimulus on temporal facilitation at corticomotoneuronal synapses. J. Physiol. (Lond.) 228: 749-763, **1973**.

Müller, E. A. Ein Leistungs-Pulsindex als Mass der Leistungsfähigkeit. Arbeitsphysiol. 14: 271-284, **1950**.
——. Physiology of muscle training. Rev. Can. Biol. 21: 303-313, **1962**.
——. How to keep fit during a voyage in space. New Scientist 17: 187-189, **1963**.

Müller, E. A. and Rohmert, W. Die Geschwindigkeit der Muskelkraft-Zunahme bei isometrischem Training. Int. Z. Angew. Physiol. 19: 403-419, **1963**.

Munscheck, H. Primary illness of the heart and sudden death by physical activity. In: Sports Cardiology. Eds.: T. Lubich and A. Venerando. Bologna: Aulo Gaggi, **1979**.

Murphy, R. J. and Ashe, W. F. Prevention of heat illness in football players. J. Amer. Med. Assoc. 194: 650-654, **1965**.

Murtomaa, M. and Korttila, K. The beginning of resuscitation by laymen: Mobile intensive care unit and emergency room. Anaesthetist 23: 398-402, **1974**.

Murukami (1960). Cited by Zachar (1971).

Mustard, J. F. and Murphy, E. A. Effect of smoking on blood coagulation and platelet survival in man. Brit. Med. J. 1: 846-849, **1963**.

Myasnikov, A. L. Influence of some factors on development of experimental cholesterol atherosclerosis. Circulation 17: 99-113, **1958**.

Myers, R. D., Bender, S. A., Krstic, M. K., and Brophy, P. D. Feeding produced in the satiated rat by elevating the concentration of calcium in the brain. Science 176: 1124-1125, **1972**.

Myrhe, L. G. and Kessler, W. V. Body density and K^{40} measurements of body composition as related to age. J. Appl. Physiol. 21: 1251-1255, **1966**.

Nadel, E. R., Holmér, I., Bergh, U., Åstrand, P. O., and Stolwyk, J. A. J. Energy exchanges of swimming man. J. Appl. Physiol. 36: 465-471, **1974**.

Nadel, E. R. and Stolwijk, J. A. J. Effect of skin wettedness on sweat gland response. J. Appl. Physiol. 33: 689-694, **1973**.

Nadel, E. R., Wenger, C. B., Roberts, M. F., Stolwijk, J. A. J., and Cafarelli, E. Physiological defences against hyperthermia of exercise. Ann. N.Y. Acad. Sci. 301: 98-109, **1977**.

Nageotte, M. E. and Kasch, F. W. Blood pressure response to continuous and intermittent exercise in hypertensive and normotensive men. Med. Sci. Sport. 10: 36, **1978**.

Nagle, F. J., Balke, B., and Naughton, J. P. Gradational step tests for assessing work capacity. J. Appl. Physiol. 20: 745-748, **1965**.

Nagle, F. J., Hagberg, J., and Kamei, S. Maximal O_2 uptake of boys and girls aged 14-17. Europ. J. Appl. Physiol. 36: 75-80, **1977**.

Naimark, A., Jones, N. L., and Lal, S. The effect of hypoxia on gas exchange and arterial lactate and pyruvate concentration during moderate exercise in man. Clin. Sci. 28: 1-13, **1965**.

Naimark, A., Wasserman, K., and McIlroy, M. B. Continuous measurement of ventilatory exchange ratio during exercise. J. Appl. Physiol. 19: 644-652, **1964**.

Nakhjavan, F. K., Natarajan, G., Smith, A. M., Drutch, M., and Goldberg, H. Myocardial lactate metabolism during isometric hand-grip test: Comparison with pacing tachycardia. Brit. Heart J. 37: 79-84, **1975**.

Nashner, L. M. Vestibular postural control model. Kybernetik 10: 106-110, **1972**.

——. Vestibular and reflex control of normal standing, pp. 291-308. In: Control of Posture and Locomotion. Eds.: R. B. Stein, K. G. Pearson, R. S. Smith, and J. B. Redford. New York: Plenum Press, **1973**.

Naughton, J. The National Exercise and Heart Disease Project. Manual of Operations. Washington, D.C.: George Washington University, **1978**.

Needham, D. M. Red and white muscle. Physiol. Rev. 6: 1-27, **1926**.

——. Biochemistry of muscle, pp. 363-415. In: The Structure and Function of Muscle, 2nd ed. Vol. 3. Physiology and Biochemistry. Ed.: G. H. Bourne. New York: Academic Press, **1973**.

Neely, J. R. and Morgan, H. E. Relationship between carbohydrate and lipid metabolism and the energy balance of heart muscle. Ann. Rev. Physiol. 31: 413-459, **1974**.

Neil, E. Afferent impulse activity in cardiovascular receptor fibers. Physiol. Rev. 40 (Suppl. 4): 201-208, **1960**.

Neil, E. and Joels, N. The carotid glomus sensory mechanism. In: Regulation of Human Respiration. Eds.: D. J. C. Cunningham and B. B. Lloyd. Oxford: Blackwell Scientific, **1963**.

Neill, W. A., Duncan, D. A., Kloster, F., and Mahler, D. J. Response of coronary circulation to cutaneous cold. Amer. J. Med. 56: 471-476, **1974**.

Neisser, U. Cognitive Psychology. New York: Appleton-Century-Crofts, **1967**.

Nelms, J. D. and Soper, J. E. Cold vasodilatation and cold acclimatization in the hand of British fish filleters. J. Appl. Physiol. 17: 444-448, **1962**.

Nelson, R., Brooks, C. M., and Pike, N. L. Biomechanical comparison of male and female distance runners. Ann. N.Y. Acad. Sci. 301: 793-807, **1977**.

Nelson, R. R., Gobel, F. L., Jorgensen, C. R., Wang, K., Wang, Y., and Taylor, H. L. Hemodynamic predictors of myocardial oxygen consumption during static and dynamic exercise. Circulation 50: 1179-1189, **1974**.

Nevison, T. O. Physical performance, total body water and monitoring of various physiological parameters at 15,000 feet and above, pp. 57-61. In: The International Symposium on the Effects of Altitude on Physical Performance. Ed.: R. Goddard. Chicago: Athletic Institute, **1967**.

Newhouse, M. T., Wright, F. J., Dolovich, M., and Hopkins, O. L. Clearance of RISA aerosol from the human lung. In: Airway Dynamics. Physiology and Pharmacology. Ed.: A. Bouhuys. Springfield, Ill.: C. C. Thomas, **1970**.

Newman, F., Smalley, B. F., and Thomson, M. L. Effect of exercise, body and lung size on CO diffusion in athletes and non-athletes. J. Appl. Physiol. 17: 649-655, **1962**.

Newmark, S. R., Himathongkam, T., Martin, R. P., Cooper, K. H., and Rose, L. I. Adrenocortical response to marathon running. J. Clin. Endocrinol. Metab. 42: 393-394, **1976**.

Newsholme, E. A. The regulation of phosphofructokinase in muscle. Cardiology 56: 22-34, **1971**.

——. The regulation of intracellular and extracellular fuel supply during sustained exercise. Ann. N.Y. Acad. Sci. 301: 81-91, **1977**.

Newsholme, E. A. and Randle, P. J. Effects of fatty acids, ketone bodies and pyruvate, and of alloxan-diabetes, starvation, hypophysectomy and adrenalectomy, on the concentrations of hexose phosphates, nucleotides and inorganic phosphate in perfused rat heart. Biochem. J. 93: 641-651, **1964**.

Newsom, Davis, J. Control of the muscles of breathing, pp. 221-245. In: M.T.P. International Review of Science. Respiratory Physiology. London: Butterworths, **1974**.

Nicholas, J. J., Gilbert, R., Gabe, R., and Auchincloss, J. H. Evaluation of an exercise therapy program for patients with chronic obstructive pulmonary disease. Amer. Rev. Resp. Dis. 102: 1-9, **1970**.

Nichols, C. W. and Lambertsen, C. J. Effects of oxygen upon ophthalmic structures, pp. 57-66. In: Underwater Physiology. Ed.: C. J. Lambertsen. New York: Academic Press, **1971**.

Nicogossian, A., Hoffler, G. W., Johnson, R. L., and Gowen, R. J. Determination of cardiac size following space missions of different durations: The second manned skylab mission. Aviation Space Environ. Med. 47: 362-365, **1976**.

Niederberger, M. Abnormal physical performance of the cardiac patient during exercise, pp. 119-130. In: Coronary Heart Disease, Exercise Testing and Cardiac Rehabilitation. Eds.: W. James and E. Amsterdam. Miami: Symposia Specialists, **1977**.

Nielsen, B. Thermoregulatory responses to arm work, leg work and intermittent leg work. Acta Physiol. Scand. 72: 25-32, **1968**.

——. Thermoregulation in rest and exercise. Acta Physiol. Scand. (Suppl.) 323: 1-74, **1969**.

——. Thermoregulation during work in carbon monoxide poisoning. Acta Physiol. Scand. 82: 98-106, **1971**.

Nielsen, M. Die Respirationsarbeit bei Körperruhe und bei Muskelarbeit. Skand. Arch. Physiol. 74: 299-316, **1936**.

——. Die Regulation der Körpertemperatur bei Muskelarbeit. Skand. Arch. Physiol. 79: 193-230, **1938**.

Niinimaa, V., Cole, P., Mintz, S., and Shephard, R. J. A head-out exercise body plethsymograph. J. Appl. Physiol. 47: 1336-1339, **1979**.

——. Oronasal distribution of respiratory airflow. Resp. Physiol. 43: 69-75, **1981**.

Niinimaa, V., Dyon, M., and Shephard, R. J. Performance and efficiency of inter-collegiate cross-country skiing. Med. Sci. Sport. 10: 91-93, **1978**.

Niinimaa, V. and Shephard, R. J. Training and oxygen conductance in the elderly. (1) The respiratory system. J. Gerontol. 33: 354-361, **1978a**.

——. Training and oxygen conductance in the elderly. (2) The cardiovascular system. J. Gerontol. 33: 362-367, **1978b**.

——. The "switch-point" for mouth breathing. Paper in preparation.

Niinimaa, V., Shephard, R. J., and Dyon, M. Determinations of performance and mechanical efficiency in nordic skiing. Brit. J. Sport. Med. 13: 62-65, **1979**.

Niinimaa, V., Woch, Z. T., and Shephard, R. J. Intensity of physical effort during a free figure skating program. Proceedings Pan American Sports Science Congress, Edmonton, **1978**.

Niinimaa, V., Wright, G., Shephard, R. J., and Clarke, J. Characteristics of the successful dinghy sailor. J. Sport. Med. Phys. Fitness, 17: 83-96, **1977**.

Nikkila, E. A. and Rissanen, A. Familial occurrence of coronary risks factors. Nordic Council for Arctic Med. Res. Rep. 19: 18-21, **1977**.

Nilsson, K. O., Heding, L. G., and Hokfelt, B. The influence of short-term maximal work on the plasma concentrations of catecholamines, pancreatic glucagon and growth hormone in man. Acta Endocrinol. 79: 286-294, **1975**.

Ninomiya, I. and Wilson, M. F. Cardiac adaptation at the transition phases of exercise in unanaesthetized dogs. J. Appl. Physiol. 21: 953-958, **1966**.

Nisbett, R. E. Determinants of food intake in obesity. Science 159: 1254-1255, **1968**.

Nishi, Y. and Gagge, A. P. Direct evaluation of convective heat transfer coefficient by naphthalene sublimation. J. Appl. Physiol. 29: 830-838, **1970**.

Niu, H., Ito, K., Takagi, K., and Ito, M. A study

of the development of cardiorespiratory function of oarsmen. In: Proceedings of International Congress of Sports Sciences, 1964. Ed.: K. Kato. Tokyo: Japanese Union of Sport Sciences, **1966**.

Noack, H. Die Sportliche Leistungsfahigkeit der Frau im Menstrualzyklus. Dtsche Med. Wschr. 79(2): 1523-1525, **1954**.

Noble, B. J. and Maresh, C. M. Effect of acute exposure to moderate altitude on the anaerobic threshold of college basketball players. Med. Sci. Sport. 10: 39, **1978**.

Noble, B. J., Metz, K. F., Pandolf, K. B., and Cafarelli, E. Perceptual responses to exercise: A multiple regression study. Med. Sci. Sport. 5: 104-109, **1973**.

Nöcker, J. Physiologie der Leibesübungen. Stuttgart: Ferdinand Enke Verlag, **1964**.

Nöcker, J. and Bohlau, V. Abhangigkeit der Leistungsfahigkeit vom Alter und Geschlecht. Munch. Med. Wschr. 97: 1517-1522, **1955**.

Nöcker, J., Låhmavin, D., and Schleusing, G. Einfluss von Training und Belastung auf den Mineralgehalt von Herz und Skelettmuskel. Int. Z. Angew. Physiol. 17: 243-251, **1958**.

Nordesjö, L. O. The effect of quantitated training on the capacity for short and prolonged work. Acta Physiol. Scand. Suppl. 405: 1-54, **1974**.

Nordin, B. E. C. Clinical significance and pathogenesis of osteoporosis. Brit. Med. J. 1: 571-576, **1971**.

Norman, R. W., Sharratt, M. T., Pezzack, J. C., and Noble, E. G. Re-examination of the mechanical efficiency of horizontal treadmill running, pp. 87-93. In: Biomechanics. Ed.: P. V. Komi. Baltimore: University Park Press, **1976**.

Norris, A. H., Shock., N. W., and Yiengst, M. J. Age differences in ventilatory and gas exchange responses to graded exercise in males. J. Gerontol. 10: 145-155, **1955**.

Norris, F. H. and Gasterger, E. L. Action potentials of single motor units in normal muscle. Electroencephalogr. Clin. Neurophysiol. 7: 115-126, **1955**.

Novak, L. P. Aging, total body potassium, fat-free mass and cell mass in males and females between ages 18 and 85 years. J. Gerontol. 27: 438-443, **1972**.

Nowacki, P., Schmidt, E., and Weist, F. The turnover of sympathicoadrenal hormones of sportsmen in training, anticipation and during competition, judged by measurements of the urinary excretion of 3-methoxy 4-hydroxymandelic acid. In: Biochemistry of Exercise. Ed.: J. Poortmans. Karger: Basel, **1969**.

Nukada, A. Hauttemperatur und Leistungsfähigkeit in Extremitaten bei Statischer Haltearbiet. Arbeitsphysiol. 16: 74-80, **1955**.

Nunn, J. F. Applied Respiratory Physiology, with Special Reference to Anaesthesia. London: Butterworths, **1969**.

Nunneley, S. A. Physiological responses of women to thermal stress: A review. Med. Sci. Sport. 10: 250-255, **1978**.

Nutrition Canada. Nutrition a national priority. A report by Nutrition Canada to Dept. of National Health and Welfare. Ottawa: Queen's Printer, **1973**.

Nutter, D. O., Schlant, R. C., and Hurst, J. W. Isometric exercise and the cardiovascular system. Mod. Concepts Cardiovasc. Dis. 41: 11-15, **1972**.

Nyboer, J. Electrical impedance plethysmography. Springfield, Ill.: C. C. Thomas, **1959**.

Nye, R. E. Influence of the cyclical pattern of ventilatory flow on pulmonary gas exchange. Resp. Physiol. 10: 321-337, **1970**.

Nygaard, E. Adaptational changes in human skeletal muscle with different levels of physical activity. Acta Physiol. Scand. Suppl. 440: 291 (Abstr.), 1976.

Nygaard, E., Bentzen, H., Houston, M., Larsen, H., Nielsen, H., and Saltin, B. Capillary supply and morphology of trained human skeletal muscle. Proceedings 27th Int. Congress Physiol. Sciences, Paris, **1977**.

Nygaard, E. and Gøricke, T. Morphological studies of skeletal muscles in women. Report 99 (in Danish). Copenhagen: August Krogh Institute. Cited by Saltin (1977), **1976**.

Nyquist, J. K. Somatosensory properties of neurons of thalamic nucleus ventralis lateralis. Exp. Neurol. 48: 123-135, **1975**.

Nyström, B. Histochemical study of end-plate bound esterases in "slow-red" and "fast-white" cat muscles during post-natal development. Acta Neurol. Scand. 44: 295-318, **1968a**.

——. Post-natal development of motor nerve terminals in "slow-red" and "fast-white" cat muscles. Acta Neurol. Scand. 44: 363-383, **1968b**.

Öberg, B. Overall cardiovascular regulation. Ann. Rev. Physiol. 38: 537-570, **1976**.

Oberholzer, R. J. H. Circulatory centers in medulla and midbrain. Physiol. Rev. 40 (Suppl. 4):

179-195, **1960**.

Oberst, F. W. Factors affecting inhalation and retention of toxic vapours, pp. 249-266. In: Inhaled Particles and Vapours. Ed.: C. N. Davis. Oxford: Pergamon Press, **1961**.

Ochwaldt, B., Bücherl, E., Kreuzer, H., and Loeschke, H. H. Beeinflussung der Atemsteigerung bei Muskelarbeit durch partiellen neuromuskulären Block (Tubocurarin). Pflüg Archiv. 269: 613-621, **1959**.

Oddershade, I., Myhre, L. G., and Dill, D. B. The effects of heat stress on cardiac output during rest and submaximal exercise. Med. Sci. Sport. 9: 52, **1977**.

Odell, W. D., Utiger, R. D., Wilber, J. F., and Condliff, P. G. Estimation of the secretion rate of thyrotropin in man. J. Clin. Invest. 46: 953-959, **1967**.

O'Donnell, T. F. The hemodynamic and metabolic alterations associated with acute heat stress injury in marathon runners. Ann. N.Y. Acad. Sci. 301: 262-269, **1977**.

Offer, G. C-protein and the periodicity in thick filaments of vertebrate striated muscle. Cold Spring Harbor Symp. 37: 87-93, **1972**.

Ogata, T. The differences in some labile constituents and some enzymatic activities between the red and white muscle. J. Biochem. (Tokyo) 47: 726-732, **1960**.

Ogata, T., Hondo, T., and Seito, T. An electron microscope study on differences in the fine structures of motor endplate in red, white and intermediate muscle fibers of rat intercostal muscle. A preliminary study. Acta Med. Okayama 21: 327-338, **1967**.

O'Hara, T. J., and Orlick, T. D. Strategies for control of competitive anxiety, pp. 305-314. In: Motor Learning, Sport, Psychology, Pedagogy and Didactics of Physical Activity. Eds.: F. Landry and W. R. Orban. Miami: Symposia Specialists, **1978**.

O'Hara, W. J. Allen, C., and Shephard, R. J. Loss of body weight and fat during exercise in a cold chamber. Europ. J. Appl. Physiol. 37: 205-218, **1977a**.

——. Treatment of obesity by exercise in the cold. Can. Med. Assoc. J. 117: 773-779, **1977b**.

O'Hara, W. J., Allen, G., Shephard, R. J., and Gill, J. W. La Tulippe: A case study of a one hundred and sixty kilometer runner. Brit. J. Sport. Med. 11: 83-87, **1977c**.

O'Hara, W., Allen, C., and Shephard, R. J. Loss of body fat during an arctic winter expedition. Can. J. Physiol. 55: 1235-1241, **1978**.

O'Hara, W., Allen, C., Shephard, R. J., and Allen, G. Fat loss in the cold: A controlled study. J. Appl. Physiol. 46: 872-877, **1979**.

Ohnishi, T. Studies on the mechanism of site I energy conservation. Europ. J. Biochem. 64: 91-103, **1976**.

Oja, P., Pyörala, K., Punsar, S., and Pekkarinen, M. L. Body weight and energy intake of middle-aged men during and after 1.5 years of exercise intervention, pp. 55-61. In: Nutrition, Dietetics and Sport. Eds.: G. Ricci and A. Venerando. Torino: Minerva Medica, **1978**.

Olafsson, S. and Hyatt, R. E. Ventilatory mechanics and expiratory flow limitation during exercise in normal subjects. J. Clin. Invest. 48: 564—573, **1969**.

Oldridge, N. The problem of compliance. Med. Sci. Sport. 11: 373-375, **1979**.

Olesen, K. H. Body composition in normal adults. In: Human Body Composition. Approaches and Applications, pp. 177-190. Ed.: J. Brožek. Oxford: Pergamon Press, **1965**.

Oliver, M. F. Metabolic response during impending infarction. Clinical implications. Circulation 45: 491-500, **1972**.

Oliver, R. M. Physique and serum lipids of young London busmen in relation to ischaemic heart disease. Brit. J. Industr. Med. 24: 181-187, **1967**.

Olson, R. E. "Excess lactate" and anaerobiosis. Ann. Int. Med. 59: 960-963, **1963**.

O'Malley, B. W. Hormonal regulation of nucleic acid and protein synthesis. Trans. N.Y. Acad. Sci. 31: 478-503, **1969**.

Oosawa, F., Kasai, M., Hatano, S., and Asakura, S. Polymerization of actin and flagellin pp. 273-307. In: Principles of Biomolecular Organization (CIBA Foundation Symposium). Eds.: G. E. W. Wolstenholme and M. O'Connor. London: Churchill, **1966**.

Opie, L. H. Metabolic response during impending infarction. 1. Relevance of studies of glucose and fatty acid metabolism in animals. Circulation 45: 483-490, **1972**.

Opit, L. J. and Charnock, J. S. A molecular model for a sodium pump. Nature (Lond.) 208: 471-474, **1965**.

Opitz, E. and Schneider, M. Über die Sauerstoffversorgung des Gehirns und den Mechanismus von Mangelwirkungen. Ergebn. Physiol. 46: 126-260, **1950**.

Orban, W. R. The Royal Canadian Air Force 5BX plan for physical fitness, 2nd ed. Ottawa:

Queen's Printer, **1962**.

Oren, J. Application of challenge methods to the evaluation of new drugs. Pediatrics 56 (Suppl.): 935-936, **1975**.

Orlander, J. E. and Kiessling, K. H. Effects of training, rest and renewed training on skeletal muscle in sedentary men, pp. 257-266. In: 3rd International Symposium on Biochemistry of Effort. Eds.: F. Landry and W. R. Orban. Miami: Symposia Specialists, **1978**.

Oscai, L. B., Babirak, S. P., McGarr, J. A., and Spirakis, C. N. Effect of exercise on adipose tissue cellularity. Fed. Proc. 33: 1956-1958, **1974**.

Oscai, L. B. and Holloszy, J. Effect of weight changes produced by exercise, food restriction, or overeating on body composition. J. Clin. Invest. 48: 2124-2128, **1969**.

Oscai, L. B. and Holloszy, J. O. Biochemical adaptations in muscle. II. Response of mitochondrial adenosine triphosphatase, creatine phosphokinase, and adenylate kinase activities in skeletal muscle to exercise. J. Biol. Chem. 246: 6968-6972, **1971**.

Oscai, L. B., Molé, P. A., and Holloszy, J. O. Effects of exercise on cardiac weight and mitochondria in male and female rats. Amer. J. Physiol. 220: 1944-1948, **1971**.

Oscai, L. B., Williams, B. T., and Hertig, B. A. Effects of exercise on blood volume. J. Appl. Physiol. 24: 622-624, **1968**.

Oscarsson, O. Functional significance of information channels from the spinal cord to the cerebellum, pp. 93-113. In: Neurophysiological Basis of Normal and Abnormal Motor Activities. Eds.: M. D. Yahr, D. P. Purpura. New York: Raven Press, **1967**.

Oscarsson, O. and Rosén, I. Projection to cerebral cortex of large muscle spindle afferents in forelimb nerves of the cat. J. Physiol. (Lond.) 169: 924-945, **1963**.

Oseid, S. Idrett og stimulerendes midler: Doping. Oslo: Norges Idrettsforbund, **1976**.

O'Shea, J. P. The effects of an anabolic steroid on dynamic strength levels of weight lifters. Nutr. Rep. Int. 4: 363-370, **1971**.

——. Scientific principles and methods of strength fitness, 2nd ed. Reading, Mass.: Addison-Wesley, **1976**.

Oshima, S. and Suzuki, S. Influence of eggs on human serum cholesterol. Jap. J. Nutr. 33: 105-112, **1975**.

Östman, I. and Sjöstrand, N. O. Effect of prolonged physical training on the catecholamine levels of the heart and adrenals of the rat. Acta Physiol. Scand. 82: 202-208, **1971**.

Östman, I., Sjöstrand, N. O., and Swedin, G. Cardiac noradrenaline turnover and urinary catecholamine excretion in trained and untrained rats during rest and exercise. Acta Physiol. Scand. 86: 299-308, **1972**.

Otis, A. B. The control of respiratory gas exchange between blood and tissues, pp. 111-119. In: The Regulation of Human Respiration. Eds.: D. J. C. Cunningham and B. B. Lloyd. Oxford: Blackwell Scientific, **1963**.

——. The work of breathing, pp. 463-476. In: Handbook of Physiology. Section 3. Respiration, Vol. 1. Eds.: W. O. Fenn and H. Rahn. Washington, D.C.: American Physiological Society, **1964**.

Otis, A. B., Fenn, W. O., and Rahn, H. The mechanics of breathing in man. J. Appl. Physiol. 2: 592-607, **1950**.

Otis, A. B., Rahn, H., Epstein, M. A., and Fenn, W. O. Performance as related to composition of alveolar air, pp. 239-253. Dayton, Ohio. Wright Field USAF. Tech. Rep. 6528, **1951**.

Outwater, J. O. On the friction of skis. Med. Sci. Sport. 2: 231-236, **1970**.

Overstall, P. W., Exton-Smith, A. N., Imms, F. J., and Johnson, A. L. Falls in the elderly related to postural imbalance. Brit. Med. J. 1: 261-264, **1977**.

Owen, O. E., Felig, P., Morgan, A. P., Wahren, J., and Cahill, G. F. Liver and kidney metabolism during prolonged starvation. J. Clin. Invest. 48: 574-583, **1969**.

Pace, H. and Rathbun, E. N. Studies on body composition. III. The body water and chemically combined nitrogen content in relation to fat content. J. Biol. Chem. 158: 685-691, **1946**.

Pacheco, B. A. Improvement in jumping performance due to preliminary exercise. Res. Quart. 28: 55-63, **1957**.

——. Effectiveness of warm-up on exercise in junior high school girls. Res. Quart. 30: 202-213, **1959**.

Pacheco, P. and Guzman, C. Intracellular recording in extensor motoneurons of spastic cats. Exp. Neurol. 25: 472-481, **1969**.

Padykula, H. and Gauthier, G. The ultrastructure of the neuromuscular junction of mammalian red, white and intermediate skeletal muscle fibers. J. Cell. Biol. 46: 27-41, **1970**.

Paez, P. N., Phillipson, E. A., Masangkay, M., and Sproule, B. J. The physiological basis of training patients with emphysema. Amer. Rev. Resp. Dis. 95: 944-953, **1967**.

Paffenbarger, R. S. Physical activity and fatal heart attack: Protection or selection? pp. 35-49. In: Exercise in Cardiovascular Health and Disease. Eds.: E. A. Amsterdam, J. H. Wilmore, and A. N. de Maria. New York: Yorke Books, **1977**.

Paffenbarger, R. S., and Hale, W. E. Work activity and coronary heart mortality. N. Engl. J. Med. 292: 545-550, **1975**.

Paffenbarger, R. S., Hale, W. E., Brand, R. J., and Hyde, R. T. Work-energy level, personal characteristics and fatal heart attack: A birth cohort effect. Amer. J. Epidemiol. 105: 200-213, **1977**.

Paffenbarger, R. S., Laughlin, M. E., Gima, A. S., and Black, R. A. Work activity of longshoremen as related to deaths from coronary heart disease and stroke. N. Engl. J. Med. 282: 1109-1114, **1970**.

Page, E. B., Hickam, J. B., Sieker, H. O., McIntosh, H. D., and Pryor, W. W. Reflex venomotor activity in normal persons and in patients with postural hypotension. Circulation 11: 262-270, **1955**.

Paillard, J. Les determinants moteurs de l'organisation de l'espace. Cah. Psychol. 14: 261-316, **1971**.

Paillard, J. and Brouchon, M. Active and passive movements in the calibration of position sense, pp. 37-55. In: The Neuropsychology of Spatially Oriented Behaviour. Ed.: S. J. Freedman. Homewood, Ill.: Dorsey Press, **1968**.

Paintal, A. S. Vagal afferent fibres. Ergebn. Physiol. 52: 74-156, **1963**.

——. The mechanism of excitation of type-J receptors and the J-reflex, p. 59. In: Breathing: Hering-Breuer Centenary Symposium. Ed.: R. Porter. London: Churchill, **1970**.

——. Vagal sensory receptors and their reflex effects. Physiol. Rev. 53: 159-227, **1973**.

Painter, D. and Green, H. Alterations in cardiac output during submaximal exercise following physical training. Canad. J. Appl. Sport. Sci. 3: 179, **1978**.

Paley, H. W., McDonald, I. G., and Peters, F. W. Abstract: Amer. Heart Assoc. 38th Scientific Meeting. Circulation 32 (Suppl. 2) 167 (Abstr.), **1965**.

Palladin, A. V. The biochemistry of muscle training. Science 102: 576-578, **1945**.

Palmer, G. J., Ziegler, M. G., and Lake, C. R. Response of norepinephrine and blood pressure to stress increases with age. J. Gerontol. 33: 482-487, **1978**.

Pandolf, K. B., Haisman, M. F., and Goldman, R. F. Metabolic energy expenditure and terrain coefficients for walking on snow. Ergonomics 19: 683-690, **1976**.

Panksepp, J. Hypothalamic regulation of energy balance and feeding behavior. Fed. Proc. 33: 1150-1165, **1973**.

Pantridge, J. F., Adgey, A. A. J., Webb, S. W., and Anderson, J. Electrical requirements for ventricular defibrillation. Brit. Med. J. 2: 313-315, **1975**.

Paoletti, R., Catapeno, A., Ghiselli, G. C., and Sirtori, C. R. Drugs and atherosclerosis, pp. 517-527. In: Atherosclerosis IV. Eds.: G. Schettler, Y. Gota, Y. Hata, and G. Klose. Berlin: Springer Verlag, **1977**.

Pařízková, J. Impact of age, diet, and exercise on man's body composition. In: International Research in Sport and Physical Education. Eds.: E. Jokl and E. Simon. Springfield, Ill.: C. C. Thomas, **1964**.

——. Nutrition and its relation to body composition in exercise. Proc. Nutr. Soc. 25: 93-99, **1966**.

——. Body Fat and Physical Fitness. The Hague: Martinus Nijhoff, **1977**.

——. The impact of ecological factors and physical activity on the somatic and motor development of preschool children. In: Physical Fitness Assessment: Principles, Practice and Application. Eds.: R. J. Shephard and H. Lavallée. Springfield, Ill.: C. C. Thomas, **1978a**.

——. Body composition and lipid metabolism in retion to nutrition and exercise, pp. 61-75. In: Nutrition, Physical Fitness & Health. Eds.: J. Pařízková and V. A. Rogozkin. Baltimore: University Park Press, **1978b**.

Pařízková, J., Eiselt, E., Sprynarová, S., and Wachtlová, M. Body composition, aerobic capacity and density of muscle capillaries in young and old men. J. Appl. Physiol. 31: 323-325, **1971**.

Pařízková, J. and Poledne, R. Consequences of long-term hypokinesia as compared to mild exercise in lipid metabolism of the heart, skeletal muscle and adipose tissue. Europ. J. Appl. Physiol. 33: 331-338, **1974**.

Pařízková, J. and Stanková, L. Influence of physical activity on a treadmill on the metabolism of adipose tissue in rats. Brit. J. Nutr. 18: 325-332, **1964**.

Park, D. M., Rennie, M. J., and Sulaiman, W. R. Uptake and release of metabolites by the working leg. J. Physiol. (Lond.) 245: 85p–86p, **1975**.

Parker, J. O., diGiorgi, S., and West, R. O. A hemodynamic study of acute coronary insufficiency precipitated by exercise. With observations on the effects of nitroglycerin. Amer. J. Med. 17: 470, **1966**.

Parmeggiani, A. and Bowman, R. H. Regulation of phosphofructokinase activity by citrate in normal and diabetic muscle. Biochem. Biophys. Res. Comm. 12: 268–273, **1963**.

Pascale, L. R., Frankel, T., Grossman, M. I., Freeman, S., Faller, I. L., and Bond, E. E. Report of changes in body composition of soldiers during paratrooper training. Denver, Col.: U.S. Army Med. Nutr. Lab. Rep. 156, **1955**.

Passmore, R. and Johnson, R. E. Some metabolic changes following prolonged moderate exercise. Metabolism 9: 452–455, **1960**.

Pate, R. R., Palmieri, P., Hughes, D., and Ratliffe, T. Serum enzyme response to exercise bouts of varying intensity and duration, pp. 193–202. In: 3rd Int. Symp. on Biochemistry of Exercise. Eds.: F. Landry and W. R. Orban. Miami: Symposia Specialists, **1978**.

Patel, R. Urinary casts in exercise. Austr. Ann. Med. 13: 170–173, **1964**.

Paterson, D. H. Methods for the calculation of cardiac output by the indirect (CO_2) Fick method in different populations. M.Sc. thesis, University of Western Ontario, **1972**.

Paterson, D. H. and Cunningham, D. A. Maximal oxygen uptake in children: comparison of treadmill protocols at varied speeds. Can. J. Appl. Sports Sci. 3: 188, **1978**.

Paterson, D. H., Shephard, R. J., Cunningham, D., Jones, N. L., and Andres, G. Effects of physical training upon cardiovascular function following myocardial infarction. J. Appl. Physiol. 47: 482–489, **1979**.

Pattengale, P. K. and Holloszy, J. O. Augmentation of skeletal muscle myoglobin by a program of treadmill running. Amer. J. Physiol. 213: 783–785, **1967**.

Patterson, G. C. and Shepherd, J. The blood flow in the human forearm following venous congestion. J. Physiol. (Lond.) 125: 501–507, **1954**.

Patterson, S. W., Piper, H., and Starling, E. H. The regulation of the heart beat. J. Physiol. (Lond.) 48: 465–513, **1914**.

Pattle, R. E. Surface lining of lung alveoli. Physiol. Rev. 45: 48–79, **1965**.

Paul, G. Insight vs Desensitization in Psychotherapy: An Experiment in Anxiety Reduction. Stanford, Calif.: Stanford University Press, **1966**.

Paul, P. FFA metabolism of normal dogs during steady state exercise at different work loads. J. Appl. Physiol. 28: 127–132, **1970**.

———. Uptake and oxidation of substrates in the intact animal during exercise, pp. 225–248. In: Muscle metabolism during exercise. Eds.: B. Pernow and B. Saltin. New York: Plenum Press, **1971**.

———. Effects of long lasting physical exercise and training on lipid metabolism. In: Metabolic Adaptation to Prolonged Physical Exercise, pp. 156–193. Eds.: H. Howald and J. R. Poortmans. Basel: Birkhauser Verlag, **1975**.

Paul, P. and Holmes, W. L. Fatty acid and glucose metabolism during increased energy expenditure and after training. Med. Sci. Sport. 7: 176–184, **1975**.

Paul, P., Issekutz, B., and Miller, H. I. Interrelationship of free fatty acids and glucose metabolism in the dog. Amer. J. Physiol. 211: 1313–1320, **1966**.

Paulev, P. E. Decompression sickness following repeated breath-hold dives. J. Appl. Physiol. 20: 1028–1031, **1965**.

———. Respiratory and cardiac responses to exercise in man. J. Appl. Physiol. 30: 165–172, **1971**.

Pauwels, J. The relationship between somatic development and motor ability, and the throwing velocity in handball for secondary school students, pp. 211–221. In: Physical Fitness Assessment. Eds.: R. J. Shephard and H. Lavallée. Springfield, Ill.: C. C. Thomas, **1978**.

Payne, A. H., Slater, W. J., and Telford, T. The use of a force platform in the study of athletic activities. A preliminary investigation. Ergonomics 2: 146–151, **1968**.

Peachey, L. D. The sarcoplasmic reticulum and transverse tubules of the frog's sartorius. J. Cell. Biol. 25: 209–231, **1965**.

Peachey, L. D. and Adrian, R. H. Electrical properties of the T-system, pp. 1–30. In: The Structure and Function of Muscle, 2nd ed., Vol. III. Physiology and Biochemistry. Ed.: G. H. Bourne. New York: Academic Press, **1973**.

Peacock, E. E. Comparison of collagenous tissue surrounding normal and immobilized joints. Surg. Forum 14: 440–441, **1963**.

Peatfield, R. C., Sillett, R. W., Taylor, D., and

McNicol, M. W. Survival after cardiac arrest in hospital Lancet 1: 1223–1225, **1977**.

Pechar, G. S., McArdle, W. D., Katch, F. I., Magel, J. R., and de Luca, J. Specificity of cardio-respiratory adaptation to bicycle and treadmill training. J. Appl. Physiol. 36: 753–756, **1974**.

Pechar, G. S. and Nelson, R. C. The kinesthetic after effects of altered resistance upon selected speed variables of a simple movement, pp. 683–692. In: Contemporary Psychology of Sport. Eds.: G. S. Kenyon and T. M. Grogg. Chicago: Athletic Institute, **1970**.

Pedersen, A. and Andersen, N. S. Work test with electrocardiogram, analysis by digital computer, pp. 202–208. In: Coronary Heart Disease and Physical Fitness. Eds.: O. A. Larsen and R. O. Malmborg. Baltimore: University Park Press, **1971**.

Pedersen, P. K. and Jørgensen, K. Maximal oxygen uptake in young women with training, inactivity, and retraining. Med. Sci. Sport. 10: 233–237, **1978**.

Pedley, F. G. Coronary disease and occupation. Can. Med. Assoc. J. 46: 147–151, **1942**.

Peleška, B. Cardiac arrhythmias following condenser discharges and their dependence upon strength of current and phase of cardiac cycle. Circ. Res. 13: 21–31, **1963**.

——. Optimal parameters of electrical impulses for defibrillation by condenser discharges. Circ. Res. 18: 10–17, **1966**.

Pelliccia, A. Influenza del lavoro muscolare sul numero e sulle funzioni delle piastrine. Med. Sport. 30: 275–282, **1978**.

Pelosi, G. and Agliate, G. The heart muscle in functional overload and hypoxia. A biochemical and ultrastructural study. Lab. Invest. 18: 86–93, **1968**.

Peñaloza, D., Sime, F., and Ruiz, L. Cor pulmonale in chronic mountain sickness: present concept of Monge's disease, pp. 41–60. In: High Altitude Physiology: Cardiac and Respiratory Aspects. Eds.: R. Porter and J. Knight. Edinburgh: Churchill-Livingstone, **1971**.

Pendergast, D. R., di Prampero, P. E., Craig, A. B., and Rennie, D. W. The influence of selected biomechanical factors on the energy cost of swimming. In: Swimming Medicine IV. Eds.: B. Eriksson and B. Fürberg. Baltimore: University Park, **1978**.

Pengelly, L. D., Alderson, A. M., and Milic Emili, J. Mechanics of the diaphragm. J. Appl. Physiol. 30: 797–805, **1971**.

Pengelly, L. D., Rebuck, A. S., and Campbell, E. J. M. Loaded Breathing. Toronto: Longman, **1974**.

Penniman, R. and Jerard, R. On the static and kinetic friction coefficients of various ski surfaces on snow, pp. 1–31. M. E. thesis, University of Vermont, **1969**.

Penpargkul, S. and Scheuer, J. The effect of physical training upon the mechanical and metabolic performance of the rat heart. J. Clin. Invest. 49: 1859–1868, **1970**.

Penrod, K. E. Nature of pulmonary damage produced by high oxygen pressures. J. Appl. Physiol. 9: 1–4, **1956**.

Pepe, F. A. Some aspects of the structural organization of the myofibril as revealed by antibody staining methods. J. Cell Biol. 28: 505–525, **1966**.

——. The structure of the myosin filament of striated muscle. Prog. Biophys. and Molec. Biol. 22: 75–96, **1971**.

——. The myosin filament: Immunochemical and ultrastructural approaches to molecular organization. Cold Spring Harbor Symp. 37: 97–108, **1972**.

Pepler, R. D. Performance and well-being in heat, p. 319. In: Temperature: Its Measurement and Control in Science and Industry, Vol. 3(3). Ed.: J. D. Hardy. New York. Reinhold, **1963**.

Peracino, E., Balzola, F., and Bruno, G. Modificazione degli elettrolite plasmatici in sogetti allenati, sottoposti a sforzi prolungati. Med. Sport. 5: 590–596, **1965**.

Perkins, J. F. Arterial CO_2 and hydrogen ion as independent, additive respiratory stimuli: Support for one part of the Gray multiple factor theory, pp. 303–317. In: The Regulation of Human Respiration. Eds.: D. J. C. Cunningham and B. B. Lloyd. Oxford: Blackwell Scientific, **1963**.

——. Historical development of respiratory physiology. In: Handbook of Physiology. Section 3: Respiration. Volume 1. Eds.: W. O. Fenn and H. Rahn. Washington, D.C.: American Physiological Society, **1964**.

Perl, W. and Cucinell, S. A. Local blood flow in human leg muscle measured by a transient response thermoelectric method. Biophys. J. 5: 211–230, **1965**.

Perley, M. M. and Kipuis, D. M. Differential plasma insulin responses to oral and infused glucose in normal weight and obese non-diabetic and diabetic subjects. J. Lab. Clin. Med.

66: 1009 (Abstr.), **1965**.

Permutt, S., Bromberger-Barnea, B., and Bane, H. N. Alveolar pressure, pulmonary venous pressure and the vascular waterfall. Med. Thorac. 19: 239–260, **1962**.

Permutt, S., Caldini, P., Maseri, A., Palmer, W. H., Sasamori, T., and Zierler, K. Recruitment versus distensibility in the pulmonary vascular bed, pp. 375–387 In: The Pulmonary Circulation and Interstitial Space. Eds.: A. P. Fishman and H. H. Hecht. Chicago: University of Chicago Press, **1969**.

Peronet, F., Cousineau, D., Nadeau, R. A., de Champlain, J., and Ferguson, R. J. Adrenal medulla activity in exercising dogs. Med. Sci. Sport. 10: 40, **1978**.

Perrault, H., Péronnet, F., Cléroux, J., Cousineau, D., Nadeau, R., Pham-Huy, H., and Tremblay, G. Electro- and echo-cardiographic assessment of left ventricle before and after training in man. Can. J. Appl. Sport. Sci. 3: 180, **1978**.

Perrine, J. J. Isokinetic exercise and the mechanical energy potentials of muscle. J. Health Phys. Educ. Res. 4: 40–44, **1968**.

Perry, S. V. The bound nucleotide of the isolated myofibril. Biochem. J. 51: 495–507, **1952**.

——. In: The Physiology and Biochemistry of Muscle as a Food. Vol. II. pp. 539–553. Eds.: E. J. Brisky, R. G. Cassens, and B. B. Narsh. Madison: Univ. of Wisconsin Press, **1970**.

Perry, S. V. and Corsi, A. Extraction of proteins other than myosin from the isolated rabbit myofibril. Biochem. J. 68: 5–12, **1958**.

Person, R. S. and Kudina, L. P. Discharge frequency and discharge pattern of human motor units during voluntary contraction of muscle. Electroencephalogr. Clin. Neurophysiol. 32: 471–483, **1972**.

Person, R. S. and Libkind, M. S. Simulation of electromyograms showing interference patterns. Electroencephalogr. Clin. Neurophysiol. 28: 625–632, **1970**.

Peter, J. B., Barnard, R. J., Edgerton, V. R., Gillespie, C. A., and Stempel, K. E. Metabolic profiles of three fiber types of skeletal muscle in guinea pigs and rabbits. Biochemistry 11: 2627–2633, **1972**.

Petersen, F. B., Graudal, H., Hansen, J. W., and Hvid, N. The effect of varying the number of muscle contractions on dynamic muscle training. Int. Z. Angew. Physiol. 18: 468–473, **1961**.

Peterson, L. H. Cardiovascular control and regulation. In: Les Concepts de Claude Bernard sur le Milieu Intérieur, p. 191. Paris: Masson et Cie, **1967**.

Peterson, R. P. Cell size and rate of protein synthesis in ventral horn neurones. Science 153: 1413–1414, **1966**.

Petit, J. M., Milic-Emili, G., and Koch, R. Le volume de reserve expiratoire pendant l'exercise musculaire chez l'homme sain. Arch. Int. Physiol. Biochim. 67: 350–357, **1959**.

Petit, J. M., Milic-Emili, G., and Sadoul, P. L'influence de la position corporelle sur le travail ventilatoire dynamique pendant l'exercise musculaire chez l'individu sain. Arch. Int. Physiol. Biochim. 68: 437–444, **1960**.

Petrén, T., Sjöstrand, T., and Sylvén, B. Der Einfluss des Trainings auf die Häufigkeit der Capillaren in Herz-und Skeletmuskulatur. Arbeitsphysiol. 9: 376–386, **1936**.

Petro, J. K., Hollander, A. P., and Bouman, L. N. Instantaneous cardiac acceleration in man induced by a voluntary muscle contraction. J. Appl. Physiol. 29: 547–551, **1970**.

Petrofsky, J. S., LeDonne, D. M., Rinehart, J. S., and Lind, A. R. Isometric strength and endurance during the menstrual cycle. Europ. J. Appl. Physiol. 35: 1–10, **1976**.

Pette, D. and Staudte, H. W. Differences between red and white muscles. In: Limiting Factors of Physical Performance, pp. 23–35. Ed.: J. Keul. Stuttgart: Thieme, **1973**.

Pew, R. W. Human perceptual-motor performance. In: Human Information Processing: Tutorials in Performance and Cognition. New York: Erlbaum, **1974**.

Phillips, C. G. Motor apparatus of the baboon's hand. Proc. Roy. Soc. B 173: 141–174, **1969**.

Phillips, C. G. and Porter, R. Cortico-Spinal Neurones. Their Role in Movement, pp. 1–450. London: Academic Press, **1977**.

Phillips, M. Effect of the menstrual cycle on pulse rate and blood pressure before and after exercise. Res. Quart. 39: 327–333, **1968**.

Phillips, W. J., Higginbotham, H. B., Frerking, H., and Paine, R. Evaluation of myocardial state by synchronized radiography and exercise. N. Engl. J. Med. 274: 826–829, **1966**.

Phillipson, E. A. Role of the vagus nerves in the control of respiration frequency in conscious dogs, pp. 130–135. In: Loaded Breathing. Eds.: L. D. Pengelly, A. S. Rebuck, and E. J. M. Campbell. Don Mills, Ont.: Longman, **1974**.

Pickering, G. The inheritance of arterial pressure,

pp. 18–27. In: The Epidemiology of Hypertension. Eds.: J. Stamler, R. Stamler, and T. N. Pullman. New York: Grune & Stratton, **1967**.

Piehl, K. Time course for refilling of glycogen stores in human muscle fibres following exercise-induced glycogen depletion. Acta Physiol. Scand. 90: 297–302, **1974**.

Pierce, A. K., Luterman, D., Loudermilk, J., Blomqvist, G., and Johnson, R. L. Exercise ventilatory patterns in normal subjects and patients with airway obstruction. J. Appl. Physiol. 25: 249–254, **1968**.

Pierce, A. K., Paez, P. N., and Miller, W. F. Exercise training with the aid of a portable oxygen supply in patients with emphysema. Amer. Rev. Resp. Dis. 91: 653–659, **1965**.

Pierson, R. N., Wang, J., Yang, M. U., Hashim, S. A., and Van Itallie, T. B. The assessment of human body composition during weight reduction: Evaluation of a new model for clinical studies. J. Nutr. 106: 1694–1701, **1976**.

Pierson, W. E. and Bierman, C. W. Free running test for exercise-induced bronchospasm. Pediatrics 56 (Suppl.): 890–892, **1975**.

Pierson, W. R. Fatigue, work decrement and endurance in a simple repetitive task. Brit. J. Med. Psychol. 36: 279–282, **1963**.

Pierson, W. R. and Lockhart, A. Effect of menstruation on simple movement and reaction time. Brit. Med. J. 1: 796–797, **1963**.

——. Fatigue, work decrement and endurance of women in a simple, repetitive task. Aerospace Med. 35: 724–725, **1964**.

Piiper, J., di Prampero, P. E., and Cerretelli, C. Oxygen debt and high energy phosphates in gastrocnemius muscle of the dog. Amer. J. Physiol. 215: 523–531, **1968**.

Pinel, J. P. G. and Schultz, T. D. Effect of antecedent muscle tension levels on motor behaviour. Med. Sci. Sport. 10: 177–182, **1978**.

Pingree, B. J. W. Acid-base and respiratory changes after prolonged exposure to 1% carbon dioxide. Clin. Sci. 52: 67–74, **1977**.

Piovano, G., Caselli, G., and Pozzilli, P. Frequency of e.c.g. abnormalities in athletes. A study of 12,000 e.c.g.s. In: Sports Cardiology. Eds.: T. Lubich and A. Venerando. Bologna: Aulo Gaggi, **1979a**.

Piovano, G., Caselli, G., and Venerando, A. Cardiopathies observed in athletes. In: Sports Cardiology. Eds.: T. Lubich and A. Venerando. Bologna: Aulo Gaggi, **1979b**.

Pipberger, H. V., Stallman, F. W., and Berson, A. S. Automatic analysis of the P-QRS-T complex of the electrocardiogram by digital computer. Ann. Int. Med. 57: 776–787, **1962**.

Pirnay, F., Deroanne, R., Marechal, R., Dujardin, J., and Petit, J. M. Consommation maximum d'oxygène dans differents types d'exercise musculaire. Arch. Int. Physiol. Biochim. 79: 319–326, **1971**.

Pirnay, F., Deroanne, R., and Petit, J. M. Maximal oxygen consumption in a hot environment. J. Appl. Physiol. 28: 642–645, **1970**.

Pirnay, F., Fassotte, A., Gazon, J., Deroanne, R., and Petit, J. M. Diffusion pulmonaire au cours de l'exercise musculaire. Int. Z. Angew. Physiol. 28: 31–37, **1969**.

Pirnay, F., Lamy, M., Dujardin, J., Deroanne, R., and Petit, J. M. Utilisation de l'oxygène par les muscles de la jambe pendant une exercise musculaire maximum. J. Physiol. (Paris) 63: 266–267, **1971**.

Pirnay, F., Petit, J. M., Bottin, R., Deroanne, R., Juchmes, J., and Belge, G. Comparaison de deux méthodes de mésure de la consommation maximum d'oxygène. Int. Z. Angew. Physiol. 23: 203–211, **1966**.

Pirnay, F., Petit, J. M., Deroanne, R., and Hausman, A. Aptitude à l'exercise musculaire sous contrainte thermique. Arch. Int. Physiol. Biochim. 76: 867–892, **1968**.

Pitts, G. C., Johnson, R. E., and Consolazio, F. C. Work in the heat as affected by intake of water, salt, and glucose. Amer. J. Physiol. 142: 253–359, **1944**.

Pitts, R. F. Physiology of the Kidney and Body Fluids. Chicago: Year Book, **1963**.

Piwonka, R. W. and Robinson, S. Acclimatization of highly trained men to work in severe heat. J. Appl. Physiol. 22: 9–12, **1967**.

Piwonka, R. W., Robinson, S., Gay, V. L., and Manalis, R. S. Preacclimatization of men to heat by training. J. Appl. Physiol. 20: 279–384, **1965**.

Planche, D. and Bianchi, A. L. Modification de l'activité des neurones respiratoires bulbaires provoquée par stimulation corticale, J. Physiol. (Paris) 64: 69–76, **1972**.

Plas, F. Electrocardiographic changes during work and prolonged effort. J. Sport. Med. Phys. Fitness 3: 131–136, **1963**.

——. Electrocardiography, pp. 61–65. In: Basic Book of Sports Medicine. Ed.: G. La Cava. Rome: International Olympic Committee, **1978**.

Plum, F. Neurological integration of behavioural

and metabolic control of breathing. In: Breathing: Hering-Breuer Centenary Symposium. Ed.: R. Porter. London: Churchill, **1970**.

Plyley, M. J. and Groom, C. Geometrical distribution of capillaries in mammalian striated muscle. Amer. J. Physiol. 228: 1376–1383, **1975**.

Pokrovsky, A. A. Aging, metabolism, motion, pp. 71–78. In: Physical Activity and Human Well-Being. Eds.: F. Landry and W. A. R. Orban. Miami: Symposia Specialists, **1978**.

Poledne, R. and Pařízková, J. Long-term training and net transport of plasma free fatty acids, pp. 201–203. In: Metabolic Adaptations to Prolonged Exercise. Eds.: H. Howald and J. Poortmans. Basel: Birkhauser Verlag, **1975**.

Pollack, G. H. Maximum velocity as an index of contractility of cardiac muscle. Circ. Res. 26: 111–127, **1970**.

Pollock, M. The quantification of endurance training. Exer. Sport. Sci. Rev. 1: 155–188, **1973**.

——. Physiological characteristics of older champion track athletes. Res. Quart. 45: 363–373, **1974**.

Pollock, M. L., Hickman, T., Kendrick, Z., Jackson, A., Linnerud, A. C., and Dawson, G. Prediction of body density in young and middle-aged men. J. Appl. Physiol. 40: 300–304, **1976**.

Pollock, M. L., Miller, H. S., Linnerud, A. C., and Cooper, K. H. Frequency of training as a determinant for improvement in cardiovascular function and body composition of middle-aged men. Arch. Phys. Med. Rehab. 56: 141–145, **1975**.

Pollock, M. S., Dimmick, J., Miller, H. S., Kendrick, Z., and Linnerud, A. C. Effects of mode of training on cardiovascular function and body composition of adult men. Med. Sci. Sport. 7: 139–145, **1975**.

Pompeiano, O. and Swett, J. E. Identification of cutaneous and muscular afferent fibres producing EEG synchronization or arousal in normal cats. Arch. Ital. Biol. 100: 343–380, **1962**.

Pool, P. E., Chandler, B. M., Seagren, S. C., and Sonnenblick, E. H. Mechano-chemistry of cardiac muscle: II. The isotonic contraction. Circ. Res. 22: 465–472, **1968**.

Poortmans, J. R. The level of plasma proteins in normal human urine, pp. 603–609. In: Proteides of the Biological Fluids. Ed.: H. Peeters. London: Pergamon Press, **1969**.

——. Serum protein determination during short exhaustive physical activity. J. Appl. Physiol. 30: 190–192, **1970**.

——. Effects of long lasting physical exercise and training on protein metabolism, pp. 212–228. In: Metabolic Adaptation to Prolonged Physical Exercise. Eds.: H. Howald and J. R. Poortmans. Basel: Birkhauser Verlag, **1975**.

——. Protein turnover during exercise, pp. 159–184. In: 3rd International Symposium on Biochemistry of Exercise. Eds.: F. Landry and W. A. R. Orban. Miami: Symposia Specialists, **1978**.

Poortmans, J. R. and Delisse, L. The effect of graduated exercise on venous pyruvate and alanine in humans. J. Sport. Med. Phys. Fitness 17: 123–130, **1977**.

Poortmans, J. R. and Jeanloz, R. W. Quantitative immunological determination of 12 plasma proteins excreted in human urine collected before and after exercise. J. Clin. Invest. 47: 386–393, **1968**.

Poortmans, J. R., Luke, K. H., Zipursky, A., and Bienenstock, J. Fibrinolytic activity and fibrinogen split products in exercise proteinuria. Clin. Chim. Acta 35: 449–454, **1971**.

Poppele, R. E. Systems approach to the study of muscle spindles, pp. 127–146. In: Control of Posture and Locomotion. Eds.: R. B. Stein, K. G. Pearson, R. S. Smith, and J. B. Redford. New York: Plenum Press, **1973**.

Portal, R. W., Davies, J. G., Robinson, B. F., and Leatham, A. G. Notes on cardiac resuscitation, including external cardiac massage. Brit. Med. J. 1: 636–641, **1963**.

Porte, D. and Williams, R. H. Inhibition of insulin release by epinephrine in man. Science 152: 1248–1250, **1966**.

Porter, K. R. and Palade, G. E. Studies on the endoplasmic reticulum. III. Its form and distribution in striated muscle cells. J. Biophys. Biochem. Cytol. 3: 269–300, **1957**.

Porter, R. and Knight, J. High Altitude Physiology: Cardiac and Respiratory Aspects. Edinburgh: Churchill-Livingstone, **1971**.

Porter, R. and Lewis, M. McD. Relationship of neuronal discharges in the precentral gyrus of monkeys to the performance of arm movements. Brain Res. 98: 21–36, **1975**.

Posner, M. I. Characteristics of visual and kinesthetic memory codes. J. Exp. Psychol. 74: 103–107, **1967**.

——. Psychobiology of attention. In: Handbook of Psychobiology. Eds.: M. Gazzaniga and C. Blakemore. New York: Academic Press, **1975**.

Posner, M. I. and Davidson, B. J. Automatic

and attended components of orienting, pp. 13-23. In: Motor learning, sport, psychology, pedagogy and didactics of physical activity. Eds.: F. Landry and W. R. Orban. Miami: Symposia Specialists, **1978**.

Poulton, E. C. Environment and Human Efficiency. Springfield, Ill.: C. C. Thomas, **1970**.

Poulton, E. C., Carpenter, A., and Catton, M. J. Mild nitrogen narcosis. Brit. Med. J. 2: 1450-1451, **1963**.

Poupa, O. Resistance of cardiac muscle to acute anoxia in high altitude adaptation, pp. 171-178. In: Exercise at Altitude. Ed.: R. Margaria. Dordrecht: Excerpta Medica Fdn. **1967**.

Pratt, P. C. and Klugh, G. H. A method for the determination of total lung capacity from postero-anterior and lateral chest roentgenograms. Amer. Rev. Resp. Dis. 96: 548-552, **1967**.

Precht, H., Christophersen, J., Hensel, H., and Larcker, W. Temperature and Life. Berlin: Springer-Verlag, **1973**.

Priban, I. P. Self-optimizing control of respiration, pp. 115-122. In: Breathlessness. Eds.: J. B. L. Howell and E. J. M. Campbell. Oxford: Blackwell Scientific, **1966**.

Prince, F. P., Hikida, R. S., and Hagerman, F. C. Human muscle fibre types in power lifters, distance runners, and untrained subjects. Pflüg. Archiv. 363: 19-26, **1976**.

Printzmetal, M., Kennamer, R., Merliss, R., Wada, T., and Bor, N. Angina pectoris. I. A variant form of angina pectoris. Amer. J. Med. 27: 375-388, **1959**.

Profant, G. R., Early, R. G., Nilson, K. L., Kusumi, F., Hofer, V., and Bruce, R. A. Responses to maximal exercise in healthy middle-aged women. J. Appl. Physiol. 33: 595-599, **1972**.

Pruett, E. D. R. Glucose and insulin during prolonged work stress in men living on different diets. J. Appl. Physiol. 28: 199-208, **1970a**.

——. Plasma insulin concentrations during prolonged work at near-maximal oxygen intake. J. Appl. Physiol. 29: 155-158; **1970b**.

Pugh, L. G. C. E. Muscular exercise on Mount Everest. J. Physiol. (Lond.) 141: 233-261, **1958**.

——. Physiological and medical aspects of the Himalayan Scientific and Mountaineering Expedition 1960-1961. Brit. Med. J. 2:621-627, **1962**.

——. Animals at high altitudes: man above 5,000 meters: Mountain exploration, pp. 861- 868. In: Handbook of Physiology, Section 4. Adaptation to the Environment. Ed.: D. B. Dill. Baltimore: Williams & Wilkins, **1964**.

——. Altitude and athletic performance. Nature 207: 1397-1398, **1965**.

——. Cold stress and muscular exercise, with special reference to accidental hypothermia. Brit. Med. J. 2: 333-337, **1967**.

——. Blood volume changes in outdoor exercise of 8-10 hour duration. J. Physiol. (Lond.) 200: 345-351, **1969a**.

——. Thermal, metabolic, blood and circulatory adjustments in prolonged outdoor exercise. Brit. Med. J. 2: 657-662, **1969b**.

——. Athletes at altitude. Lessons of the 1968 Olympic Games. Trans. Med. Soc. (Lond.). 85: 76-83, **1969c**.

——. Mean skin temperature and sweat rates of runners in Mexico, pp. 57-58. Natl. Inst. Med. Res., Scientific Report 1968-1969. London: Medical Research Council, **1970**.

——. The influence of wind resistance in running and walking and the mechanical efficiency of work against horizontal or vertical forces. J. Physiol. (Lond.) 213: 255-276, **1971**.

——. Accidental hypothermia among hill-walkers and climbers in Britain. In: Environmental Effects on Work Performance. Eds.: G. R. Cumming, D. Snidal, and A. W. Taylor. Ottawa: Canadian Association of Sports Sciences, **1972**.

——. Air resistance in sport, pp. 149-164. In: Advances in Exercise Physiology. Eds.: E. Jokl, R. L. Anand, and H. Stoboy. Basel: Karger, **1976**.

Pugh, L. G. C. E., Corbett, J. L., and Johnson, R. H. Rectal temperatures, weight losses and sweat rates in marathon runners. J. Appl. Physiol. 23: 347-372, **1967**.

Pugh, L. G. C. E., Edholm, O. G., Fox, R. H., Wolff, H. S., Harvey, G. R., Hammond, W. H., Tanner, J. M., and Whitehouse, R. H. A physiological study of channel swimming. Clin. Sci. 19: 257-273, **1960**.

Pyfer, H. R., Mead, W. F., Doane, B. L., and Frederick R. C. Group rehabilitation of cardiopulmonary patients: A five year experience. In: Proceedings of 20th World Congress of Sports Medicine. Ed.: A. Toyne. Melbourne, Australian Sports Medicine Federation, **1975**.

Pyörälä, K., Käräva, R., Punsar, S., Oja, P., Teräslinna, P., Portanen, T., Jääskeläinen, M., Pekkarinen, M. L., and Koskela, A. A controlled study of the effects of 18 months physical training in sedentary middle-aged men with high indexes of risk relative to coronary heart disease, pp. 261–265. In: Coronary Heart Disease and Physical Fitness. Eds.: O. A. Larsen and R. O. Malmbörg. Copenhagen: Munksgaard, **1971**.

Pyörälä, K., Karvonen, M. J., Taskinen, P., Takkunen, J., and Kyronseppa, H. Cardiovascular studies on former endurance athletes, pp. 301–310. In: Physical Activity and the Heart. Eds.: M. J. Karvonen and A. J. Barry. Springfield, Ill.: C. C. Thomas, **1967**.

Quest, J. A. and Gebber, G. L. Modulation of baroreceptor reflexes by somatic afferent nerve stimulation. Amer. J. Physiol. 222: 1251–1259, **1972**.

Quetelet, A. Sur l'homme et le dévelopement de ses facultés. Paris: Bachelier, **1835**.

Raab, W. Training, physical activity and the cardiac dynamic cycle. J. Sports. Med. Phys. Fitness 6: 38–47, **1966a**.

——. Prevention of Ischemic Heart Disease. Principles and Practice. Springfield, Ill.: C. C. Thomas, **1966b**.

Raab, W. and Krzywanek, H. J. Cardiac sympathetic tone and stress response related to personality patterns and exercise habit. In: Prevention of Ischemic Heart Disease. Principles and Practice. Ed.: W. Raab. Springfield, Ill.: C. C. Thomas, **1966**.

Raben, M. S. and Hollenberg, C. H. Effect of growth hormone on plasma fatty acids. J. Clin. Invest. 38: 484–498, **1959**.

Rack, P. M. H. and Westbury, D. R. The effects of length and stimulus rate on tension in the isometric cat soleus muscle. J. Physiol. (Lond.) 204: 443–460, **1969**.

Radford, P. F. and Upton, A. R. M. Trends in speed of alternated movement during development and among elite sprinters, pp. 188–193. In: Biomechanics V. Ed.: P. V. Komi, Baltimore: University Park Press, **1976**.

Radigan, L. and Robinson, S. Effects of environmental heat stress and exercise on renal blood flow and filtration fraction. J. Appl. Physiol. 2: 185–191, **1949**.

Rahe, R. H. and Arthur, R. J. Swim performance decrement over middle life. Med. Sci. Sport. 7: 53–58, **1975**.

Rahn, H. A concept of mean alveolar air and the ventilation-blood flow relationships during pulmonary gas exchange. USAF Tech. Rep. 6528: 304–313, **1951**.

——. Lessons from breath holding, pp. 293–302. In: The Regulation of Human Respiration. Eds.: D. J. C. Cunningham and B. B. Lloyd. Oxford: Blackwell Scientific, **1963**.

Rahn, H., Otis, A. B., Hodge, M., Brontman, M., Holmes, J., and Fenn. W. O. The pneumatic balance resuscitator. U.S. Air Force WADC Tech. Rep. 6528: 204–210, **1951**.

Raisz, L. G., Au, W. Y. W., and Scheer, R. L. Studies on the renal concentrating mechanism. 3. Effect of heavy exercise. J. Clin. Invest. 38: 8–13, **1959**.

Rajic, M. K., Lavallée, H., Shephard, R. J., Jéquier, J. C., LaBarre, R., and Beaucage, C. Height-weight comparison of Canadian Schoolchildren, pp. 60–74. In: Physical Fitness Assessment: Principles, Practice and Application. Eds.: R. J. Shephard and H. Lavallée. Springfield, Ill.: C. C. Thomas, **1978**.

Rakusan, K., Ost'adal, B., and Wachtlová, M. The influence of muscular work on the capillary density in the heart and skeletal muscle of pigeon. Can. J. Physiol. Pharm. 49: 167–170, **1971**.

Ralston, H. J. Recent advances in neuromuscular physiology. Amer. J. Phys. Med. 32: 85–92, **1957**.

——. Comparison of energy expenditure during treadmill walking and floor walking. J. Appl. Physiol. 15: 1156, **1960**.

Ralston, H. J., Inman, V. T., Strait, L. A., and Shaffrath, M. D. Mechanics of human isolated voluntary muscle. Amer. J. Physiol. 151: 612–620, **1947**.

Rambant, C., Leach, C. S., and Johnson, P. C. Calcium and phosphorus change of Apollo 17 crew members. Nutr. Metab. 18: 62–69, **1975**.

Ramey, M. R. Effective use of force plates for long jump studies. Res. Quart. 43: 247–252, **1972**.

——. Force plate designs and applications, pp. 303–320. In: Exercise and Sports Sciences Reviews, Vol. 3. Eds.: J. Wilmore and J. Keogh. New York: Academic Press, **1975**.

Ramsey, J. M. Carboxyhemoglobinemia in parking garage employees. AMA Arch. Environ. Health 15: 580–583, **1967**.

Randall, D. C. and Smith, O. A. Ventricular contractility during controlled exercise and emotion in the primate. Amer. J. Physiol. 226: 1051–1059, **1974**.

Randle, P. J., Garland, P. B., Hales, C. N., Newsholme, E. A., Denton, R. M., and Pogson, C. I. Interactions of metabolism and the physiological role of insulin. Recent Prog. Horm. Res. 22: 1–48, **1966**.

Rarick, G. L. Competitive sports in childhood and early adolescence, pp. 364–386. In: Physical

Activity. Human Growth and Development. Ed.: G. L. Rarick. New York: Academic Press, **1973**.

Rarick, G. L. and Larsen, G. L. The effects of variations in the intensity and frequency of isometric muscular effort on the development of static muscular strength in pre-pubescent males. Int. Z. Angew. Physiol. 18: 13-21, **1959**.

Rarick, G. L. and Thompson, J. A. J. Roentgenographic measures of leg muscle size and ankle extensor strength of seven year old children. Res. Quart. 27: 321-332, **1956**.

Rasch, P. J. and Burke, R. K. Kinesiology and Applied Anatomy. The Science of Human Movement, 4th ed. Philadelphia: Lea & Febiger, **1971**.

Rasch, P. J. and Pierson, W. R. Some relationships of isometric strength, isotonic strength and anthropometric measures. Ergonomics 6: 210-215, **1963**.

Rasmussen, R. L., Bell, R. D., and Spencer, G. D. Prepubertal exercise and myocardial collateral circulation. In: Exercise Physiology. Eds.: F. Landry and W. A. R. Orban. Miami: Symposia Specialists, **1978**.

Rathbone, J. L. Good posture, the expression of good development. In: Symposium on Posture, Phi Delta Pi. **1938**. Cited by Matthews (1963).

Rautaharju, P. M., Friedrich, H., and Wolf, H. Measurement and interpretation of exercise electrocardiograms. In: Frontiers of Fitness. Ed.: R. J. Shephard. Springfield, Ill.: C. C. Thomas, **1971**.

Rautaharju, P. M. and Karvonen, M. J. Electrophysiological consequences of the adaptive dilatation of the heart, pp. 159-183. In: Physical Activity and the Heart. Eds.: M. J. Karvonen and A. J. Barry. Springfield, Ill.: C. C. Thomas, **1967**.

Raven, P. B., Drinkwater, B. L., and Horvath, S. M. Cardiovascular responses of young female track athletes during exercise. Med. Sci. Sport. 4: 205-209, **1972**.

Rawlins, J. S. P. and Tauber, J. F. Thermal balance at depth, pp. 435-442. In: Underwater Physiology. Ed.: C. J. Lambertsen. New York: Academic Press, **1971**.

Raymond, L. W., Bell, W. H., Bondi, K. R., and Lindberg, C. R. Body temperature and metabolism in hyperbaric helium atmospheres. J. Appl. Physiol. 24: 678-684, **1968**.

Raymond, L., Sode, J., and Tucci, J. Adrenocortical response to non-exhausting muscular exercise. Acta Endocrinol. 70: 73-80, **1972**.

Raynaud, J., Varène, P., Vieillefond, H., and Durand, J. Circulation cutanée et échanges thermiques en altitude. Arch. Sci. Physiol. 27: A247-A254, **1973**.

Rechnitzer, P. A., Sangal, S., Cunningham, D. A., Andrew, G., Buck, C., Jones, N. L., Kavanagh, T., Parker, J. O., Shephard, R. J., and Yuhasz, M. S. A controlled prospective study of the effect of endurance training on the recurrence rate of myocardial infarction: A description of the experimental design. Amer. J. Epidemiol. 102: 358-365, **1975**.

Reed, J. D. Factors influencing rotary performance. J. Psychol. 28: 65-92, **1949**.

Reedy, M. K. A discussion on the physical and chemical basis of muscular contraction. Proc. Roy. Soc. B. 160: 458-460, **1964**.

Reeve, J., Swain, C. P., and Wootton, R. The acute effect of moderate exercise on blood flow to the skeleton in man. J. Physiol. (Lond.) 267: 31p-32p, **1977**.

Refsum, H. E., Meen, H. D., and Strømme, S. B. Whole blood, serum, and erythrocyte magnesium concentrations after repeated heavy exercise of long duration. Scand. J. Clin. Lab. Invest. 32: 123-127, **1973**.

Refsum, H. E. and Strømme, S. B. Urea and creatinine production and excretion in urine during and after prolonged heavy exercise. Scand. J. Clin. Lab. Invest. 33: 247-254, **1974**.

——. Relationship between urine flow, glomerular filtration and urine solute concentration during prolonged heavy exercise. Scand. J. Clin. Lab. Invest. 35: 775-780, **1975**.

Regan, T. J., Frank, M. J., McGinty, J. F., Zobl, E., Hellems, H. K., and Bing, R. J. Myocardial response to cigarette smoking in normal subjects and patients with coronary disease. Circulation 23: 365-369, **1961**.

Regan, T. J., Timmis, G., Gray, M., Binak, K., and Hellems, H. K. Myocardial oxygen consumption during exercise in lipemic subjects. J. Clin. Invest. 40: 624-630, **1961**.

Rein, H. Die Interferenz der vasomotorische Regulationen. Klin. Wschr. 9: 1485-1489, **1930**.

Reindell, H. Uber den Kreislauf der Trainierten. Uber die Restblutmenge des Herzens und über die besondere Bedeutung röntgenologischer (Kymographischer) hämodynamische Beobachtungen in Rühe und nach Belastung. Arch. Kreislaufforsch. 12: 265, **1943**.

Reindell, H., Kleipzig, H., Steim, H., Musshoff,

K., Roskamm, H., and Schildge, E. Herz Kreislaufkrankheiten und Sport. Munich: Johann Ambrosius Barth, **1960**.

Reindell, H., König, K., and Roskamm, H. Funktionsdiagnostik des gesunden und kranken Herzens. Stuttgart: Thieme, **1966**.

Reitsma, W. Skeletal muscle hypertrophy after heavy exercise in rats with surgically reduced muscle function. Amer. J. Phys. Med. 48: 237–258, **1969**.

——. Formation of new capillaries in hypertrophic skeletal muscle. Angiology 24: 45- 57, **1973**.

Remmers, J. E. and Bartlett, D. Neural control mechanisms, pp. 41–53. In: Muscular Exercise and the Lung. Eds.: J. A. Dempsey and C. E. Reed. Madison: University of Wisconsin Press, **1977**.

Remmers, J. E., Gautier, H., and Bartlett, D. Factors controlling expiratory flow and duration, pp. 122–129. In: Loaded Breathing. Eds.: L. D. Pengelly, A. S. Rebuck, and E. J. M. Campbell. Don Mills, Ont.: Longman Green, **1974**.

Renbourn, E. T. Materials and Clothing in Health and Disease. London: H. K. Lewis, **1972**.

Rendlick, R. A. and Harrington, L. A. Why the film size? Radiology 34: 536–540, **1940**.

Renneman, R. S. Cardiovascular Applications of Ultrasound. Amsterdam: North Holland, **1974**.

Rennie, D. W. The effect of very high altitudes and their associated phenomena upon muscular exercise. Actes du Congrès. Int. Union Physiol. Sci. Paris, Vol. 12: 744, **1977**.

Rennie, D. W., di Prampero, P., and Cerretelli, P. Effects of water immersion on cardiac output, heart rate and stroke volume of man at rest and during exercise. Med. Sport. 24: 223–228, **1971**.

——. Effects of water immersion on cardiac output, heart rate and stroke volume of man at rest and during exercise, pp. 159–163. Texas: Brooks USAF Base Report SAMTR 74-20, **1974**.

Rennie, D. W., Prendergast, D. R., di Prampero, P., Wilson, D., Lanphier, E. H., and Myers, C. R. Energetics of the overarm crawl. Med. Sci. Sport. 5: 65, **1973**.

Renson, R. and Van Gerven, D. Afkoelingsreakties by Zwemmen in Koud Zeewater. Hermas (Leuven) III: 191–218, **1969**.

Repke, D. I. and Katz, A. M. Calcium binding and calcium uptake by cardiac microsomes: A kinetic analyses. J. Mol. Cell Cardiol. 4: 401–416, **1972**.

Reuter, H. Exchange of calcium ions in the mammalian myocardium: Mechanisms and physiological significance. Circ. Res. 34: 599–605, **1974**.

Rewald, E. Improved work capacity at altitude by transfusional polycythemia. J. Sport. Med. Phys. Fitness 10: 96–99, **1970**.

Reynafarje, B. Myoglobin content and enzymatic activity of muscle and altitude adaptation. J. Appl. Physiol. 17: 301–305, March **1962**.

——. Haematological Changes During Rest and Physical Activity in Man at High Altitude. New York: Macmillan, **1964**.

——. Humoral control of erythropoiesis at altitude, pp. 165–170. In: Exercise at Altitude. Ed.: R. Margaria. Dordrecht: Excerpta Medica Fdn., **1967**.

Reynolds, W. J. and Milhorn, H. T. Transient ventilatory response to hypoxia with and without controlled alveolar PCO_2. J. Appl. Physiol. 35: 187–196, **1973**.

Ribisl, P. M. and Herbert, W. G. Effects of rapid weight reduction and subsequent rehydration upon the physical working capacity of wrestlers. Res. Quart. 41: 536–541, **1970**.

Ribisl, P. M. and Kachadorian, W. A. Maximal oxygen intake prediction in young and middle-aged males. J. Sport. Med. (Torino) 9: 17–22, **1969**.

Richards, D. K. A two factor theory of the warm-up effect in jumping performance. Res. Quart. 39: 668–673, **1968**.

Richardson, J. R. and Pew, R. W. Stabilometer motor performance, pp. 701–710. In: Contemporary Psychology of Sport. Eds.: G. S. Kenyon and T. M. Grogg. Chicago: Athletic Institute, **1970**.

Richardson, M. Physiological responses and energy expenditure of women using stairs of 3 designs. J. Appl. Physiol. 21: 1078–1082, **1966**.

Richardson, R. G. Altitude training. Brit. J. Sport. Med. 8: 1–63, **1974**.

Richter, G. W. and Kellner, A. Hypertrophy of the human heart at the level of fine structure: an analysis and two postulates. J. Cell. Biol. 18: 195–206, **1963**.

Ridge, B. R., Pyke, F. S., and Roberts, A. D. Responses to kayak ergometer performance after kayak and bicycle ergometer training. Med. Sci. Sport. 8: 18–22, **1976**.

Rigatto, M. Mass spectrometry in the study of the pulmonary circulation. Bull. Physio-Pathol. Resp. 3: 473–486, **1967**.

Riley, R. L. The hyperpnoea of exercise. In: The

Regulation of Human Respiration, pp. 525-535. Eds.: D. J. Cunningham and B. B. Lloyd. Oxford: Blackwell Scientific, **1963**.

——. Pulmonary function in relation to exercise, pp. 112-120. In: Science and Medicine of Exercise and Sport, 2nd ed.: Eds.: W. R. Johnson and E. R. Buskirk. New York: Harper & Row, **1974**.

Ring, G. C., Bosch, M., and Chu-Shek, Lo. Effects of exercise on growth, resting metabolism and body composition of Fischer rats. Biol. Med. 133: 1162-1165, **1970**.

Rinzler, S. H. Primary prevention of coronary heart disease by diet. Bull. N.Y. Acad. Med. 44: 936-949, **1968**.

Riva-Rocci, S. Un nuovo sfigmomanometro. Gazz. Med. Torino 47: 981-1001, **1896**.

Rivard, G., Lavallée, H., Rajic, M., Shephard, R. J., Thibaudeau, P., Davignon, A., and Beaucage, C. Influence of competitive hockey on physical condition and psychological behaviour of children, pp. 335-346. In: Frontiers of Activity and Child Health. Eds.: H. Lavallée and R. J. Shephard. Québec: Editions du Pélican, **1977**.

Robb, G. P. and Marks, H. H. Latent coronary artery disease. Determination of its presence and severity by the exercise electrocardiogram. Amer. J. Cardiol. 13: 603-618, **1964**.

Robb, G. P. and Seltzer, F. Appraisal of the double two-step exercise test. A long-term follow-up study of 3325 men. J. Amer. Med. Assoc. 234: 722-727, **1975**.

Robb, M. Feedback and skill learning. Res. Quart. 39: 175-184, **1968**.

Robbins, N., Karpati, G., and Engel, W. K. Histo-chemical and contractile properties in the cross-innervated guinea pig soleus muscle. Arch. Neurol. (Chicago) 20: 318-329, **1969**.

Roberts, D. F. The changing pattern of menarcheal age, pp. 167-175. In: Growth and Development. Physique. Ed.: O. G. Eiben. Budapest: Akadémia Kiadó, **1977**.

Roberts, J. and Alspaugh, J. Specificity of training effects resulting from programs of treadmill running and bicycle ergometer riding. Med. Sci. Sport. 4: 6-10, **1972**.

Roberts, M., Pirnay, F., Donea, B., André, A., and Petit, J. M. Tolerance à l'exercise musculaire après soustraction d'hématies. Arch. Int. Physiol. Biochim. 80: 741-747, **1972**.

Roberts, M. F., Wenger, C. B., Stolwijk, J. A. J., and Nadel, E. R. Skin blood flow and sweating changes following exercise training and heat acclimation. J. Appl. Physiol. Resp. Environ. Exer. Physiol. 43: 133-137, **1977**.

Robertson, C. H., Eschenbacher, W. L., and Johnson, R. L. Respiratory muscle blood flow distribution during expiratory resistance. J. Clin. Invest. 60: 473-480, **1977a**.

Robertson, C. H., Pagel, M. A., and Johnson, R. L. The distribution of blood flow, oxygen consumption and work output among the respiratory muscles during unobstructed hyperventilation. J. Clin. Invest. 59: 43-50, **1977b**.

Robertson, D. G., Jones, N. L., and Kane, J. W. Estimation of arterial PCO_2 from end-tidal PCO_2 in exercising adults. Cited by N. L. Jones, **1975**.

Robertson, J. D. Unit membranes: A review, with recent studies of experimental alterations and a new sub-unit structure in synaptic membranes, pp. 1-81. In: Cellular Membranes in Development. Ed.: M. Locke. New York: Academic Press, **1964**.

Robin, E. R. Editorial review: Claude Bernard's (extended) milieu intérieur revisited: Autoregulation of cell and sub-cell integrity. Clin. Sci. Mol. Med. 52: 443-448, **1977**.

Robinson, B. F., Epstein, S. E., Kahler, R. L., and Braunwald, E. Circulatory effects of acute expansion of blood volume: Studies during maximal exercise and at rest. Circ. Res. 19: 26-32, **1966**.

Robinson, S. Experimental studies of physical fitness in relation to age. Arbeitsphysiol. 4: 251-323, **1938**.

——. Temperature regulation in exercise. Pediatrics 32 (Suppl.): 691-703, **1963**.

——. Physiology of heat regulation and the science of clothing. Washington, D.C.: Division of Medical Sciences, National Academy of Sciences–National Research Council, **1957**.

Rochelle, R. H., Skubic, V., and Michael, E. Performance as affected by incentive and preliminary warm-up. Res. Quart. 31: 499-504, **1960**.

Rochelle, R. H., Stumpner, R. L., Robinson, S., Dill, D. B., and Horvath, S. M. Peripheral blood flow response to exercise consequent to physical training. Med. Sci. Sport. 3: 122-129, **1971**.

Rochmis, P. and Blackburn, H. Exercise tests. A survey of procedures, safety and litigation experience in approximately 170,000 tests. J. Amer. Med. Assoc. 217: 1061-1066, **1971**.

Rockstein, M., and Brandt, K. F. Changes in phosphorus metabolism of the gastrocnemius muscle in aging white rats. Proc. Soc. Exp. Biol. N.Y. 107: 377–380, **1961**.

Rodahl, A., O'Brien, M., and Firth, R. G. R. Diurnal variations in performance of competitive swimmers. J. Sport. Med. Phys. Fitness 16: 72–76, **1976**.

Rodahl, K. Basal metabolism of the Eskimo. J. Nutr. 48: 359–368, **1952**.

——. Nutritional requirements in the polar regions. U.N. symposium on health in the polar regions. Geneva: WHO Public Health Paper 18: 97–115, **1963**.

Rodahl, K., Åstrand, P. O., Birkhead, N. C., Hettinger, T., Issekutz, B., Jones, D. M., and Weaver, R. Physical work capacity. A study of some children and young adults in the United States. AMA Arch. Environ. Health 2: 499–510, **1961**.

Rodahl, K. and Bang, G. Thyroid activity in men exposed to cold. Arctic Aeromed. Lab. Tech. Rep., **1957**.

Rodahl, K., Birkhead, N. C., Blizzard, J. J., Issekutz, B., and Pruett, D. R. Fysiologiske forandringer under langvarig sengeleie. Nord. Med. 75: 182–186, **1966**.

Rodahl, K., Horvath, S. M., Birkhead, N. C., and Issekutz, B. Effects of dietary protein on physical work capacity during severe cold stress. J. Appl. Physiol. 17: 763–767, **1962**.

Rodahl, K., Miller, H. I., and Issekutz, B. Plasma free fatty acids in exercise. J. Appl. Physiol. 19: 489–492, **1964**.

Rodbard, S. and Pragay, E. K. Contraction frequency, blood supply, and muscle pain. J. Appl. Physiol. 24: 142–145, **1968**.

Roddie, I. C. and Shepherd, J. T. Nervous control of the circulation in skeletal muscle. Brit. Med. Bull. 19: 115–119, **1963**.

Rode, A., Ross, R., and Shephard, R. J. Smoking withdrawal program. AMA Arch. Environ. Health 24: 27–36, **1971**.

Rode, A. and Shephard, R. J. The influence of cigarette smoking upon the work of breathing in near maximal exercise. Med. Sci. Sport. 3: 51–55, **1971**.

——. Cardiac output, blood volume, and total hemoglobin of the Canadian Eskimo. J. Appl. Physiol. 34: 91–96, **1973a**.

——. Growth, development and fitness of the Canadian Eskimo. Med. Sci. Sport. 5: 161–169, **1973b**.

——. Pulmonary function of Canadian Eskimos. Scand. J. Resp. Dis. 54: 192–205, **1973c**.

Rode, A., Shephard, R. J., and Bar-Or, O. Cardiac output and oxygen conductance. A comparison of Canadian Eskimo and city dwellers, pp. 45–58. In: Pediatric Work Physiology. Proceedings of the Fourth International Symposium. Ed.: O. Bar-Or. Natanya, Israel: Wingate Institute, **1973**.

Rodgers, K. L. and Berger, R. A. Motor unit involvement and tension during maximum voluntary concentric, eccentric, and isometric contractions of the elbow flexors. Med. Sci. Sport. 6: 253–259, **1974**.

Rodstein, M. Accidents among the aged. Incidence, causes and prevention. J. Chron. Dis. 17: 515–526, **1964**.

Rodstein, M., Wolloch, L., and Gubner, R. S. A mortality study of the significance of extrasystoles in an insured population. Circulation 44: 617–625, **1971**.

Roebuck, J. A. Cited by L. E. Morehouse and A. T. Miller. In: Physiology of Exercise. St. Louis: Mosby, **1967**.

Roeske, W. R., O'Rourke, R. A., Klein, A., Leopold, G., and Karliner, J. Non-invasive evaluation of ventricular hypertrophy in professional athletes. Circulation 53: 286–292, **1976**.

Rogers, A. F. and Sutherland, R. J. Non-acclimatization of man to cold. J. Physiol. (Lond.) 277: 50p, **1978**.

Rogers, F. R. Physical capacity tests in the administration of physical education. New York: Bureau of Publications, Teachers' College, Columbia University, **1926**.

Rogers, T. A., Setliff, J. A., and Klopping, J. C. Energy cost, fluid and electrolyte balance in sub-arctic survival situations. J. Appl. Physiol. 19: 1–8, **1964**.

Rognum, T. O. and Kindt, E. Fettcellestörrelse hos 14 kvinnelige overvektige pasienter. Tid. Norske Laegeforen. 93: 1737–1739, **1973**.

Rogozkin, V. A. The effect of the number of daily training sessions on skeletal muscle protein synthesis. Med. Sci. Sport. 8: 223–225, **1976**.

Rogus, E., Price, T., and Zierler, K. L. Sodium plus potassium-activated, ouabain-inhibited adenosine triphosphate from a fraction of rat skeletal muscle, and lack of insulin effect on it. J. Gen. Physiol. 54: 188–202, **1969**.

Rohmert, W. Ermittung von Erholungspausen für statische Arbeit des Menschen. (Determination of the recovery pause for static work of man.) Int. Z. Angew. Physiol. 18: 123–164, **1960**.

——. Die Beziehung zwischen Kraft und Ausdauer bei statischer Muskelarbeit, pp. 118-136. In: Muskelarbeit und Muskeltraining. Ed.: W. Rohmert. Stuttgart: Gentner Verlag, **1968a**.

——. Rechts-Links-Vergleich bei isometrischen Armmuskeltraining mit Verschiedenem Trainingsreiz bei achtjahrigen Kindern. Int. Z. Angew. Physiol. 26: 262-393, **1968b**.

Rohrer, F. Der Strömungswiderstand in den menschlichen Atemwegen. Pflüg. Archiv. 162: 225-259, **1915**.

Rókusfalvy, P. Achievement motivation: Origin and development, pp. 397-408. In: Sport in the Modern World: Chances and Problems. Eds.: O. Grüpe, D. Kurz, and J. M. Teipel. Berlin: Springer Verlag, **1972**.

Romano, J., Engel, G. L., Webb, J. P., Ferris, E. B., Ryder, H. W., and Blankenhorn, M. A. Syncopal reactions during simulated exposure to high altitude in decompression chamber. War Med. 4: 475-489, **1943**.

Romanul, F. C. A. Capillary supply and metabolism of muscle fibers. Arch. Neurol. 12: 497-509, **1965**.

Romanul, F. C. A. and Pollock, M. The parallelism of changes in oxidative metabolism and capillary supply of skeletal muscle fibers, pp. 203-213. In: Modern Neurology. Ed.: S. Locke. New York: Little, Brown, **1969**.

Romanul, F. C. and van der Meulen, J. P. Reversal of the enzyme profiles of muscle fibres in fast and slow muscles by cross-innvervation. Nature (Lond.) 212: 1369-1370, **1966**.

Rönningen, H. Unpublished results, cited by P. O. Åstrand & K. Rodahl (1977). In: Textbook of Work Physiology. New York: McGraw Hill, **1976**.

Roos, A., Dahlstrom, H., and Murphy, J. P. Distribution of inspired air in the lungs. J. Appl. Physiol. 7: 645-659, **1955**.

Roos, A., Thomas, L. J., Nagel, E. L., and Prommas, D. C. Pulmonary vascular resistance as determined by lung inflation and vascular pressures. J. Appl. Physiol. 16: 77-84, **1961**.

Root, A. W. and Oski, F. A. Effects of human growth hormone on elderly males. J. Gerontol. 24: 97-104, **1969**.

Rose, C. P. and Goresky, C. A. Vasomotor control of capillary transit time heterogeneity in the canine coronary circulation. Circ. Res. 39: 541-554, **1976**.

Rose, D. L., Radzyminski, S. F., and Beatty, R. R. Effect of brief maximal exercise on the strength of the quadriceps femoris. Arch. Phys. Med. Rehabil. 38: 157-164, **1957**.

Rose, G. A., Holland, W. W., and Crowley, E. A. A sphygmomanometer for epidemiologists. Lancet 1: 296-300, **1964**.

Rose, G. A., Prineas, R. J., and Mitchell, J. R. Myocardial infarction and the intrinsic calibre of coronary arteries. Brit. Heart J. 29: 548-552, **1967**.

Rose, K. D. Warning for millions: intense exercise can deplete potassium. Phys. Sport. Med. 3: 26-29, **1975**.

Rose, L. B. and Press, E. Cardiac defibrillation by ambulance attendants. J. Amer. Med. Assoc. 219: 63-68, **1972**.

Rose, L. I., Bousser, J. E., and Cooper, K. H. Serum enzymes after marathon running. J. Appl. Physiol. 29: 355-357, **1970**.

Rosenzweig, D. Y., Hughes, J. M. B., and Glazier, J. B. Effects of transpulmonary and vascular pressures on pulmonary blood volume in isolated lung. J. Appl. Physiol. 28: 553-560, **1970**.

Roskamm, H. Optimum patterns of exercise for healthy adults. In: Proceedings of International Symposium on Physical Activity and Cardiovascular Health. Can. Med. Assoc. J. 96: 895-899, **1967**.

——. Myocardial contractility during exercise, pp. 225-234. In: Limiting Factors of Physical Performance. Stuttgart: Thieme, **1973a**.

——. Limits and age dependency in the adaptation of the heart to physical stress. In: Sport in the Modern World: Chances and Problems. Eds.: O. Grüpe, D. Kurz, and J. M. Teipel. Berlin: Springer Verlag, **1973b**.

Roskamm, H. and Reindell, H,. The heart and circulation of the superior athlete. In: Training: Scientific Basis and Application. Ed.: A. W. Taylor. Springfield, Ill.: C. C. Thomas, **1972**.

Roskamm, H., Weidemann, H., Samek, L., Garnandt, L., Baumann, A., Mellerowicz, H., Renemann, H., and Limon-Lason, R. Maximale Sauerstoffaufnahme, maximales Atemminutenvolumen und maximale Herzfrequenz bei Hochleistungssportlern im Verlaufe einer Akklimatisationsperiode in Font Romeu (1,800 m) und Mexico City (2,240 m). Sportartz und Sportmedizin 19: 120-132, **1968**.

Ross, J. A. Hypertrophy of the little finger. Brit. Med. J. 2: 987, **1950**.

Ross, R. S. Ischemic heart disease, pp. 1208-1226. In: Harrison's Principles of Internal Medicine. Eds.: M. M. Wintrobe, G. W. Thorn, R. D. Adams, I. L. Bennett, E. Braunwald, K. J. Isselbacher, and R. G. Petersdorf. New York:

McGraw Hill, **1970**.

Ross, W. D., Bailey, D. A., Mirwald, R., and Weese, C. H. Proportionality in the interpretation of longitudinal metabolic function data on boys. In: Frontiers of Activity and Child Health. Eds.: R. J. Shephard and H. Lavallée. Quebec City: Editions du Pélican, **1977**.

Ross, W. D., Brown, S. R., Yu, J. W., and Faulkner, R. A. Somatotype of Canadian figure skaters. J. Sport. Med. Fitness 17: 195–205, **1977**.

Rossi, F., Todaro, A., and Venerando, A. Radiological investigations in pulmonary circulation of endurance athletes. J. Sport. Med. Phys. Fitness 17: 269–274, **1977**.

Rossi, G. Asimmetrie toniche posturali, ed asimmetrie motorie. Arch. Fisiol. 25: 146–157, **1927**.

Roth, G. S. and Adelman, R. C. Age related changes in hormone binding by target cells and tissues; possible role in altered adaptive responsiveness. Exp. Gerontol. 10: 1–11, **1974**.

Roth, J., Glick, S. M., Yalow, R. S., and Berson, S. A. Secretion of human growth hormone: Physiologic experimental modification. Metab. Clin. Exp. 12: 557–579, **1963**.

Rothe, C. F. Heart failure and fluid loss in hemorrhagic shock. Fed. Proc. 29: 1854–1860, **1970**.

Rothman, R. H. and Parke, W. W. The vascular anatomy of the rotator cuff. Clin. Orthoped. 14: 176–186, **1965**.

Rothman, R. H. and Slogoff, S. The effect of immobilization on the vascular bed of tendon. Surg. Gynecol. Obstet. 124: 1064–1066, **1967**.

Rottenstein, H., Peirce, G., Russ, E., Felder, O., and Montgomery, H. Influence of nicotine on the blood flow of resting skeletal muscle and of the digits in normal subjects. Ann. N.Y. Acad. Sci. 90: 102–113, **1960**.

Roughton, F. J. W. The average time spent by the blood in the human lung capillary and its relation to the rates of CO uptake and elimination in man. Amer. J. Physiol. 143: 621–633, **1945**.

Roughton, F. J. W. and Forster, R. E. Relative importance of diffusion and chemical reaction rates in determining rate of exchange of gases in the human lung. With special reference to true diffusing capacity of pulmonary membranes and volume of blood in the lung capillaries. J. Appl. Physiol. 11: 290–302, **1957**.

Rougier, G. and Babin, J. P. Influence des activités physique sur les électrolytes sériques et musculaires. Pathol. Biol. 17: 411–427, **1969**.

———. A blood and urine study of heavy muscular work on ureic and uric metabolism in man. J. Sport. Med. 15: 212–222, **1975**.

Roush, E. S. Strength and endurance in the waking and hypnotic states. J. Appl. Physiol. 3: 404–410, **1951**.

Rowe, C. R. and Sorbie, C. Fractures of the spine in the aged. Clin. Orthoped. 26: 34–48, **1963**.

Rowe, F. A. Growth comparison of athletes and non-athletes. Res. Quart. 4: 108–116, **1933**.

Rowe, G. G., Castillo, C. A., Afonso, S., and Crumpton, C. W. Coronary flow measured by the nitrous oxide method. Amer. Heart J. 67: 457–468, **1964**.

Rowe, R. W. D. and Goldspink, G. Surgically induced hypertrophy in skeletal muscles of the laboratory mouse. Anat. Rec. 161: 69–75, **1968**.

Rowell, L. B. Visceral blood flow and metabolism during exercise, pp. 210–232. In: Frontiers of Fitness. Ed.: R. J. Shephard. Springfield, Ill.: C. C. Thomas, **1971**.

———. Human cardiovascular adjustments to exercise and thermal stress. Physiol. Rev. 54: 75–159, **1974**.

———. Regional blood flow during exercise. International Union of Physiol. Sciences Actes du Congrès Vol. 12, p. 726, **1977**.

Rowell, L. B., Blackmon, J. R., and Bruce, R. A. Indocyanine green clearance and estimated hepatic blood flow during mild to maximal exercise in upright man. J. Clin. Invest. 43: 1677–1690, **1964**.

Rowell, L. B., Blackmon, J. R., Martin, R. H., Mazzarella, J. A., and Bruce, R. A. Hepatic clearance of indocyanine green in man under thermal and exercise stresses. J. Appl. Physiol. 20: 384–394, **1965**.

———. Effects of strenuous exercise and heat stress on estimated hepatic blood flow in normal man. In: Physical Activity and the Heart. Eds.: M. J. Karvonen and A. J. Barry. Springfield, Ill.: C. C. Thomas, **1967**.

Rowell, L. B., Brengelmann, G. L., Blackmon, J. R., Twiss, R. D., and Kusumi, F. Splanchnic blood flow and metabolism in heat-stressed man. J. Appl. Physiol. 24: 475–484, **1968a**.

Rowell, L. B., Brengelmann, G. L., Blackmon, J. R., Bruce, R. A., and Murray, J. A. Disparities between aortic and peripheral pulse pressures induced by upright exercise and vasomotor changes in man. Circulation 37: 954-964, **1968b**.

Rowell, L. B., Brengelmann, G. L., Detry, J. M. R., and Wyss, C. Venomotor responses to rapid changes in skin temperature in exercising man. J. Appl. Physiol. 30: 64-71, **1971**.

Rowell, L. B., Brengelmann, G. L., and Murray, J. A. Cardiovascular responses to sustained high skin temperature in resting man. J. Appl. Physiol. 27: 673-680, **1969a**.

Rowell, L. B., Brengelmann, G. L., Murray, J. A., Kraning, K. K., and Kusumi, F. Human metabolic responses to hyperthermia during mild to maximal exercise. J. Appl. Physiol. 26: 395-402, **1969b**.

Rowell, L. B., Kraning, K. K., Evans, T. O., Ward Kennedy, J., Blackmon, J. R., and Kusumi, F. Splanchnic removal of lactate and pyruvate during prolonged exercise in man. J. Appl. Physiol. 21: 1773-1783, **1966**.

Rowell, L. B., Kraning, K. K., Kennedy, J. W., and Evans, T. O. Central circulatory responses to work in dry heat before and after acclimatization. J. Appl. Physiol. 22: 509-518, **1967**.

Rowell, L. B., Marx, H. J., Bruce, R. A., Conn, R. D., and Kusumi, F. Reductions in cardiac output, central blood volume and stroke volume with thermal stress in normal men during exercise. J. Clin. Invest. 45: 1801-1816, **1966**.

Rowell, L. B., Masoro, E. J., and Spencer, M. J. Splanchnic metabolism in exercising man. J. Appl. Physiol. 20: 1032-1037, **1965**.

Rowell, L. B. J., Murray, J. A., Brengelmann, G. L., and Kraning, K. K. Human cardiovascular adjustments to rapid changes in skin temperature during exercise. Circ. Res. 24: 711-724, **1969**.

Rowell, L. B., Taylor, H. L., and Wang, Y. Limitations to prediction of maximal oxygen intake. J. Appl. Physiol. 19: 919-927, **1964**.

Rowell, L. B., Wyss, C. R., and Brengelmann, G. L. Sustained human skin and muscle vasoconstriction with reduced baroreceptor activity. J. Appl. Physiol. 34: 639-643, **1973**.

Rowland, R. D. and Rice, R. S. Calspan Technical Report #ZS-5157-Kl. Cited by I. E. Faria and P. R. Cavanagh. In: The Physiology and Biomechanics of Cycling. New York: Wiley, **1973**.

Roy, R., Ho, K., Taylor, J., Heusner, W., and Van Huss, W. Alterations in a histochemical profile induced by weight-lifting exercise. Med. Sci. Sport. 9: 65, (Abstr.) **1977**.

Royce, J. Isometric fatigue curves in human muscle with normal and occluded circulation. Res. Quart. 29: 204-212, **1958**.

——. Oxygen consumption during submaximal exercises of equal intensity and different duration. Int. Z. Angew. Physiol. 19: 218-221, **1962**.

Rubenstein, C. J. and Summitt, J. K. Vestibular derangement in decompression, pp. 287-292. In: Underwater Physiology. Ed.: C. J. Lambertsen. New York: Academic Press, **1971**.

Rubin, G., Von Trebra, P., and Smith, K. U. Dimensional analysis of motion. III. Complexity of movement pattern. J. Appl. Psychol. 36: 272-276, **1952**.

Ruch, T. C. and Patton, H. D. Physiology and Biophysics, Circulation, Respiration and Fluid Balance, Vol. II. Philadelphia: W. B. Saunders, **1974**.

Rudjord, T. A second order mechanical model of muscle spindle primary endings. Kybernetik 6: 205-213, **1970**.

Ruff, S. and Müller, K. G. Theorie der Druckfall beschwerden und ihre Anwendung auf Tauchtabellen. Int. Z. Angew. Physiol. 23: 251-292, **1966**.

Rumball, A. and Acheson, E. D. Latent coronary heart disease detected by electrocardiogram before and after exercise. Brit. Med. J. 1: 423-428, **1963**.

Rummel, J. A., Michel, E. L., and Berry, C. A. Physiological response to exercise after space flight-Apollo 7 to Apollo 11. Aerospace Med. 44: 235-238, **1973**.

Rummel, J. A., Sawin, C. F., Buderer, M. C., Mauldin, G., and Michel, E. L. Physiological responses to exercise after space flight: Apollo 14 through Apollo 17. Aviation Space Environ. Med. 46: 679-683, **1975**.

Rumpf, E. Sportliche Höchstleistungen während der Schwangerschaft. Zentralbl. Gynaekol. 74: 870-871, **1952**.

Runcie, J. and Hilditch, T. E. Energy provision, tissue utilization and weight loss in prolonged starvation. Brit. Med. J. 2: 352-356, **1974**.

Rushmer, R. F. Anatomy and physiology of ventricular function. Physiol. Rev. 36: 400-425, **1956**.

——. Cardiovascular Dynamics, pp. 1-584. Philadelphia: W. B. Saunders, **1976**.

Rushmer, R. F., Van Citters, R. L., and Franklin, D. Definition and classification of shock. In: Shock: Pathogenesis and Therapy: An International Symposium. Berlin: Springer

Verlag, **1962**.

Rutenfranz, J. and Mocellin, R. Investigations on children and youths regarding the relationships between various parameters of the physical development and the working capacity. Second International Seminar for Ergometry, Berlin, **1967**.

Ruys, J. and Hickie, J. B. Serum cholesterol and triglyceride levels in Australian adolescent vegetarians. Brit. Med. J. 2: 87, **1976**.

Ryan, A. J. and Allman, F. L. Sports Medicine. New York: Academic Press, **1974**.

Ryan, D. E. Effects of stress on motor performance and learning. Res. Quart. 33: 111-119, **1962**.

Ryan, E. D. Relationship between motor performance and arousal. Res. Quart. 33: 279-287, **1962**.

Ryan, W. J., Sutton, J. R., Toews, C. J., and Jones, N. L. Metabolism of infused L(+) lactate during exercise. Clin. Sci. 56: 139-146, **1979**.

Ryhming, I. A modified Harvard step test for the evaluation of physical fitness. Arbeitsphysiol. 15: 235-250, **1953**.

Ryle, J. A. and Russel, W. T. The natural history of coronary disease; clinical and epidemiological study. Brit. Heart J. 11: 370-389, **1949**.

Sacdpraseuth, S. N. Contribution à l'étude de la circulation veineuse des membres inférieurs. Ph.D. thesis, Université de Lyon, **1960**.

Sackett, D. L., Anderson, G. D., Milner, R., Feinleib, M., and Kannel, W. B. Concordance for coronary risk factors among spouses. Circulation 52: 589-595, **1975**.

Sadoul, P., Heran, J., Arouette, A., and Grieco, B. Valeur de la puissance maximale supportée déterminée par des exercises musculaires de 20 minutes pour évaluer la capacité fonctionnelle des handicapés respiratoires. Bull. Physio-path. Resp. 2: 209-222, **1966**.

Sahlin, K., Alvestrand, A., Bergstrom, J., and Hultman, E. Intracellular pH and bicarbonate concentration as determined in biopsy samples from the quadriceps muscle of man at rest. Clin. Sci. Mol. Med. 53: 459-466, **1977**.

Sahlin, K., Harris, R. C., Nylind, B., and Hultman, R. E. Lactate content and pH in muscle samples obtained after dynamic exercise. Pflüg. Archiv 367: 137-142, **1976**.

Saiki, H., Margaria, R., and Cuttica, F. Lactic acid production in submaximal work. Int. Z. Angew. Physiol. 24: 57-61, **1967**.

Sakmann, B. Acetylcholine-induced ionic channels in rat skeletal muscle. Fed. Proc. 37: 2654-2659, **1978**.

Salans, L. B., Cushman, S. W., and Weisman, R. E. Studies of human adipose tissue. Adipose cell size and number in non-obese and obese patients. J. Clin. Invest. 52: 929-941, **1973**.

Salans, L. B., Horton, E. S., and Sims, E. A. H. Experimental obesity in man: Cellular character of the adipose tissue. J. Clin. Invest. 50: 1005-1011, **1971**.

Salmella, J. Psychomotor trends in International Gymnastics, p. 28. In: Abstracts of Proceedings, Annual Meeting of Canadian Association of Sports Sciences. Edmonton, Alberta. Eds.: H. Wenger and R. MacNab. Ottawa: Canadian Association of Sports Sciences, **1974**.

Salmons, S. and Vrbova, G. The influence of activity on some contractile characteristics of mammalian fast and slow muscles. J. Physiol. (Lond.) 201: 535-549, **1969**.

Salter, N. The effect on muscle strength of maximum isometric and isotonic contractions at different repetition rates. J. Physiol. (Lond.) 130: 109-113, **1955**.

Saltin, B. Aerobic work capacity and circulation at exercise in man. With special reference to the effect of prolonged exercise and/or heat exposure. Acta Physiol. Scand. 62 (Suppl. 230): 1-52, **1964a**.

——. Circulatory responses to submaximal and maximal exercise after thermal dehydration. J. Appl. Physiol. 19: 1125-1132, **1964b**.

——. Experiment cited by P. O. Åstrand and K. Rodahl (1970). In: Textbook of Work Physiology, p. 466. New York: McGraw Hill, **1970a**.

——. In: Physiological and Behavioural Temperature Regulation. Eds.: J. D. Hardy, A. P. Gagge, and A. J. Stolwijk. Springfield, Ill.: C. C. Thomas, **1970b**.

——. Muscle glycogen utilization during work of different intensities. In: Muscle Metabolism During Exercise. Eds.: B. Pernow and B. Saltin. New York: Plenum Press, **1973a**.

——. Oxygen transport by the circulatory system during exercise in man, pp. 235-252. In: Limiting Factors of Physical Performance. Ed.: J. Keul. Stuttgart: Thieme, **1973b**.

——. Fluid, electrolyte, and energy losses and their replenishment in prolonged exercise, pp. 76-97. In: Nutrition, Physical Fitness and Health. Eds.: J. Pařízková and V. A. Rogozkin. Baltimore: University Park Press, **1978**.

Saltin, B. and Åstrand, P. O. Maximal oxygen uptake in athletes. J. Appl. Physiol. 23: 353-358, **1967**.

Saltin, B., Blomqvist, G., Mitchell, J. H., Johnson, R. L., Wildenthal, K., and Chapman, C. B. Response to exercise after bed rest and after training; a longitudinal study of adaptive changes in oxygen transport and body composition. Circulation 38: VII-1 to VII-78, **1968**.

Saltin, B. and Essén, B. Muscle glycogen, lactate, ATP and CP in intermittent exercise, pp. 419-424. In: Muscle Metabolism During Exercise. Eds.: B. Pernow and B. Saltin. New York: Plenum Press, **1971**.

Saltin, B., Gagge, A. P., Bergh, V., and Stolwijk, J. A. J. Body temperature and sweating during exhaustive exercise. J. Appl. Physiol. 32: 635-643, **1972**.

Saltin, B. and Grimby, G. Physiological analysis of middle-aged and old former athletes. Comparison of still active athletes of the same ages. Circulation 38: 1104-1115, **1968**.

Saltin, B., Grover, R. F., Blomqvist, C. G., Hartley, L. H., and Johnson, R. L. Maximal oxygen uptake and cardiac output after 2 weeks at 4,300 m. J. Appl. Physiol. 25: 400-409, **1968**.

Saltin, B., Hartley, L. H., Kilbom, Å., and Åstrand, I. Physical training in sedentary middle-aged and older men. II. Oxygen uptake, heart rate and blood lactate concentration at submaximal and maximal exercise. Scand. J. Clin. Lab. Invest. 24: 323-334, **1969**.

Saltin, B., Henriksson, J., Nygaard, E., and Andersen, P. Fiber types and metabolic potentials of skeletal muscles in sedentary man and endurance runners. Ann. N.Y. Acad. Sci. 301: 3-29, **1977**.

Saltin, B. and Hermansen, L. Esophageal, rectal and muscle temperature during exercise. J. Appl. Physiol. 21: 1757-1762, **1966**.

——. Glycogen stores and prolonged severe exercise. In: Nutrition and Physical Activity, p. 32. Ed.: G. Blix. Uppsala: Almqvist & Wiksell, **1967**.

Saltin, B., Nazar, K., Costill, D. L., Stein, E., Jansson, B., Essén, B., and Gollnick, P. D. The nature of the training response; peripheral and central adaptations to one-legged exercise. Acta Physiol. Scand. 96: 289-305, **1976**.

Saltin, B. and Stenberg, J. Circulatory response to prolonged severe exercise. J. Appl. Physiol. 19: 833-838, **1964**.

Salzler, A. The influence of social factors on the physical development of young children: A contribution to the problems of acceleration, pp. 73-82. In: Growth and Development. Physique. Ed.: O. G. Eiben. Budapest: Akadémia Kiadó, **1977**.

Salzman, E. W. and Leverett, S. A. Studies in Orthostatic Venoconstriction. Dayton, Ohio: Wright Air Development Center Tech. Rep. 56-483: 1-22, **1956**.

Salzman, S. H., Hellerstein, H. K., Radke, J. D., Maistelman, H. W., and Ricklin, R. Quantitative effects of physical conditioning on the exercise electrocardiogram of middle-aged subjects with arteriosclerotic heart disease, pp. 388-410. In: Measurement in Exercise Electrocardiography. The Ernst Simonson Conference. Ed.: H. Blackburn. Springfield, Ill.: C. C. Thomas, **1969**.

Salzman, S. H., Hirsch, E. Z., Hellerstein, H. K., and Bruell, J. H. Adaptation to muscular exercise: Myocardial-epinephrine-3-H uptake. J. Appl. Physiol. 29: 92-95, **1970**.

Samek, L., Weidemann, H., Roskamm, H., Gornandt, L., Von Boroviczeny, K. G., Meyerspeer, U., Mellerowicz, H., and Limon-Lason, R. Erythrozyten, Hämoglobin und Hamatokrit bei Hochleistungsportlern im Verlaufe einer Akklimatisationsperiode in Font Romeu (1,800 m) und Mexico City (2,240). Sportarzt und Sportmedizin 19: 133-141, **1968**.

Samilson, R. L. and Morris, J. M. Surgical improvement of the cerebral palsied upper limb: electromyographic studies and results in 128 operations. J. Bone Joint Surg. 46A: 1203-1216, **1964**.

Sampson, S. R., Vidruk, E. H., and Hahn, H. L. Properties of rapidly adapting vagal receptors in intrapulmonary airways, pp. 197-208. In: Muscular Exercise and the Lung. Eds.: J. A. Dempsey and C. E. Reed. Madison: Univ. of Wisconsin, **1977**.

Sanders, A. P. and Currie, W. D. Chemical protection against oxygen toxicity, pp. 35-40. In: Underwater Physiology. Ed.: C. J. Lambertsen. New York: Academic Press, **1971**.

Sanders, T. M., White, F. C., and Bloor, C. M. Myocardial blood flow distribution in the conscious pig during steady state and exhaustive exercise. Fed. Proc. 34: 414, **1975**.

Sanders, T. M., White, F. C., Peterson, T. M., and Bloor, C. M. Effects of endurance exercise on coronary collateral blood flow in miniature swine. Amer. J. Physiol. 234: H614–H619, **1978**.

Sandler, H. and Dodge, H. T. Left ventricular tension and stress in man. Circ. Res. 13: 91–104, **1963**.

Sandler, H. and Dodge, H. T. Angiocardiographic methods for determination of left ventricular geometry and volume, pp. 141–170. In: Cardiac Mechanics: Physiological, Clinical and Mathematical Considerations. Eds.: I. Mirsky, D. Ghista, and H. Sandler. New York: Wiley, **1974**.

Sargeant, A. J., and Davies, C. T. M. Perceived exertion during rhythmic exercise involving different muscle masses. J. Human Ergol. 2: 31, **1973**.

Sargeant, A. J. and Jones, N. L. Effect of temperature on power output of human muscle during short-term dynamic exercise. Med. Sci. Sport. 10: 39, **1978**.

Sargent, F. Depression of sweating in man: So-called "sweat gland fatigue," pp. 163–212. In: Advances in Biology of Skin, Eccrine Sweat Glands and Eccrine Sweating, Vol. 3, Eds.: W. Montagna, R. A. Ellis, and A. F. Silver. New York: Pergamon Press, **1962**.

Sargent, F. and Weinman, K. P. Eccrine sweat-gland activity during menstrual cycle. J. Appl. Physiol. 21: 1685–1687, **1966**.

Sarnoff, S. J., Braunwald, E., Welch, G. H., Case, R. B., Stainsby, W. N., and Macruz, R. Hemodynamic determinants of oxygen consumption of the heart with special reference to the tension-time index. Amer. J. Physiol. 192: 148–156, **1958**.

Sarnoff, S. J. and Mitchell, J. H. The control of the function of the heart. In: Handbook of Physiology, Section 2. Circulation, Vol. I. Eds.: W. H. Hamilton and P. Dow. Washington, D.C.: Amer. Physiol. Soc., **1962**.

Sato, M. Muscle fatigue in the half rising posture. Zinruigaku Zassi 74: 195–201, **1966**.

Sato, M. and Sakate, T. Combined influences on cardiopulmonary functions of simulated high altitude and graded work loads. J. Human Ergol. 3: 55–66, **1974**.

Saumarez, R. C., Bolt, J. F., and Gregory, R. J. Neurological decompression sickness treated without recompression. Brit. Med. J. 1: 151–152, **1973**.

Saunders, N. A., Leeder, S. R., and Rebuck, A. S. Ventilatory response to carbon dioxide in young athletes. Amer. Rev. Resp. Dis. 113: 497–502, **1976**.

Saville, P. D. and Smith, R. Bone density, breaking force and leg muscle mass as functions of weight in bipedal rats. Amer. J. Phys. Anthropol. 25: 35–40, **1966**.

Saville, P. D. and Whyte, M. P. Muscle and bone hypertrophy. Clin. Orthoped. 65: 81–88, **1969**.

Sawin, C. S., Rummel, J. A., and Michel, E. L. Instrumental personal exercise during long-duration space flights. Aviation Space Environ. Med. 46: 394–400, **1975**.

Sawka, M. N., Knowlton, R. G., and Critz, J. B. Cardiovascular adaptations to repeated bouts of exercise. Med. Sci. Sport. 10: 50 (Abstr.), **1978**.

Sayed, J., Schaefer, O., Hildes, J. A., and Lobban, M. A. Biochemical indices of nutrient intake by Eskimos of Northern Foxe basin NWT. In: Circumpolar Health. Eds.: R. J. Shephard and S. Itoh. Toronto: University of Toronto Press, **1976**.

Schaefer, K. E. Carbon dioxide effects under conditions of raised environmental pressure, pp. 144–154. In: Physiology and Medicine of Diving. Eds.: P. B. Bennett and D. H. Elliott. London: Baillière, Tindall & Cassell, **1969**.

Schaefer, K. E., Bond, G. F., Mazzone, W. F., Carey, C. R., and Dougherty, J. H. Carbon dioxide retention and metabolic changes during prolonged exposure to high pressure environment. Aerospace Med. 39: 1206–1215, **1968**.

Schaefer, K. E., Carey, C. R., and Dougherty, J. H. Pulmonary function and respiratory gas exchange during saturation-excursion diving to pressures equivalent to 1,000 feet of sea water, pp. 357–370. In: Underwater Physiology. Ed.: C. J. Lambertsen. New York: Academic Press, **1971**.

Schäfer, S. S. and Schäfer, S. Die Eigenschaften einer, primären, Muskelspindel-afferenz bei rempenförmiger Dehnung und ihre mathematische Beschreibung. Pflüg. Arch. 310: 206–228, **1969**.

Schaffer, A. I. The body as a volume conductor in electrocardiography. Amer. Heart J. 51: 588–608, **1956**.

Schanberg, S. M., Schildkraut, J. J., Breese, G. R., and Kopin, I. J. NM-H^3 turnover and conversion to a diaminated conjugate as a major metabolite in rat brain. Biochem. Pharmacol. 17: 247-254, **1968**.

Scheele, K. Acute cardiac death caused by an increase of platelet aggregation during and after maximal physical stress. In: Sports Cardiology. Eds.: T. Lubich and A. Venerando. Bologna, Aulo Gaggi, **1979**.

Scheinberg, P., Blackburn, L. I., Rich, M., and Saslaw, M. Effects of vigorous physical exercise on cerebral circulation and metabolism. Amer. J. Med. 16: 549-554, **1954**.

Scher, A. M., Young, A. C., and Meredith, W. M. Factor analysis of the electrocardiogram. Test of electrocardiographic theory: Normal hearts. Circ. Res. 8: 519-526, **1960**.

Scherrer, J., Bourguignon, A., and Monod, H. La fatigue dans le travail statique. Rev. Path. Gen. Physiol. Clin. 60: 357-367, **1960**.

Scherrer, M. and Bitterli, J. Repeated estimated of D_{L,O_2} in men at different work levels. Experientia 26: 683-684, **1970**.

Scheuer, J. Physical training and intrinsic cardiac adaptations. Circulation 47: 677-680, **1973**.

Scheuer, J., Penpargkul, S., and Bhan, A. K. Experimental observations on the effects of physical training upon intrinsic cardiac physiology and biochemistry. Amer. J. Cardiol. 33: 744-757, **1974**.

Scheuer, J. and Stezoski, S. W. Effect of physical training on the metabolic, mechanical and metabolic response of the rat heart to hypoxia. Circ. Res. 30: 418-429, **1972**.

Scheuer, J. and Tipton, C. M. Cardiovascular adaptations to training. Ann. Rev. Physiol. 39: 221-251, **1977**.

Schievelbein, H. The evidence for nicotine as an etiological factor in cardiovascular disease (CVD), pp. 167-169. In: Atherosclerosis IV. Eds.: G. Schettler, Y. Goto, Y. Hata, and G. Klose. Berlin: Springer Verlag, **1977**.

Schlierf, G., Oster, P., Seidel, D., Raetzer, H., Schellenberg, B., Heuck, C. C., and Wicklein, R. L. Dietary management of atherosclerosis, pp. 511-516. In: Atherosclerosis IV. Eds.: G. Schettler, Y. Goto, Y. Hata, and G. Klose. Berlin: Springer Verlag, **1977**.

Schmalbruch, H. and Kamieniecka, Z. Fibre types in the human brachial biceps muscle. Exp. Neurol. 44: 313-328, **1974**.

Schmidt, C. F. and Comroe, J. H. Dyspnea. Mod. Concepts Cardiovasc. Dis. 13(3), **1944**.

Schmidt, D. H. The clinical and research application of nuclear cardiology, pp. 183-202. In: Heart Disease and Rehabilitation: State of the Art. Eds.: M. Pollock and D. H. Schmidt. Boston: Houghton Mifflin, **1978**.

Schmidt, R. A. A schema theory of discrete motor skill learning. Psychol. Rev. 82: 225-260, **1975**.

——. Control processes in motor skills. Exer. Sport. Sci. Rev. 4: 229-261, **1976**.

Schmidt, R. F. and Willis, W. D. Intracellular recording from motoneurons of the cervical spinal cord of the cat. J. Neurophysiol. 26: 28-43, **1963**.

Schmidt-Nielsen, K. Heat conservation in countercurrent systems, p. 143. In: Temperature: Its Measurement and Control in Science and Industry, Vol. 3, Part 3. Ed.: J. D. Hardy. New York: Reinhold, **1963**.

Schmidt-Nielsen, K. and Pennycuik, P. Capillary density in mammals in relation to body size and oxygen consumption. Amer. J. Physiol. 200: 746-750, **1961**.

Schmukler, M. and Barrows, C. H. Age differences in lactic and malic dehydrogenase in the rat. J. Gerontol. 21: 109-111, **1966**.

Schneider, E. C. A cardiovascular rating as a measure of physical fatigue and efficiency. J. Amer. Med. Assoc. 74: 1507-1510, **1920**.

Schneider, E. C. and Havens, L. C. Changes in the blood after muscular activity and during training. Amer. J. Physiol. 36: 239-259, **1915**.

Scholander, P. F. Oxygen transport through hemoglobin solutions. Science 131: 585-590, **1960**.

Scholander, P. F., Hammel, H. T., LeMessurier, H., Hemmingsen, E., and Garey, W. Circulatory adjustment in pearl divers. J. Appl. Physiol. 17: 184-190, **1962**.

Schonholzer, G. Ermüdung, Erschöpfung, Tod. Schweiz. Z. Sportmed. 5: 65-80, **1957**.

Schoop, W. Bewegungstherapie bei peripheren Durchblutungsstörungen. Med. Welt 10: 502-506, **1964**.

Schreiber, S. S., Oratz, M., Evans, C. D., Gueyikian, I., and Rothschild, M. A. Myosin, myoglobin and collagen synthesis in acute cardiac overload. Amer. J. Physiol. 219: 481-486, **1970**.

Schreiber, S. S., Rothschild, M. A., Evans, C., Reff, F., and Oratz, M. The effect of pressure or flow stress on right ventricular protein synthesis in the face of constant and restricted coronary perfusion. J. Clin. Invest. 55: 1-11, **1975**.

Schreiner, H. R. Mathematical approaches to decompression. Int. J. Biometeor. 11: 301–310, **1967**.

——. General biological effects of the helium-xenon series of elements. Fed. Proc. 27: 872–878, **1968**.

Schreiner, H. R. and Kelley, P. R. A pragmatic view of decompression, pp. 205–219. In: Underwater Physiology. Ed.: C. J. Lambertsen. New York: Academic Press, **1971**.

Schrier, R. W., Hano, J., Keller, H. I., Finkel, R. M., Gilliland, P. F., Cirksena, W. J., and Teschan, P. E. Renal, metabolic and circulatory responses to heat and exercise. Studies in military recruits during summer training, with implications for acute renal failure. Ann. Intern. Med. 73: 213–223, **1970**.

Schrier, R. W., Henderson, H. S., Tisher, C. C., and Tannen, R. L. Nephropathy associated with heat stress and exercise. Ann. Intern. Med. 67: 356–376, **1967**.

Schroter, G. Die Berufsschaden des Stutz und Bewegungsystems. Barth: Leipzig, **1958**.

Schuchhardt, S. and Losse, B. Methodological problems when measuring with PO_2 needle electrodes in semisolid media. In: Oxygen Supply. Theoretical and Practical Aspects of Oxygen Supply and Microcirculation of Tissues. Eds.: M. Kessler, D. F. Bruley, L. C. Clark, D. W. Lubbers, I. A. Silver, and J. Strauss. Baltimore: University Park Press, **1973**.

Schuck, G. R. Effects of athletic competition on growth and development of junior high school boys. Res. Quart. 33: 288–298, **1962**.

Schuerch, P. M., Dordel, H. J., Neumann, A., and Hollmann, W. The effect of ethanol on the ergometric physical performance. In: Nutrition, Dietetics and Sport. Eds.: G. Ricci and A. Venerando. Torino: Minerva Medica, **1978**.

Schultz, G. W. Effects of direct practice, repetitive sprinting and weight training on selected motor performance tests. Res. Quart. 38: 108–118, **1967**.

Schultze, H. E. and Heremans, J. F. Molecular Biology of Human Proteins, Vol. 1, pp. 719–721. Amsterdam: Elsevier, **1966**.

Schuman, S. H. Patterns of urban heat-wave deaths and implications for prevention: Data from New York and St. Louis during July, 1966. Environ. Res. 5: 59–75, **1972**.

Schwade, J., Blomqvist, C. G., and Shapiro, W. A comparison of the response to arm and leg work in patients with ischemic heart disease. Amer. Heart J. 94: 203–208, **1977**.

Schwartz, R. A. Oxygen consumption during rowing: Implications for training. Brit. J. Sport. Med. 7: 188 (Abstr.), **1973**.

Schwarz, F., ter Haar, D. J., van Riet, H. G., and Thijssen, J. H. H. Response of growth hormone, FFA, blood sugar and insulin to exercise in obese patients and normal subjects. Metab. Clin. Exp. 18: 1013–1022, **1969**.

Scoggin, C. H., Doekel, R. D., Kryger, M. H., Zwillich, C. W., and Weil, J. V. Familial aspects of decreased hypoxic drive in endurance athletes. J. Appl. Physiol. Resp. Environ. Exer. Physiol. 44: 464–468, **1978**.

Scott, G. M. Measurement of kinesthesis. Res. Quart. 26: 324–341, **1955**.

Scott, J. C. Physical activity and the coronary circulation. Can. Med. Assoc. J. 96: 853–859, **1967**.

Scrimshaw, N. S., Taylor, C. E., and Gordon, J. E. Interaction of Nutrition and Infection. Geneva: World Health Organization, **1968**.

Scrimshire, D. A., Tomlin, P. J., and Ethridge, R. A. Computer simulation of gas exchange in human lungs. J. Appl. Physiol. 34: 687–696, **1973**.

Scripture, E. W., Smith, T. L., and Brown, E. M. On the education of muscular control of power. Stud. Yale Psychol. Lab. 2: 114–119, **1894**.

Sears, T. A. Investigations on respiratory motoneurones of the spinal cord, pp. 259–274. In: Progress in Brain Research, Vol. 12. Eds.: J. C. Eccles and J. P. Schadé. Amsterdam: Elsevier, **1964**.

——. The respiratory motoneurone: integration at spinal segmental level, pp. 33–47. In: Breathlessness. Eds.: J. B. L. Howell, and E. J. M. Campbell. Oxford: Blackwell Scientific, **1966**.

Secher, N. H. Development of results in International Rowing championships, 1893–1971. Med. Sci. Sport. 5: 195–199, **1973**.

——. Isometric rowing strength of experienced and inexperienced oarsmen. Med. Sci. Sport. 7: 280–283, **1975**.

Secher, N. H., Ruberg-Larsen, N., Binkhorst, R. A., and Bonde-Petersen, F. Maximal oxygen uptake during arm cranking and combined arm plus leg exercise. J. Appl. Physiol. 36: 515–518, **1974**.

Sedgewick, A. W. Effect of actively increased muscle temperature on local muscular endurance.

Res. Quart. 35: 532-538, **1964**.

Segerra, F. and Redding, R. A. Modern concepts about drowning. Can. Med. Assoc. J. 110: 1052-1060, **1974**.

Seliger, V. Physical fitness of Czechoslovak children at 12 and 15 years of age. International Biological Programme results of investigations, 1968-1969. Acta Univ. Carol. Gymnica 5: 6-169, **1970**.

Seliger, V., Dolejš, L., Karas, V., and Pachlopniková, I. Adaptation of trained athletes' energy expenditure to repeated concentric and eccentric muscle contractions. Int. Z. Angew. Physiol. 26: 227-234, **1968**.

Seliger, V., Navara, M., and Pachlopniková, I. Der energetische Metabolismus in Verlauf des Fussballspiels. Sportarzt und Sportmedizin 54: 114-118, **1970**.

Seliger, V., Pachlopniková, I., Mann, M., Selecká, R., and Treml, J. Energy expenditure during paddling. Physiol. Bohem. 18: 49-55, **1969**.

Selvester, R. H., Collier, C. R., and Pearson, R. B. Analog computer model of the vectocardiogram. Circulation 31: 45-53, **1965**.

Selye, H. Stress and nation's health, pp. 65-74. In: Proceedings of the National Conference on Fitness and Health. Ed.: W. A. R. Orban. Ottawa: Information Canada, **1974**.

Semple, T. Acceleration of collaterals by physical activity, pp. 141-142. In: Critical evaluation of cardiac rehabilitation. Eds.: J. J. Kellerman and H. Denolin. Karger: Basel, **1977**.

Sen, R. N., Ray, G. G., and Nag, P. K. Relationship between segmental and whole body weights of some Indian subjects, pp. 384-391. In: Physical Fitness Assessment. Eds.: R. J. Shephard and H. Lavallée. Springfield, Ill.: C. C. Thomas, **1978**.

Senay, L. C. Relationship of evaporative rates to serum [Na+], [K+] and osmolarity in acute heat stress. J. Appl. Physiol. 25: 149-152, **1968**.

——. Changes in plasma volume and protein content during exposures of working men to various temperatures before and after acclimatization to heat: Separation of the roles of cutaneous and skeletal muscle circulation. J. Physiol. (Lond.) 224: 61-81, **1972**.

Senay, L. C. and Christensen, M. L. Variations of certain blood constituents during acute heat exposure. J. Appl. Physiol. 24: 302-309, **1968**.

Senay, L. C. and Fortney, S. Untrained females: Effects of submaximal exercise and heat on body fluids. J. Appl. Physiol. 39: 643-647, **1975**.

Senay, L. C. and Kok, R. Effects of training and heat acclimatization on blood plasma controls. J. Appl. Physiol. 43: 591-599, **1977**.

Senay, L. C., Mitchell, D., and Wyndham, C. H. Acclimatization in a hot, humid environment: Body fluid adjustments. J. Appl. Physiol. 40: 786-796, **1976**.

Sendroy, J. and Cecchini, L. P. Determination of human body surface area from height and weight. J. Appl. Physiol. 7: 1-12, **1954**.

Sergeant, R. L. Distortion of speech, pp. 213-225. In: The Physiology and Medicine of Diving. Eds.: P. B. Bennett and D. H. Elliott. London: Baillière, Tindall & Cassell, **1969**.

Serio, M., Piolanti, P., Romano, S., DeMagistris, L., and Guistri, G. The circadian rhythm of plasma cortisol in subjects over 70 years of age. J. Gerontol. 25: 95-97, **1970**.

Seusing, J. and Drube, H. C. The significance of hypercapnia for the occurrence of depth intoxication. Klin. Wschr. 38: 1088-1090, **1960**.

Severinghaus, J. W. Transarterial leakage: A possible mechanism of high altitude pulmonary oedema. In: High Altitude Physiology: Cardiac and Respiratory Aspects. Eds.: R. Porter and J. Knight. Edinburgh: Churchill-Livingstone, **1971**.

Severinghaus, J. W. and Mitchell, R. A. The role of cerebrospinal fluid in the respiratory acclimatization to high altitude in man, pp. 273-284. In: The Physiological Effects of High Altitude. Ed.: W. H. Weihe. London: Pergamon Press, **1964**.

Seymour, J. and Conway, N. Value of dual reports on routine electrocardiograms. Brit. Heart J. 31: 610-612, **1969**.

Shapiro, A., Shoenfeld, Y., and Shapiro, Y. Recovery heart rate after submaximal work. J. Sport. Med. Phys. Fitness, 16: 57-59, **1976**.

Shapiro, D. C. A preliminary study to determine the duration of a motor program. In: Psychology of Motor Behavior and Sport. Vol. 3. Eds.: D. Landers and R. Christina. Champaign, Ill.: Human Kinetics, **1976**.

Shapiro, V., Magazanik, A., Sohar, E., and Reich, C. B. Serum enzyme changes in untrained subjects following a prolonged march. Can. J. Physiol. Pharmacol. 51: 271-276, **1973**.

Shapiro, W., Johnston, C. E., Dameron, R. A., and Patterson, J. L. Maximum ventilatory performance and its limiting factors. J. Appl. Physiol. 19: 199–203, **1964**.

Sharkey, B. J. Intensity and duration of training and the development of cardiorespiratory endurance. Med. Sci. Sport. 2: 197–202, **1970**.

Sharma, S. D., Ballantyne, F., Goldstein, D. The relationship of ventricular asynergy in coronary artery disease to ventricular premature beats. Chest 66: 358–362, **1974**.

Sharman, I. M., Down, M. G., and Norgan, N. G. Alleged ergogenic properties of vitamin E. In: Proceedings of 20th World Congress of Sports Medicine. Ed.: A. H. Toyne. Melbourne: Australian Sports Medicine Federation, **1975**.

Sharman, I. M., Down, M. G., and Sen, R. N. The effects of vitamin E and training on physiological function and athletic performance in adolescent swimmers. Brit. J. Nutr. 26: 265–276, **1971**.

Sharp, R. H. Visual information-processing in ball games. Some input considerations, pp. 3–12. In: Motor Learning, Sport, Psychology, Pedagogy and Didactics of Physical Activity. Eds.: F. Landry and W. A. Orban. Miami: Symposia Specialists, **1978**.

Sharpey-Schafer, E. P. Effects of Valsalva's manoeuvre on the normal and failing circulation. Brit. Med. J. 1: 693–695, **1955**.

——. Effects of squatting on the normal and failing circulation. Brit. Med. J. 1: 1072–1074, **1956**.

——. Venous tone. Brit. Med. J. 2: 1589–1595, **1961**.

——. Venous tone: Effects of reflex changes. Humoral agents and exercise. Brit. Med. Bull. 19: 145–148, **1963**.

——. Effect of respiratory acts on the circulation, pp. 1875–1886. In: Handbook of Physiology, Section 2. Circulation, Vol. 3. Ed.: W. F. Hamilton. Baltimore: Williams & Wilkins, **1965**.

Sheffield, L. T. The meaning of exercise test findings, pp. 9-1 to 9-35. In: Coronary Heart Disease. Prevention, Detection, Rehabilitation with Emphasis on Exercise Testing. Ed.: S. M. Fox. Denver, Colorado: International Medical Corporation, **1974**.

Sheffield, L. T. and Reeves, T. J. Graded exercise in the diagnosis of angina pectoris. Mod. Concepts Cardiovasc. Dis. 34: 1–6, **1965**.

Sheffield, L. T. and Roitman, D. Systolic blood pressure, heart rate and treadmill work at anginal threshold. Chest 63: 327–335, **1973**.

Shen, L. C., Fall, L., Walton, G. M., and Atkinson, D. E. Interaction between energy charge and metabolite modulation in the regulation of enzymes of amphibolic sequences, phosphofructokinase and pyruvate dehydrogenase. Biochemistry 7: 4041–4045, **1968**.

Shenk, J. H., Hall, J. L., and King, H. H. Spectrophotometric characteristics of hemoglobins; beef blood and muscle hemoglobins. J. Biol. Chem. 105: 741–752, **1934**.

Shennum, P-L. and deVries, H. A. The effect of saddle-height on oxygen consumption during bicycle ergometer work. Med. Sci. Sport. 8: 119–121, **1976**.

Shepard, R. H. Effect of pulmonary diffusing capacity on exercise tolerance. J. Appl. Physiol. 12: 487–488, **1958**.

Shephard, R. J. The immediate metabolic effects of breathing carbon dioxide mixtures. J. Physiol. (Lond.) 129: 393–406, **1955a**.

——. Pneumotachographic measurement of breathing capacity. Thorax 10: 258–268, **1955b**.

——. The carbon dioxide balance-sheets of the body; their determination in normal subjects and in cases of congenital heart disease. J. Physiol. (Lond.) 129: 142–158, **1955c**.

——. Assessment of ventilatory efficiency by the single-breath technique. J. Physiol. (Lond.) 134: 630–649, **1956a**.

——. The direct interpretation of the fast vital capacity record. Thorax 11: 223–233, **1956b**.

——. A null-point discontinuous electrical pursuit meter. J. Appl. Psychol. 40: 287–294, **1956c**.

——. Physiological changes and psychomotor performance during acute hypoxia. J. Appl. Physiol. 9: 343–351, **1956d**.

——. The influence of age on the haemoglobin level in congenital heart disease. Brit. Heart J. 18: 49–54, **1956e**.

——. Some factors affecting the open-circuit determination of maximum breathing capacity. J. Physiol. (Lond.) 135: 98–113, **1957a**.

——. Changes of intramuscular blood flow during continuous high pressure breathing. J. Aviation Med. 28: 142–153, **1957b**.

——. Partitional respirometry in human subjects. J. Appl. Physiol. 13: 357–367, **1958a**.

——. Indirect method for the recording of arterial blood pressure. Porton Down, U.K.: Chemical Defence Experimental Establishment, **1958b**.

——. The timed airway resistance. J. Physiol. (Lond.) 145: 459-472, **1959**.

——. The ergonomics of the respirator, pp. 51-66. In: Design and Use of Respirators. Ed.: C. N. Davies. New York: Macmillan, **1961a**.

——. The design of a cardiac defibrillator. Brit. Heart J. 23: 7-19, **1961b**.

——. Changes in capacity of the leg veins studied by a simple counterpressure technique. Quart. J. Exp. Physiol. 46: 175-187, **1961c**.

——. The development of cardio-respiratory fitness. Med. Serv. J. Can. 21: 533-544, **1965**.

——. Oxygen cost of breathing during vigorous exercise. Quart. J. Exp. Physiol. 51: 336-350, **1966a**.

——. Dynamic characteristics of the airway and unstable breathing systems. Aerospace Med. 37: 1014-1021, **1966b**.

——. Initial "fitness" and personality as determinants of the response to a training regime. Ergonomics 9: 3-16, **1966c**.

——. Devices for the teaching of expired air resuscitation. Med. Serv. J. Can. 22: 273-284, **1966d**.

——. World standards of cardiorespiratory performance. A.M.A. Arch. Environ. Health 13: 664-672, **1966e**.

——. The maximum sustained voluntary ventilation in exercise. Clin. Sci. 32: 167-176, **1967a**.

——. The prediction of "maximal" oxygen consumption using a new progressive step test. Ergonomics 10: 1-15, **1967b**.

——. Normal levels of activity in Canadian city dwellers. Can. Med. Assoc. J. 96: 912-914, **1967c**.

——. Physical performance in Mexico City, pp. 132-134. In: International Symposium on the Effects of Altitude on Physical Performance. Ed.: R. Goddard. Chicago: Athletic Institute, **1967d**.

——. A possible deterioration in performance of short-term Olympic events at altitude. Can. Med. Assoc. J. 97: 1414, **1967e**.

——. The heart and circulation under stress of Olympic conditions. J. Amer. Med. Assoc. 205: 150-155, **1968a**.

——. Oscillations of acid-base equilibrium during maximum exercise. Int. Z. Angew. Physiol. 26: 258-271, **1968b**.

——. Intensity, duration, and frequency of exercise as determinants of the response to a training regime. Int. Z. Angew. Physiol. 26: 272-278, **1968c**.

——. Learning, habituation, and training. Int. Z. Angew. Physiol. 28: 38-48, **1969**.

——. Human endurance and the heart at altitude. J. Sport. Med. Phys. Fitness 10: 72-83, **1970a**.

——. Computer programmes for solution of the Åstrand nomogram. J. Sport. Med. Phys. Fitness 10: 206-210, **1970b**.

——. The oxygen conductance equation, pp. 129-154. In: Frontiers of Fitness. Ed.: R. J. Shephard. Springfield, Ill.: C. C. Thomas, **1971a**.

——. Prediction formulae and some normal values in pulmonary physiology. In: Handbook of Circulation and Respiration. Eds.: P. Altmann and D. S. Dittmer. Washington, D.C.: Amer. Physiol. Soc., **1971b**.

——. The working capacity of schoolchildren, pp. 319-344. In: Frontiers of Fitness. Ed.: R. J. Shephard. Springfield, Ill.: C. C. Thomas, **1971c**.

——. Standard tests of aerobic power. In: Frontiers of Fitness. Ed.: R. J. Shephard. Springfield, Ill.: C. C. Thomas, **1971d**.

——. Exercise in a cold climate. Can. Fam. Physician. 18: 44-47, **1972a**.

——. Exercise and the lungs, pp. 7-18. In: Fitness and Exercise. Eds.: J. F. Alexander, R. C. Serfass, and C. M. Tipton. Chicago: Athletic Institute, **1972b**.

——. An integrated approach to cardiorespiratory performance at sea level and at an altitude of 7,350 ft., pp. 87-100. In: Environmental Effects on Work Performance. Eds.: G. R. Cumming, D. Snidal, and A. W. Taylor. Ottawa: Canadian Association of Sports Sciences, **1972c**.

——. Alive, Man! The Physiology of Physical Activity. Springfield, Ill.: C. C. Thomas, **1972d**.

——. Some determinants of continuous and intermittent hand-grip endurance. Spor. Hekim Derg. 9: 89-103, **1974a**.

——. Men at work—Aplications of Ergonomics to Performance and Design. Springfield, Ill.: C. C. Thomas, **1974b**.

——. Sudden death: A significant hazard of exercise? Brit. J. Sport. Med. 8: 102-110, **1974c**.

——. What causes second wind? Physician Sport. Med. 2: 36-42, **1974d**.

——. The phenomenon of the second wind. Postgraduate Committee in Medicine, Univ. Sydney, Austr. 30: 84-92, **1974e**.

——. Altitude training camps. Brit. J. Sport Med. 8: 38-45, **1974f**.

——. Physical performance tests for soccer. Ottawa: Canadian Soccer Association, **1975a**.

———. Coronary artery disease: The Magnitude of the problem. In: Proceedings of International Symposium on Exercise and Coronary Artery Disease. Toronto: Toronto Rehabilitation Centre, **1975b**.

———. Physical fitness from the viewpoint of the physiologist. Keynote address. International Conference on Physical Education Research. Ed.: H. Ruskin. Natanya, Israel: Wingate Institute, **1975c**.

———. Respiratory gas exchange ratio and the prediction of aerobic power. J. Appl. Physiol. 38: 402-406, **1975d**.

———. Future research on the quantifying of endurance training. J. Human Ergol. 3: 163-181, **1975e**.

———. A new look at aerobic power. In: Medicine and Sport, Vol. 9. Eds.: E. Jokl, R. L. Anand, and H. Stoboy. Basel: Karger, **1976a**.

———. Environment, pp. 76-97. In: Sports Medicine. Eds.: J. G. P. Williams and P. N. Sperryn. London: Arnold, **1976b**.

———. Exercise and chronic obstructive lung disease, pp. 263-296. In: Exercise and Sports Sciences Reviews, Vol. 4. Eds.: J. Keogh and R. S. Hutton. Santa Barbara, Calif.: Journal Publishing Affiliates, **1976c**.

———. Endurance Fitness, 2nd ed.. Toronto: University of Toronto Press, **1977a**.

———. Exercise-induced bronchospasm: A review. Med. Sci. Sport. 9: 1-10, **1977b**.

———. Alteration in activity of vascular smooth muscle by local modulation of adrenergic transmitter release. Introductory remarks. Fed. Proc. 37: 179-180, **1978**.

———. Human Physiological Work Capacity. London: Cambridge University Press, **1978a**.

———. Physical Activity and Aging. London: Croom-Helm **1978b**.

———. The Fit Athlete. Oxford: Oxford University Press, **1978c**.

———. Vitamin E and physical performance. In: Vol. 2, Encyclopaedia of Physical Education, Fitness and Sports, **1978d**.

———. The prediction of athletic performance, pp. 113-141. In: Physical Fitness Assessment: Principles, Practice and Application. Eds.: R. J. Shephard and H. Lavallée. Springfield, Ill.: C. C. Thomas, **1978e**.

———. Fitness, obesity and health, pp. 238-261. In: Proceedings of the 1st RSG4 Physical Fitness Symposium with Special Reference to Military Forces. Ed.: C. Allen. Downsview, Ont.: Defence and Civil Institute of Environmental Medicine, **1978f**.

———. Standardization of physiological and performance test data in prepubescent children. In: 2nd International Seminar on Kinanthropometry, Louvain, Belgium, **1978g**.

———. Current status of the Canadian Home Fitness Test. Proceedings, Symposium of International Committee on Physical Fitness Research, Johannesburg. S. Afr. J. Sport. Sci., 2: 19-35, **1978i**.

———. Exercise prescription: The Canadian experience. Brit. J. Sport. Med. 12: 227-234, **1978j**.

———. Altitude, pp. 163-170. In: Basic Book of Sports Medicine. Ed.: G. La Cava. Rome: Olympic Solidarity of International Olympic Committee, **1978l**.

———. Exercise for the asthmatic patient: A brief historical review. J. Sport. Med. Phys. Fitness, 18: 301-307, **1979a**.

———. Recurrence of myocardial infarction. Observations on patients participating in the Ontario Multicentre Exercise: Heart Trial. Europ. J. Cardiol. 11: 147-157, **1979b**.

———. Current status of the Canadian Home Fitness Test. S. Afr. J. Sport. Sci. 2: 19-35, **1979c**.

———. Aging, activity and fitness: Some problems of methodology. Med. dello Sport. 32: 69-78, **1979d**.

———. Some preliminary lessons from a multicentre trial. In: Sports Cardiology Eds.: T. Lubich and A. Venerando. Bologna: Aulo Gaggi, **1979e**.

———. Effects of training upon respiratory function. Med. dello Sport. 33: 9-17, **1979f**.

———. **1979g**. See **Shephard, 1978j**.

———. Evaluation of earlier studies: Canadian study. Bethesda, Md., U.S. Public Health Service Workshop on Physical Conditioning and Rehabilitation, **1979h**.

———. Cardiac rehabilitation in prospect. In: Heart Disease and Rehabilitation: State of the Art 1978. Eds.: M. Pollock and D. H. Schmidt. Boston: Houghton Mifflin, **1979i**.

Shephard, R. J., Allen, C., Benade, A. J. S., Davies, C. T. M., di Prampero, P. E., Hedman, R., Merriman, J. E., Myhre, K., and Simmons, R. The maximum oxygen intake: An international reference standard of cardiorespiratory fitness. Bull. WHO 38: 757-764, **1968a**.

———. Standardization of sub-maximal exercise tests. Bull. WHO 38: 765-776, **1968b**.

Shephard, R. J., Allen, C., Bar-Or, O., Davies,

C. T. M., Degré, S., Hedman, R., Ishii, K., Kaneko, M., LaCour, J. R., di Prampero, P. E., and Seliger, V. The working capacity of Toronto school-children. Can. Med. Assoc. J. 100: 560–566, 705–714, **1968c**.

Shephard, R. J. and Andersen T. W. Training, work and increase of pulmonary diffusing capacity. In: Muskelarbeit und Muskeltraining. Ed.: W. Rohmert. Stüttgart: Gentner Verlag, **1968**.

Shephard, R. J. and Bar-Or, O. Alveolar ventilation in near maximum exercise. Data on preadolescent children and young adults. Med. Sci. Sport. 2: 83–92, **1970**.

Shephard, R. J. and Callaway, S. Principal component analysis of the responses to standard exercise training. Ergonomics 9: 141–154, **1966**.

Shephard, R. J., Campbell, R., Pimm, P., Stuart, D., and Wright, G. Vitamin E, exercise and the recovery from physical activity. Europ. J. Appl. Physiol. 333: 119–126, **1974a**.

Shephard, R. J., Conway, S., Thomson, M., Anderson, G. H., and Kavanagh, T. Nutritional demands of sub-maximum work: Marathon and Trans-Canadian events. In: International Symposium on Athletic Nutrition. Ed.: J. Pavluk. Warsaw: Polska Federacja Sportu, **1977**.

Shephard, R. J., Cox, M., and West, C. Some factors influencing serum lipid levels in a working population. Atheroslerosis 35: 287–300, **1980**.

Shephard, R. J. and Godin, G. Activity patterns in the Canadian Eskimo. In: Polar Human Biology. Eds.: O. G. Edholm and E. K. E. Gunderson. London: Heinemann, **1973**.

Shephard, R. J., Godin, G., and Campbell, R. Characteristics of sprint, medium and middle distance swimmers. Int. Z. Angew. Physiol. 32: 1–19, **1973**.

Shephard, R. J., Hatcher, J., and Rode, A. On the body composition of the Eskimo. Europ. J. Appl. Physiol. 30: 1–13, **1973**.

Shephard, R. J., Jones, G., Ishii, K., Kaneko, M., and Olbrecht, A. J. Factors affecting body density and thickness of subcutaneous fat. Amer. J. Clin. Nutr. 22: 1175–1189, **1969**.

Shephard, R. J. and Kavanagh, T. Biochemical changes with marathon running. Observations on post-coronary patients, pp. 245–252. In: Metabolic Adaptation to Prolonged Physical Exercise. Eds.: H. Howald and J. R. Poortmans. Basel: Birkhauser Verlag, **1975a**.

——. Mild or intensive exercise in post coronary rehabilitation, pp. 391–404. In: Proceedings 20th World Congress in Sports Medicine. Ed.: A. Toyne. Melbourne: Australian Federation of Sports Medicine, **1975b**.

——. On the stage duration for a progressive exercise test protocol, pp. 335–344. In: Physical Fitness Testing: Principles, Practice and Application. Eds.: R. J. Shephard and H. Lavallée. Springfield, Ill.: C. C. Thomas, **1978a**.

——. Predicting the exercise catastrophe in the "post coronary" patient. Can. Fam. Physician 24: 614–618, **1978b**.

——. Prognostic indices for ischaemic heart disease patients enrolled in an exercise-centred rehabilitation programme. Amer. J. Cardiol. 44: 1230–1240, **1979**.

Shephard, R. J., Kavanagh, T., and Moore, R. Fluid and mineral balance of post-coronary distance runners. Studies on the 1975 Boston Marathon. In: Nutrition, Dietetics and Sports. Eds.: G. Ricci and A. Venerando. Turin: Ediz. Minerva Medica, **1978**.

Shephard, R. J., Killinger, D., and Fried, T. Responses to sustained use of anabolic steroid. Brit. J Sport. Med. 11: 170–173, **1977**.

Shephard, R. J., Lavallée, H., Beaucage, C., Perusse, M., Rajic, M., Brisson, G., Jéquier, J. C., Larivière, G., and LaBarre, R. La capacité physique des enfants canadiens: Une comparaison entre les enfants canadiens-français, canadiens-anglais et esquimaux. 1. Consommation maximale d'oxygène et debit cardiaque. Union Méd. 103: 1767–1777, **1974**.

Shephard, R. J., Lavallée, H., Jéquier, J. C., Rajic, M., and Beaucage, C. Un programme complémentaire d'éducation physique. Etude préliminaire de l'expérience pratiquée dans le district de Trois Rivières, pp. 43–54. In: Facteurs limitant l'endurance humaine. Les Techniques d'amélioration de la performance. Ed.: J. R. LaCour. Université de St. Etienne, France, **1977**.

Shephard, R. J., Lavallée, H., Rajic, M., Jéquier, J. C., Brisson, G., and Beaucage, C. Radiographic age in the interpretation of physiological and anthropological data. In: Proceedings of International Pediatric Work Physiology Symposium. Ed.: J. Borms and M. Hebbelinck. Basel: Karger, **1978a**.

Shephard, R. J., Lavallée, H., et al. Probleme der Längsschnittuntersuchung Motorischer Entwicklung, pp. 235-251. In: Motorische Entwicklung. Probleme und Ergebnisse von Längsschnittuntersuchungen. Ed.: F. Willemczik. Darmstadt: Institut für Sportwissenschaft, **1978b**.

Shephard, R. J., Lavallée, H., Jéquier, J. C., LaBarre, R., and Rajic, M. A community approach to assessments of exercise tolerance in health and disease. J. Sport. Med. Phys. Fitness. 19: 297-304, **1979a**.

Shephard, R. J., Lavallée, H., LaBarre, R., Jéquier, J. C., Volle, M., and Rajic, M. On the basis of data standardization in prepubescent children. Proceedings of 2nd International Seminar on Kinanthropometry. Eds.: M. Ostyn, G. Beunen, and J. Simons. Basel: Karger, **1979b**.

Shephard, R. J., Lavallée, H., Rajic, M., Jéquier, J. C., LaBarre, R., and Beaucage, C. L'age radiographique. Fidélité des données et différences régionales. In press, **1979c**.

Shephard, R. J. and McClure, R. L. The prediction of cardiorespiratory fitness. Int. Z. Angew. Physiol. 21: 212-223, **1965**.

Shephard, R. J. and Olbrecht, A. J. Body weight and the estimation of working capacity. S. Afr. Med. J. 44: 296-298, **1970**.

Shephard, R. J. and Pimm, P. Physical fitness of Canadian physical education students, with a note on international differences. Brit. J. Sport. Med. Phys. Fitness 9: 165-174, **1975**.

Shephard, R. J., Rode, A., and Ross, R. Reinforcement of a smoking withdrawal program: the role of the physiologist and the psychologist. Can. J. Publ. Health 64: 542-551, **1972**.

Shephard, R. J. and Seliger, V. On the estimation of total lung capacity from chest X-rays. Radiographic and helium dilution estimates on children aged 10-12 years. Respiration 26: 327-336, **1969**.

Shephard, R. J. and Sidney, K. H. Effects of physical exercise on plasma growth hormone and cortisol levels in human subjects, pp. 1-30. In: Exercise and Sport Science Reviews 3. Eds. J. H. Wilmore and J. F. Keogh. New York: Academic Press, **1975**.

——. Exercise and Aging. Exercise and Sport Science Reviews. 6: 1-57, **1979**.

Shephard, R. J., Youldon, P. E., Cox, M., and West, C. Effect of a 6-month industrial fitness programme on serum lipid concentrations. Atherosclerosis. 35: 277-286, **1979**.

Shepherd, J. T. Physiology of the Circulation in Human Limbs in Health and Disease. Philadelphia: W. B. Saunders, **1963**.

——. Role of the veins in the circulation. Circulation 33: 484-491, **1966**.

——. Alterations in activity of vascular smooth muscle by local modulation of adrenergic transmitter release. Introductory remarks. Fed. Proc. 37: 179-180, **1967**.

Shibolet, S., Coll, R., Gilat, T., and Soher, E. Heat stroke: its clinical picture and mechanism in thirty-six cases. Quart. J. Med. 36: 525-548, **1967**.

Shibolet, S., Lancester, M. C., and Danou, Y. Heat stroke: A review. Aviation Space Environ. Med. 47: 280-301, **1976**.

Shindell, D., Pendergast, D. R., Wilson, D. W., Cerretelli, P., and Rennie, D. W. Cardiovascular and respiratory kinetics during early arm vs leg work. Fed. Proc. 36: 449 (Abstr.), **1977**.

Shock, N. W. Physiological aspects of aging in man. Ann. Rev. Physiol. 23: 97-166, **1961**.

——. Physical activity and the "rate of ageing." In: Proceedings of International Symposium on Physical Activity and Cardiovascular Health. Ed.: R. J. Shephard. Can. Med. Assoc. J. 96: 836-842, **1967**.

Shore, B. and Shore, V. Apoproteins and substructure of human serum lipoproteins, pp. 144-150. In: Atherosclerosis. Proc. 2nd Int. Symp. Ed.: R. J. Jones. Berlin: Springer Verlag, **1970**.

Shvartz, E., Meroz, A., and Birnfeld, H. $\dot{V}O_{2(max)}$ and rectal temperature in temperate and hot environments. Med. Sci. Sport. 8: 53 (Abstr.) **1976**.

Shvartz, E., Shapiro, Y., Birnfield, H., and Magazanik, A. Maximal oxygen uptake, heat tolerance, and rectal temperature. Med. Sci. Sport. 10: 256-260, **1978**.

Sidney, K. H. and Shephard, R. J. Physiological characteristics and performance of the whitewater paddler. Int. Z. Angew. Physiol. 32: 55-70, **1973**.

——. Attitudes towards health and physical training in the elderly. Effects of a physical training program. Med. Sci. Sport. 8: 246-252, **1976**.

——. Maximum testing of men and women in the seventh, eighth, and ninth decades of life. J. Appl. Physiol. 43: 280-287, **1977a**.

——. Training and e.c.g. abnormalities in the elderly. Brit. Heart J. 39: 1114-1120, **1977b**.

——. Perception of exertion in the elderly. Effects

of aging, mode of exercise and physical training. Percept. Motor Skills 44: 999–1010, **1977c**.

——. Activity patterns of elderly men and women. J. Gerontol. 32: 25–32, **1977d**.

——. Attitudes towards health and physical activity in the elderly. Effects of a physical training programme. Med. Sci. Sport. 8: 246–252, **1977e**.

——. Growth hormone and cortisol: Age differences, effects of exercise and training. Can. J. Appl. Sport. Sci. 2: 189–194, **1978a**.

——. Frequency and intensity of exercise training for elderly subjects. Med. Sci. Sport. 10: 125–131, **1978b**.

Sidney, K. H., Shephard, R. J., and Harrison, J. Endurance training and body composition of the elderly. Amer. J. Clin. Nutr. 30: 326–333, **1977**.

Siegel, W. Exercise-induced indicators of coronary atherosclerotic heart disease, pp. 3-1 to 3-21. In: Coronary Heart Disease: Prevention, Detection, Rehabilitation with Emphasis on Exercise Testing. Ed.: S. M. Fox. Denver, Colorado: International Medical Corporation, **1974**.

——. New perspectives in clinical exercise testing, pp. 3–12. In: Sports Medicine. Eds.: F. Landry and W. Orban. Miami: Symposia Specialists, **1978**.

Siegel, W., Blomqvist, G., and Mitchell, J. H. Effects of a quantitated training program on middle-aged sedentary men. Circulation 41: 19–29, **1970**.

Sikand, R. S., Magnussen, H., Scheid, P., and Piiper, J. Convective and diffusive gas mixing in human lungs: Experiments and model analysis. J. Appl. Physiol. 40: 362–371, **1976**.

Siltanen, P. and Kekki, M. Effect of exercise on the formed elements of urinary sediment. Acta Med. Scand. 164: 151–157, **1959**.

Silverberg, D. S., Shemesh, E., and Iaina, A. The unsupported arm: a cause of falsely raised blood pressure readings. Brit. Med. J. 2: 1331, **1977**.

Silverman, M. Exercise studies on asthmatic children. University of Cambridge, Ph.D. thesis, **1973**.

Simmons, R. C. G. Some effects of training upon cardiac output and its distribution M.Sc. thesis, University of Toronto, **1969**.

Simmons, R. C. G. and Shephard, R. J. Measurement of cardiac output in maximum exercise. Application of an acetylene rebreathing method to arm and leg exercise. Int. Z. Angew. Physiol. 29: 159–172, **1971a**.

——. Effects of physical conditioning upon the central and peripheral circulatory responses to arm work. Int. Z. Angew. Physiol. 30: 73–84, **1971b**.

——. Does indocyanine green obey Beer's law? J. Appl. Physiol. 30: 502–507, **1971**.

Simms, E. A. H., Danforth, E., Horton, E. S., Bray, G. A., Glennon, J. A., and Salans, L. B. Endocrine and metabolic effects of experimental obesity in man. Rec. Prog. Horm. Res. 29: 457–496, **1973**.

Simonson, E. Physiology of Work Capacity and Fatigue. Springfield, Ill.: C. C. Thomas, **1971**.

Simonson, E., Kearns, W. M., and Enzer, N. Effects of methyl testosterone treatment on muscular performance and the central nervous system of older men. J. Clin. Endocrinol. Metab. 4: 528–534, **1944**.

Simonson, E., Teslenko, N., and Gorkin, M. Einfluss von Vorübungen auf die Leistung beim 100m. Lauf. Arbeitsphysiol. 9: 152–165, **1936**.

Simoons, M. L. Computer processing of exercise electrocardiograms, pp. 155–164. In: Coronary Heart Disease, Exercise Testing and Cardiac Rehabilitation. Miami: Symposia Specialists, **1977**.

——. Computer Assisted Interpretation of Exercise Electrocardiograms. Rotterdam: Bronder-Offset B. V., **1976**.

Simoons, M. L. and Hugenholtz, P. G. Gradual changes of ECG waveform during and after exercise in normal subjects. Circulation 52: 570–577, **1975**.

——. Estimation of the probability of exercise-induced ischemia by quantitative e.c.g. analysis. Circulation 56: 552–559, **1977**.

Sinclair, D. C. Human Growth After Birth. London: Oxford University Press, **1969**.

Singer, R. N. Effect of spectators on athletes and non-athletes performing a gross motor task. Res. Quart. 36: 473–482, **1965**.

——. Balance skill as related to athletics, sex, height and weight, pp. 645–656. In: Contemporary Psychology of Sport. Eds.: G. S. Kenyon and T. M. Grogg. Chicago: Athletic Institute, **1970**.

Singh, I., Kapila, C. C., Khanna, P. K., Nanda, R. B., and Rao, B. D. P. High altitude pulmonary oedema. Lancet 1: 229–234, **1965**.

——. High altitude pulmonary oedema, pp. 165–

179. In: Exercise at Altitude. Eds.: E. Jokl and P. Jokl. Basel: Karger, **1968**.

Singh, M., Irwin, D., and Gutoski, F. P. Effect of high speed treadmill and sprint training on stride length and rate, pp. 123-134. In: Biomechanics of Sports and Kinanthropometry. Eds.: F. Landry and W. A. R. Orban. Miami: Symposia Specialists, **1978**.

Singh, M. and Karpovich, P. V. Effect of eccentric training of agonists on antagonistic muscles. J. Appl. Physiol. 23: 742-745, **1967**.

——. Strength of forearm flexors and extensors in men and women. J. Appl. Physiol. 25: 177-180, **1968**.

Sinning, W. E. Body composition assessment of college wrestlers. Med. Sci. Sport. 6: 139-145, **1974**.

Sinning, W. E., Cunningham, L. N., Racaniello, A. P., and Sholes, J. L. Body composition and somatotype of male and female nordic skiers. Res. Quart. 48: 741-749, **1977**.

Siri, W. E. In: Advances in Biological and Medical Physics. Eds.: J. H. Laurence and C. A. Tobias. London: Academic Press, **1956**.

Sjodin, B. Lactate dehydrogenase in human skeletal muscle. Acta Physiol. Scand. Suppl. 436: 1-32, **1976**.

S'Jongers, J. J., Dirix, A., Jolie, P., Borms, J., and Segers, M. Wolff-Parkinson-White syndrome and sports aptitude. J. Sport. Med. Phys. Fitness 16: 6-16, **1976**.

S'Jongers, J. J., Hebbelinck, M., Robaye, E., Vanfraechem, J., Bande, J., and Segers, M. L'influence d'un effort modéré effectué en altitude moyenne simulée en caisson, sur la fréquence cardiaque et sur la tension artérielle. Int. Z. Angew. Physiol. 24: 222-230, **1967**.

Sjöstrand, F. S., Andersson-Cedergren, E., and Dewey, M. M. The ultrastructure of the intercalated discs of frog, mouse and guinea pig cardiac muscle. J. Ultrastruct. Res. 1: 271-287, **1958**.

Sjöstrand, T. Total quantity of hemoglobin in man and its relation to age, sex, body weight, height. Acta Physiol. Scand. 18: 324-336, **1949**.

——. Volume and distribution of blood and their significance in regulating circulation. Physiol. Rev. 33: 202-228, **1953**.

——. Functional capacity and exercise tolerance in patients with impaired cardiovascular function. In: Clinical Cardiopulmonary Physiology. New York: Grune & Stratton, **1960**.

——. Blood volume, pp. 51-62. In: Handbook of Physiology. Section 2. Circulation, Vol. 1. Ed.: W. F. Hamilton. Washington, D.C.: Amer. Physiological Society, **1962**.

——. Exercise tests, pp. 515-530. In: Clinical Physiology. Ed.: T. Sjöstrand. Stockholm: Svenska Bokförlaget, **1967**.

Skerlj, B., Brozek, J., and Hunt, F. E. Subcutaneous fat and age changes in body build and body form in women. Amer. J. Phys. Anthropol. 11: 577-600, **1953**.

Skinner, J. S. The cardiovascular system with aging and exercise, pp. 100-108. In: Medicine and Sport. Vol. 4. Physical Activity and Aging. Eds.: D. Brunner and E. Jokl. Basel: Karger, **1970**.

——. Age and performance, pp. 271-282. In: Limiting Factors of Physical Performance. Ed.: J. Keul. Stüttgart: Thieme, **1973**.

Skinner, J. S., Benson, H., McDonough, J. R., and Hawes, C. G. Social status, physical activity and coronary prone-ness. J. Chron. Dis. 19: 773-783, **1966**.

Skinner, J. S., Holloszy, J. O., and Cureton, T. K. Effects of a program of endurance exercises on physical work capacity and anthropometric measurements of fifteen middle-aged men. Amer. J. Cardiol. 14: 747-752, **1964**.

Skinner, J. S., Lemieux-Searle, G., and Taylor, A. W. The effects of endurance training on the kinetics of oxygen intake at absolute and relative work loads. Med. Sci. Sport. 9: 68 (Abstr.), **1977**.

Skinner, N. S. and Costin, J. C. Role of O_2 and K^+ in abolition of sympathetic vasoconstriction in dog skeletal muscle. Amer. J. Physiol. 217: 438-444, **1969**.

Skoglund, S. Anatomical and physiological studies of knee joint innervation in the cat. Acta Physiol. Scand. 36 (Suppl. 124): 1-101, **1956**.

Skou, J. C. Enzymatic basis for active transport of Na^+ and K^+ across cell membrane. Physiol. Rev. 45: 596-617, **1965**.

Skrobak-Kaczynski, J. and Lewin, T. Secular changes in Lapps of Northern Finland. In: Circumpolar Health. Eds.: R. J. Shephard and S. Itoh. Toronto: Univ. of Toronto Press, **1976**.

Slack, J. and Mills, G. L. Anomalous low density lipoproteins in familial hyperbetalipoproteinemia, pp. 189-192. In: Atherosclerosis. Proc. 2nd Int. Symp. Ed.: R. J. Jones. Berlin: Sprnger Verlag, **1970**.

Slater-Hammel, A. T. Bilateral effects of muscle activity. Res. Quart. 21: 203-209, **1950**.

Sloan, A. W. Effect of training on physical fitness of women students. J. Appl. Physiol. 16: 167-169, **1961**.

Sloan, R. E. G. and Keatinge, W. R. Cooling rates of young people swimming in cold water. J. Appl. Physiol. 35: 371-375, **1973**.

Slonim, M. and Stahl, W. M. Sodium and water content of connective versus cellular tissue following hemorrhage. Surg. Forum 19: 53-54, **1968**.

Smiles, K. A., Elizondo, R. S., and Barney, C. C. Sweating responses during changes of hypothalamic temperatures in the Rhesus monkey. J. Appl. Physiol. 40: 653-657, **1976**.

Smiles, K. A. and Robinson, S. Sodium ion conservation during acclimatization of men to work in the heat. J. Appl. Physiol. 31: 63-69, **1971**.

Smith, C. J. V. and Britt, D. L. Obesity in the rat induced by hypothalamic implants of gold thioglucose. Physiol. Behav. 7: 7-10, **1971**.

Smith, E. B. The role of exotic gases in the study of narcosis, pp. 183-192. In: The Role of Exotic Gases in the Study of Narcosis. Eds.: P. B. Bennett and D. H. Elliott. London: Baillière, Tindall & Cassell, **1969**.

Smith, E. L. and Babcock, S. W. Effects of physical activity on bone loss in the aged. Med. Sci. Sport. 5: 68 (Abstr.), **1973**.

Smith, E. L. and Reddan, E. L. The effects of physical activy on bone in the aged. Amer. J. Roentgenol. 126: 1297, **1976**.

Smith, F. E. Indices of heat stress. London-H.M.S O. U.K. Med. Res. Council Memorandum 29, **1955**.

Smith, G. P., Gibbs, J., and Young, R. C. Cholecystokinin and intestinal satiety in the rat. Fed. Proc. 33: 1146-1149, **1973**.

Smith, H. J. and Anthonisen, N. R. Results of cardiac resuscitation in 254 patients. Lancet 1: 1027-1029, **1965**.

Smith, H. W. Salt and water volume receptors: An exercise in physiologic apologetics. Amer. J. Med. 23: 623-652, **1957**.

Smith, J. H., Robinson, S., and Pearcy, M. Re sponses to exercise, heat and dehydration. J. Appl. Physiol. 4: 659-665, **1952**.

Smith, J. L. Fusimotor loop properties and involvement during voluntary movement. Med. Sci. Sport. 4: 297-333, **1976**.

Smith, K. U. and Smith, T. J. Feedback mechanism of athletic skill and learning, pp. 83-195. In: Psychology of Motor Learning. Ed.: L. E. Smith. Chicago: Athletic Institute, **1970**.

Smith, L. E. Individual differences in strength, reaction latency, mass and length of limbs, and their relation to maximal speed of movement. Res. Quart. 32: 208-220, **1961**.

——. Relationship between muscular fatigue, pain tolerance, anxiety, extraversion-intraversion and neuroticism traits of college men, pp. 259-274. In: Contemporary Psychology of Sport. Eds.: G. S. Kenyon and T. M. Grogg. Chicago: Athletic Institute, **1970**.

Smith, L. E. and Royce, J. Muscular strength in relation to body composition. Ann. N.Y. Acad. Sci. 110 (II): 809-813, **1963**.

Smith, N. J. Gaining and losing weight in athletics. J. Amer. Med. Assoc. 236: 149-151, **1976**.

Smith, O. A. Reflex and central mechanisms in the control of the heart and circulation. Ann. Rev. Physiol. 36: 93-123, **1974**.

Smith, O. A., Jabbur, S. J., Rushmer, R. F., and Lasher, E. P. Role of hypothalamic structures in cardiac control. Physiol. Rev. 40 (Suppl. 4): 136-141, **1960**.

Smith, R. S. Properties of intrafusal muscle fibres, pp. 69-80. In: Muscular Afferents and Motor Control. Ed.: R. Granit. Stockholm: Almqvist & Wiksell, **1966**.

Smuylan, H., Cuddy, P. O., Vincent, W. A., Kashemsant, U., and Eich, R. H. Initial hemodynamic responses to mild exercise in trained dogs. J. Appl. Physiol. 20: 437-442, **1965**.

Snook, S. H. and Ciriello, V. M. The effects of heat stress on manual handling tasks. Amer. Industr. Hyg. Assoc. J. 35: 681-685, **1974**.

Snook, S. H. and Irvine, C. H. Maximum acceptable weight of lift. Amer. Industr. Hyg. J. 28: 322-329, **1967**.

Sobel, B. E. and Kaufman, S. Enhanced RNA synthesis in hypertrophied rat skeletal muscle. Physiologist 12: 360 (Abstr.), **1969**.

——. Enhanced RNA polymerase activity in skeletal muscle undergoing hypertropy. Arch. Biochem. Biophys. 137: 469-476, **1970**.

Sobel, S., Marotta, S. F., and Marbarger, J. P. Circulation plasma volume changes in anesthetized dogs during positive pressure breathing. J. Appl. Physiol. 14: 937-939, **1959**.

Soll, A. H., Kahn, C. R., Neville, D. M., and Roth, J. Insulin receptor deficiency in genetic and acquired obesity. J. Clin. Invest. 56: 769-780, **1975**.

Sonnenblick, E. H. Oxygen consumption of the

heart, pp. 89-92. In: Coronary Heart Disease and Physical Fitness. Eds.: O. A. Larsen and R. O. Malmborg. Baltimore: University Park Press, **1971**.

——. Contractility in the intact heart: Progress and problems. Europ. J. Cardiol. 13: 319-324, **1974**.

Sonnenblick, E. H., Parmley, W. W., Buccino, R. A., and Spann, J. F. Maximum force development in cardiac muscle. Nature 219: 1056-1058, **1968**.

Sonnenblick, E. H., Ross, J., and Braunwald, E. Oxygen consumption of the heart, newer concepts of its multifactorial determination. Amer. J. Cardiol. 22: 328-336, **1968**.

Sonnenblick, E. H., Spotnitz, H. M., and Spiro, D. Role of the sarcomere in ventricular function and the mechanism of heart failure. Circulation Res. 15 (Suppl. 2): 70-81, **1964**.

Southam, A. L. and Gonzaga, F. P. Systemic changes during the menstrual cycle. Amer. J. Obstet. Gynecol. 91: 142-165, **1965**.

Sowton, E. and Burkart, F. Haemodynamic changes during continuous exercise. Brit. Heart J. 29: 770-774, **1967**.

Spaulding, W. B. The clinical analysis of fatigue. Appl. Ther. 6: 911-915, **1964**.

Spence, J. T. What can you say about a twenty-year-old theory that won't die? J. Motor Behav. 3: 193-203, **1971**.

Spickard, A. Heat stroke in college football and suggestions for its prevention. Southern Med. J. 61: 791-796, **1968**.

Spiro, A. J. and Beilin, R. L. Human muscle spindle histochemistry. Arch. Neurol. Psychiatr. (Chicago) 20: 271-275, **1969**.

Spiro, D. and Sonnenblick, E. H. The structural basis of the contractile process of heart muscle under physiological and pathological conditions. Prog. Cardiovasc. Dis. 7: 295-335, **1965**.

Spiro, S. G., Juniper, E., Bowman, P., and Edwards, R. H. T. An increasing work rate test for assessing the physiological strain of submaximal exercise. Clin. Sci. 46: 191-206, **1974**.

Spitzer, J. J. and Wolf, E. H. Uptake and oxidation of FFA administered by ventriculocisternal perfusion in the dog. Amer. J. Physiol. 221: 1426-1430, **1971**.

Sprague, R. L., Barnes, K. R., and Werry, J. S. Methyl phenidate and thioridazine: Learning, reaction time, activity and classroom behavior in disturbed children. Amer. J. Orthopsychiatr. 40: 615-628, **1970**.

Sproles, C. B., Smith, D. P., Byrd, R. J., and Allen, T. E. Circulatory responses to submaximal exercise after dehydration and rehydration. J. Sport. Med. Phys. Fitness 16: 98-105, **1976**.

Sprynarová, S. Longitudinal study of the influence of different physical activity programs on functional capacity of the boys from 11-18 years. Acta Paediatr. Belg. 28 (Suppl.): 204-213, **1974**.

Spurway, N. C. Cluster analysis and the typing of muscle fibres. J. Physiol. (Lond.) 280: 39p-40p, March, **1978**.

Stacey, M. J. Free nerve endings in skeletal muscle of the cat. J. Anat. 105: 231-254, **1969**.

Stadie, W. C., Riggs, B. C., and Haugaard, N. Oxygen poisoning. Amer. J. Med. Sci. 207: 84-114, **1944**.

Staff, P. H. and Nilsson, S. Vaeske og sukkertilförsel under langvarig intens fysisk aktivitet. Tidssdrift for Den norske laegeforening 16: 1235, **1971**. Cited by P. O. Åstrand and K. Rodahl (1977).

Stahl, C. W. March hemoglobinuria. Report of five cases in students at Ohio State University. J. Amer. Med. Assoc. 164: 1458-1460, **1957**.

Stainsby, W. N. Some critical oxygen tensions and their significance. In: Proc. Int. Symp. on Cardiovascular and Respiratory Effects of Hypoxia. Eds.: I. D. Hatcher and D. B. Jennings. Basel: Karger, **1966**.

Stainsby, W. N. and Barclay, J. K. Exercise metabolism: O_2 deficit, steady level O_2 uptake and O_2 uptake for recovery. Med. Sci. Sport. 2: 177-181, **1970**.

Stålberg, E. and Ekstedt, J. Single fibre EMG and microphysiology of the motor unit in normal and diseased human muscle, pp. 113-129. In: New Developments in Electromyography and Clinical Neurophysiology, Vol. 1. Ed.: J. E. Desmedt. Basel: Karger, **1973**.

Stamford, B. A. Effects of chronic institutionalization on the physical working capacity and trainability of geriatric men. J. Gerontol. 28: 441-446, **1973**.

——. Step increment versus constant load tests for determination of maximal oxygen uptake. Europ. J. Appl. Physiol. 35: 89-93, **1976**.

Stamford, B. A. and Moffatt, R. Anabolic steroid: Effectiveness as an ergogenic aid to experienced weight trainers. J. Sport. Med. Phys. Fitness 14: 191-197, **1974**.

Stamford, B. A., Weltman, A., Moffatt, R. J., and Fulco, C. Effects of severe prior exercise on assessment of maximal oxygen uptake during one- versus two- legged cycling. Res. Quart. 49: 363-371, **1978**.

Stamler, J. Acute myocardial infarction: Progress in primary prevention. Brit. Heart J. 33 (Suppl.): 145-164, **1971**.

——. Improving life styles to control the coronary epidemic, pp. 5-48. In: Nutrition, Dietetics and Sport. Eds.: G. Ricci and A. Venerando. Torino: Minerva Medica, **1978**.

Stamler, J., Lindberg, H. A., Berkson, D. M., Shaffer, A., Miller, W., and Poindexter, A. Prevalence and incidence of coronary heart disease in strata of the labor force of a Chicago industrial corporation. J. Chron. Dis. 11: 405-420, **1960**.

Stamler, J., Stamler, R., and Pullman, T. N. The Epidemiology of Hypertension. New York: Grune & Stratton, **1967**.

Stanbrook, H. Comparison of the responses of pulmonary and systemic vessels to local hypoxia. J. Physiol. (Lond.) 284: 100p-101p, July, **1978**.

Stanesçu, N. Medical Researches in Football Game. Bucharest, Rumania: Ministry for Health, Sports Medicine Centre, **1970**.

——. Hyperfunctional Disorders of Locomotory System, as Localized Forms of Supertraining, by High Performance Sportsmen. Bucharest: Sports Medicine Center, **1971**.

Starlinger, H. and Berghoff, A. Untersuchungen über die Anwendbarkeit der endogenen Kreatinin—Clearance im Vergleich zur Inulinclearance in Rühe-und Arbeit versuchen bei gesunden Versuchspersonen. Int. Z. Angew. Physiol. 19: 194-200, **1962**.

Starlinger, H. and Lübbers, D. W. Polarographic measurements of the oxygen pressure performed simultaneously with optical measurements of the redox state of the respiratory chain in suspensions of mitochondria under steady state conditions at low oxygen tensions. Pflüg. Archiv. 341: 15-22, **1973**.

Starr, I. An essay on the strength of the heart and on the effect of aging upon it. Amer. J. Cardiol. 14: 771-783, **1964**.

Staub, N. C. and Schultz, E. L. Pulmonary capillary length in dog, cat and rabbit. Resp. Physiol. 5: 371-378, **1968**.

Staudinger, H., Krisch, K., and Loenhauser, S. Role of ascorbic acid in microsomal electron transport and the possible relationship to hydroxylation reactions. Ann. N.Y. Acad. Sci. 92: 195-207, **1961**.

Stefanik, P. A., Heald, F. P., and Mayer, J. Caloric intake in relation to energy output of obese and non-obese adolescent boys. Amer. J. Clin. Nutr. 7: 55-62, **1959**.

Stegall, H. F. Muscle pumping in the dependent leg. Circ. Res. 19: 180-190, **1966**.

Stegemann, J., Seez, P., Kremer, W., and Böning, D. A mathematical model of the ventilatory control system to carbon dioxide with special reference to athletes and non-athletes. Pflüg. Archiv. 356: 223-236, **1975**.

Stehbens, W. E. Turbulence of blood flow. Quart. J. Exp. Physiol. 44: 110-117, **1959**.

Stein, O. and Gross, J. The localization and metabolism of ^{138}I insulin in the muscle and some other tissues of the rat. Endocrinology 65: 707-716, **1959**.

Stein, R. B. Peripheral control of movement. Physiol. Rev. 54: 215-243, **1974**.

Stein, R. B. and Milner-Brown, H. S. Contractile and electrical properties of normal and modified human motor units, pp. 73-86. In: Control of Posture and Locomotion. Eds.: R. B. Stein, K. G. Pearson, R. S. Smith, and J. B. Redford. New York: Plenum Press, **1973**.

Stein, Y. and Stein, O. Biosynthesis and secretion of very low density lipoproteins, pp. 151-161. In: Atherosclerosis. Proc. 2nd Int. Symp. Ed.: R. J. Jones. Berlin: Springer Verlag, **1970**.

Steinberg, D. Drugs inhibiting cholesterol biosynthesis, with special reference to clofibrate, pp. 500-508. In: Atherosclerosis. Proceedings of 2nd Int. Symposium. Ed.: R. J. Jones. Berlin: Springer Verlag, **1970**.

Steinberg, F. U. Gait disorders in old age. Geriatrics 21: 134-142, **1966**.

Steindler, A. Kinesiology of the Human Body Under Normal and Pathological Conditions. Springfield, Ill.: C. C. Thomas, **1955**.

Steinhaus, A. H. Chronic effects of exercise. Physiol. Rev. 13: 103-147, **1933**.

——. Some factors modifying the expression of human strength. In: Toward an Understanding of Health and Physical Education. Los Angeles: Brown, **1963**.

Stellar, E. The physiology of motivation. Psychol. Rev. 61: 5-22, **1954**.

Stelmach, G. E. Retention of motor skills. Exer. Sport. Sci. Rev. 2: 1-31, **1974**.

——. Motor Control, Issues and Trends. New York: Academic Press, **1976**.

Stenberg, J., Åstrand, P. O., Ekblom, B., Royce,

J., and Saltin, B. Hemodynamic response to work with different muscle groups sitting and supine. J. Appl. Physiol. 22: 61–70, **1967**.

Stephens, J. A., Gerlach, R. L., Reinking, R. M., and Stuart, D. G. Fatiguability of medial gastrocnemius motor units in the cat, pp. 179–185. In: Control of Posture and Locomotion. Eds.: R. B. Stein, K. G. Pearson, R. S. Smith, and J. B. Redford. New York: Plenum Press, **1973**.

Stephens, J. A. and Taylor, A. Fatigue or maintained voluntary muscle contraction in man. J. Physiol. (Lond.) 220: 1–18, **1972**.

Stern, J. S. and Greenwood, M. R. C. A review of development of adipose cellularity in man and animals. Fed. Proc. 33: 1952–1955, **1974**.

Stetson, R. H. and McGill, J. A. Mechanisms of the different types of movements. Psychol. Monogr. 32: 18–40, **1923**.

Stevens, C. D., Ryder, H. W., Ferris, E. B., and Inatome, M. The rate of nitrogen elimination from the body through the lungs. J. Aviation Med. 18: 111A–132A, **1947**.

Stevens, J. C. and Krimsley, A. S. Build-up of fatigue: Role of blood flow, pp. 145–155. In: Physical Work and Effort. Ed.: G. Borg. Oxford: Pergamon Press, **1975**.

Stevens, P. M. Cardiovascular dynamics during orthostasis and the influence of intravascular instrumentation. Amer. J. Cardiol. 17: 211–219, **1966**.

Stevenson, J. A. F. Exercise, food intake and health in experimental animals. Can. Med. Assoc. J. 96: 862–866, **1967**.

Stewart, J. S. S. Management of cardiac arrest, with special reference to metabolic acidosis. Brit. Med. J. 1: 476–479, **1964**.

Stewart, K. J. and Gutin, B. The prediction of maximal oxygen uptake before and after physical training in children. J. Human Ergol. 4: 153–162, **1975**.

——. Effects of physical training on cardiorespiratory fitness in children. Res. Quart. 47: 110–120, **1976**.

Stickney, J. C., Northup, D. W., and Van Liere, E. J. Resistance of the small intestine (motility) against stress. J. Appl. Physiol. 9: 484–486, **1956**.

Stickney, J. C. and Van Liere, E. J. The effects of exercise upon the function of the gastrointestinal tract, pp. 171–179. In: Science and Medicine of Exercise and Sport. Eds.: W. R. Johnson and E. R. Buskirk. New York: Harper & Row, **1974**.

Stirling, J. L. and Stock, M. J. Metabolic origins of the thermogenesis induced by diet. Nature 220: 801–802, **1968**.

Stoboy, H., Friedebold, G., and Strand, F. L. Evaluation of the effect of isometric training in functional and organic muscle atrophy. Arch. Phys. Med. 49: 508–514, **1968**.

Stolwijk, J. A., Saltin, B., and Gagge, A. P. Physiological factors associated with sweating during exercise. J. Aerospace Med. 39: 1101–1105, **1968**.

Stolz, H. R. and Stolz, L. M. Somatic Development of Adolescent Boys. New York: Macmillan, **1951**.

Strandell, T. Circulatory studies in healthy old men. Acta Med. Scand. Suppl. 141: 1–44, **1964a**.

——. Heart rate and work load at maximal work intensity in old men. Acta Med. Scand. 176: 301–318, **1964b**.

Strandell, T. and Shepherd, J. T. The effect in humans of increased sympathetic activity on the blood flow to active muscles. Acta Med. Scand. Suppl. 472: 146–167, **1967**.

Strauss, M. B. Body Water in Man. Boston: Little, Brown, **1957**.

Streeter, D. D., Spotnitz, H. M., Patel, D. P., and Sonnenblick, E. H. Fiber orientation in the canine left ventricle during systole and diastole. Circ. Res. 24: 339–347, **1969**.

Strick, P. L. Multiple sources of thalamic input to the primate motor cortex. Brain Res. 88: 372–377, **1975**.

Ström, G. The influence of anoxia on lactate utilization in man after prolonged muscular work. Acta Physiol. Scand. 17: 440–451, **1949**.

Strömme, S. B., Andersen, K. L., and Elsner, R. W. Metabolic and thermal responses to muscular exertion in the cold. J. Appl. Physiol. 18: 756–763, **1963**.

Strömme, S. B., Meen, H. D., and Aakvaag, A. Effects of an androgenic-anabolic steroid on strength development and plasma testosterone levels in normal males. Med. Sci. Sport. 6: 203–208, **1974**.

Strömme, S. B., Stensvold, I. C., Meen, H. D., and Refsum, H. E. Magnesium metabolism during prolonged heavy exercise. In: Metabolic Adaptation to Prolonged Physical Exercise. Eds.: H. Howald and J. R. Poortmans. Basel: Birkhauser Verlag, **1975**.

Strong, C. H. Motivation related to performance of

physical fitness test. Res. Quart. 34: 497-507, **1963**.

Strughold, H. The human time factor in flight. II. Chains of latencies in vision. J. Aviation Med. 22: 100-108, **1951**.

Strydom, N. B. Environmental variables affecting fitness testing, pp. 94-101. In: Physical Fitness Assessment: Principles, Practice and Application. Eds.: R. J. Shephard and H. Lavallée. Springfield, Ill.: C. C. Thomas, **1978**.

Strydom, N. B. and Williams, C. G. Effect of physical conditioning on state of heat acclimatization. J. Appl. Physiol. 27: 262-265, **1969**.

Strydom, N. B., Rogers, G. G., Van der Walt, W. W., and Van der Linde, A. Changes in the level of serum Vitamin C in mineworkers. J. S. Afr. Inst. Mining Met. 77: 214-217, **1977**.

Strydom, N. B., Wyndham, C. H., Williams, C. G., Morrison, J. F., Bredell, G. A. G., and Von Raheden, M. K. Energy requirements of acclimatized subjects in humid heat. Fed. Proc. 25: 1366-1371, **1966a**.

Strydom, N. B., Wyndham, C. H., Van Graan, C. H., Holdsworth, L. D., and Morrison, J. F. The influence of water restriction on the performance of men during a prolonged march. S. Afr. Med. J. 40: 537-544, **1966b**.

Strydom, N. B., Wyndham, C. H., and Benade, A. J. S. The responses of men weighing less than 50 kg to standard climatic room acclimatization procedures. J. S. Afr. Inst. Mining Met. 72: 101-104, **1971**.

Stuart, D. E., Eldred, E., Hemingway, A., and Kawamura, Y. Neural regulation of the rhythm of shivering. In: Temperature: Its Measurement and Control in Science and Industry, Vol. 3(3), p. 545. Ed.: J. D. Hardy. New York: Reinhold, **1963**.

Stubbs, R. A. and Kidd, D. J. Computer analogues for decompression, pp. 300-311. In: Proceedings of Third Symposium on Underwater Physiology. Ed.: C. J. Lambersten. Baltimore: Williams & Wilkins, **1967**.

Stucke, K., Fischer, V., Feldmeier, H., and Henn, L. Fordening des Trainingniveaus durch hochkonzentrierte Eiweisszufuhr. Münch. Med. Wschr. 114: 496-503, **1972**.

Stull, G. A. and Kearney, J. T. Recovery of muscular endurance following submaximal, isometric exercise. Med. Sci. Sport. 10: 109-112, **1978**.

Styka, P. E. and Penney, D. G. Regression of carbon-monoxide induced cardiomegaly. Amer. J. Physiol.235: H516-H522, **1978**.

Sucec, A. The effects of preliminary exercise on endurance performance. Ph.D. thesis. Berkeley, Calif. University of California, **1967**.

Sugimoto, T., Sagawa, K., and Guyton, A. C. Effect of tachycardia on cardiac output during normal and increased venous return. Amer. J. Physiol. 211: 288-292, **1966**.

Sulkin, N. M. and Sulkin, D. F. An electron microscopic study of the effects of chronic hypoxia on cardiac muscle, hepatic and autonomic ganglion cells. Lab. Invest. 14: 1523-1546, **1965**.

Sulman, F. G., Pfeifer, Y., and Superstine, E. The adrenal exhaustion syndrome: An adrenal deficiency. In: The marathon, physiological, medical, epidemiological and psychological studies. Ann. N.Y. Acad. Sci. 301: 918-930, **1977**.

Summers, J. J. The role of timing in motor program representation. J. Motor Behav. 7: 229-241, **1975**.

Summitt, J. K., Kelley, J. S., Herron, J. M., and Saltzman, H. A. 1,000-foot helium saturation exposure, pp. 519-527. In: Underwater Physiology. Ed.: C. J. Lambertsen. New York: Academic Press, **1971**.

Sundsfjord, J. A., Strømme, S. B., and Aakvaag, A. Plasma aldosterone, plasma renin activity and cortisol during exercise, pp. 308-314. In: Metabolic Adaptation to Prolonged Exercise. Eds.: H. Howald and J. R. Poortmans. Basel: Birkhauser Verlag, **1975**.

Suominen, H. and Heikkinen, E. Effect of physical training on collagen. Ital. J. Biochem. 24: 64-65, **1975a**.

——. Enzyme activities in muscle and connective tissue of M. vastus lateralis in habitually training and sedentary 33 to 70 year old men. Europ. J. Appl. Physiol. 34: 249-254, **1975b**.

Suominen, H., Heikkinen, E., Liesen, H., Michel, D., and Hollmann, W. Effects of 8 weeks endurance training on skeletal muscle metabolism in 56-70 year old sedentary men. Europ. J. Appl. Physiol. 37: 173-180, **1977a**.

Suominen, H., Heikkinen, E., and Parkatti, T. Effects of eight weeks physical training on muscle and connective tissue of the M. vastus lateralis in 69 year old men and women. J. Gerontol. 32: 33-37, **1977b**.

Surawicz, B. The input of cellular electrophysiology into the practice of clinical electrocardiography. Mod. Concepts Cardiovasc. Dis. 44: 41-46, **1975**.

Surks, M. I., Chinn, K. S. K., and Matoush, L. R. O. Alterations in body composition in man after acute exposure to high altitude. J. Appl. Physiol. 21: 1741-1746, **1966**.

Suskind, M., Bruce, R. A., McDowell, M. E., Yu, P. N. G., and Lovejoy, F. W. Normal variations in end-tidal air and arterial blood carbon dioxide and oxygen tensions during moderate exercise. J. Appl. Physiol. 3: 281-290, **1951**.

Sutton, J. R. Hormonal and metabolic responses to exercise in subjects of high and low work capacities. Med. Sci. Sport. 10: 1-6, **1978**.

Sutton, J. R., Coleman, M. T., and Casey, J. H. Adrenocortical contributions to serum androgens during physical exercise. Med. Sci. Sport. 6: 72, 1974.

——. Testosterone production rate during exercise, pp. 227-234. In: 3rd International Symposium on Biochemistry of Exercise. Eds.: F. Landry and W. R. Orban. Miami: Symposia Specialists, **1978**.

Sutton, J. R., Coleman, M. J., Casey, J., and Lazarus, L. Androgen responses during physical exercise. Brit. Med. J. 1: 520-522, **1973**.

Sutton, J. R., Hughson, R. L., McDonald, R., Powles, A. C. P., Jones, N. L., and Fitzgerald, J. D. Oral and intravenous propranolol during exercise. Clin. Pharmacol. Ther. 21: 700-705, **1977**.

Sutton, J. R., Jones, N. L., and Toews, C. J. Growth hormone secretion in acid-base alterations at rest and during exercise. Clin. Sci. Mol. Med. 50: 241-247, **1976**.

Sutton, J. R. and Lazarus, L. Growth hormone in exercise: comparison of physiological and pharmacological stimuli. J. Appl. Physiol. 41: 523-527, **1976**.

Sutton, J. R., Oldridge, N. B., and Heigenhauser, G. Changes in the origin of the cardiac impulse in athletes. Med. Sci. Sport. 10: 40 (Abstr.), **1978**.

Sutton, J. R., Young, Y. D., Lazarus, L., Hickie, J. B., and Maksvytis, J. The hormonal response to physical exercise. Austr. Ann. Med. 18: 84-90, **1969**.

Suzuki, S. Experimental studies on factors in growth. Soc. Res. Child Dev. Wash., Monograph 35: 6-11, **1970**.

Svent-Gyorgi, A. Calcium regulation of muscle contraction. Biophys. J. 15: 707-723, **1975**.

Svistun, T. I. Secretion of the gastric glands on a mixed diet during a period of muscle activity of different intensities. Fiziol. Zh. 9: 215-220, **1963**. Cited by Stickney and Van Liere (1974).

Swann, H. G. Mechanism of circulatory failure in fresh and sea water drowning. Circ. Res. 4: 241-244, **1956**.

Swartman, J. R., Cook, E. A., and Taylor, P. B. Effects of exhaustive exercise and recovery on (^{3}H) PHE incorporation into contractile proteins of the rat myocardium. Can. J. Appl. Sport. Sci. 3: 194 (Abstr.), **1978**.

Swett, J. E. and Bourassa, C. M. Comparison of sensory discrimination thresholds with muscle and cutaneous nerve volleys in the cat. J. Neurophysiol. 30: 530-545, **1967**.

Swynghedauw, B., Léger, J. J., and Schwartz, K. The myosin isozyme hypothesis in chronic heart overloading. J. Mol. Cell. Cardiol. 8: 915-924, **1976**.

Sylven, B. and Malmgren, H. On alleged metachromasia of hyaluronic acid. Lab. Invest. 1: 413-431, **1952**.

Szafran, J. Decision processes and ageing, pp. 21-34. In: Behaviour, Aging and the Nervous System. Eds.: A. T. Welford and J. E. Birren. Springfield, Ill.: C. C. Thomas, **1965**.

Szent-Gyorgi, A. G. Meromyosins, the subunits of myosin. Arch. Biochem. Biophys. 42: 305-320, **1953**.

Tabakin, B. S., Hanson, J. S., and Levy, A. M. Effects of physical training on the cardiovascular response to graded upright exercise in distance runners. Brit. Heart J. 27: 205-210, **1965**.

Taccardi, B. Distribution of heat potentials on the surface of normal human subjects. Circ. Res. 12: 341-352, **1963**.

Tacker, W. A., Geddes, L. A., and Hoff, H. E. Defibrillation without a-v block using capacitor discharge with added inductance. Circ. Res. 22: 633-638, **1968**.

Tacker, W. A., Geddes, L. A., McFarlane, J., Milnor, W., Gullett, J., Havens, W., Green, E., and Moore, J. Optimum current duration for capacitor-discharge defibrillation of canine ventricles. J. Appl. Physiol. 27: 480-483, **1969**.

Takaoka, I. Effect of physical training on oxygen intake and blood lactate concentrations. Bull. Health Phys. Ed., Juntendo University, **1973**.

Talland, G. A. Effect of aging on the formation of sequential and spatial concepts. Percept. Motor Skills 13: 210, **1961**.

Tam, H. S., Darling, R. C., Cheh, H. Y., and

Downey, J. A. Sweating response: A means of evaluating the set-point theory during exercise. J. Appl. Physiol. Resp. Env. Exer. Physiol. 45: 451-458, **1978**.

Tanaka, M., Brisson, G. R., Volle, M. A., and Dion, M. Thermal responses during submaximal and maximal exercises in man. J. Sport. Med. Phys. Fitness 18: 107-116, **1978**.

Tang, S. W., Shephard, R. J., Takahashi, S., Warsh, J. J., and Stancer, H. Central and peripheral metabolites of noradrenaline and serotonin during sustained exercise. In: Proceedings of 4th International Symposium on Exercise Biochemistry, Brussels, **1979**.

Tanner, J. M. Growth at Adolescence, 2nd ed. Oxford: Blackwell Scientific, **1962**.

——. Radiographic studies of body composition in children and adults, pp. 211-236. In: Human Body Composition: Approaches and Applications. Ed.: J. Brozek. Oxford: Pergamon Press, **1965**.

——. Trend towards earlier menarche in London, Oslo, Copenhagen, the Netherlands and Hungary. Nature 243: 95-96, **1973**.

Tanner, J. M. and Israelsohn, W. Parent-child correlations for body measurements of children between the ages of one month and seven years. Ann. Human Genet. 26: 245-259, **1963**.

Tanner, J. M., Whitehouse, R. H., Marshall, W. A., Healy, M. J. R., and Goldstein, H. Assessment of Skeletal Maturity and Prediction of Adult Height (TW2 Method). New York: Academic Press, **1975**.

Tapperman, J. and Pearlman, D. Effects of exercise and anemia on coronary arteries of small animals as revealed by the corrosion cast technique. Circ. Res. 9: 576-584, **1961**.

Tartulier, M., Burret, M., and Deyrieux, F. Les pressions artérielles pulmonaires chez l'homme normal—éffets de l'âge et de l'exercice musculaire. Bull. Physiol. Pathol. Resp. 8: 1295-1321, **1972**.

Tatai, K. Comparisons of ventilatory capacities among fishing divers, nurses and telephone operators in Japanese females. Jap. J. Physiol. 7: 37-41, **1957**.

Tauchert, M., Kochsiek, K., Heiss, H. W., Strauer, B. E., Kettler, D., Reploh, H. D., Rau, G., and Bretschneider, H. J. Measurement of coronary blood flow in man by the argon method, pp. 139-144. In: Myocardial Blood Flow in Man. Methods and Significance in Coronary Disease. Ed.: A. Maseri. Torino: Minerva Medica, **1972**.

Taulbee, D. B., Yu, C. P., and Heyder, J. Aerosol transport in the human lung from analysis of single breaths. J. Appl. Physiol. Resp. Env. Exer. Physiol. 44: 803-812, **1978**.

Taylor, A. The contribution of the intercostal muscles to the effort of respiration in man. J. Physiol. (Lond.) 151: 390-402, **1960a**.

——. Some characteristics of exercise proteinuria. Clin. Sci. 19: 209-217, **1960b**.

Taylor, A. W. The effects of exercise and training on the activities of human skeletal muscle glycogen cycle enzymes, pp. 451-462. In: Metabolic Adaptation to Prolonged Physical Exercise. Eds.: H. Howald and J. R. Poortmans. Basel: Birkhauser Verlag, **1975**.

Taylor, A. W., Garrod, J., Booth, M. A., Secord, D. C., and Cary, S. M. The effects of exercise training and feeding patterns on rat adipose cell size and number. Can. J. Appl. Sport. Sci. 1: 93-97, **1976**.

Taylor, A. W., Kots, Y. M., and Lavoie, M. The effects of faradic stimulation on skeletal muscle fibre area. Can. J. Appl. Sport. Sci. 3: 185, **1978**.

Taylor, A. W., Lavoie, S., Lemieux, G., Dufresne, C., Skinner, J. S., and Vallee, J. Effects of endurance training on the fiber area and enzyme activities of skeletal muscle of French-Canadians, pp. 267-278. In: 3rd International Symposium on Biochemistry of Exercise. Eds.: F. Landry and W. A. R. Orban. Miami: Symposia Specialists, **1978**.

Taylor, A. W., Secord, D. C., Murray, P., and Bailey, G. The effects of castration and repositol testosterone treatment on exercise-induced glycogen and free fatty acid mobilization. Endocrinology 61: 13-20, **1973**.

Taylor, A. W., Stothart, J., Thayer, R., Booth, M., and Rao, S. Human skeletal muscle debranching enzyme activities with exercise and training. Europ. J. Appl. Physiol. 33: 327-330, **1974**.

Taylor, E. W. Chemistry of muscle contraction. Ann. Rev. Biochem. 41: 577-616, **1972**.

Taylor, H. J. Recent research and submarine escape. MRC (RNPRC) Report, UPS 137, **1953**.

Taylor, H. L., Blackburn, H., Puchner, T., Vasquez, C. L., Parlin, R. W., and Keys, A. Coronary heart disease in selected occupations of American railroads in relation to physical activity. Circulation 39/40 (Suppl. III): 202, **1969**.

Taylor, H. L., Buskirk, E. R., and Henschel, A. Maximal oxygen intake as an objective measure of cardiorespiratory performance. J. Appl. Physiol. 8: 73–80, **1955**.

Taylor, H. L., Henschel, A., Brozek, J., and Keys, A. The effect of bed rest on cardiovascular function and work performance. J. Appl. Physiol. 2: 223–239, **1949**.

Taylor, H. L., Henschel, A. F., and Keys, A. Cardiovascular adjustments of man in rest and work during exposure to dry heat. Amer. J. Physiol. 139: 583–591, **1943**.

Taylor, H. L., Klepetar, E., Keys, A. et al. Death rates among physically active and sedentary employees of the railroad industry. Amer. J. Pub. Health 52: 1697–1707, **1962**.

Taylor, H. L., Parlin, R. W., Blackburn, H., and Keys, A. Problems in the relationship of coronary heart disease to physical activity or its lack, with special reference to sample size and occupational withdrawal. In: Physical Activity in Health and Disease. Eds.: K. Evang and K. L. Andersen. Baltimore: Williams & Wilkins, **1967**.

Tcheng, T. J. and Tipton, C. M. Iowa wrestling study: Anthropometric measurements and the prediction of a "minimal" body weight for high school wrestlers. Med. Sci. Sport. 5: 1–70, **1973**.

Tenney, S. M. and Reese, R. E. The ability to sustain great breathing efforts. Resp. Physiol. 5: 187–201, **1968**.

Tenorová, M. and Hruza, Z. Cited by Pařízková (1973). Cs. Fysiol. 11: 485, **1962**.

Tepperman, J. and Pearlman, D. Effects of exercise and anemia on coronary arteries of small animals as revealed by the corrosion cast technique. Circ. Res. 9: 576–584, **1961**.

Terjung, R. L. Muscle fibre involvement during training of different intensities and durations. Amer. J. Physiol. 230: 946–950, **1976**.

Terjung, R. L., Baldwin, K. M., Winder, W. W., and Holloszy, J. O. Glycogen repletion in different types of muscle and in liver after exhausting exercise. Amer. J. Physiol. 226: 1387–1391, **1974**.

Terjung, R. L., Klinkerfuss, G. H., Baldwin, K. M., Winder, W. W., and Holloszy, J. O. Effect of exhausting exercise on rat heart mitochondria. Amer. J. Physiol. 225: 300–305, **1973**.

Terjung, R. L. and Tipton, C. M. Exercise training and resting oxygen consumption. Int. Z. Angew. Physiol. 28: 269–272, **1970**.

———. Plasma thyroxine and thyroid-stimulating hormone levels during submaximal exercise in humans. Amer. J. Physiol. 220: 1840–1845, **1971**.

Terjung, R. L. and Winder, W. W. Exercise and thyroid function. Med. Sci. Sport. 7: 20–26, **1975**.

Tesch, P., Larsson, L., Eriksson, A., and Karlsson, J. Muscle glycogen depletion and lactate concentration during downhill skiing. Med. Sci. Sport. 10: 85–90, **1978**.

Tesch, P., Piehl, K., Wilson, G., and Karlsson, J. Kanot. Idrottsfysiologi, Rep. 13, Trygg-Hansa, Stockholm, **1974**.

———. Physiological investigations of Swedish elite canoe competitors. Med. Sci. Sport. 8: 214–218, **1976**.

Teuber, H. L. In: The Frontal Granular Cortex and Behavior. Eds.: J. M. Warren and K. Akertz, pp. 410–444. New York: McGraw Hill, **1964**.

Thach, W. T. Discharge of cerebellar neurons related to two maintained postures and two prompt movements. I. Nuclear cell output. II. Purkinjé cell output and input. J. Neurophysiol. 33: 527–536, 537–547, **1970**.

Thadani, V. and Parker, J. Hemodynamics at rest and during supine and sitting bicycle exercise in normal subjects. Amer. J. Cardiol. 41: 52–59, **1978**.

Tharp, G. D. The role of the glucocorticoids in exercise. Med. Sci. Sport. 7: 6–11, **1975**.

Tharp, G. D. and Buuck, R. J. Adrenal adaptation to chronic exercise. J. Appl. Physiol. 37: 720–722, **1974**.

Tharp, G. D., Thadani, U., and Parker, J. O. Hemodynamics at rest and during supine and sitting bicycle exercise in normal subjects. Amer. J. Cardiol. 41: 52–59, **1978**.

Theorell, H. Kristallinsches Myoglobin, V. Die Sauerstoffbindungskurve des Myoglobins. Biochem. Z. 268: 73–82, **1934**.

Thistle, H. G., Hislop, H. J., Moffroid, M., and Lowman, E. W. Isokinetic contraction: A new concept of resistive exercise. Arch. Phys. Med. Rehabil. 48: 279–282, **1967**.

Thoden, J. S., Dempsey, J. A., Reddan, W. G., Birnbaum, M. L., Forster, H. V., Grover, R. F., and Rankin, J. A. Ventilatory work during steady-state response to exercise. Fed. Proc. 28: 1316–1321, **1969**.

Thomas, B. M. and Miller, A. T. Adaptation to forced exercise in the rat. Amer. J. Physiol. 193: 350–354, **1958**.

Thomas, C. B., Bateman, J. L., and Lindberg, E. F. Observations on the individual effects of smoking on the blood pressure, heart rate, stroke volume, and cardiac output of healthy young adults. Ann. Intern. Med. 44: 874–892, **1956**.

Thompson, H. L. and Stull, G. A. Effects of various training programs on speed of swimming. Res. Quart. 30: 479–485, **1959**.

Thompson, M. L. and Compston, M. R. Salt intake of Royal Naval Personnel. U.K. Government: MRC RNP 47/373, **1948**.

Thorén, C. Studien über die submaximale und maximale Arbeit bei Schulkindern. Z. Aertzl. Fortbild. 62: 938–942, **1968**.

Thörner, W. Trainingsversuche an Hunden. III. Histologische Beobachtungen an Herz und-Skeletmuskeln. Arbeitsphysiol. 8: 359–370, **1935**.

——. Neue Beiträge zur Physiologie des Trainings; die organentwicklung, zumal des Herzens, unter dem Einflusss anstrengender Dauerleistungen. Arbeitsphysiol. 14: 95–115, **1949-1950**.

Thorstenson, A. Muscle strength, fibre types and enzyme activities in man. Acta Physiol. Scand. Suppl. 443: 1–45, **1976**.

Thorstenson, A., Larsson, L., Tesch, P., and Karlsson, J. Muscle strength and fiber composition in athletes and sedentary men. Med. Sci. Sport. 9: 26–30, **1977**.

Thurlbeck, W. M. and Wang, N. S. The structure of the lungs, pp. 1–30. In: MTP International Review of Science. Respiratory Physiology. Ed.: J. G. Widdicombe. London: Butterworths, **1974**.

Thys, H., Farraggiana, T., and Margaria, R. Utilization of muscle elasticity in exercise. J. Appl. Physiol. 32: 491–494, **1972**.

Timiras, P. S. Developmental Physiology and Aging. New York: Macmillan, **1972**.

Tipton, C. M. Training and bradycardia in rats. Amer. J. Physiol. 209: 1089–1094, **1965**.

——. The influence of atropine on the heart rate responses of non-trained, trained and detrained animals. Physiologist 12: 376, **1969**.

Tipton, C. M., Martin, R. K., Matthes, R. D., and Carey, R. A. Hydroxyproline concentrations in ligaments from trained and nontrained rats, pp. 262–267. In: Metabolic Adaptation to Prolonged Physical Exercise. Eds.: H. Howald and J. R. Poortmans. Basel: Birkhauser Verlag, **1975**.

Tipton, C. M., Matthes, R. D., and Maynard, J. Influence of chronic exercise on rat bones. Med. Sci. Sport. 4: 55, **1972**.

Tipton, C. M., Matthes, R. D., Maynard, J. A., and Carey, R. A. The influence of physical activity on ligaments and tendons. Med. Sci. Sport. 7: 165–175, **1975**.

Tipton, C. M., Matthes, R. D., and Vailas, A. C. Influence de l'exercise sur les structures ligamentaires, pp. 103–114. In: Facteurs limitant l'endurance humaine. Les techniques d'amélioration de la performance. Ed.: J. R. LaCour. France: Université de St. Etienne, **1977**.

Tipton, C., Struck, P., Baldwin, K., Matthes, R., and Dowell, T. Response of adrenalectomized rats to chronic exercise. Endocrinology 91: 573–579, **1972**.

Tipton, C. M., Tcheng, T., and Mergner, W. Ligamentous strength measurements from hypophysectomized rats. Amer. J. Physiol. 221: 1144–1150, **1971**.

Tipton, C. M., Terjung, R. L., and Barnard, R. J. Response of thyroidectomized rats to training. Amer. J. Physiol. 215: 1137–1142, **1968**.

Tlusty, L. Physical fitness in old age. II. Anaerobic capacity, anaerobic work in graded exercise, recovery after maximum work performance in elderly individuals. Respiration 26: 287–299, **1969**.

Tomanek, R. J. Effects of age and exercise on the extent of the myocardial capillary bed. Anat. Rec. 167: 55–62, **1970**.

Tomanekh, R. H. and Colling-Saltin, A. S. Cytological differentiation of human fetal skeletal muscle. Amer. J. Anat. 149: 227–245, **1977**.

Tønnesen, K. H. Blood flow through muscle during rhythmic contraction measured by ^{133}Xenon. Scand. J. Clin. Lab. Invest. 16: 646–654, **1964**.

Tonomura, Y. Muscle Proteins, Muscle Contraction and Cation Transport. Baltimore: University Park Press, **1973**.

Tornheim, K. and Lowenstein, J. M. The purine nucleotide cycle. The production of ammonia from aspartate by extracts of rat skeletal muscle. J. Biol. Chem. 247: 162–169, **1972**.

Tornvall, G. Assessment of physical capabilities. Acta Physiol. Scand. 58, Suppl. 201, **1963**.

Torrance, R. W. Arterial chemoreceptors, pp. 247–271. In: M.T.P. International Review of Science. Respiratory Physiology. Ed.: J. G. Widdicombe. London: Butterworths, **1974**.

Tranquade, R. E. The relationship of intracellular and extracellular lactate and pyruvate concentrations. Clin. Res. 14: 179, **1966**.

Trap-Jensen, J. and Lassen, N. A. Restricted diffusion in skeletal muscle capillaries in man. Amer. J. Physiol. 220: 371-376, **1971**.

Travis, R. C. A new stabilometer for measuring dynamic equilibrium in a standing position. J. Exp. Psychol. 34: 418-424, **1944**.

Tregear, R. T. Interpretation of skin impedance measurements. Nature 205: 600-601, **1965**.

Tremblay, A., Sevigny, J., Jobin, M., and Allard, C. Diet and muscle glycogen in runners, pp. 323-326. In: 3rd International Symposium on Biochemistry of Exercise. Eds.: F. Landry and W. Orban. Miami: Symposia Specialists, **1978**.

Treumann, F. and Schroeder, W. Trainingseinfluss auf Muskeldurchblutung und Herzfrequenz. Z. Kreislauff. 57: 1024-1033, **1968**.

Triebwasser, J. H., Johnson, R. L., Burpo, R. P., Campbell, J. C., Reardon, W. C., and Blomqvist, C. G. Non-invasive determination of cardiac output by a modified acetylene rebreathing procedure utilizing mass spectrometric measurements. Aviation Space Environ. Med. 48: 203-205, **1977**.

Trivedi, B. and Danforth, W. H. Effect of pH on the kinetics of frog muscle phosphofructokinase. J. Biol. Chem. 241: 4110-4112, **1966**.

Trowell, H. Definition of dietary fiber and hypothesis that it is a protective factor in certain diseases. Amer. J. Clin. Nutr. 29: 417-427, **1976**.

Tsuji, K., Tsuji, E., Ohta, T., Oshima, S., Suzuki, H., and Suzuki, S. Effect of nutrition and physical exercise on fatty acid composition of adipose tissue in rats. Jap. J. Nutr. 30: 53-58, **1972**.

Tsukahara, N., Fuller, D. R. G., and Brooks, V. B. Collateral pyramidal influences on the corticorubrospinal system. J. Neurophysiol. 31: 467-484, **1968**.

Tucker, A. and Horvath, S. M. Regional blood flow responses to hypoxia and exercise in altitude-adapted rats. Europ. J. Appl. Physiol. 33: 139-150, **1974**.

Tunstall-Pedoe, D. S. Velocity distribution of blood flow in major arteries of animals and man. Oxford University, D.Phil. thesis, **1970**.

——. Sports cardiology, pp. 169-202. In: Sports Medicine, 2nd ed. Eds.: J. G. P. Williams and P. N. Sperryn. London: Arnold, **1976**.

Turner, D. C., Wallimann, T., and Eppenberger, H. M. A protein that binds specifically to the M-line of skeletal muscle is identified as the muscle form of creatinekinase. Proc. Nat. Acad. Sci. U.S. 70: 702-705, **1973**.

Turner, M. Faulty Posture and Its Treatment. London: Whitefriars Press, **1965**.

Turpeinen, O., Miettinen, M., Karvonen, M. J., Roine, P., Pekkarinen, M., Lehtosuo, E. J., and Alivirta, P. Dietary prevention of coronary heart disease: Long-term experiment. 1. Observations on male subjects. Amer. J. Clin. Nutr. 21: 255-276, **1968**.

Turto, H., Lindy, S., and Haline, J. Protocollagen proline hydroxylase activity in work-induced hypertrophy of rat muscle. Amer. J. Physiol. 226: 63-65, **1974**.

Tuttle, W. W. The effect of weight loss by dehydration and the with-holding of food on the physiologic responses of wrestlers. Res. Quart. 14: 158-166, **1943**.

——. Effect of physical training on capacity to do work as measured by the bicycle ergometer. J. Appl. Physiol. 2: 393-398, **1950**.

Tzankoff, S. P., Robinson, S., Pyke, F. S., and Brawn, C. A. Physiological adjustments to work in older men as affected by physical training. J. Appl. Physiol. 33: 346-350, **1972**.

Udassin, R., Shoenfeld, Y., Shapiro, Y., Birenfeld, C., and Sohar, E. Serum glucose and lactic acid concentrations during prolonged and strenuous exercise in man. Amer. J. Phys. Med. 56: 249-256, **1977**.

Ueland, K., Novy, M. J., Peterson, E. N., and Metcalfe, J. Maternal cardiovascular dynamics. IV. The influence of gestational age on the maternal cardiovascular response to posture and exercise. Amer. J. Obstet. Gynecol. 104: 856-864, **1969**.

Ufland, J. M. Einfluss des Lebensalters, Geschlechts, der Konstitution und des Berufs auf die Kraft verschiedener Muskelgruppen. I. Mitteilungen uber den Einfluss des Lebensalters auf die Muskelkraft. Arbeitsphysiol. 6: 653-663, **1933**.

Ullrick, W. C. A theory of contraction for striated muscle. J. Theor. Biol. 15: 53-69, **1967**.

Ulmeanu, F. I. C., Ciobanu, V., Clejan, L. and Moldoveanu, G. Observations concernant les modifications circulatoires chez les coureurs de Marathon. Med. Sportiva 12: 5-16, **1958**.

Ulmer, H. V. Die Abhangigkeit des Leistungsempfindens von der Tretfrequenz bei Radsportlern. Sportarzt und Sportmedizin 10: 385-390, **1969**.

Ulmer, H. V., Röske, U., and Link, K. Die Aussage fähigkeit des LPI nach EA Müller bei der

Beurteilung der körperlichen Leistungsfähigkeit von Trainierten und Untranierten. Int. Z. Angew. Physiol. 29: 343-358, **1971**.

Ultman, J. S. and Blatman, H. S. Longitudinal mixing in pulmonary airways. Analysis of inert gas dispersion in symmetric tube network models. Resp. Physiol. 30: 349-367, **1977**.

Ultman, J. S., Doll, B. E., Spiegel, R., and Thomas, M. W. Longitudinal mixing in pulmonary airways—normal subjects respiring at a constant flow. J. Appl. Physiol. Resp. Environ. Exer. Physiol. 44: 297-303, **1978**.

Urschel, C. W., Covell, J. W., Sonnenblick, E. H., Ross, J., and Braunwald, E. Myocardial mechanics in aortic and mitral valvular regurgitation; the concept of instantaneous impedance as a determinant of the performance of the intact heart. J. Clin. Invest. 47: 867-883, **1968a**.

Urschel, C. W., Covell, J. W., Graham, T. P., Clancy, R. L., Ross, J., Sonnenblick, E. H., and Braunwald, E. Effects of acute valvular regurgitation on the oxygen consumption of the heart. Circ. Res. 23: 33-43, **1968b**.

U.S. Navy. Diving Manual. Washington, D.C.: Navy Department, **1963**.

Ussing, H. H. Transport of ions across cellular membranes. Physiol. Rev. 29: 127-155, **1949**.

Uvnäs, B. Sympathetic vasodilator system and blood flow. Physiol. Rev. 40 (Suppl. 4): 69-80, **1960**.

Vaage, O. Fluorometric determination of epinephrine and norepinephrine in 1 ml. urine introducing dithiotreitol and boric acid as stability and sensitivity improving agents of the trihydroxyindole method. Biochem. Med. 9: 41-53, **1974**.

Vailas, A. C., Tipton, C. M., Matthes, R. D., and Bedford, T. G. The influence of physical activity on the strength of junctions and ligaments of diabetic rats. Med. Sci. Sport. 10: 57 (Abstr.), **1978a**.

Vailas, A. C., Tipton, C. M., Laughlin, H. L., Tcheng, T. K., and Matthes, R. D. Physical activity and hypophysectomy on the aerobic capacity of ligaments and tendons. J. Appl. Physiol. Resp. Environ. Exer. Physiol. 44: 542-546, **1978b**.

Valdivia, E. U.S.A.F. (Randolph Field, Texas) Report 55: 101, **1956**.

Vallbo, A. B. The significance of intramuscular receptors in load compensation during voluntary contractions in man, pp. 211-225. In: Control of Posture and Locomotion. Eds.: R. B. Stein, K. G. Pearson, R. S. Smith, and J. B. Redford. New York: Plenum Press, **1973a**.

——. Muscle spindle afferent discharge from resting and contracting muscles in normal human subjects, pp. 251-262. In: New Developments in Electromyography and Clinical Neurophysiology, Vol. 3. Ed.: J. E. Desmedt. Basel: Karger, **1973b**.

Van Atta, L. and Sutin, J. The response of single lateral hypothalamic neurons to ventromedial nucleus and limbic stimulation. Physiol. Behav. 6: 523-536, **1971**.

Van Beaumont, W. Red cell volume changes with changes in plasma osmolarity during maximal exercise. J. Appl. Physiol. 35: 47-50, **1973**.

Van Beaumont, W. and Bullard, R. W. Sweating: Its Rapid Response to Muscular Work. Science 141: 643-646, **1963**.

Van Beaumont, W., Greenleaf, J. E., and Julios, L. Disproportional changes in hematocrit, plasma volume, and proteins during exercise and bed rest. J. Appl. Physiol. 33: 55-61, **1972**.

Van Beaumont, W., Strand, J. C., Petrofsky, J. S., Hipkind, S. G., and Greenleaf, J. E. Changes in total plasma content of electrolytes and proteins with maximal exercise. J. Appl. Physiol. 34: 102-106, **1973**.

Van Buchem, F. S. P. and Bosschieter, E. The influence of sympathicolytics on the electrocardiogram. Koninkl. Nederl. Akademie Van Wetenschappen, Amsterdam. Proc. Ser. C: 1-14, **1970**.

Van Citters, R. L. and Franklin, D. L. Cardiovascular performance of Alaska sled dogs during exercise. Circ. Res. 24: 33-42, **1969**.

Van Dam, B. Vitamins and sport. Brit. J. Sport. Med. 12: 74-79, **1978**.

Vanden Abeele, J. Comparison of interlateral differences in gross motor skills and their relationship with handedness and footedness, pp. 57-62. In: Motor Learning, Sport, Psychology, Pedagogy, and Didactics of Physical Activity. Eds.: F. Landry and W. R. Orban. Miami: Symposia Specialists, **1978**.

Vandenberg, S. G. How stable are heritability estimates? A comparison of heritability estimates from six anthropometric studies. Amer. J. Phys. Anthropol. 20: 331-338, **1962**.

Van der Hoeven, G. M. A., Clerens, P. J. A., Denders, J. J. H., Beneken, J. E. W., and Vonk, J. T. C. A study of systolic time intervals during uninterrupted exercise. Brit.

Heart J. 39: 242-254, **1977**.

Van der Walt, W. H. and Wyndham, C. H. An equation for prediction of energy expenditure of walking and running. J. Appl. Physiol. 34: 559-563, **1973**.

Van de Woestijne, K. P. and Zapletal, A. The maximum expiratory flow-volume curve: Peak flow and effort-independent portion, pp. 61-72. In: Airway Dynamics: Physiology and Pharmacology. Ed.: A. Bouhuys. Springfield, Ill.: C. C. Thomas, **1970**.

Vaněk, M. Psychological problems of superior athletes: Some experiences from the Olympic Games in Mexico City, 1968, pp. 183-185. In: Contemporary Psychology of Sport. Ed.: G. Kenyon. Chicago: Athletic Institute, **1970**.

Vaněk, M. Psychological determinants of performance. In: Physical Activity and Cardiovascular Health. Ed.: R. Masironi. Geneva: WHO, **1979**.

Vangaard, Z. Physiological reactions to wet-cold. Aviation Space Environ. Med. 46: 33-36, **1975**.

Van Houtte, P. M. Adrenergic neuroeffector interaction in the blood vessel wall. Fed. Proc. 37: 181-186, **1978**.

Van Huss, W. D. What makes the Russians run? Nutr. Today 1: 20-23, **1966**.

Van Itallie, T. B. and Hashim, S. A. Avenues of control of serum cholesterol. In: Metabolism of Lipids as Related to Atherosclerosis. Ed.: F. A. Kummerow. Springfield, Ill.: C. C. Thomas, **1965**.

Van Itallie, T. B. and Yang, M. U. Current concepts in nutrition-diet and weight loss. N. Engl. J. Med. 297: 1158-1161, **1977**.

Van Liere, E. J., Hess, H. H., and Edwards, J. E. Effect of physical training on the propulsive motility of the small intestine. J. Appl. Physiol. 7: 186-187, **1954**.

Van Liere, E. J. and Northup, D. W. Cardiac hypertrophy produced by exercise in albino and in hooded rats. J. Appl. Physiol. 11: 91-92, **1957**.

Van Liew, H. D. Dissolved gas washout and bubble absorption in routine decompression, pp. 145-150. In: Underwater Physiology. Ed.: C. J. Lambertsen. New York: Academic Press, **1971**.

Van Liew, H. D. and Hlastala, M. P. Influence of bubble size and blood perfusion on absorption of gas bubbles in tissues. Resp. Physiol. 7: 111-121, **1969**.

Van Uytvanck, P. and Vrijens, J. Der Einfluss dynamischen und statischen Trainings auf die Entwicklung der Muskel-hypertrophie und Muskelkraft. Sportarzt und Sportmedizin 7: 149-152, **1971**.

Varène, P. and Jacquemin, C. Airways resistance: A new method of computation, pp. 99-108. In: Airway Dynamics: Physiology and Pharmacology. Ed.: A Bouhuys. Springfield, Ill.: C. C. Thomas, **1970**.

Varène, P., Timbal, J., and Jacquemine, C. Effect of different ambient pressures on airway resistance. J. Appl. Physiol. 22: 699-706, **1967**.

Varnauskas, E. and Holmberg, S. Myocardial blood flow during exercise in patients with coronary heart disease. Comments on training effects, pp. 102-104. In: Coronary Heart Disease and Physical Fitness. Eds.: O. A. Lassen and R. O. Malmborg. Baltimore: University Park Press, **1971**.

Vatner, S. F., Higgins, C. B., Franklin, D., and Braunwald, E. Role of tachycardia in mediating the coronary hemodynamic response to severe exercise. J Appl. Physiol. 32: 380-385, **1972**.

Vatner, S. F. and Pagani, M. Cardiovascular adjustments to exercise: Hemodynamics and mechanisms. Prog. Cardiovasc. Dis. 19: 91-108, **1976**.

Vávra, J. and Máček, M. Changes of the end-expiratory level and the utilization of lung reserve volumes during exercise. Int. Z. Angew. Physiol. 26: 124-130, **1968**.

Vedel, J. P. and Mouillac-Baudevin, J. Pyramidal control of the activity of dynamic and static fusimotor fibers of the cat. Exp. Brain Res. 10: 39-63, **1969**.

Vedel, J. P. and Paillard, J. Effet différential des stimulations du noyau caudé et du cortex frontal sur la sensibilité dynamique des terminaisons fusoriales primaires chez le chat. J. Physiol. (Paris) 57: 716-717, **1965**.

Vedin, J. A., Wilhelmsson, C. E., Wilhelmsen, L., Bjüre, J., and Ekstrom-Jodal, B. Relation of resting and exercise-induced ectopic beats to other ischemic manifestations and to coronary risk factors. Amer. J. Cardiol. 30: 25-31, **1972**.

Veit, H. Some remarks upon the elementary interpersonal relations with ball game teams, pp. 355-362. In: Contemporary Psychology of Sport. Eds.: G. S. Kenyon and T. M. Grogg. Chicago: Athletic Institute, **1970**.

Vellar, O. D. Nutrient Losses Through Sweating. With Special Reference to the Composition of

Whole Body Sweat During Thermally Induced Profuse Perspiration. Oslo, Norway: Universitetsforlaget, **1969**.

Venco, A., Saviotte, M., Bianchi, B., Barzizza, F., Tramarin, R., and Zolezzi, F. Electrocardiographic and echocardiographic findings in well-trained athletes. In: Sports Cardiology. Eds.: T. Lubich and A. Venerando. Bologna: Aulo Gaggi, **1979**.

Vendsalu, A. Studies on adrenaline and noradrenaline in human plasma. Acta Physiol. Scand. 49 (Suppl. 173): 1-123, **1960**.

Verhaeghe, R. H., Lorenz, R. R., McGrath, M. A., Shepherd, J. T., and Vanhoutte, P. M. Metabolic modulation of neurotransmitter release-adenosine, adenine nucleotides, potassium, hyperosmolarity and hydrogen ion. Fed. Proc. 37: 208-211, **1978**.

Vernon, H.M., Bedford, T., and Karner, C. F. The relationship of atmospheric conditions to the working capacity and the accident rate of miners. Rep. Industr. Fatigue Res. Bd. (London), 39, London HMSO, **1927**.

Vesselinovitch, D. and Wissler, R. W. Requirement for regression studies in animal models, pp. 259-263. In: Atherosclerosis IV. Eds.: G. Schettler, Y. Goto, Y. Hata, and G. Klose. Berlin: Springer Verlag, **1977**.

Viar, W. N., Oliver, B. B., Eisenberg, S., Lombardo, T. A., Willis, K., and Harrison, T. R. The effect of posture and of compression of the neck on the excretion of electrolytes and glomerular filtration. Further studies. Circulation 3: 105-115, **1951**.

Viidik, A. The effect of training on the tensile strength of isolated rabbit tendons. Scand. J. Plast. Reconstr. Surg. 1: 141-147, **1967a**.

——. Experimental evaluation of the tensile strength of isolated rabbit tendons. Biol. Med. Eng. 2: 64-67, **1967b**.

——. Elasticity and tensile strength of the anterior cruciate ligament in rabbits as influenced by training. Acta Physiol. Scand. 74: 372-380, **1968**.

——. Tensile strength properties of Achilles tendon systems in trained and untrained rabbits. Acta Orthoped. Scand. 40: 261-272, **1969**.

——. Simultaneous mechanical and light microscopic studies of collagen fibers. Z. Anat. Entwicklungsesch. 136: 204-212, **1972**.

——. Functional properties of collagenous tissues. Int. Rev. Conn. Tiss. Res. 6: 127-217, **1973**.

Viljanen, A. A. Pattern of motor activity in breathing in response to varying sensory inputs, pp. 183-193. In: CIBA Foundation Symposium on Breathing. Hering-Breuer Centenary Symposium. Ed.: R. Porter. London: Churchill, **1970**.

——. Coordination of Neuromuscular Efferent System in Breathing. London: Royal College of Surgeons, **1972**.

Viru, A. Alterations of the adrenocortical activity during exercise in relation to the training and work duration, p. 33. In: Abstracts, 2nd International Symposium on Exercise Biochemistry,, Magglingen, Switzerland, **1973**.

Viteri, F. E. Considerations on the effect of nutrition on the body composition and physical working capacity of young Guatemalan adults. In: Amino Acid Fortification of Protein Foods. Eds.: N. S. Scrimshaw and A. M. Altschul. Cambridge, Mass: MIT Press, **1971**.

Vogel, J. A., Genovese, R. L., Powell, T. L., Bishop, G. W., Bucci, T. J., and Harris, C. W. Effects of altitude on myocardium of animals. Denver, Colorado. U.S. Army Med. Res. Nutr. Lab. Labor. Rep. 322, **1969**.

Vogel, J. A. and Hansen, J. E. Cardiovascular function during exercise at altitude. In: The Effects of Altitude on Athletic Performance. Ed.: R. Goddard. Chicago: Athletic Institute, **1967**.

Vogel, J. A., Hartley, L. H., and Cruz, J. C. Cardiac output during exercise in altitude natives at sea level and high altitude. J. Appl. Physiol. 36: 173-176, **1974**.

Vogel, J. M. and Whittle, M. A. Bone mineral changes: The second manned SkyLab Mission. Aviation Space Environ. Med. 47: 396-400, **1976**.

Voigt, E. D., Engel, P., and Klein, H. Tages rhythmische Schwankungen des Leistungspulsindex. German Med. Month. 12: 394-395, **1967**.

Vokac, Z., Bell, H. J., Bautz-Holter, E., and Rodahl, K. Oxygen uptake/heart rate relationship in leg and arm exercise, sitting and standing. J. Appl. Physiol. 39: 54-59, **1975**.

Von der Groeben, J., Toole, J. G., Weaver, C. S., and Fitzgerald, J. W. Noise Reduction in Exercise Electrocardiograms by Digital Filter Techniques, pp. 41-60. I: Measurement in Exercise Electrocardiography. The Ernst Simonson Conference. Ed.: H. Blackburn. Springfield, Ill.: C. C. Thomas, **1969**.

Von Döbeln, W. Human standard and maximal

metabolic rate in relation to fat-free body mass. Acta Physiol. Scand. Suppl. 126: 1–79, **1956**.

——. Anthropometric determination of fat-free body weight. Acta Med. Scand. 165: 37–40, **1959**.

——. Kroppsstorlek, Energiomsättning och Kondition. In: Handbok i Ergonomi. Eds.: G. Luthman, U. Åberg, and N. Lundgren. Stockholm: Almqvist & Wiksell, **1966**.

Von Döbeln, W., Åstrand, I., and Bergström, A. An analysis of age and other factors related to maximal oxygen uptake. J. Appl. Physiol. 22: 934–938, **1967**.

Von Döbeln, W. and Eriksson, B. O. Physical training, growth and maximal oxygen uptake of boys aged 11–13 years. In: Pediatric Work Physiology. Proceedings of the Fourth International Symposium. Ed.: O. Bar-Or. Natanya, Israel: Wingate Institute, **1973**.

Von Döbeln, W. and Holmér, I. Body composition, sinking force and oxygen uptake of men treading water. J. Appl. Physiol. 37: 55–59, **1974**.

Von Euler, C. The control of respiratory movement, pp. 19-32. In: Breathlessness. Eds.: J. B. L. Howell and E. J. M. Campbell. Oxford: Blackwell Scientific, **1966a**.

——. Proprioceptive control on respiration. In: Muscular Afferents and Motor Control. Ed.: R. Granit. New York: Wiley, **1966b**.

——. On the role of proprioceptors in perception and execution of motor acts with special reference to breathing, pp. 139–150. In: Loaded Breathing. Eds.: L. D. Pengelly, A. S. Rebuck, and E. J. M. Campbell. Don Mills, Toronto: Longman Green, **1974**.

Von Euler, C., Herrero, F., and Wexler, I. Control mechanisms determining rate and depth of respiratory movements. Resp. Physiol. 10: 93–108, **1970**.

Von Euler, U. S. Sympatho-adrenal activity in physical exercise. Med. Sci. Sport. 6: 165–173, **1974**.

Von Euler, U. S. and Hellner, S. Excretion of noradrenaline and adrenaline in muscular work. Acta Physiol. Scand. 26: 183–191, **1952**.

Von Euler, U. S., Liljestrand, G., and Zotterman, Y. The excitation mechanism of the chemoreceptors of the carotid body. Skand. Arch. Physiol. 83: 132–152, **1939**.

Von Euler, U. S. and Lishajko, F. Improved technique for the fluorimetric estimation of catecholamines. Acta Physiol. Scand. 51: 348–356, **1961**.

Von Euler, U. S. and Luft, R. Effect of insulin on urinary excretion of adrenalin and noradrenaline. Metabolism 1: 528–532, **1952**.

Von Glutz, G., Lüthi, U., and Howald, H. Plasma growth hormone, aldosterone, cortisol and insulin changes in a 100-kilometer run, pp. 219–226. In: 3rd International Symposium on Biochemistry of Effort. Eds.: F. Landry and W. A. R. Orban. Miami: Symposia Specialists, **1978**.

Von Hoesslin, H. Ueber die Ursache der scheinbaren Abhangigkeit des Umsatzes. von der Grosse der Körperoberflache. Arch. Physiol. (Leipzig), 323–379, **1888**.

Von Pettenkofer, M. and Voit, C. Untersuchungen über dem Stoffverbrauch des normalen Menschen. Z. Biol. 2: 459, **1866**.

Vos, H. W. Human effort in the stacking of bales on a moving wagon. J. Agr. Eng. Res. 11: 238–242, **1966**.

Vranic, M., Kawamori, R., and Wrenshall, G. A. The role of insulin and glucagon in regulating glucose turnover in dogs during exercise. Med. Sci. Sport. 7: 27–33, **1975**.

Vuori, I., Pokolainen, E., and Haartiala, K. The elimination of iron in sweat produced by physical exercise and heat. Proceedings 16th World Congress of Sports Medicine, Hannover, **1966**.

Vuori, I., Saraste, M., Vihava, M., and Pakkarinen, A. Feasibility of long distance ski hikes (20–90 km) as a mass sport, pp. 530–544. In: Proceedings of 20th World Congress of Sports Medicine. Ed.: A. H. Toyne. Melbourne: Australian Sports Federation, **1975**.

Vytchikova, M. A. Increasing the vitamin B_1 content in the rations of athletes. Cited in Chem. Abstr. 52: 14787. Voprosy Pitaniya 17: 27–32, **1958**.

Wacholder, K. Weitere Untersuchungen über das Verhalten "tonischer" und "nicht tonischer" Muskeln bei ermudender Reizung. Pflüg. Archiv. 229: 133–142, **1931**.

Wachtlová, M. and Pařízková, J. Comparison of capillary density in skeletal muscles of animals differing in respect of their physical activity. Physiol. Bohemoslov. 21: 489–495, **1972**.

Wade, O. L. and Bishop, J. M. Cardiac Output and Regional Blood Flow. Oxford: Blackwell

Scientific, **1962**.

Wagman, I. H., Pierce, D. S., and Brugher, R. E. Proprioceptive influences in volitional control of individual motor units. Nature (Lond.). 207: 957–958, **1965**.

Wagner, P. D. and West, J. R. Effects of diffusion impairment on O_2 and CO_2 time courses in pulmonary capillaries. J. Appl. Physiol. 33: 62–71, **1972**.

Wahlefeld, A. W. Triglyceride determination after enzymatic hydrolysis. In: Methods of Enzymatic Analysis, 2nd English ed. Ed.: H. U. Bergmeyer. New York: Academic Press, **1974**.

Wahlund, H. Determination of physical work capacity. Acta Med. Scand. 215 (Suppl. 9): 1–127, **1948**.

Wahren, J. Quantitative aspects of blood flow and oxygen uptake in the human forearm during rhythmic exercise. Acta Physiol. Scand. 67: 1–93, (Suppl. 269): **1966**.

——. Substrate utilization during exercise, pp. 111–112. Proc. Int. Union of Physiol. Sci., Paris, **1977a**.

——. Glucose turnover during exercise in man. Ann. N.Y. Acad. Sci. 301: 45–55, **1977b**.

Wahren, J., Felig, P., and Hendler, R. Glucose and amino acid metabolism during recovery after exercise. J. Appl. Physiol. 34: 838–845, **1973**.

Wahren, J., Felig, P. H., Hagenfeldt, L., Hendler, R., and Ahlborg, G. Splanchnic and leg metabolism of glucose, free fatty acids and amino acids during prolonged exercise in man, pp. 144–153. In: Metabolic Adaptation to Prolonged Physical Exercise. Eds.: H. Howald and J. R. Poortmans. Basel: Birkhauser Verlag, **1975**.

Wahren, J., Saltin, B., Jorfeldt, L., and Pernow, B. Influence of age on the local circulatory adaptation to leg exercise. Scand. J. Clin. Lab. Invest. 33. 79–86, **1974**.

Wald, N., Howard, S., Smith, P. G., and Bailey, A. Use of carboxyhaemoglobin levels to predict the development of diseases associated with cigarette smoking. Thorax 30: 133–140, **1975**.

Walder, D. N. A possible explanation for some cases of severe decompression sickness in compressed air workers. In: The Regulation of Human Respiration. Eds.: D. J. C. Cunningham and B. B. Lloyd. Oxford: Blackwell Scientific, **1963**.

——. The prevention of decompression sickness in compressed-air workers, pp. 437–463. In: Physiology and Medicine of Diving. Eds.: P. B. Bennett and D. H. Elliott. London: Baillière, Tindall & Cassell, **1969**.

Walder, G. R., Kurtz, P., and Wilmore, J. The influence of plasma volume changes on thermoregulation during recovery from supine exercise. Personal communication. **1975**.

Walker, D. G. Animal hexokinase, p. 33. In: Essays in Biochemistry, Vol. 2. Cited by A. J. S. Benadé, C. H. Wyndham, C. R. Jansen, G. G. Rogers, and E. J. P. de Bruin. In: Plasma Insulin and Carbohydrate Metabolism After Sucrose Ingestion During Rest and Prolonged Aerobic Exercise. Pflüg. Archiv. 342: 207–218, **1966**.

Walker, I. G. The involvement of carbon dioxide in the toxicity of oxygen at high pressure. Can. J. Biochem. Physiol. 39: 1803–1809, **1961**.

Wallace, J. G. Some studies of perception in relation to age. Brit. J. Psychol. 47: 283–297, **1956**.

Waller, A. D. On muscular spasms known as "tendon-reflex." Brain 3: 179–191, **1881**.

Walløe, L. Information loss during synaptic transfer, pp. 275–276. In: Excitatory Synaptic Mechanisms. Eds.: P. Andersen and J. K. S. Jansen. Oslo: Universitetsforlaget, **1970**.

Walpurger, G. and Anger, H. Enzymatic organization of energy metabolism in rat heart after training in swimming and running. Z. Kreislaufforsch. 59: 438–449, **1970**.

Walter, H. Socioeconomic factors and human growth-observations on school children from Bremen, pp. 49–62. In: Growth and Development. Physique. Ed.: O. G. Eiben. Budapest: Akadémia Kiadó, **1977**.

Walther, E. E., Simon, E., and Jessen, C. Thermoregulatory adjustments of skin blood flow in chronically spinalized dogs. Pflüg. Arch. 322: 323–335, **1971**.

Wang, J. and Pierson, R. N. Disparate hydration of adipose and lean tissue require a new model for body water distribution in man. J. Nutr. 106: 1687–1693, **1976**.

Wang, Y. Reaction to coronary blood flow during exercise, pp. 51–54. In: Physiology of Fitness and Exercise. Eds.: J. F. Alexander, R. C. Serfass, and C. M. Tipton. Chicago: Athletic Institute, **1972**.

Wapner, S. and Witkin, H. A. The role of visual factors in the maintenance of body balance. Amer. J. Psychol. 63: 385–408, **1950**.

Warburg, O. Versuche an überlebendem Carcinomgewebe. (Methoden). Biochem. Z. 142: 317-333, **1923**.

Ward, P. The effect of an anabolic steroid on strength and lean body mass. Med. Sci. Sport. 5: 277-282, **1973**.

Warner, H. R. and Russell, R. O. Effect of combined sympathetic and vagal stimulation on heart rate in the dog. Circ. Res. 24: 567-573, **1969**.

Warnock, N. H., Clarkson, T. B., and Stevenson, R. Effect of exercise on blood coagulation time and atherosclerosis of cholesterol-fed cockerels. Circ. Res. 5: 478-480, **1957**.

Wasserman, K. Lactate and related acid base and blood gas changes during constant load and graded exercise. Can. Med. Assoc. J. 96: 775-779, **1967a**.

——. Commentary. Can. Med. Assoc. J. 96: 780, **1967b**.

Wasserman, K., Whipp, B. J., Casaburi, R., Beaver, W. L., and Brown, H. V. CO_2 flow to the lungs and ventilatory control, pp. 103-132. In: Muscular Exercise and the Lung. Eds.: J. A. Dempsey and C. E. Reed. Madison: Univ. of Wisconsin Press, **1977**.

Wasserman, K., Whipp, B. J., Koyal, S. N., and Beaver, W. L. Anaerobic threshold and respiratory gas exchange during exercise. J. Appl. Physiol. 35: 236-243, **1973**.

Watanabe, K. Running speed of skiing in relation to posture, pp. 203-210. In: Biomechanics of Sports and Kinanthropometry. Eds.: F. Landry and W. A. R. Orban. Miami: Symposia Specialists, **1978**.

Watanabe, K. and Ohtsuki, T. Postural changes and aerodynamic forces in alpine skiing. Ergonomics 20: 121-131, **1977**.

Weale, F. E. and Rothwell-Jackson, R. L. The efficiency of cardiac massage. Lancet i: 990-992, **1962**.

Wearing, M. P., Yuhasz, M. D., Campbell, R., and Love, E. I. The effect of the menstrual cycle on tests of physical fitness. J. Sport. Med. Phys. Fitness 12: 38-41, **1972**.

Wearn, J. T., Ernstene, A. C., Bromer, A. W., Barr, J. S., German, W. J., and Zschiesche, L. J. The normal behavior of the pulmonary blood vessels with observations on the intermittence of the flow of blood in the arterioles and capillaries. Amer. J. Physiol. 109: 236-256, **1934**.

Webb, P. Bioastronautics Data Book. Washington, D. C.: U.S. Natl. Aeronautics and Space Admin., **1964**.

Webb-Peploe, M. M. Effect of changes in central body temperature on capacity elements of limb and spleen. Amer. J. Physiol. 216: 643-646, **1969**.

Weber, G., Kartodihardjo, W., and Klissouras, V. Growth and physical training with reference to heredity. J. Appl. Physiol. 40: 211-215, **1976**.

Wedin, B. Cases of paradoxical undressing by people exposed to severe hypothermia, pp. 61-71. In: Circumpolar Health. Eds.: R. J. Shephard and S. Itoh. Toronto: University of Toronto Press, **1976**.

Weg, R. B. Changing physiology of aging: Normal and pathological, pp. 229-256. In: Aging: Scientific Perspectives and Social Issues. Eds.: D. S. Woodruff and J. E. Birren. New York: Van Nostrand, **1975**.

Weibel, E. R. Morphometry of the Human Lung. New York: Academic Press, **1963**.

——. Morphometric estimation of pulmonary diffusing capacity. I. Model and method. Resp. Physiol. 11: 54-75, **1970**.

——. Morphological basis of alveolar-capillary gas exchange. Physiol. Rev. 53: 419-495, **1972**.

——. A simplified morphometric method for estimating diffusing capacity in normal and emphysematous human lungs. Amer. Rev. Resp. Dis. 107: 579-588, **1973**.

Weidemann, H., Roskamm, H., Samek, L., Schneider, H. D., and Weigel, K. Sauerstoffaufnahme, Atemminutenvolumen und Saure-Basenaushalt während submaximaler und maximaler Ergometerbelastung in Freiburg (260 m) und bei akuter Höhenexposition auf Eigergletscher (2,230 m) und Jungfraujoch (3,457 m). Sportarzt und Sportmedizin 19: 147-152, **1968**.

Weihe, W. H. Time course of acclimatization to high altitude, pp. 33-36. In: The International Symposium on the Effects of Altitude on Physical Performance. Ed.: R. Goddard. Chicago: Athletic Institute, **1967**.

Weiner, J. S. and Lourie, J. A. Human Biology: A Guide to Field Methods. Oxford: Blackwell Scientific, **1969**.

Weinman, K. P., Slabochova, Z., Bernauer, E. M., Morimoto, T., and Sargent, F. Reactions of men and women to repeated exposure to humid heat. J. Appl. Physiol. 22: 535-538, **1967**.

Weis-Fogh, T. Metabolism and weight economy in migrating animals, particularly birds and insects. In: Nutrition and Physical Activity, p. 84. Ed.: G. Blix. Uppsala: Almqvist & Wiksell, **1967**.

Weiss, R. A. and Karpovich, P. V. Energy cost of exercises for convalescents. Arch. Phys. Med. 28: 447-454, **1947**.

Weissler, A. M., Kamen, A. R., Bornstein, R. S., et al. The effects of Deslanoside on the duration of ventricular systole in man. Amer. J. Cardiol. 15: 153-161, **1965**.

Weissler, A. M., Peeler, R. G., and Roehll, W. H. Relationships between left ventricular ejection time, stroke volume, and heart rate in normal individuals and patients with cardiovascular disease. Amer. Heart J. 63: 367-378, **1961**.

Welch, B., Levy, L. M., Consolazio, C. F., Buskirk, E. R., and Dee, T. E. Caloric Intake for Prolonged Hard Work in the Cold. Denver, Colorado: U.S. Army Medical Nutrition Laboratory Report 202: 1-24, **1959**.

Welford, A. T. Skill and Age. Oxford: Oxford University Press, **1951**.

——. On the sequencing of action. Brain Res. 71: 381-392, **1974**.

Welford, A. T. and Birren, J. E. Behaviour, Aging and the Nervous System. Springfield, Ill.: C. C. Thomas, **1965**.

Wells, C. L. and Horvath, S. M. Metabolic and thermoregulatory responses of women to exercise in two thermal environments. Med. Sci. Sport. 6: 8-13, **1974**.

Wells, J. G. Physical condition as a factor in pilot performance. Unpublished data, cited in Physiology of Exercise, p. 80, 4th ed. Eds.: L. E. Morehouse and A. T. Miller. St. Louis: Mosby, **1963**.

Wells, K. F. and Dillon, E. K. The sit and reach: A test of back and leg flexibility. Res. Quart. 23: 115-118, **1952**.

Weltman, A. and Katch, V. Min-by-min respiratory exchange and oxygen uptake kinetics during steady state exercise in subjects of high and low max $\dot{V}_{O_2}$. Res. Quart. 47: 490-498, **1977**.

Weltman, A., Stamford, B. A., Moffat, R. J., and Katch, V. L. Exercise recovery, lactate removal, and subsequent high intensity exercise performance. Res. Quart. 48: 786-796, **1977**.

Wen, C. P. and Gershoff, S. N. Changes in serum cholesterol and coronary heart disease mortality associated with changes in the post-war Japanese diet. Amer. J. Clin. Nutr. 26: 616-619, **1973**.

Wenger, C. B., Roberts, M. F., Stolwijk, J. A. J., and Nadel, E. R. Nocturnal lowering of thresholds for sweating and vasodilation. J. Appl. Physiol. 41: 15-19, **1976**.

Wenger, H. A. and MacNab, R. B. J. Endurance training: The effects of intensity, total work, duration, and initial fitness. J. Sport. Med. Phys. Fitness 15: 199-211, **1975**.

Wenger, N. K. Early ambulation after myocardial infarction: Grady Memorial Hospital—Emory University, School of Medicine, pp. 324-328. In: Exercise Testing and Exercise Training in Coronary Heart Disease. Eds.: J. P. Naughton, H. K. Hellerstein, and I. C. Mohler. New York: Academic Press, **1973**.

——. Does exercise training enhance collateral circulation? pp. 143-145. In: Critical Evaluation of Cardiac Rehabilitation. Eds.: J. Kellermann and H. Denolin. Basel: Karger, **1977**.

Werkö, L., Bersens, S., and Lagerlof, H. A comparison of the direct Fick and the Grollman method of determination of the cardiac output in man. J. Clin. Invest. 28: 516-520, **1949**.

Werkö, L., Varnauskas, E., Eliasch, E., Ek, J., Bucht, H., Thomasson, B., and Bergström, J. Studies on the renal circulation and renal function in mitral valvular disease. 1. Effect of exercise. Circulation 9: 687-699, **1954**.

Wessel, J. A. Fitness and physical activity for women with age. In: Physical Exercise and Activity for the Aging. Natanya, Israel: Wingate Institute, **1975**.

Wessel, J. A., Small, D. A., Van Huss, W. D., Anderson, D. J., and Cederquist, D. S. Age and physiological responses to exercise in women 20-69 years of age. J. Gerontol. 23: 269-278, **1968**.

Wessel, J. A., Ufer, A., Van Huss, W. D., and Cederquist, D. Age trends of various components of body composition and functional characteristics of women aged 20-69 years. Ann. N.Y. Acad. Sci. 110: 608-622, **1963**.

Wesson, L. G. Theoretical analysis of urea excretion by the mammalian kidney. Amer. J. Physiol. 179: 364-371, **1954**.

——. Kidney function in exercise. In: Science and Medicine of Exercise and Sport, 2nd ed.: Eds.: W. R. Johnson and E. R. Buskirk. New York: Harper & Row. **1974**.

West, J. B. Observations on gas flow in the human bronchial tree, pp. 1-7. In: Inhaled Particles and Vapours. Ed.: C. N. Davies. Oxford: Pergamon Press, **1961**.

——. Regional differences in gas exchange in the lung of erect man. J. Appl. Physiol. 17: 893-898, **1962a**.

——. Diffusing capacity of the lung for carbon monoxide at high altitude. J. Appl. Physiol. 17: 421-426, **1962b**.

——. Regional differences in blood flow and ventilation in the lung. In: Advances in Respiratory Physiology. Ed.: C. G. Caro. London: Arnold, **1966**.

——. Gas diffusion in the lung at altitude, pp. 75-83. In: Exercise at Altitude. Ed.: R. Margaria. Dordrecht: Excerpta Medica Fdn., **1967**.

——. Respiratory Physiology: The Essentials. Baltimore: Williams & Wilkins, **1974**.

——. Blood flow, pp. 85-165. In: Regional Differences in the Lung. Ed.: J. B. West. New York: Academic Press, **1977**.

West, J. C. Textbook of Servomechanisms. London: English Universities Press, **1953**.

Westbury, D. R. The response of α motoneurones of the cat to sinusoidal movements of the muscles they innervate. Brain Res. 25: 75-86, **1970**.

Wexler, B. C. and Greenberg, B. P. Effect of exercise on myocardial infarction in young vs old male rats: electrocardiographic changes. Amer. Heart J. 88: 343-350, **1974**.

Wexler, I. and Kao, F. F. Neural and humoral factors affecting canine renal blood flow during induced muscular work. Amer. J. Physiol. 218: 755-761, **1970**.

Wezler, K. and Böger, A. Über einen neuen Weg zur Bestimmung des absoluten Schlagvolumens des Herzens beim Menschen auf Grund der Windkesseltheorie und seine experimentelle Prüfung. Arch. Exp. Pathol. Pharm. 184: 482-505, **1937**.

Whalen, W. J. and Nair, P. Intracellular PO_2 and its regulation in resting skeletal muscle of the guinea pig. Circ. Res. 21: 251-261, **1967**.

——. Some factors affecting tissue PO_2 in the carotid body. J. Appl. Physiol. 39: 562-566, **1975**.

Whalen, W. J., Nair, P., and Buerk, D. Oxygen tension in the beating cat heart in situ. In: Oxygen Supply: Theoretical and Practical Aspects of Oxygen Supply and Microcirculation of Tissue. Eds.: M. Kessler, D. F. Bruley, L. C. Clark, D. W. Lübbers, I. A. Silver, and J. Strauss. Baltimore: University Park Press, **1973**.

Wheeler, E. F., El Neil, H., Wilson, J. O., and Weiner, J. C. The effect of work level and dietary intake on water balance and the excretion of sodium, potassium and iron in a hot climate. Brit. J. Nutr. 30: 127-137, **1978**.

Whipp, B. J., Torres, F., Davis, J. A., Wasserman, K., and Casaburi, R. A test to determine the parameters of aerobic function during exercise. Fed. Proc. 36: 449 (Abstr.), **1977**.

Whipp, B. J. and Wasserman, K. Efficiency of muscular work. J. Appl. Physiol. 26: 644-648, **1969**.

Whipple, G. H. Hemoglobin of striated muscle: Variations due to age and exercise. Amer. J. Physiol. 76: 693-707, **1926**.

White, H. L. and Rolf, D. Effects of exercise and of some other influences on renal circulation in man. Amer. J. Physiol. 152: 505-516, **1948**.

White, J. A., Ismail, A. H., and Bradley, C. A. Serum insulin and glucose response to graded exercise in adults. Brit. J. Sport. Med. 12: 137-141, **1978**.

White, J. R. EKG changes using carotid artery for heart rate monitoring. Med. Sci. Sport. 9: 88-94, **1977**.

White, N. M., Parker, W. S., Binning, R. A., Kimber, E. R., Ead, H. W., and Chamberlain, D. A. Mobile coronary care provided by ambulance personnel. Brit. Med. J. 3: 618-622, **1973**.

Whitelaw, A. G. L. Influence of maternal obesity on sub-cutaneous fat in the newborn. Brit. Med. J. 1: 985-986, **1976**.

Whiteside, T. C. D. Motion sickness. In: Textbook of Aviation Physiology. Ed.: J. A. Gillies. Oxford: Pergamon Press, **1965**.

Whitney, R. J. The measurement of volume changes in human limbs. J. Physiol. (Lond.) 121: 1-27, **1953**.

Whitt, F. R. Estimation of the energy expenditure of sporting cyclists. Ergonomics 14: 419-424, **1971**.

Whittaker, S. R. F. and Winton, F. R. Apparent viscosity of blood flowing in isolated hindlimb of dog, and its variation with corpuscular concentration. J. Physiol. (Lond.) 78: 339-369, **1933**.

Whittam, R. Transport and Diffusion in Red Blood Cells. London: Arnold, **1964**.

Whittenberger, J. L. Artificial respiration. Physiol. Rev. 35: 611-628, **1955**.

Wick. Cited by R. Renson. In: De regeling van de Lichaamstemperatuur bij Koude. Hermes (Leuven) III: 131-148, **1968-1969**.

Widdicombe, J. Respiratory reflexes, pp. 585–630. In: Handbook of Physiology, Section 3. Respiration. Vol. 1. Eds.: W. Fenn and H. Rahn. Washington, D.C.: American Physiological Society, **1964**.

Widdicombe, J. G. Regulation of tracheobronchial smooth muscle. Physiol. Rev. 43: 1–37, **1963**.

Widdicombe, J. G. The regulation of bronchial calibre, pp. 48–82. In: Advances in Respiratory Physiology. Ed.: C. G. Caro. London: Arnold, **1966**.

Widdicombe, J. G. MTP International Reviews of Science Physiology. Series 1. Respiratory Physiology. London: Butterworths, **1974**.

Widdowson, E. M. Mental contentment and physical growth. Lancet 1: 1316–1318, **1951**.

Wiesendanger, M., Séguin, J. J., and Künzle, H. The supplementary motor area: A control system for posture? pp. 331–346. In: Control of Posture and Locomotion. Eds.: R. B. Stein, K. G. Pearson, R. S. Smith, and J. B. Redford. New York: Plenum Press, **1973**.

Wigertz, O. Dynamics of respiratory and circulatory adaptation to muscular exercise in man. A systems analysis approach. Acta Physiol. Scand. Suppl. 363: 1–32, **1970**.

Wiggers, C. J. The mechanism and nature of ventricular fibrillation. Amer. Heart J. 20: 399–412, **1940**.

Wildenthal, K., Mierzwiak, D. S., and Mitchell, J. H. Acute effects of increased serum osmolarity on left ventricular performance. Amer. J. Physiol. 216: 898-904, **1969**.

Wildenthal, K., Mierzwiak, S., Skinner, N. S., and Mitchell, J. H. Potassium-induced cardiovascular and ventilatory reflexes from the dog hindlimb. Amer. J. Physiol. 215: 542–548, **1968**.

Wildenthal, K., Morgan, H. E., Opie, H. L., and Srere, P. A. Regulation of cardiac metabolism (symposium). Circ. Res. (Suppl. I) 38: I-1 to I-160, **1976**.

Wiley, J. F. Effects of 10 weeks of endurance training on left ventricular intervals. J. Sport. Med. Phys. Fitness 11: 104–111, **1971**.

Wiley, J. F. and Shaver, L. G. Prediction of maximum oxygen intake from running performances of untrained young men. Res. Quart. 43: 89–93, **1972**.

Wilhelmsen, L. Risk factor modification in coronary artery disease, pp. 23–35. In: Coronary Heart Disease, Exercise Testing and Cardiac Rehabilitation. Eds.: W. E. James and E. A. Amsterdam. Miami: Symposia Specialists, **1977**.

Wilhelmsen, L., Sanne, H., Elmfeldt, D., Grimby, G., Tibblin, G., and Wadel, H. A controlled trial of physical training after myocardial infarction. Prev. Med. 4: 491–508, **1975**.

Wilhelmsen, L., Tibblin, G., Aurell, M., Bjüre, J., Ekström-Jodal, B., and Grimby, G. Physical activity, physical fitness and risk of myocardial infarction. Adv. Cardiol. 18: 217–230, **1976**.

Wilkerson, J. E. and Evonuk, E. Changes in cardiac and skeletal muscle myosin ATPase ativities after exercise. J. Appl. Physiol. 30: 328–330, **1971**.

Wilkerson, J. E., Gutin, B., and Horvath, S. M. Exercise-induced changes in blood, red cell and plasma volume. Med. Sci. Sport. 9: 155–158, **1977**.

Wilkie, D. R. Man as a source of mechanical power. Ergonomics 3: 1–8, **1960**.

Willems, P. L., Zenner, R. J., and Clement, D. L. Computer assisted analysis of the Frank vectorcardiogram in top athletes and age-matched controls. In: Sports Cardiology. Eds.: T. Lubich and A. Venerando. Bologna: Aulo Gaggi, **1979**.

Williams, C., Kelman, G. R., Couper, D. C., and Harris, C. G. Changes in plasma FFA concentrations before and after reduction in high intensity exercise. J. Sport. Med. Phys. Fitness 15: 2–12, **1975**.

Williams, C. G., Bredell, G. A. G., Wyndham, C. H., Strydom, N. B., Morrison, J. F., Peter, J., Fleming, P. W., and Ward, J. S. Circulatory and metabolic reactions to work in heat. J. Appl. Physiol. 17: 625–638, **1962**.

Williams, C. G., Kok, R., Von Rahden, M., and Wyndham, C. H. Changes in the maximum oxygen intake and in anaerobic metabolism in subjects after a period of training on a bicycle ergometer. Report to Transvaal and Orange Free State Chamber of Mines Organization, **1966**.

Williams, E. S., Taggart, P., and Carruthers, M. Rock-climbing: Observations on heart rate and plasma catecholamine concentrations and the influence of oxprenolol. Brit. J. Sport. Med. 12: 125–128, **1978**.

Williams, H. G. Neurosensory development in young organisms, pp. 57–67. In: Physical Activity and Human Well-Being. Eds.: F. Landry and W. A. R. Orban. Miami: Symposia Specialists, **1978**.

Williams, L. R. T., Daniell-Smith, J. H., and

Gunson, L. K. Specificity of training for motor skill under physical fatigue. Med. Sci. Sport. 8: 162–167, **1976**.

Williams, L. R. T. and Hearfield, V. Heritability of a gross motor balance task. Res. Quart. 44: 109–112, **1973**.

Williams, M. H. Drugs and Athletic Performance. Springfield, Ill.: C. C. Thomas, **1974**.

Williams, M. H. and Ward, A. J. Hematological changes elicited by prolonged intermittent aerobic exercise. Res. Quart. 48: 606–616, **1977**.

Williams, R. D., Mason, H. L., Wilder, R. M., and Smith, B. F. Observations on induced thiamine (vitamin B_1) deficiency in man. AMA Arch. Int. Med. 66: 785–799, **1940**.

Williams, S. Underwater breathing apparatus, pp. 17–35. In: The Physiology and Medicine of Diving. Eds.: P. B. Bennett and D. H. Elliott. London: Baillière, Tindall & Cassell, **1969**.

Wilmore, J. H. Influence of motivation on physical work capacity and performance. J. Appl. Physiol. 24: 459–463, **1968**.

———. The use of actual, predicted and constant residual volumes in the assessment of body composition by underwater weighing. Med. Sci. Sport. 1: 87–90, **1969**.

Wilmore, J. H. and Behnke, A. An anthropometric estimation of body density and lean body weight in young men. J. Appl. Physiol. 27: 25–31, **1969**.

Wilmore, J. H., Royce, J., Girandola, R. N., Katch, F. I., and Katch, V. L. Physiological alterations resulting from a ten week program of jogging. Med. Sci. Sport. 2: 7–14, **1970**.

Wilmore, J. H. and Sigerseth, P. O. Physical work capacity of young girls 7–13 years of age. J. Appl. Physiol. 22: 923–928, **1967**.

Wilson, B. A., Evans, J. N., and Brantley, P. Effects of work intensity and environmental temperature on exercise-induced airway changes. Can. J. Appl. Sport. Sci. 3: 183, **1978**.

Wilson, B. A., Hermiston, R. T., Stallman, R. K., and Burton, H. Effects of pre and post O_2 breathing on performance times and ventilatory response to repeated 100-yard freestyle swims, pp. 351–356. In: Exercise Physiology. Eds.: F. Landry and W. A. R. Orban. Miami: Symposia Specialists, **1978**.

Wilson, F. N., Johnston, F. D., Rosenbaum, F. F., and Barker, P. S. On Einthoven's triangle, the theory of unipolar electrocardiographic leads, and the interpretation of the precordial electrocardiogram. Amer. Heart J. 32: 277–310, **1946**.

Wilson, J. M. G. and Jungner, F. Principles and Practice of Screening for Disease. Geneva: WHO Public Health Papers 34, **1968**.

Wilson, O. Changes in body weight of men in the Antarctic. Brit. J. Nutr. 14: 391–401, **1960**.

Wilson, R. S. and Harpring, E. B. Mental and motor development in infant twins. Developm. Psychol. 7: 277–287, **1972**.

Wilson, T. K. and Lin, K. H. Convection and diffusion in the airways and the design of the bronchial tree, pp. 5–19. In: Airway Dynamics: Physiology and Pharmacology. Ed.: A. Bouhuys. Springfield, Ill.: C. C. Thomas, **1970**.

Wilt, F. Training for competitive running. In: Exercise Physiology. Ed.: H. Falls. New York: Academic Press, **1968**.

Winder, W. W., Hagberg, J. M., Hickson, R. C., Ehsani, A. A., and McLane, J. A. Time course of sympathoadrenal adaptation to endurance exercise training in men. J. Appl. Physiol. Resp. Environ. Exer. Physiol. 45: 370–374, **1978**.

Winder, W. W. and Heninger, R. W. Effect of exercise on degradation of thyroxine in the rat. Amer. J. Physiol. 224: 572–575, **1973**.

Winslow, C. E. A., Gagge, A. P., and Herrington, L. P. Influence of air movement upon heat losses from clothed human body. Amer. J. Physiol. 127: 505–518, **1939**.

Winter, D. A. Noise measurement and quality control techniques in recording and processing of exercise electrocardiograms, pp. 159–168. In: Measurement in Exercise Electrocardiography. The Ernst Simonson Conference. Ed.: H. Blackburn. Springfield, Ill.: C. C. Thomas, **1969**.

Winter, P. M., Bruce, D. L., Bach, M. J., Jay, G. W., and Eger, E. I. The anaesthetic effect of air at atmospheric pressure. Anesthesiology 42: 658–661, **1975**.

Winters, W. G., Leaman, D. M., and Anderson, R. A. The effect of exercise on intrinsic myocardial performance. Circulation 48: 50–55, **1973**.

Wintrobe, M. M. Clinical Hematology. Philadelphia: Lea & Febiger, **1967**.

Wirth, J. C., Lohman, T. G., Avallone, J. P., Shire, T., and Boileau, R. A. The effect of physical training on the serum iron levels of college-age women. Med. Sci. Sport. 10: 223–226, **1978**.

Wissler, C. and Richardson, W. W. Diffusion of the motor impulse. Psychol. Rev. 7: 29–38, **1900**.

Withers, R. T. Anaerobic work at submaximal relative workloads in subjects of high and medium fitness. J. Sport. Med. 17: 17–24, **1977**.

Withers, R. T., Roberts, R. E. D., and Davies, G. J. The maximum aerobic power, anaerobic power and body composition of South Australian male representatives in athletics, basketball, field hockey and soccer. J. Sport. Med. 17: 391–400, **1977**.

Witkin, H. A. and Wapner, S. Visual factors in the maintenance of upright posture. Amer. J. Psychol. 63: 31–50, **1950**.

Witzleb, E. W. Chemoreceptors, hypoxia and hypercapnia, pp. 173–182. In: The Regulation of Human Respiration. Eds.: D. J. C. Cunningham and B. B. Lloyd. Oxford: Blackwell Scientific, **1963**.

Woch, Z. T., Niinimaa, V., and Shephard, R. J. Heart rate responses during free figure skating manoeuvres. Can. J. Appl. Sport. Sci. 4: 274–276, **1979**.

Wodick, R. Formal aspects of tissue spectrophotometry, pp. 72–78. In: Oxygen Supply and Physiological Function. Ed.: F. F. Jöbsis. Dallas, Texas: Professional Information Library, **1977**.

Wojtezak-Jaroszowa, J., and Banaskiewicz, A. Physical work capacity during the day and at night. Ergonomics 17: 193–198, **1974**.

Wolanski, N. An approach to the problem of inheritance of systolic and diastolic blood pressure. Genet. Pol. 10: 263–268, **1969**.

——. Genetic and ecological control of human growth, pp. 19–33. In: Growth and Development. Physique. Ed.: O. G. Eiben. Budapest: Akadémiai Kiadó, **1977**.

Woledge, R. C. The thermoelastic effect of change of tension in active muscle. J. Physiol. (Lond.) 155: 187–208, **1961**.

Wolf, A. V. Body water. Sci. Amer. 199: 125–126, **1958**.

Wolf, L. A. and Cunningham, D. A. Effects of jogging on left ventricular performance during exercise. Can. J. Sport. Sci. 3: 181, **1978**.

Wolferth, C. C. and Wood, F. C. Electrocardiographic diagnosis of coronary occlusion by use of chest leads. Amer. J. Med. Sci. 183: 30–35, **1932**.

Wolff, C. B. The effects of breathing of alternate breaths of air and a carbon dioxide rich gas mixture in anaesthetized cats. J. Physiol. (Lond.) 268: 483–491, **1977**.

Wolff, H. Physiological measurement on human subjects in the field, with special reference to a new approach to data storage. In: Human Adaptability and Its Methodology. Eds.: H. Yoshimura and J. S. Weiner. Tokyo: Japan Society for the Promotion of Sciences, **1966**.

Wollenberger, A. The role of cyclic AMP in the adrenergic control of the heart, pp. 113–190. In: Contraction and Relaxation in the Myocardium. Ed.:. W. G. Nayler. New York: Academic Press, **1975**.

Wollenberger, A. and Schulze, W. Mitochondrial alterations in the myocardium of dogs with aortic stenosis. J. Cell Biol. 10: 285–288, **1961**.

Wolthuis, R. A., Bergman, S. A., and Nicogossian, A. E. Physiological effects of locally applied reduced pressure in man. Physiol. Rev. 54: 566–595, **1974**.

Womersley, J., Durnin, J. V. G. A., Boddy, K., and Mahaffy, M. Influence of muscular development, obesity, and age on the fat free mass of adults. J. Appl. Physiol. 41: 223–229, **1976**.

Wong, A. Y. K. and Rautaharju, P. M. Stress distribution within the left ventricular wall approximated as a thick ellipsoidal shell. Amer. Heart J. 75: 649–662, **1968**.

Wong, H. Y. C. and Johnson, F. B. The effect of strenuous exercise on plasma lipids and atherosclerosis on cholesterol-fed cockerels. In: Atherosclerosis IV. Eds.: G. Schettler, Y. Goto, Y. Hata, and G. Klose. Berlin: Springer Verlag, **1977**.

Wood, E. H., Hepner, R. L., and Weidmann, S. Inotropic effects of electric currents. Circ. Res. 24: 409–445, **1969**.

Wood, J. D. Oxygen toxicity, pp. 112–143. In: Physiology and Medicine of Diving. Eds.: P. B. Bennett and D. H. Elliott. London: Baillière, Tindall & Cassell, **1969**.

——. Oxygen toxicity in neuronal elements, pp. 9–17. In: Underwater Physiology. Ed.: C. J. Lambertsen. New York: Academic Press, **1971**.

Wood, L. D. H. and Bryan, A. C. Mechanical limitations of exercise ventilation, pp. 307–316. In: Underwater Physiology. Ed.: C. H. Lambertsen, New York: Academic Press, **1971**.

——. Exercise ventilatory mechanics at increased ambient pressure. J. Appl. Physiol. Resp. Environ. Exer. Physiol. 44: 231–237, **1978**.

Wood, P. D. Effect of exercise on plasma lipids and high density lipoprotein levels. In: Topics in Ischaemic Heart Disease. An international symposium. Toronto: Toronto Rehabilitation Centre, **1979**.

Wood, P. D. S., Haskell, W. L., Lewis, S., Perry, C., and Stern, M. P. Concentration of plasma lipids and lipoproteins in male and female long-distance runners, pp. 301-306. In: 3rd International Symposium on Biochemistry of Exercise. Eds.: F. Landry and W. A. R. Orban. Miami: Symposium Specialists, **1978**.

Wood, P. D. S., McGregor, M., Magidson, O., and Whittaker, W. The effort test in angina pectoris. Brit. Heart J. 12: 363-371, **1950**.

Wood, P. D. S., Stern, M. P., Silvers, A., Reaven, G., and Von der Groeben, J. Prevalence of plasma lipoprotein abnormalities in a free-living population of the Central Valley, California. Circulation 45: 114-126, **1972**.

Woodson, R. D., Wills, R. E., and Lenfant, C. Effect of acute and established anemia on O_2 transport at rest, submaximal and maximal work. J. Appl. Physiol. Resp. Environ. Exer. Physiol. 44: 36-43, **1978**.

Woodworth, R. S. The accuracy of voluntary movement. Psychol. Monogr. 3: 1-114, **1899**.

Woolf, C. R. and Suero, J. T. Alteration in lung mechanics following training in chronic obstructive lung disease. Dis. Chest 55: 37-44, **1969**.

Woolsey, C. N., Gorska, T., Wetzel, A., Erickson, T. C., Earls, F. J., and Allman, J. M. Complete unilateral section of the pyramidal tract at the medullary level in Macaca mulatta. Brain. Res. 40: 119-124, **1972**.

Workman, R. D. Calculation of decompression schedules for nitrogen-oxygen and helium-oxygen dives. U.S. Navy Experimental Diving Unit Research Rep.: 6-65, **1965**.

——. Treatment of bends with oxygen at high pressure. Aerospace Med. 39: 1076-1083, **1968**.

——. American decompression theory and practice, pp. 252-290. In: Physiology and Medicine of Diving. Eds.: P. B. Bennett and D. H. Elliott. London: Baillière, Tindal & Cassell, **1969**.

Worth, H., Adaro, F., and Piiper, J. Penetration of inhaled He and SF_6 into alveolar space at low tidal volumes. J. Appl. Physiol. Resp. Environ. Exer. Physiol. 43: 403-408, **1977**.

Worthington, C. R. Impulsive (electrical) forces in muscle, pp. 511-519. In: Biochemistry of Muscle Contraction. Ed.: J. Gergely. Boston: Little, Brown, **1964**.

Wright, G. R., Bompa, T., and Shephard, R. J. Physiological evaluation of a winter training programme for oarsmen. J. Sport. Med. Phys. Fitness 16: 22-37, **1976**.

Wright, G. R., Jewczyk, S., Onrot, J., Tomlinson, P., and Shephard, R. J. Carbon monoxide in the urban atmosphere. AMA Arch. Environ. Health 30: 123-129, **1975**.

Wright, G. R., Randell, P., and Shephard, R. J. The influence of carbon monoxide upon selected driving skills. AMA Arch. Environ. Health 27: 349-354, **1973**.

Wright, G. R. and Shephard, R. J. Brake reaction time—Effects of age, sex and carbon monoxide. Arch. Environ. Health 33: 141-150, **1978a**.

——. Carbon monoxide, nicotine, and the "safer" cigarette. Respiration 35: 40-52, **1978b**.

——. A note on blood carboxyhemoglobin levels in police officers. Can. J. Pub. Health 40: 393-398, **1978c**.

Wright, G. R., Sidney, K. H. and Shephard, R. J. Variance of direct and indirect measurements of aerobic power. J. Sport. Med. Phys. Fitness 18: 33-42, **1978**.

Wright, P. G. A mechanism influencing venous return from a peripheral organ under cool conditions. Proc. Physiol. Soc., July **1977**.

Wright, V. and Johns, R. J. Physical factors concerned with the stiffness of normal and diseased joints. Johns Hopkins Hosp. Bull. 106: 215-231, **1960**.

Wuellner, L. H., Witt, P. A., and Harron, R. E. A method to investigate the movement patterns of children, pp. 799-808. In: Contemporary Psychology of Sport. Eds.: G. S. Kenyon and T. M. Grogg. Chicago: Athletic Institute, **1970**.

Wyatt, H. L. and Mitchell, J. H. Influences of physical training on the heart of dogs. Circ. Res. 35: 883-889, **1974**.

Wyman, R. J. Somatotopic connectivity or species recognition connectivity, pp. 45-53. In: Control of posture and locomotion. Eds.: R. B. Stein, K. G. Pearson, R. S. Smith, and J. B. Redford. New York: Plenum Press, **1973**.

Wyman, R. J., Waldron, I., and Wachtel, G. M. Lack of fixed order of recruitment in cat motoneuron pools. Exp. Brain Res. 20: 101-114, **1974**.

Wyndham, C. H. Effect of acclimatization on circulatory responses to high environmental temperatures. J. Appl. Physiol. 4: 383-395, **1951**.

——. An examination of the methods of physical classification of African labourers for manual work. S. Afr. Med. J. 40: 275–278, **1966**.

———. Physiological effects of a change in altitude from Johannesburg to sea level. S. Afr. Med. J. 46: 251–257, **1972**.

———. The physiology of exercise under heat stress. Ann. Rev. Physiol. 35: 193–220, **1973**.

———. 1973 Yant Memorial Lecture: Research in the human sciences in the gold mining industry. Amer. Industr. Hyg. J. 35: 113–136, **1974**.

———. Heat stroke and hyperthermia in marathon runners. In: The Marathon: Physiological, Medical, Epidemiological and Psychological Studies. Ed.: P. Milvy. Ann. N.Y. Acad. Sci. 301: 128–138, **1977**.

Wyndham, C. H., Bouwer, W., Van der, M., Paterson, H. F., and Devine, M. G. Practical aspects of recent physiological studies in Witwatersrand Gold Mines. J. Chem. Met. Mining Soc. S. Afr. 53: 287–313, **1953**.

Wyndham, C. H., Kew, M. C., Kok, R., Bersohn, I., and Strydom, N. B. Serum enzyme changes in unacclimatized and acclimatized man under severe heat stress. J. Appl. Physiol. 37: 695–698, **1974**.

Wyndham, C. H., Morrison, J. F., and Williams, C. G. Heat reactions of male and female caucasians. J. Appl. Physiol. 20: 357–364, **1965**.

Wyndham, C. H., Rogers, G. G., Benade, A. J. S., and Strydom, N. B. Physiological effects of the amphetamines during exercise. S. Afr. Med. J. 45: 247–252, **1971**.

Wyndham, C. H., Rogers, G. G., Senay, L. C., and Mitchell, D. Acclimatization in a hot humid environment: Cardiovascular adjustments. J. Appl. Physiol. 40: 779–785, **1976**.

Wyndham, C. H. and Strydom, N. B. The dangers of inadequate water intake during marathon running. S. Afr. Med. J. 43: 894–896, **1969**.

Wyndham, C. H. and Strydom, N. B. Körperliche Arbeit bei hoher Temperatur. In: Zentrale Themem der Sportmedizin. Ed.: W. Hollmann. Berlin: Springer Verlag, **1972**.

Wyndham, C. H., Strydom, N. B., Benade, A. J. S., and Van der Walt, W. H. The effect on acclimatization of various water and salt replacement regimens. S. Afr. Med. J. 47: 1773–1779, **1973a**.

Wyndham, C. H., Strydom, N. B., Benade, A. J. S., and Van Rensbury, A. J. Limiting rates of work for acclimatization at high wet bulb temperatures. J. Appl. Physiol. 35: 454–458, **1973b**.

Wyndham, C. H., Strydom, N. B., Maritz, J. S., Morrison, J. F., Peter, J., and Potgieter, Z. U. Maximum oxygen intake and maximum heart rate during strenuous work. J. Appl. Physiol. 14: 927–936, **1959**.

Wyndham, C. H., Strydom, N. B., Morrison, J. F., Williams, C. G., Bredell, G. A. G., Maritz, J. S., and Munro, A. Criteria for physiological limits for work in heat. J. Appl. Physiol. 20: 37–45, **1965**.

Wyndham, C. H., Strydom, N. B., Williams, C. G., and Heyns, A. An examination of certain individual factors affecting the heat tolerance of mine workers. J. S. Afr. Inst. Mining Met. 68: 79–91, **1967**.

Wyss, C.R., Brengelmann, G. L., Johnson, J. M., Rowell, L. B., and Niederberger, M. Control of skin blood flow, sweating and heart rate: role of skin vs. core temperature. J. Appl. Physiol. 36: 726–733, **1974**.

Yakovlev, N. N. Problem of biochemical adaptation of muscles in dependence on the character of their activity. J. Gen. Biol. USSR 19: 417, **1958**.

Yamabayashi, H., Takahashi, T., Tonomura, S., and Takahashi, H. Airway Dynamics. Physiology and Pharmacology. Ed.· A. Bouhuys. Springfield, Ill.: C. C. Thomas, **1970**.

Yamada, S. and Burton, A. C. Effect of reduced tissue pressure on blood flow of the fingers: the veni-vasomotor reflex. J. Appl. Physiol. 6: 501–505, **1954**.

Yamaji, K. and Shephard, R. J. Longevity and causes of death of athletes. J. Human Ergol. 6: 13–25, **1978**.

Yamaji, R. Studies on protein metabolism during muscular exercise. 1. Nitrogen metabolism in training for heavy muscular exercise. J. Physiol. Soc. Jap. 13: 476–489, **1951**.

Yamakawa, J. and Ishiko, T. Standardization of physical fitness test for oarsmen. In: Proceedings of International Congress of Sports Sciences, Tokyo. Ed.: K. Kato. Tokyo: Japanese Union of Sports Sciences, **1966**.

Yamamoto, W. S. Looking at the regulation of ventilation as a signalling process, pp. 137–148. In: Muscular Exercise and the Lung. Eds.: J. A. Dempsey and C. E. Reed. Madison: University of Wisconsin Press, **1977**.

Yamashiro, S. M., Daubenspeck, J. A., Lauritsen, T. N., and Grodins, F. S. Total work rate of breathing. Optimization in CO_2 inhalation and exercise. J. Appl. Physiol. 38: 702–709, **1975**.

Yeager, S. A. and Brynteson, P. Effects of varying training periods on the development of cardiovascular efficiency of college women. Res. Quart. 41: 589-592, **1970**.

Yeates, D. B. and Aspin, N. A mathematical description of the airways of the human lungs. Resp. Physiol. 32: 91-104, **1978**.

Yonemura, K. Resting and action potentials in red and white muscles of the rat. Jap. J. Physiol. 17: 708-719, **1967**.

Yoshimura, H. Anemia during physical training (sports anemia). Nutr. Rev. 28: 251-253, **1970**.

Yoshitake, H. Relations between the symptoms and the feeling of fatigue. Ergonomics 14: 175-186, **1971**.

Young, C. M. Body composition and body weight: criteria of overnutrition. Can. Med. Assoc. J. 93: 900-910, **1965**.

Young, D. R., Pelligra, R., Shapira, J., Adachi, R. R., and Skrettingland, K. Glucose oxidation and replacement during prolonged exercise in man. J. Appl. Physiol. 23: 734-741, **1967**.

Young, J. M. Acute oxygen toxicity in working man, pp. 67-76. In: Underwater Physiology. Ed.: C. J. Lambertsen. New York: Academic Press, **1971**.

Young, L. R. and Meiry, J. L. A revised dynamic otolith model. Aerospace Med. 39: 606-609, **1968**.

Young, R. J., Ismail, A. H., Bradley, C. A., and Corrigan, D. L. Effect of prolonged exercise on serum testosterone levels in adult men. Brit. J. Sport. Med. 10: 230-235, **1976**.

Yu, P. N. Pulmonary Blood Volume in Health and Disease. Philadelphia: Lea & Febiger, **1969**.

Yudkin, J. Diet and coronary thrombosis, hypothesis and fact. Lancet 2: 155-162, **1957**.

Zachar, J. Electrogenesis and Contractility in Skeletal Muscle Cells, pp. 1-638. Bratislava: Slovak Academy of Sciences, **1971**.

Zaharieva, E. Olympic participation by women: Effects on pregnancy and childbirth. J. Amer. Med. Assoc. 221: 992-995, **1972**.

Zak, R. Development and proliferative capacity of cardiac muscle cells. Circ. Res. (Suppl. II): 34-35, **1974**.

Zambraski, E. J., Di Bona, G. F., and Tipton, C. M. Renin-angiotensin system and autoregulation of renal blood flow in the exercising dog. Med. Sci. Sport. 9: 48, **1977**.

Zambraski, E. J., Tipton, C. M., Jordon, H. R., Palmer, W. K., and Tcheng, T. K. Iowa wrestling study: Urinary profiles of state finalists prior to competition. Med. Sci. Sport. 6: 129-132, **1974**.

Zaret, B. L., Strauss, H. W., Martin, N. D., Wells, H. P., and Flamm, M. D. Noninvasive evaluation of myocardial perfusion with potassium 43. Study of patients at rest, exercise, and during angina pectoris. N. Engl. J. Med. 288: 809-812, **1973**.

Zarrugh, M. Y. and Radcliffe, C. W. Predicting metabolic cost of level walking. Europ. J. Appl. Physiol. 38: 215-223, **1978**.

Zechman, F. W., Hall, F. G., and Hull, W. E. Effects of graded resistance to tracheal airflow in man. J. Appl. Physiol. 10: 356-362, **1957**.

Zeigler, M. G., Lake, C. R., and Kopin, I. J. Plasma noradrenaline increases with age. Nature 261: 333-335, **1976**.

Zelis, R., Mason, D. T., and Braunwald, E. Partition of blood flow to the cutaneous and muscular beds of the forearm at rest and during leg exercise in normal subjects and in patients with heart failure. Circ. Res. 24: 799-806, **1969**.

Zierler, K. L. Theory of the use of arterio-venous concentration differences for measuring metabolism in steady and non-steady states. J. Clin. Invest. 40: 2111-2125, **1961**.

——. Some aspects of the biophysics of muscle, pp. 117-183. In: The Structure and Function of Muscle, 2nd ed. Vol. III. Physiology and Biochemistry. New York: Academic Press, **1973**.

Zimmer, H. G., Steinkopff, G., and Gerlach, E. Changes of protein synthesis in the hypertrophying rat heart. Pflüg. Archiv. 336: 311-325, **1972**.

Zimmerman, E. and Parlee, M. B. Behavioural changes associated with the menstrual cycle; an experimental investigation. J. Appl. Soc. Psychol. 3: 335-344, **1973**.

Zinovieff, A. N. Heavy-resistance exercises: "Oxford technique." Brit. J. Phys. Med. 14: 129-132, **1951**.

Zirkle, L. G., Mengel, C. E., Horton, B. D., and Duffy, E. J. Studies of oxygen toxicity in the central nervous sytem. Aerospace Med. 36: 1027-1032, **1965**.

Zitnik, R. S., Ambrosioni, E., and Shepherd, J. T. Effect of temperature on cutaneous venomotor reflexes in man. J. Appl. Physiol. 31: 507-512, **1971**.

Zobl, E. G., Talmers, F. N., Christensen, R. C., and Baer, L. J. Effect of exercise on the cerebral circulation and metabolism. J. Appl. Physiol. 20: 1289–1293, **1965**.

Zocche, G. P., Fritts, H. W., and Cournand, A. Fraction of maximum breathing capacity available for prolonged hyperventilation. J. Appl. Physiol. 15: 1073–1074, **1960**.

Zohman, L. R. Early ambulation of post-myocardial infarction patients: Montefiore Hospital, p. 329. In: Exercise Testing and Exercise Training in Coronary Heart Disease. Eds.: J. P. Naughton, H. K. Hellerstein and I. C. Mohler. New York: Academic Press, **1973**.

Zoneraich, S., Rhee, J. J., Zoneraich, O., Jordan, D., and Appel, J. Assessment of cardiac function in marathon runners by graphic noninvasive techniques. Ann. N.Y. Acad. Sci. 301: 900–917, **1977**.

Zorbas, W. S. and Karpovich, P. V. The effect of weight lifting upon the speed of muscular contractions. Res. Quart. 22: 145–148, **1951**.

Zucker, R. S. Theoretical implications of the size principle of motoneurone recruitment. J. Theor. Biol. 38: 587–596, **1973**.

Zuckerman, J. and Stull, G. A. Ligamentous separation force in rats as influenced by training, detraining and cage restriction. Med. Sci. Sport. 5: 44–49, **1973**.

Zukel, W. J., Lewis, R. H., Enterline, P. E., Painter, R. C., Ralston, L. S., Fawcett, R. M., Meredith, A. P., and Peterson, B. A short-term community study of the epidemiology of coronary heart disease. A preliminary report on the North Dakota study. Amer. J. Pub. Health 49: 1630–1639, **1959**.

Zuntz, L. and Schumburg, W. E. F. Studien zu einer Physiologie des Marches. Berlin: Hirschwald, **1901**.

Zuti, W. B. and Corbin, C. B. A modified procedure for underwater weighing of children. Med. Sci. Sport. 10: 46, **1978**.

Zuti, W. B. and Golding, L. A. Comparing diet and exercise as weight reduction tools. Phys. Sport. Med. 4: 49–53, **1976**.

Zweifach, B. W. Microcirculation. Ann. Rev. Physiol. 35: 117–150, **1973**.

INDEX

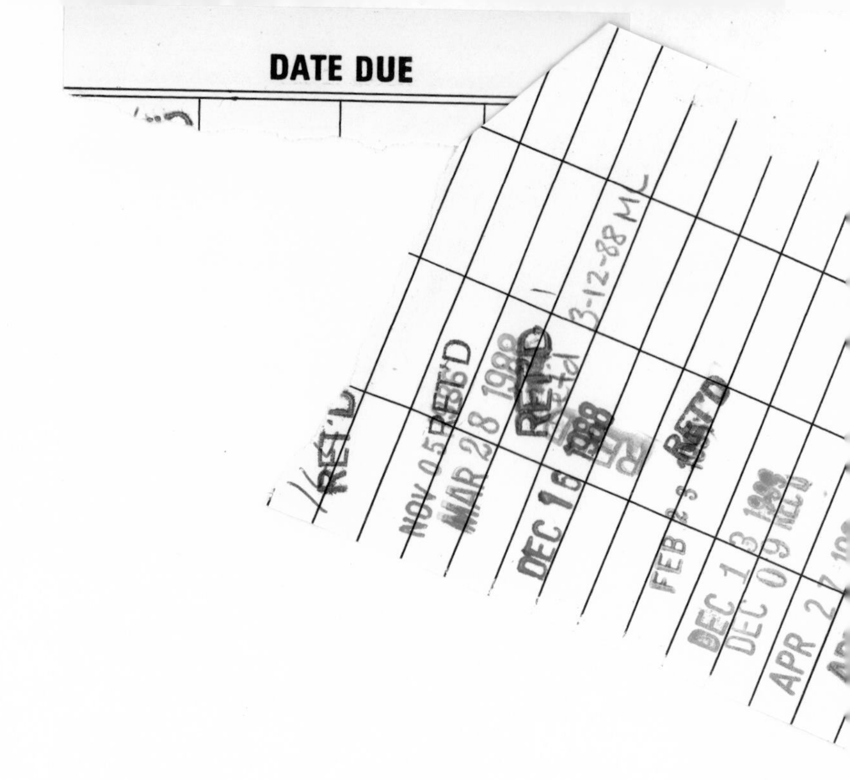